Lecture Notes in Networks and Systems

Volume 46

Series editor

Janusz Kacprzyk, Polish Academy of Sciences, Warsaw, Poland
e-mail: kacprzyk@ibspan.waw.pl

The series "Lecture Notes in Networks and Systems" publishes the latest developments in Networks and Systems—quickly, informally and with high quality. Original research reported in proceedings and post-proceedings represents the core of LNNS.

Volumes published in LNNS embrace all aspects and subfields of, as well as new challenges in, Networks and Systems.

The series contains proceedings and edited volumes in systems and networks, spanning the areas of Cyber-Physical Systems, Autonomous Systems, Sensor Networks, Control Systems, Energy Systems, Automotive Systems, Biological Systems, Vehicular Networking and Connected Vehicles, Aerospace Systems, Automation, Manufacturing, Smart Grids, Nonlinear Systems, Power Systems, Robotics, Social Systems, Economic Systems and other. Of particular value to both the contributors and the readership are the short publication timeframe and the world-wide distribution and exposure which enable both a wide and rapid dissemination of research output.

The series covers the theory, applications, and perspectives on the state of the art and future developments relevant to systems and networks, decision making, control, complex processes and related areas, as embedded in the fields of interdisciplinary and applied sciences, engineering, computer science, physics, economics, social, and life sciences, as well as the paradigms and methodologies behind them.

More information about this series at http://www.springer.com/series/15179

C. Rama Krishna · Maitreyee Dutta
Rakesh Kumar
Editors

Proceedings of 2nd International Conference on Communication, Computing and Networking

ICCCN 2018, NITTTR Chandigarh, India

Springer

Editors
C. Rama Krishna
Department of Computer Science and Engineering
National Institute of Technical Teachers Training and Research
Chandigarh, India

Rakesh Kumar
Department of Computer Science and Engineering
National Institute of Technical Teachers Training and Research
Chandigarh, India

Maitreyee Dutta
Department of Educational Television Centre
National Institute of Technical Teachers Training and Research
Chandigarh, India

ISSN 2367-3370 ISSN 2367-3389 (electronic)
Lecture Notes in Networks and Systems
ISBN 978-981-13-1216-8 ISBN 978-981-13-1217-5 (eBook)
https://doi.org/10.1007/978-981-13-1217-5

Library of Congress Control Number: 2018946585

This Springer imprint is published by the registered company Springer Nature Singapore Pte Ltd.
The registered company address is: 152 Beach Road, #21-01/04 Gateway East, Singapore 189721, Singapore

Committee Members

Chairman

I am ecstatic to learn that the Department of Computer Science and Engineering of National Institute of Technical Teaches Training and Research (NITTTR), Chandigarh is organising 2nd International Conference on "Communication, Computing and Networking (ICCCN 2018)" on 29-30 March 2018 in association with Springer.

This academic congregation will lead to deliberation on many emerging issues and possible technological solutions. The Conference is poised to present high-tech scientific outcome, promote academia-industry relations and collaborative research activities in the domain of Communication, Computing and Networking. I am sure the dynamic team of this institute will not leave any stone unturned to meet the expectations of the participants and stakeholder to update their knowledge.

I wish the Organizing Committee a grand success.

Dr. K. K. Talwar
Chairman-BOGs
NITTTR, Chandigarh

Director

I am ecstatic to learn that the Department of Computer Science and Engineering of National Institute of Technical Teachers Training & Research, Chandigarh has taken a timely initiative to address the emerging issues on Communication and Networking Technology by organising International Conference on "Communication, Computing and Networking (ICCCN-2018)" scheduled for 29-30 March, 2018.

I am made to understand that the Organising Committee has invited many speakers of international repute to make the two-day deliberations more meaningful and academically rich. I as a Director of the Institute extend a warm welcome to all the dignitaries, keynote speakers, paper presenters and delegates to the conference. I am sure the conference will provide a platform to researchers, professionals, educators and students to share innovative ideas, issues, recent trends and future directions in the fields of Communication, Computing and Networking to address the industrial and social needs.

I wish the organisers a grand success and all the participants a pleasant stay and good learning.

Prof. (Dr). S. S. Pattnaik
Director
NITTTR, Chandigarh

Conference Chairs

We are delighted to welcome delegates from India and abroad to the International Conference on "Communication, Computing and Networking (ICCCN-2018)" to be held at Department of Computer Science and Engineering of the National Institute of Technical Teachers Training & Research (NITTTR), Chandigarh in association with Springer on 29-30th March, 2018.

Undoubtedly, communication and networking technology has transformed our society in recent decades and the pace of change can only be described as disruptive. The technology itself is progressing and exploring new horizons. The conference aims at providing a platform for researchers, engineers, academicians as well as industrial professionals from all over the world to present their research results and development activities in the areas of communication, computing and networking. This conference provides opportunities for the delegates to exchange new ideas and application experiences face- to- face, to establish research relations and to find global partners for future collaboration.

It is rightly said "Alone we can do so little, together we can do so much". We are thankful to all the members of this conference who have worked hard in planning and organizing both the technical program and supporting social arrangements. In particular, we would like to take this opportunity to express our gratitude towards conference Chief Patron, Patron, members of Advisory Committee for their wise advice and brilliant suggestions on organizing the conference. Also, we would like to thank the Technical Committee and Program Committee for their thorough and timely review of the research papers and our Ph.D and M.E students for their support. We would also like to thank all the sponsors who have supported us in organizing this conference.

We wish all the success for the event.

Prof. (Dr). C. Rama Krishna
Conference Chair

Prof. (Dr). Maitreyee Dutta
Conference Chair

Conference Coordinator

I extend my most sincere welcome to all delegates of 2nd International Conference on "Communication, Computing and Networking (ICCCN-2018)" to be organized by the Department of Computer Science and Engineering, National Institute of Technical Teachers Training & Research (NITTTR), Chandigarh during 29th-30th March, 2018 in association with Springer.

Computer and Information Engineering plays an enormous role in promoting knowledge and technology which is essential for the educators, researchers, industrial and commercial concerns in the present digital age. Being a core part of this conference from the beginning, I myself feel very much enthusiastic about the conference and hope that we all will get benefit academically through mutual collaboration. This International Conference will provide an exposure to the recent advancement and innovation in the field of communication, computing and networking. With this not-to-miss conference, we are certain that we all meet to discuss the latest advances in communication, computing and networking. It is expected to be an intellectual platform to share ideas and present the latest findings and experiences in the mentioned areas.

The successful organization of ICCCN-2018 requires the talent, dedication and time of many volunteers and strong support from sponsors. My thanks goes out to the paper reviewers, the keynote speakers as well as invited speakers and authors. A special mention to our Ph.D. and ME students who have helped a lot to make this conference a successful one.

I wish the event to be a grand success.

Dr. Rakesh Kumar
Conference Coordinator

Programme Advisory Committee

Dr. Dipankar Dasgupta, University of Memphis, USA
Dr. Amitabh Trehan, Queen's University, Belfast, UK
Dr. I. Yu. Popov, St. Petersburg National Research University of Information Technologies, Russia
Dr. Abey Kuruvilla, University of Wisconsin, USA
Dr. Anu A. Gokhale, Illinois State University, USA
Dr. M. P. Poonia, Vice Chairman, AICTE, New Delhi, India
Dr. William Campbell, Birmingham City University, UK
Dr. Terry Mclvor, Academy of Technological Innovation and Research, UK
Dr. Anil Kumar Nassa, Member Secretary, NBA, New Delhi, India
Dr. Elsanosy M. Elamin, University of Kordofan, Sudan
Dr. G. Sagar, Al Ghurair University, UAE
Dr. Avichal Kapur, Joint Secretary, UGC, New Delhi, India
Dr. Fikreselam Gared, Bahir Dar University, Ethiopia
Dr. Yunfei Liu, Director, College of Information Science and Technology, Nanjing, Jiangsu, P. R. China
Dr. Dharm Singh, Professor, Computer Science, Namibia University of Science and Technology, Namibia
Dr. Manoj Singh Gaur, Director, IIT Jammu, J&K, India
Dr. Ajay K. Sharma, Director, NIT Delhi, India
Dr. Mahesh Chandra Govil, Director, NIT Sikkim, India
Dr. Lalit Awasthi, Director, Dr. B. R. Ambedkar National Institute of Technology Jalandhar, India
Prof. M. Radhakrishna, Retired Professor, IIIT Allahabad, India
Dr. Debasish Dutta, Professor, IIT Kharagpur, India
Dr. Saswat Chakrabarti, Professor, GS Sanyal School of Telecommunication, IIT Kharagpur, India
Dr. Shakti Kumar, Baddi University, India
Dr. B. N. Bhandari, Professor and Director Academic Planning, Jawaharlal Nehru Technological University, Hyderabad, India
Dr. K. Rama Naidu, Professor and Director of Evaluation, Jawaharlal Nehru Technological University, Anantapur, India
Dr. C. S. Rai, Professor and Dean, GGS Indraprastha University, New Delhi, India
Dr. Kamal Bijlani, Director, Amrita E-Learning Research Lab, Amrita University, Kerala, India
Mr. Siva G. Reddy, Founder & Chairman, Reachport, Bengaluru, Karnataka, India
Dr. Mayank Dave, Professor, NIT Kurukshetra, India
Dr. Jitender Kumar Chhabra, Professor, NIT Kurukshetra, India
Dr. H. K. Sardana, Chief Scientist, CSIR-CSIO, Chandigarh, India
Mr. Kota Srinivas, Chief Scientist, CSIR-CSIO, Chandigarh, India
Dr. Manoj Misra, Professor and Head, IIT Roorkee, India
Dr. Sanjay Agarwal, Professor, NITTTR Bhopal, India

Dr. Awadesh Kumar Singh, Professor, NIT Kurukshetra, India
Dr. Sanjeev Sofat, Professor, PEC University of Technology, Chandigarh, India
Dr. D. V. L. N. Somayajulu, Professor, NIT, Warangal, India
Dr. Ananthanarayana, Professor and Dean, NIT Suratkal, India
Dr. R. B. Patel, Professor, CCET, Chandigarh, India
Dr. Rajesh Bhatia, Professor, PEC University of Technology, Chandigarh, India
Dr. Brahmjit Singh, Professor, Department of ECE, NIT Kurukshetra, India
Dr. B. V. Ramana Reddy, Professor, GGS Indraprastha University, New Delhi, India

Technical Committee

Dr. Naveen Kumar, University of Memphis, USA
Mr. Ahmed Ali Saihood, Thi Qar University, Iraq
Dr. Upasana G. Singh, University of KwaZulu-Natal, Durban
CEng. Samudaya Nanyakkara, University of Moratuwa, Moratuwa, Sri Lanka
Er. Prasad Perera, Burapha University of Thailand, Chonburi, Thailand
Amit Kumar, College of Information Science and Technology, Nanjing, Jiangsu, P. R. China
Dr. Jagannath Aghav, Professor, College of Engineering, Pune, India
Dr. Savita Gupta, Professor, UIET, Panjab University, Chandigarh, India
Dr. Sukhwinder Singh, Professor, UIET, Panjab University, Chandigarh, India
Dr. B. C. Chaudhary, Professor, Department of Applied Science, NITTTR, Chandigarh, India
Dr. Sunil Kumar Singh, Professor and Head, CCET, Chandigarh, India
Dr. Indu Chhabra, Professor, Department of CSE, Panjab University, Chandigarh, India
Dr. Seema Verma, Professor, Department of Electronics, Banasthali Vidyapeeth, Rajasthan, India
Dr. Harish Kumar, Professor, UIET, Panjab University, Chandigarh, India
Dr. Sarabjeet Singh, Professor, UIET, Panjab University, Chandigarh, India
Dr. Shailendra Singh, Professor, PEC University of Technology, Chandigarh, India
Dr. Trilok Chand, Professor, PEC University of Technology, Chandigarh, India
Dr. Manoj Kumar, Associate Professor, Ambedkar Institute of Advanced Communication Technologies and Research, New Delhi, India
Dr. Sandeep Garg, Assistant Professor, Department of CSE, IIT Roorkee, India
Dr. S. M. Sameer, Associate Professor and Head, NIT Calicut, India
Dr. Anil Kumar Tiwari, Associate Professor, IIT Jodhpur, India
Dr. Naveen Aggarwal, Associate Professor, UIET, Panjab University, Chandigarh, India
Dr. Divya, Associate Professor, PEC University of Technology, Chandigarh, India
Dr. Lini Methew, Associate Professor and Head, NITTTR, Chandigarh, India
Dr. Rajesh Mehra, Associate Professor and Head, NITTTR, Chandigarh, India

Dr. R. C. Gangwar, Associate Professor and Head, BCET, Gurdaspur, India
Dr. Narottam Chand, Associate Professor and Head, NIT Hamirpur, India
Dr. T. P. Sharma, Associate Professor, NIT Hamirpur, India
Dr. Harsh K. Verma, Associate Professor, NIT Jalandhar, India
Mr. Rakesh Kumar Sehgal, Joint Director, C-DAC Mohali, India
Dr. Renu Dhir, Associate Professor, NIT Jalandhar, India
Dr. Vijay Laxmi, Associate Professor, MNIT Jaipur, India
Mr. Murali Krishna, Principal Engineer, Broadcom Limited, Bengaluru, Karnataka, India
Mr. Ramesh Maddara, Director Engineering, Cerium Systems Pvt. Ltd, Bengaluru, Karnataka, India
Mr. Rama Sekhar, Senior Manager, Ericsson India Pvt. Ltd., Hyderabad, India
Ms. Preeti Abrol, Principal Engineer, C-DAC Mohali, India
Dr. Balwinder Singh, Coordinator and Sr. Engineer, C-DAC Mohali, India
Mr. Rajvir Singh, Associate Professor, Chitkara University, Chandigarh, India
Dr. B. S. Bansod, Sr. Scientist, CSIR-CSIO, Chandigarh, India
Dr. Prasant K Mahapatra, Sr. Scientist, CSIR-CSIO, Chandigarh, India
Dr. Emmanuel Shubhakar Pilli, Assistant Professor, MNIT Jaipur, India
Dr. Mamta Juneja, Assistant Professor, UIET, Chandigarh, India
Dr. Anil Yadav, Assistant Professor, UIET, CSJMU Kanpur, India
Dr. Kuldeep Kumar, Assistant Professor, NIT Jalandhar, India
Dr. Poonam Saini, Assistant Professor, PEC University of Technology, Chandigarh, India
Dr. Mukesh Kumar, Assistant Professor, UIET, Panjab University, Chandigarh, India
Dr. Deepti Gupta, Assistant Professor, UIET, Panjab University, Chandigarh, India
Dr. Veenu Mangat, Assistant Professor, UIET, Panjab University, Chandigarh, India
Dr. Philemon Daniel P., Assistant Professor, NIT Hamirpur, India
Dr. Manik Sharma, Assistant Professor, DAV University, Jalandhar, India
Dr. Balwinder Raj, Assistant Professor, Dr. B. R. Ambedkar NIT Jalandhar, India
Dr. Akhilesh Kumar Sharma, Associate Professor, Manipal University Jaipur, India
Dr. Amit Jain, Assistant Professor, Sir Padampat Singhania University, Udaipur, India
Dr. Jayanti Goyal, HOD, Kanoria Girls College, Jaipur, India
Ashu Sharma, Assistant Professor, Dr. A. P. J. Abdul Kalam Technical University, UP, India
Ms. Gurpreet Kaur, Associate Professor, SGGS College, Chandigarh, India
Dr. Kavita Taneja, Assistant Professor, PU Chandigarh, India
Dr. Pankaj Kumar Sehgal, Assistant Professor, Government College, Chhachhrauli, Yamunanagar, Haryana, India
Pankaj Dev Chadha, Assistant Professor, GIMT, Kanipla, Kurukshetra, Haryana, India
Dr. Rajneesh Gujral, Professor, MM University, Mullana, Ambala, India
Dr. Devendra Parsad, Professor, CGC Landran, Mohali, India

Dr. Vikram Bali, HOD, PIET, Panipat, India
Ms. Rohini Sharma, Associate Professor, MMU Mullana, India
Dr. C Rama Krishna, Professor and Head, National Institute of Technical Teachers Training and Research, Chandigarh, India
Dr. Maitreyee Dutta, Professor and Head, National Institute of Technical Teachers Training and Research, Chandigarh, India
Dr. Rakesh Kumar, Assistant Professor, National Institute of Technical Teachers Training and Research, Chandigarh, India
Er. Shimmi S. L., Assistant Professor, NITTTR, Chandigarh, India
Mala Kalra, Assistant Professor, National Institute of Technical Teachers Training and Research, Chandigarh, India
Mr. Nitin Goyal, Assistant Professor, JMIT, Radaur, Haryana, India
Er. Anand Nayyar, Assistant Professor, KCL Institute of Management and Technology, Jalandhar, India
Roopali Punj, Assistant Professor, National Institute of Technology Hamirpur, India
Dr. Shashi Bhushan, Professor, CGC Landran, Punjab, India
Amit Doegar, Assistant Professor, National Institute of Technical Teachers Training and Research, Chandigarh, India
Nitish Devadiga, Senior Software/Research Engineer, Datarista Inc, Providence RI (Headquarters), USA
Rohit Handa, Assistant Professor, PEC, Chandigarh, India
Dr. Rajinder Sandhu, Assistant Professor, Chitkara University, Rajpura, India
Jagpreet Sidhu, Assistant Professor, Chitkara University, Rajpura, India
Shano Solanki, Assistant Professor, National Institute of Technical Teachers Training and Research, Chandigarh, India
Dr. Rupali Verma, Assistant Professor, PEC, Chandigarh, India
Poonam Singh, Lecturer, Government Polytechnic College, Behram, SBS Nagar, India
Dr. Amandeep Verma, Assistant Professor, UIET, Panjab University, Chandigarh, India
Gaurav Kumar, Director, Magma Research and Consultancy Services, Ambala, Haryana, India
Dr. Harmunish Taneja, Assistant Professor, DAV College, Chandigarh, India
Gulshan Goyal, Assistant Professor, CCET Chandigarh, India
Dr. Sandeep Kumar Goyal, Professor, MM University, Mullana, Ambala, India
Amrita Saini, Assistant Professor, Govt Women Polytechnic, Ambala City, India
Santosh Kumar Yadav, Assistant Professor, CCET, Chandigarh, India, India
Dr. Sakshi Kaushal, Assistant Professor, UIET, Chandigarh, India
Dr. Sanjeev Rana, Professor, MM University, Mullana, Ambala, India
Dr. Ajay Mittal, Associate Professor, UIET, Panjab University, Chandigarh, India
Ms. Meenu Khurana, Deputy Dean, Chitkara University, Chandigarh, India
Ms. Rishu Chhabra, Assistant Professor, Chitkara University, Chandigarh, India
Ms. Garima Saini, Assistant Professor, NITTTR, Chandigarh, India
Er. Abhishek Bajpai, Assistant Professor, SRM University, Lucknow, India

Ritula Thakur, Assistant Professor, National Institute of Technical Teachers Training and Research, Chandigarh, India
Shiv Kishan Dubey, Assistant Professor, AITH, Kanpur, UP, India
Dr. Vishal Bhatnagar, Associate Professor & Head, Ambedkar Institute of Advanced Communication Technologies and Research, Delhi
Dr. S. S. Dhami Professor, National Institute of Technical Teachers Training and Research, Chandigarh, India
Amita Jain, Assistant Professor, AIT, Delhi, India
Mr. Satish Kumar, Principal Scientist, CSIR-CSIO, Chandigarh
Vijay Kumar, Assistant Professor, Thapar Institute of Engineering and Technology, Punjab, India
Neelam Khan, Assistant Professor, NIT Srinagar, India
Amit Pratap Singh, Assistant Professor, RKGITW, Ghaziabad, India
Shekhar Singh, Maharana Institute of Professional Studies, Kanpur, India
Dr. Sunita, Assistant Professor, CCET, Chandigarh, India
Dr. Deepshikha Aggarwal, Professor, JIMS, Delhi, India
Monika Gogna, Assistant Professor, GCG-42, Chandigarh, India
Shivani Thakur, Assistant Professor, ABVGIT, Shimla, India
Shreelekha Pandey, Assistant Professor, Thapar Institute of Engineering and Technology Punjab, India
Anu Chawla, Assistant Professor, GCG-42, Chandigarh, India
Dr. Chandresh Kumar Chhatlani, Assistant Professor, JRN Rajasthan Vidyapeeth University, Udaipur, India
Dr. Sushma Bahuguna, Assistant Professor, BCIOIT, New Delhi, India
Vidhi Sood, Assistant Professor, Chanderprabhu Jain College of Higher Studies and School of Law, New Delhi, India
Minni Jain, Assistant Professor, Delhi Technological University, India
Mandeep Kaur, Assistant Professor, UIET, Chandigarh, India
Raj Kumari, Assistant Professor, UIET, Chandigarh, India
Keshav Singh, Assistant Professor, Central University of Himachal Pradesh, India
Dr. Sumit Yadav, Assistant Professor, IGDTUW, Delhi, India
Dr. Megha Gupta, Associate Professor, JIMS Rohini, Delhi, India
Rachit Garg, Assistant Professor, MPSTME, NMIMS, India
Rajeev Kumar, Assistant Professor, P.U. Swami Sarvanand Giri Regional Centre, Punjab, India
Archana B. Saxena, Associate Professor, JIMS Rohini, Delhi, India
Er. A. K. Vashishtha, IT Project Head, Vidya Knowledge Park, Meerut, Delhi-NCR, India
Amit Rathee, Research Scholar, NIT Kurukshetra, India
Yashveer Yadav, Research Scholar, National Institute of Technical Teachers Training and Research, Chandigarh, India
Pawan Kumar, Research Scholar, National Institute of Technical Teachers Training and Research, Chandigarh, India
Kamaldeep Katyal, Research Scholar, National Institute of Technical Teachers Training and Research, Chandigarh, India

Proceedings Editors

Dr. C. Rama Krishna, Professor and Head, Department of Computer Science and Engineering, National Institute of Technical Teachers Training and Research, Chandigarh, India
Dr. Maitreyee Dutta, Professor and Head, Department of Educational Television Centre, National Institute of Technical Teachers Training and Research, Chandigarh, India
Dr. Rakesh Kumar, Assistant Professor, Department of Computer Science and Engineering, National Institute of Technical Teachers Training and Research, Chandigarh, India

Local Organizing Committee

Dr. C. Rama Krishna, Professor and Head, Department of CSE, National Institute of Technical Teachers Training and Research, Chandigarh, India
Dr. Maitreyee Dutta, Professor and Head, Department of Educational Television Centre, National Institute of Technical Teachers Training and Research, Chandigarh, India
Dr. Rakesh Kumar, Assistant Professor, Department of CSE, National Institute of Technical Teachers Training and Research, Chandigarh, India
Ms. Shano Solanki, Assistant Professor, Department of CSE, National Institute of Technical Teachers Training and Research, Chandigarh, India
Mr. Amit Doegar, Assistant Professor, Department of CSE, National Institute of Technical Teachers Training and Research, Chandigarh, India
Ms. Mala Kalra, Assistant Professor, Department of CSE, National Institute of Technical Teachers Training and Research, Chandigarh, India
Mr. Pardeep Bansal, System Programmer, Department of CSE, National Institute of Technical Teachers Training and Research, Chandigarh, India
Mrs. Sangeeta Gupta, Jr. System Programmer, Department of CSE, National Institute of Technical Teachers Training and Research, Chandigarh, India
Mr. Sidharatha Nanchahal, Jr. System Engineer, Department of CSE, National Institute of Technical Teachers Training and Research, Chandigarh, India
Mr. Amrendra Sharan, Jr. System Programmer, Department of CSE, National Institute of Technical Teachers Training and Research, Chandigarh, India
Mr. Rajiv Negi, Senior Technical Assistant, Department of CSE, National Institute of Technical Teachers Training and Research, Chandigarh, India

Student Volunteers

Mr. Kamaldeep, Research Scholar, National Institute of Technical Teachers Training and Research, Chandigarh, India
Ms. Manisha Malik, Research Scholar, National Institute of Technical Teachers Training and Research, Chandigarh, India
Ms. Shweta Sharma, Research Scholar, National Institute of Technical Teachers Training and Research, Chandigarh, India
Mr. Pawan Kumar, Research Scholar, National Institute of Technical Teachers Training and Research, Chandigarh, India
Mr. Yashveer Yadav, Research Scholar, National Institute of Technical Teachers Training and Research, Chandigarh, India
Mr. Gulshan Sharma, M.E. Student, National Institute of Technical Teachers Training and Research, Chandigarh, India
Ms. Vani Kansal, M.E. Student, National Institute of Technical Teachers Training and Research, Chandigarh, India

Preface

Technology has forever changed the world we live in. We're online, in one way or another, all day long. Our phones and computers have become reflections of our personalities, our interests and our identities. They hold much that is important to us.

The connectivity that has linked individuals in the present age has become possible due to linking together of computing devices. Technically advanced devices and their increasing usability have assisted in connectivity among individuals. Communication and connectivity have never been as easy as they are today. Neither is it required to get into the hassle of arranging hardware for establishing connectivity nor is it as timely as it used to be just about a decade back.

This book contains subparts focusing on the following themes:

- Network security and privacy
- Wireless networks
- Wireless communication
- Signal and image processing
- Data science and mining
- Electronic and instrumentation
- Computing technologies

Each subpart has either survey papers highlighting the challenges or papers showing simulation or test bed-based experimental results. We hope this book will be quite useful for academicians, researchers and scientists in order to carry out further experimentation and technology enhancements.

Chandigarh
March 2018

C. Rama Krishna
Maitreyee Dutta
Rakesh Kumar

Contents

Part I Networks Security and Privacy

Experimental Analysis of DDoS Attacks on OpenStack Cloud Platform 3
Aanshi Bhardwaj, Atul Sharma, Veenu Mangat, Krishan Kumar and Renu Vig

Survey of Transport Layer Multihoming Protocols and Performance Analysis of MPTCP 15
Anurag Jagetiya, C. Rama Krishna and Yousuf Haider

Authenticating Mobile Phone Users Based on Their Typing Position Using Keystroke Dynamics 25
Baljit Singh Saini, Navdeep Kaur and Kamaljit Singh Bhatia

Differential Privacy Framework: Impact of Quasi-identifiers on Anonymization 35
Gurjeet Kaur and Sunil Agrawal

Preserving User Location Privacy in Era of Location-Based Services: Challenges, Techniques and Framework 43
Rigzin Angmo, Veenu Mangat and Naveen Aggarwal

Efficient Key Distribution and Mutual Authentication Mechanism Using Modified Needham–Schroeder and Otway–Rees Protocol for Cloud Environment 53
Ashwani Kumar Vashishtha, C. Rama Krishna and Rajiv Chechi

An Identity-Based Authentication Framework for Big Data Security 63
Vinod Kumar, Musheer Ahmad and Pankaj Kumar

Performance Analysis of Clone Node Attack in Wireless Sensor Network 73
Sachin Lalar, Shashi Bhushan and Surender

Human Electroencephalographic Biometric Person Recognition System 83
Nagsen Bansod, Siddharth B. Dabhade, Jitendra N. Dongre, K. V. Shende, S. Bhable, S. Maher, S. Thorat, K. Ankushe, S. Tharewal, V. S. Jadhav and K. V. Kale

Identifying Various Risks in Cyber-Security and Providing a Mind-Map of Network Security Issues to Mitigate Cyber-Crimes 93
Neelam Saleem Khan, Mohammad Ahsan Chishti and Mahreen Saleem

Part II Wireless Networks

Mitigating the Effect of Node Selfishness in Mobile Social Networks 107
C. C. Sobin

Energy Efficient Sector-Based Clustering Protocol for Heterogeneous WSN 117
Suniti Dutt, Gurleen Kaur and Sunil Agrawal

Real-Time Implementation of Scheduling Policies for Education Using Raspberry Pi: A Review 127
Payal Kamboj, C. Rama Krishna and S. R. N. Reddy

Performance Evaluation of LEACH Protocol Based on Data Clustering Algorithms 135
Mohit Mittal and Satendra Kumar

Performance Analysis of 3D WSN with Change in Rain Intensity 145
Raunak Monir, Ranjana Thalore and Partha Pratim Bhattacharya

A Survey on Wireless Sensor Networks Coverage Problems 153
Aastha Maheshwari and Narottam Chand

Analysis of Existing Protocols in WSN Based on Key Parameters 165
Charu Sharma and Rohit Vaid

Cellular Learning Automata-Based Virtual Network Embedding in Software-Defined Networks 173
Dipanwita Thakur and Manas Khatua

Energy-Efficient Delay-Aware Preemptive Variable-Length Time Slot Allocation Scheme for WBASN (eDPVT) 183
Tamanna Puri, Rama Krishna Challa and Navneet Kaur Sehgal

Smart Irrigation System Using Cloud and Internet of Things 195
Suresh Koduru, V. G. D. Prasad Reddy Padala and Preethi Padala

IoT-Based Condition Monitoring and Fault Detection for Induction Motor 205
R. Kannan, S. Solai Manohar and M. Senthil Kumaran

Smart Parking—A Wireless Sensor Networks Application Using IoT 217
Suman Balhwan, Deepali Gupta, Sonal and S. R. N. Reddy

Part III Wireless Communication

BATMAN: Blockchain-Based Aircraft Transmission Mobile Ad Hoc Network 233
Arushi Arora and Sumit Kr Yadav

MATLAB Simulink Modeling for Spectrum Sensing in Cognitive Radio Networks 241
Reena Rathee Jaglan, Rashid Mustafa and Sunil Agrawal

QoS-Aware Routing in Vehicular Ad Hoc Networks Using Ant Colony Optimization and Bee Colony Optimization 251
Sumandeep Kaur, Trilok C. Aseri and Sudesh Rani

Research Challenges in Airborne Ad-hoc Networks (AANETs) 261
Pardeep Kumar and Seema Verma

D-MSEP: Distance Incorporated Modified Stable Election Protocol in Heterogeneous Wireless Sensor Network 271
Anureet Singh, Hardeep Singh Saini and Naveen Kumar

Radio Environment Map Based Radio Resource Management in Heterogeneous Wireless Network 283
Rajarshi Mahapatra

Fuzzy AHP Based Technique for Handover Optimization in Heterogeneous Network 293
Riya Goyal, Tanu Goyal, Sakshi Kaushal and Harish Kumar

Improved TDMA Protocol for Channel Sensing in Vehicular Ad Hoc Network Using Time Lay 303
Ranbir Singh and Kulwinder Singh Mann

Implementation of Fractional Frequency Reuse Schemes in LTE-A Network 313
Girisha Kumar and Garima Saini

Frequency Regulation in Smart Grids Using Electric Vehicles Considering Real-Time Pricing 323
Sirat Kaur, Sukhwinder Singh and Damanjeet Kaur

Traffic Heterogeneity Analysis in an Energy Heterogeneous WSN Routing Algorithm 335
Deepak Sharma, Amarendra Goap, A. K. Shukla, Priyanka and Amol P. Bhondekar

Integration of Optical and Wireless Technology as High-Capacity Communication System: Issues and Challenges 345
Namita Kathpal and Amit Kumar Garg

Part IV Signal and Image Processing

An Analysis of Relationship Between Image Characteristics and Compression Quality for High-Resolution Satellite Images 353
Gurneet Kaur, Charu Nimesh and Savita Gupta

A Comparative Analysis of Various Image Segmentation Techniques 359
Poonam Jaglan, Rajeshwar Dass and Manoj Duhan

Deep Learning Based Shadow Detection in Images 375
Naman Bansal, Akashdeep and Naveen Aggarwal

Detection of Hemorrhagic Region in Brain MRI 383
Ujjwal Kumar Kamila, Oishila Bandyopadhyay and Arindam Biswas

A Novel Approach of Optic Disk Detection for Diagnosis of Diabetic Retinopathy 393
Aakanksha Mahajan, Sushil Kumar and Rohit Bansal

Recurrence Plot Features of RR-Interval Signal for Early Stage Mortality Identification in Sudden Cardiac Death Patients 407
Reeta Devi, Hitender Kumar Tyagi and Dinesh Kumar

Festival Framework for Synthesis of Punjabi Voice 415
Sukhpreet Kaur Gill and Parminder Singh

Medical Fusion of CLAHE Images Using SWT and PCA for Brain Disease Analysis 427
Gurpreet Kaur, Sukhwinder Singh and Renu Vig

Hybrid Classifier for Bone Age Assessment 439
Amandeep Kaur and Kulwinder Singh Mann

Recent Advances and Challenges in Automatic Hyperspectral Endmember Extraction 445
Mahesh M. Solankar, Hanumant R. Gite, Rajesh K. Dhumal, Rupali R. Surase, Dhananjay Nalawade and Karbhari V. Kale

Correction of Segmented Lung Boundary for Inclusion of Injured Diffused Regions from Chest HRCT Images 457
Rashmi Vishraj, Savita Gupta and Sukhwinder Singh

A Review of Passive Image Cloning Detection Approaches 469
Amit Doegar, Maitreyee Dutta and Gaurav Kumar

Pixel Level Image Fusion Using Different Wavelet Transforms on Multisensor & Multifocus Images 479
Harpreet Kaur, Kamaljeet Kaur and Neeti Taneja

Differentiation of Seizure and Non-seizure EEG Signals Using Analytical Approach 489
Nazia Parveen and S. H. Saeed

Part V Data Science and Mining-I

Detecting Aggressive Driving Behavior using Mobile Smartphone 513
Rishu Chhabra, Seema Verma and C. Rama Krishna

SRPF Interest Measure Based Classification to Extract Important Patterns 523
Ochin Sharma, Suresh Kumar and Nisheeth Joshi

Two-Stage Text Feature Selection Method for Human Emotion Recognition 531
Lovejit Singh, Sarbjeet Singh and Naveen Aggarwal

Authorship Analysis of Online Social Media Content 539
Ravneet Kaur, Sarbjeet Singh and Harish Kumar

Utility of OLAP and Digital ATLAS to Enhance Cross-Border Connectivity among Punjabi Sikh NRIs 551
Harkiran Kaur, Kawaljeet Singh and Tejinder Kaur

Implementing Multidimensional Analytics in NRIs Directory Databases 559
Harkiran Kaur, Kawaljeet Singh and Tejinder Kaur

Performance Evaluation of Cooperative Spectrum Sensing Over Fading Channels Based on Neural Network Learning Approach 567
Rashid Mustafa, Reena Rathee Jaglan and Sunil Agrawal

Study of NoSQL Databases for Storing and Analyzing Traffic Data 577
Mankirat Kaur, Sarbjeet Singh and Naveen Aggarwal

Applying Machine Learning Algorithms for Early Diagnosis and Prediction of Breast Cancer Risk 589
Tawseef Ayoub Shaikh and Rashid Ali

Approximation of Heaviest *k*-Subgraph Problem by Size Reduction of Input Graph 599
Harkirat Singh, Mukesh Kumar and Preeti Aggarwal

Approximate Computing for Machine Learning 607
Vinay Kumar and Ravi Kant

Experimental Evaluation of Nature-Inspired Algorithms on High Dimensions 615
Manisha Singla and K. K. Shukla

Entropy Component Correlation Analysis for Cross Pose Face Recognition 627
Kumud Arora and Poonam Garg

Diagnosis of Malignant Pleural Mesothelioma Using KNN 637
Manpreet Kaur and Birmohan Singh

Part VI Data Science and Mining-II

ECG Arrhythmia Classification Using Artificial Neural Networks 645
Saroj Kumar Pandey and Rekh Ram Janghel

An Improved Automated Question Answering System from Lecture Videos 653
Divya Mamgai, Sonali Brodiya, Rohit Yadav and Mohit Dua

A Novel Approach to Feature Hierarchy in Aspect Based Sentiment Analysis Using OWA Operator 661
Charu Gupta, Amita Jain and Nisheeth Joshi

Analyzing Behavior of ISIS and Al-Qaeda using Association Rule Mining 669
Tamanna Goyal, Jaspal Kaur Saini and Divya Bansal

A Hybrid Classification Method Based on Machine Learning Classifiers to Predict Performance in Educational Data Mining 677
Keshav Singh Rawat and I. V. Malhan

A Novel Approach for the Classification of Leukemia Using Artificial Bee Colony Optimization Technique and Back-Propagation Neural Networks 685
Rudrani Sharma and Rakesh Kumar

Improving KFCM-F Algorithm Using Prototypes 695
K. Mrudula and E. Keshava Reddy

A Time-Efficient Active Learning-Based Instance Matching System for Data Linking 703
Gulshakh Kaur, Shilpa Verma and Poonam Saini

Text Summarization Using WordNet Graph Based Sentence Ranking 711
Amita Jain, Sonakshi Vij and Devendra K. Tayal

Reckoning of Photosynthetic Pigments Using Remotely Sensed Spectral Responses of Vigna Radiata Crop for Surge Monitoring 717
Rupali R. Surase, Karbhari Kale, Amrsinh B. Varpe, Amol D. Vibhute, Hanumant Gite, Mahesh Solankar, Sandeep Gaikwad, Dhananjay Nalawade and Suresh Mehrotra

Comparative Analysis of Machine Learning Algorithms for Breast Cancer Prognosis 727
Kashish Goyal, Prakriti Sodhi, Preeti Aggarwal and Mukesh Kumar

Digital Assessment of Spatial Distribution of the Surface Soil Types Using Spatial (Texture) Features with MLC and SVM Approaches 735
Amol D. Vibhute, Karbhari V. Kale, Rajesh K. Dhumal, Ajay D. Nagne, Suresh C. Mehrotra, Amarsinh B. Varpe, Rupali R. Surase, Dhananjay B. Nalawade and Sandeep V. Gaikwad

Frequent Itemset Mining on Uncertain Database Using OWA Operator 745
Samar Wazir, M. M. Sufyan Beg and Tanvir Ahmad

Recommender System Survey: Clustering to Nature Inspired Algorithm 757
Suraj Pal Singh and Shano Solanki

Part VII Electronics and Instrumentation

Systematic Review of Dependencies in Source Code of Software and Their Categorization 771
Mrinaal Malhotra and Jitender Kumar Chhabra

Efficacy of Softwares for Generation of Dental Aligners 783
Divya Bajaj, Aditi Rawat, Deepkamal Kaur Gill, Mamta Juneja and Prashant Jindal

Craniofacial Model Generation Using CAD/CAM Software 795
Aditi Rawat, Deepkamal Kaur Gill, Divya Bajaj, Mamta Juneja, Anand Gupta and Prashant Jindal

Dimensional Accuracy of Surgical Guides Fabricated from Different Materials Using 3D Printer 805
Deepkamal Kaur Gill, Divya Bajaj, Aditi Rawat, Yash Gopal Mittal, Mamta Juneja and Prashant Jindal

Reliability-Aware Design and Performance Analysis of QCA-Based Exclusive-OR Gate 815
Gurmohan Singh, R. K. Sarin and Balwinder Raj

Design of Low Noise Amplifier for 802.16e 823
Makesh Iyer and T. Shanmuganantham

The Effect of Various Parameters on the Nozzle Diameter and 3D Printed Product in Fused Deposition Modelling: An Approach 839
Jayant Giri, Ajit Patil and Harish Prabhu

Computational Technique for Onset of P-Wave for Seismic Alert System 849
Satish Kumar, Pawan Kapur and Renu Vig

Object Oriented Dynamic Coupling and Cohesion Metrics: A Review 861
Navneet Kaur, Aradhana Negi and Hardeep Singh

IFSOM: A Two-Phase Framework for COTS Evaluation and Selection 871
Vikram Bali, Shivani Bali and Sushila Madan

HL-7 Based Middleware Standard for Healthcare Information System: FHIR 889
Meenakshi Sharma and Himanshu Aggarwal

An Artificial Fuzzy Logic Intelligent Controller Based MPPT for PV Grid Utility 901
Neeraj Priyadarshi, Farooque Azam, Akash Kumar Bhoi and S. Alam

Identification and Classification of Water Stressed Crops Using Hyperspectral Data: A Case Study of Paithan Tehsil 911
Sandeep V. Gaikwad, Amol D. Vibhute, K. V. Kale, S. C. Mehrotra, Rajesh K. Dhumal, Amarsinh B. Varpe and Rupali R. Surase

Part VIII Computing Technologies

Social Spider Foraging Based Resource Placement Policies in Cloud Environment 923
Preeti Abrol and Savita Gupta

Distributed Mutual Exclusion Algorithm with Improved Performance 933
Amit Kumar Maurya, Arvind Kumar, Akhilesh Kumar Mishra, Bharat Kumar and Priyanshu Vishwakarma

Big Data's Biggest Problem: Load Imbalancing 939
Kanak Meena and Devendra K. Tayal

Recent Trends in Green Cloud Computing 947
Shatakshi Kaushal, Dweep Gogia and Bishwesh Kumar

A FOG Computing Based Battery Swapping Model for Next Generation Transport 957
Md. Muzakkir Hussain, Mohammad Saad Alam and M. M. Sufyan Beg

A Modified FIFO with Distributed and Pipelining Scheduling in Hadoop 969
Sunil Kumar and Maninder Singh

Understanding Concepts, Future Trends, and Case Studies of Big Data Technologies 979
Khyati Ahlawat and Amit Prakash Singh

Issues in Software-Defined Networking 989
Amit Nayyer, Aman Kumar Sharma and Lalit K. Awasthi

Feasibility of Fog Computing in Smart Grid Architectures 999
Md. Muzakkir Hussain, Mohammad Saad Alam and M. M. Sufyan Beg

A Hybrid Approach for Energy-Efficient Task Scheduling in Cloud 1011
Sunil Kumar and Mala Kalra

Improving the Performance of Pre-copy Virtual Machine Migration Technique 1021
Aditya Bhardwaj and C. Rama Krishna

Improved Active Monitoring Load-Balancing Algorithm in Cloud Computing 1033
Pawan Kumar and Rakesh Kumar

Analyzing Privacy Issues in Cloud Computing Using Trust Model 1041
Parmod Kalia, Sanjeev Sofat and Divya Bansal

Author Index 1053

Editors and Contributors

About the Editors

Dr. C. Rama Krishna is Professor and Head of the Department of Computer Science and Engineering, NITTTR (Ministry of HRD, Government of India), Chandigarh. He holds a B.Tech. from JNTU, Hyderabad; an M.Tech. from Cochin University of Science and Technology, Cochin; and a Ph.D. from IIT Kharagpur in the area of mobile ad hoc networks. He is Senior Member of IEEE. Since 1996, he has been working with the Department of Computer Science and Engineering, National Institute of Technical Teachers Training and Research, Chandigarh. He has over 20 years of teaching and research experience, and his areas of interest include computer networks, wireless networks, cryptography and cyber security, E-learning, and cloud computing. He has published over 80 research papers in refereed international and national journals and conferences. He is co-inventor of one published patent. He is Associate Editor for the *International Journal of Technology, Knowledge and Society*. He is a member of advisory/technical committees of several national and international journals and conferences.

Dr. Maitreyee Dutta is Professor and Head of the Department of ETV, NITTTR (Ministry of HRD, Government of India), Chandigarh. She holds B.E. and M.E. in Electronics and Telecommunication from Guwahati University and Panjab University, respectively. She completed her Ph.D. in Engineering and Technology from Panjab University, Chandigarh. She has published 68 journal research papers and 59 conference research papers. She has conducted free courses for underprivileged students in collaboration with the social welfare organization under the UT Government, Chandigarh, and for this, she was awarded the Rota Women Distinction Award by the Rotary Club, Chandigarh, in 2013.

Dr. Rakesh Kumar is Assistant Professor in the Department of Computer Science and Engineering, National Institute of Technical Teachers Training and Research, Chandigarh. He has 14 years of teaching experience. He holds a B.Tech. (Computer Science and Engineering) from IKG Punjab Technical University, Jalandhar, and an M.Tech. (Information Technology) from GGS Indraprastha University, Delhi. He received his Ph.D. in Computer Engineering from NIT Kurukshetra, Haryana. He has published numerous papers in international journals and conferences of repute.

Contributors

Preeti Abrol STD, CDAC, Mohali, Punjab, India

Himanshu Aggarwal Computer Engineering Department, Punjabi University, Patiala, India

Naveen Aggarwal University Institute of Engineering and Technology, Panjab University, Chandigarh, India

Preeti Aggarwal Department of Computer Science and Engineering, University Institute of Engineering and Technology, Panjab University, Chandigarh, India

Sunil Agrawal Department of ECE, UIET, Panjab University, Chandigarh, India

Khyati Ahlawat University School of Information Communication and Technology, Guru Gobind Singh Indraprastha University, Dwarka, Delhi, India

Musheer Ahmad Department of Applied Sciences and Humanities, Jamia Millia Islamia, New Delhi, India

Tanvir Ahmad Department of Computer Engineering, Jamia Millia Islamia, Delhi, ND, India

Akashdeep UIET, Panjab University, Chandigarh, India

Mohammad Saad Alam Department of Electrical Engineering, ZHCET, AMU, Aligarh, India

S. Alam Millia Institute of Technology, Purnea, India

Rashid Ali Department of Computer Engineering, Aligarh Muslim University, Aligarh, Uttar Pradesh, India

Rigzin Angmo University Institute of Engineering and Technology, Panjab University, Chandigarh, India

K. Ankushe Department of Computer Science and Info Tech, Dr. Babasaheb Ambedkar Marathwada University, Aurangabad, MS, India

Arushi Arora Indira Gandhi Delhi Technical University for Women, Kashmere Gate, Delhi, India

Kumud Arora Inderprastha Engineering College, Ghaziabad, India

Trilok C. Aseri PEC University of Technology, Chandigarh, India

Lalit K. Awasthi NIT, Jalandhar, India

Farooque Azam Millia Institute of Technology, Purnea, India

Divya Bajaj University Institute of Engineering and Technology, Panjab University, Chandigarh, India

Suman Balhwan ECE, IGDTUW, Delhi, India

Shivani Bali Lal Bahadur Shastri Institute Of Management, Delhi, India

Vikram Bali J.S.S. Academy of Technical Education, Noida, Uttar Pradesh, India

Oishila Bandyopadhyay Department of Computer Science and Engineering, Indian Institute of Information Technology, Kalyani, West Bengal, India

Divya Bansal Punjab Engineering College, Chandigarh, India

Naman Bansal UIET, Panjab University, Chandigarh, India

Rohit Bansal RGPU, Noida, Uttar Pradesh, India

Nagsen Bansod Department of Computer Science and Info Tech, Dr. Babasaheb Ambedkar Marathwada University, Aurangabad, MS, India

S. Bhable Department of Computer Science and Info Tech, Dr. Babasaheb Ambedkar Marathwada University, Aurangabad, MS, India

Aanshi Bhardwaj University Institute of Engineering and Technology, Panjab University, Chandigarh, India

Aditya Bhardwaj Department of Computer Science & Engineering, NITTTR, Chandigarh, India

Kamaljit Singh Bhatia ECE, IKGPTU, Batala, India

Partha Pratim Bhattacharya College of Engineering Technology, Mody University, Sikar, Rajasthan, India

Akash Kumar Bhoi Department of EEE, SMIT, SMU, Rangpo, Sikkim, India

Amol P. Bhondekar Academy of Scientific and Innovative Research (AcSIR), CSIR-Central Scientific Instruments Organisation (CSIR-CSIO), Chandigarh, India

Shashi Bhushan CGC, Landran, Ajitgarh, Punjab, India

Arindam Biswas Department of Information Technology, Indian Institute of Engineering Science and Technology, Shibpur, West Bengal, India

Sonali Brodiya National Institue of Technology, Kurukshetra, India

Rama Krishna Challa CSE Department, NITTTR, Chandigarh, India

Narottam Chand NIT, Hamirpur, Himachal Pradesh, India

Rajiv Chechi Department of Electronics and Communication Engineering, VCE, Meerut, India

Jitender Kumar Chhabra Computer Engineering Department, National Institute of Technology, Kurukshetra, Haryana, India

Rishu Chhabra Banasthali Vidyapith, Vanasthali, Rajasthan, India

Mohammad Ahsan Chishti Department of Computer Science and Engineering, NIT Srinagar, Srinagar, Jammu and Kashmir, India

Siddharth B. Dabhade Department of Computer Science and Info Tech, Dr. Babasaheb Ambedkar Marathwada University, Aurangabad, MS, India

Rajeshwar Dass Deenbandhu Chhottu Ram University of Science and Technology, Sonipat, Murthal, India

Reeta Devi Department of Electronics and Communication Engineering, University Institute of Engineering and Technology, Kurukshetra University, Haryana, India

Rajesh K. Dhumal Geospatial Technology Research Laboratory, Department of Computer Science and IT, Dr. Babasaheb Ambedkar Marathwada University, Aurangabad, Maharashtra, India

Amit Doegar Department of Computer Science and Engineering, NITTTR, Chandigarh, India

Jitendra N. Dongre Department of Computer Science and Info Tech, Dr. Babasaheb Ambedkar Marathwada University, Aurangabad, MS, India

Mohit Dua National Institue of Technology, Kurukshetra, India

Manoj Duhan Deenbandhu Chhottu Ram University of Science and Technology, Sonipat, Murthal, India

Suniti Dutt Department of ECE, UIET, Panjab University, Chandigarh, India

Maitreyee Dutta Department of Computer Science and Engineering, NITTTR, Chandigarh, India

Sandeep Gaikwad Geospatial Technology Research Laboratory, Department of Computer Science and IT, Dr. Babasaheb Ambedkar Marathwada University, Aurangabad, Maharashtra, India

Sandeep V. Gaikwad Geospatial Technology Research Laboratory, Department of Computer Science and IT, Dr. Babasaheb Ambedkar Marathwada University, Aurangabad, Maharashtra, India

Amit Kumar Garg Department of Electronics and Communication Engineering, Deenbandhu Chhotu Ram University of Science and Technology, Murthal (Sonepat), India

Poonam Garg Insitute of Management and Technology, Ghaziabad, Uttar Pradesh, India

Deepkamal Kaur Gill University Institute of Engineering and Technology, Panjab University, Chandigarh, India

Sukhpreet Kaur Gill GNA University, Phagwara, India

Jayant Giri Department of Mechanical Engineering, Yeshwantrao Chavan College of Engineering, Nagpur, India

Hanumant Gite Geospatial Technology Research Laboratory, Department of Computer Science and IT, Dr. Babasaheb Ambedkar Marathwada University, Aurangabad, Maharashtra, India

Hanumant R. Gite Geospatial Technology Research Laboratory, Department of Computer Science and IT, Dr. Babasaheb Ambedkar Marathwada University, Aurangabad, Maharashtra, India

Amarendra Goap CSIR-Central Scientific Instruments Organisation (CSIR-CSIO), Chandigarh, India

Dweep Gogia University Institute of Engineering and Technology, Panjab University, Chandigarh, India

Kashish Goyal Department of Computer Science & Engineering, University Institute of Engineering and Technology, Panjab University, Chandigarh, India

Riya Goyal University Institute of Engineering and Technology Panjab University, Chandigarh, India

Tamanna Goyal Punjab Engineering College, Chandigarh, India

Tanu Goyal University Institute of Engineering and Technology Panjab University, Chandigarh, India

Anand Gupta University Institute of Engineering and Technology, Panjab University, Chandigarh, India

Charu Gupta Bhagwan Parshuram Institute of Technology, Delhi, India

Deepali Gupta ECE, IGDTUW, Delhi, India

Savita Gupta Department of Computer Science and Engineering, University Institute of Engineering and Technology, Panjab University, Chandigarh, India

Yousuf Haider Department of Computer Science and Engineering, Northern India Institute of Technology, Lucknow, India

Makesh Iyer Electronics Engineering Department, Pondicherry University, Puducherry, India

V. S. Jadhav Department of Computer Science and Info Tech, Dr. Babasaheb Ambedkar Marathwada University, Aurangabad, MS, India

Anurag Jagetiya Department of IT, MLV Textile & Engineering College, Bhilwara, India

Poonam Jaglan Deenbandhu Chhottu Ram University of Science and Technology, Sonipat, Murthal, India

Reena Rathee Jaglan Department of ECE, UIET, Panjab University, Chandigarh, India

Amita Jain AIACTR (Ambedkar Institute of Advanced Communication Technologies & Research), New Delhi, India

Rekh Ram Janghel National Institute of Technology, Raipur (C.G.), India

Prashant Jindal University Institute of Engineering and Technology, Panjab University, Chandigarh, India

Nisheeth Joshi Banasthali University, Vanasthali, Rajasthan, India

Mamta Juneja University Institute of Engineering and Technology, Panjab University, Chandigarh, India

K. V. Kale Geospatial Technology Research Laboratory, Department of Computer Science and IT, Dr. Babasaheb Ambedkar Marathwada University, Aurangabad, Maharashtra, India

Karbhari Kale Geospatial Technology Research Laboratory, Department of Computer Science & IT, Dr. Babasaheb Ambedkar Marathwada University, Aurangabad, Maharashtra, India

Karbhari V. Kale Geospatial Technology Research Laboratory, Department of Computer Science and IT, Dr. Babasaheb Ambedkar Marathwada University, Aurangabad, Maharashtra, India

Parmod Kalia Punjab Engineering College, Chandigarh, India

Mala Kalra Department of Computer Science and Engineering, NITTTR, Chandigarh, India

Payal Kamboj National Institute of Technical Teachers Training and Research, Chandigarh, India

Ujjwal Kumar Kamila Department of Computer Science and Engineering, Asansol Engineering College, Asansol, West Bengal, India

R. Kannan Anna University, Chennai, India

Ravi Kant Department of IT, IIIT Bhubaneswar, Bhubaneswar, India

Pawan Kapur CSIR-CSIO, Chandigarh, India

Namita Kathpal Department of Electronics and Communication Engineering, Deenbandhu Chhotu Ram University of Science and Technology, Murthal (Sonepat), India

Amandeep Kaur CSE, I.K. Gujral Punjab Technical University, Jalandhar, India

Damanjeet Kaur University Institute of Engineering and Technology, Panjab University, Chandigarh, India

Gulshakh Kaur Department of Computer Science and Engineering, PEC University of Technology, Chandigarh, India

Gurjeet Kaur Department of ECE, UIET, Panjab University, Chandigarh, India

Gurleen Kaur Department of ECE, UIET, Panjab University, Chandigarh, India

Gurneet Kaur Department of Computer Science and Engineering, UIET, Panjab University, Chandigarh, India

Gurpreet Kaur SGGS College, Chandigarh, India

Harkiran Kaur Department of Computer Science and Engineering, Thapar Institute of Engineering and Technology (Deemed to be University), Patiala, India

Harpreet Kaur CSE, Chandigarh University, Mohali, Punjab, India

Kamaljeet Kaur CSE, Chandigarh University, Mohali, Punjab, India

Mankirat Kaur University Institute of Engineering and Technology, Panjab University, Chandigarh, India

Manpreet Kaur Department of Electrical and Instrumentation Engineering, Sant Longowal Institute of Engineering and Technology, Longowal, Punjab, India

Navdeep Kaur Sri Guru Granth Sahib World University, Fatehgarh Sahib, India

Navneet Kaur Department of Computer Science, Guru Nanak Dev University, Amritsar, India

Ravneet Kaur University Institute of Engineering and Technology, Panjab University, Chandigarh, India

Sirat Kaur University Institute of Engineering and Technology, Panjab University, Chandigarh, India

Sumandeep Kaur PEC University of Technology, Chandigarh, India

Tejinder Kaur Center for Diaspora Studies, Punjabi University, Patiala, India

Sakshi Kaushal University Institute of Engineering and Technology Panjab University, Chandigarh, India

Shatakshi Kaushal University Institute of Engineering and Technology, Panjab University, Chandigarh, India

E. Keshava Reddy Department of Mathematics, Jawaharlal Nehru Technological University Anantapur, Anantapur, AP, India

Neelam Saleem Khan Department of Computer Science and Engineering, NIT Srinagar, Srinagar, Jammu and Kashmir, India

Manas Khatua Indian Institute of Technology Jodhpur, Jodhpur, Rajasthan, India

Suresh Koduru Department of Computer Science and Systems Engineering, Andhra University, Visakhapatnam, Andhra Pradesh, India

Arvind Kumar Ashoka Institute of Technology & Management, Varanasi, Uttar Pradesh, India

Bharat Kumar Ashoka Institute of Technology & Management, Varanasi, Uttar Pradesh, India

Bishwesh Kumar University Institute of Engineering and Technology, Panjab University, Chandigarh, India

Dinesh Kumar YMCA University of Science and Technology, Faridabad, Haryana, India

Gaurav Kumar Magma Research and Consultancy Services, Ambala, India

Girisha Kumar ECE Department, NITTTR, Chandigarh, India

Harish Kumar University Institute of Engineering and Technology, Panjab University, Chandigarh, India

Krishan Kumar University Institute of Engineering and Technology, Panjab University, Chandigarh, India

Mukesh Kumar Department of Computer Science and Engineering, University Institute of Engineering and Technology, Panjab University, Chandigarh, India

Naveen Kumar Thapar Institute of Engineering and Technology (Deemed University), Patiala, Punjab, India

Pankaj Kumar School of Computing Science and Engineering, Galgotias University, Greater Noida, India

Pardeep Kumar Department of Electronics, Banasthali Vidyapith, Jaipur, Rajasthan, India

Pawan Kumar NITTTR, Chandigarh, India

Rakesh Kumar Department of Computer Science and Engineering, National Institute of Technical Teachers Training and Research, Chandigarh, India

Satendra Kumar Department of Computer Science, Gurukul Kangri University, Haridwar, India

Satish Kumar CSIR-CSIO, Chandigarh, India

Sunil Kumar Directorate Livestock Farms, Guru Angad Dev Veterinary and Animal Sciences University, Ludhiana, India; Department of Computer Science and Engineering, NITTTR, Chandigarh, India

Suresh Kumar Manav Rachna International University, Faridabad, India

Sushil Kumar Amity University, Noida, Uttar Pradesh, India

Vinay Kumar Department of ECE, NIT Meghalaya, Shillong, India

Vinod Kumar Department of Applied Sciences and Humanities, Jamia Millia Islamia, New Delhi, India

Sachin Lalar CSE, IKGPTU, Jalandhar, India

Sushila Madan Lady Shri Ram College for Women, University of Delhi, Delhi, India

Aakanksha Mahajan Amity University, Noida, Uttar Pradesh, India

Rajarshi Mahapatra Dr. SPM International Institute of Information Technology, Naya Raipur, Naya Raipur, Chhattisgarh, India

S. Maher Department of Computer Science and Info Tech, Dr. Babasaheb Ambedkar Marathwada University, Aurangabad, MS, India

Aastha Maheshwari NIT, Hamirpur, Himachal Pradesh, India

I. V. Malhan Department of Computer Science & Informatics, Central University of Himachal Pradesh, Dharamshala, Himachal Pradesh, India

Mrinaal Malhotra Computer Engineering Department, National Institute of Technology, Kurukshetra, Haryana, India

Divya Mamgai National Institue of Technology, Kurukshetra, India

Veenu Mangat University Institute of Engineering and Technology, Panjab University, Chandigarh, India

Kulwinder Singh Mann Department of Information Technology, Guru Nanak Dev Engineering College, Ludhiana, India

Amit Kumar Maurya Ashoka Institute of Technology & Management, Varanasi, Uttar Pradesh, India

Kanak Meena Indira Gandhi Delhi Technical University for Women, New Delhi, India

S. C. Mehrotra Geospatial Technology Research Laboratory, Department of Computer Science and IT, Dr. Babasaheb Ambedkar Marathwada University, Aurangabad, Maharashtra, India

Suresh Mehrotra Geospatial Technology Research Laboratory, Department of Computer Science and IT, Dr. Babasaheb Ambedkar Marathwada University, Aurangabad, Maharashtra, India

Suresh C. Mehrotra Geospatial Technology Research Laboratory, Department of Computer Science and IT, Dr. Babasaheb Ambedkar Marathwada University, Aurangabad, Maharashtra, India

Akhilesh Kumar Mishra Ashoka Institute of Technology & Management, Varanasi, Uttar Pradesh, India

Mohit Mittal Department of Computer Science Engineering, Model Institute of Engineering and Technology, Jammu, India

Yash Gopal Mittal University Institute of Engineering and Technology, Panjab University, Chandigarh, India

Raunak Monir College of Engineering Technology, Mody University, Sikar, Rajasthan, India

K. Mrudula Research Scholar, Jawaharlal Nehru Technological University Anantapur, Anantapur, AP, India

Rashid Mustafa Department of ECE, UIET, Panjab University, Chandigarh, India

Md. Muzakkir Hussain Department of Computer Engineering, ZHCET, AMU, Aligarh, India

Ajay D. Nagne Geospatial Technology Research Laboratory, Department of Computer Science & IT, Dr. Babasaheb Ambedkar Marathwada University, Aurangabad, Maharashtra, India

Dhananjay Nalawade Geospatial Technology Research Laboratory, Department of Computer Science and IT, Dr. Babasaheb Ambedkar Marathwada University, Aurangabad, Maharashtra, India

Dhananjay B. Nalawade Geospatial Technology Research Laboratory, Department of Computer Science and IT, Dr. Babasaheb Ambedkar Marathwada University, Aurangabad, Maharashtra, India

Amit Nayyer HPU, Shimla, India

Nazia Parveen Department of ECE, Shri Ram Murti Smarak College of Engineering and Technology, Bareilly, UP, India

Aradhana Negi Department of Computer Science, Guru Nanak Dev University, Amritsar, India

Charu Nimesh Department of Computer Science and Engineering, UIET, Panjab University, Chandigarh, India

Preethi Padala Department of Information Technology, NITK, Surathkal, Karnataka, India

V. G. D. Prasad Reddy Padala Department of Computer Science and Systems Engineering, Andhra University, Visakhapatnam, Andhra Pradesh, India

Saroj Kumar Pandey National Institute of Technology, Raipur (C.G.), India

Ajit Patil Department of Mechanical Engineering, Yeshwantrao Chavan College of Engineering, Nagpur, India

Harish Prabhu Department of Mechanical Engineering, Yeshwantrao Chavan College of Engineering, Nagpur, India

Neeraj Priyadarshi Millia Institute of Technology, Purnea, India

Priyanka Meerut Institute of Engineering and Technology, Meerut, India

Tamanna Puri CSE Department, BML Munjal University, Gurgaon, Haryana, India

Balwinder Raj Dr. B. R, Amberdkar National Instutute of Technology, Jalandhar, Punjab, India

C. Rama Krishna Department of Computer Science and Engineering, National Institute of Technical Teachers Training and Research, Chandigarh, India

Sudesh Rani PEC University of Technology, Chandigarh, India

Aditi Rawat University Institute of Engineering and Technology, Panjab University, Chandigarh, India

Keshav Singh Rawat Department of Computer Science & Informatics, Central University of Himachal Pradesh, Dharamshala, Himachal Pradesh, India

S. R. N. Reddy CSE, IGDTUW, Delhi, India

S. H. Saeed Department of ECE, Integral University, Lucknow, UP, India

Baljit Singh Saini Sri Guru Granth Sahib World University, Fatehgarh Sahib, India; CSE, Lovely Professional University, Phagwara, India

Garima Saini ECE Department, NITTTR, Chandigarh, India

Hardeep Singh Saini Indo Global College of Engineering, Chandigarh, Punjab, India

Jaspal Kaur Saini Punjab Engineering College, Chandigarh, India

Poonam Saini Department of Computer Science and Engineering, PEC University of Technology, Chandigarh, India

Mahreen Saleem Department of Computer Engineering, A.M.U, Aligarh, UP, India

R. K. Sarin Dr. B. R, Amberdkar National Instutute of Technology, Jalandhar, Punjab, India

Navneet Kaur Sehgal Computer Science Department, BBSBEC, Fatehgarh Sahib, Punjab, India

M. Senthil Kumaran SSN College of Engineering, Chennai, India

Tawseef Ayoub Shaikh Department of Computer Engineering, Aligarh Muslim University, Aligarh, Uttar Pradesh, India

T. Shanmuganantham Electronics Engineering Department, Pondicherry University, Puducherry, India

Aman Kumar Sharma HPU, Shimla, India

Atul Sharma University Institute of Engineering and Technology, Panjab University, Chandigarh, India

Charu Sharma CSE Department, M. M. Engineering College, M. M. University, Ambala, Haryana, India

Deepak Sharma CSIR-Central Scientific Instruments Organisation (CSIR-CSIO), Chandigarh, India; Academy of Scientific and Innovative Research (AcSIR), CSIR-CSIO, Chandigarh, India

Meenakshi Sharma Computer Science and Engineering Department, GIMET, Amritsar, India

Ochin Sharma Banasthali University, Vanasthali, India; Manav Rachna International University, Faridabad, India

Rudrani Sharma Department of Computer Science and Engineering, National Institute of Technical Teachers Training and Research, Chandigarh, India

K. V. Shende Department of Computer Science and Info Tech, Dr. Babasaheb Ambedkar Marathwada University, Aurangabad, MS, India

A. K. Shukla CSIR-Central Scientific Instruments Organisation (CSIR-CSIO), Chandigarh, India; Academy of Scientific and Innovative Research (AcSIR), CSIR-CSIO, Chandigarh, India

K. K. Shukla Indian Institute of Technology (BHU), Varanasi, India

Amit Prakash Singh University School of Information Communication and Technology, Guru Gobind Singh Indraprastha University, Dwarka, Delhi, India

Anureet Singh Indo Global College of Engineering, Chandigarh, Punjab, India

Birmohan Singh Department of Computer Science and Engineering, Sant Longowal Institute of Engineering and Technology, Longowal, Punjab, India

Gurmohan Singh Centre for Development of Advanced Computing, Mohali, Punjab, India

Hardeep Singh Department of Computer Science, Guru Nanak Dev University, Amritsar, India

Harkirat Singh Department of Computer Science and Engineering, University Institute of Engineering and Technology, Panjab University, Chandigarh, India

Kawaljeet Singh University Computer Center, Punjabi University, Patiala, India

Lovejit Singh UIET, Panjab University, Chandigarh, India

Maninder Singh Department of Computer Science, Punjabi University, Patiala, India

Parminder Singh Guru Nanak Dev Engineering College, Ludhiana, India

Ranbir Singh Lovely Professional University, Phagwara, India

Sarbjeet Singh University Institute of Engineering and Technology, Panjab University, Chandigarh, India

Sukhwinder Singh University Institute of Engineering and Technology, Panjab University, Chandigarh, India

Suraj Pal Singh Department of Computer Science and Engineering, NITTTR, Chandigarh, India

Manisha Singla Indian Institute of Technology (BHU), Varanasi, India

C. C. Sobin Department of Computer Science and Engineering, MES College of Engineering Kuttippuram, Malappuram, Kerala, India

Prakriti Sodhi Department of Computer Science & Engineering, University Institute of Engineering and Technology, Panjab University, Chandigarh, India

Sanjeev Sofat Punjab Engineering College, Chandigarh, India

S. Solai Manohar Arulmigu Meenakshi Amman College of Engineering, Tiruvannamalai, Tamil Nadu, India

Mahesh Solankar Geospatial Technology Research Laboratory, Department of Computer Science and IT, Dr. Babasaheb Ambedkar Marathwada University, Aurangabad, Maharashtra, India

Mahesh M. Solankar Geospatial Technology Research Laboratory, Department of Computer Science and IT, Dr. Babasaheb Ambedkar Marathwada University, Aurangabad, Maharashtra, India

Shano Solanki Department of Computer Science and Engineering, NITTTR, Chandigarh, India

Sonal ECE, IGDTUW, Delhi, India

M. M. Sufyan Beg Department of Computer Engineering, ZHCET, AMU, Aligarh, India

Rupali R. Surase Geospatial Technology Research Laboratory, Department of Computer Science and IT, Dr. Babasaheb Ambedkar Marathwada University, Aurangabad, Maharashtra, India

Surender GTB, Bhawanigarh, Punjab, India

Neeti Taneja CSE, Chandigarh University, Mohali, Punjab, India

Devendra K. Tayal Indira Gandhi Delhi Technical University for Women, New Delhi, India

Dipanwita Thakur Banasthali University, Banasthali, Rajasthan, India

Ranjana Thalore College of Engineering Technology, Mody University, Sikar, Rajasthan, India

S. Tharewal Department of Computer Science and Info Tech, Dr. Babasaheb Ambedkar Marathwada University, Aurangabad, MS, India

S. Thorat Department of Computer Science and Info Tech, Dr. Babasaheb Ambedkar Marathwada University, Aurangabad, MS, India

Hitender Kumar Tyagi Department of Electronics, University College, Kurukshetra University, Haryana, India

Rohit Vaid CSE Department, M. M. Engineering College, M. M. University, Ambala, Haryana, India

Amarsinh B. Varpe Geospatial Technology Research Laboratory, Department of Computer Science and IT, Dr. Babasaheb Ambedkar Marathwada University, Aurangabad, Maharashtra, India

Amrsinh B. Varpe Geospatial Technology Research Laboratory, Department of Computer Science and IT, Dr. Babasaheb Ambedkar Marathwada University, Aurangabad, Maharashtra, India

Ashwani Kumar Vashishtha Department of Computer Science and Engineering, National Institute of Technical Teachers Training and Research, Chandigarh, India

Seema Verma Department of Electronics, Banasthali Vidyapith, Jaipur, Rajasthan, India

Shilpa Verma Department of Computer Science and Engineering, PEC University of Technology, Chandigarh, India

Amol D. Vibhute Geospatial Technology Research Laboratory, Department of Computer Science and IT, Dr. Babasaheb Ambedkar Marathwada University, Aurangabad, Maharashtra, India

Renu Vig University Institute of Engineering and Technology, Panjab University, Chandigarh, India

Sonakshi Vij Indira Gandhi Delhi Technical University for Women, New Delhi, India

Rashmi Vishraj University Institute of Engineering and Technology, Panjab University, Chandigarh, India

Priyanshu Vishwakarma Ashoka Institute of Technology & Management, Varanasi, Uttar Pradesh, India

Samar Wazir Department of Computer Engineering, Jamia Millia Islamia, Delhi, ND, India

Rohit Yadav National Institue of Technology, Kurukshetra, India

Sumit Kr Yadav Indira Gandhi Delhi Technical University for Women, Kashmere Gate, Delhi, India

Part I
Networks Security and Privacy

Experimental Analysis of DDoS Attacks on OpenStack Cloud Platform

Aanshi Bhardwaj, Atul Sharma, Veenu Mangat, Krishan Kumar and Renu Vig

Abstract Cloud computing has transformed the IT industry by providing convenient way of accessing resources, services and applications over the Internet. Cloud computing provides infrastructure, platform, software, and other hybrid models as pay-as-you go based services. Customers pay based on usage of particular utility. There are vulnerabilities in the cloud infrastructure which have been easily exploited by the attackers. The well-known Distributed Denial of Service (DDoS) attack is the most eminent attack in this area of computing. DDoS attacks compromise the services of the cloud servers by overwhelming it with huge volume of normal or incomplete traffic making difficult to respond further to the legitimate clients. In cloud computing, virtual machines are created which run through instances. OpenStack is one such cloud platform, which is Open Source and is mostly deployed as Infrastructure as a service (IAAS). This paper presents experimental evaluation of DDoS attack on OpenStack cloud platform. The experimentation was conducted on open source platform of OpenStack as private cloud model. AnonymousDoser, LOIC, and Slowloris were used to launch DDoS attack. These tools flood the cloud with TCP/IP and HTTP packets. Finally, different traffic analysis tools are used to analyze the traffic pattern. The analysis is done on the basis of parameters like resource utilization of CPU, memory and packet drop.

A. Bhardwaj (✉) · A. Sharma · V. Mangat · K. Kumar · R. Vig
University Institute of Engineering and Technology, Panjab University, Chandigarh, India
e-mail: bhardwajaanshi@gmail.com

A. Sharma
e-mail: atulsharma406@gmail.com

V. Mangat
e-mail: veenumangat@yahoo.com

K. Kumar
e-mail: k.salujauiet@gmail.com

R. Vig
e-mail: renuvig@hotmail.com

C. R. Krishna et al. (eds.), *Proceedings of 2nd International Conference on Communication, Computing and Networking*, Lecture Notes in Networks and Systems 46, https://doi.org/10.1007/978-981-13-1217-5_1

Keywords OpenStack · DevStack · AnonymousDoser · LOIC · Slowloris
Wireshark · HTOP · Netstat

1 Introduction

Cloud computing has altered the way of daily running facilities in IT industries [4]. It provides on demand self-service and pay on the basis of usage model. Though there are many benefits of cloud like dynamic resource provisioning, pay-as-you-go billing yet one of the disadvantage is that cloud is viable to many attacks which can cause service downtime, economic losses and degrades business reputation. One such attack on cloud is DDoS which can result in inestimable loss to cloud users and cloud providers. According to Neustar Report in 2017,[1] 849 of 1010 (84%) of the organizations experienced at least one DDoS attack in the previous 12 months as compared to 73% in 2016. In Q1 2017 Verisign,[2,3] observed an average peak attack size of 14.1 Gigabits per second, a 26% increase from Q4 2016. Verisign also stated that the industries related to Cloud/IT Services were the favorite target of DDoS attack as shown in Fig. 1.

According to cloud security Alliance [3], DDoS attack is one of the top 9 threats to cloud computing infrastructure. It has been a biggest nightmare for industrial

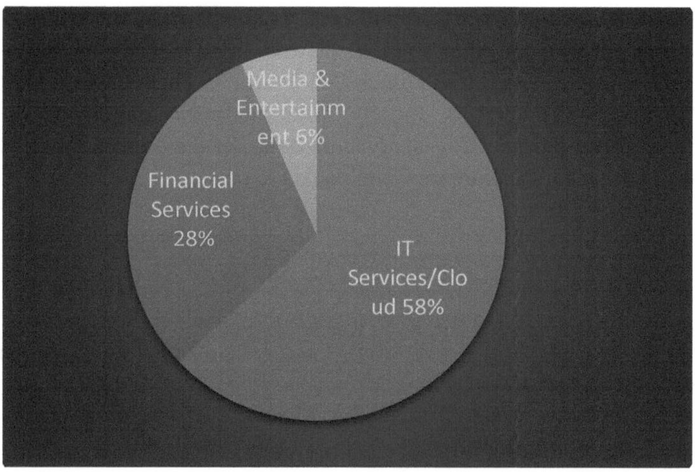

Fig. 1 DDoS attack trend in different sectors

[1]https://www.google.co.in/#q=neustar-2017-worldwide-DDoS-attacks-cyber-insights-research-report.

[2]http://www.digitalterminal.in/news/verisign-releases-q1-2017-DDoS-trends-report/9642.html.

[3]https://www.verisign.com/assets/infographic-DDoS-trends-Q12017.pdf.

activities, security and availability. DDoS is the major and largest threat to Internet and Internet of Things.

Dyn (Dynamic Network Services Company)[4] experienced the largest DDoS attack till date that shook the world on 21 October 2016. It blocked anycast server. This attack took down various big organizations like Amazon, Netflix, Reddit, Spotify, Tumblr, Twitter. The attack was launched by Mirai-based botnets. Mirai is a part of malware that look for vulnerable IoT devices like video cameras, digital video recorders, etc., and infects these devices. These IoT devices then report to command and control server which will be used as a part of large botnet. The existence of botnets that exploit inherent security weaknesses in IoT devices and the release of source code of the Mirai botnet have increased attacker's capacity to launch severe attacks. OpenStack [6] is a group of software projects that are freely available for cloud providers and enterprisers. These can be used to setup and operate computing resources, network resources and storage infrastructure. OpenStack cloud platform is scalable, deployment is easy and is plentiful of features. Due to the open nature of this cloud operating system, anyone can test and add additional components so as to meet their requirements. Even though with the help of researchers, OpenStack community has identified eleven core components for OpenStack which are officially maintained.[5] These components are Nova, Swift, Cinder, Glance, Keystone, Horizon, Ceilometers, Neutron, Heat, Trove and Sahara. Each of these components has a feature associated with it.

The first component Nova helps to provides virtual machines on request by users. Swift offers scalable storage so that one can store and retrieve lots of data through simple API. Cinder provides on demand self-service access to persistent block storage for guest Virtual machines. Glance offers repository of virtual disk images. Keystone offers service discovery, authentication and authorization for all services of OpenStack. Horizon is a graphical web-based user interface for accessing OpenStack services. Ceilometer is a single point metering component for billing utilization of resources. Neutron offers network connectivity as a service for virtualized computing platforms. Heat offers orchestration services for multiple composite cloud applications. Trove allows users to easily access relational and non-relational databases. Sahara provides data processing frameworks like hadoop, spark and storm for OpenStack-managed resources.

OpenStack[6] is used in many big organizations like Paddy Power Betfair handles more than 130 million transactions daily which is 10 times more than London stock Exchange, Verizon runs the industry's largest known network functions virtualization open stack cloud deployment across the US, Produban built a hybrid cloud infrastructure on open stack servicing Santander bank which is one of the world's largest bank. OpenStack with NASA in 2010, operated the world's largest OpenStack cloud and have experienced scaling OpenStack clouds to thousands of nodes.

---

[4]https://dyn.com/blog/dyn-analysis-summary-of-friday-october-21-attack.

[5]https://opensource.com/resources/what-is-openstack.

[6]https://www.redhat.com/en/topics/openstack.

We have seen that cloud computing environment has been one of the target of attackers. Attackers have launched DDoS attack on Linode, Rackspace, Dynamic Network Services Company, etc., leading to tremendous losses to users and company owners. Authors [1, 5] have provided various detection, prevention and mitigation techniques against DDoS attacks. This paper analysis the performance of OpenStack under DDoS attacks and till now no one has done such detailed experimentation by taking different DDoS attack tools particularly on OpenStack.

2 Implementation of DDoS on OpenStack

To implement DDoS on cloud we worked on Xen Hypervisor with OpenStack as cloud platform. First, we configured a guest operating system on Xen as Ubuntu 64 bit with 10 GB RAM. To create OpenStack based virtual machine, DevStack is used. DevStack is a series of extendable scripts used to easily and quickly create full OpenStack environment on the basis of latest Git master version. The steps[7] to be followed for the setup of the DevStack are as:

Step 1. Run the command: sudo useradd -s/bin/bash -d/opt/stack -m stack on terminal for creating a separate stack user. Since this user should have sudo privileges as it will be making many changes to the system. To provide sudo privileges run the command: echo "stack ALL = (ALL) NOPASSWD: ALL" | sudo tee/etc./sudoers.d/stack.

Step 2. Download DevStack repository from github by executing git clone https://git.openstack.org/openstack-dev/devstack.

Step 3. Create a local. conf file and preset following passwords at the root of the DevStack git repository.

ADMIN_PASSWORD = ANY_PASSWORD
DATABASE_PASSWORD = $ADMIN_PASSWORD
RABBIT_PASSWORD = $ADMIN_PASSWORD
SERVICE_PASSWORD = $ADMIN_PASSWORD

The above-mentioned steps depict the minimum configuration required to setup DevStack.

Step 4. Start the installation by running the stack script which is./stack.sh
After the installation of OpenStack 10 instances were created on it to consume some resources for checking the performance of OpenStack in a real-time scenario.

We used three attacking tools to perform DDoS on OpenStack cloud. The tools are LOIC, AnonymousDoser and Slowloris.

[7] https://docs.openstack.org/developer/devstack/.

AnonymousDoser Tool[8]: It is a tool which sends large number of TCP SYN packets which will overwhelm the server by making open connections. As a result of which the connection pool will be consumed and eventually the legitimate packets will start dropping. To launch this attack, two fields i.e. target IP and time duration (in milliseconds) for attack is filled. The intensity of the attack launched by this tool is very high which will take server down in few seconds.

LOIC Tool [2]: Low Orbit Ion Cannon (LOIC) is an open source network testing tool available at[9] developed by Praetox Technologies. It was used to attack web servers by 4chan during Project Chanology. This tool uses various flooding methods like TCP, UDP, and ICMP to launch attack against the target to exploit the resources such as CPU, storage, and bandwidth. It uses multiple threads to launch an attack. It has two forms: binary and Web-based. LOIC is one of the major DDoS attack tools which can send large number of HTTP requests to disrupt the target system. One of the main drawbacks of LOIC is that it does not spoof IP address of handlers and agents who launch the attack.

The attack was launched from three different nodes by providing the details in the GUI based interface of LOIC tool. The target IP is 192.168.51.189 which is the IP of OpenStack cloud.

Slowloris Tool [2]: It is a denial of service attack tool available at[10] and was developed by Robert "RSnake" Hansen.[11] It has both graphical and the command line user interfaces and is implemented in Perl language. The tool can exhaust machine's web server with only minimum bandwidth. It also affects unrelated services and ports. In 2009 during the presidential election of Iran, Slowloris was used as a leading tool to launch DoS attacks towards the sites of government of Iran. It creates large number of connections by sending partial requests to a victim Web server, and tries to hold these connections open for long duration. As a result, it exhausts web server's maximum concurrent connection pool, which after some time forces them to decline other legitimate connection requests from clients. The Linux command which is used to launch this attack is:

Perl slowloris.pl -dns 192.168.51.189.

3 DDoS Analysis

3.1 *Wireshark*

For AnonymousDoser: Three different nodes are used to launch DDoS attack. The IP's of these 3 nodes are 192.168.51.201, 192.168.51.61, and 192.168.51.62. The

---

[8] https://www.youtube.com/watch?v=_2_2-n_TyOk.

[9] https://sourceforge.net/projects/loic/.

[10] https://pypi.python.org/pypi/slowloris/0.1.4.

[11] https://en.wikipedia.org/wiki/Slowloris_(computer_security).

22432	135.319480	192.168.51.61	192.168.51.189	TCP	60	52881 → 80 [RST, ACK] Seq=65 Ack=498 Win=0 Len=0
22433	135.319511	192.168.51.61	192.168.51.189	TCP	60	52885 → 80 [RST, ACK] Seq=65 Ack=498 Win=0 Len=0
22434	135.319529	192.168.51.189	192.168.51.61	TCP	54	80 → 52890 [FIN, ACK] Seq=498 Ack=65 Win=29312 Len=0
22435	135.319540	192.168.51.189	192.168.51.61	TCP	54	80 → 52897 [FIN, ACK] Seq=498 Ack=65 Win=29312 Len=0
22436	135.319561	192.168.51.61	192.168.51.189	TCP	60	52886 → 80 [RST, ACK] Seq=65 Ack=498 Win=0 Len=0
22437	135.319584	192.168.51.189	192.168.51.61	TCP	54	80 → 52894 [FIN, ACK] Seq=498 Ack=65 Win=29312 Len=0
22438	135.319591	192.168.51.189	192.168.51.61	TCP	54	80 → 52887 [FIN, ACK] Seq=498 Ack=65 Win=29312 Len=0
22439	135.319592	192.168.51.61	192.168.51.189	TCP	60	52890 → 80 [RST, ACK] Seq=65 Ack=498 Win=0 Len=0
22440	135.319601	192.168.51.189	192.168.51.61	TCP	54	80 → 52889 [FIN, ACK] Seq=498 Ack=65 Win=29312 Len=0
22441	135.319622	192.168.51.61	192.168.51.189	TCP	60	52882 → 80 [RST, ACK] Seq=65 Ack=498 Win=0 Len=0
22442	135.319671	192.168.51.61	192.168.51.189	TCP	60	52895 → 80 [RST, ACK] Seq=65 Ack=498 Win=0 Len=0
22443	135.319695	192.168.51.61	192.168.51.189	TCP	60	52894 → 80 [RST, ACK] Seq=65 Ack=498 Win=0 Len=0
22444	135.319718	192.168.51.61	192.168.51.189	TCP	60	52883 → 80 [RST, ACK] Seq=65 Ack=498 Win=0 Len=0
22445	135.319756	192.168.51.61	192.168.51.189	TCP	60	52887 → 80 [RST, ACK] Seq=65 Ack=498 Win=0 Len=0
22446	135.319781	192.168.51.61	192.168.51.189	TCP	60	52884 → 80 [RST, ACK] Seq=65 Ack=498 Win=0 Len=0
22447	135.319818	192.168.51.61	192.168.51.189	TCP	60	52893 → 80 [RST, ACK] Seq=65 Ack=498 Win=0 Len=0
22448	135.319846	192.168.51.62	192.168.51.189	TCP	60	63684 → 80 [ACK] Seq=1 Ack=1 Win=65536 Len=0
22449	135.319877	192.168.51.61	192.168.51.189	TCP	60	52892 → 80 [RST, ACK] Seq=65 Ack=498 Win=0 Len=0
22450	135.319898	192.168.51.61	192.168.51.189	TCP	60	52896 → 80 [RST, ACK] Seq=65 Ack=498 Win=0 Len=0
22451	135.319920	192.168.51.61	192.168.51.189	TCP	60	52897 → 80 [RST, ACK] Seq=65 Ack=498 Win=0 Len=0
22452	135.319957	192.168.51.61	192.168.51.189	TCP	60	52889 → 80 [RST, ACK] Seq=65 Ack=498 Win=0 Len=0
22453	135.319984	192.168.51.62	192.168.51.189	TCP	117	[TCP segment of a reassembled PDU]
22454	135.320023	192.168.51.189	192.168.51.62	TCP	54	80 → 63684 [ACK] Seq=1 Ack=64 Win=29312 Len=0
22455	135.320063	192.168.51.61	192.168.51.189	TCP	60	52891 → 80 [RST, ACK] Seq=65 Ack=498 Win=0 Len=0
22456	135.322965	192.168.51.61	192.168.51.189	TCP	60	52942 → 80 [FIN, ACK] Seq=64 Ack=1 Win=65536 Len=0
22457	135.323099	192.168.51.61	192.168.51.189	TCP	60	52900 → 80 [FIN, ACK] Seq=64 Ack=1 Win=65536 Len=0

Fig. 2 Wireshark: IO packet detail of AnonymousDoser

IP address of cloud is 192.168.51.189. The Wireshark capture in Fig. 2 shows large number of reset packets being sent because of overflow of connection queue.

The IO graph in Fig. 3 shows increase in the number of packets from 0 to 1300 in just 128 s duration.

For LOIC: The IO graph in Fig. 4 shows increase in the number of packets from 0 to 2600 in just 88 s duration.

The IO graph in Fig. 5 shows increase in the number of packets from 0 to 1450 in just 16 s for Slowloris tool.

3.2 *HTOP*

It is graphically advance version of TOP. It shows the CPU utilization of all cores. Here we used 10 cores of virtual CPU's. Figure 6 shows utilization of all the ten cores separately in the case of AnoymousDoser tool. The average utilization comes out to be 30–35% for AnoymousDoser.

HTOP shows utilization of all the ten cores separately with average 60–65% CPU utilization for LOIC shown in Fig. 7 and 30–35% CPU utilization for Slowloris.

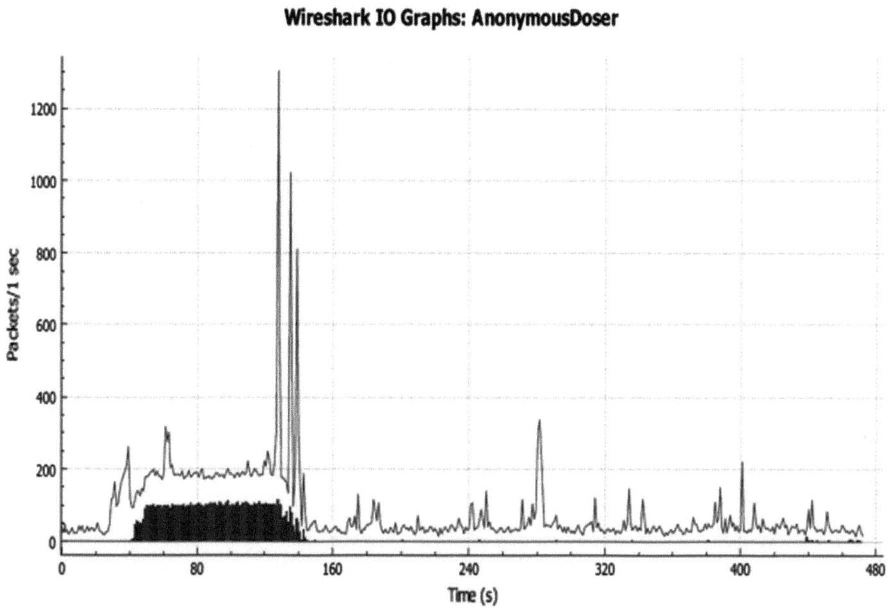

Fig. 3 Wireshark: AnonymousDoser IO graph

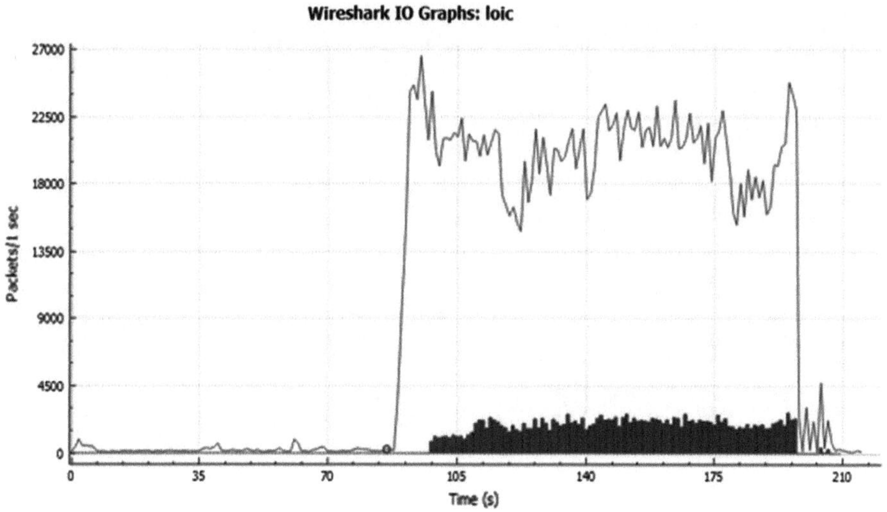

Fig. 4 Wireshark: LOIC IO graph

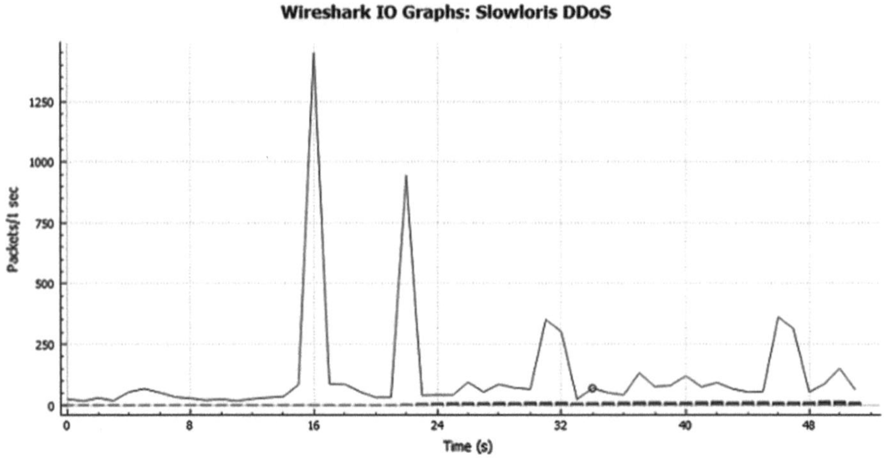

Fig. 5 Wireshark: Slowloris IO graph

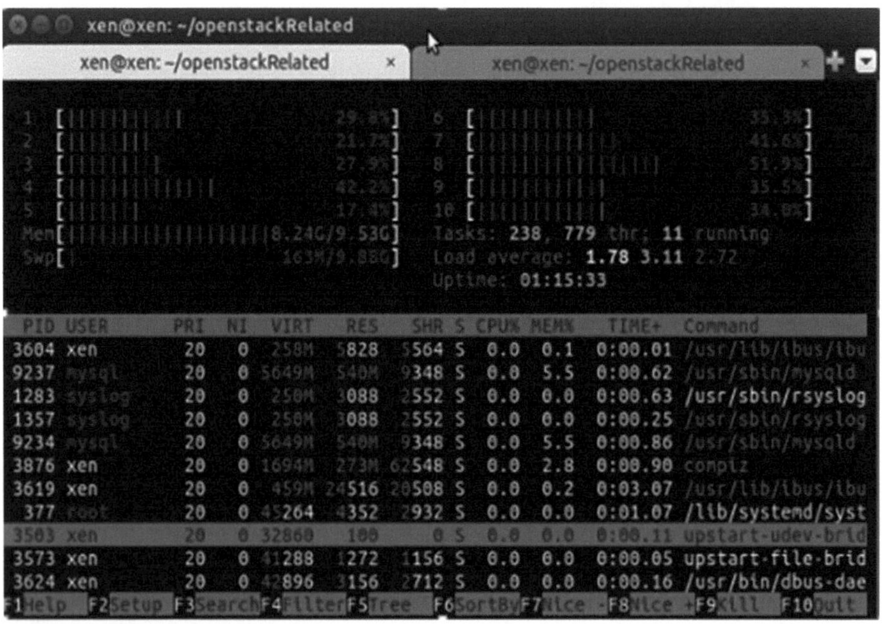

Fig. 6 AnonymousDoser CPU utilization

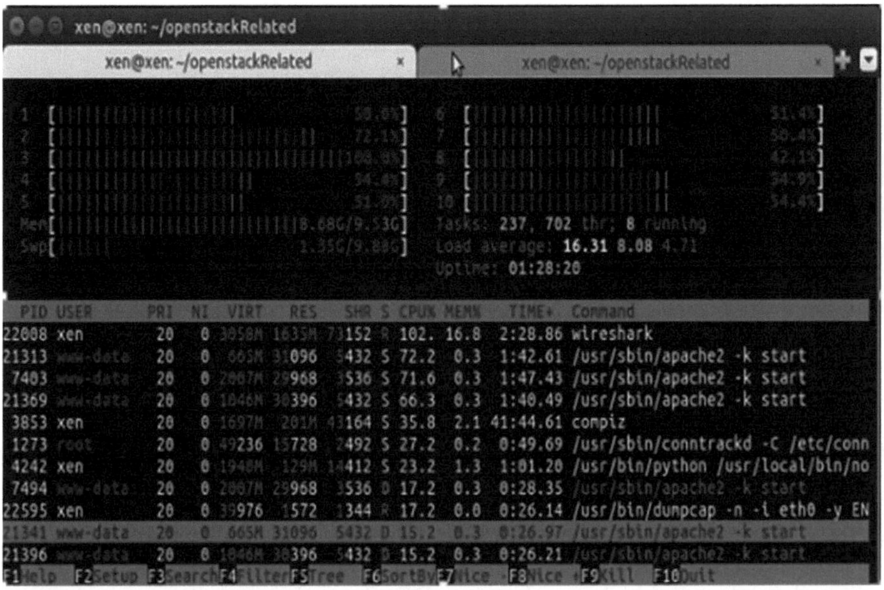

Fig. 7 LOIC CPU utilization

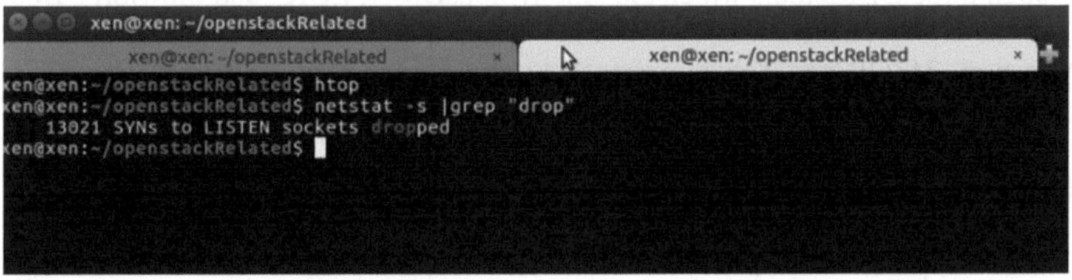

Fig. 8 AnonymousDoser packet drop

3.3 *Netstat*

Network statistics[12] is a network utility tool that has a command-line interface. It displays both incoming and outgoing TCP connections, routing tables, number of network interfaces and network protocol statistics.

Netstat shows 13021 SYN packets were lost in 128 s when attack launched through AnonymousDoser tool which is shown in Fig. 8 and 3024 SYN packets were lost in 178 s when attack launched through Slowloris shown in Fig. 9. The server becomes totally unresponsive in both the cases.

[12]https://en.wikipedia.org/wiki/Netstat.

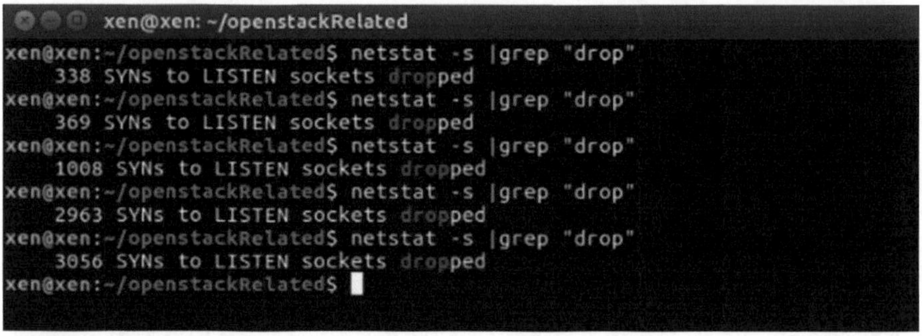

Fig. 9 Slowloris packet drop

4 Conclusion and Future Scope

Cloud Computing substituted many traditional technologies but has various security concerns as well. The frequency and size of attacks being launched on cloud have increased over years which are alarming. DDoS attack is one the major threat to cloud computing. In this work we analyzed the performance of OpenStack Cloud through three DDoS tools which are AnonymousDoser, LOIC, and Slowloris. Both TCP and HTTP packets were used for flooding which shows vulnerability in cloud. Wireshark, HTOP, and Netstat were used to analyze number of packets sent, CPU and memory utilization, packet drop, etc. Among all tools LOIC send large number of packets in few seconds. Analyses show that DDoS attacks can cripple the system resources and have highly negative impact in terms of performance and cost incurred to cloud providers and users. Our experiments have shown that the impact can be severe and spread fast in a very limited time. Thus, there is an urgent need to design and employ systematic and efficient DDoS detection and defense mechanisms on cloud platform. This work can be further extended to launch different categories of DDoS attacks in cloud in different scenarios to better understand their impact, vulnerabilities exploited by the attacks and their mechanism.

References

1. B.B. Gupta, O.P. Badve, Taxonomy of DoS and DDoS attacks and desirable defense mechanism in a cloud computing environment. J. Neural Comput. Appl. **28**, 3655–3682 (2017)
2. N. Hoque, M.H. Bhuyan, R.C. Baishya et al., Network attacks: Taxonomy, tools and systems. J. Netw. Comput. Appl. **40**, 307–324 (2014)
3. M. Jensen, J. Schwenk, N. Gruschka et al., in *On Technical Security Issues in Cloud Computing*. IEEE International Conference on Cloud Computing, Bangalore (2009), pp. 109–116

4. O. Osanaiye, K.K.R. Choo, M. Dlodlo, Distributed denial of service (DDoS) resilience in cloud: Review and conceptual cloud DDoS mitigation framework. J. Netw. Comput. Appl. **67**, 147–165 (2016)
5. G. Somani, M. Gaur, D. Sanghi et al., DDoS attacks in cloud computing: Issues, taxonomy, and future directions. J. Comput. Commun. **107**, 30–48 (2017)
6. I. Voras, B. Mihaljecić, M. Orlić et al., in *Evaluating Open-Source Cloud Computing Solutions*. MIPRO Proceedings of the 34th International Convention, Croatia (2011), pp. 209–214

Survey of Transport Layer Multihoming Protocols and Performance Analysis of MPTCP

Anurag Jagetiya, C. Rama Krishna and Yousuf Haider

Abstract Today's handheld devices like laptops, smart-phones, etc., are termed as multihomed devices as they are equipped with more than one active network interface like Ethernet, 3G, and Wi-Fi. The ability of multihoming can increase network throughput, fault recovery capabilities, and share network traffic among available paths to balance the load. However, conventional transport layer protocols like TCP were not designed to support multihoming ability. This paper explores prominent multihoming solutions suggested at the transport layer and analyzes the performance of TCP's multihoming extension MultiPath TCP under heterogeneous multiple subflows. The prime objective of this study is to find out the suitability of MultiPath TCP in scenarios where multihomed devices are using homogeneous or heterogeneous network interfaces assuming that an MPTCP-enabled client is connected to an MPTCP enabled server using two network interfaces. Two different topologies namely bottleneck and no-bottleneck are used to investigate the behavior of MPTCP in critical conditions like of bottleneck.

Keywords Multihoming · SCTP · HIP · Shim · mSCTP · MPTCP · Wi-Fi · 3G

1 Introduction

Multihoming is the ability to connect to the Internet through more than one Internet Service Provider. If one of the attachments fails due to mobility of the device or due to any other reason, the connection will be transferred to another one. Thus,

A. Jagetiya (✉)
Department of IT, MLV Textile & Engineering College, Bhilwara, India
e-mail: Anurag.mlvtec@gmail.com

C. Rama Krishna
Department of Computer Science Engineering, NITTTR, Chandigarh, India

Y. Haider
Department of Computer Science and Engineering, Northern India Institute of Technology, Lucknow, India

C. R. Krishna et al. (eds.), *Proceedings of 2nd International Conference on Communication, Computing and Networking*, Lecture Notes in Networks and Systems 46, https://doi.org/10.1007/978-981-13-1217-5_2

multihoming increases reliability in case of a network failure and enhances performance through concurrent internet connections. In conventional TCP, once the connection is established between two endpoints, during the lifetime of this connection, these endpoints can't be changed. It means that if a mobile host attaches itself to a new network, its IP address changes, but, unfortunately, endpoints cannot be changed. So, in order to establish a new connection, the old connection has to be terminated. Therefore, conventional TCP can't be used to avail the complete benefits of multihomed devices. In fact, application of multihoming can be at today's data centers, where many paths are available between two endpoints, and MultiPath routing picks one route for a particular TCP connection. It has been found that there is a non-negligible probability of multiple flows get placed on the same link and degrading the throughput [1].

2 Multihoming Protocols

Researchers have come up with many multihoming solutions on network layer and suggested protocols like Mobile IP, Level 3 Multihoming Shim Protocol for IPv6 (Shim6), etc. Later, Concept of mobility was added to transport layer that allows the modification of endpoints during the lifetime of a connection, i.e., mobility-enabled transport layer protocol to change its IP address while its end-to-end connection will remain intact. This feature has enabled hosts to manage the mobility instead of depending upon any network layer device such as a router. Therefore, this paper will cover well-known transport layer multihoming solutions submitted to IETF, like Host Identity Protocol (HIP), Stream Control Transmission Protocol (SCTP), and MultiPath TCP (MP-TCP).

2.1 Host Identity Protocol (HIP)

Internet Protocol performs dual uses as identifiers to upper layers and locator to the network layer. But, the feature needs architecture capable of changing locators without changing identifiers. Therefore, ITU-T discussed the concept of splitting ID/locator and standardized a recommendation [2]. Host Identity Protocol (HIP) is defined as a new solution for secure mobility and multihoming solutions. HIP does not use IP address' dual role feature architecture, i.e., to identify and locate hosts [3]. It separates these roles and uses a host identifier to identify a host. While IP address still preserves the role to find out the topological location of host and route the packets. Due to the addition of this new namespace, applications refer to Host Identifiers while opening connections and sending packets instead of referring to IP addresses. Transport layer generates segment that uses host identifier instead of destination IP. HIP daemon maps an IP address to host identifier at Host Identity layer. Lastly, network layer routes the packet and the socket are changed as follows:

{protocol, source HI, source port, destination HI, destination port}. In the present scenario, socket is bound to IP address, while, in HIP, a socket is bound to Host Identifier, which, in turn, dynamically connects to one or more IP address. The role of IP address is limited to find the location of the host only. Thus, an application cannot see IP address, it can only see the Host Identifier. HIP offers several benefits including mobility and multihoming support, end-to-end security, protection against processor, and memory exhausting denial-of-service attacks.

2.2 Stream Control Transmission Protocol (SCTP)

SCTP is a reliable, full duplex, message-oriented, protocol. It not only provides most of the TCP features like flow and congestion control, ordered data delivery but also provides additional services like multihoming to thwart network layer failure, partial reliability to support real-time application, multi-streaming, etc [4]. SCTP ensures strong association between two endpoints of a connection and each of that may be reached by one or more transport layer address. During initial connection setup of SCTP multihomed environment, both the end-point exchanges a list of available transport layer address (IP addresses). Then, both the endpoints define a primary path to exchange data. Besides, this primary path, SCTP endpoints also keep track of alternative paths [5]. SCTP sender usually uses the same path or destination address until being instructed by the application layer to change the path. If at any point in time, a primary destination is found inactive by SCTP's periodic heartbeat messages, SCTP may switch to the alternate destination address and retransmit the message [6]. SCTP works at the transport layer thus provides a host-centric approach. Therefore, in order to use multihoming features, both the communicating parties need to be agreed upon this service. It is found in [7] that SCTP does not support other multihoming requirements viz. load sharing, load balancing, transport layer handover, etc. Middle-boxes in modern networks doesn't recognize SCTP packets and drop them. On the other hand, applications have to explicitly choose SCTP to achieve application-level compatibility [8]. However, several APIs are defined for application to create an SCTP socket. SCTP can only provide the ability to recognize multiple paths so that connection can be shifted from one to another during a fault. But, it does not provide facility to use multiple paths simultaneously to transfer data [9]. SCTP is not interoperable with TCP. Due to this reason, developers are also less likely to develop their application that works with SCTP.

2.3 Shim Layer

A backward compatible Shim layer translates application layer system calls to TCP into corresponding calls to SCTP. This procedure remains transparent to the application layer and the Shim layer is inserted into socket layer between application and transport layers. Shim layer encourages SCTP deployment by providing a mean to translate TCP applications into SCTP applications. However, Shim is only an experimental protocol layer and not standardized yet [10].

2.4 mSCTP: SCTP with Mobility Extension

Dynamic Address Reconfiguration (DAR) extension is added to SCTP that makes it mobility friendly transport layer protocol and provides transport layer handover management in SCTP. With DAR extension SCTP endpoints can dynamically add or remove an IP address to an ongoing association and request to set the primary destination during active SCTP association [11]. This extension was required because numerous latest systems allow for dynamic addition and removal of network cards. In fact, in IPv6, a provider can renumber a network and transport association has to reinitiate to take advantage of this new configuration. This extended version of SCTP with DAR is referred as mSCTP by IETF and standardized in RFC 5061 [12]. Standard SCTP was providing fault tolerance mechanism to stationary hosts, while DAR extension made it mSCTP, i.e., mobility enable transport layer protocol. Moreover, SCTP host can choose or change its primary path of data transfer by signal its peer or server. mSCTP uses the information like signal quality from the link and physical layer to perform a better selection of new paths and make effective handover decisions [13]. The major challenge with mSCTP is to decide specific rules for changing the primary path. mSCTP does not have built-in support for location management like in Mobile IP. mSCTP does not need an additional support of network routers like Agents in Mobile IP. mSCTP does not use additional 'Binding Update' like in Mobile IP to provides the Route Optimization.

2.5 MultiPath TCP

MultiPath TCP is a backward compatible extension to TCP and is designed to use multiple interfaces concurrently so that all the pooled resources will be seen as a single logical resource to the end user. MPTCP aims to make use of available multiple paths in the network to provide better fault tolerance by switching between paths. And, increase the network throughput by using multiple paths concurrently [9]. A stable release of MultiPath TCP v0.89 in Linux kernel is available at [14]. MPTCP provides reliable, in order, byte streaming services as TCP [15]. Major design goals of MPTCP are categorized into two categories which are as follows:

(a) Functional goals: Using multiple available paths, MPTCP's performance and fault tolerance capabilities should be equal or greater than the performance of TCP over the same path.

(b) Compatible goals: MPTCP should be transparent to the application layer. and intermediate devices and middle-boxes.

In order to deal with multiple paths, several new modules are added in MPTCP like Path Management to discover Sub-flows available between MPTCP hosts and join them to existing connection, Sub-flow Management treats each sub-flow as independent TCP connection. Therefore, separate sequence number space is used per sub-flow level and a data sequence number space is used per connection level to reorder packets received from various sub-flows, Packet Scheduler assigns packets to a particular sub-flow.

3 Comparison of Multihoming Protocols

Table 1 compares SCTP, SCTP with Shim layer, mSCTP, and MPTCP on the basis of various parameters/features. It is clearly stated that SCTP is primarily designed for only fault tolerance, and, is not compatible with the application layer and middle-boxes. Therefore, Shim layer is embedded between application layer and SCTP to act as a translator. mSCTP is introduced with dynamic address reconfiguration feature to support addition and removal of any address on the fly. mSCTP has many features similar to MPTCP, but, compatibility at the application layer is still not present in it. Therefore, MPTCP is the only protocol supporting most of multihoming requirement while being compatible with existing layers and fair to regular TCP at the shared bottleneck.

4 Performance Analysis of MPTCP

It is observed that using TCP's conventional congestion control algorithms behave more aggressive to sub-flows sharing bottleneck links in MultiPath environment. Therefore, there was another challenge to design a congestion control algorithm suitable for the MultiPath environment. That algorithm should perform better or equivalent to TCP while being fair to sub-flows sharing bottleneck links [16]. Moreover, it should be effective enough to move traffic from a congested link to another. Several algorithms namely: EWTCP, Fully Coupled, Semi-Coupled, Linked Increase Algorithm (LIA), Opportunistic Linked Increase Algorithm (OLIA), BALANCED LINKED ADAPTATION ALGORITHM (BALIA), wVegas are proposed by the MPTCP working groups and other researchers to chord a balance between TCP friendliness at shared bottleneck links and effectiveness to divert traffic from a congested link. However, most of the algorithms provide better performance when sub-flows are of similar nature, e.g., both are either Wi-Fi or 3G. An experimental analysis where a wireless client uses both 3G and Wi-Fi shows different results as expected from the design goals of MPTCP. It shows that a client using MPTCP over its two

Table 1 Comparison of multihoming protocols

Multihoming protocols	SCTP	SCTP with shim layer	mSCTP	MPTCP
Application compatibility	No	Yes	No	Yes
Network compatibility	No	Yes	No	Yes
Fault tolerance capability	Yes	Yes	Yes	Yes
Transport layer handover	No	No	Yes	Yes
Simultaneous use of multiple paths	No	No	Yes	Yes
Load balancing capability	No	No	Yes	Yes
Layer(s) responsible for change of sub-flow/interface	Application	Application	Application data link, physical	Data link and physical
Backward compatibility to TCP	No	Yes	No	Yes
Stable release/ standardization	BSD OS	Undergoing	RFC 5061	RFC 6824

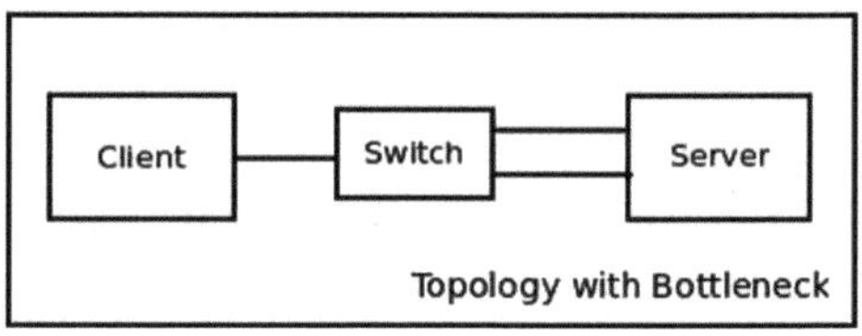

Fig. 1 Topology with bottleneck

active interfaces viz. 3G and Wi-Fi achieve lesser throughput than the same client that day achieved over Wi-Fi using conventional TCP. It is worth mentioning that the path properties of both Wi-Fi and 3G are in contrast to each other [17]. Table 2 shows a comparison of path properties of Wi-Fi and 3G Interfaces. This behavior still requires more diagnosis under different-different scenarios. To evaluate the performance of MPTCP, two topologies namely topology with bottleneck link and no-bottleneck link as shown in Figs. 1 and 2 are created using MPTCP module of NS-3 where axes indicate sub-flows [18]. Simulations composed of MPTCP enabled client and server equipped with two network interfaces are set up. Topology in Fig. 1 has bottleneck link at the switch while topology in Fig. 2 is not having any bottleneck link. Path properties of Wi-Fi and 3G which are in contrast to each other are shown in Table 3 and will be used to evaluate the performance of MPTCP in two scenarios namely homogeneous and heterogeneous interfaces.

Table 2 Comparison of the path properties of Wi-Fi and 3G interfaces

Path property	Wi-Fi	3G
Data rate	High	Low compare to Wi-Fi
Round trip time	Short	Long
Receive buffer size	Small buffer	Large buffer
Packet drop/loss rate	High	Low
Power consumption	Low	High

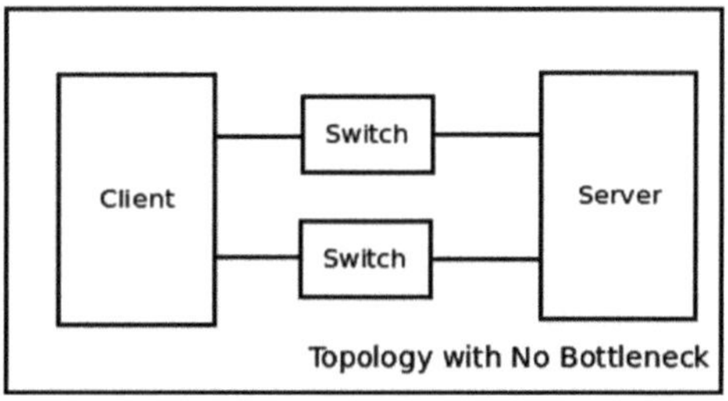

Fig. 2 Topology with no-bottleneck

Table 3 Simulation parameters

Network type	Data rate (Mbps)	Latency (ms)
Wi-Fi	10	10
3G	2	150

Throughput of these scenarios is calculated using NS-3 and shown in Figs. 3 and 4. Simulation results in Table 4 indicate that in bottleneck topology, throughput is 9.4 and 0.57 Mbps when both sub-flows are Wi-Fi and 3G respectively. When one sub-flow is 3G and another is Wi-Fi, throughput is 6 Mbps. Similarly, results are drawn for no-bottleneck scenario. In bottleneck scenario, throughput dropped to around 60% while using heterogeneous interfaces in comparison of using homogeneous interfaces. While in no-bottleneck topology throughput remains almost same in both the homogeneous and heterogeneous interfaces. Nevertheless, the number of packet drops increased substantially in heterogeneous interfaces due to the wide difference in channel properties of Wi-Fi and 3G. MPTCP is designed to balance the load among available sub-flows, but, as shown in Table 5 Wi-Fi interface was much more heavily loaded (88% of network traffic) than 3G (only 12% of network traffic) in the case of using heterogeneous interfaces with bottleneck link.

On the other side, load sharing between both the interfaces is almost equal in topology without a bottleneck. In bottleneck topology, numbers of packets drop are also much higher than no-bottleneck. Therefore, the overall, observation favors MPTCP using homogeneous interfaces in both bottleneck and no-bottleneck topologies and does not recommend the use of MPTCP in the case of bottleneck kind of topology with heterogeneous interfaces.

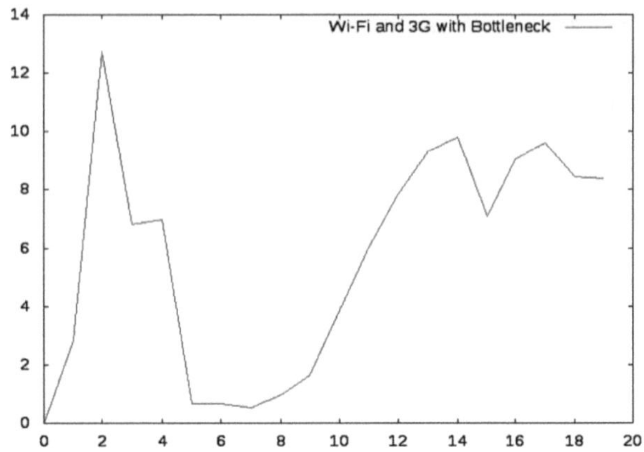

Fig. 3 MPTCP throughput in heterogeneous interfaces (bottleneck topology)

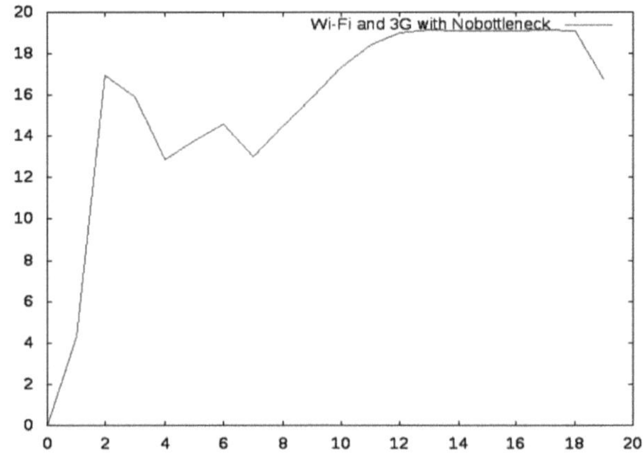

Fig. 4 MPTCP throughput in heterogeneous interfaces (no-bottleneck topology)

Table 4 Simulation results

Topology Types	(Sub-flow-I, Sub-flow-II)	Throughput (Mbps)
Bottleneck	(Wi-Fi, Wi-Fi)	9.4
	(Wi-Fi, 3G)	6
	(3G, 3G)	0.57
No bottleneck	(Wi-Fi, Wi-Fi)	16.35
	(Wi-Fi, 3G)	16.2
	(3G, 3G)	1.01

5 Conclusion

Multihoming is inevitable in order to exploit the significant potential of those interfaces. MPTCP is designed to utilize the benefits of multiple interfaces in multihomed mobile devices. But, the channel properties of available networks like Wi-Fi, 3G, etc.

Table 5 More simulation results

Network type		No. of packets drops	No. of packets transferred	
			Sub-flow Wi-Fi	Sub-flow 3G
Wi-Fi and 3G	Bottleneck	137	20,346 (88.34%)	2683 (11.64%)
	No-bottleneck	85	23,793 (54.75%)	19,658 (45.24%)

are quite contrary to each other. Therefore, MPTCP gives better throughput in networks with homogeneous sub-flows and less Bandwidth Delay Product. It is also found that in bottleneck topology, number of packets drop is much higher than no-bottleneck topology.

6 Future Scope

Simulation in this work considered data rate and delay in account, while many other parameters like buffer size, bit error rate, etc., can also be incorporated to get more precise results. This work judges MPTCP on the basis of throughput only, while, many other factors like fairness to regular TCP, power consumption in battery starved devices, extra processing overhead to rearrange out of order packets received from heterogeneous interfaces, etc., can also be evaluated before finally commenting upon the use of MPTCP. Only two network interfaces are used in this work to comment on the performance of MPTCP. The impact of more than two heterogeneous interfaces including wired and wireless can also be studied in more complex topologies [14].

References

1. B. Olivier, H. Mark, C. Raiciu, An overview of multipath TCP. IEEE **37**(5), 17–23 (2012)
2. V.P. Kafle, H. Otsuki, M. Inoue, An ID/locator split architecture for future networks. IEEE Commun. Mag. **48**(2), 138–144 (2010)
3. P. Nikander, A. Gurtov, T.R. Henderson, Host Identity Protocol (HIP): Connectivity, mobility, multi-homing, security, and privacy over IPv4 and IPv6 networks. IEEE Commun. Surv. Tutorials **12**(2), 186–204 (2010). https://doi.org/10.1109/surv.2010.021110.00070
4. T.D. Wallace, A. Sham, A review of multihoming issues using the stream control transmission protocol. IEEE Commun. Surv. Tutorials **14**(2), 565–578 (2012)
5. S. Sugimoto, B.E. Carpenter, *A Comparative Analysis of Multihoming Solution* (Ericsson Research, Nippon Ericsson K.K., Japan, 2006)
6. R. Stewart, Stream Control Transmission Protocol. RFC 4960 (2007)
7. G. Ye, T.N. Saadawi, M. Lee, in *IPCC-SCTP: An Enhancement to the Standard SCTP to Support Multi-Homing Efficiently Performance*. IEEE International Conference on Computing, and Communications (2004), pp. 523–530. https://doi.org/10.1109/pccc.2004.1395081
8. A. Ford, C. Raiciu, M. Handley, S. Barre, J. Iyengar, Architectural Guidelines for MultiPath TCP Development. RFC 6182 (2011)

9. K. Alexandros, A. Kostopoulos, H. Warma, T. Leva, B. Heinrich, A. Ford, L. Eggert, in *Towards MultiPath TCP Adoption: Challenges and opportunities, Next Generation Internet (NGI)*. 6th EURO-NF Conference (2010), pp. 1–8. https://doi.org/10.1109/ngi.2010.5534465
10. W. Ryan, *Transparent TCP-to-SCTP translation Shim layer* (Unpublished master's thesis). Delaware University, Newark Department of Computer and Information Sciences (2005)
11. L. Budzisz, R. Ferrus, F. Casadevall, On concurrent multipath transfer in SCTP-based handover scenarios
12. R. Stewart, Q. Xie, S. Maruyama, Stream Control Transmission Protocol (SCTP) dynamic address reconfiguration. RFC 5061 (2007)
13. S.J. Koh, Q. Xie, S.J. Park, Mobile SCTP (mSCTP) for IP handover support, internet-draft (2005)
14. MultiPath TCP: Linux Kernel Implementation. IP networking lab. http://www.MultiPath-tcp.org/
15. M. Scharf, A. Ford, MPTCP application interface consideration. RFC 6897 (2013)
16. C. Raiciu, M. Handly, D. Wischik, Coupled congestion control for multipath transport protocols. IETF, RFC 6356 (2011)
17. D. Wischik, C. Raiciu, A. Greenhalgh, M. Handley, in *Design, Implementation, and Evaluation of Congestion Control for MultiPath TCP*. Proceedings of the 8th USENIX Conference on Networked Systems Design and Implementation (USENIX Association, Boston, 2011), p. 8
18. M. Kheirkhah, I. Wakeman, G. Parisis, MultiPath-TCP in ns-3. Workshop on ns-3 (WNS3), Georgia Institute of Technology, Atlanta, GA (2014)

Authenticating Mobile Phone Users Based on Their Typing Position Using Keystroke Dynamics

Baljit Singh Saini, Navdeep Kaur and Kamaljit Singh Bhatia

Abstract Keystroke dynamics is an emerging biometric method for user authentication. With the increased use of mobile phones, the use of keystroke dynamics for mobiles has been evaluated by many researchers. Mostly, the research has been carried upon data which was collected from users when they were in sitting position only. This study was conducted by collecting user data in two positions, sitting and walking, and using the phone in landscape and portrait mode. The results were positive with best EER of 3.69% achieved in walking-landscape position by using standard keystroke features and random forest algorithm. The results were also better as compared to making a single user profile by combining the data from all different positions where the achieved EER was 4.12%.

Keywords Keystroke dynamics · Authentication · Mobile phones

1 Introduction

Authentication is a process to verify whether claimed physical identities of people and computers digital identity are valid [7]. Different authentication techniques provide varied levels of security but none of these technologies ensure complete security to a system. Users can be authenticated by one of the following authentication policies [7]:

B. S. Saini (✉) · N. Kaur
Sri Guru Granth Sahib World University, Fatehgarh Sahib, India
e-mail: baljitsaini28@gmail.com

N. Kaur
e-mail: drnavdeep.sggswu@gmail.com

B. S. Saini
CSE, Lovely Professional University, Phagwara, India

K. S. Bhatia
ECE, IKGPTU, Batala, India
e-mail: kamalbhatia.er@gmail.com

C. R. Krishna et al. (eds.), *Proceedings of 2nd International Conference on Communication, Computing and Networking*, Lecture Notes in Networks and Systems 46, https://doi.org/10.1007/978-981-13-1217-5_3

- Knowledge based: something known to them like password or Personal Identification Number (PIN). The problem faced by this method is that hacker can crack any password and would appear to be authorized user.
- Token: something that is under the ownership of user, for example, smart card. Such authentication methods are not so convenient because smart cards are susceptible to theft.
- Biometric: something that a user is (e.g. fingerprints, face recognition, iris scan, etc.). Here, users provide their physical attribute for authenticating themselves.

Biometrics is a strong alternative as it cannot be borrowed, stolen, or forgotten. Thus, this method is becoming globally acceptable. Biometric characteristics can be either behavioural or physiological. Physiological characteristics are the physical parameters of a certain body part, e.g. iris scanning, fingerprints, face recognition, retina scanning, etc. Behavioural characteristics are associated with the mannerisms of a person, such as signature recognition, keystroke dynamics, etc.

Keystroke dynamics being behavioural characteristic endeavours to achieve user authentication by monitoring and analysing their typing pattern through the keyboard. It takes into consideration keystroke duration, keystroke latency, force of keystrokes and other such attributes while users are typing.

Keystroke dynamics can be classified into two types: structured or static text and free or dynamic text [11]. Static analysis analyses an individual keystroke behaviour at fixed checkpoints in the system on predetermined phrases, e.g. at login time. Dynamic analysis involves periodic or constant monitoring of keystroke behaviour. For example, if a person is surfing the Internet, certain websites maybe visited frequently by the user. A list of the frequently occurring websites and the users typing behaviour, while entering the string can be stored. However, due to its intrusive nature, dynamic monitoring may lead to privacy issues.

Keystroke dynamics involve two distinctive processes: feature extraction and classification of extracted features [11]. In feature extraction, certain attributes of the user are acquired for authentication purposes. These features are extracted to represent user behaviour in keystroke dynamics. During classification, the extracted features are categorized using various algorithms such as neural networks, machine learning algorithms, etc. to determine whether the extracted features match the reference template of user or not. Based on this, user is either granted or denied access to the system.

Latency is the most frequently used feature by researchers. Latencies are of three types: release to press (RP), press to press (PP) and release to release (RR) latency [3]. Di-graph is also considered as press-to-press latency by various researchers. Some researchers call release to press time as flight time. Since the system can log the time of each key press, it is easy to extract such features from the raw information. The time between the press and release of alternate keys is called as tri-graph [5]. N-graph [18] features have also been used by some researchers to determine authentic users.

The error rates used to determine the performance of biometric authentication systems are False Acceptance Rate (FAR), False Rejection Rate (FRR) and Equal

Error Rate (EER). FAR is defined as the number of imposters those were accepted as genuine users. FRR is determined by calculating the number of genuine users who were rejected for being imposters. EER is the error value when FAR and FRR rates are same. The lower the EER, the better is the performance. FAR is preferred in the systems where the security is not paramount, while FRR is preferred in the systems where security is major concern [11].

Majority of the studies so far except for [6, 12, 16] have taken into consideration only a single position for entering user data, i.e. sitting. But a mobile phone user can be in a sitting position or in a walking position when he/she uses the mobile. In this study, we capture user data in two positions, i.e. walking and sitting, and using two orientations of mobile phone, i.e. landscape and portrait. The aim is to study whether keystroke dynamics can determine the typing pattern of the user in all situations or not. In Sect. 2, we discuss some of the previous studies, Sect. 3 discusses the methodology used in our study, Sect. 4 presents the results and Sect. 5 concludes the study by giving some prospects of the study.

2 Literature Review

In [16], the authors discussed HMOG features for user authentication in touchscreen-based systems. HMOG stands for Hand Movement, Orientation, and Grasp. HMOG features capture orientation dynamics and subtle micro-movements resulting from how a user holds, grasps and taps on the smartphone. EER of as low as 7.16% (walking) and 10.05% (sitting) was achieved when combined with tap, HMOG and keystroke features. It was observed that HMOG features reflect a better performance while walking because of the ability of HMOG features to grasp distinctive and subtle body movements caused by walking besides the hand movement dynamics caused by taps. In [1], authors presented an approach for analysis of free-text keystroke dynamics. This approach integrated analysis of di-graphs and monographs as well as it speculated the missing di-graphs depending upon monitored keystroke relations between them using neural network. The study was conducted in homogeneous and heterogeneous environments. In heterogeneous environment, 0.0152% of FAR and 4.82% FRR was achieved. Comparatively, homogeneous environment yielded FAR and FRR of 0% and 5.01%, respectively, was achieved. Mendizabal et al. [9] in their paper focused on relative study of various supervised classification methods in order to use them for biometric system construction using the already present information in the mobile phones. Some distinguished features like typing speed, finger size, pressure, time, angular and linear acceleration are extracted and processed, while the users type the 4-digit PIN. Excellent classification rates were achieved on combination of Principal Component Analysis (PCA) and Multilayer Perceptron (MLP) classifier. 80% users were correctly identified while using three samples per user, and this rate increased to 90% with the use of nine samples per user. Despite the less number of samples, the study resulted in satisfactory performance with an ERR of 20%. A new scheme to capture user typing behaviour while entering a passcode was proposed in [15]. Features like pressure applied on the screen by a user and the duration of screen press were also captured for authentication purposes. It was observed that using MLP classifier, highest accuracy results were achieved with FAR (14.06%) and FRR (14.1%). Authors in [4] authenticated users on the basis of their behaviour of typing the text messages. Features acquired from the user input were hold time and keystroke latency. Accuracy of user authentication was measured upon numeric data entry such as PIN and telephone numbers. Many neural network and pattern recognition algorithms were contrasted, and feedforward MLP proved to be most efficient resulting in EER of 10.4% and 11.3% with PIN code and telephone number, respectively, as inputs. Touch dynamic features like single and multi-touch and touch movements were used to verify mobile users identity in [10]. With an average of almost six sessions per user, data was acquired from 20 Android users and the highest performance with average error rate of 7.71% was achieved using Radial Basis Function Network (RBFN).

3 Methodology

A new dataset of 40 users was created. Data collection was carried out in five sessions. Each user was directed to type a strong password [8] tie5Raonl 80 times per session, 20 times in each posture: walking and sitting and in two different screen orientations: portrait and landscape. Thus, every user typed the input 400 times. This methodology does not deal with typing errors. If a user makes some typing error, then sample is not considered. Another reason for choosing a random input like "tie5Roanl" is that authors in [14] found out that random input yields more unique profile of a user. Data collection mechanism was carried out in uncontrolled environment. Different smartphones with Android operating system were used for the data collection to gain variability. An application was installed on the mobile phone of each user. Thus, the mobile phone itself acted as the data collection apparatus. Whenever a subject press or release a key, the software application recorded the events such as key-down, key-up time, etc. Based on these timings, the features that were acquired were dwell time, flight time, press–press and release–release time. Extracted features were then used for classification using random forest algorithm. For each user and situation, profile was created by using training set that consisted of positive samples or the samples that are of a genuine user and imposter or negative samples from other users. Figure 1 shows the flowchart for acquiring user data for building individual user profile and authenticating the user at a later stage.

4 Results

Table 1 shows the performance of random forest when the data for both the positions and orientations was considered as a single entity, i.e. one single profile was developed for each user irrespective of the position or orientation in which the user typed. The

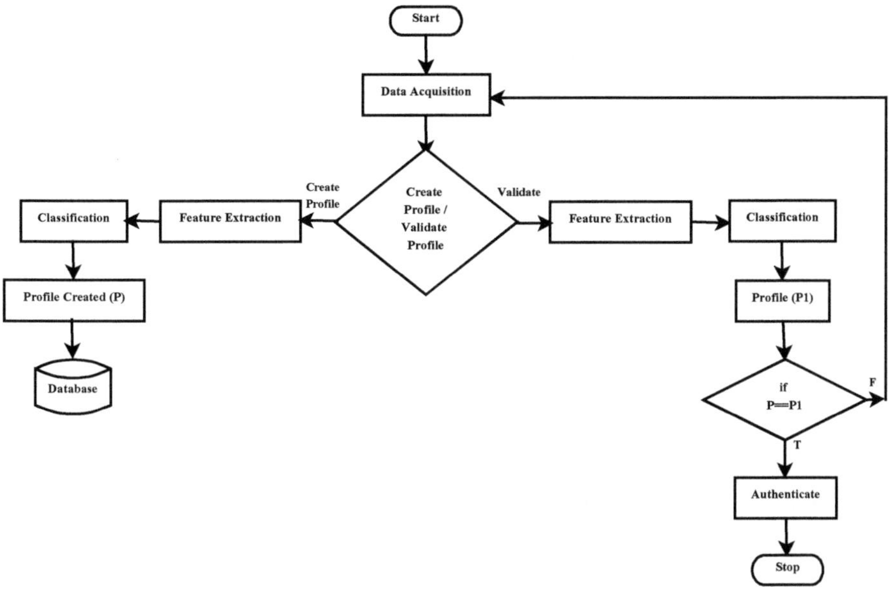

Fig. 1 User enrolment and authentication procedure

Table 1 Error rates while considering single model for each user using random forest algorithm

Posture	Screen orientation	FAR (%)	FRR (%)	EER (%)
Combined (Sitting and walking)	Combined (landscape and portrait)	8.23	3.85	4.12

average FAR achieved was 8.23%, FRR achieved was 3.85% and an EER of 4.12% was achieved.

The experiment was further expanded by taking into consideration the positions separately. Table 2 shows the results for walking and sitting positions. Results indicate that there is not much difference in the typing pattern of a user when he types in

Table 2 Error rates using random forest on walking and sitting samples without removing outliers

Posture	FAR (%)	FRR (%)	EER (%)
Walking-landscape	3.49	4.71	4.68
Walking-portrait	3.23	4.74	4.48
Sitting-landscape	2.43	4.52	4.45
Sitting-portrait	3.39	5.51	5.43

Table 3 Error rates using random forest on walking and sitting samples with outliers removed

Posture	FAR (%)	FRR (%)	EER (%)
Walking-landscape	3.63	4.67	3.69
Walking-portrait	3.23	4.74	4.48
Sitting-landscape	3.45	4.52	3.99
Sitting-portrait	4.01	5.30	5.43

walking-portrait mode or in walking-landscape mode. The EER achieved in both the cases (4.68% in walking-landscape and 4.48% in walking-portrait) was quite similar. Average FAR of 3.49% and 3.23% and average FRR of 4.71% and 4.74% in both the cases were also comparable. In case of sitting position, the results were better as compared to the walking position. An average EER of 4.45% was achieved in case of sitting-landscape mode, while an average EER of 5.43% was achieved in case of sitting-portrait mode. The possible reason for the difference in the EER rates is that in landscape mode the keyboard is bigger in size with wider and well-separated keys, thus helping the user to type more precisely. Average FAR and FRR rates are also lower in case of landscape mode as compared to portrait mode.

Also, the EER rate in sitting position is better as compared to walking position. This shows that when a user is in a still position and is doing just one thing, the typing pattern is consistent which leads to improved results because when the user is walking the typing rhythm might get affected by the uneven surface on when he/she is walking, and the user might also be concerned of being getting hit into an obstruction on the way.

The data was further analysed by removing the outliers this time. This was important to do since the experiment was conducted in an uncontrolled manner and users occasionally tend to get involved in a secondary activity like talking or looking around while typing the input text. Table 3 shows the results improved by removing the outliers. There was significant improvement in EER rate of walking-landscape position. EER of 3.69% was achieved which was the best out of all the four positions. This shows that the walking-landscape is the best position to use for user authentication provided that the user is able to concentrate solely on his typing and is not distracted by external factors. Overall, the best results were achieved when users typed in sitting positions and the phone was held in landscape mode.

We compared our results with some of the existing work done as shown in Table 4. It was observed that our results were better as compared to the existing studies.

Table 4 Comparison with previous studies

Paper	Approach	Features	Posture	Results(%)
Sitova et al. [16]	Statistical	HMOG, tap, hold time and swipe features	Sitting walking	EER: 10.05 EER: 7.16
Roh et al. [12]	Distance algorithms	Tap, motion and keystroke features	Hand table walking	EER: 7.35 EER: 10.81 EER: 6.93
Crawford et al. [6]	C 4.5-Decision tree, Logistic regression	Motion and keystroke based	Sit stand walk	FAR:1.7, FRR:6.1 FAR:1.8, FRR:5.3 FAR:1.4, FRR:5.6
Antal et al. [2]	Machine learning	Flight time, dwell time, pressure and finger area	–	EER:12.9
Rybnik et al. [13]	Statistical	Dwell time	–	EER:6.1
Systems et al. [17]	Statistical	Dwell time, flight time and pressure	–	FAR:15, FRR:0
–	Random forest	Dwell time, Flight time, press–press and release–release time	Sitting (portrait) Sitting (landscape) Walking (portrait) Walking (landscape)	FAR:4.01, FRR:5.3, EER:5.24, FAR:3.45, FRR:4.5, EER:3.99, FAR:3.32, FRR:4.65, EER:4.48, FAR:3.63, FRR:4.67, EER:3.69

5 Conclusion

This study presents an overview of research on keystroke dynamics over past few decades with emphasis on how keystroke dynamics can be used for authenticating mobile phone users. Till now, only a few studies have focused on authenticating the mobile phone users under two different conditions: walking and sitting but there was no provision of authenticating a user under different positions using a single dataset. Separate datasets have been created for a single user for different positions, thereby increasing overhead. To overcome this, we proposed a method to authenticate mobile phone users under different positions by creating a single profile of a user. This avoids the overhead included in using different datasets for the same user in walking and sitting postures. We also strived to achieve the insight of reliable posture for authentication, and it was found that sitting-landscape is the better position for authenticating a user. As a future scope, the study can be carried on increased number of users and the users belonging to different age groups. Also, motion-based features using gyroscope and accelerometer can be used to check their impact on the typing pattern of the user.

Acknowledgements The authors would like to thank all the students of Lovely Professional University who participated in the study by providing the data for analysis.

References

1. A.A. Ahmed, I. Traore, Biometric recognition based on free-text keystroke dynamics. IEEE Trans. Cybern. **44**, 458–472 (2014)
2. M. Antal, L. Szabó, I. László, Keystroke dynamics on android platform, in *8th International Conference Interdisciplinarity in Engineering, INTEr-ENG* (Tirgu-Mures, Romania, 2015), pp. 820–826
3. S.P. Banerjee, D. Woodard, Biometric authentication and identification using keystroke dynamics: a survey. J. Pattern Recognit. Res. **7**, 116–139 (2012)
4. N. Clarke, S. Furnell, B. Lines, P. Reynolds, application of keystroke analysis to mobile text messaging, in *3rd Security Conference* (Las Vegas, 2004)
5. K.R. Corpus, R. Joseph, D.L. Gonzales, L.A. Vea, A.S. Morada, Mobile user identification through authentication using keystroke dynamics and accelerometer biometrics, in *Proceedings of the International Workshop on Mobile Software Engineering and Systems* (ACM, 2016), pp. 11–12
6. H. Crawford, E. Ahmadzadeh, S. Clara, H. Crawford, R. Ohanian, Authentication on the go: assessing the effect of movement on mobile device keystroke dynamics, in *Thirteenth Symposium on Usable Privacy and Security* (2017), pp. 163–173
7. M. Karnan, M. Akila, N. Krishnaraj, Biometric personal authentication using keystroke dynamics: a review. Appl. Soft Comput. **11**, 1565–1573 (2011)
8. K.S. Killourhy, A scientific understanding of keystroke dynamics (2012)
9. I. Mendizabal-Vázquez, D. de Santos-Sierra, J. de Guerra-Casanova, S. Carmen, Supervised classification methods applied to keystroke dynamics through mobile devices, in *International Carnahan Conference on Security Technology (ICCST)* (Rome, 2014), pp. 1–6
10. Y. Meng, D.S. Wong, R. Schlegel, L.F. Kwok, Touch gestures based biometric authentication scheme for touchscreen mobile phones, in *Information Security and Cryptology Inscrypt 2012*. Lecture Notes in Computer Science vol. 7763 (2013), pp. 331–350
11. P.H. Pisani, A.C. Lorena, A systematic review on keystroke dynamics. J. Brazilian Comput. Soc. **19**, 573–587 (2013)
12. J. Roh, S. Lee, S. Kim, Keystroke dynamics for authentication in smartphone, in *Information and Communication Technology Convergence* (IEEE, 2016), pp. 1155–1159
13. M. Rybnik, M. Tabedzki, M. Adamski, K. Saeed, An exploration of keystroke dynamics authentication using non-fixed text of various length, in *International Conference on Biometrics and Kansei Engineering* (IEEE, 2013), pp. 245–250
14. B.S. Saini, N. Kaur, K.S. Bhatia, Keystroke dynamics based user authentication using numeric keypad, in *7th International Conference on Cloud Computing, Data Science and Engineering* (IEEE, Noida, 2017), pp. 25–29
15. S. Sen, K. Muralidharan, Putting 'pressure' on mobile authentication, in *Seventh International Conference on Mobile Computing and Ubiquitous Networking* (IEEE, 2014), pp. 56–61
16. Z. Sitova, J. Sedenka, Q. Yang, G. Peng, G. Zhou, HMOG: a new biometric modality for continuous authentication of smartphone users. IEEE Trans. Inf. Forensics Secur. **11**, 877–892 (2016)

17. E. Systems, D. Group, C. Science, Use of a novel keypad biometric for enhanced user identity verification, in *International Instrumentation and Measurement Technology Conference* (IEEE, Vancouver Island, Canada, 2008), pp. 12–16
18. M. Trojahn, F. Arndt, F. Ortmeier, Authentication with keystroke dynamics on touch screen keypads-effect of different N-graph combinations, in *Third International Conference on Mobile Services, Resources and Users Authentication* (2013), pp. 114–119

Differential Privacy Framework: Impact of Quasi-identifiers on Anonymization

Gurjeet Kaur and Sunil Agrawal

Abstract Due to the high volume of data available with social networking sites and companies, the privacy of the individuals is at a continuous risk. With the help of auxiliary data, the users can be tracked back. It becomes even more necessary to analyse the huge piles of data for research intentions. Hence, protection of privacy is a big concern. To deal with the privacy concerns, numerous privacy paradigms have been proposed to achieve an equilibrium between data utility and privacy. Anonymization of data before making it public for research is very important. Different privacy models include k-anonymity, l-diversity, t-closeness and differential privacy. This paper explores the role of quasi-identifiers and their roles for anonymization using differential privacy model. The research in this field can pave new ways for thinking before selecting quasi-identifiers for anonymization.

Keywords Big data · Differential privacy · Quasi-identifiers

1 Introduction

Due to escalated use of social networking sites, the rate of generation of data has grown at a very high rate and in a large volume. The swiftness of data generation has grown into the speed of petabyte's per second [1]. Amid this pile of data, information is present that directly or indirectly relates to users [2]. The variety of data has extended from the formats of spreadsheets and multimedia. Data can be categorized as semi-structured, structured and unstructured [3, 4]. Due to variability in data, there has been a mounting requisite for a revolution in the old-style systems of database management [5]. With the upturn of the data generation, sundry new issues have surfaced up, and one of them is privacy [6]. With more and more people being social, the privacy of one's life is decreasing day by day. Simple removal of distinctively identifying material from data is not adequate to avert identification. The data many

G. Kaur (✉) · S. Agrawal
Department of ECE, UIET, Panjab University, Chandigarh 160014, India
e-mail: geet.k.1212@gmail.com

C. R. Krishna et al. (eds.), *Proceedings of 2nd International Conference on Communication, Computing and Networking*, Lecture Notes in Networks and Systems 46, https://doi.org/10.1007/978-981-13-1217-5_4

companies have is publicized often for research purposes [7]. After running few algorithms, it was easily ascertained by the users. This delinquency has directed to privacy breaches people are not happy about. No one wants their browsing history to be public, or their diseases to be known by the world [8]. Hence, it is very important that the attacker does not find his way back to the users. One of the practices to avoid the violation of privacy is to anonymize the database. The anonymization manoeuvers alter the data by diverse type of modifications, by changing the structure, by clearing the values, changing the values by taxonomy and mixing the values. A method of anonymization uses set of operations to reach yearning level of concealment [9].

2 Anonymization Models

2.1 K-*Anonymity*

K-anonymity can be accomplished by generalization. An altered data set of this nature is k-anonymized if every record in the table is similar as to minimum of $k - 1$ records [10]. The elementary approach is to make certain that every QID has no less than K records in the table so as to reduce the probability of occurrence of reidentifying. Thus, chances of reidentification of the victim by the invader have the maximum probability of $1/K$ [11]. The privacy can be agreeably protected by having a higher value of K. The higher value of K means a higher loss of information [12]. The k-anonymity is prone to homogeneity attack [13] due to deficiency of variety within sensitive attributes, and the background knowledge attack [14].

2.2 L-*Diversity*

To overpower the shortcomings of the k-anonymity method, L-diversity was proposed which necessitates well representation of sensitive parameters in anonymized data sets. This technique guarantees that in each quasi-identifier group, there are minimum l distinct values for the sensitive attributes. L-diversity is not enough to avert attribute disclosure, skewness attack and similarity attack [15].

2.3 T-*Closeness*

Any equivalence class is said to have t-closeness if the distribution of the attribute in the whole table and the distance between the distribution of a sensitive attribute in this class is less than a threshold t. A table is said to have t-closeness if all equivalence classes have t-closeness [16].

2.4 *Differential Privacy Method*

This algorithm is based on the simple notion that removal of one records from the database or addition of one record into the does not affect the privacy of database. A randomized function A gives ε-differential privacy if for all data sets $M1$ and $M2$ differing on at most one element (the two data sets are called neighbouring data sets), and all $S \in$ Range (A).

$$\Pr[M(D1) \in S] \leq \exp(\varepsilon) \times \Pr[M(D2) \in S] \tag{1}$$

Having the least distinctions between the data sets, the distinction followed after the anonymization process would be less than the given value $(e\varepsilon)$[17]. Global sensitivity is:

$$\Delta g = \max||f(M1) - f(M2)||_1 \tag{2}$$

where function f: $M \rightarrow R_{\text{m}}$, $M1$–$M2$ imply they are neighbouring data sets, and $||\cdot||_1$ is the $L1$ norm. The Laplace criteria are executed to carry out the differential privacy framework. The simplified description of this algorithm is as follows:

Step 1 To accomplish differential privacy, a framework is made where the curator is between predictor and database.
Step 2 When a query is made by the predictor, keeper receives it.
Step 3 The keeper gets into the privacy impact by determining the information sensitivity and forwards the request to the database and receives the clean answer from the database.
Step 4 The keeper then mixes a suitable quantity of noise according to the privacy impact.
Step 5 This makes the result leading to embarrassing output. Hence, individual's privacy is preserved.

3 Proposed Methodology and Evaluation Metrics

To escort the development efforts, metrics are brought in use in anonymizing the records. The average risk is the arithmetic mean of all the disclosure probabilities. The lowest risk is the least risk of the whole lot of those disclosure probabilities. It presents a notion of the minimum risk to any individual whose whereabouts are recorded in the data set. The record affected by the lowest risk is the number of entries affected by the minimum risk whose whereabouts are present in our data set. For the application of the algorithms, a data set is required which is available universally. Data can be loaded into the CSV, Excel (XLS, XLSX) and JBDC format. In this paper, data is converted into CSV format so that it can be loaded into ARX

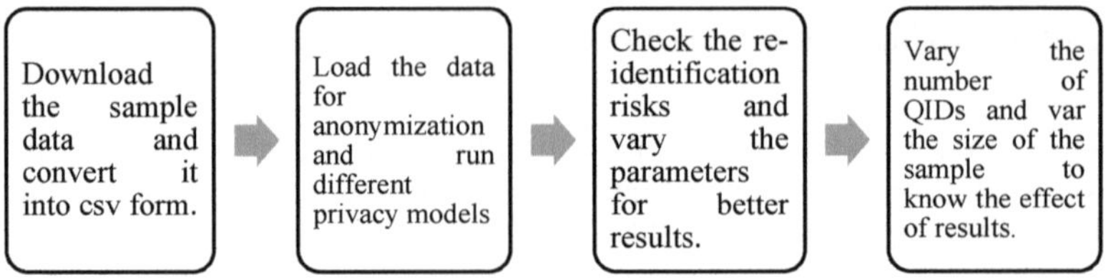

Fig. 1 Flowchart of methodology

anonymization toolbox. Various privacy models including ($\in$, d) differential privacy are present. The flowchart of the methodology is as shown in Fig. 1.

4 Simulation Results and Discussion

The simulations were run on i7 quad-core processor Windows 10 machine on ARX anonymization toolbox. The data set taken was obtained from UCI repository. This is an ADULT census data sets with 48,842 instances and 14 attributes with missing values. Missing valued tuples are removed and there are 45,222 valid tuples in total. This paper emphasizes on the study of the reidentification risks with the variation of the number and selection of quasi-identifiers. Numerous amalgamations of quasi-identifiers were taken to study the consequences (Tables 1, 2 and 3).

In Fig. 2, it is observed that the lowest risk is identified for the combinations 2 and 4. Combination 2 consists of gender, race and salary class. All of these QI's have only two classes further. Combination 4 is race, gender and marital class. There is not much of variation in their class. Since there is lesser variation in these classes, as gender is either male or female. If we do not anonymize it, it is much easier to

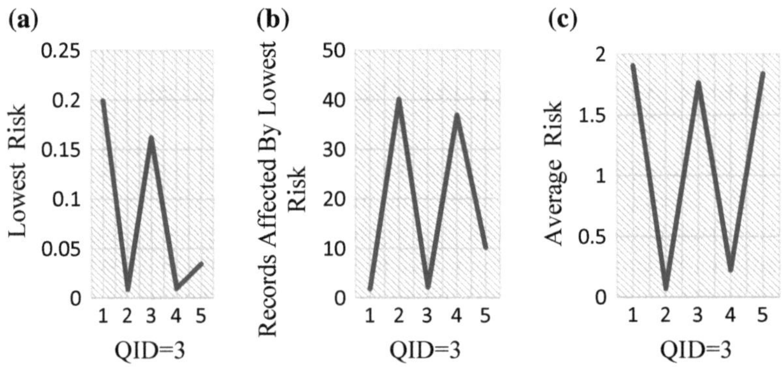

Fig. 2 Variation of **a** lowest risk, **b** records affecting lowest risk, **c** average risk with respect to the combination of three quasi-identifiers for anonymization

Table 1 Combination of three different quasi-identifiers to study the variation in reidentification risk

S. No.	QID 1	QID 2	QID 3
1	Gender (2)	Age (5)	Race (2)
2	Gender (2)	Race (2)	Salary class (2)
3	Age (5)	Race (2)	Salary class (2)
4	Race (2)	Gender (2)	Marital status (3)
5	Marital status (3)	Education (4)	occupation (14)

relate to a person. For example, if auxiliary data provides us that user has prostate cancer, it can be pointed out that user must be a man and it becomes easier to relate to the records having entries as men. Hence, it becomes important to anonymize the data entries with lesser variation as evident from the graph. The highest risk is seen in the combination 1 which includes age. As age has more classes and more variation which leads to higher risks in graph (b), we observe that the series 2 gives the best results. The combination 1 which did not give much of favourable results in the previous graph still follows the same pattern here. The records affected by lowest risk are very less for combination 1. The average risk can be seen in the next graph. For QID = 3, when the AGE is considered as quasi-identifier which has further five groups used in series 1 and 3 has a highest average risk, highest values of lowest risk and lowest percentage of records affected by lowest risk. These combinations led to highest reidentification risk. Since there is a large variation in age group, it is not that easy to point out a record (Fig. 2; Table 1).

The lowest risk depicts the lesser risk of reidentification which is our priority. Observed from the graph in Fig. 3, the combination 5 shows the lowest risk of reidentification. The combination 5 consists of race, sex, marital status and salary class. Three of the identifiers have only two classes under them. The race has black

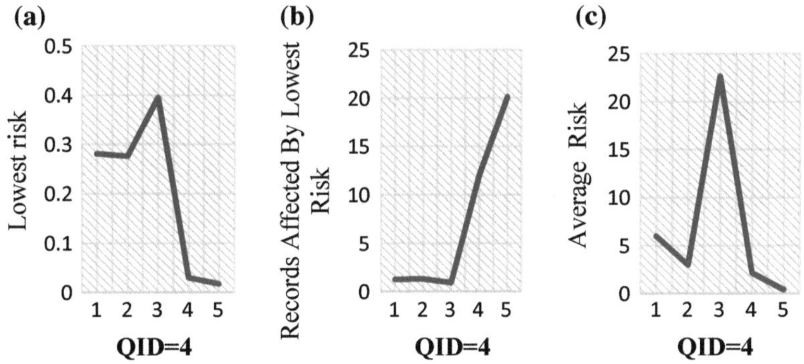

Fig. 3 Variation of **a** lowest risk, **b** records affecting lowest risk, **c** average risk with respect to the combination of four quasi-identifiers for anonymization

Table 2 Combination of four different quasi-identifiers to study the variation in reidentification risk

S. No.	QID 1	QID 2	QID 3	QID 4
1	Age (5)	Sex (2)	Race (2)	Marital status (3)
2	Age (5)	Sex (2)	Race (2)	Salary class (2)
3	Age (5)	Work class (3)	Marital status (3)	Education (4)
4	Race (2)	Sex (2)	Marital status (3)	Education (4)
5	Race (2)	Sex (2)	Marital status (3)	Salary class (2)

and white as its class. Sex has male and female as its classes. Only marital status considers three classes. But compared to other combinations considered, the variation of classes is least in the combination 5. This strongly backs our argument that the lesser variance in classes makes them more important for anonymization. The age has five classes, work class and marital status have three classes, respectively, and education has four classes. As the number of classes increases, the variation in the attribute increases. Hence, identification becomes difficult without anonymization. Hence, the combination gives the highest risk of reidentification and also the least number of records which would have the lowest risk of reidentification. As the difference between combinations 1 and 2 is the last QID, marital status in combination 1 and salary class in combination 2. And the salary class has only two classes, greater or less than 50,000 salary. It has more importance in anonymizing and protection. Hence, the combination 2 has better chances in reidentification and has a better number of records affected by lowest risk. The next graph portrays average risk of the combinations. The least reidentification risk is obtained for series 5 which uses quasi-identifiers for anonymization with least classes as other combinations have age in them which has five classes. This gives an edge to the combination 5 which is visible from the graph above (Fig. 3).

The most promising result is given by the combination 3 in Fig. 4 which has a race, sex, marital status, salary class and education as its quasi-identifiers. We observe

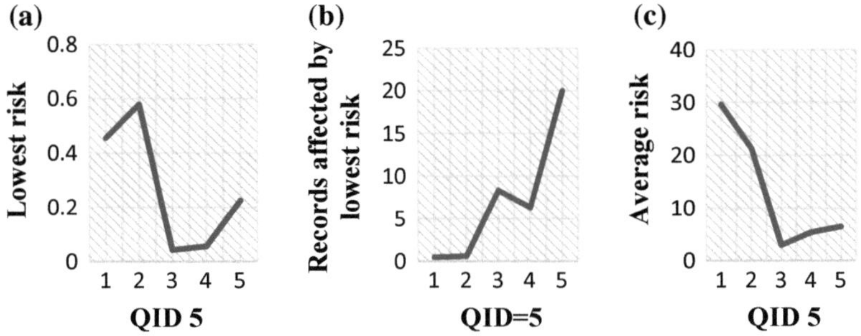

Fig. 4 Variation of **a** lowest risk, **b** records affecting lowest risk, **c** average risk with respect to the combination of five quasi-identifiers for anonymization

Table 3 Combination of five different quasi-identifiers to study the variation in reidentification risk

S. No.	QID 1	QID 2	QID 3	QID 4	QID 5
1	Age (5)	Work class (3)	Race (2)	Marital status (3)	Education (4)
2	Age (5)	Sex (2)	Race (2)	Marital status (3)	Education (4)
3	Race (2)	Sex (2)	Marital status (3)	Salary class (2)	Education (4)
4	Race (2)	Work class (3)	Marital status (3)	Salary class (2)	Education (4)
5	Race (2)	Sex (2)	Marital status (3)	Salary class (2)	Relationship (6)

from the above table that this combination has the least variation in its classes. The maximum number of classes is in the education quasi-identifier which is four. The combinations 3 and 4 differ in one quasi-identifier only. Since combination 3 has the least dispersion in its classes, it has utmost importance in anonymization for the protection of privacy. This is evident, as the combination 3 gives the best results. As the least favourable results are provided by combination 2, we observe that the pattern follows in the number of records affected by lowest risk. Average risk is least in combination 3 and worst combination is combination 1. We can conclude that this results due to the dispersion of the classes of the quasi-identifiers. It solidifies our proposition that the quasi-identifiers with a lesser dispersion of classes pays more profound blow on the reidentification risk.

5 Conclusion and Future Scope

The current analysis demonstrates that there is a probability for rupturing the discretion of users whose data has been encompassed. The unswerving link between the quasi-identifying parameters and sensitive attributes is wrecked by the anonymization. The hazard of disclosure is centred around the differential privacy. The vicissitudes in reidentification risk are observed when the variation is done in the selection of quasi-identifiers for anonymization purposes. As the number of quasi-identifiers for anonymization is increased, the reidentification risk also increases. When the quasi-identifiers with lesser number of classes are used for anonymization, the lesser risk of reidentification is obtained. This happens because of lesser the variation in data, lesser is a risk of pointing out. It is difficult to aim out a single record when there are lots of records with similar attributes. This research attempts to exhibit the line of reasoning that the selection of QI attributes can alter according to the place where they are used. Therefore, appropriate connection amongst every QI attribute values should be tried before anonymizing the data. Looking into this direction can provide immense help in looking for practical solutions for the difficulty of disclosure risk in auxiliary data. Also in actual life experiences, records in a data set are not entirely dependence free.

References

1. J. Abawajy, M.I. Ninggal, T. Herawan, Privacy preserving social network data publication. IEEE Commun. Surv. Tutorials **18**(3), 1–1 (2016)
2. A. Pervushin, A. Spivak, Determination of loss of information during data anonymization procedure. in *2016 IEEE 10th International Conference on Application of Information and Communication Technologies (AICT)*, IEEE, 2016, pp. 1–5
3. S. Zeadally, J. Contreras-Castillo, J.A. Guerrero Ibañez, Solving vehicular ad hoc network challenges with Big Data solutions. IET, Netw. **5**(4), 81–84 (2016)
4. J. Yang, B. Jiang, B. Li, K. Tian, Z. Lv, A fast image retrieval method designed for network or network big data. IEEE Trans. Industr. Inf. **13**(5), 2350–2359 (2017)
5. S. Peng, G. Wang, D. Xie, Social influence analysis in social networking big data: opportunities and challenges, IEEE Netw. 12–18 (2016)
6. S. Yu, S. Member, Big privacy: challenges and opportunities of privacy study in the age of big data. IEEE Access 2751–2763 (2016)
7. J. Qian, X. Li, C. Zhang, L. Chen, Social network de-anonymization and privacy inference with knowledge graph model. IEEE Trans. Dependable Secure Comp. **5971**(3), 1–14 (2017)
8. A. Narayanan, V. Shmatikov, Robust de-anonymization of large sparse datasets. in *2008 IEEE Symposium on Security and Privacy*, IEEE, 2008, pp. 111–125
9. A. Abdrashitov, A. Spivak, Sensor data anonymization based on genetic algorithm clustering with L-diversity. in *2016 18th Conference on Open Innovations Association and Seminar on Information Security and Protection of Information Technology*, 2016, pp. 3–8
10. R. Dewri, I. Ray, I. Ray, D. Whitley, K-anonymization in the presence of publisher preferences. IEEE Trans. Knowl. Data Eng. **23**(11), 1678–1690 (2011)
11. L. Liu, S. Member, Protecting location privacy with personalized k-anonymity : architecture and algorithms. IEEE Trans. Mobile Comput. **7**(1), 1–18 (2008)
12. J.C. Lin, Q. Liu, P. Fournier-viger, T. Hong, PTA: an efficient system for transaction database anonymization. IEEE Access **4**(4), 6467–6479 (2016)
13. Y. Wang, J. Tian, C. Yang, Y. Zhu, Research on anonymous protection technology for big data publishing. in *2016 7th IEEE International Conference on Software Engineering and Service Science (ICSESS)*, IEEE, 2016, pp. 438–441
14. S. Yu, Big privacy: challenges and opportunities of privacy study in the age of big data. IEEE Access **4**, 2751–2763 (2016)
15. J. Domingo-ferrer, D. Rebollo-monedero, J. Forne, From t-closeness-like privacy to post-randomization via information theory. IEEE Trans. Knowl. Data Eng. **22**(11), 1623–1636 (2010)
16. I. Table, A robust privacy preserving model for data publishing. in *2015 International Conference on Computer Communication and Informatics (ICCCI)*, 2015, pp. 1–6
17. K. Mohan, P. Shrivastva, M.A. Rizvi, S. Singh, Big data privacy based on differential privacy a hope for big data. in *2014 International Conference on Computational Intelligence and Communication Networks (CICN)*, November, 2014, pp. 776–781

Preserving User Location Privacy in Era of Location-Based Services: Challenges, Techniques and Framework

Rigzin Angmo, Veenu Mangat and Naveen Aggarwal

Abstract The technological advancements of mobile networks, devices, global position technology and sensors, have lead to a plethora of location-based services (LBSs) which is being offered to user. In the modern digital age, most of the automobiles use location sensing technologies and the mobile devices are equipped with GPS location tracking. In spite of the numerous potential benefits of location-based services, location awareness poses some foreseeable threats, the most important of which is location privacy. A single piece of information about location can reveal a lot about user, such as 'who', 'what', 'when' and 'where' someone is. This paper presents a comprehensive review of various aspects related to location privacy. An attempt has been made to introduce a framework to balance the trade-offs between privacy and security. The proposed framework uses privacy mechanism to protect individual's sensitive data, and a role-based access control technique to provide security of collected data through authorized user.

Keywords Location privacy · Anonymization · Obfuscation · Cryptography Location-based services

1 Introduction

The previous several years have inscribed an upswing in location-based technologies and analyses of geo-information. Technologists and naive users alike are getting interested in the spatial nature of information because of the ubiquity of geographic layer in many day-to-day activities.

R. Angmo (✉) · V. Mangat · N. Aggarwal
University Institute of Engineering and Technology, Panjab University, Chandigarh, India
e-mail: rigzin.btech9@gmail.com

V. Mangat
e-mail: veenumangat@yahoo.com

N. Aggarwal
e-mail: navagg@gmail.com

C. R. Krishna et al. (eds.), *Proceedings of 2nd International Conference on Communication, Computing and Networking*, Lecture Notes in Networks and Systems 46, https://doi.org/10.1007/978-981-13-1217-5_5

Location-based services (LBSs) are the services that take advantage of the spatial or position information of a mobile device Internet and wireless communications modes. LBS provides a wide variety of applications which transverse in different domains ranging from tracking and navigation systems that helps users reach a particular destination; directory services to find nearby businesses or events [5]; location-based tourist information; location-targeted advertising; social networking through finding friends in user area; facility to check-in at various locations; various other mobile commerce applications, entertainment services and emergency situation services. Common examples of such applications are Google Buzz, MapQuest, Yelp, Facebook, etc.

Despite the indisputable potential of location-based services, location awareness also presents inherent threats for the users, perhaps the most important of which is location privacy [10]. LBS raise the risk of privacy breaches by leaking user's sensitive information. Most users are naïve and unaware of the kind of location information service providers are storing about them, the length of time for which it is stored and the purposes it is applied to, after being processed by sophisticated analytical tools. Generally, the location data is the information which is acquired with the help of a mobile device that provides information about its current point in space. But it contains more than just location information which leads to various privacy concerns. However, not all LBS are privacy-sensitive. For example, an LBS that indicates location of public transport by answering a query like 'where is the bus number CH01GA123?' is not sensitive to location of user. But the problem arises when location is combined with user's personal identity or derivable attribute/s and this combined information gets revealed to an adversary or third party, leading to privacy threats [7] such as disclosure, tracking behaviour, identity theft, personal security, etc. A query like 'which is the nearest bus stop to catch bus to Patiala, not only accesses the user's current location but also hints at subsequent location user be at. Nowadays, the child safety concerns are also raised because the location-based mobile services are accessible to child easily. By tracking the behaviour of users, an adversary will infer the location of user with a high confidence and that information can be used for stalking or other criminal activities.

2 Related Work

Smailagic and Kogan [20] have presented CMU-TMI (Carnegie Mellon University–Triangulation Mapping and Interpolation), a new model for a location service. They have evaluated and compared CMU-TMI model implementation against other existing algorithms like CMU-PM (pattern matching), RADAR and CMU-SC (server centric) based on attributes for characterizing of training system. The attributes used are accuracy, complexity of training system, power consumption and usability.

Duckham and Kulik [8] have proposed a formal model of obfuscation that protects individual's location privacy within an insidious computing environment. The issue regarding protecting individual users' location sensitive values is addressed. Their approach focuses on balancing the level of privacy and utility for location-based services.

Puttaswamy et al. [17] have presented a prototype system named LocX for building location-based social application while preserving user location. It does not rely on any third server or components for user's location privacy, and maintains system efficiency by providing data sharing property of target applications.

Chakraborty et al. [6] have discussed about the threat of revealing private user behaviour to mobile app service that an individual is using. The issue is deciding which data should be shared so that the sensitive information of an individual cannot be inferred. The focus is not on hiding user identity rather the behaviour. They have proposed a privacy-aware framework called IpShield.

Kearns et al. [14] have introduced a design model of graph search algorithms to find and identify targeted individual while providing privacy to protected individuals. The author deals to preserve privacy rights of citizens whose essential statistics is gathered and analysed by third parties for monitoring in fields like medicine, counterterrorism or marketing.

Portnoi and Shen [16] have proposed a generic protocol named as LOCATHE, which provides a two-tier privacy authentication scheme where one is based on hiding unique identification and second deals with full individual authentication.

Rahman et al. [18] have proposed a framework called PriSens-HSAC for solution to privacy problems in the RFID tags. RFID information system can capture more sensitive type of information including time and location data as well, which may pose a threat to privacy at various stages like data collection, analysis and disclosure. So, a group-based anonymous PriSens framework is used to balance between privacy and scalability of data in healthcare.

Badar et al. [2] suggested a role-based access control (RBAC) approach which enforces the access control mechanism as soon as new user is added into the system. To recognize the user's credential for a particular role, a classification model is deployed which access user's credentials and role assignment data.

3 Location-Based Services (LBSs)

The LBS uses a mobile device's GPS equipment to trail user location, if that person has given the service requisite permissions. Once permitted, the service can identify user location down to the lowest level of hierarchy without the need for additional intervention. LBSs are services that are gaining significant attention due to their potential in delivering extra valuable information or service based on location to the user, and capabilities of using smartphones, positioning technologies and mobile networks. The main advantage of LBS for users is that the process is automated to provide personalized services using location information through various positioning technology. There is no need to enter manually again and again once user has agreed.

LBS can be thought as the confluence of numerous technologies: mobile internet, geographical information systems (GIS) and other spatial and positioning

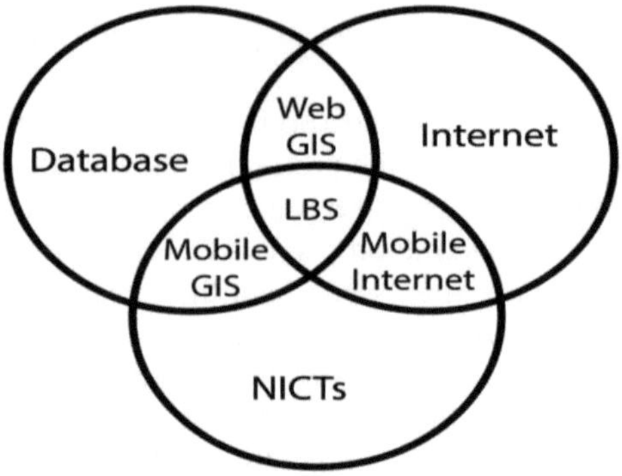

Fig. 1 LBS seen as a confluence of technologies

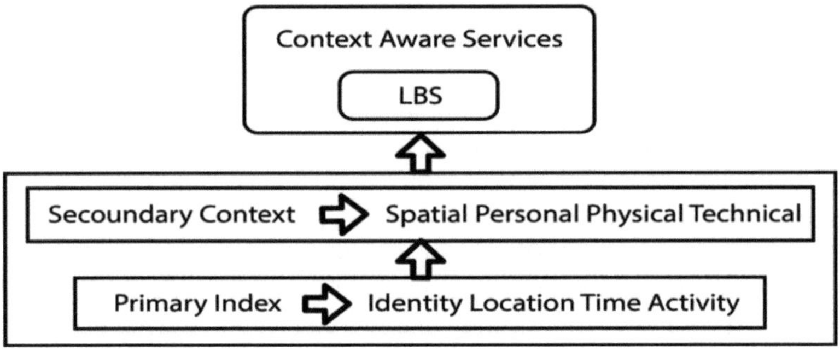

Fig. 2 Overview of LBS

technologies, the Internet and the Web and new information and communication technologies (NICTs) (Fig. 1) [4].

LBS are a type of context-aware service, wherein the service parameters and behaviour is adapted according to the user location. Various parameters determine the primary and secondary context of the task of the user as presented in Fig. 2 [4].

4 Applications of LBSs

There are various commercial applications of LBSs:

Locators for places of interest: Apps to find nearby stores for retail customers, nearby restaurants for tourists, etc., use location-based intelligence.

Proximity-based marketing: When a potential customer enters the proximity of a region, location-based mobile marketing enables targeted advertising to them.

Weather information: LBSs is used to provide real-time weather-related information and forecasts to the device used by user for future plans.

Emergency Services: In case of a calamity or accident, the LBS information can be used for tracking and locating users for rescue and assistance.

Mobile workforce management: LBS can be used by employees to check-in at field locations.

Fraud prevention: Each time a financial transaction is made using a mobile device, an LBS can provide extra security by analysing user's current location w.r.t. known locations to identify outliers or frauds.

5 Challenges in Providing Location Privacy

Lack of Awareness: The majority of users are unaware about location data and its usage. This poses a huge risk due to the scale at which location data is being collected, shared and analysed to reveal valuable private information.

Privacy is Multi-dimensional: Privacy solution has to be multi-dimensional within a social, technical, ethical and legal framework. The current policy and legal environment are not aligned with the current state of the technology.

Privacy versus utility trade-off: A balance has to be struck between protecting right of privacy of data subject with the business needs.

6 Types of Location Privacy

The two main categories of location services are push-and-pull services and sporadic and continuous services. Both types are used by location-based services and can pose a significant risk to privacy in today's era of mobile devices and IoT. It can be divided into two privacy scenarios that are in first, the location of users' remains unknown that is despite of the actual identification of user location of them remains unknown and in second, the users' identity remains unknown as in certain LBS applications only need to know users exact location despite of their actual identity like in case of emergency like road accidents. This can be done by only providing a general region to estimate of user's exact location, while still providing privacy to individual user.

7 Technical Approaches to Provide Location Privacy

There are various technical approaches which are used to provide location privacy. Some of them are as follows:

Anonymity Technologies: The concept of anonymity deals with removing link between users' real identities and their personal information. In this technique, the uniquely identifiable entity of user is replaced with temporary identifiers which can be changed periodically over time. Anonymization is a useful but not sufficient mechanism to provide privacy [21]. Also, profile-based k-anonymity has been an extension on anonymization technique Šaltenis et al. [19] for location-based user services. There are various other location anonymization techniques like perturbation, generalization, etc. Also, reidentification of users' information can be done with the help of de-anonymization techniques.

Mix Zone: As described by Beresford and Stajano [3] a mix zone is an area, wherein a mixed zone technique is used for mapping between old pseudonyms and new pseudonym of mobile users. These are changed in such a way that it does not reveal users' information to an adversary. It is an alternative solution to spatial cloaking for protecting location privacy.

Position Sharing: The idea of position sharing has been proposed to manage private location information of VANET users in a secure way in a non-trusted systems [9, 22]. First, the location information of each user is obfuscated and the obfuscated information is divided into position parts, wherein a part defines a position of strictly limited precision. Then, these parts are distributed to all the non-trusted location servers (LSs). One particular LS contains information about only a location of limited precision.

Path confusion: In the path confusion approach, the paths of two users are mixed in such a way that adversary is not able to determine which path corresponds to which user. This is similar to fake position approach, in which the fake point of users' location is transmitted. But in path confusion, the location samples can be correlated to reconstruct user trajectories. The position of target user can also be predicted based on some algorithms by known previous samples of speed, direction and next location Hoh and Gruteser [12].

Obfuscation and Coordinate Transformation: Obfuscation means to make something obscure or unclear. The method of intentionally corrupting individual's location information in order to provide privacy is known as obfuscation. A typical spatial obfuscation approach has been proposed by Ardagna et al. [1] in which instead of transmitting the exact user position, a generalized circular area is transmitted to the LBS. Duckham and Kulik [8] have proposed an alternative method of using graph obfuscation in place of circles for road networks. The obfuscation technique can be better understood with the help of Fig. 3, in which users' location query is protected.

Here, we consider the location of an institution like UIET. The precise location of a user can be obfuscated as shown in Fig. 3.

Pseudonyms: The pseudonym approach can be used to provide location privacy. It has been proposed by Jiang et al. [13] for a wireless network environment. It has been suggested not to provide permanent pseudonym rather provide temporary pseudonym that can be changed periodically to prevent reidentification and provide privacy for user's location information. Additionally, a silent period is introduced, wherein the device does not transmit information to the LBS to maintain user privacy.

Cryptography-based Approaches: In this approach, the location information is protected using encryption techniques. Yang et al. [23] have proposed anonymity preserving data collection using encryption without the help of a third party. They have also demonstrated the utility of a cryptography-based approach by inferring

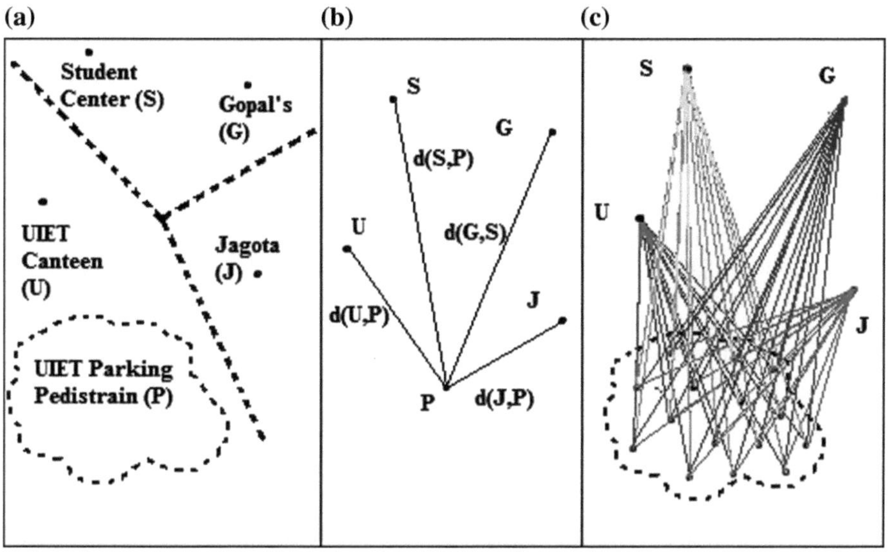

Fig. 3 By discretizing (**a**) the UIET parking region on (**b**) a collection of points (**c**) precise location no longer available after obfuscation

rules for classification for a set of data owners while protecting sensitive attributes of each individual owner.

8 Proposed Framework

As has been discussed previously, there is a strong need to address the challenge of providing location privacy to users. Alongside, there is the requirement of providing appropriate use of data to different agencies like law-and-order enforcement agencies, companies for business, emergency services for ambulance, etc., while ensuring security of data. A complex issue like privacy cannot be addressed by technology alone. However, by advancing what is technically feasible, the overall quality of the solution can be improved. So, we propose a framework that can be used to provide location privacy to users' as well as address other challenges like privacy versus utility and privacy versus security trade-off. The Proposed framework (Fig. 4) uses technical advancement with the combination of policies and regulations.

The framework utilizes hybrid privacy techniques because using only a single technique is not guaranteed to achieve desirable level of privacy for users. Since each method has its own advantages and disadvantages, so we advocate the use of hybrid privacy-preserving techniques. Furthermore, there is a need to incorporate security objectives such as mutual authentication, information confidentiality, message integrity, etc.

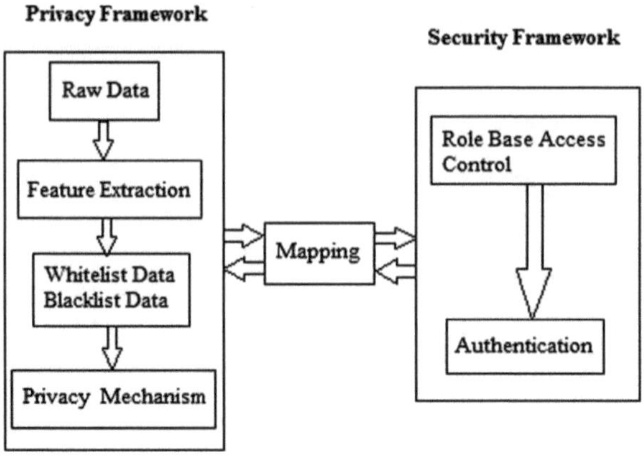

Fig. 4 Proposed privacy framework

This work proposes a multi-dimensional framework to achieve the twofold objectives of privacy and security.

1. Privacy framework provides privacy to the collected data. In this, first, the relevant features are extracted and then, a combination of privacy approaches is used to manage the user data.
2. Hierarchical security framework in which the data access is provided at various levels of generalization to authenticated users of that level only. This provides security as well as balances the utility of data.
3. Mapping technique is used to communicate between these two frameworks.

9 Conclusion

With the advancement of smart mobile devices, location-based services (LBSs) are becoming part of our day-to-day life-like location-aware emergency response, advertisements, etc. Although LBS systems have various advantages in our life, but users' privacy is threatened because their movement information can be used to selectively infer activities to reveal their personal information. Various techniques to provide location privacy to LBS users have been discussed in this paper. But users' privacy is still at risk and many open challenges are there. In this paper, we propose a framework which can be used to protect users' location privacy as well as other challenges. The proposed framework uses technical advancement with the combination of policies and regulations. In the proposed framework, we use different privacy techniques to provide user privacy and for utility and security, we use role-based access techniques at different levels so that it can fulfil the needs of businesses, government agencies, etc., while ensuring privacy of the users' sensitive data.

References

1. C. Ardagna, M. Cremonini, E. Damiani, et al., Location privacy protection through obfuscation-based techniques. in*21st Annual IFIP WG 11.3 Working Conference on Data and Applications Security*, 8–11 July 2007. Lecture Notes in Computer Science, vol. 4602 (Springer, Redondo Beach, CA), pp 47–60 (2007)
2. N. Badar, J. Vaidya, V. Atluri et al., Using classification for role-based access control management. Int. J. Technol. Policy Manage. **16**(1), 45–78 (2016)
3. A.R. Beresford, F. Stajano, Location privacy in pervasive computing. IEEE Pervasive Comput. **2**(1), 46–55 (2003)
4. A. Buczkowski, Location based services (2012), http://geoawesomeness.com/knowledge-base/. Accessed 4 Nov 2017
5. F. Calabrese, M. Colonna, P. Lovisolo et al., Real-time urban monitoring using cell phones: a case study in Rome. IEEE Trans. Intell. Transp. Syst. **12**(1), 141–151 (2011)
6. S. Chakraborty, K.R. Raghavan, M.P. Johnson, et al., A framework for context-aware privacy of sensor data on mobile systems. in*14th Workshop on Mobile Computing Systems and Applications*, Georgia, 26–27 February 2013 (ACM, USA, 2013), p. 11
7. Y. Cho, S. Lee, Detection and response of identity theft within a company utilizing location information. in*2016 International Conference on Platform Technology and Service (PlatCon)*, 15–17 Feburary 2016, IEEE, Jeju, South Korea, 2016, pp. 1–5
8. M. Duckham, L. Kulik, A formal model of obfuscation and negotiation for location privacy. in*International Conference on Pervasive Computing, February 2005*. Lecture Notes in Computer Science, vol. 3468 (Springer, Berlin, Heidelberg, 2005), pp. 152–170
9. F Durr, P Skvortsov, K Rothermel, Position sharing for location privacy in non-trusted systems. in*2011 IEEE International Conference on Pervasive Computing and Communications (PerCom)*, March 2011, IEEE, Seattle, WA, USA, 2011, pp. 189–196
10. Electronic Frontier Foundation, Location privacy (2017), https://www.eff.org/issues/location-privacy. Accessed 6 Nov 2017
11. M. Gruteser, D. Grunwald, Anonymous usage of location-based services through spatial and temporal cloaking. in*1st International Conference on Mobile Systems, Applications and Services*, 5–8 May 2003, ACM, San Francisco, California, 2003, pp. 31–42
12. B. Hoh, M. Gruteser, Protecting location privacy through path confusion. in*First International Conference on Security and Privacy for Emerging Areas in Communications Networks, 2005. Secure Communications 2005*, 5–9 September 2005, IEEE, Athens, Greece, 2005, pp. 194–205
13. T. Jiang, H.J. Wang, Y.C. Hu, Preserving location privacy in wireless LANs. in*5th International Conference on Mobile systems, Applications and Services*, June 11–13 2007, ACM, San Juan, Puerto Rico, 2007, pp. 246–257
14. M. Kearns, A. Roth, Z.S. Wu et al., Privacy for the protected (only). arXiv preprint arXiv:1506.00242
15. L. Kuang, Y. Wang, P. Ma et al., An improved privacy-preserving framework for location-based services based on double cloaking regions with supplementary information constraints. J. Secur. Commun. Netw. (2017)
16. M. Portnoi, C.C. Shen, Location-enhanced authenticated key exchange. in*2016 International Conference on Computing, Networking and Communications (ICNC)*, 15–18 February 2016, IEEE, Kauai, HI, USA, 2016, pp. 1–5
17. K.P. Puttaswamy, S. Wang, T. Steinbauer et al., Preserving location privacy in geosocial applications. IEEE Trans. Mob. Comput. **13**(1), 159–173 (2014)
18. F. Rahman, M.Z.A. Bhuiyan, S.I. Ahamed, A privacy preserving framework for RFID based healthcare systems. Future Gener. Comput. Syst. **72**, 339–352 (2017)
19. S. Saltenis, C.S. Jensen, S.T. Leutenegger et al., Indexing the positions of continuously moving objects. ACM **29**(2), 331–342 (2000)
20. A. Smailagic, D. Kogan, Location sensing and privacy in a context-aware computing environment. IEEE Wirel. Commun. **9**(5), 10–17 (2002)

21. T.M. Truta, B. Vinay, Privacy protection: P-sensitive K-anonymity property. in*Proceedings. 22nd International Conference on Data Engineering Workshops*, 2006, 3–7 April 2006. IEEE, Atlanta, GA, USA, 2006, pp. 94–94
22. M. Wernke, P. Skvortsov, F. Dürr et al., A classification of location privacy attacks and approaches. Pers. Ubiquit. Comput. **18**(1), 163–175 (2014)
23. Z. Yang, S. Zhong, R.N. Wright, Anonymity-preserving data collection, in *Eleventh ACM SIGKDD International Conference on Knowledge Discovery in Data Mining*, 21–24 August 2005, ACM, Chicago, Illinois, USA, 2005, pp. 334–343

Efficient Key Distribution and Mutual Authentication Mechanism Using Modified Needham–Schroeder and Otway–Rees Protocol for Cloud Environment

Ashwani Kumar Vashishtha, C. Rama Krishna and Rajiv Chechi

Abstract Cloud environment recent appearance radically amended the opinion regarding infrastructure, software delivery, architectures and development models. Cloud environment provides a number of services, and for which many computing environments exist on a single platform for different applications and services with the use of insecure communication channels. By keeping strike about the security issues in consideration, every cloud user is looking for strong authentic public key distribution and mutual authentication in cloud environment. The aim of this paper is to provide efficient key distribution and mutual authentication mechanism using MNSOR (Modified Needham–Schroeder and Otway Rees), between cloud service user and cloud service provider to ensure timelines of interaction without clock synchronization, man-in-the-middle attack protection and responsibility of key obtaining from trusted third-party shift from cloud service user to cloud service provider end.

Keywords Mutual authentication · Protocols · Public key infrastructure
Key distribution · Encryption and decryption · Cloud environment

1 Introduction

Among the trending technologies, cloud environment has been commercialized very rapidly in a short span of few years. Most of the time, an individual user perform computational work on their personal computer and shares it with others or want to

A. K. Vashishtha · C. R. Krishna
Department of Computer Science and Engineering, National Institute of Technical Teachers Training and Research, Chandigarh, India
e-mail: er.ashwani277@gmail.com

C. R. Krishna
e-mail: rkc_97@yahoo.com

R. Chechi (✉)
Department of Electronics and Communication Engineering, VCE, Meerut, India
e-mail: dr.rajivchechi@gmail.com

C. R. Krishna et al. (eds.), *Proceedings of 2nd International Conference on Communication, Computing and Networking*, Lecture Notes in Networks and Systems 46, https://doi.org/10.1007/978-981-13-1217-5_6

store it on cloud via an insecure channel such as Internet. Everyone wants to move on to cloud environment from personal or individual users to the users of organization. The recent surfacing of cloud environment extensively changed everyone's knowledge about infrastructure architectures, software delivery and upgrading models. Cloud environment is the infrastructure distinct as a hardware and software that provides trustworthy constant, persistent and reasonably priced access to high-end computational capabilities [1].

The migration towards cloud environment is appearing in recent years because of increasing strength of smart users day by day. Smart users are maintaining an upward number of personal data, counting bookmarks, official and personal photographs, music files, required applications and many more. Cloud environment is empowered by virtualization technology. In virtualization, an application known as hypervisor runs on host computers and number of virtual machines are created by the simulation of physical computers with required authentication. So that simulations can run any software as of operating systems and end-user applications [2]. Public key infrastructure provides the notation for forming a web trust by operationally coupled clouds through chains of trust. The mutual authentication is required to ensure that every cloud service user and cloud service provider has obtained authentic public key of each other, and they can further communicate in any of the modes of asymmetric encryption techniques. Due to the security concern and quality of service of key distribution and mutual authentication process, consistent and predictable data services are playing an important role for heterogeneous architectures accelerators in cloud environment [3].

2 Security Issues of Cloud Environment

The organization of data security and privacy in cloud environment as shown in Fig. 1 follows public cloud environment security issues as:

1. **Trust**: It is not a fresh aspect in the research field of computer science. In between a number of systems for trust, the advancement of trusted computer systems evaluation criterion (TCSEC) [4] in late 1970s and early 1980s was found most notable. The basic purpose of trust was convincing the observers about the secure functioning of the system [5]. Two parties involved in services transaction deploys the trust concept as an entity A is considered in trust with entity B, if entity A believes that entity B is behaving closely as expected and required [6]. In information society, trust is built on a variety of different foundation that is based on calculus, knowledge and on social reasons [7].
2. **Security Identification of Threats**: In security, identification of threats recognition of unique threats and challenges are handled with the implementation of fitting countermeasures. Finally, for effective integration of security control in functioning of information system and operation requirements, identified security controls are introduced in systems engineering process, as well as other

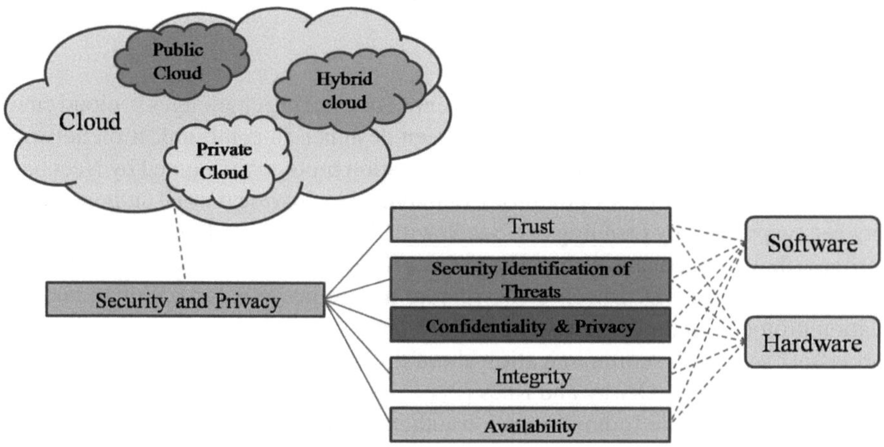

Fig. 1 Organization of data security and privacy in cloud environment

important system requirements (like reliability, maintainability and supportability) [8]. Risk evaluation is required for availability and reliability, recovery and privacy, data integrity and auditing in cloud environment [9].

3. **Confidentiality and Privacy**: Confidentiality provides authentication to the party or the system that have the capability to access sheltered data. Regarding the issues of multi-tenancy, data remanence, application security and privacy number of concerns are emerged [10]. In cloud environment, data confidentiality is interconnected with user authentication in sharing of information. For completing this task, electronic authentication is in practice for the establishment of confidentiality in user identities.
4. **Integrity**: Integrity is the key aspect of security in cloud environment. Integrity provides the guarantee of resources modification only by authorized parties in an approved way only. Integrity refers to data software and hardware both. Unauthorized deletion, modification or invention is related to data integrity in case of software as well as hardware. In cloud, the security of software integrity is the responsibility of software owner or administrator. A cloud infrastructure service provider is the trusted one who maintains data integrity and accuracy.
5. **Availability**: When an authorized party demand for system properties and required system properties are available on demand that is referred to as availability. Availability covers all data, software and hardware availability for user authorization on demand. It also increases the requirement of ubiquitous network's availability. Cloud owner responsibility is providing the guarantee of information processing, resource infrastructure and network availability to clients on demand.

3 Related Work

Authentic key distribution and mutual authentication are big challenges in cloud environment [11]. Number of protocols have been designed in communication network to address the security issues, namely authentication protocols designed by Needham and Schroeder [12], key transport protocol intended to provide mutual authentication between two parties by Denning and Sacco with the use of time stamps to remove the freshness flaw in Needham–Schroeder protocol [13], secure communication protocol designed by Andrew D. Birrell using remote procedure calls [14], mutual authentication protocol (authorized by trusted third party) that provides assurance to both parties involved in the communication about the timelines dealings without the use of clocks designed by Otway and Rees [2].

Here, we have studied communication authentication protocols designed by Needham and Schroeder [12] and mutual authentication protocol designed by Otway and Rees [2]. These protocols are performing the establishment of authenticated connections, management of cloud authenticated data transmission and signature verification with document integrity guarantee. But these protocols suffer with compromises of communication keys due to lack of mutual authentication, man-in-the-middle attack and unawareness of the availability of cloud service provider.

To solve this problem, Denning suggested the use of timestamps [13]. This would, however, further cause the problem of clock synchronization. In [2], a protocol was described for efficient mutual authentication but this protocol suffered from man-in-the-middle attack [6]. In order to overcome these problems, we have proposed a protocol, namely Modified Needham–Schroeder and Otway–Rees (MNSOR) protocol in this paper.

Proposed MNSOR is designed to solve the following two problems: authentic public key distribution and mutual entity authentication. As asymmetric key encryption technique is being used in proposed protocol, each of the cloud communicating parties has a pair of private and public key. The public key of each of the cloud communicating party is registered with a mutually trusted third party known as the authentication server [14] and named this server as MNSOR Server (MS). We have assumed that these registered public keys are always secure. Mutual entity authentication is performed by the cloud communicating parties to authenticate each other. So that each party can be assured that it has the authentic public key of the other party and that they can further communicate in any cloud environment.

- **Following conventions are used in our proposed MNSOR protocol**:
  - S and R: Cloud communicating parties where S (cloud service user) starts the conversation with R (cloud service provider).
  - MS: MNSOR Server.
  - A: Attacker.
  - A(R): Attacker impersonating R.
  - ID_S, ID_R: Identifiers of S and R respectively.

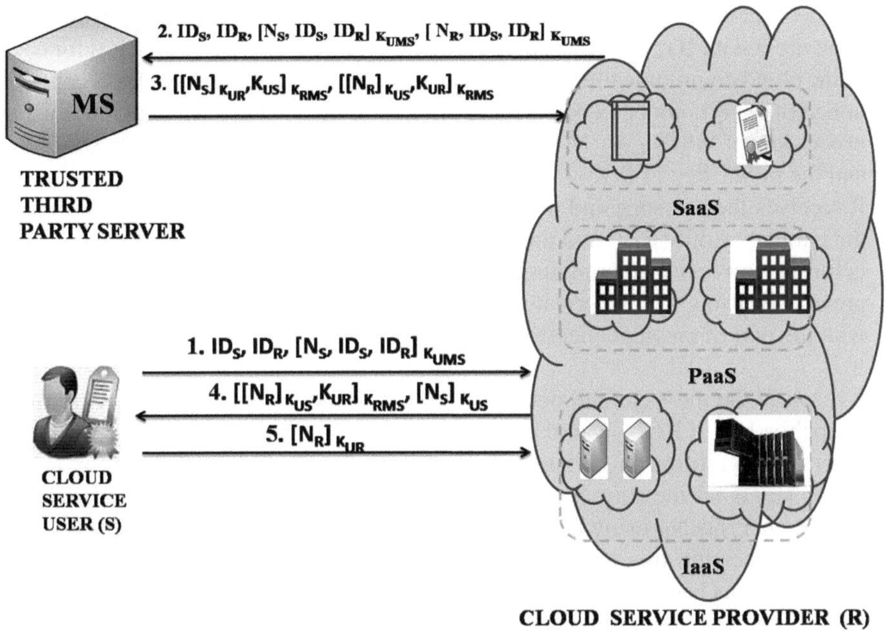

Fig. 2 Functionality of proposed MNSOR protocol

- N_S, N_R: Nonce generated by S and R, respectively.
- K_{RMS}-K_{UMS}, K_{RR}-K_{UR} and K_{RS}-K_{US}: Private–Public key pair of MS, S and R, respectively.

4 Proposed MNSOR Protocol

To overcome the security issues of Needham–Schroder and Otway–Rees protocol related to compromises of communication keys due to lack of mutual authentication, man-in-the-middle attack and unawareness about the availability of cloud service provider traced in, the functionality of proposed MNSOR protocol is shown in Fig. 2, and are as follows:

- If S wants to use of services of R, it sends a message to R, comprising of its identifier ID_S, the identifier of R, ID_R and a message encrypted using the public key of MS, K_{UMS}. The encrypted message has a nonce N_S generated by S for this communication session, ID_S and ID_R.
- R, upon receiving this message, generates a nonce N_R for this communication session, concatenates it with ID_S and ID_R and encrypts it with the public key of MS, K_{UMS}. R then concatenates this encrypted message with the message it received from R, and sends it to MS.

- MS decrypt these two encrypted parts of message using its private key K_{RMS}, and compare it with ID_S and ID_R. After verification, it prepares two messages for each of the cloud communicating parties, S and R, respectively. Message for S contains nonce of $S(N_S)$, nonce of $R(N_R)$ and the public key of $R(K_{UR})$, encrypted using private key of $MS(K_{RMS})$. Message for R contains nonce of $R(N_R)$, nonce of $S(N_S)$ and the public key of $S(K_{US})$, encrypted using private key of $MS(K_{RMS})$.
- R receives the message and decrypts the part of message using K_{UMS} to retrieve the public key of $S(K_{US})$ and the nonces. It checks N_R to verify that the public key is for the correct cloud communication session. It then encrypts the N_S, retrieved previously, using its own private key (K_{RR}). It then concatenates the encrypted N_S with the part of message received from MS, encrypted using K_{RMS} and sends it to S.
- S receives the message and decrypts the part of message using K_{UMS} to retrieve the public key of $R(K_{UR})$ and the nonce. It then uses its private key to decrypt the other part to retrieve N_S. Thus, S gets assured that the message is sent by R and public key is for the correct communication session. Now to establish its identity to R, it encrypts N_R using the public key of $R(K_{UR})$, and sends it back to R.
- R upon receiving this message decrypts it using K_{RR} and gets assured that its public key has been successfully reached the S.
- Now, S and R can communicate authentically for services over the insecure network using their private and public key pairs.

5 Analysis of Efficiency

The proposed protocol MNSOR is compared with key distribution protocol proposed by Otway and Rees in cloud computing environment in terms of efficiency.

5.1 *Standard Conventions and Assumptions*

- Time taken in encryption per block (dependence is upon algorithm used): $\mathbf{t_e}$
- Time taken in decryption per block (dependence is upon algorithm used): $\mathbf{t_d}$.
- Transmission time per block (dependence is upon broadcasting medium used): $\mathbf{t_b}$.
- Time taken in production of nonce: $\mathbf{t_p}$.
- Time taken in production of key K_{SR}: $\mathbf{t_k}$.

The running time of the key distribution protocol proposed by Otway and Rees is computed as shown in Table 1.

Table 1 Computation of running time of Otway and Rees key distribution protocol in cloud environment

Steps	Location of cloud computing parties	Time taken
1	At S	$t_p + 4t_e$
2	Transmission from S to R	$7t_b$
3	At R	$t_p + 4t_e$
4	Transmission from R to AS	$11t_b$
5	At AS	$4t_d + 4t_d + t_k + 2t_e + 2t_e$
6	Transmission from AS to R	$5t_b$
7	At R	$2t_d$
8	Transmission from R to S	$3t_b$
9	At A	$2t_d$

Hence, the total running time of the Key Distribution protocol proposed by Otway and Rees is:

$$\begin{aligned} T_1 &= t_p + 4t_e + 7t_t + t_p + 4t_e + 11t_b + 4t_d + 4t_d + t_k + 2t_e + 2t_e + 5t_b + 2t_d + 3t_b + 2t_d \\ &= (1+1)t_p + (4+4+2+2)t_e + (7+11+5+3)t_b + (4+4+2+2)t_d + t_k \\ &= 2t_p + 12t_e + 26t_b + 12t_d + t_k \end{aligned} \quad (1)$$

5.2 *Efficiency Analysis of Proposed MNSOR Over Otway and Rees Protocol in Cloud Environment*

Similarly, we can compute the running time of the proposed MNSOR protocol as shown in Table 2:

Total running time is computed as:

$$\begin{aligned} T_2 &= t_p + 3t_e + 5t_b + t_p + 3t_e + 8t_b + 6t_d + t_k + 6t_e + 6t_b + 3t_d + t_e + 4t_b + 3t_d + t_e + t_b \\ &= 2t_p + (3+3+6+1+1)t_e + (5+8+6+4+1)t_b + (6+3+3)t_d + t_k \\ &= 2t_p + 14t_e + 24t_b + 12t_d + t_k \end{aligned} \quad (2)$$

By subtracting $(2t_p + 12t_d + t_k)$ from both Eqs. (1) and (2) to eliminate common terms, we get

$$T_1 = 12t_e + 26t_b \quad (3)$$

$$\mathrm{T}_2 = 14\mathrm{t}_e + 24\mathrm{t}_b \quad (4)$$

The reduced equations, that are to be used for relative analysis, are the function of t_e and t_b only. As transmission time is a very critical factor in case of cloud

Table 2 Key distribution protocol running time of proposed MNSOR protocol

Steps	Location of cloud computing parties	Time taken
1	At S	$t_p + 3t_e$
2	Transmission from S to R	$5t_b$
3	At R	$t_p + 3t_e$
4	Transmission from R to MS	$8t_b$
5	At MS	$6t_d + t_k + 6t_e$
6	Transmission from MS to R	$6t_b$
7	At R	$3t_d + t_e$
8	Transmission from R to S	$4t_b$
9	At S	$3t_d + t_e$
10	Transmission from S to R	t_b

environment, hence the time taken for encryption would be much less as compared to the transmission time, i.e. $t_e << t_b$. So, t_e can be safely ignored from both Eqs. (3) and (4). The relationship now reduces to an expression depicting complexity as a linear function of transmission time t_b. Therefore, more the transmission time t_t, more would be the complexity of the protocol.

$$T_1 = 26t_b \tag{5}$$

$$T_2 = 24t_b \tag{6}$$

The running time efficiency of proposed MNSOR protocol over the key distribution protocol proposed by Otway and Rees in cloud computing environment can be computed as per the formula of calculating the running time efficiency of two protocols in terms of time is:

$$\eta = \left\{ \left(\frac{\boldsymbol{T}_1 - \boldsymbol{T}_2}{\boldsymbol{T}_2} \right) * 100 \right\} \%$$

After putting the values of T_1 and T_2 in the above equation, we get

$$\eta = \left\{ \left(\frac{26\boldsymbol{t}_b - 24\boldsymbol{t}_b}{24\boldsymbol{t}_b} \right) * 100 \right\} \%$$

$$\eta = \left\{ \left(\frac{2\boldsymbol{t}_b}{24\boldsymbol{t}_b} \right) * 100 \right\} \%$$

$$\eta = \left\{ \left(\frac{1}{12} \right) * 100 \right\} \%$$

Table 3 Advantages of proposed MNSOR protocol over Otway and Rees in cloud environment

Issues	Otway Rees	MNSOR protocol
Key distribution	Man-in-the middle attack is possible due to lack of mutual authentication	Secure against negotiation of key and man-in-the middle attack
Mutual authentication	Does not stay alive	Stay alive
Efficient third-party time utilization	Yes	Yes
Replay attack	No	No
Man-in-the-middle attack	Yes	No
Running time	$T_1 = 2t_p + 12t_e + 26t_b + 12t_d + t_k$	$T_2 = 2t_p + 14t_e + 24t_b + 12t_d + t_k$

$$\eta = 8.33\%$$

6 Result

Thus, it is shown that the proposed MNSOR protocol is 8.33% more efficient in running than the key distribution protocol proposed by OTWAY and REES for cloud computing environment. The proposed MNSOR protocol has following advantages over Otway and Rees protocol as shown in Table 3.

7 Conclusion

Cloud environment security requires a systematic point of view, from which security will be constructed on trust and mitigating protection to a trusted third party. Cloud environment is adopting the concept of trusted third party for key distribution and mutual authentication as the profit outweighs of its shortcomings. In this paper, disadvantages of Needham–Schroder and Otway–Rees in cloud environment has been identified and proposes an efficient key distribution and mutual authentication mechanism using MNSOR protocol with 8.33% better running time efficiency over Otway and Rees protocol in coud environment. The proposed MNSOR protocol overcomes man-in-the-middle attack possibility generated due to lack of authentication in key distribution process of Otway and Rees by taking care of mutual authentication till the end of communication in cloud environment. No timestamp is required in proposed MNSOR protocol, so it can be easily implemented in cloud environment, where clock synchronization is a most important challenge. Proposed MNSOR Server is contacted for public key distribution only when cloud service provider R is in the position of communication. The unnecessary wastage of keys can be avoided by its help. If cloud service user S gets time-stamped key but cloud service provider R is not in the position of communication then the key would get wasted. This protocol is mainly useful for a cloud service user S that wants to communicate with number of cloud service providers like R_1, R_2, …, R_n simultaneously in cloud environment for different services. The accountability of getting the public key shifts from S to R_i, in that way the traffic is reduced at S.

References

1. D. Zissis, D. Lekkas, Addressing cloud computing security issues. Future Gener. Comput. Syst. **28**(3), 583–592 (2012)
2. E. Naone, Technology overview conjuring clouds. MIT Technol. Rev. 54–56 (2009)
3. C. Rajiv, R. Khanna, Role of Swarm technology based routing in UMTS-WLAN networks. Appl. Math. Inf. Sci. Lett. **1**, 41–45 (2013)
4. L. Qiu, Y. Zhang, F. Wang, M. Kyung, H.R. Mahajan, Trusted computer system evaluation criteria. Nat. Comput. Secur. Center 525–528 (1985)
5. A. Nagarajan, V. Varadharajan, Dynamic trust enhanced security model for trusted platform based services. Future Gener. Comput. Syst. **27**(5), 564–573 (2011)
6. International Telecommunication Union, in *X-509*, https://www.itu.int/rec/T-REC-X.509. Accessed March 2016
7. D. Lekkas, Establishing and managing trust within the public key infrastructure. Comput. Commun. **26**(16), 1815–1825 (2003)
8. National Institute of Standards and Technology. in *Guide for Mapping Types of Information and Information Systems to Security Categories* (2008)
9. Gartner, *Assessing the Security Risks of Cloud Computing*, https://www.gartner.com/doc/685308/assessing-security-risks-cloud-computing. Accessed May 2016
10. Cloud Security Alliance, in *Top Threats to Cloud Computing*, https://cloudsecurityalliance.org/group/top-threats/. Accessed May 2016
11. A. Verma, S. Kaushal, Cloud computing security issues and challenges: a survey. Adv. Comput. Commun. **193**, 445–454 (2011)
12. R.M. Needham, M.D. Schroeder, Using encryption for authentication in large networks of computers. Commun. ACM **21**(12), 993–999 (1978)
13. D.E. Denning, G.M. Sacco, Timestamps in key distribution protocols. Commun. ACM **24**(8), 533–536 (1981)
14. A.D. Birrell, Secure communication using remote procedure calls. ACM Trans. Comput. Syst. (TOCS) **3**(1), 1–14 (1985)

An Identity-Based Authentication Framework for Big Data Security

Vinod Kumar, Musheer Ahmad and Pankaj Kumar

Abstract Big data raises a robust need for a network structure with the ability to support information retrieval and sharing. Many companies start to supply of big data services for Internet users and at the same time, these services also bring sever all security issues. Presently, the great part of big data set provides digital identity for users to use their services. At the moment, most of these systems use asymmetric and conventional public key cryptography (PKC) to give mutual authentication data security. However, existing authentication schemes trust commonly on the centralized servers to offer record and facilitation facilities for information retrieval. Pairing-free identity-based cryptography has some pull characteristics that gives the idea to healthy requirements in that scenario. The presented scheme is carried out in three phases and are as follows: initialization phase, registration phase and mutual authentication, and session key agreement phase. Detailed security analyzes have been made to authenticate big data server and user. Further, the paper has the resistance to possible attacks in this environment.

Keywords Big data · Elliptic curve cryptography · MapReduce
Authentication and security

V. Kumar (✉) · M. Ahmad
Department of Applied Sciences and Humanities, Jamia Millia Islamia, New Delhi, India
e-mail: vinod.iitkgp13@gmail.com

M. Ahmad
e-mail: mushr_mt@yahoo.co.in

P. Kumar
School of Computing Science and Engineering, Galgotias University, Greater Noida, India
e-mail: pkumar240183@gmail.com

C. R. Krishna et al. (eds.), *Proceedings of 2nd International Conference on Communication, Computing and Networking*, Lecture Notes in Networks and Systems 46, https://doi.org/10.1007/978-981-13-1217-5_7

1 Introduction

Big data are a word for the data collections which are high multifaceted than usual information preparing functions are insufficient to agreement through them. Disputes contain investigation, querying, capture, reposition, data continuance, psychoanalysis, updating, generosity, visualization, and data security. The term "big data" frequently conjures basically to the routine prognostication analysis, user activities analytics, or assured other highly developed information analytics techniques that exact rate from data, and infrequently to an exacting scope of data set [1]. There is small uncertainty that amount of information today, obtainable is certainly huge, although that is not the greatest right feature of this novel data environment. Big data mean collects huge data sets that cannot be handled using popular computing procedures. Big data are not purely a data; slightly it has become a whole susceptible, which occupies different tools, systems, and structures [2]. Big data engrosses the data created by different policy and purposes. Specified nearly of the area as that approach under this environment are: black box data, search large engine data, social media approach data, power grid data, stock exchange related data, transport data, structured data, and unstructured data [4, 5].

The following characteristics of big data described are as follows: (1) Volume: The amount of produced and accumulated data. The scope of data establishes the cost and probable intuition, it can preserve essentially be measured large scale. (2) Variety: The class of the environment in the data collection. It assists populace who investigates it to successfully use the subsequent penetration. (3) Velocity: In the domain, a rapidity at that the data collection are created, and managed to appropriate a requirement, and encounters that lounge in the lane of enlargement and improvement. (4) Veracity: Unpredictability of data set can hinder rules to shaft and run it. (5) Variability: Disparity of the data collection can basket progressions to manage and control. Big data have most important challenges connected with big data are as follows: curtain, apprehending data, storage, searching, transfer, analysis, sharing, and presentation [6].

Since the localities of clients are not secure in this environment, this technology offers an authentication method which offers big data of enormous quantity, and several categories to users carefully. Over this development, contributors should build data kept in assorted devices simply available without revealing the personal data of users, assures the confidentiality of users that environment. In this section, to assure the confidentiality of clients that atmosphere. In big data communication, big data technologies present everywhere, where users use them. Despite that a user placed in an unambiguous region where big data facility is affording transports to a different area, the data server offering big data technologies must relate the person statistics of the client in that environment facility region to the safety purpose, and guarantee that the private evidence of the client is not bared to a third party at some stage in the service time. At that time, there is a concern to multiple server offering big data services, and these servers cannot be placed in the similar position. Clients accepting this type of technologies should be capable to collect big data facilities at their anticipated adverts, not to a unique network server [7, 8].

In the paper, we presented an identity-based authentication framework for big data security which is secured mutual authentication mechanism to resist the confident risks. Rest of the paper is organized in the following ways: 1. Related work, 2. Preliminaries, 3. The proposed protocol, 4. Security Analysis, 5. Performance analysis, and 6. Conclusion.

2 Related Work

Apache Hadoop is the furthermost well-recognized software stages that sustenance information-exhaustive distributed relevance. It accouterments the computation model called MapReduce. Apache Hadoop policy involves of the Hadoop kernel, MapReduce, and Hadoop distributed file system (HDFS), likewise several correlated developments, together with Apache Hive, Apache HBase, etc. MapReduce, which is an encoding representation and an implementation for procedure and producing outsized capacity of data sets, established by Google, and developed by Yahoo and other network enterprises. MapReduce are foundation on the divide-and-conquer technique, and mechanism by recursively distributed a multifarious difficult into a lot of subproblems, expect these subproblems is extensible for resolving directly. Subsequently, these subproblems are allocated to a cluster of running nodes, and solved in distinct and corresponding ways. Lastly, the resolutions to the subproblems are joined to explain to the creative problem. The divide-and-conquer technique is implemented by two steps: Map step and Reduce step. In expressions of Hadoop cluster, there are two types of nodes in Hadoop organization. They are master nodes and data nodes. The master node takes the response, splits it into smaller subproblems, and allocates them to data nodes in Map step. Subsequently, the master node accumulates the counters to all the subproblems and merges them in some way to appearance the output in Reduce step [3, 9]. An authentication protocol to defend user privacy and security in big data assistances is a high stage of protection smooth when the big data user-produced arbitrary bit sign is an intermittent third party. Moreover, used in allotment users' safety consciousness communication information since it outdoes suitable random bits. Moreover, big data client's anonymity is certain because the work passes hash-chained bit series ideals, so the bit run that creates security and privacy alertness statistics is not uncovered gratuitously to a third party [7].

3 Preliminaries

3.1 Background of Elliptic Curve Cryptography

Let q be the large prime and, E indicate an elliptic curve over a prime finite field F_q, demarcated by an equation $y^2 = x^3 + cx + d$ mod q, where, $c, d \in_R F_q$, and $4c^3 + 27d^2$ mod q $\neq 0$. The additive elliptic curve group is construed as $G = \{(x, y) : x, y \in_R F_q; (x, y) \in_R E\} \cup \{\Theta\}$, where the point Θ is identified as a point at infinity that performs as the identity element in G. The point addition in elliptic curve is $P = (x_P, y_P) \in_R G$ and $Q = (x_Q, y_Q) \in_R G$, where $Q \neq P$ then $P + Q = (x_i, y_i)$, where $x_i = \mu^2 - x_P - x_Q \bmod q$, $y_i = (\mu(x_P - x_Q) - y_P)$ mod q and $\mu = (y_Q - y_P)/(x_Q - x_P)$. The scalar multiplication on the group G is explained as $nP = P + P + P + \cdots + P(n \text{ times})$. The more descriptions of elliptic curve group are given in [6].

3.2 Computational Problem

Elliptic Curve Discrete Logarithms Problem (ECDLP): For given $P, Q \in_R G$, find $t \in_R Z_q^*$ such that $P = tQ$, which is hard.
Elliptic Curve Computational Diffie–Hellman Problem (ECCDHP): For $a, b \in_R Z_q^*$ and g is the base point of G, given (g, ag, bg), then compute abg is hard to the group G.

4 The Proposed Protocol

In big data atmospheres, the services have to offer everyplace, at anytime users need them. At the same time, a user in an unambiguous place where big data service are contributed transfers to any more place, the server offering this facility has to make the individual data to client in the space provision region with safety occupation and guarantee that a separate statistics of a client is not proved to a third party throughout the maintenance duration. At the same duration, there are multiple servers, to offering this service, and the servers cannot be placed in the similar location. Client getting big data facilities must adequate to get big data facility area at their need places, not from an individual network server.

In the presented framework, the user identifiers reduced throughout the client authentication development, so that clients can offer to similar big data approach with another service networks. Now, here multiple data server is offering this facility, and they cannot be in the similar place. The server associated to a new network launches synchronization by the preceding server thus, it may accept the client identity from the earlier server and verified client. The client offered these facilities from the nearby

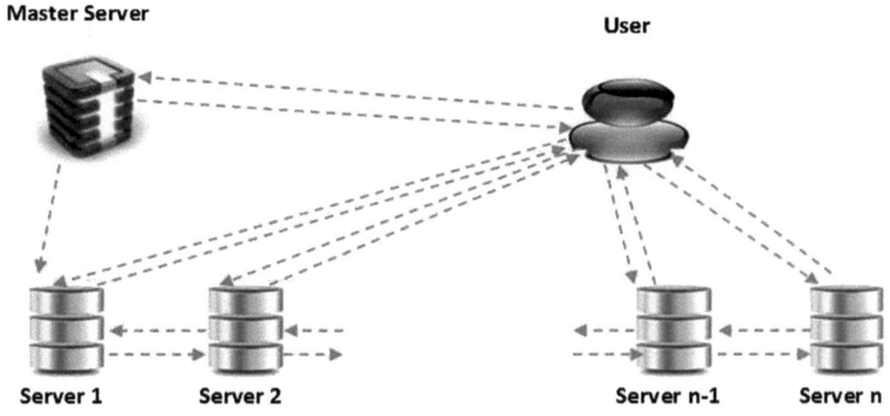

Fig. 1 The architecture for the proposed protocol

data server from the user's novel place, and not from a detailed server. Figure 1 demonstrates the complete development client authentication by the server in the proposed framework, with every server offering assistances. Although, launching synchronization to other networks. If the client requirements to offer with this type of facility in a different network, the server obtains the identity and valid data from the preceding server to validate the client earlier contributing services. This protocol scheme has the following three phases: initialization phase, user registration phase and mutual authentication, and session key agreement.

4.1 Initialization Phase

Assume that big data server BS plays the role of key distribution center (KDC) which takes a security parameter l, returns security parameter. For given l, BS takes the following steps:

Step L1: Choose an arbitrary generator $g \in_R G$.
Step L2: Select a master key $x \in_R Z_q^*$ and public key $P_{BS} = x \cdot g$.
Step L3: Choose collision free one-way hash functions

$$H_1 : \{0,1\}^* \times \{0,1\}^* \times G \times G \times G \times G \to \{0,1\}^l;$$
$$H_2 : \{0,1\}^* \times \{0,1\}^l \to \{0,1\}^l; \quad H_3 : \{0,1\}^* \to \{0,1\}^l;$$

Publish systems parameters $\langle F_q, E, G, l, g, P_{BS}, H_1, H_2, H_3 \rangle$.

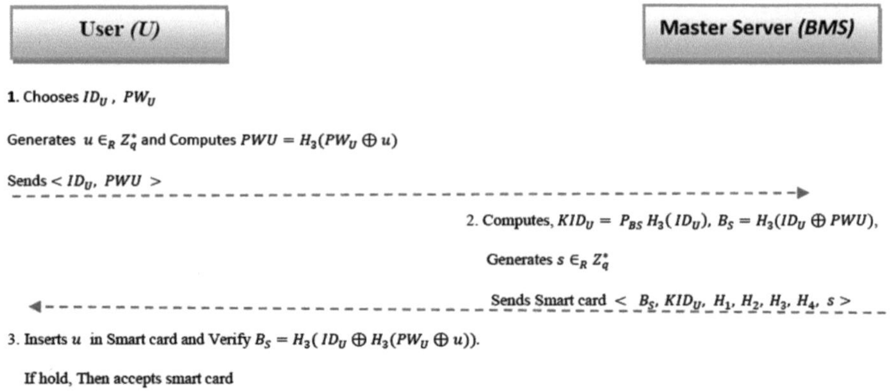

Fig. 2 User registration phase

4.2 User Registration Phase

Step R1: User U takes her/his identifier $\mathrm{ID_U}$ password $\mathrm{PW_U}$, and generates random number $u \in_R Z_q^*$ then, calculates $\mathrm{PW_U} = H_3(\mathrm{PW_U} \oplus u)$ and sends $\langle \mathrm{ID_U}, \mathrm{PW_U} \rangle$ to big data master server (BMS).

Step R2: Upon receiving massage from U at time T, then BMS computes, user authenticated key $\mathrm{KID_U} = P_{\mathrm{BS}} \cdot H_3(\mathrm{ID_U})$, $B_\mathrm{S} = H_3(\mathrm{ID_U} \oplus \mathrm{PW_U})$ and generates random number $s \in_R Z_q^*$. BMS store and sends smart $\langle B_\mathrm{S}, \mathrm{KID_S}, H_1, H_2, H_3, H_4, s \rangle$ to U, where s is secret key BMS.

Step R3: Upon getting the smart card, U inserts random number u in card. Then, the smart card is $\langle B_\mathrm{S}, \mathrm{KID_U}, H_1, H_2, H_3, H_4, s, u \rangle$.

Step R4: U enters his/her $\mathrm{ID_U}$ and $\mathrm{PW_U}$ to verify whether $B_\mathrm{S} = H_3(\mathrm{ID_U} \oplus H(\mathrm{PW_U} \oplus u))$. If it is hold, U admits the smart card (Fig. 2).

4.3 Mutual Authentication and Session Key Agreement

User U asks a service nearby big data sever BMS. U and BMS authenticate to each other and create a session key as:

Step M1: U sends his/her identity $\mathrm{ID_U}$ and password $\mathrm{PW_U}$ to login with nearby big data server *NBS* and computes $B_\mathrm{U} = H_3(\mathrm{ID_U} \oplus \mathrm{PW_U})$. Checks whether $B_\mathrm{U} =? B_\mathrm{S}$. If it is holds, then chooses a random number $R_\mathrm{U} = (x_\mathrm{U}, y_\mathrm{U}) \in_R G$ and computes $\mathrm{U}_1 = H_3(T_1)$, $M_\mathrm{U} = R_\mathrm{U} + \mathrm{U}_1 \cdot \mathrm{KID_U}$ at time stamp T_1 and anonymous identities $\mathrm{ID_U} = \mathrm{ID_U} \oplus H_2(\mathrm{ID_U} \| T_1)$. U sends $\langle T_1, \mathrm{ID_U}, R_\mathrm{U}, M_\mathrm{U} \rangle$ to NBS. Where T_1 is the current date and time of user U.

Step M2: On getting the message, NBS computes $T_2 - T_1 \leq \Delta T$, where T_2 the message getting time of NBS and ΔT is the suitable time delay in message communication. This is not hold, then NBS terminate the session otherwise computes $\text{KID}_{\text{NBS}} = P_{\text{BS}} H_3(\text{ID}_{\text{NBS}})$. Where ID_{NBS} is the identity of sever NBS. After that, NBS generates random number $R_{\text{S}} = (x_{\text{S}}, y_{\text{S}}) \in_R G$, and $x_s \in_R Z_{\text{q}}^*$, then computes $S_1 = H_3(T_3)$, $M_{\text{S}} = R_{\text{S}} + S_1 \text{KID}_{\text{NBS}}$, session key $sk_{\text{NBS}} = H_1(\text{ID}_{\text{U}} \| \text{ID}_{\text{NBS}} \| M_{\text{S}} \| M_{\text{U}} \| R_{\text{S}} || R_{\text{U}})$ and message authentication code $\text{MAC}_{\text{NBS}} = (sk_{\text{NBS}} + x_{\text{S}})g$ at time stamp T_3. NBS sends message $\langle T_3, M_{\text{S}}, R_{\text{S}}, \text{MAC}_{\text{NBS}} \rangle$ to U. Where T_3 time and date sends message to U.

Step M4: On getting the message, U verified condition $T_4 - T_3 \leq \Delta T$, where T_4 the message receiving date and time with ΔT is the suitable time delay in message communication. If condition is hold, then U generates $x_u \in_R Z_{\text{q}}^*$, then computes session key $sk_{\text{U}} = H_1(\text{ID}_{\text{U}} \| \text{ID}_{\text{NBS}} | M_{\text{S}} \| M_{\text{U}} \| R_{\text{S}} \| R_{\text{U}})$, and message authentication code $\text{MAC}_{\text{U}} = (sk_{\text{U}} + x_{\text{U}})g$. U check the condition $\text{MAC}_{\text{U}} =? \text{MAC}_{\text{NBS}}$. If the condition holds *NBS* is authenticated by *U* with session key $sk = sk_{\text{U}} = sk_{\text{NBS}}$.

Finally, session creates, where user can store/access her/his data securely over the public channel.

5 Security Evaluation

In this phase, we discussed the scheme safe against the following security properties:

- **Identity Management**: The server stores all the registered identities ID_{U} of users U in the database and verify accessibility of unique identities in each fresh registration.
- **Anonymity**: In mutual authentication and session key agreement, user sends anonymous identity $\text{ID}_{\text{U}} = \text{ID}_{\text{U}} \oplus H_2(\text{ID}_{\text{U}} || T_1)$ to big data server instead of real identity ID_{U}. The identity ID_{U} is XOR with the hashed significance of ID_{U} and T_1. Where assumed that hash function is secure in cryptosystem.
- **User Privacy**: The suggested framework never communicates user secret data in plaintext form. The messages $\langle T_1, \text{ID}_{\text{U}} = \text{ID}_{\text{U}} \oplus H_2(\text{ID}_{\text{U}} \| T_1), R_{\text{U}}, M_{\text{U}}, K_{\text{U}} \rangle$ and $\langle T_3, M_{\text{S}}, R_{\text{S}}, \text{MAC}_{\text{NBS}} \rangle$ are communicated over the public channel. Evidently, these communications cannot be interpreting easily to get identifier, password, etc. Hence, the protocol protests user privacy.
- **Replay Attack**: Replay attack is utmost common attack in authentication development. On the other hand, the mutual countermeasures are random number mechanism and time stamp. The presented protocol, adopt the random number mechanism and time stamp. The messages, mutual authentication phase U→ NBS and NBS → U are with time stamps. Hence, this protocol is strong against replay attack.
- **Mutual Authentication**: Mutual authentication is a significant attribute to a verification assistance conflicting to server spoofing attack. The proposed protocol

offers a mutual authentication for U and server BS or NBS by ECC-based private and public key exchange.

- **No Key Control**: In this protocol, user U and server NBS has a contribution into the session key neither accomplice can strength the complete session key to be a preselected significance. The session key $sk = H_1(\mathrm{ID_U}\|\mathrm{ID_{NBS}}\|M_\mathrm{U}\|M_\mathrm{S}\|R_\mathrm{U}\|R_\mathrm{S})$ depends on $M_\mathrm{U} = R_\mathrm{U} + \mathrm{U}_1 \cdot \mathrm{KID_U}$ and $M_\mathrm{S} = R_\mathrm{S} + S_1 \cdot \mathrm{KID_{NBS}}$. Thus, M_U or M_S depends on random number and hash function, therefore, any advisory cannot manage the result of the session keys or impose others.
- **Man-in-the middle Attack**: Server and user authenticate each other beyond articulate. An attacker can attempt man-in-the middle attack by conveyance the fake communication. However, to authenticate each other, server and the user interchange message authentication code (MAC). To computes MAC value, knowledge of hash value is required, then hash value does not easy to compute by everyone.
- **Key offset Attack**: In this protocol, server and user authenticated each other and established session key $sk = H_1(\mathrm{ID_U}\|\mathrm{ID_{NBS}}\|M_\mathrm{U}\|M_\mathrm{S}\|R_\mathrm{U}\|R_\mathrm{S})$. User message authentication code $\mathrm{MAC_U} = (sk_\mathrm{U} + x_\mathrm{U})g$ and server message authentication code $\mathrm{MAC_{NBS}} = (sk_\mathrm{NBS} + x_\mathrm{S})g$. If $\mathrm{MAC_U} \neq \mathrm{MAC_{NBS}}$, user U rejects the session key agreement and drives the authentication-failed message to *NBS*. Therefore, key offset attack is not possible in this protocol.
- **Session Key Agreement**: A session key $sk = H_1(\mathrm{ID_U}\|\mathrm{ID_{NBS}}\|M_\mathrm{U}\|M_\mathrm{S}\|R_\mathrm{U}\|R_\mathrm{S})$ is established between *U* and *NBS* after authentication procedure. This key is common for server and user. Adversary cannot access the session key of user.
- **Impersonation Attack**: The protocol never broadcasts user's identity $\mathrm{ID_U}$ and password PWU steadfastly through the public communication. In mutual authentication and session key agreement, phase user used $\mathrm{ID_U} = \mathrm{ID_U} \oplus H_2(\mathrm{ID_U}\|T_1)$ and $B_\mathrm{U} = H_3(\mathrm{PW_U} \oplus u)$, those are depending on hash value. Therefore, the presented scheme is strong against impersonation attack.
- **Perfect forward Secrecy**: An attacker cannot calculate session key because to calculate session key $sk = H_1(\mathrm{ID_U}\|\mathrm{ID_{NBS}}\|M_\mathrm{U}\|M_\mathrm{S}\|R_\mathrm{U}\|R_\mathrm{S})$, where to computes $\mathrm{KID_{NBS}}$ or $\mathrm{KID_U}$ is equivalent to ECCDHP in ECC and $\mathrm{M_U}$ or M_S used the XOR value which also secure.
- **PKG forward Secrecy**: An attacker cannot even calculate the user or server message authentication code $\mathrm{MAC_U} = (sk_\mathrm{U} + x_\mathrm{U})g$ or $\mathrm{MAC_{NBS}} = (sk_\mathrm{NBS} + x_\mathrm{S})g$. To computes sk_U or sk_NBS is equivalent ECCDHP in ECC and computes $\mathrm{MAC_U}$ or $\mathrm{MAC_S}$ is equivalent ECDLP in ECC.

6 Performance Analysis

In this session, the paper discussed user registration phase and mutual authentication with key agreement phase which is the main computation cost of mechanism.

User registration phase	Mutual authentication and session key agreement
4H +1PM+4XOR	7H+5PM+2PA+2XOR

H Hash function; *XOR* XOR operations; *PM* Elliptic curve point scalar multiplication operations; *PA* Elliptic curve point addition operations

7 Conclusion

As the contributions of the paper, we proposed an identity-based authentication framework for big data security which is highly secured mutual authenticated in big data environment. In this paper, user obtained the smart card in registration phase and mutually authenticated to nearby big data server, and established session key over public networks in mutual authentication and session key agreement phase. Further, this paper shows the security analysis of this protocol against several security attacks in environment. Lastly, we showed performance of the presented protocol.

References

1. J.M. Cavanillas, E. Curry, W. Wahlster, *New Horizons for a Data-Driven Economy* (Springer, 2016). https://doi.org/10.1007/978-3-319-21569-3
2. B. Dana, C. Kate, in *Six Provocations for Big Data*. Social Science Research Network. A Decade in Internet Time: Symposium on the Dynamics of the Internet and Society. September 21, 2011. https://doi.org/10.2139/ssrn.1926431
3. J. Deam, S. Ghemawat, Mapreduce: simplified data processing on large clusters. Commun. ACM **51**(1), 107–113 (2008)
4. DT&SC 7-3, *What is Big Data?* 12 Aug 2015—via YouTube
5. https://www.tutorialspoint.com/hadoop/hadoop_big_data_overview.htm
6. D. Hankerson, A.J. Menezes, S. Vanstone, *Guide to Elliptic Curve Cryptography* (Springer, 2004)
7. Y.S. Jeong, S.S. Shin, An efficient authentication scheme to protect user privacy in seamless big data services. Wireless Pers. Commun. **7–19**, 86 (2016). https://doi.org/10.1007/s11277-015-2990-1
8. H. Martin, Big data for development: a review of promises and challenges. Dev. Policy Rev. http://www.martinhilbert.net. Retrieved 07 Oct 2015
9. Y.S. Jeong, Y.T. Kim, A token-based authentication security scheme for Hadoop distributed file system using elliptic curve cryptography. J. Comput. Virol. Hack Tech. **11**, 137–142 (2015). https://doi.org/10.1007/s11416-014-0236-5

Performance Analysis of Clone Node Attack in Wireless Sensor Network

Sachin Lalar, Shashi Bhushan and Surender

Abstract Nowadays wireless sensor networks are widely used in different applications. The collection of sensor nodes consists of a Wireless Sensor Network (WSN). The sensor nodes, which have small memory, limited battery and limited processing power, sense the data and transmit to base station. Due to its characteristics, WSN can be vulnerable to different types of attacks and clone node attack is one of them. In this attack, attacker selects the few nodes and imitates them using the secret information of the selected node and then deploys in the network. In this paper, we find out the effect of clone node attack in the network. We have compared the performance of WSN in the presence of clone node attack and without clone node attack under six different scenarios in Network Simulator (NS2). We evaluate the effect of clone node attacks in term of Packet Delivery Ratio and make the comparison of packet loss % in normal and Clone node networks.

Keywords WSN · Clone node attacks · AODV · NS2

1 Introduction

WSN is the combination of thousands of sensor nodes and these nodes consists of battery, memory, power supply, transmitter, controller, and many sensors. The nodes deployed in sensor networks are categorized as base and sink nodes. Communication between these sensor nodes is possible through radio signals. These nodes sense, processes, and communicates to the base node. WSN is ad hoc in nature, so can be deployed in any area (harsh or soft) like military, environmental observation, etc. As

S. Lalar (✉)
CSE, IKGPTU, Jalandhar, India
e-mail: sachin509@gmail.com

S. Bhushan
CGC, Landran, Ajitgarh, Punjab, India

Surender
GTB, Bhawanigarh, Punjab, India

C. R. Krishna et al. (eds.), *Proceedings of 2nd International Conference on Communication, Computing and Networking*, Lecture Notes in Networks and Systems 46, https://doi.org/10.1007/978-981-13-1217-5_8

these are deployed in harsh environment, so are more vulnerable to security attacks. In the wireless sensor network, different nature of attacks on WSN are possible such as Selective forwarding attack, Acknowledgement spoofing, HELLO flood, Black hole attack, DoS attack, clone Attacks, and many more [1–3].

The purpose of the paper analyzes the performance of clone node attacks in WSN and effects of clone node attack in the network. This paper describes the effect of clone node attack in wireless sensor network. The organization of the paper is as follows: Sect. 2 gives the brief explanation of the Clone Node Attack. Section 3 represents the implementation and analysis. Section 4 concludes our work.

2 Clone Node Attack

Clone Node Attack is known as Replication attack which is identity-based attack. The attacker selects few nodes from the web and steals the secret information from the captured node. By this information the attacker can easily replicate the same nodes and deployed these clone nodes at numerous locations in the network. The clone nodes have the same identity as of authorized node. The clone nodes can communicate with their neighbor nodes as they appear in the valid nodes. A scenario of clone nodes is shown in Fig. 1 in which node A is replicated and deployed at various places. It can communicate to its neighbor nodes in the network [4–9].

With the help of clone node attack, attacker can harm the network by launching different types of attacks. The clone node can discard the valid packet which is received from the legal nodes, hence deteriorating the performance of the network. The effect of this attack will be analyzed in the NS2 simulator. We will use the AODV routing protocol in WSN and following section explains the AODV routing protocol.

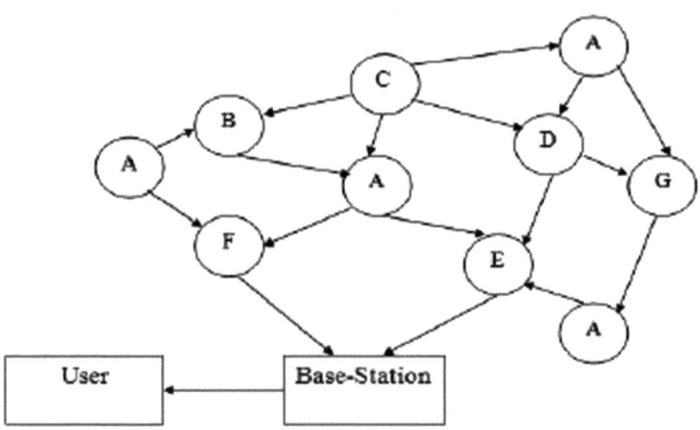

Fig. 1 Clone node attack example

3 Ad hoc on Demand Distance Vector Routing Protocol

The AODV protocol is one of the efficient protocols used for MANET where topology remains same. It finds the path to destination when required. This is the reason AODV is known as reactive routing protocol. AODV uses three types of messages: RREQ, RREP, and RERR. The RREQ message is used to find out the path to a destination. First, the source sends the RREQ message to its neighbors. The neighbor will forward that request to its neighbor until it reaches to the destination. The destination responds the request using RREP message. The source stores the information of all nodes which are in between path. If the path between source and designation is broken or damaged, then neighbor forwards the RERR message to the destination as well as source.

The method of forwarding message in the network is demonstrated in Figs. 2 and 3. The AODV routing can be vulnerable to different types of attacks. If attacker (Clone node) responds that it knows the fastest path to the receiver and that path is through clone node, then all the packets will be passed through clone node. The clone node can drop all the packets which are received from the legal node.

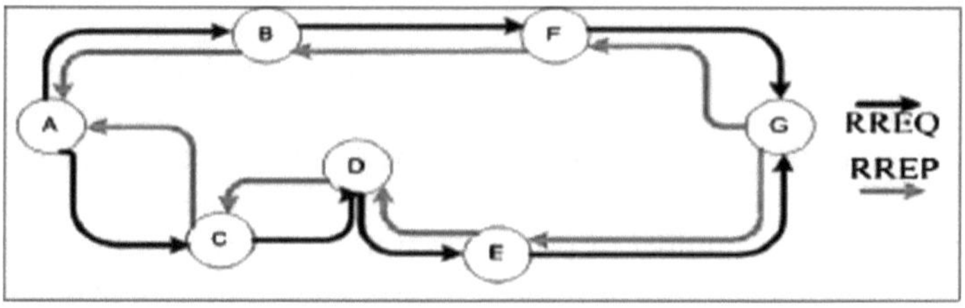

Fig. 2 Route discovery example in AODV

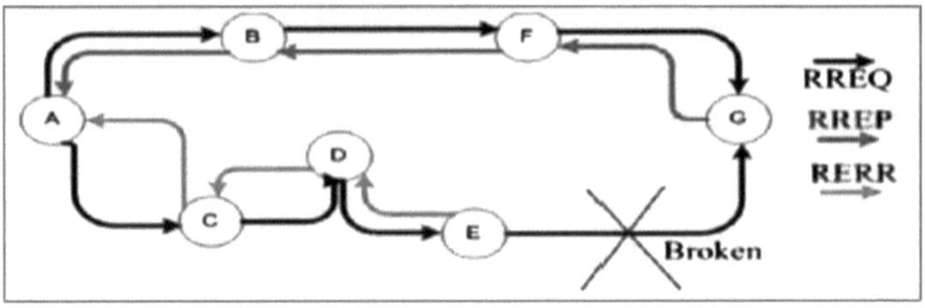

Fig. 3 AODV route error message

4 Simulation and Result

We have implemented the clone node attack in NS2 simulator. First we analyzed the performance of the wireless sensor network of 20 nodes without clone node attack in six different scenarios. In the same scenarios, we introduce the clone node in the network; then check the performance of the network. We count the number of packets sent by the sending nodes and how many of them reached the receiving nodes to get the lost packets.

In each scenario, we have found out the result from three graph/figures. The first figure in the scenario illustrates the packets forward and received in normal network when there exists no clone node.. The second figure demonstrates the forwarded packets, received packets, and drop packets in the presence of the clone node attack. The third figure shows the comparison of the loss % in the network. The loss % of the network is equal the number of packets sent to the number of packets received. In the first scenario, the number of packets received and sent is shown in Fig. 4.

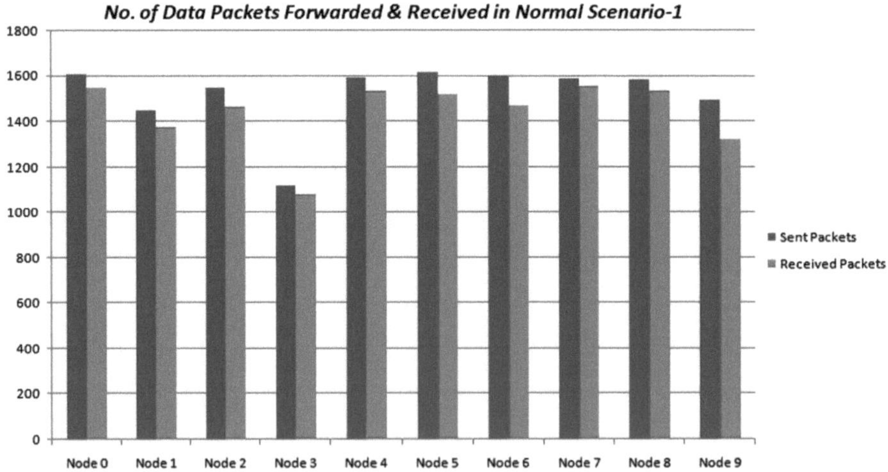

Fig. 4 Data packets forwarded and received in normal scenario-1

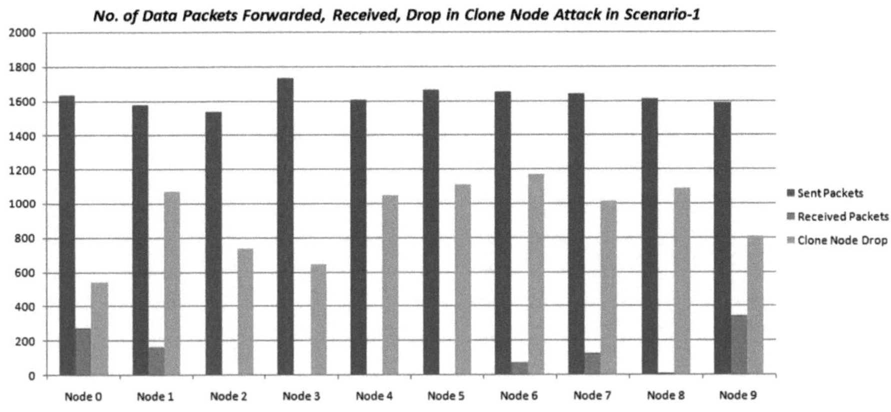

Fig. 5 Data packets in clone node attack scenario-1

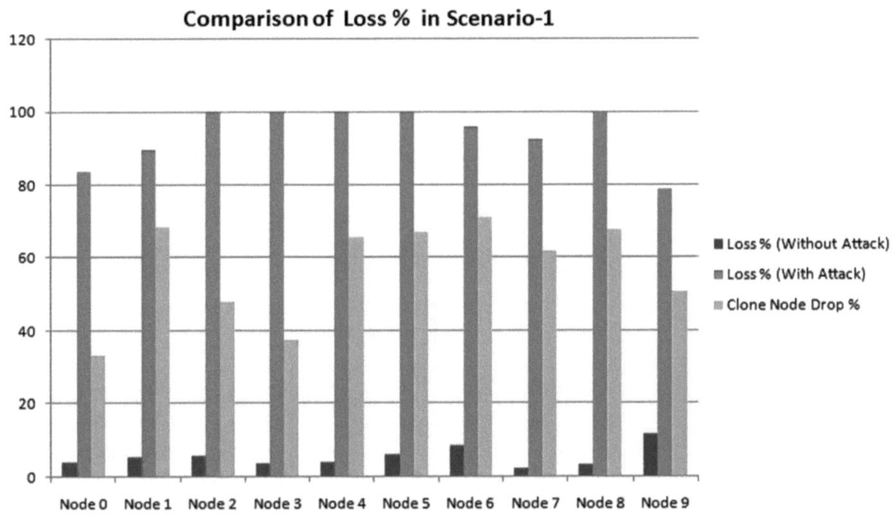

Fig. 6 Loss % comparison in scenario-1

Figure 5 shows the number of packets sent, received, and packets dropped in the same scenario when clone node is present in the network. Figure 6 shows the loss % without clone attack and clone attack in the network in scenario-1.

Figure 7 shows the loss % without clone attack and clone attack in the network in scenario-2. Figure 8 shows the loss % without clone attack and clone attack in the network in scenario-3. Figure 9 shows the loss % without clone attack and clone attack in the network in scenario-4.

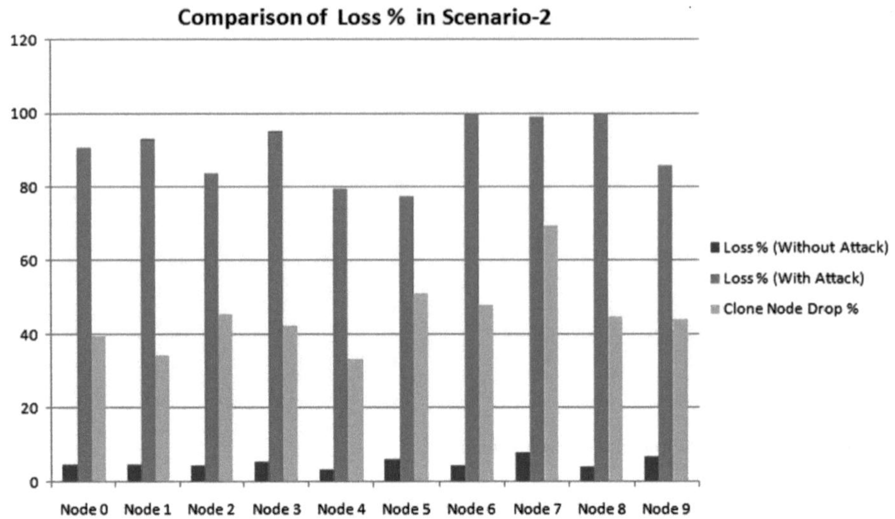

Fig. 7 Loss % comparison in scenario-2

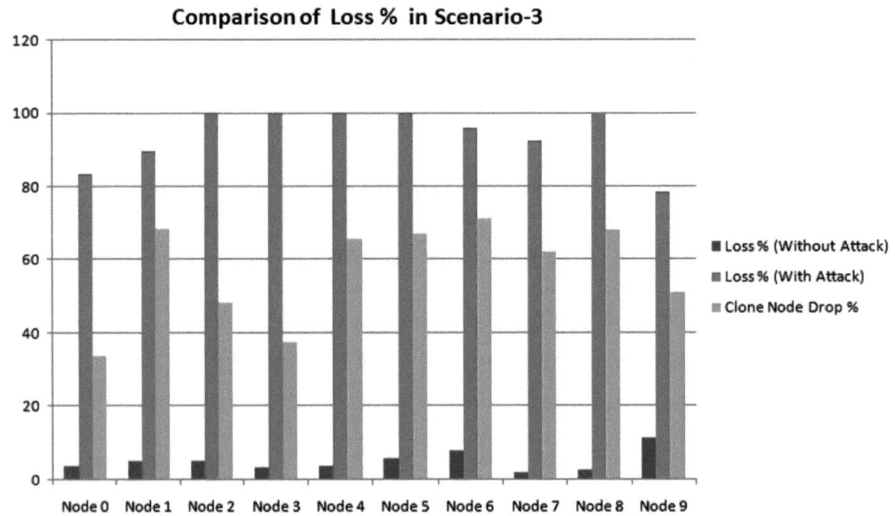

Fig. 8 Loss % comparison in scenario-3

Figure 10 shows the loss % without clone attack and clone attack in the network in scenario-5. Figure 11 shows the loss % without clone attack and clone attack in the network in scenario-6.

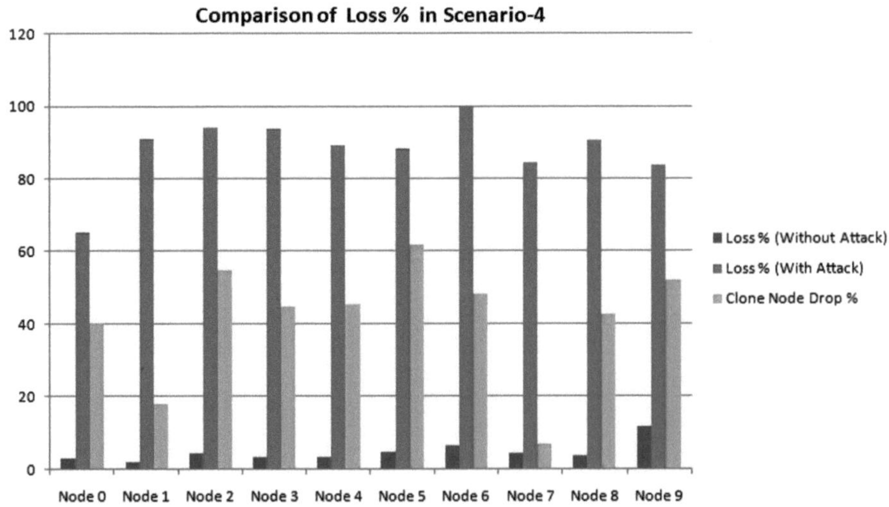

Fig. 9 Loss % comparison in scenario-4

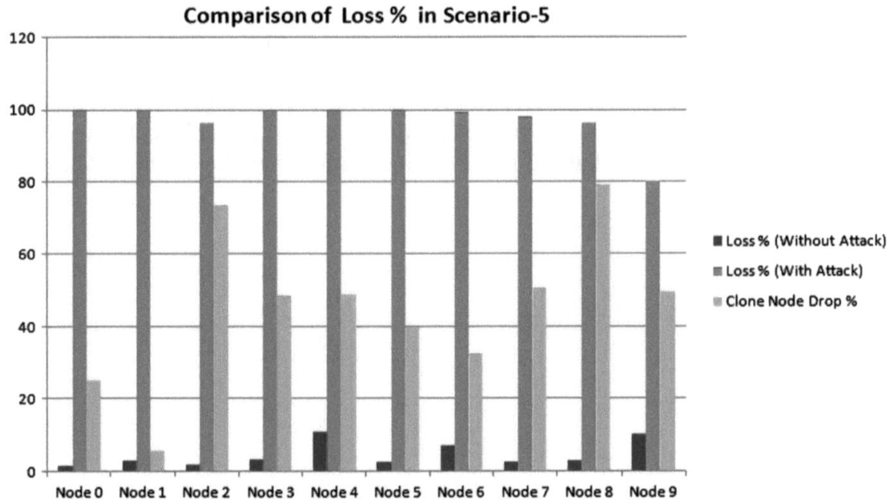

Fig. 10 Loss % comparison in scenario-5

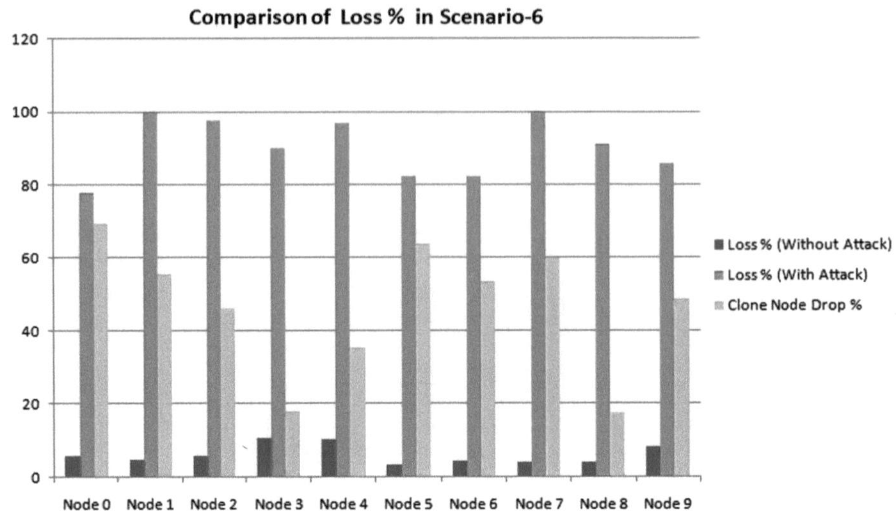

Fig. 11 Loss % comparison in scenario-6

Analysis

It has analyzed that when clone node is present in the network, then its performance will be degraded. The number of packet loss is increased in each scenario. From Fig. 12, it has been shown that the loss % is increased when the clone node attack

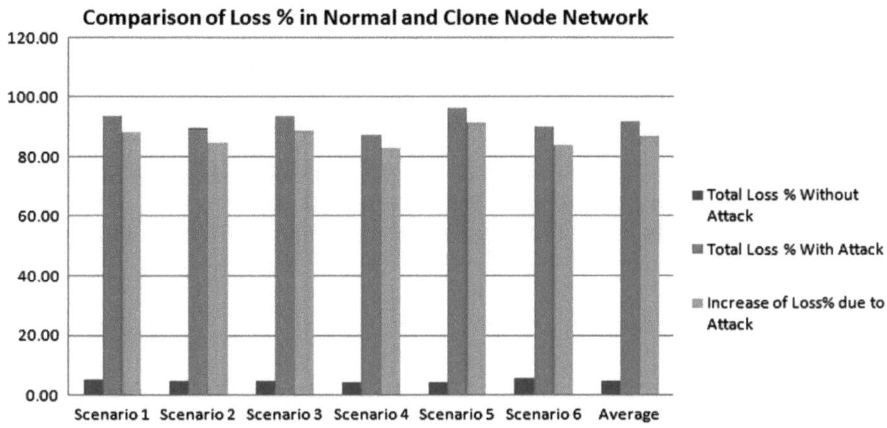

Fig. 12 Comparison of loss %

is above 80% in each scenario. The average loss % of scenarios is near to 84% due to clone node attack. The performance of the network is degraded by the clone node attack.

5 Conclusion

The wireless sensor networks are susceptible to different security attacks. In this paper, we analyzed the effect of clone node attack on the wireless sensor network. This paper analyzed that the performance of the network will be deteriorated in the presence of the clone node. The average loss % of scenarios is near to 84% due to clone node attack. The future work will be detection and prevention of replication attack in wireless sensor network.

References

1. P. Mohanty, S. Panigrahi, N. Sarma, S. Satapathy, Security issues in wireless sensor network data gathering protocols: a survey. J. Theor. Appl. Inform. Technol. **13**(1), 14–29 (2005)
2. M. Sharifnejad, M. Sharifi, M. Ghiasabadi, S. Beheshti, *A Survey on Wireless Sensor Networks Security. Fourth International Conference: Sciences of Electronic Technologies of Information and Telecommunication*, March 25–29 (2007)
3. X. Li, L. Qin, I. Liang, in *Research on Wireless Sensor Network Security*. IEEE Computer Society, International Conference on Computational Intelligence and Security (2010)
4. T. Bonact, P. Lee, L. Bushnell, R. Poovendra, in *Distributed clone detection in wireless sensor networks: an optimization approach*. 2nd IEEE International Workshop on Data security and Privacy in Wireless Networks Proceedings, Lucca, Italy (2011)

5. Y. Wang, G. Attebury, B. Ramamurthy, A survey of security networks issues in wireless sensor networks. IEEE Comm. Suv. Tutorials **8**(2), 2–23 (2006)
6. I.F. Akyildiz, S. William, S. Yogis, R. Cayirci, A survey on sensor network. IEEE Commun. Mag. **40**(8), 102–114 (2002)
7. C. Townsend, S. Arms, *Wireless Sensor Network: Principles and Applications* (Chapter 22), pp. 439–449
8. G. Padmavathi, D. Shanmuga, A survey of attacks, security mechanisms and challenges in wireless sensor networks. Int. J. Comput. Sci. Inf. Secur. **4**(1) (2009)
9. J. Ho, D. Lin, M. Wright, S. Das, *Distributed Detection of Replicas with Deployment Knowledge in Wireless Sensor Networks* (Preprint submitted Elsevier, 2009)

Human Electroencephalographic Biometric Person Recognition System

Nagsen Bansod, Siddharth B. Dabhade, Jitendra N. Dongre, K. V. Shende, S. Bhable, S. Maher, S. Thorat, K. Ankushe, S. Tharewal, V. S. Jadhav and K. V. Kale

Abstract Human head generates various signals according to the situation and activates inside the head as well as outside the head. The frequency of the Head Signal means that brain signal is different as per the level of action taken place by the person; it may be either be imaginary or motor imagery activities. From the brain signals imaginary signals are captured using MindWave Mobile Portable device. Frequency-wise channels are separated and categories as Delta, Theta, Alpha, and Beta. These channels indicated emotions, movement, sensations, vision, etc. Features are extracted of each channel using Power Spectral Density (PSD) function and Deep learning Neural Network. Feature level fusion is used for pattern matching. The Novelty of this work is a single electrode device that is used to capture

N. Bansod · S. B. Dabhade (✉) · J. N. Dongre · K. V. Shende · S. Bhable · S. Maher · S. Thorat
K. Ankushe · S. Tharewal · V. S. Jadhav · K. V. Kale
Department of Computer Science and Info Tech, Dr. Babasaheb Ambedkar Marathwada University, Aurangabad, MS, India
e-mail: dabhade.siddharth@gmail.com

N. Bansod
e-mail: nagsenbansod@gmail.com

J. N. Dongre
e-mail: jiten_7808@yahoo.com

K. V. Shende
e-mail: kanchan.shende4@gmail.com

S. Maher
e-mail: mr.santoshmaher@gmail.com

S. Thorat
e-mail: sandipthorat16@gmail.com

S. Tharewal
e-mail: sumeghtharewal@gmail.com

V. S. Jadhav
e-mail: jadhavvanita098@gmail.com

K. V. Kale
e-mail: kvkale91@gmail.com

C. R. Krishna et al. (eds.), *Proceedings of 2nd International Conference on Communication, Computing and Networking*, Lecture Notes in Networks and Systems 46, https://doi.org/10.1007/978-981-13-1217-5_9

an Electroencephalography (EEG) imaginary data from the head which is generated by brain functioning. The feature level fusion of channels and Deep learning Neural Network classification of feature give better performance. The results are proven that these EEG imaginary signals could be used as better biometrics-based authentication system.

Keywords EEG · Mindwave · Identification · Verification · Biometric

1 Introduction

An electroencephalography (EEG) is a branch of Neuroscience. Recently, researchers in a Neuroscience and computer science attracted towards novel and innovative type of biometric based on neural activity of brain signals, such as EEG signals instead of the biological traits of the human body like face, fingerprint, iris, retina, voice, etc. EEG Signal biometric trait is very difficult to duplicate, break, or guess. A novel approach is used for processing brain signal data through an EEG. The EEG gives various types of information about a person that is emotional, mediation, and sad state. We can analyze EEG Signal and find out the human Concentration, Mathematical solution power, Letter Composition method, Rotational style. These parameters are considered to person identification and verification purpose. This research work is divided into mainly Introduction, Related research work, Proposed Methodology, Experimental Result, and Conclusion.

2 Related Research Work

An EEG signals, data feature measurement through two types of algorithm these are DFT (Discrete Fourier Transform) and WPD (Wavelet packet decomposition). The distinct features of EEG signal are considered with four feature sets. The EEG signal data result was 93, 87, and 93% classification rates of three feature set. By using Multilayer Perceptron Neural Network classifies EEG signal feature data gives 100% recognition rate but limited subjects only to three. In this experiment subject has to seat normally with calm and quiet with closed eyes without any physical activity while collecting the datasets four channels are used [1].

In the recent research, identification and prediction of Motion Sickness (MS) of a driver in real life while driving the vehicle is very interesting and very important task because it can save the lives of so many peoples in traveling. MS provides one type of security to the drive as well as passengers. Prediction of emotions in real time through EEG signals is a challenging task, while performing any activity human brain produces signals and the signals are coming from various parts of our brain. In case of emotions which comes from occipital, parietal, somatosensory, etc., identification of generating signals according to the power of signals are alpha, theta bands. Identification of emotions from the signals in a certain band as per the frequency level is possible through various feature extraction techniques and classification algorithm such as PCA, LDA, BFS, FFS, KNN, SVM, NWFE, ML, etc., apart from that LDA and ML gives 95%. Therefore, it can be used more or less robust techniques for the prediction of MS very effectively [2].

This work presents HHT (Hilbert–Huang Transform) a different approach for biometric identification using electroencephalogram (EEG) signals. After the HHT amplitude and frequency were computed immediately for the classification using salient characteristics. An evaluated two publicly available databases in these scenarios, single electrode of an EEG device which used for biometric data acquisition. A first database consists of 122 subjects and second has total 109 subjects, at the time of collecting the database were subject had shown with a sequence of images on the screen and some mechanical activity or screening works got the 96 and 99% success rate respectively. These results are compared favorably with recent research articles by the various algorithm and classification [3]. A research on biometric using motor imagery EEG signals and Auto Regressive Moving Average (ARMA) are used to construct an estimated model. From that they have used ARMA-based classification system on the basis of Artificial Neural Network (ANN) approach. The extracted features are stored in the specific vector for the identification and verification on the basis of classification. Three persons, four types of the motor imagery EEG data signals were captured and perform the comparative results. Therefore, on the basis of the outcomes of [4] it shows that it can be successfully exploited for purpose of person authentication and identification.

Therefore, an EEG data signal which belongs to motor imagery strongly provides a strong biometric based authentication and identification system will be used for security purpose. At the time of EEG-based development of the system, classification played a vital role. They have compared the results of the system for the identification of imaginary movements of the persons using three different classifiers. H. Jian-Feng has compared LDA (Linear Discrimination Analysis), ANN (Artificial Neural Network) and SVM (Support Vector Machine) for classification of EEG signals, in this result LDA outperforms well as a better classifier than other algorithms [5]. The analysis of EEG data for the biometrics is concentrated on functional connectivity and measurement of time-domain statistical data which is co-dependent on each other. These two approaches are complex relations in EEG data measurement [6]. Abo-Zahad et al. [7] discusses challenges facing while practical implementation of biometric system based on the signals received from the brain for the identification of the person in a real life application.

Database acquisition is a time-consuming procedure, in device setup time is varying when selecting number of channels in the devices. In this case, 64 channels were used to collect 109 people's data; it passes the signals in the middle range called as band pass filtering for establishing functional connection among the sensors and is calculated by the Phase-based Lag Index system. From this connection data matrix is used to build the network to train the system and calculated Eigenvectors. Brain resting state in performing well, but functional connectivity gives proper results, hence it can be a next generation technique for the classification of the data. EEG-based biometric systems, biometric systems based on high-frequency scalp EEG features should be interpreted [8]. An explicitly investigated and assessed the permanence of the non-volitional EEG brainwaves over the course of time. Specifically, we analyzed how much the EEG signal changes over a period of 6 months, since any drastic change would make it unusable as an authentication method. The results are very encouraging, yielding high accuracy throughout the 6-month period [9]. The amplitude of the brain signals is the indication of circadian rhythm which is tactless of the random changes for measuring features bi-variant measure Magnitude Squared Coherence (MSC) are used and reduced the number of channels of EEG signals for identification without any affect on the accuracy of the system.

The multidimensional data classification accuracy is better for fewest numbers of samples per person by using distance-based classifier like KNN (*K*-Nearest Neighbor). In the previous literature, it is found that 64 channel data of 108 subjects gives 100% accuracy, in this case instead of 64 channels only 10 channels are used and also gives 100% recognition rate using 109 subjects' data with eye open resting position, environment for biometric identification [10].

3 Proposed Methodology

3.1 EEG Signal

In this research database is developed using a cost-effective device that is Mindwave mobile and Micromax Canvas A114 mobile phone. The aged isomer programming LLC, free downloadable software in Android OS utilized. It is a portable system used for record database of the forehead with ear reference for database developers. The imaginary activity of letter Composition is captured with five iterations of 30 s.

3.2 Feature Extraction

In this research work we acquired EEG Raw Value, eegRawValueVolts, Attention Level, meditation level, Blink strength, Delta (1–3 Hz), Theta (4–7 Hz), AlphaLow (8–9 Hz), Alpha High (10–12 Hz), BetaLow (13–17 Hz), Gamma Low (31–40 Hz), Gamma mid (41–50 Hz) these seven features.

Apart from above features five features are selected for experiment because we are dealing with normal subject database these features are accept), Gamma Low (31–40 Hz), Gamma mid (41–50 Hz).

Mean sample value (MSV)

Mean of all sample values

$$\text{MSV} = \frac{1}{N} \times \sum_{1=n}^{N} x_n \tag{1}$$

3.3 Power Spectral Density (PSD)

PSD (The Power Spectral Density) to each frequency band is extracted from EEG signals. Four levels decompose the filtered EEG data into five frequency bands, which reflect the physical activities. For each second (128 Sample's) in all channels and bands, a Fast Fourier Transform (FFT) with non-overlapping window was applied to find the PSD per band. Then the PSD is estimated as the average of the squared absolute value of the magnitude of the FFTs, as in Eq. (2)

$$\text{PSD} = \frac{1}{Nyq} \sum_{f=1}^{Nyq} |\text{FFT}|^{\wedge^2} \tag{2}$$

Nyq is the Nyquist frequency, i.e., (sampling frequency/2), and f is the frequency in Hz. The windowing with FFT investigates the influence, Hamming window with length 128 was applied before FFT producing another type of PSD named PSD with hamming because Mindwave device has one electrode wish gives Delta (1–3 Hz), Theta (4–7 Hz), AlphaLow (8–9 Hz), Alpha High (10–12 Hz), BetaLow (13–17 Hz), Gamma Low (31–40 Hz), Gamma mid (41–50 Hz) these seven features channels, obtained (5 bands * 1 electrode). The Welch PSD estimate of an input signal entailing of a discrete-time sinusoid with an angular frequency of $\pi/4$ radians/sample with additive N(0,1) white noise. Create a sine wave with an angular frequency of $\pi/4$ radians/sample with additive $N(0, 1)$ white noise. The signal is 320 samples in length. Obtain the Welch PSD estimate using the default Hamming window and DFT length. The default segment length is 71 samples and the DFT length is the 256 points, yielding a frequency resolution of $2\pi/256$ radians/sample. Because the signal is real-valued, the period gram is one-sided and there are 256/2 + 1 points [11].

3.3.1 Feature Level Fusion Σ

Concatenate the feature set of multiple channel of EEG signal. In this research work the Delta, Theta, Alpha, Beta signal channel of EEG biometric. Let $\delta = \{\delta 1, \delta 2, \delta 3, \ldots, \delta n\}$ an extracted feature of Delta signal, $\theta = \{\theta 1, \theta 2, \theta 3, \ldots, \theta n\}$ an extracted feature of Theta signal, $\alpha = \{\alpha 1, \alpha 2, \alpha 3, \ldots, \alpha n\}$ an extracted feature of Alpha signal and $\beta = \{\beta 1, \beta 2, \beta 3, \ldots, \beta n\}$ an extracted feature of Beta signal obtained by concatenating supplementing normalize feature vector and performing

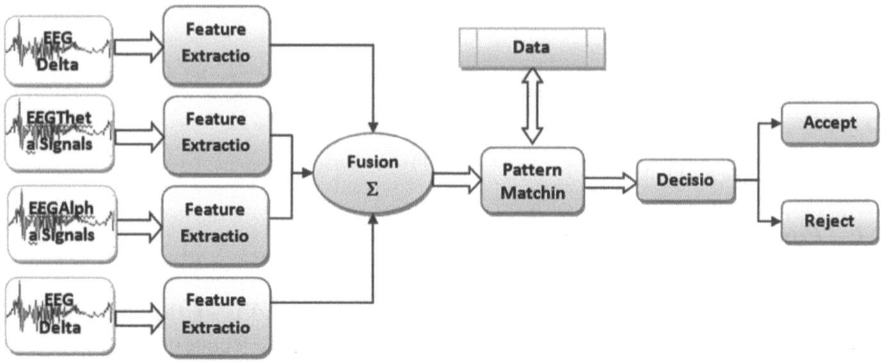

Fig. 1 Proposed methodology

feature selection on resultant fused feature vectors. We conduct extensive experiments to evaluate the efficiency and healthiness of the proposed system (Fig. 1).

3.4 Pattern Matching

3.4.1 Manhattan Distance Metric

Distance is measured of two points $X(x1, y1)$ and $Y(x2, y2)$ along with the axes of the plane with right angles, it is

$$\text{Distance} = |x1 - x2| + |y1 - y2| \tag{3}$$

3.5 Data Model

The extracted features of the data are stored in the data model using .mat file for pattern matching.

3.6 Decision

After pattern matching here, it used accept or reject decision on the basis of data available in the data model.

In this experiment, we have used 40 subjects with two sessions, but here we show only 10 subjects data in the above table because of space limitations.

4 Result and Discussion

In this experiment, the KVKRG EEG database consists of 40 subjects with five iterations and two sessions, i.e., summer and rainy. Calculated 13 EEG features are shown in Tables 1 and 2, i.e., eegRawValue, eegRawValueVolts, Attention Level, meditation level, Blink strength, Delta (1–3 Hz), Theta (4–7 Hz), Alpha Low (8–9 Hz), Alpha High (10–12 Hz), Beta Low (13–17 Hz), GammaLow (31–40 Hz) and Gamma mid (41–50 Hz), all these channels are considered as a feature. At the first time we have selected Delta (1–3 Hz) for the training of each subject and one of the single subjects to test but the result is not satisfactory. In the second experiment, the Theta (4–7 Hz) signal is considered for training features and one by one Theta signal is measured in testing but result not so good.

In the third experiment, the alphaLow (8–9 Hz) signal is considered for training features and one by one alphaLow signal is considered for testing which gives good performance. In the successive experiment, alphaHigh (10–12 Hz) used for training feature sounds better result compared to earlier experiment. BetaLow (13–17 Hz), GammaLow (31–40 Hz), GammaMid (41–50 Hz) is also exploit on same experiment but not considerable result.

In experiment no. 8, fusion of low alphaLow and alphaHigh features give better performance, therefore it is suitable for biometric.

Table 1 EEG data feature set

eegRaw Value	Delta	Theta	Alpha		Beta		gammaLow	gammaMid
			alphaLow	alphaHigh	betaLow	betaHigh		
51	16,745,434	16,750,096	10,235	9371	6354	4364	4869	3224
102	665,851	91,047	30,130	7637	7995	25,956	16,752,974	12,957
19	206,082	51,834	4173	24,355	14,860	16,748,314	24,462	18,314
28	301,085	38,758	8633	24,707	16,746,703	16,759,024	16,537	17,825
−279	1992969	99,204	15,682	68,362	26,465	27,213	16,761,920	13,203
89	483,475	46,766	68,636	70,677	16,752,442	83,017	16,597	6303
−349	16,759,164	16,756,526	16,770,661	12,770	5000	18,091	30,031	8256
−300	285723	23,618	2533	8622	4811	4421	3809	3397
24	16,765,681	28,861	16,745,500	9428	17,691	23,362	16,747,113	16,853
77	206,582	10,057	1356	5975	6416	5912	11,819	4438
48	60,769	16,746,334	16,744,346	16,767,015	15,636	26,680	23,915	6900
51	364,775	16,773,343	16,749,070	16,756,417	8894	22,048	16,496	10,577
17	1,446,025	16,769,628	24,715	14,403	16,756,180	16,766,368	22,151	8556
−264	97,468	12,890	10,171	5712	8070	6106	10,326	8207
34	273,061	158,117	83,732	26,191	8947	22,988	25,866	6478
17	616,108	19,593	14,400	16,751,418	16,755,278	30,792	15,253	12,013
39	75,894	103,275	16,749,335	23,289	21,902	22,061	17,374	8294

Table 2 Distance matrix

0	7,730,991.9491597	6,116,394.35594513	7,370,273.38984377	7,853,236.59323303	11,985,746.0169656	14,089,220.7796784	8,682,605.44068882	12,744,309.8305209	9,393,457.00000784
7,730,991.9491597	0	6,763,899.42373458	5,456,663.71187051	8,466,063.15254791	9,447,346.57628045	11,974,204.5254339	7,077,255.62712234	8,097,867.77966625	8,597,073.96610238
6,116,394.35594513	6,763,899.42373458	0	9,246,123.4406783	7,940,199.69491549	7,003,493.11864757	14,164,368.7627163	8,808,193.25423936	10620265.4067802	9,675,137.18644576
7,370,273.38984377	5,456,663.71187051	9,246,123.4406783	0	6,404,288.18644125	9,653,996.69491842	11,183,944.1694956	9,126,478.66101936	9,998,291.15254323	6,559,131.77966643
7,853,236.59323303	8,466,063.15254791	7,940,199.69491549	6,404,288.18644125	0	5,306,736.47458001	6,803,723.03390297	5,317,973.01695099	7,220,363.33898335	8,087,028.67797095
11,985,746.0169656	9,447,346.57628045	7,003,493.11864757	9,653,996.69491842	5,306,736.47458001	0	8,348,019.84745855	6,841,051.08475133	7,160,480.35593623	9,936,828.94916113
14,089,220.7796784	11,974,204.5254339	14,164,368.7627163	11,183,944.1694956	6,803,723.03390297	8,348,019.84745855	0	7,079,411.06780311	9,844,532.91525919	14,272,766.2203485
8,682,605.44068882	7,077,255.62712234	8,808,193.25423936	9,126,478.66101936	5,317,973.01695099	6,841,051.08475133	7,079,411.06780311	0	8,326,343.4067812	11,133,325.0847488
12,744,309.8305209	8,097,867.77966625	10,620,265.4067802	9,998,291.15254323	7,220,363.33898335	7,160,480.35593623	9,844,532.91525919	8,326,343.4067812	0	8,998,864.42373337
9,393,457.00000784	8,597,073.96610238	9,675,137.18644576	6,559,131.77966643	8,087,028.67797095	9,936,828.94916113	14,272,766.2203485	11,133,325.0847488	8,998,864.42373337	0

Classification of the EEG feature is the most important to biometric security. The pattern recognition is the best technique for this EEG feature classification. In this experiment, the Manhattan Distance Metric gives 61% classification and recognition rate, i.e., shown in Table 2. Distance Matrix.

5 Conclusion

The innovative is in this novel area, developed EEG data using cost-effective Mindwave mobile device for biometric purpose. We developed our own database of 40 people in two sessions. Features are extracted of EEG channels using PSD. Feature level fusions are of Delta, Theta, Alpha, and Beta channels. Manhattan distance measurement used for the classification gives 61% accuracy of classification of distinct personalities. In future we will increase the data size and one winter session data, and again we will find the unique pattern from that data to person identification.

6 Future Work

The fusion of many electrodes or features may increase the recognition rate for biometric identification and verification purposes. We can perform the fusion approach for fusion of multichannel and it may increase the recognition rate.

Acknowledgements The author would like to acknowledge and thanks to the Program Coordinator for providing me all necessary facilities and access to Multimodal Biometrics System Development Laboratory to complete this work under UGC SAP (II) DRS Phase-I and II F. No. 3-42/2009 & No. F. 4-15/2015/DRS-II (SAP-II), One Time Research Grant No. F. 19-132/2014 (BSR). We would like to acknowledge and thanks University Grants Commission (UGC), New Delhi for awarding me "UGC Rajiv Gandhi & BSR & MANF National Fellowship".

References

1. H.A. Shedeed, A new method for person identification in a biometric security system based on brain EEG signal processing. in *2011 World Congress on Information and Communication Technologies (WICT)*, vol. 2, Mumbai, 2011, pp. 1205–1210
2. Y.H. Yu, P.C. Lai, L.W. Ko, C.H. Chuang, B.C. Kuo, C.T. Lin, An EEG-based classification system of Passenger's motion sickness level by using feature extraction/selection technologies. in *The 2010 International Joint Conference on Neural Networks (IJCNN)*, Barcelona, 2010, pp. 1–6. https://doi.org/10.1109/ijcnn.2010.5596739
3. S. Yang, F. Deravi, Novel HHT-based features for biometric identification using EEG signals. in *2014 22nd International Conference on Pattern Recognition (ICPR)*, Stockholm, 2014, pp. 1922–1927
4. J.F. Hu, Biometric system based on EEG signals: a nonlinear model approach. in *2010 International Conference on Machine Vision and Human-Machine Interface (MVHI)*, Kaifeng, China, 2010, pp. 48–51

5. H. Jian-feng, Comparison of different classifiers for biometric system based on EEG signals. in *Information Technology and Computer Science (ITCS)*, 2010 Second
6. M. Garau, M. Fraschini, L. Didaci, G.L. Marcialis, Experimental results on multi-modal fusion of EEG-based personal verification algorithms. in *2016 International Conference on Biometrics (ICB)*, Halmstad, 2016, pp. 1–6. https://doi.org/10.1109/icb.2016.7550080
7. M. Abo-Zahhad, S.M. Ahmed, S.N. Abbas, State-of-the-art methods and future perspectives for personal recognition based on electroencephalogram signals. IET Biometrics **4**(3), 179–190 (2015). https://doi.org/10.1049/iet-bmt.2014.0040
8. M. Fraschini, A. Hillebrand, M. Demuru, L. Didaci, G.L. Marcialis, An EEG-based biometric system using eigenvector centrality in resting state brain networks. IEEE Signal Process. Lett. **22**(6), 666–670 (2015). https://doi.org/10.1109/LSP.2014.2367091
9. M.V. Ruiz Blondet, S. Laszlo, Z. Jin, Assessment of permanence of non-volitional EEG brainwaves as a biometric. in *2015 IEEE International Conference on Identity, Security and Behavior Analysis (ISBA)*, Hong Kong, 2015, pp. 1–6. https://doi.org/10.1109/isba.2015.7126359
10. B. Singh, S. Mishra, U.S. Tiwary, EEG based biometric identification with reduced number of channels. in *2015 17th International Conference on Advanced Communication Technology (ICACT)*, Seoul, 2015, pp. 687–691. https://doi.org/10.1109/icact.2015.7224883
11. M. Alsolamy, A. Fattouh, Emotion estimation from EEG signals during listening to Quran using PSD features. in *2016 7th International Conference on Computer Science and Information Technology (CSIT)*, 2016, pp. 1–5. https://doi.org/10.1109/csit.2016.7549457

Identifying Various Risks in Cyber-Security and Providing a Mind-Map of Network Security Issues to Mitigate Cyber-Crimes

Neelam Saleem Khan, Mohammad Ahsan Chishti and Mahreen Saleem

Abstract The aim of this review is to acquaint the researcher with the knowledge of cyber-crime. Cyber-crimes are offenses that are carried out against individuals or groups of individuals using computers and networks to commit the crimes. These crimes are carried out intentionally to damage the reputation of a victim or harm the victim physically and psychologically; using modern internet-based communication like emails, chat rooms, notice boards, websites, cell phones, etc. This paper analyzes risk perception in students and precautionary behavior in their use of the internet. It also analyzes various fields that are affected by cyber-crimes that include cyber-bullying in adolescents, cyber-crime in government organizations, and Internet of Things (IoT) and provides an insight into mitigating these hazards. It also provides a mind-map of various network security issues that needs to be considered to avoid cyber-crime.

Keywords Cyber-security · Cyber-crime · Cyber-bullying · IoT
Standard deviation · Fog

1 Introduction

Cyber-crime or computer crime is an unlawful act that includes a computer and a network. The computer may either be the target or it may have been used in the perpetration of a crime. Such crimes may threaten financial health and the nation's

---

N. S. Khan (✉) · M. A. Chishti
Department of Computer Science and Engineering, NIT Srinagar, Srinagar,
Jammu and Kashmir, India
e-mail: nlmkhan01@gmail.com

M. A. Chishti
e-mail: ahsan@nitsri.net

M. Saleem
Department of Computer Engineering, A.M.U, Aligarh, Uttar Pradesh, India
e-mail: mehkhan27@gmail.com

C. R. Krishna et al. (eds.), *Proceedings of 2nd International Conference on Communication, Computing and Networking*, Lecture Notes in Networks and Systems 46, https://doi.org/10.1007/978-981-13-1217-5_10

security. There are a number of ways a cyber-crime can take shape in which the high-profile cases include hacking, copyright infringement, theft of personal data, cyber-stalking, bullying, child pornography, and child grooming. Privacy issues arise when confidential information is intercepted or divulged, lawfully or otherwise.

When certain means are considered to enhance the cyber-security, one aspect that needs to be considered is users association with the nature of the threats and the prerequisite counter measures. Also, people acclimatize their ways on the basis of magnitude of risk they are capable to take. So, risk perception plays a very important role in improving cyber-security.

Apart from risk perception, adolescence and cyber-bullying are also factors to be considered in order to improve cyber-security. Adolescence is defined as the stage of human growth that takes place between childhood and adulthood. Its inception occurs at around age of 10 and ends around age 21. During this period, children are more vulnerable to cyber-bullying. Cyber-bullying can be defined as an antagonist, international act which is carried out by a group or individual, using electronic forms of communication, time and again; against a victim who cannot easily safeguard himself/herself. By 2020, IoT is expected to grow to 50 billion interconnected devices aggregating, and exchanging private data about their consumers, their lives, their tastes, and preferences. This will result in the increase in cyber-crimes, pertinent risks, and significant data protection issues, thereby amplifying the need to establish a greater level of security [1].

Government organizations are also vulnerable to cyber-attacks. In the past 12 months, it has been found that almost 70% of organizations claim that their security has been compromised by a successful cyber-attack. There is a shortage of proficient cyber event response professionals as reported by 65% of organizations [2].

People around the globe utilize the web and different other system models to communicate, transmit, and store data. Most of the data and information are private and very sensitive; hence, securing this data is necessary. However, these days hacking have become very common. In computer networking, hacking is any technological attempt to get hold of the victim's computer and connected systems with a malicious intent to gather relevant information for blackmailing and ruining people's lives. Hacking is historically referred to constructive, intelligent, technological act that was not necessarily binding to computer systems. So, our objective is to analyze any gaps in the existing computer network, thus enabling researchers to improve their knowledge on things to be taken care of in a computer network in order to avoid cyber-crimes.

2 Risk Perception

Risk insight and understanding plays an important role in models as predictors of precautionary measures. These precautionary measures incorporate the utilization of computer security software (e.g., antivirus s/w, financial s/w, etc.). van Schaik

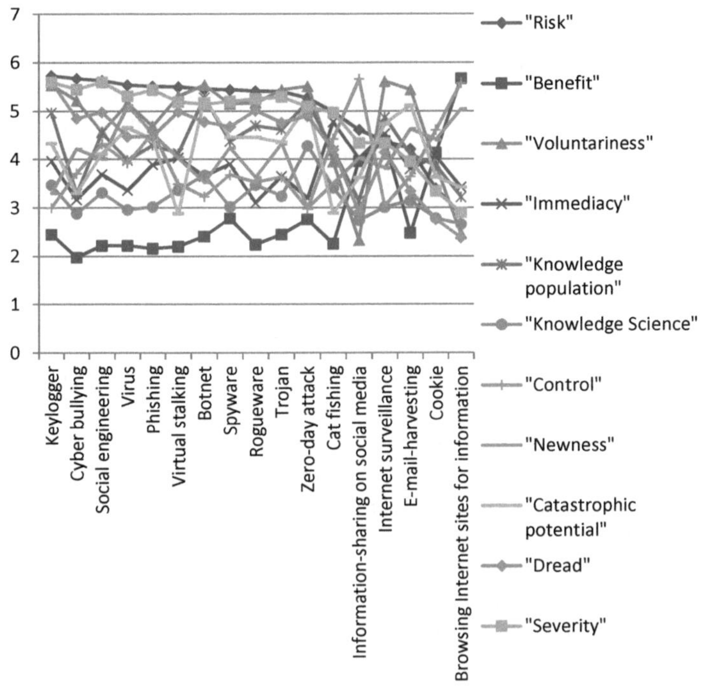

Fig. 1 Mean ratings of perceived risk and other dimensions [3]

et al. [3], in order to measure risk perception and other risk dimensions, carried out a quantitative empirical online study and analyzed an arrangement of 16 security dangers on web and 2 comparisons among 436 US and UK students. The aim of the research has three main goals: (1) Ascertain how to perceive diverse potential security affiliated hazards on the internet. (2) Exhibit the degree to which students play it safe against differing potential security affiliated dangers. (3) Determine the precursor of risk perception and discreet security behavior.

2.1 Method

An online survey was conducted [3], where 16 levels of cyber-security hazards as the independent variable were considered based on the previous research [4]. The dependent variables considered are shown in Fig. 1 along with the 16 security hazards.

The research was conducted on a total of 436 practiced web users among which 336 were female participants and 100 were male, average age = 23, SD = 3, UK: $n = 267$, USA: $n = 169$. A quantitative empirical online study using psychometric methods was conducted in which they analyzed risk perception and precautionary behavior of students related to security in their use of the internet. It became blatant by the survey that the perceived risk escalated in cases of crimes like identity theft, cyber-bullying, social engineering, and keylogger. Among the most promising predictors of perceived risk, dread, voluntariness, catastrophic potential, the immediacy, severity of consequences, and control. As well as Internet know-how and its rate of use took the top positions. Moreover, control was an imperative indicator of precautionary behavior. Students discerned identity theft as the hazardous risk.

3 Cyber-Bullying

Adolescence is a transitional stage from childhood to adulthood involving rapid physical and psychological development that is exceptionally influenced by the expanding issues of self-character, autonomy, peer kinships, and enthusiasm for popularity. During this period, children are more vulnerable to cyber-bullying. A positive school atmosphere has a positive influence on adolescent development, for example, increased confidence and school connectedness. Cyber-victimization commonly takes place outside school settings and post-school hours, henceforth scrutinizing the capacity of schools to avert it.

Holfelt and Leadbaeter [5] examined across a period of one academic year, the longitudinal and concurrent affiliations among early adolescents is roughly between grades 5 and 6, with the cyber-victimization and experiences of school atmosphere.

3.1 Method

The subsample of juveniles in their survey completed cyber-victimization questionnaires; who were in 5 and 6 grades by T1 at the starting ($n = 714$, Mage = 11.0, 52% girls) and T2 at the end (89%, $n = 638$, Mage = 11.48, 52% girls) of the academic year.

Mean, SD, and zero-order correlations among school atmosphere experiences and cyber-victimization at a specific assessment are depicted in Table 1. Zero-order correlation demonstrates that cyber-victimization, however, was adversely connected to the experience of school atmosphere within time at the starting (r range = −0.15 to −0.19) and end (r range = −0.23 to −0.29) of the academic year. However, it was seen that parent involvement and cyber-victimization was only poorly correlated across time.

The study conducted by Brett Holfelt and Bonnie J. Leadbaeter suggested that schools have a significant role in averting cyber-victimization and bullying. Attempts to build a secure and encouraging school atmosphere and to cultivate healthy relations with teachers and peers may inhibit the evolution of cyber-victimization and have a lasting impact on their use of these adverse behaviors.

Table 1 Analysis of descriptive statistics and zero-order correlations [5]

Variable	M (SD)	1	2	3	4	5	6	7	8	9	10	11	12
T1 conventional oppression	4.15 (4.00)	1											
T1 Cyber-oppression	0.52 (1.46)	0.28***	1										
T1Rationality	7.14 (2.79)	−0.39***	−0.19***	1									
T1 Equity in resource sharing	4.63 (2.60)	−0.31***	−0.15***	0.44***	1								
T1 Parent involvement	5.15 (2.70)	−0.06	−0.06	0.17***	0.05	1							
T1 Student-interpersonal rapport	9.10 (3.59)	−0.45***	−0.18***	0.64***	0.29***	0.22***	1						
T1 Student-teacher relations	15.64 (3.62)	−0.39***	−0.17***	0.56***	0.34***	0.25***	0.55***	1					
T2 Cyber-victimization	0.63 (1.44)	0.25***	0.42***	−0.18***	0.17***	−0.07	−0.19***	−0.18***	1				
T2 Rationality	7.28 (2.78)	−0.33***	−0.18***	0.63***	0.34***	0.12**	0.46***	0.39***	−0.29***	1			
T2 Equity in sharing of resources	4.31 (2.66)	−0.23***	−0.09*	0.33***	0.52***	0.05	0.24***	0.25***	−0.23***	0.42***	1		
T2 Parent involvement	4.89 (2.79)	−0.07	−0.07	0.13**	0.07	0.61***	0.15***	0.21***	−0.12***	0.18***	0.06	1	
T2 Student-interpersonal rapport	9.23 (3.68)	−0.32***	−0.15***	0.44***	0.24***	0.19***	0.63***	0.39***	−0.28***	0.62***	0.28***	0.26***	1
T2 Student-teacher rapport	15.41 (3.67)	−0.32***	−0.19***	0.40***	0.27***	0.22***	0.46***	0.62***	−0.26***	0.56***	0.35***	0.25***	0.59***

Note *$p < 0.05$, **$P < 0.01$, ***$p < 0.001$

4 Cyber-Security in IoT

Although IoT is the new hype in the digital world, nevertheless, organizations are least prepared against external threats. Privacy in data collection, sharing, and management needs to be answered as most of the IoT components are vulnerable to threats; so much so that the top concern regarding digital technologies has now been cyber warfare. There is a need to have a formal digital transformation strategy in place when the Internet of Things (IoT) comes into play. The complexity and risk increase by many folds as a greater diversity of devices are connected through IoT to a corporate network. Public infrastructures like bridges, power plants, transportation, machinery, and stoplights are under a security threat now. There has been an incredible growth curve in IOT with currently over 3 billion smartphones in use out of the total 8 billion IoT devices. These statistics are expected to climb past 25 billion by 2020. Hence, there is an immediate need to secure IoT which in effect means to enforce cyber defense effectively. This can be achieved by comprehensively understanding the underlying threats and attacks on IoT framework.

The study in [6] suggests that consumers, governments, and IoT developers actually needs to recognize the hazards and must have a response to the below-mentioned queries:

1. Who form the primary entities? 2. What constitute the assets? 3. Probable conceivable dangers? 4. Most likely danger performers? 5. Which skills and resources do malicious users have? 6. Is the current design secured against hazards? 7. Which particular threat can affect what assets? 8. What security system could be used to counter threats?

The interconnected components or devices are easily prone to cyber-attackers due to manifold reasons:

1. The attacker can easily gain physical access to most IoT devices as they function unattended by humans.
2. For most IoT components, the communication networks are wireless networks where an interloper could vindictively assemble characterized data by eavesdropping.
3. Due to limitations in power and availability of computing resources, most IoT devices cannot support sophisticated security schemes.

Three key issues were identified with IoT devices and services: confidentiality of data, trust, and privacy. Not only users but also the authorized objects are able to access the data. In order to ensure confidentiality, authorization mechanism, access control, and authentication needs to be connected over wireless networks. Trustworthiness plays an important role in such an uncertain environment of IoT where a number of nodes communicate with each other. Twofold trust needs to be ensured, first, in the interaction among entities and second, trust in the system itself [7].

4.1 Privacy Preserving in Fog Computing

Due to the tremendous increase of things in IoT, Fog computing paradigm came into existence. Although the Fog provides a scalable, decentralized solution for an increasing amount of data generated from different devices in IoT [8], the sensor data transmission among end-user devices and the Fog network needs to be secured from cyber-attacks. Kulkarni et al. [9] summarizes the following steps for privacy preserving in Fog computing.

- Sensor data collection and feature extraction.
- Introducing Gaussian noise in data to cause fuzzing of data in order to decrease the chance of spoofing and eavesdropping attacks.
- To mitigate man-in-the-middle attack: segregation (splitting of data into blocks) and then shuffling is done.
- For encryption of data blocks, public key infrastructure is implemented.
- Finally, after the transmission of separated data to Fog node, decryption and reordering of data packets is performed.

4.2 Mitigating Insider Theft

A way-out for data protection from rogue insiders is provided by StolfoS et al. [10] using Fog and Cloud computing components. To reduce security threats, it combines behavior profiling and decoy approaches. If any profile exhibits abnormal behavior, it will be tagged as suspicious by the system and the respective user will be blocked. The decoy is a fake attack employed for befuddling, detecting, and catching the rogue insider. This disinformation may include honeypots, counterfeit archives, and different sorts of goading information. This research area is huge as it demonstrates the mitigation methods and potential altering to guard against information robbery.

5 Cyber-Security in Government Organizations

Kim and Solarwinds [11] in a study mentioned that in May 2017, the UK government revealed that within the span of last 12 months, nearly two-thirds of the country's largest business had been hit by cyber-attack. Therefore, cyber-security is an important agenda for the government. A data breach or cyber-attack can cost government organizations millions of dollars. It also leads to damage to the reputation of an organization and its impact on citizens can be devastating.

Table 2 Various steps taken in order to avoid cyber-crimes

Educating the end-user	– End users should be well educated in the basics of cyber-security
Monitoring and alerts	– Monitoring the network and settings up alerts to identify suspicious activity is very critical – With comprehensive device monitoring, a malicious insider should not be capable of accessing and storing sensitive data on a peripheral device
Risk management decision-making	– Through risk assessment, one can decide where resources and limited budget should be spent – They would access technical equipment, software, systems, and assign the monetary value of the risk if an individual piece of equipment went down or hit by a virus
Up-to-date technology	– Using state-of-the-art security technology and tools is a must – Using obsolete device firmware, insecure protocols, and out-of-date security technology can leave the organization susceptible to attackers
Security team	– A well-trained team of professionals should be ready to defend critical and private data immediately – IT team should be proactive and should learn about common threats
Knowledge is power	– Keeping good knowledge of potential threats and attacks helps in preventing them

5.1 Method

Government organizations can take proper steps to protect their organization from cyber-criminals. Table 2 mentions various steps to be taken in order to protect data from cyber-crimes.

In conclusion, Joe Kim and Solarwinds mentioned that cyber-security is a journey, rather than the destination. It is an ongoing iterative process of trial and error to be successful.

6 A Mind-Map of Network Security in Mitigating Cyber-Crimes

Saad et al. [12] investigated in their review that network security in applicability at current times should inspect any voids in the present-day technology, thus facilitating researchers to intensify their knowledge while discovering answers to these potential issues. In spite of the fact that these arrangements could profit on a bigger scale, it

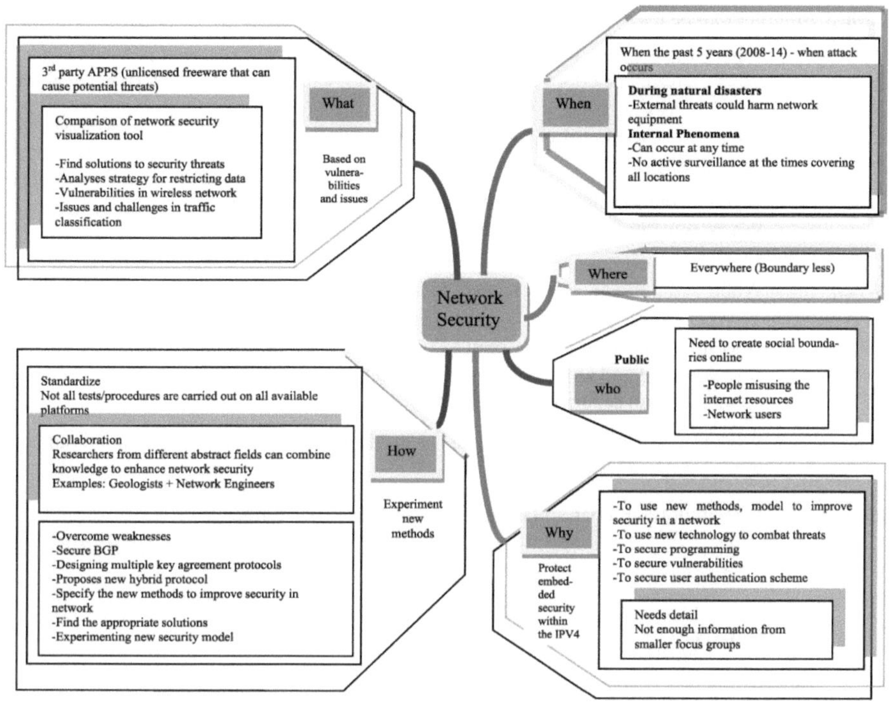

Fig. 2 Network security mind-map to mitigate cyber-crime [12]

is usually the smaller networks that are vulnerable to hackers, such as a campus network. Figure 2 depicts the mind-map of the general thought of the various factors that facilitates the heading of research in network security to mitigate cyber-crimes. In their paper, they investigated the current research area based on the 6-tuple; {what, where, when, why, who, how}, and laid down a mind-map to capture the gaps and open doors for research in the field of network security.

First, they started looking at issues and vulnerabilities. They scrutinized various tools to mitigate cyber-crime and went through the research that describes the potential threats of unlicensed freeware and third-party applications. Table 3 describes how 6-tuples can be used to mitigate cyber-crime in network security.

Amna Saad et al. concluded that it is preferable to combat the security from the source by furnishing the users with all the significant information required to prevent low-level dangers which consequently might transform into something more intricate. The general public needs to be kept well aware regarding the prevalent network security issues. It is suggested to have an "Internet task-force" or a cyber-police that can act as a monitoring team for LANs like in a campus, in order to have keen checking and perform a stern action when required. They can likewise give recommendations and tips to clients on the system keeping in mind the end goal to fortify the uprightness of the internal network.

Table 3 6-tuples can be used to mitigate cyber-crime in network security

What	– What strategies can be used to restrict data? – What consequence does modernistic technology like wireless network have on the overall network risks? – What sort of exploration was done to alleviate security threats?
Where	– Where is the location of this threat? – Where this threat usually resides?
When	– When will a threat or attack occur? E.g., No active surveillance
Why	– Why this protection is important? This protection could be possible by strengthening and amplifying the technology, approach, algorithm, model, and plan of a novel security arrangement
Who	– Who will be concerned with these issues? Research shows that the general public will be most affected
How	– How important a particular tool is to mitigate cyber-crime?

7 Conclusion

In this paper, we reviewed various research articles regarding cyber-security. We identified that risk perception in students is very important to combat cyber-crimes. We tried to analyze some of the factors responsible for cyber-bullying in adolescents and how a safe and complementary school setting, healthful kinships with teachers and peers, may prevent the arising of cyber-bullying and consequently having a long-term impact on the use of these negative practices by adolescents. Apart from students and adolescents, this paper tried to mention some of the factors that can reduce cyber-crime in government organizations. In the context of risk perception, potential security-related hazards on the internet were identified. Also, the significant predictors of perceived risk were identified. In IoT, information is very vulnerable to cyber-crimes; hence, cyber-crime has a great impact on the Internet of Things. The significant problems between IoT devices and services were identified in order to mitigate threats to data confidentiality, privacy, and trust. The significance of trustworthiness in an IoT environment was highlighted. At last, a mind-map of various security issues in network security is presented to combat the effect of cyber-attacks.

References

1. http://securityaffairs.co/wordpress/26537/cyber-crime/internet-of-things-risks.html. Last Accesses on 19 Nov 2017
2. www.dhs.gov/homeland-security-careers/dhs-cyber-security,last. Accesses on 19 Nov 2017
3. P. van Schaik et al., Risk perceptions of cyber-security and precautionary behavior. Elsevier J. Comput. Hum. Behav. **75**, 547–559 (2017)
4. V. Garg, L.J. Camp, End user perception of online risk under uncertainty, in *Hawaii International Conference on System Sciences* (Manoa, HI, 2012)
5. B. Holfelt, B.J. Leadbaeter, Concurrent and longitudinal associations between early adolescents' experiences of school climate and cyber victimization. Elsevier J. Comput. Hum. Behav. **76**, 321–328 (2017)
6. M. Abomhara, G. Koien, Cyber security and the internet of things: vulnerabilities, threats, intruders and attacks. J. Cyber Secur. **4**, 65–88 (2015) (River Publishers)
7. M. Abomhara, G. Koien, Security and privacy in the internet of things: current status and open issues, in *The International Conference on Privacy and Security in Mobile Systems* (PRISMS, Aalborg, Denmark, 2014)
8. S. Sagiroglu, D. Sinanc, Big data: a review, in *IEEE International Conference on Collaboration Technologies and Systems (CTS)* (2013), pp. 42–47
9. S. Kulkarni, S. Saha, R. Hockenbury, Preserving privacy in sensor-fog networks, in *9th IEEE International Conference on Internet Technology and Secured Transactions (ICITST)* (2014), pp. 96–99
10. J. StolfoS, M.B. Salem, A.D. Keromytis, Fog computing: mitigating insider data theft attacks in the cloud, in *IEEE Symposium On Security and Privacy Workshops (SPW)* (2012), pp. 125–128
11. J. Kim, Solarwinds, Cyber-security in government: reducing the risk. Comput. Fraud Secur. **2017**(7), 8–11 (2017)
12. A. Saad, A.R. Amran et al., Privacy and security gaps in mitigating cyber-crime: the review, in *International symposium on Agent, Multi-Agent Systems and Robotics (ISAMSR)*, 23–24 Aug 2016

Part II
Wireless Networks

Mitigating the Effect of Node Selfishness in Mobile Social Networks

C. C. Sobin

Abstract Delay-Tolerant Networks (DTNs) are challenged networks, where there may not exist an end-to-end path always. In such situations, for sending a message to destination, a source node has to store and carry it, till it finds a suitable node to forward. DTNs can be categorized as a subcategory of MANETs, based on mobility of the nodes, frequent network partitioning, etc. However, unlike in MANETs, there is no guarantee for connectivity among nodes in the network, which may cause nodes in the network to make use of encounter opportunity with other nodes for communication. In such scenarios, due to constraints on resources, some of the nodes may not forward messages further in the network and thus behave selfishly. In this paper, we have identified such selfish behavior of the nodes and proposed a heuristic-based selfishness detection and mitigation technique for DTNs. As part of simulation, we have conducted experiments using both real-world and synthetic traces and observed that DTN performance metrics were improved in terms of increasing delivery ratio, decreasing overhead ratio, and decreasing delivery delay.

Keywords Delay-tolerant networks · Selfishness

1 Introduction

Today, most of the users make use of hand-held devices like mobile phones, tablet, etc., for their daily activities. This trend creates a drift from traditional fixed infrastructure-based (wired) applications to infrastructureless (mobile-based) applications. The major difficulty in developing such mobile-based applications is

C. C. Sobin (✉)
Department of Computer Science and Engineering, MES College of Engineering Kuttippuram, Malappuram, Kerala, India
e-mail: sobincc@gmail.com

C. R. Krishna et al. (eds.), *Proceedings of 2nd International Conference on Communication, Computing and Networking*, Lecture Notes in Networks and Systems 46, https://doi.org/10.1007/978-981-13-1217-5_11

intermittent connectivity and constraints on resources such as bandwidth, storage space, etc. Delay-Tolerant Network (DTN) is a type of mobile networks, which are able to handle disruptions in connectivity and constraint on network resources. Since the connectivity among a set of DTN nodes may not be guaranteed always, store–carry and forward message forwarding strategy is used. DTNs find applications in many areas like underwater communications [1], vehicular networks [2], rural communications [3], disaster recovery management [4], etc.

The recent surge in the usage of mobile devices also highlights in the direction of a kind of DTNs, namely, Mobile Social Networks (Fig. 1), where nodes are mobile users and use multi-hop wireless communication. Ideally, in such networks, nodes are assumed to be rational in nature and cooperative to all the other nodes in message forwarding. However, in reality, not all the nodes honestly involve in message forwarding due to their constraints on resources. Such behavior of nodes, called selfishness, degrades the overall routing performance. So, while designing routing algorithm in such networks, suitable methods have to be devised to identify the selfish behavior of nodes and methods to mitigate its effect on routing performance.

In addition to the relay selection, there are other factors which affect the routing performance in DTNs. Since the resources in the network, such as available bandwidth and contact duration, as well as node resources such as buffer space and energy are limited, behavior of the nodes is an important factor to be considered while designing DTN routing algorithms. Some of the nodes may behave selfishly and do

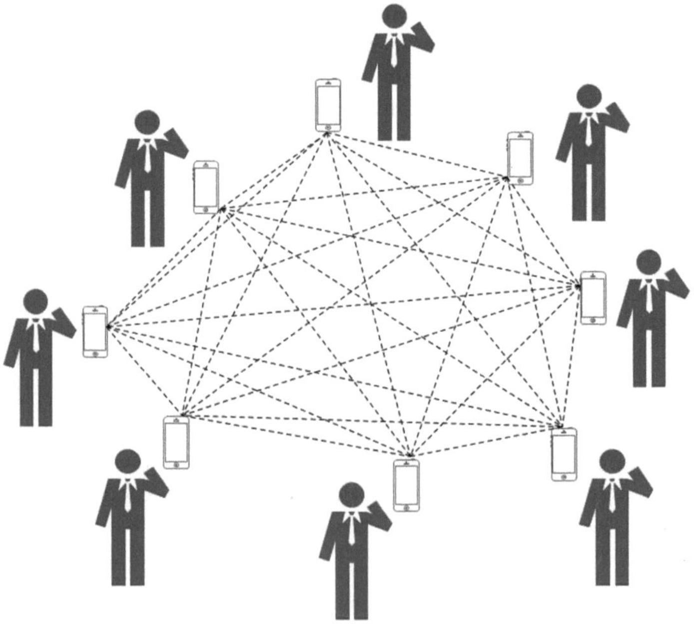

Fig. 1 Mobile social networks (MSN)

not cooperate in the message forwarding procedure, which sometimes deteriorates the routing performance. So, such behavior of the nodes was studied and proposed a heuristic-based technique to detect and improve the routing performance. The proposed method uses partial network knowledge so that it is the most suitable DTN characteristics. Apart from simulation using synthetic traces, experiments with real-world traces also guarantee the better routing performance compared to the existing methods.

2 Related Works

As relay selection is essential as well as most crucial to the performance of DTN routing schemes, a number of methods to choose a relay node have been proposed in literature. This section focuses on how different routing schemes select the relay node. A lot of information, e.g., past encounter records, social similarity, closeness, utility of the message or the node, etc., is exploited so as to choose the best node which can deliver the message to the destination timely as well as reliably. Relay selection has progressed from typical flooding-based technique to highly sophisticated prediction and social-based techniques. Epidemic routing is one of the simplest DTN routing protocols proposed by Vahdat et al. [5]. Epidemic routing adopts a flooding-based strategy, in which the source node floods the message to all the intermediate nodes, so that the message will finally reach the destination. Spray and Wait routing [6] adopts a controlled flooding strategy, in which, instead of flooding to all the neighbors, the source node floods the message to a limited set of its neighbor nodes (relays), which then directly transmit the message to the destination. Compared to epidemic routing, the number of transmissions is reduced in Spray and Wait, which in turn reduces overall resources overhead.

Lindgren et al. [7] have observed that the nodes in a network exhibit certain mobility patterns in their motion. They have used real mobility traces for their study. Based on this, they have used history of encounters as an information to predict the future contacts among nodes, particularly to the destination. A prediction value is assigned for each node in the network associated with a message for a particular destination. Nodes use the encounter history to calculate delivery predictability to all the other nodes. In relay selection, a node with highest delivery predictability is selected. In MaxProp [8], the delivery probability is used to calculate the cost of various paths from source to destination. The message is relayed on the lowest cost path. Apart from abovementioned benchmark DTN routing schemes, some of the researchers [9–12] studied node selfishness in DTN. More details regarding existing routing and data dissemination techniques can be found in [13].

3 Proposed Method

We propose a method to detect the selfish behavior of nodes in DTNs, which considers limited contact duration and limited buffer space available at every node. We use unicast communication paradigm, in which only a single source *S* and a single destination node *D* involves in the message forwarding procedure. We use a multi-copy strategy in the proposed method. The reason is that single-copy schemes have comparatively higher delivery delays and lower delivery ratio. Though replication-based schemes incur higher cost, this can be optimized, if relays are chosen intelligently. The proposed method schedules the sequence of message transfer to the encountered nodes. Since these encounters may be of short duration, choosing the relay nodes should be carefully done, so that the probability of delivering the messages to the destinations should be high. We also assume that nodes do not have any information regarding future network connectivity, or their future movements. All the messages are of the same priority initially.

To detect the selfish behavior, each node stores the message transmission history with all other nodes in the network (Fig. 2). A node is identified as a selfish node, if it is found to be not forwarding the message further in the network. This may be because of its limited resources such as buffer space and energy. When two nodes meet each other, they exchange the list of nodes from which they have received messages so far in the network. Those set of nodes, which are not in the list, are either not received a copy of message (since they are not in the communication range) or not forwarded the message (due to their selfish behavior). So, we have used a heuristic-based technique to identify selfish nodes in the network. Since the message contains the list of nodes traversed, each node once received a copy of the message, it will check the nodes at which the message is traversed and keep the information in a list. When a node has to choose a relay node from its neighbor nodes, it selects a node as a relay which is having higher value of combination of delivery predictability and message reception history. This process is repeated till the message reaches the destination. After each

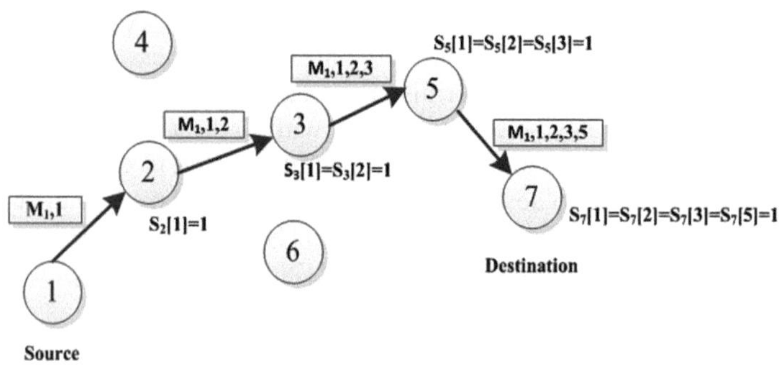

Fig. 2 Selfishness identification procedure

node executes the procedure, each node can differentiate the set of selfish nodes and set of not received the messages from source to destination and removes those nodes from prospective set of relay nodes.

4 Performance Evaluation

We have conducted simulations using opportunistic network environment [14] simulator. Map-Based Movement Mobility model (MBM) was used for the evaluation using synthetic traces. With this model, we conducted experiment with 125 nodes and varied the buffer size in nodes to see the effect on delivery ratio and the message overhead. The results of these simulations are discussed in the next section. The simulations are conducted for 12 h, with the network parameters shown in Table 1.

4.1 Routing Performance Metrics

Delivery ratio, delivery delay, and overhead ratio were used for evaluating the routing performance. The ratio of messages delivered to messages generated denotes the Delivery Ratio (DR). For calculating Overhead Ratio (OR), first the difference between relayed messages to delivered messages is calculated, which is then divided by the number of delivered messages. Delivery Delay (DD) is the time for message to reach the destination.

Table 1 Parameters used in simulation

Parameter	Value
Duration	43200 s = 12 h
Transmission speed	250 Kbps
Nodes in the network	125
TTL	300 min
Creation interval of the message	25–30 s
Size of each message	500–1024 KB
Wait Time	10–30 s
Buffer space in node	2–7 MB
P_{init}	0.75
β	0.25
γ	0.98

4.2 Experimental Results

For evaluating the routing performance of the proposed selfishness mitigation method under the presence of selfish nodes, we have used existing Prophet routing algorithm. We have compared original Prophet algorithm together with its selfishness variant and the Prophet with selfishness mitigation technique (represented as modified Prophet). For analyzing delivery ratio, overhead ratio, and delivery delay, variation is applied in buffer space of the nodes in the network from 2 MB to 7 MB. The results are summarized below.

Comparing delivery ratio (Fig. 3a), the delivery ratio of basic Prophet routing algorithm is decreased in the presence of selfish nodes in the network. Adopting our proposed method in the selfishness, variant of the Prophet algorithm identifies the selfish nodes in the network and removes them from the prospective set of relay nodes, which improves the overall delivery ratio.

Similarly, while analyzing the overhead ratio, due to the presence of selfish nodes in the network, the overhead ratio increases compared to the original Prophet algorithm, as in Fig. 3b. However, using the proposed method, the selfish nodes are bypassed in the relay selection, which decreases the overall overhead ratio of the network.

Comparing the delivery delay (Fig. 4), the selfish nature of the nodes cause them to drop the messages, which increases the overall delivery delay. Therefore, the delivery delay of selfishness variant of Prophet is higher than the original Prophet. However, the proposed selfishness mitigation method reduces the delivery delay compared to the original Prophet using its intelligent relay selection strategy.

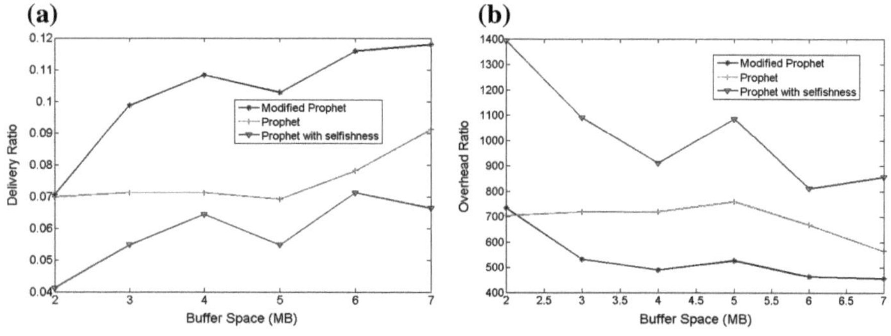

Fig. 3 Experimental results **a** Delivery ratio **b** Overhead ratio

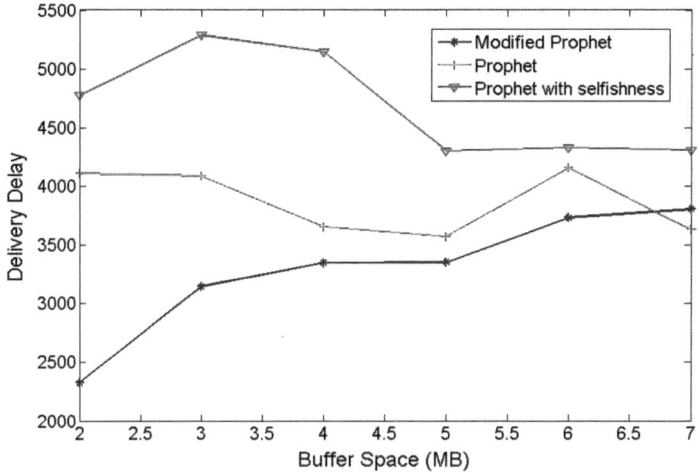

Fig. 4 Experimental results for delivery delay

4.3 *Simulation Using Real-World Traces*

We have also performed experiments using real-world traces, such as RollerNet [15]. RollerNet dataset contains Bluetooth traces groups of 5,000–15,000 Rollerblades covering a distance of 30 Km in different areas in Paris city. For comparison using RollerNet traces, we have varied the Time To Live (TTL) of the message.

Figure 5a depicts the comparison of the delivery ratio using RollerNet traces. As with the synthetic traces, the delivery ratio is decreased with the presence of selfish nodes for Prophet routing algorithm. By identifying the selfish nodes in the network and excluding them in the relay selection procedure, the delivery ratio is improved with the proposed method.

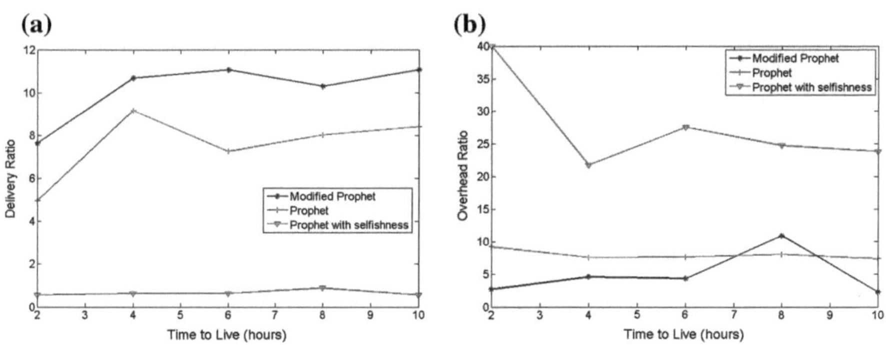

Fig. 5 **a** Delivery ratio of comparison **b** Overhead ratio comparison

The comparison of overhead ratio using RollerNet traces is listed in Fig. 5b. We can observe overhead ratio increases in Prophet, compared to the original version. However, by adopting the proposed method, the overhead ratio is decreased compared to the other variants.

5 Conclusion

DTNs find many applications in various domains, where the connectivity in the network is not guaranteed. However, the behavior of nodes is important factor to be considered while designing DTN routing algorithms. So, in this research, the selfish behavior of nodes was studied and proposed a heuristic-based technique to identify the selfish nodes and excluded them in the relay selection procedure. For analyzing the routing performance using the proposed method, we have used existing Prophet routing algorithm. Extensive experiments were conducted with synthetic traces as well as real-world traces. The experimental results indicate that the selfishness mitigation technique improves DTN routing performance metrics in terms of higher delivery ratio, lower overhead ratio, and lower delivery delay.

References

1. S.H. Bouk, S.H. Ahmed, D. Kim, Delay tolerance in underwater wireless communications (2016)
2. J. Wang, H. Zhang, X. Tang, Delay tolerant routing for cognitive radio vehicular ad hoc networks, in *2016 IEEE 22nd International Conference on Parallel and Distributed Systems (ICPADS)*, (IEEE 2016), pp. 24–31
3. C. Velásquez-Villada, Y. Donoso, Delay/disruption tolerant networking-based routing for rural internet connectivity (drinc). Int. J. Comput. Commun. Control **12**(1), 131–147 (2016)
4. S. Basu, S. Roy, S. DasBit, S. Bandyopadhyay, A human mobility based knowledge sharing approach for post disaster need assessment using dtn, in *Proceedings of the 17th International Conference on Distributed Computing and Networking*, (ACM 2016), p. 34
5. A. Vahdat, D. Becker et al. Epidemic routing for partially connected ad hoc networks. Technical report, Technical Report CS-200006, Duke University (2000)
6. T. Spyropoulos, K. Psounis, C.S. Raghavendra, Spray and wait: an efficient routing scheme for intermittently connected mobile networks, in *Proceedings of the 2005 ACM SIGCOMM workshop on Delay-tolerant networking*, (ACM 2005), pp. 252–259
7. A. Lindgren, E. Davies, S. Grasic, A. Doria, Probabilistic routing protocol for intermittently connected networks (2012)
8. J. Burgess, B. Gallagher, D. Jensen, B.N. Levine, Maxprop: Routing for vehicle-based disruption-tolerant networks. INFOCOM **6**, 1–11 (2006)
9. C. Sobin, S. Deepak, Analyzing behavioral selfishness in opportunistic environment using game theory. in *India Conference (INDICON), 2016 IEEE Annual*, (IEEE 2016), pp. 1–5
10. J. Wu, Y. Zhu, L. Liu, B. Yu, J. Pan, Energy-efficient routing in multi-community dtn with social selfishness considerations, in *Global Communications Conference (GLOBECOM), 2016 IEEE*, (IEEE 2016), pp. 1–7

11. T. Zheng, Y. Wang, X. Ma, Effects of selfishness in the opportunistic network under two different movement models, in *2016 6th International Conference on Electronics Information and Emergency Communication (ICEIEC)*, (IEEE 2016), pp. 240–244
12. C. Sobin, Analyzing the impact of selfishness on probabilistic routing algorithms in delay tolerant networks, in *2015 International Conference on Computing and Network Communications (CoCoNet)*, (IEEE 2015), pp. 186–190
13. C. Sobin, V. Raychoudhury, G. Marfia, A. Singla, A survey of routing and data dissemination in delay tolerant networks. J. Netw. Comput. Appl. (2016)
14. A. Keränen, J. Ott, T. Kärkkäinen, The one simulator for dtn protocol evaluation, in *Proceedings of the 2nd international conference on simulation tools and techniques, ICST (Institute for Computer Sciences, Social-Informatics and Telecommunications Engineering)*, p. 55 (2009)
15. RollerNet: Analysis and use of mobility in rollerblade tours. (2017 (accessed September 3, 2017))
16. Statista, report. (2017 (accessed September 3, 2017))

Energy Efficient Sector-Based Clustering Protocol for Heterogeneous WSN

Suniti Dutt, Gurleen Kaur and Sunil Agrawal

Abstract Wireless sensor networks have lately come out as a crucial computing platform. Since a WSN is composed of numerous low power, battery operated sensor nodes leading to constant energy dissipation in the network, researchers have worked on clustering techniques which tend to cope with these sensor deficiencies. In this paper, we attempt to ameliorate the lifetime, stability and throughput of the network through a new convention named Energy Efficient Sector based Clustering Protocol for heterogeneous network. In the proposed algorithm, the area of deployment is split into various sectors. The election of Cluster Head is based on maximum remaining energies of the sensors. The outcomes of the simulations establish that our proposed protocol outperforms SEP in terms of network lifetime as well as throughput by a significant amount.

Keywords WSN · Clustering · Sector formation · Residual energy · Lifetime Throughput

1 Introduction

Wireless Sensor Networks (WSNs) have received tremendous attention lately. Sensor networks are being used greatly in many fields such as military, surveillance, body sensors, structural health management, environmental monitoring etc. The sensor nodes in the network are battery operated which are non-rechargeable thus making it an important issue to avoid early node death so as to extend the lifespan of the entire network [1]. Energy resource of the sensors is thus the primary design issue and must be managed wisely. Many clustering protocols have been developed lately to combat the energy problem of WSNs. In the process of clustering [2], the network area is divided into various clusters consisting of Cluster Heads (CHs) as well as member nodes. Information sent from member nodes to the CHs is inter-cluster communi-

S. Dutt (✉) · G. Kaur · S. Agrawal
Department of ECE, UIET, Panjab University, Chandigarh 160014, India
e-mail: sunitidutt@gmail.com

C. R. Krishna et al. (eds.), *Proceedings of 2nd International Conference on Communication, Computing and Networking*, Lecture Notes in Networks and Systems 46, https://doi.org/10.1007/978-981-13-1217-5_12

cation whereas information sent from CHs to other CHs closer to the Base Station (BS) is intra-cluster communication [3]. This paper mainly focuses on reducing the energy dissipation thereby improving the network lifetime of the randomly deployed senor nodes in a WSN and also improving the rate of data packets forwarded to the BS at the same time. This is achieved by dividing the deployment region into various sectors and then selecting the CHs from each sector based on highest remaining energy of the sensors.

The remaining paper is devised as follows. Section 2 discusses the related works about various clustering protocols; Sect. 3 explains the preliminaries and problem formulation of the proposed protocol; Sect. 4 discourses the simulation outcomes and analysis of the purported scheme; and Sect. 6 concludes the findings and also presents future scope of the work.

2 Related Work

Many clustering protocols aimed at improving the energy efficiency and enhancing the network lifetime have been developed in the past. A Low-Energy Adaptive Clustering Hierarchy protocol called LEACH [4] discusses the homogeneous nature of the sensors, and is based upon arbitrary rotation of CHs to disseminate the energy load equally amongst all the sensors in the WSN. All the nodes are equally capable of being elected as CHs with probability, p. The nodes chose an arbitrary number within the range 0 and 1. If this value is less than the predefined threshold, then that node becomes the CH. The threshold measure is given as:

$$T(s) = \begin{cases} \frac{p}{1-p\left(r \mod \frac{1}{p}\right)} & \text{if } s \in G \\ 0 & \text{otherwise} \end{cases} \tag{1}$$

where, G is the set of all sensor nodes which are eligible to be elected as CHs in rth round.

Smaragdakis et al. [5] proposed a protocol called Stable Election Protocol (SEP) to examine the effect of two-level node-heterogeneity in the network. The algorithm is based upon the concept of weighted election probability of nodes to become CHs which considers the remaining amount of energy of each sensor node after every round. The threshold values for normal and advanced nodes respectively are given as:

$$p_i = \begin{cases} \frac{p_{\text{opt}}}{(1+\text{am})} & \text{if } s_i \text{ is normal node} \\ \frac{p_{\text{opt}}*(1+a)}{(1+\text{am})} & \text{if } s_i \text{ is advanced node} \end{cases} \tag{2}$$

where p_{opt} is the optimum probability, m is the fraction of advanced nodes and a is the factor by which the energy of advanced nodes exceed that of normal nodes.

The dynamically changing cluster head probability has been discussed in [6] called Enhanced Threshold Sensitive Stable Election Protocol (ETSSEP). It is a modification of SEP protocol, wherein the CHs are elected not only on the basis of remaining energies of nodes but numbers of CHs per round as well. Also ETSSEP considers three-level heterogeneity of the sensor nods which classifies nodes as normal, intermediate and advanced nodes. The threshold value is given as:

$$T(s) = \begin{cases} \frac{P}{1-P(r \cdot \bmod \frac{1}{p})} * \frac{\text{node'sresidualenergy}}{\text{network'saverageenergy} * K_{opt}} & \text{if} s \in G \\ 0 & \text{otherwise} \end{cases} \quad (3)$$

where, G is the set of all sensors eligible to become CHs, K_{opt} is the optimum value of clusters per round. The threshold values T_{nrm}, T_{int} , and T_{adv} for normal, intermediate and advanced nodes respectively are also defined in this protocol.

There are various works which include division of deployment area into zones, sectors or rings etc. for CH selection. Division of network area has been discussed in [7] in which a sector-chain based clustering protocol divides the area into various sectors in order to balance the number of nodes. A chain is constructed for each cluster, leader being the CH or the SCH, chosen on the basis of highest residual energy and minimum distance between the node and the BS. Another protocol as discussed in [8] is balanced energy efficient circular routing protocol (BEEC) in which the deployment area is assumed to be circular and is divided into ten sub-circular regions which in turn are divided into eight sectors. The transmission of information occurs between the member nodes in each sector and their respective mobile CHs based on their minimum distance from each other. Similar works have been discussed in [9] which is 3R-Reliable Rim Routing consisting of nodes in different rims with different energy levels and in [10] called Multi hop Angular routing protocol (AM-DisCNT). Both of them have been success in achieving improved throughput and stability as compared to traditional protocols.

Thus, it can be observed that enhancing network reliability, throughput, lifetime and reducing delay are main challenges faced by power constrained WSNs and are strongly influenced by using single-hop or multi-hop techniques, heterogeneous network, dividing the area of deployment into various geometrical shapes etc.

3 Proposed Scheme

The proposed work ideates designing of a clustering based hierarchical routing protocol for energy efficiency in WSN considering heterogeneity aware network, in which, the area of node deployment is divided into sectors with each sector consisting of elected CH. Location of nodes in different sectors is determined by their angle of arrivals and the probabilities of CH selection out of the normal and advanced nodes are decided based on their maximum residual energy. EESCP is based on modifications in LEACH and SEP protocols. The simulations are carried out using Matlab software. Parameters like energy exhaustion, network lifetime and number of data packets communicated to the BS also known as throughput are studied and analyzed in the proposed work.

3.1 Network Configuration

In our proposed work, we have considered first order radio energy dissipation model [4] and depending upon the distance d of each node from BS, free space path loss model (d^2 power loss) as well as multipath loss model (d^4 power loss) are also considered. Total n number of nodes have been considered in the work, out of which a fraction of nodes, m, called advanced nodes are endowed with α times additional energy than remaining $(1-m)*n$ normal nodes.

Total energy of the network is given by:

$$\mathrm{E_{total}} = \mathrm{n(1-m)*E_o + nmE_o(1+a) = nE_o(1+am)} \quad (4)$$

The work considers a rectangular 100 m × 100 m field. Assuming it to be a circular one, the field is divided into sectors, each sector corresponding to a specific angle between 0 and 360°. The number of sectors is assumed to be 10% of the total alive nodes in a given round of communication. All the sensor nodes are initially termed as NPCH i.e. No Probable Cluster Heads. The nodes which fulfill the threshold requirements and are eligible to become CHs are termed as YPCH (Yes Probable Cluster Heads). Out of all the YPCH nodes, a node is chosen in each sector to become the CH based on highest residual energy. If in a given sector, there is no YPCH, all the NPCH nodes of that sector contend to become CH based on the same criteria of highest residual energy.

The probability equations in our work have been modified considering the distance factors, distance to BS, d_i, of ith node as well as the average distance of any node from BS, d_{avg}. Depending on these distances, either the probability equation gets multiplied by the factor (d_i/d_{avg}), or it remains same as in TDEEC protocol [11].The altered probabilities for both types of nodes are given as follows:

$$\text{if } d_i \le d_{avg} \text{ then}$$

$$p_i = \begin{cases} \dfrac{p_{opt} * \text{sensor's residual energy} * d_i}{(1+am) * \text{network's average energy} * d_{avg}} & \text{if } s_i \text{ is normal node} \\ \dfrac{p_{opt} * (1+a) * \text{sensor's residual energy} * d_i}{(1+am) * \text{network's average energy} * d_{avg}} & \text{if } s_i \text{ is advanced node} \end{cases} \quad (5)$$

$$\text{if } d_i > d_{\text{avg}} \text{ then}$$

$$p_i = \begin{cases} \dfrac{p_{\text{opt}} * \text{sensor's residual energy}}{(1+\text{am}) * \text{network's average energy}} & \text{if } s_i \text{ is normal node} \\ \dfrac{p_{\text{opt}} * (1+a) * \text{sensor's residual energy}}{(1+\text{am}) * \text{network's average energy}} & \text{if } s_i \text{ is advanced node} \end{cases} \quad (6)$$

The modified equation for threshold is given as:

$$T(s) = \frac{p_i}{1 - p_i\left(r \bmod \frac{1}{p_i}\right)} * \frac{\text{Residual energy of a sensor}}{\text{Initial energy of a sensor}} \quad (7)$$

where, the probability p_i is specified separately for normal and advanced nodes according to Eq. (2).

3.2 Network Parameters

Table 1 illustrates the various parametric values assumed in the proposed work and additional settings specifically assumed considering two-level heterogeneity.

Table 1 Parameter settings

Parameters	Value
Area of network (m^2)	100 m × 100 m
Position of BS (m)	(50, 50 m)
Number of sensors	100
Length of data packet (bits)	4000 bits
Threshold distance, d_o (m)	70 m
Transmitter/receiver electronics energy	50 nJ/bit
Data aggregation energy	5 nJ/bit
Transmit amplifier energy, E_{fs}, if $d_{tobs} <= d_o$	10 pJ/bit/m^2
Transmit amplifier energy, E_{amp}, if $d_{tobs} >= d_o$(J)	0.0013 pJ/bit/m^4
Optimal probability of CH selection	0.1
Radius of two-hop, R (m)	25 m
Proportion of advanced nodes (m)	0.3
Energy factor for advanced nodes (a)	1.5
Initial energy of normal nodes (J)	0.5

3.3 Algorithm and Flowchart

The flowchart for proposed methodology is shown in Fig. 1.

Step 1: Initialize all nodes as NPCH (No Probable CH) at start of round.
Step 2: Divide the area of deployment with dimensions (100,100) into sectors where the number of sectors is computed by:
No. of sectors = 0.1 * alive nodes
Initially, we have considered 100 alive nodes in our proposed algorithm and therefore number of sectors formed will be equal to 10 as computed by the above mentioned formula. If alive nodes = 90, then number of sectors = 9 and so on.
Step 3: Taking (50, 50) and (100, 50) as reference vector, calculate the angles of all the sensor nodes in degrees from 0 to 360.
Step 4: Based on the angles, assign sector number to all the nodes.
Step 5: From each sector, select only those nodes that qualify as YPCH (Yes Probable CH) i.e. they qualify to become probable CHs according to whether they fulfill threshold conditions.
Step 6: Check the number of YPCH in each sector. We need to make only one YPCH as CH.
If YPCH = 1, make it the CH to send the data to BS.
If YPCH > 1, select the YPCH with maximum residual energy as CH and allow it to send data to the BS.

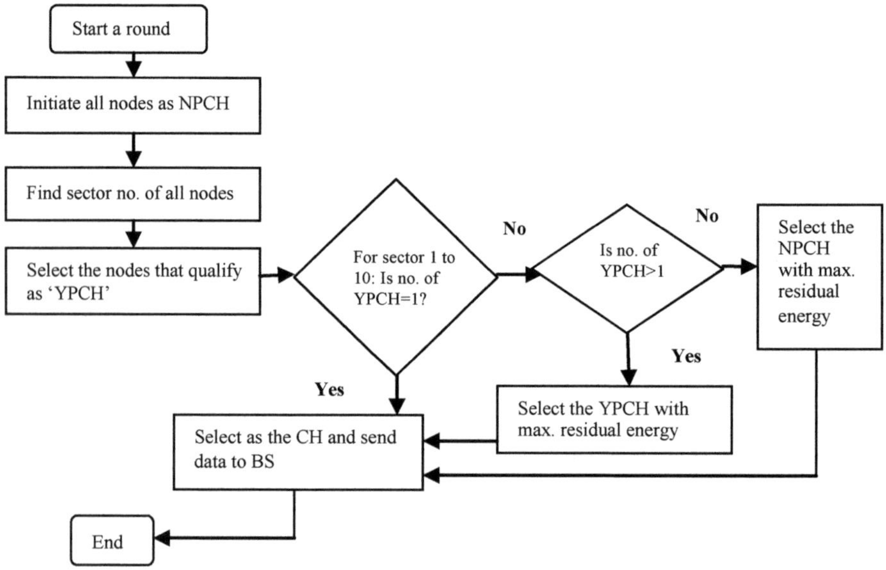

Fig. 1 Flowchart: CH selection based on maximum residual energy

Step 7: If there is no YPCH in any sector, then out of all the NPCH nodes in that sector, select the one with maximum residual energy as the CH and send data to the BS.
Step 8: Perform the usual data communication after CH selection as in LEACH.

4 Results and Discussion

Figure 2 depicts the graph comparing the network lifetime of EESCP with that of SEP and ETSSEP. The graph shows relation among the number of sensors which are not alive and the number of rounds of data communication. The First Node Death (FND) happens at round 1142 in case of SEP and in case of ETSSEP at round 1421, but it occurs at round 2433 in case of EESCP, proving the improved stability period (from beginning of network operation till death of first alive node) in EESCP. The stability period of the proposed protocol has increased by 53.06 and 41.59% as compared to SEP and ETSSEP respectively. The last sensor node in SEP dies at round 3092 and in ETSSEP the Last Node Death (LND) happens at round 5549, whereas in case of EESCP, the last sensor dies at 6051st round. The normal nodes having lesser energy as compared to advanced nodes tend to die out first. The horizontal portion of the curve in the graph shows that when all the normal sensors die out, further sensors' death does not occur for a few rounds since there is still enough energy left in the advanced sensors. The nodes again start to die out after this horizontal region.

Figure 3 shows the graph describing the number of data packets acquired at the BS versus the number of communication rounds of the purported protocol as compared to those of SEP and ETSSEP protocols.

The graph clearly shows that the proposed protocol outperforms SEP and ETSSEP here as well. The total number of data packets acquired at the BS in case of SEP is 0.020 million and in case of ETSSEP is 0.100 million whereas it is 0.133 million in

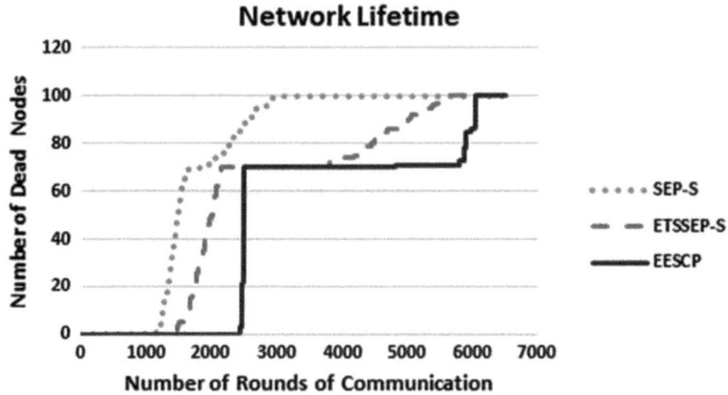

Fig. 2 Network lifetime-CH selection based on maximum residual energy

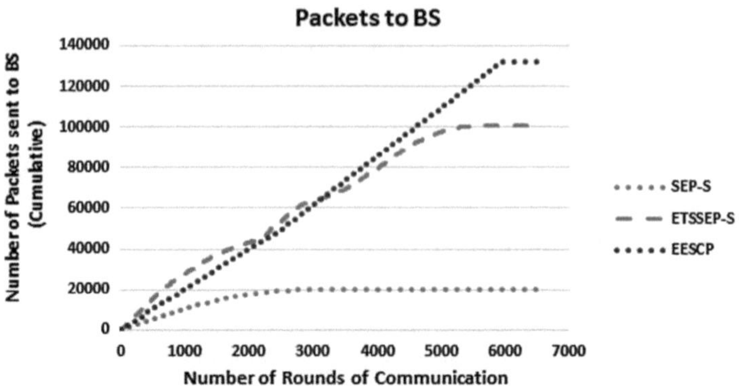

Fig. 3 Throughput-CH selection based on maximum residual energy

case of EESCP. This equals to 564% improvement with respect to SEP and 32.83% with respect to ETSSEP. We can infer from the graph that there is linear increment in the number of packets acquired at the BS followed by saturation.

5 Conclusion and Future Scope

In this paper, we have come up with a geographical location-based clustering method through sector formation in the network area and selecting the CHs from each sector based on their maximum residual energies. The network takes into consideration the two types of sensor nodes—normal and advanced—and constant bit rate traffic pattern. A significant improvement in network lifetime and throughput has been achieved with our proposed protocol as compared with SEP and ETSSEP. Further work can be carried out in the same field by considering node mobility. Also, heterogeneity may be extended by including more than two types of sensor nodes.

References

1. P. Rawat, K.D. Singh, H. Chaouchi, J.M. Bonnin, Wireless sensor networks: a survey on recent developments and potential synergies, Springer. J. Supercomput **68**, 1–48, October (2013)
2. C. Jiang, D. Yuan, Y. Zhao, Towards clustering algorithms in wireless sensor networks-a survey. IEEE Proc. WCNC (2009)
3. A.A. Abbasi, M. Younis, A survey on clustering algorithms for wireless sensor networks. Elsevier Comput. Commun. **30**, 2826–2841 (2007)
4. W.R. Heinzelman, A. Chandrakasan, H. Balakrishnan, Energy-efficient communication protocol for wireless microsensor networks, in *IEEE, Proceedings of the Hawaii International Conference on System Sciences*, January (2000)
5. G. Smaragdakis, I. Matta, A. Bestavros, SEP: A stable election protocol for clustered heterogeneous wireless sensor networks. Boston University Computer Science Department (2004)

6. S. Kumar, S.K. Verma, A. Kumar, Enhanced threshold sensitive stable election protocol for heterogeneous wireless sensor network. Springer Wirel. Pers Commun. **85**, 2643–2656 (2015)
7. N.D. Tan, N.D Viet, SCBC: Sector-chain based clustering routing protocol for energy efficiency in heterogeneous wireless sensor network. IEEE Int. Conf. Adv. Technol. Commun. (ATC) (2015)
8. A.R. Hameed, N. Javaid, S. Islam, G. Ahmed, U. Qasim, Z.A. Khan, BEEC: Balanced energy efficient circular routing protocol for underwater wireless sensor networks. IEEE Int. Conf. Intell. Networking Collab. Syst. (2016)
9. S.G. Susila, A. Vijayaselvi, Energy proficient reliable rim routing technique for wireless heterogeneous sensor networks lifespan fortification. Circ. Syst. **7**, 1751–1759 (2016)
10. M. Akbar, N. Javaid, M. Imran, A. Rao, M.S. Younis, I.A. Niaz, A multi-hop angular routing protocol for wireless sensor networks, Int. J. Distrib. Sens. Netw. **12**(9), September (2016)
11. P. Saini, A.K. Sharma, Energy efficient scheme for clustering protocol prolonging the lifetime of heterogeneous wireless sensor networks, Int. J. Comput. Appl. **6**(2), 30–36, September (2010)

Real-Time Implementation of Scheduling Policies for Education Using Raspberry Pi: A Review

Payal Kamboj, C. Rama Krishna and S. R. N. Reddy

Abstract The rise of technology has revolutionized the way we do almost anything, apart from our education system model which still remains the same. The only difference technology has made to education is that now traditional classroom lectures can be heard at a distant place as well. Other than this, there has been no contribution made to the quality of practical learning in engineering courses. In order to bridge this gap between theoretical and practical learning, we have taken one of the most important areas of Computer Science and Engineering, i.e., CPU scheduling policies in operating systems. Till now, we have just taught students about the working behavior of scheduling policies either diagrammatically or through simulation, but here we are presenting a review on the real-time implementation of scheduling policies by creating various kernels for different types of operating systems which would run first-in, first-out (FIFO), round-robin (RR), and priority scheduling policies, respectively. This helps students to understand and appreciate the behavior of these scheduling policies in a real-time environment and to analyze various parameters practically. The analyzed behavior makes students understand the advantages and drawbacks of each policy on a real-time operating system. This approach is need of the hour in order to enhance our education system.

Keywords Education · Operating system · CPU scheduling
Scheduling policies · Real-time environment · Raspberry Pi
Kernel customization · FIFO · RR · Priority

P. Kamboj (✉) · C. R. Krishna
National Institute of Technical Teachers Training and Research, Chandigarh, India
e-mail: payal_kamboj52@yahoo.com

S. R. N. Reddy
IGDTUW, New Delhi, India

C. R. Krishna et al. (eds.), *Proceedings of 2nd International Conference on Communication, Computing and Networking*, Lecture Notes in Networks and Systems 46, https://doi.org/10.1007/978-981-13-1217-5_13

1 Introduction

Education is a continuous process of gaining knowledge, skills, and values during the lifetime. Since ages, the common practice of education has been the classroom learning, which has its irreplaceable place in the education process. But this classroom learning concentrates more on theory-based lectures which are not enough for the students to get properly skilled and prepared to serve for the betterment of society. Apart from getting a professional engineering degree and paying off huge fees to the institutes, many of the engineering graduates are still unable to get employed [1]. The reason behind this lies in the focus which is being paid on the time spent to attend the lectures in the classrooms rather than the quality learning. Our schools, colleges, and universities still struggle to cope up with the rapid development in the IT industry. This alarming situation of the education system shows that the current education system does not prepare young people for the world of work and it is a well-worn complaint nowadays. Due to the existing curriculum system and lack of proper infrastructure, our education system does not provide practical learning in a real-time environment to do the experimentation on the subject learning and even if it does then it is taught in such an artificial environment that lacks many constraints of real-world programs, especially in engineering courses [2]. Practical learning helps students with critical thinking, decision-making power, creative mind development, collaborative skills, brainstorming, and ultimately establishes connections between the school, workplace, society, and the economy. The fostering practical learning needs educational institutes to ensure that the education they offer meets the deep learning and the requirements of the industry. Though many firms have initiated the practical exposure these days in the discipline of mechanical, electrical, electronics and communication, and civil engineering through the means of e-learning [3–6], still somewhere due to more complexity, Computer Science and Engineering courses lack behind when it comes to provide hands-on experience or practical learning in a real-time environment.

Computer Science and Engineering is about coming up with optimal solutions to various real-world computational problems. One of its fundamental courses is operating system, which is a system software that handles the entire functioning of the computer system which includes hardware, software, scheduling, user interaction, memory management, and many more. In practice, this subject is very complex due to its complicated processing nature which makes it even more difficult for the graduate level students to grasp it in one go. The subject is being taught in just a theoretical manner, where mere inspection of few example codes is done or simulators are used to perform the experiments which are not good enough to deepen the student knowledge on this subject. How could one understand the CPU scheduling, memory management, and deadlocks without even analyzing their working practically inside the computer system? This ultimately leads to their poor learning of the concept and failing to think any idea of inventing something new by themselves. This is the biggest reason behind lack of high-level research in many countries. Thus, the students need a way to connect the low-level details of an operating system's implementation with the high-level abstractions in the textbooks. Many efforts are being made in the past to provide a practical education through kernel means [7, 8], but somewhere they also lacked many real constraints.

In this paper, first, we would be presenting the introduction of operating system and its scheduling related concepts. After that, we will discuss the CPU scheduling in the Raspbian operating system and its source code cross compilation on the host operating system. This would be followed by the proposed work and methodology of performing the experiments and the conclusion.

1.1 Introduction of Operating System

Operating system handles almost all the functions of the computer systems by managing the available resources between hardware and software. An operating system usually can be categorized into various types on the basis of number of tasks execution allowed at a time, number of users, real-time computation, and many more parameters. In our research work, we have taken few types of operating systems into consideration which are mentioned below.

1.1.1 Types of Operating Systems

- Single-user, single-task operating system: A single user is allowed to execute one program at a time by making the kernel functions as a non-reentrant. Examples are MS-DOS and Palm OS for Palm handheld computers.
- Multitask operating system: The user is allowed to run many tasks at a given point of time by dividing the processor time between the current running tasks. The context switching time between running tasks is so less that the user has an illusion as if all programs are running concurrently. Multitasking is further categorized as preemptive and cooperative multitasking. The preemptive multitasking preempts the current running task when another task comes and gets scheduled on the processor according to the scheduling policy, whereas cooperative multitasking does not allow the current running process to get preempted before its execution. On the basis of number of users allowed, multitask operating system is divided into two types:
  - Single-user, multitask operating system: Only one user is allowed to execute multiple programs at a time.
  - Multiuser, multitask operating system: It permits multiple users to execute multiple programs in their own space simultaneously. Each of the users has sufficient resources available with them that they do not interfere in each other's performance.

1.1.2 CPU Scheduling

CPU is a processor mounted on the motherboard and responsible to take input, execute the operations on the data stream, and send the output back to the user/system. The processor is further categorized as

- Uniprocessor: The computer system with a single core is known as uniprocessor. Here, all the processing tasks in the computer system share a single processor for their execution. It is capable of executing a single thread at a given time.
- Multiprocessor: The term refers to a single computing with two or more processing units, capable of executing different instructions simultaneously. This parallel computing enables the system to execute tasks more speedily.

CPU scheduling is a process where a CPU scheduler assigns the processes to the CPU for their execution with respect to some specific scheduling policy and makes sure that there remains the high utilization of CPU, i.e., CPU does not remain idle for a long period of time. The goals are to maximize throughput, minimize response time, and maximize fairness (allocating equal processor time to available processes). The process is further categorized into two types:

- CPU-bound process: The process which spends more time in the processor and makes its utilization higher, i.e., 100% for many seconds/minutes, is known as CPU-bound process.
- I/O-bound process: The process which spends more time in waiting for input/output operations is known as input/output or I/O-bound process.

1.1.3 Scheduling Policies

The scheduling is a procedure of assigning the required resources to the processes. It is done to utilize every resource in the system efficiently and provide quality of service to the users. Scheduling policies are designed to distribute the resources among the components which request them in order to minimize the starvation of resources and increase the fairness in the system. In our research work, we have taken these three scheduling policies into consideration.

- **First-in, first-out (FIFO)**: This policy queues the processes in the order that they arrive in the ready queue to further get scheduled onto the CPU for execution. It is a non-preemptive policy.
- **Round-robin (RR)**: This policy schedules the processes on the basis of fixed time slice, which is assigned to each process in equal portions and in circular order. It is a preemptive scheduling policy.
- **Priority**: A value is assigned to all the processes and then the processes get scheduled on the basis of those values, which are nothing but priorities. It can be preemptive as well as non-preemptive.

2 CPU Scheduling in Raspbian

Raspbian is a Linux-based open-source official operating system of Raspberry Pi which can be customized according to the need. Its open source helps us to get deeper understanding and implementation details of an operating system. The reason we chose Raspberry Pi for this research work is that it enables learners to experiment over it inexpensively. Raspberry Pi also supports many Linux-based operating systems [9, 10], whose kernels can be customized according to our needs. So, this is a good way to practically demonstrate the working of scheduling policies in an operating system.

Raspbian uses a priority-based scheduling algorithm to schedule its processes on the processor. The scheduler looks through the available processes in the run queue and runs them according to some guidelines, which are different for normal processes and real-time processes. Real-time processes get higher priority over the normal processes. Every operating system including Raspbian runs on the combination of multiple scheduling policies. Raspbian implements scheduling on the basis of priority and type (normal or real-time) of a process. Here, a fixed value is assigned to all the processes and how much a task is important is determined by observing that value only. A process having higher value would run before the one with the lower value. Each task is assigned a value from the range of −20 to 19 with default being 0 [11]. This value is known as NICE value in Linux. When a process executes first time, counter (value of the process's priority) decreases by 1 with the time. NICE value describes the priority of a process. In case of normal processes, the higher the nice value, the lower the priority of a task. A real-time process has a higher priority in the system than any regular process. Its priority ranges from 0 to 99. Here, a greater number indicates higher priority. The kernel, however, has a range of 0–139 for the process priorities, where lower value signifies higher priority. The range from 0 to 99 is for the real-time processes and from 100 to 139 (mapped to nice −20 to +19) is for the normal processes. By default, this means −20 to +19 nice range maps directly onto the 100–140 priority range. Raspbian has two scheduler classes to schedule the processes. The **Completely Fair Scheduler** is a current process scheduler for normal processes, which handles the CPU resource allocation and executes all the processes fairly. It schedules the normal/regular processes with SCHED_NORMAL (Time-sharing), SCHED_BATCH, and SCHED_IDLE scheduling policies. The **Real-time scheduler** schedules the real-time processes with SCHED_FIFO (first-in, first-out) and SCHED_RR (round-robin) policies. In case of real-time processes, initially a task is scheduled with first-in, first-out scheduling policy. Here, if a higher priority process comes, the current running process gets preempted and higher priority one gets scheduled. In case any lower priority process comes then it will not get scheduled until the execution of the already scheduled higher priority process is completed. If the same priority process comes then the scheduling policy switches to round-robin policy.

Since Raspbian is an open-source operating system, its source code is easily available online [12]. The source code is written in C and assembly language. To make all the processes (normal or real-time) run in a specific desired policy, we have to dig deeper into the source code of Raspbian kernel and have to design it in such a manner that it automatically runs all the processes with one desired scheduling policy rather than the combination of many. This way we can realize the practical behavior of scheduling policy in a real-time environment. Along with the implementation of specific scheduling policy, we have to make sure that the priority of all the processes should be the same so that the priority affect can be nullified for FIFO and RR.

2.1 Cross-Compilation for Raspbian

Cross-compilation is a process of compiling an executable code for the platform other than the one in which the cross compiler is running on. There are two main ways for building up the kernel from the source code for Raspbian. One way is to build up locally on the Raspberry Pi, which usually takes more time and another is to cross-compile on a host machine, which is quicker but requires more setup on the system [13].

3 Proposed Work and Methodology

To make up the gap between the theoretical and practical aspect of learning method and to deepen student knowledge and interest in research on operating system, we have come up with an approach where we will create various kernels of Raspbian in order to show the real-time implementation of CPU scheduling policies such as first-in, first-out (FIFO), round-robin (RR), and priority. The experimental configuration required for this research work is mentioned in Table 1. This table contains the details of host system on which source code is to be customized, cross compiled, and a new kernel is to be built. It has the details of Raspberry pi as well as the details of the various experiments to be performed.

The source code of Raspbian kernel broadly contains 22 directories, out of which **kernel** directory is of our main concern. The **kernel** directory has various files, where **sched** folder contains the code files of scheduling process of the system. In **sched folder**, the main files of scheduling concern are **fair.c ,**which contains the source code of scheduling of normal processes, **rt.c** which has the source code of scheduling of real-time processes, and **core.c** which is the main file containing source code for the implementation of scheduling in the system. In the Linux kernel, there is a bandwidth cap mechanism which protects the real-time tasks to monopolize the CPU. It preserves 5% of CPU for the tasks other than the real-time ones.

Table 1 Experimental setup configuration

Host system	HP
Host system RAM	3.7 GB
Host system hard disk	310.9 GB
Host system processor	Intel Core i5 CPU 650 @ 3.20 GHz* 4
Host system OS type	64 bit OS
Host operating system	Ubuntu 16.04 LTS
Raspberry Pi	Raspberry Pi 1(Model-A) Raspberry Pi 3 (Model-B)
Processor (Raspberry Pi 1)	700 MHz single-core ARM1176JZF-S
RAM (Raspberry Pi 1)	512 MB
Processor (Raspberry Pi 3)	1.2 GHz 64-bit quad-core ARM-cortex-A53
RAM (Raspberry Pi 3)	1 GB
Raspbian kernel version	4.9

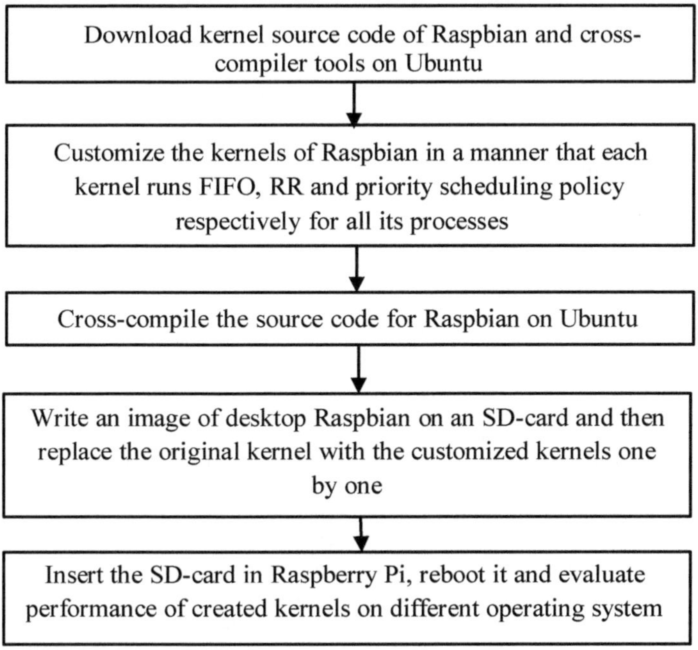

Fig. 1 A flow of the proposed methodology

3.1 Methodology

The methodology which should be followed in order to create the experiments for the practical learning is explained below in Fig. 1. These steps would create the respective kernels for each policy which could be further analyzed with respect to various parameters like response time, waiting time, turnaround time, CPU usage, etc.

4 Conclusion

The rise of technology has motivated us to customize the traditional classroom learning approach in order to provide an enhanced learning experience to students. We have presented a review on customization approach for CPU scheduling policies' practical teaching/learning in a real-time environment by customized kernel building and booting it on different types of operating systems using Raspberry Pi. This way, students would be able to analyze the parameters of scheduling policies in a real-time environment and their learning would become more impactful.

References

1. H. Atkinson, M. Pennington, Unemployment of engineering graduates: the key issues. Eng. Educ. **7,** 7–12 (2015) (Taylor & Francis Online)
2. S. Ribic, A. Salihbegovic, Tiny operating system kernel for education purposes, in *38th International Convention on Information and Communication Technology, Electronics and Microelectronics, MIPRO* (IEEE Press, Opatija, 2015), pp. 700–705
3. NPTEL CSE courses. https://nptel.ac.in/course.php?disciplineId=106
4. Course era. https://www.coursera.org
5. NITTTR Chandigarh Technology Enabled Learning Computer Science & Engineering courses. https://nitttrchd.ac.in/sitenew1/nctel/comp_sc.php
6. Skifi labs. https://www.skyfilabs.com/
7. B. Qu, Z. Wu, Kernel experiment series for operating system course teaching, in P*roceedings of IEEE International Symposium on IT in Medicine and Education (ITME)* (IEEE Press, Cuangzhou, 2011), pp. 12–16
8. B. Qu, Z. Wu, Design and Implementation of Tiny Educational OS, in *Recent Advances in Computer Science and Information Engineering* (Springer, Heidelberg, 2012), pp. 437–442
9. Raspberry Pi OS. https://en.wikipedia.org/wiki/Raspberry_Pi#Operating_systems
10. Raspberry Pi OS image. https://www.raspberrypi.org/downloads/
11. R. Love, *Linux Kernel Development*, 3rd edn. (Pearson, 2010)
12. Raspbian source code. https://www.github.com/raspberrypi/linux
13. Kernel building method and various commands using cross compilation technique. https://www.raspberrypi.org/documentation/linux/kernel/building.md

Performance Evaluation of LEACH Protocol Based on Data Clustering Algorithms

Mohit Mittal and Satendra Kumar

Abstract Sensor is a key to make our existing systems smart and easy to handle. Association of hundreds to thousands of sensors forms a huge sensor network which may have capabilities of sensing, computing, and communicating via radio frequencies. Communication technology is nowadays prominently working in the field of wireless sensor networks (WSNs). As due to the flexible nature of sensor network, it can easily connect to one another and form large network. It has a wide range of application field from indoor to outdoor environments. Besides this, it has many challenges out of which battery power is the most highlighted one. In this paper, LEACH protocol has improvised with various data clustering algorithms and developed hybrid protocols such as LEACH-FUZZY, LEACH-LVQ, and LEACH-SOFM. Simulation result shows that LEACH-FUZZY protocol performs outstandingly in comparison with LEACH, LEACH-LVQ, and LEACH-SOFM.

Keywords LEACH protocol · Learning vector quantization · SOFM Neural network · Fuzzy C-means

1 Introduction

Association of hundreds to thousands of sensor nodes or motes make wireless sensor network (WSN) [1, 2]. Individual sensors work in physical phenomena having capabilities of sensing, computing, and communicating via radio frequencies. Sensors are a simple combination of circuitry and complex usage of various protocols. The wireless sensor network is collaboration of both homogeneous and heterogeneous

---

M. Mittal (✉)
Department of Computer Science Engineering, Model Institute of Engineering and Technology, Jammu, India
e-mail: mittal.mohit02@gmail.com

S. Kumar
Department of Computer Science, Gurukul Kangri University, Haridwar, India
e-mail: satendra04cs41@gmail.com

C. R. Krishna et al. (eds.), *Proceedings of 2nd International Conference on Communication, Computing and Networking*, Lecture Notes in Networks and Systems 46, https://doi.org/10.1007/978-981-13-1217-5_14

types of sensor nodes which act according to application specific. Sensor nodes may act as sink nodes, actuator nodes, gateways, and monitoring nodes depending on the application. Wireless sensor networks are nowadays prominently in demand of usage in various fields for monitoring and tracking purposes. It expands their utilization in every field due to the connectivity of each field with the Internet. In early stage of applications, WSN is only selected for harsh environment deployment but now it is commonly used in indoors to make our working environment smart. Internet of things is the most popular emerging trend in today's scenario, where sensor network works as the backbone to this network.

Wireless sensor networks are generally deployed in different environments depending on application. Each environment has its own implementation issues but battery life is a common issue in every wireless sensor network. To resolve this issue, many routing protocols, clock synchronization algorithms, and data processing algorithms have been created but still it requires more attention.

This paper is categorized into various sections. Section 1 has explained the brief introduction of the sensor network and major challenges. In Sect. 2, LEACH protocol is elaborated. Section 3 presents clustering algorithms such as LVQ, Fuzzy C-means, and SOFM neural network. Section 4 discusses implementation and Sect. 5 includes simulation. The last section concludes the paper.

2 Low-Energy Adaptive Clustering Hierarchy Protocol (LEACH) Protocol

LEACH protocol [3, 4] is under the category of hierarchical protocol. It manages sensor nodes with different responsibilities. Some of the nodes act as cluster head (CH) nodes and rest of the sensor nodes act as non-cluster head (non-CH) nodes. LEACH protocol is a TDMA-based protocol which is collaborated with various protocols. The selection of CH nodes is depending on various parameters. After the selection of CH node, non-CH nodes sense the data and forward to CH nodes. Each non-CH node is connected to any particular CH node for particular time of span, i.e., for particular round. After aggregation of sensed data at CH node, it compresses the data and forwards to sink node. The total network lifetime is calculated on first node dead (FND), last node dead (LND), and half network alive (HNA). To reduce further energy consumption, CH rotation is done to improve overall network lifetime.

LEACH protocol executes in many rounds. These rounds are required to manage network lifetime. There are mainly two phases: cluster setup phase and steady phase. In cluster setup phase, some nodes are selected as CH node and others are non-CH nodes. These selections are done on the basis of various parameters. In this protocol, every "n" nodes compute a random number which lies in a range between 0 and 1. The threshold [T (n)] [5, 6] is calculated using

$$T(n) = \begin{cases} \frac{P}{1-P\left(r \mod \frac{1}{P}\right)} & \text{if } \in G \\ 0 & \text{Otherwise} \end{cases}, \quad (1)$$

where P denotes CH probability, r denotes number of the present round, and G represents the set of nodes for non-CH in the last $1/P$.

In the second phase, that is, steady phase, the data is transmitted from CH nodes to BS. This phase is also called data transmission phase. This phase runs for a larger period of time as compared to cluster setup phase. After completion of data transmission, next round will be started. These two phases will again execute with different sets of CH nodes and non-CH nodes.

3 Clustering Techniques

3.1 Learning Vector Quantization (LVQ)

LVQ algorithm is categorized under supervised learning technique. LVQ mainly uses class information to move Voronoi vectors in accordance to improve the quality of classified regions. Voronoi is meant to vector quantizer having least encoding distortion. LVQ provides the refined method for computing Voronoi vectors with approximations being specified by weight vectors. LVQ helps in the computation of feature map which can adaptively solve the classification complex mathematical problems.

3.2 Fuzzy C-Means

Fuzzy C-means clustering is an approach for data clustering. In this clustering algorithm, fuzzy logics are used to form clusters. It specifies some range categorization of data values and denotes some membership grade which helps to create clusters accordingly. This approach is invented by Jim Bezdek. It is highly used in distinguishing the data values into various clusters and also forms task for pattern recognition. The main aim of this algorithm is to minimize an objective function.

The objective function is computed by

$$J = \sum_{i=1}^{N} \sum_{j=1}^{C} u_{ij}^{m} \|xi - cj\|^2, \quad 1 \leq m < \infty. \tag{2}$$

Fuzzy C-means provides each data value to a membership value for each and every cluster.

3.3 Self-organizing Feature Maps (SOFM) Neural Network

SOFM neural network [7–9] falls under the category of unsupervised learning techniques and is used to optimize complex problems. SOFM neural network optimizes the problem depending on the input provided to the network. There is no target value provided for optimization. It generally does clustering of the data into various distinct clusters depending on the neighboring values. The clusters have their own minimum and maximum values. SOFM neural network [10] starts with a learning phase on input data values and after learning process completes, validation of the input values starts that validates according to the learning network which is processed.

4 Implementation

LEACH protocol is already discussed in Sect. 2. In Sect. 3, various clustering algorithms have been described. Now, in this section, flowchart of LEACH protocol using clustering algorithms is shown in Fig. 1. Process of selection of CH nodes comes under cluster setup phase in which some of the nodes act as CH node out of the total number nodes. This selection of CH node depends on various parameters. Next step is the aggregation of sensed data at CH node from various connected neighboring nodes to their corresponding CH nodes. After completion of this phase, clustering techniques one by one have been implemented to LEACH protocol.

5 Results

In the previous section, how actually LEACH protocol has been implemented with three data clustering techniques is explained with the help of flowcharts. Figure 2 shows the CRAWDAD temporal dataset. It is used for experimental analysis. Improved LEACH protocol with clustering algorithms such as LEACH-LVQ, LEACH-FUZZY (Fuzzy C-means), and LEACH-SOFM (SOFM Neural Network) is created. After this, the network lifetime for each protocol is measured. Figure 3

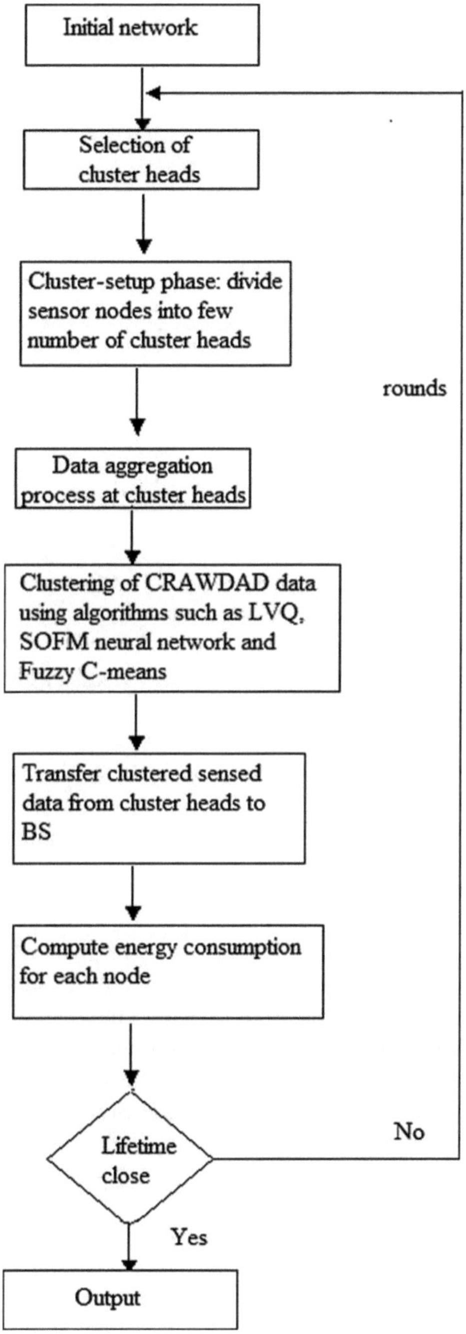

Fig. 1 Flowchart of LEACH protocol

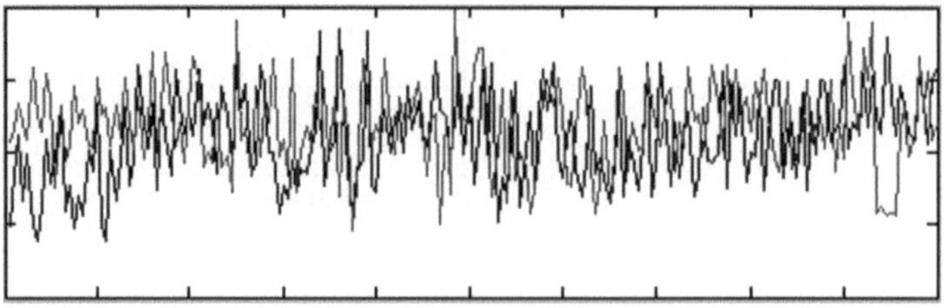

Fig. 2 CRAWDAD dataset

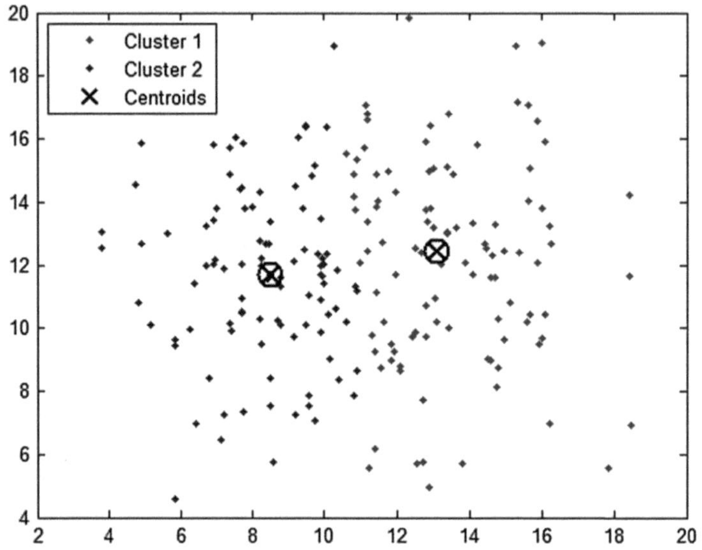

Fig. 3 Clustered data

shows the clustering result, i.e., two clusters. Figure 4a, b shows clustering using the fuzzy technique. Figure 5 shows membership function graphs for each cluster. SOFM neural network results are shown in Figs. 6 and 7. Comparative study and analysis of energy parameter among four routing protocols such as LEACH, LEACH-LVQ, LEACH-FUZZY, and LEACH-SOFM are shown in Fig. 8. Simulation is done using MATLAB (2013a).

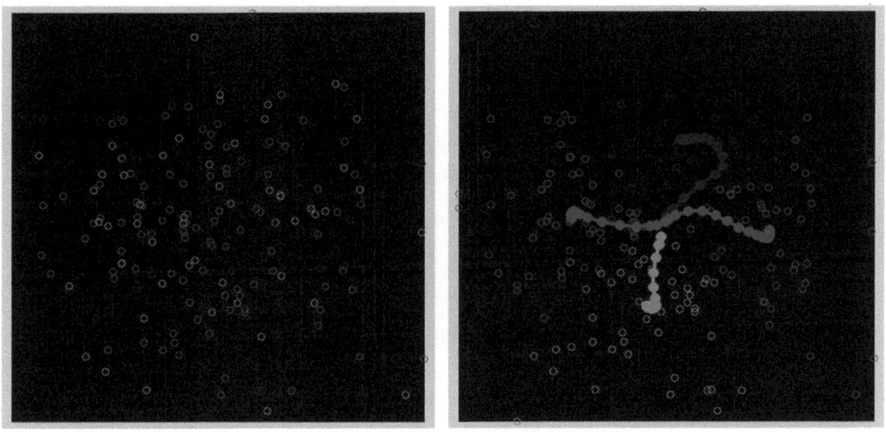

(a) Fuzzy C-means clustering before clustering

(b) Fuzzy C-means clustered representation into 4 clusters

Fig. 4 **a** Fuzzy C-means clustering before clustering. **b** Fuzzy C-means clustered representation into four clusters

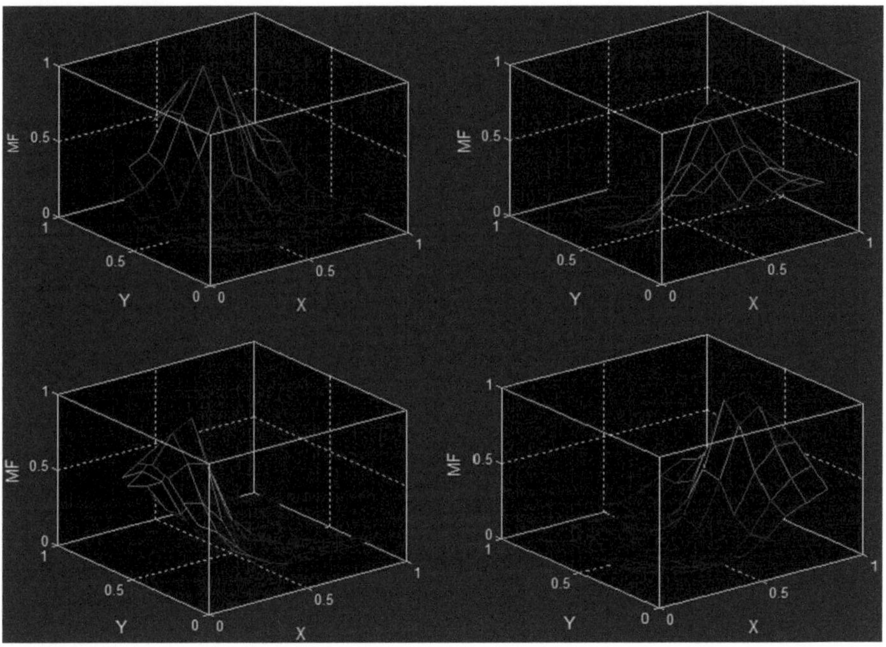

Fig. 5 Membership functions

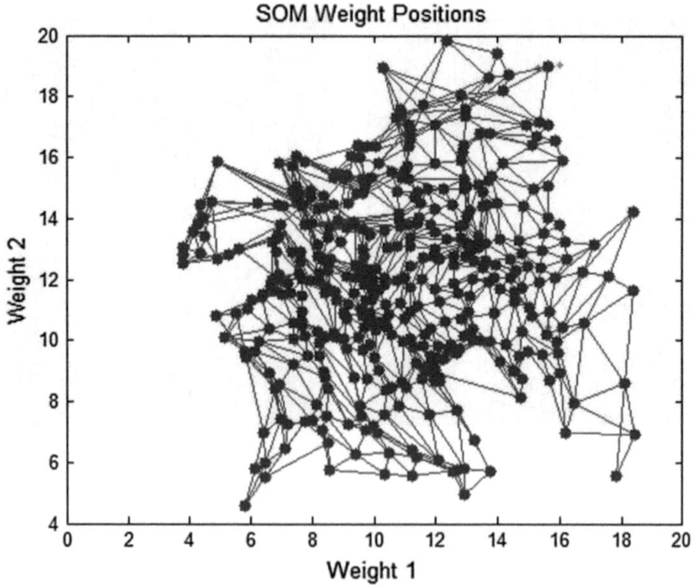

Fig. 6 Weight positions based on SOFM neural network

Fig. 7 SOFM neighbor weight distances

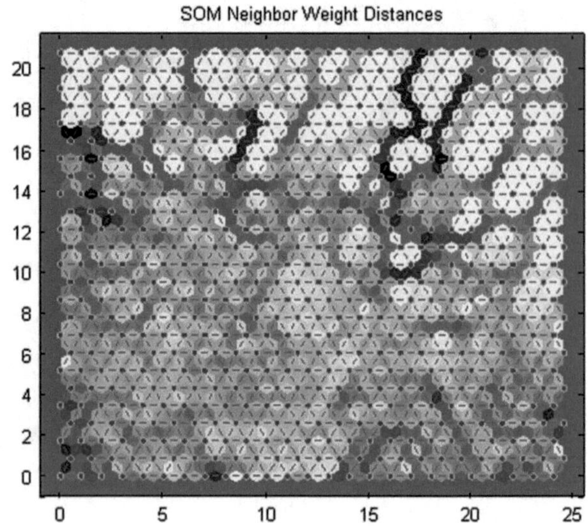

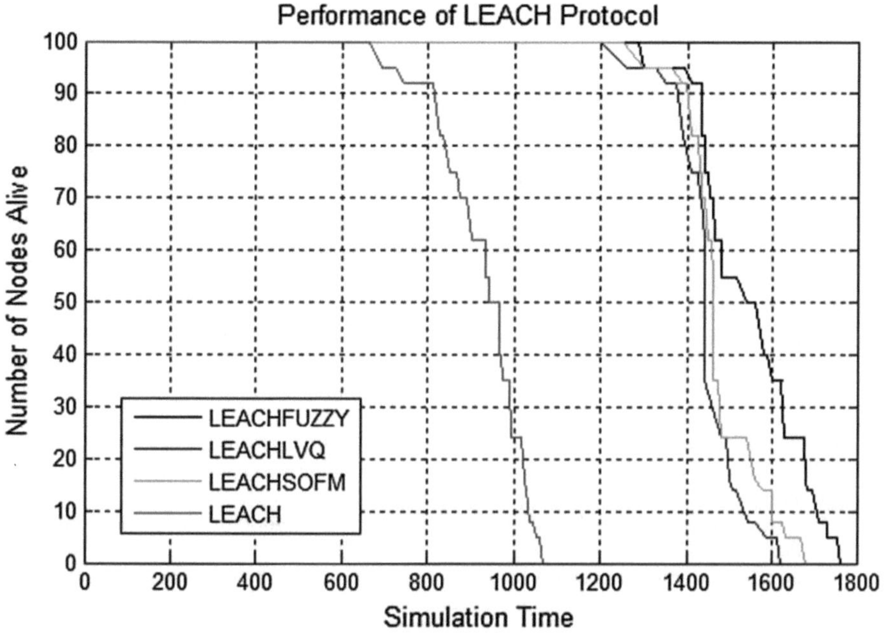

Fig. 8 Comparative analysis of network lifetime of LEACH, LEACH-LVQ, LEACH-SOFM, and LEACH-FUZZY

6 Conclusion

The wireless sensor network is a most prominent communication technology that is likely to be used in every part of real-world applications. Sensor network is the backbone for the Internet of things as it is the hot topic in nowadays scenarios. It plays a vital role in the communication field. Along with a wide range of applications, it has energy challenge to overcome this; there is a need of better routing protocol. This paper represents various data clustering algorithms such as LVQ, SOFM neural networks, and fuzzy C-means. These algorithms were created and performed well in clustering of data. Here, in this paper, these clustering algorithms are induced in LEACH protocol to improvise the network lifetime. CRAWDAD temporal dataset is taken to simulate. 100 sensor nodes have been taken for evaluation. Experimental results show that LEACH-FUZZY routing protocol is the best-performed algorithm in comparison to other data clustering algorithms.

References

1. I.F. Akyildiz, W. Su, Y. Sankarasubramaniam, E. Cayirci, Wireless sensor networks: a survey. Comput. Netw. **382**, 393–422 (2002)
2. M. Mittal, R.S. Bhadoria, *Aspect of ESB with Wireless Sensor Network*, in *Exploring Enterprise Service Bus in the Service-Oriented Architecture Paradigm* (igi-global publications, 2017), pp. 319
3. W.R. Heinzelman, A. Chandrakasan, H. Balakrishnan, *Energy-Efficient Communication Protocol for Wireless Microsensor Networks*, in*Proceeding of IEEE International Conference System Science*, Hawaii (2000), pp. 1–10
4. W.R. Heinzelman, A. Chandrakasan, H. Balakrishnan, An application-specific protocol architecture for wireless microsensor networks. IEEE Trans. Wirel. Commun. **1**, 660–670 (2002)
5. H. Oh, K. Chae, *An Energy-Efficient Sensor Routing with Low Latency, Scalability in Wireless Sensor Networks*, in*International Conference on Multimedia and Ubiquitous Engineering* (2007), pp. 147–152
6. M. Mittal, K. Kumar, *Network Lifetime Enhancement of Homogeneous Sensor Network Using ART1 Neural Network*, in*Sixth International Conference on Computational Intelligence and Communication Networks* (2014), pp. 472–475
7. L. Fausett, *Fundamentals of Neural networks: Architecture* (Algorithm and Applications, Pearson Education, 1994)
8. N. Enami, R.A. Moghadam, Energy based clustering self organizing map protocol for extending wireless sensor networks lifetime and coverage. Can. J. Multimedia Wirel. Netw. **1**, 42–54 (2010)
9. M. Mittal, K. Kumar, *Data Clustering in Wireless Sensor Network Implemented on Self Organization Feature Map (SOFM) Neural Network*, in Published in*IEEE International Conference on Computing Communication and Automation (ICCCA)* (2016)
10. M. Mittal, K. Kumar, Quality of services provisioning in wireless sensor networks using artificial neural network: a survey. Int. J. Comput. Appl. (IJCA), 28–40 (2015)

Performance Analysis of 3D WSN with Change in Rain Intensity

Raunak Monir, Ranjana Thalore and Partha Pratim Bhattacharya

Abstract Can the intensity of rainfall have major impact on terrestrial wireless communication? Using simulated results, we will look into the possibilities in this paper. The interest in developing wireless sensor network is of high importance in recent days specially looking into factors which affect the rate of data communication. Weather can be such an external factor to hamper the communication unexpectedly and at any time. In this paper, we have analyzed factors such as messages received, throughput, delay, and jitter under various intensity of rainfall in a particular scenario designed in 3D, using IEEE 802.15.4 standards. Dynamic Manet On-Demand (DYMO) was considered as the routing protocol in the simulation. Extensive simulations are done using QualNet 6.1 network simulator to validate the results and compare the graphs for better understanding. We believe this type of research work can trigger new techniques to better project the execution of WSN in its destined deployment environment.

Keywords Wireless sensor networks · Simulation · Qualnet · Rain intensity

1 Introduction

Rapid progress is being made in the field of wireless communication, and computer networks over the time, leading to the development of wireless ad hoc sensor networks, consisting of tiny, low-cost low power multifunctional sensors, which can monitor wide and remote areas with precision without the need of a human operator. Wireless Sensor Networks (WSNs) is mainly described as a self-dependent and

R. Monir (✉) · R. Thalore · P. P. Bhattacharya
College of Engineering Technology, Mody University, Sikar, Rajasthan, India
e-mail: hotandsour18_ron@yahoo.com

R. Thalore
e-mail: thalorer1603@gmail.com

P. P. Bhattacharya
e-mail: hereispartha@gmail.com

C. R. Krishna et al. (eds.), *Proceedings of 2nd International Conference on Communication, Computing and Networking*, Lecture Notes in Networks and Systems 46, https://doi.org/10.1007/978-981-13-1217-5_15

foundation-less wireless networks to help us supervise physical or environmental factors, like temperature, sound, vibration, pressure, motion, or pollutants and to mutually pass their data through the network to a main location or sink where the data can be observed and further analyzed. One can fetch necessary information from the network by placing queries and collecting results from the sink. Typically a wireless sensor network can contain up to hundreds or thousands of sensor nodes who team up to form the network. The sensor nodes can disseminate among themselves with the aid of radio signals. A wireless sensor node is equipped with sensing and computing devices, radio transceivers, and power components. WSN incorporates various sensors that are distributed around a particular node for achieving the computational operations when using IEEE 802.15.4 standards [1–3].

Rain intensity is a measure of the amount of rain fall over time. It may vary from 2 to more than 50 mm per h depending on the rate of precipitation which is namely classified as follows:

- Light rain—The rate of precipitation is below 2.5 mm per h.
- Moderate rain—Precipitation rate is between 2.5 and 10 mm per h.
- Heavy rain—Precipitation rate is between 10 and 50 mm per h.
- Violent rain—Precipitation rate is above 50 mm per h.

The sink node being above the ground may get affected due to the variation in the intensity of the rain [4]. However, it will communicate with all the sensor nodes within its range via applications.

In this paper, first section comprises of literature review where it was seen that researchers have done some studies regarding the impact of certain environmental factors on wireless communication. The next section contains the overview of the experiment conducted in 3D using Qulanet Simulator 6.1 where the simulation model and parameters have been discussed. Finally, the last section of the paper concludes with the outcome of the experiment and discussion.

2 Literature Review

Impact of environment factors on wireless links. Some researchers have worked on the effect of weather on outdoor WSNs. Boano et al. [5] introduced an extension for WSN test beds that permits to control the locally available temperature of sensor nodes and to study temperature effects on the performance of network in exact and repeatable form. Other works [6, 7] suggest that fog and rain can have a severe impact on the transmission range of WSN nodes, especially w.r.t. packet reception. Bannister et al. [8] showed that high temperature negatively affects communication, based on data from a radio survey in the desert and applied to simulations of localization and data collection. Nadeem Farukh et al. [9] analyzed the lifetime performance of hybrid and RF based WSN for ground applications and weather effect such as rain, snow, and fog on them. Ramona Marfievici et al. [10] did a case study where they experimented in three different outdoor scenarios by varying degree of vegetation along with seasonal variations.

3 Experimental Overview

In the experiment, we designed a scenario taking 100 m * 100 m of a land area and placed 100 nodes in it using Qualnet Network Simulator 6.1. The nodes were placed in five layers starting from the ground level up to 40 m below the ground, hence making it 3D. Twenty nodes were taken in each layer with 10 m spacing. The sink node was placed on the ground for necessary further communication to the base station. The weather effect was considered to be applicable to the entire area for the simulation purpose. The application used was traffic generator and using IEEE 802.15.4 the sink node was connected to 15 application for this underground communication. The sensor nodes having a battery of 200 mAhr was used and the antenna in it was omnidirectional. Weather effect was added to the entire scenario and the clouds were placed at a height of 50 m above the ground. The simulation was conducted with zero intensity of rain for a time period of 1000 s and hence the necessary result parameters were obtained. Thus the experiment was repeated 5 times with changed rain intensity, to see the variation in data communication across the network [11].

Figure 1 clearly shows how the nodes were randomly deployed and the weather effect was added to the entire network. The entire area was assumed to experience rainfall.

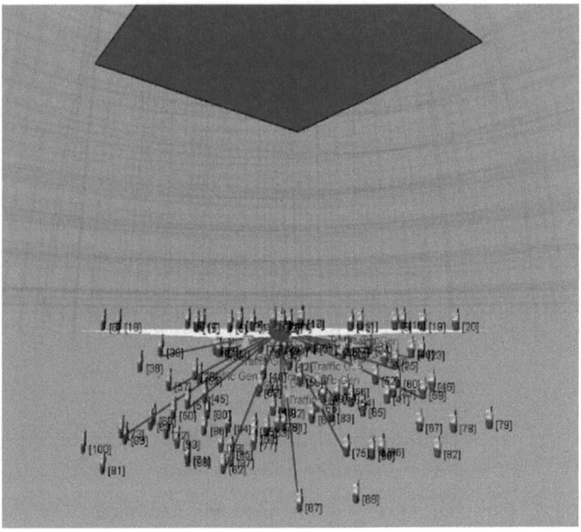

Fig. 1 3D design of the network

Table 1 show a list of parameters which were assumed for this simulation. Followed by, Fig. 2 showing the process of simulation in successive steps.

Table 1 Simulation parameters for WSN

Parameter(s)	Value(s)
Deployment strategy	Random
Total number of nodes	100
Terrain in 3D	100 m × 100 m × 40 m
Communication protocol	IEEE 802.15.4
Simulation time (sec.)	1000
Channel frequencies (Zig bee)	2.4 GHz
Routing protocol	DYMO (Dynamic Manet on-demand)
Traffic type	TRAFFIC-GEN
Rain intensity	(0, 2, 10, 40, and 60) mm
Packet size (bytes)	50
Battery capacity (mAhr)	200
Transmission range (m)	10
Antenna type	Omnidirectional

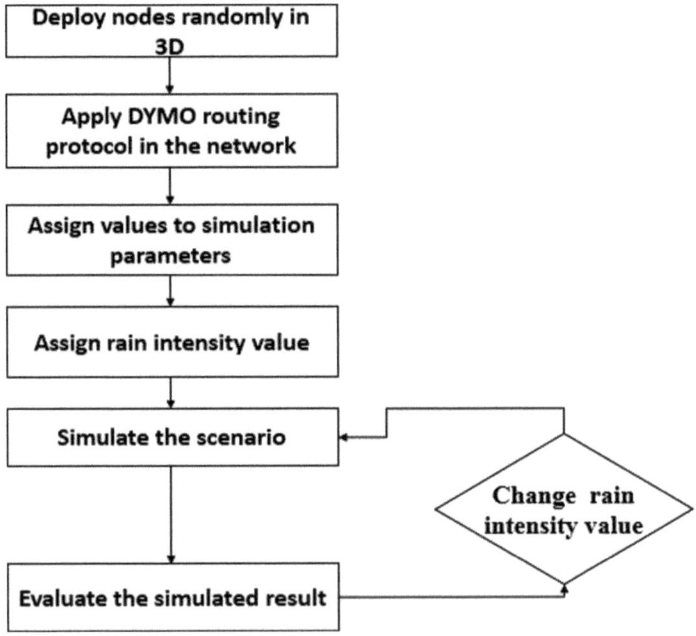

Fig. 2 Flowchart of steps during network simulation

4 Outcome of the Experiment

Using Qualnet 6.1, the scenario in 3D was simulated for one thousand seconds and parameters taken into consideration for evaluation were such as messages received, throughput, delay, and jitter, and battery power consumption. The analysis was done on the basis of simulations repeated five different times with respective value of rain intensities: no rain (0 mm), light rain (2 mm), moderate rain (10 mm), heavy rain (40 mm), and violent rain (60 mm). The graphs obtained below gives a clear idea on how it affects the network with change in the rain intensity. The only routing protocol used here was DYMO as it is among the most efficient routing protocol to be mostly preferred when using IEEE 802.15.4 standards.

Figure 3 clearly above shows a graph with a smooth decline of number of messages received in the given scenario. The data communication does not get effected to a large scale unless there is heavy rain. As the design is in 3D, hence the sink node and few other nodes being on the surface, gets affected the most.

In Fig. 4, the graph shows us the rate of throughput which indicates a smooth decline as with increase in intensity of rainfall. Although the results show that the network will still work even under heavy rainfall but with low efficiency.

Figure 5 shows lesser delay in heavy rain condition which is unexpected. But overall, the delay was acceptable and did not vary by large scale to affect packet arrival delay.

In Fig. 6, the jitter shows similar result like delay in Fig. 5, in case of heavy rainfall. However it is a good sign because in that case the packets may arrive in less number but at least the communication will take place even under such extreme weather conditions.

Fig. 3 Comparison of messages received with variation in rain intensity

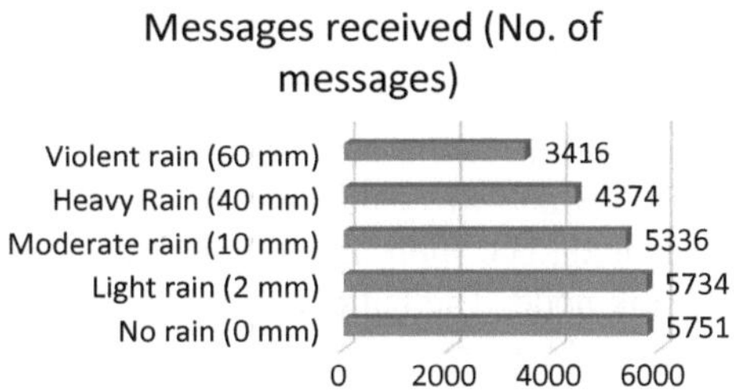

Fig. 4 Comparison of throughput with variation in rain intensity

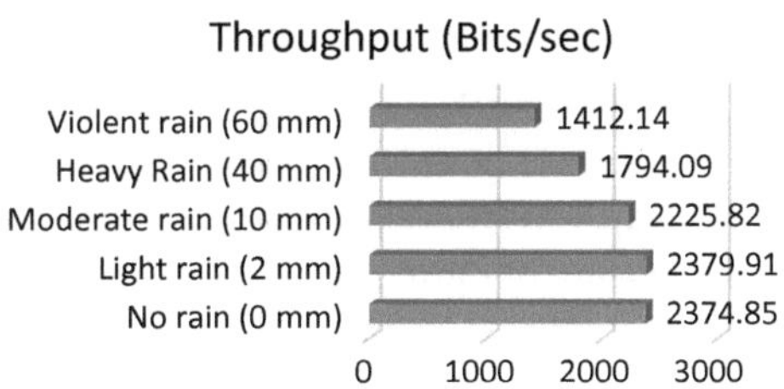

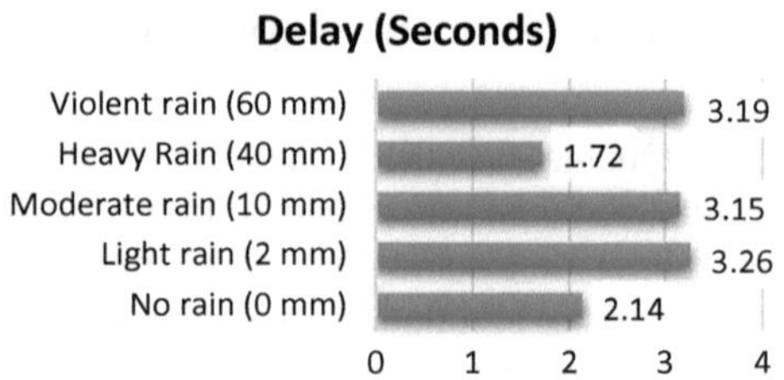

Fig. 5 Comparison of delay with variation in rain intensity

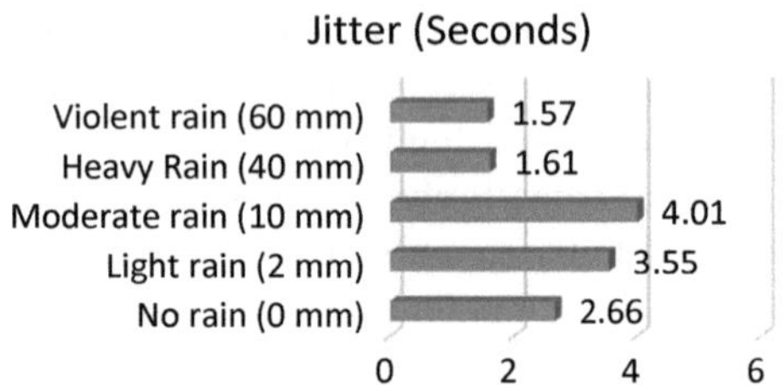

Fig. 6 Comparison of jitter with variation in rain intensity

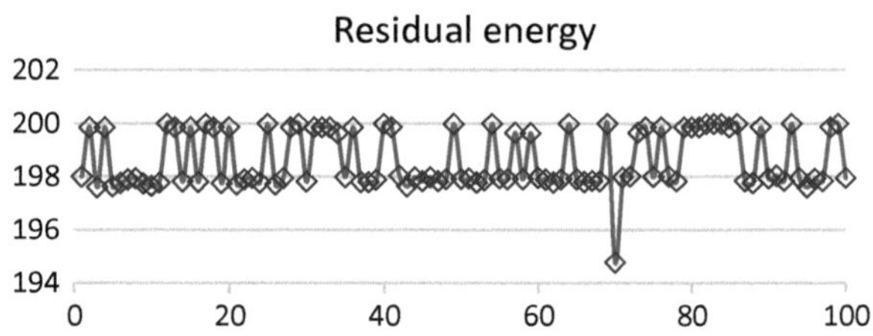

Fig. 7 Residual battery capacity in mAhr

Figure 7 shows the total residual energy left within the nodes after completion of simulation. For all the five times the simulation was conducted, the energy model was approximately same and the values varied by +0.01 only. However as expected, node number 70 consumed maximum energy, it is the sink node having to communicate the maximum in the network. Average values of 99 nodes were taken into account for E(f) which was calculated to be 198.6615 mAhr. Network lifetime of node was calculated to be approximately 41.50 h for sensor nodes in operation and 10.62 h for the sink node. The battery value needs to be increased to a higher value for longer lifetime of battery.

5 Conclusion

In this paper, the possible wireless communication using sensor network under various intensity of rainfall was explored. Using Qualnet Simulator 6.1, it was possible to obtain important parameters related to WSN and get an idea about it for future advance analysis. Rainfall seems to have impact on messages transferred and throughput in the communication with increase in intensity. However, strangely it did not affect delay and jitter much, which could be due to the use of DYMO as the routing protocol. However, it is mainly a software-based simulation conducted and further research needs to be done for obtaining practical values. Hence real-life sensor could be used and experiments can be conducted for further evaluation of this work for triggering better innovation in this field.

References

1. R.T. Manju, M.K. Jyoti, Performance evaluation of bellman-ford, AODV, DSR and DYMO protocols using qualnet in 1000 × 1000 m terrain area. Int. J. Soft Comput. Eng. (IJSCE) ISSN **2**(6), 2231–2307, Jan (2013)
2. T. Ranjana, J. Sharma, M. Khurana, M.K. Jha, QoS evaluation of energy-efficient ML-MAC protocol for wireless sensor networks. AEU-Int. J. Electr. Commun. **67**(12), 1048–1053 (2013)
3. K. Manju, J. Sharma, T. Ranjana, M.K. Jha, Direction determination in wireless sensor networks using grid topology. J. Emerg. Technol. Web Intell. **5**(2), 166–170 (2013)
4. R. Marfievici, A.L. Murphy, G.P. Picco, F. Ossi, F. Cagnacci, How environmental factors impact outdoor wireless sensor networks: a case study. In IEEE 10th International Conference on Mobile Ad-Hoc and Sensor Systems (MASS) (Oct 2013), pp. 565–573
5. C.A. Boano, M. Zúñiga, J. Brown, U. Roedig, C. Keppitiyagama, K. Römer, Templab: a testbed infrastructure to study the impact of temperature on wireless sensor networks. In Proceedings of the 13th international symposium on Information processing in sensor networks (IEEE Press, Apr 2014), pp. 95–106
6. G. Anastasi, A. Falchi, A. Passarella, M. Conti, E. Gregori, Performance measurements of motes sensor networks. In Proceedings of the 7th ACM international symposium on Modeling, analysis and simulation of wireless and mobile systems (ACM, 2004), pp. 174–181
7. N. Hameed, T. Mehmood, H.U. Manzoor, Effect of weather conditions on FSO link based in Islamabad, *arXiv preprint* arXiv:1711.10869, Nov (2017)
8. K. Bannister, G. Giorgetti, S.K.S. Gupta, Wireless sensor networking for hot applications: effects of temperature on signal strength, data collection and localization, In *Proceedings of the 5th Workshop on Embedded Networked Sensors (HotEmNets' 08)*, March (2008)
9. S. Nadeem Farukh, E. Leitgeb Chessa, S. Zaman, The effects of weather on the life time of wireless sensor networks using FSO/RF communication. radioengineering, **19**(2), 262–270 (2010)
10. R. Marfievici, A.L. Murphy, G.P. Picco, F. Ossi, F. Cagnacci, How environmental factors impact outdoor wireless sensor networks: a case study, In *Mobile Ad-Hoc and Sensor Systems (MASS), 2013 IEEE 10th International Conference on*, October 2013, pp. 565–573
11. R. Thalore, P.P. Bhattacharya, M.K. Jha, "Performance Comparison of Homogeneous and Heterogeneous 3D Wireless Sensor Networks," *Journal of Telecommunications and Information Technolog,*, 2, 32–37 (2017)

A Survey on Wireless Sensor Networks Coverage Problems

Aastha Maheshwari and Narottam Chand

Abstract All types of applications in Wireless Sensor Network (WSN) generally faces problem of coverage as per the study. Full coverage method, being simple and easy is widely adopted theoretically. To reduce the overuse of sensor, full coverage network is composed of partial ones. Full coverage network is not best for real-world problems because of a large number of restrictions. In this survey, the characteristics of partial or probabilistic coverage problems are analyzed, and compared with full coverage problems. This survey will help to overview unsettled coverage problems. Relevant models, such as detection model, network model, and deployment model based on partial and probabilistic coverage problem are also summarized. This paper discusses three main objectives; how to maximize coverage quality, how to maximize network lifetime, and how to minimize the number of sensors for coverage problems with uncertain properties by deploying, scheduling or selecting, and regulation of sensors. Future challenges are also discussed.

Keywords Wireless sensor network · Full coverage problem · Partial coverage problem · Probabilistic coverage problem

1 Introduction

Wireless Sensor Networks (WSNs) have invited attraction in industrial as well as the academic field. With the rise of microelectromechanical systems, cheap and powerful sensors and nodes are massively produced. WSNs are composed of many such nodes dynamically without the help of any predefined infrastructure. Network coverage is the central problem in WSN. Once a sensor is active, it can detect objects in detection range. If the object is detected by at least one sensor of the Region of

A. Maheshwari (✉) · N. Chand
NIT, Hamirpur, Himachal Pradesh, India
e-mail: aastha.01031993@gmail.com

N. Chand
e-mail: nar.chand@gmail.com

C. R. Krishna et al. (eds.), *Proceedings of 2nd International Conference on Communication, Computing and Networking*, Lecture Notes in Networks and Systems 46, https://doi.org/10.1007/978-981-13-1217-5_16

Interest (ROI), we say the object is covered. Consequently, we can locate objects in the ROI. The coverage problem is to determine whether the object in ROI can be covered by WSNs. Thus, the coverage quality is termed as a percentage of objects to be covered by active sensors. To check whether WSNs coverage quality is qualified, we involve measurement and adopt methods to acquire coverage quality. Major coverage problems include target, area, and barrier coverage.

In Fig. 1, the points represent the sensors and the cycles represent the detection range of sensors. Figure 1a shows working of area coverage. Figure 1b focuses on different target points (depicted by a triangle). Figure 1c studies an intruder detection problem along with a long belt region by forming a barrier. Sensor barrier detects all the intruders and their path too. To achieve coverage quality, basic requirements include full coverage that prescribes that all objects should be covered in the ROIs of WSNs. Hence, area or target coverage should cover all the points in the discrete target set, i.e., continuous region should be covered. For barrier coverage, all the intrusion points must be detected. In some applications, full coverage is unnecessary as it is too strong, thus partial coverage is taken into consideration. Partial coverage covers part of target set, subregions, and intrusion paths. Sensor detection is also a problem regarding coverage problems other than a coverage quality. Deterministic models are prepared by researchers for the same. For example, the 0/1 model that is, we use a binary variable to describe whether an object is covered by the sensor, 1 as yes and 0 as no. 0/1 model is also termed as disk model. If the object is located at the boundary of disk, we cannot assert that it is definitely covered by the sensor, so we use probabilistic approach. Probabilistic approach use random variable to show the detection information of sensor, which is based on factors as Euclidean distance. This survey mainly addresses coverage problems with partial or probabilistic properties. Probabilistic models are powerful to describe random node failures, node location inaccuracy, robotic systems [1, 2], and imperfect time synchronization [3], etc. Planned deployment is a special type of deployment technique in WSN, which is focused in [4]. It is deterministic deployment aiming at coverage-related objectives. Also, author [4] surveyed and met heuristics problem and drafted few algorithms: single-solution-based algorithm, evolutionary algorithm, swarm intelligence, and their varieties. The motivation of our survey is: (i) To discuss uncertain properties as key point coverage problems. (ii) To consider main aspects for the coverage problem, deployment, scheduling (wake-up), or movement. We strategize these motives as:

- Encapsulation of works on coverage problems with uncertain properties and create a statistical report on last decades research literature.
- Generalization of research framework on the coverage problems with uncertain properties according to their special characteristics and potential research workflow.
- Summarizing important optimization objectives and corresponding strategies for coverage problem with uncertain properties.

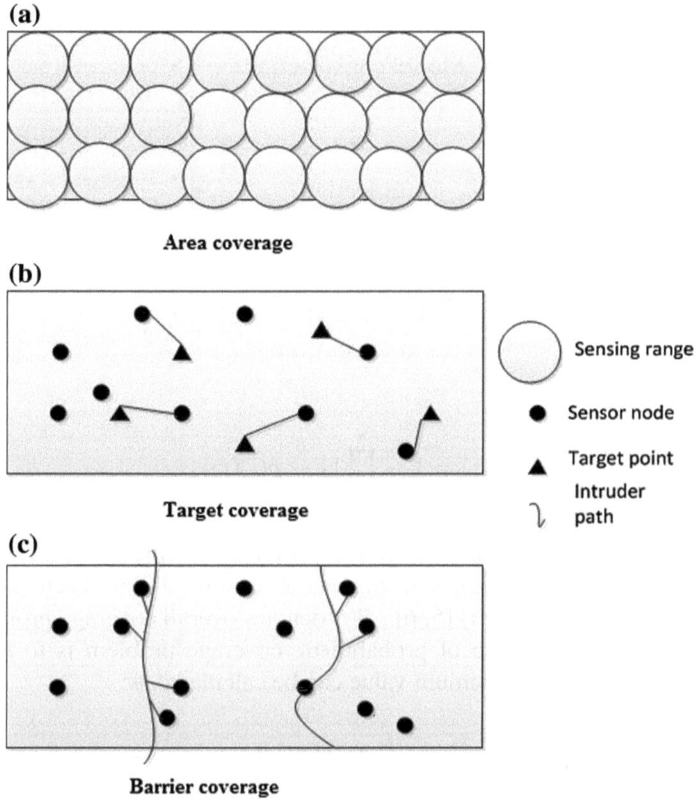

Fig. 1 **a** Area coverage. **b** Target coverage. **c** Barrier coverage

2 Coverage Properties and Problem

Full coverage Full coverage quality is the basic coverage quality and all other coverage qualities are branched from it. We use the probabilistic approach as the detection model thus defining a threshold value. If the sensors detection range probability is more than the threshold, then it is detected. Some other parameters can also be considered for the same.

Partial coverage Partial coverage weakens coverage quality in order to improve the lifetime of the network. By adjusting coverage ratio, we can achieve our purpose. The coverage ratio is a percentage of targets being covered and target coverage. Partial coverage is p-percentage coverage which adopts quantitative coverage ratio to control coverage quality, but it brings problems as detected or undetected regions are unknown and dynamic.

Probabilistic coverage In this method, the coverage is done in a probabilistic manner.

The coverage (detection) probability of a target can be expressed as:

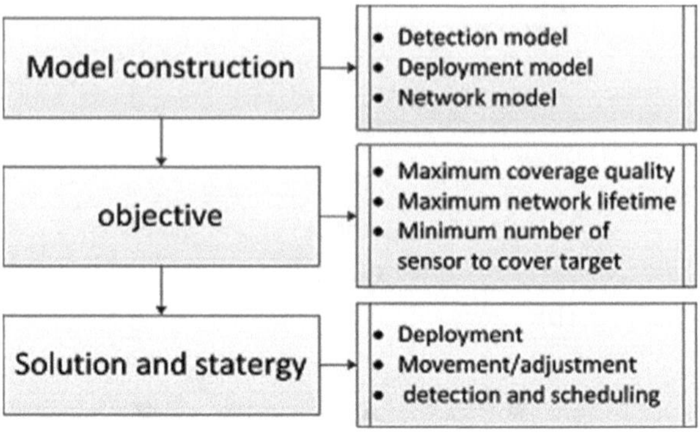

Fig. 2 Framework for solving the coverage problem

$$P(t) = 1 - \prod_{1}^{n} [1 - p(i)(t)] \tag{1}$$

where t denotes a target position in the ROI Ω, $p(i)(t)$ denotes the detection probability by the ith sensor node for, t, and n is the number of sensors. Different detection model gives different values for, $p(i)(t)$ [5, 6]., $P(t)$ denotes overall coverage probability at position, t. The main purpose of probabilistic coverage problem is to maximize coverage probability. The maximum value can be calculated as:

$$\text{Max} \int_{\Omega} R(t)P(t)\mathrm{d}t \tag{2}$$

where $R(t)$ is an object (event) density function, for some, $t \in \Omega$ [7].

3 Framework for Solving the Coverage Problem

Here, we describe a framework that uses basic research procedures for coverage problems in WSNs as shown in Fig. 2. We will follow three steps:

1. Model construction;
2. Objective(s) formalization;
3. Solution design.

3.1 Model Construction

Coverage models deal with coverage problem mathematically. A coverage model consists of detection model, network model, and deployment model which are low

Table 1 Detection and deployment model with uncertain coverage property specification

Detection model		Deployment model	
Model name	Uncertain property	Model name	Uncertain property
0/1 Model	No	Grid deployment	No
Attenuated model	Yes	Grid deployment with error	Yes
Truncate attenuated model	Yes	Random deployment	Yes
Directional model	Cannot say	Pseudo-random deployment	Yes

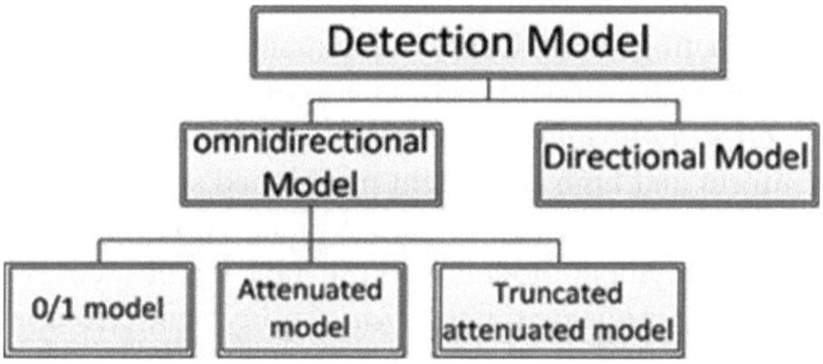

Fig. 3 Detection model

to high level. Sensor features are detected by detection model. Relation between sensors is the responsibility of deployment model. Network model is responsible for features and structure of network. If a model is probabilistic, it is marked as yes and no if it is deterministic as mentioned in Table 1.

Detection model It is based on the geometric location of the target and a sensor, which measures sensing capability and the detection quality of the sensors. Detection model is categorized as omnidirectional and directional model as shown in Fig. 3. Models property can be either deterministic or probabilistic. Euclidean distance between the sensor and target is used to depict the coverage problem. The strength of signal helps in detection of the sensor. For any sensor probability, detection function is modeled as:

$$d(s_i, t) = \sqrt{(x_i - x)^2 + (y_i - y)^2} \tag{3}$$

which denotes Euclidean distance where (x_i, y_i) and (x, y) are Cartesian coordinates of the sensor s_i and target position t.

Omnidirectional Model 0/1 model is the most widely used deterministic detection model in WSNs. It is a traditional detection model where sensor either detects the target else not. Attenuated models are better than 0/1 model as the latter ignores sensing process and utilizes more information. According to attenuated model k, is the decay factor related to the physical characteristics of the sensor. The sensor receives the signal strongest when an event occurs at the location of the sensor and vice versa. Truncated attenuated model is used when distance is too large or too small.

In this detection, probability is approximated or neglected as per the requirement. *(Rs, Ru)* is the strip detection range where *Rs* is the lower bound and *Ru* is the upper bound. *Rs* and *Ru* can be removed according to the practical situation.

Directional model From omnidirectional model, it is clear that we got to know distance is an important factor for detection function. The angle between sensor working direction and relative direction between sensor and target point are also a factor influencing detection function. Each sensor in the directional model has only one working direction, θ $(-\pi \leq \theta \leq \pi)$ defined as angle value relative to the positive x-axis. Field of view *(Fs)* of the sensor is in sector shape, i.e., the pie-shaped zone formed of effective angle of view, denoted as angle α, and two sides in length of detection range.

Deployment Model Deploying sensor in ROI is one of the fundamental steps in constructing WSN. Deployment methods are varied accordingly. Thus, it consists of deterministic model based on grid and uncertain deployment model.

Grid deployment In this model, sensor nodes are deployed on grids. They provide uniform and high consistent partitioned space. Grid deployment can be subcategorized as: Triangle [8], Square [9], and Hexagon [10]. These are the most important grid deployment models as these have the best coverage in the least sensors.

Grid deployment with error Deployment errors may occur due to some climate changes or physical phenomenon which may cause deviation in the result. This probabilistic model is introduced for node location. We can either increase the length of the grid cells so that no region is left a void or provide probabilistic coverage guarantee [8] to overcome errors.

Random deployment Practical implementation of a network for surveillance and tough environmental condition uses the random deployment of the sensor node. The deployment process is generally done by air launching the sensor nodes. The static 2D Poisson process is used to locate the sensor in the target area. It is estimated that the number of sensor nodes required for the target coverage is 3–10 times more in random deployment as compared to gird deployment mechanism [11].

Homogeneous and heterogeneous models A model where all sensors have same detection range, communication range, and computing power is called a homogeneous network. Homogeneous network properties are used as assumptions for simplification in [12–14]. But in a heterogeneous network, these may vary accordingly. The detection range of nodes varies region to region according to the target detected by nodes.

3.2 Problem Objectives

Basic entity of coverage problem of WSNs is coverage quality. Coverage ratio along with coverage probability helps to estimate coverage quality. Coverage ratio tells how much ROI is covered considered as partial coverage problem, whereas coverage probability reflects the degree of coverage for ROI discussed as probabilistic coverage

Table 2 Algorithm design for maximize coverage quality

Algorithm	Detection model			Deployment model			Type of coverage	Strategies
	0/1	Attenuated	Truncated attenuated	Grid	Random	Pseudo		
TAC [29]	Y	–	–	–	Y	–	Area	Mov
WN [30]	Y	–	–	–	Y	–	Barrier	Sch
WPC [31]	–	Y	–	–	Y	–	Area	Dep
CIAC [28]	–	–	Y	–	Y	–	Area	–
JNCA [32]	Y	–	–	–	Y	–	Area	Mov
ComNet [33]	–	Y	–	–	Y	–	Area	Mov
WCMC [34]	Y	–	–	–	Y	–	Area	Mov

problem. Here, we also cover optimization of the efficiency of WSNs as problem objectives.

Achieving coverage quality WSNs have different requirements for coverage quality. Here, each coverage quality is introduced in detail.

Achieving a certain coverage ratio It has been proved that the lifetime of WSNs will be extended greatly if the coverage ratio of the network can decrease slightly, thereby providing partial coverage rather than full coverage [15]. Full coverage is actually unnecessary in some scenarios. In such a case, partial coverage is used to prolong the lifetime of networks. Partial coverage can be defined when the coverage ratio $r < 1$. Coverage ratio and lifetime are two important but conflictive requirements in WSNs. Improving both coverage ratio and lifetime at the same time is usually difficult.

Achieving a certain coverage probability In traditional disk model, some information is lost because it considers only two cases, covered or not covered. The probability model is proposed to overcome the problem of disk model and to precisely depict the detection capabilities. In this method, the detection probability is not assigned directly to 0 if the target is not in its sensing range, unlike the 0/1 disk model. Detection probability is calculated for each sensor for a target. The probability value of each sensor is combined to result in the final decision.

Optimization objectives To achieve efficient WSN, numbers of optimization goals are proposed. The objectives of optimization process are to maximize network quality and its lifetime, and to minimize the number of sensors in the network [16].

In order to maximize the network coverage ratio, several researchers proposed a different mechanism listed in Table 2. In Wimalajeewa and Jayaweera [17], proposes a hybrid wireless network which consists of both static and mobile nodes. Mobile node moves to increase the coverage probability. In [13], effectiveness of coverage is determined by setting the desired threshold for coverage probability and heuristically Artificial Bee Colony (ABC) algorithm is used to maximize the coverage ratio of ROI.

Table 3 Algorithm design for minimize number of sensors

Algorithm	Detection model			Deployment model			Type of coverage	Strategies
	0/1	Attenuated	Truncated attenuated	Grid	Random	Pseudo		
Sensor [35]	–	Y	–	Y	–	–	Area	Sch
PMC [36]	–	–	Y	–	Y	–	Area	Sch
IS [37]	Y	–	–	–	Y	–	Barrier	Mov
TOC [38]	–	Y	–	Y	–	–	Barrier	Dep
DASFAA [39]	–	Y	–	–	Y	–	Target	Sch
ICCCN [40]	–	Y	–	Y	–	–	Target	Dep
INFOCO [41]	Y	–	–	Y	–	–	Barrier	Dep

Maximizing network lifetime Since the entire sensor network operates on battery, it is important to use the battery in an effective manner [18]. The energy consumption needs to be balanced to avoid energy wastage and make a network to last long [12]. In the target coverage network, the sleep–awake mechanism is used to increase the network life [13, 19]. Probabilistic coverage model is adopted, to increase the network life and also consider the coverage and connectivity constraint [20].

Minimizing the number of sensors In some cases, where sensors with special specification are required, to reduce the network cost, it is necessary to minimize the count of sensors in the network [21, 22]. Mobile nodes are used to provide the coverage to ROI over time with low cost in the mobile social network [3]. In this approach, minimal number of mobile nodes is required inside and outside ROI. Table 3 shows number of researchers work to reduce the count of sensors.

3.3 Solutions and Strategies

Main optimization objectives of coverage problems are classified as:

1. To maximize coverage quality;
2. To maximize network lifetime;
3. To minimize the number of sensors;

Basic strategies for different coverage objectives Deployment, scheduling, and movement are among the various solutions to achieve their objective. Deployment refers to the allocation of sensors to achieve coverage quality of the basic network. Scheduling deals with the selection of a subset of sensors to be activated for a time interval and make other sensors sleep. Selection strategy refers to one time choice for the subset of sensors for the whole time. Reallocation of sensors is dealt with movement strategy.

Deployment strategies The initial position of each sensor can be determined using this original position and the sensors can be deployed accordingly. But there is a possibility of deviation of sensors from their designated position. Some applications of deployment strategy are discussed.

- Maximizing coverage quality;
- Maximizing network lifetime;
- Minimizing number of sensors.

Basic strategies for different coverage objectives Deployment, scheduling, and movement are among the various solutions to achieve their objective. Deployment refers to the allocation of sensors to achieve coverage quality of the basic network Scheduling deals with the selection of a subset of sensors to be activated for a time interval and make other sensors sleep. Selection strategy refers to one time choice for the subset of sensors for the whole time. Reallocation of sensors is dealt with movement strategy.

Deployment strategies The initial position of each sensor can be determined using this original position and the sensors can be deployed accordingly. But there is a possibility of deviation of sensors from their designated position. Some applications of deployment strategy are discussed.

- Maximizing Coverage Quality: One of the tasks is for WSNs to maximize coverage quality. To gather more and more features of ROI as possible, limited sensors are deployed [23].
- Maximizing network lifetime: A lifetime of the individual sensor depends on battery duration, whereas lifetime of WSNs refers to the duration of time for which deployed sensors can provide adequate coverage quality. Adjusting deployment of sensors can greatly extend network lifetime [24].
- Minimizing number of sensors: Authors in [25] consider barrier coverage in bi-static radar network alike by deploying radar sensors. But, the latter one considers the cost difference between transmitter sensors and receiver sensors.

Scheduling or selection strategies If the initial positions of sensors are known, we can determine the deployment strategy. For which we use network builder can scatter a large number of sensors randomly to fulfill different coverage quality requirements. For optimized use of redundant sensors, sensors are scheduled to work alternatively. This strategy is used to maximize the network lifetime and is easily implemented.

Movement or adjustment strategies Nowadays, sensors are equipped with mobility components. Movement strategy is exploited by these mobile components. Since components are moving we can adjust the position accordingly. Energy consumption is one of the major problems regarding the same, for which minimizing movement distance becomes an important aspect. Also, adjustment of sensors parameters like the angle, detection range, etc., is important. Authors in [26] find an optimization solution with three stages, namely pattern node selection, network connectivity restoration, and coverage hole repair, by which these sensors are deployed for maximizing the covered volume ratio and keeping the connectivity. In [27], an area coverage problem with rotatable directional sensors is addressed. Two algorithms were proposed: **Concurrent Rotation and Motion Control** (CRMC), and **Staged Rotation and Motion Control** (SRMC).

4 Open Challenges

In this section, we attempt to discuss future research of coverage problems with uncertain properties, including challenging problems and new problems.

Challenging problems Here, we will deal with some difficult problems for which no perfect solution is provided.

Special K-coverage problem Some of the special K-coverage problems include 2D- and 3D K-coverage problem. 2D K-coverage problem is usually solved by Helly's theorem that states that a region is K-covered if central region of regular pentagon contains K active sensors. In case of 3D, transformed polyhedron cannot cover the space easily.

Full-view coverage problem Full-view coverage is detection range within a sector shape. One of the most famous problems is gallery camera problem. To view, an intruder sensor must face intruder from a certain angle. A full-view barrier coverage proposed by Sung and Yang [28] needs few cameras for intruder detection. Camera rotation is involved but the main problem is how to select a minimum number of cameras and rotate them so as to get full-view barrier and no perfect solution is developed for the same.

Sweep coverage problem It refers to special target coverage. Difference between sweep coverage and common target coverage is that sensors keep moving towards targets to give information. Sweep coverage requires every target to be scanned once and mobile sensors report detection record to a sink node for certain period of time. This problem is to minimize the average distance of sensors movement, which is NP-hard problem.

5 Conclusion

Rather than strict full coverage problems, a partial or probabilistic coverage problem describes real-world application more easily. We have introduced coverage problem with uncertain properties. Further, we discussed detection, network, and deployment model. In this survey, three optimization methods, i.e., maximizing coverage quality, maximizing network lifetime, and minimizing number of sensors in coverage problems with uncertain properties are covered. Some approaches for deployment, scheduling and selection, and movement or adjustment of sensors are also discussed. Finally, challenging problems for future are discussed.

References

1. F. Santoso, Sub-optimal decentralised control algorithms for blanket and k-barrier coverage in autonomous robotic wireless sensor networks. IET Commun. **4**(17), 2041–2057 (2010)
2. M. Schwager, D. Rus, J.J. Slotine, Unifying geometric, probabilistic, and potential field approaches to multi-robot coverage control. Int. J. Robot. Res. **30**(3), 371–383 (2011)
3. A. Ahmed, K. Yasumoto, Y. Yamauchi, M. Ito, Probabilistic coverage methods in people-centric sensing. J. Inf. Process. Syst. **19**, 473–490 (2011)
4. D.S. Deif, Y. Gadallah, Classification of wireless sensor networks deployment techniques. IEEE Commun. Surv. Tutorials **16**(2), 834–855 (2014)
5. J.W. Huang, C.M. Hung, K.C. Yang, J.S. Wang, Energy-efficient probabilistic target coverage in wireless sensor networks, in *IEEE International Conference on Networks* (2011), pp. 53–58
6. V. Akbarzadeh, C. Gagne, M. Parizeau, M. Argany, M.A. Mostafavi, Probabilistic sensing model for sensor placement optimization based on line-of-sight coverage. IEEE Trans. Instrum. Meas. **62**(2), 293–303 (2013)
7. G. Yang, D. Qiao, Barrier information coverage with wireless sensors, in *IEEE International Conference on Computer Communications. (INFOCOM)* (2009), pp. 918–926
8. G. Takahara, K. Xu, H.S. Hassanein, Efficient coverage planning for grid-based wireless sensor networks, in *IEEE International Conference on Communications (ICC)* (2007), pp. 3522–3526
9. D. Wang, J. Liu, Q. Zhang, Probabilistic field coverage using a hybrid network of static and mobile sensors, in *IEEE International Workshop on Quality of Service (IWQoS)* (2007), pp. 56–64
10. M. Hefeeda, H. Ahmadi, Energy-efficient protocol for deterministic and probabilistic coverage in sensor networks. IEEE Trans. Parallel Distrib Sys. (TPDS) **21**(5), 579–593 (2010)
11. P.A. Birjandi, L. Kulik, E. Tanin, K-coverage in regular deterministic sensor deployments, in *IEEE International Conference on Intelligent Sensors, Sensor Networks and Information Processing (ISSNIP)* (2013), pp. 521–526
12. W. Choi, G. Ghidini, S. K. Das, A novel framework for energyefficient data gathering with random coverage in wireless sensor networks, ACM Trans. Sens. Netw. (TOSN) **8**(4), 36 (2012)
13. R. Tan, G. Xing, B. Liu, J. Wang, X. Jia, Exploiting data fusion to improve the coverage of wireless sensor networks. ACM/IEEE Trans. Networking (TON) **20**(2), 450–462 (2012)
14. J. Zhang, R. Wang, Y. Qian, Q. Wang, A coverage control algorithm based on probability model for three-dimensional wireless sensor networks, in *IEEE International Symposium on Distributed Computing and Applications to Business, Engineering & Science (DCABES)* (2012), pp. 169–173
15. G.H. Prabhat, S.V. Rao, T. Venkatesh, Sleep scheduling for partial coverage in heterogeneous wireless sensor networks, in *IEEE International Conference on Communication Systems and Networks (COMSNETS)* (2013), pp. 1–10
16. G. Zhang, K. Sivan, F.P. S. Chin, C.C. Ko, Quality of service and coverage optimization in wireless sensor networks, in *IEEE Vehicular Technology Conference (VTC)* (2008), pp. 2824–2828
17. T. Wimalajeewa, S.K. Jayaweera, A novel distributed mobility protocol for dynamic coverage in sensor networks, in *IEEE Global Telecommunications Conference (GLOBECOM)* (2010), pp. 1–5
18. G. Xing, X. Wang, Y. Zhang, C. Lu, R. Pless, C.D. Gill, Integrated coverage and connectivity configuration for energy conservation in sensor networks, ACM Trans. Sens. Netw. (TOSN) **1**, 36–72 (2005)
19. N. Hayashi, T. Ushio, T. Kanazawa, Potential game theoretic approach to power-aware mobile sensor coverage problem. IEICE Trans. Fundam. Electron. Commun. Comput. Sci. **94**-A(3), 929–936 (2011)
20. C.M. Kim, Y. Kim, I.S. Kang, K.W. Lee, Y.H. Han, A scheduling algorithm for connected target coverage under probabilistic coverage model, in *IEEE International Conference on Information Networking (ICOIN)* (2012), pp. 86–91

21. Y. Gu, Y. Ji, J. Li, B. Zhao, An optimal algorithm for solving partial target coverage problem in wireless sensor networks. Wireless Commun. Mob. Comput. (WCMC) **13**(13), 1205–1219 (2013)
22. Z. Wang, H. Chen, Q. Cao, H. Qi, Z. Wang, Fault tolerant barrier coverage for wireless sensor networks, in *IEEE International Conference on Computer Communications (INFOCOM)* (2014), pp. 1869–1877
23. J. Zhu, B. Wang, Sensor placement algorithms for confident information coverage in wireless sensor networks, in *IEEE International Conference on Computer Communication and Networks (ICCCN)* (2014), pp. 1–7
24. L. Zhang, D. Li, H. Zhu, L. Cui, Open: an optimisation scheme of n-node coverage in wireless sensor networks. IET Wireless Sens. Syst. **2**(1), 40–51 (2012)
25. J. Chen, B. Wang, W. Liu, Constructing perimeter barrier coverage with bistatic radar sensors. J. Netw. Comput. Appl. (JNCA) **57**, 129–141 (2015)
26. X. Gong, J. Zhang, D. Cochran, K. Xing, Optimal placement for barrier coverage in bistatic radar sensor networks. ACM/IEEE Trans. Netw. (TON) **24**(1), 259–271 (2016)
27. Z. Wang, B. Wang, A novel node sinking algorithm for 3d coverage and connectivity in underwater sensor networks. Ad Hoc Netw. (AHN) **56**, 43–55 (2017)
28. T.-W. Sung, C.-S. Yang, Voronoi-based coverage improvement approach for wireless directional sensor networks. J. Netw. Comput. Appl. (JNCA) **39**, 202–213 (2014)
29. Y. Wang, G. Cao, Barrier coverage in camera sensor networks, in *ACM/IEEE International Symposium on Mobile Ad Hoc Networking and Computing (MobiHoc)* (2011), pp. 3967–3974
30. J. Habibi, H. Mahboubi, A.G. Aghdam, Distributed coverage control of mobile sensor networks subject to measurement error. IEEE Trans. Autom. Control (TAC) **61**(11), 3330–3343 (2016)
31. B. Xu, Y. Zhu, D. Kim, D. Li, H. Jiang, A.O. Tokuta, Strengthening barrier-coverage of static sensor network with mobile sensor nodes. Wireless Netw. **22**(1), 1–10 (2016)
32. S. Kumar, Impact of interference on coverage in wireless sensor networks. Wireless Pers. Commun. **74**(2), 683–701 (2014)
33. T.-W. Sung, C.-S. Yang, Voronoi-based coverage improvement approach for wireless directional sensor networks. J. Netw. Comput. Appl. (JNCA) **39**, 202–213 (2014)
34. W. Li, Y. Wu, Tree-based coverage hole detection and healing method in wireless sensor networks. Comput. Netw. (ComNet) **103**, 33–43 (2016)
35. F. Senel, Coverage-aware connectivity-constrained unattended sensor deployment in underwater acoustic sensor networks. Wireless Commun. Mobile Comput. (WCMC) **16**(14), 2052–2064 (2016)
36. A. Shan, X. Xu, Z. Cheng, Target coverage in wireless sensor networks with probabilistic sensors. Sens. **16**(9), 1372 (2016)
37. H.P. Gupta, S.V. Rao, T. Venkatesh, Analysis of stochastic coverage and connectivity in three-dimensional heterogeneous directional wireless sensor networks. Pervasive Mobile Comput. (PMC) **29**, 38–56 (2016)
38. K. Shi, H. Chen, Y. Lin, Probabilistic coverage based sensor scheduling for target tracking sensor networks. Inf. Sci. (IS) **292**, 95–110 (2015)
39. B. Wang, J. Chen, W. Liu, L.T. Yang, Minimum cost placement of bistatic radar sensors for belt barrier coverage. IEEE Trans. Comput. (TOC) **65**(2), 577–588 (2016)
40. Y. Hu, M. Xiao, A. Liu, R. Cheng, H. Mao, Nearly optimal probabilistic coverage for roadside data dissemination in urban vanets, in *International Workshops on Database Systems for Advanced Applications (DASFAA)* (2016), pp. 238–253
41. H.L. Wang, W.H. Chung, The generalized k-coverage under probabilistic sensing model in sensor networks, in *IEEE Wireless Communications and Networking Conference (WCNC)* (2012), pp. 1737–1742

Analysis of Existing Protocols in WSN Based on Key Parameters

Charu Sharma and Rohit Vaid

Abstract Wireless Sensor Networks (WSNs) is a rising technology that can be easily employed in unusual conditions such as environmental monitoring in soil, marine, earth monitoring, forest fire detection, battlefields, etc. As the nodes are randomly deployed in WSNs, security becomes one of the major issues which WSNs is facing today. The network is easily compromised due to wireless communication. The sensor nodes communication acquires different types of threats related to security. Due to limited storage and low-power of sensor nodes makes the security solutions unachievable. This paper presents the overview of WSNs, its challenges, various attacks, and issues related to security and their defensive mechanisms in WSNs.

Keywords WSNs · Sensor node · Attack · Base station · Defensive mechanisms

1 Introduction

WSNs have gained a global attention as they provide best low-cost solutions to different real-world problems. WSNs architecture is shown in Fig. 1. It consists of small-sized independent nodes or motes, which are randomly deployed in an unattended fashion. These nodes are organized in a remote area to examine environmental conditions such as pressure, temperature, etc. Sensor nodes cooperatively collect data, then process it, and interconnected nodes communicate with each other to pass the information to the base station. These nodes have limited power, bandwidth, storage, and memory space. So, the primary goal for WSNs is to utilize these resources efficiently.

C. Sharma (✉) · R. Vaid
CSE Department, M. M. Engineering College, M. M. University, Mullana, Ambala 133207, Haryana, India
e-mail: er.charusharma@mmumullana.org

R. Vaid
e-mail: rohitvaid@mmumullana.org

C. R. Krishna et al. (eds.), *Proceedings of 2nd International Conference on Communication, Computing and Networking*, Lecture Notes in Networks and Systems 46, https://doi.org/10.1007/978-981-13-1217-5_17

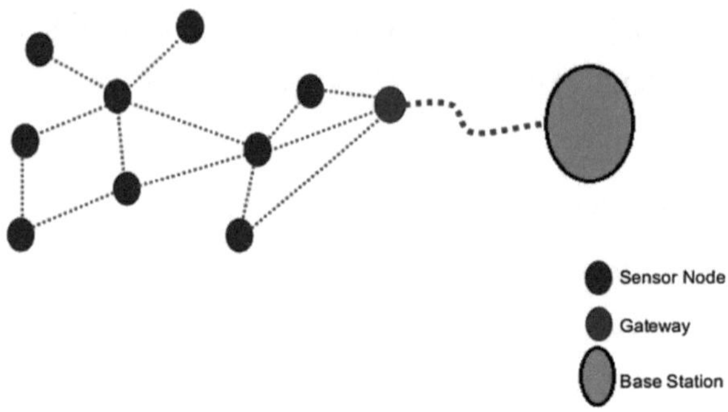

Fig. 1 Architecture of WSNs

2 Security Attacks in WSNS

- *Denial-of-Service Attack*: It makes the resources unavailable for the legal node. Thus, do not allow the victim node for accessing the resources for which it is allowed.
- *Black Hole Attack*: In this, the malicious node creates a black hole for other nodes and do not allow the data packets to escape from it.
- *Hello Flood Attack*: In this, the malicious node broadcast a HELLO message to every legal node even which is very far away from each other in the network using very high transmission power, thus breaking the security.
- *Sybil Attack*: In this type of attack, the node claims multiple identities at the same time at different locations in the network.
- *Sink Hole Attack*: In this, the malicious node consumes maximum information in the network by positioning itself where maximum traffic will stream.

3 Security Principles in WSNs

- *Confidentiality*: Confidentiality guarantees that important information is protected and only authorized parties can access the sensitive information. The loss of confidentiality is when the data is sent from sender to receiver, in-between the attacker can access the information without any permission from sender or receiver. This type of attack is known as interception.
- *Authentication*: For effective security, it is the process that ensures origin of the message is correctively recognized. The loss of authentication is when the malicious sensor had posed itself as sensor S and sends information to receiving sensor R, which shows the absence of authentication. This type of attack is called fabrication.
- *Integrity*: It means that data should be completely and accurately send in the network from source to destination without modifying it. When the attacker changes the actual route of the message and after tampering, the message sends it to the

receiver. The source and destination does not know about this change. This type of attack is called modification.

- *Availability*: Availability ensures that resources and services are available for the nodes only when they are required. The loss of availability is when the attacker intentionally stops the authorized sensor to use the services given to it. Such an attack is known as interruption.

4 Related Work

In [1], the authors reviewed the communication architecture, algorithms, different protocols, and various applications for wireless sensor networks. Various open research issues for these networks are also discussed.

In [2], the author proposed a scalable algorithm for larger networks, which helps to minimize the cost forwarding messages by passing the information along with the minimum cost path in the field but it requires regular updating of cost of each node.

The author in [3] proposed a framework to handle the problem of physical node capture attack, detection of cloned node, and removal of compromised nodes from the network. A strategy for network response is also defined in this work to ensure secure network connectivity.

Vaid et al. [4], conducted a survey on the classification of different security issues along with their remedies in Wireless Sensor Networks. The author addressed different security constraints and also various security requirements necessary to build a secure WSN framework.

Prasanna et al. [5], presented an outline of WSNs applications along with various security attacks and their countermeasures.

Araujo et al. [6], conducted a survey on the various challenges in WSNs. This work describes different types of attacks such as security attack, policy attack, high-power consumption attack, communication attack, etc., along with special mechanisms to handle these attacks.

The authors in [7], discussed various clustering protocols for WSNs and classified them based on different parameters such as cluster size, cluster density, cluster count, number of nodes deployed in the network, cluster head selection, etc., to choose the optimized clustering protocol for the network design.

The author in [8] proposed LEACH, a new clustering-based protocol which is used to minimize the energy dissipation in WSNs. It is able to uniformly allocate the energy dissipation throughout the sensors, thus doubles the network survival time. Cluster heads are randomly selected and receive data from different nodes. The received data is then aggregated, and then sends it to the base station to reduce unnecessary energy cost.

The authors in [9], reviewed the key revocation schemes and proposed a key updating scheme for removing and replacement of compromised sensor nodes for WSNs by updating the keys of all uncompromised nodes in such a way that it was unavailable for the compromised nodes without increasing communication overhead.

5 Security Models in WSNs

- *Security Protocols for Sensor Networks (SPINS)*: It has two secure building blocks: SNEP and microTESLA. SNEP provides two-party authentication, confidentiality, integrity, and data freshness. MicroTESLA provides authenticated broadcast for resource constraints environments on WSNs. SNEP and micoTESLA together provides secure communication channels using only symmetric cryptography in WSNs [10].
- *Localized Encryption and Authentication Protocol (LEAP)*: It provides an efficient key establishment and key updating procedures. It uses four types of keys—individual, pairwise, cluster, and group key. Individual key is shared with the base station, pair-wise key shared with another nodes, cluster key shared with multiple neighboring nodes, and group key shared by all network nodes. LEAP is very efficient in defending against different attacks.
- *TINYSEC*: TinySec is a lightweight link layer security architecture for WSNs. It is the alternate for the incomplete SNEP. The existing protocols are heavyweight, incomplete, and insecure so TinySec design is used to overcome the unsuitability of existing schemes. The energy cost, bandwidth, and latency are all less than 10% using TinySec.
- *LLSP*: Energy-efficient link layer security protocol which is based on the principle of TinySec has different packet format. It ensures message confidentiality, authentication, and access control. It provides early rejection capabilities. LLSP also provides low-performance overhead.

6 Comparison of Different Security Models and Routing Protocols in WSNs

A. Different Security Models

- *Based on Features*: Table 1 shows the different major features of various security protocols.
- *Based on Application Characteristics*: In Table 2, various application characteristics of different WSNs security protocols are defined. These characteristics are used to select the best security protocol for a specific application.
- *Based on Attack Protection*: An attack protection matrix of different security protocols is shown in Table 3.

Table 1 Major features

Protocol	LEDS	SPINS	LLSP	TinySec
Type	End-to-end	Node-to-base	Hop-to-Hop	Hop-to-Hop
Key management	Yes	Yes	No	No
Location awareness	Partial	No	No	No
Scalable	Partial	Low	Low	Partial

Table 2 Application characteristics

Protocol	Application characteristics
LEDS	It provides end-to-end secure applications It provides physical attack protection
SPINS	Best suited for small-size network Communication pattern is node-to-base
LLSP	Best for in-network processing applications Resource constraints environment
TinySec	Best for in-network processing and local broadcast Can be combined with high-level protocols

Table 3 Attack protection

Protocol	LEDS	SPINS	LLSP	TinySec
Replay	–	Yes	Yes	No
Injection	Strong	Partial	Maybe	Maybe
Alternation	Strong	Partial	Maybe	Maybe
DOS	Medium	–	Low	Low

B. **Routing Protocols**

- *Based on different parameters*: Table 4 shows the comparison of routing protocols based on various parameters such as data aggregation, reliability, processing overhead, etc.

7 Defensive Mechanisms

In WSNs, as the network grows, it is very difficult to identify and authenticate each and every mote in the network. The risk of secure transmission of information over network also increases as malicious node and can easily modify the information in the network. To achieve the security in WSNs, various mechanisms have been proposed as given below:

Table 4 Routing protocols comparison

Routing protocol	Data aggregation	Delay (in terms of time)	Processing overhead	Data delivery model	Type of network	Reliability
LEACH	Yes	Low	High	Cluster based	Fixed	High
PEGASIS	No	High	Low	Chain based	Fixed	High
TEEN	Yes	High	High	Threshold value driven	Fixed	Medium
APTEEN	Yes	High	High	Threshold value driven	Fixed	High
HEED	Yes	High	Very high	Cluster based	Fixed	High

- In existing schemes, there is a risk factor of path key exposure problem so as to avoid compromised nodes from modifying the sensitive information in WSNs, an efficient multi-path key establishment solutions need to be investigated.
- To lengthen the network's survival time better than the existing systems, an efficient and secure data aggregation techniques need to be proposed for cluster-based WSNs.

8 Conclusion

In this paper, we have provided the general outline of WSNs which includes characteristics, challenges, and various security attacks. We also analyzed various issues related to security different security models with comparison and proposed their defensive mechanisms, which help us to achieve the desired security goals in WSNs. The proposed solutions make the WSNs more secure and efficient to fight against malicious nodes.

References

1. I.F. Akyildiz, W. Su, Y. Sankarasubramaniam, E. Cayirci, Wireless sensor networks: a survey. Comput. Netw. **38**, 393–422 (2002) (Elsevier)
2. F. Ye, A. Chen, S. Lu, L. Zhang, A scalable solution to minimum cost forwarding in large sensor networks, in *Proceedings of the IEEE International Conference on Computer Communication and Networks (ICCCN)*, 15–17 Oct 2001 (Phoenix, USA, 2001), pp. 304–309
3. T. Bonaci, L. Bushnell, R. Poovendran, Node capture attacks in wireless sensor networks: a system theoretic approach, in *49th IEEE Conference on Decision and Control (CDC)*, vol. 1 (Atlanta, Georgia, USA, 2010), pp. 6765–6772

4. R. Vaid, V. Kumar, Security issues and remedies in wireless sensor networks—a survey. Int. J. Comput. Appl. **79** (4), 31–39 (2013)
5. S. Prasanna, S. Rao, An overview of wireless sensor networks applications and security, Int. J. Soft Comput. Eng. (IJSCE) **2** (2). ISSN: 2231-2307 (May 2012)
6. A. Araujo, J. Blesa, E. Romero, D. Villanueva, Security in cognitive wireless sensor networks—Challenges and open problems. EURASIP J. Wirel. Commun. Netw. **2012**, 48 (2012)
7. Sanjeev Kumar Gupta, Neeraj Jain, Poonam Sinha, Clustering protocols in wireless sensor networks: a survey. Int. J. Appl. Inf. Syst. **5**(2), 41–50 (2013)
8. H. Wendi Rabiner, A. Chandrakasan, H. Balakrishnan, Energy-efficient communication protocol for wireless microsensor networks, in *Proceedings of the 33 Annual Hawaii International Conference on System Sciences-2000* (IEEE, 2000), p. 10
9. R. Vaid, V. Kumar, KURCS: key updating for removing & replacement of compromised sensor nodes from wireless sensor networks. Int. Organ. Sci. Res. J. Comput. Eng. (IOSR-JCE), **17** (3), 57–67 (2015) (Ver. V)
10. A. Perrig, R. Szewczyk, J.D. Tygar, V. Wen, D.E. Culler, SPINS: security protocols for sensor networks. Wirel. Netw. **8**(5), 521–534 (2002)

Cellular Learning Automata-Based Virtual Network Embedding in Software-Defined Networks

Dipanwita Thakur and Manas Khatua

Abstract Software-defined networking (SDN) is a propitious technology for achieving network virtualization by decoupling the control and data planes of a network. SDN hypervisor supports multiple virtual SDN-based networks logically isolated from each other. Each virtual SDN has its own controller and allocated resources over physical network. For achieving optimal resource allocation, there is a need of efficient virtual network embedding (VNE) approach in multidomain virtual SDN-based network. In this paper, we propose a self-adjusted, online, distributed virtual network mapping strategy based upon the idea of irregular cellular learning automata. We consider two aspects of the network during the execution of VNE in SDN—node and link mapping, and optimal placement of SDN controller. We evaluate the proposed scheme vSDN-CLA using Mininet. The simulation results show significant performance improvement in terms of throughput and end-to-end delay. Considering a substrate network of 100 nodes, we observed that the proposed scheme achieved 23.72 and 10.55% higher throughput, and 28.13 and 42% lesser end-to-end delay compared to that in two benchmark schemes DM-vSDN and CO-vSDN, respectively.

Keywords Virtual network embedding · SDN · Cellular learning automata

1 Introduction

Emerging applications in information and communication technologies (ICTs) are imposing new directions to future Internet. Those directions create different challenges for providing profound services, capabilities, and facilities. To overcome those challenges, we need heterogeneous network architecture on a common physical

D. Thakur (✉)
Banasthali University, Banasthali, Rajasthan, India
e-mail: tdipanwita@banasthali.ac.in

M. Khatua
Indian Institute of Technology Jodhpur, Jodhpur, Rajasthan, India
e-mail: manaskhatua@iitj.ac.in

C. R. Krishna et al. (eds.), *Proceedings of 2nd International Conference on Communication, Computing and Networking*, Lecture Notes in Networks and Systems 46, https://doi.org/10.1007/978-981-13-1217-5_18

network. In this regard, network virtualization comes into existence by providing overlay on physical networks. Software-defined networking (SDN) [1] realizes the network virtualization by creating multiple logical networks upon a physical network and, at the same time, managing the network resources depending on users requirements.

Through the virtualization, multiple tenants can share the same underlying SDN infrastructure. FlowVisor [2] is one of the popular tools of network virtualization in SDN. FlowVisor slices the network considering some quality-of-service (QoS) parameters such as topology, bandwidth, switch, CPU, and flow space. Each slice has its own policy that defines its resources, and a controller associated with it. OpenFlow protocol is widely used in between data plane and control plane to run the APIs.

1.1 Motivation

For achieving network virtualization in traditional networks, multiple challenges are addressed due to its vendor-specific design. However, in SDN, creation of virtual network using software program is easy, and it provides strong QoS and service-level agreement (SLA) control. During the creation of virtual networks (VNs), it is required to map the VNs onto underlying physical networks. This process of mapping is called virtual network embedding (VNE). Several approaches of VNE (e.g., [3, 4]) were proposed for achieving efficient resource utilization through virtualization using either *path splitting* or *integer linear programming (ILP)* concepts. In path splitting, different sub-flows are created from a single virtual flow. Each and every sub-flow needs flow rules in each of the switches along the substrate path it supports. Hence, this approach is expensive due to the use of more content-addressable memory. On the other hand, ILP is nondeterministic polynomial time algorithm, and it suffers from many other disadvantages. Therefore, there is a need of low-cost VNE scheme.

1.2 Contribution

Unlike the existing schemes, in this paper, we use the idea of "irregular cellular learning automata" ICLA [5] to formulate the VNE problem and solution. In brief, we summarize our contributions as follows:

- We formulate the virtual SDN-based network embedding problem as the combination of two distinct problems—VNE and optimal placement of the controller.
- We design a coordinated online virtual SDN embedding algorithm, namely vSDN-CLA, to minimize the controller-to-switch delay and maximize the throughput.
- We perform extensive experiments to evaluate the proposed scheme with respect to the end-to-end delay and throughput followed by a comparative analysis with the benchmark schemes.

2 Related Work

We outline some current research on the area of VNE [6] and virtualization in SDN [1, 7] environment. An SDN-based virtualized network has been considered for flow migration based resource management in [8]. However, they did not consider the problem of controller placement in SDN-based network. Considering the selection of SDN controller, Demirci and Ammar [9] designed a virtual link and node mapping algorithm for balancing the nodes of the substrate network, and to minimize the controller-to-switch delay. However, in their proposed scheme, they did not provide any coordination between the controller placement and node mapping followed by link mapping. Zhou et al. [10] proposed a multi-domain VNE strategy for SDN for improving the scalability without considering the controller selection problem and flow entry resource allocation in node mapping. Chowdhury et al. [11] framed the VNE problem as a mixed integer program, and later suggest two virtual embedding algorithms by initiating an interrelation between the phases of node and link mapping. Recently, Gong et al. [12] proposed a novel heuristic-based online virtual SDN mapping algorithm using ILP, namely, *Co-vSDN*, to minimize the controller-to-switch delay for each virtual network. In brief, most of the VNE algorithm ensures optimal utilization of overall substrate network resources mainly in traditional substrate network but not in SDN-based substrate network.

3 Proposed Method

3.1 Problem Description

FlowVisor [2] slices substrate SDN to create multiple logically isolated virtual SDNs (vSDNs). The slicing is done based upon the availability of resources such as CPU capacity of switches, link bandwidth, and flow tables. When a request of vSDN arrives, it is either accepted or postponed based upon the availability of the resources. The allocated resources are released by the hypervisor when the vSDN request expires. Since each vSDN has its own controller, the controller placement problem should be addressed along with the VNE problem. Further, the embedding is done based upon the resource capacity and bandwidth constraints.

3.2 Modeling Using ICLA

At the outset, it is mentioned that throughout the paper the keywords "vertex", "node", and "cell" are used to represent the similar entity but in different contexts. Similarly, for the keywords "edge" and "link" too. Now, the SDN-based substrate network is defined as an ICLA [13], $L = \{G^s\langle V^s, E^s\rangle, \phi, A, F, P, Q\}$, where

- V^s is the finite set of substrate nodes (e.g., OpenFlow switches and controller nodes) represented as cells in A. E^s is the finite set of substrate links.
- ϕ is the finite set of states denoted by ϕ_i associated with each cell based upon the action probability vector $p = \{p_1, p_2, \ldots, p_n\}$. In this work, we have only two states—either a physical node is available to map to the virtual node or it is not available to map till now.
- A is the set of n learning automata (LA) as each node contains one LA independently. Let at time t, α_i is the action chosen by the automata. The action probability p is updated based on the following recurrence equations:

$$p_i(t+1) = p_i(t) + a(1 - p_i(t)) \tag{1}$$

$$p_j(t+1) = p_j(t) + a(1 - p_j(t)) \quad \forall j, j \neq i \tag{2}$$

$$p_i(t+1) = (1 - b)p_i(t) \tag{3}$$

$$p_i(t+1) = b/(r-1) + (1 - b)p_j(t) \quad \forall j, j \neq i \tag{4}$$

  The parameters a and b represent the reward and penalty values, respectively. r is the total number of actions.
- $F : \phi_i \rightarrow \beta$ is the function of the "local rule" in each node where β is the set from which reinforcement signal takes its value. In this work, β has two values—either reward a or penalty b.
- $P : \phi \times \beta \rightarrow \phi$ is the learning algorithm for the LA. The learning algorithm updates the probability vector p on the basis of supplied reinforcement signal and its selected action α_i.
- $Q : \phi \rightarrow \alpha$ is the probabilistic decision function, based on which each LA chooses its action. In this paper, we have only two actions—either the request is accepted or postponed.

Note that the superscript "s" indicates substrate network and "v" indicates virtual network. Each node $i \in V^s$ has a certain switching capacity S_i^s which defines the maximum number of flow information a switch can store in its flow table. At a particular time t, the flow table size of node i is denoted by $S_i(t)$. Each link $e_{ij} \in E^s$ of the substrate SDN connects the substrate nodes i and j, and it has bandwidth B_{ij}^s. Each node i has some content-addressable memory capacity M_i^s and CPU capacity CPU_i^s. In this network configuration, a requested vSDN network is defined as $G^v\langle V^v, E^v\rangle$. Each vSDN request consists of a tenant's resource requirement including the virtual network topology. The probabilistic decision function Q considers few node and link mapping constraints which are as follows:

- *Memory Capacity*: Requested virtual node capacity is always less than or equal to the substrate node capacity. This capacity is defined as content-addressable memory.

$$M_l^v \leq M_j^s; \quad l \in V^v, \quad j \in V^s \tag{5}$$

- *CPU Capacity*: Requested virtual node CPU capacity is always less than equal to the total substrate node capacity.

$$CPU_l^v \leq CPU_j^s; \quad l \in V^v, \quad j \in V^s \tag{6}$$

- *Switching Capacity*: At a particular time t, the flow table size of a virtual node is less than or equal to the switching capacity of the corresponding substrate node.

$$S_l^v(t) \leq S_j^s; \quad l \in V^v, \quad j \in V^s \tag{7}$$

- *Link Bandwidth*: Requested bandwidth in virtual link should be less than or equal to the substrate link bandwidth.

$$B_{kl}^v \leq B_{ij}^s; \quad e_{kl} \in E^v, e_{ij} \in E^s \tag{8}$$

If the conditions represented by Eqs. (5)–(8) are satisfied, then the selected action of the LA is "Accepted" and then the reinforcement signal is to *reward the action*, otherwise it *penalizes the action* and updates the state of ϕ and to map the virtual SDN request to the substrate network. Equations (1–4) represent the rewarding and penalizing functions for the selected actions, respectively.

3.3 Controller Placement

Let $D(l_{ij})$ be the transmission delay of the substrate link $l_{ij} \in E^s$ which is calculated by the total delay. The average controller-to-switch delay is defined as D, which is formulated as

$$D = \sum\nolimits_{l_{ij} \in E^s} D(l_{ij})/|E_c| \tag{9}$$

where $|E_c|$ is the total numbers of virtual control link. If n_c is the controller node, then it must be attached with one of the switch nodes of the substrate SDN network. The virtual nodes and the controller node cannot be mapped with the same substrate node. To select the optimal location of the controller node, embedding nodes of the virtual links l_{uv} are calculated by $(\sum_{ij \in E^s} f_{ij}^{uv} - 1)$, where $f_{ij}^{uv} \in \{0, 1\}$ is a binary variable to indicate the mapping between a virtual link and substrate link.

Algorithm 1 ICLA-based Coordinated VNE Algorithm in vSDN

Inputs:$(G^s\langle V^s, E^s\rangle, \phi, A, F, P, Q), (G^v\langle V^v, E^v\rangle)$
Output: VNE

Establish an ICLA.
Initialize action probability by 0.5
for Each node i in the ICLA **do**
    Get the requested node and link capacity
    Node i selects an action α_i.
    if $\alpha_i ==$ 'Accepted' **then**
        $\beta = 1$
    else
        $\beta = 0$
    end if
    if $\beta == 1$ **then**
        Reward node i
    else
        Penalize node i
    end if
    for Each rewarded node's link e_{lk} **do**
        Find minimum bandwidth path
        Update state of node i.
        Update action probability vector of i.
    end for
end for

Algorithm 2 ICLA-based Controller Placement Algorithm in vSDN

Inputs: VNE result, Permissible Delay
Output: Placement of the Controller Node

/*Generate a set S^c containing substrate controller nodes.*/
Create a set Z with all rewarded substrate nodes.
for Each node i in Z **do**
    Calculate delay d in between i and its neighbor nodes.
end for
for $i = 1$ to $|V^v|$ **do**
    Find all the substrate nodes satisfying $d <$ Permissible Delay, then put them in a set S^c
end for
for $i = 1$ to $|S^c|$ **do**
    Compute total controller-to-switch hops
end for
Select the S^{ci} with minimum hop count as the controller
Higher probability node is selected if there are more than one substrate nodes with minimum hops.
Update state of node S^{ci}.
Update the action probability vector

3.4 vSDN-CLA Algorithm

The proposed VNE algorithm is shown in Algorithm 1. The reinforcement signal value will be 1 if the substrate node capacity is greater than the requested node capacity. If the reinforcement signal value becomes 1, then the node is awarded, otherwise the node is penalized. If the node is rewarded, then the successful node mapping is done. After the completion of successful node mapping, if more than one path is found in between two rewarded nodes that can be satisfied with the requested bandwidth, then the minimum bandwidth path will be selected for link mapping. After the successful VNE, the controller of each and every virtual network is selected by Algorithm 2.

4 Performance Evaluation

For evaluating the performance of the proposed algorithm, we use Mininet emulator [14]. We consider the metrices controller-to-switch delay, end-to-end delay, and the average throughput for comparison study of the proposed scheme with the benchmark schemes CO-vSDN [12] and DM-vSDN [9].

4.1 Experimental Setup

In Mininet, we have different topologies like minimal, single, reversed, linear, tree, and so on. Using these topologies, we evaluate the proposed scheme with various network properties. The substrate network is configured with 100 nodes, and the virtual SDN network consists of 5–20 nodes. For each substrate node, the capacity of the content-addressable memory, capacity of CPU, and switching capacity are varied from 50 to 100 following the uniform distribution. For each substrate link, the delay varies from 5 to 20 ms. The available bandwidth capacity of each of the substrate links varies from 50 to 100 in uniformly distributed manner. We took virtual SDN request of different sizes such as 2–10 nodes represented as "small", 10–20 nodes represented as "medium", and 20–50 nodes represented as "large", respectively. We use FlowVisor [7] as an SDN hypervisor which separates the substrate network in different isolated virtual networks. Each virtual network has its own controller. In other words, we have different controllers for each and every virtual SDN network, and we set up NOX [7] as a virtual SDN controller.

4.2 Simulation Results

At the outset, we evaluate the performance of our proposed scheme vSDN-CLA in terms of average controller-to-switch delay and compare the results with that in the benchmark schemes. The received results are plotted in Figs. 1 and 2, respectively. In Fig. 2, we measured the average delay under different sizes of request for virtual SDN. In the experiment corresponding to Fig. 1, we took the tree topology with the

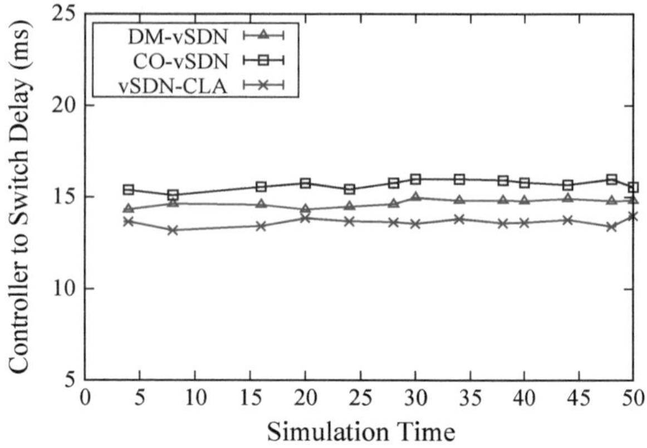

Fig. 1 Average delay from controller to switch

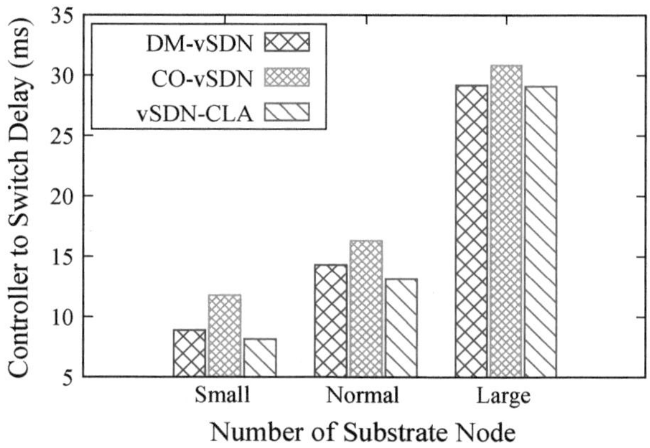

Fig. 2 Average delay from controller to switch under different sets of vSDN

"Large" number of nodes. In this experiment, 20 different virtual SDN networks are embedded on 6 substrate networks sliced by the FlowVisor. We have measured end-to-end delay using "ping" command, and the throughput is measured by executing file transfer between each pair of nodes via the TCP with the Iperf tool. We took the arithmetic means to show the results. The received result is plotted in Figs. 3 and 4, respectively. We observed that, for both the metrices, the proposed scheme vSDN-CLA outperforms the benchmark schemes. We have compared the received results of the proposed scheme with two benchmark schemes namely CO-vSDN [12] and DM-vSDN [9]. Both the schemes have used ILP formulation, whereas we have used ICLA-based formulation. The principal aspect of any learning system is its rate of learning or identically rate of convergence. Learning system is useful because learning process should be completed before the necessary changes take place in the environment or we can say that learning process takes place in slowly changing environment. That is why learning is effective. So in case of the learning automata, the parallel operation is done to increase the rate of convergence. As we are talking about the SDN network and the virtual network, request will be multiple at any point

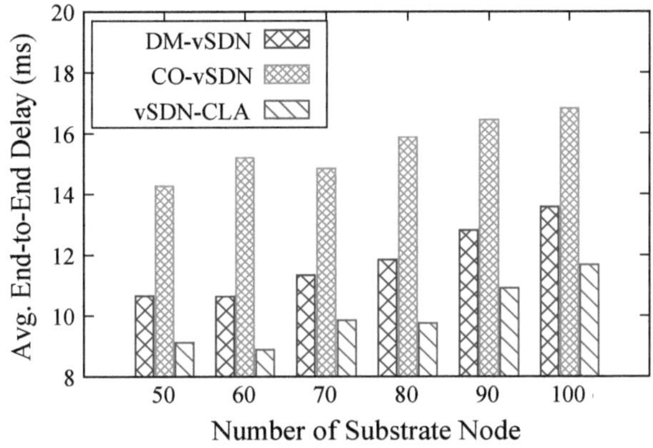

Fig. 3 Average end-to-end delay

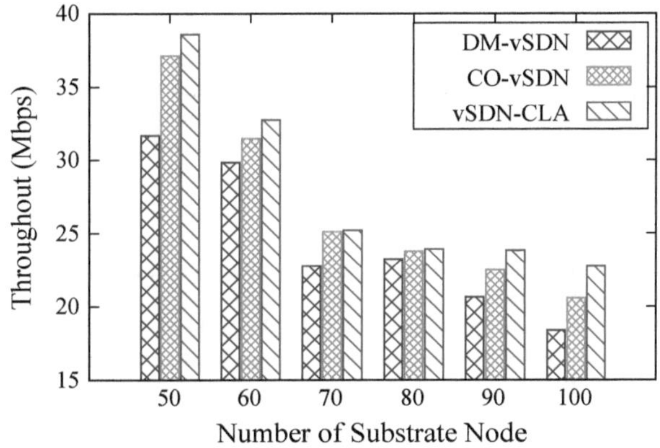

Fig. 4 Throughput

of time over a single substrate network, and each node in the substrate network runs their corresponding learning algorithm in parallel, and hence gives better rate of convergence.

5 Conclusion

This paper presents an ICLA-based VNE and controller placement algorithm, namely, vSDN-CLA in SDN. The aim of this solution is to achieve an efficient mapping between virtual network and substrate network in SDN domain. At the same time, the proposed mapping satisfies the QoS requirements in terms of memory, CPU, bandwidth, and switching capacity of the nodes. We performed extensive simulations using Mininet. The simulation results show that the vSDN-CLA schemes significantly outperform the existing benchmark schemes in terms of controller-to-switch delay, end-to-end delay, and throughput. For example, considering a substrate network of 100 nodes, the proposed scheme achieved 23.72 and 10.55% higher throughput, and 28.13 and 42% lesser end-to-end delay compared to that in two benchmark schemes, namely, DM-vSDN and CO-vSDN, respectively. In this work, we restricted our analysis by not considering many limiting factors such as energy consumption, and fault tolerance, flow migration, and unavailability of path. In future, we have planned to design a more flexible scheme considering those aspects. For instance, using dynamic flow migration technique, we can achieve optimal resource routing.

References

1. D. Kreutz, F.M. Ramos, P. Verissimo, C. Rothenberg, S. Azodolmolky, S. Uhlig, Software-defined networking: a comprehensive survey. Proc. IEEE **103**(1), 14–76 (2015)
2. R. Sherwood, G. Gibb, K.K. Yap, G. Appenzeller, M. Casado, N. McKeown, G. Parulkar, Can the production network be the testbed? in *Proceedings of the Ninth USENIX Conference on Operating Systems Design and Implementation* (2010), pp. 1–6
3. M. Yu, Y. Yi, J. Rexford, M. Chiang, Rethinking virtual network embedding: substrate support for path splitting and migration. SIGCOMM Comput. Commun. Rev. **38**(2), 17–29 (2008)
4. M. Capelle, S. Abdellatif, M.J. Huguet, P. Berthou, Online virtual links resource allocation in software-defined networks, in *Proceedings of the 2015 IFIP Networking Conference* (May 2015), pp. 1–9
5. M. Esnaashari, M.R. Meybodi, Irregular cellular learning automata. IEEE Trans. Cybern. **45**(8), 1622–1632 (2015)
6. A. Fischer, J.F. Botero, M.T. Beck, H. de Meer, X. Hesselbach, Virtual network embedding: a survey. IEEE Commun. Surv. Tutor. **15**(4), 1888–1906 (2013)
7. A. Blenk, A. Basta, M. Reisslein, W. Kellerer, Survey on network virtualization hypervisors for software defined networking. IEEE Commun. Surv. Tutor. **18**(1), 655–685 (2016)
8. R. Mijumbi, J. Serrat, J. Rubio-Loyola, N. Bouten, S. Latre, F.D. Turck, Dynamic resource management in SDN-based virtualized networks, in *Proceedings of IEEE International Workshop on Management of SDN and NFV Systems* (2014)
9. M. Demicri, M.H. Ammar, Design and analysis of techniques for mapping virtual networks to software-defined network substrates. Comput. Commun. **45**, 1–10 (2014)
10. B. Zhou, W. Gao, S. Zhao, X. Lu, Z. Du, C. Wu, Q. Yang, Virtual network mapping for multi-domain data plane in software-defined networks, in *Proceedings of the International Conference on Wireless Communications, Vehicular Technology, Information Theory and Aerospace Electronic Systems* (May 2014), pp. 1–5
11. N.M.M.K. Chowdhury, M.R. Rahman, R. Boutaba, Virtual network embedding with coordinated node and link mapping, in *Proceedings of the IEEE INFOCOM* (April 2009), pp. 783–791
12. S. Gong, J. Chen, Q. Kang, Q. Meng, Q. Zhu, S. Zhao, An efficient and coordinated mapping algorithm in virtualized SDN networks. Front. Inf. Technol. Electr. Eng. **17**(7), 701–716 (2016)
13. M. Esnaashari, M.R. Meybodi, A cellular learning automata based clustering algorithm for wireless sensor networks. Sens. Lett. **6**(5), 723–735 (2008)
14. R.L.S. de Oliveira, A.A. Shinoda, C.M. Schweitzer, L.R Prete, Using mininet for emulation and prototyping software-defined networks, in *Proceedings of 2014 IEEE Colombian Conference on Communications and Computing* (2014), pp. 1–6

Energy-Efficient Delay-Aware Preemptive Variable-Length Time Slot Allocation Scheme for WBASN (eDPVT)

Tamanna Puri, Rama Krishna Challa and Navneet Kaur Sehgal

Abstract The recent evolution of wireless sensor technology is providing unlimited opportunities to remote monitoring and health care applications using Wireless Body Area Sensor Networks (WBASNs). While using WBASN for patient monitoring, emergency situations can appear anytime that requires immediate action without much delay. This leads to the requirement for a new energy-efficient and delay-aware scheme for WBASNs. eDPVT is fulfilling such requirements by proposing a scheme for the Media Access Control (MAC) layer of WBASN. eDPVT is designed to reduce the packet delivery delay for emergency data, which is being transmitted from heterogeneous biosensor nodes to eDPVT base station node in WBASN. In addition to the delay, this scheme also reduces energy consumption with the introduction of variable-length time slots in the context-free period of the super frame. eDPVT handles emergency situations with least delay with the help of preemption mechanism used in this scheme. For performance evaluation, eDPVT is compared with eMC-MAC protocol. ns3 simulation of eDPVT results show that eDPVT is energy-efficient scheme, and there is a significant reduction in average packet delivery delay.

Keywords WBASN · MAC layer · Slot allocation · Energy-efficient Delay-aware · MAC protocol · QoS

T. Puri (✉)
CSE Department, BML Munjal University, Gurgaon, Haryana, India
e-mail: tamannasehgal81@gmail.com

R. K. Challa
CSE Department, NITTTR, Chandigarh, India

N. K. Sehgal
Computer Science Department, BBSBEC, Fatehgarh Sahib, Punjab, India

C. R. Krishna et al. (eds.), *Proceedings of 2nd International Conference on Communication, Computing and Networking*, Lecture Notes in Networks and Systems 46, https://doi.org/10.1007/978-981-13-1217-5_19

1 Introduction

Latest trends in IoT and sensor networks are giving directions to the high-quality medical services for normal as well as disabled persons. Normal routine of the monitored person is not affected by the use of wireless sensor nodes [1–3]. A typical WBASN is implemented using low-power wearable or implantable biosensor nodes in the body of a patient. Any smartphone or a similar wireless device can act as a base station for this network, which can receive biomedical readings collected by biosensor nodes from patient's body, and send these readings to the monitoring devices like hospital's patient monitoring system. The base station may use fastest possible Internet service to communicate this information.

WBASN is challenging due to its challenging requirements viz. energy minimization, low-power consumption, low delay, and QoS. Energy efficiency is very important aspect to be covered in WBASNs because sensor node's increased energy consumption may lead to heating up of the sensor node and may damage human body tissues [4, 5]. The biggest challenge is how to handle emergency situations. In case of an emergency situation, patient's information must reach monitoring station in least possible time so that patient can be diagnosed as soon as possible. Therefore, the delay for the data packets containing emergency data needs to be reduced. Time slot allocation and preemption methods help in reducing this delay. In literature, various researchers have suggested different methods of slot allocation and preemption.

Existing literature covers a number of MAC protocols which are designed keeping in mind traffic load, energy efficiency, delay, and QoS [6–11]. A number of MAC protocols consider QoS requirements of biosensor nodes while assigning time slots in CFP period, but none of the MAC protocols worked on the heterogeneity of sensor node based on type of data or size of the payload to be transmitted by these nodes. A slot allocation scheme based on heterogeneity of data packets is proposed in [12] but this scheme does not consider emergency data packets. Therefore, the concept of preemption to give priority to emergency data traffic is not implemented in this scheme.

eDPVT assumes that heterogeneous biosensor nodes may have different payload sizes and exploits this fact to design eDPVT scheme for WBASNs. Priority of data packets is taken into account while allocating time slots. The priorities are set according to QoS of data packets. Very few MAC protocols are using preemption scheme to handle emergency data packets. eDPVT uses the concept of preemption in order to communicate emergency data immediately by halting the communication of lower priority data.

The performance of eDPVT is compared with eMC-MAC [13] protocol. eMC-MAC protocol considers QoS requirements of different data packets and assigns the time slots accordingly. It also uses a preemption scheme to handle emergency data. But the time slots are of fixed length and not based on the payload size of the data packets. Parameters used for the performance evaluation include Average Packet Delivery Delay (APDD) and total energy consumption. The results of performance evaluation indicate that eDPVT performs better as compared to eMC-MAC while taking energy efficiency and APDD of different data packets into account.

2 Related Works

In IEEE 802.15.4 MAC protocol [14, 15], the super-frame structure is having only seven GTSs (Guaranteed Time Slots) to be assigned in CFP (Contention-Free Period) period. Slot allocation scheme is facing much delay. LDTA-MAC [16] overcomes the limitations of IEEE 802.15.4 MAC protocol and reduces the delay by modifying the slot allocation scheme. LDTA-MAC has reduced delay and energy consumption in comparison to IEEE 802.15.4 MAC protocol but QoS requirements of different types of traffic generated by heterogeneous sensor nodes are not considered. Fixed size of time slots is allocated assuming biosensor nodes having same payload size which is not true in a real scenario. Emergency handling is not supported by this protocol.

ATLAS [17] is designed for multi-hop communication in WBASN. The super-frame structure considered in this protocol is traffic adaptive and has been changed based on the observed traffic load. Four different super-frame structures have been defined based on traffic load conditions. The protocol is traffic adaptive in terms of traffic load but considers fixed period of the time slots allocated in CFP period and QoS. There is no provision for emergency data packet handling.

Time slot allocation in PNP-MAC protocol [18] is preemptive but emergency data is transmitted in a non-preemptive fashion. The protocol is not QoS-aware. Heterogeneity of sensor nodes is not considered in this protocol. Traffic is categorized into two categories only which is not the case in WBASNs. Further categorization of traffic is required in order to utilize the channel properly. Different types of sensors may have different payload sizes but in this protocol, all the sensor nodes are assumed to have same payload size. The time slots used in the protocol are of fixed length. Therefore, these slots are not utilized properly and nodes are not allowed to be in sleep mode until that slot is completed. This condition leads to energy consumption due to idle listening.

McMAC [19] highlights different QoS requirements of a variety of traffic classes in WBASNs. The strength of the protocol lies in handling emergency data in each and every period of the super frame in a very beautiful manner by giving it the highest priority and reducing the delay for emergency data transmission to the minimum. Super-frame structure satisfies multi-constrained QoS requirements. The drawback of this protocol is that it uses FCFS-based slot allocation and does not take remaining lifetime of packets into consideration, which leads to dropping of packets if they are served later according to the scheme.

eMC-MAC proposed by Pandit et al. [13] uses priority-based slot allocation and considers remaining lifetime of data packets while assigning a slot so that the number of packet drops could be reduced further. The use of preemptive slots helps in transmission of emergency data with least possible delay. Traffic is classified based on different QoS requirements of data packets and data packets are handled accordingly in different periods. Fixed-length slot allocation is used, which leads to energy wastage due to idle listening within individual slots. Payload size of all eMC-MAC sensor nodes is assumed to be the same which is actually not the case in a real scenario.

Another traffic-adaptive scheme similar to eMC-MAC [12] allocates variable-length time slots to the sensor nodes in the CFP period based on their payload sizes, and also considers the priority of the data packets based on their QoS. This scheme has no provision for handling emergency data packets in CFP period. In other words, this scheme does not perform well in case of emergency situations.

3 Materials and Methods

3.1 Network Model and Assumptions

In proposed scheme eDPVT, single-hop star topology as shown in Fig. 1 is considered with a base station equipped with fast processor and large memory. The base station invokes eDPVT time slot allocation algorithm as soon as it receives all the requests from different sensor nodes for time slots, and sends a notification to the sensor nodes regarding their slot allocation instantly. These sensor nodes sense data readings from human body and transmit these readings to base station in the form of data packets. eDPVT considers sensor nodes with different payload sizes. These sensor nodes are assumed to be clock synchronized with the base station.

3.2 Design Model of eDPVT

Data traffic in a WBASN is classified into five different categories. So, five different types of Data Reading packets are taken into consideration which are Emergency Data reading Packets (EDRPs), Critical Data Reading Packets (CDRPs), Reliability-Constrained Data Reading Packets (RDRPs), Delay-Constrained Data Reading Packets (DDRPs), and Normal Data Reading Packets (NDRPs).

3.2.1 Super-Frame Structure

The super-frame structure of eDPVT is defined as shown in Fig. 2. In this super-frame structure, the advertisement is used only for synchronization purpose. Slot allocation information is not broadcasted in this period. In CAP 1 period, CDRPs and RDRPs can transmit slot allocation request for guaranteed time slots to the base station. NDRPs and DDRPs do not have any reliability constraints so these types of

Fig. 1 eDPVT topology

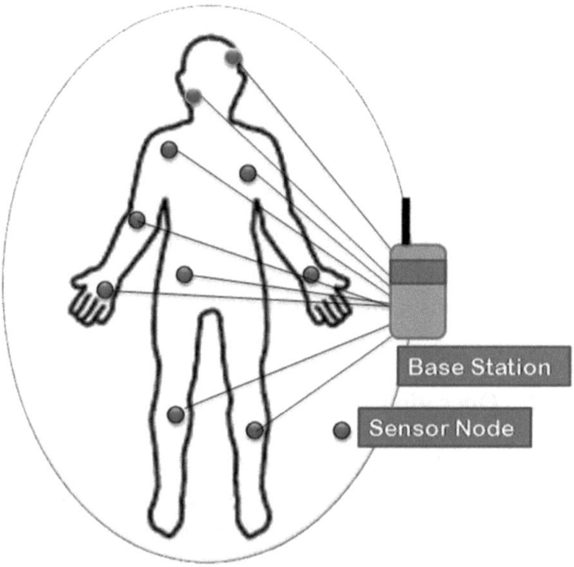

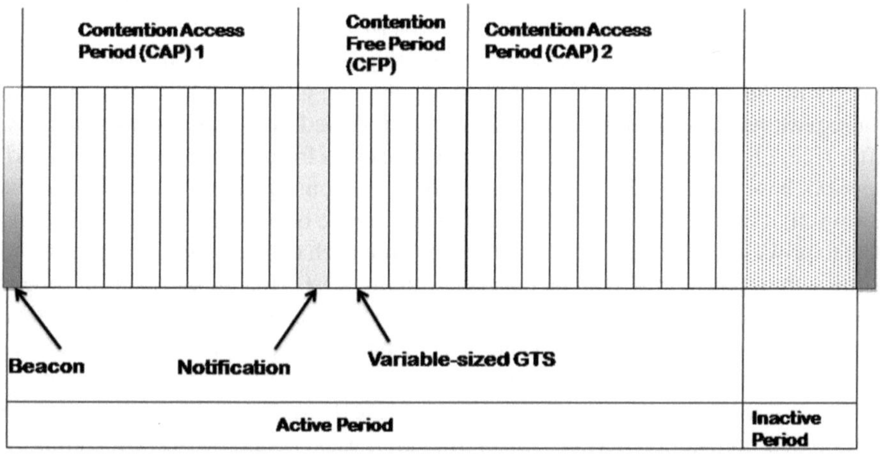

Fig. 2 Super-frame structure of eDPVT

packets do not request for GTS in CAP 1 period. NDRPs do not require guaranteed resources since the data is routine data. DDRPs are loss-tolerant, therefore these packets do not request for guaranteed slot.

After CAP 1 period, a notification period is introduced to broadcast slot allocation information to sensor nodes. The notification period is followed by CFP period. This CFP period contains a number of timeslots allocated to sensor nodes. Slot length for different sensor nodes is decided based on the payload size shared by these sensor nodes with the base station. Emergency time slots are also introduced in the CFP period to handle emergency situations. These time slots are kept to be of short length to receive slot requests from sensor nodes having EDRPs. CFP period is followed by CAP 2 period to accommodate normal traffic and delay-constrained traffic so that sensor nodes having these types of traffic can transmit the sensed data to eDPVT base station.

As soon as CAP period is over, the inactive period of super frame is started in which all the sensor nodes remain in sleep mode. Base station may transmit the collected data to the monitoring station during this period. As sensor nodes are in the sleep mode and wake up only in the next super frame to receive the beacon, energy consumption is reduced by keeping the sensor nodes in low power mode.

3.2.2 Operation

The communication process in WBASN is controlled by the base station. Base station in WBASN broadcasts advertisements to sensor nodes. After receiving advertisement from base station, these sensor nodes synchronize themselves and transmit data according to the periods defined by base station in the super-frame structure. The data packet containing information about duration of each period of super frame is broadcasted by eDPVT base station in the advertisement.

In CAP 1 period, the sensor nodes generating emergency data transmit EDRPs directly to the base station. The sensor nodes having CDRPs and RDRPs transmit slot request packets to the base station using prioritized random back-off and CCA. After receiving all the EDRPs and slot request packets from sensor nodes in the CAP 1 period, base station calls eDPVT time slot allocation algorithm. According to the algorithm, it ensures that CDRPs are given preference over RDRPs while allocating slots to provide QoS aware slot allocation. Within each category of packets, a sorted queue is maintained in which data packet with the least remaining packet lifetime is kept first in the queue so that it can be allowed to send its data earlier than other data packets. This method reduces the number of packet drops, which further reduces energy consumption due to retransmissions and makes eDPVT energy-efficient.

The base station prepares a notification packet and broadcasts this packet to sensor nodes in the notification period. This notification packet contains slot allocation list for the CFP period of the super frame. This list indicates time slot during which a particular sensor node can transmit its CDRPs and RDRPs to base station. As soon as sensor nodes receive this notification packet, they set their wakeup schedule accordingly and go to sleep mode. Then, each sensor node allocated time slot in the CFP period, wake up in its dedicated time slot and check whether this time slot is preempted or not. If this time slot is not preempted, sensor node transmits the data packet. Otherwise, it changes the mode to sleep mode and transmits this data packet in the next super frame.

In the notification packet, base station also defines slot numbers for the Emergency Time Slots (ETSs). ETS slots are kept in CFP period for emergency data handling. If there is any emergency event at any sensor node during CFP, then this sensor node can request for data transmission of EDRPs using ETS slots. During ETS, sensor node can send emergency time slot request packet to the base station. After receiving emergency slot request packets during ETS, the base station calls proposed time slot preemption algorithm. According to this algorithm, the base station then broadcasts emergency notification packet to the sensor nodes. This packet is transmitted at the end of ETS slot.

After receiving emergency notification packet, sensor nodes readjust their wakeup schedules and go back to sleep mode. This notification packet contains time slot information about preempted and privileged sensor nodes. Privileged sensor node can transmit its EDRPs in the allocated time slot, which is preempted by base station to give priority to EDRPs over lower priority data packets. As the time slots allocated by the base station are of variable size, preemption is more complex as compared to eMC-MAC which has fixed size time slots. Preemption of time slots reduces the packet delay for Emergency Data Packets.

After CFP period is complete, CAP 2 period allows the sensor nodes having DDRPs and NDRPs to transmit their data packets. As DDRPs and NDRPs can tolerate packet losses and these packets are not reliability-constrained, dedicated time slots are not required for these types of packets. Sensor nodes transmit these reading packets to the base station using prioritized random back-off and CCA. DDRPs are given priority over NDRPs to reduce their packet delay. Base station receives all DDRPs and NDRPs during CAP 2. At the end of CAP 2 period, all the nodes go back to sleep mode.

eDPVT is designed to employ following energy-saving approaches. Sensor nodes wake up only when necessary and remain in sleep mode rest of the time, saving their energy. For example, sensor nodes having only DDRPs or NDRPs will wake up only in CAP 2 period. Variable-length time slots further reduce energy consumption by adjusting the slot size according to sensor node's payload size. Sorted list of data reading packets helps in reducing energy consumption which may occur due to retransmission of dropped packets.

4 Performance Evaluation

For performance evaluation, time slot allocation and preemption schemes of eDPVT are compared with that of eMC-MAC protocol [13]. eDPVT time slot allocation and preemption scheme are implemented in ns-3.19 [20] on Ubuntu 14.04 LTS. Wireshark tool is used to read the trace files. The performance of eDPVT is analyzed and compared with eMC-MAC protocol. Different parameters used for the simulation of the schemes are shown in Table 1.

Table 1 Simulation parameters

Parameter	Value
Total number of sensor nodes	10
Initial energy value at each sensor node	100 J
Radius of data transmission	2 m
Payload size	16 bytes, 32 bytes
Super-frame period	1 s
CAP size	76.8 ms
Channel data rate	250 kbps
Simulation time	100–1000 s

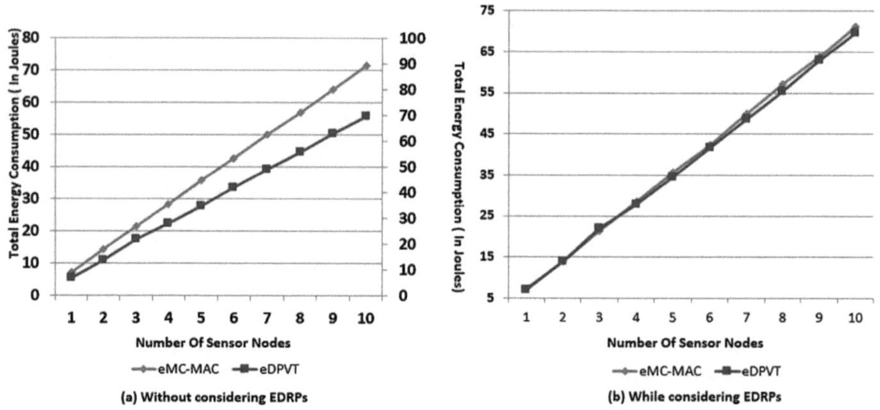

Fig. 3 Effect of varying the number of sensor nodes on total energy consumption

4.1 Simulation Results

The simulation results of eDPVT and eMC-MAC for two different scenarios illustrate the effectiveness of eDPVT in terms of APDD and energy consumption. To analyze the performance of eDPVT and eMC-MAC without considering EDRPs, emergency time slot request packets are not transmitted in ETS slots during simulation. As there is no emergency time slot request in ETS slots, there will be no preemption. To analyze the effect of varying the number of sensor nodes on eDPVT, simulation results are calculated by varying the number of sensor nodes from one to ten. Figure 3 presents energy consumption analysis of eDPVT and eMC-MAC with considering and without considering EDRPs.

Figure 4a shows that APDD has reduced value in eDPVT as compared to eMC-MAC at each data point while no EDRP is considered in traffic. The schemes are also compared while considering EDRPs in the traffic in Fig. 4b. The performance comparison of eDPVT and eMC-MAC in terms of APDD for different data packets is presented in Figs. 5, 6, and 7. It can be seen that number of sensor nodes is making

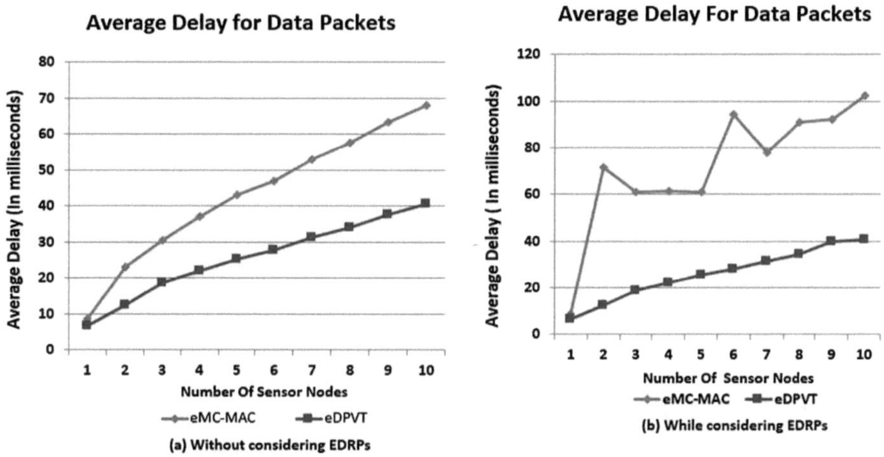

Fig. 4 Effect of varying the number of sensor nodes on average delay for all data packets

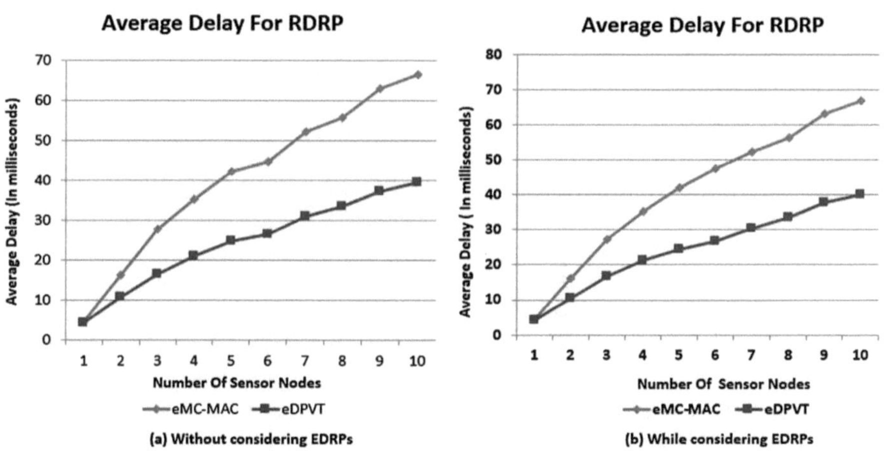

Fig. 5 Effect of varying the number of sensor nodes on average delay for RDRPs

very little effect on APDD for EDRPs in case of eDPVT, achieving the objective of the research to handle the emergency situation with least delay. Varying the number of sensor nodes shows significant impact on the results obtained for CDRP packets.

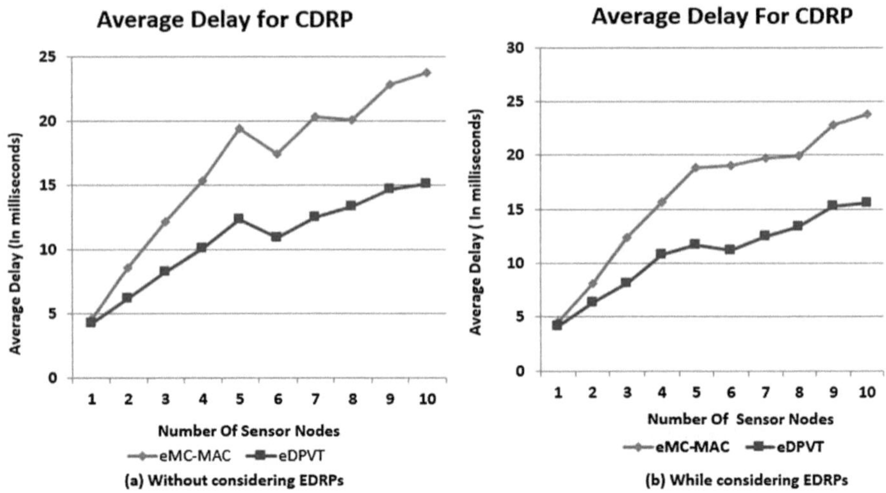

Fig. 6 Effect of varying number of sensor nodes on average delay for CDRPs

Fig. 7 Effect of varying number of sensor nodes on average delay for EDRPs

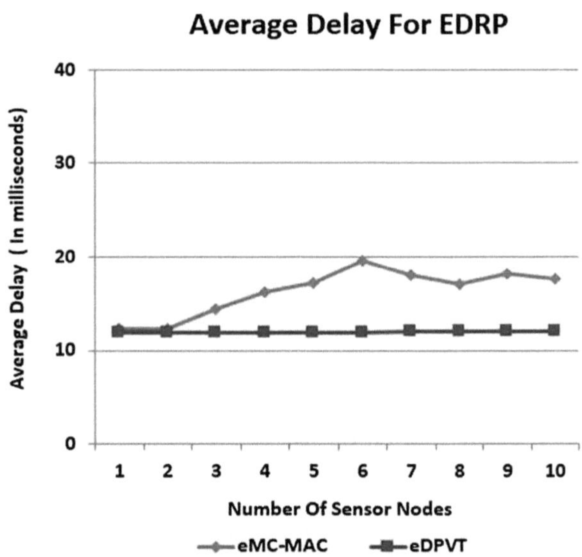

5 Conclusion

In this paper, eDPVT scheme is presented which reduces the APDD for different data packets considerably when compared with eMC-MAC and also reduces energy consumption. The results for APDD for data packets generated by sensor nodes of WBASN are improved by 60% in eDPVT when compared to eMC-MAC. APDD for EDRPs is reduced by 31% when compared with eMC-MAC. APDD for CDRPs is considerably reduced by 35% in eDPVT as compared to eMC-MAC. APDD for RDRPs is reduced by 40% in comparison to eMC-MAC. Overall energy consumption is reduced in eDPVT by saving the energy wastage due to idle listening.

References

1. J. Ahmad, F. Zafar, Review of body area network technology and wireless medical monitoring. Int. J. Inf. **2**(2), 186–188 (2012)
2. S. Movassaghi, J. Lipman, Wireless body area networks: a survey. IEEE Commun. Surv. Tutorials (2013)
3. S. Ramli, R. Ahmad, Surveying the wireless body area network in the realm of wireless communication, in *7th International Conference on Information Assurance and Security (IAS)* (2011), pp. 58–61
4. N. Javaid, S. Hayat, M. Shakir, M.A. Khan, S.H. Bouk, Z.A. Khan, Energy efficient MAC protocols in wireless body area sensor networks-a survey. J. Basic Appl. Sci. Res. (JBASR) **3**(4), 770–781 (2013)
5. R. Sruthi, Medium access control protocols for wireless body area networks : a survey. Procedia Technol. **25**, 621–628 (2016) (RAEREST)
6. S. Ullah, S.M. Bin Shen, R. Islam, P. Khan, S. Saleem, K.S. Kwak, A study of medium access control protocols for wireless body area networks. Sensors **10**(1), 128–145 (2010)
7. F. Masud, A.H. Abdullah, G. Abdul-salaam, F. Ullah, Traffic adaptive MAC protocols in wireless body area networks. Hindawi Wirel. Commun. Mob. Comput. Article ID 8267162, 14 pages. https://doi.org/10.1155/2017/8267162 (2017)
8. A. Rahim, N. Javaid, M. Aslam, Z. Rahman, U. Qasim, Z.A. Khan, A comprehensive survey of MAC protocols for wireless body area networks, in *2012 Seventh International Conference on Broadband, Wireless Computing, Communication and Applications (BWCCA)* (12–14 Nov 2012), pp. 434, 439. https://doi.org/10.1109/bwcca.2012.77
9. O. Bouachir, A.B. Mnaouer, F. Touati, PEAM: a polymorphic, energy-aware MAC protocol for WBAN, in *2016 23rd International Conference on Telecommunications, ICT 2016*. https://doi.org/10.1109/ict.2016.7500491 (2016)
10. M.B. Rasheed, N. Javaid, A. Haider, U. Qasim, Z.A. Khan, T.A. Alghamdi, An energy consumption analysis of beacon enabled slotted CSMA/CA IEEE 802.15.4, in *28th International Conference on Advanced Information Networking and Applications Workshops (WAINA)* (May 2014), pp. 372–377
11. V. Mainanwal, Comparison, evaluation and characterization of MAC protocols in wireless body area network, in *1st International Conference on Next Generation Computing Technologies (NGCT-2015)* (2015), doi:978-1-4673-6809-4, pp. 175–180
12. T. Puri, R.K. Challa, N.K. Sehgal, Energy efficient QoS aware MAC layer time slot allocation scheme for WBASN, in *Proceedings of Fourth International Conference on Advances in Computing, Communications and Informatics (ICACCI) Special Session on Energy Efficient Wireless Communications and Networking*, Kochi, India (IEEE, 11th Aug 2015), pp. 966–972 (ISBN No.: 978-1-4799-8792-4/15)
13. S. Pandit, K. Sarker, M.A. Razzaque, A.M.J. Sarkar, An energy-efficient multiconstrained QoS aware MAC protocol for body sensor networks. J. Multimedia Tools Appl. (MTAP) **73**(3), 1–22 (April 2014) (Springer US)
14. Wireless Medium Access Control (MAC) and Physical Layer (PHY), Specification for low-rate wireless personal area networks (WPANs). IEEE Std. **802**(4) (15-2006) (2003)
15. N.F. Timmons, W.G. Scanlon, Analysis of the performance of IEEE 802.15.4 for medical sensor body area networking, in *First Annual IEEE Communications Society Conference on Sensor and Ad Hoc Communications and Networks* (2004), pp. 16–24

16. C. Li, B. Hao, K. Zhang, Y. Liu, J. Li, *A Novel Medium Access Control Protocol with Low Delay and Traffic Adaptivity for Wireless Body Area Networks*, vol. 35 (Springer Science+ Business Media, LLC 2011, 2011), pp. 1265–1275
17. M.O. Rahman, C.S. Hong, S. Lee, Y.-C. Bang, ATLAS: a traffic load aware sensor MAC design for collaborative body area sensor networks. Sensors **11**(12), 11560–11580 (2011)
18. J. Yoonm, G. Ahn, PNP-MAC: preemptive slot allocation and non-preemptive transmission for providing QoS in body area networks, in *7th IEEE Consumer Communications and Networking conference* (2010), pp. 1–5
19. M.M. Monowar, M.M. Hassan, F. Bajaber, M. Al-Hussein, A. Alamri, McMAC: towards a MAC protocol with multi-constrained QoS provisioning For Diverse Traffic In Wireless Body Area Networks. Sensors **12**(11), 15599–15627 (2012)
20. The Network Simulator Version 3, ns-3. www.nsnam.org

Smart Irrigation System Using Cloud and Internet of Things

Suresh Koduru, V. G. D. Prasad Reddy Padala and Preethi Padala

Abstract An intensive utilization of water resources in industries, agriculture, and groundwater consumption by humans for various purposes has degraded the water levels. A focus on effective utilization of water resources with simplified irrigation across different agricultural farms is required with the advancement of technology. This paper presents a framework based on cloud and Internet of things for implementing a smart irrigation system. Based on the defined framework, a use case for automated smart irrigation system is developed and a competent mechanism is defined for effective utilization of excessive water generated from showers to increase the groundwater levels. The use case provides flexibility to farmers for monitoring the farms in real time using the farmer's cockpit. Here, heterogeneous devices are firmly integrated to empower smart irrigation and to monitor the system in real time. The use case actuation and automation are done based on certain imposed constraints to respond as per inputs and outputs generated by various devices installed in smart irrigation system.

Keywords Internet of things · Smart irrigation · Cloud computing

1 Introduction

Today, with the growing population, the consumption of natural resources is increasing exponentially across different industries. It is expected that the world population will increase to 9.8 billion by 2050 which is currently 7.6 billion as per official United Nations estimates [1] and due to this, freshwater and food consumption will also increase exponentially. In countries like India, 89% of groundwater is used in

S. Koduru (✉) · V. G. D. P. R. Padala
Department of Computer Science and Systems Engineering, Andhra University, Visakhapatnam, Andhra Pradesh, India
e-mail: koduru112@gmail.com

P. Padala
Department of Information Technology, NITK, Surathkal, Karnataka, India

C. R. Krishna et al. (eds.), *Proceedings of 2nd International Conference on Communication, Computing and Networking*, Lecture Notes in Networks and Systems 46, https://doi.org/10.1007/978-981-13-1217-5_20

irrigation and the sources for groundwater supply are shallow tube wells and dug wells [2]. Due to this, there is groundwater depletion where farmers are forced to use expensive deepwater equipment which has led to their hardship and cost [3]. It is imperative that cautious measures need to be taken to increase the groundwater levels and to provide food safety for the growing population. The solution for this is to modernize the existing irrigation mechanisms and agriculture systems using cloud and Internet of things [4].

2 Literature Survey

With the changes in technologies and rapid consumption of natural resources by humans, it is vital to implement smart irrigation techniques to overcome the scarcity of water resources in near future. Smart irrigation using wireless sensors for precision agriculture, drip irrigation automation system for water level monitoring, are discussed by considering various factors like soil moisture, humidity, and temperature in [5]. In [6], couple of irrigation tools, SmartLine and Hunter Pro-C2, are used to achieve water savings and better crop yield. Here, Hunter Pro-C2 has realized better economic benefits and large amount of irrigation water is saved. Unplanned usage of water and irrigation with automatic valve mechanism based on the soil moisture is illustrated in [7, 8] which minimizes manual intervention of farmers for crop irrigation. In [9], a framework for efficient water distribution management is defined using wireless sensor network (WSN) and field-programmable gate array (FPGA), where data is transmitted using ZigBee and based on which control of motor on/off conditions is determined.

The abovementioned research efforts majorly focused on illustrating smart irrigation techniques by using various irrigation tools, automating irrigation systems using actuators, and sensors for minimizing the water consumption in crop irrigation and to achieve better crop yield. But it is vital that these smart irrigation techniques must be extended for preserving water generated from rainfall which will help to increase the groundwater levels in agricultural farms and to enable farmers to save crops from unforeseen rainfall by integrating farms with real-time weather conditions. This paper proposes a framework based on cloud and Internet of things to implement a smart irrigation system that saves crops during unforeseen rainfall, increases groundwater levels with a competent mechanism, and to reuse excessive water generated during rainfall for crop irrigation. Based on the defined framework, a use case for smart irrigation system is developed which extends the features depicted in [7–9] where agricultural farms are closely monitored based on real-time weather conditions and soil moisture. Also, a competent mechanism is defined to avoid rainwater depletion, utilize excessive water generated from rainfall for crop irrigation, and save farms during unforeseen rainfall.

3 Framework

The framework proposed in Fig. 1 is based on cloud and Internet of things where agricultural farms can be monitored by farmers anytime anywhere. The framework comprises two levels—control system and farmer's cockpit which are reusable for diverse applications.

Control system consists of three segments—Application interface, database segment, and Internet of things service cockpit. The devices that are connected to real world are registered in IoT services cockpit to capture data in real time and to update it in database. The database segment stores the weather forecast data and real-time data that comes from sensors and actuators installed in farms. In application interface segment, a refined business logic is defined to control the sensors, actuators, and motors installed in farms for updating real-time data from farms. The data is further consumed in farmer's cockpit to take decisions based on the environmental conditions, soil moisture, and water availability for irrigation in farms. Farmer's cockpit is a sophisticated user interface for the farmers to monitor the farms in real time and to control the irrigation system based on the environmental conditions. A competent mechanism is also defined in application interface segment where the irrigation system in farms is automated based on certain imposed constraints and rules.

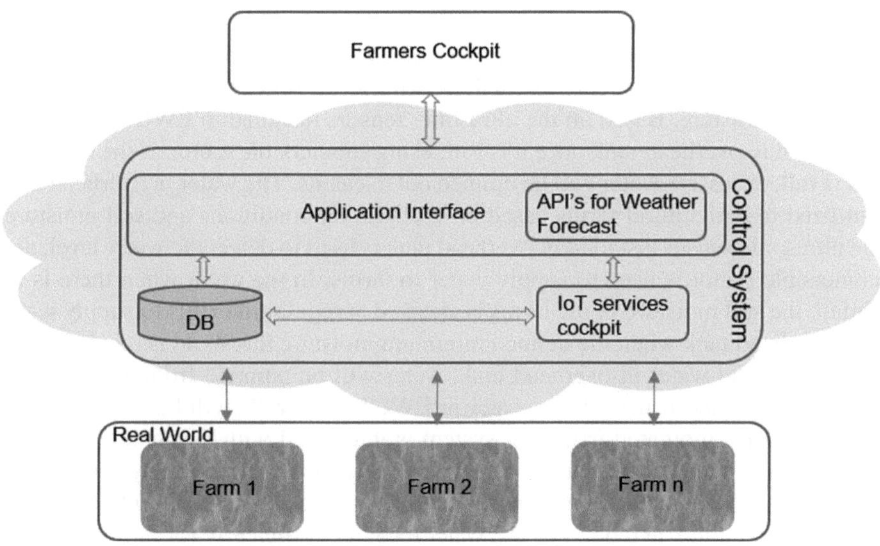

Fig. 1 Framework for smart irrigation system

4 Use Case Implementation

In this section, a use case with process flow is devised based on the defined framework that establishes a seamless integration between the devices installed in farms. The smart irrigation system is automated for efficient utilization of water resources to increase the groundwater levels during showers and to save crops from unforeseen rainfall.

4.1 Use Case Design for Smart Irrigation System

Smart irrigation use case is designed as shown in Fig. 2 where water resources are utilized efficiently and help to increase groundwater levels. The components used for the use case are overhead tank (OT), excess water collection tank (EWCT), submersible motors, relay module to control the motors, ultrasonic sensors to measure water level in tanks, and soil moisture sensors to measure the soil moisture in crops. The use case is automated based on the environmental conditions and soil moisture. Supply of water to farms is done based on the soil moisture values and weather forecast as per the defined process flow shown in Fig. 3. In case of heavy rains and when the soil moisture reaches to the defined maximum threshold, excessive water will be collected in rainwater trenches (RWT), where it will be an aid to increase the groundwater levels. Once the rainwater trench is full, excessive water will be drained to excess water collection tank that will be utilized for further farm irrigation based on the soil moisture. Based on the ultrasonic sensors installed in EWCT, water will be pumped to overhead tank once it is full, using submersible motor. If the overhead tank is full, excessive water will be drained out to canals. The water in overhead tank is utilized in agricultural farms based on the weather conditions and soil moisture. The ultrasonic sensor installed in overhead tank is used to detect the water level, and submersible motor is used to supply water to farms. In the event when there is no rainfall, the soil moisture in the farms is checked at regular intervals to supply water from overhead tank when the defined minimum moisture threshold is reached.

If there is no water in overhead tank, water will be pumped from excess water collection tank and if there is no water in EWCT, borewells will be the source for irrigation. The automated irrigation system is integrated with farmer's cockpit for the farmers to monitor the farms in real time. The smart irrigation system aids to decrease the groundwater depletion, saves crops during unforeseen rainfall, reduces manual intervention, and helps to use water resources efficiently for farm irrigation.

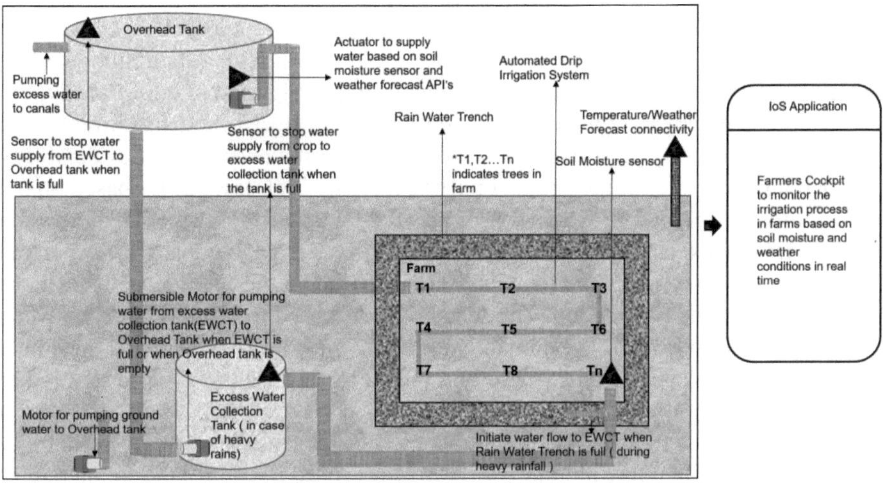

Fig. 2 Proposed smart irrigation system

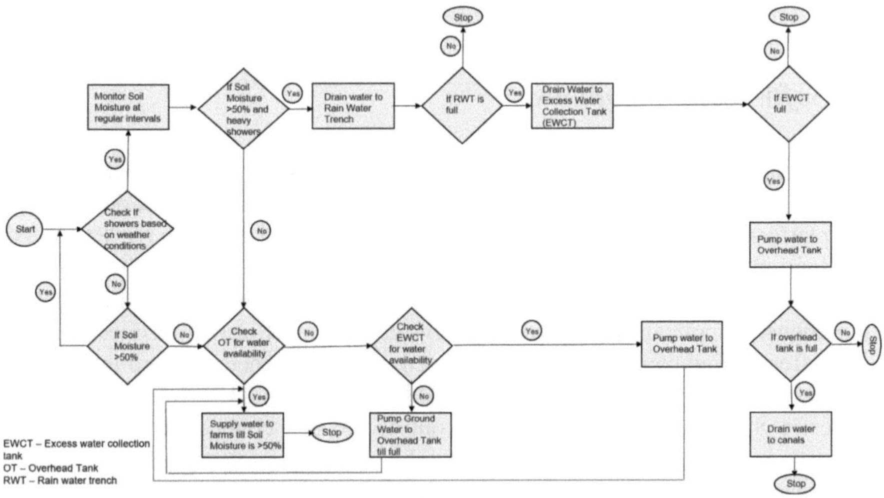

Fig. 3 Process flow for smart irrigation system

4.2 *Hardware Components*

Currently, there are several platforms and plug-and-play devices available for programming Internet of things. To implement the prototype, Raspberry PI 3 module is used which is a high-performance low-cost computer. Raspberry PI 3 enables to interact with real-world objects using sensors and actuators with ease and flexibility. The main features of Raspberry PI 3 are plug-and-play capabilities, high processing speed, and performance which uses Quadcore 1.2 GHz Broadcom BCM2837 64bit

CPU, adequate RAM, option to choose overclock presets to maximize performance based on CPU load [10]. Soil moisture sensor, ultrasonic sensor, and relay module are tightly coupled with Raspberry PI to enable smart irrigation system and to monitor the soil moisture in real time based on weather forecast. Soil moisture sensor HC-SR501 is used to measure the soil moisture in farms and it provides flexibility to adjust onboard potentiometer sensitivity to indicate high/low values based on the defined threshold [11]. Ultrasonic sensor HC-SR-04 is used to trigger an alert when overhead tank or excess water collection tank is full [12], and it consists of two parts: a transducer which produces an ultrasonic sound based on the water level in the tanks and the other listens to the echo. The sensor is configured to produce an echo when the tank is about to be full. Submersible motor [13] installed in overhead head tank is used to supply water to farms based on the soil moisture threshold and weather conditions. The motor installed in excess water collection tank is used to pump water to overhead tank based on the ultrasonic sensor values and water availability in overhead tank. The relay module 5V10A2 [14] is used to control the submersibles motors where the motors will be switched on/off based on the soil moisture and water availability in tanks.

4.3 *Software Components*

It is important that the farmers need to monitor the farms anytime anywhere in real time and the solution for this is cloud platform. Today, there are a lot of open-source cloud platforms and to implement the use case, SAP cloud platform (SCP) is used. SAP cloud platform helps to build the desired applications economically with ease and flexibility [15]. It provides an agile platform as a service (PaaS) which enables comprehensive application development services to connect various business systems and to offer myriad capabilities like enhanced user experience, secured integration, analytics, Internet of things services, and HANA database [15]. Internet of things services in SAP cloud platform provide out of the box connectivity between the devices to establish seamless integration where business-relevant data is derived using broad variety of protocols [16]. Raspberry PI, soil moisture sensor, ultrasonic sensor, and relay module are registered using Internet of things services, and the data that comes from these devices is stored in HANA database which is a relational database management system that combines online analytical processing (OLAP) and online transactional processing (OLPT) into a single in-memory database [17]. Here, the communication between the real-world objects and cloud is established using JSON that is lightweight data interchange format and language independent [18]. Application development is done using SAP cloud platform Web IDE which is a web-based development environment [19]. Here, a refined application logic is defined for smart irrigation system and data that comes from real-world objects if validated based on the predefined conditions.

A mobile application is used for farmer's cockpit to monitor the farms in real time anytime anywhere and it is developed using Apple's integrated development

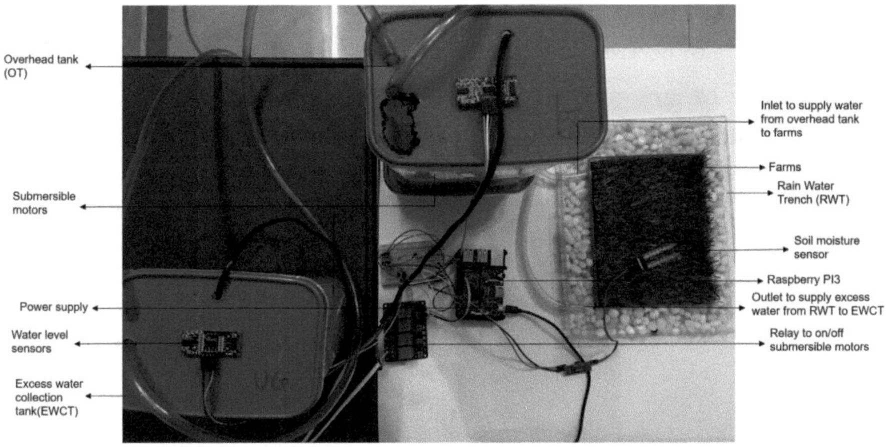

Fig. 4 Setup of smart irrigation system

environment which provides flexibility to establish a seamless integration to develop native applications swiftly using Apples modern programming language [20]. Using the developed mobile application, farmers can monitor the soil moisture in farms and can check the real-time weather forecast using yahoo weather APIs [21] to achieve a competent irrigation mechanism.

5 Results and Discussions

This section presents the use case results for smart irrigation system where tests were carried out based on the defined process flow. An algorithm is defined for smart irrigation use case depicted in Fig. 4, and it is monitored using the developed farmer's cockpit mobile application as shown in Fig. 5a, b. The application provides flexibility to farmers for monitoring the soil moisture, weather forecast, and to effectively manage the irrigation system in real time. The use case is tested successfully in multiple cycles based on two-step process—Irrigation process during showers and irrigation process without showers.

To test the irrigation mechanism, a threshold of 50% is defined for soil moisture and the cutoff water level for the tanks is defined as one liter when it is about to be full. In the event of showers and when the soil moisture is greater than 50% as shown in Fig. 5a, excess water in the farms is drained out to rainwater trench (RWT) and later to excess water collection tank (EWCT). An event is triggered successfully from water level sensor that is installed in EWCT when the tank is full, and the submersible pump in EWCT is successfully started to pump water to overhead tank. Once the overhead tank is full, water is drained to canals. In the event of no showers, soil moisture is checked at regular intervals and when it is reached to minimum

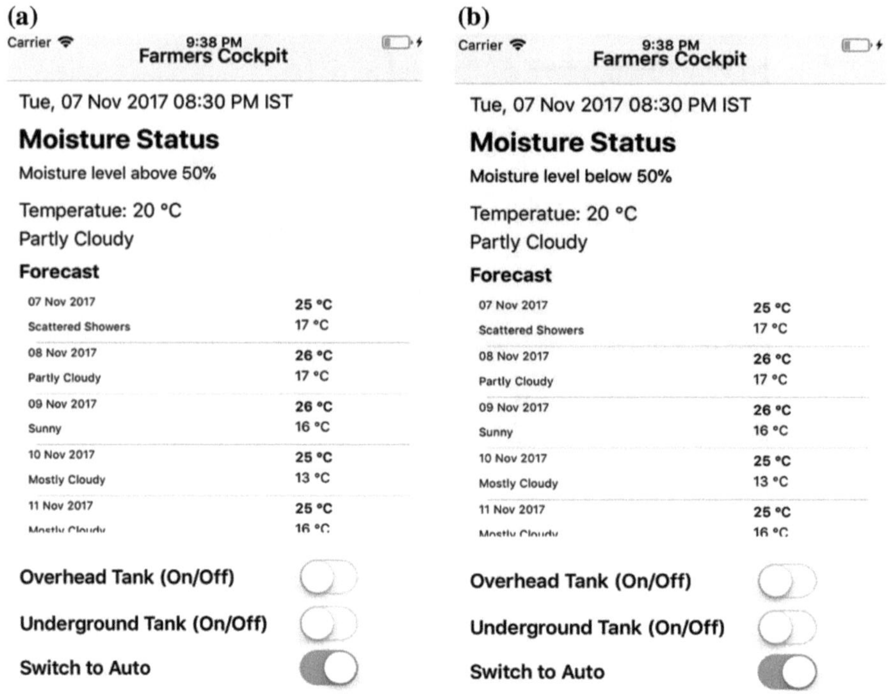

Fig. 5 **a** Soil moisture above 50% with scattered showers. **b** Soil moisture below 50%

threshold of less than 50% as shown in Fig. 5b, the submersible pump in overhead tank is automatically started to supply water to farm. If the overhead tank is empty, water from EWCT is pumped automatically till the tank is full and later water is pumped to farms till a threshold of soil moisture greater than 50% is reached. If the excess water collection tank is empty, irrigation of the farms is dependent on canals and borewells. The use case provides a flexibility to decide whether the irrigation is automatic or manual. Tests were carried out for automatic irrigation in the event of failure of motors where the farmer can switch off the motors from the mobile application as shown in Fig. 5a, b.

6 Conclusion

In this paper, a seamless integration of devices is enabled to achieve a smart irrigation system. Here, irrigation of farms is done using a competent water preservation mechanism, soil moisture, and weather forecast to effectively utilize the water resources and to avoid groundwater depletion. The smart irrigation in farms is monitored in real time using farmer's cockpit and it is tested successfully in several cycles based on the defined framework considering water supply from borewells and weather conditions as an assumption. The use case can be further extended by integrating with solar equipment to empower an independent smart irrigation system without a need to depend on power supply to the integrated devices. The defined infrastructure can also be reused in manufacturing and chemical industries with few customizations.

References

1. United Nations Department of Economic and Social Affairs. https://www.un.org/development/desa/en/news/population/world-population-prospects-2017.html
2. R. Suhag, *Overview of Ground Water in India*. No. id: 9504. (2016)
3. P. Singh. India's groundwater crisis. http://www.livemint.com/Opinion/v4nXpXNxSJtxQNlEbvtJFL/Indias-groundwater-crisis.html
4. F. TongKe, Smart agriculture based on cloud computing and IOT, J. Convergence Inf. Technol. **8**(2) (2013)
5. E. Gajendran, S.R. Boselin Prabhu, M. Pradeep, An analysis of smart irrigation system using wireless sensor, Multi-disc. J. Sci. Res. Educ. **3** (2017)
6. H.M. Al-Ghobari, F.S. Mohammad, M.S.A. El Marazky, Evaluating two irrigation controllers under subsurface drip irrigated tomato crop, Spanish J. Agric. Res. **14**(4), 1206 (2017)
7. A. Tyagi, et al. Smart irrigation system (2017)
8. V.L. Akubattin, et al. Smart irrigation system (2016)
9. P. Ranade, S.B. Takale, Smart irrigation system using FPGA based wireless sensor network. IEEE Proc. **15**, 223–224 (2016)
10. R. Pi, https://www.raspberrypi.org/products/raspberry-pi-3-model-b/
11. KitsGuru, http://www.kitsguru.com/soil-moisture-sensor
12. Rees52, http://rees52.com/65-ultrasonic-sensor-module-hc-sr-04.html
13. Primeoffers, http://www.primeoffers.in/B072P6TX9W-globuspelto-submersible-water-pump-blackwhite-orangeblue-for-desert-air-cooler-aquarium-fountains.html
14. PaisaWapas, https://www.paisawapas.com/p-5v-10a-2-channel-relay-module-shield-for-arduino-arm-pic-avr-dsp-electronic-8865719
15. SAP Cloud Platform, https://cloudplatform.sap.com/capabilities.html
16. SCP IoT Services, https://cloudplatform.sap.com/capabilities/iot/iot-service.html
17. SAP HANA database, https://www.sap.com/india/products/hana/features/in-memory-database.html
18. JSON, http://www.json.org/
19. SAP Web IDE, https://cloudplatform.sap.com/capabilities/devops/web-ide.html
20. Apple Developer, https://developer.apple.com/library/content/referencelibrary/GettingStarted/DevelopiOSAppsSwift/
21. Yahoo Developer, https://developer.yahoo.com/weather/

IoT-Based Condition Monitoring and Fault Detection for Induction Motor

R. Kannan, S. Solai Manohar and M. Senthil Kumaran

Abstract The Internet of Things (IoT) has enormous development in recent trends of industrial, environmental, and medical applications. The availability of massive amount of processing power in the cloud, new opportunities have emerged for complete automation of industrial devices along with Industrial Wireless Sensor Networks (IWSNs). The continuous monitoring and earlier fault detection of the machines builds the efficient process control in the industrial automation. The detection and classification of faults in the machine-learning techniques depends on the number of features involved in it. Increasing the dimensionality of the features will affect the classification accuracy adversely. To overcome these issues, the proposed method of Minimal Relevant Feature Extraction-based Class-Specific Support Vector Machine (CS-SVM) algorithm introduces the suitable technique to extract the relevant features and classify the faults accordingly. The relevant feature prior to the classification increases the accuracy effectively, and also the less dimensionality of the proposed algorithm is more suitable for time-consuming nature in cloud. The implementation of the above-proposed system provides an online machine condition monitoring and accurate fault prediction in the cloud platform using IoT service along with IWSNs.

Keywords Internet of things (IoT) · Industrial wireless sensor networks (IWSNs) · Fault detection · Classification of faults and cloud computing

R. Kannan (✉)
Anna University, Chennai, India
e-mail: rpsskannan@gmail.com

S. Solai Manohar
Arulmigu Meenakshi Amman College of Engineering, Tiruvannamalai, Tamil Nadu, India
e-mail: ssmanohar76@gmail.com

M. Senthil Kumaran
SSN College of Engineering, Chennai, India
e-mail: senthilkumarnm@ssn.edu.in

C. R. Krishna et al. (eds.), *Proceedings of 2nd International Conference on Communication, Computing and Networking*, Lecture Notes in Networks and Systems 46, https://doi.org/10.1007/978-981-13-1217-5_21

1 Introduction

The Wireless Sensor Networks (WSNs) are more flexible communication system for transmitting the data to the desired location. In recent trend, IoT-based condition monitoring with WSNs is highly utilized in the automation industries due to its enormous quality of detecting objects [1]. The Wireless Sensor Network needs more concentration in the scenario of energy consumption, cost, and reliable communication to obtain the better performance. The most common task of a IoT-based applications is transmitting huge sensed data to a specific node.

The diagnosis misclassifications and the incorrect maintenance scheduling are the major deficiencies in the conditional monitoring that requires the novel operating scenarios are called as novelty detection [2]. The standard machine-learning models such as Support Vector Machine (SVM) [3] is based on the pattern recognition and are more suitable for data classification. It has high dimensionality in nature which reduces the accuracy level.

The rest of the paper is ordered as follows. Section 2 deals about related works. Design methodology for condition monitoring, fault detection, and classification are presented at Sect. 3. In Sect. 4, hardware setup and outputs are described. Finally, conclusion of the proposed method with its applications in IoT scenario is conferred in Sect. 5.

2 Related Work

This section presents a detailed related review of proposed work for industrial automation applications using IoT service.

2.1 IWSN-Based Approaches

Lu and Gungor [4] explored nonintrusive diagnostic algorithm for remote energy monitoring using WSNs. Hou and Neil [5] discussed the detailed analysis of the fault diagnosis using IWSN and experimentally verified the two-step classifier method.

2.2 IoT-Based Approaches

Chi et al. [6] proposed the usage of Complex Programming Logic Device (CPLD)-based core controller for the sensor interface in IoT platform. Salvadori [7] proposed the concept of Dynamic Power Management (DPM) in IoT scenario for the condition monitoring system.

2.3 *Fault Diagnosis-Based Approaches*

Lie et al. [8] reviewed and summarized the research publications regarding the conditional monitoring. The comprehensive review stated that the characteristics and unique behaviors of fault were identified and analyzed. Prieto et al. [9] proposed two types of methods in order to diagnosis the fault based on the feature behavior in conditional monitoring applications.

3 Real-Time Online Machine Condition Monitoring and Fault Detection

The fault detection is an essential process for preventing further damage to the machines in the automation industries. Hence, the real-time condition monitoring of the machine is necessary for prediction of fault, and diagnosing the fault classes helps to avoid the fault occurrence in future. This section discusses the implementation details of real-time online condition monitoring systems and the methodology used for the fault detection and classification using Industrial Wireless Sensor Network (IWSN) along with IoT service in the cloud platform. The Fig. 1 represents the structure of the proposed system.

The proposed system is classified into three subsystems: (1) Industrial Wireless Sensor Network System; (2) Motor Fault Diagnosing System; and (3) Cloud Platform System.

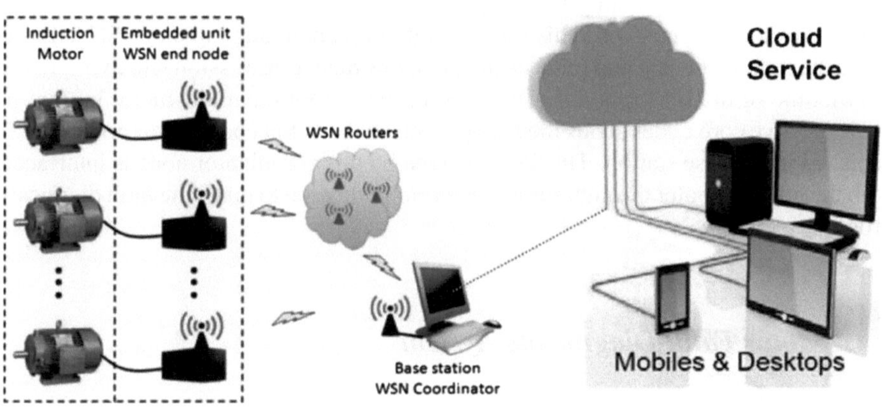

Fig. 1 Structure of the proposed system

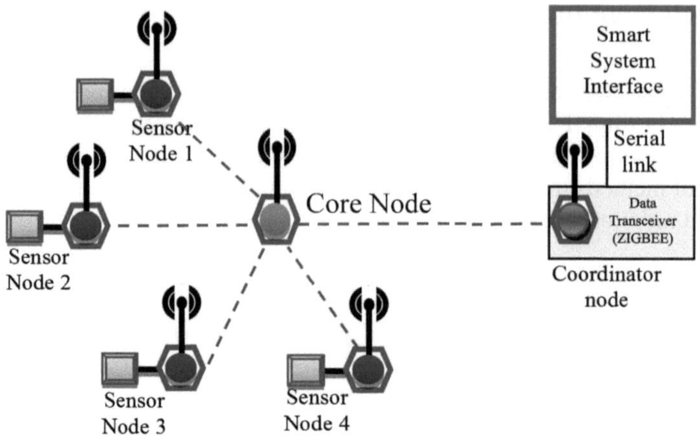

Fig. 2 Functional diagram of IWSN

3.1 Industrial Wireless Sensor Network System

The wireless communication technologies have become increasingly popular for emerging applications in the industrial automation domain. The industrial wireless network standard IEEE 802.15.4 (ZigBee) is utilized for this proposed system. The proposed IWSN functional diagram is shown in Fig. 2.

This system has four sensors of voltage, current, vibration, and speed. Among these sensor nodes, a core node is implemented, which is used to take care of the resource allocation, channel scheduling, and data fusion based on the priority condition of the data transmission. This functionality implementation in the WSN helps to avoid the redundant data and reduces the large raw data transmission which improves the lifetime of the sensor nodes. The data received from the multi-motor system in the respective core nodes sends the data packets to the WSN coordinator node, which is placed in the base station. The data received by this coordinator node is interfaced with the host computer through serial port communication to make the fault diagnosis process.

3.2 Motor Fault Diagnosing System

In this system, the data received from the above energy-efficient IWSN is processed for diagnosing the fault in the industrial motor systems. With the increase in number of features, the detection and classification of the faults are limited and hence, the dimensionality of the feature set needs to be reduced. The present work discusses the implementation details of proposed Minimal Relevant Feature-based Fault Classification (MRFC) using Class-Specific Support Vector Machine (CS-SVM) algorithm. The proposed system consists of the following process: preprocessing, features extraction, and classification of faults.

3.2.1 Preprocessing

The data set that denotes the locations of fault conditions in voltage, current, vibrations, and speed are based on the mechanical and electrical analysis with the historical data set. After the data acquisition process is carried out by the IWSN systems, the unfilled entries of the real-time input data set are removed by the preprocessing in the host computer.

3.2.2 Feature Extraction

The nature of the fault is either individual or combinational. In this work, the proposed system is analyzed the four parameters (voltage, current, speed, and vibration) of the motor under various fault conditions. Initially, the list of records for each class in the preprocessing section is made, and the features regarding the minimum and maximum are extracted as follows:

$$\begin{aligned} Ft_1 &= \min(L_{Vt}); \quad Ft_2 = \max(L_{Vt}); \\ Ft_3 &= \min(L_{Cr}); \quad Ft_4 = \max(L_{Cr}); \\ Ft_5 &= \min(L_{\mathrm{Rpm}}); \quad Ft_6 = \max(L_{\mathrm{Rpm}}); \\ Ft_7 &= \min(L_{Vb}); \quad Ft_8 = \max(L_{Vb}): \end{aligned}$$

With these features, the comparison of training input parameters regarding the fault diagnosis process classifies the fault.

3.2.3 Classification of Faults

The data to be classified or tested is initialized as Ts_n. The test values of fault conditions in voltage, current, vibration, and speed are regarded as Vt_T, Ct_T, Vb_T and rpm_T. The Class-Specific Support Vector Machine (CS-SVM) is modified in this section to classify the faults based on the relevant feature set extracted.

The modified formula for classification is expressed as:

$$Ts_C = \operatorname{argmin} \frac{1}{2} \|w\|^2 + C_n(\text{corresponding to fault}) - \sum_{j:y_j} \tau_j \tag{1}$$

where

w Weight corresponding to the feature set
C_n Class
τ Slack variable

The classification algorithm using MRFC method is shown below.

Classification Algorithm

For x = 1 to size of Ts_n
For y = 1 to Size of C_n
If $(Vt_T >= Ft_1 \&\& Vt_T <= Ft_2)$
If $(Ct_T >= Ft_3 \&\& Ct_T <= Ft_4)$
If $(\text{rpm}_T >= Ft_5 \&\& \text{rpm}_T <= Ft_6)$
If $(Vb_T >= Ft_7 \&\& Ct_T <= Ft_8)$
Then
$Ts_C = \text{argmin} \frac{1}{2} w^2 + C_n (\text{corresponding to fault}) - \sum_{j:y_j} \tau_j$
Else
Then $Ts_C = 0$
End if;
End for y;
End for x;

The performance analysis of fault diagnosis is presented in Fig. 3.

The simulation environment is constructed using MATLAB and validated with the fault data set (historical) and real-time data set based on motor specifications. The relevant min-max fault range-based feature extraction improved the specificity, accuracy, and precision by 12.5, 65.7, and 60.3%, respectively, compared with the SVM algorithm (shown in Fig. 3a, b). The accuracy of the fault classification obtained by the proposed learning algorithm is around 98%.

3.3 *Cloud Platform System*

As per the discussion in the previous subsystem, the performance analysis proved that the proposed MRFC machine-learning algorithm produced 98% of accuracy in classification of fault class. This light weighted learning algorithm is well suited for cloud platform to predict the fault of the real-time data under IoT scenario. The cloud vendors currently provide state-of-the-art services for IoT operations. In the cloud platform, the various services can be utilized based on the application requirements such as remote control, data analysis, and machine learning, etc.

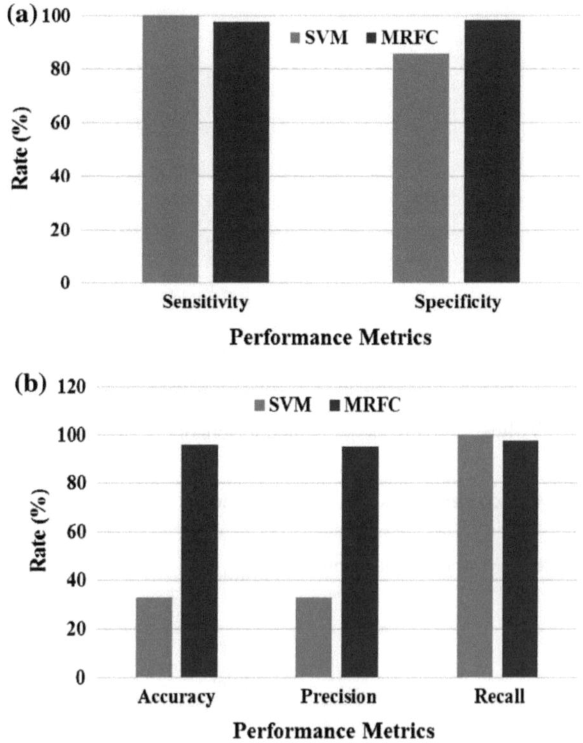

Fig. 3 **a**, **b** Performance analysis of fault diagnosis

3.3.1 Creation of Instances of Run-Time Environment in Cloud Platform

The IBM Bluemix is a cloud computing platform, which is used to implement the present work in the internet circumstance. The creation of the required instances of run-time environment in the IBM cloud platform is a foundation work of implementation, which is used to incorporating the IoT and data storage services and interlinked with each others. The Node-RED is a programming tool in IBM cloud platform, which is associated with the run-time application environment.

The present MRFC-based fault prediction algorithm developed by the Python script is deployed into the Node-RED flow editor function block. This function block is connected to the IoT service platform, which is receiving the real-time data from the sensor devices using MQTT protocol. The data received by the MQTT protocol is in the form of JSON message payload. This message payload is extracted in the function block using Python script, and insisted to the learning process and then prediction of the fault classes accordingly.

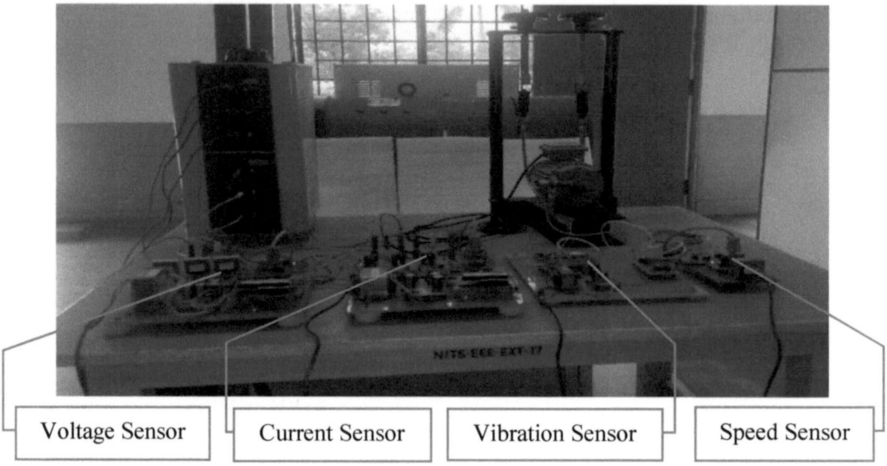

Fig. 4 Experimental setup (on sensor node)

Table 1 Motor specifications

Parameters	Value
Type	three-phase induction motor
Frequency	50 Hz $\pm$5%
Voltage	415 V (Y) $\pm$10%
Power	0.37 kW
Current	1.05 A
Speed	1380 rpm
Power factor	0.74

3.3.2 Implementation of IoT Service and Data Storage Service

The Internet of Things (IoT) service establishes the communication bridge between the sensor devices and the run-time environment application in the cloud platform. In the IoT platform, the interconnected devices need to communicate using lightweight protocol. In the present implementation, the MQTT protocol with JSON message payload format is used for the data communication.

4 Experimental Setup and Results

The experimental setup of on sensor node and WSN coordinator node is shown in Figs. 4 and 5, respectively. Table 1 shows the specification of motor used for the present work.

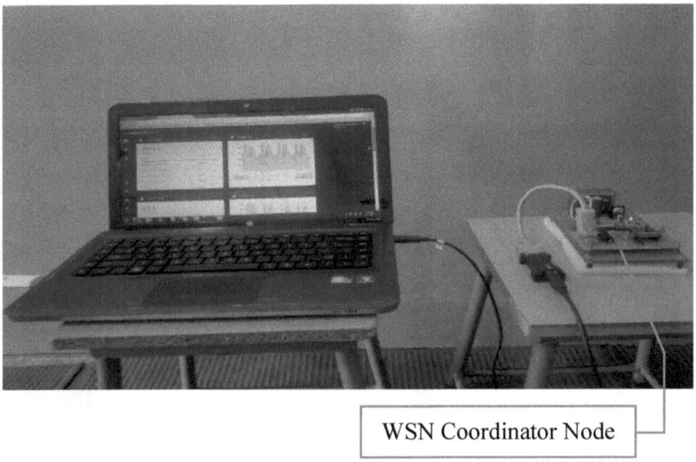

Fig. 5 Experimental setup (WSN coordinator node)

4.1 Sensors and Communication Devices

The various sensors and communication devices are used for implementation of the present work. A 230 V/9 V potential transformer along with the dual operational amplifier (TL082ACN) performed as voltage sensor. Hall Effect sensor (HE025T01) is used for measuring current. Vibration is measured by MEMS accelerometer (ADXL335) and IP18305DF is used to measure the speed of the motor. Zigbee is used as the wireless communication device and has a defined rate of 250 kb/s.

4.2 Experimental Result

The seven classes of fault conditions including normal operation condition are classified in the learning algorithm. The fault classes due to the individual and combined parameters are represented as: class 0—Normal Operation, class 1—voltage and current fault, class 2—vibration fault, class 3—speed fault, class 4—voltage–current–vibration fault, class 5—voltage–current–speed fault, class 6—vibration–speed fault, and class 7—voltage–current–vibration–speed fault.

The various fault conditions are injected into the motor based on above mentioned fault classes with respect to changes in the parameter values of voltage, current, vibration, and speed correspondingly. The experimental results proved that the MRFC algorithm exactly predicts the fault classes and normal operating class with respect to the fault data injected in the motor under real-time condition.

The fault classes and the real-time parameter values can be viewed and is plotted with respect to time in the IoT service platform of the cloud presented in Fig. 6.

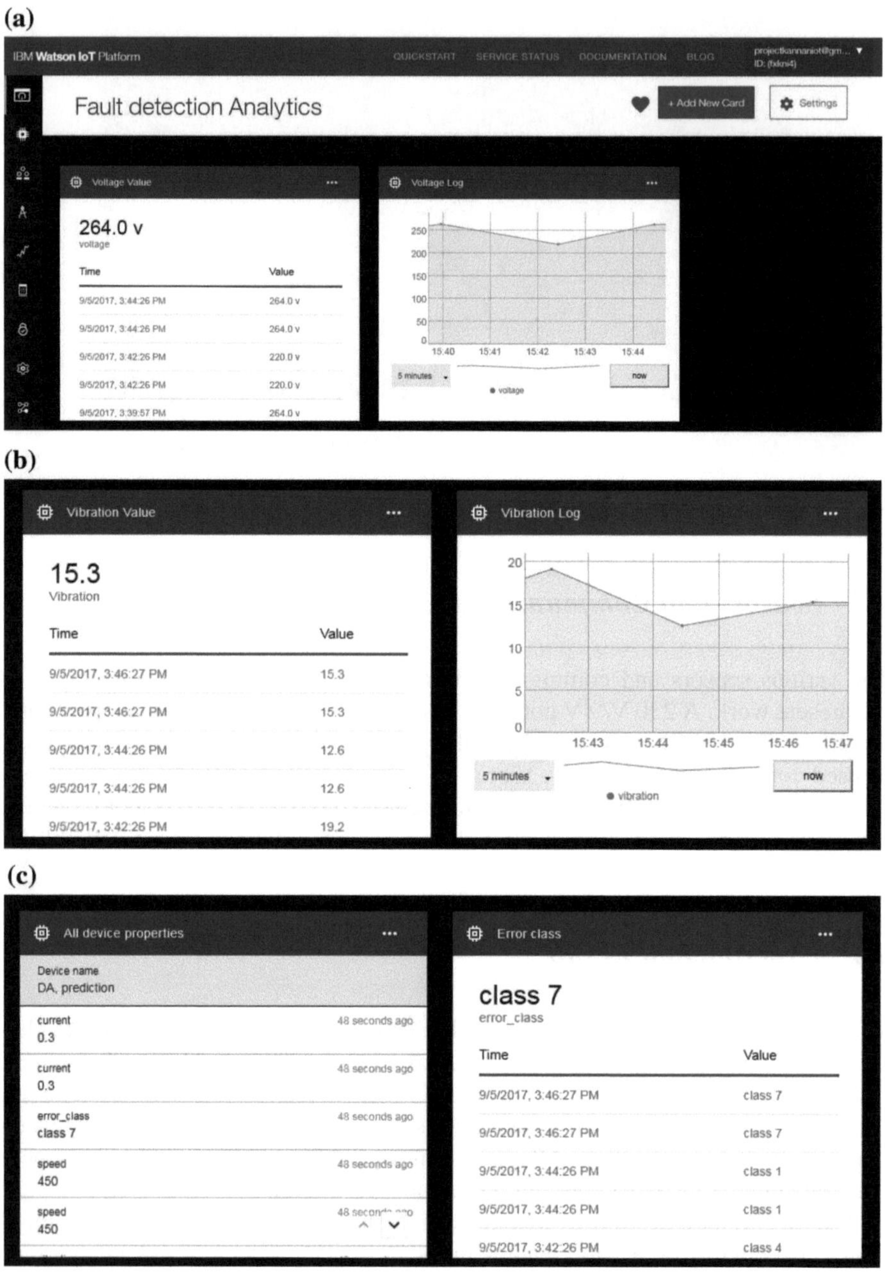

Fig. 6 **a–c** Online condition monitoring and fault prediction of real-time data

5 Conclusion

The fault detection is an essential process for preventing further damage to the machines in the automation industries. The simple control algorithm is developed for fault diagnosis of the induction motor under the IoT environment along with industrial wireless sensors network. The implementation of core node among the various sensor nodes improved the lifetime of each sensor nodes and minimized the data transmission delay by priority-based resource allocation and data fusion techniques. With the utilization of the above communication protocol, the fault diagnosis of the motor is achieved with MRFC algorithm. The performance analysis shows that the present MRFC algorithm improved the specificity, accuracy, and precision with the suitable feature dimensionality reduction technique. The advancement of minimum features extraction, the MRFC algorithm is deployed into the cloud platform using IoT service to predict the fault class. The output confirmed that various fault classes are predicted exactly with the minimum time interval. Hence, the entire process of the present implemented system improved the functionalities of the fault diagnosing of machines in the industries automation applications.

References

1. M. Shailaja, G. Nikkam, V.R. Pawar, Water parameter analysis for industrial application using IoT, in *Proceedings of 2nd International Conference on Applied and Theoretical Computing and Communication Technology* (*iCATccT*) (2016), pp. 703–707
2. J.A. Carino, Enhanced industrial machinery condition monitoring methodology based on novelty detection and multi-modal analysis. IEEE Access **4**, 7594–7604 (2016)
3. Z. Yin, J. Hou, Recent advances on SVM based fault diagnosis and process monitoring in complicated industrial processes. Neurocomputing **174**, 643–650 (2016)
4. B. Lu, C. Vehbi Gungor, Online and remote motor energy monitoring and fault diagnostics using wireless sensor networks. IEEE Trans. Ind. Electron. **56**, 4651–4659 (2009)
5. L. Hou, W. Bergmann Neil, Novel industrial wireless sensor networks for machine condition monitoring and fault diagnosis. IEEE Trans. Instrum. Meas. **61**, 2787–2798 (2012)
6. Q.P. Chi, H. Yan, C. Zhang, Z. Pang, L. Da Xu, A reconfigurable smart sensor interface for industrial WSN in IoT environment. IEEE Trans. Industr. Inf. **10**, 1417–1425 (2014)
7. F. Salvadori, Monitoring in industrial system using wireless sensor network with dynamic power management. IEEE Trans. Instrum. Meas. **58**, 3104–3111 (2009)
8. Y. Lei, J. Lin, M.J. Zuo, Z. He, Condition monitoring and fault diagnosis of planetary gearboxes: A review. Measurement **48**, 292–305 (2014)
9. M.D. Prieto, G. Cirrincione, A.G. Espinosa, J.A. Ortega, H. Henao, Bearing fault detection by a novel condition-monitoring scheme based on statistical-time features and neural networks. IEEE Trans. Industr. Electron. **60**, 3398–3407 (2013)

Smart Parking—A Wireless Sensor Networks Application Using IoT

Suman Balhwan, Deepali Gupta, Sonal and S. R. N. Reddy

Abstract A smart parking system is for the purpose of finding vacant parking slot in parking zones without involving manual efforts and thus avoidance of the need to spend fuel and time efforts. Various sensors deployed in the parking zone helps in determining the parking slot availability, and the information can be easily accessed using the internet by the users. In developed prototype of smart parking—a wireless sensor network application integrates various sensors, which senses if parking zone is free or occupied, and the information/data measured by the sensors is uploaded to the cloud from which the user can access the information using an Android app. The sensors data is also updated in the repository database at the master node, and some of the quality of service parameters are also evaluated based on the valuable data.

Keywords Wireless sensor network · Raspberry pi · IR sensor · Ultrasonic sensor · Parking system · Thing speak · Quality of service · Heterogeneous network · IoT

1 Introduction

To get parked is an important component in a vehicle's lifecycle. With the advent of technology and along the development of civilization, the population of vehicles kept on increasing at a very fast pace. But the parking infrastructure has not kept pace with the rate of increase of vehicles. It has led to congestion of vehicles on roads and also vehicles are parked in unauthorized spaces. A research about vehicular traffic has brought to light a startling fact that around 30% of the traffic congestion is due

S. Balhwan · D. Gupta (✉)
Mtech, ECE, IGDTUW, Delhi, India
e-mail: deepalig92@gmail.com

Sonal
ECE, IGDTUW, Delhi, India

S. R. N. Reddy
CSE, IGDTUW, Delhi, India

C. R. Krishna et al. (eds.), *Proceedings of 2nd International Conference on Communication, Computing and Networking*, Lecture Notes in Networks and Systems 46, https://doi.org/10.1007/978-981-13-1217-5_22

to the ignorance of the vehicle's driver about a nearby available parking space [1]. One major cause is the mismanagement of the parking spaces. This problem can be ameliorated to a significant extent using Wireless Sensor Network. It will lead to savings in terms of human efforts, fuel consumption, and optimum use of the parking space.

2 Smart Parking System

This system utilizes various techniques to efficiently manage the current parking system [1]. Earlier, variety of technologies have been used in mitigating problems which arise in parking [2] with communication elements like Bluetooth, Zigbee, GSM, GPS,Wi-Fi, RFID, and applications like Short Message Service (SMS), image processing, cloud-based server as well as Android with the help of SoCs like Arduino, Raspberry Pi [3]. Pros and Cons of different smart parking systems have been given in [4].

3 Proposed Work

This paper aims to introduce a smart parking system using Internet of Things (IoT), Raspberry Pi3 Board, IR sensors, and ultrasonic sensors. The information from the sensors will be sent wirelessly to the cloud using a processing module from where the users with the help of their Android smartphone app can determine about the availability of the parking space without the need to manually search for the same and hence, saving efforts and resources like fuel and time. So, we have tried to build a prototype which aims to notify the users about vacant parking slot using sensors, which senses the presence or the absence of vehicles and the measured information is then uploaded to a cloud application from where the connected users can access the sensed information via mobile app and thus can benefit from it (Fig. 1). Our prototype can be further expanded and realized for practical scenarios where the application can be developed, which senses information of several parking slots spread out geographically and depending on the location information of the mobile users, the relevant information for them could be provided.

The work carried out introduces a wireless sensor network for a smart parking system based on IoT platform which uses sensors to sense the data and a Raspberry Pi (RPi), which transports the data over Wi-Fi to the cloud. Also, an heterogeneous network is proposed to implement different Wireless Sensor Networks (WSNs) where the different wireless sensor node send their data to a master or centralized node implemented by use of a Raspberry Pi (RPi). Database is created at all nodes (master and slaves). Various Quality of Service (QoS) parameters are also evaluated for the proposed system.

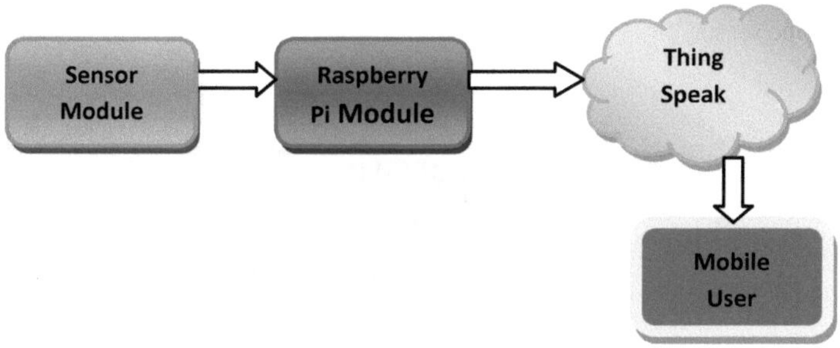

Fig. 1 Block diagram

Fig. 2 IR sensor

4 Architecture

4.1 Hardware Architecture

The hardware consists of the device used for the detection of vehicles in the parking slot, the communication system which interconnects it to the web server and the user end device for accessing this data. The hardware encompassing the above functionalities is composed of the Raspberry Pi module, IR sensor, and ultrasonic sensor.

4.1.1 Sensor Module: This module is installed in the parking place and there will be a sensor node for each parking space. The sensors used are—infrared sensor and ultrasonic sensor. Both these sensors are used for the same purpose, which is to sense the presence/absence of a vehicle in the parking slot and thereby determine the parking status of a parking slot, i.e., whether it is vacant or not.

4.1.1.1 Infrared Sensor: It is an electronic device which determines the presence/absence of an obstacle by sensing the light wavelength in its nearby area. It emits or detects infrared spectrum (Fig. 2). Using the IR sensors in each parking slot, the presence or absence of a vehicle is detected and accordingly, message is thus received by the user.

Specification of the IR sensor [5]:

(i) Operating Voltage: 5 V DC
(ii) Sensitivity up to 30 cm

Fig. 3 Ultrasonic sensor

4.1.1.2 Ultrasonic sensor: The ultrasonic sensor (Fig. 4) is placed in each parking slot. It can thus, tell about the occupancy of a parking slot by sending out sound waves at a frequency of 40 kHz. The sound waves will return back after encountering an obstacle. The distance between the sensor and the obstacle is calculated by measuring the round trip time of the sound waves and the speed of the sound waves. The calculated distance is then used to tell about the presence/absence of the vehicle in the parking slot.

Specification of ultrasonic sensor [6]:

(i) Operating Voltage: 5 V DC
(ii) Operating Current: 15 mA
(iii) Working Frequency: 40 Hz
(iv) Maximum Range: 4 m
(v) Minimum Range: 2 cm (Fig. 3)

4.1.2 Raspberry Pi module: The Raspberry Pi (RPi) is an ultra-low-cost and a very small-sized (size of a credit card) computer that can support a number of peripheral devices like keyboard, mouse, computer monitor, laptop, or TV [7]. RPi is used as an SoC (System on Chip) with the capability of having CPU, GPU, USB ports, audio, video, HDMI ports, and various other interfaces. It has slot for SD card. The OS image is burned on the SD card from which the Rpi boots. It has GPIO pins to which the sensors are connected and the processing is within the RPi. The RPi has inbuilt Wi-Fi module and Bluetooth module for the purpose of communication. Various programming languages can be used with RPi like Python (Fig. 4).

4.2 Software Architecture

The feasibility of hardware usage for all requirements is reinforced with the efficacy of the supporting software. The primary theme of "IoT" utilized in the implementation of this solution, makes the need for software a primary essentiality. The architecture of smart parking system has been shown in Fig. 5.

Fig. 4 Raspberry Pi board

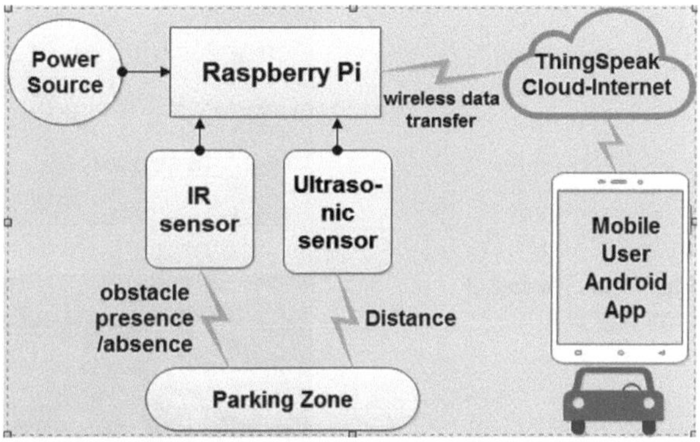

Fig. 5 Architecture of the smart parking system

4.2.1 The interfacing of the cited hardware with each other and the web server, and the accessing of data by the end user has been built upon a common programming platform. The Raspberry Pi module and the sensors communicate by means of the Python script run on the RPi.

4.2.2 ThingSpeak Cloud: It is used as the repository/database where the sensors data regarding the occupancy of the parking slot in the parking zone is uploaded. The information stored at ThingSpeak can be accessed using mobile app and by connecting to the channels where the information is displayed in graphical format. ThingSpeak is an IoT application to store, retrieve, and analyze data from sensors [8].

4.2.3 Mobile/Android module: This module is installed as an Android app in the user's phones and displays the status of parking lot by connecting to the ThingSpeak Cloud.

5 Implementation

Smart parking—a proposed IoT framework has been developed and implemented as shown in the Figs. 6, 7, 8, and 9 and the results have been carried out based on the developed environment.

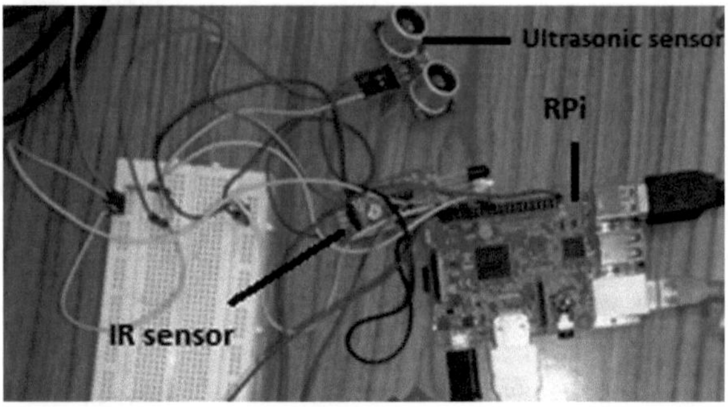

Fig. 6 Interfacing diagram of overall system

Fig. 7 Interfacing of raspberry pi

Fig. 8 Interfacing of IR sensor

Fig. 9 Interfacing of ultrasonic sensor

6 Mote Development

The mote has been developed and deployed using the 3D printing environment so that it can work out in every environment. The mote developed has been shown in the Fig. 10.

The occupancy of a parking space is measured using the sensors (ultrasonic Sensor and IR Sensor) interfaced with Raspberry Pi. The following steps are performed:

6.1 After measuring data, i.e., presence/absence of an obstacle in case of IR sensor and distance in case of ultrasonic sensor, the data is sent to "ThingSpeak" using internal Wi-Fi of RPi.

6.2 Registration is done on the ThingSpeak Cloud for uploading of the measured data. The data is displayed graphically.

6.3 A mobile app in the end user's mobile device then talks to the cloud by connecting to it and the information regarding the vacancy of the parking space is fetched on mobile app and hence is known to the user.

Fig. 10 Smart parking Mote

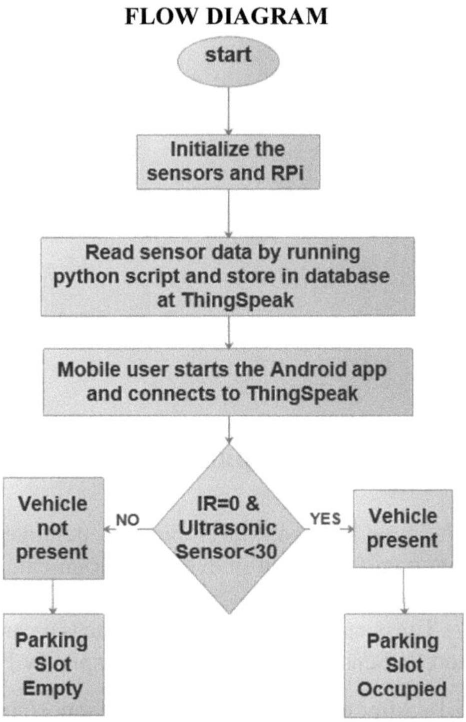

Fig. 11 Flow diagram of the smart parking system

Flow diagram for the developed smart parking system is shown in Fig. 11.

7 Results

The results have been evaluated on the display console as shown in the Fig. 12.

When the developed prototype will be on, it gives the measured distance in case of ultrasonic sensor in cm and the presence/absence of an object in case of IR sensor. The measured data from the sensors is sent to ThingSpeak (Fig. 13) from the Raspberry Pi using the Wi-Fi communication module. The data is displayed graphically for the two fields (distance and the obstacle's presence/absence) corresponding to the data sensed by the two sensors (Fig. 14).

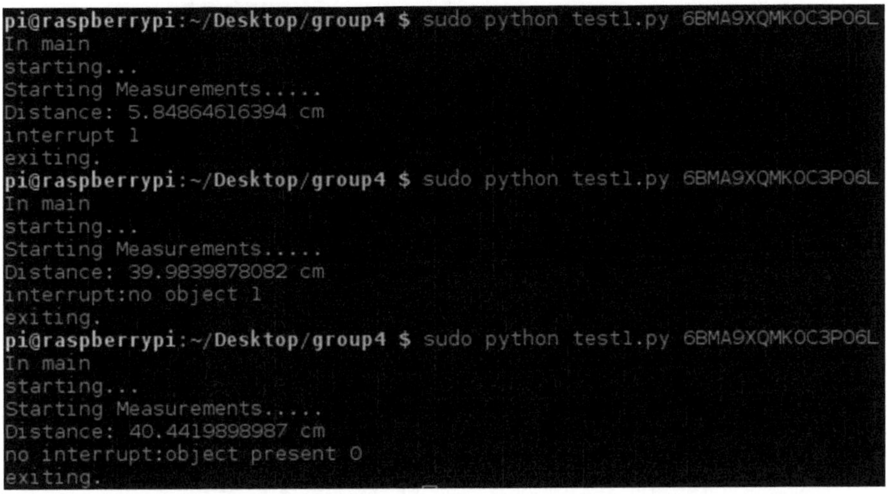

Fig. 12 Console showing output of sensors

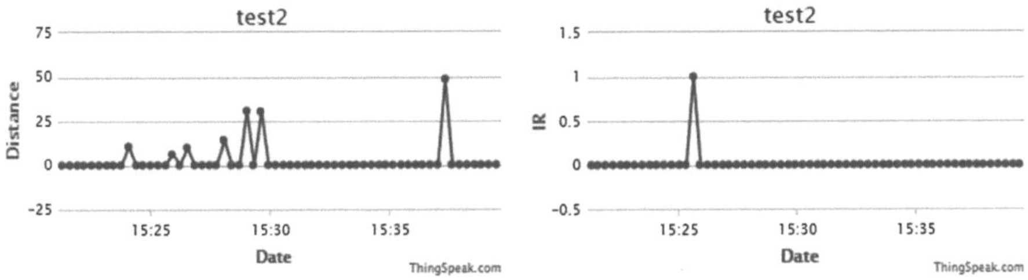

Fig. 13 Data of sensors from Raspberry Pi 3 to ThingSpeak

8 Heterogeneous Smart WSN

An heterogeneous WSN is made by connecting together different WSNs like smart parking system, LPG and fire detection system, intrusion detection system, and so on. One of the RPi is made as the master node and others as the slave nodes. The slave nodes communicate to the master RPi using inbuilt Wi-Fi.

The send and receive codes are written in Python script (Figs. 15 and 16).

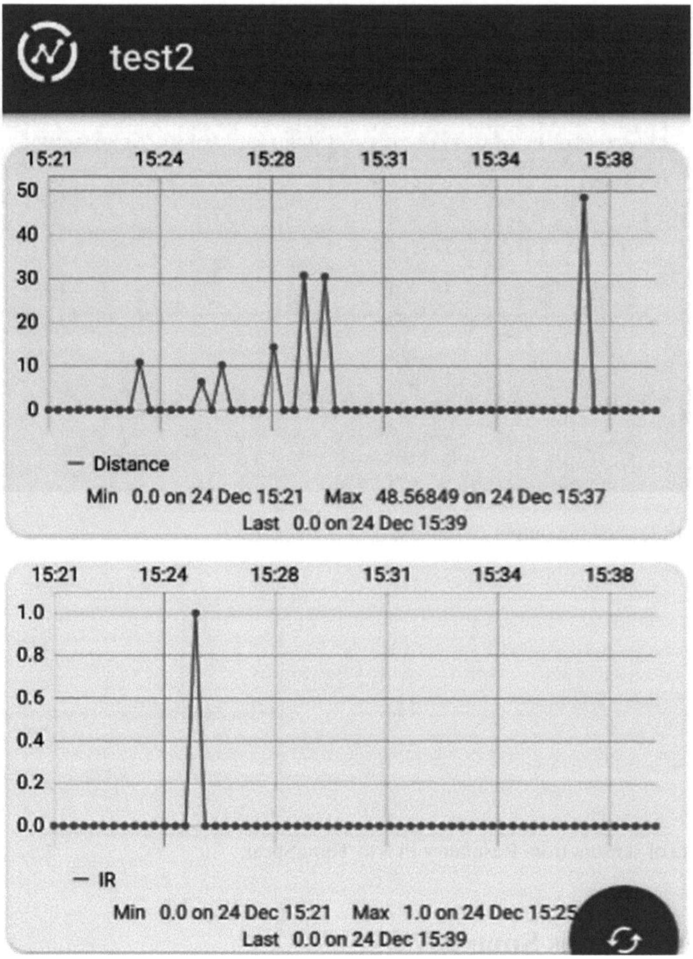

Fig. 14 Android app

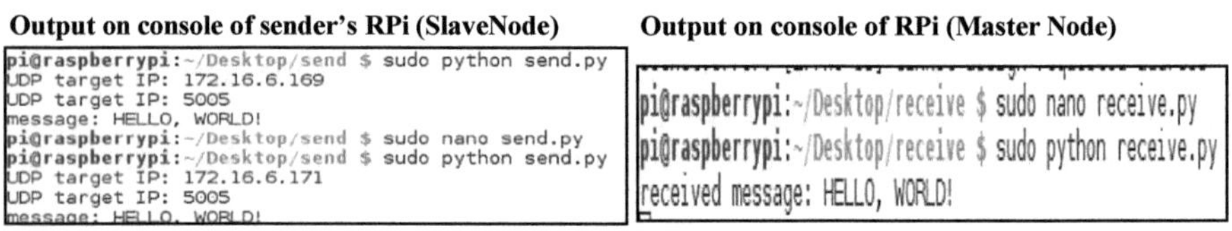

Fig. 15 Console showing execution of sample codes (slave and master node), output on console of sender's RPi (slave node), and output on console of RPi (master node)

```
pi@raspberrypi: ~
File  Edit  Tabs  Help
pi@raspberrypi ~ $ sudo apt-get install mysql-server python-mysqldb
Reading package lists... Done
Building dependency tree
Reading state information... Done
mysql-server is already the newest version.
python-mysqldb is already the newest version.
0 upgraded, 0 newly installed, 0 to remove and 5 not upgraded.
```

Fig. 16 Setting up MySQL server

```
mysql> select * from node;
+--------------------------------+-----------+----------+
| details                        | date1     | time1    |
+--------------------------------+-----------+----------+
| testdata                       | 2017-11-18 | 10:46:04 |
| Node 4: Distance= 3274.24883842 | 2017-11-18 | 10:53:24 |
| Node 4: Distance= 14.6195888519 | 2017-11-18 | 10:54:19 |
| Node 4: Distance= 3313.13037872 | 2017-11-18 | 10:54:22 |
| Node 4: Distance= 14.6195888519 | 2017-11-18 | 10:54:25 |
| Node 4: Distance= 3316.96867943 | 2017-11-18 | 10:54:28 |
| Node 4: Distance= 14.6358013153 | 2017-11-18 | 10:54:31 |
```

Fig. 17 Slave node data stored at slave node

```
pi@raspberrypi:~/Desktop $ sudo python receive.py
Welcome to SmartWSN
Node 1: Theft Detected, mail sent
receive.py:17: Warning: Data truncated for column 'Date' at row 1
  curs.execute("INSERT INTO wsnstatus(Status, Date, Time) VALUES(%s, now(),now())",data)
Node 1: Theft Detected, mail sent
Node 4: Distance= 3222.33247757
Node 4: Object Status0
Node 4: Distance= 3259.13476944
Node 4: Object Status0
Node 4: Distance= 3248.93712997
Node 4: Object Status0
Node 4: Distance= 3235.4888916
Node 4: Object Status0
Node 4: Distance= 3228.38377953
Node 4: Object Status0
Node 4: Distance= 3240.74578285
Node 4: Object Status0
```

Fig. 18 Slave node data at master node

8.1 Creation of Database and Storing the Readings of Multiple Sensor Nodes

The database is created using MySQL (Fig. 16) as MySQL is scalable, can be tuned easily, supports user management and permissions, and supports client–server architecture where the client accesses the database remotely.

MySQL is installed with Python bindings for MySQL.

The database table is created where the sensors readings from the client or slave node are stored along with date and time (Figs. 17 and 18).

9 Evaluation of QoS Parameters

Quality of Service (QoS) is the guarantee provided by a service provider, i.e., a quality measure that is provided with a condition, for example, maximum bandwidth, minimum delay/latency, etc. QoS parameters are met by the network while delivering data flow [9]. QoS is characterized by bandwidth, packet loss probability, end-to-end delay, jitter, throughput, etc. Depending upon the requirements, the applications may have varying QoS demands like real-time applications may tolerate packet loss, however, the latency/delay encountered should be limited. Similarly, applications involving banking transactions may tolerate delay but no packet loss. QoS is a term which can be defined differently depending upon whether it is required by an application or for a network [10].

QoS requirements in case of WSN is quite challenging due to the various issues encountered in WSN. The resource constraints like memory, bandwidth, and power constraints along with the decentralized operation of the WSN with capability of autonomous/self-configuration, the limited battery life and in some cases, the mobility of the nodes in the WSN make the QoS requirements of WSN different from those with the traditional. QoS in WSN is mostly application specific and depends on sensor node measurements, deployment of the sensor nodes like random, uniform, or grid, the various coverage scenarios of the sensor nodes and the quantity of the sensor nodes that are active. QoS metrics of a WSN are measured and given in [11].

In our paper, we have evaluated the delay encountered by the packets from the sender nodes to a centralized node in case of an heterogeneous network (Fig. 19). The packets are of a standard length format. The jitter received at the master node is also measured. The impact of the increase in the length of the packets on the delay and jitter was also observed (Fig. 20).

10 Conclusion and Future Work

We have designed and developed a wireless sensor node, i.e., a smart parking system using IoT, employing two sensors and a Raspberry Pi for detection of parking slot availability. Inbuilt Wi-Fi (IEEE standard 802.11) is used as the communication module. The data sensed by the sensors and processed by the RPi is uploaded on the cloud from where the user can determine the availability of the parking space using an Android app. The sensor node along with other sensor nodes is connected and thus, an heterogeneous wireless sensor network is created. One of the nodes is made as the master node and the rest acts as slave nodes and send their data to the master node. Various QoS parameters are observed and calculated when packets are transmitted from slave nodes and received at the master node in an ethernet connected heterogeneous network. Thus, the efficiency of the wireless sensor network is known.

In future, we would be evaluating further QoS parameters like minimum bandwidth consumption when dense traffic is communicated from all the nodes to the

Status	Date	Time
test	2017-11-12	06:46:04
Node 4: Distance= 3249.05467033	2017-11-12	06:49:16
Node 4: Object Status0	2017-11-12	06:49:16
Node 4: Distance= 3235.28623581	2017-11-12	06:49:19
Node 4: Object Status0	2017-11-12	06:49:19
Node 4: Distance= 3225.6155014	2017-11-12	06:49:22
Node 4: Object Status0	2017-11-12	06:49:22
Node 4: Distance= 3236.68456078	2017-11-12	06:49:26
Node 4: Object Status0	2017-11-12	06:49:26
Node 4: Distance= 3244.40169334	2017-11-12	06:53:07
Node 4: Object Status0	2017-11-12	06:53:07
Node 4: Distance= 3255.0573349	2017-11-12	06:53:11
Node 4: Object Status0	2017-11-12	06:53:11
Node 4: Distance= 3219.41018105	2017-11-12	06:53:14
Node 4: Object Status0	2017-11-12	06:53:14
Node 4: Distance= 3236.83047295	2017-11-12	06:53:17
Node 4: Object Status0	2017-11-12	06:53:17
Node 4: Distance= 3242.2413826	2017-11-12	06:53:20
Node 4: Object Status0	2017-11-12	06:53:20
Node 4: Distance= 3221.83799744	2017-11-12	06:53:23
Node 4: Object Status0	2017-11-12	06:53:23
Node 4: Distance= 3203.85432243	2017-11-12	06:53:27
Node 4: Object Status0	2017-11-12	06:53:27

Fig. 19 Delay in receiving packets from a particular sensor

Node 4: Object Status0	2017-11-12	06:53:07
Node 4: Distance= 3255.0573349	2017-11-12	06:53:11
Node 4: Object Status0	2017-11-12	06:53:11
Node 4: Distance= 3219.41018105	2017-11-12	06:53:14
Node 4: Object Status0	2017-11-12	06:53:14
Node 4: Distance= 3236.83047295	2017-11-12	06:53:17
Node 4: Object Status0	2017-11-12	06:53:17
Node 4: Distance= 3242.2413826	2017-11-12	06:53:20
Node 4: Object Status0	2017-11-12	06:53:20
Node 4: Distance= 3221.83799744	2017-11-12	06:53:23
Node 4: Object Status0	2017-11-12	06:53:23
Node 4: Distance= 3203.85432243	2017-11-12	06:53:27
Node 4: Object Status0	2017-11-12	06:53:27

Fig. 20 Delay and jitter observed

master node and the power consumption efficiency using sampling and packetization. Also, the security of the wireless sensor node in terms of the confidentiality of the data, integrity of the data, and the availability of the data when connected to the cloud which is in turn connected to the internet is to be considered, and steps to be taken to mitigate the various security attacks that are mounted by exploiting the vulnerabilities when WSN node is connected to the internet.

References

1. V.N. Chokshi, A survey: smart parking system using internet of things (IoT). Int. J. Adv. Eng. Res. Dev. **4**(4), 847–849 (2017)
2. I.F. Akyildiz, W.S. Sankarasubramaniam, Wireless sensor networks: a survey. Comput. Netw. **38**, 393–422 (2002)
3. A. Khanna, R. Anand, IoT based smart parking system, in *International Conference on Internet of Things and Applications (IOTA)*, Maharashtra Institute of Technology, Pune, India (2016)
4. N. Moses, Y.D. Chincholkar, Smart parking system for monitoring vacant parking. Int. J. Adv. Res. Comput. Commun. Eng. **5**(6), 717–720 (2016)
5. Ultrasonic Ranging Module HC—SR04. Retrieved (November 27, 2017) from www.micropik.com
6. IR Obstacle Sensor. Retrieved (November 27, 2017) from forum.researchdesignlab.com
7. Raspberry Pi. Retrieved (November 27, 2017) from www.raspberrypi.org/help/what-is-a-raspberry-pi/
8. IoT Analytics—ThingSpeak Internet of Things. Retrieved (November 27, 2017) from https://thingspeak.com/
9. E. Crawley et al., in *A Framework for QoS-Based Routing in the Internet RFC 2386*, Internet Eng. Task Force (1997). http://fttp.ietf.org/internet-draftsdraft-ietf-qosr-framework-02.txt
10. D. Chen, P.K. Varshney, QoS support in wireless sensor networks: a survey, in *Proceedings of the International Conference on Wireless Networks (ICWN 04)* (Las Vegas, 2004), pp. 227–233
11. J.E. Mbowe, G.S. Oreku, Quality of service in wireless sensor networks scientific research. Wireless Sens. Netw. **6**(2), 19–26 (2014). https://doi.org/10.4236/wsn.2014.62003

Part III
Wireless Communication

BATMAN: Blockchain-Based Aircraft Transmission Mobile Ad Hoc Network

Arushi Arora and Sumit Kr Yadav

Abstract Automatic Dependent Surveillance–Broadcast (ADS-B) systems is a recent aviation technology, which has established its standing in the domain of communication between aircrafts and has almost replaced the conventional radar system for transmission of messages. Various loopholes related to privacy and security has been found in ADS-B, by researchers such as message broadcast tampering. In this paper, we propose a methodology based on BATMAN algorithm, which secures the communication between aircrafts and ground stations. This framework for authentication is a novel network topology based on a decentralized blockchain. The authentication is done using a three-way handshake between On-Air Unit (OAU) and Ground Station (GS). The proposed technology involves a group formation of authenticated mobile nodes for communication using the distributed public ledger and further utilizing its security features. The impact of BATMAN algorithm is that it further secures AANET (Aircraft Ad-hoc Network) for ADS-B system.

Keywords ADS-B · Blockchain · ATC · Privacy · Security · Ledger · HashMaps

1 Introduction

The future of transportation system beholds Air Traffic Control (ATC) as its cornerstone. Hartsfield–Jackson Atlanta International Airport has been the world's busiest airport by passenger traffic since 1998 handling a total of 882,497 aircraft movements in 2015 [24]. This augmentation of the air traffic load is advancing at a tremendous pace. Automatic Dependent Surveillance–Broadcast (ADS-B) is a genre of Air Traffic Management and Control (ATM/ATC) Surveillance system. It is a technology that will soon replace that the conventional radar-based systems [18]. Along with

A. Arora (✉) · S. K. Yadav
Indira Gandhi Delhi Technical University for Women, Kashmere Gate, Delhi, India
e-mail: arushi1250@gmail.com

S. K. Yadav
e-mail: sumit@igdtuw.ac.in

C. R. Krishna et al. (eds.), *Proceedings of 2nd International Conference on Communication, Computing and Networking*, Lecture Notes in Networks and Systems 46, https://doi.org/10.1007/978-981-13-1217-5_23

ADS-B, there exists two broadcast protocols Flight Information Service (FIS-B) and Traffic Information Service (TIS-B), which are used to send the reply back to the On-Air Unit (OAU) from the Ground Station (GS). New threats are emerging along with this technology which forms an integral component of Next Generation Air Transportation System. There is no methodology for entity authentication which advents one of the serious peril. This does not safeguard immunization from illegitimate messages advancing from third parties or uncertified units making it hackable [4]. Also, these messages can be effortlessly tinkered as they lack the use of data signatures for authentication. One serious repercussion of message encryption dearth is eavesdropping. As a consequence of these momentous privacy and security vulnerabilities, messages can be easily spoofed putting many lives at risk [19]. For example, an aircraft can be forced to reroute its path when a spoofed message informs it of another aircraft flying nearby at the same altitude. Scrutinizing of these messages is possible by conferring the backup radar signal. But this approach manifests to be unfeasible as it consumes plenty of time and energy in a situation wherein a decision has to be made within seconds. Also, there have been some instances of hijacked emergency signals [14]. Federal Aviation Administration (FAA) has also asserted the need of the hour to avert hackers targeting aircraft and air traffic control systems [12]. Another safety alert was issued by Britain's Civil Aviation Authority [21]. Previously, an unknown voice broke into the communication system of US Air flight approaching Washington's Reagan National Airport putting the lives of passengers at risk. The message caused confusion for the pilot of this plane along with the pilots of two aircrafts in the airspace [13].

The blockchain is one of the most ingenious and promising technologies of the future which can overcome the above-stated encumbrances [6]. It is effectively replacing the current transaction system. The idea of blockchain was pioneered by a researcher with a pseudonym "Satoshi Namakato" who applied this technology for implementing the cryptocurrency—Bitcoin [15]. The technology is decentralized and uses a distributed public ledger in which blocks are encrypted and chained together in a chronological order. All the nodes in this framework have a copy of the public ledger and are connected on a peer-to-peer basis [8]. A pair of a public and private key is associated with each node of the network. A block, which is the basic building unit of the chain, stores transactions at a particular timestamp. It encompasses the transactions, its hash value, timestamp, a signature of the block, and nonce [3, 20]. The blockchain is a decentralized which is verifiable and transparent [17]. The records can be altered only if more than 51% of the nodes are in control of the hacker which is unsustainable. In [1, 2, 22, 23] algorithm based on machine learning were used to solve various existing real-world problems. The technology is autonomous and it maintains the anonymity of the sender and receiver in the transaction by utilizing public and private keys of the nodes [16].

In this paper, BATMAN, a scheme to secure communication between aircrafts and ground stations is proposed. Using a public ledger, i.e., a blockchain, authenticated nodes form a group and are allowed for communication. BATMAN prevents any foreign entity which is not a part of that group, to send false messages or to tamper them. The security features of blockchain are further utilized in this mobile network.

The approach is discussed in detail under Sect. 3 of this paper. In Sect. 2 below, we discuss the related work.

2 Related Work

The blockchain is one of the most propitious technologies of the future. In its germinal state itself, the technology has successfully replaced economic transaction system in various organizations. The technology beholds the power to revolutionize the way data is stored or exchanged using an incorruptible public ledger. In addition to its salient features which include immutability, validation, decentralization, and transparency, blockchain promises to provide privacy and security at all points of time. Automatic Dependent Surveillance–Broadcast (ADS-B) systems is a recent aviation technology, which has established its standing in the domain of communication between aircrafts and has almost replaced the conventional radar system for transmission of messages [10]. The ADS-B system uses the standard Global Navigation Satellite System (GNSS) and the traditional communication system between the aircraft and the fixed reference, i.e., Air Traffic Control (ATC). These figures are the broadcasted to the open space. Other ADS-B enabled and the ground ATC are capable to use this information for traffic control in real time. There are various ambiguities that come into consideration in a full-scale implementation of ADS-B. The loopholes that exist are defined in the research paper [25]. This study aims at discovering the drawbacks of implementation of ADS-B and raising awareness regarding the security to oscillate between the technological advancement and the increasing threat magnitude. There are various ambiguities that come into consideration in a full-scale implementation of ADS-B. The paper [5] explains the breakdown of ADS-B protocol in terms of security and attacks on these devices. Ground reality of ADS-B with all its pros and cons is described in [19], but it could not suggest a feasible solution to the problems risen by the ADS-B. In this paper, we propose a synthesis of ADS-B technology and the blockchain protocol so as to overcome all the drawbacks of the ADS-B technique. A blockchain protocol can be defined as a continually expanding of the chain of blocks which are linked and secured with the ciphertext [9, 11]. Each block contains transaction information and is linked to a hash function of the previous block. The blockchain protocol finds its implementation presently in the Bitcoin concept which is a peer-to-peer transaction of digital money. The paper [7] explains how blockchain is implemented.

3 Proposed Work

In this section, we propose a scheme BATMAN, to secure the broadcast of ADS-B messages using blockchain. The concept is explained in detail using three algorithms where all nodes of the group are interconnected on a P2P basis. In the first algorithm, the aircraft registers with the Air Traffic Control (ATC) for the first time. The aircraft submits its identification details, such as FNo_i Flight no, Airlines ID AID_i, Electronic License No., i.e., ELN_i, Country ID, i.e., CID_i, and other required identification details. The ATC assigns a Pseudo ID (PID_i), Public–Private key pair, namely PK_i and SK_i to the aircraft and stores the mapping of the actual identity with the assigned PID_i, and the corresponding public key for that PID_i.

This entry forms a transaction of the $Block_i$ digitally signed by the ATC and added to the identity ledger β with other necessary descriptors. The hash is computed and result forms the HPrev for the next block. $Block_i$ is uploaded to ledger β. The issued PID_i, a copy of the identification ledger entry in β and a pointer PTR_i to the $Block_i$ along with the Private Key SK_i for the issued PID_i are stored with GS. The structure of public ledger β is shown in Fig. 1. Table 1 lists the keys used in the algorithms.

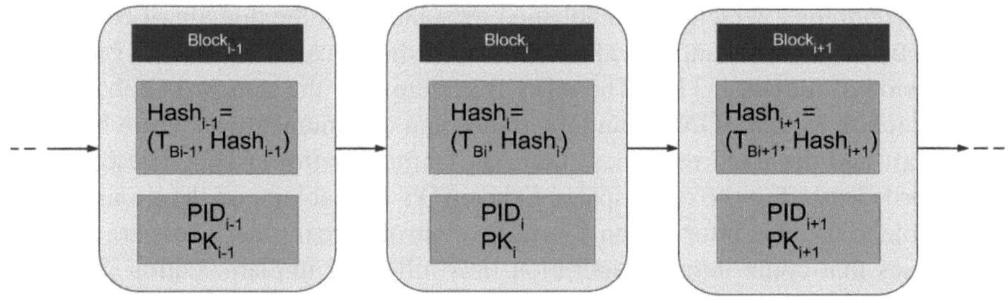

Fig. 1 The structure of public ledger β

Table 1 Keys used in the algorithm

Key	Description	Key	Description
OAU_i	ith On-Aircraft Unit	PID_i	Pseudo ID assigned by ATC
GS	Ground Station	DS_{ATC}	Digital Signature of ATC
FNo_i	Flight Number of OAU_i	H()	Hashing function
AID_i	Airlines ID of OAU_i	$Block_i$	Block of the public ledger
G_k	Group public key	$HPrev_i$	Hash of the previous block
CER_{GS}	Certificate given to GS by the ATC	T_{Bi}	Transaction added to the ith block
ID_i	Original Identity given to the ATC	PTR_i	Pointer to the ledger copy
G_m	Group of nodes	CID_i	Country ID of OAU_i
G_n	Group of nodes	ID_{GS}	ID of Ground Station
SK_i	Private key of ith OAU	PK_i	Public key of ith OAU
S_{GS}	Private Key of the GS	DS_{GS}	Digital Signature of GS
B	Public Ledger for authentication; resides with all GSs	ELN_i	Electronic License Number of OAU_i
HM	HashMap	M_i	Message
K_{AOUi}	Key stored in AOU_i	GS_k	kth Ground Station

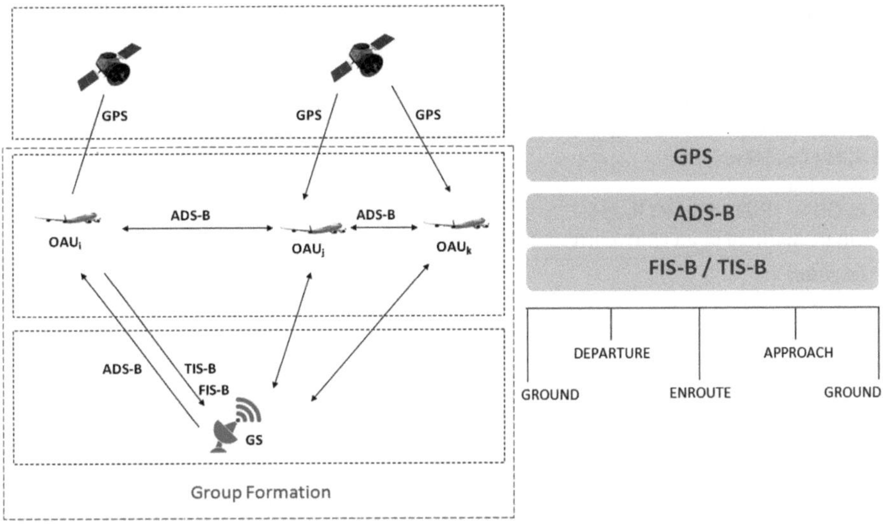

Fig. 2 Group formation after authentication of AOUs

The second algorithm explains the authentication of OAU with GS when they encounter for the first time. When the OAU_i is on the ground (before the flight), it authenticates itself with the GS with a three-way handshake and becomes part of the group of aircrafts in range of the GS. For this, some initial formalities are performed. The OAU_i receives a message containing the GS's certificate CER_{GS} from an in-range GS, it sends a three-tuple message M_i containing its PID_i, ledger copy, and pointer details encrypted with ID_{GS}. The GS upon receiving the message decrypts it using its private key S_{GS}, and gets the PID_i, corresponding to the ledger entry and pointer to the block. It searches for the block decrypts the transaction entry with PK_{ATC} and obtains the corresponding public key PK_i. To verify the identity of the OAU_i, as it is who it claims to be, it encrypts a challenge using the PK_i which can only be decrypted if it has the corresponding private key SK_i. When the OAU_i is able to decrypt using the SK_i, now its sends a response to the challenge after which it completes the mutual authentication step. Upon confirming the identity and authentication of the AOU_i, it now stores the PID_i and other necessary details, in a HashMap to convey to other GSs for authenticating this PID to avoid the above steps for authentication. It simultaneously sends the copy of mapping entry digitally signed with its private key S_{GS} which gets stored in AOU_i. Also, GS sends the group public key G_k, which the GS uses to broadcast messages to G_n. The group formation is shown in Fig. 2.

The OAU will encounter various GS right from its departure to landing. In order to tackle the situation of repeated authentication with simultaneously changing GSs after a period of time, we define the third algorithm. When the aircraft moves from the coverage of one GS to another GS, it need not perform the above handshake as it has already been authenticated by one GS. As the aircraft AOU_i enters the coverage

Algorithm 1:	*Registration of the aircraft (AOUi)*
Input: Aircraft's identification information, public ledger β	
1.1. $AOU_i \rightarrow ATC\{Details: FNo_i, AID_i, ELN_i, CID_i\}$	
1.2. $MAP(ID_{i(ATC)}, \{(PID_i, PK_i)DS_{ATC}\})$	
1.3. T_{Bi}: $\{(PID_i, PK_i$, other descriptors$)DS_{ATC}\}$	
1.4. $H (T_{Bi}, HPrev)$	
1.5. $Upload(T_i, \beta)$	
1.6. $GS \leftarrow \{PID_i, SK_i, PTR_i, \beta\}$	
Output: Addition of an entry corresponding to a new aircraft in authentication ledger β	

Algorithm 2:	*Authentication: OAUs first encounter with a GS*
Input: Aircraft's identification information, public ledger β	
2.1. $GS \rightarrow OAU$: $\{CER_{GSi}\}$	
2.2. $M_i = \{(PID_i, \beta, ID_{GS}, PTR_i)\}$	
2.3. Encrypt (ID_{GS}, M_i)	
2.4. $OAU \rightarrow GS$: $\{M_i\}$	
2.5. $GS \rightarrow Decrypt(S_{GS})$	
2.6. $GS \rightarrow Search(\beta, PK_{ATC})$	
2.7. $GS \rightarrow Encrypt(PK_i$, Challenge$)$	
2.8. $OAU \rightarrow Decrypt(SK_i$, Challenge$)$	
2.9. $K_{AOUi} = DS_{GS} (S_G)$	
2.10. $HM = (PID_i, K_{AOUi})$	
2.11. $AOU_i \rightarrow Store (K_{AOUi})$	
2.12. $GS \rightarrow Store (PID_i)$	
2.13. $GS \rightarrow AOU_i (G_k)$	
2.14. $G_n = G_n \, U \, (OAU_i)$	
Output: First time authentication AOU_i and membership with Ground Station GS	

Algorithm 3:	*Authentication with changing GS (GS_k)*
Input: Aircraft's identification information, public ledger β	
3.1. $OAU_i \rightarrow \{(PID_i, ID_{GSi}, CER_{GSi}) DS_{GSi}\}$	
3.2. $G_m = G_m \, U \, (OAU_i)$	
3.3. $GS_k \rightarrow AOU_i (G_k)$	
Output: Authentication AOU_i and membership with changing Ground Station GS	

of another GS, it sends the authentication copy by previous GS, which is digitally signed by the previous GS. Since all GSs are assumed to form a network, thus upon receiving a network joining request from PID_i, it just verifies with previous GS, which searches the HashMap and returns a true value for this PID_i, which is then made part of the group in-range G_m of new GS_k and given new Group key G_k.

4 Conclusion and Future Work

As there has been always debate and news related to the security of communication link in the discipline of aviation, it becomes evident that there is a need for more secure forms of communication. ADS-B is the latest technique developed to overcome the conventional radar communication, though it has its own share of disadvantages and loose ends. Blockchain protocol was initially developed to ensure security in the P2P transfer of Bitcoins, but its startling performance has opened doors for implementation in relatively less secure areas. In this paper, we proposed a scheme BATMAN, to secure the broadcast of ADS-B messages using blockchain technology. Formation of a group by allowing authenticated nodes to enter secures the network from various malicious activities like tampering with messages or sending false signals. Various security issues related to ADS-B can be successfully overcome using BATMAN. From the above-proposed algorithm, we can conclude that an approval and implementation of ADS-B technique along with the infusion of blockchain concept for security management can lead to a significant measure of security and safety. Blockchain indeed is the technology of the future. A lot of financial multinationals and banks are working to handle transactions using blockchains. The blockchain technology is on its way reinventing the way we work and live. It eliminates the use of centralized devices in IoT (Internet of Things), cybersecurity, and networking. The technology provides a new way of managing trust and can be effectively applied in insurance and domains like finance. It eliminates the involvement of third party, hence finding its effective utilization in private transport and ride sharing.

References

1. Y. Ahuja, S.K. Yadav, Multiclass classification and support vector machine. Global J. Comput. Sci. Technol. Interdisc. **12**(11), 14–20 (2012)
2. F.S Al-Anzi, S.K. Yadav, J. Soni, Cloud computing: security model comprising governance, risk management and compliance, in*2014 International Conference on Data Mining and Intelligent Computing (ICDMIC)* (2014, September), pp. 1–6
3. G. Brambilla, M. Amoretti, F. Zanichelli, *Using Block Chain for Peer-to-Peer Proof-of-Location*. Arxiv Preprint arXiv:1607.00174
4. Cable News Network (CNN), http://edition.cnn.com/2012/07/26/tech/web/air-traffic-control-security/index.html
5. A. Costin, A. Francillon, Ghost in the air (Traffic): on insecurity of ADS-B protocol and practical attacks on ADS-B devices.*Black Hat* USA, 1–12 (2012)
6. A. Dorri, M. Steger, S.S. Kanhere, R. Jurdak, *BlockChain: A Distributed Solution to Automotive Security and Privacy*. Arxiv Preprint arXiv:1704.00073 (2017)
7. I. Eyal, A.E. Gencer, E.G. Sirer, R. Van Renesse, *Bitcoin-NG: a scalable blockchain protocol* (NSDI, New Delhi, 2016), pp. 45–59
8. M. Fukumitsu, S. Hasegawa, J. Iwazaki, M. Sakai, D. Takahashi, A proposal of a secure P2P-type storage scheme by using the secret sharing and the blockchain, in *IEEE 31st International Conference on Advanced Information Networking and Applications (AINA)* (2017, March), pp. 803–810
9. A. Kosba, A. Miller, E. Shi, Z. Wen, C. Papamanthou, Hawk: the blockchain model of cryptography and privacy-preserving smart contracts, in *IEEE Symposium on Security and Privacy (SP)* (2016, May), pp. 839–858
10. E.A. Lester, Benefits and Incentives for ADS-B Equipage in the National Airspace System, Doctoral dissertation, Massachusetts Institute of Technology, 2007
11. X. Li, P. Jiang, T. Chen, X. Luo, Q. Wen, A survey on the security of blockchain systems. Future Gener. Comput. Syst. (2017)
12. D. McCallie, J. Butts, R. Mills, Security analysis of the ADS-B implementation in the next generation air transportation system. Int. J. Crit. Infrastruct. Prot. **4**(2), 78–87 (2011)

13. J.A. McCartin, *Collision Course: Ronald Reagan, the Air Traffic Controllers, and the Strike That Changed America* (Oxford University Press, Oxford, 2011)
14. N. Moran, G. De Vynck, in *Westjet Hijack Signal Called False alarm*. Bloomberg (Jan 2015) [Online]
15. S. Nakamoto, in *Bitcoin: A Peer-to-Peer Electronic Cash System* (2008)
16. M. Pilkington, in *Blockchain Technology: Principles and Applications* (2015)
17. C.L. Reyes, in *Moving Beyond Bitcoin to an Endogenous Theory of Decentralized Ledger Technology Regulation: An Initial Proposal* (2016)
18. R. Sittler, J. Ru, S. Sivananthan, D. Vader, A new algorithm of radar to ADS-B registration. Air Traffic Control Q. **19**(3), 191 (2011)
19. M. Strohmeier, M. Schäfer, R. Pinheiro, V. Lenders, I. Martinovic, On Perception and Reality in Wireless Air Traffic Communication Security. IEEE Trans. Intell. Transp. Syst. **18**(6), 1338–1357 (2017)
20. M. Swan, *Blockchain: Blueprint for a New Economy* (O'Reilly Media Inc, Newton, 2015)
21. K. Sweet, *Aviation and Airport Security: Terrorism and Safety Concerns* (CRC Press, Boca Raton, 2008)
22. D.K. Tayal, S.K. Yadav, Sentiment analysis on social campaign "Swachh Bharat Abhiyan" using unigram method. AI Soc. **32**(4), 633–645 (2017)
23. D.K. Tayal, S. Yadav, K. Gupta, B. Rajput, K. Kumari, Polarity detection of sarcastic political tweets, in *2014 International Conference on Computing for Sustainable Global Development (INDIACom)* (2014, March), pp. 625–628
24. A. Unal, Y. Hu, M.E. Chang, M.T. Odman, A.G. Russell, Airport related emissions and impacts on air quality: Application to the Atlanta International Airport. Atmos. Environ. **39**(32), 5787–5798 (2005)
25. M. Wolf, M. Minzlaff, M. Moser, Information technology security threats to modern e-enabled aircraft: A cautionary note. J. Aerosp. Inf. Syst. **11**(7), 447–457 (2014)

MATLAB Simulink Modeling for Spectrum Sensing in Cognitive Radio Networks

Reena Rathee Jaglan, Rashid Mustafa and Sunil Agrawal

Abstract A simulation model of spectrum sensing detector based on energy is developed using MATLAB Simulink in this paper. The model is designed considering optimal threshold and Signal-to-Noise Ratio (SNR) conditions. Users own block is designed for optimal threshold calculation, where Matlab algorithm is written in the background (MATLAB editor window). Real and estimated Primary User (PU) activities at different time intervals are investigated and tabulated from the simulation results. Further, the designed model is extended for Cooperative Spectrum Sensing (CSS).

Keywords Spectrum sensing (SS) · Cognitive radio network (CRN)
Probability of detection (Pd)

1 Introduction

Research in wireless communication has become an increasingly important topic in twenty-first century with aggravated demand for high data rates, electronic, and communication gadgets across the globe. However the available spectrum is limited due to the static assignment policy. It has been observed by the telecom regulatory bodies that the spectrum has been greatly underutilized most of the time [1].

CR is an existing technology that has sparked intense interest in recent years among research communal. It has a profound impact in wireless communication with the increase in competition for usage of available spectrum. Finding the vacant spectrum band and its usage by SUs while providing adequate protection to PUs is the chief concept of this technology [2, 3]. Thus, Spectrum Sensing (SS) plays a key role. It is one of the most important functionalities of CR. Research has been done on SS techniques by various researchers [4, 5] studying pros and cons of each technique. However, Energy Detection (ED) technique is used in this paper due to its simplicity

R. R. Jaglan (✉) · R. Mustafa · S. Agrawal
Department of ECE, UIET, Panjab University, Chandigarh 160014, India
e-mail: reenarathee5@gmail.com

C. R. Krishna et al. (eds.), *Proceedings of 2nd International Conference on Communication, Computing and Networking*, Lecture Notes in Networks and Systems 46, https://doi.org/10.1007/978-981-13-1217-5_24

and ease. Moreover, it is the most commonly used and popular technique with no prior information requirement of PU. Liang et al. [6] in formulated sensing-tradeoff problem using this technique. In CRNs, main focus is to protect PU, i.e., maximize Pd with constraint on Pf. The authors' further investigated tradeoff problem for low SNR regimes of-15 dB.

An improved ED spectrum sensing scheme based on past history of sensing results (average of test statistics) while preserving same level complexity and application as that of conventional scheme has been proposed by M. L. Benitez and F. Casadevall in [7]. The authors evaluated performance improvement and observed lesser time between two consecutive sensing events for same Pd and Pf. There are three main factors of sensing performance, i.e., channel between PU transmitter and SU, sensing time and detection threshold that have been considered in [8]. The expression for average Pd is derived which is further used by authors' to optimize detection threshold in terms of SNR. However, the local SS is deteriorated with practical wireless phenomenon like multipath fading and shadowing. Hence, an active technique for improving detection performance by exploiting spatial diversity comes into consideration and has been beautifully presented by I. F. Akyildiz in [9]. In practice, detection performance needs consideration of these practical phenomenon and Cooperative Spectrum Sensing (CSS) takes the power of mitigating these effects.

In this presented paper, simulation model in ED is developed and studied using MATLAB Simulink environment, taking into account optimal threshold and SNR conditions. These are important parameters which needs to be taken care for efficient sensing unlike [6, 7]. Further, the designed model has been extended to CSS with cooperation of five SUs. The paper deals with PU activities at different time intervals for these techniques.

2 The Energy Detection Study

Signal detection in CRN is a binary hypothesis problem and can be represented

$$Y(n) = W(n) : H_0. \tag{1}$$

$$Y(n) = hX(n) + W(n) : H_1, \tag{2}$$

where $Y(n)$ is signal samples received at the detector deciding presence or absence of PU, $W(n)$ is AWGN, h is the channel gain, $X(n)$ is PU signal samples at the detector. The detector has to choose one of the hypothesis depending on the test statistics. Energy of the sensed signal is estimated, termed as test statistics using [3].

$$T(Y) = \frac{1}{N}\sum_{n=1}^{N}(Y[n])^2. \tag{3}$$

where $N = f_s$ is sample size,is sample time and f_s is sampling frequency. $T(Y)$ is compared with λ to conclude a decision. The threshold [6] and decision statistics can be given as

$$\lambda = \frac{\frac{Q^{-1}(Pf)}{\sqrt{N}} + 1}{\text{SNR}}. \tag{4}$$

$$Z = \begin{cases} T(Y) \leq \lambda : H_0 \\ T(Y) > \lambda : H_1 \end{cases}. \tag{5}$$

Each SU will either detect presence of PU signal or will detect its absence. If decision statistics, i.e., Z is larger than λ PU exists else does not exist.

3 Simulink

Simulink has been used by numerous researchers for designing and simulating models of multidomains. It is assimilated with Matlab, allowing incorporation of MATLAB algorithms into models [10, 11].

3.1 Vital Subsystem for Spectrum Sensing

The hierarchy of a model includes subsystems. A subsystem contains a subset of blocks within the main model. The subsystems that are vital for SS Simulink model are PU and SU signal [12, 13] (Fig. 1).

White Gaussian noise is added to signal by AWGN channel. The input signal may be real or complex and accordingly produces output signal. S-function blocks (User-Defined Functions block) have been used for spectrum energy and optimal threshold calculation. Matlab algorithm has been written in these blocks and has been

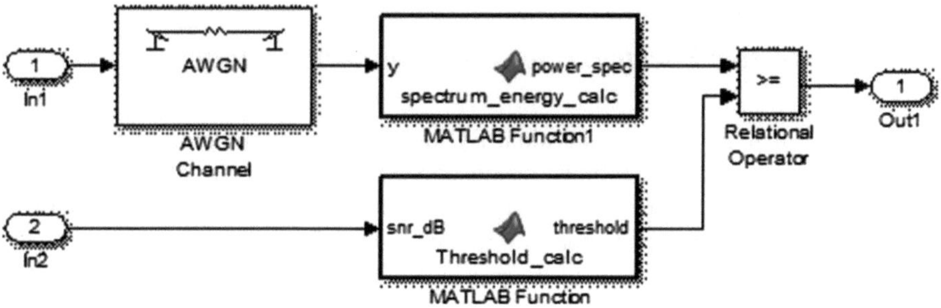

Fig. 1 Subsystem for SU signal

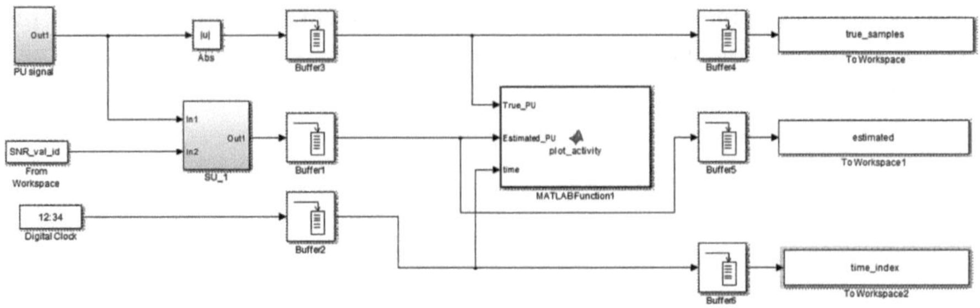

Fig. 2 Simulink model for ED technique

further added to the Simulink model. This is how own function blocks are created in Simulink with MATLAB code in the background. S-functions are compiled as MEX files using mex utility. Further, relational operator block has been used to compare signal's energy with threshold.

3.2 Energy Detection Simulation Model

The model has been developed using blocks of Simulink and by creating subsystems as shown in Fig. 3. Digital clock block is needed when current simulation time is required within a discrete system. The sample time needs to be adjusted. The Abs block outputs absolute value of input. The Buffer block adjusts input sequence to smaller or larger frame size, frame-based processing is performed. Input's data of each column is redistributed producing different frame size's output. Slower frame rate output means an input signal is buffered to a larger frame size or vice versa (Fig. 2).

In ED, sampling frequency needs to be at least twice the center frequency in order to satisfy Nyquist criteria. Here, real and estimated activity of PU have been determined by using user-defined block.

3.3 Cooperative Spectrum Sensing Simulation Model

Performance of local SS by individual SU degrades in attendance of fading and shadowing. Thus, researchers have proposed CSS to combat the effects of fading and shadowing [14, 15] in practical wireless scenarios. In his Simulink model five SUs have been taken into consideration to conclude a global decision.

Here, individual decision about the presence/absence of PU is taken by each SU while the global or final decision is taken at the fusion center. In this paper, AND

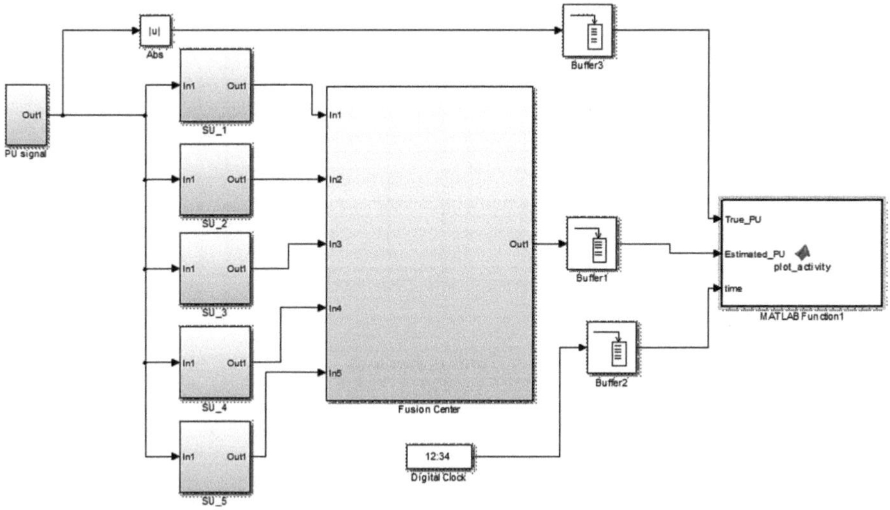

Fig. 3 Simulink model for CSS

Table 1 PU real and estimated activity at 10 dB SNR for ED

Time (samples) [$\times 10^{-6}$]	0	0.5	1	2	2.5
Real PU activity	1	1	0	1	1
Estimated PU activity	1	1	0	1	1

decision rule has been used at the fusion center. CSS takes into account spatial characteristics of each SU.

4 Results and Investigations

The simulation results shown below quantify the real and estimated PU activity for the used techniques, i.e., ED and CSS. The Simulink models have been designed considering optimal threshold and SNR conditions. The PU activity have been modeled randomly as ON/OFF activity corresponding to 1 (presence) and 0 (absence) respectively.

The white gaps within blue bands is when the algorithm missed to detect PU activity due to noise. The performance has been evaluated at 10 dB SNR in Figs. 4 and 5.

The results are fine as ED performs well for high and moderate SNR conditions as shown in Table 1. So, low SNR condition has been taken for further investigation and the results have been demonstrated in Figs. 6, 7 and Table 2.

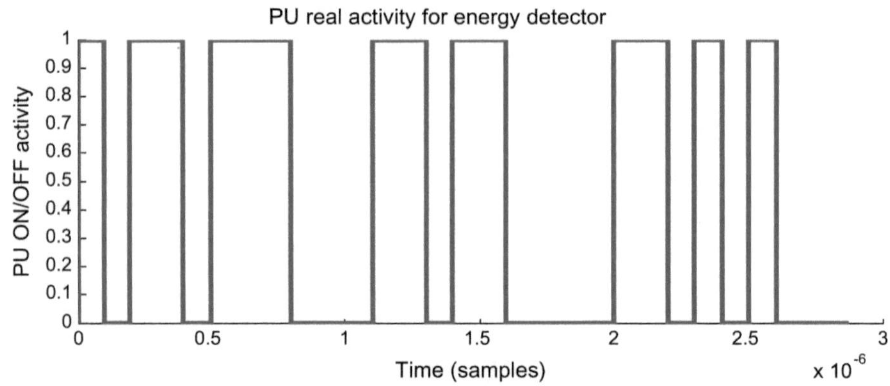

Fig. 4 PU real activity for energy detection technique at 10 dB SNR

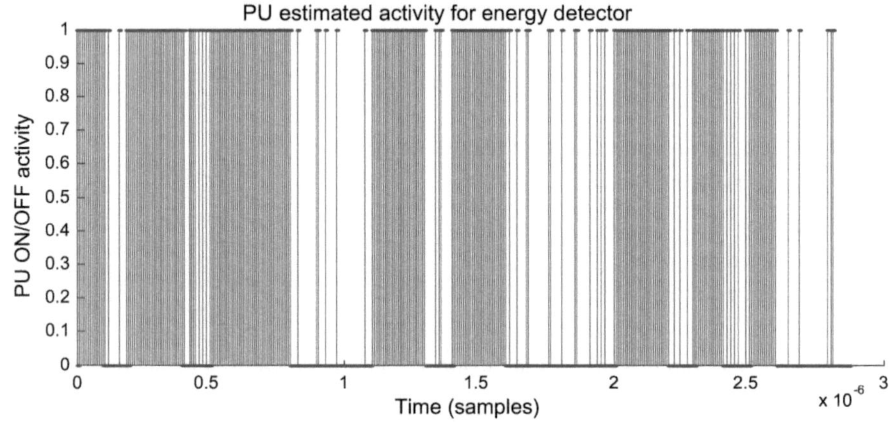

Fig. 5 Estimated PU activity for energy detection technique at 10 dB SNR

Table 2 PU real and estimated activity at −10 dB SNR for ED

Time (samples) [$\times 10^{-6}$]	4.8	4.9	5.0	5.1	5.2
Real PU activity	1	1	1	1	1
Estimated PU activity	0	0	0	0	0

However, from Figs. 6, 7 and Table 2, it becomes clear that this detection algorithm fails at low SNR, thus giving rise for need of more accurate detectors like cyclostationary- or entropy-based detectors to perform well in low SNR conditions.

The designed Simulink model refreshes after every 3 s for each value of SNR. Afterwards, a new SNR floor is selected and probability of detection is estimated for the previous SNR value.

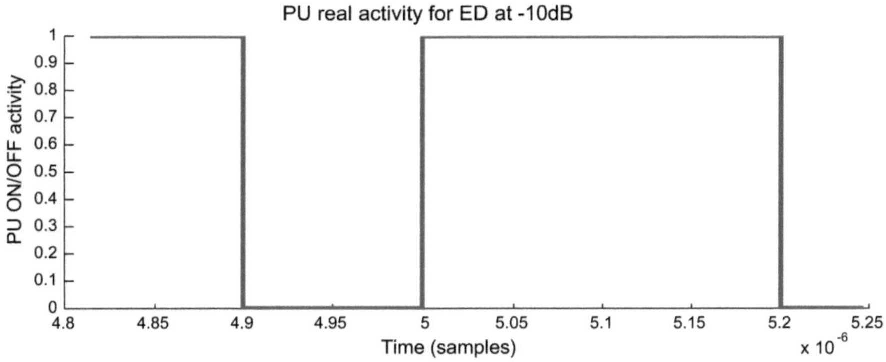

Fig. 6 PU real activity for energy detection technique at −10 dB SNR

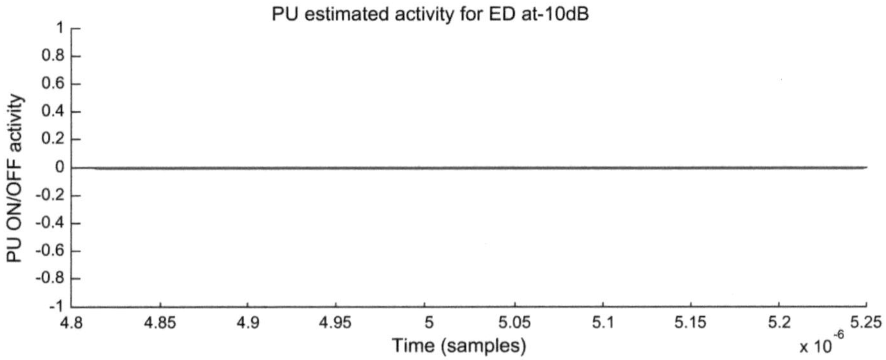

Fig. 7 Estimated PU activity for energy detection technique at −10 dB SNR

From Fig. 8 it can be clearly observed that the technique works well for moderate and high SNR conditions and for low SNR values it is deteriorating constantly. Further, CSS simulation results for PU activity considering five SUs have been shown in Figs. 9, 10. For simplicity, AND decision rule has been used at fusion center to reach a global decision.

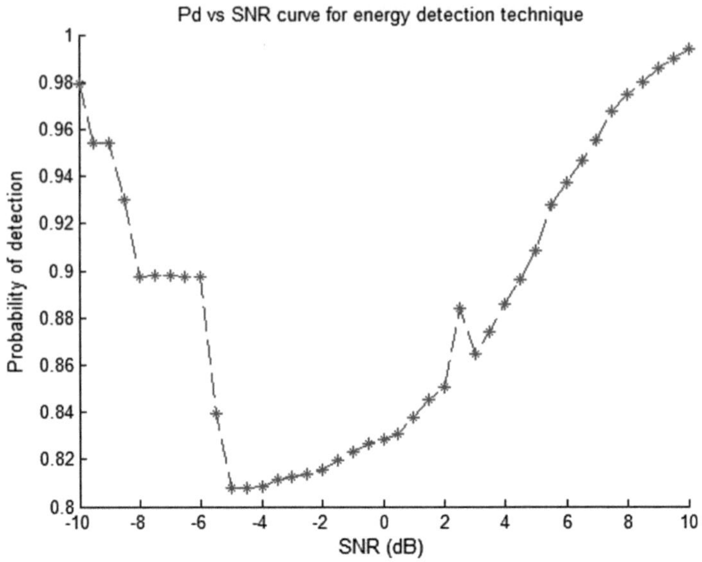

Fig. 8 Pd versus SNR curve for energy detection technique

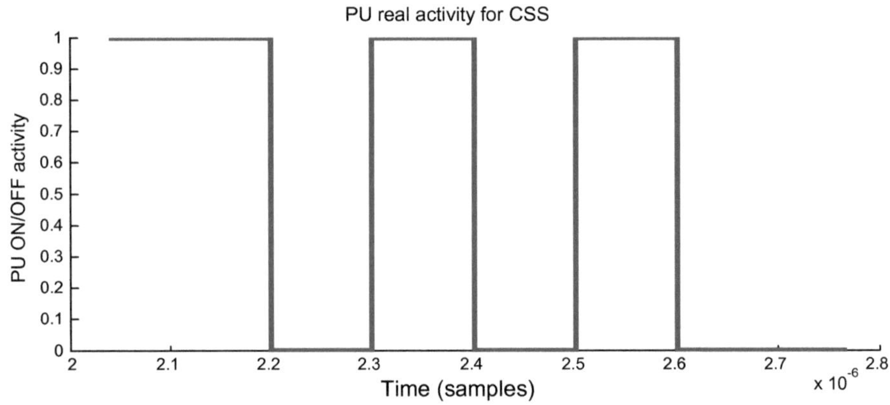

Fig. 9 PU real activity for cooperative spectrum sensing at 10 dB SNR

CSS aids in accurate detection of PU signal considering spatial characteristics. The above Table 3 shows the real and estimated PU activities at 10 dB SNR.

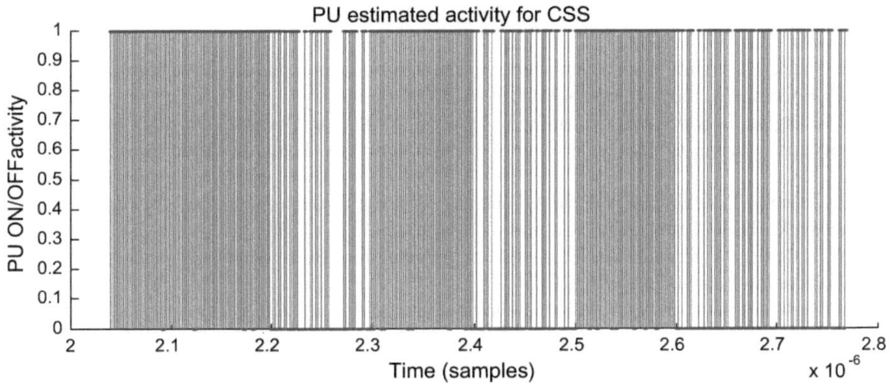

Fig. 10 Estimated PU activity for cooperative spectrum sensing at 10 dB SNR

Table 3 PU real and estimated activity for CSS

Time (samples) [$\times 10^{-6}$]	2	2.1	2.2	2.3	2.4	2.5	2.6
Real PU activity	0	1	1	1	1	1	1
Estimated PU activity	0	1	1	1	1	1	1

5 Conclusions

In this paper, Simulink model for ED and CSS scheme have been designed. Users own block has been designed for optimal threshold calculation, where MATLAB algorithm has been written in the background (MATLAB editor window). Thus, making it convenient for implementing whole concept of the specified technique. Further, investigations have been performed for high and low SNR conditions. Optimal threshold and Signal to Noise Ratio (SNR) conditions have been considered providing more accurate results. It can be concluded that ED technique works well for high SNR conditions while for low SNR conditions it fails. Moreover, the model has been extended for CSS with cooperation of five SUs. Matlab Simulink can be used to simulate modules of CRN and further the concept can be implemented for real scenarios.

References

1. E. Tugba, L. Mark, P. Ken, in *Spectrum Occupancy Measurements, Limestone*. (Loring Commerce Centre, Maine, 18–20 Sept 2007). http://www.sharedspectrum.com/papers/spectrum-reports/
2. M. Song, C. Xin, Y. Zhao, X. Cheng, Dynamic spectrum access: from cognitive radio to network radio. IEEE Wirel. Commun. **19**(1), 23–29 (2012)

3. B. Wang, K.J.R. Liu, Advances in cognitive radio networks: A Survey. IEEE J. Sel. Top. Sign. Proces. **5**(1), 5–23 (2011)
4. T. Yucek, H. Arslan, A survey of spectrum sensing algorithms for cognitive radio applications. IEEE Commun. Surv. Tutorials **11**(1), 116–130 (2009)
5. R.R. Jaglan, S. Sarowa, R. Mustafa, S. Agrawal, N. Kumar, Comparative study of single-user spectrum sensing techniques in cognitive radio networks. Procedia Comp. Sci. **58**, 121–128 (2015)
6. Y.C. Liang, Y. Zeng, E.C.Y. Peh, A.T. Hoang, Sensing-throughput tradeoff for cognitive radio networks. IEEE Trans. Wirel. Commun. **7**(4), 1326–1337 (2008)
7. M.L. Benitez, F. Casadevall, Improved energy detection spectrum sensing for cognitive radio. IET Commun. **6**(8), 785–796 (2012)
8. E. Chatziantoniou, B. Allen, V. Velisavljevic, Threshold optimization for energy detection based spectrum sensing over hyper-rayleigh fading channels. IEEE Commun. Lett. **19**(6), 1077–1080 (2015)
9. I.F. Akyildiz, B.F. Lo, R. Balakrishnan, Cooperative spectrum sensing in cognitive radio networks: A survey. Phys. Commun. **4**, 40–62 (2011)
10. http://in.mathworks.com/help/simulink/
11. N. Lorvancis, D. Jalihal, Performance of p-norm detector in cognitive radio networks with cooperative spectrum sensing in presence of malicious users. Wirel. Commun. Mob. Comput. **17**, 1–8 (2017)
12. R.R. Jaglan, R. Mustafa, S. Agrawal, Performance evaluation of energy detection based cooperative spectrum sensing in cognitive radio network. Proc. First Int. Conf. Inf. Commun. Technol. Intell. Syst. **2**, 585–593 (2016)
13. R. Bouraoui, H. Besbes, Cooperative spectrum sensing for cognitive radio networks: fusion rules performance analysis, in *Proceedings of International IEEE Wireless Communications and Mobile Computing Conference (IWCMC)* (2016)
14. M. Moradkhani, P. Azmi, M.A. Pourmina, Optimized energy limited cooperative spectrum sensing in cognitive radio networks. Comput. Electr. Eng. **42**, 1–11 (2014)
15. H.M. Farag, E.M. Mohamed, Soft decision cooperative spectrum sensing with noise uncertainty reduction. J. Pervasive Mob. Comput. **35**, 146–164 (2017)

QoS-Aware Routing in Vehicular Ad Hoc Networks Using Ant Colony Optimization and Bee Colony Optimization

Sumandeep Kaur, Trilok C. Aseri and Sudesh Rani

Abstract Vehicular Ad Hoc Networks (VANETs) are the kind of mobile Ad Hoc networks (MANETs) which consist of mobile nodes (vehicles) and fixed nodes (road-side units). Various applications such as road safety, comfort of passenger, entertainment, public safety, sign extension, and intelligent transportation require the communication between the vehicles. VANETs have special characteristics like frequent topology changes and high mobility which disconnect the network very frequently. Therefore, the highly dynamic nature of vehicles and the rapid changes in network topology are the most challenging issues in VANETs. In these situations, the most important requirement is the transfer of data from source to goal with a certain level of quality of service (QoS). In this paper, we propose QoS-Aware Routing Protocol for VANETs called QoS-Aware Routing in VANETs (QARV) in which packets reach the destination while satisfying the QoS. This protocol works in highway scenario. The new concept called Terminal Intersection concept is used in these two protocols in order to reduce the congestion and lessen the time for exploring the route. In this paper, two bio-inspired algorithms, Ant Colony Optimization (ACO) and Bee Colony Optimization (BCO), are used to achieve the results. This paper provides the comparative analysis of various performance parameters of both the algorithms.

Keywords Routing · QoS · Ant colony optimization · Bee colony optimization

S. Kaur (✉) · T. C. Aseri · S. Rani
PEC University of Technology, Chandigarh, India
e-mail: Brarsumandip92@gmail.com

T. C. Aseri
e-mail: A_trilok_aseri@yahoo.com

S. Rani
e-mail: jsudesh@gmail.com

C. R. Krishna et al. (eds.), *Proceedings of 2nd International Conference on Communication, Computing and Networking*, Lecture Notes in Networks and Systems 46, https://doi.org/10.1007/978-981-13-1217-5_25

1 Introduction

Due to the development of wireless technologies and short-range communication technologies, VANETs are quite popular these days [1]. VANETs consist of vehicles and roadside units. In intelligent transportation system (ITS), vehicles play various roles such as source vehicle, destination vehicle, and router to send the packets to other vehicles. The motivation to develop VANETs is collecting road information and disseminating this information to various vehicles to improve the road safety applications [2]. VANETs are a key technology in ITS due to the availability of navigation system, global positioning system (GPS), and other sensors that can collect the vehicle speed, location [3, 4]. The various ITS applications are vehicle safety, public safety, intersection collision avoidance, vehicle diagnostics, traffic monitoring. VANETs have special characteristics like predictable mobility, highly mobile, self-organizing, and no energy constraints. Although there is remarkable achievement in the field of VANETs, there are still many challenges such as routing, scalability, signal fading, bandwidth limitations, privacy, security, and QoS. Unfortunately, the traditional wireless technologies are not applicable to VANETs directly, because they have characteristics like frequent topology changes and high mobility which disconnects the network very frequently. Therefore, the highly dynamic nature of vehicles and the rapid changes in network topology are the most challenging issues in VANETs [5]. In these situations, the most important requirement is the transmission of data from source to goal with a certain level of QoS. The QoS means the transmission of data from source to goal with minimum delay and minimum overhead. The various QoS parameters are connectivity probability, reliability, availability, link duration, hop count, end to end delay, and stability [6, 7].

The existing QoS routing protocols have various limitations like: (1) Some protocols use extra control messages to estimate the connectivity of the vehicles. (2) The techniques of calculating the QoS are not effective, (3) The routing algorithms are not scalable and adaptive. (4) The existing protocols suffer from increased complexity, lack of stability in networks. (5) Due to fast-paced property of vehicles in VANETs, every node often escapes from current access purpose and breaks into covering field of another access purpose. (6) Multi-Constrained Path Selection problem is a major issue in VANETs.

In this paper, QoS-aware routing protocol called QARV is proposed to deal with these limitations. We propose an ACO-based QARV algorithm and BCO based QARV algorithm to resolve this QoS issue. Both the algorithms use the opportunistic method to find the optimal routes. Once the optimal route is established, the data transfer takes place between the vehicles. This paper compares the results of both ACO-based QARV algorithm and BCO based QARV algorithm and validates the results.

2 Related Work

VANETs are featured with exchanging information amongst vehicles in real time. It needs knowledge packets ought to travel through the transport network from supply nodes to the destiny nodes. The highly dynamic nature of vehicles and the rapid changes in network topology are the most challenging issues in VANETs. Therefore, routing protocol is crucial within the operation of VANETs. In these situations, the most important requirement is the transfer of data from source to goal with a certain level of QoS. Various researchers attempted to handle this QoS routing issue by applying various approaches.

Sun et al. proposed a protocol that is QoS routing protocol and deals with dynamic layout and keeps the balance between stability and potency of the rule. However, whether the link is connected between the vehicles is calculated by transmitting beacon packets which can cause the channel crowding within the case of more traffic density [8]. Namboodiri et al. proposed Prediction Based Routing for VANETs (PBR) predicts the time of the link exists between the source and the goal [9]. It predicts the route lifetimes of the existing routes and finds the new routes from source to goal. This protocol is applicable in mobile gateway scenario. It reduces the route failures while increasing packet delivery ratio. Zhao et al. proposed Vehicle-Assisted Data Delivery Protocol in which route is decided on the basis of historical route QoS parameters. These are not valid in VANETs because it has dynamic topology [10].

Zhang et al. proposed an artificial bee colony based multicast routing in VANETs that finds the Steiner minimum tree (SMT) of graph G from the source to the goal [10]. The original artificial bee colony (ABC) is applied to the SMT problem. Jerbi et al. proposed efficient geographical routing use only local traffic parameters to find the route which is not suitable in large traffic conditions [11]. Eiza et al. proposed Situation-Aware QoS Routing Algorithms for Vehicular Ad Hoc Networks that uses the awareness of situation and ACO for developing the situation-aware multicast routing algorithm for VANETs [12]. Situational awareness means getting the knowledge from the current status of the vehicles, integrating this knowledge with the previous knowledge, and using the resulting knowledge to get decisions for the future events. This protocol finds the best route between the vehicles by considering various QoS constraints using Ant Colony Optimization [13].

Naumov et al. proposed Connectivity-Aware Routing in Vehicular Ad Hoc Networks; a source transmits packets towards its destination. After getting the request packets, the destination finals the best routing path and so acknowledges this route's data to the source [14]. This paper attempts to tackle the above limitations and proposes QoS-aware routing protocol using ACO and BCO.

3 QARV Protocol Using ACO

In this section, we propose a QoS-aware routing protocol using Ant Colony Optimization.

3.1 *Main Idea*

Our main idea is to provide QoS in VANET environment using bio-inspired algorithms such as ACO. First, we have a tendency to style a Terminal Intersection (TI) concept using ACO to ascertain the simplest routing path satisfying the various QoS constraints.

3.2 *Proposed Protocol*

We propose QARV routing protocol. It is an intersection routing protocol that looks the best QoS route. The QoS is measured in terms of Delay, PC, and PDR. Every vehicle foremost transmits a routing appeal to its TIS (terminal intersection for a supply vehicle) for routing data before redirecting knowledge packets. If this connection information is available, a favorable acknowledgement is shipped back to supply automobile S, that immediately begins knowledge packets redirecting. Else, TIS gives contrary acknowledgement to S.

3.3 *Assumptions*

Every automobile is provided with GPS, digital chart, and navigation system which offer the automobiles data, places of crossings and also the road segment length. We tend to suppose that the source vehicles will acquire the geographic places of their various target location services and the totally dissimilar transmission pairs have the same QoS. A fixed node is placed on every crossing to assist packet redirecting and routing data storage.

3.4 *Protocol Description*

A terminal intersection is set by two parameters: The movable inclination means the moving direction of the transmission station. The distance from itself to the next crossings Whoever supported these variables, a score is allotted to every candidate intersection, and then the one owing the very best score is chosen as the TI.

The optimal route establishment. *Candidate route derivation* After the selection of terminal intersection for the supply vehicle S and the target vehicle D, S first initiates a routing appeal to TIS. If the information of routing towards TID is already present at TIS and it is not expired, TIS sends a favorable response to S; otherwise, TIS sends a contrary message to S so initiates to look the optimum route.

To find the candidate paths from TIS to TID, TIS sends a gaggle of forward ants around TID. Once the forth ant reaches at crossing I_i, it, first of all, saves I_i's ID and its delay, then it uses a random call to pick consecutive crossing supported each both global and native pheromone keep regionally at $I_{i.}$ Let there are K next crossings of I_i, namely I_1, I_2, …, I_{K-1}, I_K. The probability p_{ij} with which the forth ant means forward ant back ant means backward ant I_j as the next crossing is given by Eq. (1).

$$p_{ij} = \frac{\mathrm{LP}_{ij^{\propto}} \cdot \mathrm{GP}_{ij^{\beta}}}{\sum_{m=1}^{K} \mathrm{LP}_{im^{\propto}} \cdot \mathrm{GP}_{im^{\beta}}} \quad (1)$$

where LP and GP are the pheromone value locally and globally respectively.

Whenever the forth ant reaches TID, and if its delay is a smaller amount, then $\mathrm{D_{th}}$ the route is taken as the candidate path. The problem occurs when all the coming ants do not satisfy the demand delay. Here, the D_{th} is the delay threshold.

Optimal route selection this selection is done using the reverse ants. The forth ant is converted into back ant if it satisfies the delay threshold value.

Once all the reverse ants gain TIS, we tend to equate fitness value of all the on the routes, and also the route with the highest worth of fitness value is chosen as the best one. Then, TIS transmits a favorable acknowledgement to supply vehicle to begin the transfer.

4 QARV Using BCO

In this protocol, the colony consists of three teams of bees: Employed, onlooker, and scout bees. It is presumed that there is just single employee bee for every food source [15]. The amount of utilized bees means employed bees. There are three type of bees i.e. employed bees, onlooker bees and scout bees within the swarm is capable the amount of food origins round the swarm. Utilized bees attend their food supply and are available back to swarm and whirl on this space. The utilized bees attend their food supply and are available back to swarm and whirl on this space. The utilized bee whose food supply has been forsaken makes a scout and initiates to go looking for locating a replacement food supply. Onlookers see the dances of utilized bees and select food origins betting on whirls. The basic of the algorithm means the basic idea of the algorithm are described below.

Begin food origins means when the bees start searching food from their source made for whole used bees ITERATE every used bee moves to a food supply in her remembrance and discovers a next supply, then calculates its nectar quality and whirl within the swarm.

Every looker sees the whirls of used bees and picks one in all their supplies counting on the whirls, so moves there to supply. Then it calculates its nectar quality.

Food origins are discovered and are exchanged with the new food origins determined by scouts. The optimal food supply launch thus far is considered. In this algorithm, the place of food supply describes resolution to the best downside and also the nectar quantity of a food supply matches to the standard (fitness). The quantity of the utilized bees is adequate within the population. At the primary step, random startup population (food supply positions) is produced. Here the population means the food source positions iterate the steps of the finding operations of the utilized, onlooker and scout bees severally, the utilized bee generates an alteration on the supply place in her remembrance and determines a replacement food supply place only if the nectar quality of the new one is more than that of the previous supply, the bee remembers the new supply place and deletes the position of origins.

When all bees finish the method, all distribute the position data of the origins. Every viewer calculates the nectar data taken from all used bees and so picks a food supply betting on the nectar quantity of origins. As within the study of the used bee, it creates an alteration on the supply place in her remembrance and calculates its nectar quantity. The bee remembers the new place and leaves the recent one. The origins forsaken are discovered and new source are created to get exchanged with the forsaken ones by false scouts.

5 Experimental Setup

Various parameters taken for simulation are described in Table 1.

Table 1 Simulation parameters

Name of parameter	Value
Simulator used	NS-3
Simulation area	5000 m * 5000 m
Simulation time	150 s
Communication range	150–350 m
Packet size	1024 bytes
D_{th}	80 s

6 Simulation Results

6.1 Simulation Parameters PDR

Connectivity Probability. It is the probability of how long the route exists between the vehicles.

Delay. The normal time taken from source to the goal for the exchange of packets.

6.2 Simulation Results

PDR. Figure 1 shows that the green line in the graph shows the BCO PDR with time and the blue line shows the ACO PDR with time. It can be inferred from the figure that the packet delivery ratio increases with time in both the cases but BCO PDR is better than the ACO PDR. The BCO packet delivery ratio is more effective than ACO PDR because the speed of bees is faster than ants as the bees fly in the sky while ants move slowly on ground. Using bees, the transfer of packets is faster The second reason is the direction, i.e., ants move only in one direction while the bees move in different directions, so it is easy to transfer packets in BCO than in case of ACO

CP. It refers to the probability of how long the route exists between the vehicles. Figure 2 shows that the green line in the graph shows the BCO CP with vehicle space density and the blue line shows the ACO connectivity probability with vehicle space density. It can be inferred from the figure that the CP increases with increase in vehicle space density in both the cases but BCO CP is better than the ACO CP. If the vehicle space density is low, then it is difficult to maintain the connectivity between

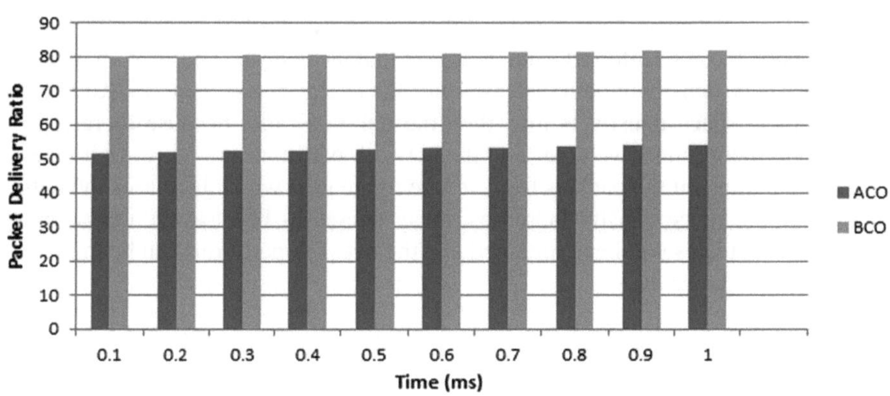

Fig. 1 Comparison of PDR of QARV using ACO and BCO

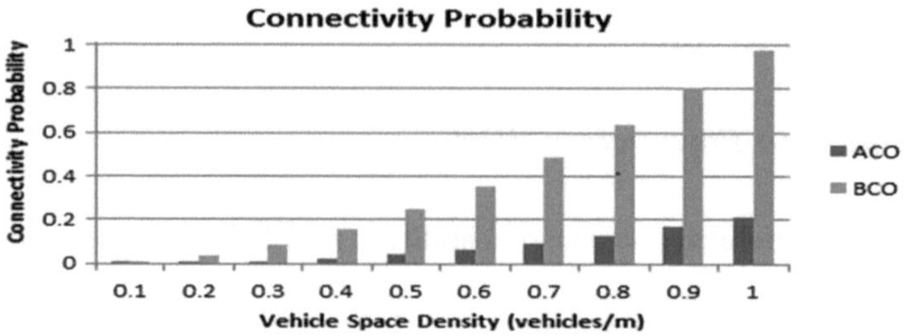

Fig. 2 Comparison of connectivity probability of QARV using ACO and BCO

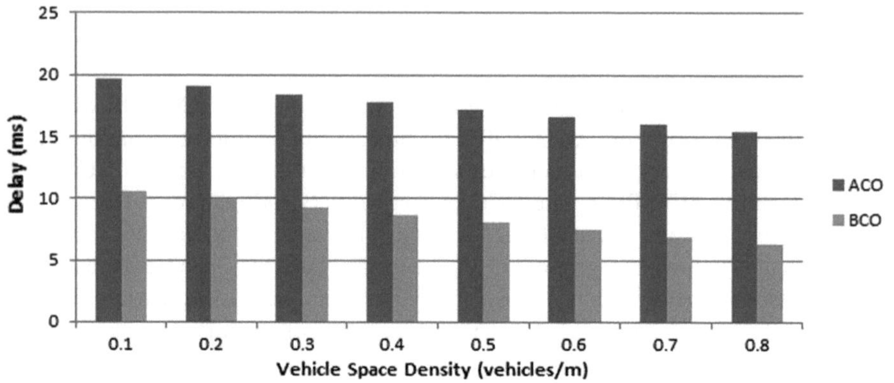

Fig. 3 Comparison of delay QARV using ACO and BCO

the vehicles. As the vehicle space density increases, the network road segments can be improved, this leads to increase in connectivity.

Delay. Delay refers to the normal time taken from source to the goal for the exchange of parcels. Figure 3 shows that the green line in the graph shows the BCO delay with vehicle space density and the blue line shows the ACO delay with vehicle space density. It can be inferred from the figure that the delay decreases with increase in vehicle space density in both the cases but BCO delay is less than the ACO delay. If the vehicle space density is low, then the delay is more because there is less connectivity between the vehicles and the network partitions are more, so it takes a long time to transmit the packets from source to the goal which increases the delay. As the vehicle space density increases, delay decreases because the connectivity between the vehicles will be maintained and the network partitions can be improved which reduces the delay.

7 Conclusion

This paper presents a comparison of two bio-inspired algorithms such as ACO and BCO which provide an efficient mechanism for maintaining QoS routing. The aim of both the protocols is to maintain QoS in VANET routing in terms of PDR, CP, and Delay. The QARV using BCO performs better than QARV using ACO because of two reasons. The first reason is the speed of the bees is faster than the ants. The second reason is the direction of the bees. The bees move in different directions. If the connection breaks, it is easy to maintain the connectivity between the bees faster than maintaining the connectivity in case of ants. Simulation results show that the QARV using BCO is better than the QARV using ACO in terms of packet delivery ratio, connectivity probability, and delay.

8 Future Scope

The future scope is to check the QoS in Vehicular Ad Hoc Networks using combinatorial algorithms such as Particle Swarm Optimization, Tabu search, etc.

References

1. S. ur Rehman, M. Arif Khan, T.A. Zia, L. Zheng, Vehicular ad-hoc networks (VANETs): an overview and challenges. J. Wirel. Netw. Commun. **3**(3), 29–38 (1997)
2. S. Al-Sultan, M.M. Al-Doori, A.H. Al-Bayatti, H. Zedan, A comprehensive survey on vehicular Ad Hoc network. J. Netw. Comput. Appl. **37**, 380–392 (2014)
3. J. Blum, A. Eskandarian, L. Hoffmman, Challenges of inter vehicle Ad Hoc networks. IEEE Trans. Intell. Transp. Syst. **5**, 347–351 (2004)
4. G.M.T. Abdalla, M.A. Abu-Rgheff, S.M. Senouci, in *Current Trends in Vehicular Ad Hoc Networks*, University of Plymouth-School of Computing, Communications and Electronics, UK and France Telecom-Research and Development CORE, France (2007), pp. 1–9
5. G.M.T. Abdalla, M.A. Abu-Rghef, S.M. Senouci, Current trends in vehicular Ad Hoc networks, in *1st International Workshop UBIROADS*, Marrakech, Morocco (2007)
6. G. Li, L. Boukhatem, J. Wu, Adaptive quality of service based routing for vehicular Ad Hoc networks with ant colony optimization. IEEE Trans. Veh. Technol. **66**, 1–15 (2004)
7. S. Ian, O. Tonguz, Enhancing Vanet connectivity through roadside units on highways. IEEE Trans. Veh. Technol. **60**, 3586–3602 (2011)
8. Y. Sun, S. Luo, Q. Dai, Y. Ji, An adaptive routing protocol based on QoS and vehicular density in urban VANETs. Int. J. Distrib. Sens. Netw. **11**, 1–5 (2015)
9. V. Namboodiri, L. Gao, Prediction-based routing for vehicular Ad Hoc networks. IEEE Trans. Veh. Technol. **56**, 2332–2345 (2007)
10. J. Zhao, G. Cao, VADD: vehicle assisted data delivery in vehicular Ad Hoc networks. IEEE Trans. Veh. Technol. **57**, 1910–1922 (2008)
11. M. Jerbi, T. Rasheed, Y. Doudane, Towards efficient geographic routing in vehicular Ad Hoc networks. IEEE Trans. Veh. Technol. **58**, 5048–5058 (2009)
12. X. Zhang, X. Zhang, C. Gu, A micro-artificial bee colony for multicast routing in vehicular Ad Hoc networks. Elsevier J. Ad-Hoc Netw. **58**, 1–9 (2016)

13. M.H. Eiza, T. Owens, Q. Ni, Q. Shi, Situation-aware QoS routing algorithm for vehicular Ad Hoc networks. IEEE Trans. Veh. Technol. **64**, 5520–5535 (2016)
14. V. Naumov, T.R. Gross, Connectivity-aware routing (CAR) in vehicular Ad Hoc networks, in *IEEE INFOCOM 26th IEEE International Conference on Computer Communications* (2007), pp. 1919–1927
15. S. Batim, A.H. Mellouk, Bee life based multi constraints multicast routing optimization in vehicular Ad Hoc networks. J. Netw. Comput. Appl. **36**, 981–991 (2011)

Research Challenges in Airborne Ad-hoc Networks (AANETs)

Pardeep Kumar and Seema Verma

Abstract Wireless Ad-hoc Networks have been one of the major research issues from last one and half decade. However, the primary researches have focused on ground vehicles and random mobility scenarios. But due to recent advancements in communication technologies and signal processing techniques, nowadays, we have a new and special kind of kind of ad-hoc networks called "Airborne Ad-hoc Networks (AANETs)" in which aircrafts are implemented as nodes traveling with significant speeds. Because of unique characteristics like higher mobility, highly dynamic aeronautical environment and time-varying inter-aircraft radio link quality, there are new research challenges in AANETs. These networks can be used for enhancing the situational awareness, flight efficiency, and flight coordination in military and civilian applications. In this paper, we have presented the theoretical review of research challenges in AANETs which have been found after study of previous research papers. It has been found that current routing protocols and 2-D mobility models which are being used for typical Ad-hoc networks are not able to cope with AANETs environment. There is a requirement to implement the physical movement of aircrafts in more realistic manner. For this, AANETs need domain-specific routing protocols, geographical protocols, and hierarchical protocols along with memory-based three-dimensional mobility models to deal with the challenges of highly dynamic aeronautical environment.

Keywords AANETs · Highly dynamic aeronautical environment · Flight efficiency · Flight coordination

---

P. Kumar (✉) · S. Verma
Department of Electronics, Banasthali Vidyapith, Jaipur, Rajasthan, India
e-mail: pardeep.lamba@gmail.com

S. Verma
e-mail: seemaverma3@yahoo.com

C. R. Krishna et al. (eds.), *Proceedings of 2nd International Conference on Communication, Computing and Networking*, Lecture Notes in Networks and Systems 46, https://doi.org/10.1007/978-981-13-1217-5_26

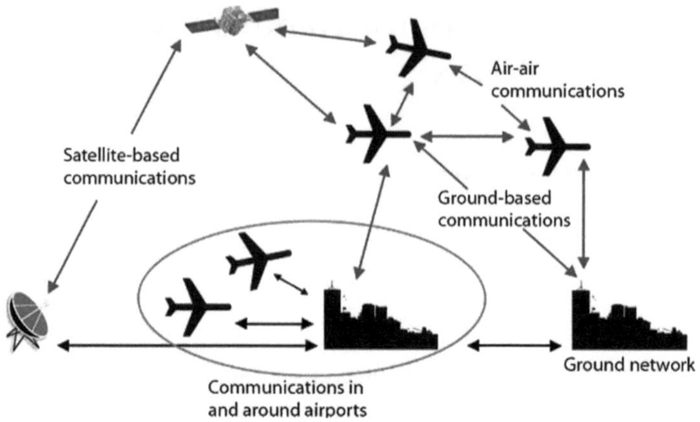

Fig. 1 Airborne ad-hoc network with both air-to-air communication and air-to-ground communications [14]

1 Introduction

Wireless ad-hoc networks are categorized into Mobile Ad-hoc Networks (MANETs) and Vehicular Ad-hoc Networks (VANETs). VANETs are used for communication among vehicles or communication among vehicles and roadside equipment. With the use of new advanced technologies in wireless communication, the latest wireless network family has been among us known as Airborne Ad-hoc Networks (AANETs). AANETs consist of aircrafts as flying nodes. These are special class of VANETs [13]. AANETs are also known as Flying Ad-hoc Networks (FANETs).

Airborne Ad-hoc Networks can be established between aircrafts for civilian applications while for military application, UAVs (Unmanned aerial vehicles) can be used as mobile nodes. In AANETs, there are two types of communication [2].

1.1 Aircraft-to-Aircraft Communication

Different aircrafts can communicate with each other directly for different application areas like target tracking and path planning. There can be a long or short range of communication between aircrafts. It is also called air-to-air communication as shown in Fig. 1.

1.2 Aircraft-to-Infrastructure Communication

Here, aircrafts may communicate with ground stations, satellites, and warships to provide information services. This technique is also known as air-to-ground communication as shown in Fig. 1.

The major features of these networks are very high speed, three-dimensional movement in the air [2, 5], lesser node density compared to traditional MANETs and VANETs due to larger distance between nodes up to several kilometers [10], limited available bandwidth, frequent link and route breaks due to aircraft banking, coordinated movement (which cannot be implemented with random mobility models), etc.

The uniqueness of AANETs has presented a lot of challenges for researchers. The speed of aircrafts (around 300–400 km per hour) is the major cause of difference between original MANETs and AANETs. Traditional routing protocols are not able to cope with these highly dynamic networks. So, Airborne Ad-hoc Networks need geo-location assisted, hierarchical and domain-specific routing protocols to deal with the challenges of highly dynamic aeronautical environment. Till now, there have been different implementations of AANETs but mostly they have been done using random mobility models (synthetic models or memoryless models) which cannot trace aircraft movement as in real environment. So there is a strong need of implementing these networks in 3D mobility models. Along with this, appropriate simulator design is also required to do research in this field.

The rest of the paper is organized as follows: Sect. 2 explains the related work which has been done so far in this area. Various research challenges with implementation of AANETs are formulated in Sect. 3. Finally, concluding remarks are given in Sect. 4.

2 Related Research Work

Aircraft communication creates a very challenging scenario for Mobile Ad-hoc Networks. Despite research achievements and increasing interest in this airborne networking technology, high mobility, dynamic topology changes and low node density create a highly challenging scenario for researchers to be able to provide reliable communication solutions. With these challenges, many research works have been performed on this emerging topic.

Despite the various challenges, the researchers have developed many routing protocol solutions for MANETs. They are mainly categorized as topology-based and geographical-based protocols [7]. Topology-based scheme has been further subdivided into three types of protocols: proactive, reactive, and hybrid protocols. Hybrid protocols are a combination of proactive and reactive protocols. They make use of distance vectors to establish best paths to the destination. The routing information is flooded only when there is any change in the network topology. Every node has its own routing zone of fixed size known as number of hops [7]. These protocols are not able to cope with highly mobile Aircraft Ad-hoc Networks.

The geographical routing protocols have been proposed to handle large control overhead, solve scalability and low packet delivery ratio issues. Here, to maintain routing information, an intelligent forwarding scheme is used instead of topology information as used in previous protocols. Every intermediate node needs to find one neighbor from its neighbor table based on destination's location which is contained in packet. This selected node should be nearest to destination node to forward the packet. This packet sending procedure in GPSR is known as greedy forwarding. This type of geographical protocols can be very effective for these highly dynamic Airborne Ad-hoc Networks.

Different fundamental issues and design considerations with development of airborne networks are presented by Wang et al. [25]. The authors have addressed physical layer, data link layer, and network layer aspects for AANETs design along with use of airborne networks in civil aviation for safety and efficiency improvement. Kiwior et al. have compared Ad-Hoc on Demand Distance Vector protocol (AODV), Temporally Ordered Routing Algorithm (TORA) and Open Shortest Path First-version3 (OSPFv3-MANET), and Optimized Link State Routing Protocol (OLSR) for intermittent links in highly dynamic airborne networks [9]. Here, OSPFv3-MANET and OLSR outperformed traditional MANET protocols. Rafols Ramirez has described the link management method for airborne networks [17]. This paper identifies various issues with link management like criteria for node admission, conditions to indicate need of hand-off; corrective actions needed to improve degraded performance of the network, efficient algorithms to detect link degradations. Rohrer et al. [18] have defined implementation of a 3-D Gauss–Markov mobility model using NS-3 simulator. The authors have also presented a domain-specific routing protocol AeroRP which mainly focused on routing data packets efficiently among airborne nodes. It overcomes the transmission range problem in airborne networks. Abdel Ilah Alshabtat proposed a routing protocol for these networks [1]. This protocol is named as Directional Optimized Link State Routing protocol (DOLSR) which is equipped with directional antenna. The proposed protocol outperformed AODV, OLSR and Dynamic Source routing (DSR) protocol with reduced overhead packets and minimized end-to-end delay. Yegui Cai et al. have studied the MAC issue in UAV ad-hoc networks with multi-packet reception capability and full-duplex radios [4]. This paper formulated MAC schemes with perfect and imperfect channel state information as combinatorial optimization and discrete stochastic optimization problem. They have used a token-based technique for information updating in the network to overcome the contention based MAC schemes. Abhishek Tiwari et al. have discussed the issue of interoperability between different networks [22]. The authors have proposed the use of airborne nodes to solve this issue because they have potential to provide large area coverage, low latency as compared to satellite communication. Cheng et al. have presented a comparison of reactive and proactive MANET routing protocols [5]. This paper shows that Optimized Link State Routing Protocol (OLSR) performs much better in comparison with AODV and Open Shortest Path First with MANET Designated Router (OSPF-MDR). OLSR provides higher delivery rate at higher speeds, lesser average end-to-end delay and higher delivery rate for increased transmission radius and low traffic loads. Bekmezci et al. have presented new network family Flying Ad-hoc Networks (FANETs) [2]. Comparison of MANET and VANET has been done here. Ehssan Sakhaee et al. have introduced the concept of ad-hoc networking among aircrafts [20]. This novel approach can be used to increase the data rate and practicality of future in-flight broadband Internet access. This technology can be used to decrease the internet traffic load on existing satellite nodes. Kevin Peters et al. have presented a protocol for packet delivery in highly dynamic and multi-Mach speed environment which is named as AeroRP [15]. The analysis result shows that the proposed protocol has advantage over MANET routing protocol in terms of delay, overhead, accuracy etc. Shangguang et al. presented aided geographical routing protocol (A-GR) for AANETs which is based upon Automatic Dependent Surveillance-Broadcast (ADS-B) system [23]. The proposed protocol uses the velocity and position of aircraft which is provided by airborne ADS-B system. Ki-II Kim has presented a performance evaluation of usual routing algorithms in high speed and new mobility pattern of aircraft ad-hoc networks [8]. O. K. Sahingoz has presented different networking models for Flying Ad-hoc Networks [19]. This paper presents routing, path planning, and quality of services issues at different networking layers. Chao Yin et al. [26] have proposed a new routing protocol for Flying UAV networks. To decrease the transmission delay, an enhanced routing protocol named "Fountain-code based Greedy Queue and Position Assisted (FGQPA)" routing protocol is presented here. First, this protocol used Power Allocation and Routing (PAR) technique to remove the effect due to node queue backlog on transmission delay of the network. Then, a "Nearest Span" policy was integrated into the PAR technique to reduce the delay further. An adaptive framework is designed and implemented by Weijun et al. [24] which allows many MAC protocols to combine and then switch mutually for UAV ad-hoc networks. The UAVs can switch to the most suitable MAC protocol. The concept of opportunistic routing in wireless sensor networks assisted by UAVs is presented by Ma et al. [12]. Due to dynamic behavior of UAVs, sensor nodes have different opportunities to be in communication range of UAVs. This paper has introduced two protocols: All Neighbors Opportunistic Routing (ANOR) and Highest Velocity Opportunistic Routing (HVOR). In ANOR protocol, the source node shares its data packets to every neighbor that are within range. In HVOR protocol, the source node transmits packets to a node moving with highest speed.

The importance of mobility models in AANETs is also explained by Jean-Daniel in [3]. To implement the realistic behavior of UAVs, the simulation environment should be able to replicate the components of the movement scenario of UAVs. One such major component is the mobility model which must be able to trace the almost real movement patterns of UAVs. After implementing four protocols AODV, OLSR, Reactive-Geographic hybrid Routing (RGR) and geographical routing protocol (GRP) over these mobility models, the result proves that the protocol performance varies with the change of mobility models. The analysis also indicates that Enhanced Gauss–Markov (EGM) mobility model and RGR protocol are best suited for UAANET.

3 Research Challenges in AANETs

Airborne Ad-hoc Networks have the traditional problems of wireless communications, in addition, their high mobility and multi-hop nature with lack of fixed infrastructure create a number of design constraints and complexities [2, 4, 6, 18, 19, 21].

The various research challenges with AANETs are listed as follows.

3.1 Physical Layer Issues

At this layer, antenna structures and radio propagation models are the key factors. Because of the highly dynamic environment, node movements and terrain structures, UAVs or aircrafts do not maintain the link among each other. To solve the problem of scalability, new protocols and algorithms are needed to be designed so that any number of flying nodes can be connected with minimum performance degradation.

3.2 MAC Layer Issues

These networks have link quality fluctuations due to high mobility and distance variation between nodes. This link outage and quality fluctuation affect the design of MAC layer.

3.3 Network Layer Issues

At this layer, development of new routing protocol to improve peer-to-peer communication for coordination and collision avoidance in multi-UAV environment is required.

3.4 Transport Layer Issues

The main responsibility of transport layer has been reliability. Congestion control algorithms are needed to be implemented for reliable and efficient AANET design.

3.5 *Cross-Layer Architectures*

This architecture can be used to enhance the interaction among different layers to share the information about network. A cross-layer can be designed to combine all layers into a single protocol for more efficient AANET architectures in multi-aircraft systems.

3.6 *Hierarchical Routing*

To solve the problem of scalability, clustering or hierarchical routing concept can be used [11].

After a vast literature survey, it is found that variable link quality due to very high mobility of aircraft nodes presents many challenges to Quality of Service (QOS) implementation in AANETs using different routing protocols. The existing routing protocols may not be as effective as in traditional MANETs because they have not been tested for high mobility and link variability conditions. So, there is strong need for designing suitable protocols for these highly dynamic Ad-hoc networks.

The challenges due to limited bandwidth may be overcome by using directional antenna. The congestion due to transmitting nodes can be avoided by use of multiple access scheme, i.e., STDMA (spatial Reuse TDMA). But at the same time, generation of STDMA schedule to effectively utilize the network resources is a challenging task. A layer integration approach is also proposed for AANETs [6]. This new approach consists of a communication layer which connects the physical and the application layers directly and bypasses other layers. Different functions like routing, clustering, and MAC are coordinated at the communication layer (Table 1).

The designing of the MAC layer for AANETs is a major challenge for guaranteeing low packet error rate, high throughput, and low network latency [4].

To represent the accurate physical movement of nodes in highly dynamic AANETs, mobility models are also crucial factors. The previous research of traditional ad-hoc networks had incorporated the synthetic mobility models which are memoryless and random in nature. The realistic movement of the aircraft nodes can be represented by using memory-based 3-D mobility models. Because the movement of an aircraft node will not be random all time, the present position may be found out by its previous velocity and position at any point of time [18]. So, the mobility models must be having memory.

4 Conclusions and Future Scopes

This review paper gives an overview about the uniqueness of the emerging AANETs technology. We provide a recent literature review and related issues and challenges in layered approach for implementing these highly dynamic networks. But communication is a major issue so far faced by the previous researches in this emerging

Table 1 Comparison of protocols used in recent researches in airborne networks

Protocol	Type	Method of neighbor discovery	Simulator used	Mobility model	PDR	E2E delay	Overhead	Reference
DOLSR	Proactive	Broadcasting Hello messages to one-hop neighbors and then select MPR nodes	OPNET 4.5	Random Waypoint Mobility model	More than AODV, DSR and OLSR	Less than AODV, DSR and OLSR	Lesser than OLSR, AODV and DSR due to reduction in no. of MPRs	[1]
A-GR	Geographic	It makes use of location and mobility information	Qualnet 5.0	Gauss–Markov Model	A-GR provides more PDR with little degradation as the no. of nodes increases to 50 or more	At high node density, A-GR provides lesser delay compared to GPSR and GRAA	For GPSR and GRAA, overhead increases exponentially	[23]
Improved reactive and geographic (IRG)	Geographic	Reactive, Greedy Forwarding	NS-2.35	Random waypoint model	IRG outperform the AODV and GPSR with increase of speed because selection of next hop is based upon relative velocity between nodes	IRG provides slightly lesser delay than GPSR. AODV has highest delay with increase in node velocity	Routing overhead increases for all three, i.e., IRG, GPSR and AODV when speed is increased. GPSR performs better at low velocity compared to IRG and AODV. But at high speeds, IRG overcomes GPSR	[16]
Aero RP	Geographic	The packet forwarding decisions are made hop by hop. A velocity dependent parameter called Time to intercept (TTI) provides an idea about relative speed of the potential neighbor w.r.t receiving node.	NS-3	Random Waypoint Model	The PDR increases for Aero RP as the no. of nodes are increased	Aero RP in ferry or buffer modes holds the packet for some specified time which at the same time ensures more packet delivery. But the other two modes of Aero RP, i.e., Drop Beacon and Drop Beaconless have least delay	Aero RP and OLSR creates less no. control packets	[15]

(continued)

Table 1 (continued)

Protocol	Type	Method of neighbor discovery	Simulator used	Mobility model	PDR	E2E delay	Overhead	Reference
Geographic Routing Protocol for Aircraft Ad-Hoc network (GRAA)	Geographical	Three-dimensional up-to-date information about the location of aircraft and its direction of movement to calculate Euclidean distance between nodes	Qualnet	Random waypoint Mobility model and Two Ray Ground radio propagation model	For predetermined route and increase in speed, PDR is higher for GRAA than GPSR	Due to its hybrid approach, GRAA provides faster delivery of packets to destination than GPSR. Because in case of GPSR, once the packet is delivered to unexpected zone, it increases end-to-end delay		[7]

field. Various types of protocols are also discussed here. There are needs to design novel geographical and hierarchical protocols to deal with these networks. Because they will be able to route the packets even with frequent node movements and link quality variations due to the varying distance between communicating nodes. They make use of node's location information to select the best route in terms of distance and reliability. This study also reveals that proactive protocols may perform better as compared to reactive protocols. We also found that to study the aeronautical environment in a more realistic way, there is a need to implement them in 3-D mobility models.

References

1. A.I. Alshabtat, L. Dong, J. Li, F. Yang, Highly-dynamic cross-layered aeronautical network architecture. IEEE Trans. Aerosp. Electron. Syst. **47**(4), 2742–2765 (2011). https://doi.org/10.1109/TAES.2011.6034662
2. I. Bekmezci, O.K. Sahingoz, S. Temel, Flying ad-hoc networks (fanets): a survey. Ad Hoc Netw. **11**(3), 1254–1270 (2013)
3. J.D.M.M. Biomo, T. Kunz, M. St-Hilaire, Y. Zhou, Unmanned aerial ad hoc networks: simulation- based evaluation of entity mobility models impact on routing performance. Aerospace **2**(3), 392–422 (2015)
4. Y. Cai, F.R. Yu, J. Li, Y. Zhou, L. Lamont, Medium access control for unmanned aerial vehicle (UAV) ad-hoc networks with full-duplex radios and multipacket reception capability. IEEE Trans. Veh. Technol. **62**(1), 390–394 (2013). https://doi.org/10.1109/TVT.2012.2211905
5. B.N. Cheng, S. Moore, A comparison of Manet routing protocols on airborne tactical networks, in *MILCOM 2012–2012, IEEE Military Communications Conference* (2012), pp. 1–6, https://doi.org/10.1109/milcom.2012.6415798

6. W. Huba, N. Shenoy, A manet architecture for airborne networks with directional antennas. Int. J. Adv. Netw. Serv. **5**, 333–345 (2012)
7. S. Hyeon, K.I. Kim, A new geographic routing protocol for aircraft ad hoc networks, in *29th Digital Avionics Systems Conference* (2010), pp. E.1–E.2, https://doi.org/10.1109/dasc.2010.5655476
8. K.I. Kim, A simulation study for typical routing protocols in aircraft ad hoc networks. Int. J. Softw. Eng. Appl. **7**(2), 227–234 (2013)
9. D. Kiwior, L. Lam, Routing protocol performance over intermittent links, in *MILCOM 2007—IEEE Military Communications Conference* (2007) pp. 1–8
10. D. Kiwior, E.G. Idhaw, S.V. Pizzi, Quality of service (QoS) sensitivity for the OSPF protocol in the airborne networking environment, in *MILCOM 2005 IEEE Military Communications Conference*, vol. 4 (2005), pp. 2366–2372 https://doi.org/10.1109/milcom.2005.1606022
11. C. Lee, S. Kang, K.I. Kim, Design of hierarchical routing protocol for heterogeneous airborne ad hoc networks. Int. Conf. Inf. Netw. **2014**(ICOIN2014), 154–159 (2014). https://doi.org/10.1109/ICOIN.2014.6799683
12. X. Ma, S. Chisiu, R. Kacimi, R. Dhaou, Opportunistic communications in WSN using UAV, in *2017 14th IEEE Annual Consumer Communications Networking Conference (CCNC)* (2017), pp. 510–515, https://doi.org/10.1109/ccnc.2017.7983160
13. H. Meka, L.M. Chidambaram, S.K. Madria, M. Linderman, M. Kumar, S. Chakravarthy, Roman: routing and opportunistic management of airborne networks, in *2011 International Conference on Collaboration Technologies and Systems (CTS)* (2011), pp. 555–562
14. K. Namuduri, in *Electrical Engineering* (2017), URL https://electrical.engineering.unt.edu/research/laboratories/autonomoussystems-laboratory
15. K. Peters, A. Jabbar, E. K. Etinkaya, J.P.G. Sterbenz, A geographical routing protocol for highly-dynamic aeronautical networks, in *2011 IEEE Wireless Communications and Networking Conference* (2011), pp. 492–497, https://doi.org/10.1109/wcnc.2011.5779182
16. P. Qu, Z. Zhu, Z. Yan, X. Hu, X. Tan, Integrate advantage of geographic routing and reactive mechanism for aeronautical ad hoc networks. Int. J. Hybrid Inf. Technol. **8**(7), 85–94 (2015)
17. R. Ramirez, Link management in the air force airborne network, in *MILCOM 2005 IEEE Military Communications Conference*, vol. 4 (2005), pp. 2488–2493, https://doi.org/10.1109/milcom.2005.1606041
18. J.P. Rohrer, E.K. Etinkaya, H. Narra, D. Broyles, K. Peters, J.P.G. Sterbenz, Aerorp performance in highly-dynamic airborne networks using 3d Gaussmarkov mobility model, in *2011—MILCOM Military Communications Conference* (2011), pp. 834–841, https://doi.org/10.1109/milcom.2011.6127781
19. O.K. Sahingoz, Networking models in flying ad-hoc networks (fanets): concepts and challenges. J. Intell. Rob. Syst. **74**(1–2), 513 (2014)
20. E. Sakhaee, A. Jamalipour, N. Kato, Aeronautical ad hoc networks, in *IEEE Wireless Communications and Networking Conference WCNC* 2006, vol. 1 (2006), pp. 246–251
21. B. Shah, K.I. Kim, A survey on three-dimensional wireless ad hoc and sensor networks. Int. J. Distrib. Sens. Netw. **10**(7), 660–701 (2014)
22. A. Tiwari, A. Ganguli, A. Sampath, Towards a mission planning toolbox for the airborne network: optimizing ground coverage under connectivity constraints, in *2008 IEEE Aerospace Conference* (2008), pp. 1–9, https://doi.org/10.1109/aero.2008.4526375
23. S. Wang, C. Fan, C. Deng, Sun Q. GuW, F. Yang, A-GR: A novel geographical routing protocol for aanets. J. Syst. Architect. **59**(10), 931–937 (2013)
24. W. Wang, C. Dong, H. Wang, A. Jiang, Design and implementation of adaptive MAC framework for UAV ad hoc networks, in *2016 12th International Conference on Mobile Ad-Hoc and Sensor Networks (MSN)* (2016), pp. 195–201, https://doi.org/10.1109/msn.2016.039
25. Y. Wang, Y.J. Zhao, Fundamental issues in systematic design of airborne networks for aviation, In *IEEE Aerospace Conference, IEEE* (2006) 8 p
26. C. Yin, Z. Xiao, X. Cao, X. Xi, P. Yang, D. Wu, Enhanced routing protocol for fast flying UAV network, in *2016 IEEE International Conference on Communication Systems (ICCS)* (2016) pp. 1–6, https://doi.org/10.1109/iccs.2016.7833587

D-MSEP: Distance Incorporated Modified Stable Election Protocol in Heterogeneous Wireless Sensor Network

Anureet Singh, Hardeep Singh Saini and Naveen Kumar

Abstract The nodes in Wireless Sensor Networks are communicating wirelessly and their energies are depleted in the communication among themselves or with the Base Station (BS). In order to save their energies, various heterogeneous routing protocols have been discovered. In this paper, we have proposed D-MSEP (Distance incorporated Modified Stable Election Protocol) which is an improved version of M-SEP (Modified Stable Election Protocol) as it incorporates the distance factor in probabilistic formula of Cluster Head selection in each cluster. The inclusion of distance factor helps in avoidance of Cluster Head selection farther from the BS. This leads to saving of huge amount of energy leading to the escalated stability period by 32 and 173.38% as compared to the M-SEP protocol. In addition to this, the network lifetime is enhanced by 56.20% as compared to M-SEP. The proposed protocol is best suitable for large area applications to disseminate the information from the network.

Keywords Heterogeneous wireless sensor network · M-SEP; D-MSEP
Distance incorporated

1 Introduction

Wireless sensor network (WSN) has been promising in providing the accessibility to the remote areas where the human intervention was never feasible. With the tremendous growth of sensing technology in almost every field, WSN has emerged as the

A. Singh (✉) · H. S. Saini
Indo Global College of Engineering, Chandigarh, Punjab, India
e-mail: oyeluckykuckyoye@gmail.com

H. S. Saini
e-mail: hardeep_saini17@yahoo.co.in

N. Kumar
Thapar Institute of Engineering and Technology (Deemed University), Patiala, Punjab, India
e-mail: n.kumar.1987@ieee.org

C. R. Krishna et al. (eds.), *Proceedings of 2nd International Conference on Communication, Computing and Networking*, Lecture Notes in Networks and Systems 46, https://doi.org/10.1007/978-981-13-1217-5_27

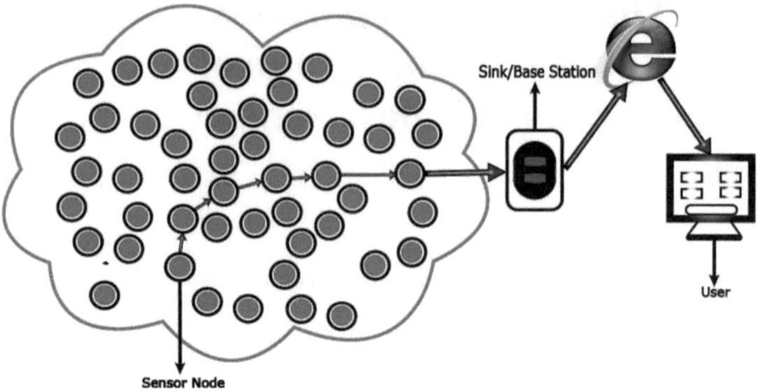

Fig. 1 WSN architecture [1]

one of the important support for exploiting the non-attended applications [1]. In fact, it has created an unparalleled interest among the researchers because of sheer number of applications. Designed mainly for battlefield surveillance, it is now used in almost every field like agricultural, habitat monitoring, environmental operations including forest fire detection, flood detection, detection of volcanic eruption, etc. [2]. WSN consists of various small-sized, battery-operated nodes which help in monitoring the attributes, deployed in large-scale areas. The information is disseminated from the network and is passed on to the Base Station or sink. WSN is totally application specific, so the nature of nodes and nature of sink in terms of being static and mobile depends on the application for which it is designed. The data collected at the sink is further communicated to the end user with the use of Internet. The architecture of WSN is shown as in Fig. 1.

The sensor node is equipped with four basic main components namely battery, microprocessor, sensing unit, and transceiver. There are other application-dependent units like mobilizer and GPS device. Battery resources have been the most important part of a sensor node due to its irreplaceable nature, once it is exhausted. Therefore, the main objective in WSNs has been the conservation of energy of nodes. For achieving this goal of energy conservation, various routing protocols which are energy efficient have been developed. To perform the energy-efficient balanced routing, the sensor nodes are grouped which are termed as clusters. Clustering not only provides scalability to the network, instead it also makes the network much more load balanced leading to the enhancement in the network lifetime [3]. The selection of Cluster Head (CH) is a research topic being worked on by many WSN researchers. CH works in aggregation of data from the cluster members and to send it to the sink. WSN is made to work in two types of network, heterogeneous and homogeneous network. Homogeneous network deals with the same types of nodes in terms of energy, computation capabilities, connectivity range, whereas the heterogeneous network posses nodes with different afore-mentioned parameters [4].

The paper is organized in the following manner: The related literature work is discussed in Sect. 2. Section 3 explains briefly about heterogeneous routing protocols: Sect. 4 states the main contribution to the research work. Section 5 consists of proposed protocol's explanation. Section 6 presents and discusses simulation results. In the end, there is conclusion and future scope.

2 Related Work

Since the development of various applications facilitated by WSN, the one of the major concern has been in the conservation of energy while the data transmission is in progress. Homogeneity in sensor network does not exist ideally. So, introduction of heterogeneity in the network provides much stability to the network. SEP which introduced the concept of different levels of energy nodes in the network provided much enhanced network lifetime to the network. It failed when it was being applied for more than two levels [5]. DEEC [6] introduced the factor of energy ratio, i.e., ratio of average and residual energy. However, it made the advanced nodes suffering from penalized effect as it kept on making advanced nodes as Cluster Heads more frequently irrespective of their energy stock. DDEEC [7] worked in improving the DEEC protocol by declaring a threshold value, which decides the criteria of selection of Cluster Heads by treating all nodes equally. After the development of two-level routing protocols, EEHC [8] was the first one who introduced three levels of heterogeneity, containing supernodes, advanced nodes, and normal nodes. The performance evaluation of EEHC has shown the improvement of 10% over the LEACH [9] protocol. However, results shows that EEHC suffered from the drawback that it could not handle the penalizing effect as happened in DEEC protocol. EDDEEC [10] worked similar to the DDEEC but on the three levels. It introduced the threshold concept to avoid the penalizing of high energy-rich nodes. BEENISH [11] and D-BEENISH [16] worked for the four levels of heterogeneity by introducing the ultra-nodes along with super, advanced, and normal nodes. Load balancing is being discussed in EEZECR [12] protocol. M-SEP [13] is the modified version of SEP protocol which modifies the threshold formula for Cluster Head selection. The simulation results of M-SEP protocol gives 40 and 55%, respectively, longer than LEACH and SEP protocol. A balanced energy-efficient protocol proposed in [14] and a stable election-based protocol is proposed in [15] and gives an insight on how to achieve an efficient network.

3 Heterogeneous Routing Protocols

Since the heterogeneous network has come into limelight, the much of the research has been focusing towards achieving the network stability while developing the routing protocols. While clustering is performed in WSN, the much of research is focused on the energy-efficient Cluster Head selection. Cluster Head selection is discussed in the following two protocols with their threshold and probabilistic formulae are as follows.

3.1 Stable Election Protocol (SEP)

It was the first protocol to exploit the heterogeneity at the two levels by considering the normal nodes and advanced nodes in the network. CH selection has been based on the weighted probability which is given as in Eq. 1, but it suffered from instability when made to work for more than two levels.

$$P_{\text{nrm}} = \left\{ \frac{P_{\text{opt}}}{1 + am} \text{for normal nodes; } \frac{P_{\text{opt}}\,(1 + a)}{1 + am} \text{for advance nodes;} \right. \tag{1}$$

In Eq. (1), P_{opt} is the optimum Number of Cluster Heads in the network where "a" is the fraction of amount of energy and "m" is the fraction of advanced nodes.

3.2 Modified Stable Election Protocol (M-SEP)

This protocol is a modified version of SEP which enhances the network lifetime by considering the energy of nodes in current ongoing round and average energy of the whole network.

$$P_{\text{nrm}} = \frac{P_{\text{opt}}}{1 + am * Ei(n)} \tag{2}$$

In above Eq. (2), $Ei(n)$ is the current nodal energy and P_{nrm} is the normal nodes probability. The threshold formula for the Cluster Head selection is given as in Eq. (3).

$$T(S_{\text{nrm}}) = \begin{cases} \frac{3 * P_{\text{nrm}}(i) * E_{\text{avg}}(r)}{1 - P_{\text{nrm}}\left(r \mod \left(\frac{1}{3 * P_{\text{nrm}}(i)}\right)\right) * Ei(n)} & \text{if } S(i) \in G \\ 0 & \text{otherwise} \end{cases} \tag{3}$$

E_{avg} is the average energy of the network. CH selection equations are based on the threshold-based probabilistic equations for advanced nodes is given by Eqs. (4) and (5)

$$\text{P}_{\text{adv}} = \frac{P_{\text{opt}}(1 + a)}{(1 + am) * Ei(n)} \tag{4}$$

$$\text{T}(S_{\text{adv}}) = \begin{cases} \frac{3 * P_{\text{adv}}(i) * E_{\text{avg}}(r)}{1 - P_{\text{adv}}\left(r \mod \left(\frac{1}{3 * P_{\text{adv}}(i)}\right)\right) * Ei(n)} & \text{if } S(i) \in G' \\ 0 & \text{otherwise} \end{cases} \tag{5}$$

G' is the set of nodes which are eligible to become Cluster Heads.

4 Proposed Protocol: Distance Incorporated Modified Stable Election Protocol (D-MSEP)

In this work, two levels of energy heterogeneity is considered; normal nodes and advanced nodes. Equation (6) gives the probability for normal nodes to become Cluster Head. The proposed work introduces the distance factor in the threshold-based equation with the condition if the nodes are located within the average distance to the sink as in Eq. (7) which determines the Cluster Head selection among the normal nodes. However, the threshold equations remain the same as that of M-SEP if the distance to the sink is more than the average distance from the sink.

The novelty of the proposed work is the selection of CH in M-SEP protocol is made energy efficient by incorporating distance factor. Also, the performance of proposed protocol (D-MSEP) is evaluated by validating its performance with SEP and M-SEP protocols dealing with the sink located outside the network which is intended to make it applicable for the unattended applications.

4.1 Normal Cluster Head Selection

Among normal nodes, the Cluster Head is selected using Eqs. (6), (7), and (8).

If distance $\leq Davg$

$$P_{\mathrm{nrm}} = \frac{P_{\mathrm{opt}}}{1 + am * Ei(n)} \tag{6}$$

$$T(S_{\mathrm{nrm}}) = \begin{cases} \frac{3 * P_{\mathrm{nrm}}(i) * E_{\mathrm{avg}}(r)}{1 - P_{\mathrm{nrm}}\left(r \bmod \left(\frac{1}{3 * P_{\mathrm{nrm}}(i)}\right)\right) * Ei(n)} & \text{if } S(i) \in G' \\ 0 & \text{otherwise} \end{cases} \tag{7}$$

If distance $> D_{\mathrm{avg}}$

$$T(S_{\mathrm{nrm}}) = \begin{cases} \frac{3 * P_{\mathrm{nrm}}(i) * E_{\mathrm{avg}}(r) * D(i)}{1 - P_{\mathrm{nrm}}\left(r \bmod \left(\frac{1}{3 * P_{\mathrm{nrm}}(i)}\right)\right) * Ei(n) * D_{\mathrm{avg}}} & \text{if } S(i) \in G' \\ 0 & \text{otherwise} \end{cases} \tag{8}$$

4.2 Advanced Cluster Head Selection

Among normal nodes, the Cluster Head is selected using Eqs. (9), (10), and (11).

$$\mathrm{P_{adv}} = \frac{P_{\mathrm{opt}}}{(1+am) * Ei(n)}(1+a) \tag{9}$$

If distance $\leq Davg$

$$\mathrm{T}(S_{\mathrm{adv}}) = \begin{cases} \frac{3*P_{\mathrm{adv}}(i)*E_{\mathrm{avg}}(r)}{1-P_{\mathrm{adv}}\left(r \mod \left(\frac{1}{3*P_{\mathrm{adv}}(i)}\right)\right)*Ei(n)} & \text{if } S(i) \in G' \\ 0 & \text{otherwise} \end{cases} \tag{10}$$

If distance $> Davg$

$$T(S_{\mathrm{adv}}) = \begin{cases} \frac{3*P_{\mathrm{adv}}(i)*E_{\mathrm{avg}}(r)*D(i)}{1-P_{\mathrm{adv}}\left(r \mod \left(\frac{1}{3*P_{\mathrm{adv}}(i)}\right)\right)*Ei(n)*D\mathrm{avg}} & \text{if } S(i) \in G' \\ 0 & \text{otherwise} \end{cases} \tag{11}$$

4.3 Radio Energy Consumption (REC) Model

The basic model for REC is given by the set of Eqs. (12), (13), (14), and (15). There is standard energy consumption that a node encounters while transmitting data to the sink.

$$E_{tx}(l,d) = lE_{\mathrm{elc}} + lE_{\mathrm{efs}}d^2 \text{for } d < d_{\mathrm{o}} \tag{12}$$

$$E_{tx}(l,d) = lE_{\mathrm{elc}} + lE_{\mathrm{efs}}d^4 \text{for } d > d_{\mathrm{o}} \tag{13}$$

where d is the distance between the two nodes or between node and sink, and the threshold distance is represented by d_{o}. The reception of message also consumes energy by the following equation:

$$E_{rx}(l) = lE_{\mathrm{elec}} \tag{14}$$

During data aggregation process, there is some amount of energy consumption given by the following equation:

$$E_{dx}(l) = mlE_{da} \tag{15}$$

Table 1 Specifications of parameters for simulating the network

Parameter	Value
Network coverage	(100, 100) m
BS location	(50, 175) m
Number of nodes	100
Initial energy	0.1 J
Eelec	50 nJ per bit
Efs	10 pJ/bit/m^2
Emp	0.0013 pJ/bit/m^4
d_0	87 m
Eda	5 nJ/bit/signal
Packet size	2000 bit

4.4 Network Area and Simulation Parameters

The network area is 100 m $\times$ 100 m with total number of nodes deployed randomly is 100. Sink is placed outside the network at (50, 175) m. All normal nodes are equipped with energy of 0.1 J and advance fraction $a = 1$ and $m = 0.1$. The parameters used for simulating the environment of WSN are presented in Table 1.

5 Simulation and Discussions

D-MSEP protocol is simulated in MATLAB 2013a. Advanced nodes are 10 in numbers and remaining 90 nodes out of 100 total nodes are normal nodes. The energy of advanced nodes are 0.2 J as compared to 0.1 J of normal nodes.

5.1 Stability Period

The period of completing number of rounds before the first node dies is stability period. After simulation, it is observed that stability period in D-MSEP is enhanced by 31.90% as compared to M-SEP and 173.38% as compared to SEP protocol. It is the result of the proposed approach being implemented to make the Cluster Head selection energy efficient. It is because the depletion of energy is decreased due to the avoidance of Cluster Head selection at the farther distance from the sink. SEP has 124 rounds, M-SEP covers 257 rounds, and D-MSEP covers 339 rounds until the first node is dead as shown in Fig. 2.

Fig. 2 Stability period of network

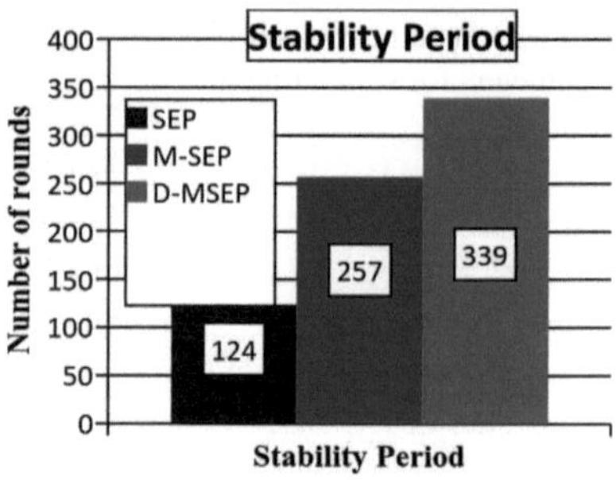

Fig. 3 Lifetime comparison

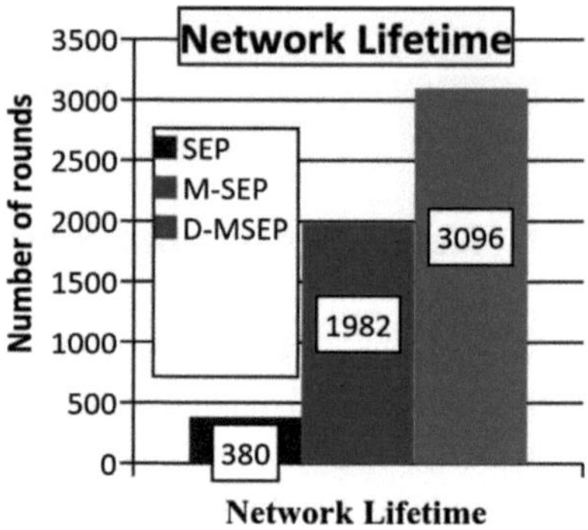

5.2 *Network Lifetime*

It is the important parameter to validate the performance of D-MSEP. It is defined as the number of rounds covered from the moment when the network starts functioning and the time it stops operating. It can be termed as last node dead. As shown in Fig. 3, the last node dead is at 3096 rounds in case of D-MSEP as compared to 1982 rounds and 380 for M-SEP and SEP protocols, respectively, leading to the enhancement by 56.20% as compared to M-SEP.

5.3 *Number of Dead Nodes Versus Rounds*

The graph in Fig. 4 shows that the number of dead nodes versus rounds is much better in case of D-SEP as compared to M-SEP and SEP.

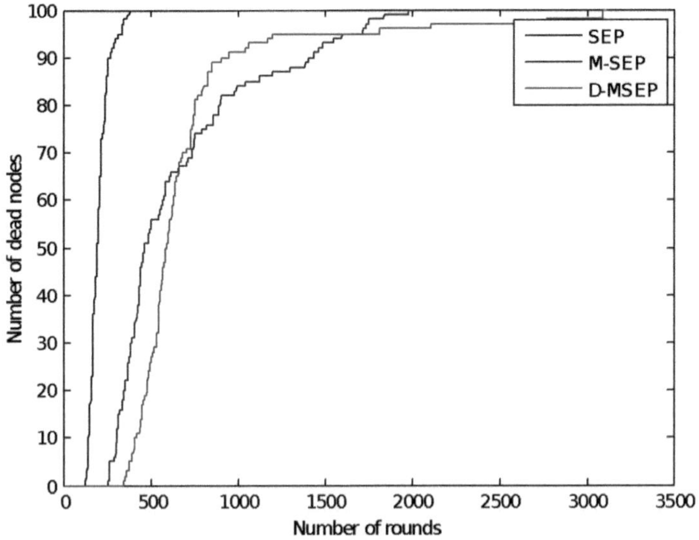

Fig. 4 Dead nodes comparison

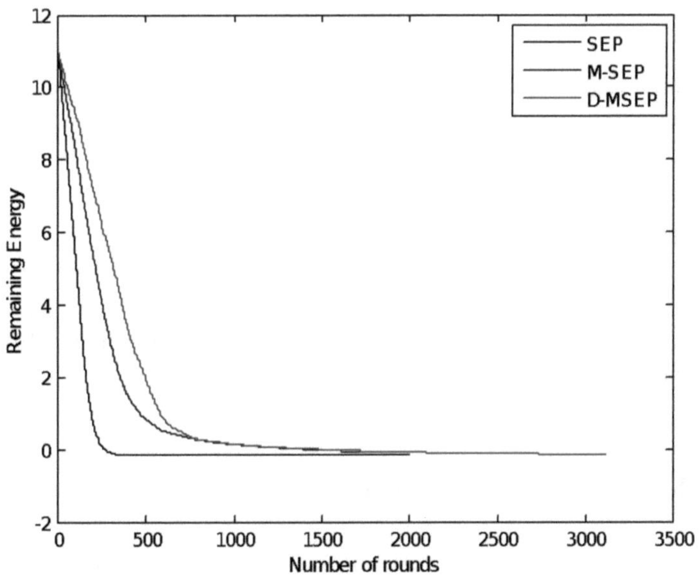

Fig. 5 Remaining energy of network

5.4 Remaining Energy of Network

It is the one another performance metric which depicts the rate of energy depletion in the network as the data transmission proceeds. The remaining energy of the D-MSEP

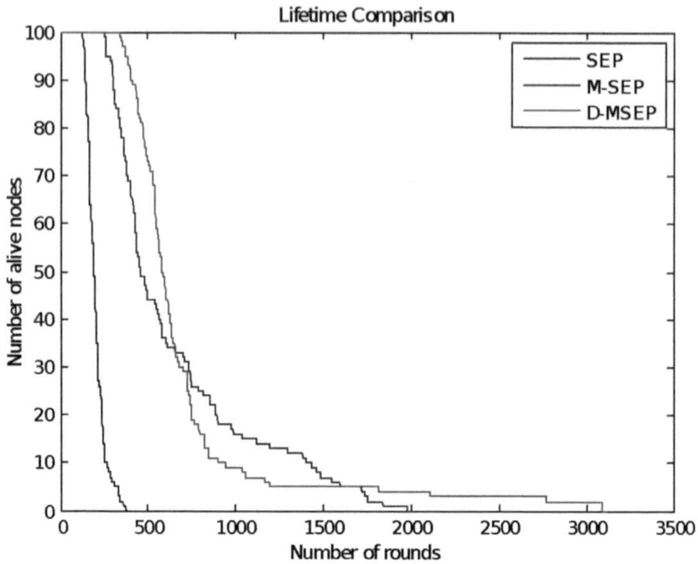

Fig. 6 Number of alive nodes versus rounds

covers much more rounds as compared to the M-SEP and SEP protocols as shown in the Fig. 5.

5.5 *Number of Alive Nodes Versus Rounds*

The load balancing in the network is ensured by this parameter. As it can be observed from Fig. 6, the graph of alive nodes versus rounds is steeper than the M-SEP accounting to the load balancing in the network.

6 Conclusion and Future Work

Heterogeneous routing makes the network much more stable and energy balance thereby covering much more rounds as compared to heterogeneous protocols. A D-MSEP protocol is proposed which incorporates the distance factor for Cluster Head selection. It performs on two level of heterogeneity which makes it more economical as compared to other heterogeneous routing protocols. It is clear from the simulation results that the proposed D-SEP shows better performance than M-SEP and SEP protocols in terms of stability period, remaining energy of the network, network lifetime, and number of alive nodes versus rounds accounting to the network load balancing. The future work will be increasing throughput by incorporating moving sink in the network.

References

1. I.F. Akyildiz, W. Su, Y. Sankarasubramaniam, E. Cayirci, A survey on sensor networks. IEEE Commun. Mag. **40**, 102–114 (2002)
2. J.N. Al-Karaki, A.E. Kamal, Routing techniques in wireless sensor networks: a survey. IEEE J. Wirel. Commun. **11**, 6–28 (2004)
3. S. Bandyopadhyay, E.J. Coyle, An energy efficient hierarchical clustering algorithm for wireless sensor networks, in *Proceedings of Twenty-Second Annual Joint Conference of the IEEE Computer and Communications, IEEE Societies*, vol. 3 (2003), pp. 1713–1723
4. M. Yarvs, N. Kushalnagar, H. Singh, Exploiting heterogeneity in sensor networks, in *Proceedings of 24th Annual Joint Conference of the IEEE Computer and Communications Societies (INFOCOM), IEEE Societies*, vol. 2 (2005), pp. 878–890
5. G. Smaragdakis, I. Matta, A. Bestavros, SEP: A stable election protocol for clustered heterogeneous wireless sensor networks, in *Proceedings of the Second International Workshop on SANPA* (2004), pp. 1–11
6. L. Qing, Q. Zhu, M. Wang, Design of a distributed energy-efficient clustering algorithm for heterogeneous wireless sensor networks. Comput. Commun. J. **29**, 2230–2237 (2006)
7. B. Elbhiri, R. Saadane, S. El Fkihi, D. Aboutajdine, Developed distributed energy efficient clustering (DDEEC) for heterogeneous wireless sensor networks, in *5th International Symposium on I/V Communications and Mobile Network (ISVC)* (2010), pp. 1–4
8. D. Kumar, T.C. Aseri, R.B. Patel, EEHC: energy efficient heterogeneous clustered scheme for wireless sensor networks. Comput. Commun. J. **32**, 662–667 (2009)
9. W.R. Heinzelman, A. Chandrakasan, H. Balakrishnan, Energy-efficient communication protocol for wireless microsensor networks, in *Proceedings of the 33rd Hawaii International Conference on System Sciences*, vol. 8 (2000), p. 8020
10. N. Javaid, T.N. Qureshi, A.H. Khan, A. Iqbal, E. Akhtar, M. Ishfaq, EDDEEC: enhanced developed distributed energy-efficient clustering for heterogeneous wireless sensor networks. Procedia Comput. Sci. **19**, 914–919 (2013)
11. T.N. Qureshi, N. Javaid, A.H. Khan, A. Iqbal, E. Akhtar, M. Ishfaq, BEENISH: balanced energy efficient network integrated super heterogeneous protocol for wireless sensor networks. Procedia Comput. Sci. **19**, 920–925 (2013)
12. S. Verma, K. Sharma, Energy efficient zone divided and energy balanced clustering routing protocol (EEZECR) in wireless sensor network. Int. J. Circ. Syst. (CSIJ) **1**, 19–29 (2014)
13. D. Singh, C.K. Panda, Performance analysis of modified stable election protocol in heterogeneous WSN, in *International Conference on Electrical, Electronics, Signals, Communication and Optimization (EESCO)* (2015), pp. 1–5
14. A. Preethi, E. Pravin, D. Sangeetha, Modified balanced energy efficient network integrated super heterogeneous protocol, in *International Conference on Recent Trends in Information Technology (ICRTIT)* (2016), pp. 1–6
15. P.G.V. Naranjo, M. Shojafar, A. Abraham, E. Baccarelli, A new stable election-based routing algorithm to preserve aliveness and energy in fog-supported wireless sensor networks, in *IEEE International Conference on Systems, Man, and Cybernetics (SMC)* (2016), pp. 002413–002418
16. R. Kumar, R. Mathur, D-BEENISH: distance incorporated balanced energy efficient network integrated super heterogeneous protocol for WSN. Int. J. Syst. Control Commun. **8**, 187–203 (2017)

Radio Environment Map Based Radio Resource Management in Heterogeneous Wireless Network

Rajarshi Mahapatra

Abstract This work uses radio environment map (REM) to allocate spectrum dynamically among the heterogeneous base stations (BSs) within a particular geographical area. With the knowledge of BSs location and their power models, this work first forms a REM of the corresponding coverage area by considering serving BS and interfering BSs, then uses this REM for radio resource management (RRM) among different BSs to minimize outage probability and reducing service blocking to the users. The simulation result shows that the REM-based resource allocation performed better with increasing complexity.

Keywords Radio Environment Map (REM) · Radio Resource Management (RRM) · Heterogeneous wireless network

1 Introduction

Heterogeneous wireless technologies deployed in environment support QoS requirements with increasing number of users of diverse applications. In cellular, the cell dimension is moving from macrocell to microcell to picocell to femtocell to accommodate more number of subscribers [2]. Spectrum allocation among the coexisting different tier cells is a challenging task to reduce interference while supporting user QoS [5]. Among many techniques, radio environment maps (REM) can be used for efficient radio resource management (RRM) mechanism in wireless network [6]. REM refers to a database that dynamically stores information about the environment. REM can be formed in two ways; local REM databases will be formed with radio-propagation-related information, such as propagation losses, signal strengths, and the locations of individual BSs or BSs in close proximity and the global REM stores the less dynamic parameters, such as QoS metric from the local REM [7, 8].

R. Mahapatra (✉)
Dr. SPM International Institute of Information Technology, Naya Raipur,
Naya Raipur, Chhattisgarh, India
e-mail: rajarshim@ieee.org

C. R. Krishna et al. (eds.), *Proceedings of 2nd International Conference on Communication, Computing and Networking*, Lecture Notes in Networks and Systems 46, https://doi.org/10.1007/978-981-13-1217-5_28

This work proposes a REM-based RRM mechanism among the coexisting heterogeneous BSs, while causing no (or insignificant) interference to the other. At first, local and global REM databases have been constructed in step and a central controller allocates the spectrum with the help of REM databases.

The proposed technique uses simple mechanism to allocate radio resource among BSs appropriately as per traffic generation to minimize outage. This spectrum assignment to BSs (i.e., eventually to UE) for satisfying UE's QoS requirement is to reduce service blocking. This allocation needs to be controlled dynamically due to dynamic traffic generation & requirements, which are based on BS's locations, their characteristics, UE generation, UE's location, and their requirements. In particular, the proposed allocation technique aims to minimize outage probability and minimum service blocking of UE within heterogeneous network (HetNet). In addition to this, the RRM technique can also activate and deactivate the BSs as and when required which help to reduce network power consumption. The contributions in this paper as follows:

- To propose a REM-based RRM techniques in HetNet environment.
- To minimize outage probability and service blocking of UE's within HetNet.
- To reduce network power consumption while satisfying UE's requirements.

2 System Model and Utility Measure

In this work, HetNet consists of femtocell BSs within the coverage of macrocell, microcell and picocell BSs with randomly deployed. The system model of the proposed REM-based RRM technique is shown in Fig. 1. As shown in figure, local REM database formed with the help of locations of individual BSs and their power models and the global REM store the less dynamic parameters, such as QoS metric or the parameters that may affect a high number of network nodes. After forming the REM databases, the central controller placed along with global REM assigns the radio resource among the BSs to minimize outage probability and service blocking (not satisfy QoS of requested service).

2.1 Outage Probability

The probability of coverage in a downlink cellular network at decreasing levels of generality. The signal to interference and noise ratio (SINR) of the k-th UE at a distance r from its serving BS m can be expressed as $\gamma = \frac{G_k^{mm} P^m}{\sigma^2 + I_r}$, where P^m is the transmit power of m-th BS, and I_r is the interference on the UE and σ^2 is the noise PSD. In this work, the received signal power at the UE k is $G_k^{mm} = L_k^h(r_k^m) \times \eta_k^m$, where L and η are the distance-dependent channel gain and the shadowing component for k-th UE of m-th BS. Then, the average outage probability is defined as $p(\gamma_T) = prob[\gamma < \gamma_T]$ [9].

2.2 Service Blocking Outage

Service blocking means that QoS of the particular UE is not supported by the network (bandwidth in the case). This work considered a Markov queue model to analyze the service blocking. Let c_m, c_n, c_p and c_f be the wireless channels allocated to macro, micro, pico, and femtocell, respectively. The traffic is generated by Poisson process, with rate λ. The service blocking probability of UE b for "case m, n, h, f, v" is $S^b_{mnhfv} = \frac{a^c/c!}{\sum_{j=0}^{c} a_j/j!}$, where a is the traffic offered to a group in Erlang B, which is defined by λ/μ, $1/\mu$ is the average duration of traffic, and $c = mc_m + nc_n + pc_p + fc_f$.

Outage of signal strength and blocking of service occurs independently with each other. Signal strength outage occurs when $SINR < \gamma_T$, whereas service blocking outage occurs when UE, which is having $SINR > \gamma_T$ cannot serve by BS mainly due to limited bandwidth, restriction to number of supported UE, and high throughput requirement. Thus, UE at position $r = (r_1, r_2)$ suffers outage with probability of $p(\gamma_T))$. Then the total outage probability along with service blocking can be expressed as $1 - (1 - p(\gamma_T))(1 - S^b_{mnhfv})$.

In general, UEs are randomly distributed within the specified coverage area and they communicate with the serving BS while all other BSs act as interferers. This work considered that the UEs are connected to nearest BSs, which are associated using Voronoi tessellation; namely the UEs in the Voronoi cell of a particular BS of its tier are served by it. Here, each UE has the capability to connect to any tier BS. Since, the UEs are in the heterogeneous environment, the cross-tier connection, cross-tier handover, etc. can pose as a serious bottleneck to connect nearest BS. However, several technologies (regarding vertical handover) as well as standards (IEEE 802.21) came up to support vertical handover seamlessly.

After defining the Voronoi tessellation, now it is the turn for UE association with nearest BS of suitable tier. This work categorizes UEs into four types; (A) high throughput of static indoor UE, (B) high throughput of static outdoor UE, (C) low mobile outdoor UE, and (D) high mobile outdoor UE. During association, a preference list of each type of UE has been considered. For *Type A*, UE prefers femtocell, *Type B* UE prefers picocell, *Type C* UE prefers microcell, and *Type D* UE prefers macrocell [10]. After the association, REM plays a significant role to allocate radio resource to the BSs, i.e., users.

3 REM-Based RRM Technique

At the beginning of REM-based RRM techniques, the REM has been formed for the area under observation in two layers, local REM and global REM with the knowledge of BS's locations and their power model. Here, 3GPP path loss models for different BSs are used. Each BS stores their own REM in local REM database. Once constructed, UE association will start with knowledge of their location, type, and preference.

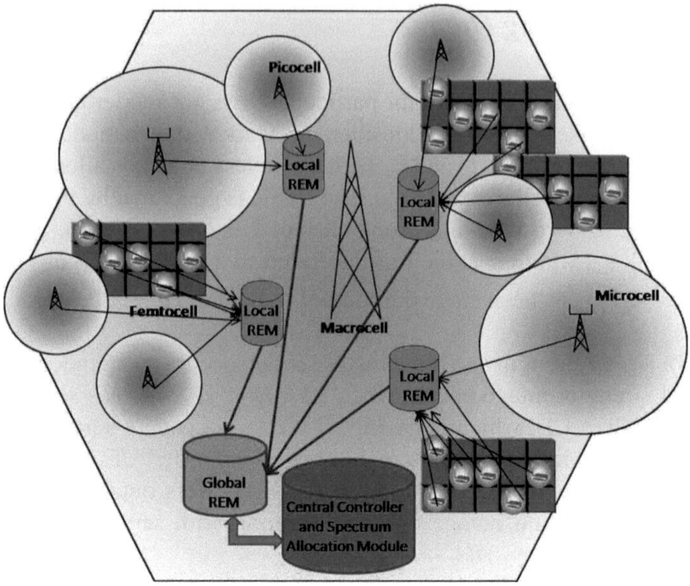

Fig. 1 System model to the proposed work

For *Type A*, UE femtocell is considered. *Type A* UE is associated with nearest femtocell BS as preferences. Similarly, *Type B* and *Type C* UEs are associated with picocell and microcell, respectively. *Type D* UEs are associated with macrocell only. Here, each BS can serve a limited number of UE (*e.g.*, maximum 5 UEs for femtocell). This parameter plays an important role during the association. If any BS cannot serve a particular UE due to this, then UE can associate any nearest BS of other tier or same tier if available.

The associated BS will be considered as serving BS of that particular UE and all other BSs are considered as interfering BSs. After association, the received SINR will be calculated from the REM database. Having knowledge of UE association, spectrum allocation started from the cell of higher coverage area, i.e., from macrocell to femtocell. There are two reasons behind this; first, it overlaps more number of smaller cells, the value of total interference can be minimized by using farthest spectrum allocation for overlapping smaller cell and second, there is a low BW requirements for high mobile UEs in general. This allocation procedure depends on activation/deactivation of BSs in the area. After the assignment, network can find the number of UE, which are mainly two types; one whose SINR is below threshold, causing outage and other whose SINR above the threshold but not enough to satisfy particular service.

Selection of appropriate BSs depends on several factors, such as, UE's location, UE's requirement, UE mobility, traffic load of the network and traffic distribution within this region, etc. With the help of these information, BS can determine the required transmit parameters to satisfy the UE's requirements. Local REM algorithm provides the few options for possible communication scenario and forwards them to the all BSs, which are present within its communication range. These information are then forwarded by each BSs to the central controlling center for global REM. Upon receiving these, the controlling center takes the decision on the BSs access based on the global spectrum available, which is at network level, i.e., number of UEs, number of activated BSs and network power consumption. Finally, the controlling center acts on UE request and assigns the radio resource to UE.

Among these, few parameters have an impact locally, which are considered as local variables and others are considered as global variables [7]. Local variables are used in local energy efficiency (EE) to optimize the energy consumption within limited geographical region or UE equipment specific. Whereas global EE parameters are used to optimize global, i.e., network energy consumption. Local EE provides the *local snapshot* of radio environment at UE level, i.e., BS coverage, throughput requirements, supported mobility. These factors influence the BS activation/deactivation to minimize energy consumption metric. In this allocation process, some BSs are not being utilized, which can be put into sleep mode to minimize network energy consumption.

4 Calculation of Interefence and Energy Consumption in REM-Based RRM Technique

After allocating spectrum, the interference and energy consumption of the network have been estimated. In order to minimize outage and service blocking with REM-based RRM technique, it is essential to estimate the interference on each UE to satisfy UE's requirements. REM provides information about SINR at a particular location, i.e., UE's location. Since, the entire geographical area covers by four types of cells (as shown in Fig. 2) and at any moment, UE connects to a particular tier BSs, which is nearest (using Voronoi tessellation) and supports its type. In HetNet environment, different tiers BSs overlap each other and contribute interference on the UEs. Since, the entire geographical area covers by four types of cells, among these only macrocell is always active, rest are activated and deactivated appropriately as and when required.

In this proposed scheme, generalized fractional frequency reuse scheme has been modified, where some portion of frequency always is allocated to macrocell, say f_M and $f_B - f_M$ are allocated to other BSs according to their activation, where f_B is the total amount of frequency available for communication [6].

The total interference (I^{Tot}) can be expressed as

$$I^{Tot} = \sum_{l=1}^{f_B} \sum_{i=1}^{N_{sub}} \left(\alpha_{i,k} I_i^{lh} + \sum_{d=1}^{N_{tot}} J_i^{dl} \right) \tag{1}$$

where $\alpha_{i,k}$ is the binary assignment variables $\{0, 1\}$, I_i^{lh} is mutual interference described in [6], and N_{sub} is the number of RBs.

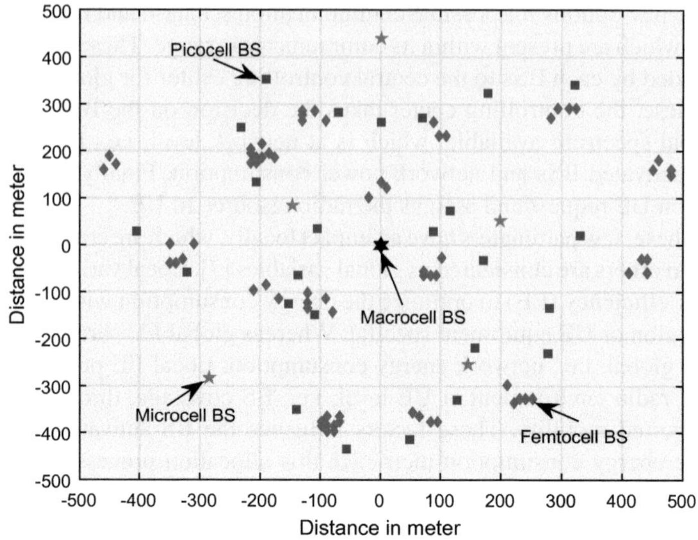

Fig. 2 Position of different BSs in HetNet environment

Due to the characteristic of linear combination, E^{v}_{mnhf} [9] is expressed as

$$E^{v}_{mnhf} = mE^{macro}_{active} + nE^{micro}_{active} + hE^{pico}_{active} + fE^{femto}_{active} \quad (2)$$

The EE of the system is defined by E^{v}_{mnhf}/N_c.

5 Results and Discussion

The proposed REM-based RRM technique has been simulated in wireless environment over the entire service area of 500 × 500 m. In this simulation, 20 blocks of femtocell (5 × 5 femto in each blocks, as per the BeFEMTO specification [3]), are deployed with 20% activation. The microcell (5 in numbers) and picocell (25 nos) are randomly deployed inside macrocell coverage of 500 × 500 m. Figure 2 shown position of different tier BSs in HetNet environment. With this all active BSs, Fig. 3 shows the REM of SINR of the coverage area under consideration. In this figure, each BS is active and during the REM of each BSs are formed by considering it as serving BS and all other BSs as interfering station.

As per the recent study of traffic distribution [4], considered in this analysis as 55% of indoor UE, 18% high mobile UE, 15% of outdoor low mobile UE and 12% static outdoor UE. In this algorithm, UEs are associated with BS based on their characteristics. This work also considers the maximum number of UE association to a particular BSs, like femtocell, it is limited to 5, for picocell the number is 10, and

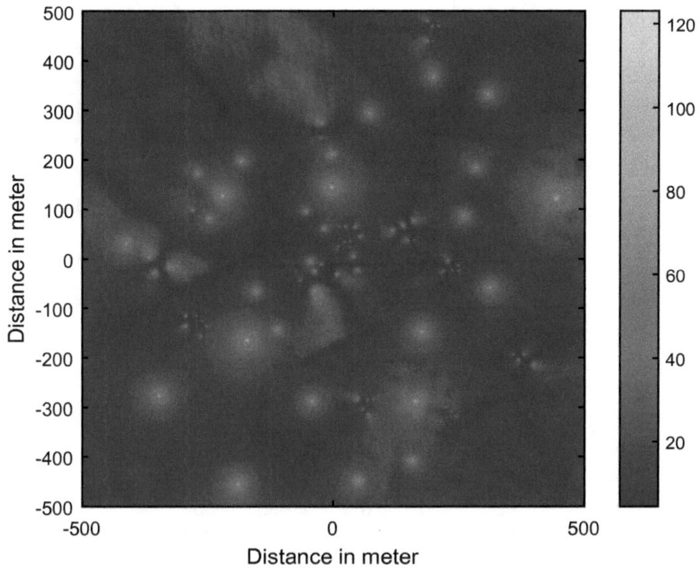

Fig. 3 REM of SINR with all BSs activated

microcell can support maximum of 30 UE and macrocell supports maximum of 50 UEs per sector. Different types of UEs are considered to understand the effectiveness of the proposed RRM algorithm. Indoor and outdoor UEs have been considered appropriately within the path loss models. In case of mobile UE, the effect of Doppler spread has been taken into consideration during spectrum allocation. The mobile with 100 kmph at 2 GHz frequency, the Doppler spread is approximately 180 KHz, i.e., one RB. Using the REM-based RRM technique, Fig. 4 shows the spectrum allocation scenario in a typical deployment of BSs. This allocation satisfies interference threshold and service blocking criteria. The different color represents the different frequencies. As shown in figure, the neighboring cells are assigned with different frequency band (more separation) to minimize interference. Figure 5 shows the outage of the proposed scheme in comparison with the past SINR-based scheme.

However, the service blocking and the probability of outage decrease with more activation of BSs. More number of activated BSs increase the network power consumption and also support improved network throughput. Table 1 provides the detailed numbers of mobile of each type, which falls in outage category. Figure 6 shows the normalized network energy efficiency with number of activated BSs. As evident from the figure, the EE increases upto a certain number of BSs activation. The EE decreases with more BS activation, which eventually increases network power consumption, whereas the throughput will not increase significantly due to increase in interference.

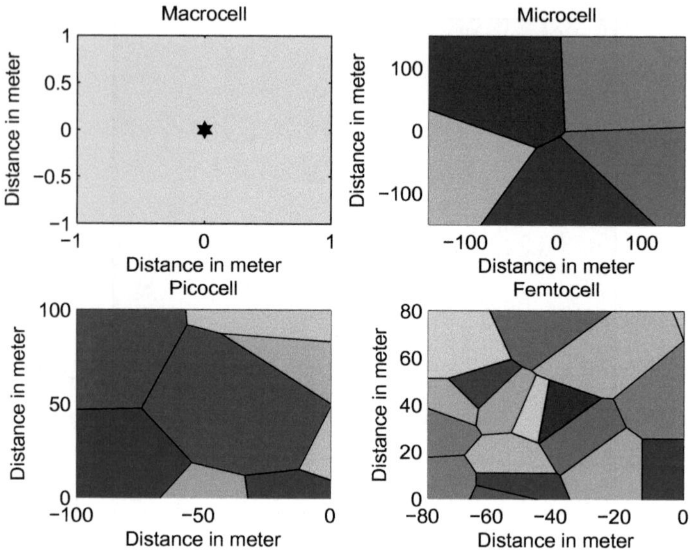

Fig. 4 Radio spectrum allocation across heterogeneous BSs

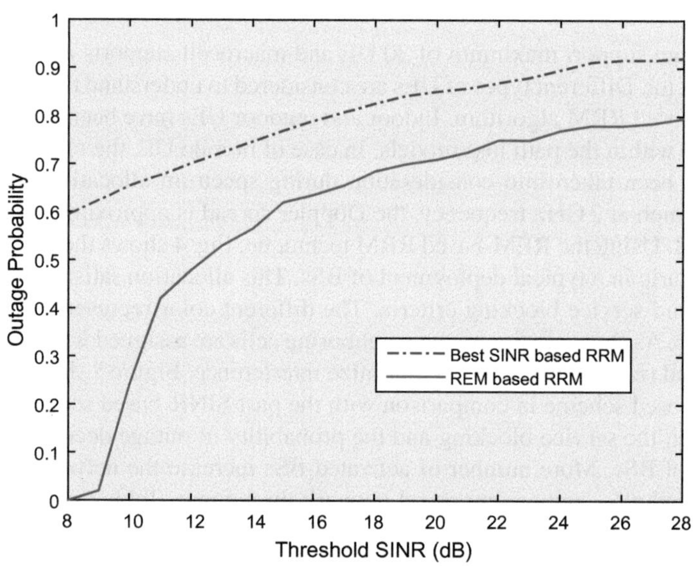

Fig. 5 Comparison of outage probability

Table 1 Number of UE with outage and service blocking at 20 dB.

UE type	Number of UE	Outage UE	SE (bps/Hz)	Data rate (Mbps)	Service blocked UE
Type A	165	58	5	10	30
Type B	36	10	5	2	17
Type C	45	25	3	2	13
Type B	54	25	4	1	10

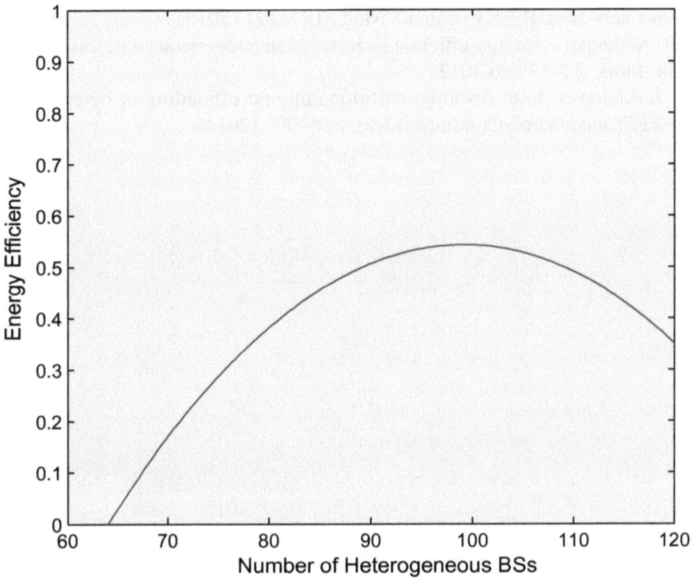

Fig. 6 Normalized EE in activated BS in HetNet

6 Conclusion

This paper proposed an REM-based RRM technique in HetNet wireless environment. This proposed scheme minimizes outage and service blocking. Simultaneously, it minimizes network energy consumption by appropriately activated and deactivated BSs. The simulation results suggest that a tradeoff is needed between EE, number of activated BSs and outage probability.

References

1. G. Bansal, M.J. Hossain, V.K. Bhargava, Optimal and suboptimal power allocation schemes for OFDM-based cognitive radio systems. IEEE Tran. Wirel. Commun. **7**, 4710–4718 (2008)
2. A. Damnjanovic et al., A survey on 3GPP heterogeneous networks. IEEE Wirel. Commun. **18**(3), 10–21 (2011)

3. Deliverable D2.3, EU-FP7 Project, Broadband evolved FEMTO networks (2012)
4. Femto Forum, Femtocells? Natural solution for offload, a Femto Forum topic brief (June 2010), www.femtoforum.org
5. M. Ho et al., Power control for interference management and QoS guarantee in heterogeneous networks. IEEE Commun. Lett. **19**(8), 1402–1405 (2015)
6. R. Mahapatra, E.S. Strinati, Interference-aware dynamic spectrum access in cognitive radio network, in *IEEE PIMRC,* Toronto, Canada (2011)
7. R. Mahapatra et al., Energy-efficient green framework of future heterogeneous wireless network. Comput. Netw. Elsevier **57**(6), 1518–1528 (2013)
8. J. Perez-Romero et al., On the use of radio environment maps for interference management in heterogeneous networks. IEEE Commun. Mag., 184–191 (2015)
9. A. Singh, R. Mahapatra, Energy efficient analysis of heterogeneous wireless network, in *IEEE IACC,* Delhi, India, 22–23 Feb 2013
10. S. Singh, J. Andrews, Joint resource partitionaing and offloading in heterogeneous cellular network. IEEE Trans. Wirel. Commun. **13**(2), 888–901 (2014)

Fuzzy AHP Based Technique for Handover Optimization in Heterogeneous Network

Riya Goyal, Tanu Goyal, Sakshi Kaushal and Harish Kumar

Abstract With the exponential growth of the telecommunication networks high data rate, low latency become highly demandable. In wireless communication, User Equipment (UE) demands always be on the network so that packet losses are minimum during mobility in the network and cater the better Quality of Service (QoS). Handover is the mechanism which transfers the control of UE from one network to another without interruption. If the target network is not preeminent, then it may cause the handover failure and handover ping-pong effect. The selection of the network should be optimal so as to preserve the overall QoS of the network. The proposed scheme selects the optimal network from all the available networks. The results are analyzed numerically and show the best network in terms of less handover failure rate that can help for seamless connectivity.

Keywords Handover · RSS · Dwell time · Fuzzy-AHP · 4G · WLAN · WiMAX

1 Introduction

With the growth in a telecommunication network, a new variety of smart and powerful devices are developed. All these devices demand high data rate and low latency to support multimedia traffic. Wireless networks are generally implemented for the multimedia services as it avoids the installation cost incumbent in the wired networks [4]. It introduces new wireless telecommunication technologies like Universal

R. Goyal (✉) · T. Goyal · S. Kaushal · H. Kumar
University Institute of Engineering and Technology Panjab University, Chandigarh, India
e-mail: riya.goyal111@gmail.com

T. Goyal
e-mail: tanugoyal27@gmail.com

S. Kaushal
e-mail: sakshi@pu.ac.in

H. Kumar
e-mail: harishk@pu.ac.in

C. R. Krishna et al. (eds.), *Proceedings of 2nd International Conference on Communication, Computing and Networking*, Lecture Notes in Networks and Systems 46, https://doi.org/10.1007/978-981-13-1217-5_29

Mobile Telecommunication System (UMTS), Fourth Generation (4G), Wireless Local Area Network (WLAN), WiMAX, etc.

4G systems supply all-IP solution to the users where data, voice and multimedia traffic is used on "anytime and anywhere" basis. 4G system has two networks: Mobile WiMAX and Long-Term Evolution (LTE). An extremely versatile radio interface was deployed in 2009 referred to as 3rd Generation Partnership Project (3GPP). 3GPP introduces the LTE networks to support multimedia traffic as it offers high data rate and low latency [1].

LTE provides advancement in the High Speed Packet Access (HSPA). Swedish telecom operator Telia Sonera has first deployed the LTE network in December 2008, in capitals of Oslo, Norway, Stockholm, and Sweden [3]. From the WiMAX standard used for the fixed access, IEEE created IEEE 802.16e, which is called as Mobile WiMAX. A large number of WiMAX networks are industrially conveyed over the world. WLAN is described as a network of wireless devices that possess high-frequency radio waves and furthermore incorporates an access point to the Internet. The WLAN system acts as a bridge between User Equipment (UE) and wired structure of computers, servers, and routers. This standard defines the specification of the physical layer and MAC layer. Due to the availability of WLAN in each and every smart device and low-cost Wi-Fi networks, interworking between LTE and WLAN has become the affordable and fruitful solution to address increasing demand for data customers [15].

UE demands always be on the network so as to minimize the losses and provide the better Quality of Service (QoS) in the network. Handover is the process of transferring an outgoing call between the evolved nodes (eNB) without handover failure. Handover can be horizontal handover and vertical handover. It consists of three phases: handover initialization, network selection, and handover execution [2]. With the estimation reports of a handover, UE starts the handover to the serving eNB. UE plays out the downlink radio channel estimations in light of the Received Signal Strength (RSS), periodically. When the necessary conditions for the network are convinced, then the UE delivers the corresponding report [7], but if the selected network is not optimal, then it may breach the QoS of the network. Conventional method uses RSS for the selection of best network [5]. Along with this, other parameters are also required to select a network from available networks. Hussein et al. [10] used RSS, the load on eNB and uplink Signal-to-Noise Ratio (SINR) for the selection of a cell in the homogenous network but authors do not consider the other essential parameters such as dwell time, bandwidth, power transmission, etc. Goyal et al. [8] used AHP method for the selection of suitable network and results show that delay is the main criteria. The author does not consider other essential parameters and also not calculated the handover failure rate. In this paper, a scheme is proposed for the selection of best network in heterogeneous network scenario so as to maintain overall QoS of the network. The network is selected by considering criteria such as RSS, velocity, bandwidth, load on eNB, power transmission, dwell time, and cell radius. By applying fuzzy AHP method, best ranked network is selected based on minimum handover failure.

The rest of the paper is organized as follows: Sect. 2 describes the literature survey. Section 3 presented the proposed scheme. In Sect. 4, results are numerically analyzed and Sect. 5 concludes the paper.

2 Literature Survey

In this section, basics of handover in the network and fuzzy AHP method is discussed.

2.1 Basics of Handover in Networks

The coverage region of the eNB is limited and UE is moving from one place to another. To overcome the call interruption problem, the control of UE is to be transferred from one eNB to another. This is called as handover. The signal strength of the UE decreases as it moves outward the serving eNB. During the selection of eNB, when there is lack of resources, it results in handover failure. There are three types of handover failures, i.e., too early, too late, and false handover. It is easy to handle them too early and too late kind of handover by taking one or two threshold values of RSS. But the problem arises when false handover will occur [13].

2.2 Fuzzy AHP Method

The Fuzzy Analytical Hierarchy Process (Fuzzy AHP method) is an extensive multi-attribute decision-making method which is deployed to solve the hierarchy fuzzy problems. The extent fuzzy AHP method [6] is used, which was originally introduced by Chang. In this, the pair-wise comparisons matrix have fuzzy numbers. The fuzzy AHP method is implemented with the accompanying six stages [16]:

- Develop the structure of the matrix with various criteria and design the pair-wise comparison matrix $\left(A_{ij}\right)$. *It represents the examination framework with the 9-point scale.*
- Examine the consistency of pair-wise comparison matrix. If the resultant matrix is consistent, it would fulfill Eq. (1).

$$s_{ij} + s_{ji} = 1 \tag{1}$$

- Change the output of the comparison matrix into fuzzy variables that possess the values ranging from zero to one, for the computation of positive fuzzy matrix that satisfies Eq. (2).

$$s_{ij} = \frac{a_{ij}}{a_{ij} + 1} \tag{2}$$

- Compute the fuzzy weights of the comparison matrix with the help of Eq. (3).

$$w_i = \frac{z_i}{\sum_{i=1}^{n} z_i}, \text{ where } z_i = \frac{1}{\sum_{j=1}^{n} \frac{1}{s_{ij}} - n} \tag{3}$$

- Take the mean of the weighted matrix and combine the results of the criteria.
- Calculate the Consistency Index (CI) using Eq. (4), and obtain the resultant scores. If CR $\leq$ 0.01, then the matrix is consistent otherwise choose the weights again.

$$\mathrm{CR} = \frac{\mathrm{CI}}{\mathrm{RI}}, \quad \text{where } RI \text{ is Random Consistency Index.} \tag{4}$$

The next section presents the proposed approach.

3 Proposed Scheme

Based on the reasons of the handover failure as discussed in Sect. 2.1, there is need to further optimize the process of network selection during handover. Various schemes and parameters are taken differently to optimize the selection process for the network. The major objective of the proposed work is to reduce the handover failure. For the calculation of handover failure rate according to the parameters described below, an objective function (F_n) has to be computed. The focus of the scheme proposed in [11] is on velocity, RSS, and their major goal is to minimize the false handover, but the author does not consider the other essential parameters. In this paper, handover failure rate is calculated by considering velocity, RSS, and other essential parameters such as current load on eNB, power transmission, dwell time, cell radius, etc. These parameters are discussed below:

- **RSS (Received Signal Strength)**—Handover is initiated only when the RSS value of the current network drops below the predefined threshold. A minimum of −20 dB Signal-to-Noise Ratio (SINR) is needed to detect RSS [11]. RSS is calculated as shown in Eq. (5).

$$\mathrm{RSS} = \mathrm{pt} - \mathrm{p}\log(d_k) + X_\sigma \tag{5}$$

  where RSS is the received signal strength by the UE from eNB, and pt is the transmitted power from eNB, respectively, p is the path loss of the propagation environment. X_σ in dB is attenuation due to shadowing with zero mean and standard deviation.

- **Velocity**—Depending on the speed of the UE, the UE will select cellular or WiMAX networks, respectively. Quality of handover is much affected when the velocity is high, which would cause different types of handover failure rates [12]. Equation (6) computes the velocity using monitoring index:

$$\alpha = v\left(\frac{\omega}{\delta}\right)^{1/\beta} \tag{6}$$

where ω represents the RSS level at current location of UE, δ is a predefined threshold, v is the velocity, and α is the monitoring of signal.

- **Dwell time**—It is the time spent by the UE in the same state. When dwell time is less than the handover time, there is a handover failure due to insufficient time for handover completion. By estimating the dwell time in a region, UE can limit unnecessary handover and handover failure [9]. It is calculated from Eq. (7)

$$T = \left(\frac{2a\ \sin(\emptyset)}{v}\right);\quad 0 \leq T \leq \frac{2a}{v} \tag{7}$$

where T is the dwell time, v is the velocity of the UE, a is the radius of the cell, and $\emptyset = \arctan\left(\frac{m_l - m_t}{1 + m_l m_t}\right)$, m_l and m_t are the slopes of the different cells.

- **Load on eNB**—When eNB broadcasts, the Master Information Block (MIB) signals contain the information of available resource block.
- **Power Transmission**—It directly depends on the checking of the available networks amid handover which is shown in Eq. (8) [14].

$$\mathrm{PL}(d)_{\mathrm{db}} = S + 10\mathrm{n}\log(d) + (X_\sigma) \tag{8}$$

where PL is the power transmission and S is the path loss constant.
Based on all these parameters, an objective function (F_n) is derived as shown in Eq. (9) that minimizes the handover failure rate according to the function (F_n), handover failure rate is computed, and the network that has minimum value of (F_n) is selected as the best network among the available networks.

$$\mathrm{F}_n = \frac{\mathrm{RSS} \times \mathrm{Velocity} \times \mathrm{Power\ Transmitted} \times \mathrm{Load\ on\ eNB}}{\mathrm{Cell\ Radius} \times \mathrm{Dwell\ Time}} \tag{9}$$

To implement the objective function, fuzzy AHP method is used where LTE, WiMAX, and WLAN are the different networks available as the alternatives and RSS, velocity, power transmission, current load on eNB, cell radius, and dwell time are the criteria. Thus, the proposed scheme minimizes the handover failure rate and chooses the network, accordingly. The process to optimize the handover for the selection of network is shown in Fig. 1. RSS value is compared with the predefined threshold, and if RSS value is lesser, then the timer is increased and that timer will be compared with the dwell time of the network. Only when the timer is greater than the dwell time, extended fuzzy AHP is applied to compute the objective function and it results in the selection of the best network among the available networks.

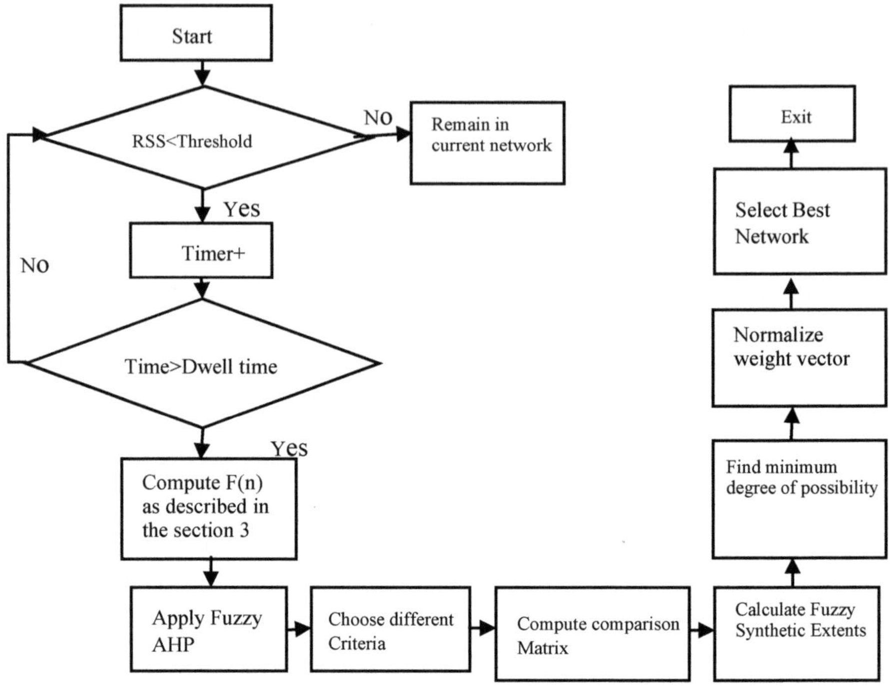

Fig. 1 Flowchart of the proposed scheme

4 Analysis of Proposed Method

Numerical analysis has been done for the validation of the proposed method and also to highlight the benefits of it. The result is analyzed by considering three access network interfaces called as alternatives, i.e., LTE, WLAN, and WiMAX. RSS, velocity, the power transmitted, the load on eNB, bandwidth, cell radius, and dwell time are considered as criteria, as input to the proposed scheme which is implemented in Matlab. Table 1 represents the normalized matrix of the criteria.

In fuzzy AHP method, the comparison ratios are used to be able to tolerate vagueness. By applying fuzzy extend analysis, it is easy to compare the alternatives with different criteria. Table 2 represents the alternative matrix that shows the characteristics of different networks. The values of the matrix are normalized in the form of 0's and 1's.

From the above analysis, CR is calculated by Eq. (4). The RI for our case is 1.32, and it provides CR < 0.01, so the matrix is said to be consistent.

With the use of Triangular Fuzzy Numbers (tfn), the ordinal number matrix is converted to a tfn using tfn numbers, and then the minimum degree of possibility is evaluated. Table 3 represents the normalized weight matrix by averaging the weights of the parameters.

Table 1 Aggregated fuzzy comparison matrix

Parameter	RSS	Velocity	Power transmitted	Load on eNB	Bandwidth	Cell radius	Dwell time
RSS	1	3	3	5	7	5	7
Velocity	0.33	1	9	7	7	5	3
Power transmitted	0.33	0.11	1	0.20	0.20	0.14	0.20
Load on eNB	0.20	0.14	5	1	5	0.33	0.33
Bandwidth	0.14	0.14	5	0.20	1	7	0.20
Cell radius	0.20	0.20	7	3	0.14	1	7
Dwell time	0.14	0.33	5	3	5	0.14	1

Table 2 Alternative matrix

Available networks	RSS	Velocity	Power transmitted	Load on eNB	Bandwidth	Cell radius	Dwell time
LTE	1	1	1	1	0	1	0
WiMAX	0	0	0	1	1	1	1
WLAN	1	1	0	1	1	0	1

Table 3 Weight vectors

Criteria	Weights
RSS	0.29
Velocity	0.35
Power transmitted	0.00
Load on eNB	0.09
Bandwidth	0.05
Cell radius	0.14
Dwell time	0.08

Table 4 Fuzzy vector matrix

Networks	Rank analyzed
LTE	0.87
WiMAX	0.37
WLAN	0.86

Finally, a final score is obtained by taking the geometrical mean of the criteria for each network [6]. Table 4 represents the overall weights of different networks.

For this case, based on the final score, LTE network got selected as the most reliable network during mobility since it has minimum handover failure rate. Analysis of Eq. (9) illustrates that if cell radius and dwell time are less as compared to other parameters, the probability of handover failure increases and vice versa. The proposed scheme helps to select the best network out of all available networks.

5 Conclusion

In this paper, the technique is proposed to select the best network out of all available network that results in minimum handover failure rate. It has been observed that UE is moving from one network to another network freely at high speed. If the target network is not optimal, then it will degrade the overall QoS of the network. To overcome this problem, extended fuzzy AHP method based scheme is proposed by considering the RSS, velocity, cell radius, dwell time, bandwidth, the power transmitted, the load on eNB criteria, and hence, optimized the network. Results are analyzed numerically and show that LTE network is the best network among WLAN and WiMAX in terms of handover failure rate. In future, we will focus on other factors that affect UE and network during handover like energy utilization of UEs, packet loss, etc.

References

1. I.F. Akyildiz, D.M. Gutierrez-Estevez, E.C. Reyes, The evolution of 4G cellular systems: LTE-Advanced. Phys. Commun. **3**(4), 217–244 (2010)
2. A.D. Assouma, R. Beaubrun, S. Pierre, Mobility management in heterogeneous wireless networks. IEEE J. Sel. Areas Commun. **24**(3), 638–648 (2006)
3. D. Astély, E. Dahlman, A. Furuskär, Y. Jading, M. Lindström, S. Parkvall, LTE: the evolution of mobile broadband. IEEE Communications magazine **47**(4), 44–51 (2009)
4. P. Datta, S. Kaushal, Exploration and comparison of different 4G technologies implementations: a survey, in *Engineering and Computational Sciences (RAECS), 2014 Recent Advances in IEEE* (2014), pp. 1–6
5. K. Dimou, M. Wang, Y. Yang, M. Kazmi, A. Larmo, J. Pettersson, ... Y. Timner, Handover within 3GPP LTE: design principles and performance, in *70th IEEE Conference on Vehicular Technology Conference Fall (VTC 2009-Fall)* (2009), pp. 1–5
6. I. Ertuğrul, N. Karakaşoğlu, Comparison of fuzzy AHP and fuzzy TOPSIS methods for facility location selection. Int. J. Adv. Manuf. Technol. **39**(7), 783–795 (2008)
7. T. Goyal, S. Kaushal, A study of energy efficiency techniques using DRX for handover management in LTE-A networks, in *International Workshop on Multiple Access Communications* (Springer, Cham, Sept 2015), pp. 33–44
8. T Goyal, S. Kaushal, Optimized network selection during handover using analytic hierarchy process in 4G network, in *IEEE International Conference on Advances in Computing, Communication, and Automation (ICACCA)* (Spring, 2016), pp. 1–4
9. R. Hussain, S.A. Malik, S. Abrar, R.A. Riaz, H. Ahmed, S.A. Khan, Vertical handover necessity estimation based on a new dwell time prediction model for minimizing unnecessary handovers to a WLAN cell. Wirel. Pers. Commun. **71**(2), 1217–1230 (2013)
10. Y.S. Hussein, B.M. Ali, M.F.A. Rasid, A. Sali, A.M. Mansoor, A novel cell-selection optimization handover for long-term evolution (LTE) macrocellusing fuzzy TOPSIS. Comput. Commun. **73**, 22–33 (2016)

11. A. Jain, S. Tokekar, Optimization of vertical handoff in UMTS_WLAN heterogeneous networks, in *IEEE International Conference on Emerging Trends in Communication, Control, Signal Processing and Computing Applications* (*C2SPCA*) (2013), pp. 1–5
12. M. Khan, K. Han, An optimized network selection and handover triggering scheme for heterogeneous self-organized wireless networks. Math. Probl. Eng. **2014**, 11 (2014)
13. T. Komine, T. Yamamoto, S. Konishi, A proposal of cell selection algorithm for LTE handover optimization, in *IEEE Symposium on Computers and Communications* (*ISCC*) (2012), pp. 000037–000042
14. S. Kunarak, R. Suleesathira, Vertical handoff decision and network merit for integrated wireless and mobile networks, in *Telecommunication Networks and Applications Conference* (*ATNAC*), *IEEE Australasian* (2011), pp. 1–6
15. L. Luan, M. Wu, Y. Chen, X. He, C. Zhang, in *Handover Parameter Optimization of LTE System in Variational Velocity Environment* (2011), pp. 395–399
16. C. Singla, S. Kaushal, Cloud path selection using fuzzy analytic hierarchy process for offloading in mobile cloud computing, in *2nd International Conference on Recent Advances in Engineering and Computational Sciences* (*RAECS*) (2015), pp. 1–5

Improved TDMA Protocol for Channel Sensing in Vehicular Ad Hoc Network Using Time Lay

Ranbir Singh and Kulwinder Singh Mann

Abstract VANET is a decentralized type of network in which vehicle acts like mobile nodes, can join together, and start communicating with each other. The reactive routing protocol is the type of protocol which establishes a path by gathering network information at the time of path establishment. The TDMA is the medium access control protocol, which assigns channels to each vehicle and vehicles can communicate with each other on the given time slots. Due to self-configuring nature of the network, clocks are weekly synchronized, which leads to packet collision in the network. In this research, time lay technique has proposed in which whole network has divided into zones and zone heads are selected from each zone, which are responsible to synchronize clocks of the mobile vehicle. The simulation of the proposed model has done in NS2, and it has been analyzed that proposed technique performs well in terms of NRL, route lifetime, and PDR.

Keywords VANETs · TDMA · MAC · Reactive · NRL · PDR

1 Introduction

The VANETs are one of the most prominently researched areas of today's technology which help in determining the communication of vehicles and the roadside units with each other. VANETs are deployed to decrease the probability of road accidents and to improve passenger comfort by allowing vehicles to exchange different kinds of data messages between the mobile vehicles and the installed infrastructure. If the vehicles are communicating among themselves, it is known as Vehicle-to-Vehicle (V2V) communication, whereas if it happens between vehicles and roadside infrastructure

R. Singh (✉)
Lovely Professional University, Phagwara, India
e-mail: ranbir.21123@lpu.co.in

K. S. Mann
Guru Nanak Dev Engineering College, Ludhiana, India
e-mail: mannkulvinder@yahoo.com

C. R. Krishna et al. (eds.), *Proceedings of 2nd International Conference on Communication, Computing and Networking*, Lecture Notes in Networks and Systems 46, https://doi.org/10.1007/978-981-13-1217-5_30

it is called as Vehicle-to-Roadside (V2R) communication. Through VANETs, vehicles can easily share the present status of the traffic flow or a perilous circumstance, for example, a mishap. There is a proper connection provided to the all the nodes present in the network among each other and also the elements of roadside network. There is no need to determine the base or foundation of the creation of the network. The important data is sent and received throughout the network. There are warnings given to the required vehicles regarding any kind of situations [1]. There is a growth in utilization of the Wi-Fi IEEE 802.11 technology, which helps with the deployment of the VANETs. There are two standards basically involved here, which are the 802.11b and 802.11g. The vehicles can choose any of these standards for providing a connection with the wireless network interface. The requirements of the highly dynamic networks are, however, not fulfilled by these standard because they are general-purpose standards. VANET is a highly dynamic network, and so these standards are of no much use for it due to their small sizes. There are different behaviors and characteristic properties of VANETs which help them be different from the other networks. There are various routing protocols introduced within the VANETS. The protocols are classified on the basis of various parameters [2].

The data packets are transmitted from a source to a particular destination by the unicast routing protocols. The personalized comfort applications and commercial applications are supported by these protocols. Within the ad hoc environment, the unicast routing protocols are used in a wide range. The virtual partitioning of dynamic nodes among different groups is known as the clustering of VANETs. The nodes are identified to be part of specific clusters. For routing, relaying of inter-cluster traffic, scheduling of intra-clusters traffic, and for channel assignment of cluster members, a special node is selected which is known as the cluster head. Within the routing process, no cluster members involve. The advantages of proactive and reactive techniques are combined to generate the reactive routing protocols [3]. There are two levels into which the hybrid protocols are divided. The information related to routing of all the nodes is maintained and updated to rest of the nodes of the network, by the inner layer which is proactive. The most commonly utilized routing protocols in VANETs are broadcast routing protocols. They are widely deployed within the safety-related applications. The packet is transmitted to all the nodes present in the network within the broadcast mode. This message is re-broadcasted to all the other nodes of the network by each node. Within the broadcast routing protocols, flooding is a prominent technique utilized.

1.1 MAC Protocols in VANETs

Vehicular Ad Hoc Networks (VANETs) have emerged as another new class of efficient information dissemination technology that can meet various requirements of Intelligent Transportation System (ITS) applications which aim to improve traffic safety and efficiency. Each mobile node in VANETs works intelligently to assist other nodes with/without the guidance of the available infrastructure. VANETs offer a wide area of applications, but due to several exclusive properties and highly dynamic nature, this network emerged many challenges that encourage the researchers to investigate their research in an attempt to trim down these issues. In VANETs, an efficient MAC protocol helps in achieving reliable communication, but have strict requirements in terms of QoS. The huge networks can be managed very easily as hierarchical structured with the help of clustering mechanism. A single hierarchy is generated by partitioning nodes among the clusters. The medium access delay is to be minimized through the VANET MAC protocols. These protocols are widely utilized within the various safety applications due to such characteristics. The access to the shared medium is arbitrated with the help of a medium access control protocol [5]. The nodes within the transmission range of each other are restricted from transmitting the data on the similar duration by the MAC protocol. In order to improve the transportation system, the safety and non-safety-related services are provided by VANETs. The channel is to be accessed in an efficient manner and should be collision free for each vehicle. The manner in which all the vehicles can be accessed within VANETs is provided by MAC protocol. There are numerous MAC protocols designed within these networks. Various issues are also being faced during the designing of MAC protocols such as the high mobility of nodes; the change is the topology of networks, the hidden/exposed node related issue, and so on [6]. In order to provide MAC layer implementation within the Wireless Access in Vehicular Environment (WAVE) standard, the IEEE 802.11p is introduced. A TDMA-based contention-free MAC protocol is referred to as VeMAC. The TDMA frames are generated by dividing VeMAC time. There are numerous time slots present within each of the frames. A contention-free Self-Organizing MAC Protocol for DSRC-based Vehicular Ad Hoc Networks (VeSOMAC) aids in exchanging TDMA slot detail during the MAC scheduling.

2 Literature Review

Hu et al. [1] have proposed the multi-hop service process into a single virtualized service. This will help in analyzing the impact of multi-hop transmission on delay performance. They have also applied martingale theory to test FIFO and earliest EDF scheduling policies [1]. To analyze the MAC sub-layer access performance, an IEEE 802.11p EDCA mechanism has adopted. The simulation results show that the proposed model is better in terms of super-martingale end-to-end backlog and delay bound has compared to existing standard bounds. They have also investigated the delay end backlog performance to test the effect of number of vehicles.

Rossi et al. [2] presented that there is a huge utilization of MAC protocols in time-critical applications that arise concerns regarding the robustness and adaptiveness of protocols. In this paper, the authors have investigated that how vehicular density performance can be enhanced using maximum CW size [2]. They have developed a stochastic model in order to obtain optimal maximum CW that is integrated into amended CSMA/CA protocol. This will result in maximizing the single-hop throughput among adjacent vehicles. The simulation results show that the proposed optimized protocols are effective in terms of channel throughput and transmission delay performance as compared to standard CSMA/CA.

Khan et al. [3] have presented an advanced MAC scheme especially for the emergency systems. The CCH utilization uniformly distributes the load on SCHs which has improved by the presented scheme. In order to achieve messages on time, the mode of the network switches to emergency from general that help in utilizing the common service channel effectively. This will increase the probability for timely reception of messages and reduces the rate of transmission collisions and latency. This will increase the probability of message delivery. The simulation results show that it performs better in contrast to other and thus validates its performance as compared with existing MAC schemes.

Ndih et al. [4] have presented a contention-free MAC protocol for fast and reliable multi-hop data dissemination in vehicular ad hoc networks which takes advantage of two-packet collisions to improve network capacity. There are mainly two operations included in MAP, such as first, a dynamic subdivision of the space into multiple sub-spaces using one relay node and then each performs a deterministic medium access that is free of control messages [4].

Gawas et al. [5] proposed a cross-layer approach for efficient dissemination of emergency messages in VANETs (CLDEM). The message redundancy and maintaining low end-to-end communication delays of this approach will minimize the output. They have proposed a scheme that will help in selecting a one-hop neighbor relay as a potential forwarder that relays the broadcast messages and improves the transmission reliability in a platoon of vehicles. There is need of minimum overhead and minimum bandwidth consumption to control the broadcast messages by selected relay [5]. They have adopted 802.11e MAC to provide service differentiation in different traffic classes. The simulation results show that the proposed cross-layer scheme effectively propagates the critical broadcast messages with minimum latency as compared to existing schemes.

Hadded et al. [6] concluded that ADHOC MAC provides an efficient broadcast service for inter-vehicle communications and solves MAC issues such as the hidden-exposed terminal problem and QoS provisioning. The authors have also presented a comparison between simulation and analytical results in order to validate the mathematical model and protocol [6]. The network simulator NS-2 and the realistic road traffic simulator SUMO is also being used for analysis purposes.

3 Research Methodology

The vehicular ad hoc network is the decentralized type of network in which no central controller is present and due to non-presence of central controller, the routing, time synchronization, and security are the three major issues of the network. To reduce packet loss in the vehicular ad hoc, the whole network is divided into fixed size zones and zones heads are selected in each zone. The zone heads are selected in a zone on the basis of speed and distance. In this work, the novel technique has been proposed which synchronize the clocks of the vehicle nodes. The proposed technique is based on the MAC time for the clock synchronization in the network. The node which aggregated the data to zone head will also the sent the current time header. The zone head when receives the packet will check the time header and when the time at the vehicle node and zone head gets mismatched, then the zone head will adjust the clock according to the current time. In the last step, clock synchronization of the zone heads will adjust its clocks according to the received time from their adjacent cluster heads. When the clocks of the zone nodes get synchronized, then the modes which are applied at each node will work efficiently and energy consumption in the network get reduced, and also throughput increased at a steady rate (Fig. 1).

3.1 Proposed Algorithm

Input: vehicle nodes with week clock synchronization
Output: vehicle nodes with synchronized clocks

1. Deploy the vehicle ad hoc network with finite number of vehicle nodes
2. Divide whole network into fixed size zones
3. While (zone head selected)
   Repeat for loop for each node in the network
   If (speed (i) > speed (i + 1) && distance (i) < distance(i + 1))
   Zone head = node (i)
   End
4. The vehicle nodes transmit data to the zone head
5. Check (MAC)
   If (Mac time of the vehicle node! = zone head time)
   Zone head adjusts its clock according to vehicle node time
   Else
   Communication starts in the network
   End

4 Experimental Results

This work has been implemented in NS2, and the results have been compared in terms of various parameters such as NRL, the lifetime of the route, and PDR Table 1.

In Fig. 2, normalized routing load increases with increasing the number of nodes and the proposed technique shows the minimum normalized route load as compared to others. This may be explained by the fact that its route discovery process is decreased as compared to the others. Similarly, in Fig. 3, the path is established from source to destination through the centralized party. Route Lifetime (RLT) (time of

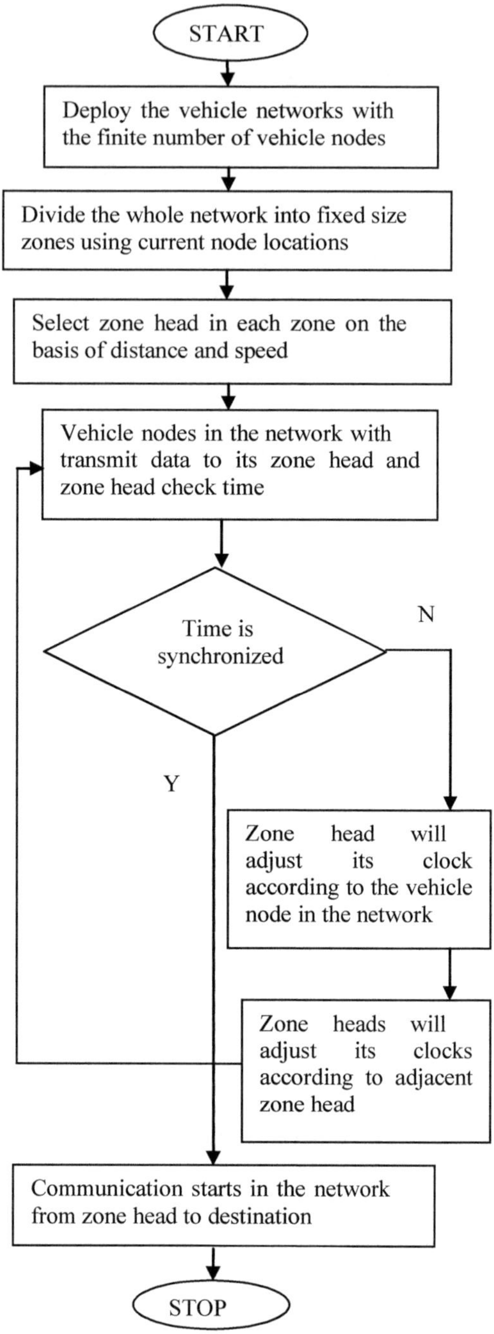

Fig. 1 Proposed flowchart

Table 1 Simulation parameters

Parameters	Values
Antenna type	Omnidirectional
Link layer type	LL
MAC type	802.11
Channel frequency	2.4 GHz
Routing protocol	LAR
Area	800*800 m
Range	100 m
Number of vehicles	20–80
Traffic type	CBR

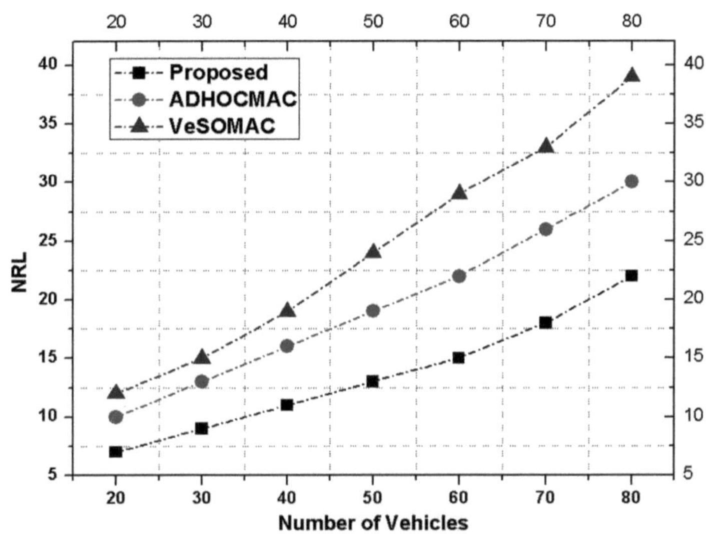

Fig. 2 Graph of normalized route load versus number of vehicles

a connection between the source and destination) is the delay between the time of arrival of the message RREQ to the destination and the breaking time on the road created by the same message RREQ. The route lifetime in the proposed technique is more as compared to existing which leads to reduction that is link failure.

The packet delivery fraction is defined as the ratio of number of data packets received at the destinations over the number of data packets sent by the sources, as shown in Fig. 4; the performance of the proposed technique is compared with existing in terms of PDR. The packet delivery ratio of the proposed technique is increased as compared to existing techniques.

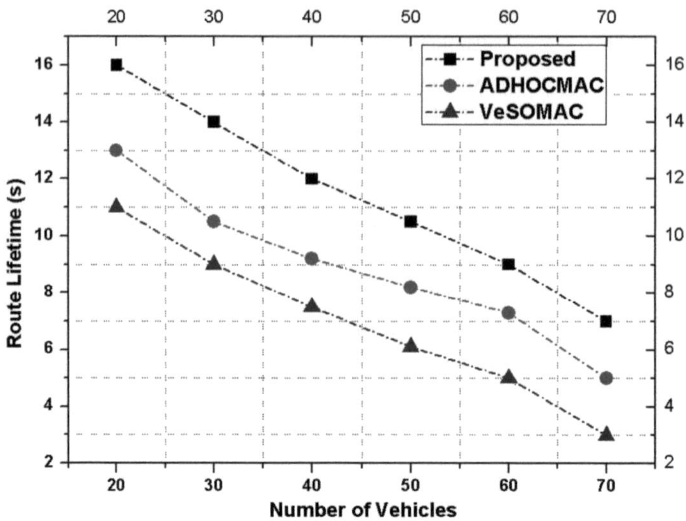

Fig. 3 Graph of route lifetime versus number of vehicles

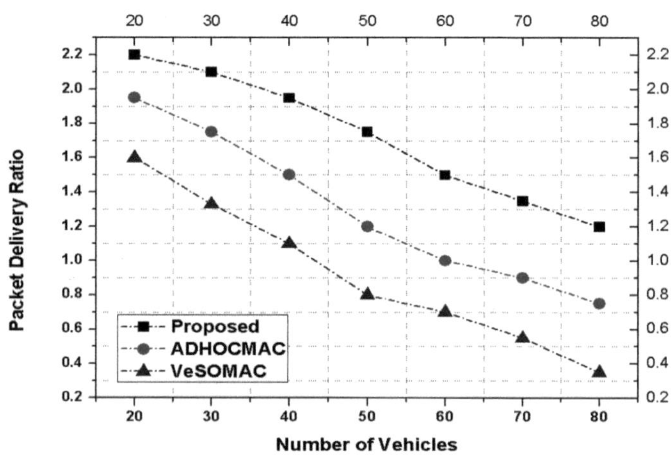

Fig. 4 SPacket delivery ratio versus number of vehicles

5 Conclusion

In this work, it has been concluded that the path from source to destination will be established using reactive routing protocols. The TDMA protocol is used for the channel sensing. Due to non-clock synchronization, TDMA protocol does not work fine and it leads to packet loss in the network. In this research, time lay technique of vehicular ad hoc network is proposed which synchronize clocks of the vehicle nodes. The simulation is performed in NS2, and the results are improved up to 20%

in terms of various parameters. In future, the mutual authentication technique will be applied, which increase the security of the network.

References

1. Y. Hu, H. Li, Z. Chang, Z. Han, End-to-end backlog and delay bound analysis for multi-hop vehicular ad hoc networks. IEEE Trans. Wireless Commun. **16**, 6808–6821 (2017)
2. G.V. Rossi, K.K. Leung, Optimised CSMA/CA protocol for safety messages in vehicular ad-hoc networks, in *2017 IEEE Symposium on Computers and Communications* (*ISCC*), vol. 5 (2017), pp. 121–128
3. S. Khan, M. Alam, M. Fränzle, A hybrid MAC scheme for wireless vehicular communication smart technologies, in *17th International Conference on IEEE EUROCON 2017*, vol. 7 (2017), pp. 889–895
4. E.D.N. Ndih, S. Cherkaoui, MAP: contention-free MAC protocol for VANETs with PLNC, in *IEEE ICC 2017 Mobile and Wireless Networking*, vol. 7 (2017), pp. 1001–1008
5. M.A. Gawas, P. Hurkat, V. Goyal, L.J. Gudino, Cross layer approach for efficient dissemination of emergency messages in VANETs, in *Ninth International Conference on IEEE Ubiquitous and Future Networks* (*ICUFN*), vol. 5 (2017), pp. 206–211
6. M. Hadded, P. Muhlethaler, A. Laouiti, Performance evaluation of a TDMA-based multi-hop communication scheme for reliable delivery of warning messages in vehicular networks, in *13th International Conference on Wireless Communications and Mobile Computing Conference* (*IWCMC*), vol. 8 (2017), pp. 1029–1034

Implementation of Fractional Frequency Reuse Schemes in LTE-A Network

Girisha Kumar and Garima Saini

Abstract This paper presents an enhancement of spectral efficiency in Long-Term Evolution Advanced (LTE-A) network through Fractional Frequency Reuse (FFR) scheme. The 3GPP LTE-A standard for wireless communication came with a solution to increase the network performance specially for users in cell edge region. The most popular Inter-Cell Interference Coordination (ICIC) technique is fractional frequency reuse. In FFR, the total cell region is separated as two regions such as cell interior region and edge region. The entire accessible bandwidth is divided among cell interior and edge regions. This paper presents the different FFR schemes, mainly strict FFR and Soft Frequency Reuse (SFR). It also focuses on comparison of FFR and SFR techniques with conventional frequency reuse technique. The simulation results show that using Fractional Frequency Reuse schemes, there is an improvement in performance of cell edge user in terms of Signal-to-Interference-Plus-noise Ratio (SINR), throughput, and spectral efficiency as related to conventional frequency reuse method.

Keywords Fractional frequency reuse · Inter-cell interference coordination · Long-term evolution advanced · Soft frequency reuse

1 Introduction

The evolution of number of mobile users since last decade increases rapidly. Due to multimedia applications, the requirement of data rate of network also increases. The rapid growth of data rate of user causes an increase in capacity of system. Due to high system capacity, the need of wireless spectrum also increases. Regrettably, the spectrum of wireless network is very limited and too costly. So, a proper technique

G. Kumar (✉) · G. Saini
ECE Department, NITTTR, Chandigarh, India
e-mail: girish.smp66@gmail.com

G. Saini
e-mail: garima@nitttrchd.ac.in

C. R. Krishna et al. (eds.), *Proceedings of 2nd International Conference on Communication, Computing and Networking*, Lecture Notes in Networks and Systems 46, https://doi.org/10.1007/978-981-13-1217-5_31

Fig. 1 Strict FFR

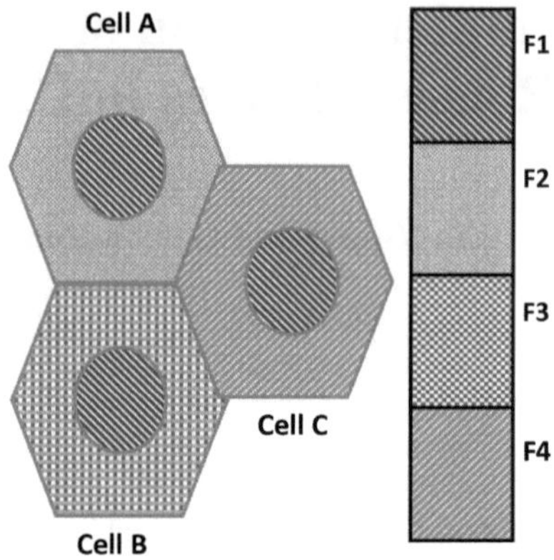

is needed to satisfy high requirement of system capacity with limited spectrum in improved wireless communication networks like LTE and LTE-A [1–4].

To achieve high requirement of system capacity, the frequency reuse techniques are used. In conventional frequency reuse technique with reuse factor 1, the cells are grouped into cluster and the frequency allotment to each cell within the cluster is not repetitive. In the conventional frequency reuse scheme, inter-cell interference occurs. In case of LTE/LTE-A networks, the spectrum management and ICI coordination are very important [5]. FFR and SFR are two methods to increase the efficiency of spectrum by reducing the inter-cell interference in LTE/LTE-A networks. In FFR scheme, the cell is parted into two areas such as cell center and edge region. The total accessible bandwidth is partitioned as subparts. The quantity of sub-bands for cell center and edge regions is allotted individually. The main aim of FFR schemes is that the users in adjacent cell edge region should use distinct frequencies and reduce the interference for cell edge user from cell center user. This increases the user probability of coverage in the region of cell edge [6].

In strict FFR scheme, the total frequency spectrum is partitioned into primary segments and secondary segments. The primary segment is allotted to cell center user and secondary segments are allotted to users in the region of cell edge. The entire cell center user uses common frequency band, whereas cell edge region user utilizes the part of frequency segments with a reuse factor α. So in strict FFR, the number of sub-band Radio Frequency (RF) is RF + 1. The cell center user can use either cell center or edge sub-band but cell edge user has to use only cell edge sub-band so that there is no interference between users of edge and center region [7, 8]. As shown in Fig. 1 for three-cell strict FFR, the total spectrum is partitioned into four sub channels $F1$, $F2$, $F3$, and $F4$. The center regions of all the cells are allotted using same frequency band $F1$, and the three cell edge regions are allotted using three different frequency bands such as $F2$, $F3$, and $F4$. The cell edge frequencies are allotted with a reuse factor of Δ. So, the total frequency band required for three-cell cluster is 4.

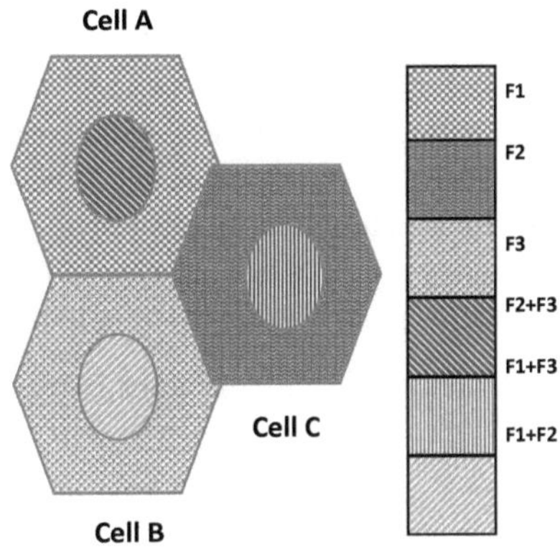

Fig. 2 Soft frequency reuse

In SFR, there is a share of frequency sub-bands for cell edge user with a reuse factor α. The cell center user shares the sub-band with edge user of other cell in such a way that there is no interference between them. The sub-band which is allotted to the cell edge user can also be used by cell center user if it is not used by edge user. The Shannon capacity for the cell edge user is reduced due to lack of availability of spectrum. To overcome this, high carrier power is allotted at the edge user in order to improve the SINR and finally increases the Shannon capacity [9–11].

As shown in Fig. 2, for three-cell SFR, the total spectrum is divided into six sub-band frequencies such as $F1$, $F2$, $F3$, $F4$, $F5$, and $F6$. The center regions of three cells are allotted using different frequency band such as $F4$, $F5$, and $F6$. The edge regions are allotted using frequencies such as $F1$, $F2$, and $F3$ in such a way that there is no interference between them.

The remaining section of the paper is structured in the following manner, Sect. 2 describes the network model procedure along with simulation parameters, simulation outcomes are discussed in Sect. 3, and lastly concluding the paper in Sect. 4.

2 Network Model

In the proposed model considered M cells, in each cell K user are uniformly distributed. In conventional reuse technique, the whole frequency spectrum is used in a cell with reuse factor 1. In case of FFR and SFR, the cell is separated into two portions such as cell interior and outer regions as mentioned in Sect. 1. Generally, the radius of interior region of cell may be selected as two-third of total radius of cell that is $\left(r_{\text{inner}} = \left(\frac{2}{3}\right) * R\right)$ where R denotes the radius of cell. First, the SINR value for different reuse scheme is calculated. Later, using 500 Monte Carlo realizations average values of fading and shadowing is estimated.

2.1 Working Principle of Network Model

After calculating SINR's values of each user, these SINR values are mapped to CQI entity [12], to decide the type of modulation used. The modulation type gives information about number of bits/symbol. Finally, using Shannon's formula, spectral efficiency is calculated. In case of resource allocation algorithm, subcarrier allotment is a major part. The resource allocation algorithm works as follows:

1. Calculate number of subcarrier per user using Eq. 1

$$s_u = \frac{R_\mathrm{u}}{b\Delta f} \tag{1}$$

   where R_u is required data rate of user in bps, Δf is subcarrier spacing, and b is number of bits/symbol.
2. Randomly allot a specific subcarriers to each user based on service class.
3. If subcarriers are free after step 2, then remaining carriers are allotted to each user based on round-robin fashion.
4. The condition for allotment of subcarrier in step 3 is that the allotted subcarrier per user should be less than maximum subcarriers required per user.
5. If the total required subcarrier is less than total available subcarrier after step 2, then subcarriers are subtracted from each user in a round-robin fashion.
6. The condition for subtraction of subcarrier from each user is in step 5 is the allotted subcarrier per user should be greater than minimum subcarriers required per user.

After allotment of subcarriers to user throughput is calculated. Further using 500 Monte Carlo simulations, same procedure is repeated to estimate average per user parameters such as SINR, throughput, and spectral efficiency and these parameters are then plotted with respect to user distance from BS.

2.2 Simulation Parameters

Table 1 provides all the simulation parameters required for simulations. The network used for simulation consists of 19 cell clusters and 48 users per cell which are uniformly distributed within a cell. In this paper, the LTE-A network is simulated using the parameters like cell radius, noise power density, and base station transmission power all are according to 3GPP standard.

Table 1 Simulation parameters

Sl. no.	Parameter	Value
1	Cluster size	19 cells
2	Cell radius	100 m
3	Users per cell	48
4	Channels	1200
5	System bandwidth (after carrier aggregation)	60 MHz
6	Channel bandwidth	15 kHz
7	Carrier frequency	2.3 GHz
8	Base station transmission power	23 dbm
9	Noise power spectral density	−173 dbm/Hz

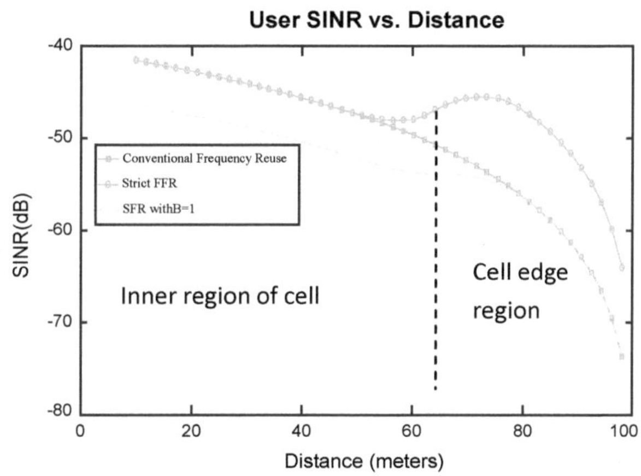

Fig. 3 Plot of SINR of user versus distance

3 Results and Discussion

The simulation of LTE-A network for distinct frequency reuse methods like conventional frequency reuse, strict FFR, and SFR scheme is done in this paper. The network efficiency is analyzed by SINR, spectral efficiency, and throughput of user.

Figure 3 shows the plot of SINR of user with reference to user distance to the base station in distinct frequency reuse methods. It justifies that in case of inner region (up to cell radius of 67 m), both conventional and strict FFR reuse schemes provide same SINR value. The transmission power and interference levels in the interior region of cell are same in both conventional frequency reuse and strict FFR scheme. Both of these schemes will provide best SINR value near to base station and it decreases as the user move away from base station. In the outer region of cell (from cell radius of 66–100 m), the strict FFR performance is better in terms of SINR as compared to conventional frequency reuse method.

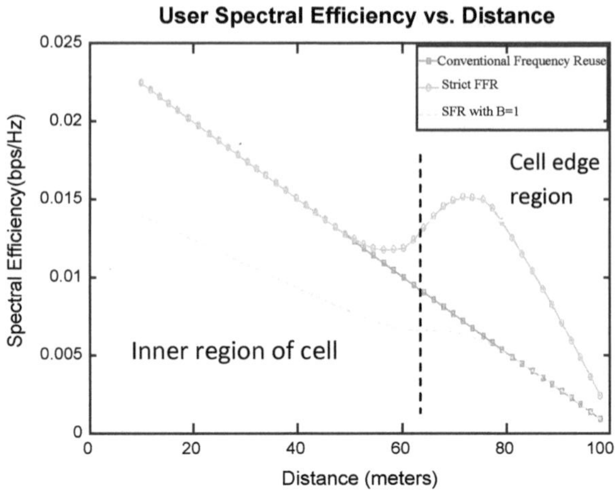

Fig. 4 Plot of spectral efficiency of user versus distance

The variation of spectral efficiency per user with respect to user distance to base station is shown in Fig. 4. According to Shannon's formula, the spectral efficiency depends on SINRs so it will observe the same effect in spectral efficiency as in case of SINR.

Figure 5 shows the plot of user throughput with reference to user distance to the base station in distinct frequency reuse methods. In conventional frequency reuse method, the throughput is higher in central region (up to cell radius of 67 m) of cell and it decreases as user move towards the cell edge region (From cell radius of 67–100 m). Only in the interior region of cell the throughput is high for conventional method as compared to strict FFR and SFR. In case of cell edge region, the strict FFR provides better throughput than conventional reuse schemes.

The CDF plot for SINR per user is shown in Fig. 6. The main purpose of using CDF is to analyze the total efficiency of distinct frequency reuse methods irrespective of position of user in a cell. The strict FFR has best SINRs value at cell edge region as compared to conventional reuse schemes due to less interference effect.

Figure 7 shows CDF plot for spectral efficiency per user in conventional frequency reuse, Strict FFR, and SFR schemes. According to Shannon's formula, spectral efficiency is a function of SINRs values so similar effect is observed as in case of SINR.

The CDF plot for average throughput per user is shown in Fig. 8. The plot indicates that conventional frequency reuse scheme provides better throughput after certain limit (approximately greater than 7 kbps) as compared to strict FFR and SFR. The conventional frequency reuse scheme provides the best throughput only in interior region of cell, and it provides worst throughput performance in the edge region of cell.

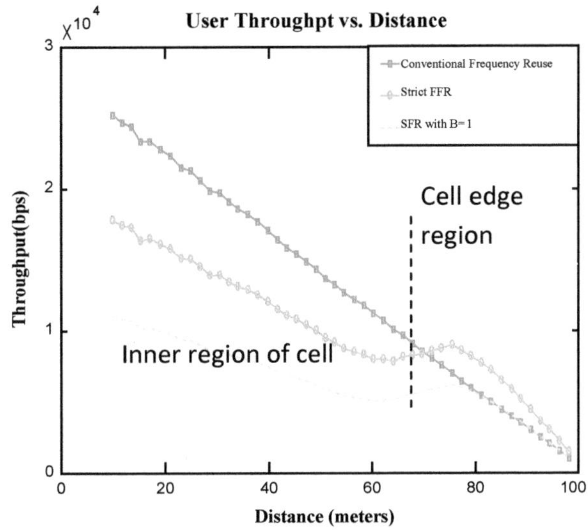

Fig. 5 Plot of user throughput versus distance

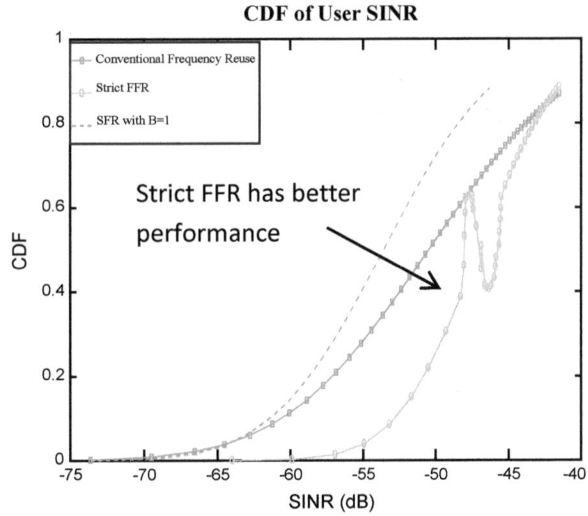

Fig. 6 CDF plot for SINR per user

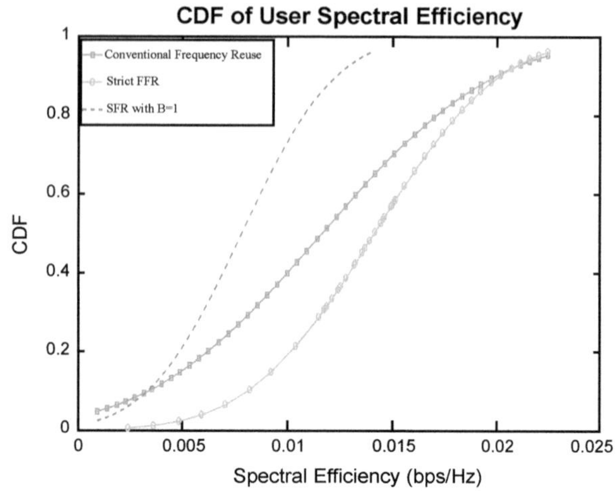

Fig. 7 CDF plot for average spectral

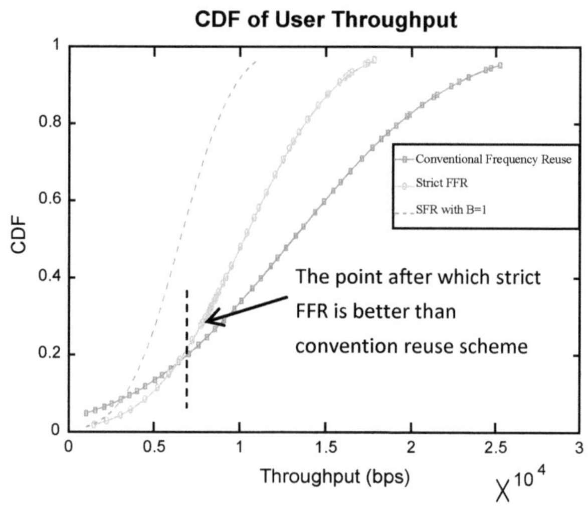

Fig. 8 CDF plot of average throughput efficiency per user throughput per user

4 Conclusion

In LTE-A advanced networks, the problem of increasing the cell edge performance is a major research area. The ICI coordination technique came with the solution of FFR schemes to increase the performance of user specially for cell edge users. The results of simulation justify that strict FFR has improved efficiency in terms of SINRs values and spectral efficiency at cell edge region as compared to conventional reuse technique. The throughput for cell edge user is best in strict FFR due to availability of more number of subcarriers reserved for user in edge of cell. In case of central region of cell conventional reuse technique, it provides better SINR and throughput. In SFR schemes, throughput and SINR value can be improved using proper power control factors.

References

1. S. Ohmori, Y. Yamao, N. Nakajima, The future generations of mobile communications based on broadband access technologies. IEEE Trans. Commun. **38**(12), 134–142 (2000)
2. Y. Zhou, J. Wang, T.S. Ng, K. Higuchi, M. Sawahashi, OFCDM: a promising broadband wireless access technique. IEEE Commun. Mag. **46**(3), 39–49 (2008)
3. E-UTRA and E-UTRAN overall description; stage 2 (Rel. 8), 3rd Generation Partnership Project, Cedex, France, Tech. Spec. TS 36.300 V8.12.0 [Online]. Available http://www.3gpp.org
4. Requirements for future advancement for evolved universal terrestrial radio access (E-UTRA) (Rel. 9), 3rd Generation Partnership Project, Cedex, France, Tech. Spec. TS 36.913 V9.0.0 [Online]. Available http://www.3gpp.org
5. N. Himayat, S. Talwar, A. Rao, R. Soni, Interference management for 4G cellular standards [WIMAX/LTE UPDATE]. IEEE Commun. Mag. **48**(8), 86–92 (2010)
6. T.D. Novlan, R.K. Ganti, A. Ghosh, J.G. Andrews, Analytical evaluation of fractional frequency reuse for OFDMA cellular networks. IEEE Trans. Wirel. Commun. **10**(12), 4294–4305 (2011)
7. F. Baccelli, M. Klein, M. Lebourges, S. Zuyev, Stochastic geometry and architecture of communication networks. J. Telecommun. Syst. **7**(1), 209–227 (1997)
8. T. Brown, Cellular performance bounds via shotgun cellular systems. IEEE J. Sel. Areas Commun. **18**(11), 2443–2455 (2000)
9. M. Qian, Y. Li, B. Vucetic, X. Yang, J. Shi, Adaptive soft frequency reuse schemes for wireless cellular networks. IEEE Trans. Veh. Technol. **64**(1) (Jan 2015)
10. S. Gupta, S. Kumar, R. Zhang, S. Kalyani, K. Giridhar, L. Hanzo. Resource allocation for D2D links in the FFR and SFR aided cellular downlink. IEEE Trans. Commun **64**(10) (Oct 2016)
11. G. Giambene, V. Le, T. Bourgeau, H. Chaouchi, Iterative multi-level soft frequency reuse with load balancing for heterogeneous LTE-a systems. IEEE Trans. Wirel. Commun. **16**(02) (Feb 2017)
12. S. Pietrzyk, G. Janssen, Radio resource allocation for cellular networks based on OFDMA with QoS guarantees, in *Proceedings of IEEE Global Telecommunications Conference* (Dec 2004), pp. 2694–2699

Frequency Regulation in Smart Grids Using Electric Vehicles Considering Real-Time Pricing

Sirat Kaur, Sukhwinder Singh and Damanjeet Kaur

Abstract The main ancillary service which is the focus of the research is frequency regulation. The real-time pricing is considered corresponding to the load division introduced as peak, off-peak, and mid-peak load. An algorithm is devised that incorporates load levelling and charging/discharging costs of Electric Vehicles (EVs). It gives an account of the number of EVs coming for charging/discharging at the Smart Grid (SG) for each hour. Results of the proposed algorithm provide the trend that is favourable to EV consumer and beneficial to the SG. It has been observed from the results that the frequency is regulated and the cost function adds an additional parameter to increase the efficiency. Load levelling is achieved which helps to deduce that the frequency at the SG is regulated.

Keywords Electric vehicles (EVs) · Frequency regulation · Smart grids (SGs) Vehicle-to-grid (V2G) technology

1 Introduction

Nowadays, Electric Vehicles (EVs) are seeing a continuous rise in their number. These EVs are useful for achieving the various ancillary services such as frequency regulation, voltage profile maintenance, and power factor correction, etc., in the Smart Grid (SG) environment. To meet ancillary services of Smart Grid, EVs can play a vital role by supplying power during peak hours. EVs can regulate frequency by charging their batteries from Smart Grid during off-peak hours and discharging it during peak hours. The integration of EVs in the Smart Grid is via bidirectional power flow and bidirectional communication links. This bidirectional flow of energy between SG and EVs is referred to as Vehicle-to-Grid (V2G) technology. EVs have been implemented in road transportation as a part of V2G technology but there are various

---

S. Kaur (✉) · S. Singh · D. Kaur
University Institute of Engineering and Technology, Panjab University, Chandigarh, India
e-mail: siratosan@icloud.com

C. R. Krishna et al. (eds.), *Proceedings of 2nd International Conference on Communication, Computing and Networking*, Lecture Notes in Networks and Systems 46, https://doi.org/10.1007/978-981-13-1217-5_32

challenging issues like degradation of EV battery, communication overhead between an EV and the SG, changes in the whole infrastructure of distribution network, charging/discharging price demand of EV owners amongst others which needs to be addressed.

Kempton and Tomic [1] state that they expect 'the 21st century will see fossil fuels displaced by intermittent renewable energy' in 2005. They propose that since EVs have rechargeable batteries, they can balance the frequency deviations either by supplying or withdrawing energy from the grid. Base load market is the least beneficial for V2G, whereas regulation market is the most beneficial for V2G technology. This paper is amongst the first of its kind to establish V2G technology as a landmark that could change the power system management.

Aggregation of EVs is introduced by Guille and Gross [2]. They state that connecting a single EV to the Smart Grid (SG) will have no meaningful impact on the SGs. Hence, connecting thousands of EVs, also known as aggregation of EVs, to the SG will impact the grid not only as a load but as a storage/generation device. A mechanism to optimally schedule real-time electricity consumption was proposed by Papavasiliou et al. [3]. The main focus is on a supply bidding function to incorporate charging of EVs in residential area. A demand response scenario is presented where the residential area gives a certain electricity consumption capacity corresponding to taking part in providing ancillary services to the grid. But there was an inherent assumption regarding the supply bidding function which would require prior knowledge so that the highest bidder could be selected.

Coordinated charging is the best kind of charging which will not only stabilize the SG but also utilize the EV to their full potential is the conclusion that Habib et al. [4] provide after an extensive survey on the impact of V2G technologies and charging strategies implemented in the electric environment. The roles of aggregators pertaining to the aggregation of EVs in the SG is further discussed and explained in [5]. Hui Liu et al. suggest that aggregators need to calculate not only the total Frequency Regulation Capacity (FRC) but also communicate with the EV Charging Station (CS) or individual EVs. This scheme did not support simultaneous charging and discharging of EVs. It neglects EVs demands and concentrates more on the frequency regulation itself. The study on battery degradation of EVs was done in [6] by Fabian Rücker et al. According to the authors, the EV life depends heavily on the battery it is using. Therefore, various charging strategies are studied which helps to restore the EV battery to lessen its degradation. A novel charging method for the lithium-ion batteries is presented in [7] by Wang Zhifu et al.

There is a huge burden on the SG due to supply–demand gap as highlighted in [8] by Kuljeet Kaur et al. providing a systematic framework to establish the actors/entities in the grid environment along with their respective roles which in turn became a background for the presented research. They have two cases established for the working of the proposed algorithm. The first case study is for the database corresponding to one-hour time interval given in the Pennsylvania–New Jersey–Manhattan (PJM) dataset, the second is the database with the time interval of 15 min between each frequency interval taken from Electric Reliability Council of Texas (ERCOT) dataset.

Both case studies took into account the different databases in order to widen the research perspective.

Real-world EV charging data was used in order to develop an algorithm to study demand charging at commercial sites in [9] by Guanchen Zhang et al. The concept of Distribution Locational Marginal Price (DLMP) and Real-Time (RT) market was utilized in [10] by Pierluigi Siano and Debora Sarno. An algorithm that provided cost savings for residential users was developed in order to study the impact of various parameters involved in the algorithm through Monte Carlo Simulations (MCS) in order to better tabulate their results.

The various algorithms that are implemented in order to provide effective charging were discussed in [11] by Qinglong Wang et al. who gave a detailed survey on the strategies. The authors have stated that the EVs are a trending topic in the world with their number increasing to 5 million per year by 2020. EVs are seen from their Load Characteristics and Statistical Charging aspects. The survey also states that the current charging schemes are not capable of providing continuously controllable charging power.

A survey on economic-driven approach for charging of EVs was presented in [12] by Wenjing Shuai et al. who shed the light on techno-economic environment for the EVs. According to the survey, EVs are not only a participant in the SGs but also a member of the Electricity market calling the EVs prosumers. The comparison of 'Unidirectional charging mechanisms' with the 'Bidirectional charging mechanisms' are done and models regarding the two are provided. In case of bidirectional charging, V2G technology is stated to be used as storage for renewable energy. But there are limited amount of analytical results for V2G due to economic constraints. A roadmap to integrate the EVs into the SG using a smart battery charger in the EVs was presented in [13] by Tugrul U. Daim et al. The proposed algorithm charged the EVs in a way that the battery lifespan of the EVs was increased corresponding to the current scenario of EV charging.

A survey on dispatching strategies opted by the EVs in regulating frequency in the SGs is provided in [14] by Chao Peng et al. An algorithm that took demand dispatch management as a means to regulate charging for EVs is proposed in [15] by Sang-Keun Moon and Jin-O Kim. The main focus is to determine the mismatch between loads and charging costs. A bidding strategy algorithm to study aggregator behaviour in real-time day-ahead market was given by Baringo and Amaro [16]. A stochastic optimization model in order to achieve scenarios of uncertainties to calculate prices for charging the EVs through aggregators is used. Austrian year - head market data in order to balance grid using decentralized pricing is used in [17] by B. Fassler et al. The authors study the day-ahead market scenario with the given data and conclude that primary frequency control market is better suited for EVs.

A survey on the recent studies which used EVs as a means of charging the SG and using as a transportation medium is given in [18] by Nicolò Daina et al. Yao et al. [19] propose a real-time charging scheme for the demand response market catering to EV parking station. The authors also compare their own work with existing strategies in order to provide a full-scale study of what lacked in previous studies. The authors' specific focus was on the parking station where maximizing the charging will provide operational profits to the consumer. The scheduling algorithm proposed by the authors used State of Charge (SOC) of EV as a dynamic constraint which updated only after cost is calculated. A decentralized algorithm that carefully performed in order to achieve customer satisfaction along with operator profitability was given by Omran and Filizadeh [20]. Wenzel et al. [21] propose an algorithm for the aggregation of EVs in the SG. Their main focus is on the EV aggregator and its real-time charging scenario. They provide an algorithm which uses future information for charging and discharging and handle this information for regulating the frequency. Diana et al. [22] provided a survey on various EV models that have been given. They have also stated that short time periods pertaining to the given models scales down the annual use of time periods to only a few minutes. Frequency regulation is an important ancillary service provided by SGs and EVs are an integral part of this service. EVs act as mediators along with aggregators to balance the load on the SG. This study focuses on balancing the net load on the SG using charging load and discharging load through the number of EVs coming for charging or discharging depending on the hour of the day. In the current paper, division of time is essential to understand when discharging and when charging can take place without overloading the SG. Load balancing is an important aspect to SG. And hence, EVs are used to balance the actual load to avoid sudden blackouts and valleys and peaks in the readings of the load. However, many methods do not include discharging cost into the algorithm as a first-hand parameter. The proposed method provides a way to divide the time such that charging and discharging rates cater to those times.

2 Approach and Proposed Method

Much of the research work has been focused on the charging of the EVs through the SGs. The proposed algorithm incorporates the discharging of the EVs in the SG in order to stabilize the SG. This research focuses on using EVs as a means for balancing the frequency in the Smart Grids. The first step takes all the 24 values of the regulation signal from the database corresponding to each hour. Then, the Coloured Petri Net (CPN) algorithm is implemented that uses this signal to give the number of EVs based on whether they approach for charging or discharging. The proposed work can be represented as a sixfold as shown in Fig. 1.

To validate the proposed approach, the PJM regulation signal data of June 2014 is used. The regulation signal is the key factor in deciding the charging and discharging of EVs. The load is divided into three categories: off-peak load, mid-peak load, and peak load. The peak load is defined as the load when demand is nearly the capacity of the system, and off-peak load is the load when demand is quite low as compared to the rating of the system.

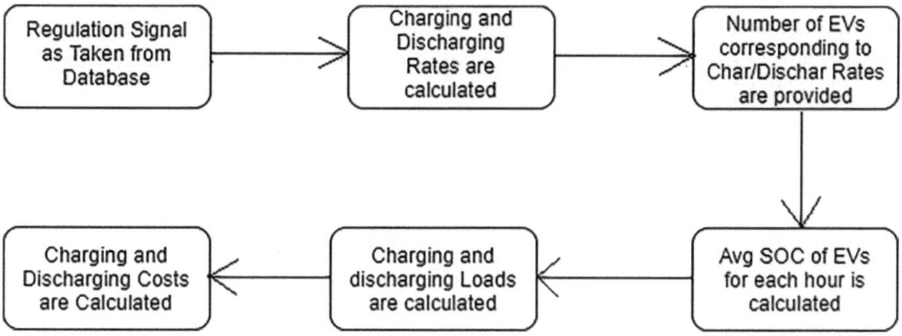

Fig. 1 The proposed method

Table 1 Charging and discharging rates

Type of hour	Time	charging rate	Discharging rate
Off-peak hours	23:00–09:00	3.155 Rs./kW h	12.749 Rs./kW h
Peak hours	10:00–12:00	No charging permitted	3.155 Rs./kW h
Mid-peak hours	09:00–10:00, 12:00–23:00	8.241 Rs./kW h	8.241 Rs./kW h

2.1 *Charging and Discharging Rates*

In the algorithm, it can be seen that the discharging rate will be highest during off-peak hours while the charging rate will be highest during peak hours. The various charging/discharging rates during peak, off–peak, and mid-peak hours are mentioned in Table 1 which will directly effect the charging/discharging costs of EVs.

Number of EVs: The number of EVs is one of the important parameters that defines not only the EVs arrival/departure for charging/discharging, but also defines the time at which EV arrives. The newly arrived EVs, operating EVs and departing EVs are calculated as given in Eqs. (1)–(4)

$$\mathrm{EV}_{\mathrm{C}}^{\mathrm{op}} = \mathrm{EV}_{\mathrm{C}}^{n}/E_{\mathrm{C}}. \tag{1}$$

$$\mathrm{EV}_{\mathrm{D}}^{\mathrm{op}} = \mathrm{EV}_{\mathrm{D}}^{n}/E_{\mathrm{D}}. \tag{2}$$

$$\mathrm{EV}_{\mathrm{C}}^{\mathrm{dp}} = \left(\mathrm{EV}_{\mathrm{C}}^{\mathrm{T}} - \left(\mathrm{EV}_{\mathrm{C}}^{n} + \mathrm{EV}_{\mathrm{C}}^{\mathrm{op}}\right)\right)/E_{\mathrm{C}}. \tag{3}$$

$$\mathrm{EV}_{\mathrm{D}}^{\mathrm{dp}} = \left(\mathrm{EV}_{\mathrm{D}}^{\mathrm{T}} - \left(\mathrm{EV}_{\mathrm{C}}^{n} + \mathrm{EV}_{\mathrm{D}}^{\mathrm{op}}\right)\right)/E_{\mathrm{D}}. \tag{4}$$

where $\mathrm{EV}_{\mathrm{C}}^{n}$ is the number of newly arrived EVs for charging, $\mathrm{EV}_{\mathrm{D}}^{n}$ is the number of newly arrived EVs for discharging, $\mathrm{EV}_{\mathrm{C}}^{\mathrm{op}}$ is the number of operating EVs for charging, $\mathrm{EV}_{\mathrm{D}}^{\mathrm{op}}$ is the number of operating EVs for discharging, $\mathrm{EV}_{\mathrm{C}}^{\mathrm{dp}}$ is the number of departing EVs after charging, $\mathrm{EV}_{\mathrm{C}}^{\mathrm{T}}$ is the total number of EVs coming for charging, $\mathrm{EV}_{\mathrm{D}}^{\mathrm{T}}$ is the total number of EVs coming for discharging, EV_D^{dp} is the number of departing EVs after discharging, E_D is the discharging energy, and E_C is the charging energy.

Average SOC of EVs: State of Charge (SOC) of each EV is calculated using Eqs. (5)–(8)

$$SOC_{BC} = SOC * Char\_rate. \tag{5}$$

$$SOC_{BD} = SOC * Dis\_rate. \tag{6}$$

$$SOC_{AC} = 100 - (100. * (E\_new\_charg/E\_rated)). \tag{7}$$

$$SOC_{AD} = 100 * (E\_new\_discharg/E\_rated). \tag{8}$$

where SOC_{BC} is the State of Charge(SOC) before charging of an individual EV, SOC is the SOC of an EV pertaining to its arrival at the SG, Char_rate is the Charging Rate for that hour, SOC_{BD} is the SOC before discharging of an individual EV, Dis_rate is the discharging rate for that hour, SOC_{AC} is the SOC after charging of an EV, E_new_charg is the new capacity of an EV pertaining to its charging, E_rated is the rated capacity of the electric vehicle's battery, SOC_{AD} is the SOC after discharging of an EV, and E_new_discharg is the capacity of EV pertaining to its discharging.

After the calculation of individual SOCs of EVs, the number of EVs corresponding to a signal hour is accumulated and average of SOCs before charging/discharging and after charging/discharging are calculated.

2.2 Vehicle-to-Grid (V2G) and Grid-to-Vehicle (G2V) Technology

According to the algorithm, the actual load signifies the regulation signal provided in the database. Load weights (w_t) are defined for peak, off-peak, and mid-peak hours which decides the charging and discharging of loads.

$$TCLd = BaseLd(t) + \sum_{EV=1}^{n} Ch\_Dmd(t). \tag{9}$$

Here, TCLd is the total charging load, BaseLd is the base load for each hour, and Ch_Dmd is the charging demand for each EV at that hour. The power available for charging during *t* hour is given in Eq. 10.

$$CLDev = TCLd(t) * P_t * EV_C(t). \tag{10}$$

where CLDev is the charging load deviation for each hour, P_t is the regulated power for charging, and EV_C is the number of EVs arriving for charging for each hour.

The G2V discharging load deviation is calculated in a similar manner considering the number of EVs coming for discharging. Charging and discharging of each hour will occur simultaneously. Whether the user profits or the grid stabilizes depends on the amount of revenue generated for each hour.

To verify whether the load is levelled, a simple mathematical formula is used:

$$L_{\mathrm{Net}} = L_{\mathrm{A}} - L_{\mathrm{C}} + L_{\mathrm{D}} \tag{11}$$

where L_{Net} is the net load, L_{A} is the preliminary load, L_{C} is the charging load, L_{D} is the discharging load

2.3 Charging and Discharging Cost

After the load calculation, charging and discharging cost are calculated using charging price/discharging price and load values during that hour.

An objective function is calculated beforehand to accumulate all the required parameters as given in Eq. 12.

$$\mathrm{ObjFn}_{\mathrm{Char}} = W_{\mathrm{t}} * (\mathrm{Pt}_{\mathrm{reg}} + \mathrm{Pt}_{\mathrm{unreg}}) * \mathrm{Char\_Rate}. \tag{12}$$

where $\mathrm{ObjFn}_{\mathrm{Char}}$ is the objective function pertaining to calculating the charging cost at hour t, W_{t} is the load Weight at that hour, $\mathrm{Pt}_{\mathrm{reg}}$, $\mathrm{Pt}_{\mathrm{unreg}}$ are the regulated power and unregulated power for charging, and Char_Rate is the charging rate.

$$\mathrm{CC} = (\min(\mathrm{ObjFn}_{\mathrm{Char}}) * (\mathrm{SOC}_{\mathrm{BC}} - \mathrm{SOC}_{\mathrm{AC}}) * \mathrm{EV}_{\mathrm{C}})/\mathrm{SOC}_{\mathrm{AC}}. \tag{13}$$

where CC is the charging cost, $\mathrm{SOC}_{\mathrm{BC}}$ is the SOC before charging, $\mathrm{SOC}_{\mathrm{AC}}$ is the SOC after charging, and EV_{C} is the number of EVs for charging.

Charging cost is a direct function of the load at that point along with the magnitude of SOC updated as given in Eq. 13. The objective function of discharging cost differs in the type of parameters is used as follows:

$$\mathrm{ObjFn}_{\mathrm{Dis}} = W_{\mathrm{t}} * \mathrm{Pt}_{\mathrm{reg}} * \mathrm{Dis\_Rate}. \tag{14}$$

where $\mathrm{ObjFn}_{\mathrm{Dis}}$ is the objective function for calculating the discharging cost at time t, W_{t} is the load weights pertaining to each hour for discharging, $\mathrm{Pt}_{\mathrm{reg}}$ is the regulated power for discharging, and Dis_Rate is the discharging rate pertaining to that hour. In Eq. 14, the regulated power is only used for discharging where in the case of charging cost, the regulated and unregulated power for charging is used. This is because the unregulated discharging will ultimately go back to the grid, while the unregulated charging will disrupt the grid and will lead to sharp valleys and peaks.

$$\mathrm{DC} = \left(\min(\mathrm{ObjFn}_{\mathrm{Dis}}) * (\mathrm{SOC}_{\mathrm{BD}} - \mathrm{SOC}_{\mathrm{AD}}) * \mathrm{EV}_{\mathrm{D}}(t)\right)/\mathrm{SOC}_{\mathrm{AD}}. \tag{15}$$

where DC is the discharging cost, SOC_{BD} is the SOC before discharging, SOC_{AD} is the SOC after discharging, and EV_D is the number of EVs arriving for discharging at that hour as given in Eq. 15.

3 Results

The various regulation signal parameters are calculated to stabilize the SG so as to achieve its full potential when compared to conventional grids which break down when load increases or decreases drastically. The specified parameters are obtained using the (Pennsylvania–New Jersey–Manhattan) PJM dataset constituting regulation data for 1 June 2014 and 5 June 2014 is used. The results are tabulated for frequency support over one-hour interval.

3.1 Number of Electric Vehicles (EVs) for Charging and Discharging

The number of EVs is calculated in three parts—newly arriving EVs, operating EVs, and departing EVs. These are then added together to get the number of EVs coming for charging/discharging pertaining to each hour. Figure 2 represents the number of EVs coming for charging and discharging. The number of EVs for discharging is higher during peak hours (10:00–12:00) and those for charging is higher during off-peak hours. The number of EVs coming for charging during peak hours (10:00–12:00) is 0 (zero). This is because charging is not permitted for those hours.

3.2 Charging Cost and Discharging Cost

As seen in Fig. 3, the charging is not permitted for the load at the peak hours (10:00–12:00). This is because the SG at that point is at a stage, where the demand is at the rate of supply and charging would lead to sudden blackouts. Hence to avoid grid failures, Grid-to-Vehicle (G2V) charging is not permitted at peak hours. Charging cost depends on the charging demand at any hour as well as number of EVs at that hour.

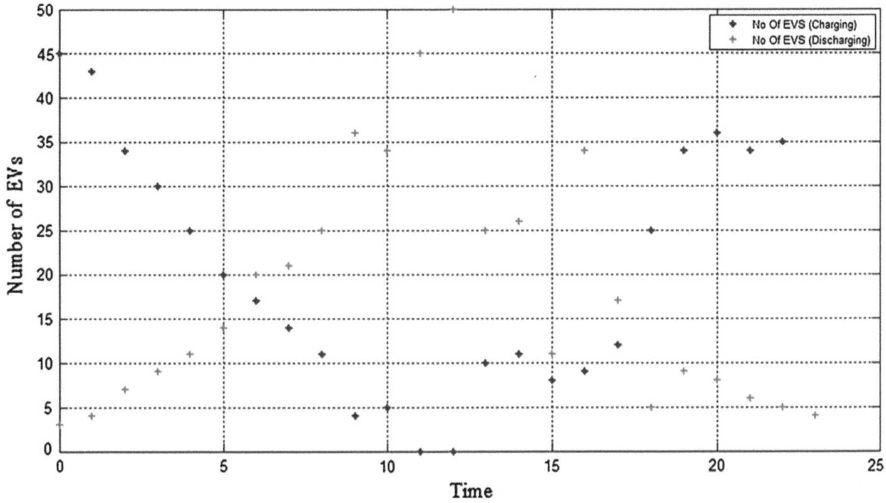

Fig. 2 Number of EVs coming for charging/discharging for 1 June 2014

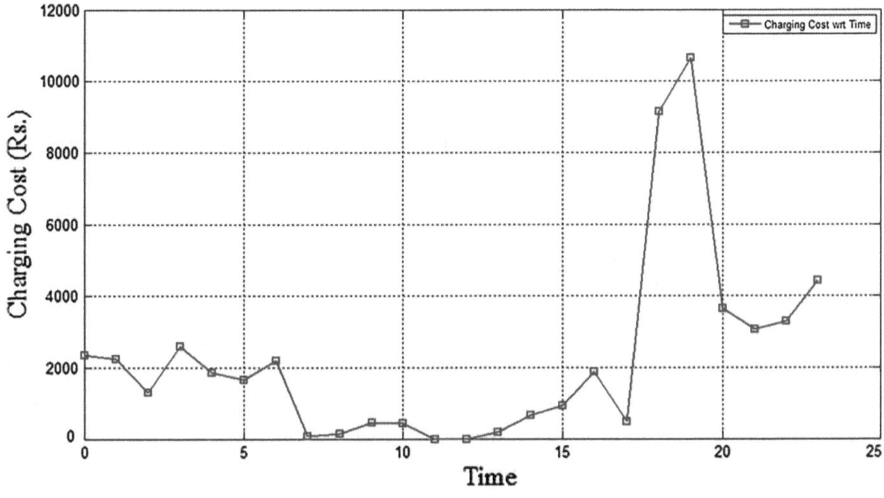

Fig. 3 Charging cost w.r.t. time for 1 June 2014

The EV paid the cost by the SG operators according to the time at which the EV arrives at the SG for discharging. According to the results, it is beneficial for the consumer to arrive at the peak hours for discharging. Though discharging rate will be less, the time for discharging will be more and hence more revenue will be generated. As seen in the graph given in Fig. 4, the peak hours (10:00–12:00) give the highest discharging cost of 19518.00 (11:00) and 35695.00 (12:00) for 1 June 2014.

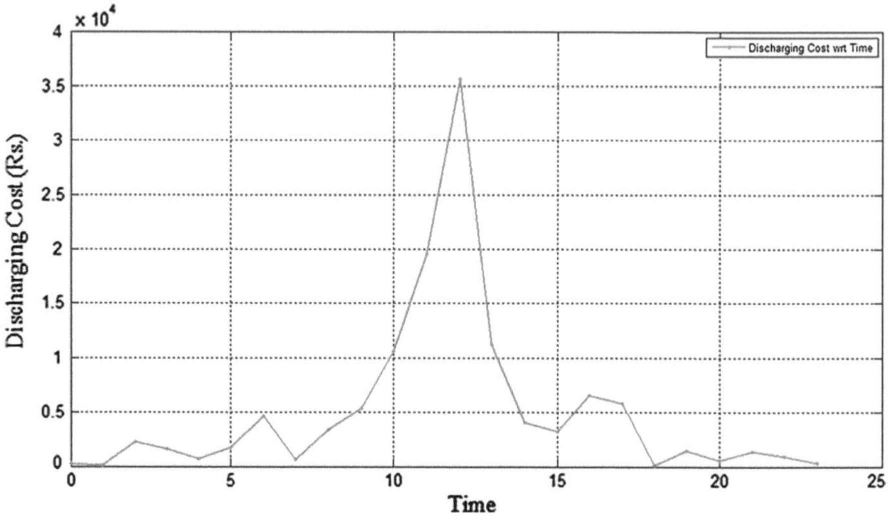

Fig. 4 Discharging cost w.r.t. time for 1 June 2014

3.3 Charging and Discharging Load

Discharging at peak hours when demand is near its rated capacity, regulates the frequency, and provides a way for EVs to utilize their battery power. Contrary to discharging where the EV owner is paid for discharging, here the grid generates the revenue for charging the EVs.

The net load gives the levelled load which tells that the frequency at the SG is regulated. According to the graphs given in Fig. 5, the frequency was regulated. The main focus of this study is to regulate frequency at the grid, using EVs to cater to the ancillary service resulting in load levelling. The load during G2V exchange takes away the energy from the grid, while the load during V2G exchange gives back to the grid. Here, the net load leads to load levelling where there are no valleys and peaks after the charging and discharging for each hour takes place. The actual load has irregularities in energy while the net load is almost a straight line as a result of regulating the frequency.

4 Conclusion

This work has proposed a novel technique to leverage the Electric Vehicle's (EVs) charging and discharging abilities in order to regulate frequency at the Smart Grid (SG). The actual load helps to regulate the frequency at the SG by incorporating it to the net load, thereby providing load levelling at the SG. The traffic at the grid also helps to detect whether the SG can cater to that many number of EVs in a particular

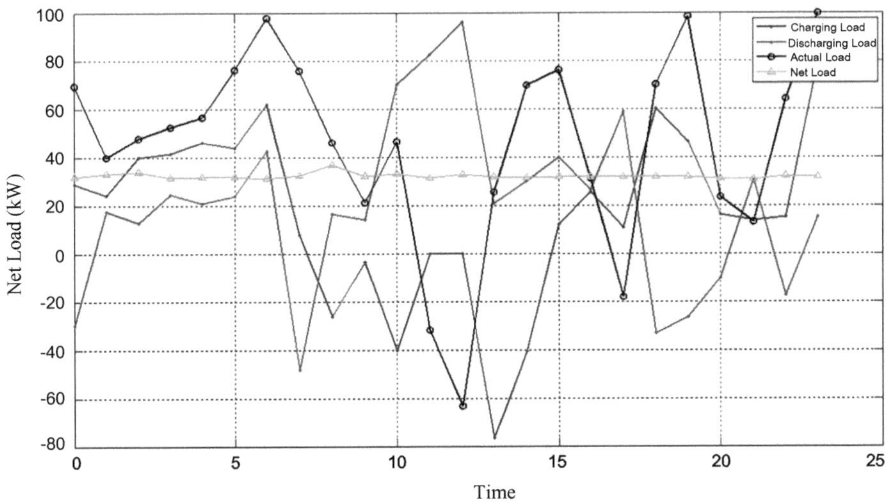

Fig. 5 Net load w.r.t. time for 1 June 2014

hour. The effectiveness of the proposed scheme has been studied through the real-time data taken from the PJM database corresponding to the regulation signals. The experimental results show that discharging data is equally important as the charging data. This method is quantifiably better as it incorporates discharging into the already existing methods along with better division of time, which was not seen in the existing methodologies. Therefore, this paper provides a study for load levelling along with real-time pricing in the SG using EVs for frequency regulation.

5 Future Scope

The peak load caters to only discharging at the SG while charging is not permitted during these hours. But in a real-time scenario, it can be seen that many EVs will come for charging even at the peak hours. Therefore, the EVs pertaining to charging can implement (Vehicle-to-Vehicle) V2V charging.

V2V charging corresponds to the Vehicle-to-Vehicle charging where an EV with a high State of Charge (SOC) can transfer some of its battery capacity to the EV with lower SOC. This will entail the EV-to-EV revenue generation along with the existing revenue that is being generated.

References

1. W. Kemptom, J. Tomic,Vehicle-to-grid power implementation: from stabilizing the grid to supporting large scale renewable energy. J. Power Resour. **144** (1), 280–294 (2005)
2. C. Guille, G. Gross, A conceptual framework for the vehicle-to-grid (V2G) implementation. Energy Policy Elsevier **37**(11), 4379–4390 (2009)
3. A. Papavasiliou, H. Hindi, D. Greene, Market-based control mechanisms for electric power demand response, in *IEEE Conference on Decision and Control* (ISSN No. 0191–2216) (2010)
4. S. Habib, M. Kamran, U. Rashid, Impact analysis of vehicles-to-grid technology and charging strategies of electric vehicles in distribution networks: a review. J. Power Sources Elsevier **277**, 205–214 (2015)
5. H. Liu, Z. Hu, Y. Song, J. Wang, X. Xie, Vehicle-to-grid control for supplementary frequency regulation considering charging demands. IEEE Trans. Power Syst. **30**(6), 3110–3119 (2015)
6. F. Rücker, I. Bremera, S. Linden, J. Badeda, D.U. Sauer, Development and evaluation of a battery lifetime extending charging algorithm for an electric vehicle fleet. Sci. Direct Energy Procedia **99**, 285–291 (2016)
7. W. Zhifu, W. Yupu, L. Zhi, S. Qiang, R. Yinan, The optimal charging method research for lithium-ion batteries used in electric vehicles. Energy Procedia **104**, 74–79 (2016)
8. K. Kaur, R. Rana, N. Kumar, M. Singh, S. Mishra, A coloured petri net based frequency support scheme using fleet of electric vehicles in smart grid environment. IEEE Trans. Power Syst. **31**(6), 4638–4649 (2016)
9. G. Zhang, S.T. Tan, G.G. Wang, Real-time smart charging of electric vehicles for demand charge reduction at non-residential sites. IEEE Trans. Smart Grids **PP**(99) (ISSN No. 1949–3053) (2016)
10. P. Siano, D. Sarno, Assessing the benefits of residential demand response in a real time distribution energy market. Appl. Energy **161**, 533–551 (2016)
11. Q. Wang, X. Liu, J. Du, F. Kong, Smart charging for electric vehicles: a survey from the algorithmic perspective. IEEE Commun. Surv. Tutorials **18**(2), 1500–1518 (2016)
12. W. Shuai, P. Maille, A. pelov.: charging electric vehicles in the smart city: a survey of economy-driven approach. IEEE Trans. Intell. Transp. Syst. **17**(8), 2089–2106 (2016)
13. T.U. Daim, X. Wang, K. Cowan, T. Shott, Technology roadmap for smart electric V2G of residential chargers. J. Innov. Entrepreneurship **15**(5) (2016)
14. C. Peng, J. Zou, L. Lian, Dispatching strategies of electric vehicles participating in the frequency regulation on power grid: a review. Renew. Sustain. Energy Rev. Elsevier **68**(Part 1), 147–152 (2017)
15. S.-K. Moon, J. Kim, Balanced charging strategies for electric vehicles on power systems. Appl. Energy **189**, 44–54 (2017)
16. L. Baringo, R.S. Amaro, A stochastic robust optimization approach for the bidding strategy of an electric vehicle aggregator. Electr. Power Syst. Res. **146**, 362–370 (2017)
17. B. Faessler, P. Kepplinger, J. Petrasch, Decentralized price-driven grid balancing via repurposed electric vehicle batteries. Elsevier Energy **118**, 446–455 (2017)
18. N. Daina, A. Sivakumar, J.W. Polak, Modelling electric vehicles use: a survey on the methods. Renew. Sustain. Energy Rev. **68**(Part 1), 447–460 (2017)
19. L. Yao, W.H. Lim, T.S. Tsai, A real-time charging scheme for demand response in electric vehicle parking station. IEEE Trans. Smart Grid **8**(1), 52–63 (2017)
20. N.G. Omran, S. Filizadeh, A semi-cooperative decentralized scheduling scheme for plug-in electric vehicle charging demand. Electr. Power Energy Syst. **88**, 119–132 (2017)
21. G. Wenzel, M. Negrete-Pincetic, D.E. Olivares, J. MacDonald, D.S. Callaway, Real time charging strategies for an electric vehicle aggregator to provide ancillary services. IEEE Trans. Smart Grids **PP**(99) (2017)
22. N. Daina, A. Sivakumar, J.W.Polak, Electric vehicle charging choices: modelling and implications for smart charging services. Transp. Res. Part C: Emerg. Technol. Elsevier **81**, 36–56 (2017)

Traffic Heterogeneity Analysis in an Energy Heterogeneous WSN Routing Algorithm

Deepak Sharma, Amarendra Goap, A. K. Shukla, Priyanka and Amol P. Bhondekar

Abstract Wireless Sensor Network (WSN) is an important element of Internet of Things arena. Energy efficiency is the prime factor for attaining the long life of WSN systems, which are resource constrained especially in energy. During the WSN operations, the communication activities consume maximum energy share of the WSN nodes. As routing algorithms are associated with the communication activities of the network, the energy efficiency of the routing algorithms become a critical factor in WSNs. The early work in WSN routing considers the deployment of homogeneous nodes. In practical scenarios, the WSN nodes are not homogeneous in their configuration, capabilities and/or behavior. The heterogeneity of WSN nodes has been broadly classified into three major categories, viz. energy, computation, and link. The heterogeneity of traffic/data generation rate is another important aspect, which considers nodes with heterogeneous data transmission requirements. This paper considers traffic heterogeneity aspect along with the energy heterogeneity for improved routing decision in the WSN. The paper discusses how the performance of the system is affected under different heterogeneity scenarios. It considers a two-level energy and traffic heterogeneous scenario in a clustering-based WSN. It also proposes an improved method of cluster head selection for the given scenario. The work comes under multi-heterogeneity consideration in WSN routing algorithms, which is an important aspect for designing efficient routing algorithms for realistic WSNs.

Keywords Internet of things · Wireless sensor networks · Routing protocols
Energy heterogeneity · Traffic heterogeneity

D. Sharma (✉) · A. Goap · A. K. Shukla · A. P. Bhondekar
CSIR-Central Scientific Instruments Organisation (CSIR-CSIO), Chandigarh, India
e-mail: deepakskc@yahoo.com

D. Sharma · A. K. Shukla · A. P. Bhondekar
Academy of Scientific and Innovative Research (AcSIR), CSIR-CSIO, Chandigarh, India

Priyanka
Meerut Institute of Engineering and Technology, Meerut, India

C. R. Krishna et al. (eds.), *Proceedings of 2nd International Conference on Communication, Computing and Networking*, Lecture Notes in Networks and Systems 46, https://doi.org/10.1007/978-981-13-1217-5_33

1 Introduction

The early work in Wireless Sensor Networks (WSNs) routing algorithms does not give emphasis to nodes' heterogeneity aspects. However, in practical WSN deployments, nodes homogeneity is difficult to attain and maintain. Even during homogeneous WSN operations, the energy consumed by different nodes on wireless communications is different, which creates an energy heterogeneous scenario over its operations. The WSN is an element of Internet of Things (IoT) ecosystem, where an interoperability of various heterogeneous subsystems and devices is expected. The disparity in nodes' configurations, capabilities and/or functional behavior creates heterogeneities in WSNs. A constructive consideration of these disparities (i.e., heterogeneity) can improve the performance of WSNs. Researchers have considered routing algorithms for energy, computation and link heterogeneous WSN scenarios [1]. SEP (Stable Election Protocol) [2] routing protocol is among the early work for energy heterogeneous WSNs, which is still used to analyze the performance improvement by the newly developed algorithms in the area. SEP proposes a new cluster head selection mechanism to improve the energy efficiency in energy heterogeneous WSN. It shows that the routing algorithm for homogeneous WSNs (LEACH [3, 4]) does not judiciously utilize the extra energy of energy heterogeneous nodes. Consideration of traffic heterogeneity (based on disparity of packet length) in WSN routing algorithms is relatively less explored area [5–9]. The traffic heterogeneity is an important aspect for WSNs with heterogeneous sensing requirements, where different sensors have different amount of data to be communicated to the base station.

For designing the routing algorithms for more realistic WSNs, consideration of multiple heterogeneities is an important aspect. This paper presents the performance analysis of a routing algorithm (SEP) for energy heterogeneous WSN under heterogeneous traffic scenarios. It also presents an improved cluster head selection method for the given/proposed multi-heterogeneity scenario.

The rest of this paper is arranged as follows: Sect. 2 introduces the system model used for the analysis. In Sect. 3, we present a performance analysis of the SEP algorithm under the proposed multi-heterogeneous environment. An improved routing algorithm is also proposed for the scenario in this section. In Sect. 4, the results of this work are discussed. Finally, the conclusions and some perspectives to extend this work are presented in Sect. 5.

2 System Model

To analyze the effect of traffic heterogeneity in an energy heterogeneous WSN routing scenario, SEP [2] has been considered as the base algorithm. SEP was developed with an aim to effectively utilize the extra energy of the energy heterogeneous nodes during the routing process. SEP considers LEACH-like clustering approach with improved cluster head selection mechanism to cater energy heterogeneity. It considers LEACH-

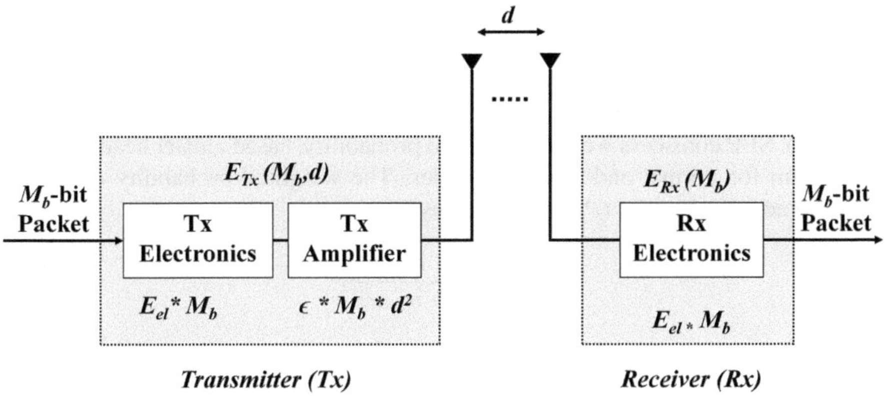

Fig. 1 Radio model

like simulation environment with base station/sink located at the center of the square area. The first-order radio model (Fig. 1) has been considered, where the energy spent by the radio to transmit M_b-bits message to a distance d is given by:

$$E_{Tx} = \begin{cases} M_b \cdot E_{el} + M_b \cdot \in_{fs} \cdot d^2 \text{ if } d < d_0 \text{ where } d_0 = \sqrt{\frac{\in_{fs}}{\in_{mp}}} \\ M_b \cdot E_{el} + M_b \cdot \in_{mp} \cdot d^4 \qquad \text{if } d \geq d_0 \end{cases} \tag{1}$$

where E_{el} is the receiver or the transmitter circuit's per bit dissipated energy and the distance between the sender and the receiver is d. The $\in_{fs}$ and the $\in_{mp}$ are transmitter amplifier losses for the free space and the Multipath scenarios, respectively. The energy dissipated in receiving an M_b-bits message is given by:

$$E_{Rx} = M_b.E_{el} \tag{2}$$

Assuming 100 numbers of nodes (n) deployed uniformly over a 100 m × 100 m square area. The sink node is placed at the center of the field and the distance of any node to the sink is $\leq d_0$, then the free space model can be considered for the intra-cluster and the inter-cluster communications. The energy spent by the cluster head nodes (E_{CH}) is given by:

$$E_{CH} = M_b.E_{el}\left(\frac{n}{k} - 1\right) + M_b.E_{DA}\left(\frac{n}{k}\right) + M_b.E_{el} + M_b. \in_{fs} .d^2_{toBS} \tag{3}$$

where E_{DA} is the data aggregation energy spent per bit by cluster heads, k is the number of clusters, and d_{toBS} is the distance between the cluster head and the sink. The energy dissipated by the non-cluster head nodes (E_{nonCH}) is given by:

$$E_{nonCH} = M_b.E_{el} + M_b. \in_{fs} .d^2_{toCH} \tag{4}$$

where d_{toCH} is the distance between the member nodes and their cluster head.

SEP considers that m_e percent of nodes (called advance nodes) have α_e (additional energy factor) times more energy in comparison to normal nodes. In comparison to a uniform cluster head selection criteria of routing algorithm for homogeneous WSNs (LEACH), SEP considers weighted optimal probability-based cluster head selection mechanism for normal and advanced nodes. The weighted probability of normal (P_{nr}) and advanced nodes (P_{ad}) are given by:

$$P_{\mathrm{nr}} = \frac{P_{\mathrm{op}}}{1 + \alpha_e.m_e} \tag{5}$$

$$P_{\mathrm{ad}} = \frac{P_{\mathrm{op}}}{1 + \alpha_e.m_e} \times (1 + \alpha_e) \tag{6}$$

where P_{op} is the optimal probability to become cluster head. Similar to LEACH, a node becomes a cluster head in the current round if the random number selected by the node is less than the threshold. The threshold functions for the normal nodes (T_{nr}) and the advanced nodes (T_{ad}) are given by:

$$T_{\mathrm{nr}}(n) = \begin{cases} \frac{P_{\mathrm{nr}}}{1-P_{\mathrm{nr}}.\left(r \bmod \frac{1}{P_{\mathrm{nr}}}\right)} & \text{if } n \in G' \\ 0 & \text{otherwise} \end{cases} \tag{7}$$

$$T_{\mathrm{ad}}(n) = \begin{cases} \frac{P_{\mathrm{ad}}}{1-P_{\mathrm{ad}}\cdot\left(r \bmod \frac{1}{P_{\mathrm{ad}}}\right)} & \text{if } n \in G'' \\ 0 & \text{otherwise} \end{cases} \tag{8}$$

where G' and G'' are the set of nodes that have not become cluster heads within the last $1/P_{\mathrm{nr}}$ and $1/P_{\mathrm{ad}}$ rounds of the epoch, respectively; and the current round is represented by r. Similar to LEACH, nodes other than the cluster head nodes join the cluster considering the signal strength of the cluster heads' advertisement message. Each member node is assigned a time slot by its cluster head node for data communication. The cluster head aggregates the data collected from the member nodes before sending it to base station. After a certain time period, the complete process is repeated.

3 Traffic Heterogeneity Consideration in Routing Algorithm for Energy Heterogeneous WSN

To consider the heterogeneity in terms of traffic generation from the different nodes of the network, we have considered that the packet length of the nodes is different for different nodes. For simplicity, we have considered two types of traffic heterogeneous nodes, which is similar to the two type of energy heterogeneous nodes scenario in SEP. Further, the same heterogeneous nodes are having energy and traffic heterogeneity.

Table 1 System parameters

Parameter	Value
Area/network size	100 m × 100 m
Number of sensor nodes (n)	100
Base station/sink location	Center of the field
Normal node's initial energy	0.5 J
Energy consumed in transmit/receive electronics (E_{el})	50 nJ/bit
Energy consumed in data aggregation by cluster head (E_{DA})	5 nJ/bit/signal
Energy consumed by transmit amplifier in free space scenario ($\in_{\text{fs}}$)	10 pJ/bit/m^2
Energy consumed by transmit amplifier in multipath scenario $\left(\in_{\text{mp}}\right)$	0.0013 pJ/bit/m^4
Normal node's packet length (M_{b})	4000 bits

The analysis has been carried out in MATLAB and the parameters considered for simulation environment are shown in Table 1.

3.1 Analysis of Traffic Heterogeneity in Energy Heterogeneous WSN Routing

In this section, we have considered the effect of traffic heterogeneity in SEP algorithm. We have considered that the energy-rich nodes may have higher traffic generation rate in terms of their packet size. We have considered $m_e = m_t$ percent of randomly selected nodes (traffic heterogeneous nodes) with $(1 + \alpha_t)$ time longer packet length in comparison to normal nodes. It is assumed that the network can sustain the heterogeneous traffic for the given simulation environment.

Figure 2 shows four different scenarios for four different values of traffic load (α_t), where each scenario considers three different values of m_e (=m_t). The four results show the performance for α_t = 0, 1, 2, 4 with α_e = 1 and $m_e = m_t$ = 0.1, 0.2, 0.3 in each scenario. It shows that the stable period for higher m_t is penalized more than the lower values of m_t, with an increase in traffic load (α_t). Further, the stability period is not affected much with an increase in only m_t under traffic heterogeneous scenario (α_t>0).

Figure 3 shows that similar to homogeneous WSN case [9], the stable period of SEP deteriorates with an increase in traffic heterogeneity. The $m_e = m_t = 0.1$; $\alpha_e =$ 1; $\alpha_t = 0$ shows the performance of original SEP algorithm without traffic heterogeneity. An increase in packet length (α_t), while maintaining energy heterogeneity,

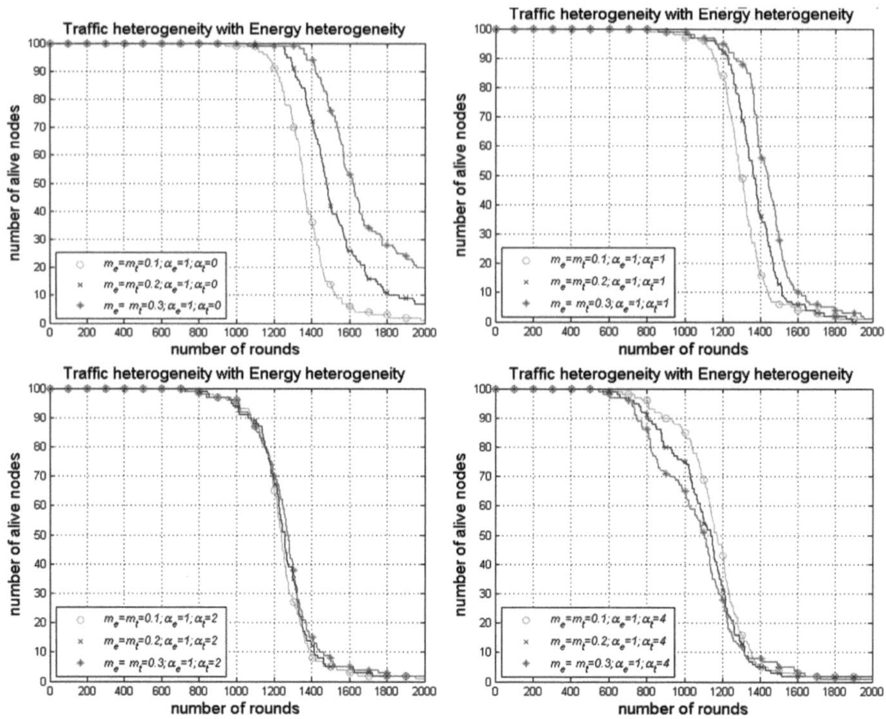

Fig. 2 Traffic heterogeneity in SEP: varying m_t for α_t (four different scenarios for α_t = 0, 1, 2, 4)

deteriorates the performance (stable period) of SEP protocol significantly. The results follow the pattern even for different percentage of heterogeneous nodes.

3.2 An Improved Routing Algorithm Considering Traffic Heterogeneity Along with Energy Heterogeneity

In this section, we present a traffic heterogeneity aware protocol (SEP-T) for energy and traffic heterogeneous scenario. Based on the analysis in the last section, an efficient cluster head selection mechanism is proposed, which considers both type heterogeneities in a constructive manner. The new weighted probabilities for normal nodes ($P_{\text{nr\_th}}$) and advances nodes ($P_{\text{ad\_th}}$) (more energy with longer packet length) are given by:

$$P_{\text{nr\_th}} = \frac{P_{\text{op}}}{1 + \alpha_e . m_e} \times (1 + \alpha_t \cdot m_t) \tag{9}$$

$$P_{\text{ad\_th}} = \frac{P_{\text{op}}}{1 + \alpha_e . m_e} \times (1 + \alpha_e - \alpha_t \cdot m_t) \tag{10}$$

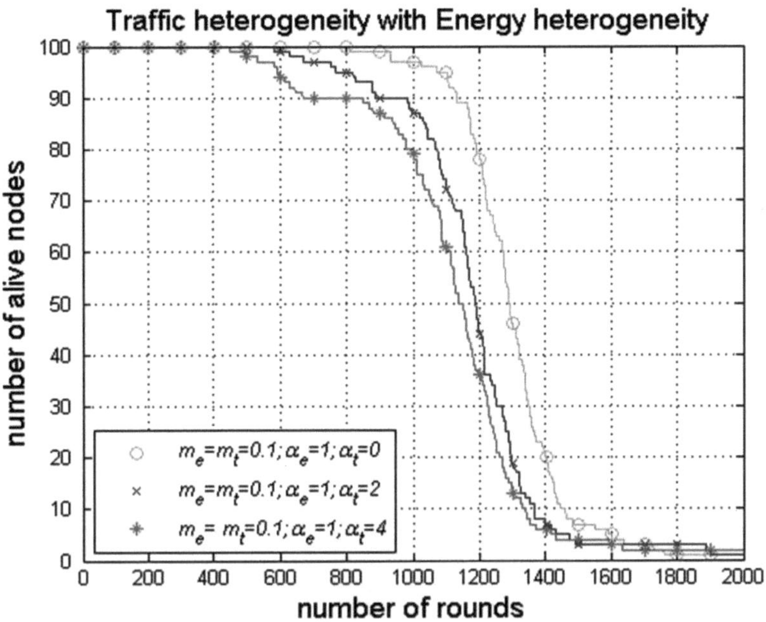

Fig. 3 Traffic heterogeneity in SEP: varying α_t for $m_t = 0.1$

The remaining mechanism is same as discussed in system model section for SEP protocol. The performance of new algorithm is evaluated by replacing P_{nr} and P_{ad} with $P_{\text{nr\_th}}$ and $P_{\text{ad\_th}}$, respectively, in the system model.

Figure 4 shows the performance of proposed algorithm under different heterogeneous scenarios. The stable period of SEP deteriorate under traffic heterogeneity, which is improved by the proposed mechanism (SEP-T).

4 Result and Discussion

The improved cluster head selection method (SEP-T) shows improvement in stable period under different heterogeneity scenarios (Fig. 4). For a particular deployment, an increase in α_t from 0 to 1 in SEP ($m_e = m_t = 0.1$; $\alpha_e = 1$) decreases the stable period from 808 to 677 rounds; however, the SEP-T improves the stable period to 761 rounds. An increase in α_t from 0 to 2 in SEP ($m_e = m_t = 0.1$; $\alpha_e = 1$) decreases the stable period from 808 to 581 rounds; however, the SEP-T improves the stable period to 660 rounds. An increase in α_t from 0 to 2 in SEP ($m_e = m_t = 0.2$; $\alpha_e = 1$) decreases the stable period from 959 to 604 rounds; however, the SEP-T improves the stable period to 676 rounds. An increase in α_t from 0 to 1 in SEP ($m_e = m_t = 0.2$; $\alpha_e = 2$) decreases the stable period from 1059 to 857 rounds; however, the SEP-T improves the stable period to 925 rounds.

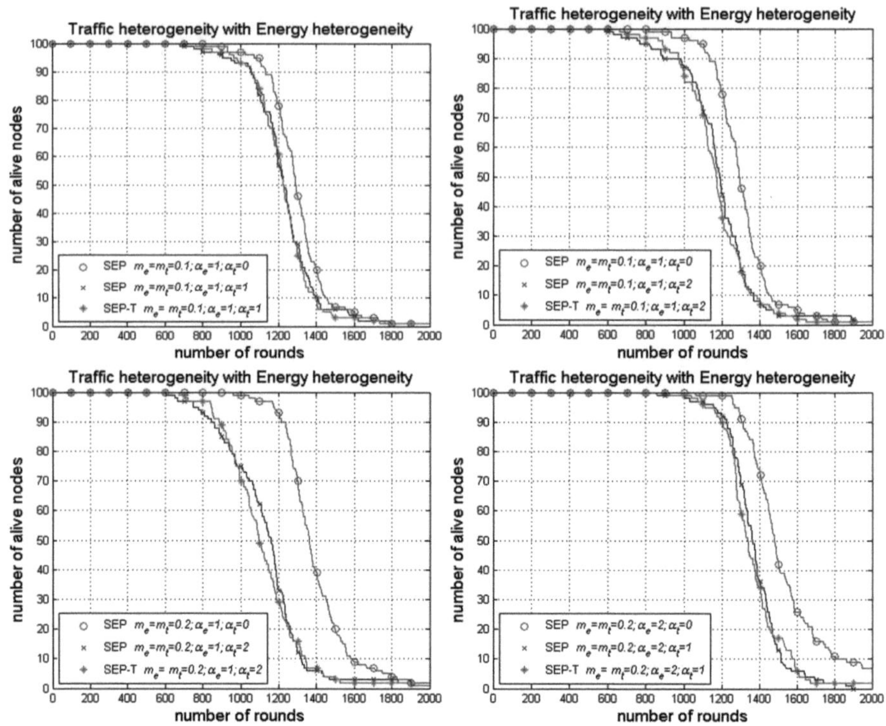

Fig. 4 Considering traffic and energy heterogeneity in cluster head selection

5 Conclusion

This paper analyzes the effect of traffic heterogeneity in a routing algorithm for energy heterogeneous WSN. The results show that the stable period of SEP deteriorates significantly with an increase in traffic heterogeneity. Further, the stable period for WSN with higher percentage of heterogeneous nodes is penalized more with an increase in traffic load of heterogeneous nodes. The proposed improved cluster head selection method (SEP-T), based on the analysis results, shows better energy efficiency in terms of improved stability period under different heterogeneous scenarios. The work is a step towards consideration of multiple heterogeneities in WSN, which can help in realizing more realistic WSN deployments. The future work includes analysis of random/different levels of multiple heterogeneities at node level along with consideration of latest routing techniques for comparisons.

References

1. S. Tanwar et al., A systematic review on heterogeneous routing protocols for wireless sensor network. J. Netw. Comput. Appl. **53**, 39–56 (2015)
2. G. Smaragdakis et al., SEP: a stable election protocol for clustered heterogeneous wireless sensor networks, in *Second International Workshop on Sensor and Actor Network Protocols and Applications (SANPA 2004)* (2004)
3. W.R. Heinzelman et al., Energy-efficient communication protocol for wireless microsensor networks, in *2000. Proceedings of the 33rd annual Hawaii international conference on System Sciences*, vol. 2 (2000), p. 10
4. W.B. Heinzelman et al., An application-specific protocol architecture for wireless microsensor networks. Wireless Commun. IEEE Trans. **1**, 660–670 (2002)
5. L. Xiaoya et al., Energy efficient routing protocol based on residual energy and energy consumption rate for heterogeneous wireless sensor networks, in *2007 Chinese Control Conference* (2007), pp. 587–590
6. X. Li et al., A routing protocol for balancing energy consumption in heterogeneous wireless sensor networks, in *International Conference on Mobile Ad-Hoc and Sensor Networks* (2007), pp. 79–88
7. A. Ojha et al., Traffic heterogeneity consideration in clustering based WSN routing algorithms, in *IETE Proceedings of 47th Mid Term Symposium on Modern Information and Communication Technologies for Digital India (MICTDI), Chandigarh* (2016)
8. H. Zhou et al., A novel stable selection and reliable transmission protocol for clustered heterogeneous wireless sensor networks. Comput. Commun. **33**, 1843–1849 (2010)
9. D. Sharma et al., A traffic aware cluster head selection mechanism for hierarchical wireless sensor networks routing, in *Presented at the IEEE International Conference on Parallel, Distributed & Grid Computing*, India (2016)

Integration of Optical and Wireless Technology as High-Capacity Communication System: Issues and Challenges

Namita Kathpal and Amit Kumar Garg

Abstract Radio over Fiber (also referred as Radio on the Fiber, Hybrid Fiber Radio and Fiber Radio Access) is a desegregated technology that merges wireless system with optical fiber for the imminent distribution of broadband services. RoF significant feature is to transmit radio frequency signals (20 Hz–300 GHz) over optical fiber to fulfill the demand of large data traffic by using the merits of optical fiber which gives large bandwidth (50 THz), less attenuation (0.2 dB/km), and low electromagnetic interference. In the present work, various issues and challenges related to RoF system architecture concerned with the efficient utilization of centralized control capability have been compared along with the current state of art of RoF system.

Keywords Radio frequency over fiber (RFoF) · Intermediate frequency over fiber (IFoF) · Base band over fiber (BBoF) · Central station (CS) · Base station (BS)

1 Introduction

To fulfill the high bit rate communication services demand and to distribute the wireless mm wave signal over the large geographical area, hybrid system consisting of fiber and radio technologies called RoF are proposed. RoF consociates the effectiveness of optical fibers for the transportation of radio signals to various wireless access points situated in different area with the main advantage of wireless system, i.e., mobility and ubiquity. The exemplar provided by RoF system is beneficial as it transmits radio signal in their original analog form. RoF system architecture fundamentally consists of CS, BS, and optical fiber which interconnect CS and BS. The basic elements of CS are Modulators, Optical Sources, Optical

N. Kathpal (✉) · A. K. Garg
Department of Electronics and Communication Engineering, Deenbandhu Chhotu Ram University of Science and Technology, Murthal (Sonepat), India
e-mail: namitakathpal2016@gmail.com

A. K. Garg
e-mail: garg_amit03@yahoo.co.in

C. R. Krishna et al. (eds.), *Proceedings of 2nd International Conference on Communication, Computing and Networking*, Lecture Notes in Networks and Systems 46, https://doi.org/10.1007/978-981-13-1217-5_34

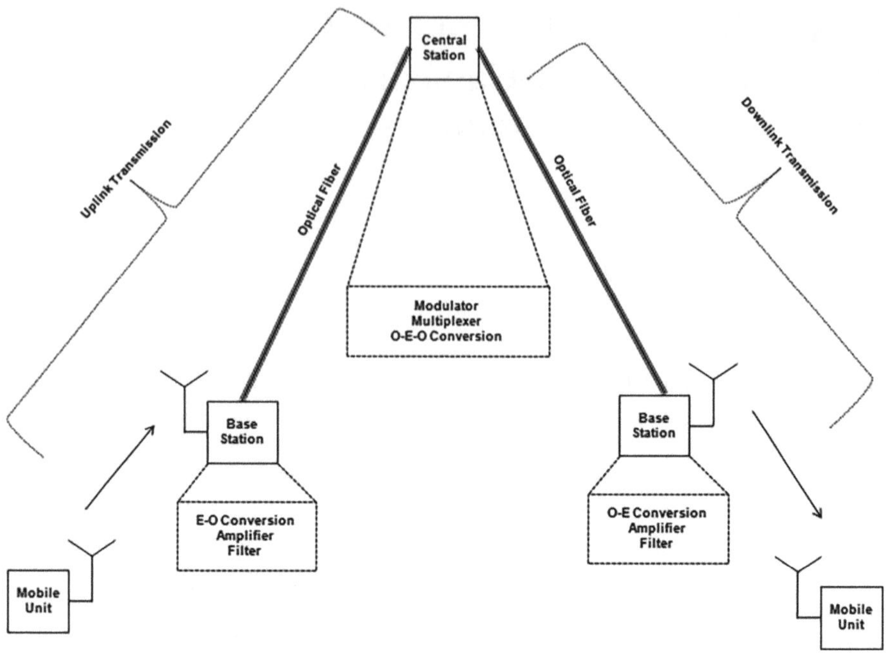

Fig. 1 RoF system architecture

Detectors, and Multiplexers, i.e., all the equipment mandatory to perform RF signal processing and BS comprises of O–E/E–O converters, filters, and amplifiers. This architecture allows cost-effective BS, as they only need to perform O–E/E–O conversions, filtering and amplification functions whereas CS has to perform RoF control functions such as modulation, demodulation and frequency allocation thus CS are expensive in comparison to BS. Figure 1 shows the general RoF architecture.

In the uplink transmission from Mobile Unit to Central Station, the RF signals received at BS are amplified and converted into optical signal using optical source and then, this light signal is transferred to CS via fiber. In the downlink transmission from CS to Mobile Unit, data from the user modulates RF source which in turn modulates optical carrier from light source. The modulated optical signal is carried to BS via optical fiber and at the BS this optical signal is converted into RF signal and this wireless signal is transmitted to Mobile Unit using BS antenna.

2 RoF System Architectures and Related Issues

RoF system architecture is classified into three categories depending on the frequency range of the Radio Signal to be transported

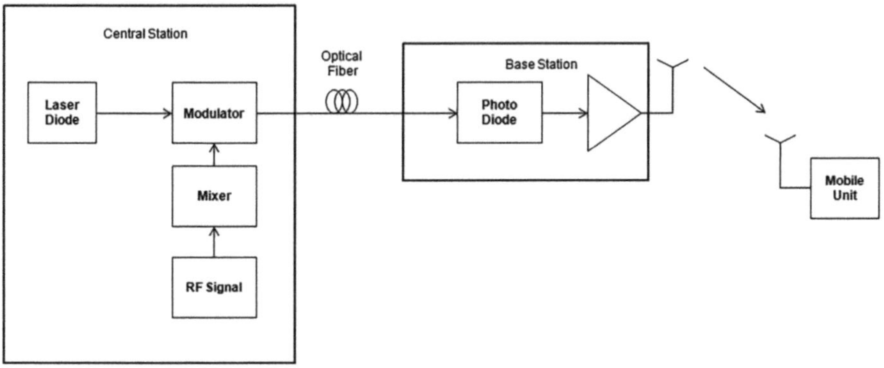

Fig. 2 RFoF downlink system architecture

1. RFoF (Radio Frequency over Fiber)
2. IFoF (Intermediate Frequency over Fiber)
3. BBoF (Baseband over Fiber)

1. RFoF architecture

In RFoF architecture (also known as Analog RoF) shown in Fig. 2, the electrical radio signals in the range of 10–40 GHz are optically modulated in CS and transmitted to BS through optical fiber without altering the radio center frequency. Therefore, frequency up/down conversion is not required at BS this make RFoF architecture the most simplest and cost-effective architecture but the link performance is affected by chromatic dispersion [1] and nonlinear distortion which in turn limits the transmission of signal over long distance.

Thus, various linearization techniques using analog or digital signal processing [2] was proposed to mitigate these impairments.

2. IFoF architecture

In IFoF architecture (also known as Digitized RoF or Digitized IoF) depicted in Fig. 3, the electrical signals in the range of 10–100 MHz are transposed onto an optical carrier at CS and the modulated light signal is propagated through optical fiber to BS. At the BS, the transmitted optical signal is recovered by optical detector and the detected electrical signal is upconverted to RF which in turn transmitted to MU. The BS in this architecture required additional components for frequency up and down conversions. Thus, this RoF system architecture is expensive but it reduces chromatic dispersion effects by employing band pass signaling technique [3]. Band pass signaling technique has certain limitations which affect the data transmission rate, therefore Delta Sigma Modulator (DSM) based DIoF system had been proposed [4] which reduces the requirements of active devices for recovering the transmitted signal at Base Station. Further continuous time low pass DSM has proposed [5] which shows better results in terms of SNR.

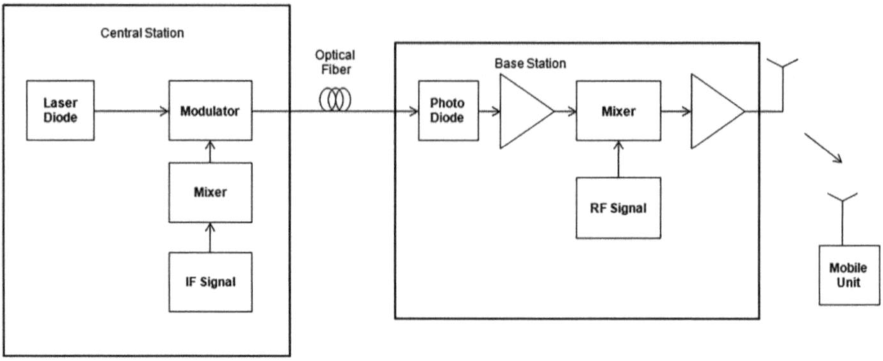

Fig. 3 IFoF downlink system architecture

3. BBoF architecture

In BBoF architecture (also known as Digital RoF) shown in Fig. 4, the radio signals in the range of 20 Hz–20 kHz are modulated at CS and transmitted to BS. At BS, the electrical signal is detected by the photodetector which is then upconverted to RF through IF or directly and transmitted to MU. In BBoF, the effect of chromatic dispersion is negligible but BS configuration is most expensive.

In this architecture at CS, the RF signal is converted into digital signal by ADC and at the BS received digital signal is converted into analog by DAC but this conversion produces jitter [6] which limits the link performance. In [7], all-photonic DRoF architecture has been proposed which employs electro-optical modulator to perform optical sampling and modulation at CS which reduces the jitter to femtosecond and also produces less BER (10^{-62}) as compared to ARoF. As BRoF requires up converters at BS and demands of up converters can be reduced by employing heterodyne photo-detection [12].

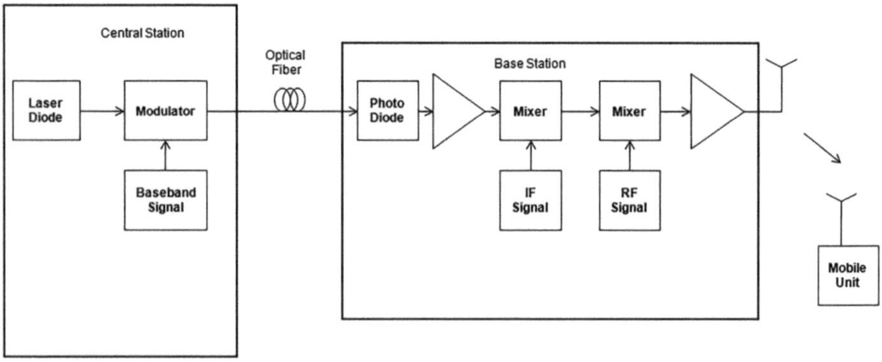

Fig. 4 BBoF downlink system architecture

Table 1 Comparison of various RoF architectures for wireless communication

S. no.	Parameters	RFoF	IFoF	BBoF
1.	Transmission distance [7]	15 km	65 km	65 km
2.	BER [8]	10^{-9}	10^{-58}	10^{-62}
3.	BS designing	Least complex as frequency conversion is not required	Complex due to requirement of frequency conversions	Complex due to requirement of frequency conversions
4.	Immunity to interference	Less	High as laser and photodiode are operating at low frequency	High as laser and photodiode are operating at low frequency
5.	High bit rate transmission [9]	<1 Gbps	<1 Gbps	1 Gbps to 2.5 Gbps
6.	Rayleigh scattering (σ_s) [10]	High as σ_s is directly proportional to frequency (frequency range in RFoF is from 10–40 GHz)	Medium (as frequency range is from 10–100 MHz)	Least (as frequency range is from 20 Hz–20 kHz)
7.	Hardware requirement [11]	Least due to BS designing	More as additional RF components are required	More as additional RF components are required

3 Conclusion

In this paper, an overview of various RoF system architectures has been provided along with current state of art. Table 1 shows the advantage and disadvantage of various RoF architectures.

Based on the literature survey, it has been concluded that the reduction in transmission frequency leads to the transmission of signal over long distance (65 km) with less BER (10^{-62}).

References

1. A. Ng'oma, P. Shih, J. George, F. Annunziata, M. Sauer, C. Lin, W. Jiang, J. Chen, S. Chi, 21 Gbps OFDM wireless signal transmission at 60 GHz using a simple IMDD radio-over-fiber system, in *Optical Fiber Communication Conference (OFC), San Diego (CA), USA* (2010) (ISBN: 9781557528858)
2. H. Chen, J. Li, K. Xu, Y. Dai, F. Yin, J. Lin, *Experimental Investigation on Multi-Dimensional Digital Predistortion for Multi-Band Directly-Modulated Radio-Over-Fiber Systems, Optical*

Fiber Communication Conference (OFC) (California United States, San Francisco, 2014). ISBN 9781557529930
3. P.A. Gamage, A. Nirmalathas, C. Lim, D. Novak, R. Waterhouse, Design and analysis of digitized RF-over-fiber links. J. Lightwave Technol. **27**(12), 2052–2061 (2009)
4. S. Jang, G. Jo, J. Jung, B. Park, S. Hong, A digitized IF-over-fiber transmission based on low-pass delta-sigma modulation. IEEE Photonics Technol. Lett. **26**(24), 2484–2487 (2014)
5. S. Jang, S. Hyun, K. Kim, J. Jung, B. Park, Design of CMOS continuous-time low-pass delta-sigma modulator for digital distributed antenna system based on IF-over-fiber transmission, in *19th International Conference on Advanced Communication Technology (ICACT)* (2017), pp. 835–838
6. G.C. Valley, Photonic analog-to-digital converters. Opt. Express **15**(5), 1955–1982 (2007)
7. D. Novak, R.B. Waterhouse, A. Nirmalathas, C. Lim, P.A. Gamage, T.R. Clark, M.L. Dennis, J.A. Nanzer, Radio-over-fiber technologies for emerging wireless systems. IEEE J. Quantum Electron. **52**(1), 1–12 (2016)
8. S.R. Abdollahi, H.S. Al-Raweshidy, R. Nilavalan, A. Ahmadinia, An integrated transportation system for baseband data, digital and analogue radio signals over fibre network, in *7th International Wireless Communications and Mobile Computing Conference* (2011), pp. 2003–2008
9. L. Chen, Y. Shao, X. Lei, H. Wen, S. Wen, A novel radio-over-fiber system with wavelength reuse for upstream data connection. IEEE Photonics Technol. Lett. **19**(6), 387–389 (2007)
10. J. Ko, S. Kim, J. Lee, S. Won, Y.S. Kim, J. Jeong, Estimation of performance degradation of bidirectional WDM transmission systems due to rayleigh scattering and ASE noise using numerical and analytical models. IEEE J. Lightwave Technol. **21**(4), 938–946 (2003)
11. A. Bogoni, L. Poti, R. Proietti, G. Meloni, F. Ponzini, P. Ghelfi, Regenerative and reconfigurable all optical logic gates for ultra fast applications. Electron. Lett. **41**(7), 435–436 (2005)
12. U. Gliese, S. Norskov, T.N. Nielsen, Chromatic dispersion in fiberoptic microwave and millimeter-wave links. IEEE Trans. Microw. Theory Tech. **44**(10), 1716–1724 (1996)

Part IV
Signal and Image Processing

An Analysis of Relationship Between Image Characteristics and Compression Quality for High-Resolution Satellite Images

Gurneet Kaur, Charu Nimesh and Savita Gupta

Abstract Prediction of compression quality of an image at a pre-encoding stage plays a vital role in any compression model because not only does it greatly reduce the associated costs, but also provides a basis for further optimization of the compression algorithm employed. The present work aims to facilitate this prediction by establishing a relationship between the inherent features of an image and its quality post compression. Since the former varies significantly for different categories of images such as nature scene, medical, remote sensing, etc., results obtained for high-resolution satellite images are largely different from similar analyses typically conducted for nature scene images. We establish the effects of gradient-based Image Activity Measure, average gray level and image contrast on Gradient Magnitude Similarity Deviation, which has recently emerged as a highly efficient Full Reference Image Quality Assessment model.

Keywords DWT compression · Image activity · Gradient magnitude similarity deviation · Full reference image quality assessment

1 Introduction

The last few decades have witnessed an exponential rise in the amount of image and video data being utilized and transmitted worldwide. Consequently, the demand for efficient compression algorithms is also increasing in order to bring down storage costs, bandwidth required for transmission and transmission time at given transfer rates. A variety of transforms have been developed for image compression, such as the Discrete Cosine Transform and Discrete Wavelet Transform. Of these, DWT is a highly efficient method for sub-band decomposition of an image and has replaced DCT methods earlier used in compression payloads of remote sensing satellites [1].

G. Kaur (✉) · C. Nimesh · S. Gupta
Department of Computer Science and Engineering, UIET,
Panjab University, Chandigarh, India
e-mail: gurneetkaur.28@gmail.com

C. R. Krishna et al. (eds.), *Proceedings of 2nd International Conference on Communication, Computing and Networking*, Lecture Notes in Networks and Systems 46, https://doi.org/10.1007/978-981-13-1217-5_35

The Consultative Committee for Space Data Systems recently published an image data compression standard which describes an algorithm comprising of DWT and bit-plane encoder. The said algorithm has much lower complexity as compared to other DWT-based standards, such as JPEG2000, and is hence more practical for use onboard a spacecraft [2]. In the present work, we employ this standard to encode and decode high resolution satellite images.

The quality analysis models in lossy coding can broadly be divided into three categories—full reference (FR) ones—where the original image is available as reference, no reference ones—which work only with the degraded image and partial reference models—which use attributes of the original image as reference. Of the FR-IQA models, metrics such as RMSE and PSNR have been used extensively in the past. Although their mathematical tractability is quite appealing, these metrics suffer from a serious drawback—they do not correlate well with the Human Vision System (HVS). In an effort to establish IQAs which are closer to subjective human perception, several measures such as Structural Similarity [3] and the related MS-SSIM, Visual Information Fidelity [4], Universal Image Quality Index [5], Gradient Magnitude Similarity Deviation (GMSD) [6], etc., were developed. Of these, GMSD has emerged as a highly efficient yet effective model for post-compression quality analysis.

The Gradient Magnitude Similarity Deviation model, like many other FR-IQAs, computes image quality in two steps. In the first step, a Local Quality Map is computed by locally comparing original and degraded images. Overall quality is then determined from the LQM by using a pooling strategy. Several choices exist for this pooling strategy—the simplest and the most popular being average pooling. However, since different parts of the image contribute differently to image quality, a weighted pooling strategy yields better results. GMSD employs standard deviation for obtaining overall quality score from the similarity map and is considerably faster than the currently available state-of-art FR-IQAs.

Prediction of quality at pre-encoding stage is vital to every compression model as it cuts down costs and enables optimization of coding parameters. Gradient-based Image Activity Measure (IAM) [7] is widely used in prediction models as it captures the fine details to which HVS is sensitive. Apart from image activity, we use average luminance and contrast as predictors for GMSD for high-resolution remote sensing images compressed using the aforementioned CCSDS standard.

2 Image Characteristics and Compression Quality

Remote sensing images are typically characterized by varied textures, strong edges, and fine detail. In lossy wavelet based compression methods, most errors are observed at locally active regions. Hence, image activity can give an accurate estimate of distortion post encoding. Spatial characteristics average luminance (I_g), image contrast (I_c), and gradient based Image Activity Measure (I_a) used as quality predictors are described as follows:

$$I_{\text{g}} = \frac{1}{W * H} \sum_{j=1}^{H} \sum_{i=1}^{W} x_{ij},$$

$$I_{\text{c}} = \sqrt{\frac{1}{W * H} \sum_{j=1}^{H} \sum_{i=1}^{W} \left(x_{ij} - I_l\right)^2}$$

$$I_{\text{a}} = \frac{1}{W * (H - 1)} \sum_{j=1}^{W} \sum_{i=1}^{H-1} \left|x_{ij} - x_{i(j+1)}\right| + \frac{1}{(W - 1) * H} \sum_{i=1}^{W-1} \sum_{j=1}^{H} \left|x_{ij} - x_{(i+1)j}\right|$$

where w and h represent width and height of the image respectively, and x_{ij} is pixel intensity.

3 Results and Analysis

For the present work, satellite images were obtained from earthexplorer.gov.us, which permits access to HR orthogonal aerial imagery data taken over US for free. Figure 1 shows some images from the dataset. Metadata of the dataset is given in Table 1.

Fig. 1 High-resolution satellite images

Table 1 Metadata of training set

Number of images	70
Type	Digital grayscale
Bits per pixel	8
Resolution of image	30 cm
Format	TIFF

In order to prepare the images for compression as per BPE input specifications, basic pre-processing operations were performed:

- Conversion of RGB images to grayscale
- Conversion from TIFF to RAW image format

Images in training dataset were compressed to 0.5 bpp using 9/7 Integer DWT provided by CCSDS Image Data Compression Standard. Post decompression, GMSD was calculated using the original images as reference. Image activity, average gray level and contrast were computed for each image. Figure 2 shows parts of the original image, decompressed image and the similarity map generated prior to GMSD calculation. Visible distortions in object boundaries, blurring of edges and loss of fine detail can be observed in the decompressed images.

To determine whether spatial features of the image have any impact on post-compression quality, scatter graphs were plotted and analyzed. Figure 3 shows the relationship between GMSD and the image gray level (I_{g}), Contrast (I_{c}), gradient-based image activity (I_{a}) and the products $I_{\mathrm{a}} * I_{\mathrm{c}}$, $I_{\mathrm{a}} * I_{\mathrm{g}}$ and $I_{\mathrm{a}} * I_{\mathrm{g}} * I_{\mathrm{c}}$. While no correlation was observed between quality and average gray level or quality and image contrast when taken independently, these two quantities when combined with I_{a}, displayed good correlation with objective quality score. Further, some degree of correspondence was observed between quality values and $I_{\mathrm{a}} * I_{\mathrm{g}} * I_{\mathrm{c}}$ but a relatively higher degree of scattering was present in this plot. Image Activity Measure is proportional to the occurrences of edge transitions and amount of fine detail present in an image. Since degradations caused by DWT compression are more prominent in edges and fine detail, it is not unexpected that GMSD, which quantifies degradation present in the image, corresponds well with IAM.

4 Future Work

This paper examines the relationship between GMSD and spatial image characteristics. In order to construct a model for estimation of GMSD based on image features for HR satellite images, regression can be employed to identify a suitable subset of image characteristics. The objective of such a model would be to minimize the error between actual and predicted GMSD values for a test dataset of images. This model can be employed to determine compression quality at pre-encoding stage which will

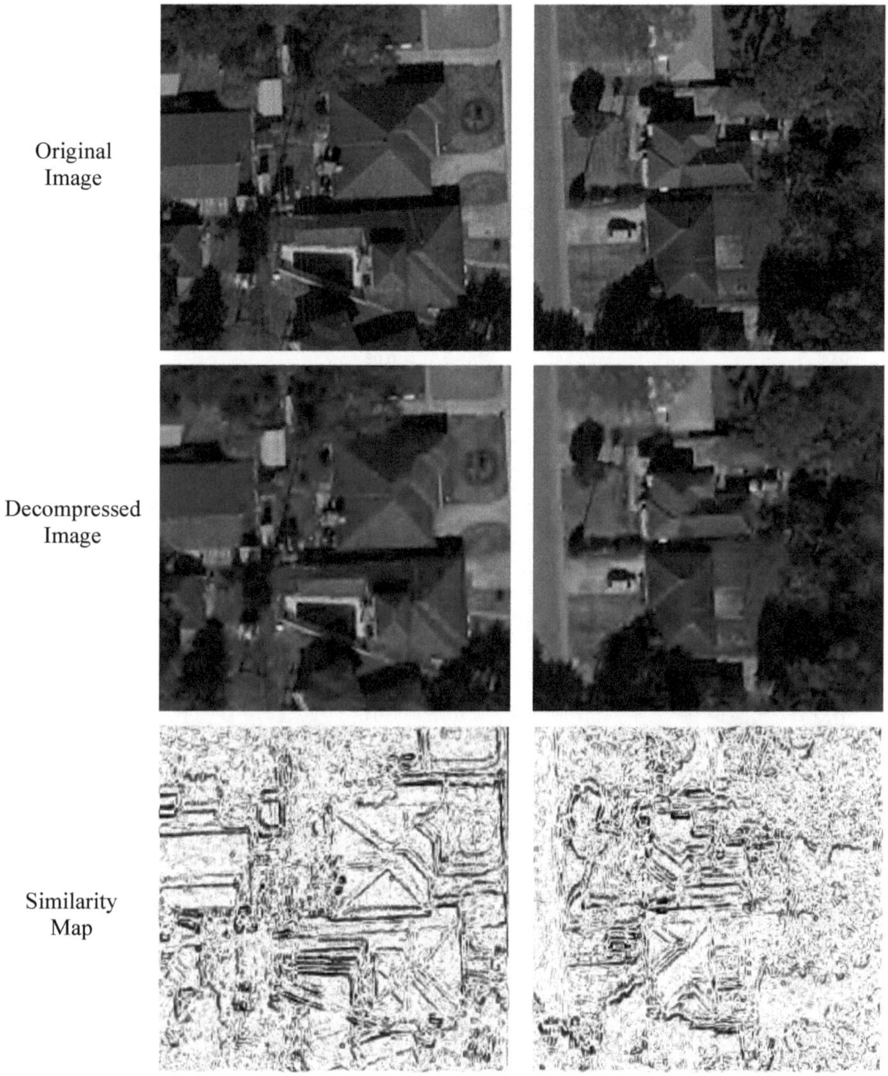

Fig. 2 A comparison between original and decompressed images

reduce costs associated with encoding and can also be used to optimize the compression algorithm. Further, the model can be refined by incorporating frequency domain image characteristics.

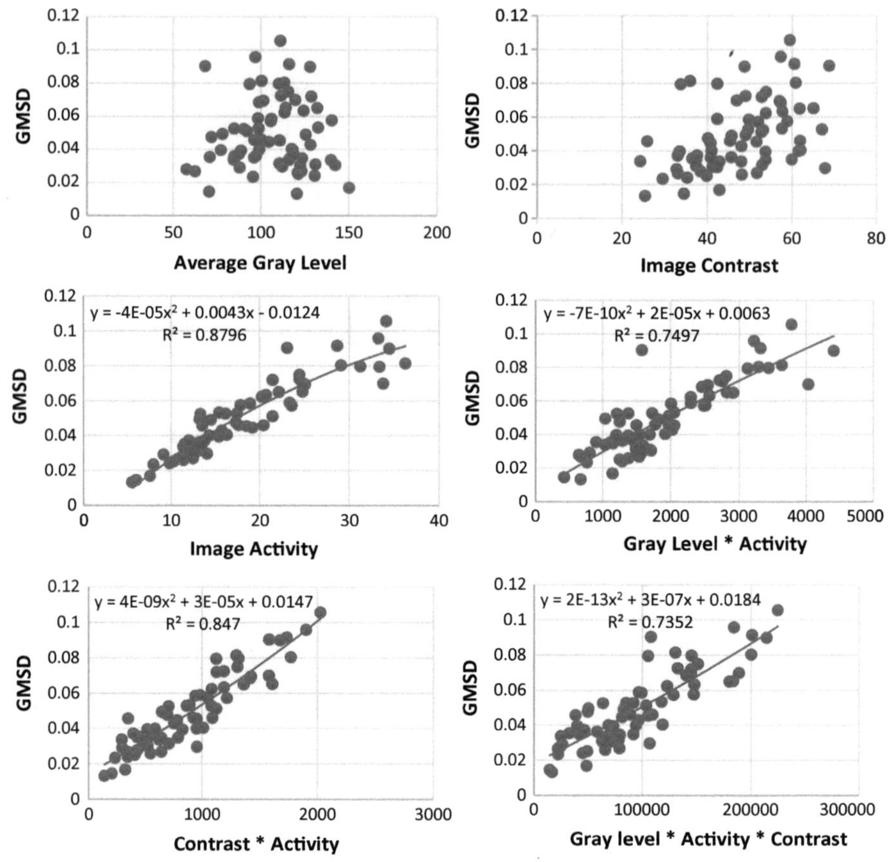

Fig. 3 Scatter plots of image characteristics against GMSD

References

1. H. Jiang, K. Yang, T. Liu, Y. Zhang, Quality prediction of DWT-based compression for remote sensing image using multiscale and multilevel differences assessment metric. Math. Probl. Eng. (2014). https://doi.org/10.1155/2014/593213
2. Consultative Committee for Space Data Systems, *Image Data Compression. Recommendation for Space Data System Standards* (2005)
3. Z. Wang, A.C. Bovik, H.R. Sheikh, E.P. Simoncelli, Image quality assessment: from error visibility to structural similarity. IEEE Trans. Image Process. **13**(4), 600–612 (2004)
4. H.R. Sheikh, A.C. Bovik, Image information and visual quality. IEEE Trans. Image Process. **15**(2), 430–444 (2006)
5. Zhou Wang, A.C. Bovik, A universal image quality index. IEEE Signal Process. Lett. **9**(3), 600–612 (2002)
6. W. Xue, L. Zhang, X. Mou, A.C. Bovik, Gradient magnitude similarity deviation: a highly efficient perceptual image quality index. IEEE Trans. Image Process. **23**(2), 684–695 (2014)
7. S. Saha, R. Vemuri, An analysis on the effect of image features on lossy coding performance. IEEE Signal Process. Lett. **7**, 104–107 (2000)

A Comparative Analysis of Various Image Segmentation Techniques

Poonam Jaglan, Rajeshwar Dass and Manoj Duhan

Abstract In this ascension era of technology, Magnetic resonance imaging (MRI) emerges as the utmost clinically acceptable imaging modality for detection and diagnosis of tumors. The Breast tumor is leading scrupulous diseases among women. In last two decades, image segmentation has got a high boost and attention from the researchers across the globe. To represent the image in such a way which is easy to analyze and more meaningful, the process of segmentation is used. It is the primal step in processing images of different types. Therefore, the image is sectioned into desirable building blocks. Basically, it provides the meaningful objects of the image. Literature provides a variety of image segmentation algorithms even though there is a requirement of an efficient segmentation technique which can work efficiently on all sorts of images. The key extract of an algorithm lies within the superiority of segmentation performed by a particular method. The availability of segmentation algorithms is quite large, so the analysis of these algorithms might be interesting to the researchers. This paper reviews segmentation techniques such as theory-based, region-based, thresholding, edge-based, Neural Network-based, Model-based, and Partial differential equation based on the basis of their functioning, utility, advantages, disadvantages, and applications.

Keywords Magnetic resonance imaging · Partial differential equation
Neural network · Markov random field · Fuzzy C-means

P. Jaglan (✉) · R. Dass · M. Duhan
Deenbandhu Chhottu Ram University of Science and Technology,
Sonipat, Murthal, India
e-mail: jaglanpoonam@gmail.com

R. Dass
e-mail: rajeshwardas10@gmail.com

M. Duhan
e-mail: duhan_manoj@rediffmail.com

C. R. Krishna et al. (eds.), *Proceedings of 2nd International Conference on Communication, Computing and Networking*, Lecture Notes in Networks and Systems 46, https://doi.org/10.1007/978-981-13-1217-5_36

1 Introduction

In the year 2016, approximately 1.5 lac new cases of breast cancer (over 10% of all cancers) have been registered in India [1]. Breast MRI becomes a clinically applicable imaging modality for breast cancer screening due to its high sensitivity and moderate specificity as compared to others [2]. MRI proves excellent at surgical implants imaging or the augmented breast [3]. Medical investigation of breast cancer is well recognized but finding out the proper technique for early detection is still the challenging one [4]. Initially, preprocessing should be considered for contrast enhancement before further processing [5]. Thus, the next step in image analysis is segmentation. Patterns color, shapes, and texture are the basic features to segment the complex image into the simple objects in the human vision; the image segmentation follow the same process for the computer vision. The process to part the image into distinct regions is so-called image segmentation. By and large, the image representation is so refined that the image outcomes are more functional for ongoing researches. This paper is further categorized as: In Sect. 1 introduction is presented, Sect. 2 gives a brief description of image segmentation, Sect. 3 explains the subsisting techniques of segmentation and Sect. 4 demonstrates the conclusion and future scope.

2 Image Segmentation

This process gets the higher attention in medical image processing. It aspires to section a digital image into a set of regions where each pixel is uniform as well as visually distinct with respect to characteristics like boundary, intensity, texture and color, etc., to identify objects/boundaries in an image. This partitioning is domain independent. It provides the significant information which can be easily analyzed.

From last many years, this area of research is speeded up at an extent. The researchers came up with enormous segmentation algorithms time to time because of its prodigious significance. Thus, an algorithm created for a particular image's type may generally not acceptable by the other class of images, as each doing the segmentation but not in the same way as another [6]. So from this point of view, there is a need to develop a specific algorithm which may fulfill every objective and must be effectively applicable to each kind of digital image. The process selection depends upon the problem formulation, type of image and output required. The field of image segmentation deserves the tremendous researches and hence a protean image segmentation technique needs to be developed on fast pace.

3 Image Segmentation Techniques

A distinct number of segmentation methods are proposed by the researches carried out in this field in the past decades. Segmentation techniques are initially classified as object, layer and block-based methods [7]. Where object based methods segment an image into regions in which the exact boundaries of the object are found. The object can be any text, photo or a graphical object, etc. Many researchers found this technique quite complex than the others. Layer Based method divides the image into layers, i.e., foreground, back ground, and mask layers. The mask layer decides the reconstruction of the final image from the other two layers. Object-based- and layer-based techniques are not much applicable to medical imaging so did not explain in detail. Whereas in the block-based approaches partition the image into rectangular blocks based on various image attributes, i.e., histogram, color, gradient, wavelet coefficients, etc. It is the most simplified segmentation technique. The hybridization of the above three categories also developed by the researchers. Block-Based Segmentation approach is further divided into two types as per the fundamental traits of pixels: Discontinuity and Similarity are as follows [8]:

a. **Boundary-based detection methods**

These methods rely on some discontinuity feature of the pixel. The images are segmented into regions by the unforeseen changes in the gray level/intensity of the image. Generally, it can identify various corners, edges, points as well as lines in the image. Pixel misclassification error is the prominent limitation. Edge detection technique is the main example of this type.

b. **Region-based detection methods**

These depend on similarity properties like lines, points and edges. The segmentation is done on the premise of similitude in intensity levels in according to a set of prerequisite criterion. Algorithms like region merging and splitting, thresholding, and region growing, etc., comes under this category.

Figure 1 shows the categorization of segmentation algorithms on the basis of various image attributes and Table 1 gives the relative comparison of different segmentation algorithms.

3.1 Region-Based

In this, the object's related pixels are grouped for segmentation. An image is segmented into the homogenous regions. In this, grouping of regions typically based on their anatomical or functional roles. The detected area for segmentation should be closed. It can also be termed as "Continuity Based". The patterns with same intensities inside the cluster formed by neighboring pixels make different sub-regions of the input image. There must be a pixel relevant to region at every step and that particular

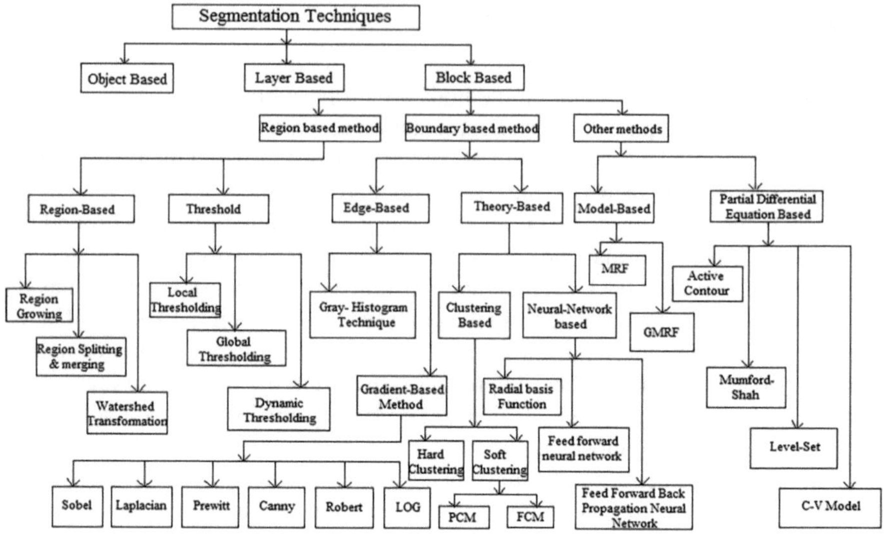

Fig. 1 Classification of image segmentation techniques

Table 1 Comparative analysis of various image segmentation techniques

S. no.	Segmentation techniques	Description	Types	Advantages	Disadvantages	Applications
1	Region-based	Sub-regions formation based on the patterns having same intensity values within a cluster of neighboring pixels	1. Region growing 2. Region splitting or merging 3. Watershed transformation	Simple and more immune to noise	It requires lots of computational time	Pixel aggregation, piecewise constant radiance, 3D reconstruction
2	Edge-based	Detection of pixels having abrupt intensity transition is done and then the extracted ones are joined altogether to form the closed object boundaries	1. Gray histogram technique 2. Gradient-based method	It can retrieve the information even through weak boundaries. Upgrade the positional accuracy	Image with too many edges is problematic. Less immune to noise than other techniques	Face identification, medical image processing, biometrics

(continued)

Table 1 (continued)

S. no.	Segmentation techniques	Description	Types	Advantages	Disadvantages	Applications
3	Thresholding based	It comes with a threshold value *T* which sectioned the entire image into only two intensities so-called binary image	1. Global thresholding 2. Local thresholding 3. Dynamic thresholding	Fewer computations. Easily understandable. Intrinsic and simple	Noise-sensitive. Not suitable for complex images. Not for multi-channel images	Locate cysts, tumors and distinct pathologies, medical image analysis
4	Theory-based	The derivatives from different regions taken into consideration in this technique	1. Clustering based 2. Neural network-based	Straightforward for classification. Implementation is quite simple	Computation time is high. Features are often image dependent. Feature selection criterion is uncertain. Spatial information is less utilized	Volume of tissues is accurately measured
5	Model-based	For object reproduction, specific learning regarding the geometrical state must be compared with the local information. Works well only when precise shape of the object is known	1. Markov random field 2. Gaussian Markov random field	Minimizes the number of misclassified pixels. Faster convergence. Fewer tendencies to be trapped in local minima	Hard to calculate. Expensive. Large storage required	Remote sensing and texture segmentation
6	PDE-based	Particularly the models are designed to solve problems where the approximate shape of the boundary is known. Active-contour model or snakes is the prominent one	1. Active contour 2. Level set 3. Mumford-shah model 4. C. V Model	Fast and efficient. Implicit, geometric characteristics of the evolving structures are straightway reckoned. Intrinsic	Highly immune to noise. Computational complexity is high	Computer vision and medical image analysis, stereo reconstruction, object extraction and tracking

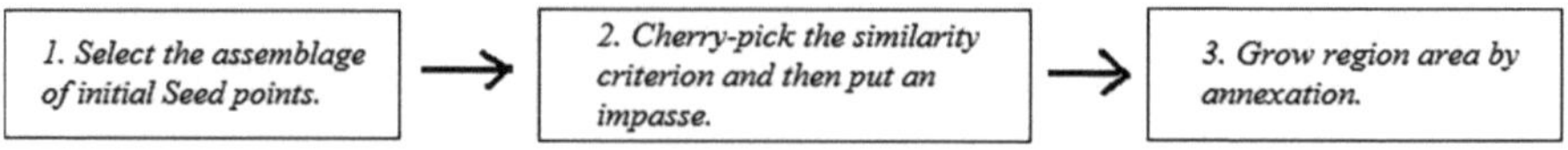

Fig. 2 Process of region growing

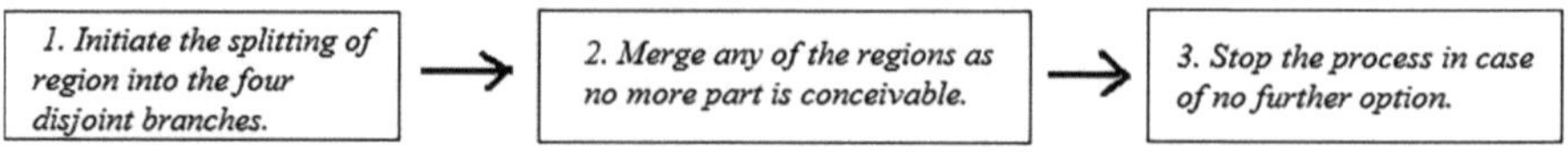

Fig. 3 Three steps of region splitting and merging

pixel is considered. These techniques are regularly utilized for intuitive methods for segmentation because of their high computational efficiency [9] but at the cost of high computational time [10]. Further classification is as below.

3.1.1 Region Growing

The accumulation process of pixels with comparable properties is done for the formation of a region. It bunches the pixels of the whole image into subparts or huge areas based on some standard criterion for growth. Region growing can be processed as shown in Fig. 2 [11].

3.1.2 Region Splitting and Merging

First the image is partitioned into a set of random detached regions and then merging and/or splitting can be done to satisfy some specific conditions. In fact, splitting has more impact on the overall process. The specific one can be accessible in the form of quad-trees as the name suggests that each node has exactly four branches. This technique is complex and time-consuming. For this the steps are as shown in Fig. 3 [12].

3.1.3 Watershed Transformation

It basically done the intensity-based assemblage of pixels as each pixel contains a different intensity value. It is a mathematical morphological operation reckoned imperative to solve formidable problems of image segmentation specifically breast images [13, 14]. Its textured patterns sometimes cause over-segmentation so researchers still finding the remedy. This is a puissant segmentation technique due to its processing speed, simplicity and entire image partition [13].

3.2 *Edge-Based*

Techniques based on finding discontinuities in the gray scale values are simply known as edge or boundary-based methods. Edges can be defined as the discontinuity of gray level intensity value at the boundaries [15]. The basic types of edges are step, point and line edges out of which step and line edges are infrequent for real-time images as sharp changes rarely subsist in the real images [16]. All kinds of edges can be detected mainly by spatial masks. It detects first and then extracting the objects that have a quick transition in the gray value. And finally these objects are joined to form the closed object boundaries. However, it may be classified as.

3.2.1 Gray Histogram Technique

This method is quite efficient comparatively. First, a histogram is formed on the basis of intensity level of each pixel so that edges and valleys are easily located in the image. If significant edges and valleys were identified, it is difficult to use this method. The specific value of threshold T gives the proper result, and to find out the limits of gray level intensity is quite difficult because the impact of noise on gray histogram is uneven.

3.2.2 Gradient-Based Method

Gradient can be considered as the first derivative of an image. This method responds well at points having rapid transition among two regions. These points also called edge pixels have high value of gradient and must be joined to build closed boundaries. Typically, gradient operators convolve with the image.

All the edge detection operators are listed below:

Sobel Operator

This is the most widely used technique introduced by Sobel in 1970 [6]. Edges can be finding out by the approximate sobel value to the derivative. The point with highest gradient value termed as edges. It is very much similar to the Roberts operator.

Prewitt Operator

It is introduced by Prewitt in 1970 [6] and most suitable to estimate the orientation as well as magnitude of an edge. It is very simple in operation as well as the edge detection and their orientation is quite easy. This operator is less precise and also sensitive to noise.

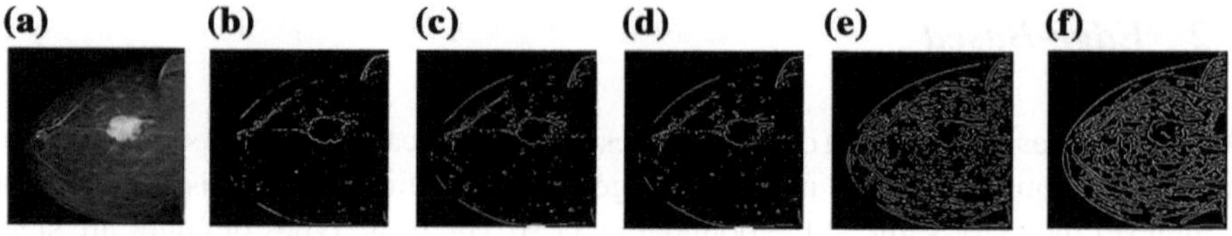

Fig. 4 Qualitative results of different edge detection operators. **a** Breast MR image [18]. **b** Robert. **c** Sobel. **d** Prewitt. **e** LOG. **f** Canny

LOG Operator

This can be termed as Marr-Hildreth edge detector. The pixels having gradient value at its maximum are considered as edges. At the zero-crossing, edge direction can prevail.

Canny Operator

It is one of the best edge detectors which possesses minimum error while detecting true edge points and also prevails good noise immunity. Also known as an optimal edge detector due to finding the edges optimally without emotive the features.

Laplacian Operator

In this, the second-order derivative expression is computed to detect the edges of the image as this particular expression has zero crossings at the prime edges within the image [17].

Robert Operator

Lawrence Roberts introduced this technique in 1965. It is quite simple and computes fast. The magnitude of the spatial gradient of the breast MR image at a particular point can be represented by pixel values at that point in the output. It accentuates the high spatial frequency regions considered as edges.

Figure 4a depicts the breast MR image [18] Fig. 4b–f shows the qualitative results of different edge detection operators. The Canny is most promising one as per the qualitative analysis shown in Fig. 4, yet it is comparatively time-consuming. Practically, it is necessary to maintain a balance in between accurate edge detection and noise reduction [6]. Some additional or fake edges will be detected in the case of high noise density. Hence, these techniques are effective for the noise-free images.

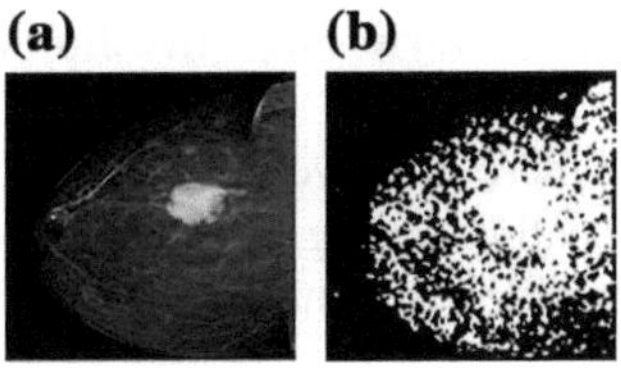

Fig. 5 **a** Breast MR image [18]. **b** Image after thresholding

3.3 *Threshold*

This is the easiest as well as efficient methodology for image segmentation based on the qualities of the picture. The values used as threshold must be acquired by the histogram of the input image's edges. So, the threshold gives an accurate value only if the edge detections are accurate. More logically, it converts a multi-level image into two-level image, i.e., only a specific threshold value T is selected first, any pixel having value equal or greater than T is considered as foreground and else pixel having value less than T is taken background [19]. Figure 5a shows the original breast MR image [18] and Fig. 5b shows the thresholded image. MRI images generally suffer from noise which corrupt the histogram of the image so that thresholding process also disrupts [20]. It does not perform well for complex images. The three thresholding strategies are.

3.3.1 Global Thresholding

If extremely large intensity distinction exists among the object's background and foreground, a sole value of threshold can essentially be utilized to differentiate both objects and this process is so-called Global thresholding. Hence, in this kind of thresholding, the threshold value estimation must depend solely on the property of picture intensity [21]. Otsu, entropy-based, object attribute-based thresholding are some examples of thresholding strategies [22].

3.3.2 Local Thresholding

This technique firstly segments the whole image into sub-regions and then assigns distinct threshold values to each one. Therefore, this type of thresholding process is called local thresholding. The filtering process should take place to remove spasmodic gray levels after thresholding. It takes more time for image segmentation in comparison to global thresholding. It is more applicable to images with varying backgrounds [21].

3.3.3 Dynamic Thresholding

If there are several objects in the image having distinct intensity regions then some locally varying threshold values (T_1, T_2, ... T_n) for each pixel should be used to partition the image which depends upon point values of gray level image.

3.4 Theory-Based

In this kind of segmentation technique, a number of algorithms were outgrowth by distinct works, which plays a vital role for segmentation approach. In general, they include algorithms based on wavelet, clustering, genetic and neural network.

3.4.1 Clustering Techniques

In this, a finite set of categories are identified and then clusters are formed by grouping them together. It is an unsupervised learning task. It relies on the basic principle to maximize the intra-class similarity as well as the inter-class similarity is minimized. The imaging traits, i.e., color, size, intensity, texture, etc., impinge the clustering algorithms. The similarity measures affect the outcomes. It is generally classified as:

Hard Clustering

Each pixel is related to only one cluster and each cluster differentiated by sharp boundary lines. *k*-means algorithm [23] is the prominent one. The number of pixels *p* is assigned to *k* different clusters as per the closeness measured by the Euclidian distance. Initially, the centroids must be chosen arbitrary. The mean must be recalculated of every cluster when each pixel is clustered and this is a repetitive process to get some significant results. The main drawback is that the different results occurred at every execution, the number of clusters must be determined [6]. The essential steps are [24]:

(i) The numbers of cluster are randomly selected.
(ii) Compute the histogram of pixel intensities and certain random pixels are chosen as centroids amongst the *k* pixels.
(iii) The mean of a certain region is considered as the centroid and there should be a huge gap between each centroid.
(iv) Now do the comparison of every pixel to each centroid and assigned it to the specified one.
(v) Then reconsider the mean of each cluster and also the location of every centroid is recalculated.

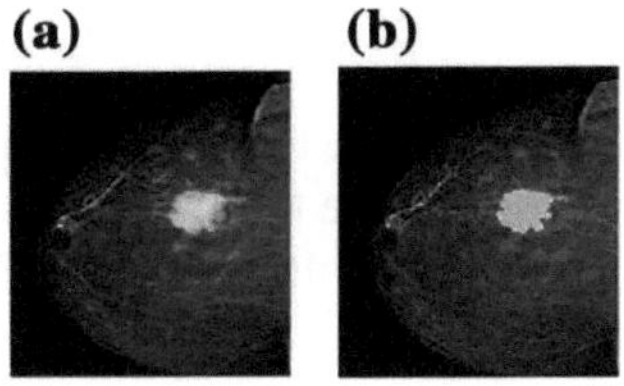

Fig. 6 **a** Breast MR image [18]. **b** K-means segmented image

(vi) Repeat step 4 and 5, until centroids stop moving. Therefore, Image separated into clusters.

Figure 6a depicts the breast MR image [18] and Fig. 6b shows the K-means segmented image.

Soft Clustering

It is the efficient algorithm because of its feature of maximum information preservation [25]. Recently, fuzzy clustering algorithms are integrated with object shape which can be considered as a unique feature [26]. Therefore, it is suggested to extract constant geometric contours to avoid the segmentation of random objects. It further categorized as Possibilistic c-means (PCM) and Fuzzy c-means (FCM), etc.

FCM: According to this, every pixel of the image assigned to more than one cluster as per some membership function. It may reproduce the brittle representation of the system's behavior. The outcome is restricted by spherical clusters even then it is the promising one [27]. Generally, it provides more flexibility than the corresponding hard clustering algorithm.

3.4.2 Neural Network-Based Segmentation

In this, first neural network mapping is done in which each neuron is identified as a pixel [28]. Then, by the use of dynamic equations, the image edges are extracted to command that every neuron must be stated towards minimum energy defined by neural network [29]. Generalization, learning ability and reminiscence are the three basic properties of neural network [30]. Generally, neural networks are processed in layers. A large number of interconnected "nodes" containing an "activation function" forms the layers. The "input layer" transfers the particular patterns to the "hidden layer" where the real processing is done by the weighted connections network. [30]. Then, a connection must be formed in between "hidden layer" and "output layer". It is basically classified as:

Radial Basic Function

It is one of the techniques of the neural network. The preprocessed image is required to upgrade the training process. This method gives highly accurate results as well as the time complexity is less [31].

Feed Forward Neural Network

The information flow is unidirectional starts from the input layer and reaching the output layer via hidden layer without following any cycles [32]. The main examples are Hebb, Perceptron, Ada-line and Madaline networks.

Feed Forward Back Propagation Neural Network

In this, a specific set of training data fed in the forward direction via the network for each training iteration. Then the error calculation is done at the output layer on the basis of some target information consequently the required changes of weights has been taken place which is then implemented throughout the network to begin the next iteration. Furthermore, repeat the entire procedure by the use of the next pattern of training [33].

The major issue in the case of breast MR image is to train the neural network model and only then one can identify the tumor as benign or malignant. All the three neural networks stated above gives the different accuracy even if trained on the same set of inputs. The back propagation neural network proves better in terms of accuracy as compared to the others [31].

3.5 *Model-Based*

Generally, local information is the essential one in each algorithm [34] while in this; the specific learning of the geometrical state of the object must have a contrast with the local information to make a representation of the objects. The exact shape of the object must be known to make this system pertinent.

3.5.1 Markov Random Field

MRF-based segmentation comes under this. To identify the edges accurately, MRF must be combined with edge detection [10]. The process performs in an iterative fashion as coarse resolution is segmented initially and finer resolution gets the secondary attention.

3.5.2 Gaussian Markov Random Field

Another method includes Gaussian Markov Random Field (GMRF) in which consideration of spatial dependencies between pixels takes place. In region growing, Gaussian Markov Model (GMM) based segmentation is taken into consideration. The GMM extended version also identifies the region, edge cues as well as feature space [35].

3.6 Partial Differential Equation Based

It is the finding of Kass et al. in 1987 [36]. Kass invented this for finding regions those are familiar even in the presence of noise and other uncertainty. PDE-based methods are mainly conveyed by snakes also called active-contour model. This model describes the approximate shape of the boundary. Other techniques are described below.

3.6.1 Active Contours

The basic idea behind this is to elaborate a curve, subject to refrainment from a given image for detecting objects. These are computer generated curves to find object boundaries under the influence of internal and external forces while moving within the image [37]. It has multiple advantages as compared to classical feature attraction techniques. They usually track dynamic objects while autonomously searching for the minimum state. Main drawbacks are noise sensitivity and high computational complexity [38]. Various improvements have been made by the researchers to the basic model, but still not overcome fundamentally.

3.6.2 Level Set Model

The method developed by Osher and Sethian was much influential [39]. This method easily follows shapes having topological changes. Therefore, it becomes a great tool for modeling time-varying objects. The problems of corner point producing, curve joining and breaking due to its topological irrelevancy and stability may be solved. The disadvantage lies with the fact that the image gradient is the fundamental rule for edge-stopping function [40] and the curve may pass the object boundaries as it never level at zero at the edges.

3.6.3 Mumford-Shah Model

The stopping criterion taken into consideration is global information of the entire image for segmentation purpose. This feature of globalization results in the best image segmentation [41].

3.6.4 C–V Model

In this, image must be partitioned into two regions out of which one can represents the objects needs to be detected and the other one shows the background. C–V model is not based on edge function, to stop the evolving curve on desired boundary [42]. It can detect objects even with smooth boundaries.

4 Conclusion

All the applicative techniques of image segmentation are briefed in this paper on the basis of their functioning utility, advantages, disadvantages, and applications. The description of every method shown in Table 1 helps in the selection of appropriate technique as per the various image attributes. Since a number of parameters like color, intensity, noise, etc., affect these algorithms, this emphasizes the necessity of appropriate image segmentation technique. Thresholding and region-based are less immune to noise; so least applicable to MR images. Theory-based techniques are time-consuming as it is required to train the data first. But FCM behaves well for medical imaging. Active-contour and level set methods finds their way in the field of MR images as it is fast enough as well as efficient. Still, application-based image segmentation algorithms developed on daily basis by the researchers but needs to come up with the protean one that is independent of the input image and accurate enough. Therefore, the researchers can come up with the hybrid approach of two or more segmentation techniques (as per the area of application given in Table 1) to give the better segmentation results. Also the researches may carried out by modify some of the existing techniques. Any optimization technique can be combined with the ongoing techniques of segmentation to present the good outcome [43]. Researchers still need to design a universal as well as effective algorithm for image segmentation.

References

1. S. Dey, Cancer cases in India likely to soar 25% by 2020, in *Indian Council of Medical Research* (May 2016) (http://timesofindia.indiatimes.com/india/Cancer-cases-in-India-likely-to-soar-25-by-2020-CMR/articleshow/52334632.cm8s)
2. A. Qusay, Al-Faris et al., Computer-aided segmentation system for breast mri tumour using modified automatic seeded region growing (BMRI-MASRG). J. Digit. Imaging **27**, 133–144 (2014)
3. A.E. Hassanien, T. Kim, Breast cancer MRI diagnosis approach using support vector machine and pulse coupled neural networks. J. Appl. Logic **10**(4), 277–284 (2012)
4. B.K. Gayathri, P. Raajan, in A survey of breast cancer detection based on image segmentation techniques, in *International Conference on Computing Technologies and Intelligent Data Engineering (ICCTIDE)* (Jan 2016), pp. 1–5
5. A. Ella, Hassanien et al., MRI breast cancer diagnosis hybrid approach using adaptive ant-based segmentation and multilayer perceptron neural networks classifier. Appl. Soft. Comput. **14**, 62–71 (2014)
6. R.C. Gonzalez, R.E. Woods, S.L. Eddins, *Digital Image Processing Using MATLAB*, 2nd edn. (Tata McGraw Hill, Education Private Limited, 2010)
7. D. Banupriya, M. Sundaresan, in Enhanced Hybrid algorithms for compound image segmentation, in *2nd International Conference on Computing for Sustainable Global Development (INDIACom)* (Mar 2015), pp. 672–676
8. R. Dass, Priyanka, S. Devi, in Image segmentation techniques. IJECT **3**(1) (Jan–Mar 2012)
9. M.A. Aswathy, M. Jagannath, Detection of breast cancer on digital histopathology images: present status and future possibilities. Inf. Med. Unlocked **8**, 74–79 (2017)
10. J. Acharya, S. Gadhiya, K. Raviya, in Segmentation techniques for image analysis: a review. IJCSMR **2**(1) (Jan 2013)
11. A. Javadpour, A. Mohammadi, Improving brain magnetic resonance image (MRI) segmentation via a novel algorithm based on genetic and regional growth. J. Biomed. Phys. Eng. **6**(2) (2016)
12. M. Ahlem et. al., Comparison of automatic seed generation methods for breast tumor detection using region growing technique, in *IFIP International Conference on Computer Science and its Applications, Springer International Publishing* (2015), pp. 119–128
13. S.R. Shareef, Breast cancer detection based on watershed transformation. Int. J. Comput. Sci. **11**(1) (Jan 2014)
14. A. Mustaqeem, A. Javed, T. Fatima, An efficient brain tumor detection algorithm using watershed and thresholding based segmentation. Image, Graph. Signal Process. **10**, 34–39 (2012)
15. N.M. Zaitoun, M.J. Aqel, Survey on image segmentation techniques, in *International Conference on Communication, Management and Information Technology*, (2015), pp. 797–806
16. N. Senthilkumaran, R. Rajesh, Edge detection techniques for image segmentation—a survey of soft computing approaches. Int. J. Recent Trends Eng. Inf. Pap. **1**(2) (May 2009)
17. R. Muthukrishnan, M. Radha, Edge detection techniques for image segmentation. Int. J. Comput. Sci. Inf. Technol. (IJCSIT) **3**(6) (Dec 2011)
18. http://www.google.co.in/breastmri
19. S. Saini, K. Arora, A study analysis on the different image segmentation techniques. IJICT **4**(14), 1445–1452 (2014)
20. N. Tirpude et. al., A study of brain magnetic resonance image segmentation techniques. Int. J. Adv. Res. Comput. Commun. Eng. **2**(1) (Jan 2013)
21. A. Norouzi et al., Medical image segmentation methods, algorithms, and applications. IETE Tech. Rev. **31**(3), 199–213 (2014)
22. N. Senthilkumaran et al., Efficient implementation of niblack thresholding for MRI brain image segmentation. Int. J. Comput. Sci. Inf. Technol. **5**(2), 2173–2176 (2014)
23. R. Yogamangalam, B. Karthikeyan, Segmentation techniques comparison in image processing. IJET **5**(1) (Mar 2013)
24. H.M. Moftah et al., Adaptive k-means clustering algorithm for MR breast image segmentation. Neural Comput. Appl. **24**, 1917–1928 (2013)

25. S. Chebbout, H.F. Merouani, Comparative study of clustering based colour image segmentation techniques, in *Eighth International Conference on Signal Image Technology and Internet Based Systems* (2012), pp. 839–844
26. P. Singh, R.S. Chadha, A novel approach to image segmentation. IJARCSSE **3**(4) (Apr 2013)
27. N. Kaur, J. Singh, V. Sharma, Analysis and comprehensive study: image segmentation techniques. IJRASET **3**(I) (Jan 2015)
28. A.S. Manraj, Current image segmentation techniques-a review. IJCSIT **6**(2), 1940–1942 (2015)
29. A.E. Hassanien et. al., A Novel hybrid perceptron neural network algorithm for classifying breast MRI tumors, in *AMLTA 2014, CCIS 488, Springer International Publishing Switzerland* (2014), pp. 357–366
30. A. Raad et.al., Breast cancer classification using neural network approach: MLP and RBF, in *The 13th international Arab conference on Information Technology* (Dec 2012), pp. 15–19
31. S. Hallad, H. Roopa, Comparing three neural network techniques in the classification of breast cancer. Int. J. Adv. Res., Ideas Innov. Technol. **3**(4), 198–203 (2017)
32. S. Matta, Review: various image segmentation techniques. IJCSIT **5**(6), 7536–7539 (2014)
33. P. Sudharsan, C.L.Y. sivakumari, A comparitive analysis of segmentation techniques in mammogram images. IJERT **2**(12) (Dec 2013)
34. N.R. Pal, S.K. Pal, A review on image segmentation techniques. Pattern Recogn. Elsevier **26**(9), 1277–1294 (1993)
35. R. Azmi, N. Norozi, A new markov random field segmentation method for breast lesion segmentation in MR images. Med. Signals Sens. **1**(3), 156–164 (2011)
36. X. Jiang, R. Zhang, S. Nie, Image segmentation based on PDEs Model: a Survey, in *3rd International Conference on Bioinformatics and Biomedical Engineering* (2009), pp. 1–4
37. P. Karch, I. Zolotova, An experimental comparisons of modern methods of segmentation, in *IEEE 8th International Symposium on SAMI* (2010), pp. 247–252
38. D. Baswaraj et.al., Active contours and image segmentation: the current state of the art. Glob. J. Comput. Sci. Technol. Graph. Vis. **12**(11) (2012)
39. R. Dass, Vikash, Comparative analysis of threshold based, K-means and level set segmentation algorithms. IJCST **4**(1) (Mar 2013)
40. J. Patra et al., Segmentation techniques used image recognization and SAR image processing. IOSR J. Comput. Eng. (IOSR-JCE) **16**(2), 37–43 (2014)
41. A. Taneja et. al., A performance study of image segmentation techniques, in *4th International Conference on Reliability, Infocom Technologies and Optimization (ICRITO)* (Trends and Future Directions) (2015), pp. 1–6
42. P. Getreuer, Chan vese segmentation. Image Process. On Line (IPOL) **2**, 214–224 (2012)
43. R. Dass, S. Devi, Priyanka, Effect of wiener-helstrom filtering cascaded with bacterial foraging optimization to despeckle the ultrasound images. Int. J. Comput. Sci. Issues **9**(4), 372–380 (2012)

Deep Learning Based Shadow Detection in Images

Naman Bansal, Akashdeep and Naveen Aggarwal

Abstract Various computer vision applications required Shadow detection and removal for example object tracking and recognition, scene interpretation, and video supervision. Since shadows have similar characteristics as that of the objects shadow pixels can be classified as part of object. This may cause problems such as merging or loss of object, alternation, and misinterpretation of object shape. To deal with this problem, we represent a deep learning based framework to automatically detect shadows in images. Our method learns many significant features automatically using supervised approach in Convolutional Neural Networks (ConvNets). The approach also makes use of drop out to improve results. The proposed methodology is tested on SBU dataset and results are promising.

Keywords Convolutional neural network · Shadow detection · Feature extraction

1 Introduction

In general, a shadow is formed as soon as a light source is blocked by an opaque item. In other words, when an opaque object is positioned between background surface and a light source, it results in variation of illumination in that area, as it does not allow the light to access the adjoining regions of foreground object. Change of illumination is less significant at the outer boundaries as compared to the center region of shadows. Hence, shadow can be classified as: penumbra and umbra region. Umbra can be defined as darker regions in the shadow in which direct light is completely obstructed whereas penumbra are the lighter regions of shadow.

N. Bansal (✉) · Akashdeep · N. Aggarwal
UIET, Panjab University, Chandigarh, India
e-mail: namanbansal07@gmail.com

Akashdeep
e-mail: akashdeep@pu.ac.in

N. Aggarwal
e-mail: navagg@gmail.com

C. R. Krishna et al. (eds.), *Proceedings of 2nd International Conference on Communication, Computing and Networking*, Lecture Notes in Networks and Systems 46, https://doi.org/10.1007/978-981-13-1217-5_37

In various computer vision applications, object detection is an essential phase. A precise evaluation of object shape and size is required so that object tracking or recognition can be performed. The problem gets more complex when shadows are also part of video or image. Because shadow shows similar characteristics as that of the object therefore the shadow region is classified as part of object. This leads to reduction in performance of various computer vision applications for example segmentation, image recognition, and object tracking as shadows cause object merging or occlusion.

Although shadows have many disadvantages, they can be helpful to find the information such as object shape, size and orientation as well as direction and intensity of light source. Shadow detection is required to overcome misclassification problem and to extract useful features.

This paper presents a deep CNN-based approach for detecting shadows in images. CNN automatically extracts features from the input image and uses them to detect shadows. The proposed CNN architecture consists of six layers organized as two pair of convolutional and ReLU layers, followed by a dropout layer and at end a convolutional layer [Conv-ReLU-Conv-ReLU-Dropout-Conv]. It is trained over 2700 images of size 256 * 256.

2 Related Work

Several shadow detection techniques have been presented in the literature. Prati et al. [1] conducted a study on moving shadow detection techniques in which classification was done on the basis of algorithm-based taxonomy. Sanin et al. [2] observed that choosing features has more impact in shadow detection than the selection of any algorithm.

On the basis of [1, 2] shadow detection techniques are divided into major categories of: Spectrum-based and Geometry-based models (Fig. 1).

Geometry-based methods detect shadows by taking advantage of geometric information such as camera location, background scene, and location of light source.

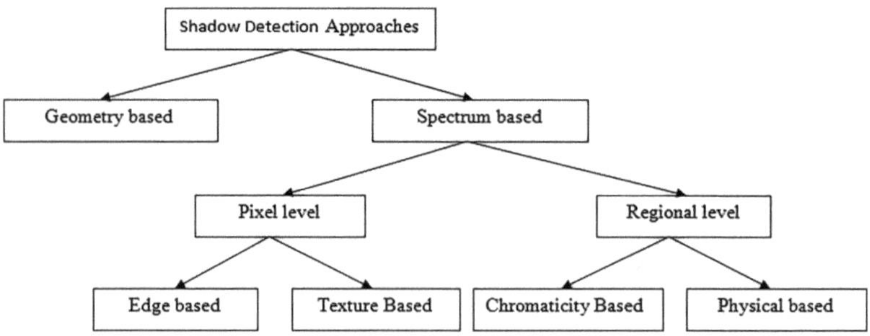

Fig. 1 Classification of shadow detection methods

A three-stage algorithm presented by Chen et al. [3] is used in shadow detection of pedestrians posing vertically. First, to detect shadow features Support Vector Machine was trained and applied on foreground camouflage. Then, the foreground camouflage is partitioned into shadow sub-regions and humans by a linear classifier. A 2D Gaussian filter was used in the final stage to reconstruct the shadow area with the help of background region. A coarse to fine method for shadow removal is presented by Hsieh et al. [4]. A moment-based technique was used at the coarse stage, for estimating rough boundaries among moving object and the shadows. By Gaussian shadow modeling approximation of shadow region is further computed at the fine stage. Algorithms using **Chromaticity based Methods** attempt to use appropriate color spaces to depict the difference in pixel value and appearance when shadow appears. Cucchiara et al. [5] used Hue-Saturation-Value (HSV) color space in shadow recognition. Shadows were detected by calculating rate of change in HSV component of the frame and background referred. Chen et al. [6] proposed YUV color space method to obtain luminance information and to keep chrominance component intact for shadow detection. These techniques are easy to execute but are prone to noise and do not work with strong shadows. **Physical-based Methods** were proposed to model the particular appearance of shadow points according to reflections and illumination. Physical-based color features were used to detect shadows by Huang et al. [7] in outdoor and indoor environments. When object have chromaticity similar to background, physical methods does not give good results. **Edge-based Methods** are very useful in detecting shadows as edges do not vary under changing illumination. Panicker et al. [8] extracted foreground and background mask using Sobel operator. Then these two edge maps were correlated to preserve internal edges of the object. Finally, vertical and horizontal operations were applied to reconstruct object shape. For detecting shadows in indoor sequences, Xu et al. [9] proposed a static edge correlation method. In this method Change Detection Mask (CDM) was generated and canny edge detection was applied to detect shadow regions. **Texture based Methods** usually apply two steps: (i) selection of shadow points with the help of weak shadow detector and then (ii) categorizing these candidates as object or shadow on the basis of texture correlation. The technique presented by Sanin et al. [10] makes use of color features to recognize shadow regions and then applied gradient texture correlation to discriminate them from object. Zhang et al. [11] presented a different algorithm for detection of shadow established on ratio edge, providing that ratio edge is illumination unvaried. As texture based methods have to compute different neighborhood evaluations for every pixel, they are usually slow.

In literature along with the handcrafted features which are used to extract shadows, many approaches based on automatic feature extraction have significant weightage. Shen et al. [12] presented an effective learning based algorithm for detecting shadows in particular image. Local structure of shadow edges was extracted using Convolutional Neural Network (CNN), to increase local consistency in pixel labels. Network architecture consists of seven layers with multiple filters of size 5 * 5. The shadow and bright measures were calculated from the detected shadow edges. Khan et al. [13] proposed a framework which learns the features automatically with multiple convolutional deep neural networks. The architecture consists of a seven-layer network which has alternate convolutional and sub-sampling layers. The given method extracts elements alongside of the object boundaries and at the super-pixel level. Features were learned using a window which is focused at interest points in both the cases. The calculated posteriors were then given to a Conditional Random Field (CRF) model for generating efficient shadow outlines for shadow detection. Author attained highest accuracy of 93.16% on UIUC dataset. Khan et al. [14] presented an algorithm for automatically detecting and removing shadows. The algorithm automatically extracted relevant features using convolutional neural networks in a supervised manner. The extracted features were given to a CRF model for generating efficient mask of shadow. A Bayesian formulation was proposed by using these shadow masks, for correctly extracting shadow matte and removing the shadows. Proposed framework gives good result for both umbra and penumbra regions.

The above approaches require user assistance, or assumptions regarding scene properties or object surface. To overcome these problems, we present a deep learning architecture for shadow detection independent of user assistance or any similar assumptions.

3 Proposed Scheme

For detection of shadows in images we propose a CNN-based framework for automatic detection of required features. The CNN architecture used consists of alternating Convolutional and ReLU layers (Fig. 2) followed by a dropout layer. The CNN layer will help to extract detailed features from raw images which are then fed to Relu layer for activation map. Different filters are used in convolutional layer, which are then convolved with input feature map. The convo layer uses filters of size 5 * 5 for extracting features. Number and dimensions of filters have been set after experimentation. The sequence of convo-relu is repeated twice and its output is given to drop out layer. Dropout layer has been added to improve accuracy within obtained results.

The proposed technique uses 32 filters of size 5 * 5 at each conv layer. ReLU layer provides an activation function, proposed framework uses $f(x) = \mathrm{Max}(0, x)$ activation. The reason for using Max Activation function is to suppress impartial results. Dropout layer randomly ignores nodes to prevent interdependencies between different nodes. This layered architecture allows CNN to learn features at multilevel hierarchy. A dropout of 0.2 has been used in the proposed framework. The final layer of network is convolutional layer in which 1 filter of dimensions 5 * 5 * 32 is used to output the required shadow in image. During training process, framework uses stochastic gradient descent to automatically learn features in supervised manner and uses back-propagation for gradient computation. Once the network is trained, it takes 256 * 256 grayscale image as an input and processes it to provide shadow in that image.

Input: Image of size [$256 \times 256 \times 1$] is provided to proposed architecture

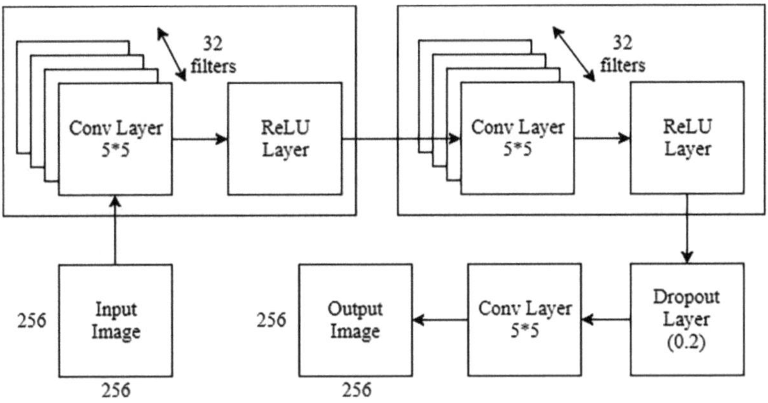

Fig. 2 Proposed CNN architecture for shadow detection

Table 1 Details of each layer in proposed CNN architecture

Layer type	Input size	Filter size	No. of filters	Stride	Padding value
conv	256 * 256 * 1	5 * 5 * 1	32	1	2
Relu	256 * 256 * 32	–	–	–	–
conv	256 * 256 * 32	5 * 5 * 32	32	1	2
Relu	256 * 256 * 32	–	–	–	–
dropout	256 * 256 * 32	–	–	–	–
conv	256 * 256 * 32	5 * 5 * 32	1	1	2

Conv layer: It will calculate a dot product between filter weights and region of input image. Output of this layer is [256 * 256 * 32] feature map.

ReLU layer: It will apply an activation function, max (0, x) thresholding at zero. This will not change the size of feature map.

Dropout layer: It will randomly select the nodes to be dropout with a probability of 0.2.

In this architecture, CNN is trained over 2700 images for batch size 1 with 50 epochs and learning rate of 0.01 and. Table 1 provides details of each layer used in the architecture.

4 Results and Discussion

4.1 Dataset

SBU Shadow Dataset: The dataset consists of 4089 images along with their ground truth. Only 3600/4089 images are used. For training purpose 2700 images are used and 900 images are used for the purpose of testing. The dataset contains variable size RGB images which are resized to 256 * 256 grayscale images.

4.2 Results

To evaluate the superiority and effectiveness of the proposed method, Qualitative results obtained by proposed framework are shown on three different types of images in Figs. 3, 4 and 5. First column of Figs. 3, 4 and 5 consists of input image having shadows, second column consists of the expected shadow results, and column third, shows the obtained results after the implementation of proposed CNN framework. The results show that our proposed framework detects shadows successfully. Results are equally good on satellite images (Fig. 3), outdoor scenes (Fig. 4) and images having shadows of humans (Fig. 5). As can be seen in Fig. 3c the obtained results are very close to the expected results showing high accuracy in shadow detection by proposed framework. The achieved results as shown in Fig. 4, depicts the presence of some white spots along with the shadow. This is due to the similar characteristics of the region and shadow. Some shadow information is lost in Fig. 5, but result are parallel to the expected results of the image consisting shadow.

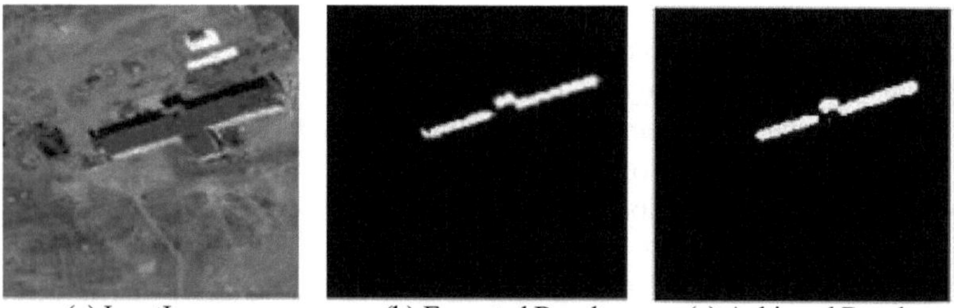

(a) InputImage (b) Expected Result (c) Achieved Results

Fig. 3 Results of proposed approach on image#10 of SBU dataset

(a) Input Image (b) Expected Result (c) Achieved Results

Fig. 4 Results of proposed approach on image#17 of SBU dataset

(a) Input Image (b) Expected Result (c) Achieved Results

Fig. 5 Results of proposed approach on image#35 of SBU dataset

5 Conclusion

This research paper has proposed a deep learning based architecture for solving the problem of shadow detection in images. An automatic feature learning CNN approach is presented to extract the most significant features from a single image. CNN architecture consists of multiple hidden layers along with input layer and an output layer. Hidden layers are either convolutional, ReLU or dropout layer. The approach uses a 6-layer CNN framework [Conv-ReLU-Conv-ReLU-Dropout-Conv] on single images for detection of shadows. 32 filters of size 5 * 5 are used in each convolutional layer and a dropout of 0.2 is applied in the framework. Qualitative evaluation of results obtained through proposed method shows that it performed well in various types of single images such as in outdoor and aerial images. The approach is required to be tested quantitatively for further approval. The proposed approach can be further improved by developing more deep CNN architectures for precise results. The approach can be further tested on videos and other benchmark datasets for further testing.

References

1. A. Prati, I. Mikic, M.M. Trivedi, R. Cucchiara, Detecting moving shadows: algorithms and evaluation. IEEE Trans. Pattern Anal. Mach. Intell. **25**(7), 918–923 (2003)
2. A. Sanin, C. Sanderson, B.C. Lovell, Shadow detection: a survey and comparative evaluation of recent methods. Pattern Recogn. **45**(4), 1684–1695 (2012)
3. C.-C. Chen, J.K. Aggarwal, Human shadow removal with unknown light source, in *IEEE 2010 20th International Conference on Pattern Recognition (ICPR)* (2010), pp. 2407–2410
4. J.-W. Hsieh, W.-F. Hu, C.-J. Chang, Y.-S. Chen, Shadow elimination for effective moving object detection by Gaussian shadow modeling. Image Vis. Comput. **21**(6) (2003), 505–516
5. R. Cucchiara, C. Grana, M. Piccardi, A. Prati, Detecting moving objects, ghosts, and shadows in video streams. IEEE Trans. Pattern Anal. Mach. Intell. **25**(10), 1337–1342 (2003)
6. C.-T. Chen, C.-Y. Su, W.-C. Kao, An enhanced segmentation on vision-based shadow removal for vehicle detection, in *IEEE 2010 International Conference on Green Circuits and Systems (ICGCS)* (2010), pp. 679–682
7. J.-B. Huang, C.-S. Chen, Moving cast shadow detection using physics-based features, in *IEEE Conference on Computer Vision and Pattern Recognition, 2009. CVPR 2009* (2009), pp. 2310–2317
8. J. Panicker, Vinitha, M. Wilscy, Detection of moving cast shadows using edge information, in *IEEE 2010 The 2nd International Conference on Computer and Automation Engineering (ICCAE)*, vol. 5, pp. 817–821 (2010)
9. D. Xu, X. Li, Z. Liu, Y. Yuan, Cast shadow detection in video segmentation. Pattern Recogn. Lett. **26**(1), 91–99 (2005)
10. A. Sanin, C. Sanderson, B.C. Lovell, Improved shadow removal for robust person tracking in surveillance scenarios, in *IEEE 2010 20th International Conference on Pattern Recognition (ICPR)* (2010), pp. 141–144
11. W. Zhang, X.Z. Fang, X.K. Yang, Q.MJ. Wu, Moving cast shadows detection using ratio edge. IEEE Trans. Multimed. **9**(6), 1202–1214 (2007)
12. L. Shen, T.W. Chua, K. Leman, Shadow optimization from structured deep edge detection, in *Proceedings of the IEEE Conference on Computer Vision and Pattern Recognition* (2015), pp. 2067–2074
13. S.H. Khan, M. Bennamoun, F. Sohel, R. Togneri, Automatic feature learning for robust shadow detection, in *2014 IEEE Conference on Computer Vision and Pattern Recognition (CVPR)* (2014), pp. 1939–1946
14. S.H. Khan, M. Bennamoun, F. Sohel, R. Togneri, Automatic shadow detection and removal from a single image. IEEE Trans. Pattern Anal. Mach. Intell. **38**(3), 431–446 (2016)

Detection of Hemorrhagic Region in Brain MRI

Ujjwal Kumar Kamila, Oishila Bandyopadhyay and Arindam Biswas

Abstract Accidental brain injury causes life-threatening situations due to acute bleeding inside skull or brain. Automated detection of such hemorrhage from brain1 MRI can help doctors and medical staffs to plan for the treatment and save patient's life. In this paper, we have proposed a brain MRI analysis approach to detect the presence of epidural or subdural hemorrhage in brain matter. The proposed approach integrates *k*-means clustering, adaptive thresholding, and curvature analysis to segment the intracranial region, identify the mid-line shift of horn region, and detect the presence of hemorrhage in brain MRI. On detection of hemorrhage, the process segment the hemorrhagic region and identify its location in the brain.

Keywords Hemorrhage · *k*-means clustering · Adaptive thresholding
Curvature · Midline shift

1 Introduction

Brain hemorrhage is the consequence of bleeding within the brain substance from the ruptured blood vessels. Accidental brain injury (results in epidural and subdural hematoma) and hemorrhagic stroke (intracerebral hemorrhage) result in severe damage and leakage of blood vessels. Computer aided analysis of brain MR images

---

U. K. Kamila
Department of Computer Science and Engineering, Asansol Engineering College, Asansol, West Bengal, India
e-mail: ujjwal.kamila@gmail.com

O. Bandyopadhyay (✉)
Department of Computer Science and Engineering,
Indian Institute of Information Technology, Kalyani, West Bengal, India
e-mail: oishila@gmail.com

A. Biswas
Department of Information Technology, Indian Institute of Engineering Science and Technology, Shibpur, West Bengal, India
e-mail: barindam@gmail.com

C. R. Krishna et al. (eds.), *Proceedings of 2nd International Conference on Communication, Computing and Networking*, Lecture Notes in Networks and Systems 46, https://doi.org/10.1007/978-981-13-1217-5_38

can help the doctors in faster diagnosis and treatment planning for brain hemorrhage patients. Segmentation of hemorrhagic clot (hematoma) in the MRI is a challenging task as the components of brain such as gray matter (GM), horns, cerebrospinal fluid (CSF), white matter (WM), and hematoma appear in MRI with several overlapping intensity ranges. The variety of shapes and location of hematoma also makes it difficult to identify among brain matters.

Different segmentation approaches such as region growing, thresholding, morphological operations, and active-contour based approaches are used for lesion detection in brain MRI and CT images. An integrated approach of dual-tree complex wavelet transform (DT-CWT) using k-means algorithm is proposed by Zhang et al. [13] for brain MRI segmentation. Saad et al. [11] have applied region based histogram analysis followed by Gamme-law transformation to compute the optimum thresholding value for segmentation of brain lesion. Chen et al. [5] have proposed coherent local intensity clustering formulation using non-local regularization mechanism for brain image segmentation. In many cases, traumatic brain injury results in deformation of brain mid-line. Liu et al. [7] have introduced a method of tracing the brain mid-line shift by establishing the relationship between the hemorrhage and the mid-line deformation using linear regression model. The region-based active contour model uses this initialized contour to segment the ROI. This unsupervised technique segment the image into different objects by grouping the similar pixels in the feature space. Fuzzy C-means clustering is also applied to initialize the contour of hemorrhagic region in brain CT image [1]. The computer-aided estimation of ideal mid-line (IML) has attracted researchers in the recent years. Researchers have used the segmented contour of the intracranial region of brain CT images for ideal mid-line detection [10, 12].

In this paper we have proposed an integrated approach to detect intracranial hemorrhage and segment the hemorrhagic region from T2-weighted brain MRI. The proposed approach uses k-means clustering followed by morphological operations to segment the intracranial region from brain MRI. Adaptive thresholding is applied to generate the horn contour. Analysis of contour curvature of intracranial region and horn region are applied to determine the mid-line shift in the hemorrhage affected brain MRI. We have evaluated the quality of segmentation of the hemorrhagic region using Jaccard index. The rest of the paper has been organized as follows: Sect. 2 explains different phases of the proposed approach. The experimental results and efficiency of the method are discussed in Sect. 3.

2 Proposed Approach

The proposed approach uses the T2-weighted brain MR images. The entire process comprises five main phases which include intracranial area extraction, horn contour generation, computation of mid-line shift, hemorrhage detection, and segmentation of hemorrhagic region in the input MR image. Figure 1 shows different phases of the proposed approach.

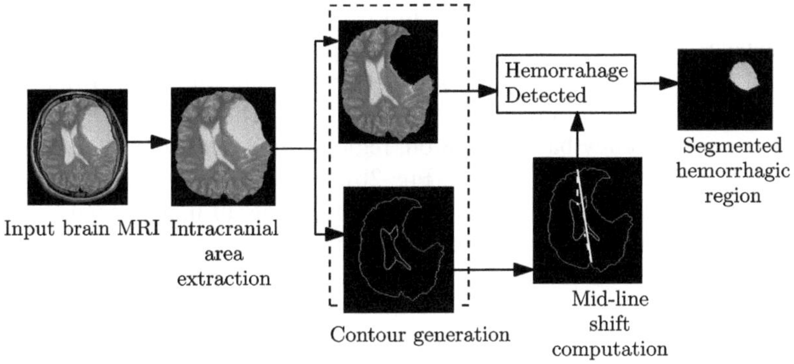

Fig. 1 Main components of the proposed approach

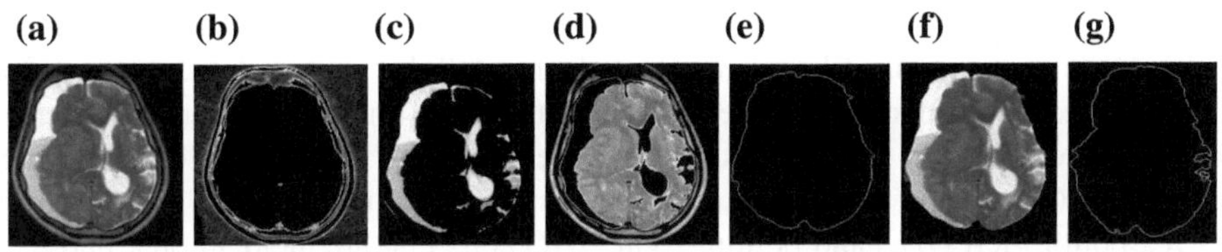

Fig. 2 k-means results: **a** Input brain MRI, **b** cluster 1 (centroid at lower intensity), **c** cluster 2 (centroid at highest intensity range), **d** cluster 3 (centroid at medium intensity range), **e** intracranial region contour (after dilation, region filling, and connected component analysis on cluster 1 image), **f** extracted intracranial region, **g** intracranial region contour (without hemorrhagic region)

2.1 Intracranial Area Analysis

The brain MR image appears with the outermost skull boundary, dura-matter, and intracranial region with GM, WM, CSF, and horns. Gray scale image (with intensity range 0–255) of brain MRI consists of dark background with skull and intracranial boundary, bright horn region (and hemorrhagic region for image with hemorrhage), and gray WM, GM, and CSF region. We have applied k-means clustering, morphological operations, and adaptive thresholding to generate the contour of intracranial region and horns.

Intracranial region extraction using K-means clustering:

k-means clustering partitioned data into k mutually exclusive clusters by minimizing the Euclidean distance metrics. Each cluster has a center point which has minimum sum of distances from all other points in that cluster [4]. As gray scale (0–255) brain MRI appears with dark background, bright horn and hemorrhagic region, and gray region with WM, GM, and CSF, we have applied k-means clustering approach on brain MR image to generate three clusters. The clusters appear with centroids at three different intensity ranges. One cluster (cluster 1 in Fig. 2b with centroid at lowest intensity range) appears with dark background and skull, dura-matter, and intracranial region boundary, one cluster (cluster 2 in Fig. 2c with centriod at highest intensity range) appears with bright horn and hemorrhagic region, and one cluster (cluster 3 in Fig. 2d with centriod at medium intensity range) appears with gray region i.e. WM, GM, and CSF area. Intracranial region contour is extracted (shown in Fig. 2e) by applying morphology based dialation, region filling, and connected component analysis [3] on lower intensity cluster (Fig. 2b). Contour of intracranial area without hemorrhagic region (Fig. 2g) is generated by masking all higher intensity regions (hemorrhage and horn region) of the extracted intracranial area (Fig. 2f) and applying region filling on the masked image.

2.2 Horn Contour Generation Using Adaptive Thresholding

Adaptive thresholding [2] is applied on the segmented intracranial image (Fig. 2f) to generate the contour of horn area. It performs cellular analysis of the image. In this approach, each pixel (k) of the image J is traversed and a small window around it with 8-neighboring pixels is examined (see Fig. 3b). The method compute the maximum (max_i) and minimum (min_i) intensity values of top-left (TL), top-right (TR), bottom-left (BL) and bottom-right (BR) cells that are incident on k and use these values in adaptive thresholding calculation. In each move, the next pixel is selected whenever the method identifies a pixel with an intensity value higher than the adaptive-threshold value of the present pixel (Fig. 3a). The adaptive threshold τ_{apt} is computed as:

$$\tau_{apt} = \gamma\tau_l + (1-\gamma)\left(\tau_l + \tau_{pr}\right) \tag{1}$$

where $\gamma \in [0,1]$ represents the smoothing factor for the exponential moving average, $\tau_l = \frac{1}{2}(max_i - min_i)$, and τ_{pr} represents the adaptive threshold calculated for the previous pixel on the cellular region contour. For computational simplicity, we consider $\gamma = \frac{1}{2}$ [2]. A cell is said to contain a part of intracranial boundary, provided $max_i - min_i \geqslant \tau_{apt}$. Figure 3b shows an object with high- and low-intensity pixels. Blue circles represent the contour pixels that are identified using the adaptive thresholding approach. In each traversal, the object remains on the right side (left side) of the contour for a counter-clockwise (clock-wise) move.

2.3 Midline Shift Analysis

The severity of brain injury is detected from the midline shift in brain MRI. Clinicians depends on the midline shift to analyse the change of symmetry for diagnosis of abnormalities in brain MRI. In this paper, we have proposed a novel approach for the midline shift detection by analyzing the curvature of brain contour and horns. The proposed method - (a) analyse upper and lower part of intracranial region contour to identify the points with maximum curvature as mid-points, (b) perform connected

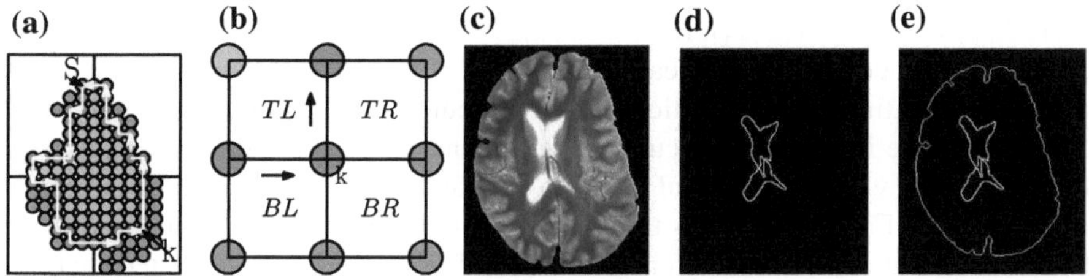

Fig. 3 **a** Object-contour pixel selection, **b** 8-neighbours of pixel k, **c** intracranial region, **d** contour of horn (using adaptive thresholding), **e** intracranial contour with horn contour

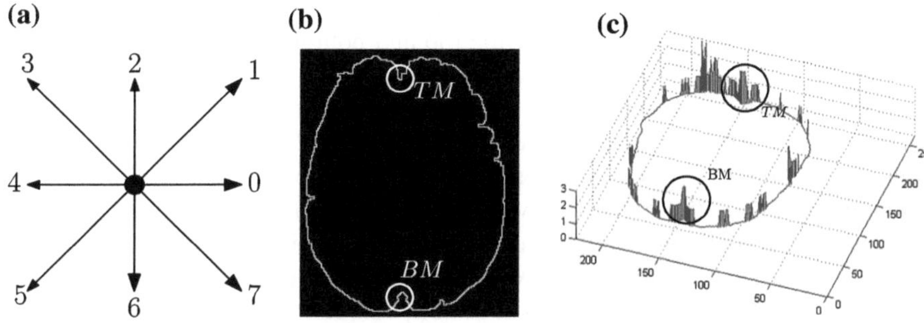

Fig. 4 **a** Chain code directions, **b** intracranial contour, **(c)** discrete curvature of intracranial contour

component analysis (CCA) [3] on intracranial region contour to identify horn components (horn segments), and (c) analyse horn component (each horn segment, in case of multiple segments) to detect the mid-points (top and bottom point with maximum curvature).

Contour Curvature Analysis

To determine the midline which divide the brain into two hemisphere, we have used discrete curvature estimation based on chain code sequence.

Chain-code analysis of digital curve: A digital curve (R) can be defined as a sequence of digital points (with integer coordinates) in which two points $(x_1, y_1) \in R$ and $(x_2, y_2) \in R$ are neighbours if $\max(|x_1 - x_2|, |y_1 - y_2|)$=1. Thus, the chain code of a point $k_i \in R$ w.r.t. its previous point $k_{i-1} \in R$ is given by $c_i \in 0, 1, 2, \ldots, 7$ [8] (Fig. 4a).

The *discrete curvature* (κ) at a point k_i on digital curve R is computed as the *sum of differences* between the mean angular direction of C (> 1) vectors (i.e., chain codes) of the proceeding curve segment of k_i and that of C vectors on the succeeding curve segment of k_i as follows [9].

$$\kappa(p_i, C) = \frac{1}{C}\sum_{j=1}^{C} min \begin{Bmatrix} min(H_{i+j}, 8 - H_{i+j}), \\ min(H^{+}_{i+j}, 8 - H^{+}_{i+j}), \\ min(H^{-}_{i+j}, 8 - H^{-}_{i+j}) \end{Bmatrix} \qquad (2)$$

where $H_{i+j} = |c_{i+j} - c_{i-j+1}|$, $H^{+}_{i+j} = |c_{i+j+1} - c_{i-j+1}|$, and $H^{-}_{i+j} = |c_{i+j} - c_{i-j}|$; c_{i+j} is the chain code of the jth leading point and c_{i-j} is that of the jth trailing point w.r.t. p_i. Under the above definition of discrete curvature, a digital curve will show a high curvature in a relatively-curved region, and a low curvature in a relatively-straight region, whereas a digital arc or a digital straight line segment will show zero-curvature. Figure 4b shows the contour of the intracranial region and Fig. 4c shows the variation of discrete curvature (κ) of contour. It can be noted that the zero-curvature appears in relatively straight and arc region, whereas the regions with sharp bends (such as region 'A' and 'B') exhibit high peaks. We have computed the discrete curvature values for all the pixels belonging to the top and bottom region of intracranial contour. The pixels with maximum curvature value are identified ('TM' and 'BM' of Fig. 4c) as midpoint for the top and bottom part of intracranial contour. If multiple pixels appear with high curvature value, then for every pixel p_i, the chain code pattern (as shown in Fig. 4a) of leading and trailing pixels are analyzed for a window of 10 pixels. Ideally, the chain code string (c_is, e.g {2,3}) for trailing pixels of p_i should have the complementary chain code string (c_i's, {6,7}) in the leading pixels of p_i. The highest curvature pixel with maximum number of (c_i, c_i') pair present in leading and trailing pixels is selected as midpoint. Same process is repeated for horn contour also and pixels with maximum curvature are identified as horn mid-points for top and bottom part of horn contour.

Midline Shift Computation

We calculate the slope (m_1) of the straight line between top and bottom mid points ('A' and 'D' of Fig. 5) of the intracranial contour. This line is used as a reference line for midline shift detection. In next step, the slope (m_2) between intracranial contour top midpoint ('A') and horn contour top midpoint ('B') is computed. Similarly, the slope (m_3) between intracranial contour bottom midpoint ('D') and horn contour bottom midpoint ('C') is also calculated. In order to detect the midline shift, we calculate the angle between intracranial top midpoint and horn top midpoint (∠BAD) using slopes m_1 and m_2. Angle between intracranial bottom midpoint and horn bottom midpoint (∠CDA) is also computed using same approach. Midline shift is detected when any of these two angles (∠BAD or ∠CDA) exceeds 1.5°. Figure 5a, b shows two cases of midline shift in hemorrhagic brain MRI. Figure 5c shows a healthy brain MRI where both top and bottom angles are very small.

2.4 Hemorrhage Location Identification

The contour of intracranial region (S_1) (after skull removal (Fig. 2e) is compared with the contour of segmented intracranial region (S_2) (Fig. 2g) to detect the location and area of the hemorrhagic region. Upper and lower midpoints of intracranial contour (TM and BM of Fig. 4a) are used to divide each intracranial contour image into two halves (LS_1 and RS_1 for S_1 and LS_2 and RS_2 for S_2). The area difference (in pixels) between each respective half of S_1 and S_2 ((LS_1 - LS_2), (LR_1 - LR_2))

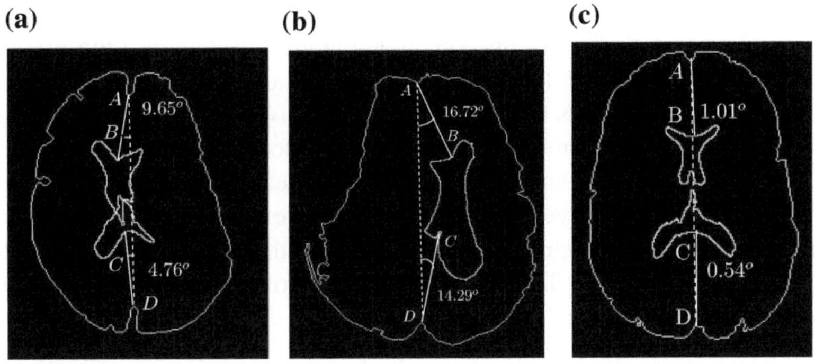

Fig. 5 **a** Midline shift (left side), **b** Midline shift (right side), **c** Healthy image

Input MRI	Intracranial region	Contour	Comments	Input MRI	Intracranial region	Contour	Comments
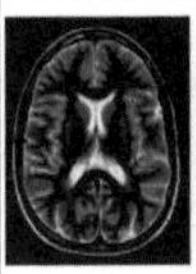 (a1)	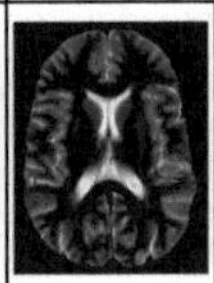 (a2)	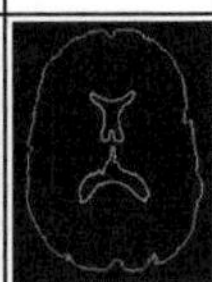 (a3)	ΔArea: 54(pix), Md.shift: 0.6°, Status: Healthy	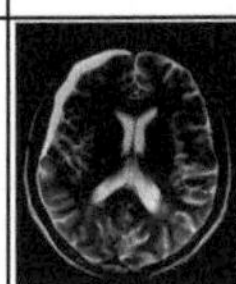 (a4)	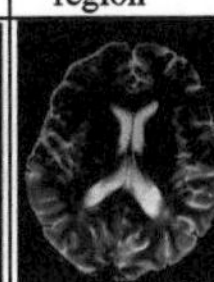 (a5)	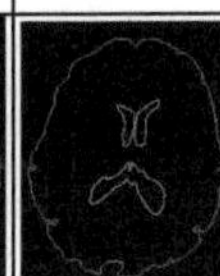 (a6)	ΔArea: 207(pix), Md.shift: 1.6°, Status: Hemorrhage (left side) Jaccard-Index: 0.92
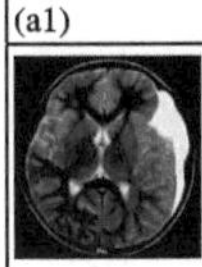 (b1)	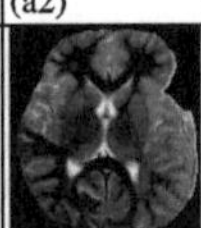 (b2)	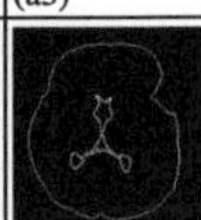 (b3)	ΔArea: 213(pix), Md.shift: 1.5°, Status: Hemorrhage (right side) Jaccard-Index: 0.97	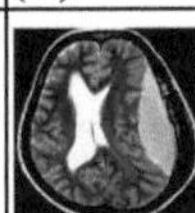 (b4)	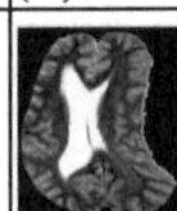 (b5)	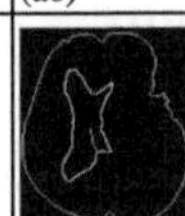 (b6)	ΔArea:257(pix), Md.shift:1.8°, Status: Hemorrhage (right side) Jaccard-Index: 0.94

Fig. 6 Different phases of hemorrhage detection

is computed to detect the location of hemorrhage. It is observed that a high area difference ($\Delta Area > 200 pixels$) in left or right hemisphere of MRI indicates the presence of hemorrhagic region in that portion (Fig. 6).

3 Results

We have used a dataset of 30 T2-weighted brain MRI to implement and test the proposed method. Experimental results (some are shown in Fig. 6) show satisfactory outcome as the presence of hemorrhage and its location are correctly identified in all the cases. We have applied intracranial region and horn contour generation, contour curvature analysis, midline shift computation to diagnose the presence of hemorrhage in the MRI and detect hemorrhage location. It is observed that midline shift (Md. shift in comments of Fig. 6) calculation gives clear indication of hemorrhagic cases. However in some cases the top and bottom part of intracranial contour region appears with low curvature change. This may leads to erroneous result in midline shift calculations. To overcome this situation we have used both midline shift calculation and hemorrhagic region segmentation with area computation to identify the hemorrhagic situations. To evaluate the quality of the segmented hemorrhagic region, we have used Jaccard similarity indexes [6]. Jaccard index is computed by comparing the region A segmented by the proposed approach with region B marked by the expert. Jaccard index J(A,B) can be defined as -

$$J(A, B) = \frac{A \cap B}{A \cup B} \tag{3}$$

The value of $J(A, B)$ lies between 0 and 1. In ideal situation where the two region matches perfectly, $J(A, B) = 1$ and for entirely different region, it becomes 0. Comments of Fig. 6 shows the Jaccard index value for few hemorrhage affected brain MRI.

4 Conclusion

The proposed approach analyses brain MR images and extracts the hemorrhage affected region successfully. The location of hemorrhage (left or right side of MRI) is also identified correctly in each test image. This paper focuses on accidental brain hemorrhage scenarios such as epidural and subdural hemorrhage. Diagnosis and segmentation of subarachnoid and cerebral hemorrhage can be proposed as an extension of this work.

Acknowledgements Some MRI test-images used in this work are taken from the website of Radiopaedia (https://radiopaedia.org/articles/ageing-blood-on-mri) and Open Access Biomedical Image Search Engine (https://openi.nlm.nih.gov/gridquery.php?simCollection=PMC4556906_JMedLife-08-278-g004&query=&req=3&m=1&n=100). We acknowledge all of them with thanks. The MRI images used for this work are uploaded in our drive (https://drive.google.com/drive/folders/1SbHipIztu9WPBHLHFP7RK-wuhqTism0a?usp=sharing).

References

1. H.S. Bhadauria, A. Singh, M.L. Dewal, An integrated method for hemorrhage segmentation from brain CT imaging. Comput. Electr. Eng. **39**, 15271536 (2013); Elsevier **39**(5), 15271536 (2013). https://doi.org/10.1016/j.compeleceng.2013.04.010

2. A. Biswas, S. Khara, P. Bhowmik, B.B. Bhattacharya, Extraction of region of interest from face images using cellular analysis, in *Proceedings of ACM Compute* (2008), pp. 1–8
3. B. Chanda, D.D. Majumdar, *Digital Image Processing and Analysis* (Printice Hall of India, New Delhi, 2007)
4. T. Chen, Y. Chen, S. Chien, Fast image segmentation based on K-means clustering with histograms in HSV color space. J. Sci. Res. **44**(2), 337–351 (2010)
5. Z. Chen, J. Wang, D. Kong, F. Dong, A nonlocal energy minimization approach to brain image segmentation with simultaneous bias field estimation and denoising. Mach. Vis. Appl. **25**, 529–544 (2014)
6. W. Hsu, Improved watershed transform for tumor segmentation: application to mammogram image compression. Expert Syst. Appl. **39**, 3950–3955 (2012)
7. R. Liu, S. Li, C.L. Tan, B.C. Pang, From hemorrhage to midline shift: a new method of tracing the deformed midline in traumatic brain injury CT images, in *Proceedings of IEEE International Conference on Image Processing (ICIP)* (2009), pp. 2637–2640. https://doi.org/10.1109/ICIP.2009.5414092
8. S. Pal, P. Bhowmick, Estimation of discrete curvature based on chain-code pairing and digital straightness, in *Proceedings of IEEE International Conference on Image Processing (ICIP)* (2009), pp. 1097–1100. https://doi.org/10.1109/ICIP.2009.5413475
9. S. Pal, R. Dutta, P. Bhowmick, Circular arc segmentation by curvature estimation and geometric validation. Int. J. Image Graph. **12**(4), 1250,024–1–1250,024–24 (2012)
10. X. Qi, A. Belle, S. Shandilya, W. Chen, C. Cockrell, Y. Tang, K.R. Ward, R.H. Hargraves, K. Najarian, Ideal midline detection using automated processing of brain CT image. Open J. Med. Imaging Sci. Res. **3**(2), 51–59 (2013). https://doi.org/10.4236/ojmi.2013.32007
11. N.M. Saad, S.A.R. Abu-Bakar, S. Muda, M. Mokji, Segmentation of brain lesions in diffusion-weighted MRI using thresholding technique, in *Proceedings of IEEE International Conference on Signal and Image Processing Applications* (2011), pp. 249–254. https://doi.org/10.1109/ICSIPA.2011.6144092
12. H.L. Tong, M.F.A. Fauzi, S.C. Haw, Automated hemorrhage slices detection for CT brain images. Proc. Second Int. Visual Inf. Conf. **7066**, 268–279 (2011). https://doi.org/10.1007/978-3-642-25191-7-26
13. J. Zhang, W. Jiang, R. Wang, L. Wang, Brain MR image segmentation with spatial constrained K-mean algorithm and dual-tree complex wavelet transform. J. Med. Syst. (2014). https://doi.org/10.1007/s10916-014-0093-2

A Novel Approach of Optic Disk Detection for Diagnosis of Diabetic Retinopathy

Aakanksha Mahajan, Sushil Kumar and Rohit Bansal

Abstract Diabetic retinopathy (DR) is emerging technology in the field medical imaging. Because of diabetic symptom in body, loss of vision can happen and it is mainly due to Diabetic Macular Edema (EMD) and Retinopathy Proliferative Diabetic (RPD). Recent studies claims the new treatment methods and prevention. There are very effective treatments and they are optimal when the DR is detected in early stages, even when the patient has no symptoms. This paper deals with development of a computational oriented system to help in the pre-diagnosis of diabetic retinopathy, a progressive disease that develops in patients with diabetes mellitus and is characterized by the appearance of vascular lesions in the retina as increase in permeability and intra-retinal haemorrhages, intra-retinal accumulation of fluids and fluid, closure of blood vessels, capillaries and retinal arterioles and growth of new vessels blood inside and on the retinal surface. Some of these injuries are visible in fundus images, which are basically the images of the retina illuminated with white light and taken with a camera. The main objective of this paper is to develop a recognition algorithm of images that can automatically detect the optic disk. It is first proposed to correct distortions of luminosity produced by nature intrinsic optics of the image acquisition method. Then it is proposed to perform the detection of the optical disk.

Keywords Diabetic retinopathy · Diabetic macular edema
Lighting correction algorithm · Optic disk detection · Retinopathy proliferative diabetic

---

A. Mahajan (✉) · S. Kumar
Amity University, Noida, Uttar Pradesh, India
e-mail: er.aakanksha@yahoo.co.in

S. Kumar
e-mail: skumar21@amity.edu

R. Bansal
RGPU, Noida, Uttar Pradesh, India
e-mail: rohitbansaliitr@gmail.com

C. R. Krishna et al. (eds.), *Proceedings of 2nd International Conference on Communication, Computing and Networking*, Lecture Notes in Networks and Systems 46, https://doi.org/10.1007/978-981-13-1217-5_39

1 Introduction

Diabetic retinopathy (DR) is defined by the presence of one or more lesions vascular diseases in patients with diabetes. DR is a progressive disease, characterized by secondary clinical features to capillary changes with closure of blood vessels, growth of new vessels in the retina and iris, increased permeability and accumulation of acids and lipids [1]. The changes in early stages are loss of pericytes (one of the cells contra of the connective tissue layer surrounding the capillaries [2]), thinning of the basement membrane and appearance of microaneurysms. There are changes in the blood ointment in the retina. The internal barrier breaks showing itself as increased capillary permeability and intraregional hemorrhage. The capillaries and the retinal arterioles close (retinal no perfusion). In the most advanced stages, new blood vessels grow in the iris. Microaneurysms are almost always the first clinical sign of diabetic retinopathy, and they look like deep intraregional points with diameters between 15 and 60 [3]. The rupture of microaneurysms and an increase in permeability capillary results in intra-retinal hemorrhages. The small hemorrhages intraretinals are a typical sign of diabetic retinopathy. Increased capillary permeability results in intra-retinal accumulation of fluids and lipids seen as well accumulating zones of color between white and yellow called hard exudates [4].

Areas of capillaries are generated that, when merged, generate areas of no-perfused retina (non-perfused retina) [5]. Capillary closure is not visible without angiography uoresceinica. Diabetic Macular Edema (EMD) is characterized by intra-retinal accumulation of liquids and lipids within the macula and can occur at any stage of the illness [6, 7].

Many type of check-ups have been proven by the digital photography of the retina which is currently preferred by the community. The prevalence of DR has recently been studied in different regions of geographies and different ethnic groups [8].

2 Proposed Methodology

It is proposed to divide the system into three fundamental stages. First the problem of the non-uniform illumination present in the images will be considered, then the detection of the optical disk and the macula will be made. Finally, the segmentation will be done automatically for hard and soft exudates. It is proposed to model non-uniform lighting based on two main sources.

Optical aberrations of the image acquisition system together with the geometry of the human eye and the effects of the nature of visible light cross different optical media, specifically, refraction, and reflection. With corrected non-uniform illumination, it is proposed to detect the optical disk and the macula by means of a pattern recognition with learning supervised algorithm which consists of two stages, a training stage which generates an average sub-image of the structures of interest and a test stage where the best match of the sub-image with the new one is searched exhaustively image to be detected.

Finally for the detection of exudates, it is proposed to apply an improved algorithm of contrast, based on operations of mathematical morphology and then apply a threshold that allows finding the brightest regions in the image. It is important to highlight that for a correct evaluation of the performance of the system, it has a set of digital fundus images taken from patients and selected by the means of a protocol of nest in clinical bases that were manually segmented by an expert ophthalmologist in the area. These set of images constitutes our gold standard or Ground Truth on the basis of which we will make the comparisons with the results obtained from the proposed system.

2.1 Correction of Non-uniform Illumination

Inadequate lighting of the scene as well as inappropriate conditions while capturing the image, for example, an unsettled position of the capture system of an image, can introduce severe distortions in the resulting digital image. These distortions are usually perceived as slow intensity variations in the whole image or sudden variations in the edges of the image. This effect is commonly called as Inhomogeneous intensity (intensity inhomogeneity), Illumination non-uniform, shading (shading) or bias field.

Retinal images are acquired with a fundus camera that receives the light ejected from the surface of the retina and is recorded by a sensor, for example, CCD. Often the images obtained have a non-uniform illumination and therefore, there is variability in luminosity and local contrast. This problem seriously affects the diagnostic process and consequently due to that, injuries in some areas can become difficult to recognize for a human observer. On the other hand, different methods of automatic analysis based on the computer system have been proposed to help the medical specialist to obtain the useful information of the images, for example, calculation of the tortuosity of the veins or detection of microaneurysms and exudates. Images with high variability in contrast and illumination, intra and inter image are extremely difficult to analyze with the proposed automatic methods and the results can be disastrous, therefore, the first step is the standardization of images, which consists in taking all the images to be analyzed to a standard value in contrast and luminosity to correctly perform its analysis and comparison [8].

To be able to perform a correct normalization in the luminosity of the image, an appropriate model is required that represents the bias or non-uniformity present. The hypothesis that is accepted in general about no homogeneity in the intensity of the image supposes that this was shown as a function that varies smoothly in space and that alters the intensities in each point in the image that would otherwise be constant for the same type of tissue regardless of its position. In the simplest form, the model assumes that the Non-homogeneity of intensity is multiplicative or additive, that is, the noun-noun field it is multiplied or added to the real image. In the field of images, Magnetic Resonance is often used the multiplicative model because of its consistency with the physical process of acquisition and with the sensitivity in the coil of reception [9].

In Magnetic Resonance, the different models of image formation have been proposed in the literature, depending on how they interact with the bias-free image, $u(x)$, the non-homogeneous intensity field, $b(x)$ and noise $n(x)$ [9]. One of the proposed models assume that the noise is approximated by a Gaussian probability density function itself of the scanner and therefore independent of the bias in the intensity, therefore, the image acquired $v(x)$ is modeled as

$$v(x) = u(x)b(x) + n(x) \tag{1}$$

In another model, only the biological noise is scaled in intensity by the bias field $b(x)$ so that the signal-to-noise ratio (SNR) is preserved

$$v(x) = (u(x) + n(x))b(x) \tag{2}$$

The third model of imaging is based on the log transformation rhythmic intensities where the multiplicative non-homogeneous intensity field it becomes additive:

$$\log(v(x)) = \log(u(x)) + \log(b(x)) + n(x) \tag{3}$$

In the last decade, various lighting correction methods have been proposed. In [9] a macro classification is proposed in two groups called prospective (future, possible) and Retrospective (retrospective, retroactive). The first one refers to the calibration and improvement of the physical process of acquisition of the image while the second one focuses exclusively on the information which provides the acquired image and sometimes some type of apriori information. The complete classification is [9] [10]:

- Prospective.
- Ghost.
- Multiple coils.
- Special sequences.
- Retrospective.
- Filtered out.
- Surface adjustment.
- Intensity.
- Gradient.
- Segmentation.
- Maximum likelihood (ML).
- Algorithm with fuzzy logic of C-medias.
- Nonparametric
- Histogram.
- High frequency maximization.

- Minimization of information.
- Check histograms.
- Others.

A non-uniform lighting correction operator or shadow tries to remove the background information of the image. This can be done by calculating an approximation of the background and then make a subtraction between the original image and the approximation found (when it is assumed that non-uniform lighting has an additive nature in the image). To avoid negative values usually a constant is added

Shape

$$[\mathrm{SC}(f)]_{(x)} = f(x)[A(f)\}_x + c, \tag{4}$$

where $[SC(f)]_{(x)}$ is the corrected image in luminosity, $[A(f)\}_x$ corresponds to the image approximation of the background and c a constant.

The shading effect due to non-uniform illumination is a function that varies slowly with spatial coordinates [11].

In the past, it was preferred to use the medium filter as the medium-sized filter shows better results in the conservation of edges and generally produce better results in retinal images. Now, as there are different forms of efficient implementation of this algorithm, the median has been used more in recent years [12].

The size of the kernel or filter window must be determined by the factors such as the resolution of the image, the size of the structures present in the image (vessels blood, macula, optic disk) and the size of the lesions present (exudates, microaneurysms, etc.). The size of the kernel is important because in general, it will always be bigger than the injuries, but if the lesion is of comparable size with the size of the kernel, a hemorrhage that occupies an important region in the image of the retina, the size of the kernel becomes a crucial variable for the later automatic detection of injuries because the liposuction can confuse the injury with non-uniform lighting due to its size. For this reason, in general, the lighting correction methods are more sophisticated than those described to obtain a more reliable result.

In [13] a non-uniform polynomial illumination model is proposed that is of multiplicative nature. An iterative process is carried out in the columns and then the polynomial is adjusted that best fits a set of points on a line, equally spaced in the image obtained from a filter of the medium. These points are updated in each step, eliminating the points that present an error greater than a pre-established maximum. After the elimination of the points of greater error, the adjustment cycle is repeated. In [14] a modification to the algorithm is proposed by Foracchia et al. [15] which it is based on an additive-multiplicative model of non-uniform illumination.

The model of the resulting image is modeled as: $I = U$ SM + SA where U is the ideal image, I is the observed image, and SM and SA are the additive components and multiplicative of non-uniformity in illumination. They have proposed a model to calculate the distance of Mahalanobis for each pixel in the image taking values of mean and standard deviation calculated in a window of size w to segment the image into two parts: bottom and non-bottom by thresholding with a value. After, the first segmentation, the average and the standard deviation of the background image are calculated, which are the estimators of the SA and SM components respectively, to get the corrected image U.

2.2 Optical Disk Detection

The optical disk is presented in the fundus image as a circular structure having a bright high contrast with the background. Various methods have been proposed for optical disk detection in digital images of fundus. In [16] the optical disk is located, making use of its variation in levels of gray. This approach has shown to have good results when there are very few pathologies in the image (or there are not). Since some exudates appear bright, the contour of the optical disk is not found in said work. In [17] and in [18] the outline of the optical disk is detected using the Hough transform, the gradient of the image is calculated and it determines the circle that best fits him, this approach requires considerable processing time and also requires that certain conditions to be met in the geometry of the optical disk that does not always occur. In [19] it is located tracking the blood vessels at the origin, the method is certainly one of the safest and logical methods to locate the optical disk, but it depends to a larger extent, on the detection of blood vessels. In this case, it is advisable to separate the tasks of segmentation so as not to fall into accumulation of errors and reduce the processing time. In [20] a threshold of area is used to locate the optical disk and the watershed transformation is used to find the outline of it. Most methods do not work well if the contrast of the image is poor [18].

In [21] a method is proposed using morphological alteration and the watershed transformation.

In [22] a method of automatic segmentation of the optical disk and the fovea defining the problem as a regression problem is proposed. The KNN regression method is used to predict the distance in pixels to the object of interest in any position of the image based on a set of characteristics measured in that position, mixing contributions of the properties of the image and derived contributions of the segmentation of the vascular network in the retina. A prediction of the distance to the optical disk is the minor one selected as the center of it.

The method is trained with 500 images and was evaluated in 500 test images. A successful detection rate of 99. 4% was achieved.

2.3 Segmentation of Exudates

In [23] images are acquired in RGB mode and the green channel is used to apply a morphological lock using a structural element octagonal to increase the contrast between exudates and the background of the image. For extraction of exudates, a method is proposed to make a quotient between the addition and subtraction of green channels (which emphasizes exudates) and the red channel (which emphasizes the bottom). The exudates can then be selected from the resulting image through thresholding. This approximation depends to a great extent on the value of the selected threshold. In [24] a statistical solution is proposed based on mixed models (mixture models, MM) to perform a thresholding dynamic that allows separating hard exudates from the bottom. The MMs provide a statistical technique to estimate probability densities in the same way as the multimodal histograms. The standard method to obtain the estimate of the maximum likelihood is through the iterative algorithm expectation-maximization (EM).Considering the MM parameters converge, the first Gaussian component is associated with the dark-red elements, the second is associated with the background and the third with light yellowish regions. The threshold value obtained is used to separate hard exudates from the bottom.

In [25] a solution is proposed using the grouping method (or clustering) fuzzy c-means. Initially, the pre-processing is done to the image performing color normalization and contrast increase, then a classification is made of the whole image in two classes: exudate and non-exudate. For this, a set is used for initial characteristics such as color, size, edge intensity and texture to feed a genetic algorithm to classify the characteristics and identify the subset of better sorting results [26]. The feature vectors selected are classified using a multilayer neural network. The algorithm was tested with a database of 300 retinal images manually labeled and a sensitivity of 96% and a specificity of 94.6% was reached. In [27] a method is proposed to use the watershed method with controlled markers. Initially, an average filter to blur small objects is performed with low variations intensity and objects of interest is highlighted. The green channel is removed in space of RGB color and a gamma transformation of contrast enhancement is applied. Then the gradient of the magnitude of the image for the watershed algorithm is calculated and the concept of markers to control over-segmentation is used. Two types of markers are used, internal (associated with the object of interest) and external (associated at the bottom). The internal markers are obtained from the gradient image using the technique extended minima transformation, which computes the regions with brighter pixels and external markers partition the image into regions in where each region contains an internal marker. The method was validated on the basis of DRIVE and STARE data, obtaining a sensitivity of 94% and a predictive value of 91.9%. In [21] an algorithm based on transformations is proposed morphologically. It can be divided into two parts, first, using the green channel of the RGB space are candidates that are regions that possibly contain exudates, and then morphological techniques are applied to find the exact contours.

2.4 Proposed Lighting Model

It is assumed that the luminous eye that arrives electronic sensor coming from the retina can be separated, as if it came from two different sources of light, one that provides an image with only the objects of interest and another with the noun-lighting pattern, so that the following lighting model proposed in [15] where the observed retinal image I can be modeled as:

$$I = f\left(I^0\right) = f(I_b^0 + I_f^0), \tag{5}$$

where I^0 is the original image, I_b^0 is the original background image (Background), I_f^0 is the original foreground image (Foreground) and $f\left(I^0\right)$ is the transformation of acquisition. I_f^0 is the image with all the structures (blood vessels, optical disk and macula) and lesions (microaneurysms, exudates, etc.) and I_b^0 is the image without any structure or injury but with the non-uniform lighting pattern. In this way, assuming $f\left(I^0\right)$ as a linear function, the image of interest or corrected may be calculated by means of

$$I^0 - I_b^0 = I_f^0 \tag{6}$$

2.5 *Optical Disk Detection and Macula*

The optical disk appears in the fundus images as a region normally clearer than its surroundings and represents the optic nerve. From there, veins emerge and arteries of the central retina that cover most of the retina after a series of bifurcations. Set the position of the optical disk in the images Fundus is important for several reasons. Many important pathologies in the retina can have affectation in the optic nerve. Also, given that the Optical Disk can easily be confused with lesions such as large exudates, its detection is important to exclude it from the set of detected lesions in automatic systems. In addition, the detection of the optical disk is fundamental for establishing a reference frame within the fundus image [15].

The application of two different proposed strategies, the first using the morphology of the blood vessels and the second using the template matching algorithm are described.

2.5.1 Detection Using Blood Vessel Morphology

This detection proposal is based on the morphological characteristics of the network capillary in the retina image and its branching pattern. From the optical disk, the main blood vessels are detached forming an arch of the temporal side, enclosing the area where the macula is. For this purpose, the algorithm for detection of blood vessels RISA was used (Retinal Image Multiscale Analysis) proposed in [27]. With this tool, the thickest blood vessels were segmented, then the object with the largest area in 8-connectivity was selected. The search for the center was followed using a simple iterative process that consists of finding the center of mass of the blood vessel object and eliminate the farthest pixel using the distance Euclidean as a metric, within the set of pixels of the object. The last pixel to be removed was selected as the chosen pixel belonging to the optical disk.

In the first approach, although based on a morphological criterion robust, the results obtained were not entirely satisfactory because in the implementation phase, several problems were found. The first one being, the center of mass of the blood vessel network does not always lead to the optical disk space center and the second is that the results of blood vessel extraction using RISA are slightly affected in images that have a lot of injuries (exudates and microaneurysms).

2.5.2 Detection Methodology Using Comparison of Templates

The strategy used in this methodology is innovative and simple. It will apply a template matching algorithm using the template as not only the optical disk region but the optical disk pair—macula, as a single structure. This strategy is based on the fact that the optic disk and the macula, despite being present in small variations between different patients is assumed in practice. This search will be done using the correlation as a measure of similarity using the Fast Fourier transform as an algorithmic means to accelerate the process search The result will be influenced by a spatial probability function that proposes that the optical disk presents a greater probability of localization towards right and left ends of the image and a lower probability of location towards the central (horizontal) area of the image. This is because the protocols of acquisition of images for the type of pathologies that was a major concern establishes that the main diagnostic image must be a posterior pole image which clearly visualize the main structures (optic disk, blood vessels and macula) with the macula towards the center of the image and the optical disk towards the horizontal ends.

2.5.3 Training

The training process begins by obtaining the limits (Lower, Upper, Right, and Left) and the center of mass of each one of the structures (Optical Disk and macula) as well as the midpoint between the two center of mass in each of the images from the manual segmentation or Ground Truth. This midpoint between the center of mass (which will be referred to as the reference point) will be subsequently used as the reference in the registration of the images. Taking into account, that, usually the macula appears below the optic disk, between the two structures an angle is generated, that varies slightly among different patients To reduce this variability, the rotation of each image was performed on the reference point degrees, so that the optic disk and the macula were at the same vertical level. The size of the average image (or pattern image) was defined from the most external coordinates (from the limits of manual segmentation) to the reference point of all the images, that is, the limits were selected that contained all the optical disk—macula boxes in each image. From this distribution the mean values, and standard deviation that will be used in the process of next search.

$$\theta_{\text{mean}} = \frac{1}{N}\sum_{i=1}^{N}\theta_i,\ \theta_{\text{std}} = \sqrt{\frac{1}{N-1}\sum_{i=1}^{N}(\theta_i - \theta_{\text{mean}})^2} \tag{7}$$

In algorithm two, a pseudo-code of the training process is shown.

2.6 Classification

The basic idea of the search process is to correlate the pattern image with the classification image, making use of the fast Fourier transform as an algorithmic medium, both for a right eye and for a left eye and perform subtraction between two, for three different angles which give 6 different cases: Normal and mirror image to look for right eye and left eye.

Three different angles

$$\theta_{\text{mean}} - \theta_{\text{std}},\ \theta_{\text{mean}},\ \theta_{\text{mean}} + \theta_{\text{std}}.$$

The correlation between two images is defined as

$$f[m,n]og[m,n] = \sum_{z=-\infty}^{\infty}\sum_{\omega=-\infty}^{\infty} f[z,\omega]g^*[m+z,n+\omega] \tag{8}$$

The convolution between two images is defined as

$$f[m,n] * g[m,n] = \sum_{z=-\infty}^{\infty}\sum_{\omega=-\infty}^{\infty} f[z,\omega]g^*[m-z,n-\omega] \tag{9}$$

The relationship between correlation and convolution is given by

$$f[m,n]og[m,n] = f[m,n] * g^*[-m,-n] \tag{10}$$

And the relationship between the convolution and the Fourier transform is given by the convolution theorem:

$$x[n] * g[n] \leftrightarrow X\left(e^{jw}\right)G\left(e^{jw}\right) \tag{11}$$

Given three images i [m; n] the search image, pd [m; n] the eye pattern image right and pi [m; n] the left eye pattern image, and given a rotation function spatial which applies a rotation of certain degrees relative to the center of the image i, and given $I(jw) = F_{fi}$ [m; n] g, where F denotes the Fourier transform, the correlation between the two images i [m; n] and p [m; n] as shown in Eq. (12).

$$
\begin{aligned}
i[m,n]^{\circ} R(p[m,n],\theta) &= \sum_{z=-\infty}^{\infty} \sum_{\omega=-\infty}^{\infty} i[z,\omega] p_{\theta}[m+z, n+\omega] \\
&= i[m,n] * p_{\theta}^{*}[-m,-n] \rightarrow \mathcal{F}\{i[m,n]\}\mathcal{F}\{p_{\theta}^{*}[-m,-n]\} \\
&\rightarrow I(j\omega) \cdot p_{\theta}^{*}(j\omega)
\end{aligned} \tag{12}
$$

The position search will be carried out in the most likely location of the optical disk in the images, as shown in Eq. (13).

$$
\begin{aligned}
C_{rr1}[m,n] &= \left(i[m,n]^{\circ} p_{\theta 1}^{d}[m,n]\right) - \left(i[m,n]^{\circ} p_{\theta 1}^{i}[m,n]\right) \\
C_{rr2}[m,n] &= \left(i[m,n]^{\circ} p_{\theta 2}^{d}[m,n]\right) - \left(i[m,n]^{\circ} p_{\theta 2}^{i}[m,n]\right) \\
C_{rr3}[m,n] &= \left(i[m,n]^{\circ} p_{\theta 3}^{d}[m,n]\right) - \left(i[m,n]^{\circ} p_{\theta 3}^{i}[m,n]\right)
\end{aligned} \tag{13}
$$

The result of the correlation for each of the proposed angles can be interpreted as a probability function where a local/absolute maximum corresponds with a high probability value of finding the pattern image at that point. Clinically, it can be established that when photographs are taken of the posterior pole, there is a greater chance of finding the optical disk at the horizontal ends, right and left of the image, being less likely to find it the horizontal center of the image, where there is probability of finding the macula Because of this, a mask image r $[m; n]$ that will weigh the result obtained in the subtraction of correlations, making its (absolute) value increase towards the horizontal ends and decrease toward the center, The best chance of finding a right eye in the position $[m; n]$ is given by Eq. (14).

$$
\left[m^{d}, n^{d}\right] = arg \underset{[m,n]}{\max} r[m,n] * C_{rrk}; \quad k = 1, 2, 3 \tag{14}
$$

The highest probability of finding a left eye in the position $[m; n]$ is given by Eq. (15).

$$
\left[m^{i}, n^{i}\right] = arg \underset{[m,n]}{\max} r[m,n] * C_{rrk}; \quad k = 1, 2, 3 \tag{15}
$$

And finally the choice of search between a right eye and a left eye will be given by

$$
[x_{\text{od}}, y_{\text{od}}] = \left\{ \begin{array}{c} \left[m^{d}, n^{d}\right], si\left|C_{rrk}\left[m^{d}, n^{d}\right]\right| \geq \left|C_{rrk}\left[m^{i}, n^{i}\right]\right| \\ \left[m^{i}, n^{i}\right], i\text{n another also} \end{array} \right\} \tag{16}
$$

As a final step, the optical disk area was adjusted using a basic shift algorithm of stockings. Iteratively, following the procedure of sliding of stockings, the position of the optical disk was adjusted until close of the real center.

3 Result Analysis

The images obtained in [28] were analyzed as (Fig. 1; Table 1):

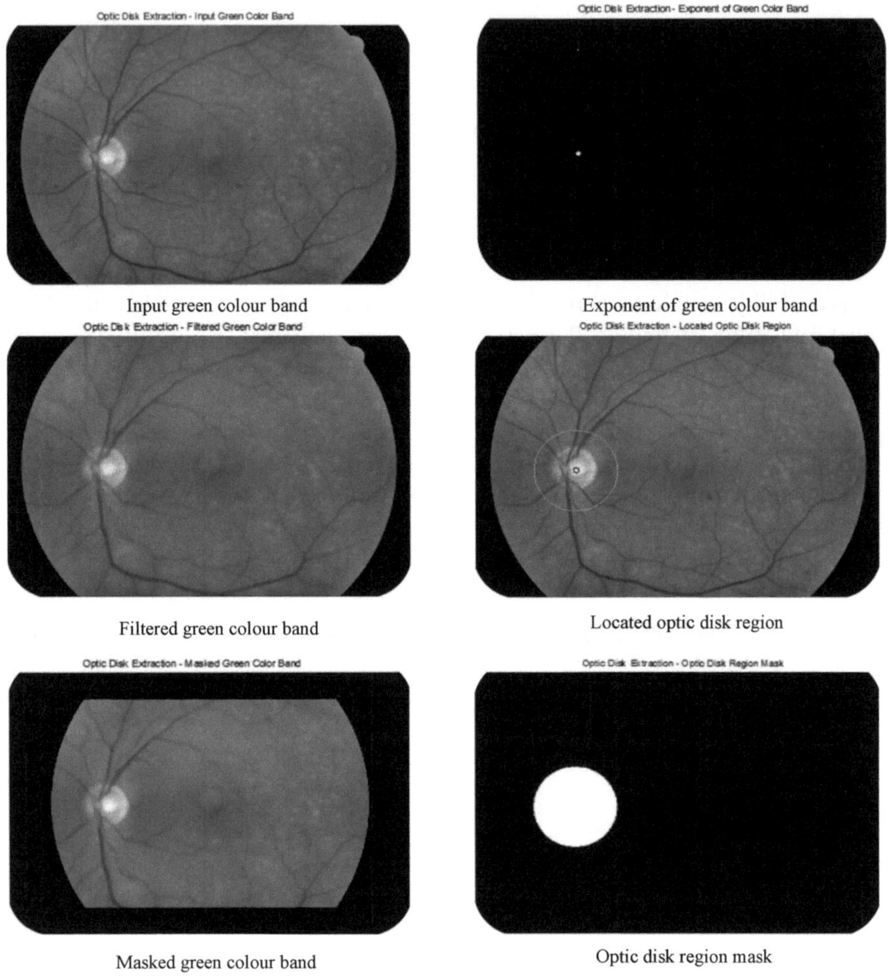

Fig. 1 Optic disk region mask

Table 1 Result analysis

Image database	Precision rate (%)
HRF database	98.5
DRIVE	96.9

4 Conclusion

In this paper, the lighting correction algorithm and optic disk detection algorithm were proposed using segmentation having template based approach. It is necessary to note that the basic idea of the search process is to correlate the pattern image with the classification image, making use of the Fast Fourier transform as an algorithmic medium. The observed results claims 96.9% accuracy in DRIVE database when compared with ground truth values.

References

1. D.J. Hwang, K.M. Lee, M.S. Park, S. Hee Choi, J.I. Park, J.H. Cho, K.H. Park, S.J. Woo, Association between diabetic foot ulcer and diabetic retinopathy. PLoS One **12**(4), e0175270 (2017)
2. American Diabetes Association, Diagnosis and classification of diabetes mellitus. Diabet. Care **37**(Supplement 1), S81–S90 (2014)
3. H. Pratt, F. Coenen, D.M. Broadbent, S.P. Harding, Y. Zheng, Convolutional neural networks for diabetic retinopathy. Procedia Comput. Sci. **90**, 200–205 (2016)
4. Diabetic Retinopathy Clinical Research Network, Aflibercept, bevacizumab, or ranibizumab for diabetic macular edema. N. Engl. J. Med. **2015**(372), 1193–1203 (2015)
5. J.A. Wells, A.R. Glassman, A.R. Ayala, L.M. Jampol, N.M. Bressler, S.B. Bressler, A.J. Brucker, F.L. Ferris, G.R. Hampton, C. Jhaveri, M. Melia, Aflibercept, bevacizumab, or ranibizumab for diabetic macular edema: two-year results from a comparative effectiveness randomized clinical trial. Ophthalmology **123**(6), 1351–1359 (2016)
6. H.J. Jelinek, M.J. Cree, D. Worsley, A. Luckie, P. Nixon, An automated microaneurysm detector as a tool for identification of diabetic retinopathy in rural optometric practice. Clin. Exp. Optom. **89**(5), 299–305 (2006)
7. S. Guigui, T. Lifshitz, J. Levy, Diabetic retinopathy in Africa: Advantages of screening. Postgrad. Med. **123**(4), 119–125 (2011)
8. M. García, C.I. Sánchez, J. Poza, M.I. López, R. Hornero, Detection of hard exudates in retinal images using a radial basis function classifier. Ann. Biomed. Eng. **37**(7), 1448–1463 (2009)
9. U. Vovk, F. Pernus, B. Likar, A review of methods for correction of intensity inhomogeneity in MRI. IEEE Trans. Med. Imaging **26**(3), 405–421 (2007)
10. L. Kubecka, J. Jan, R. Kolar, Retrospective illumination correction of retinal images. J. Biomed. Imaging **2010**, 11 (2010)
11. J.S. Suri, D. Wilson, S. Laxminarayan (eds.), *Handbook of Biomedical Image Analysis*, vol. 2 (Springer Science & Business Media, Berlin, 2005)
12. A.R. Doherty, N. Caprani, C.Ó. Conaire, V. Kalnikaite, C. Gurrin, A.F. Smeaton, N.E. O'Connor, Passively recognising human activities through lifelogging. Comput. Hum. Behav. **27**(5), 1948–1958 (2011)
13. T. Tasdizen, E. Jurrus, R.T. Whitaker, Non-uniform illumination correction in transmission electron microscopy. MICCAI Workshop on Microscopic Image Analysis with Applications in Biology (2008), pp. 5–6
14. G.D. Joshi, J. Sivaswamy, in Colour Retinal Image Enhancement Based on Domain Knowledge. IEEE Sixth Indian Conference on Computer Vision, Graphics & Image Processing, ICVGIP'08 (2008, December), pp. 591–598
15. M. Foracchia, E. Grisan, A. Ruggeri, Luminosity and contrast normalization in retinal images. Med. Image Anal. **9**(3), 179–190 (2005)

16. R.S. Maher, S.N. Kayte, S.T. Meldhe, M. Dhopeshwarkar, Automated diagnosis non-proliferative diabetic retinopathy in fundus images using support vector machine. Int. J. Comput. Appl. **125**(15) (2015)
17. J.F. Korobelnik, D.V. Do, U. Schmidt-Erfurth, D.S. Boyer, F.G. Holz, J.S. Heier, E. Midena, P.K. Kaiser, H. Terasaki, D.M. Marcus, Q.D. Nguyen, Intravitreal aflibercept for diabetic macular edema. Ophthalmology **121**(11), 2247–2254 (2014)
18. A. Pinz, S. Bernogger, P. Datlinger, A. Kruger, Mapping the human retina. IEEE Trans. Med. Imaging **17**(4), 606–619 (1998)
19. K.S. Argade, K.A. Deshmukh, M.M. Narkhede, N.N. Sonawane, S. Jore, in *Automatic Detection of Diabetic Retinopathy Using Image Processing and Data Mining Techniques*. IEEE 2015 International Conference on Green Computing and Internet of Things (ICGCIoT) (2015, October), pp. 517–521
20. T. Walter, J.C. Klein, Segmentation of color fundus images of the human retina: Detection of the optic disc and the vascular tree using morphological techniques. Med. Data Anal. 282–287 (2001)
21. S.D. Walter, Properties of the summary receiver operating characteristic (SROC) curve for diagnostic test data. Stat. Med. **21**(9), 1237–1256 (2002)
22. M. Niemeijer, M.D. Abràmoff, B. Van Ginneken, Fast detection of the optic disc and fovea in color fundus photographs. Med. Image Anal. **13**(6), 859–870 (2009)
23. H. Mir, H. Al-Nashash, U.R. Acharya, in *Assessment of Retinopathy Severity Using Digital Fundus Images*. IEEE 2011 1st Middle East Conference on Biomedical Engineering (MECBME) (2011), pp. 200–203
24. C.I. Sánchez, M. García, A. Mayo, M.I. López, R. Hornero, Retinal image analysis based on mixture models to detect hard exudates. Med. Image Anal. **13**(4), 650–658 (2009)
25. A. Osareh, B. Shadgar, R. Markham, A computational-intelligence-based approach for detection of exudates in diabetic retinopathy images. IEEE Trans. Inf Technol. Biomed. **13**(4), 535–545 (2009)
26. J. Eswaran, M. Soundararajan, R. Kumar, S. Knapp, UnPAKing the class differences among p21-activated kinases. Trends Biochem. Sci. **33**(8), 394–403 (2008)
27. M.E. Martinez-Perez, A.D. Hughes, S.A. Thom, A.A. Bharath, K.H. Parker, Segmentation of blood vessels from red-free and fluorescein retinal images. Med. Image Anal. **11**(1), 47–61 (2007)
28. https://www5.cs.fau.de/research/data/fundus-images/

Recurrence Plot Features of RR-Interval Signal for Early Stage Mortality Identification in Sudden Cardiac Death Patients

Reeta Devi, Hitender Kumar Tyagi and Dinesh Kumar

Abstract Recurrence is a basic property of scattering dynamical systems. Recurrence plots visualize the recurrent behavior of these systems. The cardiovascular system in sudden cardiac death (SCD) patients appeared to be such dynamical system as the patient meets sudden death within few minutes or an hour after onset of the symptoms. Therefore, the present study was designed to explore an algorithm to predict the SCD 1 h before the "SCD Onset" using recurrence plots of the RR-intervals. The MIT-BIH database was used for designing the samples of the study. The 5-min ECG signals of the SCD patients which were 1 h before the SCD onset and just 5 min before the SCD onset were preprocessed for noise removal and RR-interval extraction. The RR-intervals were then processed to obtain a set of nonlinear features using recurrence plots. These features were checked for correlation among the SCD patients using Spearman's function of correlation. The classification of the optimal features into normal control subjects and SCD was implemented using classifier k-Nearest Neighbor or k-NN classifier. A maximum accuracy of 98.44% was obtained using k-NN. The results indicate that, the proposed algorithm can efficiently identify the person at risk of progressive SCD 1 h before, which would be helpful in supporting sufficient time for the medical staff attending the patient to respond with preventive treatment.

Keywords RR-interval · Recurrence plots · Spearman's correlation k-NN · SCD

R. Devi (✉)
Department of Electronics and Communication Engineering, University Institute of Engineering and Technology, Kurukshetra University, Haryana, India
e-mail: reetakuk@gmail.com

H. K. Tyagi
Department of Electronics, University College, Kurukshetra University, Haryana, India
e-mail: hitender.tyagi@gmail.com

D. Kumar
YMCA University of Science and Technology, Faridabad, Haryana, India
e-mail: dineshelectronics@gmail.com

C. R. Krishna et al. (eds.), *Proceedings of 2nd International Conference on Communication, Computing and Networking*, Lecture Notes in Networks and Systems 46, https://doi.org/10.1007/978-981-13-1217-5_40

1 Introduction

The cardiovascular diseases which are non-communicable diseases are a leading cause of death worldwide. Sudden cardiac death among the cardiovascular diseases particularly put a large burden because it cuts short the potential years of contribution of the affected persons to their family and the society. It is defined by W.H.O criteria as the natural sudden death due to cardiac causes occurring within 24 h. But out of witnessed deaths nearly 80% occurred within 2 h of the symptoms onset [1]. In the context of time, sudden is defined for most clinical purposes as one hour between a change in clinical status announcing the onset of the life threatening cardiac arrest [2]. Despite significant advances in the medical science, there is no standard procedure adopted to identify the sudden cardiac death due to this type of cardiac arrest at an early stage. Therefore, identification of the patients at risk of sudden cardiac death is a great challenge for the medical fraternity. Many scientists and engineering research groups are trying to validate different signal processing techniques to solve the problem. Patil et al. and Jilani et al. [3, 4] developed methods considering age, sex, race, history of smoking and illness and so on for modeling of an SCD pattern so prediction of SCD can be related to that particular pattern in future. On the other hand, Lammert et al. [5] developed methods based on feature extraction techniques from a patient's ECG signal. These features were used as electrophysiological marker of SCD. Apart from that, statistical analysis based SCD identification methods also were proposed by some research groups [6–9]. In spite of producing good prediction results, all these methods could not be made into healthcare products due to their own limitations. In this paper, we focus on another type of measure of complexity, which is based on the method of recurrence plots. This nonlinear approach along with quantitative features has been introduced for the analysis of chaos–chaos transition in the dynamical systems [10]. Therefore, in this paper we introduced recurrence rate, maximum length of diagonal line, entropy, trend, laminarity and trapping time as the six measures of complexity based on recurrence plots for identification of sudden cardiac death at an early stage of 1 h before "SCD Onset". We used the Spearman's correlation to find out correlation among the features of the recurrence plots in SCD patients. We found that four features out of six are significantly correlated to each other in the SCD patients at 1 h before "SCD Onset" and 5 min before "SCD Onset". These features are then subjected to k-NN classifier for classification into normal or the SCD subjects. K-fold cross validation is used to validate the training (70%) and testing samples (30%). An accuracy of 98.44% is obtained as the classification rate of k-NN classifier for SCD patients.

2 Materials and Methods

2.1 *Dataset*

ECG signals were downloaded from MIT/BIH database for normal control subjects (Normal Sinus Rhythm Database) and sudden cardiac death patients (Sudden Cardiac Death Holter Database) [11]. In this work, total 55 signals consisting of 40 SCD ECG signals and 15 normal ECG signals of lead MLII were studied. The samples of SCD patients were divided into two groups of 20 samples each. The first group consisted of 5 min ECG signals for each patient at 1 h before SCD. The second group consisted of last 5 min ECG signals to the SCD. The ECG signals of the SCD patients who were continuously paced for their heart beat were not included in the sample design process of the study.

2.2 *Preprocessing and RR-Interval Extraction*

The sampling frequency used for the ECG signals under study was 125 Hz for SCD patients and 128 Hz for normal subjects. In this study, 5 min duration sampled ECG signals were used. An illustration of the dataset taken under consideration for SCD patients which were one hour before and 5 min before the SCD incidence is shown in Fig. 1. Then the signals were subjected to IIR high pass filter with cut off frequency at 1 Hz and IIR notch filter for the removal of baseline wandering noise and power line interference, respectively. The QRS peaks from the filtered ECG signals were detected using standard Pan-Tompkins algorithm [12]. This way preprocessed ECG signals were used to extract the RR-intervals. Figure 2 shows plot of RR-intervals as a heart rate variability signal of an SCD patient. The processing of RR-intervals for obtaining the nonlinear features of recurrence plots [13, 14] was performed using computational software MATLAB 8.1.

2.3 *Recurrence Plots (RP)*

Recurrence plots are used for the applications of nonlinear data analysis. The phase space of a signal using time delay method [15] is reconstructed first to analyze the signal using RP theory. In this method, an m dimensional space time series vector x_i where $i = 1, 2, 3, \ldots, N$ is embedded using time delay factor (τ)

$$A_i = \left[x_i x_{i+\tau} \ldots x_{i+(m-1)\tau}\right]; \quad i = 1, 2, 3, \ldots, N - (m-1)\tau \tag{1}$$

The false nearest-neighbors algorithm [15] for the estimation of the smallest sufficient embedding dimension as well as the mutual information function [16]

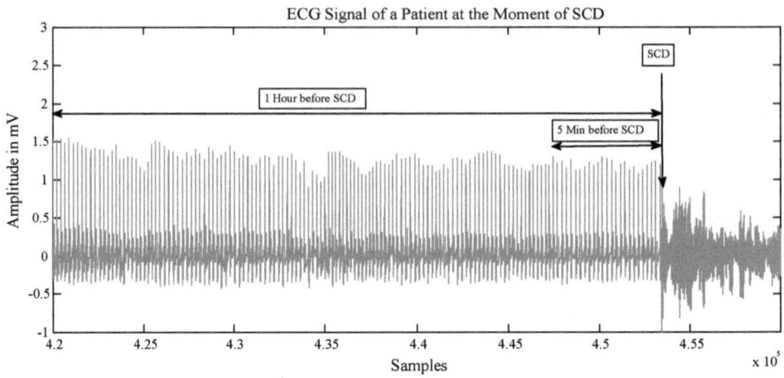

Fig. 1 ECG signal of a patient at the moment of SCD and few seconds afterwards

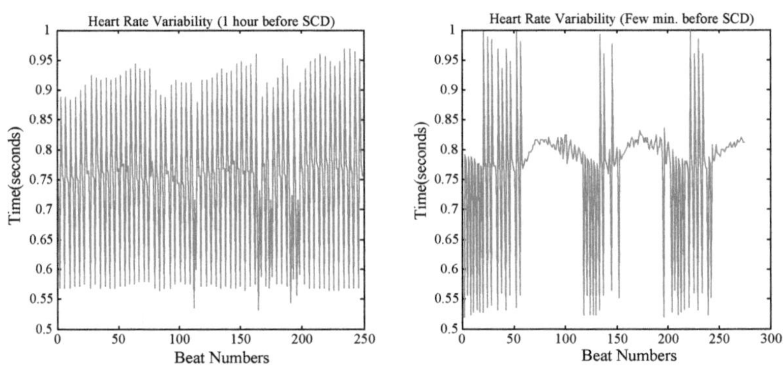

Fig. 2 Plot of RR-intervals (heart rate variability) of an SCD patient

for an appropriate time delay have been used in this work. After reconstructing the state space, the recurrence plot of a signal can be obtained by plotting the elements of an $M \times M$ recurrence matrix, given in the following equation:

$$R_{i,j} = \Theta\big(\varepsilon - A_i - A_j\big); \quad i, j = 1, 2, \ldots, M \tag{2}$$

where

$$M = N - (m - 1)\tau,$$

$\varepsilon =$ a threshold distance,
$\|.\| =$ the Euclidean norm, and
$\Theta(x)$ the Heaviside function

2.4 Features Extracted from RP

In this work, six features from the recurrence plot of each 5 min ECG signal were extracted. The related extracted features [13] are as follows.

Recurrence rate (REC): It measures the percentage of recurrent points in the M dimension recurrence matrix given as

$$\text{REC} = \frac{1}{M^2} \sum_{i,j=1}^{M} R_{i,j} \tag{3}$$

Lmax: Excluding the length of the main diagonal line, it measures the longest length of the diagonal line.

ENTR: The diagonal line segments were quantified using Shannon entropy of their probability distribution function $P(L)$

$$\text{ENTR} = -\sum_{L=L_{\min}}^{M} P(L)\ln P(L) \tag{4}$$

Trend: The linear regression coefficient is measured using the quantitative measure "trend" over the recurrence point density

$$\text{TREND} = \frac{\sum_{i,j=1}^{M}\left(1-\frac{M}{2}\right)\left(RR_{i,j}-\langle RR_{i,j}\rangle\right)}{\sum_{i,j=1}^{M}(i,j-M/2)^2} \tag{5}$$

Laminarity (LAM): Excluding diagonal line segments, it quantifies the percentage of recurrent points forming vertical line segments

$$\text{LAM} = \frac{\sum_{l_v=l_{v\min}}^{M} l_v p(l_v)}{\sum_{i,j}^{M} R_{i,j}}, \tag{6}$$

where $p(l_v)$ is the number of recurrent points forming vertical line segments.

Trapping time (TT): The average length of vertical structures $(p(l))$ is measured by trapping time

$$\text{TT} = \frac{\sum_{l=l_{\min}}^{M} l p(l)}{\sum_{l=l_{\min}}^{M} p(l)} \tag{7}$$

Table 1 Statistical significance of the recurrence plots features using Spearman's correlation

Recurrence plots quantification features		Spearman's correlation	
Sr. No.	Feature	RHO	*p*-value
1	**Recurrence rate**	**0.2476**	**0.2925**
2	LMAX	0.4726	0.0354
3	ENT	0.5092	0.0219
4	TND	0.4569	0.0428
5	**LAM**	**0.4111**	**0.0718**
6	TT	0.4714	0.0359

3 Results

The present work aims to investigate the nonlinear features of recurrence plots of 5 min duration SCD ECG signal which is 1 h earlier to the SCD of the patient. In total, there were 6 features for each of 55 subjects (15 normal subjects, 20- patients of 5 min before SCD and 20- patients of 1 h before SCD). Out of total 6 features per subject, the best and optimal features were selected using the forward sequential feature selection algorithm. The simple machine learning algorithm k-NN classifier (with 70% samples for training and 30% samples for testing) was used to classify the SCD patients and normal control subjects. The random selection of the training and testing samples was done using k-fold cross validation. The extracted features were validated statistically using Spearman's function for correlation with a maximum significance level of 0.05 for its significance on SCD classification. Results of correlation between two groups of SCD patients are listed in Table 1.

Except two (shown in bold fonts), out of total six features, all the basic set of features in nonlinear processing of RR-intervals using recurrence plots, showed significant low *p*-values between SCD patients, thereby, accepting the null hypothesis of two samples being correlated with some form of correlation. The results ensured that SCD patients can be identified at high risk of SCD at the stage of one hour earlier to their SCD as they hold correlation with the SCD patients of 5 min earlier to the SCD. In predicting the SCD one hour before to the actual incident the calculated classifier's performance measures are reported in Table 2. The measured values for accuracy, sensitivity, specificity and precision are 98.44, 100, 97.91 and 95% respectively for k-NN classifier. A comparison of the results obtained in present study with other similar studies is presented in Table 3. As evident from Table 3, compared to other published results for the prediction of sudden cardiac death at an early stage, the results obtained in present study are more useful. For prediction of SCD, the database used, features domains and number of features used are almost comparable in all the studies presented here. However, in this work, the authors tried to explore the possibility of predicting the SCD much earlier of 1 h before SCD than 1–5 min of the actual incidence onset as shown in Table 3. A much early prediction period such as one hour is necessary to alarm the clinicians attending the patient, so that

Table 2 Classifier performance measures

Classifier	AC (%)	SN (%)	SP (%)	P (%)
k-NN	98.44	100	97.91	95

Table 3 Comparison of present study with other similar research studies

Reference	Database	Prediction duration before SCD	Features extracted from	No. of Features used in identification of SCD	Accuracy (%)
Ebrahimzadeh et al. [6]	MIT/BIH	1 min	HRV	20	91.23
Murukesan et al. [9]	MIT/BIH	2 min	HRV	7	96.36
Murukesan et al. [8]	MIT/BIH	5 min	HRV	7	93.71
Fujita et al. [7]	MIT/BIH	4 min	HRV	4	94.7
Present study	MIT/BIH	1 h	Recurrence plots	6	98.44

preventive treatment could be initiated at an early stage which would effectively save the precious life.

4 Discussion

The present study performed for prediction of sudden cardiac death one hour before to the actual incidence of SCD is very useful to warn the concerned clinicians attending the patient. Compared to other similar studies, therefore, the present technique would provide sufficient precious time for preventive treatment action to be taken by the concerned medical staff. The designed methodology for nonlinear processing of RR-intervals effectively removed the noises and other interferences from the ECG signals downloaded from MIT/BIH database. In total six nonlinear features of recurrence plots were extracted. Using Spearman's correlation and forward sequential search machine learning algorithm for feature selection, features of non-relevance were dropped. The reduced dimension dataset was then applied to k-NN classifier for classification into normal control subjects or SCD. In predicting the SCD cases at the stage of one hour before the actual incidence, the maximum values of the classifiers' performance measures obtained using k-NN are as 98.44% accuracy, 100% sensitivity, 97.91% specificity and 95% precision values. Among different types of features, four out of six features were found highly correlated ($p < 0.05$) between two groups of the SCD patients which implies that features extracted from recurrence plots can predict SCD at an early stage of one hour before it actually occurs. For clinical relevance of the study, ECG signals selected for simulation, materials and methodology adopted and extraction of different features of RR-intervals using recurrence plots are in confirmation to the published guidelines. However, generalization of the presented algorithm requires more number of datasets of diverse SCD patients to be tested. Future studies of the present work would thereby, focus on modeling the type of correlation between two groups of the SCD patients.

References

1. E. Ladich, R. Virmani, A. Burke, Sudden cardiac death not related to coronary atherosclerosis. Toxicol. Pathol. **34**, 52 (2006)
2. S.S. Chug, Early identification of risk factors for sudden cardiac death. Nat. Rev. Cardiol. **7**, 318–326 (2010)
3. S. Patil et al., Intelligent and effective heart attack prediction system using data mining and artificial neural network. Eur. J. Sci. Res. **31**(4), 642–656 (2009)
4. T. Jilani et al., Acute coronary syndrome prediction using data mining techniques-an application. Int. J. Inf. Math. Sci. **5**(4), 295–299 (2009)
5. M.E. Lammert et al., Electrocardiographic predictors of out-of hospital sudden cardiac arrest in patients with coronary artery disease. Am. J. Cardiol. **109**(9), 1278–1282 (2012)
6. E. Ebrahimzadeh, M. Pooyan, Early detection of sudden cardiac death by using classical linear techniques and time-frequency methods on electrocardiogram signals. J. Biomed. Sci. Eng. **4**, 699–706 (2011)
7. H. Fujita, U.R. Acharya, Vidya K. Sudarshan, D.N. Ghista, S.V. Sree, W.J.E. Lim, J.E.W. Koh, Sudden cardiac death (SCD) prediction based on non-linear heart rate variability features and SCD index. Appl. Soft Comput. **43**, 510–519 (2016)
8. L. Murukesan, M. Murugappan, I. Omar, S. Khatun, S. Murugappan, Time domain features based sudden cardiac arrest prediction using machine learning algorithms. J. Med. Imaging Health Inf. **5**, 1267–1271 (2015)
9. L. Murukesan, M. Murugappan, M. Iqbal, M. Saravanan, Machine learning approach for sudden cardiac arrest prediction based on optimal heart rate variability features. J. Med. Imaging Health Inf. **4**, 1–12 (2014)
10. F. Taken, in *Detecting Strange Attractors in Turbulence Dynamical Systems and Turbulence*. Lecture Notes in Mathematics, vol. 898 (Springer, Berlin, 1981), pp. 366–381
11. A.L. Goldberger, L.A.N. Amaral, L. Glass, J.M. Hausdorff, P.C. Ivanov, R.G. Mark, J.E. Mietus, G.B. Moody, C.K. Peng, H.E. Stanley, PhysioBank, PhysioToolkit and PhysioNet: Components of a new research resource for complex physiologic signals. Circulation **101**, e215–e220 (2000)
12. J. Pan, W. Tompkins, A real time QRS detection algorithm. IEEE Trans. Biomed. Eng. **32**, 230–236 (1985)
13. H. Yang, Multiscale recurrence quantification analysis of spatial vectorcardiogram (VCG) signals. IEEE Trans. Biomed. Eng. **58**(2), 339–347 (2011)
14. Y. Chen, H. Yang, Multiscale recurrence analysis of long-term nonlinear and nonstationary time series. Chaos Solitons Fractals **45**(7), 978–987 (2012)
15. M.B. Kennel, R. Brown, H.D. Abarbanel, Determining embedding dimension for phase-space reconstruction using a geometrical construction. Phys. Rev. A **45**, 3403–3411 (1992)
16. A.M. Fraser, H.L. Swinney, Independent coordinates for strange attractors from mutual information. Phys. Rev. A **33**, 1134–1140 (1986)

Festival Framework for Synthesis of Punjabi Voice

Sukhpreet Kaur Gill and Parminder Singh

Abstract Speech generation is process of converting the text to speech called TTS system. This paper presents the generation of Punjabi waveform by using Festival framework. Building of new voice is done through support of Festvox, Festival and Speech Tools. Festival acts as an Engine for waveform synthesis and statistical parametric synthesis method is implemented for building Punjabi voice. The required data for building the voice is recorded in the noiseless environment and maximum Punjabi valid phonemes are covered in the corpus. Text Processing is done to collect the nice prompts to build the accurate voice Model. The accuracy factor is calculated through Mel-cepstral distortion (MCD) parameters.

Keywords CLUSTERGEN synthesizer · Festival · Speech synthesis · Festvox

1 Introduction

In modern era of technological world, speech synthesis is one of the major research areas, as speech has an indispensable role in the human life. Its major applications are in reading electronic mails, display reading by blind persons and to train robots. A synthesized voice is produced from the small elements and it is further trained by passing through different modules.

Speech synthesis is process in which sequence of input text is converted into synthesized waveform as output. In synthesis system when text is converted into speech that is known as Text-to-Speech (TTS) system. There is various Speech Synthesis system used to build the new voice by researchers. Festival Framework has multilingual Support [1]. Festival is implemented in two languages, C++ and

S. K. Gill (✉)
GNA University, Phagwara, India
e-mail: gillsukhi92@gmail.com

P. Singh
Guru Nanak Dev Engineering College, Ludhiana, India
e-mail: parminder2u@gmail.com

C. R. Krishna et al. (eds.), *Proceedings of 2nd International Conference on Communication, Computing and Networking*, Lecture Notes in Networks and Systems 46, https://doi.org/10.1007/978-981-13-1217-5_41

scheme [2]. Festival Framework offers different sort of methods for synthetic speech generation like unit selection approach, diphone based and Statistical Parametric Synthesis approach. For different Indian Languages, the voice models are developed using various techniques and quality of such synthesized speech is estimated by doing its similarity with natural tested speech unit. Statistical Parametric synthesized method is used named as CLUSTERGEN method.

This paper focuses on developing Punjabi voice using Festival framework. From Brahmi script, Indian languages script has been developed [3]. The Punjabi is language of north India. It is mother tongue of Punjab. In this paper Phoneme is chosen as basic unit for speech synthesis. For Punjabi language, there is need for lexicon list along with phoneme symbol on the basis of which we can build the language specific rules. Festival offer support for various waveform synthesis Technique: Unit selection database, diphone Database and Statistical Parametric synthesis. The paper use the waveform synthesis technique that is statistical parametric speech synthesis technique also called CLUSTERGEN voice model generation. This is easy method to build good quality of synthesized output voice or waveform as compare to unit selection approach. This method uses the Hidden Markov Model synthesis Technique [4]. In this frequency parameters (F0 contour) and Mel-cepstral (MCEP) are extracted. These parameters are extracted from the information collected at the time of labelling module, such extracted parameters from labelled data acts as base for synthesized voice. So, to get the accurate label values when labelled movement is done through 10 passes to achieve the good quality of speech. Then at synthesis time these parameters are selected to generate the voice model. This method used the maximum likelihood criteria to select the parameters to build the model. At time of waveform synthesis generation for inputted sequence of text, the features for sequence is selected from already build model on the basis of extracted parameters from corpus, these are used to reconstruct the waveform for inputted string of text.

2 Related Works

Festival also provides support for Indic languages. Festival is an open source. Festival uses three basic modules as text processing, prosodic processing and waveform synthesis [5]. Currently, various researchers applied the different waveform synthesis technique and different basic units (phoneme, syllable etc.) were experimented and quality has been compared. Mostly, Unit selection and diphones was used to produce speech waveform by various researchers. In open-Domain unit selection speech synthesis minimum user experience and knowledge of speech synthesis is required [6]. In unit selection waveform synthesis different basic units (syllable, phonemes etc.) were tested by researchers. As text to speech synthesis in Bengali has been generated using syllable as basic unit [7]. Whereas, for Malayalam language generation the phoneme as basic unit have been used [8]. Statistical Parametric synthesis is one of the methods which is easiest method to reconstruct waveform from extracted parameters. In Unit selection quality is depend upon quality and database size. Also, the cost for building voice using this technique is expensive and specialization is required [9]. Statistical Parametric synthesis CLUSTERGEN voice is developed. To run CLUSTERGEN voices the new version of Festival and Festvox was updated. CLUSTERGEN voices are released in new version beta 1.96 of Festival. In hybrid Technique the unit selection Technique is applied on parameters extracted by the CLUSTERGEN method [10]. Researchers concluded that Hybrid technique is better than Clunits and Statistical parametric Method. Currently, Festival has only support for Hindi, Gujarati, Bengali and Rajasthani [14]. But no work is done for Punjabi language in Festival Framework. Concatenative method is used to developed Text to speech synthesis system and syllables act as basic unit for speech generation [11].

In this paper the voice model is developed using CLUSTERGEN Technique for Punjabi language. According to the analysis of Punjabi phonemes, phonemes are of three types: V (total valid phonemes are 20), CV (Non-Nasalized valid phonemes are 313) and CV (Nasalized valid phonemes are 324) [12, 13].

3 Proposed Work

3.1 Punjabi TTS Synthesis System

Steps and methods required for building TTS system is difficult Task. For waveform generation various modules processing is performed. To build a voice five main steps are required: defining phoneme lay out, lexicon creation, analysis of training text, voice model development, create synthesizer (waveform) [14]. For Punjabi language Grapheme based synthesis method is easy method for developing good quality of voice [15]. It uses the CLUSTERGEN voice generation method. Fewer resources are available for building new Indic voice; Grapheme to phoneme synthesizer is best method for such languages. The various modules of processing are shown in Fig. 1 for Festival framework.

3.2 Training Corpus Development

For the complete coverage of Punjabi phoneme, the corpus of around 500 sentences from newspaper is collected and moulded them in required festival format as shown in Fig. 2. The words in sentences are selected in such a way that there should be accurate prompts which contain the nice prompts to build the voice model.

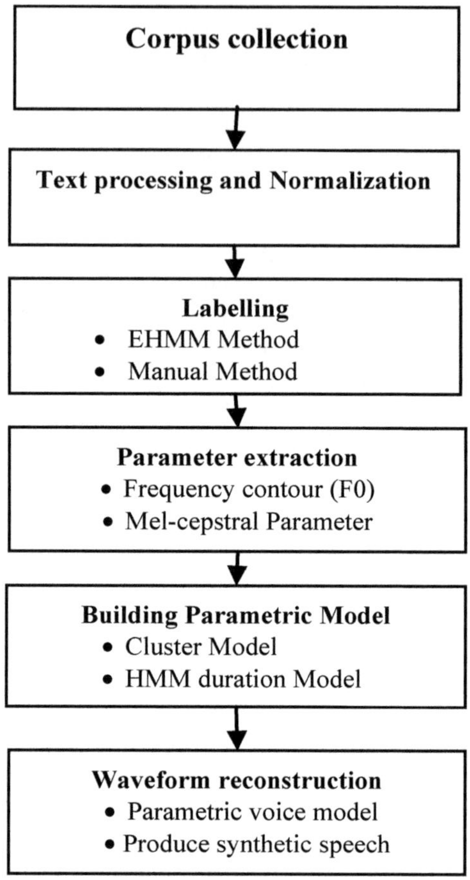

Fig. 1 Punjabi speech TTS synthesis system

```
( punjabi_a001 "ਮੁਖਤਾਰ ਨਕਵੀ ਮਹਾਨ ਸੁਪਨਸਾਜ਼ ਗੁਰਬਖਸ਼ ਚਾਤਰਿਕ ਰਾਹੀਂ ਲੌਪੋਕੇ ਦੌਲਤ ਥਿਘੇ ਬੰਜਰ ਜ਼ਮੀਨ ਗਜ਼ ਦੇ ਹਿਸਾਬ" )
( punjabi_a002 "ਫਸੀਲ ਖੂਹ ਬਾਗ਼ ਨੇਤਾ ਕੇਬੀ ਧੀਰ ਹੈਪੀ ਪਿਤਾ ਵੀ ਸਨ" )
( punjabi_a003 "ਅਪਣਾਉਣ ਸਬੰਧੀ ਲੋੜੀਂਦੀ ਵਿਦਿਆ ਹਾਸਲ ਸਕਦੇ " )
(punjabi_a004 "ਦੇਖਣ ਵਿਚ ਤੇ ਲਾਗਲਾ ਖੇਤਰ ਕਰਕੇ ਤਾਰਾ ਆਰਟ ਕਰਾਫ ਘਰ ਬਹੁਤ ਹੀ ਆਦਿ ਸਥਾਨ ਹੈ" )
```

Fig. 2 Festival prompt format

The corpus should be accurate and recording should be done in noise free environment. Proper pruning should be done to do the accurate labelling. The corpus development and normalization is main base for building the new voice, because on the basis of this proper labelling can be done. Information of labelled data is used to synthesis the speech of good quality. So, there should not be poor alignment between the transcript and the acoustic, there should not be too much silence in the prompts and silence should not be longer than 200 ms.

Table 1 Mapping of independent vowels

Character or ordinal	Unicode character	SAMPA phones
ਅ	2565	A
ਆ	2566	A:
ਇ	2567	I
ਈ	2568	i:
ਉ	2569	U
ਊ	2570	u:
ਏ	2575	E
ਐ	2576	Ai
ਔ	2579	O
ਓ	2580	Au

3.3 Adding Language Specific Rules

For the new language development, we required the knowledge of Language exact rules. To complete the lexicon list entries, we required the complete knowledge of ordinals or characters used in the Punjabi language and Unicode character for respective ordinal. In this we perform the mapping, firstly mapping of Unicode character to the corresponding ordinal, secondly mapping of Unicode character to the corresponding phoneme means we required to map the ordinals generated earlier to SAPMA phones. Now, we required to add the Language exact rules. The characters or ordinals of Punjabi language divided into independent vowels, vowels, consonants etc. Table 1 represents the Independent vowels used during implementation.

3.4 Labelling

Labelling is the most important part for the synthesis. It assigns the waveform to phoneme. The data building is done before labelling for prompts and the alignment is performed. We represent the chunk of text and relationship which is basic object for synthesis and assigns the symbols to those chunks. The valid and needed information regarding these chunks are collected in different files and going to act as milestone for further modules of synthesis process. Manual labelling is best method for accurate labelling of phones. The information extracted at labelled time is used at the time of voice model building. Wavesurfer tool is used for manual labelling of SAMPA phonemes "v+aI+nB+A+k+u:+v+A+9r" in word "ਵੈਨਕੂਵਰ" is shown in the Fig. 3.

The phoneme is selected as main basic component for synthesis for avoiding large amount of data space and for the superb quality of synthesized voice statistical method is used. The data is divided into two sets: training set and testing set, the sentences under testing modules are used check the quality of synthesized data. The

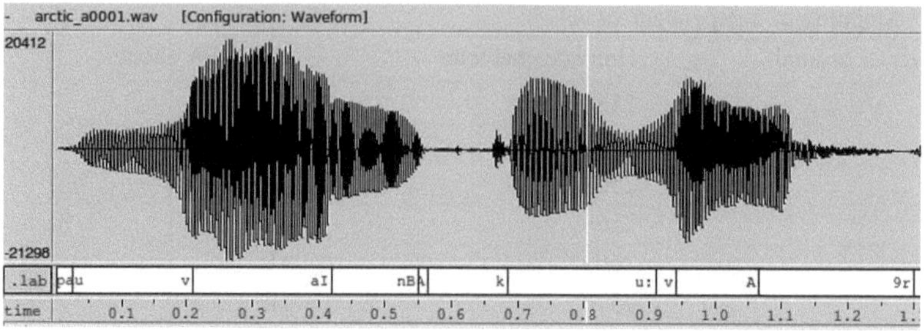

Fig. 3 Labelled position of phonemes “ਵ+ਐ +ਨ+ਅ +ਕ+ਊ +ਵ+ਅ +ਰ” in word “ਵੈਨਕੂਵਰ”

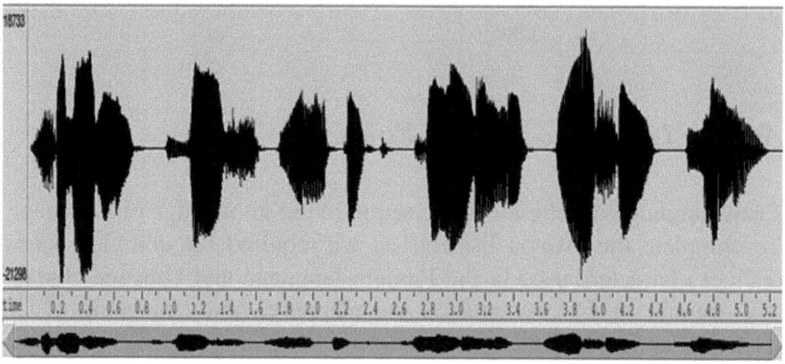

Fig. 4 Original waveform recording for sequence of text “ਸਵੇਰੇ ਦੇਸ਼ ਨੈਤਿਕ ਖੁਨਦਾਨ ਨਸ਼ੇ ਬੋਲਣ”

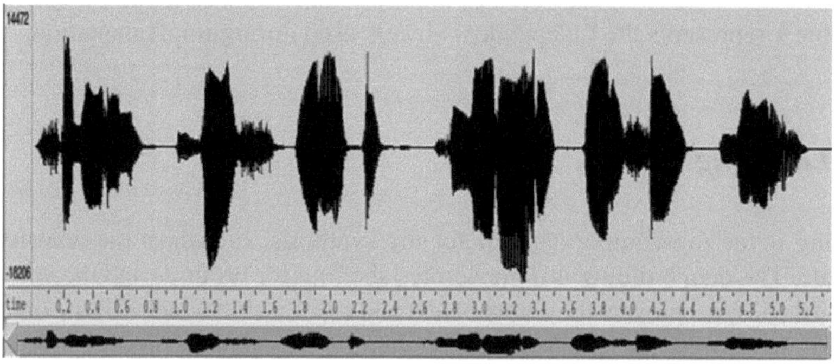

Fig. 5 Waveform synthesized by CG method for text “ਸਵੇਰੇ ਦੇਸ਼ ਨੈਤਿਕ ਖੁਨਦਾਨ ਨਸ਼ੇ ਬੋਲਣ”

one of such testing module contain prompt “ਸਵੇਰੇ ਦੇਸ਼ ਨੈਤਿਕ ਖੁਨਦਾਨ ਨਸ਼ੇ ਬੋਲਣ” and its original waveform implementation is presented in Fig. 4.

The generation of CG (CLUSTERGEN) cluster and models is done which further leads to the regenerate of this test related to their technologically advanced model, the waveform generated is presented in Fig. 5.

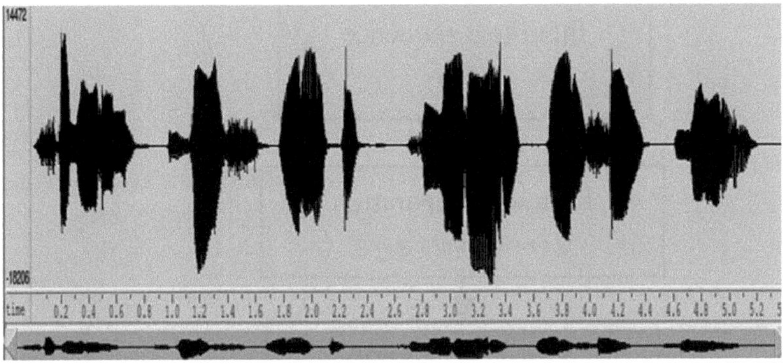

Fig. 6 Synthesized waveform generated for text "ਸਵੇਰੇ ਦੇਸ਼ ਨੈਤਿਕ ਖੂਨਦਾਨ ਨਸ਼ੇ ਬੋਲਣ" by moving labels

Label Movements are done in 10 passes for improvement in CG sounds. The test data also regenerated the test data for checking quality, one such test utterance "ਸਵੇਰੇ ਦੇਸ਼ ਨੈਤਿਕ ਖੂਨਦਾਨ ਨਸ਼ੇ ਬੋਲਣ" is re-generated and its re-generated waveform is shown in Fig. 6.

3.5 Parameter Extraction and Model Development

From labelled data features are extracted these extracted features parametric are used to build the voice model. In feature extraction, there are many features that required extracting but it depend upon the choice of basic unit and wave synthesis technique. We required the extraction of Frequency parameters (F0 contours) and Mel—cepstrum. Maximum likelihood criteria are used to select the parameters features. The maximum likelihood criteria method ignores all possible value except the most likely occur and select the parameters that could produce the best possible result in given condition. The selected parameters or values of features are used to build the voice model. By combined coefficients of both parameters frequency contours and Mel-cepstral are used to build the parameter model. Firstly, generation of cluster model is done from extracted parameters on the basis of maximum likelihood criteria. Secondly, duration Model is generated for the new voice model.

3.6 Reconstruction of Waveform

In this we reconstruct the waveform for input text sequence. It extracts waveform for input text by selecting features of corresponding phoneme from generated voice model in previous module. The Reconstruction of input text sequence "ਪਰਚੇ" is shown in Fig. 7. In this the text Sequence is converted into separated phonemes called char-

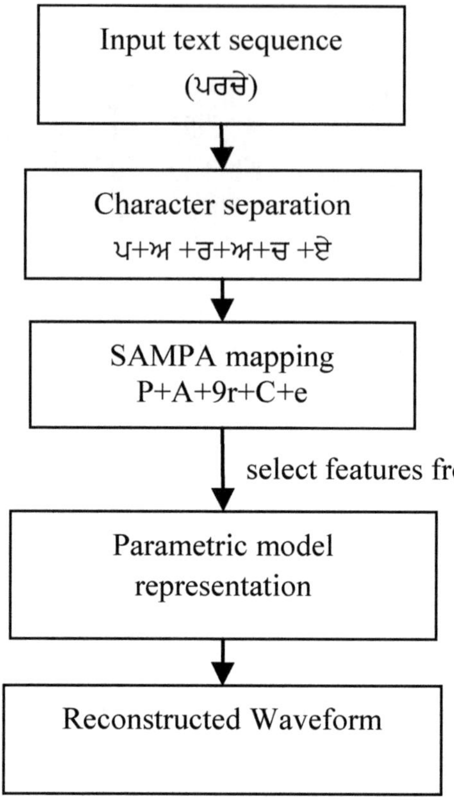

Fig. 7 Reconstruction of waveform

acter Separation as "ਪ+ਅ +ਰ+ਅ+ਚ+ਏ", then mapped to SAMPA phones, these SAMPA phones select their corresponding features parameter from already generated voice model and finally at end reconstruct the waveform for inputted text sequence. Lower twist or lower digression or twist of synthesized voice in comparison to original voice the it shows that synthesized voice is of higher quality, Fig. 6 represents that re-generated voice has less alteration in comparison to original one as in Fig. 5 one, it means that it is of superior quality. Maximum likelihood criteria is used for selecting features to build the tree in CLUSTERGEN (CG) method then with improving labels movements we can achieve the superior quality of voice [16].

4 Results

Experimentation is done to get the higher quality of synthesized speech and quality of synthesized voice is measured through Mel-cepstral distortion (MCD) scores. If

Table 2 Mel-cepstral distortion (MCD) scores values

Synthesized speech	MCD scores	MCD scores after moving labels	Gain
Punjabi language	5.39	5.13	0.37
Awk database voice	5.25	5.01	0.23

Table 3 Comparative analysis of various languages against MCD scores

Synthesized speech	MCD scores
Punjabi	5.13
Bengali	4.96
Hindi	5.24
Kannada	5.01
Tamil	5.30
Malayalam	5.1
Telugu	4.39

there is less Mel-cepstral distortion (MCD) scores then quality of speech is improved. Table 2 shows the improvement of Mel-cepstral distortion (MCD) scores by moving labels.

The Table 3 given below shows the comparison of Punjabi synthesised speech with the synthesized speech of other languages done by [3].

The value of Mel-Cepstral distortion (MCD) scores for Punjabi is 5.13 which is quite lower than Hindi, Tamil and Malayalam. But it is higher than Bengali, Telugu and Kannada. These variations in the scores are due to labelling techniques. In our experimentation we have used the mixture of automatic labelling and manual labelling. The different passes of the labels have been done according to the exactness and correctness of sound tags for the respective Punjabi utterances. The comparative analysis of the various Mel-Cepstral distortion (MCD) scores of different languages has been shown graphically in the following Fig. 8.

5 Conclusion

Development of Punjabi speech is described in this paper. The Grapheme based speech generation which uses statistical parametric approach is used which was easy to implement. The speech corpus of 500 sentences is used to cover maximum valid phonemes. The Recording in noiseless environment is done and complete analysis of Punjabi phonemes is done to build language rules. Also, at each module all issues are taken into consideration that can affect the quality of output voice and mainly focused on text processing, manual labelling which is main base to synthesized the voice. The output waveform for the given input text using generated CLUSTERGEN voice

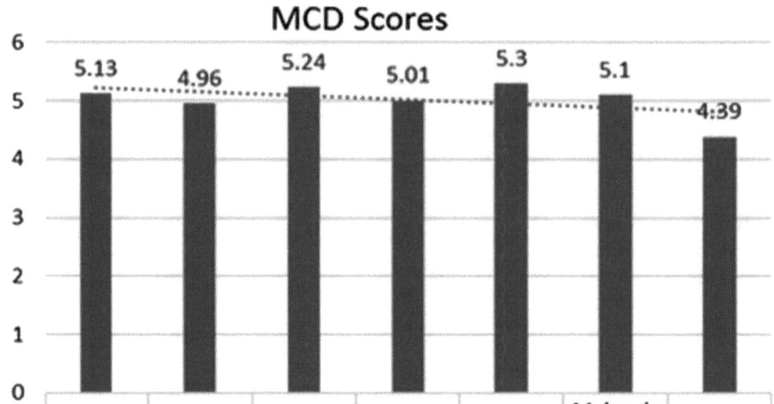

	Punjabi	Bengali	Hindi	Kannada	Tamil	Malayalam	Telugu
■ MCD Scores	5.13	4.96	5.24	5.01	5.3	5.1	4.39

Fig. 8 Graph showing the comparative analysis of Mel-cepstral distortion (MCD) scores

model is natural in sounding and very smooth. For storing voice models generated by this method takes the less memory.

References

1. B. Kumar, B. Chettri, Currents trends, frameworks and techniques used in speech synthesis-a survey. Int. J. Soft Comput. Eng. **2**, 2231–2307 (2012)
2. P. Taylor, A.W. Black, R. Caley, The architecture of the festival speech synthesis system, in Proceedings of ESCA workshop in Speech synthesis (1998), pp. 147–151
3. K. Prahallad, N. Kumar, S. Rajendran, V. Keri, The-IIT-H indic speech databases, in Proceedings of Interspeech (Portland, Oregon, USA 2012)
4. S. Roy, A technical guide to concatenative speech synthesis for hindi using festival. Int. J. Comput. Appl. **86**(8), 30–34 (2014)
5. A.W. Black, K. Lenzo, Building voices in the festival speech synthesis system (Cambridge, 1999), pp. 23–93 (for Festival version 1.3.1)
6. K. Richmond, S. King, Multisyn: open-domain unit selection for the festival speech synthesis system. Sci. Direct **49**(4), 317–330 (2007)
7. N.P. Narendra, S.K. Rao, K. Ghosh, R.R. Vempada, S. Maity, Development of syllable-based text to speech synthesis system in Bengali. Int. J. Speech Technol. **14**, 167–181 (2011)
8. B.K. Rajan, V. Rijoy, D.P. Gopinath, N. George, Duration modeling for text to speech synthesis system using festival speech engine developed for malayalam language, in Proceedings of International Conference Circuit, Power and Computing Technologies (2015), pp. 1–5
9. A.W. Black, CLUSTERGEN: a statistical parametric synthesizer using trajectory modeling, in Proceedings of Interspeech (Pittsburgh, A, USA, 2006), pp. 1762–1765
10. A.W. Black, C.L. Bennett, B.C. Blanchard, J. Kominek, B. Langner, K. Prahallad, A. Toth, CMU blizzard 2007: A hybrid acoustic unit selection system from statistically predicted parameters, in Proceedings of Blizzard challenge Workshop (Pittsburgh, PA, 2007)

11. P. Singh, L. Singh, text to speech synthesis system for Punjabi language, in Proceedings of COLING 2012 (Mumbai, India 2012)
12. S. Luthra, P. Singh, Punjabi speech generation system based on phonemes. Int. J. Comput. Appl. **49**(13), 40–44 (2012)
13. V. Goyal, G.S. Lehal, Evaluation of Hindi to Punjabi machine translation system. Int. J. Comput. Sci. Issues **4**(1), 36–39 (2009)
14. A.W. Black, K. Lenzo, Multilingual text–to-speech synthesis, in Proceedings of the ICASSP (Montreal, Canada, 2004)
15. A.W. Black, K. Lenzo, Building synthetic voices, 7th edn. (Cambridge, 2014), pp. 147–148 (for FestVox 2)
16. G.K. Sukhpreet, Generation of Punjabi speech using festival framework. M.Tech, Thesis submitted in Guru Nanak Dev Engineering College, Ludhiana, Punjab, India, 2016

Medical Fusion of CLAHE Images Using SWT and PCA for Brain Disease Analysis

Gurpreet Kaur, Sukhwinder Singh and Renu Vig

Abstract The imaging technologies have become an essential component of medical care with faster scanning time and improved spatial resolution. Image fusion is a intricate technique to coalesce vital details from images and produce a single output. To unfold the modality aspects, its strength, limitation and compare the performance of single modal and multimodal image fusion, a contrast enhanced multiresolution fusion framework techniques is proposed using stationary wavelets with restrictive coefficient spreading, in order to precisely transfer the salient features into the final image. On applying CLAHE, multiscale decomposition is applied to extort approximation and details using stationary wavelets. The detail diagonal, vertical and horizontal constituents are combined using absolute maximum selection criteria. Principal Component Analysis (PCA) combines approximation constituents of the medical images. Inverse stationary transform is applied on all the components. The two clinical brain analysis cases of degeneration and Neoplastic disease well illustrate the performance using statistical assessment evaluation techniques PSNR and MSE for comparisons. The plausible method efficiently compares the essence of single and multimodal image fusion and delivers superior result using proposed algorithms revealing imperative diagnostics details.

Keywords Single modal · Multimodal · Image fusion · Stationary wavelets CLAHE · Neoplastic disease · Degenerative · Principle components

G. Kaur (✉)
SGGS College, Chandigarh, India
e-mail: gurpretkg@gmail.com

S. Singh · R. Vig
University Institute of Engineering and Technology,
Panjab University, Chandigarh, India

C. R. Krishna et al. (eds.), *Proceedings of 2nd International Conference on Communication, Computing and Networking*, Lecture Notes in Networks and Systems 46, https://doi.org/10.1007/978-981-13-1217-5_42

1 Introduction

Medical image fusion [1] combines multimodal patient imagery into a combined fused image. It aims to cautiously select and combine extract patient anatomy, structural abnormalities, physiological functions, delineate the precise position of tissue damage or tumour. The degree of medical information can be well interpreted for detection, diagnosis or further analysis. Medical images may be selected from either single modal or multimodal inputs from pre- or post-treatment times for fusion [2]. The fusion process is not trivial and unless executed diligently the original diagnostic information is endangered. Medical images are non-invasive and possess critical information for various medical reasons [3].

This paper presents a novel method to assess single modal and multi-modal medical image fusion technique. The Contrast Limited Adaptive Histogram Equalization (CLAHE) method is applied for image enhancement to augment the structural features. The resultant images are decomposed using stationary wavelets. As levels of decomposition increase, it results in coefficient expansion. At first level of decomposition, PCA is applied on the low sub bands and for high, max rule is used, followed by inverse stationary wavelets. The proposed process can well presume diagnostic information from medical images, with few artifacts using single and multimodal image fusion.

This manuscript has five main sections. Clinical aspects of modalities and data set illustrations in Sect. 2. Section 3 encompasses the proposed technique and its implementation details, Sect. 4 present performance evaluation based on statistical experiments and discussion, Sect. 5 confers conclusion.

2 Clinical Potential of Modalities

Medical imaging modalities contain vital information for clinical processes. Each imaging techniques encompass the organ or the tumor in a limited preview. Ultrasonography (USG), Computed Tomography (CT), Magnetic Resonance (MR), Magnetic Resonance Angiography (MRA), depict the high decree structural data while functional information is represented adequately using the low-spatial resolution images acquired from functional MRI (fMRI), PET, SPECT [4, 5]. Hence, multiple medical images can be integrated to obtain added critical information, accurate diseases diagnosis and reduce the storage volume. Single Modal Image Fusion is useful when a lone scan may not sufficiently reveal underlying tissue, tumor or organ. It aids in improving the image quality, perform accurate guidance during intervention procedures or biopsy, to precisely localize suspected cancers with precision, and other clinical practices [3]. It is limited by the fact that lighter cancers or function information may be recorded and refined from the image sets and thus it is not possible to capture all aspects pertaining to diagnosis under a single scan. Single Modal Image Fusion have better correlation amongst input images with lesser registration errors still whole body anatomical region matching is far from achieved. Medical problems are heterogeneous in nature with most of the diagnostic information kept hidden from the underlying tissue or organ [3]. Also one specific modality may be limited in its capabilities and may not adequately represent sufficient diagnostic component. Multimodal Image Fusion combines images from more than one imaging modality. It enhances the chances of improved diagnosis of malignancies by overcoming the limitations due to dependency on a single modality. Moreover it brings forth physiological, structural and functional patient data forming a single fused output confirming its supremacy over single modal fusion. Each type of data can further reinforce the presence of an abnormally, correct errors resultants of alternating modality and fill the gap of missing information. Multimodal fusion is responsible for overcoming imaging limitations and resulting in accurate and of superior quality fused image. The performance of multimodal techniques are limited by increased complexity, higher registration errors, less correlated inputs, and added artifacts. Only few instances for single modal medical fusion are quoted in literature, pertaining to temporal aspects or for low quality images. It acquires only a single type of information anatomical or functional, which remain insufficient to provide a reasonable diagnostic content [3].

3 Proposed Fusion Framework

Spatial Domain and Transform Domain techniques are broad fusion categorizations [6]. Various scientific literatures have been categorized into pixel, feature, decision level fusion [7]. Proposed work is implemented under pixel-level as it contains original measure quantities and is thus the most preferred technique. Multiresolution analysis is adapted to represents the localization of features at proper scale. It executes fusion independent of orientations, scale and exhibits similarity with human visual system [8]. Multiresolution transforms are listed under pyramid-based and wavelet-based techniques. Pyramid transforms in literature are Laplacian Pyramid [9], Gradient Pyramid [10], Ratio-to-low pass Pyramid [11], FSD Pyramid, Contrast Pyramid and Morphological Pyramids [12]. The pyramid-based techniques suffer from blocking effects when the input images are considerably different [13]. These result in poor signal-to-noise ratio, provide no directional information and are limited in their ability to represent diagnostic features in multiple scales [14, 15] while superior results are achieved using wavelets. Discrete wavelet transform (DWT) [16–21] perform sub sampling and have three orientations, horizontal, diagonal and vertical. These are computationally efficient and result in non redundant image representation. Wavelets lag shift dependency and are unable to preserve segment information, leading to lost edge information, block artifacts and result in poor fusion results [22, 23]. Stationary Wavelet transforms (SWT) is a member of the wavelet family and is able to overcome these drawbacks. In the proposed work CLAHE is applied on the one of the input image pair. SWT computes approximation and detail components. Detailed components are fused at first level in order to restrict the coefficient expansion using

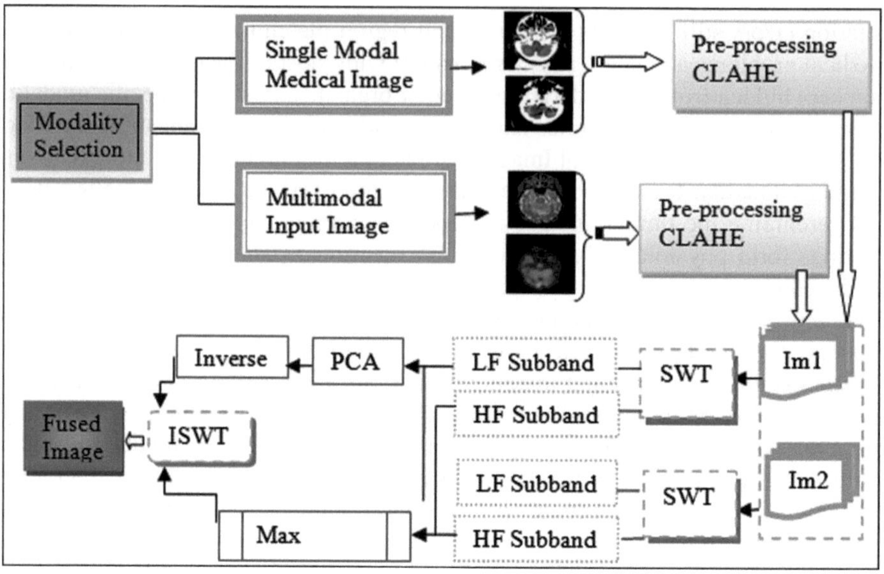

Fig. 1 Projected medical fusion framework

maximum selection rule. PCA is executed on the approximation coefficients obtained from SWT. All the coefficients are processed using inverse SWT to obtain the fused diagnostic image. Figure 1 illustrates the proposed framework.

3.1 Database

The medical images of different modalities have been downloaded from the website of The Harvard Brain Atlas http://med.harvard.edu/AANLIB/ [24]. Cross-sectional neuro-imaging cases of degenerative and neoplastic brain tumor cases have been used in this study. Case1 depicts a 75 year old man suffering from Meningioma, with left lower extremity weakness, difficulty with memory and concentration. He underwent craniotomy and tumor resection.

Multi-temporal 26 slices of CT/CT were acquired to perform single modal image fusion.

Case 2 illustrates the case of 73-year-old woman suffering from Alzheimer disease, with elevated arterial blood pressure documented on several occasions and had a hysterectomy used to execute multimodal image fusion using 46 slices of MRI T2 and SPECT images. The test data details are as depicted in Table 1 and sample of each dataset in Fig. 2.

Table 1 Test dataset details

Medical	Case 1	Case 2
Organ	Brain	Brain
Disease	Neoplastic tumor	Degeneration
Study	Meningioma	Alzheimer
Modality	Single modal	Multi modal
Type	CT/CT	MRI T2/SPECT
Image slices	26	46

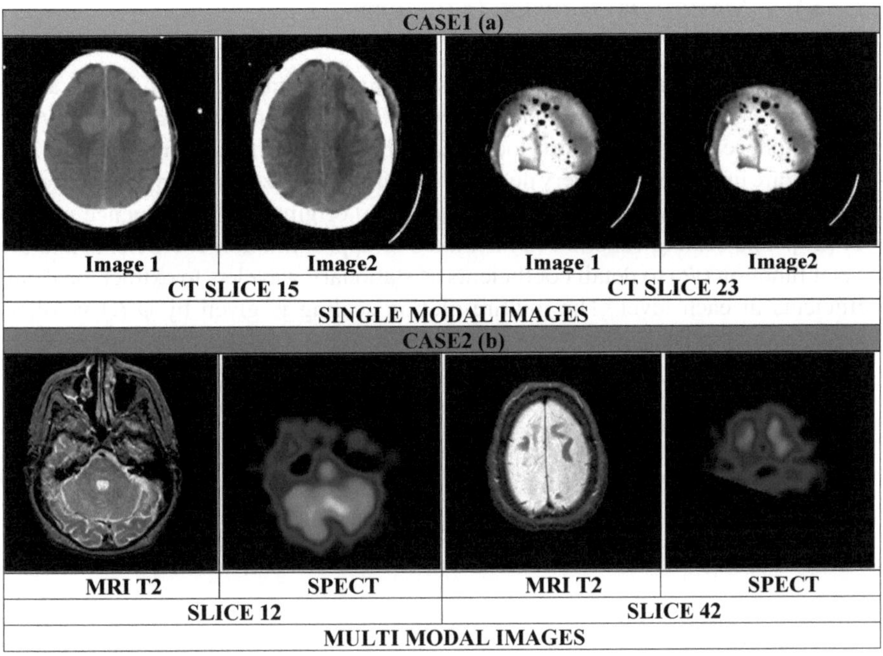

Fig. 2 Sample of each dataset **a** CT-CT **b** MRI T2-SPECT

3.2 Image Enhancement Based on CLAHE

Various global and local contrast enhancement techniques are proposed in literature. Global techniques produce poor results as the local details are poorly defined and thus are unable to represent the extent of spatial variation from medical images as in Contrast stretching and histogram equalization. The visual human system is sensitive to intensity contrast and is prejudiced by the pixel intensity values and of its neighbours [6], the local contrast enhancement techniques such as adaptive histogram equalization(AHE), CLAHE better define the local details [25]. AHE divides the image into tiles, combines neighbouring tiles by bilinear interpolation. It amplifies the noise levels while eliminating boundaries and enhancing edges. Histogram is cropped at a level and noise amplification is confined by CLAHE. Successful experiments by [26] conclude CLAHE based detection on dense mammographic backgrounds. In CLAHE the contrast of small regions called the tiles is enhanced. The neighbouring tiles were combined using bilinear interpolation. In this work, CLAHE transform in applied to medical input and augment contrast locally. This facilitates better visual output and reduced artifacts for further processing.

3.3 SWT Stationary Wavelet Transform

Stationary transform is a spectral method superfluous in nature, is also known as the 'algorithme a trous'. SWT is preferred over DWT as it is time invariant and provides improved phase information **and** localization by suppressing down sampling [23]. The filters are up sampled at every decomposition level and zeros are inserted between the filter taps, preserving time information of the original signal sequence at each level [27]. The scaling function (φ) is linked to approximation coefficients and the wavelet function (Ψ) to detail coefficients of stationary wavelets to extract the detail coefficients at each level. The multiresolution scaling is given by $\varphi\ (t)$ wavelets represented as $\psi_j\ (t)$

$$\varphi(t) = \sqrt{2} \sum_m ho(m)\varphi(2t - m) \tag{1}$$

$$\psi_j(t) = \sqrt{2} \sum_m hi(m)\varphi(2t - m), \quad j = 1, 2 \tag{2}$$

and $f(t)$ is as below

$$f(t) = \sum_a^b y = c(y)\varphi \,.\, x(t) + \sum_{i=0}^{2} \sum_{x=0}^{\infty} \sum_{y=-\infty}^{\infty} d_i(x, y)\varphi i, x, y(t), \tag{3}$$

where $c(y)$ and $d_i\ (x, y)$ are the expansion coefficients in the integral.

This undecimated version of DWT does not down sample the input coefficients, having same number of pixels at each decomposition level, it notably ensures abrupt changes.

The L_j and H_j are low pass and high pass filters at level zero. $I_m A_{j+1}$ is the resultant approximation coefficient from previous level, detail components are given by $I_m H_{j+1}$ for horizontal, $I_m V_{j+1}$ for vertical and $\mathrm{I}_m D_{j+1}$ for diagonal component, resulting from the upsampling of original coefficients.

$$L_j + 1 = \uparrow 2(L_j) \text{ and} \tag{4}$$

$$H_j + 1 = \uparrow 2(H_j) \tag{5}$$

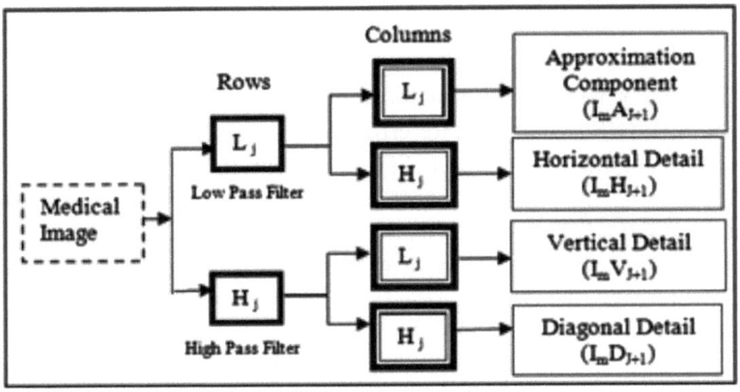

Fig. 3 Stationary wavelet transform decomposition

The decomposition for SWT is depicted in Fig. 3 and is given as follows:

$$I_m A_{J+1,a1,a2} = \sum_{b1} \sum_{b2} L_j^{\uparrow 2j}(b_1 - 2a_1) L_0^{\uparrow 2j}(b_2 - 2a_2) I_m A_{j,a1,a2} \quad (6)$$

$$I_m H_{J+1,a1,a2} = \sum_{b1} \sum_{b2} L_j^{\uparrow 2j}(b_1 - 2a_1) L_0^{\uparrow 2j}(b_2 - 2a_2) I_m H_{j,a1,a2} \quad (7)$$

$$I_m V_{J+1,a1,a2} = \sum_{b1} \sum_{b2} H_j^{\uparrow 2j}(b_1 - 2a_1) H_0^{\uparrow 2j}(b_2 - 2a_2) I_m V_{j,a1,a2} \quad (8)$$

$$I_m D_{J+1,a1,a2} = \sum_{b1} \sum_{b2} H_j^{\uparrow 2j}(b_1 - 2a_1) H_0^{\uparrow 2j}(b_2 - 2a_2) I_m D_{j,a1,a2} \quad (9)$$

Medical fusion is a gradual procedure, with diverse procedures and diverse methods [28]. As diagnostic information is vital, SWT well preserves the detailed information with better computational complexity than most transform domain techniques including NSCT [29].

3.4 PCA Principal Component Analysis

The statistical PCA procedure converts the linked variables and produce principal components linearly unlinked in nature, sorted in sequence with highest variance first. The resulting vectors are uncorrelated. PCA can be implemented for dimension reduction of image data, preserving important disparity [30]. PCA orthogonalizes input vectors components to uncorrelated them, and sort them as per relevance. It forms eigenvectors and eigen values in the eigenvector directions. PCA reveals the internal structure of input so as to best explain the variance in the underlying data [31].

4 Objective Experiment Criteria

There are various objective criteria for evaluation of fusion results as peak signal–noise ratio (PSNR), mean square error (MSE), signal–noise ratio (SNR), cross correlation (CR), standard deviation (SD) and many more. The metrics PSNR and MSE are used in the study as these realize well with meticulous informative under medical images and substance [32].

PSNR is defines input quality with gold standard image. As no such image is available for reference, the computation indicates the fusion results and is defined by

$$\text{PSNR} = 10\log_{10}\frac{L^2}{\text{MSE}} \tag{10}$$

MSE corresponds to the error in original with final fused image. The high registration accuracy gives small output value. Mean sq error = 0 for equivalent images and is given by

$$\text{MSE}(X_f) = \frac{1}{I \times J}\sum_{i=1}^{I}\sum_{j=1}^{J}[X_f(i,j) - X_o(i,j)]^2 \tag{11}$$

4.1 Results and Discussions

Objective performance evaluation by PSNR and MSE are discussed to validate results. Each modality works in its own preview under varied conditions, providing specific insight into an organ, lesions or functionality [33]. It is difficult to attain all the requisite diagnostic details from single modality as it may either be void of functional or anatomical information. When the different input modalities are in agreement or of complementary nature, medical image fusion provides improved diagnostic information. Image fusion using the proposed technique hold superior results with complimentary multimodal data enhancing the spatial resolution and sharpening, for improved diagnosis. The resultant image of multimodal fusion is as depicted in Fig. 4 and Table 2 summarizes the outcome below.

5 Conclusion

Though massive technological advancements have been through, there still is need to suitably decipher them into clinical benefits with excellence. Also the fusion results are mainly dependent on medical interpretations to extract clinical conclusions. Single modality may scarcely engross the possible mechanism with respect to medical diagnosis. Multimodal techniques have superior outcomes in providing diagnostic

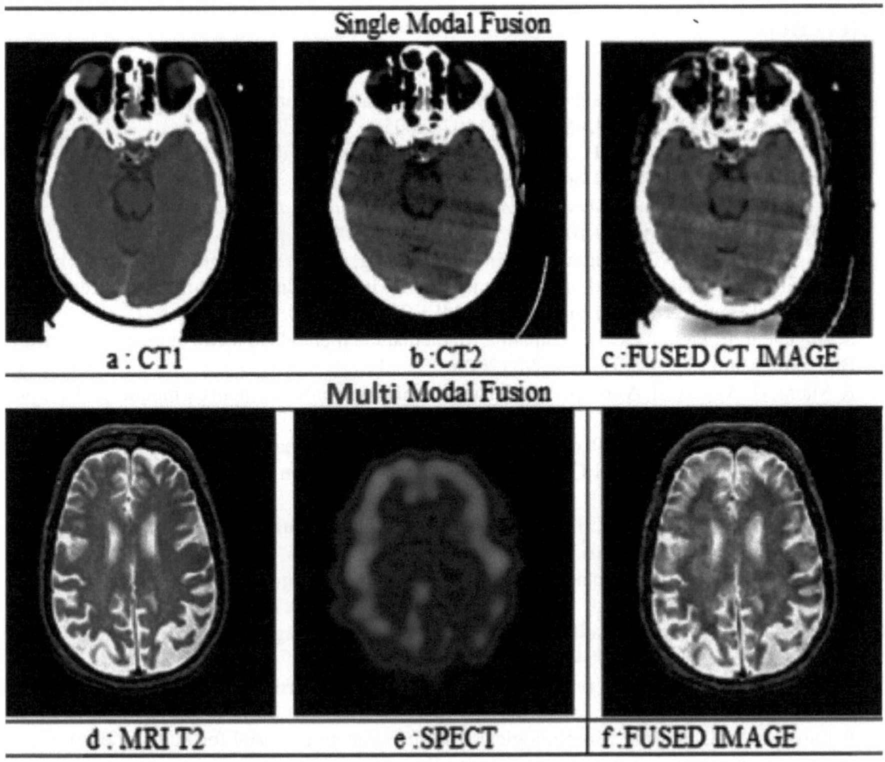

Fig. 4 **a** CT Image1 **b** CT Image2 **c** Fused CT Image **d** MRI T2 **e** SPECT **f** Fused image

Table 2 Results

Technique				
	Single modal fusion		Multi modal image fusion	
	PSNR	MSE	MSE	PSNR
AVG	14.06	0.039267	0.012513	17.03
PCA	14.68	0.034031	0.016269	17.89
DWT	13.40	0.091152	0.042992	13.67
SWT	6.58	0.219866	0.081121	13.02
Proposed	18.12	0.015433	0.004720	20.09

information with increased computation complexity. Under Single modal 26 slices of multi-temporal CT are fused and 46 slices of multimodal MRI T2 and SPECT are fused using the proposed technique. The results have been computed based on statistical evaluation using PSNR and MSE in comparison with standard state of art techniques. The results clearly indicate the demarcation amongst single and multimodal fusion results. The proposed fusion process can well presume diagnostic information from medical images, with fewer artifacts.

References

1. Z. Wang, D. Ziou, C. Armenakis, D. Li, Q. Li, A comparative analysis of image fusion methods. IEEE Trans. Remote Sens. **43**(6), 1391–1402 (2005)
2. Z.H. Xu, L. Liu, L. Tong, L.H. Zhou, C. Chen, Wavelet medical image fusion algorithm based on local area feature. Biol. Biomed. Rep. **2**(1), 25–31 (2012)
3. B.H. Hasegawa, H. Zaidi, *Dual-Modality Imaging: More Than the Sum of its Components*, pp. 35–81. https://doi.org/10.1007/0-387-25444-7_2 (2006)
4. Z. Jingzhou, Z. Zhao, T. Jionghua, L. Ting, M. Zhiping, Fusion algorithm of functional images and anatomical images based on wavelet transform. in IEEE 2nd International Conference on Biomedical Engineering and Informatics BMEI'09 (2009), pp. 1–5
5. S. Angenent, E. Pichon, A. Tannenbaum, Mathematical methods in medical image processing. Bull. Am. Math. Soc. **43**(3), 365–396 (2006)
6. R. Maini, H. Aggarwal, A comprehensive review of image enhancement techniques. J. Comput. **2**(3), 8–13 (2010)
7. C.K. Solanki, N.M. Patel, Pixel based and wavelet based image fusion methods with their comparative study, National Conference on Recent Trends in Engineering and Technology (2011)
8. G. Bhatnagar, Q.M.J. Wu, Z. Liu, Directive contrast based multimodal medical image fusion in NSCT domain. IEEE Trans. Multimedia **15**(5) (2013)
9. P. Burt, E. Adelson, The Laplacian pyramid as a compact image code. IEEE Trans. Commun. **31**(4), 532–540 (1983)
10. V. Petrovic, C. Xydeas, Gradient -based multiresolution image fusion. IEEE Trans. Image Process. **13**(2), 228–237 (2004)
11. A. Toet, Image fusion by a ratio of low pass pyramid. Pattern Recogn. Lett. **9**(4), 245–253 (1989)
12. F. Laporterie, G. Flouzat, Morphological pyramid concept as a tool for multi resolution data fusion in remote sensing. Int. Comput-Aided Eng. 63–79 (2003)
13. D.L. GHill, D.J. Hawkes, Medical image registration using voxel similarity measures, in Applications of Computer Vision in Medical Image Processing. AAAI spring symposium series (1994), pp. 34–37
14. A. Toet, A morphological pyramidal image decomposition. Patt. Recognit. Lett. **9**(4), 255–261 (1989)
15. R. Redondo, F. Sroubek, S. Fischer, G. Cristobal, Multifocus image fusion using the log-gabor transform and a multisize windows technique. Inf. Fusion **10**, 163–171 (2009)
16. H. Li, B.S. Manjunath, S.K. Mitra, Multisensor image fusion using the wavelet transform. Graph. Models Image Process. **57**(3), 235–245 (1995)
17. Y. Yang, D.S. Park, S. Huang, N. Rao, Medical image fusion via an effective wavelet-based approach, EURASIP J. Adv. Sig. Process. (2010)
18. D.J. Field, Scale-invariance and self-similar 'Wavelet' transforms: an analysis of natural scenes and mammalian visual systems, in *Wavelets, Fractals and Fourier Transforms: New Developments and New Applications* (Oxford University Press, 1993), pp. 151–193
19. I. De, B. Chanda, A simple and efficient algorithm for multifocus image fusion using morphological wavelets. Sig. Process. **86**(5), 924–936 (2006)
20. J. Teng, X. Wang, J. Zhang, S. Wang, Wavelet-based texture fusion of CT/MRI images, in 3rd International Congress on Image and Signal Processing (CISP), vol.6 (2010), pp. 2709–2713
21. V.D. Calhoun, Feature-based fusion of medical imaging data. IEEE Trans. Inf. Technol. Biomed. **13**(5), 711–720 (2009)
22. M. Beaulieu, S. Foucher, L. Gagnon, Multispectral image resolution refinement using stationary wavelet transform, in Proceedings of 3rd IEEE International Geoscience and Remote Sensing Symposium (2003), pp. 4032–4034
23. G.P. Nason, B.W. Silverman, *The Stationary Wavelet Transform and Some Statistical Applications* (Technical Report BS8 1Tw, University of Bristol, 1995)

24. The Whole Brain Atlas at http://www.med.harvard.edu/AANLIB/home.html
25. E.D. Pisano, S. Zong, B.M. Hemminger, R. Marla Deluca, E. Johnston, M. Keith Muller, P. Braeuning, S.M. Pizer, Contrast limited adaptive histogram equalization image processing to improve the detection of simulated spiculations in dense mammograms. J. Digit. Imaging **11**(4), 193–200 (1998)
26. S.M. Pizer, E.P. Amburn, J.D. Austin et al., Adaptive histogram equalization and its variations. Comput. Vision, Graph. Image Process. **39**, 355–368 (1987)
27. M. Holschneider, R. Kronland-Martinet, P. Tchamitchian, *A Real-Time Algorithm for Signal Analysis with the Help of the Wavelet Transform* (Springer-Verlag 1989), pp. 289–297
28. Y. Dai1, Z. Zhou, L. Xu, The application of multi-modality medical image fusion based method to cerebral infarction. EURASIP J. Image Video Process. (2017)
29. Q. Jiang, X. Jin, S.-J. Lee, S. Yao, A novel multi-focus image fusion method based on stationary wavelet transform and local features of fuzzy sets in IEEE Access (2017)
30. W. Hao-quan, X. Hao, Multimode medical image fusion algorithm based on principal component analysis, in IEEE symposium on Computer Network and Multimedia Technology (2009), pp. 1–4
31. S. Guruprasad, M.Z. Kurian, H.N. Suma, Fusion of CT and PET medical images using hybrid algorithm DWT-DCT-PCA. 978-1-4673-8611-1/15/$31.00 IEEE (2015)
32. S. Hariharasudhanl, B. Raghu, An investigative approach and analysis on fusion techniques of images for medical beneficial applications. Ind. J. Sci. Technol. **9**(48) (2016)
33. G. Kaur, R. Vig, S. Singh, A review on medical image fusion aspects and techniques, in Proceedings of 1st International Conference on Communication, Computing and Networking—ICCCN 2017 (NITTTR Chandigarh, India, 2017)

Hybrid Classifier for Bone Age Assessment

Amandeep Kaur and Kulwinder Singh Mann

Abstract Bone age assessment (BAA) is a task performed on radiographs by the pediatricians in hospitals to predict the final adult height and to diagnose growth disorders by monitoring skeletal development. Typically, the height and diameter of the wrist were the main indicators considered for judging the bone age. In this research work, it is established that by conducting one-way ANOVA along Procrustes on the shape data of hand X-rays of the persons it is possible to create a discriminate function that would help to classify the geometrical mean shape between a male and a female hand for bone age assessment. The approach explained in this paper also helps in building a better shape model for further extraction of bone parts and at the same time, it helps to build a classifier feature set for supervised hybrid classification for bone age assessment. Primarily this algorithm works on the basis of computing variability of the total shape, between the group's variability and Procrustes distance. The two-way ANOVA was also conducted to achieve bone shape separations for evaluation purposes. The performance evaluation of the methods was done and its outcome shows a fair degree of accuracy. The work automatically helps to reduce the number of procedures or steps involved in bone age assessment.

Keywords X-rays · Bone age · Shape analysis · Classification

1 Introduction

Bone age assessment (BAA) is a task performed on radiographs by the pediatricians in hospitals to predict the final adult height and to diagnose growth disorders by monitoring skeletal development. Before the advent of radiographs [1], it was a

A. Kaur (✉)
CSE, I.K. Gujral Punjab Technical University, Jalandhar, India
e-mail: amandeep76@gmail.com

K. S. Mann
Department of Information Technology, Guru Nanak Dev Engineering College, Ludhiana, India

C. R. Krishna et al. (eds.), *Proceedings of 2nd International Conference on Communication, Computing and Networking*, Lecture Notes in Networks and Systems 46, https://doi.org/10.1007/978-981-13-1217-5_43

simple visual assessment or the measurement of bones was taken manually. Typically, the height and the diameter of the wrist were the main indicators considered for judging the age. The parents are normally worried about the physical growth of their kids [2], so they invent emotion and energies in finding the ways to predict or check the maturity of their wards. Today, bone age assessment has many other dimensions and a fully [3] automated bone age system is a need of the hour.

Not so recently, in the year 2006, India hosted AFCU-20 championship and in one of the games, the national team coach Bob Houghton expressed his astonishment at the number of senior Indian squad leading the U-20 front line in the football team. The age of many players was the issue of controversy and most of the sports fraternity feels appalling at such age controversies. The menace of age fraud has come into sharper focus once again following recent media allegations surrounding India's 21-member squad for next month's FIFA U-17 World Cup. The All India Football Federation (AIFF) has denied such allegations but Indian football history is unfortunately decorated with such incidents. The Mohun Bagan Academy team was banned from the U-18 I-League, while Gurgaon-based Conscient FC was disqualified from the Delhi zone of the U-16 I-League and Bengaluru-based Ozone FC from the national league of the same tournament– all for having fielded overage players. Ozone played the final—where they lost to Minerva Academy—following a stay order from the Karnataka High Court. Not just in football, in fact in many sports the age assessment remains a burning issue.

Mass migration due to the spread of the Syrian crises triggered a renewed interest of the radiologist and the stakeholders involved in providing the shelter, refuge, and care to the migrants. The radiologists were engaged in doing the assessment of the bone of the people directly receiving the benefits of the aid or in cases where the child needs to be segregated from the adults. This gave birth to a new kind of demand for medical gadgets that can perform the bone age assessment on the borders sites, refugee camps, etc. This medical equipment should be connected to the database servers online for further processing and storage of bio-physical records of the migrants. The governments of Ireland, Britain, Greece, and many other countries faced the same issues.

Bone age assessment can help not only finding the growth patterns of various people but also help in identifying many primary and secondary diseases. Diseases related to malfunctioning of bones, bone shape or formation, bone density and bone tissue damage, etc. The dimensional and shape analysis of the bones is correlated with other clinical studies and then the identification of the bone diseases can be performed. These developments are however dependent upon few classical methods such as Greulich Pyle (GP) [3] and Tanner Whitehouse (TW) [4] of bone age assessment but also on the capability of the person taking measurements of bones. Today, bone age assessments are done with help of computer-aided solutions and applications. The use of artificial intelligence, probability models, and machine learning methods are already making waves in automating this problem. But, intermediate steps may require additional improvements that can help to reduce the procedures for building such systems. The statistical analysis can help in finding procedures that can work on differentiation or variation of the bone shapes of male or female for example, especially those methods that use active contour [5–7].

2 Problem Statement

The convention methods of measuring the bone maturity using Atlas will require the development of many atlases which would be based on different demographics because it is not only age that affects the bone formation and its shape. The gender and ethnicity of the personal also affects the size, dimensions, and shape of the bones. This makes the age prediction process more cumbersome and prone to errors. To overcome this issue the classification algorithms needs to be fed with respective data for it to become accurate in classification. Hence, to get accurate data feed the segmentation methods that rely on the shape analysis of the bone need to incorporate methods that define the shape of the bone invariantly and a mean shape of the bone is computable mathematically. Moreover, this bone shape representation as a matrix must be qualified for shape classification. The scope of this work is to build bone shape analysis algorithm that helps in predicting the age of a person so that even issues relating to the fusion of bones is also addressed as an independent matrix and an additional feature such as the degree of fusion would not be required.

3 Methodology

This section discusses the steps involved in constructing the bone shape classification algorithm that can help in bone age assessments.

Step 1: The initial process is to build a dataset of all the landmarks that seem to represent the shape of the object of interest. As shown in Fig. (1) the major landmarks of the Distal Phalanges and metacarpals are marked and the dataset of such landmarks for each image is created. This dataset is processed further so that supervised shape learning can be as per the age, gender and ethnicity of the subject. The following description helps in understanding the procedure

(a) Let "lmp.n" a number of major landmark points "lmp" manually collected.
(b) Let (lmp.x, lmp.y) be the location of the landmark points collected.
(c) Let Imp.I be variable representing the Image of Bone Part.
(d) Remove repeating data points or duplicate points.
(e) Interpolate to get more pairs of points (lmp.x, lmp.y).
(f) Mark the Landmark points with 1, other points as zero.
(g) Interpolate new set of landmark points and evenly space them, so that we final shape matrix.

Step 2: Supervised Classification: The shape matrix will have slight changes as per the gender and other factors. The female gender shape matrix will have borders

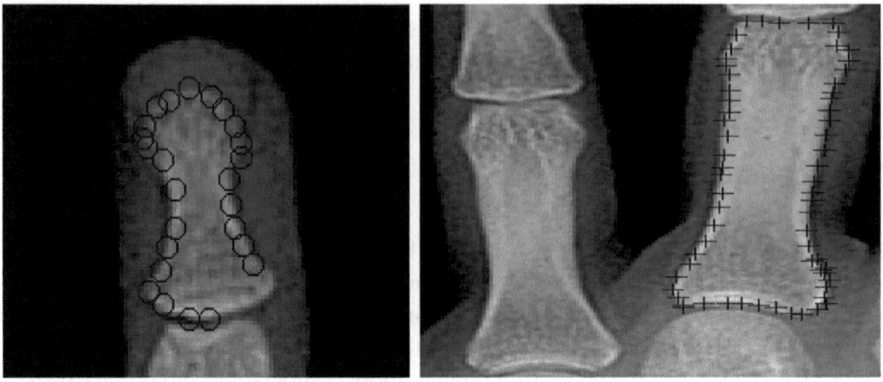

Fig. 1 Major landmarks of distal phalanges and meta carpals bones

data points that are closer to their centroid where else in case of a male of the same age may have typical border data points which are far from their centroid as compared to the female counterparts. Then if an input bone shape has close space mapping with respect to the reference bone shape then we can safely infer that distance measurement is helpful in identifying the similarity. For computing the similarity, we need to define that landmarks point are invariant in nature even if these data points are transformed, scaled and rotated. This would complete the definition of the landmarks as shapes. Hence, the following three sub-steps are done.

(a) Translation: In this process, the image is translated so that the mean of the objects points lies on the centroid of the object.
Landmarks_data_points = ({lmn.x1, lmny1}, {lmn.x2, lmny2}…),
The mean of these points is Xmean = (lmn.x_1 + lmn.x_2 + lmn.$x_{i)}$/I, Ymean = (lmn.y_1 + lmn.y_2 + lmn.$y_{i)}$/i
Translate, such that the mean is translated into origin (lmn.x.lmn.y) -> (lmn.x_{1-} Xmean.$x_{1,}$ lmn.y_{1-} Ymean.$y_{1)}$
(b) Scaling: Likewise, the scale is removed by re-scaling the object so that root mean square distance from the point to be translated to origin 1.
S = mean (lmn.x − Xmean.x_1) / (lmn.y − Ymean.y_1), this division will make the scale 1 when the point coordinates are divided by the Bone part image object scale.
(c) Rotation: Initially it involves computing the angle to the center of all points and then subtracting the mean angle to build new data points that have same rotation.

Step 3: ANOVA Test: In this step, try to find the whether the variance of landmarks "l" have similar variance.

(a) One-Way ANOVA:
(b) Two-Way ANOVA:

Table 1 Runs conducted on the proposed algorithm

Test runs		
	Predicted by algorithm (a) male distal bone	Predicted by algorithm (b) male distal bone
Actual (a)	4	0
Actual (b)	1	3

Step 4: Conduct Procrustes Distance Analysis: After confirming that the variance between the classes (e.g., distal bone shape) of data points is similar, we conduct the Procrustes distance analysis

Step 5: Bone Part Classifier:

Pseudo logic:

Step 1: Evaluate the mean bone shape for each class defined by age, gender, and ethnicity.
Step 2: For new shape matrixes, Obtain a matrix, after satisfying the non-centralize and centralized mean shape condition.
Step 3: The algorithm's allocation rule is to assign any new object to the sup-population in which its shape features (or landmarks) shows the smallest distance "d". If the distance of an object to class "***g***" is equal to its distance to class g (class belonging to the reference image feature set).

4 Evaluation and Results

A major step in constructing a classification rule is to assess its accuracy. The reason for that is that it is not always possible for a classifier to perfectly distinguish new objects. In particular, in some cases, the features of whole classes are mathematically indistinguishable. In other words, high overlapping/repetition of the available measurements between regions of a sub-population can result in poor prediction. According to this, it is extremely convenient to measure the performance of a classification rule. To achieve this, the comparison between actual classes with corresponding predicted classes is carried out in order to count misclassified objects. The following are the performance outcomes.

Table 1 reflects the condition of 8 test runs were conducted on the proposed algorithm. It is apparent from the above matrix that is having two conditions; one for each condition (a) and (b) that failed and was predicted wrongly in case of condition "a male distal bone":

$$\text{Accuracy} = \frac{\text{TP} + \text{TN}}{P + N} \tag{1}$$

$$\text{i.e.}\quad \text{Accuracy} = (3+3)/8 * 100 = 75\%.$$

The accuracy in Eq. 1 shows that the proposed algorithms perform fairly enough to distinguish between the classes based on the shapes of the bone parts, it assigned each object to a particular class if its distance value to that class was smaller than the values for the rest of classes.

5 Conclusion

It is apparent that an object in an image can be defined by its geometric shape obtained from the landmarks. These landmarks form the basis of defining the various kinds of shapes a bone may make take especially, in stages when one bone fuses into the other distinct bone part. Mathematically, this shape must be maintained even if its landmarks data point are put under rotation and is scaled to smaller or larger shapes. For assessing the age of a person, the separation of bone shapes, bone shape classification can help in building accurate systems. The validity of the model of shape classification is a necessary step for the selection of the landmarks. In this paper, we have followed a matching approach, with which distance between corresponding landmarks was computed so that maximum possible agreement on the shape of the bone could be achieved. In this process, we used invariance analysis such as Euclidean distance based similarity computation (scaling, rotation, and translation) and we found a better evaluation of the shapes by using Procrustes distance.

References

1. M.A. Breen, A. Tsai, A. Stamm, P.K. Kleinman, Bone age assessment practices in infants and older children among society for pediatric radiology members. Pediatr. Radiol. 1–6 (2016) (Bone Age Assessment)
2. M.F. Darmawan, S.M. Yusuf, M.R. Abdul Kadir, H. Haron, Age estimation based on bone length using 12 regression models of left-hand X-ray images for Asian children below 19 years old. Leg. Med. **17**(2), 71–78 (2015)
3. M. Haktan, *Computer Assisted Bone Age Assessment: Biomedical Engineering-Bone Age Assesment* (2005)
4. M. Urschler et al., Applicability of Greulich-Pyle and tanner-Whitehouse grading methods to MRI when assessing hand bone age in forensic age estimation: a pilot study. Forensic Sci. Int. **266**, 281–288 (2016)
5. J. Zhang, F. Lin, X. Ding, Automatic determination of the greulich-pyle bone age as an alternative approach for chinese children with discordant bone age, Horm. Res. Paediatr. **86**(2), 83–89 (2016) (Neural Network, 1,2)
6. G. Gao, C. Wen, H. Wang, X. Lizhong, Fast multiregion image segmentation using statistical active contours. IEEE Signal Process. Lett. **24**(4), 417–421 (2017)
7. Y. Lu, T. Zhou, Lip segmentation using localized active contour model with automatic initial contour. Neural Comput. Appl. 1–8 (2017)

Recent Advances and Challenges in Automatic Hyperspectral Endmember Extraction

Mahesh M. Solankar, Hanumant R. Gite, Rajesh K. Dhumal, Rupali R. Surase, Dhananjay Nalawade and Karbhari V. Kale

Abstract The advancements in hyperspectral remote sensing are increasing continuously and recording a wealth of spatial as well as spectral information about an object, but resulting high volume of data. Analysis and classification of this high volume hyperspectral data needs a ground truth data or spectral library or image based endmembers which assist to unmix the mixed pixels and map their spatial distribution. Till date, though several hyperspectral endmember extraction algorithms have been proposed, every algorithm has its own limitations. The perfect endmember extraction algorithm would find unique spectra with no prior knowledge. This paper discusses the recent improvements and challenges in hyperspectral endmember extraction. The algorithms evaluated includes PPI, NFINDR, FIPPI and ATGP. The experiments are performed on the subset of Hyperion and AVIRIS_NG datasets.

Keywords Hyperspectral endmember extraction · PPI · FIPPI · NFINDR ATGP · AVIRIS-NG

M. M. Solankar (✉) · H. R. Gite · R. K. Dhumal · R. R. Surase · D. Nalawade · K. V. Kale
Geospatial Technology Research Laboratory, Department of Computer Science and IT, Dr. Babasaheb Ambedkar Marathwada University, Aurangabad 431004, Maharashtra, India
e-mail: csit.mms@bamu.ac.in

H. R. Gite
e-mail: csit.hrg@bamu.ac.in

R. K. Dhumal
e-mail: dhumal19@gmail.com

R. R. Surase
e-mail: rupalisurase13@gmail.com

K. V. Kale
e-mail: kvkale.csit@bamu.ac.in

C. R. Krishna et al. (eds.), *Proceedings of 2nd International Conference on Communication, Computing and Networking*, Lecture Notes in Networks and Systems 46, https://doi.org/10.1007/978-981-13-1217-5_44

1 Introduction

The recent revolutions in hyperspectral imaging has enriched with both the spatial and spectral information about a surface, but increases the challenges in analysis due to large volume of the data. The supervised analysis of the hyperspectral data is comprehensively dependent on the reference information, it may be the spectral library, ground truth information or image-derived endmembers [1]. It is very challenging to collect ground truth information or get a reference spectra of every element within the scene in spectral database, because of large spatial coverage of the data [2]. The image derived endmembers solves these issues with its scene dependency only [3].

The endmembers in hyperspectral images are the spectral signatures corresponding to the unique materials within the scene [4, 5]. Till date, various algorithms have been developed to find the hyperspectral endmembers. Every algorithm is having its own limitations either at the algorithms initialization phase or at algorithm termination phase [6, 7]. This paper attempt to evaluate the performance of four popular End Member Extraction (EME) algorithms namely Pixel Purity Index (PPI), NFINDR, Fast Iterative Pixel Purity Index (FIPPI) and Automatic Target Generation Process (ATGP) on two different hyperspectral datasets. Every algorithm is tested four times on a single dataset with varying its input parameters. The experimental results explored the efficiency of the every algorithms with respect to the hypothetical range of input parameter values. Before proceeding to the experimental work, all of the four algorithms are described below.

1.1 PPI

The PPI by Boardman et al. [8] finds a set of vertices of a convex hull of an image and as an endmember. The MNF transform is used to decrease data dimensionality. The PPI begins with large number of randomly generated N-dimensional vectors also knows as skewers $\{\mathrm{Skewer}_i\}$. For every Skewer_i, each and every pixel vectors is projected on Skewer_i to identify pixel vectors having extreme positions and produce extrema set of this particular Skewer_i, indicated by $S_{\mathrm{extrema}}(\mathrm{Skewer}_i)$. Every Skewer_i produces a diverse extrema sets, some of the pixel vectors may present in more than one extrema sets. Indicator function for S is defines as,

$$I_S(x) = f(x) = \begin{cases} 1; \text{ if } x \in S \\ 2; \text{ if } x \notin S \end{cases} \quad \text{and} \quad N_{\mathrm{PPI}}(x) = \sum_j I_{S_{\mathrm{extrema}}(\mathrm{Skewer}_i)}(x) \qquad (1)$$

Here, N_{PPI} is PPI count for pixel vector x. Find the PPI counts for every pixel vectors given by Eq. (**1**). Define the threshold T, so that all pixel vectors with $N_{\mathrm{PPI}} > T$ can be considered as pure.

1.2 NFINDR

The NFINDR by Winter [9] is fully unsupervised approach. It takes MNF transformed data as an input. It starts with random selection of user defined number of pixels and consider those pixels as endmembers. Suppose, "M" is the matrix containing pure pixels augmented by row of ones

$$M = \begin{bmatrix} 1 & 1 & \cdots & 1 \\ \underset{m_1}{\rightarrow} & \underset{m_2}{\rightarrow} & \cdots & \underset{\mathrm{m_i}}{\rightarrow} \end{bmatrix}, \tag{2}$$

where, m_i is spectra for endmember i. Now the simplex volume (V) made by the endmembers is proportional to the det. of M.

$$V(M) = \frac{1}{(1-1)}\mathrm{Abs}(|M|) \tag{3}$$

Here, 1–1 indicated data dimensions. To upgrade the estimation of endmembers, every pixel vector should be assessed with possibility of being pure or closely pure. For this, a sample volume is assessed for each and every pixel in each pure pixel location by substituting that pure pixel and recalculating the volume. If the replacement gives rise in volume, the endmember gets replaced by pixel. This procedure is continued till there are no further substitutions of the endmembers.

1.3 FIPPI

The FIPPI by Chang [10] uses VD to define the number of pure signatures ("p") required to extract. MNF transformation is applied to retain the first "p" components. It generates "p" number of skewers $\left\{\mathrm{skewer}_i^{(0)}\right\}_{i=1}^{p}$ formed by electing those pixel vectors that are corresponding to target pixel vectors produced using ATGP. In case of iterative rule $k \geq 0$, for every $\mathrm{skewer}_i^{(k)}$, every pixel vector is projected on this individual skewer to identify extrema set by Eq. (1). Then it finds the pixel vectors that gives largest NPPI $\left(r_i^{(k)}\right)$ by Eq. (1) say $\left\{r_i^{(k)}\right\}$. The stopping rule forms the union $\mathrm{skewer}_i^{(k+1)} = \left\{r_i^{(k)}\right\}_{\mathrm{NPPI}\left(r_i^{(k)}\right)>0}\left\{\mathrm{skewer}_i^{(k)}\right\}$. If $\mathrm{skewer}_i^{(k+1)} = \mathrm{skewer}_i^{(k)}$, then new pixels are not added to the skewer set.

FIPPI has enhanced the PPI in few aspects. It make use of VD to calculate the count of endmembers supposed to be pulled out, which is one of the key problem in the PPI.

1.4 ATGP

The ATGP by [11] developed to find possible target signatures with no prior knowledge. It is an orthogonal subspace projection based mechanism to find pure spectral signatures.

The remainder of this paper is structured as below. Section 2 describes the datasets used for experiments. Section 3 gives methodology followed for the study. Section 4 presents the comparative experimental observations, performance criteria and test sets on both the datasets. Section 5 summarizes the main concluding observations derived from the experiments and results.

2 Datasets

Two real hyperspectral datasets (i.e., AVIRIS-NG and Hyperion) are used for experimental analysis.

2.1 AVIRIS-NG

The AVIRIS-NG image is having 2680×719 spatial size with 425 spectral channels with a spectral coverage from 380 to 2500 nm. The spatial and spectral resolutions of the dataset are 5 m and 5 nm respectively [12]. The subset (300 × 300 × 320) of it, shown in Fig. 1 is taken for EME. The scene contains the vegetation and urban areas.

2.2 Hyperion

The Hyperion image is having 3471×1001 spatial size with 242 spectral channels with 10 nm spectral bandwidth and covers 4.0–2.5 μm range of electromagnetic spectrum. Its spatial resolution is 30 m [13]. The subset (300 × 250 × 155) of it, shown in Fig. 2 is taken for EME. The scene contains vegetation, urban and water bodies.

Fig. 1 AVIRIS-NG FCC

Fig. 2 Hyperion FCC

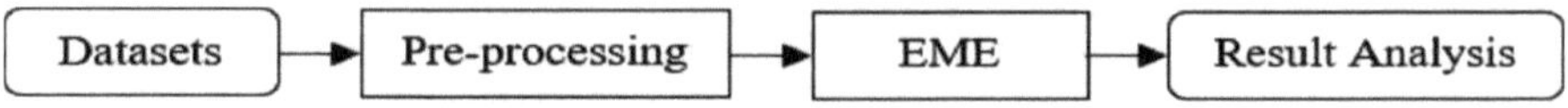

Fig. 3 Methodology

3 Methodology

The experiments are performed in Python 2.7.14 (64 bit) on Intel CORE i5 7th Generation system with 16 GB RAM. Figure 3 shows the methodology followed for experimental work.

3.1 Preprocessing of both the datasets involves atmospheric correction, bad-band removal and subset the original image data. The RTM-based ATREM and Scene-based QUAC correction mechanism are used for atmospheric corrections of dataset 2.1 and dataset 2.2 respectively. In bad-band (i.e., bands containing noise or no data) removal, 105 bad bands from dataset 2.1 and 87 bad bands from dataset 2.2 are removed. The final spatial and spectral subsets of dataset 2.1 (300 × 300 × 320) and dataset 2.2 (300 × 250 × 155) are taken for EME.

3.2 The EME algorithms discussed in Sect. 1 are implemented on dataset 2.1 and dataset 2.2 in strategic way by uniformly varying their input parameters.

3.3 The experimental outcomes and comparative observations of all the four algorithms are briefly discussed in Sect. 4.

4 Results and Discussion

This segment elaborates the successive experiments that use hyperspectral image data to make the widespread evaluation of EME algorithms. Before detailed discussion and analysis of results, it is essential to highlight input parameters used for EME. As per the requirements of discussed algorithms, the number of pure signatures to be extracted is manually defined. In practice, it is very tough to extract the accurate endmembers within a single run. To examine the association between input parameters and their effects on final outcomes, every algorithm is evaluated four times on both the datasets 2.1 and 2.2. Table 1. Gives the experimental outcomes and comparative observations of all the four algorithms on dataset 2.1 (Test-A) and dataset 2.2 (Test-B).

The PPI is application dependent and very sensitive to parameters "*I*" and "Th" [14]. As the value of "*I*" goes beyond the thousands of iterations, it maximizes the computational complexity of the system [15]. All the counts having pixel purity greater than zero are considered to be pure [16]. The experimental results shows that the random nature of initial skewer generation might produce slightly different set of endmember during every run. The value of "Th" is strongly associated with the value of "*I*". The probability of getting more pure pixels increases with increase in the threshold value. Figures 4a and 5a shows the endmembers extracted using PPI from dataset 2.1 and dataset 2.2 respectively.

The major issue with NFINDR is to identify how many endmembers to be extracted [17]. The iterations given for NFINDR are three times the number of endmembers. The NFINDR can be developed both in parallel and sequential mode of implementation. The parallel implementation outperforms the sequential implementation but raises the computational complexity [18, 19]. Figures 4b and 5b plots the NFINDR outcomes on dataset 2.1 and dataset 2.2 respectively.

Table 1 reveals that, in case of FIPPI, the endmembers supposed to be extracted (Ea) and the actually extracted endmembers (Er) are varies. FIPPI returns only most possible endmembers, rather than satisfying the **Ea**. The FIPPI works significantly faster than PPI and returns better results.

Table 1 Comparative observations of PPI, NFINDR, FIPPI and ATGP on both datasets

	PPI				NFINDR			FIPPI				ATGP	
Test	*I*	Th	*E*	*T*	*E*	*I*	*T*	*Ea*	E_r	*I*	*T*	*E*	*T*
A-1	500	10	16	58	7	21	5	10	7	30	3	7	12
A-2	500	20	10	58	8	24	10	20	8	60	6	8	14
A-3	2000	10	25	251	9	27	11	30	9	90	9	9	16
A-4	2000	20	21	248	9	50	11	50	9	150	16	9	16
B-1	500	10	7	4	4	12	3	10	4	30	2	4	2
B-2	500	20	4	4	5	15	4	20	5	60	4	5	3
B-3	2000	10	34	32	6	18	4	30	6	90	5	6	3
B-4	2000	20	14	32	11	33	6	50	11	150	10	11	6

A Dataset 2.1, *B* Dataset-2.2, *I* Iterations, *Th* Threshold, *E* Endmembers, E_a Endmembers to extract, E_r Extracted endmembers, *T* Time in seconds

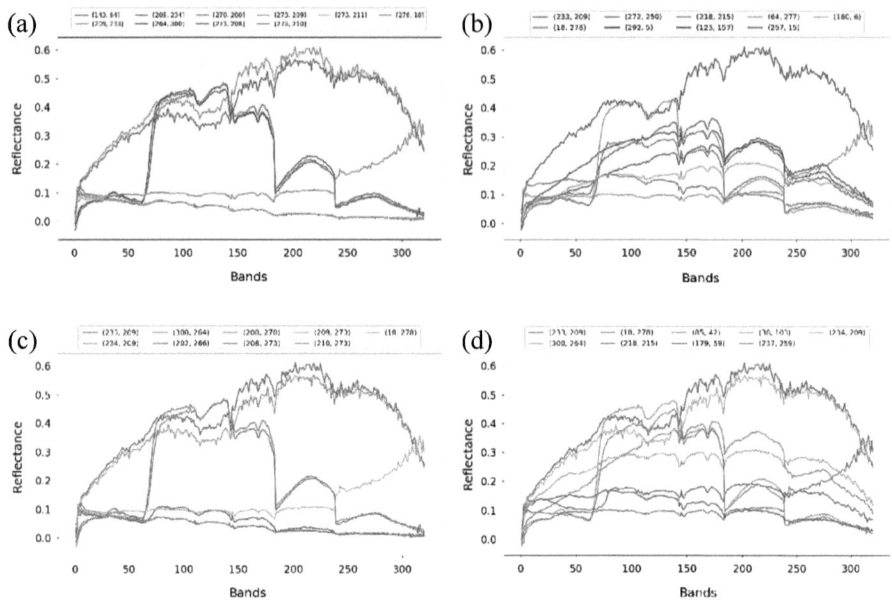

Fig. 4 **a** PPI, **b** NFINDR, **c** FIPPI, **d** ATGP

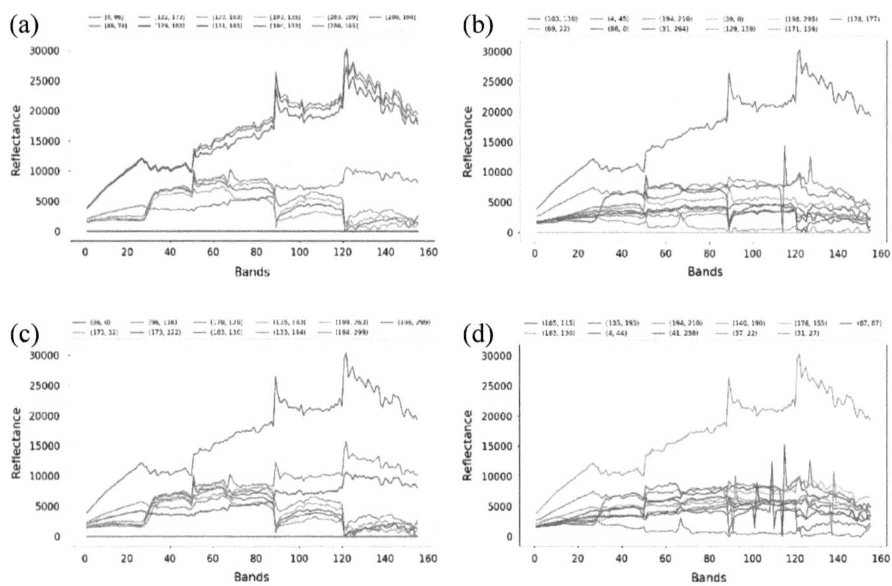

Fig. 5 **a** PPI, **b** NFINDR, **c** FIPPI, **d** ATGP

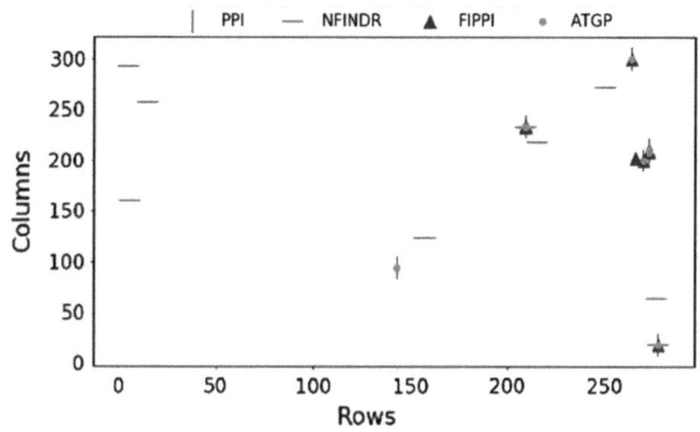

Fig. 6 AVIRIS-NG endmembers

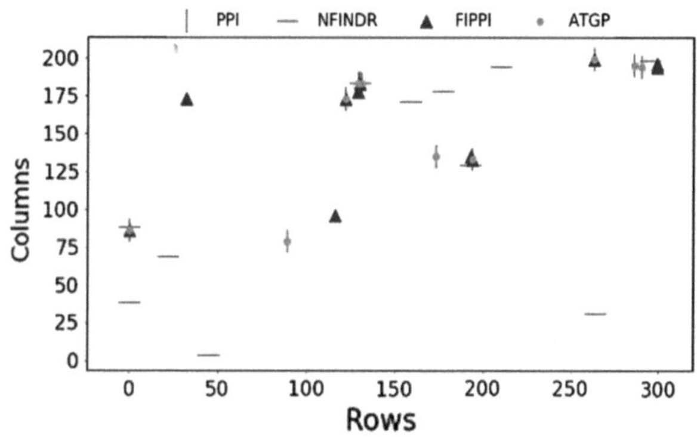

Fig. 7 Hyperion endmembers

Table 1 tells that ATGP works with approximately equal speed as compared to NFINDR and FIPPI. The only matter is to decide how many number of endmembers to be identified [20].

Figures 6 and 7 scatters the comparative endmember coordinates extracted by all the four algorithms using both the datasets 2.1 and 2.2 respectively.

As we can see in Figs. 6 and 7, most of the endmember extracted by PPI, FIPPI, and ATGP overlaps. NFINDR gives varying results as compared to other three algorithms. All of the four algorithms uses only spectral information for endmember identification. The fusion of both spatial and spectral information improves the endmember extraction as compared to using only spectral information [21].

5 Conclusion

This paper compares the four popular endmember extraction algorithms on two different hyperspectral datasets. The experimental results reveals that NFINDR algorithms give diverse results as compared to PPI, FIPPI, and ATGP. The PPI, FIPPI, and ATGP algorithm produces approximately similar results. The PPI gives noticeable results with higher threshold and higher iterations, but takes much more time as compared to other three algorithms. The execution time of the NFINDR, FIPPI and ATGP is much less as compared to PPI. The major issue in automatic extraction of the hyperspectral endmembers (i.e. how many number of endmembers are supposed to be extracted) is still not appropriately addressed. The PPI and NFINDR algorithms starts with random initial vector generation, which creates inconsistent output in every run of algorithm on the same dataset with same input parameters. The overall observations tell that the FIPPI algorithm works well for both the datasets.

Acknowledgements The Authors thankfully acknowledge to DST, GOI, for financial assistance [MRP No. BDID/01/23/2014-HSRS/35 (ALG-V)] and giving AVIRIS-NG data. The authors also extend sincere thanks to UGC SAP and DST-FIST for providing lab facilities to Department of CSIT, Dr. B. A. M. University, Aurangabad-(MS), India. We also express our gratitude towards USGS for providing EO-1 Hyperion data.

References

1. K.V. Kale et al., A research review on hyperspectral data processing and analysis algorithms. Proc. Natl. Acad. Sci., India, Sect. A **87**(4), 541–555 (2017)
2. A. Plaza, et. al., A quantitative and comparative analysis of endmember extraction algorithms from hyperspectral data. IEEE Trans. Geosci. Remote Sens **42**(3), 650–663 (2004)
3. C.I. Chang, Q. Du, Estimation of number of spectrally distinct signal sources in hyperspectral imagery. IEEE Trans. Geosci. Remote Sens **42**(3), 608–619 (2004)
4. A. Plaza, et. al., An experimental evaluation of endmember generation algorithms, in*Proceedings of SPIE*, vol. 5995 (2005), pp. 599501-1-9)
5. P.J. Martínez et al., Endmember extraction algorithms from hyperspectral images. Ann. Geophys. **49**(1), 93–101 (2006)
6. C.I. Chang et al., A new growing method for simplex-based endmember extraction algorithm. IEEE Trans. Geosci. Remote Sens. **44**(10), 2804–2819 (2006)
7. J. Plaza, E.M. Hendrix, I. García, G. Martín, A. Plaza, On endmember identification in hyperspectral images without pure pixels: a comparison of algorithms. J. Math. Imaging Vision **42**(2), 163–175 (2012)
8. Boardman, et. al.,*Mapping Target Signatures Via Partial Unmixing of AVIRIS Data* (1995)
9. M.E. Williams, N-FINDR: an algorithm for fast autonomous spectral end-member determination in hyperspectral data, in*SPIE Conference on Imaging Spectrometry* (1999), pp. 266–275
10. C.I. Chang, A. Plaza, A fast iterative algorithm for implementation of pixel purity index. IEEE Geosci. Remote Sens. Lett. **3**(1), 63–67 (2006)
11. A. Plaza, C.I. Chang. Impact of initialization on design of endmember extraction algorithms. IEEE Trans. Geosci. Remote Sens **44**(11), 3397–3407
12. R.O. Green, et. al., *Airborne Visible/Infrared Imaging Spectrometer Next Generation (AVIRIS-NG) Data User's Guide—India Campaign 2015* (2015). Available at: http://vedas.sac.gov.in:8080/aviris/pdf/20150726_AVRISINGDataGuide_v4.pdf, pp. 1–13

13. J. Pearlman, et. al, Overview of the hyperion imaging spectrometer for the NASA EO-1 mission. in *IEEE 2001 International Geoscience and Remote Sensing Symposium*, vol. 7 (2001), pp. 3036–3038
14. A. Plaza, C.I. Chang, Fast implementation of pixel purity index algorithm. in *Proceedings of the SPIE Conference on Algorithms and Technologies for Multispectral, Hyperspectral, and Ultra spectral Imagery XI*, vol. 5806 (Mar 2005), pp. 307–317)
15. Wu et al., Real-time implementation of the pixel purity index algorithm for endmember identification on GPUs. IEEE Geosci RS Lett **11**(5), 955–959 (2014)
16. Y. Li, C. Gao, S.Y. Chen, C.I. Chang, Endmember variability resolved by pixel purity index in hyperspectral imagery. in *Satellite Data Compression, Communications, and Processing X*, vol. 9124 (May 2014), p. 91240I (International Society for Optics and Photonics)
17. A. Plaza, C.I. Chang, An improved N-FINDR algorithm in implementation, in *Proceedings of SPIE*, vol. 5806 (July 2005), pp. 298–304
18. Q. Du, N. Raksuntorn, N.H. Younan, R.L. King, Variants of N-FINDR algorithm for endmember extraction. in *Proceedings of SPIE*, vol. 7109 (Oct 2008), pp. 71090G1-8
19. L. Ji et al., Modified N-FINDR endmember extraction algorithm for remote-sensing imagery. Int. J. Remote Sens. **36**(8), 2148–2162 (2015)
20. C.I. Chang et al., Comparative study and analysis among ATGP, VCA, and SGA for finding endmembers in hyperspectral imagery. IEEE J. Sel. Top. Appl. Earth Observations Remote Sens. **9**(9), 4280–4306 (2016)
21. D.M. Rogge et al., Integration of spatial–spectral information for the improved extraction of endmembers. Remote Sens. Environ. **110**(3), 287–303 (2007)

Correction of Segmented Lung Boundary for Inclusion of Injured Diffused Regions from Chest HRCT Images

Rashmi Vishraj, Savita Gupta and Sukhwinder Singh

Abstract Lung field extraction has received considerable attention during the last four decades. Because, this is one of the important step in radiologic pulmonary image analysis. Various methods have been developed for accurate lung extraction. However, those methods work well for homogenous regions but often get failed to detect diffused regions. Here, the proposed algorithm is evaluated on Lung Tissue Research Consortium (LTRC) to include homogenous and injured diffused regions. Automated methods for segmenting several anatomical structures in chest CT images are also proposed: namely thorax extraction, large airway elimination, lung identification and boundary correction. Proposed approach is compared and validated with the existing Region-based Level Set Method (RbLSM).

Keywords Lung tissue research consortium (LTRC) · Lung field extraction
Diffused parenchymal lung diseases

1 Introduction

Lung is an important organ in human body. It performs exchange of oxygen and carbon dioxide. However, prolonged habit of smoking, increase of tobacco use, exposure to pollution and other carcinogenic chemicals effect the normal functionality of lungs which leads to lung cancer and other lung diseases commonly called Diffused Parenchymal Lung Diseases (DPLDs) [1]. Such diseases are one of the primary reason of deaths universally. Accounting the latest statistics of WHO report, the deaths owing to lung diseases in India were increasing accounting to 11% of the total deaths. A lot of around 142.09 in every one lakh, expire due to some type of lung disease

R. Vishraj (✉) · S. Singh
University Institute of Engineering and Technology, Panjab University, Chandigarh, India
e-mail: rashmi.cse.uiet@gmail.com

S. Gupta
Department of Computer Science and Engineering, University Institute of Engineering and Technology, Panjab University, Chandigarh, India

C. R. Krishna et al. (eds.), *Proceedings of 2nd International Conference on Communication, Computing and Networking*, Lecture Notes in Networks and Systems 46, https://doi.org/10.1007/978-981-13-1217-5_45

or the other giving India the unconvinced ranking first in lung disease deaths in the world [2]. The American Lung Association approximated more than 400,000 deaths in every year in the United States linked only to lung diseases [3]. The number of lung disease patients is rapidly increasing despite the advancement in diagnosis and treatment. Some revolutionary efforts need to be made to decrease the rate of morbidity and premature deaths associated to such diseases.

Earlier, biopsies were done to inspect the inside part of the human body. But, with the advancement in digital imaging technology, images of the pathology bearing region inside the human body are taken by clinicians to visualize the disease and provide safer diagnosis. To inspect lung related diseases, X-ray scans are done followed by Computed Tomography (CT) scans [4]. Therefore, CT imaging is used to detect lung diseases. According to the report of the Early Lung Cancer Action Project (ELCAP) a low dose of a Thoracic Computed Tomography (CT) images contain more information about tissue of body part than X-ray images [5]. This helps in detection of a wide range of health problems which include cysts, infections, appendicitis and cancer. As, Computed Tomography (CT) is highly preferred because it is non-invasive diagnosis of lung diseases. However, some limitations are also associated with the use of CT images. A single CT scan examination generates large number of slice images. Thus, radiologists and surgeons have to outlook series of CT slices on a screen or films, which is actually a tedious job, consumes a lot of time and often lead to subjectivity issues. To overcome these issues, extensive research is going on for the development of Computer Aided Diagnosis (CAD) system for analyzing bio-medical imaging. As, CAD systems provide the radiologists with numerical evaluation in order to assist in the diagnosis of various diseases on medical images [6].

A lot of research has been detailed in the literature for lung segmentation. In which, the lung region is extracted from the CT scan image so called the preprocessing step [7]. Various state-of-the-art methods of lung segmentation confide on a large gray value disparity among parenchyma region and neighboring tissues. But, these methods usually fail on scans with lungs that contain dense pathologies, and such scans occur frequently in clinical practice [8, 9]. Existing fully automated lung segmentation methods often do not succeed because of their lack of ability to proficiently include human input to deal misclassifications errors. This paper presents a lung extraction method for CT images that is highly proficient and robust [10]. The identification of DPLDs from CT is not an easy assignment for radiologists as the complication and discrepancy between the visual patterns in the diseased images.

Moreover, the presence of disease in the lungs hampers the performance of the software attempting to locate lung margins. As, consolidation pattern along the pleural margin of the lungs lead an inaccurate boundary delineation in which the consolidation is treated as outside the lung boundary. In the literature, almost all image segmentation techniques for lung extraction function well only with absent or minimal lung pathologic conditions [11].

2 Related Work

Human lung has complex anatomy, not any particular algorithm is proficient enough to segment out the pulmonary regions. Such methods are region growing, thresholding, morphology and clustering methods. Such methods are generally combined to make more efficient and robust methods. These are usually classified into two groups, i.e., 2D methods and 3D methods.

In 2D methods, lung region is segmented slice by slice. Antonelli fused traditional methods such as thresholding and morphological operations to extract lung region from CT images [12]. Shojaii et al. [13] used 2D marker based watershed transform to detect the lung borders [13]. Li et al. [14, 15] discussed optimal thresholding and mathematical morphology operations for lung extraction [14, 15]. Their method removed Partial Volume Effect (PVE) and also separates left and right lung. Chama et al. [16] used mean shift segmentation followed by geometrical properties for lung field extraction on LIDC dataset [16]. Wu et al. [17] used a graph cut based method for lung extraction in endoscopic videos [17]. Vishraj et al. [18] developed an Intuitionistic Fuzzy domain Region-based Level Set Method (IFRbLSM) to detect pleura attached lung nodules in HRCT images. As, automated delineation of exact boundaries of lung region in Thoracic Computed Tomography (CT) images is one of the difficult tasks owing to the weak edges. Because the presence of disease at pleura region merges the boundary of lung tissue with the chest wall. They used Intuitionistic fuzzy sets exploiting the inherent capability to handle fuzzy boundaries [18].

Some of the 3D segmentation methods are discussed in this section. De et al. [19] developed a framework by combining 3D binary morphological operations and 3D region growing to extract lung region from the CT images [19]. Wei et al. [20] presented a fully automatic method for extraction of lung field region. In this, several methods were combined i.e. optimal iterative thresholding, 3D connectivity labeling and 3D region growing method for the extraction of lung parenchyma [20]. Choi et al. [21] developed a method for extraction of lung boundary by including the juxta-pleural nodules. In this, thresholding and 3D connected component analysis methods were used to extract lung tissue [21].

3 Proposal Overview

In this research work, an easy and accurate lung field segmentation technique is proposed and detailed in the following subsection. Proposed strategy consists of four phases, i.e., thorax extraction, large airway elimination, lung identification, and lung boundary correction.

3.1 Thorax Extraction and Large Airway Elimination

In this module, Level Set Method followed by thresholding is used for rough extraction of thorax. The objective of thorax extraction is to remove the regions other than the human body. Expecting accurate thorax extraction by using only thresholding technique is not a right approach because to the subsequent reasons. First, Hounsfield unit (HU) of the human fat and the patient bed are very close to each other. Second, even in HRCT images PVE is seen because of limited resolution which makes segmentation more difficult. So to keep it simple, we have cropped the image by covering the thorax region within a window. This window is computed by taking ± 150 along y axis around centroid of the image. It has been observed that window size works well for all the images.

In Large airway elimination module, trachea is removed using connected component analysis, where disconnected components are identified and given different labels. The following constraints are then applied on connected components to identify the trachea and non-trachea regions, i.e., area and circularity [22].

- *Area*: Is computed of an individual connected component by analyzing region properties.
- *Circularity*: It is assumed that trachea is similar to circular shape. So any component having circularity less than 0.8 and minimum and maximum radius between 5 and 22 mm is identified as trachea. For computing circularity (i.e., $c_{\max}$) we compute the estimated radius from the area parameter of the connected components by using Eq. 1, where r_{est} is the estimated radius of the trachea computed using Eq. 2

$$c\max = 1 - \left(\frac{r\max}{r_{\text{est}}}\right) \tag{1}$$

$$r_{\text{est}} = \sqrt{\frac{\text{Area}}{2\pi}} \tag{2}$$

For trachea detection among connected components we check condition mentioned in Eq. 3. Here, $r_{\min}$ and $r_{\max}$ are assumed as the minimum radius and maximum radius of the trachea.

$$r_{\min} > 5 \;\&\; r_{\max} < 22 \;\&\; (\text{abs}(c_{\max}) < 0.8) \tag{3}$$

3.2 Lung Identification and Boundary Correction

Lung field region is extracted from real time CT images to speed up the computation process. This work is an extension to our previous work where Region-based Level Set Method (RbLSM) is used for the extraction of the lung field region from the HRCT images. The technique [18] used worked efficiently on LIDC dataset but would not work so well on LTRC dataset. Because, it ignores various small and diffused lung regions which leads to wrong results. So, RbLSM is fused with concavity patching method which is earlier used to fill the concavities due to juxta-pleural nodules [23]. In this work, morphological closing is used to correct the lung contour. In morphological closing, dilation operation is performed followed by erosion having same structuring element for both i.e. disk having radii 10 which was empirically found suitable for proposed method. In this way, new boundary is constructed, which detects the concavities and segregated and diffused lung regions.

4 Results and Discussion

The proposed methods are executed and tested on Lung Tissue Research Consortium (LTRC) [24]. Repository contains 128 patients affected with 13 ILDs having 108 image series with 411 annotated lung tissue patterns. The research work is implemented on MatLab 2016b platform and run on PC, having an Intel core i5-5200U processor at 2.20 GHz with 8 GB RAM.

As described in previous section it is a difficult task to include pleura injured diffused lung regions. So, to include those regions concavity patching method is fused with the RbLSM and their results are presented in Figs. 1, 2, 3 and 4. In all figures (a) represents original image, (b) represents the extracted lung boundary, using RbLSM (c) shows the corrected lung boundary using the proposed approach (d) represents the lung field region obtained using RbLSM, where (e) shows the corrected lung field region using the proposed method. From Fig. 1b, we can easily visualize the concavity regions and the diffused lung regions. By using proposed method lung contour is corrected (c) and an accurate lung field region is extracted and represented in Fig. 1e.

Here in results we have shown the results on four image slices where Fig. 1a represents the Original image of Patient 1231 Series 2 slice 150, in Fig. 2a Original image of Patient 8843 Series 4 slice 99, in Fig. 3a Original image of Patient 9395 Series 2 slice 350 and in Fig. 4a Original image of Patient 14902 Series 2 slice 100 is represented.

5 Conclusion

In this research paper, a method is proposed for correct lung boundary extraction to include diffused lung regions. Here, we have used morphological and Level Set Methods on individual slices to extract the lung field region. Later, stack of slices is formed to visualize the whole lung organ. From the literature, we have deduced that Level Set methods perform better than traditional methods. So, in this work,

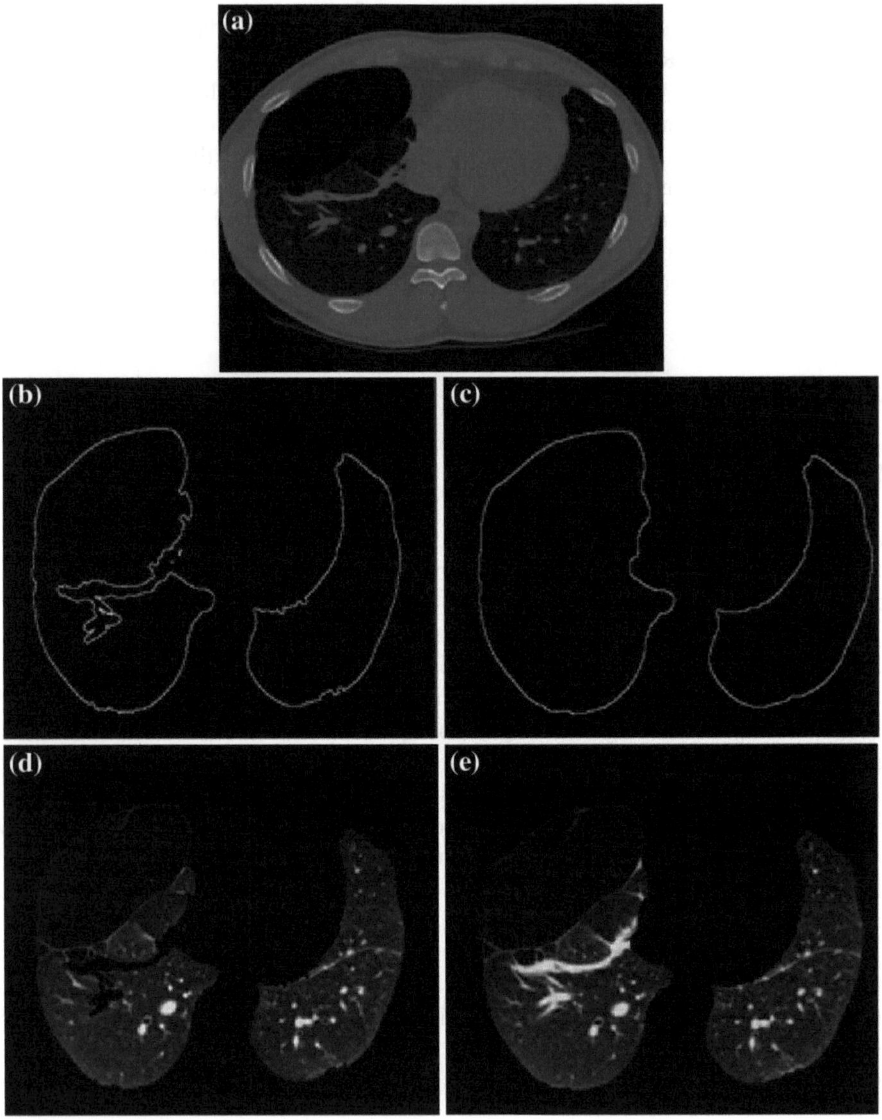

Fig. 1 Result of patient 1231 series 2 slice 150

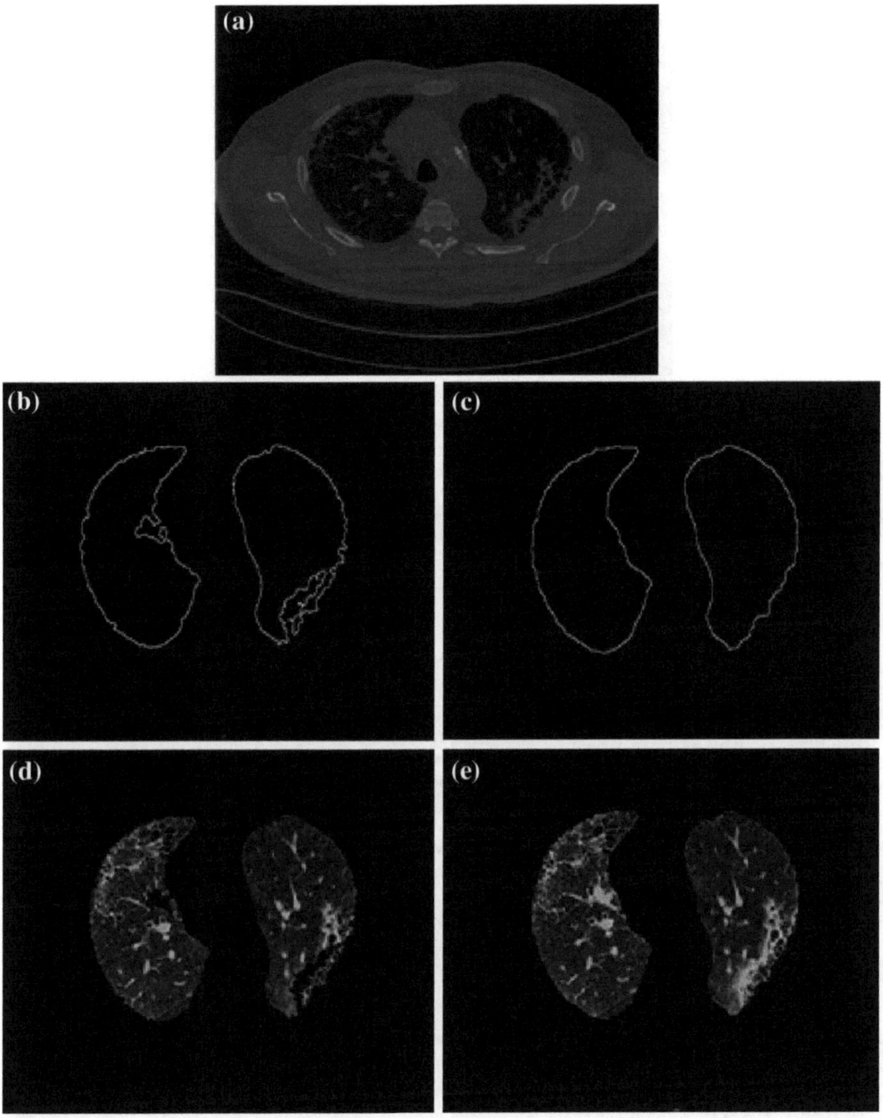

Fig. 2 Result of patient 8843 series 4 slice 99

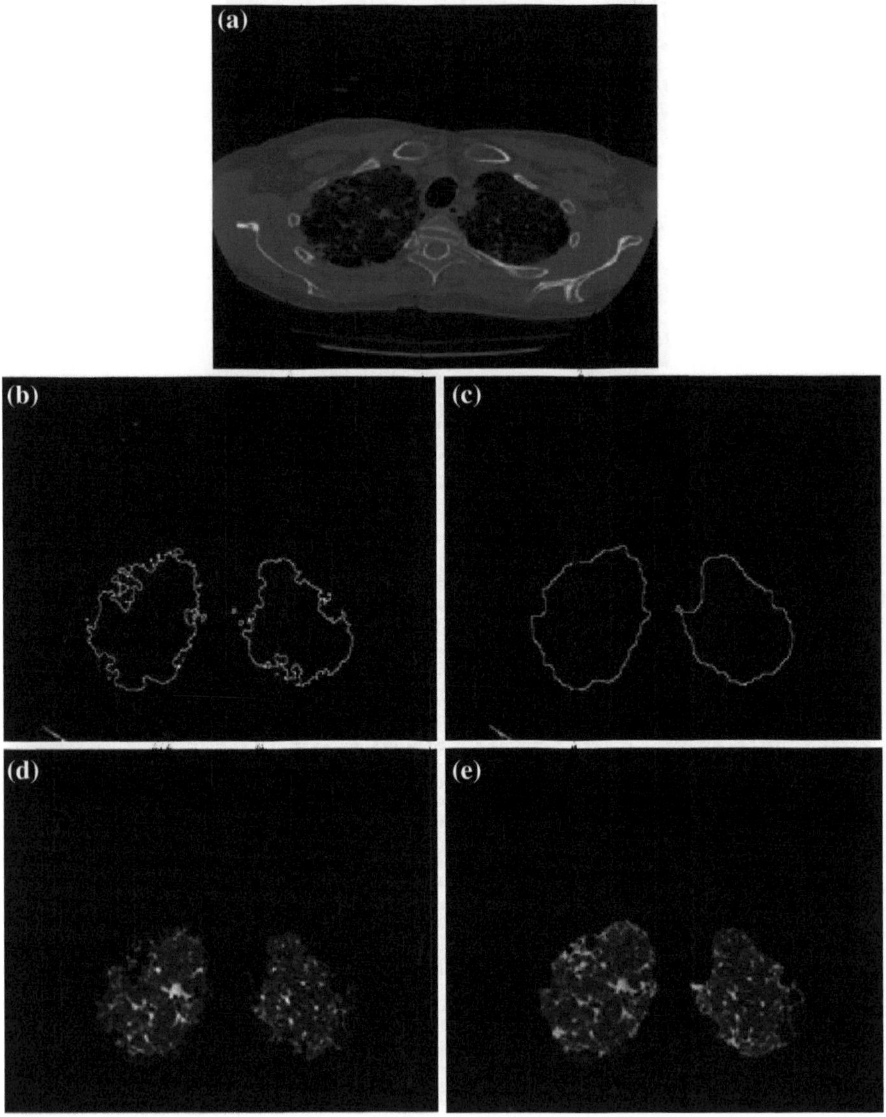

Fig. 3 Result of patient 9395 series 2 slice 350

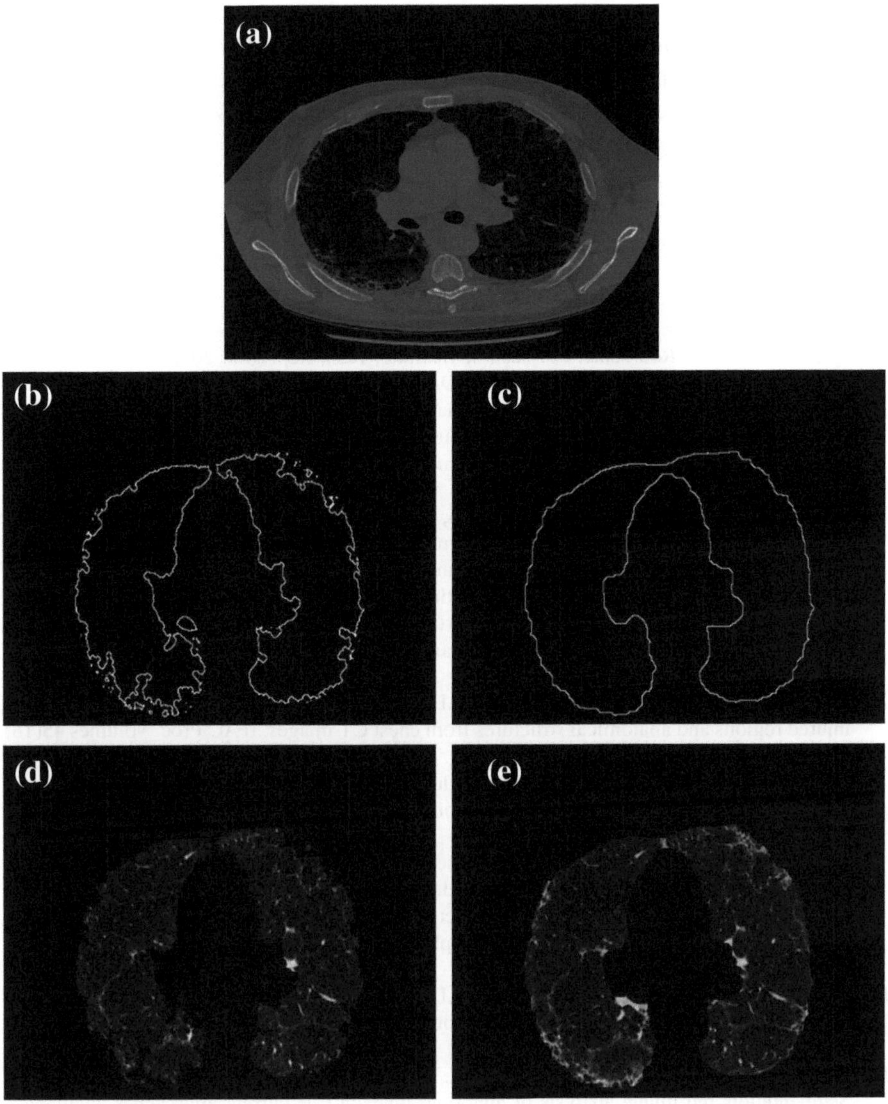

Fig. 4 Result of Patient 14902 Series 2 slice 100

proposed technique is compared with RbLSM. By analyzing results we can conclude that proposed technique extracts the accurate lung contour than the RbLSM.

In the future, we plan to incorporate the 3D segmentation methods for volumetric extraction of lung filed region. The proposed method will also be tested on other databases of parenchymal diseases.

References

1. C. Demir, B. Yener, *Automated Cancer Diagnosis Based on Histopathological Images: A Systematic Survey*. Rensselaer Polytechnic Institute, Tech. Rep. (Mar 2005)
2. G. Rao, *India Tops World in Lung Disease Deaths* [Online]. The Hindu. Available: http://www.thehindu.com/news/national/andra-pradesh/India-tops-world-in-lung-disease-deaths/article7372468.ece. Accessed 17 Jan 2017
3. American Lung Association, *Estimated Prevalence and Incidence of Lung Disease* (Tech. Rep, Washington, DC, Apr 2013)
4. A. Depeursinge, J. Iavindrasana, G. Cohen, A. Platon, P.A. Poletti, H. Müller, Lung tissue classification in 21st IEEE International Symposium on HRCT data Integrating the Clinical Context, in Computer-Based Medical Systems, 2008. CBMS'08 (17 June 2008), pp. 542–547
5. H.K. Weir, M.J. Thun, B.F. Hankey, L.A. Ries, H.L. Howe, P.A. Wingo, A. Jemal, E. Ward, R.N. Anderson, B.K. Edwards, Annual report to the nation on the status of cancer, 1975–2000, featuring the uses of surveillance data for cancer prevention and control. J. Nat. Cancer Inst. **95**(17), 1276–1299 (3 Sept 2003)
6. Y. Iwao, T. Gotoh, S. Kagei, T. Iwasawa, M.D. Tsuzuki, Integrated lung field segmentation of injured regions and anatomical structures from chest CT images. IFAC Proc. Volumes **45**(18), 85–90 (1 Jan 2012)
7. S. Zhou, Y. Cheng, S. Tamura, Automated lung segmentation and smoothing techniques for inclusion of juxta-pleural nodules and pulmonary vessels on chest CT images. Biomed. Signal Proc. Control. **13**, 62–70 (30 Sept 2014)
8. Y. Zhou, J. Bai, Multiple abdominal organ segmentation: an atlas-based fuzzy connectedness approach. IEEE Trans. Inf. Technol. Biomed. **11**(3), 348–352 (2007)
9. J.K. Udupa, S. Samarasekera, Fuzzy connectedness and object definition: theory, algorithms, and applications in image segmentation. Graphical Models Image Proc. **58**(3), 246–261 (1 May 1996)
10. Y. Iwao, T. Gotoh, S. Kagei, T. Iwasawa, M.D. Tsuzuki, Integrated lung field segmentation of injured regions and anatomical structures from chest CT images. IFAC Proc. Volumes **45**(18), 85–90 (1 Jan 2012)
11. A. Mansoor, U. Bagci, B. Foster, Z. Xu, G.Z. Papadakis, L.R. Folio, J.K. Udupa, D.J. Mollura, Segmentation and image analysis of abnormal lungs at CT: current approaches, challenges, and future trends. Radio Graph. **35**(4), 1056–1076 (14 July 2015)
12. M. Antonelli, B. Lazzerini, F. Marcelloni, Segmentation and reconstruction of the lung volume in CT images, in Proceedings of the 2005 ACM Symposium on Applied computing, ACM (13–17 Mar 2005), pp. 255–259
13. R. Shojaii, J. Alirezaie, P. Babyn, Automatic lung segmentation in CT images using watershed transform, in IEEE International Conference on Image Processing, 2005. ICIP 2005, vol. 2 (11–14 Sept 2005), pp. II–1270
14. W. Li, S.D. Nie, J.J. Cheng, A fast automatic method of lung segmentation in CT images using mathematical morphology, in World congress on medical physics and biomedical engineering (Springer, Berlin, Heidelberg, 2007, 2006), pp. 2419–2422
15. C. Lei, L. Xiaojian, Z. Jie, C. Wufan, Automated lung segmentation algorithm for CAD system of thoracic CT. J. Med. Coll. PLA. **23**(4), 215–222 (1 Aug 2008)

16. C.K. Chama, S. Mukhopadhyay, P.K. Biswas, A.K. Dhara, M.K. Madaiah, N. Khandelwal, Automated lung field segmentation in CT images using mean shift clustering and geometrical features, in *Medical Imaging 2013: Computer-Aided Diagnosis, International Society for Optics and Photonics*, vol. 8670 (28 Feb 2013), p. 867032
17. S. Wu, M. Nakao, T. Matsuda, Automatic GrabCut based lung extraction from endoscopic images with an initial boundary, in 2016 IEEE 13th International Conference on Signal Processing (ICSP) (6 Nov 2016), pp. 1374–1378
18. R. Vishraj, S. Gupta, S. Singh, Intuitionistic fuzzy domain level set method for automatic delineation of juxta-pleural pulmonary nodules in thoracic CT images. Curr. Med. Imaging Rev. **14**(2), 280–288 (1 Apr 2018)
19. G. De Nunzio, E. Tommasi, A. Agrusti, R. Cataldo, I. De Mitri, M. Favetta, S. Maglio, A. Massafra, M. Quarta, M. Torsello, I. Zecca, Automatic lung segmentation in CT images with accurate handling of the hilar region. J. Digit. Imaging. **24**(1), 11–27 (1 Feb 2011)
20. Y. Wei, G. Shen, J.J. Li, A fully automatic method for lung parenchyma segmentation and repairing. J. Digit. Imaging. **26**(3), 483–95 (1 June 2013)
21. W.J. Choi, T.S. Choi, Automated pulmonary nodule detection based on three-dimensional shape-based feature descriptor. Comput. Methods Programs Biomed. **113**(1), 37–54 (31 Jan 2014)
22. P.C. Lo, *Segmentation of Lung Structures in CT* (Doctoral dissertation, Faculty of Science, University of Copenhagen)
23. R.N. Jorge, J.M. Tavares, M.P. Barbosa, A.P. Slade (eds.), *Technology and Medical Sciences*. (CRC Press 2011)
24. B. Bartholmai, R. Karwoski, V. Zavaletta, R. Robb, D.R. Holmes, The Lung Tissue Research Consortium: An extensive open database containing histological, clinical, and radiological data to study chronic lung disease, In *The Insight Journal-2006 MICCAI Open Science Workshop*. (17 Aug 2006) Available: http://hdl.handle.net/1926/221.

A Review of Passive Image Cloning Detection Approaches

Amit Doegar, Maitreyee Dutta and Gaurav Kumar

Abstract Image Cloning is also known as copy-paste/copy-move image forgery where a fragment of the image or object is copy and paste into some other area of the same image. It is a type of image tampering with a motive either to hide the object or to falsify the information of the image. Thus it makes difficulty in the trustworthiness of the images in various real-time applications. With the easy accessibility of image manipulation software, the number of cases of image tampering is increasing. Hence there is a growing need for robust, accurate and efficient digital image forgery detection approaches. This review presented a brief discussion of various approaches for image cloning detection and will be useful for the researchers as a future direction in the area of image forensics and image forgery detection approaches implementation.

Keywords Image forgery · Image cloning detection · Image forensics

1 Introduction

Images become a significant resource of information in the digital world as they are the fastest means of information and medium of communication. Images are being used in various spheres of real-time applications like science, law, education, politics, media, military, medical imaging and diagnosis, art piece, digital forensics, intelligence, sports, scientific publications, journalism, photography, social media and business. In recent years, forged images have affected the above-mentioned

A. Doegar (✉) · M. Dutta
Department of Computer Science and Engineering, NITTTR, Chandigarh, India
e-mail: amit@nitttrchd.ac.in

M. Dutta
e-mail: maitreyee@nitttrchd.ac.in

G. Kumar
Magma Research and Consultancy Services, Ambala, India
e-mail: kumargaurav.in@gmail.com

C. R. Krishna et al. (eds.), *Proceedings of 2nd International Conference on Communication, Computing and Networking*, Lecture Notes in Networks and Systems 46, https://doi.org/10.1007/978-981-13-1217-5_46

Fig. 1 Example of image cloning that appeared in press in July 2008 [7, 8]

application areas [1, 2]. Digital image acts a significant part of different technologies and fields. The use of digital cameras, personal computers, and sophisticated image processing software are available for modification and for manipulation of images. These tools are scalable and provide user interface features. Manipulating and tampering the images today can be effectively accomplished not only by specialists but also by novice users [3]. These tampered images are not recognizable and so real in perception in a way that authenticity is lost [4]. Therefore, integrity and authenticity verification of images has gained researchers attention in image processing field.

The basic aim of tampering or image cloning is the modification or falsification of images to distort some data in the image [5, 6]. The cloned image as shown in Fig. 1 shows the four missiles but only three of them are actual: two other sections (encircled) imitated other image sections by exploiting the image cloning.

In Sect. 2, various types of image forgery have been described. The approaches to detect any type of forgery are classified into two categories namely, active and passive approaches.

In active approaches images needs to protect through a digital signature or through watermarking techniques whereas passive approaches do not require any kind of pre-embed operation of digital signature or watermark. The drawback of active approaches is that it needs to pre-embed either with a digital signature or with watermarking, whereas a large number of images present today on web, social media and other applications are not active in nature. Thus we have focused on the review of various passive techniques for image cloning detection which are reviewed in Sect. 3.

2 Types of Image Tampering

The tampering or simple manipulation of the digital images is having a specific taxonomy with the following categories:

A. Image Cloning or Copy-Move Tampering: In this method, cloning represents the false or tampered region which is not present in the source image. This type of image comes from copy-move image forgery in which few or various portions of images can be moved or replaced within the same image and a cloned image is produced [7].

B. Image Splicing: In this method, a distinct fake image is produced by combining two or more images. This type of tampering is also known as photo-montage which uses image splicing so that two images can be sticked together using various image editing tools [9].
C. Image Retouching: In this method, the image is less modified or slight changes are made or some features of the image are enhanced [5].
D. Image Resampling (Mirroring, Stretching, Rotation and Resize): For making a composite of two objects it is achievable that one object might be resized, rotated, stretched, or mirrored to match the relative dimensions or height of another object. This scheme introduces non-negligible changes which can be utilized to identify image tampering created by resampling [4].

3 Related Work

In this section, the literature is reviewed to identify the various approaches for image cloning detection:

Costanzo et al. [10] reviewed the attacks which are capable to remove shift key-points to compromise the SIFT-based detection of copy-move forgery and proposed three approaches to identify the images whose SIFT key-points were globally or locally removed. Ye et al. [11] proposed the natural authentication code for "blocking artifact caused by JPEG image compression". Authors also described a technique based on an estimated quantized table using the DCT coefficient histogram. Amerini [7] proposed a SIFT-based approach to deal with multiple cloning in the tampered image. Gomase and Wankhade [12] introduced that detecting forged images is one of the challenging tasks and proposed a method which combines operations of both copy-move and fast copy-move techniques to detect the forgery. Huang et al. [13] introduced a technique which makes use of quantization matrix to recognize double JPEG compression in Image Tampering. Bianchi and Piva [14] suggested a strategy to detect the non-aligned double JPEG compression in the light of perception using SVM classifier for the dataset of 100 uncompressed TIFF images. Hsu et al. [15] proposed a histogram of orientated Gabor magnitude (HOGM) based scheme for the detection of copy-move forgery on CoMoFoD database and applied Gabor filter and sorted feature vector with lexicographical sorting to obtain the similar features from different blocks. Ardizzone et al. [16] implemented a hybrid approach that compares triangles instead of single points or blocks. Triangles are matched based on the shapes and color information. Ansari et al. [17] mentioned the overview of various techniques for image forgery detection, that are format based, physical environment based, geometry-based, camera-based and reviewed the approaches for pixel-based image forgery detection. Ardizzone et al. [18] presented an approach based on SIFT that detect the alterations in the image and compared the proposed method, with respect to keypoint clustering, texture analysis, and cluster matching method.

Ryu et al. [19] explored the method based on "features based on Zernike moments of circular blocks" to detect the copy-move forged regions. Bayram et al. [20] presented Fourier-Mellin Transform (FMT) and counting bloom filter based approach for image forgery detection. Li et al. [21] proposed Discrete Wavelet Transformation (DWT) based approach for image tampering detection. Ghorbani and Faraahi [22] presented the improved algorithm based on DWT and DCT Quantization Coefficient Decomposition to detect the image cloning tampering. Sadeghi et al. [23] divided image into blocks and applied Fourier Transform and calculated each block's statistical characteristics and find similar blocks to detect the duplicated areas. Muhammad et al. [24] proposed two methods namely Local Binary Pattern (LBP) and steerable pyramid transform (SPT) to extract features and applied Support Vector Machine to detect the passive image forgery. Jaberi et al. [25] proposed a Mirror Reflection Invariant Feature Transform (MIFT) based approach. Authors used dense features rather than pixel correlation and concluded that MIFT gives higher accuracy for short duplicate region. Kang and Wei [26] proposed a robust and efficient method based on SVD for dimension reduction and image feature vector extraction. In the proposed approach authors applied lexicographical sorting on rows and column vectors to detect the tampered image. Chihaoui et al. [27] proposed an automatic hybrid method using the SIFT and SVD for detecting the similar regions which are duplicated. Features are extracted using SIFT descriptors and use similarity matrix calculation and proximity matrix calculation for feature matching. Cao et al. [8] divide the image into fixed-size blocks and implemented DCT based approach to detect the duplicated area. Fridrich et al. [28] developed two algorithms for detection for copy-move forgery, one is on the basis of exact match and other on the basis of the approximate match for DCT coefficients. Liu et al. [29] developed an integrated approach based on local features and SIFT. This approach is robust to complex geometrical transform rotation, scaling, and illumination changes. Huan et al. [30] proposed the DCT based approach to obtain the features and applied the package clustering algorithm to improve the detection precision. Zhang et al. [31] focused on post-processed forgery operations, mainly geometric distortions. Authors introduced Analytic FMT (AFMT) and focused on its discretization. AFMT is described in polar coordinates, authors converted the coordinate system from polar to cartesian coordinates. Huang et al. [32] proposed approaches based on FFT, SVD, and PCA for copy-move tampering detection and for feature matching. Gan [33] proposed an approach which is robust against scaling and rotation based on AFMT. Correlated features were sorted using the lexicographic method and to identify similar continuous regions correlation coefficient was applied. Yuan et al. [34] proposed a Log-Polar domain using band limitation based approach for duplication detection. Bo et al. [35] proposed a SURF (Speed up Robust Features) based approach to detect copy- move image forgery, which is invariant to scaling and rotation.

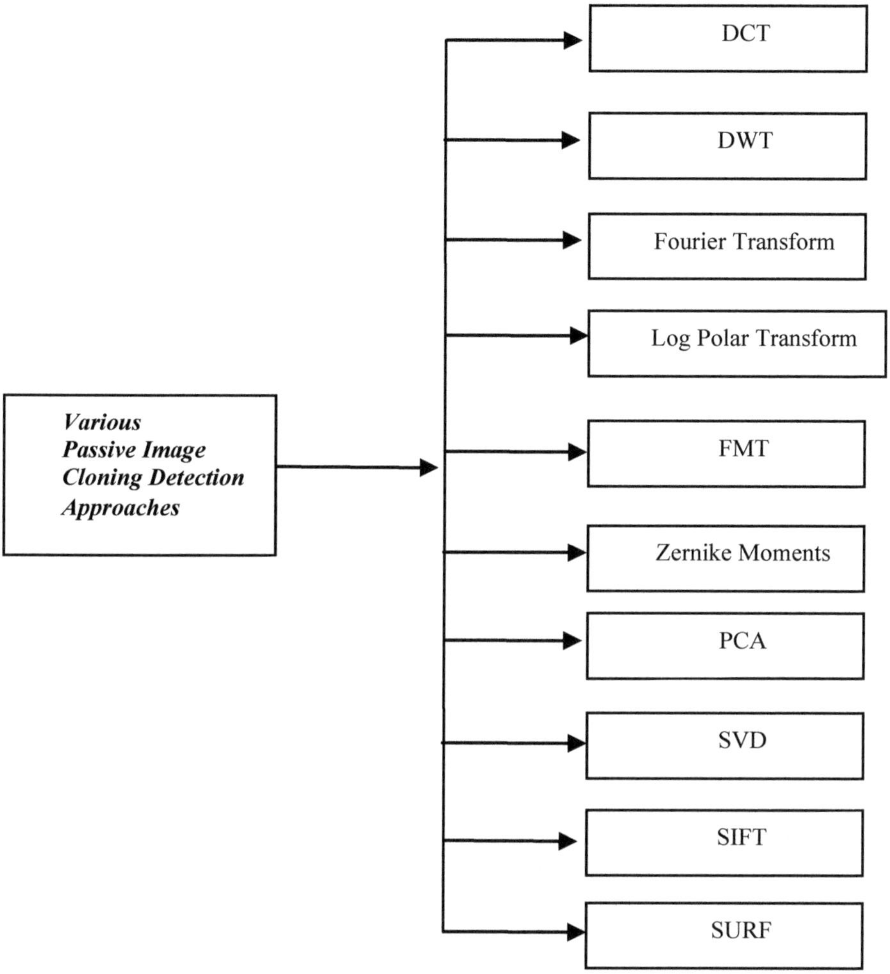

Fig. 2 Approaches for passive image cloning detection

4 Approaches for Passive Image Cloning Detection

The various approaches as mentioned in Fig. 2 for Image cloning detection are summarized in this section.

DCT is a transformation that does not actually perform compression. It is a preparation stage for coefficient quantization stage. This approach exploits DCT coefficients as features. DWT decomposes the signal into an orthogonal set of wavelets [36]. One of the considerable characteristics of Fourier transform is that irrespective of the type of signal, it is possible to describe the signal as a collection of sine and/or cosine waves multiplied by the weighting functions. Log-Polar resamples an image from a conventional grid to log-polar grid, and back [36]. Fourier-Mellin Transform (FMT) performs radial projection on the log-polar coordinate and are invariant to rotation, scale and translation [36]. Zernike Moments are characterized over the unit circle and are known for its computational speed. This technique is invariant to rotation up to 15 degrees. PCA aims to minimize the dimensionality of data into reduced dimensional feature space that describes the image data effectively and economically. Singular value decomposition (SVD) is another popular image transform that re-factors the image into three matrices. SIFT extracts stable points from images and attaches to them robust features [37]. SURF is a local feature detection algorithm that is based on the use of Hessian Matrix. SURF performs well as compared to the other algorithms because of the use of the integral image. The advantages and disadvantages of all the approaches are summarized in Table 1.

Table 1 Description of various Passive Image Cloning Approaches

Image cloning approaches	Authors	Advantages	Disadvantages
DCT	Fridrich et al. [28], Cao et al. [8], Zhao et al. [38], Wang et al. [30]	Robust against blur and noisy images	Fails in case of rotation and scaling, high computational complexity
DWT	Khan et al. [39], Li et al. [21]	Invariant to JPEG compression and additive noise. Reduce time complexity and robust to JPEG compression	Fails in case of rotation and scaling
Fourier transform	Sadeghi et al. [23], Ye et al. [11]	Robust to compression, scaling, blurring, and noise	Limited to rotation invariant up to 10° and scaling up to 10%
Log-polar transform	Bravo-Soloria and Nandi [40], Yuan et al. [34], Myna et al. [41]	Invariant to reflection and rotation	Performance of the approach degrades when tampering consists of blur and scaling
FMT	Bayram et al. [20], Guo et al. [42]	Capable to identify the tampered images which are post-processed with geometric distortions like translation, rotation and gaussian noise	FMT features are insensitive to scaling up to 10%
Zernike moments	Ryu et al. [19], Liao et al. [43]	Robust against noise, JPEG compression, blurring and rotation	Cannot handle tampering against scaling or affine transformation

(continued)

Table 1 (continued)

Image cloning approaches	Authors	Advantages	Disadvantages
PCA	Popescu and Farid [44]	Efficient method, low false positives rate computational time	Sensitive to noise or lossy compression, high computational complexity
SVD	Zhao et al. [38], Chihaoui et al. [27]	Robust to JPEG compression up to QF 70, less time complexity	Not invariant to rotation and scaling
SIFT	Saleem et al. [45], Amerini et al. [7], Huang et al. [46], Chihaoui et al. [27]	Robust against noise attack, JPEG compression, rotation and scaling.	Partially invariant to illumination changes
SURF	Bo et al. [35]	Robust against geometric transformations like scaling and rotation	Generally doesn't work well with lighting changes and blur

Most of these approaches have been implemented uses the benchmark datasets as shown in Table 2. The image resolution in these datasets varies from 128×128 to 3888×2592 pixels.

Table 2 Existing datasets for Image forgery detection [7, 15, 34, 47–49]

Name	Total images	Image size
Columbia university	1845	128×128
Image forensics	10	200×200
CASIA v1.0	1725	374×256
CASIA v2.0	12614	240×160 to 900×600
Image manipulation	48	420×300 to 3888×2592
MICC-F220	220	722×480 to 800×600
MICC-F600	600	800×533 to 3888×259
MICC-F2000	2000	2048×1536
CoMoFoD	260	512×512 to 3000×200
CMFD_db	160	768×1024

5 Conclusion and Future Scope

The image cloning is applied to hide or falsify the important information. There are various approaches associated with the detection of forgery or cloning of images and failure rate is one of them that has an impact on the detection of image forgery. Some proposed algorithms are not able to detect actual cloned region, some have high time complexity and some are failed to detect cloned areas because of various postprocessing operations.

Thus a singular approach to detect image cloning is not sufficient to detect or deal with multiple types of postprocessing operations such as scaling, rotation, brightness adjustment, contrast enhancement or degradation, sharpening, blurring, additive noise, histogram equalization, cropping, recompression and geometric operations and it is a challenging task. Thus there is a need to develop a feature fusion based approach and machine learning based approach to deal with all such kind of post-processing operations in image cloning or tampering detection.

References

1. S. Murali, B.S. Anami, G.B. Chittapur, Detection of copy-create image forgery using luminance level techniques, in Proceedings of 3rd National Conference on Computer Vision, Pattern Recognition, Image Processing and Graphics, NCVPRIPG (2011), pp. 215–218
2. O.M. Al-Qershi, B.E. Khoo, Passive detection of copy-move forgery in digital images: state-of-the-art. Forensic Sci. Int. **231**(1–3), 284–295 (2013)
3. M.A. Quereshi, M. Deriche, A review on copy move image forgery detection techniques, in Proceedings of IEEE 11th International Multi-Conference on Systems, Signals and Devices (SSD) (2014), pp. 1–5
4. H. Farid, *Image Analysis*. [Online]. Available: http://www.cs.dartmouth.edu/farid. Accessed 22 Oct 2017
5. D. Chauhan, D. Kasat, S. Jain, V. Thakare, *Survey on Keypoint Based Copy-Move Forgery Detection Methods on Image*, vol. 85 (Elsevier, Science Direct, Procedia Computer Science, 2016), pp. 206–212
6. C.N. Bharti, P. Tandel, A survey of image forgery detection techniques, in Proceedings of IEEE International Conference on Wireless Communications, Signal Processing and Networking, (WiSPNET 2016) (2016), pp. 877–881
7. S.I. Amerini, L. Ballan, R. Caldelli, A. Del Bimbo, A SIFT-based forensic method for copy-move attack detection and transformation recovery. IEEE Trans. Inf. Forensics Secur. **6**(3), 1099–1110 (2011)
8. Y. Cao, T. Gao, L. Fan, Q. Yang, A robust detection algorithm for copy-move forgery in digital images. Forensic Sci. Int. **214**(1–3), 33–43 (2012)
9. G.B. Chittapur, B.S. Anami, S. Murali, H. Prabhakara, B.S. Anami, Comparison and analysis of photo image forgery detection techniques. Int. J. Comput. Sci. Appl. **2**(6), 45–56 (2012)
10. A. Costanzo, I. Amerini, R. Caldelli, M. Barni, Forensic analysis of SIFT keypoint removal and injection. IEEE Trans. Inf. Forensics Secur. **9**(9), 1450–1464 (2014)
11. S. Ye, Q. Sun, E. Chang, Detecting digital image forgeries by measuring inconsistencies of blocking artifact, in Proceedings of IEEE International Conference on Multimedia and Expo (July 2007), pp. 12–15
12. P.G. Gomase, N.R. Wankhade, Advanced digital image forgery detection: a review. IOSR J. Comput. Sci. (IOSR-JCE) **2014**, 80–83 (2014e-ISSN: 2278-0661, p-ISSN: 2278-8727)

13. F.J. Huang, J.W. Huang, Y.Q. Shi, Detecting double JPEG compression with the same quantization matrix. IEEE Trans. Inf. Forensics Secur. **5**(4), 848–856 (2010)
14. T. Bianchi, A. Piva, Detection of nonaligned double JPEG compression based on integer periodicity maps. IEEE Trans. Inf. Forensics Secur. **7**(2), 842–848 (2012)
15. C.M. Hsu, J.C. Lee, W.K. Chen, An efficient detection algorithm for copy-move forgery, Proceedings—2015 10th Asia Joint Conference on Information Security (2015), pp. 33–36
16. E. Ardizzone, A. Bruno, G. Mazzola, Copy-move forgery detection by matching triangles of keypoints. IEEE Trans. Inf. Forensics Secur. **10**(10), 2084–2094 (2015)
17. M.D. Ansari, S.P. Ghrera, V. Tyagi, Pixel-based image forgery detection: a review. IETE J. Educ. **55**(1), 40–46 (2014)
18. E. Ardizzone, A. Bruno, G. Mazzola, Detecting multiple copies in tampered images, in Proceedings of IEEE International Conference on Image Processing, (ICIP) (2010), pp. 2117–2120
19. S. Ryu, M. Lee, H. Lee, Detection of copy-rotate-move forgery using zernike moments, in Proceedings of Springer International Workshop on Information Hiding, Berlin Heidelberg, vol. 1 (2010), pp. 51–65
20. S. Bayram, H. Sencar, N. Memon, An efficient and robust method for detecting copy-move forgery, in Proceedings of IEEE International Conference on Acoustics, Speech and Signal Processing (2009), pp. 1053–1056
21. K.-F. Li, T.-S. Chen, S.-C. Wu, Image tamper detection and recovery system based on discrete wavelet transformation, in IEEE Pacific Rim Conference on Communications, Computers and Signal Processing, vol. 1 (2001), pp. 164–167
22. M. Ghorbani, A. Faraahi, DWT-DCT (QCD) based copy-move image forgery detection, in IEEE 18th International Conference on Systems, Signals and Image Processing (IWSSIP) (June 2011)
23. S. Sadeghi, H.A Jalab, S. Dadkhah, Efficient copy-move forgery detection for digital images, in Proceedings of World Academy of Science, Engineering and Technology(WASET), vol. 71 (2012), p. 755
24. G. Muhammad, M.H. Al-Hammadi, M. Hussain, A.M. Mirza, G. Bebis, Copy move image forgery detection method using steerable pyramid transform and texture descriptor. IEEE EuroCon **2013**(July), 1586–1592 (2013)
25. M. Jaberi, G. Bebis, M. Hussain, G. Muhammad, Accurate and robust localization of duplicated region in copy-move image forgery. Mach. Vis. Appl. **25**(2), 451–475 (2014)
26. X.K.X. Kang, S.W.S. Wei, Identifying tampered regions using singular value decomposition in digital image forensics, in Proceedings of International Conference on Computer Science and Software Engineering, vol. 3 (2008), pp. 926–930
27. T. Chihaoui, S. Bourouis, K. Hamrouni, Copy move image forgery detection based on SIFT descriptors and SVD matching, in Proceedings of the International Conference on Advanced Technologies for Signal and Image Processing (2014), pp. 125–129
28. A. Fridrich, B. Jessica, D. Soukal, A.J. Lukas, Detection of copy-move forgery in digital images, in Proceedings of Digital Forensic Research Workshop (2003)
29. B. Liu, C. Pun, A SIFT and local features based integrated method for copy-move attack detection in digital image, in Proceedings of IEEE International Conference on Information and Automation (2013), pp. 865–869
30. H. Wang, H.-X. Wang, X.-M. Sun, Q. Qian, A passive authentication scheme for copy-move forgery based on package clustering algorithm. Multimedia Tools Appl. **76**(10), 12627–12644 (2017)
31. J. Zhong, Y. Gan, Detection of copy–move forgery using discrete analytical Fourier-Mellin transform. Nonlinear Dyn. **84**(1), 189–202 (2016)
32. D.-Y. Huang, C.-N. Huang, H. Wu-Chih, C.-H. Chou, Robustness of copy-move forgery detection under high JPEG compression artifacts. Multimedia Tools Appl. **76**(1), 1509–1530 (2017)
33. Y. Gan, J. Zhong, Application of AFMT method for composite forgery detection. Nonlinear Dyn. **84**(1), 341–353 (2016)
34. Y. Yuan, Y. Zhang, S. Chen, H. Wang, Robust Region duplication detection on log-polar domain using band limitation. Arab. J. Sci. Eng. **42**(2), 559–565 (2017)

35. B. Xu, W. Junwen, L. Guangjie, D. Yuewei, Image copy-move forgery detection based on SURF, in IEEE International Conference on Multimedia information networking and security (MINES) (2010), pp. 889–892
36. [Online]. Available: http://klapetek.cz/wdwt.html. Accessed 2 Nov 2017
37. M. Sonka, V. Hlavac, R. Boyle, *Image Processing, Analysis and Machine Vision* (Cengage Publisher, 2017)
38. J. Zhao, J. Guo, Passive forensics for copy-move image forgery using a method based on DCT and SVD. Forensic Sci. Int. **233**(1–3), 158–166 (2013)
39. S. Khan, A. Kulkarni, Robust method for detection of copy-move forgery in digital images, in Proceedings of IEEE International Conference on Signal and image processing (ICSIP 2010) (2010), pp. 69–73
40. B.-S. Sergio, A.K. Nandi. Exposing duplicated regions affected by reflection, rotation and scaling, in IEEE International Conference on Acoustics, Speech and Signal Processing (ICASSP) (2011), pp. 1880–1883
41. A.N. Myna, M.G. Venkateshmurthy, C.G. Patil, Detection of region duplication forgery in digital images using wavelets and log-polar mapping, in International Conference on Computational Intelligence and Multimedia Applications (ICCIMA 2007) (2007), pp. 371–377
42. X. Guo, Z. Xu, Y. Lu, Y. Pang, An application of Fourier-Mellin transform in image registration, in Proceedings of the 5th International Conference on Computer and Information Technology (CIT) (2005), pp. 619–623
43. S.X. Liao, M. Pawlak, On the accuracy of zernike moments for image analysis. IEEE Trans. Pattern Anal. Mach. Intell. **20**(12), 1358–1364 (1998)
44. A.C. Popescu, H. Farid, *Exposing Digital Forgeries by Detecting Duplicated Image Regions*, vol. 2000 (Department Computer Science, Dartmouth College, Technology Report TR2004-515, 2004), pp. 1–11
45. S. Mariam, A key-point based robust algorithm for detecting cloning forgery, in IEEE International Conference on Control System, Computing and Engineering (ICCSCE), vol. 4 (2014), pp. 2775–2779
46. H. Hailing, W. Guo, Y. Zhang, Detection of copy-move forgery in digital images using SIFT algorithm, in IEEE Pacific-Asia Workshop on Computational Intelligence and Industrial Application, PACIIA'08, vol. 2 (2008), pp. 272–276
47. N.B.A. Warif, A.W.A. Wahab, M.Y.I. Idris, R. Ramli, R. Salleh, S. Shamshirband, K.-K.R. Choo, Copy-move forgery detection: survey, challenges and future directions. J. Netw. Comput. Appl. **75**, 259–278 (2016)
48. J. Zhong, Y. Gan, J. Young, L. Huang, P. Lin, A new block-based method for copymove forgery detection under image geometric transforms. Multimedia Tools Appl. **76**(13), 14887–14903 (2017)
49. [Online]. Available: http://forensics.idealtest.org/. Accessed 7 Nov 2017

Pixel Level Image Fusion Using Different Wavelet Transforms on Multisensor & Multifocus Images

Harpreet Kaur, Kamaljeet Kaur and Neeti Taneja

Abstract Image fusion is performed after a supreme step of image registration and it is a consolidative display method of two or more images which can improve the quality of an image. This paper presents the implementation and comparison among the image fusion discrete wavelet transform techniques that are SWT, DDB2WT, SIDWT, and DT-CWT. The performance comparison of these techniques are evaluated with the help of PSNR, RMSE, PFE, CF, RF, and SF metrics and concluded that DT-CWT gives better results among them.

Keywords Image fusion · SWT · SIDHWT · DDB2WT · DT-CWT
Quality metrics

1 Introduction

An Image Fusion is a process of obtaining a single fused informative resultant image by combining a set of that is more inclusive rather than any of the input images [1]. It provides the reliability and high accuracy by eliminating the duplicate data. Its extensive benefit in the field of medical, scrutiny, military applications, etc. [2, 3]."Wavelet Transform is a mathematical tool for seismic wave analysis" [4, 5]. The detail decomposition of an input image with wavelet transform is shown below in Fig. 1.

The Comparison of various techniques likes SWT (Stationary wavelet transform) [6, 7], DDB2WT (Discrete Daubechies (2,2) wavelet transform) [8, 9], Shift invari-

H. Kaur (✉) · K. Kaur · N. Taneja
CSE, Chandigarh University, Mohali, Punjab, India
e-mail: harpreet8307@gmail.com

K. Kaur
e-mail: kamaljeet.samra9@gmail.com

N. Taneja
e-mail: neeti.taneja30@gmail.com

C. R. Krishna et al. (eds.), *Proceedings of 2nd International Conference on Communication, Computing and Networking*, Lecture Notes in Networks and Systems 46, https://doi.org/10.1007/978-981-13-1217-5_47

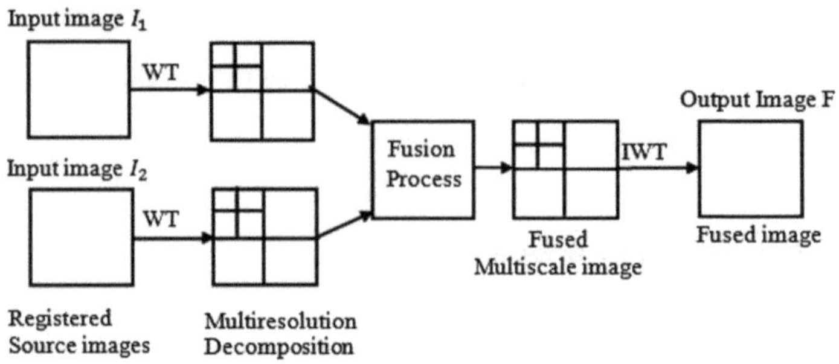

Fig. 1 General block diagram of an image fusion base on WT

ant Discrete Haar Wavelet Transform [10], DTCWT ('Dual tree complex wavelet transform') [11–14] of wavelet transform are implemented in this paper.

Pre requesting step for implementing the fusion techniques is to adequately align and register the source input images that are attained by a camera at disparate view-points focus levels, angles view and by different sensors using image registration technique [15–17]. In this paper, it is supposed that the source input images are already registered. Second to obtain the maximum information and better contrast Image fusion techniques are applied on the registered input images.

2 Quality Metrics for Result Assessment

2.1 RMSE (Root Mean Square Error)

This metric evaluate the difference of pixels between the fused and reference image [18]. Its value increases when images are different types. RMSE smaller value means better the fusion algorithm.

$$\mathrm{RMSE} = \sqrt{\frac{1}{MN}\sum_{y=0}^{N-1}\sum_{x=0}^{M-1}[I(x, y) - F(x, y)]^2} \tag{1}$$

Here $M, N =$ size's of images,
$(x, y) =$ image co-ordinate, $I =$ input image, $F =$ Output image

2.2 *PFE (Percentage Fit Error)*

Minimum the value of the PFE, better the performance of the fusion algorithm is counted.

$$\text{PFE} = 100 * \text{Norm}\big((I(:) - F(:))/\text{Norm}(I(:))\big) \tag{2}$$

Here Norm = Norm operator, I = input image, F = Output image

2.3 *PSNR (Peak Signal-to-Noise Ratio)*

Greater value of PSNR is considered better. For fused image PSNR is defined below:-

$$\text{PSNR} = \left(10\log_{10}\left(\frac{L^2}{\text{RMSE}}\right)\right) \tag{3}$$

Here L = Gray level in the image [19, 20].

2.4 *SF (Spatial Frequency)*

Spatial frequency is evaluated using horizontal frequency (RF) and vertical spatial frequency (CF) [21]. Its higher value preferred.

3 Experiment and Results

This section presents the implementation and the relative performance of different discrete wavelet transforms based on fusion methods. The performance of SWT, DDB2WT, SIDHWT, DT-CWT fusion techniques are evaluated based on PSNR, RMSE, PFE, CF, RF and SF metrics on different eight pairs of input images in which three pair sets are of Multisensor images like CT (computed tomography) as shown in Figs. 2, 3 and 4 and five pair sets are of Multifocus images as shown in Figs. 5, 6, 7, 8 and 9.

Wavelets based image fusion methods are implemented in MATLAB2010. The final results of quality metrics are obtained on an output images (that obtained after applying different fusion methods) is shown in Fig. 10.

On the bases of visual perception all the four methods gives good result. But visual perception is not an accurate way to judge any of the fusion methods so we use quality metrics to obtain correct and accurate results.

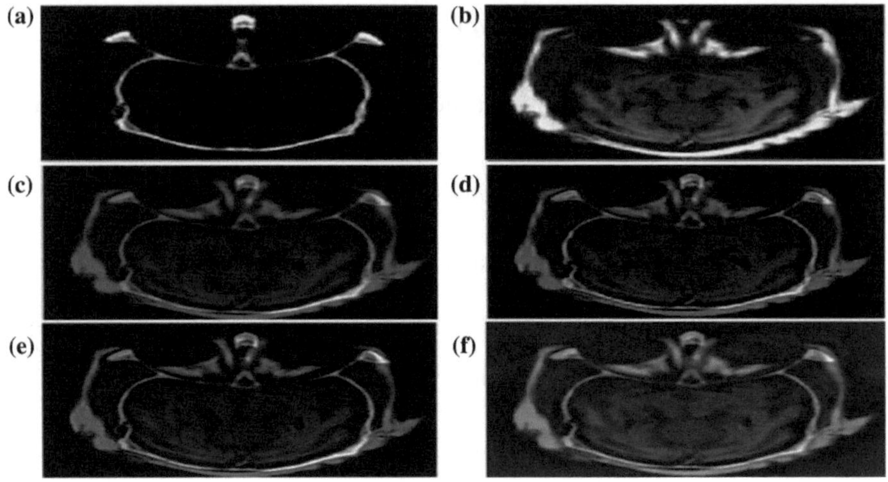

Fig. 2 **a** Input image1 (ct.png) **b** Input image2 (mri.png) **c** SWT **d** DDB2WT **e** SIDHWT **f** DT-CWT

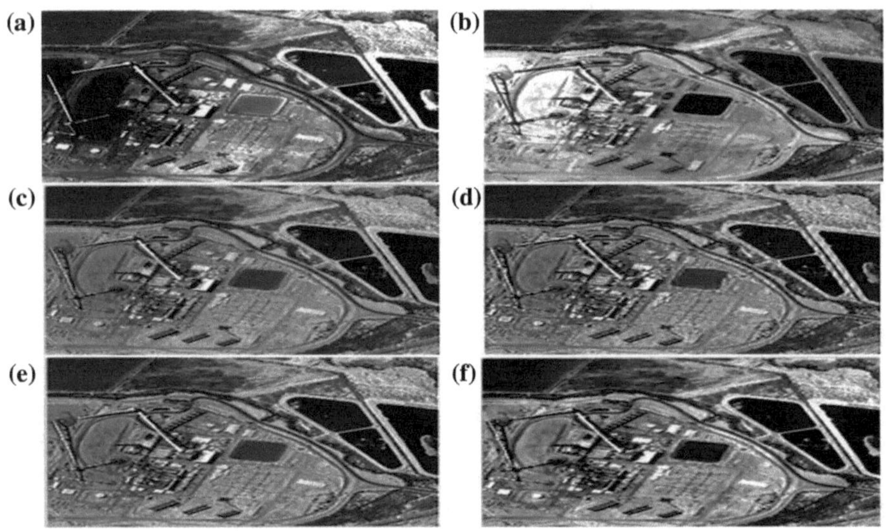

Fig. 3 **a** Input image1 (sensor1.png) **b** Input image (sensor2.png) **c** SWT **d** DDB2WT **e** SIDHWT **f** DT-CWT

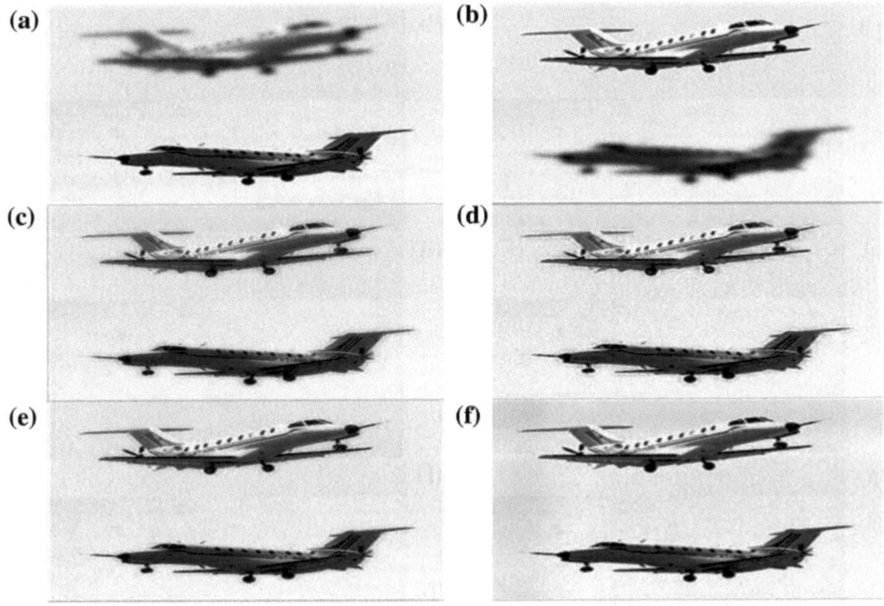

Fig. 4 **a** Input image1 (sensor1.png) **b** Input image2 (sensor2.png) **c** SWT **d** DDB2WT **e** SIDHWT **f** DT-CWT

DDB2WT method have no shift variant property and not detect the information from diagonal sides. Shift invariant Discrete Haar Wavelet Transform method is a modified Haar wavelet transform method which have a shift invariant property but does not have a directionality property. It gives a better result when information is distributed mostly in horizontal and vertical directions but not diagonally, SF metric gives good result.

At higher decomposition levels above level 4, DT-CWT provides better result than SWT but at level 2 it provides almost same results. SWT provides better result on those images in which more information is concentrated at horizontal and vertical directions than diagonal directions. In case, when information is concentrated at diagonal directions than DT-CWT based fusion method provides better results.

Result of RMSE and PFE metrics are shown in combined Fig. 10 which indicated that DT-CWT gives lowest value of RMSE and PFE for every fused image and DDB2WT gives highest values which indicates that DT-CWT provides better results.

(a) (b) (c) (d) (e) (f)

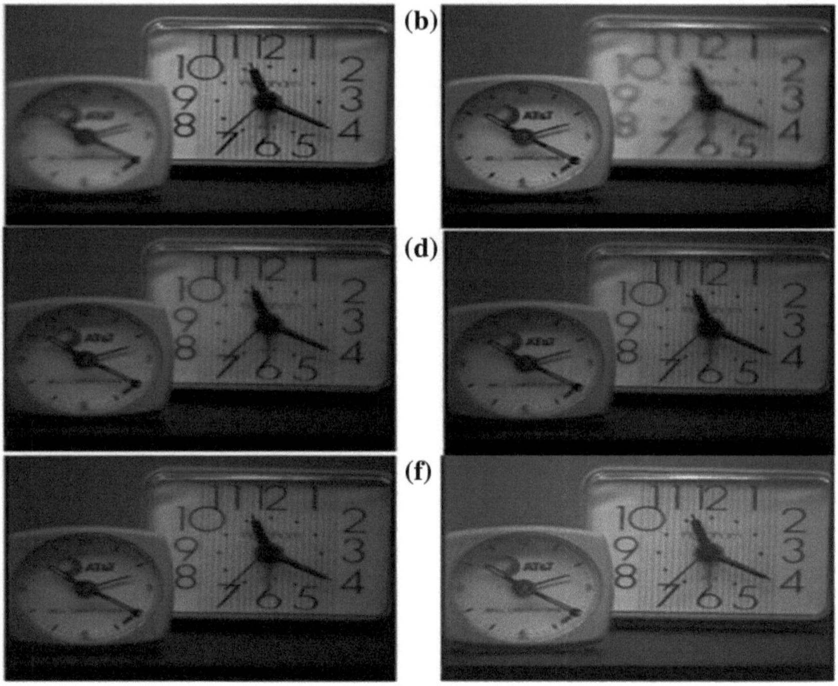

Fig. 5 **a** Input image1 (focus1.png) **b** Input image2 (focus2.png) **c** SWT **d** DDB2WT **e** SIDHWT **f** DT-CWT

4 Conclusion and Future Scope

In this paper, the comparative study has been done among the different discrete wavelet transform based fusion methods named SWT, DDB2WT, SIDHWT, DT-CWT. From the results, it is clear that fusion based on the DT-CWT method gives the better result not only on the base of visual perception but also on the quality performance metrics that are PSNR, RMSE, PFE, RF, CF, SF. And also SWT has a shift invariant and directionality properties which provide its results better as compare to DDB2WT that have no shift invariant property and SIDHWT. But all these methods preserve only the image information and loss the edge information so in future to preserve the edge information combine DT-CWT with some segmentation technique.

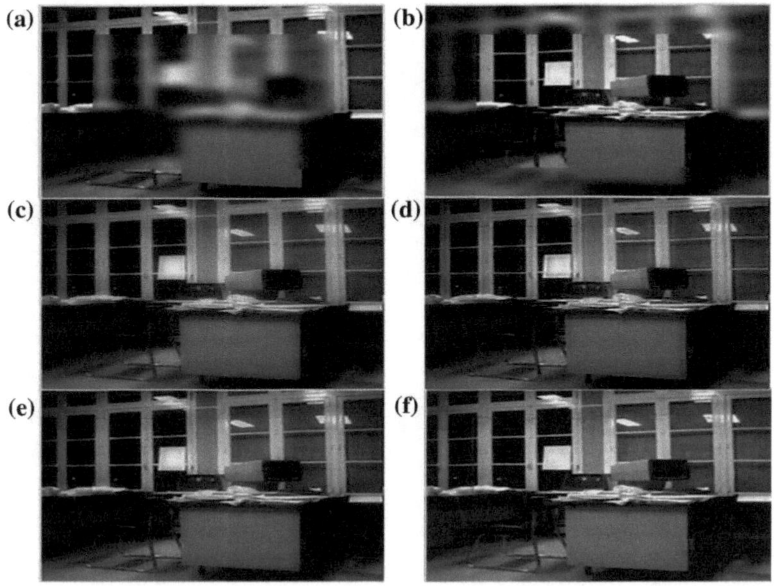

Fig. 6 **a** Input image1 (focus1.png) **b** Input image2 (focus2.png) **c** SWT **d** DDB2WT **e** SIDHWT **f** DT-CWT

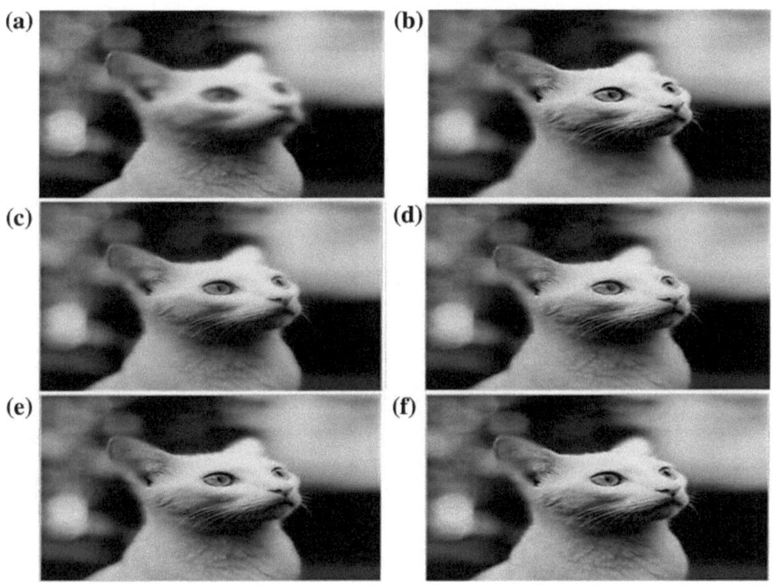

Fig. 7 **a** Input image1 (focus1.png) **b** Input image2 (focus2.png) **c** SWT **d** DDB2WT **e** SIDHWT **f** DT-CWT

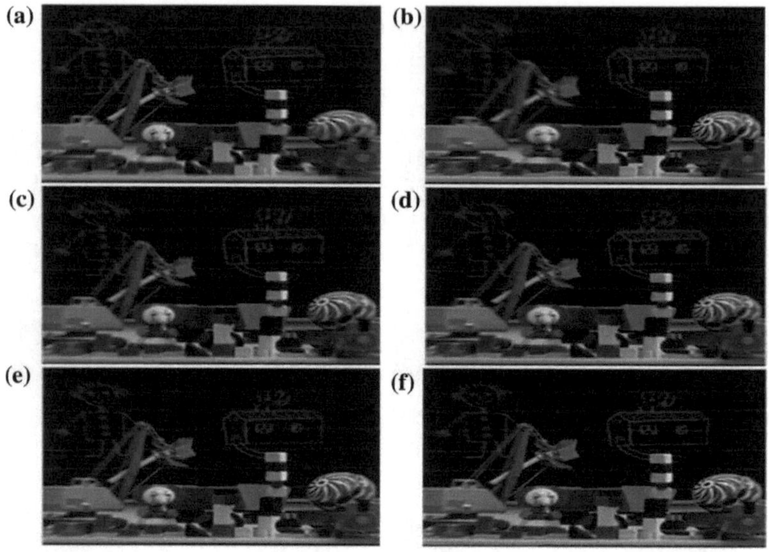

Fig. 8 **a** Input image1 (focus1.png) **b** Input image2 (focus2.png) **c** SWT **d** DDB2WT **e** SIDHWT **f** DT-CWT

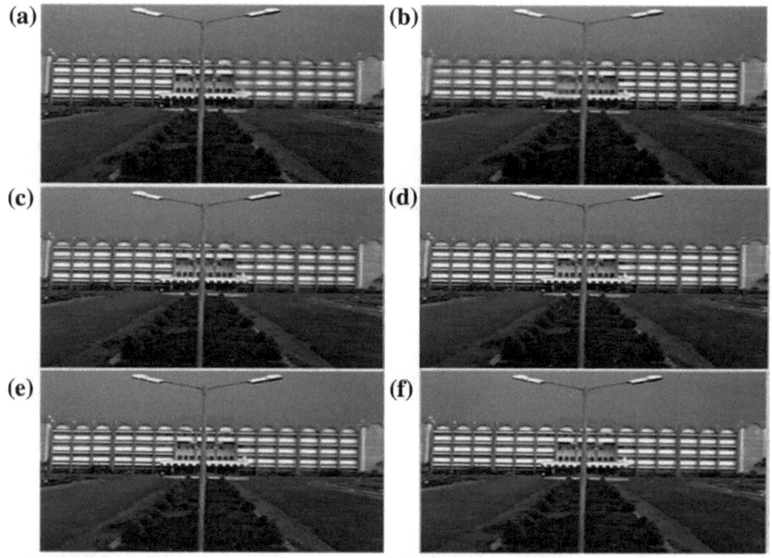

Fig. 9 **a** Input image1 (focus1.png) **b** Input image2 (focus2.png) **c** SWT **d** DDB2WT **e** SIDHWT **f** DT-CWT

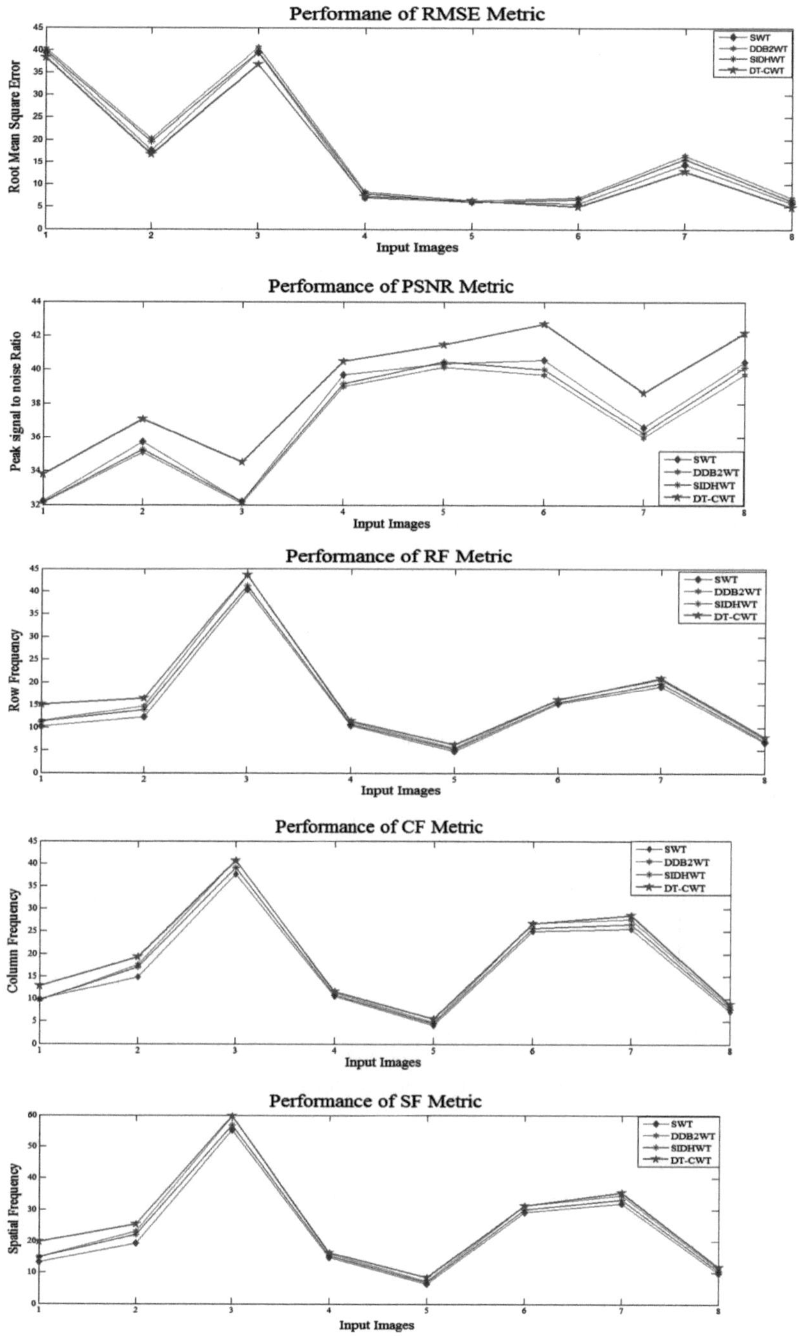

Fig. 10 Various metrics result

References

1. D.K. Sahu, M.P. Parsai, Different image fusion techniques–a critical review. Int. J. Mod. Eng. Res. (IJMER) **2**(5) (Sept–Oct 2012)
2. G. Pajares, J.M. de la Cruz, A wavelet-based image fusion tutorial. Pattern Recogn. (2004)
3. P.M. de Zeeuw, E.J.E.M. Pauwels, J. Han, multimodality and multiresolution image fusion. Int. Conf. Comput. Vis. Theor. Appl. (VISAPP 2012)
4. S. Mallat, A theory for multi-resolution signal: the wavelet representation, in *IEEE Transactions on Pattern Analysis and Machine Intelligence*, vol. 11 (July 1989)
5. M. Sifuzzaman, M.R. Islam, M.Z. Ali, Application of wavelet transform and its advantages compared to fourier transform. J. Phys. Sci. **13**, 121–134 (2009)
6. F. Lalibert´e, L. Gagnon, Y. Sheng, Registration and fusion of retinal images: a comparative study, in *International Conference on Pattern Recognition 2002*, Québec (11–15 Aug 2002), pp. 715–718
7. P.P. Mirajkar, D.R. Sachin, Image fusion based on stationary wavelet transform. Int. J. Adv. Eng. Res. Stud. E-ISSN2249–8974 (July–Sept 2013)
8. D.A. Yocky, Image merging and data fusion by means of the discrete two-dimensional wavelet transform. J. Opt. Soc. Amer. A **12** (1995)
9. M.J. Shensa, The discrete wavelet transform: wedding the à Trousand Mallat algorithms. IEEE Trans. Signal Process. **40**(10), 2464–2482 (1992)
10. M. Veterli, C. Herley, *Wavelets and Filter Banks: Theory and Design*
11. N.G. Kingsbury, The dual-tree complex wavelet transform: a new technique for shift invariance and directional filters, in *Proceedings of 8th IEEE DSP Workshop*, Bryce Canyon, UT, USA, paper no. 86 (1998)
12. S. Sruthy, L. Parameswaran, A.P. Sasi, *Image Fusion Technique using DT-CWT* (IEEE, 2013) (978-1-4673-5090)
13. V.P.S. Naidu, Block DCT based image fusion techniques. e-J. Sci. Technol. (e-JST)
14. R.P.S. Chauhan, R. Dwivedi, S. Negi, Comparative evaluation of DWT and DT-CWT for image fusion and De-noising. Int. J. Appl. Inf. Syst. (IJAIS) **4**(2) (ISSN: 2249-0868) (Foundation of Computer Science FCS, New York, USA, Sept 2012)
15. B. Zitova, J. Flusser, Image registration methods: a survey. Image Vis. Comput. **21**, 977–1000 (2003)
16. S.S. Bisht, B.Gupta, P. Rahi, Image registration concept and techniques: a review. Int. J. Eng. Res. Appl. **4**(4)(Version 7), 30–35 (Apr 2014)
17. F. Lalibert´e, L. Gagnon, Y. Sheng, Registration and fusion of retinal images: a comparative study, in *International Conference on Pattern Recognition 2002*, Québec (11–15 Aug 2002), pp. 715–718
18. V.P.S. Naidu, J.R. Raol, Fusion of out of focus images using principal component analysis and spatial frequency. J. Aerosp. Sci. Technol. **60**(3), 216–225 (2008)
19. Z. Wang, A.C. Bovik, A universal image quality index. IEEE Signal Proc. Lett. **9**(9), 81–84 (2002)
20. P.S. Fisher, Image quantity measures and their performance. IEEE Trans. Commun. **43**(12), 2959–2965 (1995)
21. R. Dhawan1, N.K. Garg, A comprehensive review on different techniques of image fusion. Int. J. Eng. Comput. Sci. **3**(3), 5098–5101 (Mar 2014)

Differentiation of Seizure and Non-seizure EEG Signals Using Analytical Approach

Nazia Parveen and S. H. Saeed

Abstract One of the most preferred approaches to analyze seizures and non-seizures is wavelet analysis of electroencephalogram (EEG) spectrum. In this study, recording of EEG data (using internationally recognized "10–20" system of electrode placement) has been used to study features of normal subjects and seizures. The data obtained has been analyzed using wavelet transform tool. The analysis involves three steps: preprocessing, feature extraction, and classification. The preprocessing involves noise removal and decomposition of EEG signal into different bands. The feature extraction includes computation of energy, mean, variance, etc. using wavelet transform. Other parameters such as interquartile range (IQR) and mean absolute deviation are also computed. A normal EEG can be identified on the basis of normal rhythmic pattern in different bands. The deviations from the normal pattern have been identified by time–frequency analysis of the pattern.

Keywords Wavelet · Seizures · Spectral analysis

1 Introduction

65 million people have been suffering from epilepsy worldwide. 35 million could be cured by seizure control by giving medication and surgery. 15 million are still incurable [1]. Seizures are "hyper synchronous" activity of the neuron's cell which results in sudden disorder in the brain [2]. These seizures directly effect on the consciousness of the person. The correct prediction of this epilepsy will enable doctors to take prior action so as to avoid the injury caused by sudden seizures.

Nazia Parveen (✉)
Department of ECE, Shri Ram Murti Smarak College of Engineering and Technology, Bareilly 243202, Uttar Pradesh, India
e-mail: er.nazia123@gmail.com

S. H. Saeed
Department of ECE, Integral University, Lucknow 226021, Uttar Pradesh, India
e-mail: ssaeed@iul.ac.in

C. R. Krishna et al. (eds.), *Proceedings of 2nd International Conference on Communication, Computing and Networking*, Lecture Notes in Networks and Systems 46, https://doi.org/10.1007/978-981-13-1217-5_48

The electroencephalogram is one of the indispensable tools for diagnosing such type of neural disorders [3, 4]. It could record spikes [5], sudden occurrences, complexities of waves, sharp wave, etc. With the use of EEG recording epilepsy can be identified even in short period of EEG recording but to get unclear nature of seizure long-term video monitoring is necessary. It is a time-consuming and costly endeavour. Neurophysiologist can identify the seizure while monitoring but it was not found to be very efficient. So, there is a need of an automatic seizure detection scheme to facilitate the prediction of seizures [6]. A number of methods have been adopted for the analysis of these disorders primarily based [7] on time domain, frequency domain, and time–frequency domain analysis of EEG spectrum [4]. On decomposition, EEG consists of five sub-bands [8]: delta (0–4 Hz), theta (4–8 Hz), alpha (8–12 Hz), beta (13–30 Hz), and gamma (30–60 Hz). To study the brain dynamics, entire sub-band gives more accurate information than individual bands. By amplifying sub-band of interest or each sub-band, neuronal activities could be identified. This is a basis of this paper.

Earlier approaches which are based on Fourier transform are the witness of the fact that the epileptic seizure makes the changes in frequency bands. So this is not appropriate to study the nonstationary nature of EEG by frequency decomposition only. Time–frequency analysis could give better picture and outperforms the conventional methods. Wavelet is a standout among the latest and mainstream strategies which provide a unified framework [9] for a number of techniques developed for various signal-processing applications [10]. Particularly, it is of immeasurable interest for the analysis of nonstationary signals like EEG [11], because it is an alternate technique to the classical short-time Fourier transform (STFT) or Gabor transform.

Unlike STFT, which uses a single analysis window, the wavelet transform (WT) uses short windows at high frequencies and long windows at lower frequencies. There is another approach based on fast wavelet decomposition method known as harmonic wavelet packets transform (HWPT) [12]. This approach can be used for detection seizures for long as well as short EEG signals. This method can achieve higher frequency resolutions without recursive calculation making the computation faster. One more method suggested [13] that automated diagnosis of epilepsy framework involves a combination of feature extraction in multi-domain and nonlinear analysis of EEG signals. Chen et al. [14] developed a framework for optimal DWT settings which helps in improving accuracy and reducing computational cost incurred in the detection of seizures.

In this paper, a seizure detection scheme is discussed where EEG data is used to detect generalized seizure. The wavelet transform [15] is used for decomposition of signals which provides both time and frequency information. Feature extraction takes place using different parameters like energy, mean variance, interquartile range [16], and absolute deviation. On the basis of above parameters, seizures and non-seizures would be distinguished.

2 Methodology

The EEG data of normal subjects and humans suffering from generalized epilepsy has been taken CHB-MIT database. For recording of EEG data, "10–20" system [7, 17, 18] of electrode placement is being used with 23 channels, 256 Hz sampling frequency, and 19 electrodes and a ground electrode attached to the surface of scalp [19]. This study involves analysis of data using wavelet transform [20] tool. Using wavelet transform, the signal components in time–frequency space can be better localized [21]. Some wavelet-based features [1, 22] (such as energy, mean, variance, etc.) and some features without wavelet decomposition (such as interquartile range, mean absolute deviation, etc.) have been computed. The value of the parameters obtained for seizures and non-seizures are quite different which helps in distinguishing healthy subjects from seizure patients. The study involves detection of generalized seizures.

Out of 100 datasets, preprocessing and feature extraction for seizures and non-seizures is done for 10 datasets.

2.1 *Preprocessing*

The preprocessing refers to any transformations or reorganizations that occur between collecting the data and analyzing the data [9]. The preprocessing may be done to merely organize the data to facilitate analysis or to remove the artifacts due to eyeball movements, eyes blinks, improper placement of electrodes, etc. The artifacts are potential fluctuations of nonneural origin [6].

The analysis done on raw data suffers from problems such as poor resolution, low SNR, and artifacts results [23]. The preprocessing involves noise removal from the EEG signal. In this method, moving average filter of order 3 and 5 is used to remove the noise. The higher order filter may also be used but they resulted in distortion of the signal (Fig. 1).

During preprocessing, the decomposition of EEG [11, 24, 25] dataset into different bands (delta, theta, alpha, beta, and gamma) has been done both for seizures and non-seizures (Figs. 2 and 3).

In Tables 1 and 2, power [26] corresponding to different bands of frequencies is computed for seizures and non-seizures.

2.2 *Feature Extraction*

In this stage, parameters computed include energy, mean, variance, standard deviation [27], interquartile range, mean absolute deviation, etc. The plots for histogram, Fourier transform, power spectral density, scalogram, spectrogram, etc. have been obtained using wavelet.

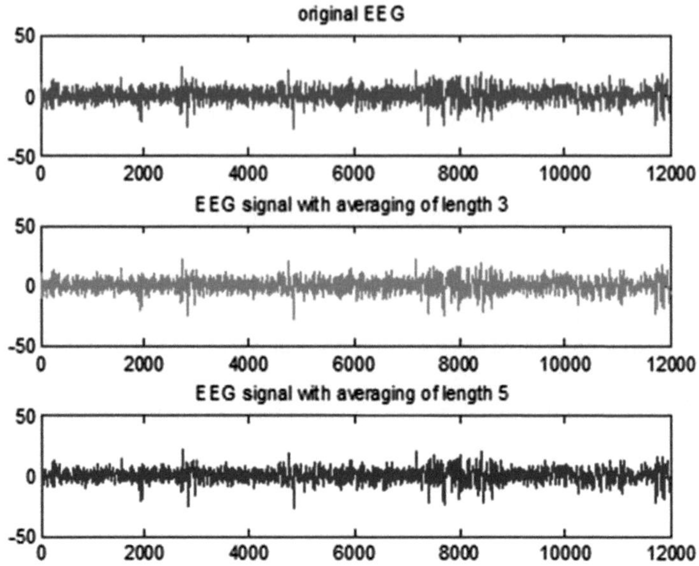

Fig. 1 Moving average filter applied to EEG signal

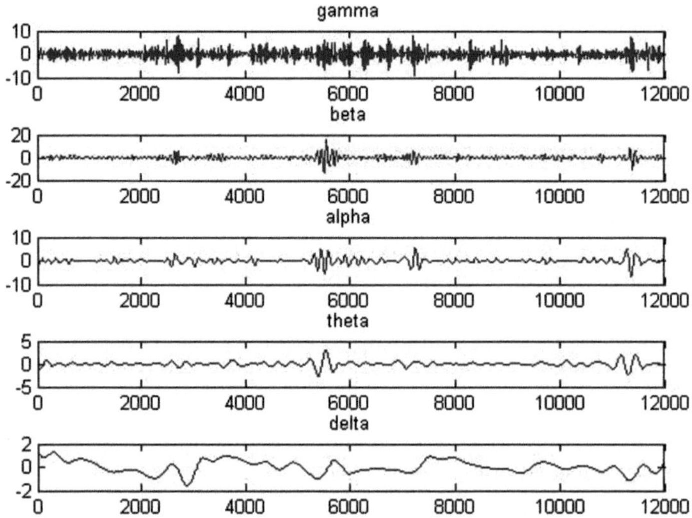

Fig. 2 Frequency bands in seizure data set

Table 1 Power in different bands of EEG for seizures datasets

Parameter	Sei1	Sei2	Sei3	Sei4	Sei5	Sei6	Sei7	Sei8	Sei9	Sei10
Δ Power	0.2919	0.1793	0.0764	0.3904	0.9360	0.1504	0.2965	0.8616	0.8304	0.2175
θ Power	0.3048	0.4669	0.1898	0.3806	0.7140	0.3814	0.5046	0.7672	0.5091	0.6500
α Power	1.2190	1.7175	0.8667	3.0815	4.1217	1.0388	2.0457	4.5865	2.2153	1.0636
β Power	3.0788	4.1082	1.7947	2.4217	12.9629	4.6567	8.9947	9.1923	4.8714	5.1277
Total power	4.8945	6.4718	2.9276	6.2743	18.7346	6.2274	11.8414	15.4076	8.4262	7.0589
Relative Δ	0.0596	0.0277	0.0261	0.0622	0.0500	0.0242	0.0250	0.0559	0.0985	0.0308
Relative θ	0.0623	0.0721	0.0648	0.0607	0.0381	0.0612	0.0426	0.0498	0.0604	0.0921
Relative α	0.2491	0.2654	0.2961	0.4911	0.2200	0.1668	0.1728	0.2977	0.2629	0.1507

Table 2 Power in different bands of EEG for non-seizures datasets

Parameter	Nonsei1	Nonsei2	Nonsei3	Nonsei4	Nonsei5	Nonsei6	Nonsei7	Nonsei8	Nonsei9	Nonsei10
Δ Power	0.7506	1.1067	0.7506	0.1643	0.5292	0.7237	0.2987	0.1513	0.5909	0.5445
θ Power	0.6414	0.8373	0.6414	0.2024	0.4845	0.4890	0.1708	0.0965	0.3443	0.2854
α Power	3.6942	2.1877	3.6942	0.5855	1.3288	1.5971	0.9870	0.5028	0.7539	0.7495
β Power	4.6885	5.1644	4.6885	1.6861	3.8196	3.9856	2.5205	2.2295	1.9764	1.5862
Total power	9.7747	9.2961	9.7747	2.6384	6.1621	6.7954	3.9770	2.9802	3.6655	3.1656
Relative Δ	0.0768	0.1191	0.0768	0.0623	0.0859	0.1065	0.0751	0.0508	0.1612	0.1720
Relative θ	0.0656	0.0901	0.0656	0.0767	0.0786	0.0720	0.0429	0.0324	0.0939	0.0901
Relative α	0.3779	0.2353	0.3779	0.2219	0.2156	0.2350	0.2482	0.1687	0.2057	0.2368

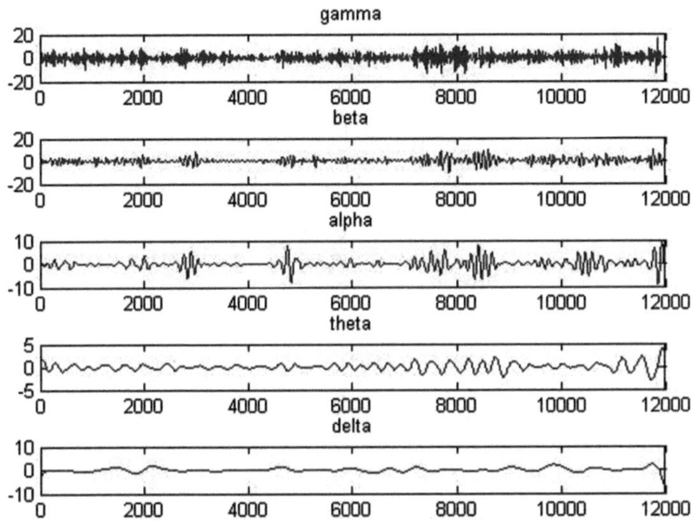

Fig. 3 Frequency bands in non-seizure data set

The next step in feature extraction is decomposition [28] of seizure and non-seizure data using wavelet transform to obtain detail and approximate coefficients [29] (Table 3). The energy of detail and approximate coefficients [30, 31] is then computed using db4 (level 8). The statistical features computed include mean, variance, standard deviation, interquartile range, and mean absolute deviation (Table 4). These features are computed for seizures in addition to non-seizures and assist in discriminating seizures and non-seizures.

The energy at each decomposition level [32] has been calculated using

$$\mathrm{ED}_i = \sum_{j=1}^{N} \left|\mathrm{D}_{ij}\right|^2, \quad i = 1, 2, 3 \ldots 1 \tag{1}$$

$$\mathrm{EA}_i = \sum_{j=1}^{N} \left|\mathrm{A}_{ij}\right|^2, \quad i = 1, 2, 3 \ldots 1 \tag{2}$$

where $i = 1, 2, 3 \ldots 1$ and N is the number of coefficients of detail or approximation at each decomposition level.

In Tables 5 and 6, features without wavelet decomposition such as MAD, IQR, mean, variance, etc. have been computed for seizures and non-seizures datasets.

The Fourier transform breaks up any time-varying signal into its frequency components of varying magnitude [33] and is defined as

$$F(k) = \int_{-\infty}^{\infty} f(t)\mathrm{e}^{-2\pi ikx}\mathrm{d}x \tag{3}$$

Table 3 Energy of non-seizure datasets

	D1	D2	D3	D4	D5	D6	D7	D8	A8
S1	0.1059	0.9137	5.8822	20.0132	18.7746	10.2814	9.6669	2.3816	31.9806
S2	0.0645	0.6137	5.9463	26.0028	17.2767	9.2390	6.8590	3.0418	30.9562
S3	0.0085	0.2006	3.7595	23.2148	22.1134	11.8638	7.0866	1.4503	30.3025
S4	0.0088	0.3405	13.2984	67.9993	11.8117	3.9020	1.5062	0.6207	0.5123
S5	0.0293	0.3278	4.2460	17.8813	26.4399	12.6754	7.6012	3.2434	27.5557
S6	0.0074	0.1760	4.0882	18.7226	16.3345	8.8537	7.1376	2.4833	42.1967
S7	0.0119	0.3474	9.0866	46.7944	14.5909	6.8046	4.0732	1.0189	17.2721
S8	0.0101	0.2970	9.3048	56.1317	13.5074	5.5307	2.0644	0.4111	12.7429
S9	0.2439	2.1584	8.1153	17.2650	21.5573	9.4741	5.3915	2.7390	33.0556
S10	0.1788	1.6439	11.1807	36.1166	13.9270	4.3845	4.9873	1.7251	25.8561

Table 4 Energy of seizure dataset

	D1	D2	D3	D4	D5	D6	D7	D8	A8
S1	0.0146	1.0119	18.2969	12.2965	23.6568	18.5115	7.4594	2.4956	16.2569
S2	0.0109	0.8646	10.2217	22.0838	17.9279	22.0010	10.4371	3.1387	13.3143
S3	0.0139	0.8499	10.5247	33.9542	21.4962	19.2847	9.5761	2.4156	1.8847
S4	0.0077	0.4183	7.9864	32.2080	12.4556	9.1295	15.1947	1.9178	20.6820
S5	0.0062	0.5560	7.7546	12.3726	19.0775	21.1855	9.8147	1.6754	27.5574
S6	0.0101	0.6338	7.6445	22.5786	20.8681	23.0300	7.0112	2.8896	15.3342
S7	0.0113	0.9073	9.7598	23.8259	22.1901	27.7581	6.9744	1.8501	6.7229
S8	0.0069	0.3978	7.5679	22.4734	20.3374	18.0633	10.4773	1.9904	18.6856
S9	0.0082	0.8277	11.1504	14.8506	15.2786	18.2349	10.3882	2.3141	26.9474
S10	0.0097	0.7887	10.6209	27.5564	19.5173	21.1642	5.4624	3.1422	11.7382

Table 5 Tabulation of statistical features for seizure datasets

Features	Sei1	Sei2	Sei3	Sei4	Sei5	Sei6	Sei7	Sei8	Sei9	Sei10
Mean absolute deviation	2.7075	2.8816	2.2112	3.4507	4.6525	2.9072	4.0255	4.8443	3.1898	3.4696
Interquartile range	4.14	4.32	3.42	5.37	6.64	4.33	5.995	7.3400	4.7600	5.3700
Mean	0.0358	0.0149	−0.0438	−0.0122	0.0113	0.0137	0.1151	−0.1455	0.0260	0.0039
Variance	14.0737	16.418	9.2123	21.0688	43.8337	16.9449	30.9667	42.8605	19.9409	21.7797
Standard deviation	3.7515	16.418	3.0352	4.5901	6.6207	4.1164	5.5648	6.5468	4.4655	4.6669

Table 6 Tabulation of statistical features for non-seizures datasets

Features	NonSei1	Non Sei2	NonSei3	NonSei4	NonSei5	NonSei6	NonSei7	NonSei8	NonSei9	NonSei10
Mean absolute deviation	4.2912	4.6543	4.6543	4.9723	3.5732	3.7236	4.3110	4.6191	2.9386	4.0208
Interquartile range	6.86	6.9	6.9	7.545	5.85	5.8500	6.7900	6.9950	4.7850	6.5700
Mean	0.0526	−0.0179	−0.0179	0.0163	−0.0353	−0.0062	−0.0311	0.0030	−0.0415	−0.0326
Variance	31.5293	40.6908	40.6908	43.3274	21.0291	24.0590	31.4127	37.7025	14.3133	26.3913
Standard deviation	5.6151	6.3789	6.3789	6.5824	4.5857	4.9050	5.6047	6.1402	3.7833	5.1372

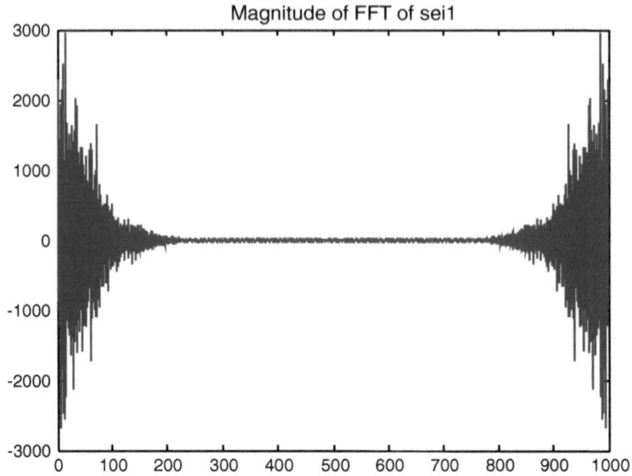

Fig. 4 Fourier transform of seizure

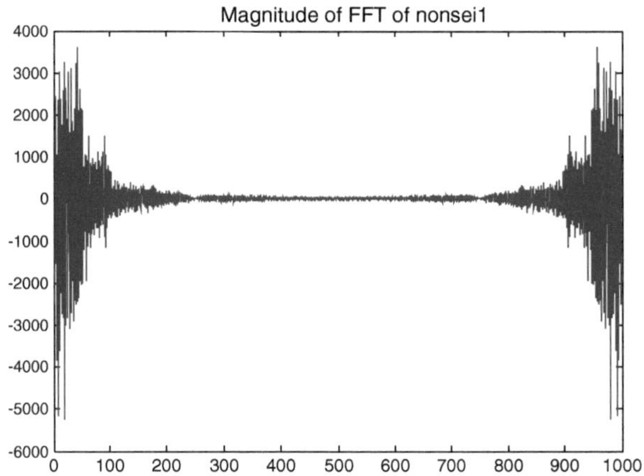

Fig. 5 Fourier transform of non-seizure

The Fourier transform is represented with the power spectrum. The frequency components present in the EEG signal can be identified for seizures and non-seizures. The Fourier transform obtained for seizure dataset is shown in Fig. 4 and that of non-seizure is represented in Fig. 5.

One of the most widely used techniques for detecting an epileptic seizure is power spectral analysis [33, 34]. It involves partitioning the EEG signal into its periodic components to identify the most prominent frequencies (in amplitude) using power spectral density of seizure and non-seizure dataset.

3 Results

On comparison of Figs. 2 and 3, it can be observed that there are several peaks which have abnormal amplitudes in the $\pm$ and μ frequency bands in Fig. 3. These peaks may be due to the abnormalities in brain such as epilepsy, tumors, and traumas. Moreover, different frequency components present in the signal can be identified using Fourier transform. In Figs. 6, 7, 8, and 9, wavelet transform (db7) has been applied to seizure data to obtain approximate and detail coefficients and the histograms. The scalogram in Figs. 12 and 13 has been shown for dataset of seizures and non-seizures, where x-axis represents time [23, 27], y-axis is the scale factor (reciprocal of frequency), and magnitude of the correlation is represented by pixel intensity. The continuous variation in the intensity of color can be observed in scalogram of seizure and non-seizures. Scalogram plots are exclusive, and the abrupt changes in the epileptic EEG signal can be observed through changes in the color. There is continuous variation in the color in the plot depicting the randomness as well chaotic nature of the EEG signal. On decomposing seizure and non-seizure data, the spectrogram for wavelet coefficients [35] has been plotted in Figs. 14 and 15. The variation in intensity of different colors can be easily visualized in seizure spectrogram with respect to non-seizure spectrogram.

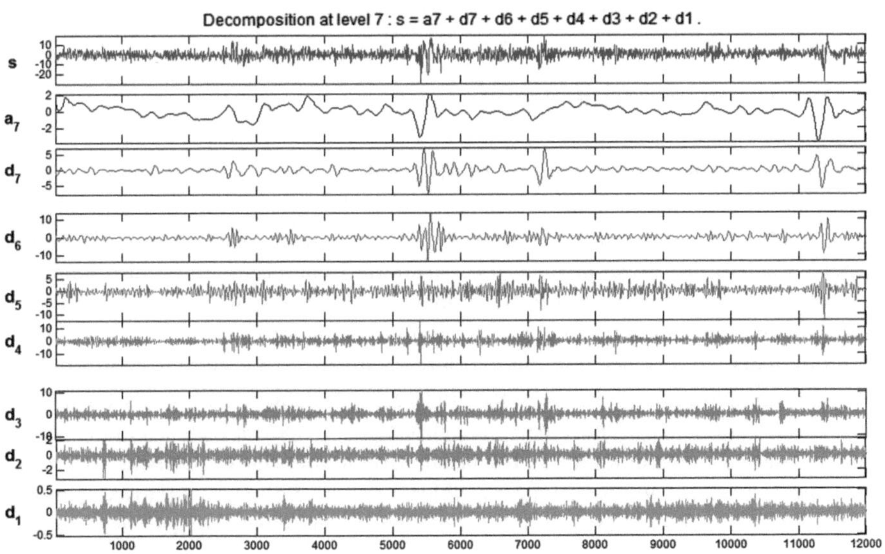

Fig. 6 Application of db7 wavelet transform to seizure dataset

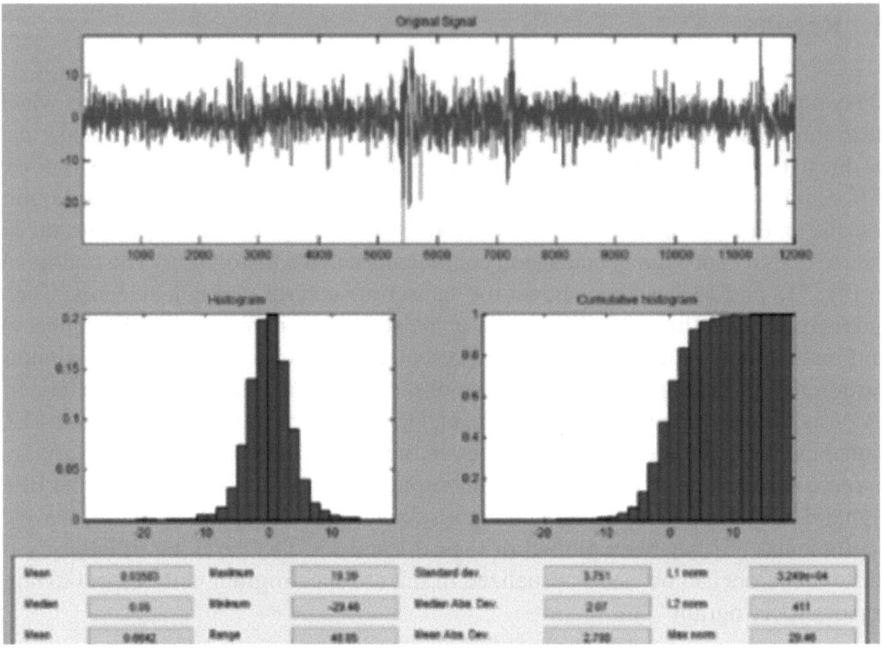

Fig. 7 Histogram of db7 wavelet transform for seizure dataset

4 Discussion and Conclusion

Epileptic seizure detection using an EEG signal is a rich area for signal analysis. Seizure detection and prediction refer to features that are capable of distinguishing seizure and non-seizure states. A number of approaches have been adopted for this purpose. Each of the procedures has its own strengths and limitations, attributable to the complexities of seizure waveforms, trade-offs between sensitivity and specificity, trade-offs among accuracy, latency time, and platform constraints.

In this work, a very accurate and widely accepted technique has been used for statistical computation of seizures and non-seizures data. EEG signals for seizure sufferers and non-seizures have been decomposed into special bands, and the energy of each band has been computed . It has been found that power in distinctive bands of seizure is higher than non-seizures (Tables 3 and 4). Similarly, exclusive parameters were computed, namely, mean, standard deviation, interquartile range (Tables 5 and 6), and so on, which indicates an exceptional deviation in seizure values in comparison to non-seizures.

Additionally, it has been found from results that within the case of seizures there was a transfer of energy from lower coefficients to higher coefficients. The energy spectral density of seizure shows off negative peak at 250 Hz while that of non-seizure exhibits positive peak at the same frequency (Figs. 10 and 11). Further, the

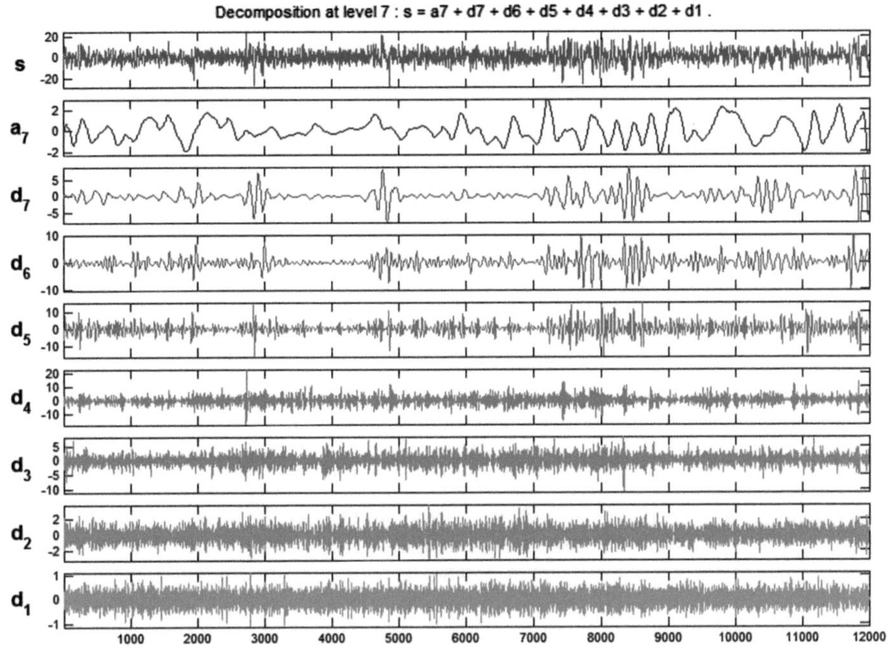

Fig. 8 Application of db7 for non-seizure dataset

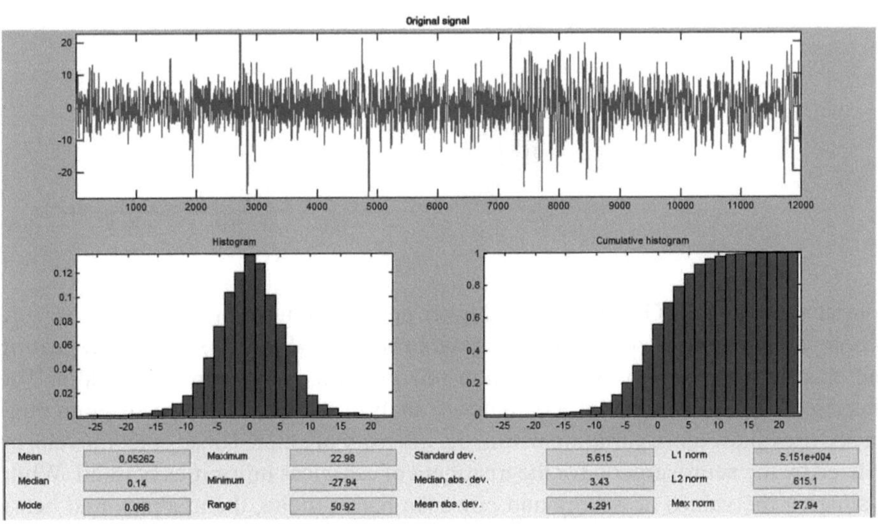

Fig. 9 Histogram of db7 for non-seizure dataset

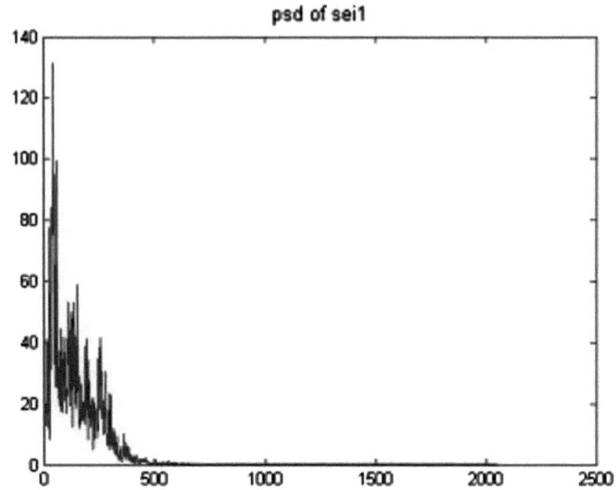

Fig. 10 PSD of seizure

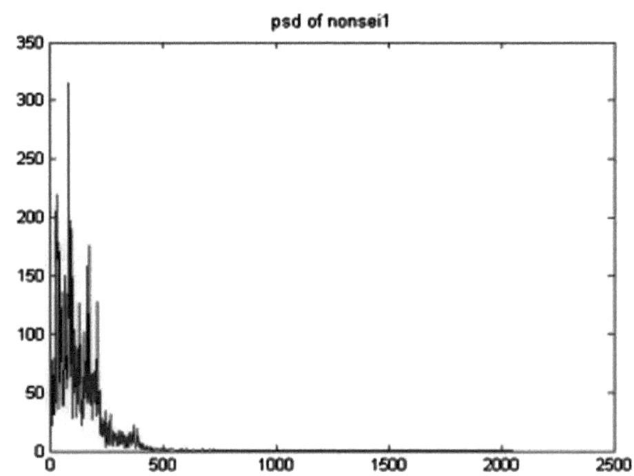

Fig. 11 PSD of non-seizure

Fourier spectra of EEG signals show higher peaks and more rhythms in the seizure as compared to non-seizure. The scalogram of seizures (Fig. 12) showcase repeating and chaotic pattern even as that of non-seizures (Fig. 13) seems as random. The aftereffect of this examination facilitates to analyze seizures. The spectrogram plots for seizures show more variation within the intensity of color. Those outcomes can be utilized by the neurosurgeon for the treatment of ceaseless infirmities of mind. While evaluating the values of energy and power in bands (delta, theta, alpha, and beta), it could be inferred that energy and power act as a good indicator to differentiate mind states of seizures subjects and non-seizure subjects. A decrease in the value of power in the EEG signals of non-seizure subjects indicates more rhythmic and ordered behavior.

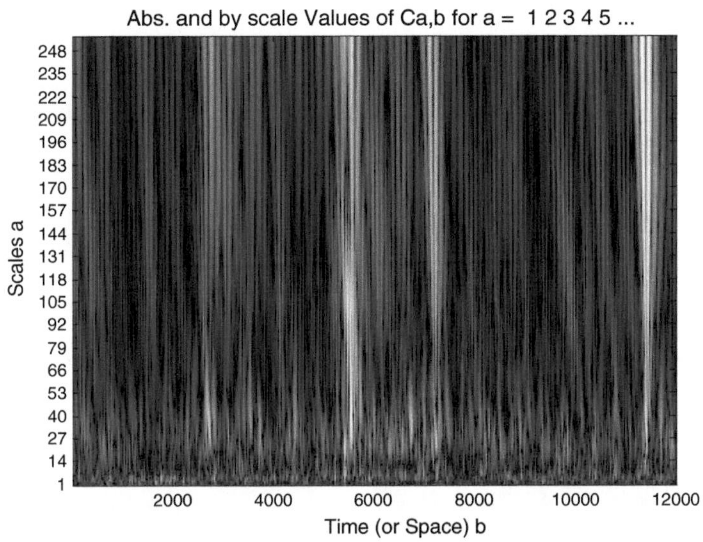

Fig. 12 Scalogram of seizure

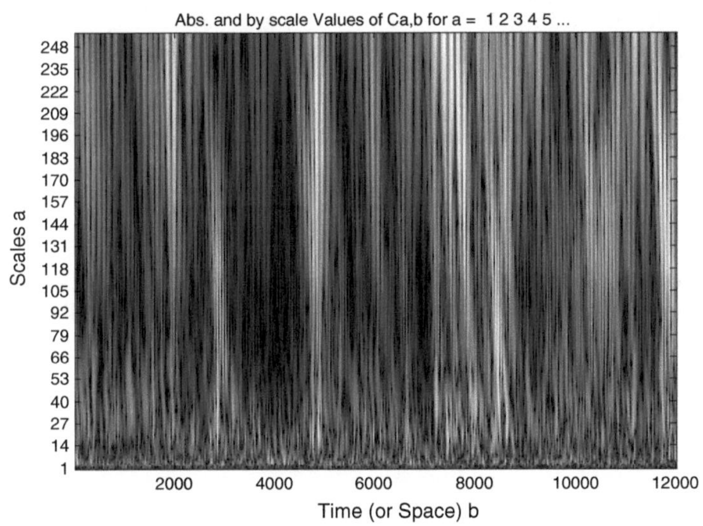

Fig. 13 Scalogram of non-seizure

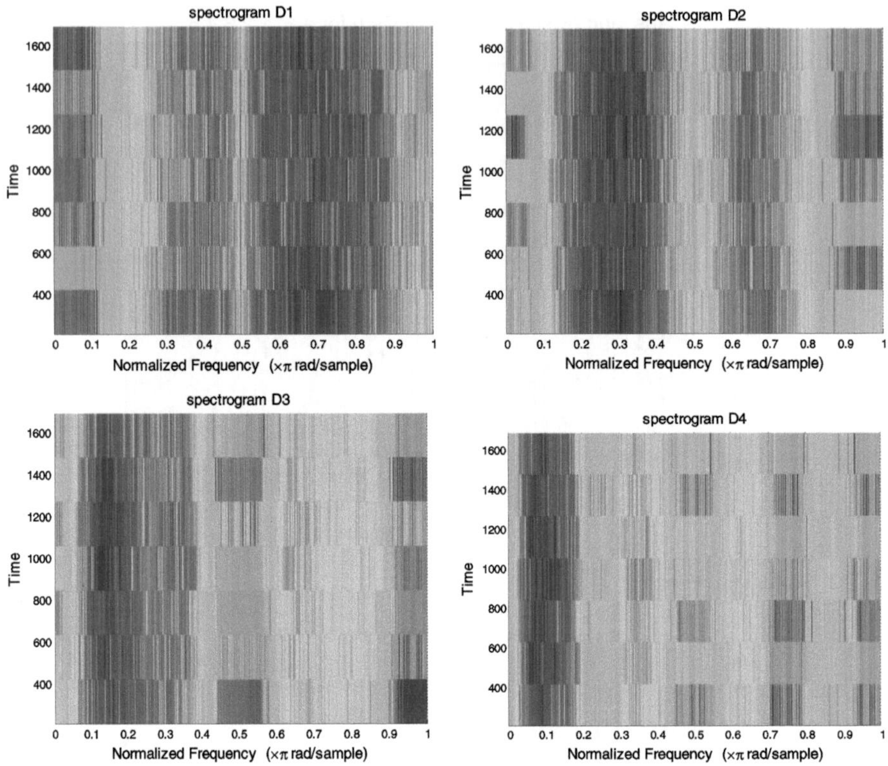

Fig. 14 Spectrogram of D1-D4 for seizure

Acknowledgements The author is thankful to MIT open source for the data of seizures and non-seizures without which this study would not have been possible. The author is also thankful to all colleagues who have directly or indirectly assisted in this work.

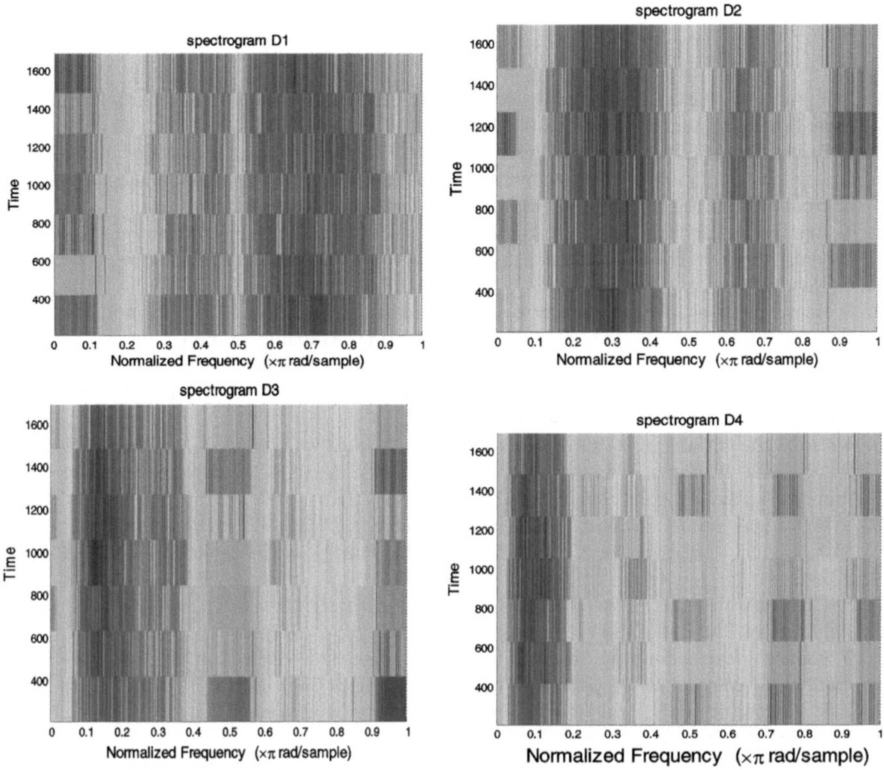

Fig. 15 Spectrogram of D1-D4 for non-seizure

References

1. U.R. Acharya, Automated diagnosis of epileptic EEG using entropies. Res. Article, Biomed. Sign. Process. Control **7**(4), 401–408 (July 2012)
2. R. Fisher, W. van Emde Boas, W. Blume, C. Elger, P. Genton, P. Lee, J. Engel, Epileptic seizures and epilepsy: definitions proposed by the international league against epilepsy (ILAE) and the international bureau for epilepsy (IBE). Epilepsia **46**(4), 470–472 (2005)
3. V. Bajaj, R.B. Pachori, Classification of seizure and non seizure signals using empirical mode decomposition, in *IEEE Transactions on Information Technology in Biomedicine*, vol. 16, No. 6 (Nov 2012)
4. S. Ramgopal et al., Seizure detection, seizure prediction, and closed-loop warning systems in epilepsy. Epilepsy Behav. **37**, 291–307 (29 Aug 2014) (Available online)
5. T.A. Tzallas et al., Automated epileptic seizure detection methods: a review study, in *Epilepsy—Histological, Electroencephalographic and Psychological Aspects*, ed. by D. Stevanovic (InTech, 2012) (ISBN: 978–953-51-0082-9)
6. K. Nidal, A.S. Malik, *EEG/ERP Analysis: Methods and Applications* (CRC Press, 2014)
7. R.A.S. Ruiz, R. Ranta, V. Louis-Dorr, EEG montage analysis in the Blind Source Separation framework. Biosignal Process. Control **6**(1), 77–84 (2010)
8. R.M. Rangayyan, *Biomedical Signal Analysis*, 2nd edn. (Wiley-IEEE Press, 2015)
9. M.X. Cohen, *Analyzing Neural Time Series Data: Theory and Practice* (MIT Press (17 Jan 2014)
10. S. ChavanArun et al., *Signal Pre-processing using Wavelet Transform*, vol. 3, No. 1 (Serial Publications Pvt. Ltd., 2011)
11. Y. Kumar, Wavelet Entropy Based EEG Analysis for Seizure Detection, in *2013 IEEE International Conference on Signal Processing, Computing & Control* (ISPCC) (26–28 Sep 2013)
12. L.S. Vidyaratne et al., Real-time epileptic seizure detection using EEG. IEEE Trans. Neural Syst. Rehabil. Eng. **25**(11) (Nov 2017)
13. L. Wang et al., Automatic epileptic seizure detection in EEG signals using multi-domain feature extraction and nonlinear analysis. Entropy **19**, 222 (2017). https://doi.org/10.3390/e19060222
14. D. Chen, S. Wan, J. Xiang, F.S. Bao, A high-performance seizure detection algorithm based on discrete wavelet transform (DWT) and EEG. PLoS ONE **12**(3) (9 Mar 2017) (e0173138). https://doi.org/10.1371/journal.pone.0173138
15. N.E. Zahra, H.A. Sevindir, Z. Aslan, A.H. Siddiqi, Wavelets in medical imaging, in *AIP Conference Proceedings* (18–22 July 2011)
16. O. Farooq et al, Automatic seizure detection using inter quartile range. Int. J. Comput. Appl. (Apr 2012)
17. D. Coyle, T.M. Mc Ginnity, G. Prasad, Improving the separability of multiple EEG features for a BCI by neural-time-series-prediction-preprocessing. Biosignal Process. Control **5**(3), 196–204 (2010)
18. N.F. Ince, F. Goksu, A.H. Tewfik, S. Arica, Adapting subject specific motor imagery EEG patterns in space–time–frequency for a brain computer interface. Biosignal Process. Control **4**(3), 236–246 (2009)
19. K. Lehnertz, Epilepsy and nonlinear dynamics. J. Biol. Phys. **34**, 253–266 (2008)
20. H. Adeli et al., Analysis of EEG records in an epileptic patients using wavelet transform. J. Neurosci. Methods (Oct 2002)
21. S. Sanei, J.A. Chambers, *EEG Signal Processing, Centre of Digital Signal Processing* (Wiley, Cardiff University, UK, 2007)
22. U.R. Acharya, Application of non-linear and wavelet based features for the automated identification of epileptic EEG signals. Int. J. Neural Syst (2012)
23. O. Faust et al., *Wavelet-Based EEG Processing for Computer-Aided Seizure Detection and Epilepsy Diagnosis* (Elsevier Seizure, 2015)
24. H. Adeli et al., A wavelet chaos methodology for analysis of EEGs and EEG sub-bands to detect seizure and epilepsy. IEEE Trans. Biomed. Eng. **54**(2) (Feb 2007)

25. M.Z. Parvez et al., EEG signal classification using frequency band analysis towards epileptic seizure prediction, in *16th International* Conference *on Computer and Information Technology* (Khulna, Bangladesh, 8–10 Mar 2014)
26. A.H. Siddiqi et al., *Relevance of Wavelets and Inverse Problems to Brain* (Mathematics in Science and Technology, World Scientific Press, 2011)
27. T. Helen et al., Efficient EEG analysis for seizure monitoring in epileptic patients, in *2013 IEEE Systems Applications and Technology Conference* (LISAT)
28. A. Varsavsky et al., *Epileptic Seizures and the EEG: Measurement, Models, Detection and Prediction* (CRC Press, 2011)
29. Y. Meyer, *Wavelets and Application* (Paris: Masson, 1992)
30. H.R. Mohseni et al., Seizure detection in EEG signals: a comparison of different approaches, in *Proceedings of the 28th IEEE EMBS Annual International Conference* (New York, USA, 2006)
31. M.Z. Parvez, M. Paul, Epileptic seizure detection by analyzing EEG signals using different transformation techniques. Elsevier J.of Neurocomputing **145** (2014)
32. N. Ahammad et al., Detection of epileptic seizure event and onset using EEG. BioMed Res. Int. (Annual Issue, 2014)
33. L.J. Greenfield, J.D. Geyer, P.R. Carney, *Reading EEGs: A Practical Approach* (Lippincott Williams and Wilkins, 2009)
34. R. Bilas et al., Analysis of normal and epileptic seizure EEG signal using empirical mode decomposition. Elsevier J. Comput. Methods Programs Biomed. (2011)
35. R. Sharma et al., *Biomedical Signal and Image Processing in Patient Care edited by Kolekaretal.* IGI Global Series (2010)

Part V
Data Science and Mining-I

Detecting Aggressive Driving Behavior using Mobile Smartphone

Rishu Chhabra, Seema Verma and C. Rama Krishna

Abstract In the course of most recent couple of years, there has been an increasing interest to monitor driver's behavior, road and traffic conditions in order to prevent accidents and save lives of the commuter. Different methods are available in the literature like special mounted vehicle cameras, sensors, Advanced Driver Assistance Systems (ADAS) etc. to monitor road, traffic conditions, and driver's behavior. The use of mobile smartphone inbuilt sensors to monitor driving behavior has been one of the most cost-effective methods in developing countries like India. Earlier work in this paradigm mainly focuses on the use of mobile smartphone's inbuilt accelerometer as a monitoring device for various driving maneuvers like sudden acceleration, sudden braking, and sharp turns. In this paper, we design and implement a system which uses an inbuilt accelerometer to detect sudden changes in acceleration; however, the system detects unsafe/sharp turns using inbuilt gyroscope in an efficient manner. The system further categorizes the driver as an aggressive or nonaggressive driver based on the observed driving pattern.

1 Introduction

Intelligent Transportation Systems (ITS) focuses on adding intelligence to the transportation system with the aim of making the commute safe and easy. Monitoring driver's behavior is considered as a complex paradigm and various methods like surveillance videos, simulators, specialized sensors, vehicle mounted cameras, ADAS and inbuilt mobile smartphone sensors have been specified in the literature

R. Chhabra (✉) · S. Verma
Banasthali Vidyapith, Vanasthali, Rajasthan, India
e-mail: rishuchh@gmail.com

S. Verma
e-mail: seemaverma3@yahoo.com

C. Rama Krishna
NITTTR, Chandigarh, India
e-mail: rkc_97@yahoo.com

C. R. Krishna et al. (eds.), *Proceedings of 2nd International Conference on Communication, Computing and Networking*, Lecture Notes in Networks and Systems 46, https://doi.org/10.1007/978-981-13-1217-5_49

to generate an alert in case of improper driving. It has been concluded that mobile smartphones are an efficient and cost-effective technique that is owned by the majority of people worldwide and in India [29]. In this paper, we use an inbuilt mobile smartphone accelerometer to detect sudden acceleration and sudden braking. Contrary to the previous work, the proposed system uses an inbuilt gyroscope to detect sharp turns. The proposed system does not make use of magnetometer; instead it uses accelerometer and gyroscope to detect driving behavior. The fused sensor data obtained from an inbuilt magnetometer, accelerometer, and gyroscope have been used in [30] to detect various driving maneuvers. However, most of the low budget mobile smartphones do not include a magnetometer sensor and moreover; gyroscope is efficient than magnetometer to detect device motion [1].

The paper is organized as follows: Sect. 2 briefly discusses the related research, Sect. 3 gives the proposed system design and implementation, Sect. 4 discusses different patterns analyzed using the accelerometer and gyroscope and Sect. 5 presents the conclusion and future directions.

2 Related Work

Monitoring driver behavior has become one of the key research areas under the umbrella of ITS. The researchers have employed various techniques like specialized simulators [5, 9, 11, 25, 31], specialized vehicle mounted sensors and devices [14, 21, 23, 24, 26, 27], ADAS [3] and inbuilt mobile phone sensors [10, 13, 15, 17, 28] to accomplish this task. A comparative analysis of different techniques to detect driver's behavior has been given in [8]. In [18], different methods for detecting driver behavior like specialized sensors in a vehicle, roadside sensors, and inbuilt sensors have been discussed. A system Nericell has been designed and implemented for observing road and traffic conditions by means of smartphones in [22]. In [13], road conditions and driver behavior has been detected using an inbuilt accelerometer. Thresholds has been set for various driving maneuvers like safe acceleration, safe deceleration, lane changing, etc. and high pass filter has been used to filter the values. In [17], a system MIROAD has been proposed to detect various driving maneuvers using fused sensor data and Dynamic Time Warping (DTW) algorithm for end point detection. In [10], drunk driver behavior has been detected using an accelerometer and orientation sensor. [15] proposed a system that uses inbuilt sensors of a mobile smartphone and focuses on erratic driving behavior caused while overtaking. In [4], along with fuzzy inference system, inbuilt mobile phone sensors have been used to detect driver's behavior using dynamic thresholds obtained by statistical analysis. Inbuilt smartphone sensors have been used in [2], for accident detection and reporting. In [7], a system has been proposed to detect abnormal driver behavior like fast turns, weaving, etc. using accelerometer and orientation sensor. Support Vector Machine (SVM) has been trained and then used to classify driver behavior. In [19], the researchers have used the inbuilt accelerometer readings to detect sudden acceleration changes, sharp turns, road bumps, potholes, etc. In [12], fused sensor data from mobile smartphone accelerometer, gyroscope, and magnetometer have been obtained. DTW algorithm and Baye's classification have been applied to classify the driver's behavior as unsafe or safe. In [30], an android application has been developed to recognize driver behavior. The algorithm guides the user to choose between pure accelerometer readings or fused sensor data. However, the proposed algorithm cannot use fused sensor data and data from the accelerometer at the same time. In [6], a system V-Sense has been developed. The system detects, differentiates, and alerts the driver about various steering maneuvers like driving on curvy roads, lane changes and turns using the inbuilt accelerometer, magnetometer, gyroscope, and GPS. In [32], the system uses inbuilt accelerometer and gyroscope to obtain the centripetal acceleration and angular speed to detect driver phone usage. The system uses a gyroscope to detect the turn directions and the angular speed of the vehicle while turning. In [20], a system SenSafe has been proposed that senses the behavior of surrounding vehicles using smartphone inbuilt sensors; and alerts the driver and pedestrians. Another system SenSpeed, to sense the speed of the vehicle without GPS has been developed in [16]. It uses an accelerometer to detect vehicle acceleration and gyroscope to detect the angular velocity. The proposed work uses inbuilt smartphone sensors to detect the driver behavior. It is distinct in the way it detects turns using a gyroscope by establishing thresholds, and not using an accelerometer or fused sensor data for the detection of turns.

3 System Design and Implementation

The device used is an Android smartphone Xiaomi Redmi Prime 2 with API level 19. The device has an inbuilt accelerometer, GPS, gyroscope, and magnetometer. Figure 1 describes the proposed approach. For changes in longitudinal acceleration, the Y-axis readings of the accelerometer have been observed and for changes in angular velocity, the Z-axis readings of gyroscope have been observed. The gyroscope Z-axis values decrease when the vehicle turns right and increase on turning left [20]. The readings from the accelerometer are filtered using a high pass filter in order to remove high frequency noise and then the low pass filter is used as a smoothing filter. For smoothening of accelerometer and gyroscope values, exponential moving average has been used as a low pass filter [33] and the algorithm for the same has been written in JAVA.

3.1 Data Collection and Data Processing

Data is collected using an Android App which registers the accelerometer sensor and the gyroscope sensor in the beginning and the delay is set to 200 ms. The phone is fixed at the vehicle's dashboard in portrait mode. Figure 2 shows the interface of the Android App "Driver Behavior Detection .apk". The app starts recording and

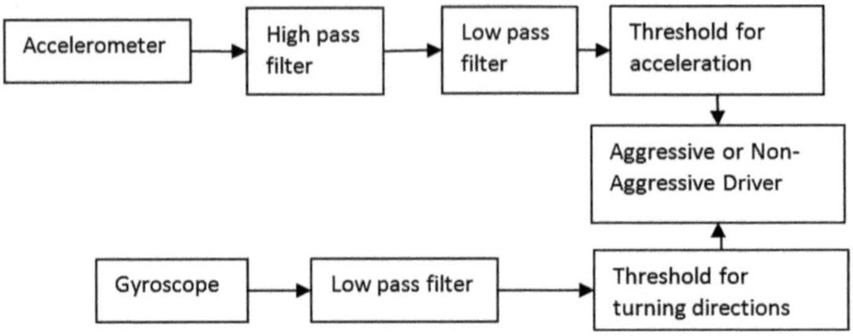

Fig. 1 Proposed methodology

Fig. 2 Android application interface to detect aggressive and non-aggressive driver behavior

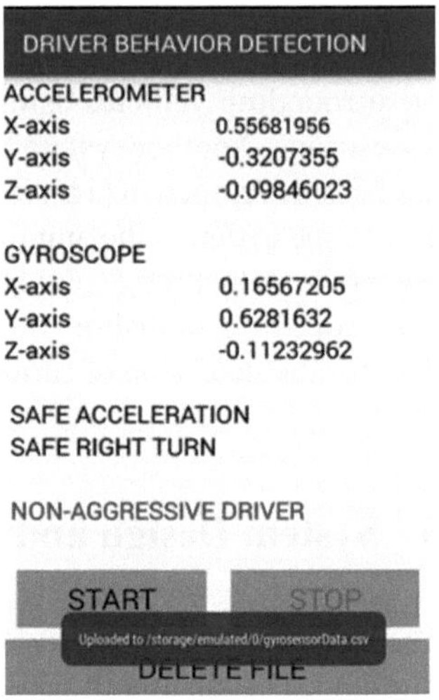

displaying the data when we tap the START button. On tapping the STOP button, the data recorded for the accelerometer is stored in sensorData.csv file and data from the gyroscope is stored in gyrosensorData.csv file. These .csv data files can be deleted from the app by tapping the DELETE FILE button; otherwise, whenever we run the app next time, the files get overwritten. On tapping the STOP button, the overall driving status is displayed as aggressive or non-aggressive on the basis of how the driver drove the car for a particular time duration under consideration (between start and stop). If the count of safe events encountered (safe acceleration, safe deceleration, safe left turn, and safe right turn) is more than the unsafe events (sudden acceleration, sudden deceleration, sharp left turn, and sharp right turn) then the overall status displayed is "NON-AGGRESSIVE DRIVER", otherwise, the status is "AGGRESSIVE DRIVER".

Table 1 shows data from .csv file for filtered accelerometer readings using high pass and low pass filter and filtered gyroscope readings using a low pass filter. The data still contain certain false positives and false negatives. Algorithm 1 presents the procedure to compute the overall driving status when the STOP button is tapped.

Algorithm 1 Algorithm to display the overall driving status

Begin
Initialize $safe = 0, unsafe = 0$
if Driving Event is safe acceleration or safe deceleartion or safe turn **then**
  $safe++$
else
  $unsafe++$
end if
if $safe >= unsafe$ **then**
  Output: NON-AGGRESSIVE DRIVER
else
  Output: AGGRESSIVE DRIVER
end if
End

Table 1 Filtered accelerometer and gyroscope data respectively for driver X

Accelerometer readings

X	Y	Z
0.812113	5.064216	5.19446
0.806699	5.030454	5.15983
0.756859	4.981942	5.101583
0.711573	4.90937	5.020355
0.688572	4.794312	4.917678
0.664758	4.668505	4.800908

Gyroscope readings

X	Y	Z
0.174464	0.108677	−0.16182
0.279257	0.212499	−0.05011
0.185826	−0.34917	−0.34816
−0.03561	−0.22494	−0.41088
0.21537	0.080754	−0.33758
−0.03779	−0.33747	−0.32694

4 Patterns Analyzed

A number of test drives have been conducted to obtain the patterns for different driving events. After analyzing the processed data, thresholds determined are given in Table 2. Accelerometer threshold values have been verified by the results obtained in [19, 30]. To obtain the threshold values for the gyroscope readings, the real-time dataset has been collected by driving a car with the smartphone mounted in the front of vehicle; and recording the data for safe and unsafe/sharp (left and right) turns. It has been analyzed that the values recorded for the turns are synchronous with the results and data obtained by [20].The following patterns have been observed for different driving events:

Table 2 Threshold and data used for driving events

Event	Sensor used for detection	Axis used for detection	Threshold values
Safe acceleration	Accelerometer	Y-axis	1–3 (m/s^2)
Hard acceleration	Accelerometer	Y-axis	$>$3(m/s^2)
Safe deceleration	Accelerometer	Y-axis	-1 to -3 (m/s^2)
Hard deceleration	Accelerometer	Y-axis	<-3(m/s^2)
Safe left turn	Gyroscope	Z-axis	0.4 to 1.2(rad/s)
Sharp left turn	Gyroscope	Z-axis	$>$1.2(rad/s)
Safe right turn	Gyroscope	Z-axis	-0.4 to -1.2 (rad/s)
Sharp right turn	Gyroscope	Z-axis	<-1.2(rad/s)

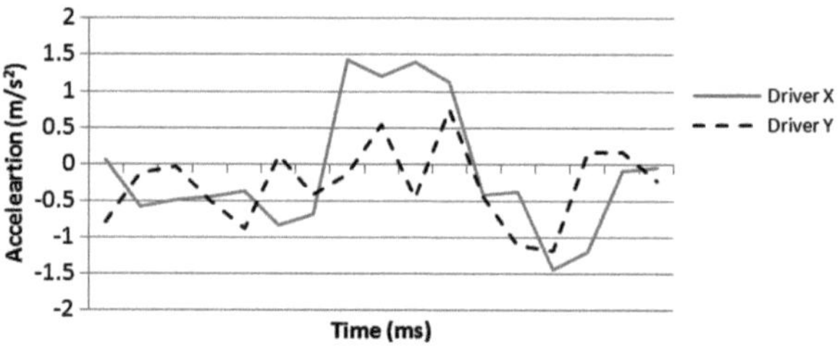

Fig. 3 Safe acceleration and deceleration driving pattern for Driver X and Driver Y

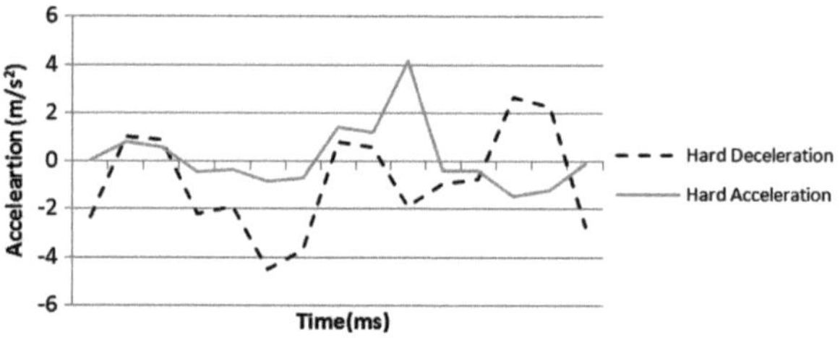

Fig. 4 Hard deceleration and hard acceleration driving pattern for Driver X

4.1 *Safe and Unsafe/Hard Acceleration, Deceleration*

Figure 3 shows the driving pattern involving safe acceleration and safe deceleration. When the value of the Y-axis of the accelerometer is lying between 1 and 3(m/s^2) or between -1 and -3(m/s^2), it is considered as safe acceleration and safe deceleration respectively. Figure 4 shows the driving pattern for hard acceleration and sudden braking i.e. hard deceleration. This is implied by a sudden increase or decrease in the value of accelerometer Y-axis data. When the value of Y-axis is >3 or < -3(m/s^2), it is considered as hard acceleration and deceleration respectively.

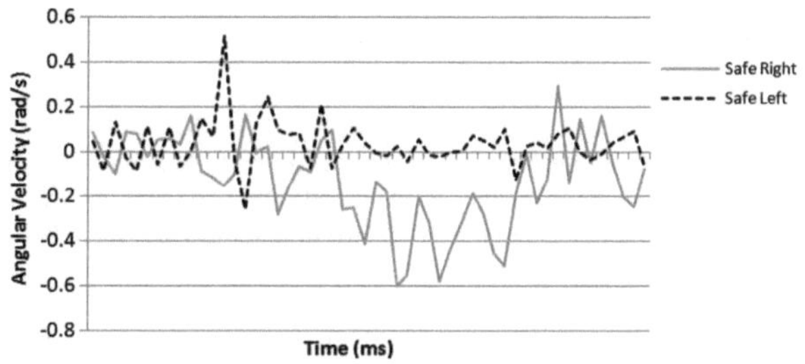

Fig. 5 Safe left turn and Safe right turn driving pattern for Driver X

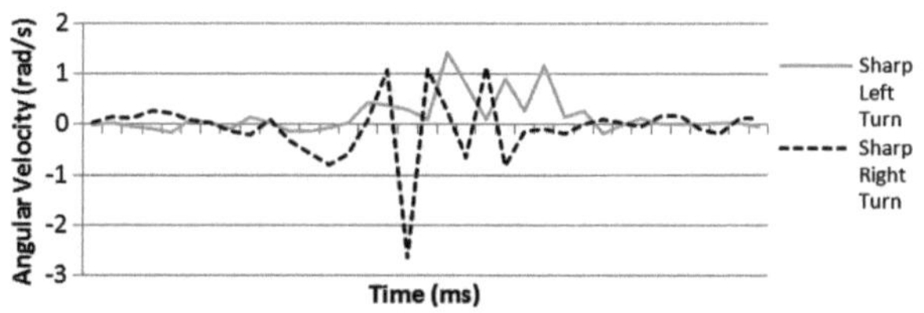

Fig. 6 Sharp left turn and Sharp right turn driving pattern for Driver X

4.2 Safe and Unsafe/Sharp Turns

Figure 5 depicts the driving pattern for the safe left turn and safe right turn. When the value of Z-axis of the gyroscope is lying between 0.4 and 1.2 (rad/s), it is considered as safe left turn and when it is between −0.4 and −1.2 (rad/s), it is considered as a safe right turn. Figure 6 represents an unsafe/sharp left turn when the value of the Z-axis of the gyroscope exceeds 1.2(rad/s) and an unsafe/sharp right turn is detected by the application when the value of Z-axis of gyroscope is less than −1.2(rad/s).

5 Conclusion and Future Directions

This work shows that the driver's behavior can be detected on the basis of the data obtained from the smartphone's accelerometer and a gyroscope sensor. The Android application developed uses the data from the accelerometer to detect acceleration changes and data from the gyroscope to detect turns. On the basis of analysis of driving data, a driver has been categorized as an aggressive or non-aggressive driver. The future plan is to extend this work to detect road conditions and considering various other factors that contribute to the driving style of a driver. We also plan to make the application independent of orientation and movement of the mobile smartphone.

References

1. Accelerometer and Gyroscope in Android Motion Sensors GameLab. https://gamelab.university-id.net/2017/02/accelerometer-and-gyroscope-in-android-motion-sensors/
2. F. Aloul, I. Zualkernan, R. Abu-Salma, H. Al-Ali, M. Al-Merri, IBump: smartphone application to detect car accidents. Comput. Electr. Eng. **23**, 66–75 (2015). https://doi.org/10.1016/j.compeleceng.2015.03.003
3. T. Bar, D. Nienhüser, R. Kohlhaas, J.M. Zöllner, Probabilistic driving style determination by means of a situation based analysis of the vehicle data, in *Proceedings of the IEEE Conference on Intelligent Transportation Systems, ITSC* (2011), pp. 1698–1703. https://doi.org/10.1109/ITSC.2011.6082924
4. G. Castignani, T. Derrmann, R. Frank, T. Engel, Driver behavior profiling using smartphones: a low-cost platform for driver monitoring. IEEE Intell. Transp. Syst. Mag. **7**(1), 91–102 (2015). https://doi.org/10.1109/MITS.2014.2328673
5. R. Cervero, University of California transportation university of California. Transp. Res.-D **2**(2), 107–123 (1997). https://doi.org/10.1068/a201285
6. D. Chen, K.T. Cho, S. Han, Z. Jin, K.G. Shin, Invisible sensing of vehicle steering with smartphones, in *Proceedings of the 13th Annual International Conference on Mobile Systems, Applications, and Services - MobiSys '15* (2015), pp. 1–13. https://doi.org/10.1145/2742647.2742659, http://dl.acm.org/citation.cfm?doid=2742647.2742659
7. Z. Chen, J. Yu, Y. Zhu, Y. Chen, M. Li, D 3: abnormal driving behaviors detection and identification using smartphone sensors, in *2015 12th Annual IEEE International Conference on Sensing, Communication, and Networking (SECON)*. IEEE (2015), pp. 524–532
8. R. Chhabra, S. Verma, C.R. Krishna, A survey on driver behavior detection techniques for intelligent transportation systems, in *2017 7th International Conference on Cloud Computing, Data Science & Engineering-Confluence*. IEEE (2017), pp. 36–41
9. G. Christian, M. Quintero, J.O. Lopez, A.C.C. Pinilla, Driver behavior classification model based on an intelligent driving diagnosis system, in *2012 15th International IEEE Conference on Intelligent Transportation Systems* (2012), pp. 894–899. https://doi.org/10.1109/ITSC.2012.6338727, http://ieeexplore.ieee.org/document/6338727/
10. J. Dai, J. Teng, X. Bai, Mobile phone based drunk driving detection. ...), in 2010 4th International (2010), pp. 1–8. https://doi.org/10.4108/ICST.PERVASIVEHEALTH2010.8901, http://ieeexplore.ieee.org/xpls/abs_all.jsp?arnumber=5482295
11. I.G. Daza, N. Hernandez, L.M. Bergasa, I. Parra, J.J. Yebes, M. Gavilan, R. Quintero, D.F. Llorca, M.A. Sotelo (2011) Drowsiness monitoring based on driver and driving data fusion, in *Proceedings of the IEEE Conference on Intelligent Transportation Systems, ITSC* (2011), pp. 1199–1204. https://doi.org/10.1109/ITSC.2011.6082907
12. H. Eren, S. Makinist, E. Akin, A. Yilmaz, Estimating driving behavior by a smartphone. Proc. IEEE Intell. Veh. Symp. **254**, 234–239 (2012). https://doi.org/10.1109/IVS.2012.6232298
13. M. Fazeen, B. Gozick, R. Dantu, M. Bhukhiya, M.C. González, Safe driving using mobile phones. IEEE Trans. Intell. Transp. Syst. **13**(3), 1462–1468 (2012). https://doi.org/10.1109/TITS.2012.2187640
14. R. Feng, G. Zhang, B. Cheng, An on-board system for detecting driver drowsiness based on multi-sensor data fusion using dempster-shafer theory, in *Proceedings of the 2009 IEEE International Conference on Networking, Sensing and Control, ICNSC* (2009), pp. 897–902. https://doi.org/10.1109/ICNSC.2009.4919399
15. E.H. Gazali, Monitoring Erratic Driving Behaviour caused by Vehicle Overtaking using Off-the-shelf Technologies. October (October) (2010)
16. H. Han, J. Yu, H. Zhu, Y. Chen, J. Yang, Y. Zhu, G. Xue, M. Li, SenSpeed: sensing driving conditions to estimate vehicle speed in urban environments. Proc. IEEE INFOCOM (2014). https://doi.org/10.1109/INFOCOM.2014.6847999
17. D.A. Johnson, M.M. Trivedi, Driving style recognition using a smartphone as a sensor platform, in *Proceedings of the IEEE Conference on Intelligent Transportation Systems, ITSC*, pp. 1609–1615 (2011). https://doi.org/10.1109/ITSC.2011.6083078

18. N. Kalra, D. Bansal, Analyzing driver behavior using smartphone sensors: a survey. Int. J. Electron. Electr. Eng. **7**(7), 697–702 (2014)
19. N. Kalra, G. Chugh, D. Bansal, Analyzing driving and road events via smartphone. Int. J. Comput. Appl. **98**(12), 5–9 (2014). https://doi.org/10.5120/17233-7561. http://research.ijcaonline.org/volume98/number12/pxc3897561.pdf
20. Z. Liu, M. Wu, K. Zhu, L. Zhang, SenSafe: a smartphone-based traffic safety framework by sensing vehicle and pedestrian behaviors. Mob. Inf. Syst. (2016). https://doi.org/10.1155/2016/7967249
21. D. Mitrovic, Reliable method for driving events recognition. IEEE Trans. Intell. Transp. Syst. **6**(2), 198–205 (2005). https://doi.org/10.1109/TITS.2005.848367
22. P. Mohan, V.N. Padmanabhan, R. Ramjee (2008) Nericell: rich monitoring of road and traffic conditions using mobile smartphones, in *Proceedings of the 6th ACM Conference on Embedded Network Sensor Systems - SenSys '08* (2008), p. 323. https://doi.org/10.1145/1460412.1460444
23. K. Oguchi, D. Weir, System for predicting driver behavior. US Patent App. 11/968,864 (2008)
24. W.Q.W. Qing, Y.W.Y. Weiwei, A driver abnormality recondition model based on dynamic Bayesian network for ubiquitous computing, in *2010 3rd International Conference on Advanced Computer Theory and Engineering ICACTE*, vol. 1 (2010), pp. 320–324. https://doi.org/10.1109/ICACTE.2010.5579007
25. G.C.M. Quintero, J.A.O. Lopez, J.M.P. Rua, Intelligent erratic driving diagnosis based on artificial neural networks, in *2010 IEEE Andescon* (2010), pp. 1–6. https://doi.org/10.1109/ANDESCON.2010.5631576, http://ieeexplore.ieee.org/document/5631576/
26. M. Sakairi, M. Togami, Use of water cluster detector for preventing drunk and drowsy driving, in *Proceedings of IEEE Sensors* (2010), pp. 141–144. https://doi.org/10.1109/ICSENS.2010.5690673
27. H. Singh, J.S. Bhatia, J. Kaur, Eye tracking based driver fatigue monitoring and warning system, in *India International Conference on Power Electronics, IICPE 2010* (2011). https://doi.org/10.1109/IICPE.2011.5728062
28. P. Singh, N. Juneja, S. Kapoor, Using mobile phone sensors to detect driving behavior, in *Proceedings of the 3rd ACM Symposium on Computing for Development - ACM DEV '13* (2013), p. 1. https://doi.org/10.1145/2442882.2442941, http://dl.acm.org/citation.cfm?doid=2442882.2442941
29. Smartphone users in India 2015-2021 | Statistic. https://www.statista.com/statistics/467163/forecast-of-smartphone-users-in-india/
30. P. Vavouranakis, S. Panagiotakis, G. Mastorakis, C.X. Mavromoustakis, J.M. Batalla, Recognizing driving behaviour using smartphones, in *Beyond the Internet of Things* (Springer, Berlin, 2017), pp. 269–299
31. T. Wakita, K. Ozawa, C. Miyajima, K. Igarashi, K. Itou, K. Takeda, F. Itakura, Driver identification using driving behavior signals. IEEE Conf. Intell. Transp. Syst. Proc. ITSC **2005**, 907–912 (2005). https://doi.org/10.1109/ITSC.2005.1520171
32. Y. Wang, J. Yang, H. Liu, Y. Chen, M. Gruteser, R.P. Martin, Sensing vehicle dynamics for determining driver phone use, in *Proceeding of the 11th Annual International Conference on Mobile Systems, Applications, and Services - MobiSys '13* (2013), p. 41. https://doi.org/10.1145/2462456.2464447, http://dl.acm.org/citation.cfm?doid=2462456.2464447
33. M. Xie, D. Pan, Accelerometer gesture recognition. Stanf. Lect. **1**, 1–5 (2014). http://cs229.stanford.edu/proj2014/MichaelXie,DavidPan,AccelerometerGestureRecognition.pdf

SRPF Interest Measure Based Classification to Extract Important Patterns

Ochin Sharma, Suresh Kumar and Nisheeth Joshi

Abstract Associative classification is a field of data mining that deals with the mining of the associations among different data variables and further classifying these variables. There are various techniques, explored by different researchers to classify association rules. Most of these classification techniques use confidence interest measure in scrutiny and ranking the rules. But confidence measure itself produces many times inaccurate results as investigated in various literature. So, in this paper, classifier based on a recent interest measure named Significant Rule Power Factor (SRPF) has been explored. The experiments are conducted through implementation in Weka tool. Results show that SRPF-based classifier is a new-generation classifier with light in weight and better results in accuracy.

Keywords Association rule mining · RPF · SRPF · Classifier · Data mining
Data analysis · Statistical measure · Machine learning

1 Introduction

Data mining is a field that focuses on how to extract valuable hidden and significant patterns from various data repositories. Associative classification is a branch of data mining that is used to classify the various related items and to find the valuable patterns or relationship among them. Some essential definitions related to associative classification are as follows:

Definition 1 Association rule mining is a technique for extracting relationship among variables. A rule is defined as T->U, where T is called antecedent part of a rule and U is consequent part of the rule.

O. Sharma (✉) · N. Joshi
Banasthali University, Vanasthali, India
e-mail: ochinsharma2@gmail.com

O. Sharma · S. Kumar
Manav Rachna International University, Faridabad, India

C. R. Krishna et al. (eds.), *Proceedings of 2nd International Conference on Communication, Computing and Networking*, Lecture Notes in Networks and Systems 46, https://doi.org/10.1007/978-981-13-1217-5_50

Definition 2 An itemset represented by attribute–value pairs as $\{(a_i, b_i), v_i\}$ and $\{(c_j, d_j), v_j)$ where a_i, b_i, c_j, d_j are related attributes and v is the value of item set.

Definition 3 Associative classification: In associative classification, rule's antecedent part represents an itemset and consequent represents a class label. Classifier is built from a set of association rules. It is in the form of $\{(a_i, b_i\ldots) \text{->} c_i\}$, where $(a_i, b_i\ldots)$ represents an itemset and c_i represents a class label such that $c_i \in C$.

Definition 4 Interest measure: It is a statistical constraint while knowledge extraction. It is useful in rule pruning, important rule extraction, and also helpful in rule ranking. Depending upon requirements, different interest measures can be used [1]. A few of well-known interest measures are support, confidence [2], conviction [3], leverage [4], and RPF [5].

Class-based association rules (CARs) were first introduced by Ma and Liu in [6]. Different methods have been used to mine CARs, for example, CBA (2) [7], MMAC [8], CPAR [9], and CMAR [10]. CAR-based classifiers could show more accurate results as compared to other classifiers.

Nguyen et al. said in the future scope of [11] that interest measure can help in building an accurate classifier and also helpful in rule ranking. Mostly, classifiers such as CBA [6], CBA (2) [7], MMAC [8], and CMAR [10] used "Confidence" measure to classify the important rules and building the classifier afterward. But many researchers, Brin et al. [3], Piatetsky [4], Ochin et al. [5, 12], and Neda et al. [13], have discussed "Confidence" interest measure and its various associated drawbacks. Hence, the accuracy of confidence-based classifiers is a subject of investigation.

Neda in her thesis [13] said that it would be important to classify rules based on the lift interest measure to improve associative classification as lift is a better interest measure than confidence but it has never done. However, in the proposed work, classification of the rules is based on SRPF interest measure that is better than the lift measure. Significant rule power factor interest measure [12] has been used to build a classifier by using various concepts of existing classifiers and also discussing the experimental outcome of this.

2 Why SRPF Classifier

SRPF-based classifiers have used some important essentials from associative classification theory, including association rule mining, rule pruning, and rule ranking. It classifies the important rules to keep the classifier light and accurate. For this, SRPF classifier has also used features of existing classifiers such as CBA (2), C4.5 [14], CMAR [10], and MCAR [15]. The framework and essential features of SRPF-based classifier contain the following:

Table 1 Important rule identification based on confidence, lift, and SRPF

Scenario	*P* (T)	*P* (U)	*P* (TU)	Confidence	Lift	SRPF
I	0.2	0.5	0.15	0.15/0.2 = 0.75 = 75%	0.15/(0.2*0.5) = 0.15/0.10 = 1.15	0.02/0.1 = 0.22
II	0.3	0.5	0.2	0.2/0.3 = 0.66 = 66%	0.2/(0.3*0.6) = 0.2/0.18 = 1.11	0.04/0.1 = 0.40

- Categorization of attribute-wise classes to narrow the search space.
- Class-based association rule generation [6].
- It provides provision for associated class-wise rule generation. This means two or more classes can be combined to generate rules.
- A newly developed interest measure (SRPF) is being used instead of confidence measure.
- Post-rule pruning and ranking.
- Categorizing rules as high priority, medium priority, and low priority, on the basis of SRPF values of rules [10].
- To build a classifier, only high ranked rules are to be considered to keep the classifier light and more accurate [15].

The detailed framework is explained here:

2.1 SRPF Interest Measure

SRPF-based classifier generates association rules same as CBA-2 [6]. CBA uses “Confidence” measure to identify valuable rules. But many researchers have criticized that “Confidence” measure has flaw full results [3–5, 12, 13]. On the other hand, a recent interest measure named SRPF is based on Shapiro’s principle [3] and proven to be an effective interest measure theoretically.

SRPF is being calculated as SRPF (TU) = (*P* (TU)/*P* (T) *P* (U)) * *P* (TU), where *T* and U are the correlated items. *P* (T) is the chances of occurrence of item T and *P* (U) is the chances of occurrence of item B. *P* (TU) is the chances of occurrence of both T and U items together. Classifier like CBA has a constraint of a single support to generate the rules, whereas SRPF classifier facilitates multiple support systems like in [7, 8]. So, this classifier is also perfect to handle rare item problem [16]. Further, SRPF classifier provides a recent provision to generate rules by combining few classes, if required. For example, diary class items (milk…) and confectionary class items (bread…) might be dependent on each other. Then, rules can be generated by combining both of these classes.

In Table 1, scenario 2 contains greater occurrences of antecedent part *P* (T) items and also greater occurrences of rule association *P* (TU). Only SRPF is able to predict that rule associated with the second scenario is more important.

Table 2 SRPF obeying Shapiro and Tan's principles

Interest measure	P1	P2	P3	O1	O2	O3	O4	O5
SRPF	Y	Y	Y	Y	N	N	N	Y

Shapiro [4] and Tan [17] have given some desired principles that an interest measure should follow. Table 2 shows the rules followed by SRPF measure. SRPF follows all the three principles of Shapiro and two principles of Tan.

2.2 SRPF Based Classification

Brute force rule generation method produces a huge number of rules as compared to class-based association rules. For example, in a transactional database, if a number of items are 30. Number of association rules by brute force $= 3^d - 2^{d+1} + 1 =$ 205888984611000 rules. But by categorizing attribute-wise classes to narrow the search [18], rules generated are $\sum_{i=1}^{TC} 2^{Ci}$ where TC = Total number of Classes, C_i = items for ith class, where $i = 1$ to total number of classes. With a class-based approach, if there are eight classes and different items associated with respective classes are 2, 6, 4, 3, 3, 4, 3, and 5, then total numbers of generating rules are $3^d - 2^{d+1} + 1 => 2 + 602 + 50 + 12 + 12 + 50 + 12 + 180 = 920$ rules. Here is the algorithm to generate class-based association rules [6] and prune the rules on the basis of support and SRPF interest measures.

SRPF_Classifier (Frequent itemsets, Total_class, Classwise_attributes)

Description: This module will generate class-wise association rules and discard those rules that have less than the class-wise support threshold and SRPF min threshold:

1. L_1 = {frequent itemsets}, I = 1, p = 1.
2. For (class = I; class <= Total_class; i++)//For each class generate rules
3. Begin
4. For (k = 2; k <= Items_class [p]; k++)//Candidates need to be generated up to the items for a class
5. C_k = Candidate_generation ($L_{k-}1$);//Generate candidates by self-JOIN ($L_{k-}1$)
6. Prune C_k based on pessimistic error rate and as described in Sect. 2.3.
7. For all transactions t∈D do begin
8. C_t = Subset (C_k, t)//Candidates present in transaction
9. For all transactions c∈C_t do begin
10. c.count ++
11. End of For loop step 8
12. L_k = {c ∈ C_k & c.count >= minSupport (class[p])}
13. If (L_k .SRPF >= minSRPF)
14. End of For loop of step 6

Fig. 1 Successful compilation after re-coding and recompiling weka jar files

15. return $U_k\ L_k$;
16. End of For loop of step 4
17. $CAR_I = L_k / CAR_I$ store class-wise generated rule
18. p++;//Repeat rule generation for next class
19. End of For loop of step 2
20. End of algorithm

The total classes are predefined. As algorithm receives the frequent itemsets, class-wise items generate the class-wise association rules (steps 2–5). After rules generation, the system verifies rules against the minimum support and also checks rules with minimum SRPF threshold (steps 11, 12). Those rules having less than the support and SRPF threshold will be pruned.

2.3 *Rule Pruning and Ranking*

Rule pruning is an essential step in building a classifier. It is done through the database coverage [7] and pessimistic error technique used in C4.5 [14]. It prunes a rule "*R*" as follows: If pessimistic error rate of rule R_A is higher than the pessimistic error rate of rule R_B, then rule R_A will be pruned. On the other side, if pessimistic error rate is same for both the R_A and R_B rules. R_A is said to have higher precedence if

1. SRPF value of R_A is greater than R_B, but if both rules have same values of SRPF then
2. Check whose support is higher, if both are having same support value then
3. Check which rule has more itemset, if both rules having same number of items in the rule
4. Check timestamp, which rule is generated first.

3 Experiments and Discussion

Weka tool [19] is an open source and coded in Java. JEdit tool [20] has been used to edit Weka files. ANT tool [21] has been used to recompile the Weka .jar files (Refer Fig. 1).

Table 3 Generation of CAR rules based on SRPF values

Data Set	No. of instances	No. of attributes	Support	SRPF	Without class association rules	Class-based association rules (CAR)
Weather	14	5	0.1	0.1	1410	41
Train	10	4	0.1	0.1	320	42
Breast cancer	286	10	0.01	0.1	34452	6402

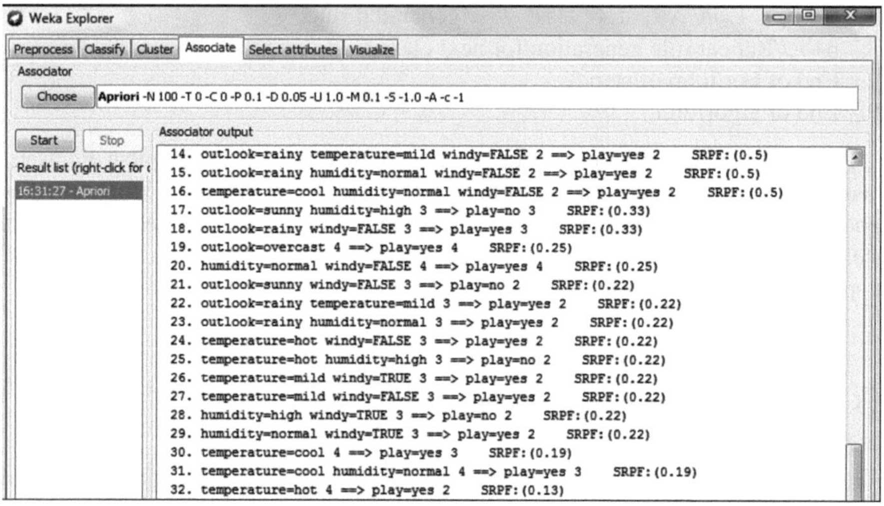

Fig. 2 Experimental values of SRPF for CAR

Three different datasets are used from UCI repository [22] and experiments are conducted; refer Table 1 and Fig. 3. As found, datasets produced less numbers of rules in class-wise rule generation as compared to without class-based rules. Table 3 is showing the detailed summary of datasets used, support, and SRPF minimum thresholds while rule generation (Fig. 2).

Refer to Table 4, in which comparison of the accuracy of different classifiers has been shown, such as trees (J48, Random Forest), Bayes classifier, rule-based classifiers (OneR, JRIP), and associative classifiers (CBA, MMAC). Results show that SRPF classifier is more accurate over existing classifiers. For Weather, tic-tac-toe, and breast cancer datasets, the accuracies reported are 87.3, 99, and 79%, respectively.

Table 4 Accuracy comparison of SRPF classifier

Classifier	Weather	Tic-tac-toe	Breast cancer
ZeroR	64.2	65.34	70.2
J48	50	84.55	75.5
Random forest	71.4	96.34	69.5
Naïve Bayes	57.14	70.04	71.6
MMAC	71.66	99.29	72.1
CBA	85	100	69.66
OneR	42.85	69.93	65.73
JRIP	64.2	97.8	70.9
SRPF classifier	87.3	100	79

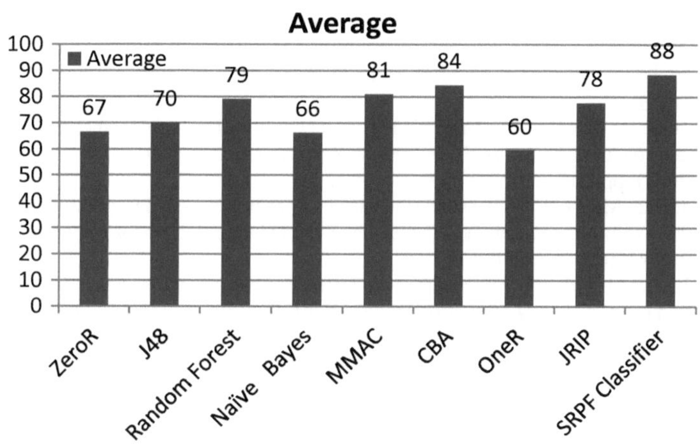

Fig. 3 Average accuracy comparison of various classifiers

Refer Fig. 3, the average accuracy of SRPF classifier is reported as 88%, greater than many existed classifiers.

4 Conclusion

Associative classification has proven its worth with greater accuracy. Mostly, associative classifiers are based on the confidence interest measure. This work has examined classification based on a recent interest measure. Experiments conducted on Weka showed that by using SRPF interest measure, we can achieve more accurate results with no overhead.

References

1. M. Hahsler, *A Probabilistic Comparison of Commonly Used Interest Measures for Association Rules* (2015). Available online: http://michael.hahsler.net/research/association_rules/measures.html
2. R. Agrawal, R. Srikant, Fast algorithms for mining association rules, in *Proceedings of the 20th International Conference on Very Large Data Bases*, VLDB, vol. 1215 (12 Sept 1994), pp. 487–499
3. S. Brin, R. Motwani, J.D. Ullman, S. Tsur, Dynamic itemset counting and implication rules for market basket data. ACM SIGMOD Rec. **26**(2), 255–264 (1 June 1997)
4. G. Piatetsky-Shapiro, Discovery, analysis, and presentation of strong rules. Knowl. Discovery Databases 229–238 (1991)
5. Ochin, S. Kumar, N. Joshi, Rule power factor: a new interest measure in associative classification. Procedia Comput. Sci. **93**, 12–18 (31 Sept 2016)
6. B.L. Ma, Integrating classification and association rule mining. in *Proceedings of the Fourth International Conference on Knowledge Discovery and Data Mining* (Aug 1998)
7. B. Liu, Y. Ma, C.K. Wong, Classification using association rules: weaknesses and enhancements, in *Data Mining for Scientific and Engineering Applications* (2001), pp. 591–605 (Springer US)
8. F.A. Thabtah, P. Cowling, Y. Peng, MMAC: a new multi-class, multi-label associative classification approach. in *Fourth IEEE International Conference on Data Mining, 2004. ICDM'04* (1 Nov 2004), pp. 217–224
9. X. Yin, J. Han, May. CPAR: classification based on predictive association rules, in *Proceedings of the 2003 SIAM International Conference on Data Mining. Society for Industrial and Applied Mathematics* (2003), pp. 331–335
10. W. Li, J. Han, J. Pei, CMAR: accurate and efficient classification based on multiple class-association rules, in *ICDM 2001, Proceedings IEEE International Conference on Data Mining* (2001), pp. 369–376
11. L.T. Nguyen, B. Vo, T.P. Hong, H.C. Thanh, Interestingness measures for classification based on association rules. in *International Conference on Computational Collective Intelligence* (28 Nov 2012), pp. 383–392 (Springer Berlin Heidelberg)
12. O. Sharma, S. Kumar, N. Joshi, Significant rule power factor—an algorithm for new interest measures. Smart Innov. Syst. Technol–SIST **77**, 289–298 (2017). Springer
13. N. Abdelhamid, *Deriving Classifiers with Single and Multi-Label Rules using New Associative Classification Methods* (2013).
14. J.R. Quinlan, *C4.5: Program for Machine Learning* (Morgan Kaufmann, 1992)
15. F. Thabtah, P. Cowling, Y. Peng, MCAR: multi-class classification based on association rule, in *The 3rd ACS/IEEE International Conference onComputer Systems and Applications* (2005), p. 33
16. R.U. Kiran, P.K. Reddy, Mining rare association rules in the datasets with widely varying items' frequencies, in *International Conference on Database Systems for Advanced Applications* (Springer Berlin Heidelberg, 1 Apr 2010), pp. 49–62
17. Pang-Ning Tan, Vipin Kumar, Jaideep Srivastava, Selecting the right objective measure for association analysis. Inf. Syst. **29**(4), 293–313 (2004)
18. Z. Tang, Q. Liao, A New class based associative classification algorithm. IMECS **2007**, 685–689 (2007)
19. R.R. Bouckaert, E. Frank, M. Hall, R. Kirkby, P. Reutemann, A. Seewald, D. Scuse, *WEKA Manual for Version 3-7-3* (The University of WAIKATO, 2010)
20. S. Pestov, *jEdit Open Source Programmer's Text Editor* (2002)
21. S. Loughran, H. Erik, *Ant in Action: of Java Development with Ant* (Dreamtech Press, 2007)
22. http://www.ics.uci.edu/~mlearn/MLRepository.html

Two-Stage Text Feature Selection Method for Human Emotion Recognition

Lovejit Singh, Sarbjeet Singh and Naveen Aggarwal

Abstract In this paper, two-stage text feature selection method is proposed to identify significant features to effectively recognize the human emotions from the unstructured text documents. The proposed method employs two-stage feature filtering mechanism, namely, semantic, and statistical stage. The first stage consists of semantic-based method which extracts the meaningful words from the unstructured text data using parts of the speech (PoS) tagger. It identifies the noun, verb, adverb, and adjective as prospective words for detecting text-based human emotions. The second stage employs chi-square (χ^2) method to remove the weak semantic features with lower statistical score. The effectiveness of the two-stage feature selection method is evaluated and compared with existing methods with support vector machine (SVM) classifier on the publically available and widely accepted ISEAR dataset. The results obtained from the analysis indicate that the SVM classifier with two-stage method has achieved 10.6, 15.46, and 34.45% improvement in emotion recognition rate as compared with the single-stage methods such as PoS method, χ^2 method, and baseline.

1 Introduction

The social media started gaining attention two decades ago by permitting the humans to share their expressions in the form of ideas, information, interest related to career, incident, or the other ways such as virtual communities and networks [5]. The most popular social media platforms are Twitter, Facebook, LinkedIn, Whatsapp, Google Plus, and YouTube. These social platforms commonly allow people to express their

L. Singh (✉) · S. Singh · N. Aggarwal
UIET, Panjab University, Chandigarh, India
e-mail: pu.lovejitjhajj@gmail.com

S. Singh
e-mail: sarbjeet@pu.ac.in

N. Aggarwal
e-mail: navagg@gmail.com

C. R. Krishna et al. (eds.), *Proceedings of 2nd International Conference on Communication, Computing and Networking*, Lecture Notes in Networks and Systems 46, https://doi.org/10.1007/978-981-13-1217-5_51

feelings in the form of unstructured words. There is a need of text emotion recognition system to extract the essential information from the user unstructured text data [4]. The statistical methods have been developed to improve the performance text emotion classification models [6, 9, 12]. The statistical methods did not consider the semantic meaning of the word while constructing text feature subset. In this paper, the two-stage text feature selection method is proposed to select the words with their semantics and statistics significance. The proposed method employs parts of the speech tagger to extract the emotion potential words, namely, nouns, verbs, adverbs, and adjectives. It filters out the other insignificant words such as pronouns, prepositions, conjunctions, determinants, articles, and numbers. The chi-square method is applied to compute the significance score of each semantic word and filter out those words with lower χ^2 score than desired threshold. The main contributions of this paper are as follows: First, the proposed two-stage feature selection method selects the words which are semantically reliable, and statistically significant to effectively recognize the human emotions from unstructured text. The two-stage method is robust to represent the text documents. It has achieved 99.5% feature subset hit rate and only 0.5% feature subset miss rate in representing the ISEAR dataset [8] unstructured text instances. Second, the experimental results indicate that the feature subset selected with two-stage method has improved 10.6, 15.46, and 34.45% emotion recognition accuracies of SVM classifier as compared with the model using PoS method, χ^2 method, and baseline features, respectively. The other sections of the paper are presented as follows: prior work is briefly discussed in Sect. 2. The proposed method is presented in Sect. 3. The results and discussion about the effectiveness and comparison of proposed method with existing approaches are presented in Sect. 4. Last, the proposed work is concluded in Sect. 5.

2 Prior Work

In the prior work, the statistical text feature selection methods are employed to improve the effectiveness of text classification models. In [12], a comparative analysis of term strength (TS), chi-square (χ^2), information gain (IG), document frequency (DF), and mutual information (MI)-based statistical feature selection approaches have been presented for the text categorization on the Reuters corpus. In [9], an empirical study of χ^2, MI, IG, and DF feature selection approaches have been presented to analyze the human sentiments on Chinese documents. In [6], enhanced IG feature selection technique has been proposed for sentiment analysis using movie review corpus. In [11], a comparative analysis of DF, χ^2, MI, and IG feature selection methods have been carried out for Chinese hotel online review sentiment analysis. At present, majority of the existing work is performed using statistical feature selection methods to enhance the recognition rate of the human sentiment classification model [6, 9, 11, 12]. At present, there is no existing study performed with the statistical methods for text human emotion recognition. The main problem of statistical methods is that they are not considering the semantics of the words while constructing text feature subset [6, 9, 12]. To overcome this limitation, two-stage text feature selection method is proposed to consider semantics and statistics of the words during the selection of significant features to construct feature subset.

3 Proposed Method

This section provides the illustration of the two-stage method with parts of the speech tagger (POST) at first stage to semantically select the text features, and χ^2 method at the second stage to statistically reduce the semantic feature subset, and the performance of proposed method is assessed with support vector machine (SVM)-based classification model. The block diagram of the proposed method is presented in Fig. 1.

3.1 First-Stage Text Feature Selection

The unstructured text data is a collection of words, numbers, and special symbols. The parts of the speech tagger (POST) tokenized the unstructured text into the slices known as tokens $w_1, w_2, \ldots, w_n$. The POST predicts the part of the speech tags

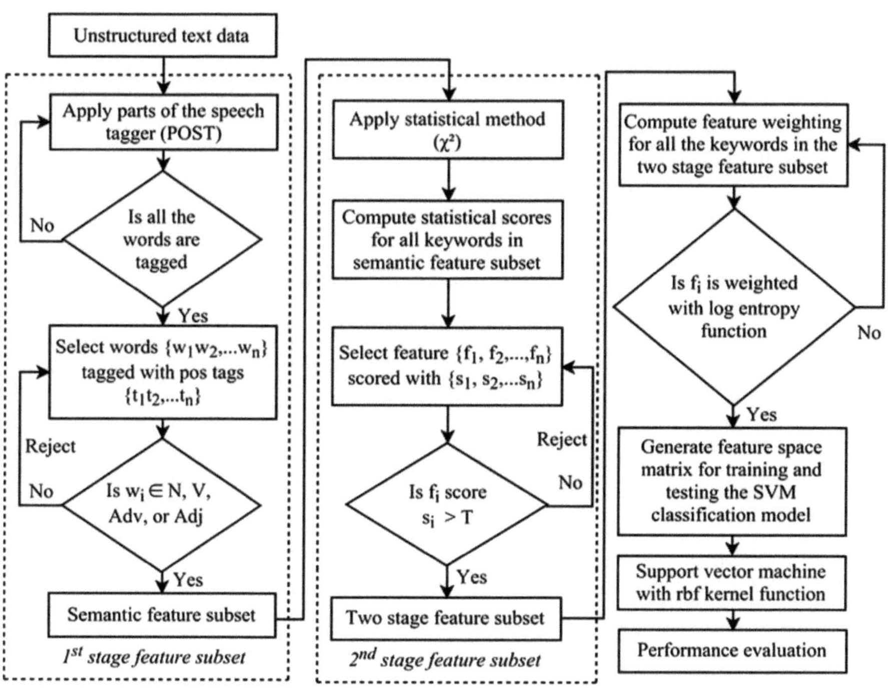

Fig. 1 Proposed method for human emotion recognition

$t_1, t_2, \ldots, t_n$ for every token extracted from unstructured text data. The words w_i, $i = 1, 2, \ldots, n$ are tagged with t_i, $i = 1, 2, \ldots, n$ from a set of tags T which includes noun (NN, NNS, NNP, NNPS), pronoun (PRP, PRP$, TO, WP, WP$), verb PRP, PRP$, TO, WP, WP$), adverb (RB, RBR, RBS), adjective (JJ, JJR, JJS), preposition (IN), conjunction (CC), interjection (UH), and symbol (SYM). The Stanford POST [10] log-linear bidirectional dependency network based maximum entropy model is employed to tag the current token w_i with parts of the speech tag t_i. The two other online parts of the speech taggers are applied to achieve maximum accuracy in the predictions of parts of the speech tags for each token. The online learning algorithm based on sparse network of linear separator (SNOW) architecture has been employed in the identification of parts of the speech tags for tokens [6]. Further, the probability based online parts of the speech tagger has been also applied to utilize the Bayes' theorem, a Markov assumption, and the Viterbi algorithm for tagging the tokens in the unstructured text data [1]. The majority-based voting scheme is performed to identify the parts of the speech tags for every tokens and to generate the semantic-based feature subset which consists of nouns, verbs, adverbs, and adjectives.

3.2 Second-Stage Text Feature Selection

The chi-square statistical method is applied at second stage to remove the insignificant words in the semantic feature subset. The chi-square method [12] determines the independence relationship between feature f_i and the class e_j (joy, anger, disgust, fear, sadness, guilt, and shame, j = 7), and the χ^2 score of the feature is compared to the χ^2 distribution with six degrees of freedom at 10% level of significance which means feature f_i and class e_j are independent with 90% of confidence when score is less than chi-square distribution threshold T. The higher χ^2 score than T indicates that the hypothesis of independence is incorrect and there is a dependency relationship between a feature f_i and emotion class e_j. The χ^2 score is computed with the help of two-way contingency of feature f_i with emotion class e_j using Eq. 1:

$$\chi^2(f_i, e_j) = \frac{N \times (O(f_i, e_j) \times O(\neg f_i, \neg e_j) - O(f_i, \neg e_j) \times O(\neg f_i, e_j))^2}{(O(f_i, e_j) + O(\neg f_i, e_j)) \times (O(f_i, e_j) + O(f_i, \neg e_j)) \times (O(f_i, \neg e_j) + O(\neg f_i, \neg e_j)) \times (O(\neg f_i, e_j) + O(\neg f_i, \neg e_j))} \tag{1}$$

In Eq. 1, $O(f_i, e_j)$ describes the frequency of feature f_i occurred in the emotion class e_j, $O(f_i, \neg e_j)$ describes the frequency of feature f_i occurred without emotion class e_j, $O(\neg f_i, e_j)$ describes the frequency of feature, f_i is not occurred in the class e_j, $O(\neg f_i, \neg e_j)$ describes the frequency at which neither f_i occurred and nor e_j occurred. The value of N is computed as $O(f_i, e_j) + O(\neg f_i, \neg e_j) + O(f_i, \neg e_j) + O(\neg f_i, e_j)$. The χ^2 value in Eq. 1 is calculated for a feature over a single class.

However, the final chi-square score is assigned to particular feature through finding the maximum χ^2 value over all the classes using Eq. 2:

$$\chi^2(f_i) = \max_{j=1}^{m}[\chi^2(f_i, e_j)] \quad (2)$$

3.3 SVM Classification Model

The supervised support vector machine classification model is deployed to recognize human emotion with proposed feature subset fs_{ij}. The fs_{ij} space matrix is labeled with class c_k for training the SVM classifier model. The formal definition for the input training data in set theory is presented as follows.

$$fs = \{(fs_i, c_i) | fs_i \in \Re^p, c_i \in \{joy, anger, disgust, fear, sadness, guilt, shame\}\} \quad (3)$$

The one-against-all support vector machine architecture is adopted to deploy the multi-classification model [2]. The SVM with radial basis function (RBF) kernel is adopted to build the classification model for text emotion recognition. The aim of support vector machine is to locate the best hyperplane with maximum separating margin among the emotion classes in the training space fs_{ij}. The kernel function maps the feature subset space into the dot product space using Eq. 4.

$$(fs, fs') \mapsto k(\mathbf{fs}, \mathbf{fs}') \quad (4)$$

where k is the kernel function in dot product feature space and the rbf kernel in vector space is defined as

$$k_{rbf}(\mathbf{fs}, \mathbf{fs}') = exp[-\gamma \parallel \mathbf{fs} - \mathbf{fs}' \parallel^2], \quad \gamma > 0 \quad (5)$$

where $\parallel \mathbf{fs} - \mathbf{fs}' \parallel^2$ is squared euclidean distance among the feature vectors **fs** and $\mathbf{fs}'$. The γ parameter sets the width of the bell-shaped curve function. It is defined as $\gamma = \frac{1}{2\sigma^2}$, where σ^2 is the tunning parameter. The bell-shaped curve become wider with large σ^2 value and it narrows with small σ^2 value. The training error is minimized with selecting optimal constant parameter cost c, which force the support vector machine to reduce the misclassification data points as much as possible with cost parameter c = 50. The performance of the SVM classifier is measured with standard metrics such as accuracy and f-measure.

4 Results and Discussion

The effectiveness of the proposed method is assessed on the ISEAR dataset [8]. The International Survey on Emotion Antecedents and Reactions (ISEAR) contains more than 7000 emotional instances in the form of text which were collected from various

Table 1 Emotion detection accuracies of the SVM classifier with proposed two-stage method in comparison with POS method, statistical method, and baseline features on ISEAR dataset

Method	Acc (%)	FSMR (%)
Baseline	37.98	0
χ^2 method	56.97	2.60
POS method	61.83	0.1
Two-stage method	72.43	0.5

Acc: Accuracy, FSMR: Feature subset miss rate

regional people with different cultural backgrounds. The acquired instances were annotated by 1096 annotators with diversified cultural backgrounds. The ISEAR dataset was annotated with seven discrete emotion labels, namely, joy, anger, disgust, fear, sadness, guilt, and shame. There are 9200 total number of baseline features in the ISEAR dataset. The two-stage feature selection method has extracted 3000 (excluding plural and proper noun) noun, 2487 verb, 501 adverb, and 1340 adjective features with parts of the speech tagger using ISEAR dataset. The first stage consists of 7328 (noun ∪ verb ∪ adverb ∪ adjective) number of features in the semantic feature subset. The second stage employed the χ^2 method which computed the statistical significance score for each word in the semantic feature subset. The $f_i, i = 1, 2, \ldots, n$, where n is number of features in the semantic feature subset are selected with $\chi^2 score > T$, where T (threshold) value which is decided with χ^2 distribution using six degrees of freedom at 10% level of significance. The χ^2 method has selected top 900 features from the semantic feature subset for text human emotion recognition. The performance of the SVM classifier with proposed two-stage feature selection method is evaluated and compared with the model using statistical method (χ^2), parts of the speech method (POS), and the baseline on the ISEAR dataset with 10-fold cross-validation approach. The performance is shown in Table 1, which indicates that the two-stage text feature selection method has improved 10.6, 15.46, and 34.45% emotion recognition accuracies of SVM classifier as compared with the model using POS method, χ^2 method, and baseline, respectively. The two-stage feature subset has 0.5% feature subset miss rate which means 99.5% of documents in the ISEAR datasets are represented by proposed feature subset. It is acceptable and improved over the χ^2 feature selection method. The confusion matrices for the detail results with proposed method are presented in Table 2. The f-measure of the SVM classifier for detail comparison with two-stage method, POS method, χ^2 method, and baseline is presented in Fig. 2.

The two-stage method has improved 15.26, 19.85, and 31.83% f-measure of SVM classifier in recognizing the anger emotion over the POS method, χ^2 method, and baseline, respectively. Similarly, the two-stage method indicates $\approx$19.74 $\pm$ 9.55%, $\approx$16.89 $\pm$ 9.59%, $\approx$24.84 $\pm$ 13.94%, $\approx$13.65 $\pm$ 12.48%, $\approx$16.42 $\pm$ 6.97%, and $\approx$26.15 $\pm$ 11.13% improvement in f-measure of the SVM classifier in predicting the disgust, fear, guilt, joy, sad, and shame emotions, respectively, over the POS method, χ^2 method, and baseline.

Table 2 Confusion matrix of recognizing seven emotions with proposed two-stage feature selection method on ISEAR datset

(%)	Anger	Disgust	Fear	Guilt	Joy	Sad	Shame
Anger	**71.43**	5.95	1.79	10.71	4.76	2.38	2.98
Disgust	7.50	**74.38**	6.25	2.50	1.88	3.13	4.38
Fear	6.67	5.45	**78.79**	1.21	0.61	1.21	6.06
Guilt	13.82	4.61	2.63	**68.42**	4.61	3.29	2.63
Joy	7.01	1.91	3.18	4.46	**71.97**	5.73	5.73
Sad	15.85	4.68	4.68	7.02	6.43	**69.01**	2.34
Shame	6.92	5.03	5.66	3.14	1.89	4.40	**72.96**

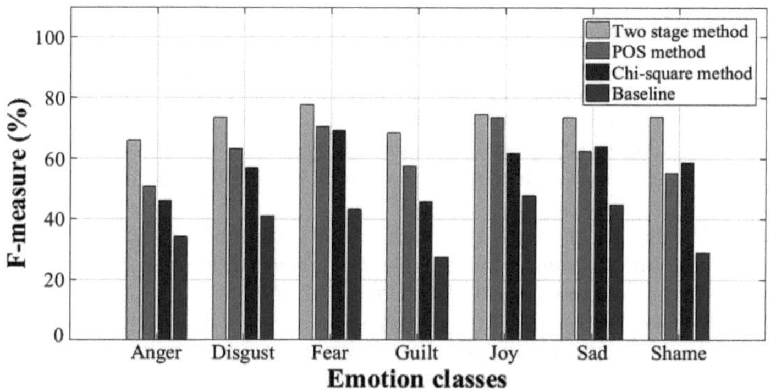

Fig. 2 F-measure of text emotion recognition model using proposed two-stage feature selection method in comparison with POS method, χ^2 method, and baseline features

5 Conclusion

In the prior work, the statistical methods have been developed to identify the optimal text feature subset without considering the semantics of the words while constructing text features [6, 9, 12]. In this paper, the proposed method considers the semantics and statistical significance of words during the selection of text features. The proposed method employs parts of the speech tagger to extract the emotion potential words, namely, nouns, verbs, adverbs, and adjectives. It employs χ^2 method to select most significant semantic features to construct optimal text feature subset. The effectiveness of the two-stage approach is assessed and compared with the existing approaches using SVM on ISEAR dataset. The results obtained through the analysis clearly shows that the proposed method significantly improved the classifier's emotion recognition rate, and f-measure with acceptable feature subset miss rate as compared with only parts of the speech method, statistical method, and baseline. In the future work, efforts will be made to construct a knowledge base on the basis of relationship among the features in two-stage feature subset.

Acknowledgements This work was supported by University Grant Commission (UGC), Ministry of Human Resource Development (MHRD) of India under Basic Scientific Research (BSR) fellowship for meritorious fellows vide UGC letter no. F.25-1/2013-14(BSR)/7-379/2012(BSR) Dated 30.5.2014.

References

1. X. Carreras, I. Chao, L. Padro, M. Padro, FreeLing: an open-source suite of language analyzers. in *Proceedings of the 4th International Conference on Language Resources and Evaluation (LREC)* (2004), pp. 239–242
2. C.C. Chang, C.J. Lin, LIBSVM: a library for support vector machines. ACM Trans. Intell. Syst. Tech. (TIST) **2**(3), 27 (2011). https://doi.org/10.1145/1961189.1961199
3. P. Ekman, *Basic Emotions*, Handbook of Cognition and Emotion Psychology Department (Wiley, UK, 1999)
4. D.M.E.D.M. Hussein, A survey on sentiment analysis challenges. J. King Saud Univ.-Eng. Sci. (2016). https://doi.org/10.1016/j.jksues.2016.04.002
5. A.O. Jonathan, S.W. Steve, Social media definition and the governance challenge: an introduction to the special issue. Telecommun. Policy **39**, 745–750 (2015). https://doi.org/10.2139/ssrn.2647377
6. P. Konez, J. Paralic, An approach to feature selection for sentiment analysis. in *Proceedings of the 15th IEEE International Conference on Intelligent Engineering Systems (INES)* (2011), pp. 357–362. https://doi.org/10.1109/INES.2011.5954773
7. D. Roth, D. Zelenko, Part of speech tagging using a network of linear separators. in *Proceedings of the 17th International Conference on Computational Linguistics (ICCL)*, vol. 2, (1998). pp. 1136–1142. https://doi.org/10.3115/980691.980755
8. K.R. Scherer, H. Wallbott, International survey on emotion antecedents and reactions (ISEAR) (1990). http://www.affective-sciences.org/home/research/materials-and-online-research/research-material/. Accessed 5 June 2017
9. S. Tan, J. Zhang, An empirical study of sentiment analysis for chinese documents. Expert Syst. appl. **34**(4), 2622–2629 (2008). https://doi.org/10.1016/j.eswa.2007.05.028
10. K. Toutanova, D. Klein, C.D. Manning, Y. Singer, Feature-rich part-of-speech tagging with a cyclic dependency network. in *Proceedings of the 2003 Conference of the North American Chapter of the Association for Computational Linguistics on Human Language Technology*, vol. 1 (2003), pp. 173–180. https://doi.org/10.3115/1073445.1073478
11. H. Wang, P. Yin, J. Yao, J.N. Liu, Text feature selection for sentiment classification of Chinese online reviews. J. Exp. Theor. Artif. Intell. **25**(4), 425–439 (2013). https://doi.org/10.1080/0952813X.2012.721139
12. Y. Yang, J.O. Pedersen, A comparative study on feature selection in text categorization. in *Proceedings of the 14th International Conference on Machine Learning (ICML)*, vol. 97 (1997), pp. 412–420

Authorship Analysis of Online Social Media Content

Ravneet Kaur, Sarbjeet Singh and Harish Kumar

Abstract Authorship analysis is used to analyze different excerpts of text by unknown authors. Commencing from literary works in the early eighteenth century, authorship analysis is a most sought out option for forensic investigations till date. Throughout the literature, authorship analysis tasks have been performed on lengthy literary documents to small social network messages. With a steep drift of communication preferences from offline to online media, authorship works dedicated toward online media need a thorough investigation. In this paper, we have provided generalized methodologies adopted for different authorship analysis tasks followed by the review of prominent articles in each domain. Works applicable to online content have been reviewed considering the features used, approaches deployed, and statistics related to taken data samples. A tabular comparison concerning different data details has been given for better understanding and in-depth analysis.

Keywords Authorship analysis · Attribution · Verification · Profiling

1 Introduction

Social media has overpowered the clout of traditional communication measures such as letter writing, telegrams, or other offline practices. Internet being an essential part of the human life has made people rely more on online communication sources such as instant messaging, blog writing, social networking websites, etc. for communicating and information spreading. With the huge amount of data present online nowadays,

R. Kaur (✉) · S. Singh · H. Kumar
University Institute of Engineering and Technology, Panjab University, Chandigarh, India
e-mail: ravneets48@gmail.com

S. Singh
e-mail: sarbjeet@pu.ac.in

H. Kumar
e-mail: harishk@pu.ac.in

C. R. Krishna et al. (eds.), *Proceedings of 2nd International Conference on Communication, Computing and Networking*, Lecture Notes in Networks and Systems 46, https://doi.org/10.1007/978-981-13-1217-5_52

there has been a sudden inclination in the misuse of the present information such as to spread spam, abusive content, and threatening. Hence, it has become essential to know the authenticity of the message posted on diverse online platforms such as blogs, social media sites, news forums, or other such platforms. Anonymous spread of some sensitive perilous activities such as trolling, sarcastic comments, rumors, dispersion of negative information, threatening, rebuking, and obscene postings usually involves criminal investigations wherein it becomes highly relevant to know the original author of the post.

The term coined as authorship analysis involves the investigation of various tracts of text to determine its authorship. Though authorship analysis has a broad application base such as in literary works, de-anonymization tasks, source code author identification, and much more, recently, it has got wide attention in social media analysis where the primary challenge is working with short texts. Also, short messages are commonly deployed in criminal activities and hence application of authorship analysis in crime investigations seems to be prolific. This may include detection of hackers and hacked profiles, phishing posts, illegal activities, etc.

2 Authorship Analysis

According to Koppel et al. [11], authorship analysis falls into three categories, specifically, authorship attribution (identification), authorship verification, and authorship characterization (profiling) (Fig. 1).

2.1 Authorship Attribution

Authorship attribution, often studied as authorship identification, is a many-to-one classification problem where among a given set of users and an unknown user, the

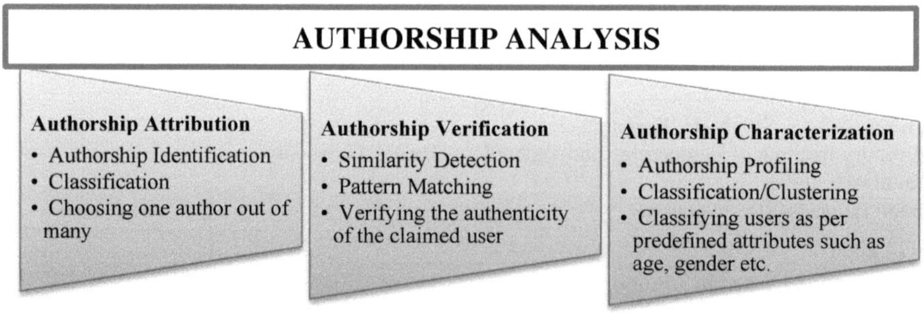

Fig. 1 Categories of authorship analysis

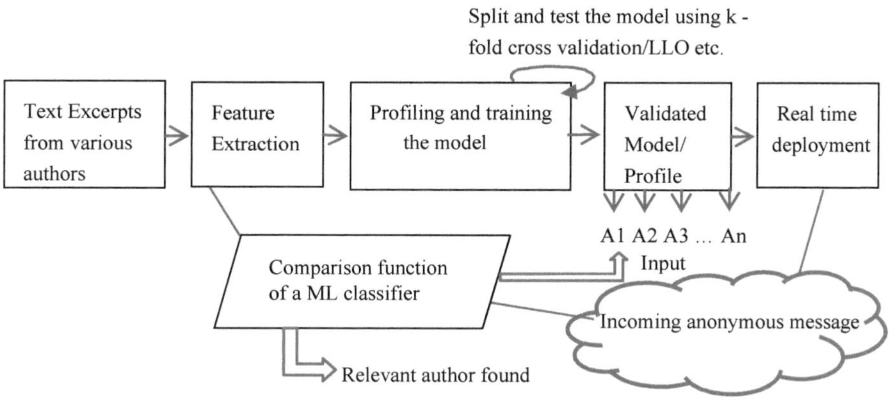

Fig. 2 A simplified authorship attribution framework

task is to determine a correct user for the given record. With "n" users in place, the closest counterpart is yielded as the label of the unknown sample. Unlike the previous tradition of studying the problem in a closed fashion where even in the worst scenario assuredly a class is returned as a label, recent works have studied the problem as an open-set problem where the unlabeled text may not necessarily be the one by the given users. In short, the former scenario assumes the author to be present in the training set and hence calls for the decisive match, whereas in the latter case if no close match is found, a rejection policy is followed stating the author to be someone outside the given set. Researchers have used varying excerpts of text ranging from long documents in literary works to the tweets limited to just 140 characters for the task of authorship identification. A general methodology adopted for the authorship attribution is shown in Fig. 2.

Figure 2 reflects generalized steps followed for the task of authorship identification. For an unknown sample, the job is to determine who among a given set of users (A_1, A_2, …, A_n) is the actual author of the anonymous text. The task is accomplished by independently modeling the profiles of referenced users and comparing the unknown data against each profile to determine the probable author. The problem if studied as a closed set does return the most probable match. But for the open-set scenario, if no probable match is found beyond a threshold, an author outside the given set is alarmed.

2.2 *Authorship Verification*

Authorship verification, also studied as similarity detection, is the one-to-one binary classification task and could be considered as a subset of the open-set authorship identification problem where for a given anonymous message one has to determine

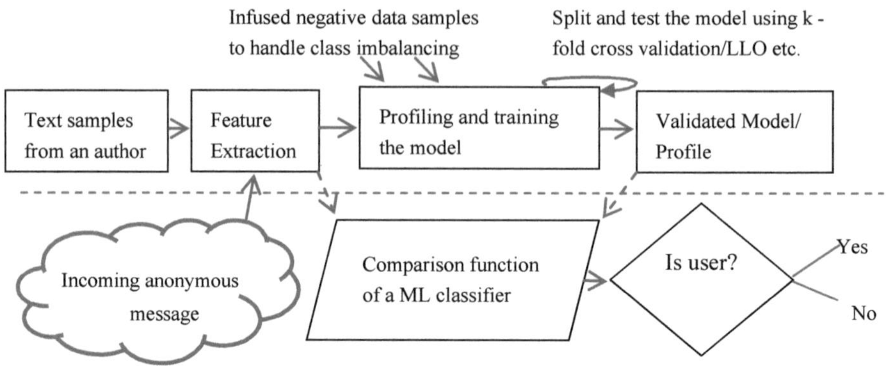

Fig. 3 A generalized authorship verification process

whether it is written by the suspected user or not. The basic methodology followed for the authorship verification task is shown in Fig. 3.

A behavioral profile is built by selecting and extracting the relevant features from a user's data samples. The same sets of features obtained from the unknown sample are compared against the learned patterns. After comparing with the already learned patterns of both positive and negative classes, a class with which the feature values of the unknown sample match the most is predicted as the label.

2.3 *Authorship Characterization*

Authorship profiling/characterization focuses on determining the characteristics such as age, gender, linguistic information, etc. of the authors of the written text. Given an anonymous text and a set of predefined categorizations, the task is to determine the required demographic characteristics of a user.

As with other authorship analysis tasks, features extracted from the labeled data samples for each demographic characteristic are independently analyzed and profiled. Machine or other self-learning techniques are used to analyze the efficiency of different classifiers or cluster models. Accuracy estimates of different classifiers reflect the efficiency of each method. Likewise, weighing the features as per their importance gives an estimate of the significance of each feature to distinguish different authors (Fig. 4).

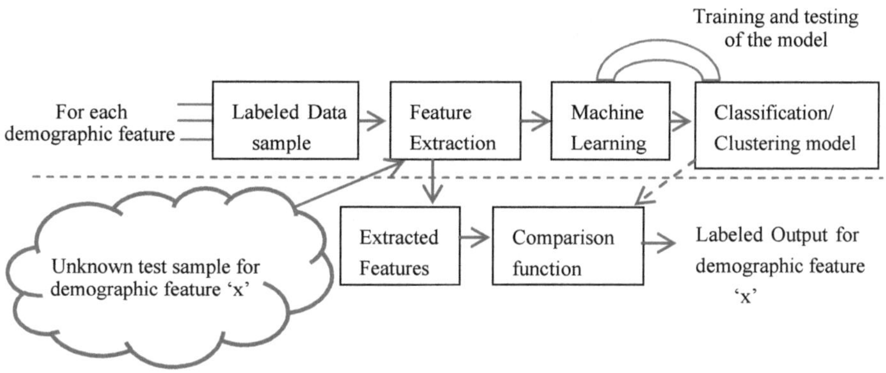

Fig. 4 A generalized authorship characterization process

3 Review of Relevant Literature

This section covers some of the prominent works in each of the authorship analysis domains to give some deep insights. A detailed description of works is followed by a tabular comparison to distinguish works on the basis of some useful parameters.

3.1 Authorship Attribution

Though work on authorship attribution dates back to eighteenth century with a prime focus on literary works, more recently it has raised the attention of researchers for short online content. Zheng et al. [23] proposed a framework for the identification of authors of online messages by working on lexical, syntactic, structural, and content-specific features. It was experimentally evaluated that lexical and syntactic features were highly beneficial for longer messages but failed to act as good discriminators for shorter online messages. Also, use of structure-based and content-specific features helped to accurately classify and discriminate different authors by attaining an accuracy of 97.69 and 88.33% on English and Chinese datasets, respectively. Among the different classifiers, the best performance was achieved by SVM followed by neural networks and decision trees (C4.5).

Abbasi and Chen [1] developed a K-L transform-based stylometric technique called Writeprints for authorship identification and similarity detection of both asynchronous and synchronous online messages. Their work extended the existing feature set by incorporating Idiosyncratic (i.e., misspelled words) features which proved to be highly beneficial. Both static (content-free) and dynamic (content-dependent) features were used with information gain as a feature selection measure. Authorship identification was studied as a classification problem and in addition to Writeprints, classifiers such as SVM and ensemble SVM were also used.

Layton et al. [12] checked the efficiency of a popular authorship attribution technique, SCAP on short messages and inferred that it was able to achieve a maximum accuracy of 72.9% on tweets. Character-based n-grams were used to build behavioral writing profiles of users, and top L words for each user were matched against the test cases. Simplified Profile Intersection(SPI) was used as a distance metric to evaluate the authorship analysis technique, SCAP. Furthermore, an unsupervised learning measure called Silhouette Coefficient (SC) was used along with the creation of sub-profiles using k-means++ clustering algorithm.

Unlike existing approaches restricted to a limited set of users (usually up to 300 authors), Narayanan et al. [16] used stylometric techniques for authorship identification on a larger dataset. Some distinctions in the use of feature set include deployment of linguistic instead of context-specific features, avoidance of "bag-of-words" model, use of only a limited set of words in case of word-based features, use of only single-character function frequencies instead of bigrams and trigrams, and the introduction of syntactic features. Among 1188 features analyzed, the ones showing good performance include the length of posts (both in terms of number of words and characters), frequency of punctuation marks, patterns of upper and lower case letters, and syntactic features. Among different classifiers, performances of RLSC and nearest neighbor were found to be better.

Johansson et al. [9] focused on author identification and alias matching using timeprint features such as month, hour, period, and type of the day. Experimental results revealed that such features were able to predict and identify the users better than stylometric features as temporal characteristics reflect the chronotype behavior of individuals which help to analyze the characteristic behavior of an individual. A supervised approach was used for author identification, while an unsupervised method was employed for alias matching.

Peng et al. [19] studied text analysis as (i) n-grams analysis, which used simple text features such as lexical and syntactic features and (ii) forensic linguistics with complex features such as syntactic, structural, and content based. Researchers proposed a multi-bit n-gram method where with each step only a single bit was moved. As a classification problem, Euclidean distance was used as a distance measure for kNN classifier and Interquartile Range (IQR) for outlier detection attaining an average accuracy rate of 90 and 85.7%, respectively.

Rocha et al. [22] performed forensic authorship attribution by utilizing the bag-of-words stylometric features, such as char n-grams, word n-grams, POS n-grams, and other lexical and syntactic features. Additionally, researchers also presented a comprehensive review of the existing authorship attribution techniques. A number of generalized stylometric features and classifiers along with the ones specific to social network scenario and small-scale data were enlightened. Both closed and open-set recognitions were performed.

3.2 Authorship Verification

Brocardo et al. [5] applied the supervised machine learning technique on short e-mail messages for authorship verification. Instead of analyzing the frequency distribution of n-grams in a text, a binary classification approach was used. Extending the work further, a hybrid approach was followed in Brocardo et al. [6] to combine the outputs of SVM and logistic regression classifiers with mutual information as a feature selection method.

PAN datasets,[1] also explored in Potha and Stamatatos [21], were probed for the authorship verification task by Mayor et al. [14] using techniques such as n-gram based Naïve Bayes, imposter, and sparse representations. Use of bag-of-words model was supplemented by other characteristics such as punctuations, prefixes, suffix, stop words, etc. It was found that sparse representation method performed exceptionally well in comparison to n-grams and imposter methods.

Barbon et al. [4] implemented an n-gram based authorship verification approach setting its suitability for the detection of compromised accounts. A three-step approach was followed, beginning from the creation of baseline profiles to matching and updating with Simplified Profile Intersection (SPI) as a similarity measure. Use of kNN as a classifier helped to attain an accuracy of 93%.

Li et al. [13] examined the competence of stylometric features on short social network messages for authorship verification highlighting its importance for the detection of compromised accounts. Investigating and profiling the writing patterns of normal and fraudulent users helped to supplement the existing login time authentication schemes. Around 233 stylometric and social network-based features were examined achieving an overall accuracy of 79.6%.

Similarly, stylometric features were also examined by Iqbal et al. [8] in different classification (AdaBoost, Naive Bayes, and DMNB) and regression techniques (linear regression, SVM with RBF, and SMO). To analyze both false positives and negatives, DET curve was used and regression analysis gave comparatively better performance than classification.

Kocher and Savoy [10] implemented an unsupervised authorship verification technique using a distance measure called SPATIUM-L1 in k-most frequent words selection. Instead of machine learning classification, statistical analysis was performed and performance comparable to that of a meta-classifier was achieved.

Brocardo et al. [7] analyzed the hitch of machine learning architectures for the task of authorship verification and examined how deep belief networks helped to attain more promising results (error rate of 5–12%). Gaussian–Bernoulli deep belief network (GB-DBN) with a layered framework of Restricted Boltzmann Machines (RBMs) succeeded by a machine learning classifier were used for classification. In addition to lexical, syntactic, and application-specific features, different combinations of n-gram features were analyzed. Information Gain (IG) and Mutual Inclusion (MI) were used to select relevant features for each respective user.

---

[1] http://pan.webis.de/.

3.3 *Authorship Characterization*

Argamon et al. [2] profiled the documents by gender, age, language, and personality using Bayesian Multinomial Regression (BMR). BMR being impervious to overfitting is a multivariate logistic regression model, and hence was hypothesized to give effective performance. Experiments were performed using frequency of occurrence of a feature value normalized over the total count of words. Accuracy values of 76.1, 77.7, 82.3, and 65.7% were achieved for gender, age, language, and personality using combination of content- and style-based features.

Similarly, Pennacchiotti and Popescu [20] performed the authorship profiling of Twitter users by their profile information, tweeting behavior, linguistic, and social networking characteristics. Use of such features helped to achieve F-score values of 88.9, 76.1, and 65.5, respectively, for the classification of political affiliation (democrats/republic), ethnicity (African-Americans or not), and Starbucks fans or not. Gradient Boosted Decision Trees (GBDT) was used as a classification algorithm as it had a faster decoding rate in contrast to its competitors.

To identify age and gender from the given excerpts of chat conversations, Meina et al. [15] profiled different style, structure, and semantic patterns of users. Ensemble-based classification was used with random forest giving 60% accuracy for both gender and age which was higher than that achieved with individual classifiers, viz., Naïve Bayes, SVM, and classifier committees.

Patra et al. [18] used the classification approach using stylometric features (lexical, syntactic, and structural) for profiling authors by gender and age group. Apart from the *n*-gram feature analysis which failed to give good performance, different word class frequency lists, positive and negative word class lists, list of stop words, smiley words, foreign words, pronouns, and punctuations were also analyzed. Using decision trees as a classification algorithm, accuracies of 56.83 and 28.95% were achieved for the classification of gender and age, respectively.

Palomino-Garibay et al. [17] applied both regression and classification tasks using random forest algorithm for the detection of age and gender of tweet authors using lexical, statistical, and word-specific features. Averaging process was used to draw conclusions from independent classifiers for better outcomes with reduced variance, otherwise expected through imbalanced selection of features. Overall, 70 and 61% accuracies were achieved for gender and age, respectively.

Table 1 summarizes the experimental details of the textual data and features used for different authorship analysis tasks.

4 Conclusion

Increase in the use of social media platforms for both authentic and spurious purposes has augmented the choice of authorship analysis task for forensic investigations. Considering the topic to be subtle but important, this paper highlighted the

Table 1 Statistics related to data and features used in different authorship analysis tasks

Refs.	Category	Number of users	Number of documents/articles/post	Size of 1 document/article/post	Features used	Classifier	Performance accuracy
Zheng et al. [23]	Attribution	20 English/20 Chinese	48 English/37 Chinese	169 words English/807 words	Lexical, syntactic, structural, content based	SVM, NN, C4.5	97.6% (English) 88.33% (Chinese)
Abbasi and Chen [1]	Attribution	400 (100 for each dataset)	NA	27,774; 23,423; 43,562; 1422 words/author resp. for each data	Lexical, syntactic, structural, content-specific and idiosyncratic	Writeprint, Ensemble SVM	92, 96, 88.8, 50.4%
Layton et al. [12]	Attribution	50	120 tweets	Maximum 140 char	*n*-grams	SCAP method	70% (max. 72.9%)
Narayanan et al. [16]	Attribution	3628	1,00,000 blog posts	305 words per post	Linguistic stylometric features	SVM, NN, NB, RLSC	Varying precision (max 80%)
Johansson et al. [9]	Attribution	1000	NA	Each with at least 200 posts	Timeprints (day, month, hour, period, and type of the day)	SVM, NB	99% accuracy (100 users) and 75% (1000 users)
Peng et al. [19]	Attribution	40	Total 94,274 comments	Variable length (Concatenated text)	*n*-grams	kNN, outlier classifier	90% (kNN) 85.7% (Outlier)
Rocha et al. [22]	Attribution	10,000 (Testing: 50 users)	10 M tweets (Testing: 500 tweets/user)	Maximum 140 characters	Bag-of-words (*n*-grams)	RF, PMSVM, PPM-5, SCAP	70% accuracy for 50 authors
Azarbonyad et al. [3]	Attribution	133 Twitter users	Per person Avg: Tweets: 1820 E-mail: 3200	Average char per: Tweet: 101 E-mail: 648	Unigrams made of 4-gram characters	SCAP	Tweets: 80% E-mail: 94%
Iqbal et al. [8]	Verification	158	Each user with 200 e-mails	200 e-mails	Lexical, syntactic, structural, and topic specific	AdaBoost, NB DMNB	82.9%
Brocardo et al. [5]	Verification	87	Enron e-mails (50 blocks per author)	500 characters per block	Stylometry, mainly focused on *n*-grams	Threshold setting	EER-14.35%
Li et al. [13]	Verification	30	9259 posts	Average 20.6 words	Lexical, syntactic, structural, and social network specific	kNN, SVM, NB, Dec. Trees, NN	79.6%
Barbon et al. [4]	Verification	1000 Twitter users	Max 500 tweets/user	Maximum 140 characters	*n*-grams ($n = 6$)	kNN	Accuracy: 93% Precision: 99%

(continued)

Table 1 (continued)

Refs.	Category	Number of users	Number of documents/articles/post	Size of 1 document/article/post	Features used	Classifier	Performance accuracy
Argamon et al. [2]	Profiling	Gender/Age: 19,320 Language: 258	NA	Gender/Age: 7250 Language: 580–845 Personality: 250–1950	Content based and style based	BMR	76.1, 77.7, 82.3 and 65.7% for gender, age, language, personality
Pennacchiott and Popescu [20]	Profiling	6000	5000 positive/5000 negative	Maximum 140 characters	Profile, tweeting behavior, linguistic, and social networking	Threshold setting	88.9, 76.1 and 65.5% (categorically)
Meina et al. [15]	Profiling	0.5 M English and 0.12 M Spanish conversations	0.23 M English and 76 K Spanish files	Total 180 M English and 21 M Spanish words	Structural, style, and semantic features	kNN, NB, SVM	Gender: 63.2%, Age: 61.1%, Gender+Age: 65.3%

generalized methodologies adopted for different authorship analysis tasks (attribution, verification, and characterization) followed by some prominent works done by various researchers in each domain. Though each analysis domain has its own application base, authorship verification appears to be the most sensitive and crucial topic related to problems such as breaching, obfuscation, account compromisation, anonymous rebuking, and forgery. Second, it was analyzed that most of the studies relied on machine learning classification techniques (few recent works also utilized deep learning concepts) laying focus on various content-specific and content-independent features. In the near future, with the size of online communications getting reduced day by day, there is a need to examine the effectiveness of existing approaches for reduced size messages. Moreover, forthcoming techniques need to implement some incremental learning methods to incorporate the dynamism and enormous volume of data getting generated.

Acknowledgements This publication is an outcome of the R&D work undertaken project under the Visvesvaraya PhD Scheme of Ministry of Electronics & Information Technology, Government of India, being implemented by Digital India Corporation.

References

1. A. Abbasi, H. Chen, Writeprints: a stylometric approach to identity-level identification and similarity detection in cyberspace. ACM Trans. Inf. Syst. **26**, 7 (2008)
2. S. Argamon, M. Koppel, J.W. Pennebaker, J. Schler, Automatically profiling the author of an anonymous text. Commun. ACM **52**, 119–123 (2009)
3. H. Azarbonyad, M. Dehghani, M. Marx, J. Kamps, Time-aware authorship attribution for short text streams, in *Proceedings of the 38th International ACM SIGIR Conference on Research*

and Development in Information Retrieval (2015), pp. 727–730
4. S. Barbon, R.A. Igawa, B.B. Zarpelão, Authorship verification applied to detection of compromised accounts on online social networks. Multimed. Tools Appl. **76**, 3213–3233 (2017)
5. M.L. Brocardo, I. Traore, S. Saad, I. Woungang, Authorship verification for short messages using stylometry, in *2013 International Conference on Computer, Information and Telecommunication Systems (CITS)* (2013), pp. 1–6
6. M.L. Brocardo, I. Traore, I. Woungang, Authorship verification of e-mail and tweet messages applied for continuous authentication. J. Comput. Syst. Sci. **81**, 1429–1440 (2015)
7. M.L. Brocardo, I. Traore, I. Woungang, M.S. Obaidat, Authorship verification using deep belief network systems. Int. J. Commun. Syst. **30**, 12 (2017)
8. F. Iqbal, L.A. Khan, B. Fung, M. Debbabi, E-mail authorship verification for forensic investigation, in *Proceedings of the 2010 ACM Symposium on Applied Computing* (2010), pp. 1591–1598
9. F. Johansson, L. Kaati, A. Shrestha, Time profiles for identifying users in online environments, in *2014 IEEE Joint Intelligence and Security Informatics Conference (JISIC)* (2014), pp. 83–90
10. M. Kocher, J. Savoy, A simple and efficient algorithm for authorship verification. J. Assoc. Inf. Sci. Technol. **68**, 259–269 (2017)
11. M. Koppel, J. Schler, S. Argamon, Computational methods in authorship attribution. J. Assoc. Inf. Sci. Technol. **60**, 9–26 (2009)
12. R. Layton, P. Watters, R. Dazeley, Authorship attribution for twitter in 140 characters or less, in *2010 Second Cybercrime and Trustworthy Computing Workshop (CTC)* (2010), pp. 1–8
13. J.S. Li, L.-C. Chen, J.V. Monaco et al., A comparison of classifiers and features for authorship authentication of social networking messages. Concurrency Comput. Pract. Experience **29** (2017)
14. C. Mayor, J.G.G. Hernández, A.I.T. Castro et al., A single author style representation for the author verification task. in *CLEF* (Working Notes) (2014), pp. 1079–1083
15. M. Meina, K. Brodzinska, B. Celmer et al., Ensemble-based classification for author profiling using various features. Noteb. Pap. CLEF (2013)
16. A. Narayanan, H. Paskov, N.Z. Gong et al., On the feasibility of internet-scale author identification, in *2012 IEEE Symposium on Security and Privacy (SP)* (2012), pp. 300–314
17. A. Palomino-Garibay, A.T. Camacho-González, R.A. Fierro-Villaneda et al., A random forest approach for authorship profiling, in *Proceedings of CLEF* (2015)
18. B.G. Patra, S. Banerjee, D. Das et al., Automatic author profiling based on linguistic and stylistic features. Noteb. PAN CLEF (2013)
19. J. Peng, K.-K.R. Choo, H. Ashman, Bit-level n-gram based forensic authorship analysis on social media: identifying individuals from linguistic profiles. J. Netw. Comp. Appl. **70**, 171–182 (2016)
20. M. Pennacchiotti, A.-M. Popescu, A machine learning approach to twitter user classification. Icwsm **11**, 281–288 (2011)
21. N. Potha, E. Stamatatos, A profile-based method for authorship verification, in *SETN* (2014), pp. 313–326
22. A. Rocha, W.J. Scheirer, C.W. Forstall et al., Authorship attribution for social media forensics. IEEE Trans. Inf. Forensics Secur. **12**, 5–33 (2017)
23. R. Zheng, J. Li, H. Chen, Z. Huang, A framework for authorship identification of online messages: writing-style features and classification techniques. J. Am. Soc. Inf. Sci. Technol. **57**, 378–393 (2006)

Utility of OLAP and Digital ATLAS to Enhance Cross-Border Connectivity among Punjabi Sikh NRIs

Harkiran Kaur, Kawaljeet Singh and Tejinder Kaur

Abstract Data analytics is one of the prominent research areas and finds its application in various domains. By expending this concept, this paper presents a case study of Punjabi Sikh NRIs that are one of the largest Diasporas in the world today. Diaspora has acted as a great reservoir for financial, political, and moral support for their homelands. Therefore, Diaspora has become a prominent part of human enquiry and research. However, with the invention and proliferation of Information and Communication Technologies (ICT), Diaspora people/communities connect effectively with the people of their community living in different countries. The proposed paper discusses the two main techniques, e.g., Online Analytical Processing (OLAP) and digital ATLAS, which can be used for the purpose of enhancing cross-border connectivity among Diasporas. These techniques provide a two-segment support. The first segment (OLAP) assists the development of a framework of database of profiles of the Diasporas of a particular community. These profiles include the prominent figures working in the mainstream political associations, in the fields of education, defense, entertainment, technology and economics, and various organizations of a few countries such as UK, Malaysia, USA, Canada, and Australia. The second segment (Digital ATLAS) facilitates the development of a digital ATLAS composed of detailed statistical profiles of the Diaspora community. This digital ATLAS draws upon the available official data in various locations of the countries, using visualizations and dashboards. The implementation details of five phases considered for these techniques have been elaborated in this paper.

H. Kaur (✉)
Department of Computer Science and Engineering, Thapar Institute of Engineering and Technology (Deemed to be University), Patiala, India
e-mail: harkiran.kaur@thapar.edu

K. Singh
University Computer Center, Punjabi University, Patiala, India
e-mail: singhkawaljeet@pbi.ac.in

T. Kaur
Center for Diaspora Studies, Punjabi University, Patiala, India
e-mail: tejinder_sp@yahoo.co.in

C. R. Krishna et al. (eds.), *Proceedings of 2nd International Conference on Communication, Computing and Networking*, Lecture Notes in Networks and Systems 46, https://doi.org/10.1007/978-981-13-1217-5_53

Keywords Intelligent system · Intelligent databases · Visualizations · Digital ATLAS · Cultural computing · Intelligent directory · OLAP

1 Introduction

Diaspora (a Greek word meaning "dispersion") denotes the movement or migration of a group of people, sharing a national/ethnic identity, away from an established or ancestral homeland [1]. As per this definition, since the Diasporas of particular communities have dispersed in various countries crossing the national and territorial borders, in the matters of culture, language, religion, custom, folklore, emotion, and loyalty, they feel connected with their original homelands and they believe that their homelands too have claim on them [2].

Since very limited opportunities are available to Diaspora scattered around the world to explore and experience their history and culture, hence, it is very important to devise a way for them to build a sense of collective identity and community consciousness, in order to live peacefully inside their adopted host lands, with a difference.

Presently, Diaspora communities remain connected with each other via traditional directories, Internet forums, and social networking sites using Internet and other technological gadgets as an extension to it. The traditional NRI directories, used by the Diaspora for connectivity, exhibit some flaws including the following:

i. Records of NRI profiles are static and non-scalable in nature.
ii. Difficult search utilities provided by these traditional directories.
iii. Maximum information provided by these directories is the contact information of the NRI profiles and no analytical results are generated for the same.

For this purpose, cultural computing comes to the rescue of Diaspora, which helps them to experience the connectivity with the peers of their homeland in closer terms. The umbrella approach cultural computing is composed of intelligent databases as its subset. The proposed techniques OLAP and digital ATLAS (an Intelligent Database approach) in this paper create a virtual homeland for the Diaspora on cyberspace, which furnishes a sense of connectivity with their respective homelands.

1.1 Cultural Computing

Culture computing (Culture and Computing) is an emerging research area of Human–Computer Interaction (HCI), which aims to overcome different cultural barriers and create cultural bonding and awareness in international communities using Information and Communication Technologies (ICT) [3]. Cultural computing helps to learn and experience the culture of homeland in an interactive way by using the following mechanism:

a. Expounding the use of Internet and other technological gadgets as an extension to the social activities with an aim to gain a proper understanding of the history and personal identities [4];
b. Digitization and preservation of digital cultural objects and their dissemination through Internet forums and social network sites;
c. Providing intelligent access to cultural information;
d. Development of efficient tools to facilitate understanding of even non-tech savvy people, for their respective cultures and to connect with their respective cultural communities with improved efficiency [3].

This paper aims to enlighten one of the contemporary areas of cultural computing: "augmenting cross-border connectivity among cultural communities" (here Sikh Diaspora communities). In the present scenario, cube technology or OLAP is utilized for Big Data applications such as Facebook data analysis, cricket matches data analysis, weather data predictions, etc. This paper presents the application of intelligent database approach (the subset of cultural computing), that is, Online Analytical Processing (OLAP) on the traditional NRIs directory. This directory also records and analyses employment record of Diasporas in their respective domains, various events conducted by Sikh organizations located in foreign countries, problems faced by Diaspora, gadgets used by Diasporas belonging to different age groups, and providing knowledge related to women's rights, to Diaspora.

Corresponding to the results retrieved, it plots the visualizations including digital ATLAS on the dashboards for the said purpose. OLAP is a business intelligence tool that inculcates data detection, in addition to performing complex analytical calculations, report viewing, and predicting "what if" scenarios. OLAP differs from traditional databases, and as in OLAP, data is accumulated and analyzed. These systems provide speedy results for queries that summate large volumes of detailed data. This step further explores significant trends in data warehouses or data marts.

2 Literature Survey

Literature survey carried out confirmed the following facts about the Punjabi Sikh Diaspora, which made the authors to conceptualize that Digital Diaspora could prove to be a great reservoir for moral and cultural support to the Sikh Diaspora from their homeland.

Fact #1: The Indian Diaspora has tremendously migrated to different countries in the world [5].
Fact #2: *Naturalized citizenship*: Majority of Punjabi Sikh Diaspora are living as naturalized citizens in their host countries, which means, these people need not surrender their cultural symbols along with their ethnic identities [6].
Fact #3: It has been observed that technology can be used for gaining a proper understanding of their histories and personal identities by Diaspora since most of

them use Internet and other technological gadgets as an extension and not as a replacement of their social activities.
Fact #4: Like other diasporic groups, Indian/Punjabi migrants intensively invest in remittances to the home space for altruism, development, self-interest, family links, and social purposes. This benefits both the place where they currently exist as well as propel the economic growth and other forms of advancements back home. By this way, the Diasporic communities bring many changes in their countries of origin for modernization and better quality of life [7].

Thus, it can be said that Punjabi Sikh Diaspora has proved to be a significant source of support for the financial, political, and moral purposes for their homeland.

There is Diaspora directory already available, namely, *International NRIs Directory—Indians Abroad and Punjabi Impact* (used for communication), which was compiled and published by Nirpal Singh Shergill in 2014. This directory contains the profiles of prominent Indian Diaspora residing overseas, their contact details, Indian/Punjabi newspapers, Sikh Organizations, and many other details pertaining to their work areas in different countries [8].

Omar Boutkhoum et al. in 2013 anticipated a multidimensional model, to assimilate multi-criteria analysis in OLAP systems [9]. Since OLAP by then was used only for archiving, organizing, analysis, and multidimensional modeling, it was limited in consideration of multi-criteria and quality aspect of decision system.

In 2014, the e-Diasporas ATLAS was developed which is the reinstitution of scientific findings and presentation of Diasporas. It assembled the archive of 8000 migrant websites and produced a vast and moving e-corpus [10].

3 Proposed Case Study of Punjabi Sikh Diaspora Using OLAP and Digital ATLAS

The proposed study explains and implements the significant gears of Cultural Computing—OLAP and digital ATLAS.

a. As the first step, a multidimensional (OLAP) directory database framework/structure of prominent Diaspora profiles is created. This OLAP framework is efficient enough to answer multidimensional analytical (MDA) queries. At the same time, this technique is certainly effective to process user requests and respond immediately using Online Transaction Processing (OLTP). The proposed directory will overcome the drawbacks of the NRI Directory that is available in the form of hard copy text file that was published in 2014. The key feature of OLAP technique is that the proposed multidimensional NRIs directory in this paper is ever expanding and supports continuous evolution.
b. In the second step, a digital ATLAS containing statistical data profiles of the Diaspora is developed for various countries using visualizations and dashboards. This serves as an improved version of existing e-Diaspora ATLAS available for the same purpose.

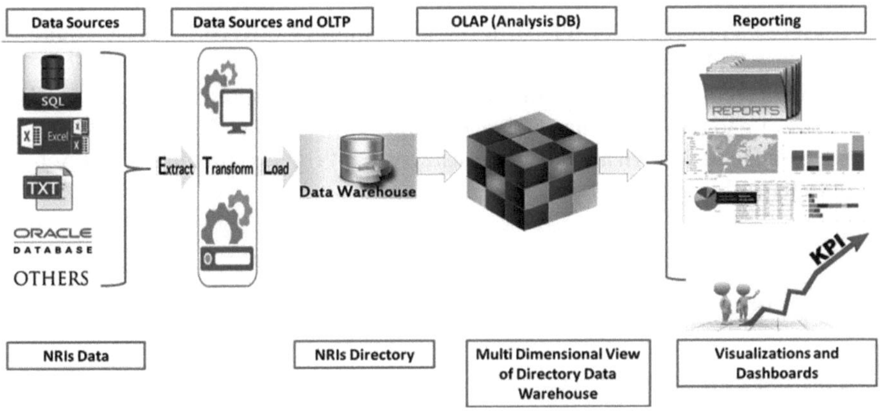

Fig. 1 Online analytical processing for Sikh NRIs directory data warehouse (DW). *Source* own research

This technique enables the Diasporas to connect with other people of their communities that reside in the same or different host countries, belong to the same domain or area of interest, such as education, defense, entertainment, technology and economics, various organizations, and so on. Figure 1 describes the steps for online analytical processing of Sikh NRIs directory.

The task for online analytical processing has been accomplished in five phases:

Phase 1: Diaspora data sources are available in numerous forms including SQL database, EXCEL sheets, ORACLE database, and text files. These heterogeneous data sources are provided as input to the Extract–Transform–Load (ETL) process.
Phase 2: The phase 2 is the ETL Process. It extracts the heterogeneous data from SQL, ORACLE, and EXCEL database and transforms it into a single (homogeneous) form of database (here ORACLE database) and loads it into the Data Warehouse (DW).
Phase 3: In this phase, the DW is divided into n number of manageable data marts (subsets of Data Warehouse).
Phase 4: OLAP Cube is created in this phase, which provides multidimensional views to the different NRI users. Multiple factors are associated with the Diaspora studies. The present study focusses upon the subset of these factors, which are mentioned as Key Performance Indicators (KPIs).
Phase 5: Multidimensional reports are generated that incorporate the operations including slicing, dicing, roll-up, roll-down, and pivoting. Further, a digital ATLAS is being formed using these views.

This prototype addresses the following Key Performance Indicators (KPIs) across the aforementioned five countries:

a. Employment in accordance with their domain (Leading KPI).
b. Number of events (Quantitative KPI) and type of events (Qualitative KPI) conducted by Sikh institutions.

c. Diaspora satisfaction in line with the opportunities offered by these countries (Leading KPI).
d. Problem types (Lagging KPI) and count reported by Diaspora (Quantitative KPI).
e. Types of gadgets compliant with age groups of Diaspora (Qualitative KPI).
f. Knowledge to women about various categories of rights (Directional and Actionable KPI).

4 Implementation of Proposed Case Study

To apply this case study, the authors have utilized the software, NET framework as a front end, and ORACLE AWM (Analytical Workspace Manager) as a back end. Power BI desktop software has been used for developing the visualizations and dashboards. Figure 2 presents the star schema implemented using the said approach. Star schema is the data warehousing schema that comprises huge central virtual table (fact table) with no redundancies. All the dimension tables refer fact table. Furthermore, star schema is very efficient in managing ad hoc queries [11]. Here, "Profiles", "Sikh_Organizations", "Events", and "Activities" are the dimensions connected to fact table, that is, cube designed for the said purpose.

This cube will compute the measures that are selected as KPIs. The cube in this diagram is the fact table that includes all the measures (or KPIs) addressed in Sect. 3 of this paper.

Figure 3 elaborates the visualizations, depicting the KPIs enlisted above, in the form of a Dashboard. The first user control is the slicer. From the slicer, the user can select the domain of the Diaspora. The map in this dashboard describes the digital ATLAS. It highlights the geographical area in the map in which Diaspora related to the selected domain is working. This map also shows the count of Diaspora in a particular domain, on mouse hover event. The third visualization is the count of

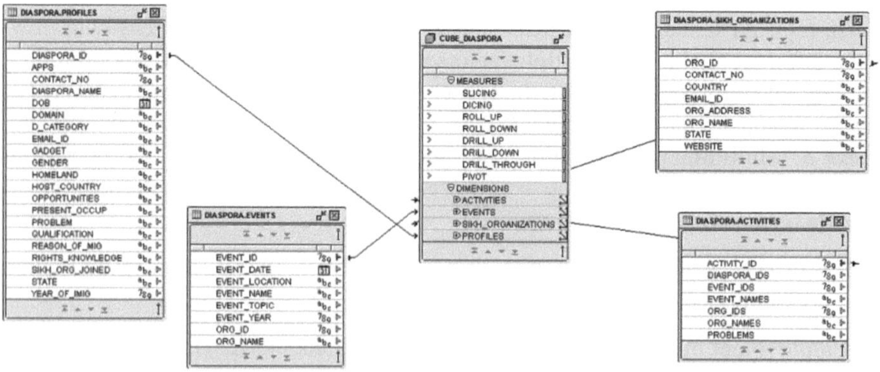

Fig. 2 Star schema in AWM for Sikh NRIs directory. *Source* own research

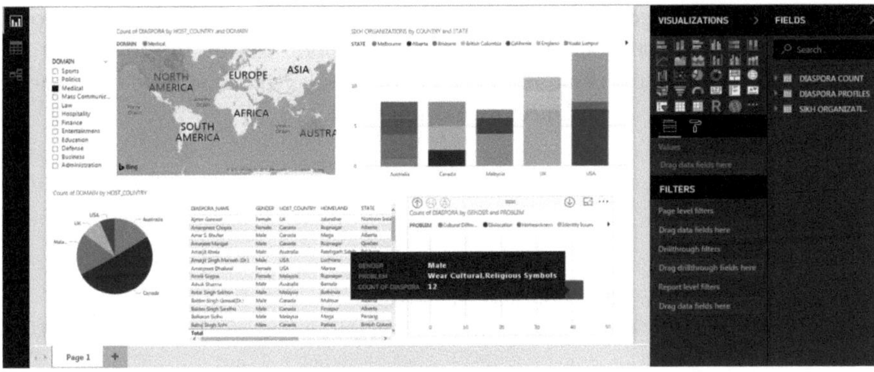

Fig. 3 Visualizations in dashboard for Sikh NRIs ATLAS. *Source* own research

Sikh Organizations in different countries, state-wise. This dashboard also presents the record of the data through which these results are obtained. The last visualization in this dashboard describes the gender-wise count of problems and types of problems faced by Diaspora.

These dashboards act as the virtual world to the Diasporas, as they can visualize the possible connections to which they can connect and escalate their circle and hence enhance their connectivity. These connections are generated based on common homelands, host countries, work domain, and so on.

5 Conclusion

The study proposed and described in this paper analyzes and visualizes Punjabi Sikh profiles of Diaspora. This technique provides an interface to the Diaspora communities. OLAP technology in the current era is utilized for big data analysis applications. However, as concluded from the literature studied, OLAP has never been utilized for the analysis of Diaspora data. The OLAP element of the proposed study computes the user's area of interest (or working or domain) or problems and establishes the connectivity among the Diaspora that shares common area of working or facing common set of problems in the foreign countries. Further, a digital ATLAS is provided that is composed of detailed statistical profiles of the Punjabi Sikh community drawing upon official data in various locations of the countries using visualizations and dashboards. To realize these objectives, set of Key Performance Indicators (KPIs) have been identified. In this paper, a prototype implementation of these techniques has been showcased. This prototype is implemented in five phases. In Phase 1, various types of data sources such as Excel sheets, SQL, and Oracle database have also been considered. In Phase 2, the ETL process has been implemented. Phase 3 marks the creation of Data Warehouse of NRIs directory. Phase 4 converts the DW into a multidimensional cube to provide n number of views to n types of different users.

Phase 5 is the final stage that supplies the user with sliced, diced, and pivot reports. In future, Diasporas, other than Punjabi Sikh Diaspora, can be considered for the implementation of these techniques. Further, this study can be enriched by covering more host countries to which Diaspora has migrated.

References

1. (n.d.). Retrieved 21 Oct 2017, from https://en.wikipedia.org/wiki/Diaspora
2. B. Ashcroft, G. Griffiths, H. Tiffin (eds.), *Key Concepts in Postcolonial Studies* (Routledge, London, 1998)
3. M. Rauterberg, *From Personal to Cultural Computing: How to Assess a Cultural Experience* (Information nutzbar machen, 2006)
4. M. Rauterberg, HCI as an engineering discipline: to be or not to be!? Afr. J. Inf. Commun. Technol. **2**(4), 163–184 (2006)
5. Non-resident Indian and person of Indian origin (n.d.). Retrieved 3 Jan 2017, from https://en.wikipedia.org/wiki/Non-resident_Indian_and_person_of_Indian_origin
6. Generations and their gadgets (3 Feb 2011). Retrieved 25 Sept 2017, from http://www.pewinternet.org/2011/02/03/generations-and-their-gadgets
7. R. Cohen, *Global Diaspora: An Introduction* (University of Washington Press, Seattle, 2003)
8. NS Shergill (ed.), International directory of Punjabi NRIs released at House of Commons, London (2014). Retrieved from https://www.theindianpanorama.news/shergills-international-directory-indians-abroad-unjab-impact-released-in-london/
9. MH Omar Boutkhoum, Integration approach of multicriteria analysis to OLAP systems: multidimensional model, in *IEEE 2013 ACS International Conference on Computer Systems and Applications (AICCSA)*, Ifrane, Morocco (2013)
10. e-diasporas: the concept. (n.d.). Retrieved 3 Oct 2017, from http://www.e-diasporas.fr/
11. K. Dhanasree, C. Shobabindu, A survey on OLAP, in *2016 IEEE International Conference on Computational Intelligence and Computing Research (ICCIC)*, Chennai, India (2016), p. 9

Implementing Multidimensional Analytics in NRIs Directory Databases

Harkiran Kaur, Kawaljeet Singh and Tejinder Kaur

Abstract NRIs Directory/Database is the best available opportunity in today's world, to locate the NRIs on globe. This directory can be utilized by several organizations to expand their businesses overseas. It can also be used by NRIs to locate persons common to their homeland or persons having same work area, situated at various geographical locations. Having these directories leads the users to crack right business deals, direct and speedy communications with potential businesses, individuals related to common domains or common homelands. Presently, several NRI directories are available both online and offline directories, e.g., *International NRIs Directory—Indians Abroad and Punjabi Impact*, NIIR Project Consultancy Services, E-mail Database Store that offers NRI E-mail Database Pack, and many others. In general, these directories contain the profiles of prominent Indian Diaspora residing overseas, their personal contact details, Indian/Punjabi newspapers, Sikh organizations, business contact numbers, business e-mail ids, and many other details pertaining to their work areas in different countries. However, there are several problems associated with these traditional directories. This paper addresses the problems of handling large amounts of data stored in traditional NRIs directories and further proposes the cube technique to overcome these problems and enhance the utilities of these directories.

Keywords Intelligent databases · Multidimensional model · Intelligent directory OLAP · Cube technology

---

H. Kaur (✉)
Department of Computer Science and Engineering, Thapar Institute of Engineering and Technology (Deemed to be University), Patiala, India
e-mail: harkiran.kaur@thapar.edu

K. Singh
University Computer Center, Punjabi University, Patiala, India
e-mail: singhkawaljeet@pbi.ac.in

T. Kaur
Center for Diaspora Studies, Punjabi University, Patiala, India
e-mail: tejinder_sp@yahoo.co.in

C. R. Krishna et al. (eds.), *Proceedings of 2nd International Conference on Communication, Computing and Networking*, Lecture Notes in Networks and Systems 46, https://doi.org/10.1007/978-981-13-1217-5_54

1 Introduction

Presently, Diaspora data is recorded in the form of NRIs directory, as a hard copy, where data is neither safe nor used for enhancing connectivity among Punjabi Sikh Diaspora. The proposed work is projected to enhance the utilities of traditional NRIs directory by:

a. Locating a Diaspora in a foreign country, or more specifically in a foreign state, that too by searching his profile domain-wise, homeland-wise, gender-wise, and so on.
b. Generating reports of Diaspora's progression at each level such as country level, domain level, and gender level.
c. Exploring the events organized by Sikh organizations at year-wise level.

The objective of this research study is to provide Diasporic information accessible means for answering ad hoc queries. These include broad categories of Diasporic progression such as Diasporic information, their current work areas, their qualification, information about their roots in India, and their work experience history in their respective fields (based on the key descriptors) and build connections among Punjabi Sikh Diaspora scattered in the five countries UK, Malaysia, USA, Canada, and Australia.

In addition to this, the following Key Performance Indicators (KPIs)/Key descriptors augment the efficacy of the traditional NRIs directory databases:

i. Host country-wise count of Diaspora migrations since year 2000 {Leading KPI}.
ii. Qualification-wise count of Diaspora migrations in host countries since year 2000 {Quantitative KPI}.
iii. Occupation of Diaspora as per their qualification {Leading KPI}.
iv. Host country-wise count of Sikh organizations established overseas {Leading KPI}.
v. Number of events {Quantitative KPI} and type of events {Qualitative KPI} conducted by Sikh organizations overseas.
vi. Number of events participated in by Diaspora host country-wise {Directional and Actionable KPI}.
vii. Yearly participation count of Diaspora in events across the host countries {Actionable KPI}.
viii. Diaspora satisfaction in accordance with the opportunities offered by these countries {Leading KPI}.
ix. Problem types {Lagging KPI} and count reported by Diaspora {Quantitative KPI}.
x. Types of gadgets compliant with and used by various age groups (e.g., youth, adult, teens, middle age, and senior citizens) of Diaspora {Qualitative KPI}.
xi. Knowledge to women about various categories of rights {Directional and Actionable KPI}.
xii. Employment of Diaspora across different countries overseas {Leading KPI}.

xiii. Count of occupations obtained by Diaspora according to their work areas or domains {Leading KPI}.
xiv. Count of opportunities {Quantitative KPI} and type of opportunities {Qualitative KPI} presented to Diaspora, host countries' wise.
xv. Types of gadgets used by Diaspora host country-wise {Qualitative KPI}.
xvi. Number of demographic distribution (Youth, adult, middle age, senior citizens, men, and women) across different domains in the host countries {Directional KPI}.

The proposed study concentrates upon the development of framework of multidimensional directory database of Punjabi Sikh Diaspora profiles of significant figures. In order to answer Multidimensional Analytical (MDA) queries, Online Analytical Processing (OLAP) approach has been utilized. Therefore, the proposed technique follows Kimball's approach to multidimensional modeling including the following steps [1]:

i. Select the business process: For the proposed application, the business process is selected by

   a. Developing framework of database of Sikhs/Punjabi profiles specifically of emerging community leaders who have become significant figures in the mainstream political associations/education/technology/economic/organizations of UK, Malaysia, USA, Canada, and Australia.
   b. Developing an ATLAS composed of detailed statistical profiles of the Punjabi Sikh community drawing upon official data in various locations of the country using visualizations and dashboards.

ii. Declare the grain: This step includes the following:

   a. Integrating data from different types of sources: NRIs data sources are available in several forms, e.g., SQL database, EXCEL database, ORACLE database, text files, and many more. The historical data is merged with current data: These heterogeneous datasets are converted into homogeneous form (ORACLE database in the proposed work).
   b. To improve quality of data: The data now obtained is cleansed to be loaded into the database.

iii. Identify the dimensions: In the proposed techniques, dimensions across which multidimensional views are generated including Diaspora profiles, activities, Sikh organizations, and events.
iv. Identify the facts

   a. To generate the dashboard for the evaluation of the results of data aggregation regarding Diaspora and provide summarized reports based upon selected KPIs or measures.

Dimensional models emphasize upon measurements by partitioning data into the context of measures or Key Performance Indicators (KPIs). These models are initialized in relational databases, namely, multidimensional databases (STAR schemas). STAR schemas comprise fact table(s) connected to related dimension tables through primary keys/foreign key relationships.

2 Literature Survey

Literature survey elaborates various research papers and articles on cube technology, online analytical processing, multidimensional modeling, mapping of relational databases to multidimensional cubes, and several comparative readings and pioneering ideas for creating hypercube and querying this cube of NRIs directory databases.

According to Cabibbo and Torlone [2], multidimensional design approach produces a logical schema from "Entity-Relationship (ER) diagram" and it may further generate multidimensional schemas in terms of relational databases.

Boehnlein and Ulbrich-vom Ende [3] introduced a hybrid approach for deriving logical schemas from *Structured Entity Relationship (SER) diagrams*. SER is an extension of ER diagram that envisages the presence of dependencies among the classes/objects. This is why SER is reportedly considered as an improved substitute to categorize multidimensional database objects.

The authors At-taibe and El Mohajir [4] applied a technique to convert OLTP mechanism for Decision Support Systems (DSS) to OLAP method using Moody and Kortink technique. This DSS was supposed to plan educational capacity of a University.

Chandwani and Breja [5] implemented an approach to convert an XML schema into STAR schema. Using this technique, the data could be retrieved from the XML schema, and further it can be converted to a homogeneous form, which makes it compatible for loading into the centralized data warehouse.

OLAP provides different views to different users. For the top officials (Overseas Ministries), summary information can be made accessible to them, by applying OLAP operations such as drill up is applied to the Diaspora directories. Whereas for enhancing the connectivity among the Diaspora, Diaspora residing at different geographic locations can view the information about other Diaspora within their countries/domains/cultures/religions, by applying OLAP operations such as drill down to the Diaspora directories [6].

In data warehouse, the data is set up in hierarchical model. As the first step, data is put into tables, which are also known as dimensions, then into facts and further into aggregate facts. The simplest form of the combination of fact table and dimension tables is called STAR schema. The access layer helps the user to retrieve the data [7].

A Diaspora directory is already available, namely, International NRIs Directory—Indians Abroad and Punjabi Impact (in hard copy), which has been compiled and published by Nirpal Singh Shergill in 2014. It supplies information (including names, addresses, contact numbers, or e-mail ids) about religious and cultural institutions, Sikh Gurudwaras, radios, TV channels, and Punjabi Newspapers established by NRIs overseas in more than 40 countries [8]. Another NRI directories published include one under NIIR Project Consultancy Services, and other is e-mail database store that offers NRI e-mail database pack. This database incorporates the NRI mailing list and contacts. Hence, such sources that provide Diaspora information either are paid and charge high costs, or are in the form of hard copies, and hence does not facilitate simple searching or search using filters or data analysis.

Many research studies undertaken till date are based on multidimensional modeling to augment basic acquaintance in the area of multidimensional modeling. These have been described briefly in above section. This analysis enabled the authors to discover the "multidimensional integrity constraints" along with maintenance of correct data transformations.

3 Proposed Methodology and Research Outcomes Using Multidimensional Analytics on NRIs Directory Databases

i. Design and develop data mart that helps to perform thorough data analysis of Diaspora information, for augmenting cross-border connectivity.
ii. The proposed framework comprises an integrated platform for complete data analysis, which determines Diaspora's progression over the years (since they have migrated to host countries), as per different KPIs/Key descriptors.

This research work proposes statistical analysis with a number of techniques that are suitable for the examined tasks such as each yearly progression report of Diaspora in foreign countries. The proposed work utilizes **ETL (E**xtraction ➔ Cleansing➔ **T**ransformation ➔ **L**oading) process. Data from same or different types of resources will be extracted and integrated into data warehouse periodically. The main focus of this research is dimension modeling as shown in Fig. 1. OLTP is an entity model, which is meant for the up-to-date schemas that characterize only the minor online transactions. Directory databases store historical data, which is least supported by OLTP. So, in order to solve these problems, OLAP is being widely used for handling data warehouses such as directories. OLAP includes the aggregated, historical data, and stored in multidimensional schemas [9]. The traditional ETL (Extract–Transform–Load) DW multidimensional approach to business intelligence is one of the most effective approaches for dealing with big data applications such as directory databases.

3.1 Implementation Steps

The development environment for the proposed research utilized the following tools that support the building and querying hypercube of NRIs directory database: Java Development Kit 1.8.0_20, Oracle Database 10g Enterprise Edition Release2,

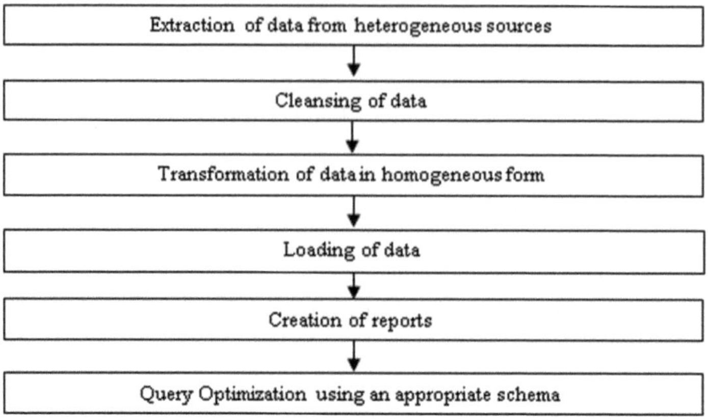

Fig. 1 Dimensional modeling process

Analytic Workspace Manager (AWM) Version 11.2.0.4.0B, and Oracle SQL developer Version 4.0.3.16 Build MAIN-16.84.

Past data is unstructured, and there is no proper mechanism to retrieve or search the desired information based on the user query. The proposed framework/system encompasses an integrated platform for detailed analysis of Diaspora's past data. On applying the OLAP operations, data analysis of Diaspora information can be achieved. The entities in the proposed work include Diaspora profiles, activities, events, and Sikh organizations. Entities utilized in ER diagram have been entirely mapped with the dimensions of proposed STAR schema.

In the proposed study, AWM tool has been utilized for creating, developing, and managing multidimensional data in an Oracle data warehouse. The container of OLAP information, that is, an analytic workspace (AW) is formed, then OLAP dimensions, levels of dimensions, hierarchies of levels, and cubes are created. The detailed set of steps for the said purpose includes the following [1]:

i. Create analytic workspace: Analytical workspace contains dimensions, levels, hierarchies, mappings, the cube, and cube mappings.
ii. Create dimensions: Dimensions—ACTIVITIES, EVENTS, PROFILES, and SIKH_ORGANIZATIONS.
iii. Create levels and hierarchies of dimensions. Example: ACTIVITY_ID and DIASPORA_IDS are the levels, and H_ACTIVITIES is the hierarchy of ACTIVITY dimension.
iv. STAR schema design based on cube's structure based on the requirements. Figure 2 presents the mapping of dimensions with the imported relational sources.
v. Defining cube and measures: In this process, cube "CUBE_DIASPORA" is created as the fact table and DIASPORA_PROFILES, DIASPORA_EVENTS, DIASPORA_SIKH_ORGANIZATIONS, and DIASPORA_ACTIVITIES as

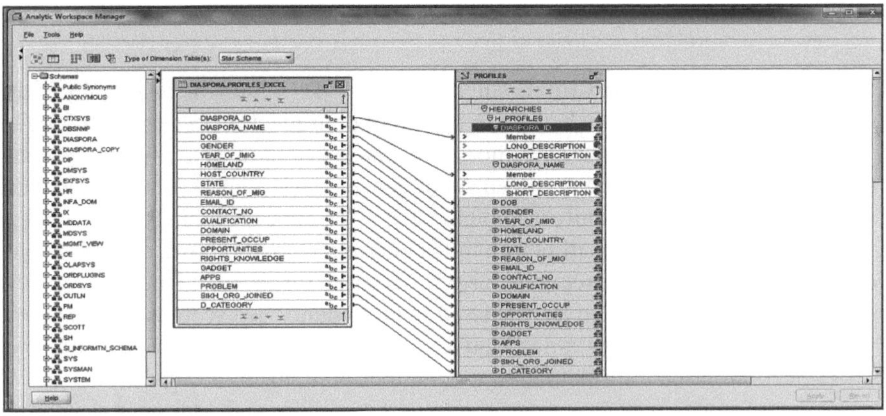

Fig. 2 Mapping of hypercube dimensions with the imported relational source

the dimension tables. Measures include the calculations/operations mentioned as KPIs.

vi. Relevant dimensions are assigned to the cube. The hypercube is loaded with records present in relational source by using "Maintain Cube" function in AWM. Now, the cube is available for query by using Oracle SQL developer and user-defined slices, dices, and pivots. A multidimensional database offers the analyst to perform following queries on the cube:

   a. Slicing: When a single value in any dimension is selected, this leads to reduction in dimensionality of the cube.
   b. Dicing: The dice operation presents a subcube by selecting two or more dimensions.
   c. Roll-up/Roll-down: This function enables the analyst to perform (move up) aggregations on a data cube hierarchy for a dimension or by moving down in a dimension level's hierarchy.
   d. Pivoting or rotation: It enables variations in the dimensional placement of the cube, that is, rotates the axis of data to observe the data from user-defined viewpoints [10].

With these operations, multidimensional views of the NRIs directory database are generated, which is presently missing in the traditional directories. This also enhances the utility as well as the performance of the NRIs directory databases, as instead of joins, the fact table (using aggregations in cube) performs the said task. In addition to this, the aforesaid traditional directory is static in nature, whereas the proposed technique will augment this NRIs directory to be scalable and modifiable. Using the proposed technique, modifiability and scalability are improved as the database entries of NRIs profiles can be updated via SQL developer and these changes are reflected in the cube by "Maintain Cube" procedure of AWM.

4 Conclusion

The traditional NRI directory offers single end user 2-D (2-Dimensional) view of the data in the tabular form. This had led to a number of drawbacks in terms of usage of these directories. In this research, authors have developed multidimensional structure of NRIs directory database that is well suited for dimensioning of summarized data. This multidimensional structure is implemented using cube technology, which offers 3-D view/n-D view of NRIs directory to the end user, based on user's requirement. For the said purpose, the aggregations are created from the fact table (cube), using the measures (or KPIs) identified as discussed in aforementioned sections of the paper. OLAP cube acts as the base for descriptive analytical processing, thus providing end user's selection-based information access. Furthermore, the multidimensional structure proposed in this paper supports continuous evolution and addresses scalability issue of the directory database.

References

1. M.R. Ralph Kimball, *The Data Warehouse Toolkit: The Complete Guide to Dimensional Modeling* (Wiley, USA, 2000)
2. L. Cabibbo, R. Torlone, A logical approach to multidimensional databases, in *International Conference on Extending Database Technology* (Springer, Berlin, 1998), pp. 183–197
3. M. Böhnlein, A. Ulbrich-vom Ende, Business process oriented development of data warehouse structures, in *Data Warehousing 2000* (2000), pp. 3–21
4. A. At-taibe, B. El Mohajir, From ER models to multidimensional models: the application of moody and Kortink technique to a university information system, in *2014 Third IEEE International Colloquium in Information Science and Technology (CIST)* (2014), pp. 155–162
5. M.B. Gunjan Chandwani, Implementation of XML schema to star schema. Int. J. Latest Trends Eng. Technol. (IJLTET) **5**(2) (2015)
6. K.S. Ravinder Pal Singh, Design and research of data analysis system for student education improvement (Case study: student progression system in university), in *2016 International Conference on Micro-Electronics and Telecommunication Engineering (ICMETE)* (2016), pp. 508–512
7. D.M. West, *Big Data for Education: Data Mining, Data Analytics, and Web Dashboards*. Governance Studies at Brookings (2012)
8. N.S. Shergill (ed.), International directory of Punjabi NRIs released at House of Commons, London. (2014). Retrieved from https://www.theindianpanorama.news/shergillsinternational-directory-indians-abroad-punjab-impactreleased-in-london/
9. R.A. Acharya, *Basics of Enterprise Reporting, Fundamentals of Business Analytics* (Wiley, New Delhi, 2014)
10. R.S. Ephraim Turban. *Business Intelligence: A Managerial Approach*, 2nd edn. (Pearson, USA, 2010)

Performance Evaluation of Cooperative Spectrum Sensing Over Fading Channels Based on Neural Network Learning Approach

Rashid Mustafa, Reena Rathee Jaglan and Sunil Agrawal

Abstract Cognitive Radio (CR) is a promising technique for effective utilization of the radio spectrum. Detection accuracy is compromised when a network user experiences fading and shadowing effects. To solve this problem cooperative spectrum sensing (CSS) scheme is used. Energy detection (ED) technique is used by each Secondary Users (SUs) to make a correct decision about primary user's (PU) presence or absence. Finally, all SUs communicate the PU information to fusion center (FC) using Phase Shift Keying (PSK) Modulation scheme. Hard fusion rules such as AND, OR, and Majority are used at the FC for determining the global decision about PU presence or absence. To meet a balance between performance and complexity, an artificial neural network (ANN) based CSS with reliable FC decision is discussed in the present study. It is observed that detection accuracy of the proposed ANN model is much better as compared to conventional hard fusion rules in fading environments.

Keywords Artificial neural network · Cooperative spectrum sensing
Probability of missed detection · Probability of false alarm

1 Introduction

Recently cognitive radio (CR) is proposed as an efficient technique for improving the spectrum scarcity problem by permitting secondary user (SU) access the license band opportunistically for a significant portion of time without causing any interference to primary user (PU). In cognitive radio technology, spectrum sensing (SS) plays a key role. Among SS techniques such as Match Filter Detection, Cyclostationary Feature Detection, Energy Detection (ED) is the mostly used technique in the literature [1], due to its simplicity and efficiency. In fading and shadowing environment, SS performance is degraded. So sensing from single user

R. Mustafa (✉) · R. R. Jaglan · S. Agrawal
Department of ECE, UIET, Panjab University, Chandigarh 160014, India
e-mail: mustafarashid80@gmail.com

C. R. Krishna et al. (eds.), *Proceedings of 2nd International Conference on Communication, Computing and Networking*, Lecture Notes in Networks and Systems 46, https://doi.org/10.1007/978-981-13-1217-5_55

cannot be reliable hence multiple user scenario or cooperative spectrum sensing (CSS) is proposed to overcome this problem [2, 3].

In CSS methods multiple SUs share incorporate information to the common base station or fusion center (FC) through control channels for PU status detection [4]. CSS can be employed either in a centralized way or in a distributed fashion. In the first scheme, the CR common base station collects all detecting results from the cognitive radio users (CRs) and detects the vacant band [5]. The distributed scheme requires exchange of observations between CRs. A key portion of a CSS model design is the data fusion. There are two main ways of combining the sensing results: hard combining and soft combining. In soft combining the CRs send the entire local sensing observation to the FC. Then shared data is combined using diversity techniques classified as maximum ratio combining (MRC) or equal gain combining (EGC). Soft combining scheme brings the best sensing performances and accuracy since there is multi bit information to process by the FC, however, it also suffers in reporting time and overhead to the control channel in terms of bandwidth requirement. The hard combining rule is the method that incurs less overhead to the system because of one bit information. Each CR makes a local individual decision and sends it as one bit information to the FC or other CRs. The FC, then, combines the shared information using linear fusion rules (OR, AND, and Majority). In the AND rule, the spectrum is considered occupied if all CRs have considered it occupied. In the OR rule, the spectrum is considered occupied if at least one CR have decided so. The l out of K is a middle term of the last two.

There are many researchers who investigated the performance of CSS in cognitive radio network (CRN) over different configurations [6–8]. The performance of CSS in different fading channel has been studied [9–12].

All the schemes which applied at FC for decision-making provide possible methods to implement CSS. However these schemes increase complexity of CRs and degrade the performance of CRN. In literature Artificial neural network (ANN) is found to be performing well in cases of pattern recognition and classification etc.

In CRN, ANN is considered as an efficient tool for providing learning ability as it can be applied for predicting the future outputs in reference to given inputs. ANN in CRN is employed in different scenarios such as spectrum sensing [13], channel state selection [14] and learning [15–18].

In [19] a new fusion rule based on Back Propagation (BP) in neural network depending on reliability is studied. Present paper deals with an ANN based CSS with reliable FC decision to meet a balance between performance and complexity.

Remaining paper is sectioned as follows: System model along with notations is discussed in Sect. 2. Simulation model algorithm is explained in Sect. 3. In Sect. 4 simulation results are discussed. Conclusion is drawn in Sect. 5.

2 System Model

We considered a CSS scenario of 'K' SUs, one PU and one FC in the proposed system. Each SU performs energy detection for making their own local decision about availability of PU. These individuals' local decisions are transmitted to FC by each SU through Additive White Gaussian Noise (AWGN) and Rayleigh fading channel in order to conclude a global decision.

2.1 Energy Detection Model

PU sends signal through digital modulation technique and it is transmitted through fading and noise environment. The energy detection (ED) block diagram is represented in Fig. 1. In which received signal has been processed via band pass filter with bandwidth W. Then output is squared through nonlinear device and the integral is obtained over the given interval. Finally, the received energy in terms of statically data (test statistics) is compared to the threshold value (λ) to mark a final decision on primary user absence or presence.

The PU detection using ED follows a binary hypothesis test, represented as

$$y(n) = \begin{cases} w(n), & H_0 \\ hx(n) + w(n), & H_1 \end{cases}, \tag{1}$$

where H_0 indicates absence of the PU signal and H_1 indicates present of the PU signal, $y(n)$ is the sample signal obtained by SU and $x(n)$ is transmitted PU signal, $w(n)$ is AWGN (with respect to zero-mean and signal variance) and h represents the gain in the channel bandwidth between both the channels. Energy of the signal in terms of test statistics can be given as

$$E = \frac{1}{N}\sum_{n=1}^{N}[y(n)]^2, \tag{2}$$

where N is the sample size. The test statistics is compared with predefined threshold then we are applying some decision rules which decide the correct status of PU signal. Energy detection process done with all SU and determine the final status of PU present or absent. Then all SU send the final data to FC. In FC different fusion

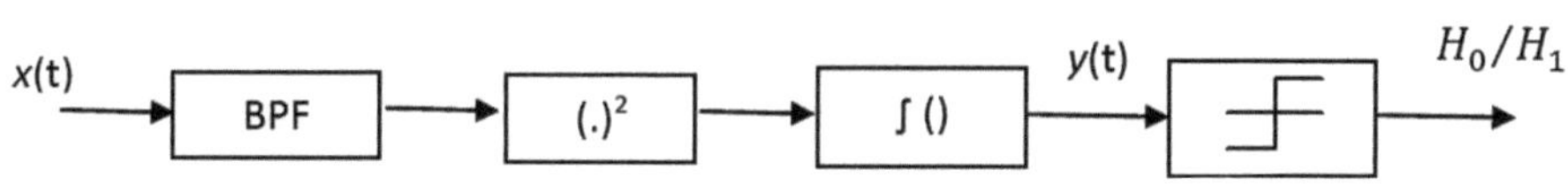

Fig. 1 Energy detection diagram

techniques are applied such as OR, AND, Majority and NN in order to conclude a global decision.

2.2 Cooperative Spectrum Sensing

Let l samples of the received signal is collected by K, SU in CSS scenario and sensed information sent to the base station or FC in the form of one bit binary information. Then FC makes a final decision in accordance to the following rule:

$$T = \sum_{i=1}^{K} Y_i \begin{cases} < l \; H_0 \\ \geq l \; H_1 \end{cases}, \tag{3}$$

where T is decision statistics and Y_i is the local decision statistics of ith SU.

The FC receives the information from K users decides H_1 when at least l out of K, SU sense results inferring H_1. Otherwise, the FC decides H_0. Different rules are summarized below:

OR decision rule can be evaluated by $l = 1$ in the above equation. In the OR rule, the spectrum is considered occupied if at least one CR has decided so.

AND rule can be evaluated by $l = K$ in the above equation. In the AND rule, the spectrum is considered occupied if all CRs have considered it occupied.

The Majority rule is a middle term of the last two. It can be evaluated by taking $l = \frac{K}{2}$ in the above equation.

2.3 Artificial Neural Network Based Cooperative Spectrum Sensing

Learning is a necessary characteristic of CR technique. With learning capability, CR technology can make a knowledge base and apply its behavior and actions according to the ambient radio environment. In ANN systems, a set of input-output data of the real process is resolved through measurement. Later, these arrays are used as an example to draw a conclusion for future situation.

The main attraction of using ANN-based CSS is that a cognitive radio system can learn and train itself from past data of previous scenario and ultimately make global decision without the need of current scenario.

In this paper ANN approach is implemented through training and testing, where all SUs data were used as NN input and PU actual status was taken as NN target. After neural network training the performance is measured through confusion matrix. ANN model is created and employed in our proposed scheme as a fusion tool.

3 Simulation Model

Proposed model is developed using Simulink in MATLAB version R2015a and the steps are as follows:

(a) Generate primary user signal with N samples using random integer generator and Bernoulli generator.
(b) The modulation scheme taken PSK.
(c) Generate AWGN channel with zero-mean and variance.
(d) Then received signal is conducting energy detection process.
(e) Then energy of each SU compares with predefined threshold (λ) to makes a binary decision (1 or 0) about PU activity. Then one bit decision sends to FC.
(f) The FC makes a global decision, i.e., H_1 or H_0 by using different hard fusion rules.
(g) Steps 1–6 are repeated N times to consistently estimate the detection probability (P_d) false alarm probability (P_{fa}) and missed detection probability (P_{md}).
(h) Next add Rayleigh faded channel after step 3.
(i) Steps 4–7 are repeated N times to consistently estimate the (P_d), (P_{fa}) and (P_{md}).
(j) The simulation data are saved in workspace for further investigations.
(k) For Proposed NN fusion steps carried out after step 10.
(l) Take all SUs data as a NN input.
(m) Take PU actual status as NN target.
(n) Train the network using NN pattern recognition tool box.
(o) Also creating trained ANN model which is used in our proposed scheme as a fusion tool.
(p) Then proposed ANN model performance is evaluated.

4 Results and Discussion

We considered a CRN, with one PU, one FC and number of SUs = 14 situated at different distance from the base station. The ED takes 1000 samples of PU signal and false alarm probability is set to $P_{\text{fa}} = 0.01$.

All simulation was carried out on MATLAB version R2015a over an AWGN and Rayleigh fading channel. We described the CSS performance in various scenarios through complementary receiver operative characteristic (CROC) curves.

CROC curves of the 14 SUs spectrum sensing in AWGN and Rayleigh channel following different hard fusion rules are represented in Figs. 2 and 3. Average SNR (γ) is taken as 10 dB.

From Figs. 2 and 3 it has been observed that SS is harder in the presence of fading and the ED performance degrades more in Rayleigh channel than in the AWGN channel and OR fusion rule has better performance than other fusion rules such as AND rule and MAJORITY fusion rules, because OR decision fusion rule includes

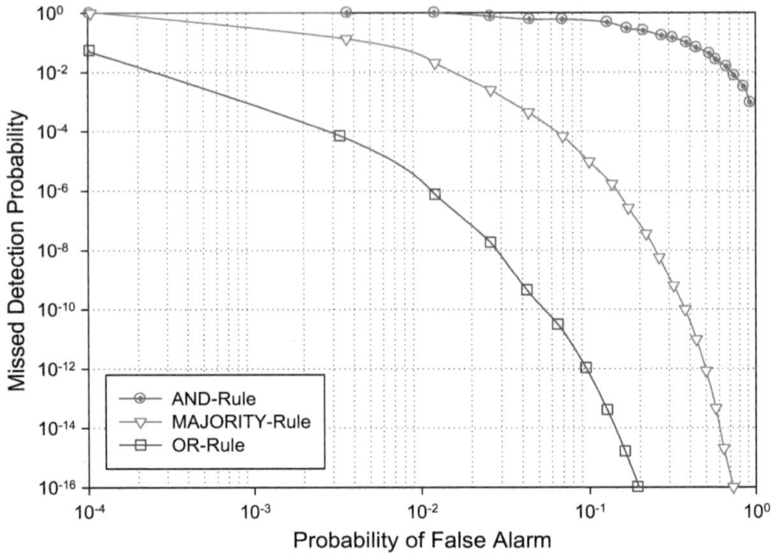

Fig. 2 CROC curves of hard fusion rule in AWGN channel

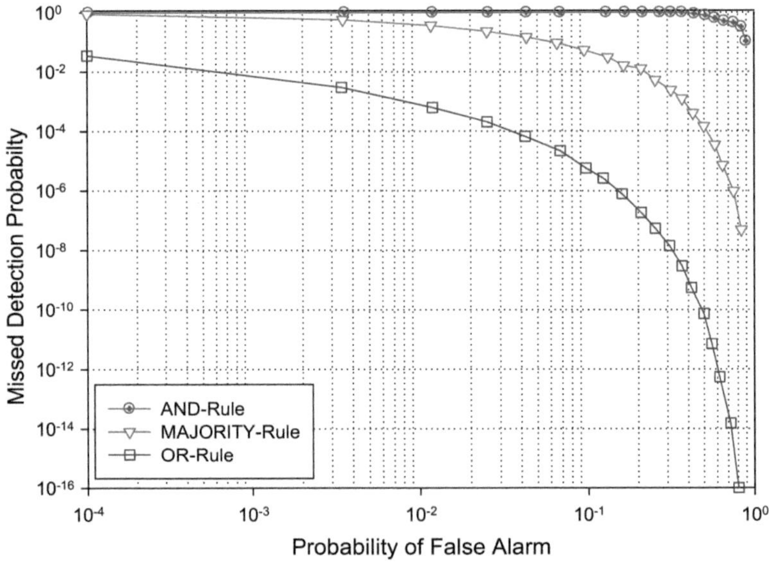

Fig. 3 CROC curves of hard fusion rule in Rayleigh channel

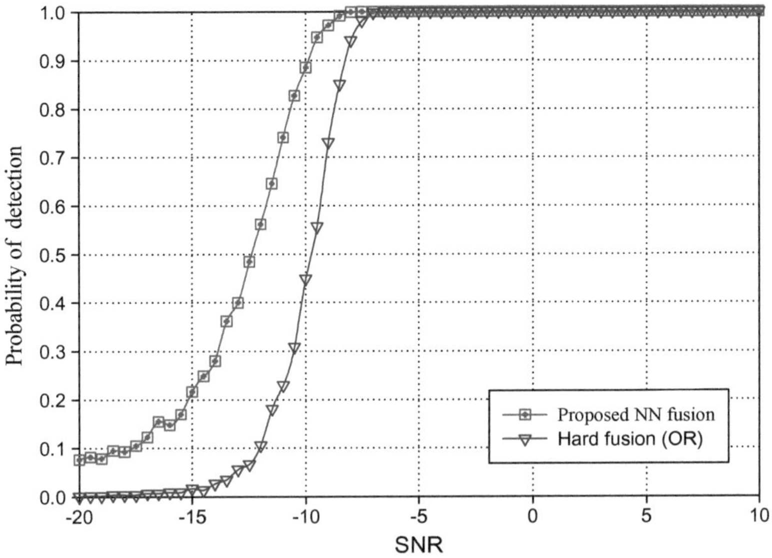

Fig. 4 Detection accuracy curves for different fusion rules for different SNR values

result of a single secondary user to decide the PU status. The results suggest that detection accuracy is high in non-fading AWGN channel when compared to Rayleigh fading channel.

Figure 4 shows the detection probability (P_d) at different SNR values ranging from -20 to 10 dB and fixed probability of false alarm $P_{\text{fa}} = 0.01$ in CSS. This figure compares detection accuracy curves of proposed neural network fusion scheme with conventional hard fusion rule. It has been noted that detection accuracy of the proposed NN fusion scheme is much better as compared to conventional hard fusion rules.

Figure 5 represents the comparative performance of the proposed neural network fusion scheme with related kind of work in the literature. From this figure it has been observed that ROC of the proposed neural network scheme is much better than BP neural network scheme in [19].

5 Conclusion

In the present paper performance of CSS scheme in CRN is investigated. We proposed a novel ANN based fusion model for detection of PU status in two different channels. AWGN and Rayleigh channel were considered and CSS performance was compared under different fusion rules and proposed ANN fusion rule. Proposed ANN fusion rule was found to be much better in detection accuracy as compared to conventional

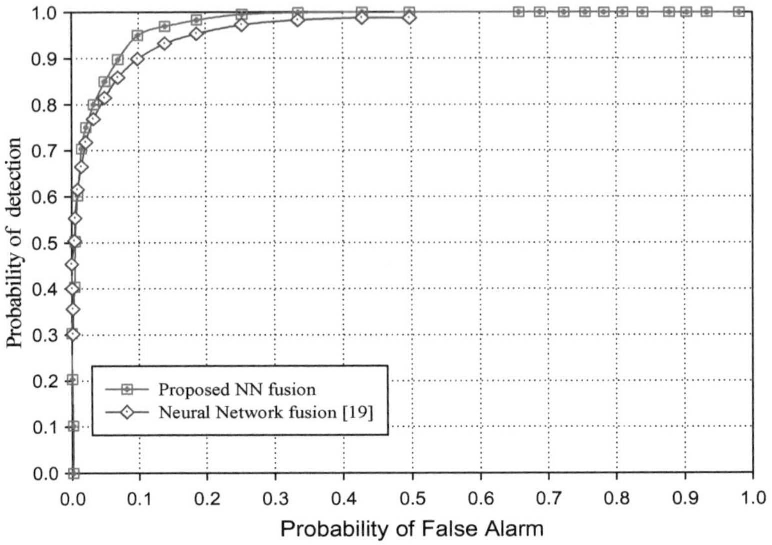

Fig. 5 Comparative performance

hard fusion rules. This study proves that ANN has great potential in recognition of physical channel attributes. Furthermore ANN increases the effectiveness and accuracy of CRN. In future work we will simulate other digital modulation schemes in order to widen the scope of application, also other fading channel like nakagami, rician channel also be used for further investigation.

References

1. R.R. Jaglan, S. Sarowa, R. Mustafa, S. Agrawal, N. Kumar, Comparative study of single-user spectrum sensing techniques in cognitive radio networks. Procedia Comput. Sci. **58**, 121–128 (2015)
2. I.F. Akyildiz, W.Y. Lee, M.C. Vuran, S. Mohanty, Next generation/dynamic spectrum access/cognitive radio wireless networks: a survey. Comput. Netw. **50**(30), 2127–2159 (2006)
3. X. Chen, H.-H. Chen, W. Meng, Cooperative communications for cognitive radio networks-from theory to applications. IEEE Commun. Surv. Tutorials **16**, 1180–1192 (2014)
4. G. Ganesan, Y. Li, Cooperative spectrum sensing in cognitive radio networks, in *First IEEE International Symposium on New Frontiers Dynamic Spectrum Access Networks*, USA (2005), pp. 137–143
5. Y. Zhang, J. Xiang, Q. Xin, G.E. Fien, Optimal sensing cooperation for spectrum sharing in cognitive radio networks, in *European Wireless Conference*, Denmark (2009), pp. 216–221
6. R.R. Jaglan, R. Mustafa, S. Sarowa, S. Agrawal, Performance evaluation of energy detection based cooperative spectrum sensing in cognitive radio network, in *Proceedings of First International Conference on Information and Communication Technology for Intelligent Systems*, vol. 2, no. 51 (2016), pp. 585–593

7. A.A. Alkheir, M. Ibnkahla, A selective decision-fusion rule for cooperative spectrum sensing using energy detection. Wirel. Commun. Mob. Comput. **16**(12), 1603–1611 (2016)
8. E.C.Y. Peh, Y.C. Liang, Y.L. Guan, Y. Zeng, Optimization of cooperative sensing in cognitive radio networks: a sensing-throughput trade off view. IEEE Trans. Veh. Technol. **58**, 5294–5299 (2009)
9. F.B.S. De Carvalho, W.T.A. Lopes, M.S. Alencar, Performance of cognitive spectrum sensing based on energy detector in fading channels. Procedia Comput. Sci. **65**,140–147 (2015)
10. S. Nallagonda, S.D. Roy, S. Kundu, Performance evaluation of cooperative spectrum sensing scheme with censoring of cognitive radios in Rayleigh fading channel. Wirel. Pers. Commun. 1409–1424 (2012)
11. M. Ranjeeth, S. Anuradha, Performance of fading channels on energy detection based spectrum sensing. Procedia Mater. Sci. **10**, 361–370 (2015)
12. J. Li, B. Li, M. Liu, Performance analysis of cooperative spectrum sensing over large and small scale fading channels. AEU Int. J. Electron. Commun. **78**, 90–97 (2017)
13. P. Giribone, R. Revetria, M. Antonetti, R. Tabolacci, Use of artificial neural networks as support for energy saving procedures, in *Telecommunications, Telecomm Energy Conference*, USA (2000), pp. 159–162
14. S. Dhanjal, Artificial neural networks in speech processing: problems and challenges, in *IEEE Pacific Rim Conference on Communication Computer Signal Processing*, Canada (2001), pp. 510–513
15. Z. Zhang, X. Xie, Intelligent cognitive radio: research on learning and evaluation of CR based on neural network, in *IEEE 5th International Conference on Information and Communication Technology*, Cairo (2007), pp. 33–37
16. C. Zhang, S. Wang, D. Li, J. Yang, J. Zhang, Semi-supervised behavioral learning and its application. Optik (Stuttg) **127**, 376–382 (2016)
17. A. Al-Dulaimi, L. Al-Saeed, An intelligent scheme for first run cognitive radios, in *4th International Conference on Next Generation Mobile Applications, Services and Technologies (NGMAST)*, Jordan (2010), pp. 158–161
18. X. Dong, Y. Li, C. Wu, Y. Cai, A learner based on neural network for cognitive radio, in *IEEE 12th International Conference on Communication Technology*, Nanjing (2010), pp. 893–896
19. Y. Chen, H. Zhang, H. Hu, Q. Wang, An efficient cooperative spectrum sensing algorithm based on BP neural network, in *Proceedings of IEEE on Wireless Communication and Sensor Network (WCSN)*, China (2014), pp. 297–301

Study of NoSQL Databases for Storing and Analyzing Traffic Data

Mankirat Kaur, Sarbjeet Singh and Naveen Aggarwal

Abstract Traffic congestion is an omnipresent problem which is faced by every big city where irregular traffic incidences increase the misery of road users in a disproportionately large manner. Traffic congestions can be regulated through simulating and enhancing traffic control with improved traffic management which is also known as traffic monitoring. Traffic-related Big Data gathered from sensors reflect some new characteristics such as multisource, high volume, heterogeneity, high velocity, continuity and spatio-temporal. Such type of datasets generated in real time need to be stored in databases in a scalable manner in order to provide congestion status of traffic to commuters timely. In this paper, NoSQL databases are highlighted as a successor to the traditional databases followed by enlisting their features and types. A tabular comparison concerning different data stores in terms of their features is given for better understanding and in-depth analysis of each type for their possible application in concerned research area.

Keywords NoSQL database · Relational database · Traffic congestion
Big data · ITS

1 Introduction

Urban population is increasing at a staggering rate and it directly leads to the rapid growth of a number of private vehicles in metropolitan cities, resulting in travel time delays, time loss, long tailbacks, user stress, fuel consumption, and pollution. Traffic congestion has drained money of largest economies very progressively [1, 2]. As two

M. Kaur (✉) · S. Singh · N. Aggarwal
University Institute of Engineering and Technology, Panjab University, Chandigarh, India
e-mail: mankirat@pu.ac.in

S. Singh
e-mail: sarbjeet@pu.ac.in

N. Aggarwal
e-mail: navagg@gmail.com

C. R. Krishna et al. (eds.), *Proceedings of 2nd International Conference on Communication, Computing and Networking*, Lecture Notes in Networks and Systems 46, https://doi.org/10.1007/978-981-13-1217-5_56

billion vehicles are circulating on road, there is an urgent need to identify congested areas in cities and re-route vehicles in case of sporadic traffic incidences. Traffic congestions can be regulated through simulating and enhancing traffic control with improved traffic management, which is also known as traffic monitoring. With the advancement of traffic monitoring system, a term ITS was coined, which means Intelligent Transportation Systems. ITS is a broader term encompassing sensors, communication technologies, and advanced applications which help in managing existing infrastructure and provide innovative services to transport systems leading to the reduction in traffic congestions.

The major challenge in the field of ITS is the overwhelming growth in size, variety, and velocity of traffic data collected from a large number of installed sensors [3]. Due to the widespread increase in the accessibility of the internet and mobile applications, the traffic state can be regulated by spreading messages on traffic-related apps and thus major logjams can be avoided. But these activities have greatly aroused a need to efficiently store and process a large amount of structured, semi-structured, and unstructured real time as well as historic data generated from numerous sensor devices. The traditional data models such as hierarchical, network and relational focuses on storing only structured data in form of trees, graphs and tables, respectively. However, NoSQL databases handle unstructured data by partitioning and replicating data over multiple clusters. This paper focuses on comparing the NoSQL and relational databases followed by discussion of various features and types.

2 Data Models for Storing Traffic Data

The traditional data models comprised of Hierarchical, Network and Relational models. Hierarchical model cannot handle sophisticated relationships, whereas Network model maintains all records using pointers. These models are efficient for linear storage mediums such as tapes but insertion, deletion and update operations are difficult to process queries in these models as they require a large number of pointer adjustments. The shortcomings of these models are addressed by Relational model. It stores data in form of tables which are linked with each other using key fields. Relational databases show compliance with ACID properties and are suitable for all types of applications. The traffic data collected from the web and mobile applications have following characteristics:

1. High volume: data about traffic conditions captured through sensor devices is far beyond the limits of megabytes or gigabytes and reaches terabytes.
2. High Velocity: the frequency of data generation is increasing day by day.
3. High Variety: the traffic data is usually multimodal in nature which includes images, audio, video, text files etc.

The above characteristics satisfy all the three paradigms of Big Data [4]. In order to store such a large amount of unstructured data, it poses a question whether relational databases can cope up with the changing storage requirements of advanced applications. In next section, the inability of relational databases to store Big Data is described, followed by the emergence of NoSQL database and its various types.

3 Storing Big Data in Relational Databases

Today's web applications are very agile and consist of different types of attributes such as text, image, video, and audio. Storing data of every such application in relational databases, even if their data model does not match, leads to increased complexity through expensive mapping into multiple tables and use of complex algorithms. Although, the data model matches with the relational store, the huge feature set that may not be required by application further incurs additional overhead with increased cost and complexity. On the similar note, inserting and removing features in web application supported by SQL database cannot be possible without making system unavailable for a small period of time. The famous web apps such as Facebook, Amazon need to handle preponderance of read and write requests on a very large scale along with managing the increasing amount of data. Dealing with such agile requirements and data without any evident delay requires a flexible schema, speed and distributed databases [5].

According to CAP theorem, any shared system can satisfy at most two properties at a time from the three, viz. Consistency (C), Availability (A) and Partition Tolerance (P). Due to the normalized data model and compliance with ACID properties, the relational databases are built to support consistency and availability. To accomplish this, they have to relinquish partition tolerance which makes relational databases difficult to scale horizontally [6, 7]. SQL databases have a predefined layout of tables and to query these tables structured query language is used. Using join operation multiple tables can be queried. Typically, the SQL databases scale well in a vertical manner, i.e., by running it on a more powerful computer by upgrading its processing power and memory. Vertical scaling cannot be done beyond a certain point and after that, the data need to be distributed across clusters. In a distributed environment, relational stores cannot work, as joining and lock operations across multiple computers become very expensive. The high performance and availability is the prime requirement of many web companies, therefore, the database should be integrated with high replication ability, data partitioning, and failover mechanisms. *Since relational databases are not suitable to handle these functionalities, hence NoSQL databases have emerged to cater the changing requirements of advanced applications* [6].

4 Storing Traffic Data in NoSQL Databases

The emergence of growing unstructured data and infrastructure needs from organizations, e-Commerce, location-based and social media applications has led to the development of non-relational or NoSQL databases. The applications which require anomaly detection, compromised account detection and predicting trends in shopping by customers need a large amount of data to be fused from multiple sources in real time [7]. To cater these requirements, NoSQL (means Not Only SQL) has been developed that do not have properties of traditional SQL stores and has replication and partitioning as their inbuilt features. Usually, the relational stores are developed by first deciding the conceptual and logical models then the physical model is prepared where the type of queries addressed by the model is decided. But in case of NoSQL databases, the queries that application will be addressing is decided first, then the data model is constructed to support listed queries. These databases promote the horizontal scaling of data by constructing clusters where data is partitioned and replicated over multiple systems. The key features of these databases are as follows [8]:

1. Big data Features: It satisfies the three V's of Big Data by enabling the capturing, storing, processing and analyzing the large volumes of data, heterogeneous in nature (i.e., acquired from different sources) arriving at very high frequency.
2. Flexible Schemas: In RDMS databases, changing schema undergoes rigorous testing framework and schema rarely evolves over time. But in NoSQL databases, the schema is changing throughout the life of application to cater the changing requirements with time. There is no restriction of putting data in fixed table columns and unique data per row id, therefore, the data model is designed to satisfy the need of the individual application.
3. Eventual Consistency: The CRUD (i.e., creates, read, update, and delete) operations are expected from every database to execute efficiently. In course of providing fast and concurrent insert and read operations in NoSQL database, the updates are guaranteed to be propagated to all partitions eventually. This means that data fetched from the database at any point of time are not guaranteed to be up-to-date. This is known as Eventually Consistency. This is provided to ensure high availability and partitioning of data by relaxing consistency.
4. Horizontal Scaling: In order to provide the faster insertions and reads from the database without any processing delay, the data need to be accessed quickly from the database. In RDMS, this is facilitated through vertical scaling without partitioning and distributing data over multiple nodes as it increases the overhead of searching table in each partition. But in NoSQL there is no such constraint of complying ACID, scalability is provided through adding new complete systems with their own processors and disks.
5. Shared Nothing Architecture: According to this architecture, each node is self-sufficient having its own memory and storage and scalability is provided through linearly adding a new node using inexpensive commodity hardware. It criticizes the concept of centralized storage of data on a single machine. Data is partitioned

into smaller subsets and distributed over multiple nodes; this process is known as Sharding. NoSQL databases are built on this architecture and sharding is used for partitioning of data.

6. Compliance with BASE properties: NoSQL databases do not comply with ACID properties of relational data stores and provide a better performance through partitioning data. It satisfies another set of properties known as BASE properties which are Basic Availability, Soft State and Eventually Consistent. Basic availability means that the system is always available to the client but with delayed synchronization and it may also undergo temporary inconsistency. Soft State means that the state of the database changes without any input or updates. Eventual consistency means that if no updates are made in the database for a long span of time then all clients will see same value of database items.
7. CAP Theorem: This theorem says that any database system satisfies at most any two properties at a time from Consistency, Availability and Partition tolerance. All NoSQL systems comply partition tolerance but from consistency and availability, one is chosen according to the requirements of the application. Some databases choose consistency while other NoSQL databases prefer availability at the cost of compromising the other one.

Various studies have shown that NoSQL databases perform better than relational data stores by comparing them on the basis of read and write operations.

Van der Veen et al. [6] compared PostgreSQL, a variant of SQL database and two NoSQL databases viz. MongoDB and Cassandra on the physical server and on a virtual machine to assess the impact of virtualization on the performance of databases. They concluded that SQL database performs best when multiple clients issue single read requests and when single client issue multiread requests to the database. NoSQL databases perform best in all other situations of single and multiple clients issuing multi- read-and-write requests to the database. They inferred that NoSQL databases are the best choice for small or large critical sensor applications when write performance is important, whereas SQL databases are the best choice when flexible query capabilities are needed.

Li and Manoharan [8] evaluated the performance of SQL database, SQL Express and five NoSQL databases on simple CRUD operations on key-value pairs. They concluded that except Couchbase and MongoDB (NoSQL databases) which performed best in all operations, not all NoSQL databases perform better than SQL databases. They carried out experiments on a single client and multiple operations but still, NoSQL databases outperformed the SQL one.

Boicea et al. [9] compared NoSQL database against SQL database. They used Oracle DBMS as the implementation of SQL database and MongoDB as the implementation of NoSQL. They reported that for 100,000 records, the insertion time in Oracle is higher as compared to MongoDB and in case of deletion and update operations, MongoDB performed much better than Oracle. There is a huge gap in time taken by both databases in performing delete and update operations.

Working on a similar note, Indrawan-Santiago [10] differentiated ten NoSQL databases with SQL databases on the basis of the data model, indexing, sharding, transaction model and support for ad hoc queries. She compared quantitatively SQL with NoSQL and reported that NoSQL outperforms the SQL databases and is likely to enhance the database management capability of enterprises.

5 Comparison of Types of NoSQL Databases

NoSQL databases are broadly categorized into four types, namely, Key-Value Stores, Extensible Record or Column data store, Document data store and Graph databases. These are briefly described as follows:

Key-Value Store: This is the simplest database which is very similar to maps and dictionaries with a single index of key values over whole data. These data stores are schema-free and the only way to provide some structure to data is by assembling matching key-value pairs into collections. They are suitable for basic query operations which are based on the primary key attribute and its query speed is higher than SQL databases [11]. They hold most of its data in memory and thus can be used for caching intensive relational queries. Since the values are in form of bytes which are difficult to understand, key indices are used for retrieving data items from the database. As values are independent and unlinked, the relationships need to be modeled at application logic [5]. The examples of key-value stores are Project Voldemort, Redis, Riak.

Document Store: In these databases, data is stored in form of document and collections. A document consists of key-value pairs and keys have to be unique in each document, like JSON document. Each document has special unique key-ID within a collection of documents. As it is stored in interpretable JSON type format, complex data structures can be handled very conveniently [5]. These data stores are also schema-free and adding new documents in the collection or in another document with a different type of attributes can be done very easily, similar to inserting new attributes to documents. They allow developers to create secondary indexes on documents [12]. These stores have the ability to store objects in a serialized format (XML, JSON, BSON). Values are transparent to a system and thus the relationship between documents can be modeled easily through embedding or referencing styles. Complex data structures can be handled more conveniently, like nested-objects. Therefore, they can model relationships between different values and thus, can address complex queries very easily without incurring any overhead of querying at application logic. These data stores are suitable for applications that require aggregations across collections, data integration, and schema migration tasks. The application areas include real-time analytics such as games, small blog websites, etc. [13–15]. Examples are CouchDB, SimpleDB, MongoDB, Couchbase.

Column Family Stores: These stores are motivated by Google Bigtable. It is conceptually similar to RDBMS, where data model consists of rows and columns. Scalability is assured through splitting rows and columns over multiple nodes. Rows are split through sharding on the primary key [12] and columns of a table are distributed over multiple nodes by using column-groups (column family). A column family is set of related columns which are physically stored together. In order to achieve performance gain, similar access columns are grouped in a single column family. Set of column families are further grouped into super columns [15]. In these databases, column families must be predefined which leads to less flexibility of schema design. Column family stores are very efficient in domains where a huge amount of data need to be stored having a varying number of attributes. Both horizontal and vertical partitioning can be used simultaneously on the same table. A random number of key-value pairs can be stored in the rows. The values in key-value pairs are not interpreted by the system, therefore, relationships are handled at application logic. Multi-version concurrency control is used for maintaining concurrency. The data model is sparse in terms of handling null values as it stores only the key-value pairs that are required by the system in a row. These databases are suitable for storing a large amount of data with a varying number of attributes, as the data model can be partitioned efficiently [5]. The examples are Cassandra, HBase.

Graph Databases: It is used for efficient management of data when the applications contain heavily linked data and the database needs cost-intensive operations like recursive joins to query them. In this database, the objects with embedded key-value pairs are modeled in form of nodes and edges. Partitioning graph databases is very tedious work which requires domain-specific knowledge and complex algorithms. Information is not gained by simple key lookups but by analyzing relationships between entities. There is often a trade-off between distributing nodes on multiple servers or placing heavily linked nodes at a closer distance in order to avoid large traversal leading to huge performance penalty [5, 15]. Examples are Neo4j, Sesame, GraphDB.

The above four categories of NoSQL databases differ in the approach followed by them along with the principal set of features such as partitioning, replication, concurrency model and so on. It is very difficult to compare all the four types of data stores on the basis of these features. Therefore, the popular variants of column and document stores chosen for comparison purpose are

- Column or Extensible record databases (Cassandra, HBase)
- Document databases (MongoDB, CouchDB)

From the evaluation of four databases in Table 1 namely, Cassandra, HBase, CouchDB and MongoDB it can be observed that though three of them can store large amount of data by providing same features and equivalent performance in different applications but MongoDB is preferable for storing and analyzing traffic data due to the following reasons:

- It has special support for geospatial processing.
- It stores received data in JSON format which is compatible with many programming language constructs. This helps in easy import/export over multiple platforms.
- The data stored in JSON format is easily visualized in form of trajectories on any Maps API.

Table 1 Comparison of NoSQL databases

Features	NoSQL			
	HBase	Cassandra	CouchDB	MongoDB
Concurrency control	Optimistic concurrency control through locking mechanisms	Weak concurrency through multi-version concurrency control	Multi-version concurrency control	Locks are provided for atomic level operations
Consistency	Strong consistency through locks and logging	Provides eventual consistency	Provides eventual consistency	Provides atomic consistency
Partitioning	Range-based partitioning	Consistent hashing partitioning	Consistent hashing partitioning	Range-based partitioning
Replication	Asynchronous replication	Optimistic replication	Asynchronous replication	Optimistic replication
Secondary index support	Built-in support for secondary indexes not available	Secondary indexes not built-in	Secondary indexes available	Secondary indexes available
Data structure used for building indexes	B-tree indexes supported	Ordered-hash indexes	B-tree indexes supported	B-tree indexes supported
CAP theorem	It qualifies consistency and partition tolerance	It qualifies availability and partition tolerance	It qualifies availability and partition tolerance	It qualifies consistency and partition tolerance
Automatic failure detection support	Automatic failure detection	Fully automatic failure detection	Failure detection not automatic	Fully automatic failure detection
Automatic recovery support	Recovery process is done by administrator	Fully automatic recovery process	Recovery process is automatic	Recovery is automatic
Map-reduce support	Supported	Supported	Supported	Supported
Data storage	Data is stored in HDFS	Data is stored on disk	Data is stored on disk	Data is stored on disk
Storage format	HDFS	SSTable storage format	Data is stored in serialized form, JSON	Data is stored in binary serialized form, BSON

(continued)

Table 1 (continued)

Features	NoSQL			
	HBase	Cassandra	CouchDB	MongoDB
Transaction support	Local transactions possible only in single shard	Local transactions possible only in single shard	Transactions are not supported	Transactions are not supported
Automatic sharding	Automatic sharding possible	Automatic sharding possible	Automatic sharding not available	Automatic sharding possible
Type of applications supported	Suitable for applications dealing with huge amounts of data	It is popularly used in developing financial service applications	Real-time analytics, logging, small and flexible websites like blogs	Department of motor vehicle application with vehicles and drivers
Architectures used	Master-slave architecture used	Ring-architecture used here	It uses multi-master architecture	Master-slave architecture used here
Remarks	It provides excellent consistency in column databases through locks and logging. Therefore, it is being used as an efficient data storage method for home energy management (HEM) mobile applications where consistency is vital that is it should always return the latest written value	Best choice for large critical applications as it was built to scale horizontally EPC(electronic product code) information services have used it for storage as they require write operations more frequently than read and support flexible schema with extension fields that can vary according to contents of user	Provide ACID semantics at the document-level Applications which need to respond too many parallel read and write requests and which only have to provide a certain level of consistency It does not provide declarative query language and puts burden on programmer for distributing map-reduce tasks	Best choice for a small or medium-sized non-critical sensor application when write performance is important It also allows easy scaling of databases for large number of sensors. It is used by Aydin et al. [16] for high performance write support for QuickServer and by Boulmakoul et al. [17] for visualizing trajectories on a map using JSON

- It offers a number of indexes and query mechanisms to handle geospatial information.
- Most of the sensor-based applications have used MongoDB for storing data in conjunction with Hadoop clusters to facilitate fast analysis and processing.

6 Conclusion

As the traffic in cities and towns is increasing at unprecedented rates, it has highly increased the misery of commuters. The only way to get rid of mile long tailbacks is the prior delivery of information related to current traffic state through mobile or web applications. Therefore, real-time streaming as well as historical traffic data storage and processing is required to satisfy the demands of ITS applications for providing real-time traffic monitoring and management. The traditional databases, such as Hierarchical, Network and Relational have become incapable in handling the changing storage requirements and complex queries of advanced applications. This lead to the evolution of databases, known as NoSQL, which have replication and partitioning data as their inbuilt features. These databases support flexible schema and data model is designed according to the user queries addressed by the application. In this paper, the features and types of NoSQL databases have been discussed and the comparison between prominent databases has been carried out on the basis of sixteen different features. It has been observed that MongoDB is most preferable for handling and storing the subtle characteristics of traffic data.

References

1. INRIX, Americans will waste $2.8 trillion on traffic by 2030 if gridlock persists [Online] (2014). http://inrix.com/press-releases/americans-will-waste-2–8-trillion-on-traffic-by-2030-if-gridlock-persists/. Accessed 27 Jan 2017
2. J. Manyika et al., Big data: the next frontier for innovation, competition, and productivity [Online] (2011). http://www.mckinsey.com/business-functions/digital-mckinsey/our-insights/big-data-the-next-frontier-for-innovation. Accessed 27 Jan 2017
3. J. Gantz, D. Reinsel, Extracting value from chaos state of the universe, in *Proceedings of International Data Corporation (IDC)*, June 2011
4. P. Warden, *Big Data Glossary* (O'Reilly Media, USA, 2011)
5. R. Hecht, S. Jablonski, NoSQL evaluation: a use case oriented survey, in *Proceedings of International Conference Of Cloud Computing (CSC 2011)*, pp. 336–341
6. J.S. Van der Veen et al., Sensor data storage performance: SQL or NoSQL, physical or virtual, in *Proceedings of IEEE Fifth International Conference of Cloud Computing* (2012), pp. 431–438
7. N. Leavitt, Will NoSQL live up to their promise? Computer **43**(2), 12–14 (2010)
8. Y. Li, S. Manoharan, A performance comparison of SQL and NoSQL databases. in *2013 IEEE Pacific Rim Conference Communications, Computers and Signal Processing (PACRIM)*, pp. 15–19
9. A. Boicea et al., MongoDB vs Oracle–database comparison, in *2012 Third International Conference on Emerging Intelligent Data and Web Technologies (EIDWT)* (2012), pp. 330–335

10. M. Indrawan-Santiago, Database research: are we at a crossroad? reflection on NoSQL, in *IEEE 2012 15th International Conference Network-Based Information Systems (NBiS)*, Sept 2012, pp. 45–51
11. J. Han et al., Survey on NoSQL database, in *IEEE 2011 6th International Conference Pervasive Computing and Applications (ICPCA)*, Oct 2011, pp. 363–366
12. R. Cattell, Scalable SQL and NoSQL data stores. ACM SIGMOD Rec. **39**(4), 12 (2013)
13. Y. Gu et al., Analysis of data replication mechanism in NoSQL database MongoDB, in *2015 IEEE International Conference Consumer Electronics-Taiwan (ICCE-TW)* (2015), pp. 66–67
14. Y. Gu et al., Application of NoSQL database MongoDB, in *2015 IEEE International Conference Consumer Electronics-Taiwan (ICCE-TW)* (2015), pp. 158–159
15. V.N. Gudivada et al., NoSQL systems for big data management, in *Services, 2014 IEEE World Congress*, 27 June 2014, pp. 190–197
16. G. Aydin et al., Architecture and implementation of a scalable sensor data storage and analysis system using cloud computing and big data technologies. J. Sens. **2015** (2015)
17. A. Boulmakoul et al., Towards scalable distributed framework for urban congestion traffic patterns warehousing. Appl. Comput. Intell. Soft Comput. **2015** (2015)

Applying Machine Learning Algorithms for Early Diagnosis and Prediction of Breast Cancer Risk

Tawseef Ayoub Shaikh and Rashid Ali

Abstract With the advancement in the technological age, the deadly diseases threatening human survival also increase at the same pace. Breast cancer being at number two in causing the deaths among women is equally among the most curable type of cancer if diagnosed prior to time. There is an utmost thirst for diagnosis of breast cancer through an automation system in everyday health applications. This paper uses dimensionality reduction technique offered by Weka tool called WrapperSubsetEval on two benchmark cancer datasets of Wisconsin and Portuguese "Breast Cancer Digital Repository" (BCDR), on top four data mining algorithms available in literature. The final experiments carried in MATLAB and Weka demonstrated that Naive Bayes, J48, k-NN and SVM got an improvement in accuracy from 92.6186, 92.9701, 96.1336, 97.891 to 97.0123, 96.8366, 97.3638, 97.9123% in case of Wisconsin dataset and an improvement from 87.4126, 80.4196, 93.7063, 91.6084 to 89.5105, 90.9091, 97.9021, 95.1049% in case of BCDR-D01_Dataset.

Keywords Machine learning algorithms · Breast cancer · Naïve Bayes · SVM MATLAB · Weka

1 Introduction

Breast cancer is rising at an alarming rate by being the number two among the fatal diseases in the world in women with over 1.5 million foreseen cases spotted in 2010 and triggering a threat to human survival in causing demises to half a million per year, according to a report of World Health Organization [1]. Its effects are dangerously increasing where it is responsible for one in every six deaths among women in the

T. A. Shaikh (✉) · R. Ali
Department of Computer Engineering,
Aligarh Muslim University, Aligarh, Uttar Pradesh, India
e-mail: tawseef37@gmail.com

R. Ali
e-mail: rashidaliamu@rediffmail.com

C. R. Krishna et al. (eds.), *Proceedings of 2nd International Conference on Communication, Computing and Networking*, Lecture Notes in Networks and Systems 46, https://doi.org/10.1007/978-981-13-1217-5_57

European Union [2]. A hot research oriented topic in the modern times, breast cancer continuously makes its victims at an alarming rate where in 2012, likely 1.67 million fresh cancer cases were spotted. India is also not opaque from its deadly shadow. With a frequency of almost 1,44,000 fresh cases of breast cancers per annum, it is emerging as number one female cancer in metropolitan India. A likely probability of 100,000 new breast cancer patients is spotted in India per annum [3]. From 5 per 100,000 female population per year in rural areas, its range enlarges to 30 per 100,000 female population per year in urban areas in India [4]. With a trivial share of the world population in developed countries, still it accounts for 50% of breast cancers identified worldwide [5]. This whole scenario compels to come up with new techniques and methods in order to fight against this threatening disease, thus making entry of Information and Communication Technologies (ICT) an ideal choice for the same. Twenty-first century witnessed rapid growth of ICT age and there is hardly any sphere of human life where it has not laid its foot prints. Since the modern healthcare is getting shifted from cure based to care based evidence medicine. The priority is given to early diagnosis and detection of the diseases when they are in initial phase of their development. The latest next generation human genome sequencing is such an example where at early advanced stage could it be possible to see which base pair in the DNA has mutated and can lead to cancer on later stages. The same mutation could be reversed at the initial stage, thus nipping the roots in the bud.

So a woman can get a better chance of complete regaining from the cancer if diagnosed at prior stage, thus firing the need of development of efficient diagnosis techniques. A decrease will occur in the connected disease and mortality rates if timely revealing of breast cancer can become a possibility. Radiologists use the technique of screening mammography as the chief imaging modality for prior breast cancer detection because it has got the credit of being the only method of breast imaging in shrinking mortality rates related to breast cancer [6]. Normally in traditional cases, double-reading (same mammograms are read by two radiologists individually) has been encouraged in decreasing the percentage of overlooked cancers and it is at present the supreme technique incorporated in most of the screening programs instead of the fact that it earns in surplus workload and costs [7]. So a platform for the computer-aided detection/diagnosis (CADe/CADx) systems is established for backing up a single radiologist reading mammograms providing sustenance to her/his decisions [8]. ICT can show impending roles in combating this anti life threat. In fact, big data discussion has made its entry into city's talk nowadays in being a promising dimension that is expected to leave its hall mark on all major fields. Its spectrum in healthcare domain are expanding rapidly because of its increased performance in saving costs, predicting aftermaths, optimal cure within budgets and nurturing quality of health care to protect people's survival.

The suspicious lesions identified by the radiologist are sorted out by the CADx systems and lesions detection is focused by CADe systems [9]. The presented work concentrates on CADx systems. Since CADx systems archetypally has their base on machine learning classifiers (MLC) for affording diagnosis well advanced in time. A combination of forecasters is prerequisite for pronouncing the observation in order to train an MLC for breast cancer diagnosis. In order to make the inference whether a certain surveillance is from a cancerous finding or not, a high discriminant power should be possessed by the classifier [10]. This not only being an opportunistic but also a challenging theme that has congregated the concentration of research of quite a lot of sciences, from computer vision, artificial intelligence, mathematics and statistics to medicine. Thus, an assembly of related predictors may be used for diagnosis inferring [11].

Remaining paper is fashioned as. Section 2 concentrates on the description of the datasets used in this study. It also throws a brief light on the methodology of extracting the productive feature vector from large high dimensional feature space. The brief information about the mining algorithms which serviced the present work are discussed in Sect. 3. It also brings about the classification parameters used in the evaluation of the classification accuracy of the selected classifiers. Experimental results are drawn both in tabular and graphical form and explored in Sect. 4. The results are discussed in Sect. 5 in discussion part and finally the paper is concluded with Conclusion as Sect. 6.

2 Materials and Methods

This slice designates the assessment procedure of image descriptors for breast cancer verdict. It enlightens about the extraction of the feature vector from the image mamma graphs for training machine learning classifiers to envisage the pinpointing of a lesion (Fig. 1). This section unfolds the data sets upon which the experiments were carried on, shadowed by an ephemeral enlightenment of the image descriptors that were evaluated.

2.1 Data Sets

In this study, two breast cancer datasets are castoff in order to pinpoint the general best method and classifier.

Wisconsin breast cancer diagnosis (WBCD): It is developed by the University of Wisconsin Hospital grounded exclusively on an FNA (Fine Needle Aspiration) test for breast masses finding [12]. A total of 699 clinical instances is present in this dataset, possessing 458 (65.52%) benign and 241 (34.48%) malignant. Every clinical instance is described by a set of 9 attributes having assigned integer values whose range lies from 1 to 10 and one class has a binary value of either 2 or 4 as output as a convenience for representing benign and malignant cases respectively. From this dataset 16 missing occurrences are detached in order to gain high accuracy framing out the final dataset possessing 683 clinical occurrences, with 444 (65.01%) benign and 239 (34.99%). malignant cases.

Breast Cancer Digital Repository (BCDR): It is collected form Portuguese female patients using an average age of 54.4 years old, fluctuating in the range from

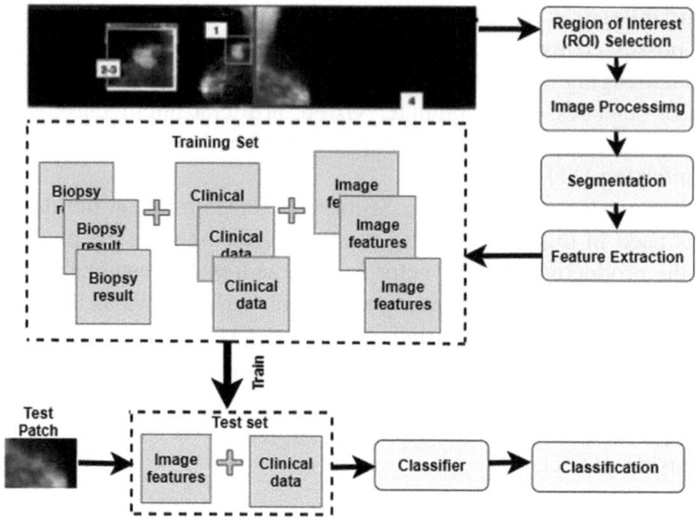

Fig. 1 Extracting image features from the image mammography lesions and training a classifier based upon same features

28 to 82 [13]. BCDR-F01 is the head data set free for public. From 362 segmentation, 187 (51.66%) share is employed by benign findings and the residual 175 (48.35%) go to malignant findings.

3 Evaluation

This chapter deals with the brief defining and working of selected data mining algorithms fruitful for our work in this paper. It also engulfs the metrics which are used as a measuring parameter for calculating the classification accuracy of the selected data mining algorithms on both the datasets. It also adds some more classification measuring parameters even if they were not directly showed in this work.

3.1 Algorithms Used

Four famous and most common used data mining algorithms in studies are used in this paper.

Naive Bayes: It is a special algorithm whose background lies in the famous foundation laid down by Bayes theorem and belongs to probabilistic method of classifiers. Bayes Classifier also known as generative model has its secret in computing class conditional probability in terms of posterior and prior probabilities. Considering a testing instance possessing 'd' different features and having values $X = <x_1, x_2, \ldots, x_d>$ respectively. For determining posterior probability $P(Y(T) = i|x_1, x_2, \ldots, x_d)$ that the class $Y(T)$ of test instance T is I, the Bayes rule results in:

$$P(Y(T) = i|x_1, x_2, \ldots, x_d) = i).\frac{P(x_1, x_2, \ldots, x_d|Y(T) = i}{P(x_1, x_2, \ldots, x_d)} \quad (1)$$

k-NN: K nearest neighbor is an instance base learning (IBK) which is a type of lazy learning. In it the stage of training model building is often dispensed and test instance has a direct linkage with the training instances for producing a classification model. This approach crafts locally optimized model precise to the test instance.

J48: It is Decision Tree algorithm that uses a split criteria for splitting the data into the corresponding labels. The splitting condition can be single attribute known as Univariate or multiple featured known as Multivariate. The goal here is to recursively make splitting of training data for maximizing the discrimination of different classes over different nodes. Gini-index and Entropy are used to quantify the same. If $p_1, \ldots, p_k$ is the portion of the records fitting to k different classes in a node N, then Gini-index $G(N)$ of a node N is

$$G(N) = 1 - \sum_{i=1}^{k} p_i^2 \quad (2)$$

Corresponding Entropy $E(N)$ is:

$$E(N) = -\sum_{i=1}^{k} p_i \cdot \log(p_i) \quad (3)$$

The objective is to always minimalize the weighted sum of Gini-index or entropy for splitting while developing the training model.

SVM: Support Vector Machine use linear conditions to make the classification. An SVM classifier is equivalent to a single level decision tree with a very sensibly preferred multivariate split condition. Weka uses a specific efficient optimization algorithm inside Sequential Minimal Optimization (SMO) for SVM. The goal is to increase the margin of the separating hyper plane:

$$\text{Objective function} = \frac{||\overline{W}||^2}{2} + C \cdot \sum_{i=1}^{n} \xi_i \quad (4)$$

ξ_i is a Slack parameter whose purpose is to incorporate soft margins and C adjusts the importance of margin and slack necessities. Nonlinear SVM are focus of the present era of research which are learned using kernel methods. Here the pair wise dot product between different training instances and between different test instances are used as similarity values, which in turn open the gates for transformations of data into multidimensional space. Kernel function (dot product) is:

$$K(\overline{X}, \overline{Y}) = \phi(\overline{X}) \cdot \phi(\overline{Y}) \quad (5)$$

Performance evaluation of classifiers are evaluated from two different perspectives, i.e., Visualization Techniques (ROC analysis and Reject Curves) and Statistical techniques (Confusion Matrix, Precision, Recall, Sensitivity, specificity and F-Measure).

3.2 Classification Metrics

Consists a list of parameter normally used for finding out the classification accuracy of the classifier.

Sensitivity (also called *Recall sensitivity, recall, hit rate or true positive rate* (*TPR*)]: Sensitivity is the share of genuine positives that are properly acknowledged as positives by the classifier.

$$\text{Sensitivity} = \text{TPR} = \text{TP}/(\text{TP} + \text{FN}) \tag{6}$$

Specificity (also called *True Negative Rate*): Specificity is capability of classifier to isolate negative results.

$$\text{Specificity} = \text{TNR} = \text{TN}/(\text{TN} + \text{FP}) \tag{7}$$

Precision [positive predictive value (PPV)]: Measure of relevant retrieved instances.

$$\text{Precision} = \text{PPV} = \text{TP}/(\text{TP} + \text{FP}) \tag{8}$$

Accuracy: Gives the share of correctly classified instances.

$$\text{Accuracy} = (\text{TP} + \text{TN})/(\text{TP} + \text{TN} + \text{FP} + \text{FN}) \tag{9}$$

F_1 score (also F-score or F-measure or balanced F-score): It is the harmonic mean of recall and precision:

$$F_1 = 2 * \frac{1}{\frac{1}{\text{recall}} + \frac{1}{\text{precision}}} = 2 * \frac{\text{precision} * \text{recall}}{\text{precision} + \text{recall}} \tag{10}$$

3.3 Regression Metrics

Consists of statistical parameters like MAE, MSE and RMSE, etc.

4 Results and Discussion

In this section the outputs from experiments are calculated in the mathematical terms and the same are plotted in both tabular and graphical form.

In this paper, four most widely used data mining algorithms available from the literature are applied on the two benchmark cancer datasets using all time WEKA toolkit [14]. The algorithms used are Naive Bayes, J48, k-NN, and SVM in Weka tool and corresponding classification parameters are noted down. Weka offers a special type of Wrapper feature selection facility known as WrapperSubsetEval [15] which outputs the optimal feature subset from feature space. Two types of feature selection are widely used in the literature [16–18]:

Filter Models: Independent of the specific algorithm being used, a hard principle on a feature or set of features is the trademark of this model that has the usability in evaluating the suitability of classification [19–21].

Wrapper Models: Here algorithm and feature selection process is packed together which makes the feature selection process algorithmic specific. This technique believes on the fact that different algorithms may exertion better with different feature vectors [22–24].

The same four algorithms are applied on the modified datasets of the original datasets and corresponding classification parameters are noted down again. The results when compared showed an increase in the classification accuracy of all four algorithms. An increase from 92.6186 to 97.0123 on Wisconsin dataset and from 87.4126 to 89.5105 on BCDR-D01 dataset in case of Naive Bayes occurred. In the same way an increase from 92.9701 to 96.8366 in case of Wisconsin dataset and from 80.4196 to 90.9091 on BCDR-D01 dataset in case of J48 occurred. Similarly an increase from 96.1336 to 97.3638 on Wisconsin dataset and from 93.7063 to 97.9021 on BCDR-D01 dataset in case of k-NN occurred. Finally an increase from 97.891 to 97.9123 on Wisconsin dataset and from 91.6084 to 95.1049 on BCDR-D01 dataset in case of SVM occurred as visible from the above Table 1 and Fig. 2. The corresponding misclassification of all the four algorithms also decreased accordingly.

Table 1 Accuracy improvement on Wisconsin and BCDR-D01 datasets on four algorithms

Accuracy				
	Initial Wisconsin dataset (%)	Modified Wisconsin dataset (%)	Initial BCDR-D01_Dataset (%)	Modified BCDR-D01_Dataset (%)
Naive Bayes	92.6186	97.0123	87.4126	89.5105
J48	92.9701	96.8366	80.4196	90.9091
k-NN	96.1336	97.3638	93.7063	97.9021
SVM	97.891	97.9123	91.6084	95.1049

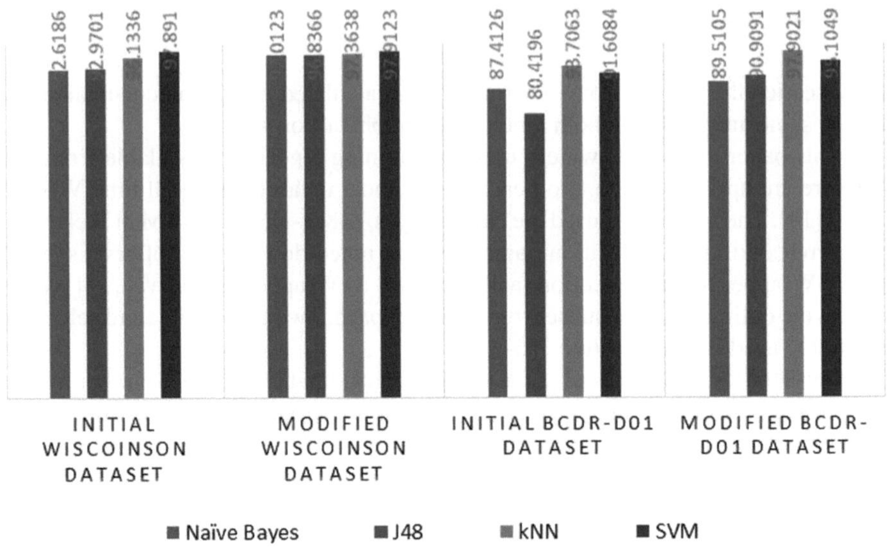

Fig. 2 Graphical representation of accuracy improvement on the modified datasets

5 Discussion

Mining of Data and Big data are the hot research topics of the last few decades and Machine Learning is the backbone for framing up a practical shape to all the concepts related to these. After the industrial revolution was witnessed by human history where the muscular energy of living beings was transformed into the moving engines, these mechanical systems now do the tons of amount of work in a very short amount of time, for which manually it was taking both a lot of effort and time. Carrying on these foundations, human brain was always in search of bringing about a new revolution where they were thirsty not only to produce motion in machines but also impart them with a sense to make judgment, take reasonable decisions and solve complex computational problems by using the prior experiences. This all gave birth to Artificial Intelligence where data is the biggest asset and to mine it for drawing conclusive results is done by different Classifiers.

In this work, we used the four most used data mining classifiers on the two medical science cancer datasets and corresponding results are noted down. The experiments are carried out in famous Weka tool. A Wrapper kind of dimensional reduction technique of WrapperSubsetEval is applied on both the datasets and again the same parameters are noted down. Initially Naive Bayes got a classification accuracy of 92.6186%, which after dimensional reduction reached up to 97.01235 on Wisconsin dataset. Same way Naive Bayes got classification accuracy of 87.4126%, which after dimensional reduction reached up to 89.5105%, on BCDR-D01_ Dataset. Carrying the same calculation on same datasets using J48, the results showed an increase from 92.9701 to 96.8366% in case of Wisconsin dataset and increase from 80.4196 to 90.9091% in case of BCDR-D01_ Dataset. Same way k-NN got 96.1336% as initial classification accuracy which got improved to 96.8366% in case of Wisconsin dataset and same way got initial accuracy of 93.7063% which upon modifying dataset reached up to 97.9021%, % in case of BCDR-D01_Dataset. Finally the SVM got 97.891% as initial classification accuracy which got improved to 97.9123% on Wisconsin dataset. Lastly, the SVM got 91.6084% as initial classification accuracy which got improved to 95.1049% on BCDR-D01_ Dataset.

6 Conclusions

The techniques of dimensional reduction has been of widely use in the literature. It has been proving as the promising result oriented treasure making the field of Machine Learning more mature. Lot of techniques of reducing the number of dimensions does exist and each one is showing good results on different type of Classifiers. Ranging from Linear Discriminant Analysis (LDA), Idempotent Component Analysis (ICA), Principal Component analysis (PCA), Generalized discriminant analysis (GDA), Backward Feature Elimination (BFE), Forward Feature Construction (FFC), etc. it has got a wide range in the IT world nowadays. In future, more dimensional reduction techniques using soft computing concepts are coming as these are the bridges for making the mining of data smooth.

Acknowledgements This work is partly supported by Research Fellowship of the "Visvesvaraya Ph.D. Scheme for Electronics & IT", Ministry of Electronics & Information Technology (MeitY), Government of India (GoI), Vide Grant no. PHD-MLA/4(39)/2015-16.

References

1. B.R. Matheus, H. Schiabel, Online mammographic images database for development and comparison of CAD schemes. J. Digital Imaging **24**(3), 500–506 (2011)
2. I.C. Moreira, I. Amaral, I. Domingues, A. Cardoso, M.J. Cardos, J.S. Cardoso, INbreast: toward a full-field digital mammographic database. Acad. Radiol. **19**(2), 236–248 (2012)
3. A. Nandakumar, N. Anantha, T.C. Venugopal, R. Sankaranarayanan, K. Thimmasetty, M. Dhar, Survival in breast cancer: a population-based study in Bangalore, India. Int. J. Cancer **60**(5), 1–5 (2006)
4. S. Swaminathan, *Consensus Document for Management of Breast Cancer* (Indian Council of Medical Research, New Delhi, 2016), pp. 12–20
5. D.M. Parkin, Global cancer statistics in the year 2000. Lancet Oncol. **2**, 533–543 (2001)
6. H.D. Nelso, K. Tyne, A. Naik, C. Bougatsos, B.K. Chan, L. Humphrey, Screening for breast cancer: systematic evidence review update for the US Preventive Services Task Force. Ann. Intern. Med. **151**(10), 727, 1–22 (2009)
7. L. Tabar, L. Vita, T.H.H. Chen, A.M.F. Yen, A. Cohen, T. Tot, S.Y.H. Chiu, S.L.I.S. Chen, J.C.Y. Fann, J. Rosell, H. Fohlin, R.A. Smith, S.W. Duffy, Swedish two-county trial: impact of mammographic screening on breast cancer mortality during 3 decades. Radiology **260**(3), 658–663 (2011)

8. R.R. Pollán, G.M. López, C.S. Ortega, D.G. Herrero, F.J. Valiente, R.M. Solar, P.N. González, M. Vaz, J. Loureiro, I. Ramos, Discovering mammography-based machine learning classifiers for breast cancer. J. Med. Syst. **36**(4), 2259–2269 (2012)
9. J. Diz, G. Marreiros, A. Freitas, Using data mining techniques to support breast cancer diagnosis. *New Contributions in Information Systems and Technologies*, vol 1 (Springer, Berlin, 2015), pp. 689–700
10. K. Rodenacker, A feature set for cytometry on digitized microscopic images. Anal. Cell. Pathol. **25**(1), 1–36 (2001)
11. J.S. Suri, D.L. Wilson, S. Laxminarayan, *Handbook of Biomedical Image Analysis*, vol 2 (Springer Science & Business Media, Germany, 2005)
12. Data repository for machine learning. http://archive.ics.uci.edu/ml/datasets.html. Last visited 28-10-2017
13. M.A.G. López, N. Posada, D.C. Moura, R.R. Pollán, J.M.F. Valiente, C.S. Ortega, M. Solar, D.G. Herrero, I. Ramos, J. Loureiro, T.C. Fernandes, B.M.F. Araújo, BCDR: a breast cancer digital repository, in *15th International Conference on Experimental Mechanics, FEUP-EURASEMAPAET*, Porto/Portugal, ISBN: 978-972-8826-26-02, 22–27 July (2012)
14. M. Hall, The WEKA data mining software: an update. SIGKDD Explor. **11**(1) (2009)
15. R. Kohavi, G.H. John, Wrappers for feature subset selection. Artif. Intell. **97**, 273–324 (1997)
16. S. Jeyasingh, M. Veluchamy, Modified bat algorithm for feature selection with the Wisconsin Diagnosis Breast Cancer (WDBC) dataset. Asian Pac. J. Cancer Prev. **18**, 1257–1264 (2017)
17. S.A. Josephine, K. Shannon, *Application of Data Mining Techniques in Improving Breast Cancer Diagnosis*, vol 9420 (2016), pp. 1–10
18. S. Sasikala, S.A. Balamurugan, S. Geetha, Multi Filtration Feature Selection (MFFS) to improve discriminatory ability in clinical data set. Appl. Comput. Inform. **12**, 117–127 (2016)
19. S. Sasikala, S.A. Balamurugan, S. Geetha, A novel feature selection technique for improved survivability diagnosis of breast cancer, in *2nd International Symposium on Big Data and Cloud Computing (ISBCC'15)*, vol 50. Procedia Computer Science, Elsevier (VIT University Chennai, India, 2015), pp. 16–23
20. L.T. Vinh, S. Lee, Y. Park, B.J. Auriol, A novel feature selection method based on normalized mutual information. Int. J. Appl. Intell. **37**(1), 100–120 (2011)
21. S. Lin, S. Chen, Parameter determination and feature selection for C4.5 algorithm using scatter search approach. Int. J. Soft Comput. **16**(1), 63–75 (2011)
22. X. Lu, X. Peng, P. Liu, Y. Deng, B. Feng, B. Liao, A novel feature selection method based on CFS in cancer recognition, in *IEEE 6th International Conference on Systems Biology (ISB)* (IEEE Computer Society, China, 2012), pp. 226–231
23. M.D. MonirulKabi, M.D. Shahjahan, M. Kazuyuki, A new local search based hybrid genetic algorithm for feature selection. Int. J. Neuro Comput. **74**(17), 2914–2928 (2011)
24. T. Ruckstieb, C. Osendorfer, P.V.D. Smagt (2012) Minimizing data consumption with sequential online feature selection. Int. J. Mach. Learn. Cybern. **4**(3), 235–243 (2012)

Approximation of Heaviest k-Subgraph Problem by Size Reduction of Input Graph

Harkirat Singh, Mukesh Kumar and Preeti Aggarwal

Abstract Heaviest k-Subgraph problem is to detect a subgraph of k vertices from a given undirected weighted graph G such that the sum of the weights of the edges of k vertices is maximum. Finding heaviest k-subgraph is a NP-hard problem in the literature. We have proposed an approach for approximating the solution of heaviest k-subgraph in which greedy approach is used to reduce the size of a graph which is used as input for branch and bound implementation of the heaviest k-subgraph problem.

Keywords Heaviest k-subgraph · Approximation algorithm · Branch and bound algorithm

1 Introduction

The density of a subgraph on vertex set S in a given undirected graph $G = (V, E)$ is defined as $d(S) = \frac{|E(S)|}{|S|}$, where $E(S)$ is the set of edges in the subgraph induced by S.

The problem of finding the densest subgraph of a given graph G can be solved optimally in polynomial time [1]. When we require the subgraph to have a specified size, the problem of finding a maximum density subgraph becomes NP-hard.

Massive graphs constructed with data retrieved from the web and various social networks have received remarkable interest in analyzing sub structures with the use of densest subgraph problems. Algorithms for finding dense subgraphs have been used

H. Singh (✉) · M. Kumar · P. Aggarwal
Department of Computer Science and Engineering, University Institute of Engineering and Technology, Panjab University, Chandigarh 160014, India
e-mail: harkiratsingh.tu@gmail.com

M. Kumar
e-mail: mukesh_rai9@yahoo.ac.in

P. Aggarwal
e-mail: pree_agg@pu.ac.in

C. R. Krishna et al. (eds.), *Proceedings of 2nd International Conference on Communication, Computing and Networking*, Lecture Notes in Networks and Systems 46, https://doi.org/10.1007/978-981-13-1217-5_58

in various application related to community detection, detection of events in social media data, pattern analyses in graphs, spam detection, etc. In above applications, the graph constructed from data become massive and due to this, for extracting dense subgraphs, fast algorithms are required.

In this paper, we used the greedy technique to approximate the solution of heaviest k-subgraph. The main intuition of greedy method is to always select a choice that looks best among the available options at that moment. This heuristic might not always give an optimal solution, but can be helpful in giving a near-optimal solution. In this paper greedy approach is used while constructing the graph from twitter data. We reduce the size of the graph which acts as input to branch and bound implementation [2] of heaviest k-subgraph.

2 Related Work

Heaviest k-subgraph problem is a NP-hard and there does not exist polynomial time algorithm to find an exact solution. Still there exists some of the researches on approximating heaviest k-subgraph problem. Asahiro et al. [3] describe a greedy algorithm for the heaviest k-subgraph problem. They assign weight to each of the vertices. This is equal to the total sum of the weight of the edges connecting to that vertex known as a weighed degree. They again and again remove a minimum weighted degree vertex in the currently remaining graph, until exactly k vertices are left.

Feige et al. [4] present an approximation algorithm for the heaviest k-subgraph problem using semi-definite programming. Papailiopoulos et al. [5] present an algorithm that combines spectral and combinatorial techniques. This algorithm uses vectors lying in a low-dimensional subspace of the adjacency matrix of the graph to obtain candidate subgraphs. Depend upon the spectrum of graph this algorithm comes with novel performance. Khuller et al. [6] give a 1/2-approximation algorithm for (DalkS) and show that DamkS is as hard as DkS within a constant factor. Charikar [7] describes a simple heuristic that has a 2-approximation guarantee. Nicolas et al. [8] present approximation algorithms with having time complexity moderately exponential or parameterized which relate trade-offs between approximability and complexity.

3 Proposed Technique

In this section, we presented a proposed technique for approximating the solution of heaviest k-subgraph. Flow diagram of various steps is shown in Fig. 1.

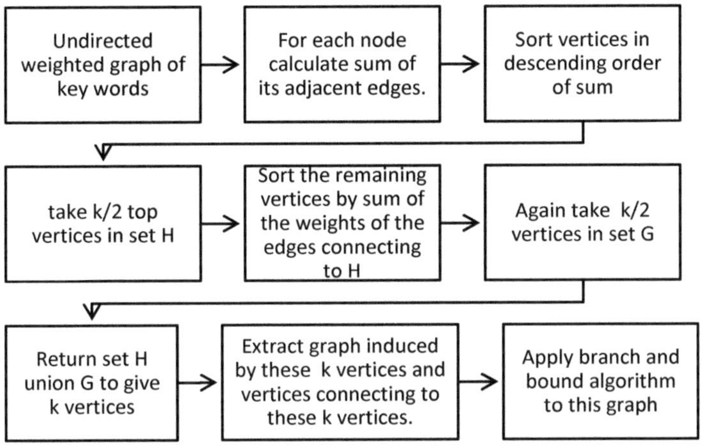

Fig. 1 Flow diagram of proposed technique

3.1 Preprocessing of Data

In this step, preprocessing of Twitter data is done. Each tweet is first tokenized and then Stanford POS Tagger is used to remove the stop words. Remaining words in the particular tweet are stored in the same sequence as in the original tweet.

Let each tweet be represented by $T_i = \{st_1, x_1, st_2, x_2, x_3, st_3 \ldots\ldots st_n, x_m\}$ which contains stop words st_n and other words x_n. After preprocessing stopwords are removed and tweet is represented by $T_i = \{x_1, x_2, x_3, \ldots, x_m\}$.

3.2 Graph Formation

Let $T = \{T_1, T_2, T_3, \ldots, T_n\}$ be the set of tweets and $T_i = \{x_1, x_2, x_3, \ldots, x_m\}$ be one of the tweets where set of x_m is the sequence of words in tweet T_i.

A graph $G = (V, E)$ be the structure specifying relationships between the words of a collection, where V corresponds to the set of words, called vertices or nodes and E is the set of relations among words. Each word x in T_i represents a node in G and an edge between any two nodes corresponds to the simultaneous occurrence of the two words in a tweet. The weight denotes the number of such instances.

Let Edge(x, y) be the boolean function which tells whether their exists edge between node x and y.

$$\text{Edge}(x, y) = \begin{cases} \text{true}, & x \text{ and } y \text{ co-occur in atleast one tweet} \\ \text{false}, & \text{otherwise} \end{cases}$$

For undirected graph, let it be G, (x, y) is an unordered pair and edge between x and y is bidirectional, thus $\text{Weight}(x, y)$ is same as $\text{Weight}(y, x)$

$$\text{Weight}(x, y) = \{w, \text{ where } w \text{ is the number of co-occurences of } x \text{ and } y\}$$

.

3.3 *Heaviest* k-*Subgraph*

Problem definition: Heaviest k-Subgraph problem is to detect a subgraph of k vertices from a given undirected weighted graph G such that the sum of the weights of the edges of k vertices is maximum.

3.4 *Approximation of Heaviest* k-*Subgraph*

The main idea of this method is to reduce the size of the graph which acts as input to branch and bound method. We used a greedy approach to select vertices of the graph. We did not pick directly top k weighted degree vertices rather we used some technique to reduce the size of the graph. We first calculate the weighted degree of each of the vertices. Then we take top $k/2$ weighted degree vertices in set H and sort the remaining vertices by the sum of the weights of the edges connecting to set H. From that sorted vertices, again we take top $k/2$ vertices in set G. In next step, we extract graph induced by these k vertices of set H and G and vertices connecting these k vertices. According to our approach large graph is reduced to a small size on which branch and bound implementation of heaviest k-subgraph can give an approximate solution. In next section, we have presented pseudocode of the above approach.

3.5 *Proposed Algorithm*

1. Input: A graph $G(V, E)$ and integer k
2. Output: Approximate heaviest k-subgraph of $G(V, E)$
3. Let S is an array that contains sum of the weights of the edges connecting each node
4. for each node u in V do
5. S[u] = sumOfAdjacentEdges(u);
6. $S = \textit{sortDescendingOrderByWeight}(S)$;
7. Set H;
8. //remove top $k/2$ from S and put in set H
9. for i = 1 to k/2

```
10.         H[i] = removeTop(S);
11. //Put remaining element from in Set R
12. R = V − H;
13. //for each node in R calculate sum of weight of edges connecting to set H
14. For each node u in R
15.         SumR[u] = calculateSumOfNeighbourInH(u);
16.         SumR = sortDescendingOrderBySum(SumR);
17. /remove top k/2 from SumR and put in set G
18. for i = 1 to k/2
19.         G[i] = removeTop(SumR);
20. Set K = H ∪ G;
21. //Extract graph induced by vertices of set k and vertices connecting to these k
    vertices.
22. Graph G′ = inducedGraph(G, K);
23. //apply branch and bound on graph G′
24. k_subGraph = branchAndBound(G′, k);
25. Return k_subGraph;
```

4 Experiments and Results

We have used Twitter Streaming API and Twitter4J (an unofficial Java library for Twitter API) to collect a set of tweets. We use these samples for our proposed algorithm for approximating heaviest k-subgraph problem. The algorithms were implemented in Java and were run on a machine with 2.53 GHz clock.

4.1 Results

In this section, we compare the result of approximation algorithm of a heaviest k-subgraph problem with exact branch and bound algorithm of heaviest k-subgraph problem. We compare running time and solution of the algorithm, i.e., weight of heaviest k-subgraph.

Figure 2 shows the running the time of approximate algorithm with exact branch and bound. It shows that exact branch and bound runs faster when the value of k is smaller, i.e., up to 15. When the value of k get increase more than a certain value (here more than 15), the approximate algorithm runs faster. Running time of exact branch and bound algorithm increase exponentially after a certain value of k. After the value of $k = 20$, the time of exact branch and bound explode. But on the other hand approximate algorithm Twitter able to solve the problem of large k without exploding.

In Fig. 3 we compared solution of the heaviest k-subgraph problem, i.e., the weight of k-subgraph returned by both of the algorithms. Branch and bound algorithm [1]

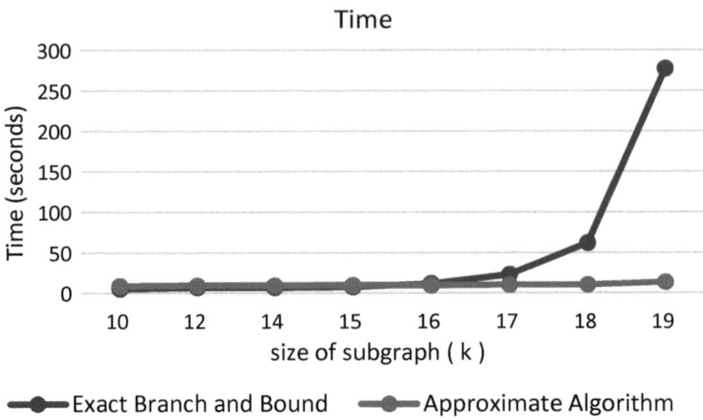

Fig. 2 Time comparison

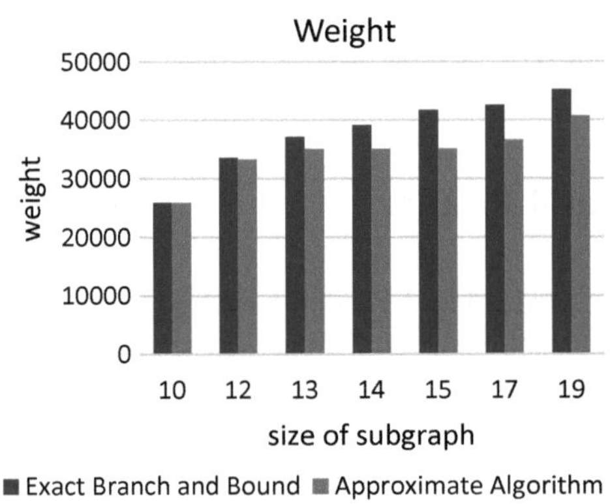

Fig. 3 Weight comparison

gives the exact solution of the heaviest k-subgraph problem whereas approximate algorithm approximates this value. According to results if we compare the ratio of the approximate solution with the exact solution, it shows that most of the time, the ratio is above 0.9.

After analyzing results of Figs. 2 and 3, we can say that approximate algorithm can run efficiently for a large value of k while maintaining the ratio of approximate solution to exact solution more than 0.9 most of the times. After analyzing results of Figs. 2 and 3, we can say that approximate algorithm can run efficiently for large value of k while maintaining the ratio of approximate solution to exact solution more than 0.9 most of the times.

5 Conclusion

We have proposed an approximating algorithm for a heaviest k-subgraph problem where the main idea is to reduce the size of the graph which acts as input for branch and bound algorithm. Results show that as the value of k is increased approximate algorithm runs faster with a ratio of approximate solution to exact solution above 0.9 most of the times. Future work includes improvement of this ratio so that approximate algorithm give the more accurate result.

References

1. A.V. Goldberg, Finding a maximum density subgraph. Technical report (1984)
2. M. Letsios, O.D. Balalau, M. Danisch, E. Orsini, M. Sozio, Finding heaviest k-subgraphs and events in social media, in *2016 IEEE 16th International Conference on Data Mining Workshops (ICDMW)*, Barcelona (2016), pp. 113–120
3. Y. Asahiro, K. Iwama, H. Tamaki, T. Tokuyama, Greedily finding a dense subgraph. J. Algorithms **34**(2), 203–221 (2000)
4. U. Feige, *The Dense k-Subgraph Problem 1 Introduction* (1999)
5. D. Papailiopoulos, I. Mitliagkas, A. Dimakis, C. Caramanis, Finding dense subgraphs via low-rank bilinear optimization, in *Proceedings of the 31st International Conference on Machine Learning (ICML-14)* (2014), pp. 1890–1898
6. S. Khuller, B. Saha, On finding dense subgraphs, in *Icalp*, vol 5555 (2009), pp. 597–608
7. M. Charikar, Greedy approximation algorithms for finding dense components in a graph, in *Approximation Algorithms for Combinatorial Optimization* (2000), pp. 84–95
8. N. Bourgeois, A. Giannakos, G. Lucarelli, I. Milis, V.T. Paschos. Exact and superpolynomial approximation algorithms for the densest k-subgraph problem. Eur. J. Oper. Res. **262**, 894–903

Approximate Computing for Machine Learning

Vinay Kumar and Ravi Kant

Abstract Inexact computing is an attractive paradigm when it comes to trading off computation quality with effort expended. Many applications are error-tolerant and can produce results with a slight shift in accuracy and a large reduction in computations. In this article, we present the techniques for Approximate Computing. Also, strategies have been discussed to find approximation amenable code and maintain the tolerance level of a machine learning algorithm. This article immediately aims to provide insights to research scientists about Approximate Computing techniques and motivate more works in this area in near future. Approximate computing is all about adapting to application accuracy needs and providing savings over power and energy.

Keywords Inexact computing · Approximation · Machine learning
Data mining · Error-tolerant · Accuracy

1 Introduction

There is an enormous amount of data to be processed. On the contrary, we have limited resources to make the same possible. Inexact computing is one mechanism that can be used to sacrifice minimal accuracy and gain tremendous energy savings. Having an answer with a precision of eight to nine decimal places is not always profitable. It is equivalent to a solution correct to three or four decimal places. It is very crucial to realize the fact that there are costs, in terms of expended energy to come at the more accurate answer. At the same time, special care should be taken for the tolerance level of the application. In previous researches, it is shown how

V. Kumar (✉)
Department of ECE, NIT Meghalaya, Shillong, India
e-mail: Vinay.kumar@nitm.ac.in

R. Kant
Department of IT, IIIT Bhubaneswar, Bhubaneswar, India
e-mail: B414016@iiit-bh.ac.in

C. R. Krishna et al. (eds.), *Proceedings of 2nd International Conference on Communication, Computing and Networking*, Lecture Notes in Networks and Systems 46, https://doi.org/10.1007/978-981-13-1217-5_59

Isaac Newton's numerical methods could help us achieve the same. Using Newton's calculations, one can save energy by calculating the necessary precision instead of the ultimate precision. There are many machine learning algorithms, which can be applied using inexact computing approach. This can reduce enormous amount of power usage by sacrificing a slight accuracy. This paper is inspired from the idea that the learning model's intrinsic robustness to noise may be leveraged to relax certain constraints on the underlying hardware.

1.1 Overview of Approximate Computing

The idea of approximate computing is simple: deliberately reduce accuracy to save energy, memory, and/or time. It is paradigm to take reducing accuracy as an opportunity not a loss. Different techniques of implying approximate computing have emerged over the past few years, exploring the space of possibilities: numerical approximations, neural network accelerators, unreliable hardware, etc. The biggest challenge is to bring this concept down to every down programmers in a principled way while ensuring a certain level of required correctness.

Approximate computing is different from conceptually related paradigms like probabilistic computing or stochastic computing. Approximate computing does not involve assumptions on the stochastic nature of any process implementing the system. However, it uses the statistical nature of data and algorithms to trade quality for power savings. Probabilistic computing is briefly explained in [1].

1.2 Areas of Implementation

Approximate computing is highly used in neural network accelerators. Neural network is generally used in error-tolerant applications. There are some researches that propose a number of techniques to approximate them. Neural network shows parallelism and can be accelerated using dedicated hardware. There are many quality metrics like pixel difference in images, classification and clustering of data, ranking accuracy, etc. which can be a subject to inexact computing. Universal image quality index and satisfiability check are among other quality metrics. For several applications, there are multiple performance metrics that can be used for calculating quality loss, for example, both accuracy of clustering and mean centroid distance can be used as a performance metrics for k-means clustering. Other areas of implementation include image processing and MapReduce. Approximate computing also contributes in the field of various devices and components like simulators, analytical models, CPUs, and GPUs. Shoushtari et al. [2] presented an inexact computing technique for cache memory, ultimately SRAM cells.

2 Insights of Approximate Computing in Machine Learning

There are many implications of inexact computing in the huge domain of machine learning where one can trade off a slight accuracy to huge energy savings. The concept of machine learning in itself is a probabilistic model which leads to approximate results. Trading accuracy in here is havoc but when it comes to processing of huge amount of data, such measures save a lot of energy at the cost of sacrificing slight accuracy.

Several algorithms in machine learning work on the principle of approximation and fetch efficient results over any dataset.

2.1 *Algorithms and Usage*

There are many applications like data compression, data mining, pattern classification, or pattern recognition which involves clustering. There are many proposed methods for the same including one on isodata [3, 4], and some randomized approaches like in Clara [5], Clarans [6], and modules based on neural nets [7].

***K*-means Clustering**: *It is one of the most famous clustering algorithms in use and its most famous heuristic is called Lloyd's algorithm* [8–10]. In a loop, it computes the neighborhood of a center point and then migrates to the centroid of its neighborhood, until some given convergence criterion is satisfied. Ultimately, it converges to local optimum [11]. The computation of neighbors in Lloyd's algorithm is the most expensive step but a lot of implementation for the same have been found lately [12–16] (Fig. 1).

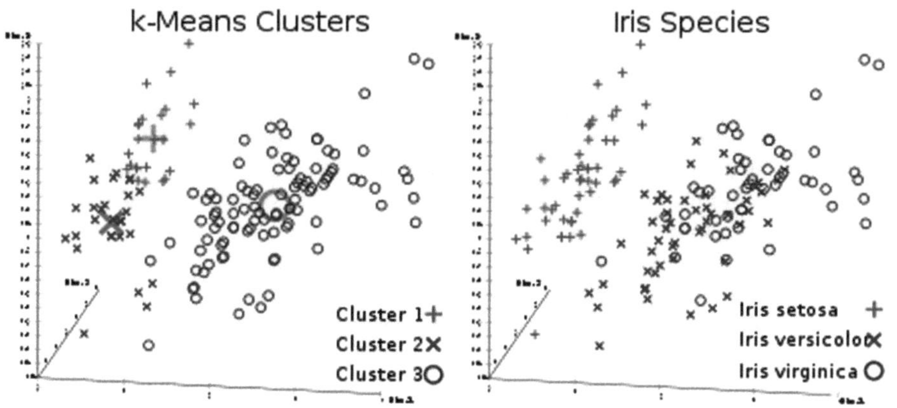

Fig. 1 *K*-means clustering

There has always been a classic tradeoff between approximation and execution time. There exist many clustering algorithms that are able to produce outputs nearer to optimal solution. Arora et al. and Kolliopoulos in their paper mentioned about $(1+\varepsilon)$ approximation algorithm for the Euclidean space problem [17, 18]. The running time of the approximation algorithm in the second paper was found to be $O\ (2^{1/}\varepsilon^{d}\ n \log n \log k)$ with a fixed dimension d. Agarwal and Procopiuc gave another $(1+\varepsilon)$-approximation algorithm which runs with a complexity of

$$O(n \log k) + (k/\varepsilon)^{O(k^{l-1/d})}.$$

An important breakthrough was achieved by Matousek [19] when he proposed an asymptotically efficient approximation algorithm which runs with a complexity of

$$O(n(\log\ n)^{k}\varepsilon^{-2k^{2}d}) \text{ time for fixed } k \text{ and } d.$$

Other approaches are out there which involve approximation algorithms to develop more practical and efficient algorithms having weaker, but still constant factor of approximation. Thorup worked on solving location problems in sparse graphs using similar metrics. Therefore, the analysis is different and it relies on geometric properties very particular to k-means problem. With an accuracy loss of 5%, nearly 550 times energy savings could be achieved [20]. A k-means clustering algorithm, up to 50× energy saving, can be achieved by allowing a classification accuracy loss of 5% [20]. Similarly, incurring an error up to 5%, using neural approximation technique, an inverse kinematic application could accelerate up to 26 times compared to GPU execution [21].

Neural Networks: It is a common observation that neural networks are used in error-tolerant applications, and there are some research papers that propose techniques to approximate them. Using backpropagation as a technique, Venkataranami et al. tried to approximate neural network which is used for training neural networks and quantifying the impact of approximation over any neuron. The neurons having the least impact were further replaced by their approximate versions to create the approximate model of the neural network.

Zhang et al. also gave a technique for approximating the neural networks. They defined the neurons as resilient and a small deviation from its original computation would lead to a drastic change in the output quality. A theoretical approach was presented to find the resilience factor in each hidden or output layer. The neurons with the least resilience are subject to approximation.

Du et al. proposed a hardware neural network accelerator which was based on the fact that neural networks allow iterative training which further allows suppressing the impact of the all the neurons producing large error. Their inexact neural network accelerators provided significant savings in energy, delay, and area.

3 Strategies of Approximate Computing

A number of strategies exist in order to approximate any model which majorly includes precision scaling, loop perforation, load value approximation, voltage scaling, memory access skipping, lossy compression, using memorization, use of neural networks, data sampling, etc. [22]. Some of these strategies are explained in brief.

Precision Scaling: There are several approximate computing techniques that change the precision width of input data to reduce storage as a requirement.

Loop Perforation: There are techniques that use loop perforation as an approach, which is basically skipping some looping steps to reduce computational overheads.

Load Value Approximation: A load miss in the cache leads to the data being fetched from the main memory or the next level cache which incurs high latency. Load value approximation leverages the approximable nature of the applications to approximate load values, thus hiding the cache miss latency.

Using Memoization: The process of storing the results of functions for later use and reuse with same input is termed as memoization.

There could be several strategies to execute inexact computing and use approximation in hardware design and software modules. Several researches are being done to bring about the change in computation using the principle of approximation.

4 What Else to Approximate?

There are ways to find approximable code segments or critical sections, i.e., control flow in a program in order to make the approximate model. Identification of approximation amenable code is important and tolerance level of the application should be analyzed before implying inexact computing. Approximation of one code segment always affects the other. Overall benefits of approximation are as follows:

- Have high performance/energy savings.
- Be flexible to adapt to application accuracy needs.
- Have minimal hardware overhead.

Approximate computing is further implemented in the following areas:

1. Image processing and multimedia,
2. Signal processing,
3. Machine learning,
4. Scientific computing,
5. Database search, and
6. MapReduce.

This can further be extended to many other streams with extensive research and experimental efforts.

5 Conclusion

The opportunities and obstacles in the field of approximate computing have been highlighted in the article. Thus, this article can be concluded with a brief presentation of the challenges underlying in this field. The existing approximate computing techniques have focused on multimedia genre or iterative algorithms. These error-resilient loads comprise only a part of the computational workloads. When these "low-hanging fruits" gradually vanish, researchers must turn their attention to general-purpose problems and optimal solutions, thus extending the scope of approximate computing to the complete spectrum of computing theory and applications.

References

1. K. Palem, A. Lingamneni, What to do about the end of Moore's law, probably!, in *Proceedings DAC* (2012), pp. 924–929
2. M. Shoushtari et al., Exploiting partially-forgetful memories for approximate computing. IEEE Embed. Syst. Lett. (2015)
3. G.H. Ball, D.J. Hall, Some fundamental concepts and synthesis procedures for pattern recognition preprocessors, in *International Conference on Microwaves, Circuit Theory, and Information Theory*, Tokyo, Japan, Sept 1964
4. A.K. Jain, R.C. Dubes, *Algorithms for Clustering Data* (Prentice Hall, Englewood Cliffs, NJ, 1988)
5. L. Kaufman, P.J. Rousseeuw, *Finding Groups in Data: An Introduction to Cluster Analysis* (Wiley, New York, 1990)
6. R.T. Ng, J. Han, Efficient and effective clustering methods for spatial data mining, in *Proceedings of the Twentieth International Conference on Very Large Databases*, Santiago, Chile, Sept 1994, pp. 144–155
7. T. Kohonen, *Self-Organization and Associative Memory*, 3rd edn. (Springer, New York, 1989)
8. E. Forgey, Cluster analysis of multivariate data: efficiency vs. interpretability of classification. Biometrics **21**, 768 (1965)
9. S.P. Lloyd, Least squares quantization in PCM. IEEE Trans. Inf. Theor. **28**, 129–137 (1982)
10. J. MacQueen, Some methods for classification and analysis of multivariate observations, in *Proceedings of the Fifth Berkeley Symposium on Mathematical Statistics and Probability*, vol. 1. Berkeley, CA (1967), pp. 281–296
11. S.Z. Selim, M.A. Ismail, K-means-type algorithms: a generalized convergence theorem and characterization of local optimality. IEEE Trans. Pattern Anal. Mach. Intell. **6**, 81–87 (1984)
12. K. Alsabti, S. Ranka, V. Singh, An efficient k-means clustering algorithm, in *Proceedings of the First Workshop on High Performance Data Mining*, Orlando, FL, Mar 1998
13. T. Kanungo, D.M. Mount, N.S. Netanyahu, C. Piatko, R. Silverman, A.Y. Wu, An efficient k-means clustering algorithm: analysis and implementation. IEEE Trans. Patt. Anal. Mach. Intell. **24** (2002) (to appear)
14. S.J. Phillips, Acceleration of k-means and related clustering problems, in *Algorithm Engineering and Experiments* (*Proceedings ALENEX '02*), ed. by D.M. Mount, C. Stein. Lecture notes computer science, vol. 2409 (Springer, Berlin, 2002) (to appear)
15. D. Pelleg, A. Moore, Accelerating exact k-means algorithms with geometric reasoning, in *Proceedings of the ACM SIGKDD International Conference on Knowledge Discovery and Data Mining*, San Diego, CA, Aug 1999, pp. 277–281
16. D. Pelleg, A. Moore, k-means: extending k-means with efficient estimation of the number of clusters, in *Proceedings of the Seventeenth International Conference on Machine Learning*, Palo Alto, CA, July 2000

17. S. Arora, P. Raghavan, S. Rao, Approximation schemes for Euclidean k-median and related problems, in *Proceedings of the Thirtieth Annual ACM Symposium on Theory of Computing*, Dallas, TX, May 1998, pp. 106–113
18. S. Kolliopoulos, S. Rao, A nearly linear-time approximation scheme for the Euclidean k-median problem, in *Proceedings of the Seventh Annual European Symposium on Algorithms*, ed. by J. Nesetril. Lecture notes computer science, vol. 1643 (Springer, Berlin, 1999), pp. 362–371
19. J. Matousek, On approximate geometric k-clustering. Discrete Comput. Geom. **24**, 61–84 (2000)
20. V.K. Chippa, S. Venkataramani, S.T. Chakradhar, K. Roy, A. Raghunathan, Approximate computing: An integrated hardware approach, in *Asilomar Conference on Signals, Systems and Computers*, Pacific Grove, CA, 2013, pp. 111–117
21. B. Grigorian, N. Farahpour, G. Reinman, BRAINIAC: Bringing reliable accuracy into neurally-implemented approximate computing, in *IEEE 21st International Symposium on High Performance Computer Architecture (HPCA)*, Burlingame, CA, 2015, pp. 615–626
22. S. Mittal, A survey of techniques for approximate computing. ACM Comput. Surv. **48** (2016)

Experimental Evaluation of Nature-Inspired Algorithms on High Dimensions

Manisha Singla and K. K. Shukla

Abstract This paper concentrates on four very similar metaheuristic optimization algorithms: Differential Evolution (DE), Genetic Algorithm (GA), Particle Swarm Optimization (PSO), and Cuckoo Search (CS) algorithm. These optimization algorithms are used to solve optimization problems with real parameters having real parametric functions. This paper gives a brief discussion of these algorithms followed by the experiment over various benchmark functions. Many researchers have attempted to compare these algorithms on various benchmark functions. This work compares these algorithms on high dimensions over benchmark functions like Ackley's function, Alpine function, Brown function, Deb function, and Powell sum function. These above algorithms are compared on the basis of time required to converge on various benchmark functions. Our experiments indicate that the CS algorithm outperforms others when the dimensions are high, whereas in some cases, it is comparable to DE.

Keywords Metaheuristic optimization · Benchmark functions · Genetic algorithm · Differential evolution · Particle swarm optimization and cuckoo search

1 Introduction

Optimization problem deals with optimization algorithms by minimizing or maximizing the objective function in the given problem. Population-based metaheuristics methods and the numerical methods are the two categories of optimization algorithms where numerical-based optimization algorithms solve the problems having smaller search space and also in a finite amount of time [1]. This limitation of numerical-based optimization is solved by population-based optimization algorithms. These algorithms can also deal with the problems having complex and comparatively larger

M. Singla (✉) · K. K. Shukla
Indian Institute of Technology (BHU), Varanasi, India
e-mail: manishasigla.rs.cse17@itbhu.ac.in

K. K. Shukla
e-mail: kkshukla.cse@iitbhu.ac.in

C. R. Krishna et al. (eds.), *Proceedings of 2nd International Conference on Communication, Computing and Networking*, Lecture Notes in Networks and Systems 46, https://doi.org/10.1007/978-981-13-1217-5_60

search space. These are having high potential to solve harder and complex real-world problems. It is also demonstrated earlier that population-based algorithms provide much better and accurate results than numerical methods.

Earlier researchers have compared some of the algorithms on the basis of their accuracy and time taken by the algorithm. By modifying some of the parameters in the algorithm, this work produces a comparison among the abovementioned optimization algorithms on the basis of the time taken by the algorithms over various benchmark functions *when the solution space is high dimensional.*

2 Description of Optimization Algorithms

Various population-based optimization algorithms are discussed here which minimize or maximize the objective function by applying various functional constraints. Introduction to these algorithms is as shown below:

2.1 Genetic Algorithm (GA)

Genetics and biology are the reasons for the development of genetic algorithms (GA). The element x_k, $k = 1, \ldots, D$, is considered as a gene with a candidate solution x_j, $j = 1, \ldots, N$ and the set of candidate solution is considered as the population in the language of GA algorithm. This algorithm works in such a way that the candidates are randomly chosen at the beginning of the algorithm and the new candidate solutions or the child are generated by modifying the parent. The basic steps followed for implementing genetic algorithm are described in Algorithm 1. GA is the evolutionary algorithm which is invented by John Holland in 1975 [2]. The original GA comprises three genetic operators, i.e., selection, crossover, and mutation. First, parents are selected in the selection stage. The selection operator randomly selects the individuals and then compares these individuals on the basis of their fitness values. From these selected individuals, new individuals or chromosomes are generated with the help of crossover and mutation operations.

Algorithm 1

1. Randomly select individuals of a population.
2. Evaluate fitness value of the population.
3. while(!termination condition) do
4. Select the individuals with best fitness value.
5. Perform crossover and mutation to generate new individuals.
6. Evaluate fitness values of new individuals.
7. Least-fit individuals are replaced by new individuals.

2.2 *Differential Evolution (DE)*

DE is a considered as another important and powerful real-parameter based optimization algorithm which is first introduced by price and Storn and described in [3]. Algorithm of DE is as shown below in Algorithm 2 [4].

Algorithm 2

```
Randomly select individuals (chromosomes) of the population.
Evaluate fitness value of the population.
while(!termination condition) Do
  for (P[i]∈ Population)
        A[i]=New(P[i], Population, popSize, weight, CR)
        if (fitness(A[i]<fitness(P[i])
            NewPopulation=A[i]
        else
            NewPopulation=P[i]
          endif
    end
    Population=NewPopulation
    fitness(Population)
end
```

2.3 *Particle Swarm Optimization (PSO)*

PSO is the algorithm which is first proposed by Eberhart and Kennedy in 1995 [5]. PSO algorithm consists of the set of particles and these particles move around their search space on the basis of their past (best) location [6]. The algorithm of the PSO algorithm is as shown in Algorithm 3 [7].

2.4 *Cuckoo Search (CS)*

Cuckoo search algorithm gets its inspiration from the activities of the cuckoo bird [8]. This algorithm uses the following representations: Each egg of a nest is a solution, and a cuckoo egg is considered as a new solution. The main aim of this algorithm is to make use of the better solutions (cuckoos) so as to replace other solutions in the nests. This algorithm is based on the below-described rule:

1. At a time only one egg is laid by Cuckoo bird and dumped in a randomly chosen nest.
2. The nest which is having best quality eggs is taken to the next generation.
3. Host nests are fixed in number and if the host bird finds the egg of cuckoo bird, then that egg is not considered in further calculations.

Algorithm 3

```
Randomly select individuals (chromosomes) of the population.
Initialize P_gbest=0
for(i=1 to popSize)
    Assign p_velocity to each chromosome randomly.
    P_position=Random(popSize)
    P_lbest=P_position
    if(fitness(P_lbest)<fitness(P_gbest))
            P_gbest=P_lbest
    endif
end
while(!terminationCondition) do
    for(chromosome ∈ Population)
            P_velocity=update_velocity (p_velocity, P_gbest, P_lbest)
            P_position=update_position (P_position, P_velocity)
            if (fitness(p_position)<fitness(P_lbest))
                    P_lbest=p_position
                    if(fitness(P_lbest)<fitness(P_gbest))
                        P_gbest=P_lbest
                    endif
            endif
    end
end
return(p_gbest)
```

Deb and Yang also discovered that rather than random walk, search style by levy flights is much better [8]. The complete pseudocode of CS algorithm is described below [9].

Algorithm 4

```
begin
Choose an objective function f(x)
Randomly generate initial population of n hosts nest.
Evaluate fitness value and rank them.
```

```
while(k>MaxGen) or Stop criterion
        k=k+1
    Obtain a cuckoo randomly or generate new solution by Levy flights.
Evaluate fitness_val (i).
        Choose a random nest j
        if(fitness_val(i)>fitness_val(j))
            Replace j by new solution.
        end if
Evaluate fitness value, rank the eggs and find the current best solution.
end while
end
```

3 Experiments

Experiments are performed on various dimensions over abovementioned functions. These are performed in Anaconda using Python language over system with 4 GB RAM. Here, these algorithms are compared for dimensions like $D = 5$, 20, 50, and 100. These benchmark functions are defined below with their results over above-mentioned metaheuristic algorithms. In the below-described experiments, tuning parameters considered in case of GA are crossover rate (CR) and mutation rate (MR) equals 0.5 and 0.5, respectively. In case of DE, CR considered is 0.9 and value of inertia is 0.5. In case of PSO, acceleration constants carry a value of 1.5, inertia weight taken is 0.8, velocity lower bound is −1, and velocity upper bound is +1. In case of CS, discovery rate considered during experiments is 0.25.

Ackley function

This function is generally used to solve evolutionary algorithms, and this n-dimensional function is having more than one local minima and only one global minima at $x = 0$ and having value = 0. The Ackley's function is shown by Eq. 1 [10]:

$$f(x) = 20 * e^{-0.02\sqrt{\frac{1}{D}\sum_{i=1}^{D} x_i^2}} - e^{\frac{1}{D}\sum_{i=1}^{D}\cos(2\pi x_i)} + 20 + e \tag{1}$$

The two-dimensional (for two variables) representation of the Ackley function is as shown in Fig. 1 [11]. This benchmark function is implemented by abovementioned metaheuristic algorithms over various dimensions ($D = 5$, 20, 20, and 100). Table 1 shows the evaluation time taken by these algorithms, and Fig. 2 represents the graphical comparison of these algorithms on different dimensions (with x-axis representing various dimensions $D = 5$, 20, 50, and 100). It is observed that with the abovemen-

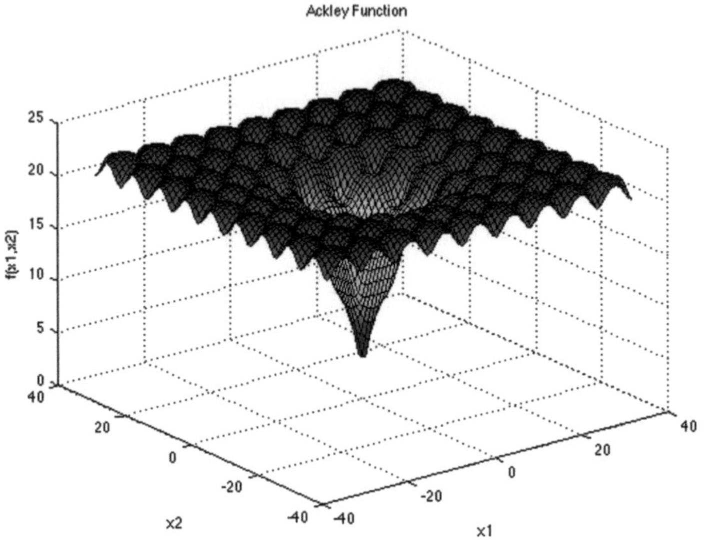

Fig. 1 Ackley function for two variables

Table 1 Evaluation time (in sec) needed over Ackley function for various dimensions

Algorithm	Dim = 5	Dim = 20	Dim = 50	Dim = 100
GA	6.31	6.16	12.08	21.25
DE	1.55	6.37	6.73	12.85
PSO	1.83	6.2	12.04	20.03
CS	2.078	3.375	7.094	11.964

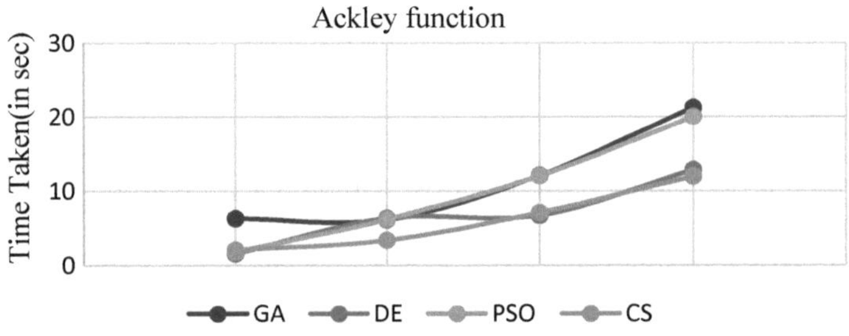

Fig. 2 Comparison of various algorithms over Ackley function

tioned parameters of these algorithms, cuckoo search algorithm outperforms other algorithms except DE at some points.

Alpine function

Alpine is a differentiable function and is defined over *n*-dimensional space. Also, this function is not convex. Pictorial representation of Alpine function is shown in

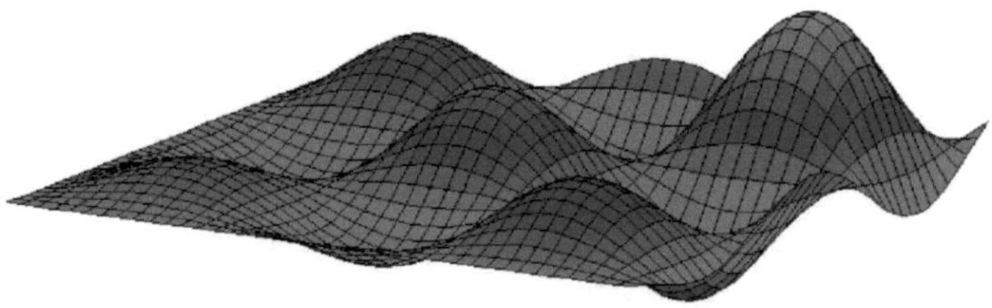

Fig. 3 Alpine function for two variables

Table 2 Evaluation time (in sec) needed over Alpine function for various dimensions

Algorithm	Dim = 5	Dim = 20	Dim = 50	Dim = 100
GA	6.4	6.33	12.4	19.36
DE	1.5	6.39	6.77	12.07
PSO	1.67	6.31	11.73	17.87
CS	1.703	3.157	6.172	11.261

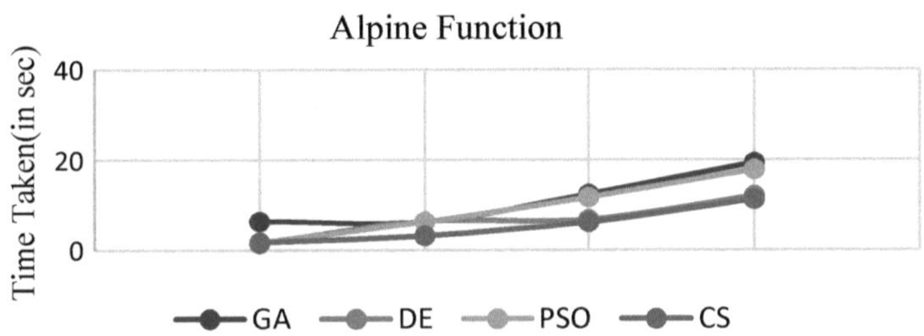

Fig. 4 Comparison of various algorithms over Alpine function

Fig. 3 [12]. Alpine function is defined in Eq. 2 [12], and its results with different dimensions corresponding to various metaheuristic algorithms are shown in Table 2.

The corresponding graphical comparison of these algorithms over alpine function is shown in Fig. 4 (with x-axis representing various dimensions $D = 5$, 20, 50, and 100). Alpine function has its global minimum at (0, 0) having its global minimum value 0.

The equation for the alpine function is as shown below:

$$f(x) = \sum_{i=1}^{D} |x_i \sin(x_i) + 0.1x_i| \tag{2}$$

It can be observed that here also cuckoo search algorithm outperforms other algorithms.

Brown function

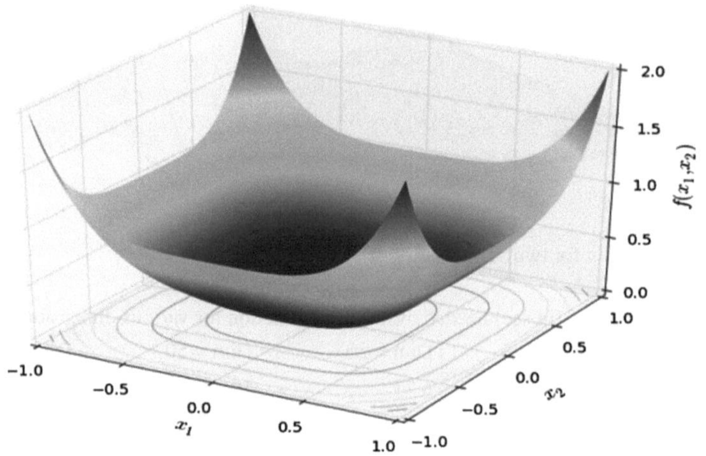

Fig. 5 Brown function for two variables

Table 3 Evaluation time (in sec) needed over Brown function for various dimensions

Algorithm	Dim = 5	Dim = 20	Dim = 50	Dim = 100
GA	6.28	7.02	13.72	14.1
DE	1.81	6.32	10.31	10.79
PSO	1.86	6.17	11.6	12.09
CS	2.641	6.844	15.251	25.808

The equation for brown function is as shown below and pictorial representation of Brown function is shown in Fig. 5 [13]:

$$f(x) = \sum_{i=1}^{D-1} \left(x_i^2\right)^{\left(x_{i+1}^2+1\right)} + (x_{i+1}^2)^{(x_i^2+1)} \quad (3)$$

Table 3 shows the evaluation time needed by various algorithms over Brown function for various dimensions. Cuckoo search works better for certain dimensions but not for all the dimensions here. Its graphical comparison is shown in Fig. 6 (with x-axis representing various dimensions $D = 5$, 20, 50, and 100).

Deb function

The equation for the Deb function is as shown below [14]:

$$f(x) = \frac{-1}{D} \sum_{i=1}^{D} \sin^6(5\pi x_i) \quad (4)$$

Table 4 represents the evaluation time needed by various algorithms over Deb function, and Fig. 7 shows the comparison of various algorithms over Deb function

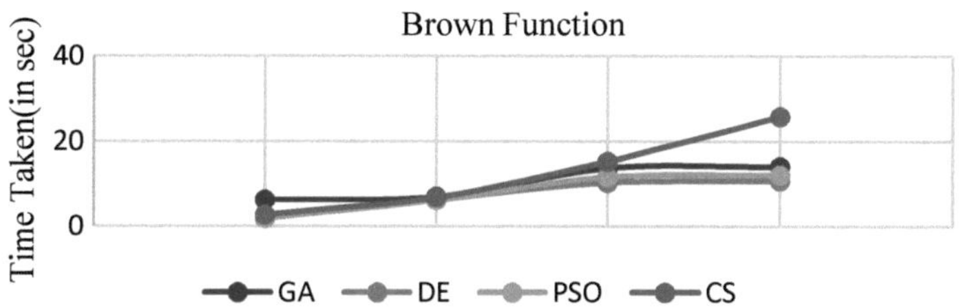

Fig. 6 Comparison of various algorithms over Brown function

Table 4 Evaluation time (in sec) needed over Deb function for various dimensions

Algorithm	Dim = 5	Dim = 20	Dim = 50	Dim = 100
GA	6.374	6.21	12.05	19.62
DE	1.54	6.38	10.59	6.69
PSO	1.56	6.25	12.21	17.89
CS	1.713	2.635	4.563	7.856

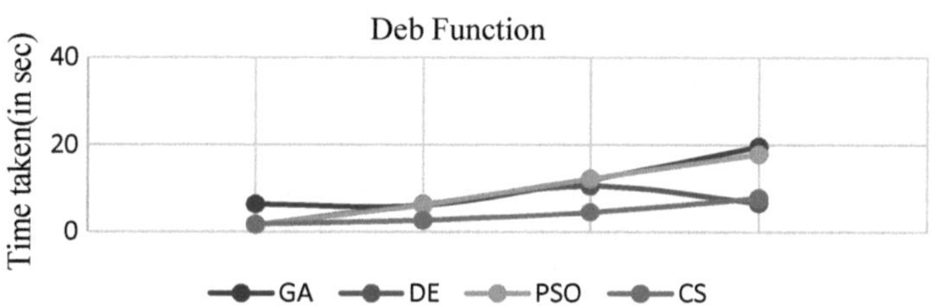

Fig. 7 Comparison of various algorithms over Deb function

graphically (with x-axis representing various dimensions D = 5, 20, 50 and 100). Here, also CS outperforms other algorithms.

Powell sum function

The equation for the Powell sum function is as shown below [14]:

$$f(x) = \sum_{i=1}^{D} |x_i|^{i+1} \tag{5}$$

Figure 8 shows the graphical representation of the comparison of various algorithms over Powell sum function (with x-axis representing various dimensions D = 5, 20, 50, and 100) shown in Table 5.

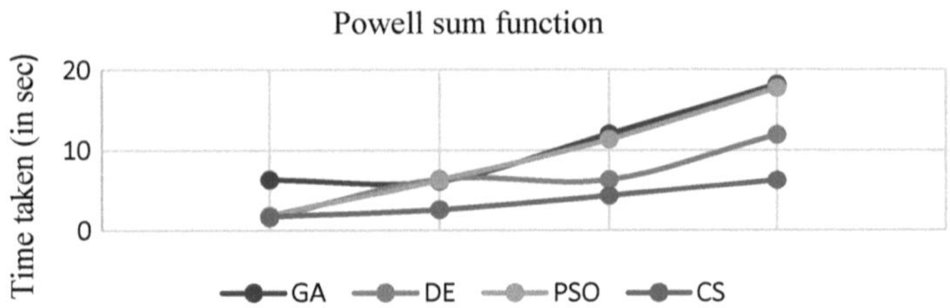

Fig. 8 Comparison of various algorithms over Powell sum function

Table 5 Evaluation time (in sec) needed over Powell sum function for various dimensions

Algorithm	Dim = 5	Dim = 20	Dim = 50	Dim = 100
GA	6.34	6.2	12.03	18.16
DE	1.59	6.42	6.35	11.92
PSO	1.9	6.31	11.36	17.73
CS	1.703	2.578	4.328	6.262

4 Conclusions and Future Scope

From the above experiments, it is observed that CS algorithm works better than other algorithms for higher dimensions (in some cases it is comparable to DE) except when it is implemented for Brown function. Here, these comparisons are made on certain parameters of these algorithms. These comparisons are made on the standard metaheuristic algorithms but their modified versions can also be compared or standard algorithms with some modifications in parameters can also be compared.

References

1. E.K. Nyarko, R. Cupec, D. Filko, A comparison of several heuristic algorithms for solving high dimensional optimization problems. Int. J. Electr. Comput. Eng. Syst. **1**, 1–8 (2014)
2. D. Whitley, A genetic algorithm tutorial. Stat. Comput. 65–85 (1994)
3. Differential evolution. Available at: https://en.wikipedia.org/wiki/Differential_evolution [Online]. Accessed 9 Nov 2017
4. R. Storn, K. Price, Differential evolution—a simple and efficient heuristic for global optimization over continuous spaces. J. Global Optim. **4**, 341–359 (1997)
5. Particle swarm optimization. Available at: https://en.wikipedia.org/wiki/Particle_swarm_optimization [Online]. Accessed 10 Nov 2017
6. H. Singh, B. Singh, A comparison of optimization algorithms for standard benchmark functions. Int. J. **7** (2017)
7. J. Kennedy, Particle swarm optimization, in *Encyclopedia of Machine Learning* (Springer, US, 2011), pp. 760–766
8. "Cuckoo Search". Available at: https://en.wikipedia.org/wiki/Cuckoo_search [Online]. Accessed 10 Nov 2017

9. A.H. Gandomi, X.S. Yang, A.H. Alavi, Cuckoo search algorithm: a metaheuristic approach to solve structural optimization problems. Eng. Comput. **29**(1), 17–35 (2013)
10. S. Surjanovic, Ackley function. Available at: https://www.sfu.ca/~ssurjano/ackley.html [Online]. Accessed 10 Nov 2017
11. MathWorks. Ackley function. Available at: https://in.mathworks.com/matlabcentral/fileexchange/37000ackleyfunction?focused=5234563&tab=function [Online]. Accessed 10 Nov 2017
12. M. Clerc, Alpine function. Available at: http://clerc.maurice.free.fr/pso/Alpine/Alpine_Function.htm [Online]. Accessed 10 Nov 2017
13. Brown function. Available at: http://mathworld.wolfram.com/BrownFunction.html [Online]. Accessed 11 Nov 2017
14. X.S. Yang, *Nature-Inspired Optimization Algorithms* (Elsevier, New York, 2014)

Entropy Component Correlation Analysis for Cross Pose Face Recognition

Kumud Arora and Poonam Garg

Abstract This paper aims to shelve into correlation of entropy components from cross poses of the face. The quadratic entropy components in correspondence with the high frequencies are extracted. In this paper, we formulate CCA in semi supervised manner for finding correlations between quadratic entropy features. Correlation of components via canonical correlation analysis helps in setting up better correspondence between a discrete set of non-frontal pose and the frontal pose of same subject. Both linear and nonlinear version of CCA is explored to find correlations among cross pose features. Due to correlation finding, we can consider joint feature combinations and evaluations rather than considering each of the images individually. Experimental findings demonstrate the canonical correlation of unsupervised spectral feature selection based on information theoretic concepts performs better than correlation of Eigen faces. Also nonlinear version of canonical correlation finds correlation better. This paper majorly confines to 2D cross face recognition.

Keywords Renyi entropy · CCA (Canonical correlation Analysis) · KCCA (Kernel canonical correlation Analysis) · KECA (Kernel entropy component Analysis)

1 Introduction

Face recognition of a person across the pose different from the enrolled one still poses bottleneck for accurate recognition. A nonlinear transformation of the 2D face image happens when the face pose changes. Change in the viewpoints lead to self-occlusion of distinguishing face features. Recognition accuracy falls steeply when the probe

K. Arora (✉)
Inderprastha Engineering College, Ghaziabad, India
e-mail: Kumud.arora76@gmail.com

P. Garg
Insitute of Management and Technology, Ghaziabad, Uttar Pradesh, India
e-mail: pgarg@imt.edu

C. R. Krishna et al. (eds.), *Proceedings of 2nd International Conference on Communication, Computing and Networking*, Lecture Notes in Networks and Systems 46, https://doi.org/10.1007/978-981-13-1217-5_61

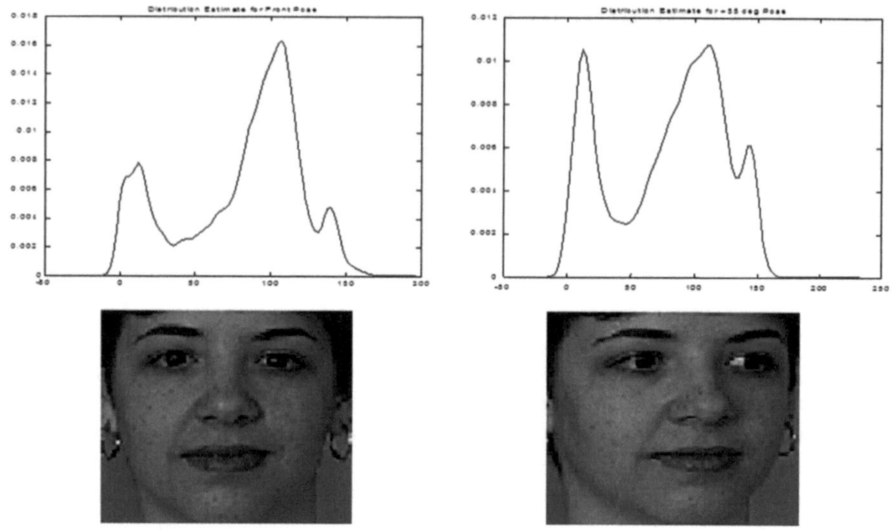

Fig. 1 Pixel distribution estimates under two face poses

face pose is different from the one that is enrolled in the database. Modeling of the relations between view subspaces of an individual from different viewpoints can be done by learning mappings from one pose to another. View subspaces from different viewpoints of the same individual are basically correlated. Figure 1 shows that the slight change in the pose causes change in the distribution estimate of the intensity values and pixel correlation reduces to 49%.

The problem of recognizing cross pose faces can be formulated as minimizing the dissimilarity measure between two vectors from different subspaces. Hoetlling [1] proposed theoretically, by using CCA method, correlations between two different sets of variables can be maximized. Thus, performing CCA on the identity coupled face poses; the latent correlation over two poses can be maximized. But due to the fact that cross pose face coupling has nonlinear relationship, the effectiveness of relation analysis using raw pixel intensities with CCA remains a bit limited. Entropy, as per information theory [2], is a nonlinear function used to measure the uncertainty or randomness of data flow. It can be used to capture the latent information hidden in the data even when the data distribution source is not completely known. In information theory, Shannon entropy is the most famous measure of randomness out of many kind of entropies proposed [3]. Apart from Shannon entropy other key entropy is Rényi entropy [4]. Renyi's quadratic entropy is appealing because it can be used as a measure of entanglement and can be estimated directly from input data set. Jenssen [5] proposed Kernel Entropy Component Analysis (KECA), an approach that uses kernel based estimator in the input space for Renyi's entropy estimation and optimum entropy projections.

In case of face, the entropy of high-frequency component describes its detail structure. Thus the problem of finding relationship between the images of two cross pose variations reduces to finding high-frequency overlap. A least square formulation of CCA [6] is used for finding inter-view supervised correlations. In our work we explore to find out inter-view semi-supervised correlations from the training samples via Renyi's entropy estimate. Section 2 gives a brief review of cross pose face recognition and Kernel Entropy Component Analysis. Section 3 describes the semi-supervised formulation of CCA to extract entropy components as features from identity coupled frontal and pose dataset. Subsequently in Sect. 4 experimental results are presented and lastly conclusions and direction for future works are presented in Sect. 5.

2 Background and Related Work

In this section, we give a brief overview of least square formulation of CCA and Kernel Entropy and Component Analysis.

2.1 *Least Square Formulation of CCA*

The Canonical Correlation—a multivariate analysis of correlation [1] is one of the extensively used methods for two view subspace learning. CCA is desirable in analyzing the strength of association between two constructs. The CCA provides useful information about the linear correlation between two interrelated constructs, while PCA fails to expose this information [7]. A special property of canonical correlations is that they do not change when subjected to affine transformations of the variables, i.e., they remain invariant to affine transformations. For two views of the same subject, $X \in R^{dXn}$ and $Y \in R^{kXn}$, CCA calculates the projection of each representation $Wx \in R^{d}$ and $Wy \in R^{d}$ in such a way that they remain maximally correlated even in the reduced subspace with correlation coefficient

$$\rho = \frac{w_x^T XY^T w_y}{\sqrt{\left(w_x^T XXY^T w_x\right)\left(w_y^T XXY^T w_y\right)}} \tag{1}$$

Su et al. [6] reported Least Square formulation of CCA in multi-class label classification. Their work utilized four matrices

$$C_{XX} = XX^T \in R^{d\times d} \tag{2}$$

$$H = Y^T\left(YY^T\right)^{-1/2} \in R^{n\times k} \tag{3}$$

$$C_{HH} = XHH^T X^T \in R^{d\times d} \tag{4}$$

$$C_{DD} = C_{XX} - C_{HH} \in R^{d \times d} \tag{5}$$

The solution to CCA is expressed as Eigen vectors corresponding to top Eigen values of $(C_{XX} \dagger C_{HH})$ subject to the condition that original data points are made linearly independent before centering, i.e., rank $(X) = n - 1$ and rank $(Y) = k$. Intuitively, for a given input of high-frequency patch, there will be high correlation with similar corresponding cross pose patch. Li et al. [8] in their work measured the similarities of the local patches by correlations in a subspace constructed by Canonical Correlation Analysis. Cross pose recognition was performed by directly summing up the similarity of all patch pairs. Recently Li et al. [9] used both holistic and local face features for maximizing the intra-subject across-pose correlations via CCA. Nadile et al. [10] used KCCA for learning 2D correspondence mapping between front face and profile one.

2.2 *Feature Extraction and Dimensionality Reduction*

Principal component analysis (PCA)—one of the classical methods for dimensionality reduction has been widely used for face recognition [11]. It can extract representative features from high dimensionality, noisy and linear correlated data. It can ensure that the extraction of linear features while it may not be able to extract useful nonlinear features in case of nonlinear data. Nonlinear methods are required to handle the nonlinear data, among which kernel principal component analysis (KPCA) [12] is the most prominent one. KPCA is an extension of traditional linear PCA and by using kernel trick it implicitly maps the original features into a high-dimensional feature space. Mapping to high-dimensional feature space makes the feature space linearly separable and then the linear PCA is conducted. Principal components analysis (PCA), kernel PCA (KPCA) based feature extraction methods gauge the significance of components according to their covariance contribution. They do not consider the entropy contribution, which vestiges important supplementary information for the covariance. Both PCA and KPCA are typical spectral dimensionality reduction methods which mine features by selecting the top eigenvalues and then selecting their corresponding eigenvectors [9]. Kernel entropy component analysis (KECA) is an information theory-based dimensionality reduction method, first proposed and employed in pattern recognition by Jenssen [5]. This method attempts to extract Renyi quadratic entropy of the input data set by means of a kernel-based estimation.

In KECA foundation is laid on information theoretic learning based on the quadratic information measures, combined with Parzen windowing followed by Eigen decomposition of un-centered kernel matrix and sorting of Eigen vectors according to information of projections. It is fundamentally different from other methods in two ways: on the one hand, the selection of top eigenvalues and corresponding eigenvectors is not necessary; on the other hand, the dimension reduction reveals the intrinsic structure related to the Renyi entropy of the input data [13].

KECA has been applied in face recognition to feature extraction and pattern recognition successfully, showing superior performance over PCA and KPCA [14, 15]. Recently Ruan and Wang [16] utilises kernel entropy component analysis (KECA) with weighted multiresolution face features to improve face recognition accuracy. Verdiguier et al. [17] proposed an extension of the KECA method, named Optimized KECA (OKECA), that directly extracts the optimal features retaining most of the data entropy by using Independent Component Analysis (ICA) framework.

3 Proposed Approach

In this paper semi-supervised CCA is performed on identity coupled frontal and pose dataset. CCA uses entropy components as features with a goal to get the maximum information out of the two feature extractors. Kernel entropy component analysis can be considered as an alternative to Kernel PCA for face representation where the projections are characterized from information theoretic perspective. Following are the steps of the approach used:

1. Input frontal database and identity coupled pose dataset.
2. Perform Canonical Correlation Analysis with least square formulation [6] to find the correlation between frontal and pose features in latent space.
3. Instead of choosing top Eigen vectors corresponding to top Eigen values KECA is applied, which reveals the structure related to Renyi entropy. Renyi entropy given by

$$Z = -\int \log p(x)^2 \mathrm{d}x \tag{6}$$

4. Renyi entropy is estimated from given dataset as $-\log(V)$ through Kernel Density Estimation [5], where V is the information potential associated with number of retained components.
5. By using kernel decomposition V is given by

$$\sum\nolimits_{i=1}^{Nc} \left(\lambda_j^{\frac{1}{2}} 1_N^T e_j\right)^2, \tag{7}$$

where e_j are the eigenvectors in columns, E =[e1; e2; : : : ; en], and D is a diagonal matrix containing the eigenvalues $[\lambda_1, \lambda_2, \lambda_3, \lambda_4 \ldots, \lambda_n]$ of K, and $Nc \le n$, the number of retained components. The term $\left(\lambda_j^{\frac{1}{2}} 1_N^T e_j\right)^2$ denotes quadratic entropy values. Nc, number of retained components is determined by retaining only those KECA Eigen vectors that have non zero sum of their corresponding components and positive Eigen value.

6. $L-1$ norm is applied to select the well-aligned features.

7. To discriminate all correlated pairs, correlation value is taken factor as similarity measure.

4 Experimental Results

For experimental evaluations, 2D images from FERET Database [18] is used. Identity coupled images of the database are divided into two groups front and pose set. Only positive Yaw angles are taken for experiments. Each image of the database is cropped to 80×80 sizes. We present the Canonical correlation performance of the KECA (quadratic entropy features) feature descriptor which also acts as dimensionality reduction approach.

As seen from the sample figure higher frequency blocks corresponds to higher entropy and its found to be maximum for the combined blocks showing left eye, right eye and nose tip (Fig. 2). In spite of easy computation, Shannon Entropy measures cannot be used as reliable feature due to their high sensitivity for noise perturbations. Also due to low correlation factor (0.5992) between front database and pose database, the error with PCA reach 40% and with KPCA it reaches to 42.4% (Fig. 3).

Analyzing canonical correlation on the training dataset (Front & Pose Dataset) with inbuilt matlab command "canoncorr" results in warning "Matrix is close to singular or badly scaled. Results may be inaccurate". Two factors accounting for this: Small sample size problem arising out of high dimensionality of face data as compared to number of training samples (200 images of 80×80 dimension) and the failure of CCA to extract linear relation from face database which has inherent non-linear nature. Figure 4 shows the canonical correlation plot of entropy feature values.

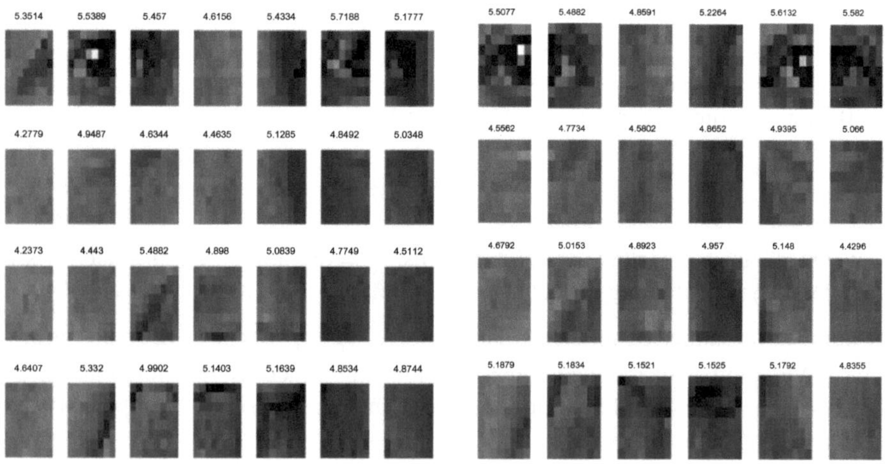

Fig. 2 Shannon entropy features from 8×8 blocks of front and posed image

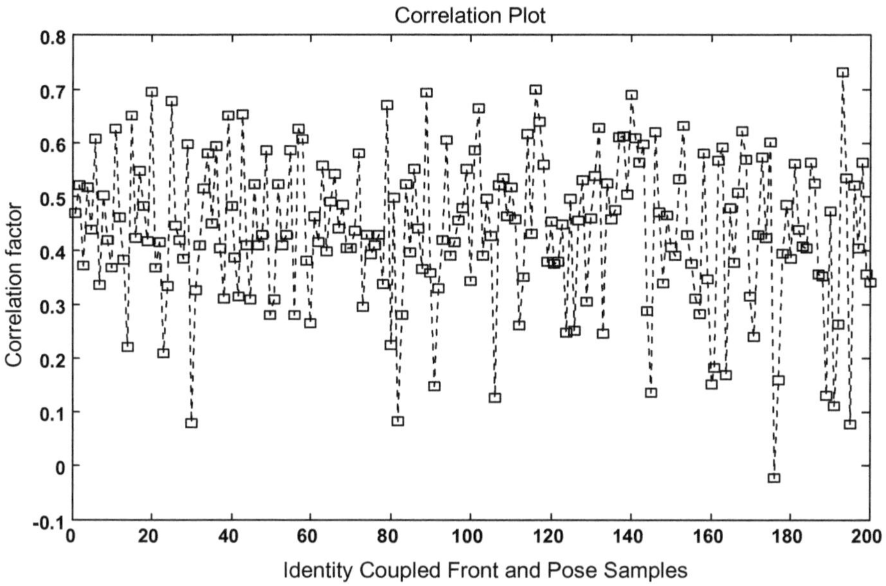

Fig. 3 Correlation plot between identity coupled samples

Correlation coefficient increases among the identity related samples as the canonical dimension increases. Figure 5 shows the canonical correlation plot calculated by using kernels on raw pixel values. In spite of increase in computational complexity, performance of quadratic features is better than intensity features. In context of the retained information potential, quadratic entropy features fared better even when the kernel form of CCA is used on the raw intensity values.

5 Conclusions and Future Scope

In this paper we evaluate the proposed formulation of CCA in semi supervised manner for finding correlations between entropic kernel components in the latent space. While KECA sort the kernel eigenvectors by entropy, proposed formulation of CCA explicitly searches for the maximization of the correlated features in order to retain most informative content. Correlation of components via canonical correlation analysis helps in setting up better correspondence between a discrete set of non-frontal pose and the frontal pose of same subject. For the future works we will explore partial least squares method for correlating front and posed dataset. Also we will explore the effectiveness of using weights for multi-resolution components of kernel entropy in order to maximize the correlations in latent space.

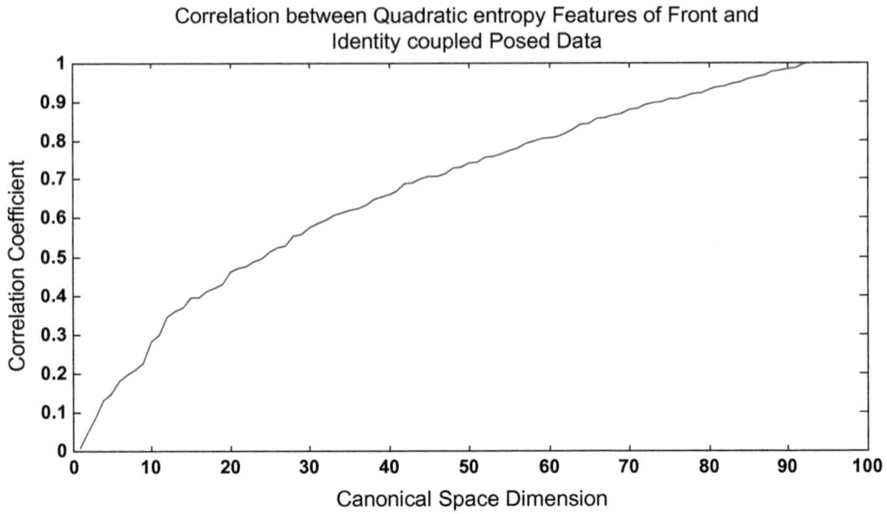

Fig. 4 CCA plot of correlations between quadratic entropy features

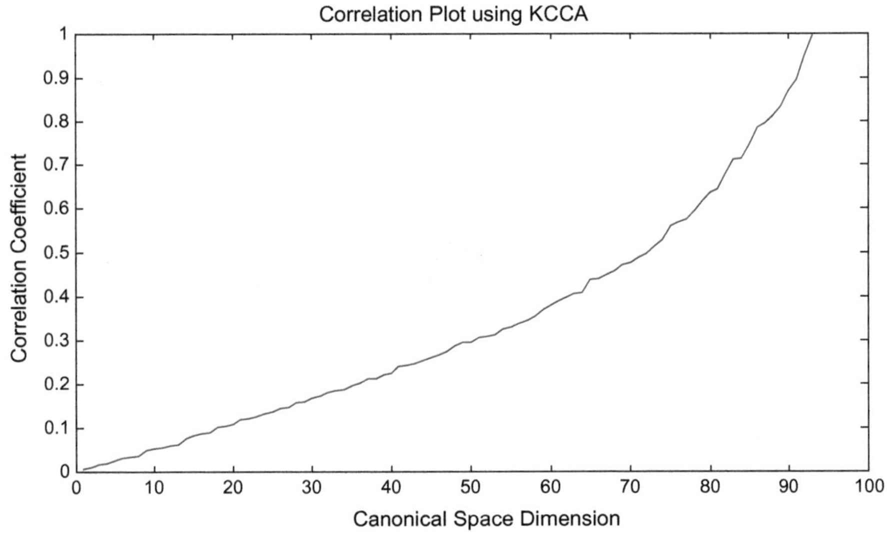

Fig. 5 KCCA plot of raw intensity features

References

1. H. Hotelling, Relations between two sets of variates. Biometrika **28**(3/4), 321–377 (1936)
2. N.R. Pal, S.K. Pal, Entropy: A new definition and its applications. IEEE Trans. Syst. Man Cybern. **21**(5), 1260–1270 (1991)

3. C. Shannon, A mathematical theory of communication. Bell Syst. Tech. J. **27**, 379–423, 623–656; Math. Rev. (MathSciNet) **10**, 133e (1948)
4. M. Skórski, in *Shannon Entropy Versus Renyi Entropy from a Cryptographic Viewpoint*. IMA International Conference on Cryptography and Coding (Springer, Cham, 2015), pp. 257–274
5. R. Jenssen, Kernel entropy component analysis. IEEE Trans. Pattern Anal. Mach. Intell. **32**(5), 847–860 (2010)
6. L. Sun, S. Ji, J. Ye, in *A Least Squares Formulation for Canonical Correlation Analysis*. Proceedings of the 25th International Conference on Machine Learning (ACM, New York, 2008), pp. 1024–1031
7. D.R. Hardoon, S. Szedmak, J. Shawe-Taylor, Canonical correlation analysis: An overview with application to learning methods. Neural Comput. **16**(12), 2639–2664 (2004)
8. A. Li, S. Shan, X. Chen, W. Gao, in *Maximizing Intra-Individual Correlations for Face Recognition Across Pose Differences*. IEEE Conference on Computer Vision and Pattern Recognition, CVPR 2009, pp. 605–611
9. A. Li, S. Shan, X. Chen, B. Ma, S. Yan, W. Gao, Cross-pose face recognition by canonical correlation analysis. arXiv preprint arXiv:1507.08076 (2015)
10. M. Nadil, F. Souami, A. Labed, H. Sahbi, KCCA-based technique for profile face identification. EURASIP J. Image Video Process. **1**, 1–13 (2017)
11. I.T. Jolliffe, J. Cadima, Principal component analysis: A review and recent developments. Phil. Trans. R. Soc. A **374**(2065), 20150202 (2016)
12. B. Schölkopf, A.J. *Smola, Learning with Kernels: Support Vector Machines, Regularization, Optimization, and Beyond* (MIT press, Cambridge, 2002)
13. D. Xu, D. Erdogmuns, in *Renyi's Entropy, Divergence and Their Nonparametric Estimators*. Information Theoretic Learning (Springer, New York, 2010), pp. 47–102
14. B.H. Shekar, M.S. Kumari, L.M. Mestetskiy, N.F. Dyshkant, Face recognition using kernel entropy component analysis. Neurocomputing **74**(6), 1053–1057 (2011)
15. A. Kar, D. Bhattacharjee, D.K. Basu, M. Nasipuri, M. Kundu, A face recognition approach based on entropy estimate of the nonlinear DCT features in the logarithm domain together with kernel entropy component analysis. arXiv preprint arXiv:1312.1520 (2013)
16. X. Ruan, S. Wang, in *Face Recognition Based on Weighted Multi-Resolution Kernel Entropy Component Analysis*. IEEE 29th Chinese Control and Decision Conference (CCDC), pp. 6118–6123 (2017)
17. E. Izquierdo-Verdiguier, V. Laparra, R. Jenssen, L. Gómez-Chova, G. Camps-Valls, Optimized kernel entropy components. IEEE Trans. Neural Netw. Learn. Syst. **28**(6), 1466–1472 (2017)
18. P.J. Phillips, H. Wechsler, J. Huang, P.J. Rauss, The FERET database and evaluation procedure for face-recognition algorithms. Image Vis. Comput. **16**(5), 295–306 (1998)

Diagnosis of Malignant Pleural Mesothelioma Using KNN

Manpreet Kaur and Birmohan Singh

Abstract Malignant pleural mesothelioma is a major health issue and is a cause of the concern as the number of cases are increasing constantly. The early diagnosis is necessary for the survival of the affected. An approach for the diagnosis of Mesothelioma has been proposed in this paper, which distinguishes normal person and the patient. The proposed approach uses a Differential Evolution Based Feature Selection Method, the input to which is a combination of the results from Mutual Information Maximization and Sequential Backward Selection. For classification K-Nearest Neighbor has been used and the average classification accuracy achieved is 99.07%.

Keywords Mesothelioma · Sequential backward selection and mutual information maximization · K-nearest neighbor

1 Introduction

Malignant pleural mesothelioma is a cancer that affects the people in the industrialized countries. The persons who are exposed to a high level of asbestos are having more chances of getting affected by this either by inhaling or by ingesting it. Asbestos exposure attributes to 86–95% contribution in the development of this disease [4].

Peritoneal mesothelioma starts showing its effect in the abdomen with abdominal pain, then results in weight loss, nausea, and vomiting. The other symptoms include anemia, chest pain, and hemoptysis (https://blogs.biomedcentral.com/on-health/2017/03/21/latest-mesothelioma-life-expectancy-stats-and-survival-rates/). This cancer

M. Kaur
Department of Electrical and Instrumentation Engineering, Sant Longowal Institute of Engineering and Technology, Longowal 148106, Punjab, India

B. Singh (✉)
Department of Computer Science and Engineering, Sant Longowal Institute of Engineering and Technology, Longowal 148106, Punjab, India
e-mail: birmohans@gmail.com

C. R. Krishna et al. (eds.), *Proceedings of 2nd International Conference on Communication, Computing and Networking*, Lecture Notes in Networks and Systems 46, https://doi.org/10.1007/978-981-13-1217-5_62

propagates in the thin tissue membrane of lubricating cells, which wall around to protect the heart, lungs, and abdominal. Tiny fibers of asbestos enter the body through tracheal or esophageal pathways (http://markets.businessinsider.com/news/stocks/Malignant-Mesothelioma-Pipeline-Review-H2-2017-1002380105).

Malignant pleural mesothelioma is not a common disease the incidents of which are in 1–2 per million/year, however, in the developed countries, these are 10–30 per million/year for men and 1–5 per million/year for women [6]. In Italy, there were 3.6 cases in men and 1.3 cases in women per lac persons per year, for the year 2008 [10]. The number of cases in the last years has globally increased slightly. The lag time of 30–50 years after exposure to asbestos is the reason [1]. It is predicted that the expected number of cases will be at an estimated peak in 2020. It will be a significant health issue even after the estimated peak in some of the countries, and India is one of these [2].

In Australia, there were 575 deaths, out of which 465 males and 110 females, in the year 2013. In Great Britin, 2515 people lost their lives due to this disease, out of which there were 2101 males and 414 females in the year 2014 (https://blogs.biomedcentral.com/on-health/2017/03/21/latest-mesothelioma-life-expectancy-stats-and-survival-rates/).

The survival period depends upon the stage of diagnosis. This is mostly diagnosed in the third stage of cancer and the survival period is up to one year. Since it provides a less survival time after prognosis, diagnosis at the early stage is a major concern [13].

Between in 1990–2001, in Italy, 9.8 months was survival period, once it is diagnosed, and after 3 years, less than 10% could survive. The numbers of cases with this disease are increasing and there is a fear of about 800 deaths in 2012–2025 per year in males [10].

A significant overlap between malignant and benign mesothelial cells as well as between malignant mesothelioma and adenocarcinoma cells, make the diagnosis difficult [3]. Different researchers have proposed diagnosis systems for classification of Mesothelioma disease. Er et al. [6] proposed a probabilistic neural network based system for diagnosis of normal and patient instances in Mesothelioma disease dataset. Er et al. [5] used Artificial Immune Systems (AIS) for malignant Mesothelioma disease diagnosis. Nilashi et al. [9] used EM, PCA, Classification and Regression Trees (CART) and Fuzzy rule based technique for classification of the disease. Ushasukhanya and Sridhar [13] proposed a technique for analysis of malignant Mesothelioma using artificial intelligence. The best results were achieved by probabilistic neural network and Random Search method. Tutuncu and Cataltas [12] proposed a method for diagnosis of Mesothelioma using different classifiers; J48, Bayes Net, Sequential Minimal Optimization (SMO), Logistics Model Tree (LMT), Logistic regression, Multi class classifier, Random Committee PART and Artificial Neural Networks. The best results were achieved with artificial neural networks.

In this work, a methodology has been proposed for the diagnosis of pleural malignant mesothelioma that uses Differential Evolution Based Feature Selection Method, in combination with Sequential Backward Selection and Mutual Information Maximization. The reduced set of feature is used for classification with K-Nearest Neighbor to distinguish between normal data and malignant mesothelioma.

2 Database

The database considered in this work is "Mesothelioma disease dataset" which was prepared at Dicle University Faculty of Medicine. This is a dataset publicly available at UCI (University of California, Irvine) Machine Learning Repository. In this dataset, there are 324 instances indicating different patient records. Each instance contains 34 features; Age, gender, city, asbestos exposure, malignant mesothelioma type of Malignant Mesothelioma, duration of asbestos exposure, diagnosis method, side the duration of symptoms, respiratory distress (dyspnea), ache on chest, weakness, habit of cigarette, performance status (WBC count), hemoglobin (HGB), platelet count PLT, sedimentation, blood lactic dehydrogenase LDH, alkaline phosphatase ALP, (Total protein), albumin (albumin), glucose, pleural lactic dehydrogenase, pleural protein, pleural albumin, pleural glucose, Pleural effusion, pleural thickness on tomography, pleural level of acidity pH and C-reactive protein (CRP). This is followed by the class of the instances; which belongs to either healthy patient records (228 records) or cancer patient records (96 patients) [6].

3 Feature Selection

In medical classification applications, the numbers of features are very large, which may include irrelevant and redundant features. The presence of irrelevant and/or redundant features may result in over-fitting as well as increases the processing time. The feature selection procedure is a process of selecting an optimal set of features in order to improve the classification accuracy [11].

There are three classes of feature selection techniques. Wrapper methods, make use of some classifiers to score a given subset of features; filter methods, which study inherent properties of the data; and hybrid methods, which combine the properties of these two methods [12].

In this paper, a procedure proposed by Tiwari et al. [12] has been used for selection of non-redundant and relevant set of features. They proposed a methodology in which they used a combination of Mutual Information Maximization algorithm and Sequential Backward Selection to generate a set of features which is input to Differential Evolution based Feature Selection [8] technique. The final set achieved consists of 10 features. The computation time and the consistency of the results has been improved.

4 Results and Discussion

The methodology proposed for analysis of malignant Mesothelioma disease has been evaluated on Mesothelioma disease dataset available publicly. K-Nearest Neighbor, Support Vector Machine and Artificial Neural Networks have been used for classification. The best accuracy has been achieved by K-Nearest Neighbor (KNN), with 10 numbers of nearest neighbors, and with Euclidian Distance, where all the features are given equal weights.

KNN is a non parametric algorithm which has been used since 1970 in many applications of pattern recognition, classifications, etc., classifies the cases on the basis of similarity measures. It is mainly used when all the attributes are continuous [7].

The performance of the system has been measured with sensitivity, specificity, and accuracy. In the proposed system, False Positive (FP) is a negative label to a positive point, False Negative (FN) is a positive label to a negative point, True Positive (TP) and True Negative (TN) are the correct classification of the labels. In this proposed system, TP is taken as number of correctly classified patient cases, TN value is number of normal cases duly recognized, FP is quantity of normal cases recognized as patient cases and FN is a quantity of patient cases identified as normal cases. The results achieved from this work, show a sensitivity of 98.96%, specificity of 99.12% and average accuracy of 99.07% with 10 fold cross validation procedure.

A comparison of the results achieved has been made with the work of the other researchers as shown in the Table 1. In the year 2012 Er et al. [6] achieved an average accuracy of 96.30% with the probabilistic neural network. In the year 2015 Er et al. [5] used Artificial Immune Systems and reported an accuracy of 97.74%. In the year 2017 Nilashi et al. [9] reported the prediction accuracy of 93.07% by EM, PCA, CART and Fuzzy rule based technique. In the year 2017 Ushasukhanya and Sridhar [13] reported an accuracy of 96.3% with probabilistic neural network and Random Search method. In the same year, Tutuncu, and Cataltas, achieved the best results with ANN claiming an accuracy of 99.07%. The average accuracy achieved by the proposed system is also 99.07%.

Table 1 A comparison of the results with the other research works

Paper and author	Year	Accuracy
Er et al.	2012	96.30
Er et al.	2015	97.74
Nilashi et al.	2017	93.60
Ushasukhanya and Sridhar	2017	96.30
Tutuncu and Cataltas	2017	99.07
Proposed methodology	2018	99.07

5 Conclusions

In this paper, a methodology for analysis of malignant pleural mesothelioma has been proposed. An optimal set of features has been selected by eliminating redundant and non-relevant features in the feature set. K-Nearest Neighbor with the 10 numbers of nearest neighbors, and with Euclidian Distance, where all the features are given equal weights, has been used for classification. The average accuracy achieved is 99.07% with 98.96% sensitivity. The results are quite comparable with the other approaches available in the literature. The proposed approach may be significant in the detection of disease at the early stage, and reduce the death rates.

References

1. P. Baas, D. Fennell, K.M. Kerr, P.E. Van Schil, R.L. Haas, S. Peters, Malignant pleural mesothelioma: ESMO clinical practice guidelines for diagnosis, treatment and follow-up. Ann. Oncol. **26**(suppl_5), v31–v39 (2015)
2. A.C. Bibby, S. Tsim, N. Kanellakis, H. Ball, D.C. Talbot, K.G. Blyth, N.A. Maskell, I. Psallidas, Malignant pleural mesothelioma: an update on investigation, diagnosis and treatment. Eur. Respir. Rev. **25**(142), 472–486 (2016)
3. L. Chen, S.G. Caldero, S. Gmitro, M.L. Smith, G. Petris, M.A. Zarka, Small orangiophilic squamous-like cells: An under-recognized and useful morphological feature for the diagnosis of malignant mesothelioma in pleural effusion cytology. Cancer Cytopathol. **122**(1), 70–75 (2014)
4. M. Corfiati, A. Scarselli, A. Binazzi, D. Di Marzio, M. Verardo, D. Mirabelli, C. Negro et al., Epidemiological patterns of asbestos exposure and spatial clusters of incident cases of malignant mesothelioma from the Italian national registry. BMC Cancer **15**(1), 286–300 (2015)
5. O. Er, A.Ç. Tanrikulu, A. Abakay, Use of artificial intelligence techniques for diagnosis of malignant pleural mesothelioma. Dicle Tıp Dergisi **42**(1), 5–11 (2015)
6. O. Er, A.C. Tanrikulu, A. Abakay, F. Temurtas, An approach based on probabilistic neural network for diagnosis of Mesothelioma's disease. Comput. Electr. Eng. **38**(1), 75–81 (2012)
7. M.A. Jabbar, B.L. Deekshatulu, P. Chandra, Classification of heart disease using K-nearest neighbor and genetic algorithm. Procedia Technol. **10**, 85–94 (2013)
8. R.N. Khushaba, A. Al-Ani, A. Al-Jumaily, Feature subset selection using differential evolution and a statistical repair mechanism. Expert Syst. Appl. **38**(9), 11515–11526 (2011)
9. M. Nilashi, O. Bin Ibrahim, H. Ahmadi, L. Shahmoradi, An analytical method for diseases prediction using machine learning techniques. Comput. Chem. Eng. (2017). https://doi.org/10.1016/j.compchemeng.2017.06.011
10. C. Pinto, S. Novello, V. Torri, A. Ardizzoni, P.G. Betta, P.A. Bertazzi, D. Mirabelli, Second Italian consensus conference on malignant pleural mesothelioma: State of the art and recommendations. Cancer Treat. Rev. **39**(4), 328–339 (2013)
11. B. Surendiran, A. Vadivel, A new feature reduction method for mammogram mass classification. Control Comput. Inf. Syst. **140**, 303–311 (2011)
12. S. Tiwari, B. Singh, M. Kaur, An approach for feature selection using local searching and global optimization techniques. Neural Comput. Appl. **28**(10), 2915–2930 (2017)
13. S. Ushasukhanya, S.S. Sridhar, Survey on artificial intelligence techniques in the diagnosis of pleural mesothelioma. Int. J. Pharm. Technol. **9**(1), 29042–29051 (2017)

Part VI
Data Science and Mining-II

ECG Arrhythmia Classification Using Artificial Neural Networks

Saroj Kumar Pandey and Rekh Ram Janghel

Abstract Electrocardiogram (ECG) arrhythmia is referred to as a change in human heart rhythm, and it becomes either too slow or very large compared to normal heart rhythms. This may cause disease affecting cardiac. Early correct identification of arrhythmia is important in the detection of cardiac disease and getting the better treatment of a patient. Numerous classifiers are present for ECG diagnosis. Artificial neural network (ANN) is one of the more popular and very widely utilized models for ECG diagnosis. In this paper, we introduced three different ANN models, which are classified as healthy and arrhythmia classes and using UCI repository ECG 12 lead signal feature extracted data. This particularly uses ANN models that are trained as well as tested on back-propagation feedforward neural network, recurrent neural network (RNN), and radial basis function (RBF) networks. We evaluated the diagnosis testing result in the form of classification accuracy, sensitivity, and specificity. Among these three contrast ANN models, RNN models have shown better diagnosis result up to obtained 83.1% testing classification accuracy with selected attributes.

Keywords Arrhythmia · Electrocardiogram · Recurrent neural network · Radial basis function · Diagnosis · Back-propagation algorithm

1 Introduction

Diseases affecting cardiac have become usual in day by day. Diseases affecting cardiac is growing because of unhealthy living style increase these days and in medical conditions like diabetes, hypertension, and tobacco smoking. The heart can be affected due to multiple conditions. Among different sickness individual has experienced, heart infections are as yet one of the vital issues in the world. Early discovery

S. K. Pandey (✉) · R. R. Janghel
National Institute of Technology, Raipur (C.G.), India
e-mail: sarojpandey23@gmail.com

R. R. Janghel
e-mail: rrj.iiitm@gmail.com

C. R. Krishna et al. (eds.), *Proceedings of 2nd International Conference on Communication, Computing and Networking*, Lecture Notes in Networks and Systems 46, https://doi.org/10.1007/978-981-13-1217-5_63

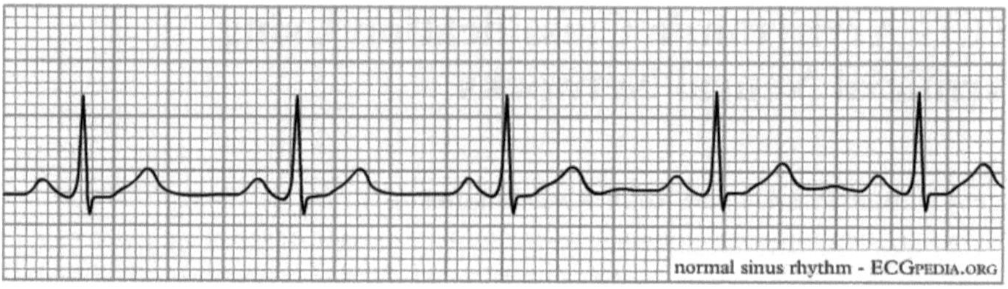

Fig. 1 Normal heart pulse

and legitimate medicinal treatment of maladies relating to the cardiac can spare existences of the unhealthy person in instances of unexpected death [1]. One of the critical classification devices of heart working is ECG or EKG (Electrocardiogram) which makes a realistic record of the heart's electrical impulses. Figure 1 appears the ECG examples of normal heart pulse and abnormal heart pulse. An arrhythmia is an abnormal heart pulse. There are two fundamental sorts of arrhythmias. Bradycardia is the time when the heart rate is too low under 60 beats for consistently. Tachycardia is the point at which the heart rate is too quick more than 100 pulse rates every minute. Whenever arrhythmias are extreme, the cardiac capacity to pump blood might be lessened causing shortness of breath, chest torment, experiencing tired, and loss of consciousness. On the off chance is that more serious, it can cause heart assault or death [2]. In this way, to maintain a strategic distance from any awful happenings, we are to recognize and classify the heart arrhythmia. Methodologies have as of now been created for diagnosing cardiovascular arrhythmias in view of ECG signal information, yet at the same time indicate poor execution. Different machine learning and data mining techniques are being conveyed to enhance the recognition of ECG arrhythmia. Yet, in any case, there are contrasts between the cardiologs and the program diagnosis. Taking the cardiologist's as best quality level, we are to limit the distinction by methods for machine learning apparatuses [3].

In this paper, we used six different models of artificial neural networks (ANN) and focused on the best simulation results to differentiate between normal and abnormal cardiac activities. In the following section, this paper is grouped as follows. Section 2 reports the associated works of various classification techniques. Section 3 describes the methodology description, and Sect. 4 explains the simulation results. Conclusions and further work are presented in Sect. 5.

2 Survey of Related Works

Distinctive methods have been available to develop robotized recognition and finding of ECG. Self-organizing maps (SOM), support vector machines (SVM), multilayer perceptron (MLP), Markov models, and fuzzy or neuro-fuzzy systems and of various procedures have been recommended to enhance execution [4–6]. The examination of

ECG is primarily identifying its example and diagnosing arrhythmia persistently. To date, a few analysts have made undertakings to apply SVM and distinctive another classifier to diagnosis cardiac beats. Various strategies have been introduced over before years for working up the motorized structures to absolutely order the ECG information. These include wavelet transform [7, 8], direct vector quantization [9], probabilistic neural network [10], and fuzzy crossover neural system algorithms [11]. Silipo et al. [12] proposed a differentiation task for characterization of ECG applying two arrangement strategies: one with administered learning strategy and other learning with unlabeled information. Sugiura et al. [13] built up a fuzzy rationale-based methodology for recognizing ECG arrhythmias and segregating ventricular arrhythmias. Acharya et al. [4] utilized cardiac rate changeability (HRV) as the base flag and executed ANN and fuzzy proportionality connection for the grouping of four ECG arrhythmias. Kohli et al. [14] suggest SVM-based arrhythmia arrangement associated with three strategies: one against one, one against all, and fuzzy choice capacity. In this paper, one against all strategy gives the better exactness contrast with different techniques. Jadhav et al. [15] developed three diverse models of ANN for the determination of heart arrhythmia. In this paper, ANN models are prepared to utilize static back-engendering with force figuring out how to grouping of arrhythmia.

3 Methodology

In this task, ANN models have been used for the diagnosis of ECG arrhythmia dataset. The training takes place through the repetitive process until of weight adjustments used to its beginning weight after epoch iteration of the training process. Figure 2 presents the overall used methodology. As found in the tests results in our ANN models, classify ECG signal as either normal or arrhythmic. In this article, we used

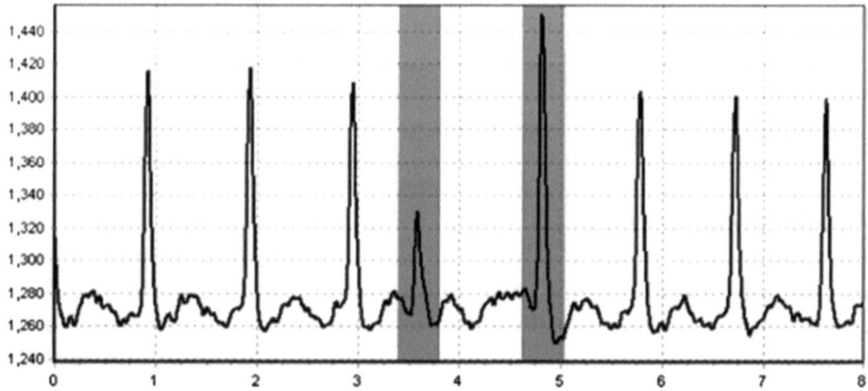

Fig. 2 Abnormal heart pulse

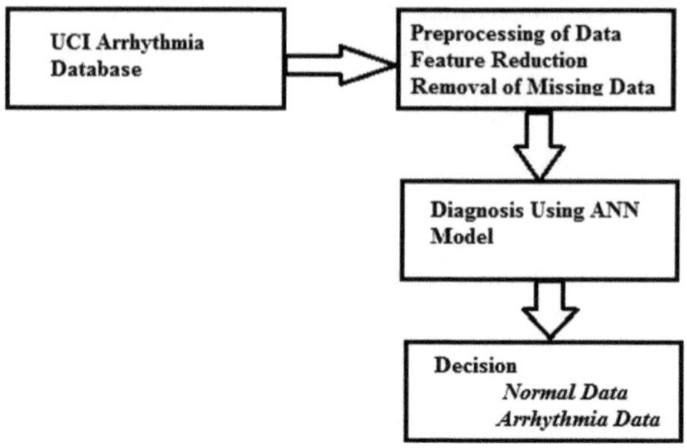

Fig. 3 Block diagram of entire methodology

three ANN models, namely, back-propagation network (BPA), radial basis function (RBF) network, and recurrent neural network (RNN). The dataset contains several missing attribute values which are deleted the entire subjects. Various methods of neural networks are taken as input data values in the normalized form; hence, obtain the normalized data value first, select the highest value from the given attributes, and then divide all the data values of given attribute by this highest value. Thus, we get all values of an attribute in interval zero and one. The BPA is applied to the gradient descent to calculate the new weights and biased values. For better performance, BPA is able to adjust the network weight quickly. The graph representing the error is known as error space of the model for each association of weight and biases. The main objective of anyone training model is to obtain the global optima in given search space. Multiple times BPA may obtain confined to local minima because of the unavailability of somewhat global leading procedure or the attempt to investigate the whole error margin which is of large dimensionality and very complex. The ANN model applied in this investigation existed in the input layer, one hidden layer, and output layers. The sigmoid and purelin transfer function is used in both hidden and output layers of neurons, respectively. The performance measure used was mean sum-squared error (MSE) and confusion matrix [16] (Fig. 3).

4 Simulation Results

4.1 Database Used

This dataset was donated by Dr. H. Altay Guvenir in 1998. This database includes the total number of subjects as 452. The number of features is 279 in which 206 features are linear valued and the remaining are nominal. There are 32 subjects with missing information so we discard these instances and use only 392 instances. We also deleted columns with all 0s and also deleted columns of which most of the entries are 0s. We obtained 182 columns of which 173 are numerical and 9 are categorical attributes [17]. We used only 161 attributes for training and testing of ANN models. The objective is to differentiate between the healthy and unhealthy of cardiac and to classify it in one of the two groups. Class 1 mentions the normal ECG data and class 0 mentions the arrhythmia.

4.2 Performance Evaluation

We have examined the three main measurement performances on various machine learning classification algorithms that are sensitivity, specificity, and classification accuracy. These measurements are described by using confusion matrix as true positive (TP), true negative (TN), false positive (FP), and false negative (FN). TP observation happens when the observation of arrhythmia occurs by the classifier and matches with the physician's observation, and TN observation happens when classifier and physician both detect the not present of arrhythmia. FP occurs when the classifier observed an arrhythmia but physician observed not an arrhythmia case, and FN occurs when classifier does not observe an arrhythmia case but physician observed arrhythmia. Classification accuracy refers to the total number of correctly classified instances that is the addition of true positive and true negative divided by the total number of instances (TP, TN, FP, and FN). For binary classification,

$$Accuracy = \frac{T_P + T_N}{Totalnumberofinstances} \tag{1}$$

Sensitivity refers to the ratio of correctly classified true positive (TP) divided by the sum of true positive (TP) and false negative (FN). It is also called the true positive rate.

$$Sensitivity = \frac{T_P}{T_P + F_N} \tag{2}$$

Classification specificity is the rate of correctly classifying the negatives referred to as specificity. It is equal to the ratio of TN to the sum of FP and TN.

$$Specificity = \frac{T_N}{F_P + T_N} \tag{3}$$

In this work, first we have used 100% dataset for training purpose and then different datasets for testing of model for obtaining trained accuracy. For finding the testing accuracy, we have used five different datasets for trained model used for obtaining better tested accuracy. We have used fivefold cross-validation of testing results to obtain better accuracy.

4.3 Experiment Results

Tables 1, 2, 3, 4, 5, and 6.

5 Conclusions and Further Work

This task introduced a powerful artificial intelligence-based diagnosis model for ECG arrhythmia diagnosis using 12 lead UCI ECG signal databases. We have applied three well-known ANN models to the contrast between the present normal signal data and arrhythmic signal data. Based on the contrast analysis of the obtained results, RNN is the strongest ANN model for the specified ECG arrhythmia diagnosis work. It is shown that the testing accuracy of RNN method is better from other two ANN methods. In RNN method, we obtained classification testing accuracy 83.1% and sensitivity 86.67% and specificity 66.67%, respectively. This work gives the better result in terms of accuracy and compared to the results of some recent task using the same database. Our further work of research will introduce some more hidden layers (deep learning) for ECG arrhythmia classification and try to improve the performance of diagnosis results. We will also be using multiclass classification techniques to classify the different types of arrhythmias.

Table 1 Classification result of ANN models for using 50–50 training and testing dataset

ANN Model	Sensitivity %	Specificity %	Accuracy %
RNN	72.15	68.42	70.91
RBF	70.03	58.10	65.13
BPA	68.35	60.71	64.28

Table 2 Classification result of ANN models for using 60–40 training and testing dataset

ANN Model	Sensitivity %	Specificity %	Accuracy %
RNN	75.96	73.58	75.15
RBF	69.83	57.20	63.33
BPA	69.16	59.23	64.68

Table 3 Classification result of ANN models for using 70–30 training and testing dataset

ANN Model	Sensitivity %	Specificity %	Accuracy %
RNN	73.68	64.29	70.33
RBF	74.10	57.15	68.50
BPA	63.50	50.40	60.16

Table 4 Classification result of ANN models for using 80–20 training and testing dataset

ANN Model	Sensitivity %	Specificity %	Accuracy %
RNN	74.07	58.33	71.23
RBF	77.28	63.49	70.40
BPA	68.12	62.22	64.10

Table 5 Classification result of ANN models for using 90–10 training and testing dataset

ANN Model	Sensitivity %	Specificity %	Accuracy %
RNN	**86.67**	**66.67**	**83.05**
RBF	83.28	63.66	75.25
BPA	77.10	61.18	74.35

Table 6 Comparison of this work with other existing works

Literatures	Models	Classes	Accuracy %
Polat et. al [19]	Fuzzy weighted AIRS	2	76.2
Uyar et.al [20]	SVM and LR	2	74.9
Elasyad [21]	LVQ 2.1	2	74.12
Zuo et. al [22]	KDF-WKNN	2	70.66
This work	**RNN**	**2**	**83.1**

References

1. Y. Ozbay, B. Karlik, A recognition of ECG arrhytihemias using artificial neural networks, in *2001 Conference Proceedings of the 23rd Annual International Conference of the IEEE Engineering in Medicine and Biology Society*, vol. 2 (2001), pp. 1680–1683
2. M. Mitra, R.K. Samanta, Cardiac arrhythmia classification using neural networks with selected features. Procedia Technol. **10**, 7684 (2013)
3. Q. Li, C. Rajagopalan, G.D. Clifford, A machine learning approach to multi-level ECG signal quality classification. Comput. Methods Programs Biomed. **117**(3), 435447 (2014)
4. U. Rajendra Acharya, P. Subbanna Bhat, S.S. Iyengar, A. Rao, S. Dua, Classification of heart rate data using artificial neural network and fuzzy equivalence relation. Pattern Recognit. **36**(1), 6168 (2003)
5. R.D. Raut, S.V. Dudul, Arrhythmias classification with MLP neural network and statistical analysis, in *Proceedings of - 1st International Conference on Emerging Trends in Engineering Technology ICETET 2008* (2008), pp. 553–558
6. B.M. Asl, S.K. Setarehdan, M. Mohebbi, Support vector machine-based arrhythmia classification using reduced features of heart rate variability signal. Artif. Intellect. Med. **44**(1), 5164 (2008)
7. G. Selvakumar, K.B. Bagan, B. Chidambararajan, Wavelet Decomposition for Detection and Classification of Critical ECG Arrhythmias (2007), pp. 8084
8. L. Khadra, A. Fraiwan, W. Shahab, Neural-wavelet analysis of cardiac arrhythmias
9. P. Melin, J. Amezcua, F. Valdez, O. Castillo, A new neural network model based on the LVQ algorithm for multi-class classification of arrhythmias. Inf. Sci. (Ny) **279**, 483–497 (2014)

10. S. Yu, Y. Chen, Electrocardiogram beat classification based on wavelet transformation and probabilistic neural network **28**, 1142–1150 (2007)
11. M. Engin, ECG beat classification using neuro-fuzzy network **25**, 1715–1722 (2004)
12. R. Silipo, G. Bortolant, C. Marchesi, Supervised and unsupervised learning for ciiagnos t ic E, Evaluation (1996), pp. 931–934
13. T.K. Toshifumi Sugiural, H. Hiratal, Y. Harada2, Automatic discrimination of arrhythmia **20**(I), 108–111 (1998)
14. N. Kohli, N.K. Verma, A. Roy, SVM based Methods for Arrhythmia Classification in ECG (2010), pp. 486–490
15. S.M. Jadhav, S.L. Nalbalwar, A.A. Ghatol, Artificial neural network based cardiac arrhythmia disease diagnosis, in *2011 International Conference Process Automation and Control Computing* (2011), pp. 16
16. R. Tiwari, A. Shukla, P. Kaur, Diagnosis of Thyroid Disorders using Artificial Neural Networks (2009), pp. 67
17. http://archive.ics.uci.edu/ml/machine-learning-databases/arrhythmia/
18. S.M. Jadhav, Artificial neural network models based cardiac arrhythmia disease diagnosis from ECG signal data. Int. J. Comput. **44**, 813 (2012)
19. K. Polat, S. ahan, S. Gne, A new method to medical diagnosis: artificial immune recognition system (AIRS) with fuzzy weighted pre-processing and application to ECG arrhythmia. Expert Syst. Appl. **31**(2), 264269 (2006)
20. A. Uyar, F. Gurgen, Arrhythmia classification using serial fusion of support vector machines and logistic regression. Appl. 2007. 4th IDAACS 2007 (2007), pp. 560–565
21. A.M. Elsayad, Classification of ECG arrhythmia using learning vector quantization neural networks, in *Proceedings - 2009 International Conference on Computing Engineering System ICCES09* (2009), pp. 139–144, 2009
22. W.M. Zuo, W.G. Lu, K.Q. Wang, H. Zhang, Diagnosis of cardiac arrhythmia using kernel difference weighted KNN classifier. Comput. Cardiol. **35**, 253256 (2008)

An Improved Automated Question Answering System from Lecture Videos

Divya Mamgai, Sonali Brodiya, Rohit Yadav and Mohit Dua

Abstract Learning is important for students for individual as well as nation development. However, due to the ever increasing student base and lack of sufficient number of mentors all over the world, one-to-one student-teacher interaction in real life has become a tough task. Question Answering (QA) systems reduce the requirement for physical interaction between students and teachers. QA systems allow students to post their queries and get answers for the same. The paper discusses implementation of an automated QA system that uses a knowledge base constructed from the video lectures recorded in the classrooms as well as the ones that are available online. Two approaches to generate the transcript from the video lectures have been implemented: (1) with use of Online Speech Recognition APIs (Application Programming Interfaces) and (2) with use of offline method. CMU sphinx has been automatically trained using predefined dataset in offline method. Transcripts from both approaches, i.e., online and offline are then compared on the basis on semantic similarity and results are presented.

Keywords Question answering system · NLP · Transcript · Language model

1 Introduction

Learning is an important part of every student's life. Not only a student gets knowledge about a subject matter but also acquires skills to use them. Therefore, an effective

D. Mamgai (✉) · S. Brodiya · R. Yadav · M. Dua
National Institue of Technology, Kurukshetra, India
e-mail: divyamamgai21@gmail.com

S. Brodiya
e-mail: sonalibrodiya@gmail.com

R. Yadav
e-mail: rohity2304@gmail.com

M. Dua
e-mail: er.mohitdua@nitkkr.ac.in

C. R. Krishna et al. (eds.), *Proceedings of 2nd International Conference on Communication, Computing and Networking*, Lecture Notes in Networks and Systems 46, https://doi.org/10.1007/978-981-13-1217-5_64

learning becomes crucial in the knowledge-driven world today. The process of education was limited to a classroom with a teacher leading the process until beginning of twenty first century. The concept of online learning was questionable. However, with the advent of Internet, e-learning came into picture and a number of affordable solutions were available with plethora of e-learning systems such as online courses and video lectures [1]. Recently, a number of QA systems have been proposed where users enter the query in their native language and the system gives the closely matched output [2, 3].

A QA system enables the student to engage in the learning process anytime and anywhere, at his or her own pace and leisure. This kind of online learning has been found effective. During the academic year 2002–2003, independent researchers at Eduventures found that 30,000 American high school students took classes online [4]. Through vivid resources, a student can get hands-on in any area of interest. The online learning world is not only limited to students but also helps the adults as well.

According to [5], the following improvements have been noticed by the top companies in the learning field

- Reduction in terms of time competence
- Increment in productivity through learning interventions
- Improvement in staff satisfaction/engagement
- Decrement in delivery time
- Improved ability to change products or procedures.

Online learning is not only diverse but it is also beneficial for the environment. Britain's Open University's study found that producing and providing e-Learning courses consumes an average of 90% less energy and produces 85% fewer CO_2 emissions per student than conventional face to face [6].

However, this medium of instruction fails when it comes to interaction. Online lectures and courses are one-way sources of information and fail to address the provision to answer the doubts and queries by the students. For example, a student can only watch the lecture but not ask questions that may arise. These learning interactions are missing, and hence the efficacy of online video lectures becomes questionable [7]. The research work in [7, 8] intends to eliminate this limitation by using these video lectures to produce a knowledge base that will serve as a source to answer the queries of the students. Various parameters such as use of presentations and error correction have been considered to generate accurate transcripts. The results obtained with and without including these parameters have been compared with human generated transcripts to see which option gives more accurate answers. This is done considering two approaches: (1) Natural Language Processing [9] and (2) Pattern Matching, to fulfill this intent.

Although transcript generation has been discussed in the previous literature, the transcripts are still not accurate and can be improved. Accuracy of transcript is a major factor because it forms the basis of further query extraction and answer fetching as well. If the transcript generated is itself not accurate, it will certainly lead to incorrect and irrelevant results. In the meanwhile, offline speech to text [10] options have been developed where they may be trained for personal uses. One such

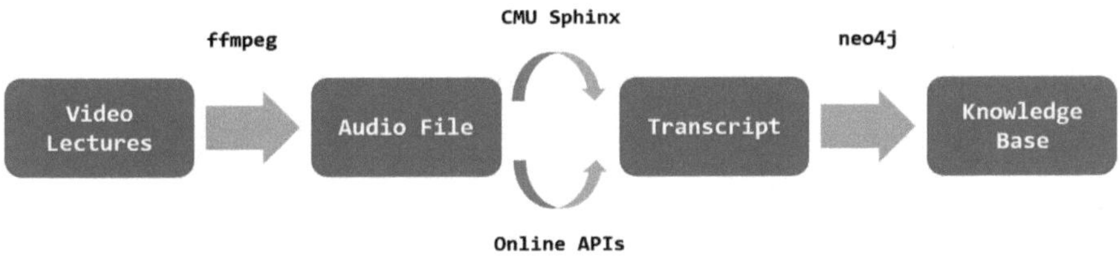

Fig. 1 The overall flow of the system

option is CMU Sphinx [11, 12], an open source for speech to text which needs to be trained first. However, manual training may consume many hours and also requires lot of efforts and is also prone to errors. Further, it is equally important that transcript is grammatically correct because grammar gives sense and meaning to a sentence. Wrong grammar may change the meaning of the sentence.

This paper works on improving the accuracy of the transcripts generated from the video lectures. Plethora of speech to text APIs are available, which convert speech input into text output [8]. Using these APIs to generate transcripts would indubitably provide much better results. In this paper, three such APIs are considered: (1) Google Cloud Speech Recognition [13], (2) Wit.ai [14] and (3) Microsoft Bing [15]. Second, a method to automate the training of Sphinx has been proposed to reduce the overhead involved in manual training [11]. This trained source is then used to generate transcripts which are further improved by including grammar check and grammar correction [16].

2 System Overview

In this section, the overall system design is discussed in detail. The flow of the system is shown in Fig. 1. The application takes video lectures as input and feeds them into knowledge base after converting these videos to transcripts. The system developed takes a video lecture as an input to generate a transcript which is fed into the knowledge base. Secondly, relevant answers are fetched from the knowledge base in case a query is made.

2.1 Technologies Used

2.1.1 Ffmpeg 2.8.4 -Python Package Index

Ffmpeg [17] is a software package that contains libraries and utilities to encode, decode and convert videos and audios of almost any format at a very fast rate. It reads from an arbitrary number of input files such as regular files, network streams, grabbing devices, etc. and writes to an arbitrary number of output files.

2.1.2 Speech to Text APIs

API stands for Application Program Interface. APIs are basically tools for building software applications. An API acts like a software intermediary that allows two applications to talk to each other. The APIs utilised in converting audios to text are called Speech to Text APIs. Various APIs such as Google Speech Recognition, Wit.ai, Microsoft Bing and CMU Sphinx are integrated with the application to compare results.

2.1.3 Neo4j

It is a Graph Database Management System. It has been used to create knowledge base which is in the form of graph. All the transcribed data goes into this knowledge base and it also acts as a source for answers for the queries [18].

The application requires that video lectures be saved locally on the system. The path to the video file is first required by ffmpeg to convert it into audio. This audio is taken through two routes. One route uses online APIs to generate notes from the audio whereas the other passes the audio through an offline source and puts grammar check and grammar correction afterwards to generate notes. The accurate notes out of the two are further added into the knowledge base using neo4j [18].

2.2 Architecture

The application mainly focuses on one module of the system described in the existing paper, which is the generation of the transcript. The working of transcription module is shown in Fig. 2. The intent to improve accuracy of transcripts is done taking two approaches and results from both of these approaches are compared. The two approaches discussed will be: (1) Transcription using online APIs and (2) through CMU Sphinx.

2.2.1 Transcription Using Online APIs

Online APIs such as Google Speech Recognition, Wit.ai, and Microsoft Bing have already been developed and can be accessed using authentication keys. Such APIs take audio file as input; audio being in particular formats, and converts it into transcripts, specifically lecture notes here. These notes are locally saved by the application and also fed into knowledge base and are accessible for evaluation.

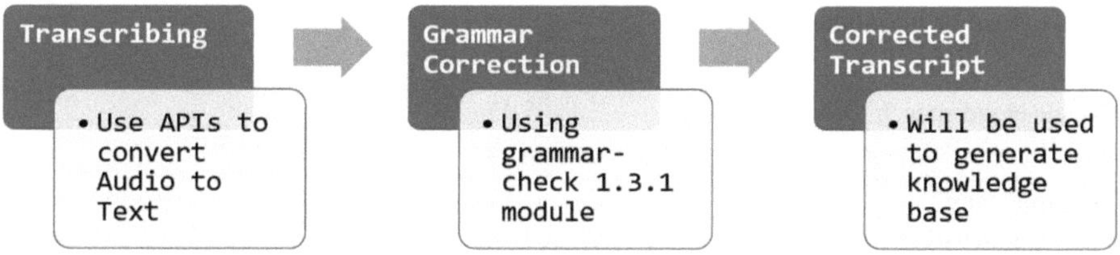

Fig. 2 Transcription module

2.2.2 Transcription Using CMU Sphinx

The sphinx was trained on a set of videos so as to include the domain specific terms into the dictionary. The videos used for training the sphinx were taken from Youtube along with their closed captions (CC), if possible. The audio part of the videos was downloaded using the youtube-dl along with the closed caption as only the audio is relevant for training. The downloaded audio files were converted into a wav file of 16,000 Hz sample rate, mono channel and normalized audio so as to enhance the audio for better clarity. The converted audio files and downloaded subtitle files or closed captions were then further processed to generate the .fileids and .transcription file. The wav files were then converted to the corresponding mfc files. The timestamp was removed from the closed captions file and the entire text was converted to lowercase after removing the punctuations. The .fileids file contains the list of all the audio files, without the extensions, that will be used for training in a line by line where each line contains name of a single audio acoustic file (.mfc file). The .transcription file contains the corresponding transcription, generated from closed captions. These files are then used to include the words or terms that were originally not present in the dictionary, basically the technical or domain specific terms.

The output generated through the APIs is further processed through grammar check and correct to detect grammatical flaws that may have occurred during transcription process and correct them accordingly. This is achieved using Language check module in python.

3 Evaluation Experiment and Results

Evaluation of transcripts generated from two sources is carried out and both the results are compared for accuracy. Online APIs have been built on the top of machine learning algorithms but unlike them, Sphinx is explicitly trained and it has been found that transcripts generated from Sphinx followed by processing from grammar check and correction give the best results. The results obtained by performing semantic analysis between actual and generated transcripts is shown in Table 1.

Table 1 Results from both approaches

API	Accuracy (%)
Google cloud speech recognition	85.47
Wit.ai	80.23
Microsoft bing	51.57
CMU sphinx (trained)	92.36

4 Conclusions and Future Research

The discussed work uses automatically trained CMU Sphinx for generating video transcripts, followed by grammar check and correction. The results show more accuracy than other standard proposed APIs. This paper considers the possibilities of training the CMU Sphinx on videos that are in English language. The work can further be extended by training CMU Sphinx for other languages and multilingual videos. Processing of videos in terms of filtering noise or improving the sound quality can be done before transcription process and it can also be automated to reduce human effort and improving video quality.

References

1. H. Yang et al., in *VideoQA: Question Answering on News Video*. Proceedings of the Eleventh ACM International Conference on Multimedia (ACM, New York, 2003)
2. Z. Virk, M. Dua, in *An Advanced Web-Based Bilingual Domain Independent Interface to Database Using Machine Learning Approach*. Proceedings of International Conference on Communication and Networks (Singapore, 2017), pp. 581–589
3. R. Devi, M. Dua, Performance evaluation of different similarity functions and classification methods using web based hindi language question answering system. Procedia Comput. Sci. **92**, 520–525 (2016)
4. J. Price, Definition of Online Education|Synonym. http://classroom.synonym.com/definition-online-education-6600628.html
5. Synergy Learning: 5 Interesting facts about eLearning. http://www.synergylearning.com/5-interesting-facts-elearning/
6. Talent LMS: 20 Interesting Facts About eLearning & LMS [Infographic]. https://www.talentlms.com/blog/20-facts-elearning-infographic/
7. K.M. Alhawiti, *Natural Language Processing and its Use in Education* (Computer Science Department, Faculty of Computers and Information technology, Tabuk University, Tabuk, Saudi Arabia, 2014)
8. J. Cao et al., in *Automated Question Answering from Lecture Videos: NLP vs. Pattern Matching*. Proceedings of the 38th Annual Hawaii International Conference on System Sciences, 2005, HICSS'05 (IEEE, Big Island, 2005), p. 43a
9. R. Collobert et al., Natural language processing (almost) from scratch. J. Mach. Learn. Res. **12**, 2493–2537 (2011)
10. P. Khilari, A Review on speech to text conversion methods (2015)
11. C.F. Liao, *Understanding the CMU Sphinx Speech Recognition System* (Department of Computer Science, National Chengchi University, 2002)

12. PocketSphinx: PocketSphinx API Documentation. https://cmusphinx.github.io/doc/pocketsphinx/index.html
13. Google cloud Platform: google-cloud. https://googlecloudplatform.github.io/google-cloud-python/
14. Wit.ai: Natural Language for Developers. https://wit.ai/
15. Microsoft Azure: Cognitive Service Try experience. https://azure.microsoft.com/enin/try/cognitive-services/?api=speech-api
16. Github, Inc., myint/language-check. https://github.com/myint/language-check
17. FFmpeg: FFmpeg. https://ffmpeg.org/
18. Neo4j Inc.: Neo4j Documentation. https://neo4j.com/docs/

A Novel Approach to Feature Hierarchy in Aspect Based Sentiment Analysis Using OWA Operator

Charu Gupta, Amita Jain and Nisheeth Joshi

Abstract In today's digital age opinions enable effective decision-making ability of both the customer and manufacturer. These opinions can be analysed using aspect-based sentiment analysis. The analysis uses the features of the product for determining the final polarity score. Researchers have focussed on extracting the features and giving them optimum weights (selected top k features) for final classification. A major drawback with these approaches is that they follow a uniform weight assignment scheme for all the features. It is observed that in any domain there can be a 'three-class-bifurcation' (either implicit or explicit)—low level, medium level, high level of features. In real-life the features belonging to high level are sometimes adapted at either low or medium level. This can be termed as overlapping of features. This encourages us to design a new feature hierarchy for effective weight distribution of overlapped features. The proposed methodology comprises of two steps: lookup and update. The resultant updated weights of the features can then be used for further processing through classification in aspect-based sentiment analysis.

Keywords Aspect-based sentiment analysis · Feature selection · Feature hierarchy · OWA operator · Web mining

1 Introduction

Sentiment Analysis [1] is a popular research topic in today's age. Today opinions and sharing of user's experience have become very popular. Sentiment analysis (docu-

C. Gupta
Bhagwan Parshuram Institute of Technology, Delhi, India

A. Jain (✉)
AIACTR (Ambedkar Institute of Advanced Communication Technologies & Research), Delhi, India
e-mail: amita_jain_17@yahoo.com

N. Joshi
Banasthali University, Vanasthali, Rajasthan, India

C. R. Krishna et al. (eds.), *Proceedings of 2nd International Conference on Communication, Computing and Networking*, Lecture Notes in Networks and Systems 46, https://doi.org/10.1007/978-981-13-1217-5_65

ment, sentence or aspect level) aims at extracting useful information at both micro and macro level [2, 3]. Aspect-based sentiment analysis (ABSA) [4, 5] has now being seen as having a high potential in determining the effective polarity [6]. This carefully extracted information (in terms of features) in the data is an asset to the decision makers.

With the growing market of electronic commerce the purchasing patterns have drastically changed. The ease to decide (by looking at online reviews) what to buy while sitting in one's living room is a depiction of the strength of technology we have today. Not only for individuals but for companies working in electronic commerce domain (B2C), it is essential to find out the performance of their product in online market. The internet is flooded with reviews, opinions, about different products or services offered by various companies. Looking at this scenario we can understand the need of analysis at two ends—customer and the company/manufacturer. First, at the customer end the need is to finely analyse what is being said over the internet about the products and services they plan to buy. Second on the company/manufacturer end where they analyse what is actually being said about their products.

It is observed that the features of the products define their characteristics. A careful analysis of the features reveal the most appropriate importance of the feature. Feature selection is an exhaustive and highly explored research area [7]. Features decide the decision maker's opinion. But a major drawback of existing approaches is that they do not consider the overlapping of features in two different segments (An exemplified illustration is given in the next section of problem statement). With the present work, this limitation of other existing approaches is overcome by using a feature hierarchy for the domain.

The rest of this paper is organized as follows: Section 2 discusses the problem statement with an illustrative example. Section 3 is devoted to preliminaries which have been used in developing the proposed methodology. Sections 4 and 5 give an insight to the proposed method and conclusion respectively.

2 Problem Statement

Sentiment analysis works on different levels of details (granularity). Aspect-based analysis focuses on extracting and classifying aspects (features) for effective determination of sentiment polarity.

Nowadays users write different types of reviews. One of the review by a user from "carwale.com" [8] on Renault Kwid[1] automobiles presented in Fig. 1. The review shows a clear fetching of feature payload from a higher segment car (Kwid compared

[1]Renault Kwid: The **Renault Kwid** is an entry-level crossover produced by the French car manufacturer **Renault**, intended for the Indian market.

Fig. 1 Renault Kwid (Review by Sandeep Nangia, 11 months ago on "carwale.com")

> With these exteriors Mini Duster has born. Engine performance is good not excellent if you compare with AltoK10 or WagonR Renault Engines are much superior in this segment.

in looks with Mini Duster[2]). This feature overlapping is essential to consider and this is the idea on which the proposed feature hierarchy is built.

3 Materials and Methods

In this section we define the concepts of major components which are used in developing the proposed methodology.

3.1 Sentiment Analysis

Web mining refers to the techniques used to scrape the internet to uncover key data points that can be analysed to provide intelligence relevant to the task at hand [9].

Web Mining trends include Social Network Mining, Sentiment Analysis, Web Semantics (refer Fig. 2). Sentiment analysis is a field of study that analyses people's opinion, sentiments, organizations, individuals, appraisals, issues, events, topics and their attributes [10]. Finding Sentiment can be formally defined as finding the quadruple (s, g, h, t) where, 's' represents the sentiment, 'g' represents the target object, 'h' represents the holder and 't' represents the time at which the sentiment was expressed.

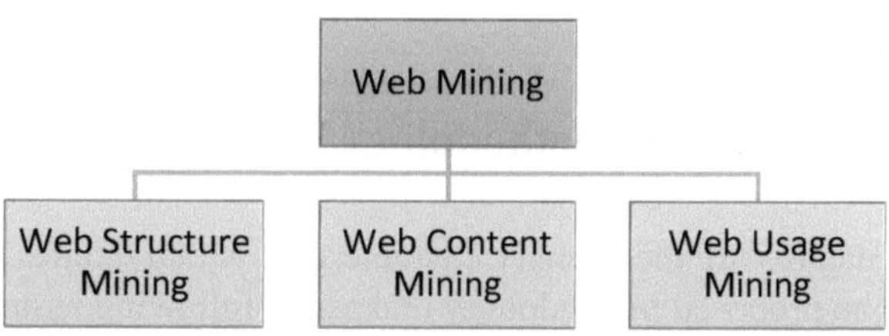

Fig. 2 Web mining classification

[2]Duster: The **Dacia Duster** is a compact sport utility **vehicle** (SUV) by the French manufacturer **Renault** since 2010.

As in sentiment analysis we are modelling human emotions that are too large, sentiment polarity is used. Polarity describes the direction of the sentiment. It can be positive, negative or neutral [11]. Sentiments in rich corpora can be analysed at different levels. These different levels of Analysis are

Document level: To classify whether an opinion document expresses a positive or negative sentiment polarity.

Sentence Level: To determine whether each sentence expressed a positive, negative or neutral (subjective classification) opinion.

Entity and Aspect Level: It is a fine-grained analysis called feature level or feature based opinion mining. This level computes the sentiment by looking at the opinion directly. An opinion consists of a pair of Sentiment and a Target.

At both the Document Level and Sentence Level, the values of the sentiment which are computed are not associated with the topics directly [12]. In contrast, aspect level analysis finds sentiment-target pairs in a given text. This encourages a more specific analysis at aspect level.

3.2 OWA Operator

Ordered Weighted Average operator, considers decision-making under ignorance as opposed to belief structures [13]. It is a weighted average of ordered values of a variable. Given a set of values C_1, C_2,..., C_n; OWA consists of choosing a normalized set of weighting factor $W = [W_1, W_2, \ldots, W_n]$ where,

$$W_{(i)} \in [0, 1] \quad \text{and} \quad \sum_i W_{(i)} = 1 \tag{1}$$

Also Compute OWA (C_1, C_2, ..., C_n) as follows:

$$\text{OWA}(C_1, C_2, \ldots, C_n) = \sum_i W_{(i)} \cdot B_{(i)}, \tag{2}$$

where, $B_{(i)}$ is the ith largest element in (C_1, C_2, ..., C_n). A detailed description of the proposed methodology is given in Sect. 4.

4 Proposed Methodology

Features are the primary characteristic which defines the productivity of a product. The proposed methodology makes the following assumptions:

1. From the feature list of the domain, one of the feature is considered for defining the level boundaries. The domain under consideration has three levels—low, medium and high for optimum classification of other features.

Table 1 Construction of feature hierarchy

Level boundaries	Features	Weights
High level	f_1	a
	f_2	b
	f_3	c
	f_4	..
Medium level	f_5	..
	f_6	..
	f_7	..
	f_8	..
Low level	f_9	..
	f_{10}	..

2. For a particular level, the features are arranged in an increasing weight order. Approaches like information gain cab be used as weighting criteria of the features.

After considering these assumptions, the proposed system constructs a feature hierarchy. The structure of this hierarchy is given in Table 1. It consists of a feature set (f_1, …, f_n) for each of the level boundaries- High, Medium and Low. Weights are given to the features using information gain [14].

The significance of the proposed feature hierarchy is that the features are arranged in decreasing order of the class (level) quality and not based on decreasing/increasing feature weight.

Automotive domain is selected for the proposed architecture. It is observed that the parameters can be easily understood in this domain. Automobiles have various features like, mileage, turning radius, boot space, infotainment, pop-up tweeters, champagne flutes and so on. The level can be decided on the basis of the feature price. Through careful analysis, it is observed that some features which belong to the High-Range automobiles are adapted in new models of Medium-Range- or Low-Range automobiles.

In this case, the feature weight for that particular Level becomes more. For this, the proposed methodology employs two steps.

4.1 Lookup

Let us consider a feature f_z, mostly found in High-Range automobiles. But a user (who reviewed a medium-range automobile $M_{x,}$ where 'x' refers to the segment) has given a positive adaptation of feature f_z in M_x.

Now the weight of overlapped f_z (W_{fz}) should be added and updated for medium-range automobiles.

4.2 Update

The update action triggers updating the weight of f_z (W_{fz}) for M_x and thus W_{fz} is updated using OWA operator in the following Eq. 3:

$$\text{New } W_{f_z}(M_x) = \text{OWA}\left(w_{f_1}, w_{f_2}, \ldots, w_{f_n}\right) \tag{3}$$

The OWA operator enhances the process of updating the feature weight which helps in effective weight evaluation of the aspects in aspect based sentiment analysis.

5 Conclusion

In this paper, a model of feature hierarchy is built for aspect based sentiment analysis. This feature hierarchy will enable effective approximation of sentiment polarity, determination of implicit aspects and can be applied on any domain of interest. Aspect extraction is not considered here. The proposed method uses lookup and update method to evaluate the final weight of the overlapping feature. Lookup saves the time of searching for feature weight because the feature hierarchy is segmented from high level to low level. In continuation to the proposed method, the classification of aspects can be effectively enhanced.

References

1. B. Liu, *Sentiment Analysis and Opinion Mining*, 1st edn. (Morgan & Claypool, San Rafael, 2012)
2. W. Medhat, A. Hassan, H. Korashy, Sentiment analysis algorithms and applications: A survey. Ain Shams Eng. J. **5**, 1093–1113 (2014)
3. S.M. Mohammad, Challenges in sentiment analysis, in *A Practical Guide to Sentiment Analysis*, ed. by D. Das, E. Cambria, S. Bandyopadhyay (Springer, Berlin, 2016)
4. K. Schouten, F. Frasincar, Survey on aspect level sentiment analysis. J. Knowl. Data Eng. **28**(3), 813–830 (2016)
5. Y. Jo, A.H. Oh, in *Aspect and Sentiment Unification Model for Online Review Analysis*. Proceedings of the Forth International Conference on Web Search and Web Data Mining (2011), pp. 815–824
6. O.D. Clercq, in *The Many Aspects of Fine Grained Sentiment Analysis: An Overview of the Task and its Main Challenges*. Second International Conference on Human and Social Analytics: HUSO (2016), pp. 23–28
7. C. Nicholls, F. Song, Comparison of feature selection methods for sentiment analysis, in *Advances in Artificial Intelligence. AI 2010. Lecture Notes in Computer Science*, vol. 6085, ed, by A. Farzindar, V. Kešelj (Springer, Berlin, 2010)
8. https://www.carwale.com/renault-cars/kwid/userreviews-p2/. Last Accessed 28 Sept 17
9. K. Sharma, G. Shrivastava, V. Kumar, Web mining: Today and tomorrow, in *ICECT*, vol. 1 (2011)

10. K. Schouten, F. Frasincar, Survey on aspect –level sentiment analysis. IEEE Trans. Knowl. Data Eng. **28**(3), 813–830 (2015)
11. M. Hu, B. Liu, in *Mining and Summarizing Customer Reviews*. Proceedings of 10th ACM SIGKDD International Conference on Knowledge Discovery and Data Mining (2004), pp. 168–177
12. H. Tang, S. Tan, X. Cheng, A survey on sentiment detection of reviews. Expert Syst. Appl. **12**(3), 10760–10773 (2009)
13. R.R. Yager, On ordered weighted averaging aggregation operators in multicriteria decision making. IEEE Trans. Systems Man Cybernet. **18**, 183–190 (1998)
14. K. Schouten, F. Frasincar, R. Dekker, An information gain driven feature study for aspect based sentiment analysis, in *Natural Language Processing and Information Systems* (2016), pp. 48–59

Analyzing Behavior of ISIS and Al-Qaeda using Association Rule Mining

Tamanna Goyal, Jaspal Kaur Saini and Divya Bansal

Abstract Social media have been exploited by terrorist groups to share their massacres plans, recruitment, buying weapons, or propagating their violent plans. Terrorist groups named ISIS and Al-Qaeda are the most active and well known for using social media to propagate their violent intents over online discussion forums. It becomes necessary to study the behavior of these terrorist groups over online social media. In this paper, we present association rule mining based approach to extract a feature set for terroristic groups named ISIS and Al-Qaeda. We used the Global Terrorism Dataset which contains systematic information on terrorist attacks worldwide since 1970. Entropy-based feature extraction technique is used to extract top features which are then further used to find association rules. Eclat (Equivalence Class Transformation) and Apriori algorithms are used to mine association rules from prepared data. Rules for ISIS and Al-Qaeda are computed separately, and are then further classified using machine learning classification algorithms. Our research contributes to the smart and novel application of data mining algorithms and computational intelligence to study the behavior of the most popular and active terrorist groups over social media.

1 Introduction

Recent years have witnessed exploitation of online social media by terrorist groups in many ways. Networking, sharing information, planning, recruitment, and mobilization are set of terroristic activities which have done over online social media. The Irish Republican Army, Naxalites, Boko Haram, Hezbollah, ISIS (Islamic States of

T. Goyal · J. K. Saini (✉) · D. Bansal
Punjab Engineering College, Chandigarh, India
e-mail: sainijassi87@gmail.com

T. Goyal
e-mail: goyaltamanna161@gmail.com

D. Bansal
e-mail: divya@pec.edu.in

C. R. Krishna et al. (eds.), *Proceedings of 2nd International Conference on Communication, Computing and Networking*, Lecture Notes in Networks and Systems 46, https://doi.org/10.1007/978-981-13-1217-5_66

Iraq and Syria), Tehrik-i-Taliban Pakistan, Lashkar-e-Taeba, Taliban, and Al-Qaeda are terrorist groups which are getting active over online social networks such as Twitter or dedicated dark web forums [1, 2]. ISIS and Al-Qaeda are two most popularly known terrorist groups which are well known for using social media to propagate their agenda and violent activities. It becomes a challenging task for security agencies to explore and mine the behavioral features of these terrorist groups. Several research organizations have taken charge to examine trends and patterns of terrorists to study behavioral models of terrorists.

Figure 1a shows the word cloud of top 50 terrorist groups as reported by Global Terrorism Dataset (GTD). GTD is open source dataset which contains information about various terroristic incidents in a form of 134 attributes [3]. The font size of each terrorist group name is proportional to number of attacks done by terrorist group as reported by GTD. It is evident from the figure that Boko Haram, Al-Qaeda, ISIS, and Al-Shabaab are the most active terrorist group since 1970, which in terms is an open research challenge to researchers in the field to study their behavior and attack patterns. Figure 1b depicts the number of attacks done by ISIS and Al-Qaeda as per reported by GTD. It is seen that a number of attacks are increasing exponentially over years which opens an era for researchers to study the behavior and analyze the patterns of attacks by ISIS and Al-Qaeda in order to counter terrorism.

In this paper, we analyzed the behavior of two terrorist groups, i.e., ISIS and Al-Qaeda, by understanding the patterns of attacks done by both the groups in different countries. The data of these groups is compiled and preprocessed from GTD and analyzed further. We applied Apriori and Eclat algorithms to mine the association rules and classified these rules using naive Bayes, SVM, and decision tree algorithms to make predictions for the future. This paper is further organized as follows: Sect. 2 elaborates literature survey in the same domain. Section 3 explains the detailed proposed methodology. Experimental results are illustrated in Sect. 4, and finally, Sect. 5 concludes the paper and lists the future scope of the work done.

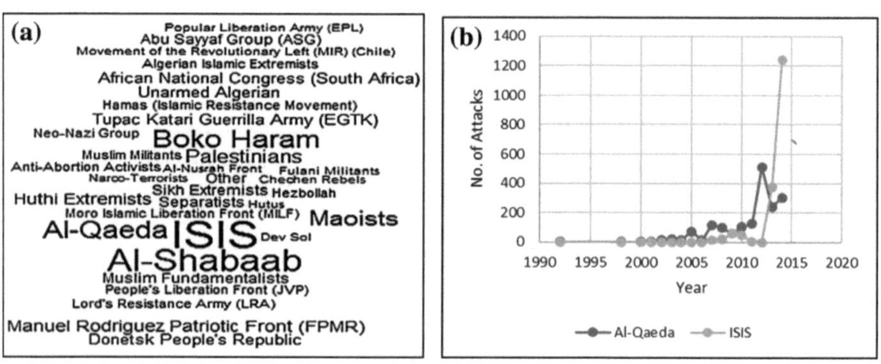

Fig. 1 **a** Top 50 Terrorist Groups Worldwide as reported by GTD **b** No. of Attacks since 1990 by ISIS and Al-Qaeda as reported by GTD

2 Related Work

Computational techniques from machine learning, data mining, and social network analysis are studied in literature to highlight which techniques are used by researchers to study behavior of terrorist groups [4–11]. Literature survey is divided into two parts. First, we studied what are techniques used in all research evidences as listed in Table 1. Second, we looked at which terroristic organizations have been targeted by research agencies. Table 2 describes work done on various terrorist groups.

Table 1 Different techniques used for the study of different terrorist groups

Technique used	Citation
Data preprocessing techniques, data mining, and classifiers	[12]
COSM (Crime ontology similarity measure)	[13, 14]
Natural Language Processing (NLP), event clustering, event trending, and narrative generation	[15]
Stochastic Opponent Modeling Agents (SOMA)	[16, 17]
Naive Bayes and Apriori algorithm	[18]
Class Association Rule mining (CPAR and CMAR)	[19]
Classification techniques (Naive Bayes, SVM, decision tree, and multilayer perceptron)	[20]
Data mining techniques, Classifiers—Naive Bayes, SVM, and AdaBoost	[21]
Clustering techniques	[22]
Logistic regression, Support Vector Machines (SVM), boosting, naive Bayes models, and classification trees	[23]
STONE	[24]
Rule mining techniques	[25]

Table 2 Research work done on different terrorist groups

Terrorist group	Work Done	Citation
Lashkar-e-Taiba (LeT)	LeTs activities, behavior, and environment are studied in Pakistan and J&K using SOMA	[4, 16, 17]
ISIS	Analyzed and classified the tweets of ISIS using machine learning	[5, 10, 21]
Al-Qaeda	The magazines produced by Al-Qaeda are examined to estimate the similarity and difference among ISIS, Al-Qaeda, and Taliban	[5]
Terrorist groups of Istanbul	On the basis of criminology perspective, COSM (ontology-based similarity measure) is used to classify them into three groups: extreme left, separatism, and extreme right groups	[13]
Terrorist groups of Turkey	COSM is used to classify the terrorist network on the basis of their similarity	[14]
Hezbollah	SOMA is used to predict the behavioral model of this terrorist behavior	[17]

Related work done in the field illustrates that there is no prior work done to extract features from GTD about popular terrorist groups ISIS and Al-Qaeda, and hence, demands to extract a feature set of these violent extremist groups which can assist any security agency worldwide to perform behavioral rule-based predictions about future attacks.

3 Proposed Methodology

For a given item set, association rule mining algorithms can be utilized to discover frequent patterns and draw inferences based on the discovered patterns. We used Eclat and Apriori algorithms which are described further. Eclat (Equivalence Class Transformation) works on the basis of a depth-first search, whereas Apriori works on the basis of a breadth-first search. Eclat uses simple intersection operations for equivalence class clustering along with bottom-up lattice traversal to mine frequent itemsets and works well for smaller datasets. Apriori counts the support of itemsets and uses a candidate generation function which exploits the downward closure property of support and works well for large datasets. Our proposed behavioral rule-based prediction algorithm can be described step by step:

1. Extract and prepare data about ISIS and Al-Qaeda from Global Terrorism Dataset.
2. Extract useful features from this data using entropy and information gain as mentioned below:

$$Entropy(set) = \sum_{i=1}^{k} -P(value_i)log_2P(value_i) \quad (1)$$

   where P(value)$_i$ is the probability of getting the ith value when randomly selecting one from the set.
3. Association rules are calculated with extracted features using Apriori and Eclat algorithms by setting different support and confidence levels.
4. Classify the rules using different classification algorithms.
5. Compute different performance metrics accuracy, sensitivity, specificity, etc.
6. Verify and validate the results.

4 Experimental Results

By using information gain and entropy, 37 attributes are selected out of 134 attributes from Global Terrorism Dataset. The top seven out of 37 variables that contributed the most are as follows:

Table 3 Sample extracted rules using Apriori algorithm

Rule No	Rule (lhs=> rhs)
1	attacktype1_txt=Bombing/Explosion=> country_txt=Iraq
2	weaptype1_txt=Explosives/Bombs/Dynamite, weapsubtype1_txt=Vehicle=>attacktype1_txt= Bombing/Explosion
3	attacktype1_txt=Bombing/Explosion=> weaptype1_txt=Explosives/Bombs /Dynamite
4	targtype1_txt=Military,natlty1_txt=Yemen =>country_txt=Yemen
5	attacktype1_txt=Bombing/Explosion, natlty1_txt=Iraq=>weaptype1_txt =Explosives /Bombs/Dynamite

1. Target1: Target of incident like military, public place, government, transportation, etc.
2. City: It defines name of the city, village, or town in which the incident occurred.
3. Location: Information about location of incident.
4. Weapsubtype1_txt: Type of weapon used for the attack like bomb, firearms, biological, etc.
5. Nalty1_txt: Nationality of target of attack.
6. Country_txt: Name of country attacked.
7. Attacktype1_txt: Type of attack like assassination, hijacking, kidnapping, bombing/ explosion, etc.

Feature names used above are same as used in Global Terrorism Dataset Codebook [3]. The rules are extracted by Apriori algorithm by considering different supports and confidences. Sample extracted rules are shown in Table 3.

Rule (1): When the type of attack is bombing or explosion, then the country is Iraq.

Rule (2): When the main weapons used are explosives, bombs, or dynamite and a vehicle is used as another weapon, then attack tends to be bombing or explosion.

Rule (3): When a bombing or explosion is used for attack, then the main weapon used for attack would be explosives, bombs, or dynamite.

Rule (4): When the target is military and nationality of attack is Yemen, then country would be Yemen.

Rule (5): When the type of attack is bombing or explosion and nationality of attack is Iraq, then weapon used for the attack would be explosion, bombs, or dynamite.

After associative rule mining, these rules are classified using different classifiers including Naive Bayes, SVM, and decision tree. In the classification method, to construct a model, a training set is required which comprises a set of attributes with one attribute being the attribute of the class. Thus, a random sample of one-third of the dataset of rules is considered for training and the trained model is used for classification on the basis of the terrorist group name and by using this trained model the instance is classified. The remainder of the dataset is used for testing the model.

Table 4 Performance measures

S. No.	Performance measure	Naive Bayes	Decision tree	SVM
1	Specificity	0.98	0.95	0.99
2	Sensitivity	0.96	0.80	0.99
3	Positive predictive value	0.95	0.63	0.99
4	Negative predictive value	0.99	0.98	0.99
5	Kappa measure	0.91	0.69	0.98
6	Accuracy	0.98	0.93	0.99

We used R programming to perform all experiment. The output of these classifiers is compared on the basis of different parameters, as shown in Table 4.

It can be inferred from Table 4 that SVM emerged as the best classifier for the classification of our association rules of ISIS and Al-Qaeda.

5 Conclusion

Social media emerges to give all of us a platform to share and discuss any information or opinions. But violent extremists have exploited these online social networks for their malicious intent. We proposed a novel approach to find associations among two popularly known terrorist groups: ISIS and Al-Qaeda. We further automated the process to identify whether attacks were done by ISIS or Al-Qaeda using machine learning classification algorithms. SVM is the best classifier for our case as compared to naive Bayes and decision tree. We discovered seven critical feature sets named as target, city, location, weapon type, nationality, country, and attack type description, to find association rules. It is seen that when type of attack is bombing or explosion and nationality of attack is Iraq, then the attacking group is ISIS, and when the main weapons used are explosives, bombs, or dynamite and vehicle are used as another weapon, then the attacking group is Al-Qaeda.

In future, this work can be extended to study more terrorist groups. Different features can be accommodated from social media also to study behavior of terrorist groups.

References

1. Accessed on 15 Aug 2017, http://listovative.com/top-10-most-dangerous-terrorist-organizations-in-the-world/
2. V.S. Subrahmanian ed., Handbook of Computational Approaches to Counterterrorism (Springer Science & Business Media, 2012)
3. Global Terrorism Database. Accessed 19 May 2016, https://www.start.umd.edu/gtd/
4. A. Mannes et al., A computationally-enabled analysis of Lashkar-e-Taiba attacks in Jammu and Kashmir. 2011 European Intelligence and Security Informatics Conference (EISIC) (IEEE, 2011)

5. T. Mahmood, K. Rohail, Analyzing terrorist events in Pakistan to support counter-terrorism-Events, methods and targets, in *2012 International Conference on Robotics and Artificial Intelligence (ICRAI)* (IEEE, 2012)
6. Jacob R. Scanlon, Matthew S. Gerber, Automatic detection of cyber-recruitment by violent extremists. Sec. Inf. **3**(1), 1–10 (2014)
7. F. Spezzano, V.S. Subrahmanian, A. Mannes, Reshaping terrorist networks. Commun. ACM **57**(8), 60–69 (2014)
8. P. Su et al., Mining actionable behavioral rules. Decis. Support Syst. **54.1**, 142–152 (2012)
9. U.K. Wiil, J. Gniadek, N. Memon, Measuring link importance in terrorist networks, in *2010 International Conference on Advances in Social Networks Analysis and Mining (ASONAM)* (IEEE, 2010)
10. D.B. Skillicorn, Empirical assessment of al qaeda, ISIS, and taliban propaganda, in *2015 IEEE International Conference on Intelligence and Security Informatics (ISI)* (IEEE, 2015)
11. D.B. Skillicorn et al., Understanding South Asian Violent Extremist Group-group interactions, in *2014 IEEE/ACM International Conference on Advances in Social Networks Analysis and Mining (ASONAM)* (IEEE, 2014)
12. WITS - Home Page. Accessed 19 May 2016, http://wits.worldbank.org/WITS/WITS/Default-A.aspx?Page=Default
13. International Terrorism: Attributes of Terrorist Events (ITERATE) | Duke University Libraries. Accessed 19 May 2016, http://library.duke.edu/data/collections/iterate
14. S. Ginsberg, 2012 Database Spotlight: Minorities at Risk Organizational Behavior (MAROB) | START.umd.edu. Accessed 19 May 2016, https://www.start.umd.edu/news/database-spotlight-minorities-risk-organizational-behavior-marob
15. E. Spertus, M. Sahami, O. Buyukkokten, Evaluating similarity measures: a large-scale study in the orkut social network, in *Proceedings of the Eleventh ACM SIGKDD International Conference on Knowledge Discovery in Data Mining (KDD '05)* (ACM, New York, NY, USA, 2005), pp. 678–684
16. J Pagn, Improving the classification of terrorist attacks a study on data pre-processing for mining the Global Terrorism Database, in *2010 2nd International Conference on Software Technology and Engineering (ICSTE)*, vol. 1 (IEEE, 2010)
17. E. Serra, V.S. Subrahmanian, A survey of quantitative models of terror group behavior and an analysis of strategic disclosure of behavioral models. IEEE Trans. Comput. Soc. Syst. **1.1**, 66–88 (2014)
18. F. Ozgul, C. Atzenbeck, Z. Erdem, How much similar are terrorists networks of istanbul?, *2011 International Conference on Advances in Social Networks Analysis and Mining (ASONAM)* (IEEE, 2011)
19. F. Ozgul et al., *Intelligence and Security Informatics*, Specific similarity measure for terrorist networks: how much similar are terrorist networks of Turkey? (Springer, Berlin, 2011), pp. 15–26
20. L. Guohui, et al. Study on correlation factors that influence terrorist attack fatalities using Global Terrorism Database. Procedia Eng. **84**, 698–707 (2014)
21. J. Lauten chlager, et al., Group Profiling Automation for Crime and Terrorism (GPACT). Procedia Manuf. **3**, 3933–3940 (2015)
22. A.R. Kulkarni, V. Tokekar, P. Kulkarni, Identifying context of text documents using Nave Bayes classification and Apriori association rule mining, in *2012 CSI Sixth International Conference on Software Engineering (CONSEG)* (IEEE, 2012)
23. Bangaru Veera Balaji, Vedula Venkateswara Rao, Improved classification based association rule mining. Int. J. Adv. Res. Comput. Commun. Eng. **2**(5), 2211–2221 (2013)
24. A. Al Deen, M. Nofal, S. Bani-Ahmad, Classification based on association-rule mining techniques: a general survey and empirical comparative evaluation. Ubiquitous Comput. Commun. J. **5.3** (2010)
25. M. Ashcroft et al., Detecting jihadist messages on twitter, in *EISIC 2015*, Manchester, UK. IEEE Computer Society (2015)
26. RAND Corporation, RAND HRS Data File (v.N) | RAND (1994). Accessed 19 May 2016, http://www.rand.org/labor/aging/dataprod/hrs-data.html

A Hybrid Classification Method Based on Machine Learning Classifiers to Predict Performance in Educational Data Mining

Keshav Singh Rawat and I. V. Malhan

Abstract Machine learning algorithm can be applied in education data mining (EDM) to extract knowledge. Educational data mining is an important practice of automatic extraction and segmentation of useful information from the education data sources. This paper is focused on comparison and study of hybrid model of classification and machine learning algorithms based on decision tree, clustering, artificial neural network, Naïve Bayes, etc. This paper introduces concepts of popular algorithm for new researchers of this area. The paper discusses hybrid classification model using machine learning algorithms using voting that can be used to analyze the performance of students. We have used open source data mining tool Weka for a practical experiment on data set of students that serve the purpose of prediction, classification, visualization, etc. The findings of this paper reveal that hybrid method of classification are more efficient for prediction of student-related data.

Keywords Machine learning · Educational data mining · Decision tree · Hybrid classification · NB · ANN

1 Introduction

Data mining is a process of automatics extraction of useful information from the large data sets repositories, etc. Data mining helps to find relationships and discover patterns by using machine learning algorithm, statistics, and visualization. Data mining application in the field of education facilitates better discovery of the learning resources, and improves efficiency of learning process. The main objective of educational data mining is to improve the learning quality and understanding that helps us

K. S. Rawat (✉) · I. V. Malhan
Department of Computer Science & Informatics, Central University of Himachal Pradesh, Dharamshala, Himachal Pradesh, India
e-mail: keshav79699@gmail.com

I. V. Malhan
e-mail: imalhan_47@rediffmail.com

C. R. Krishna et al. (eds.), *Proceedings of 2nd International Conference on Communication, Computing and Networking*, Lecture Notes in Networks and Systems 46, https://doi.org/10.1007/978-981-13-1217-5_67

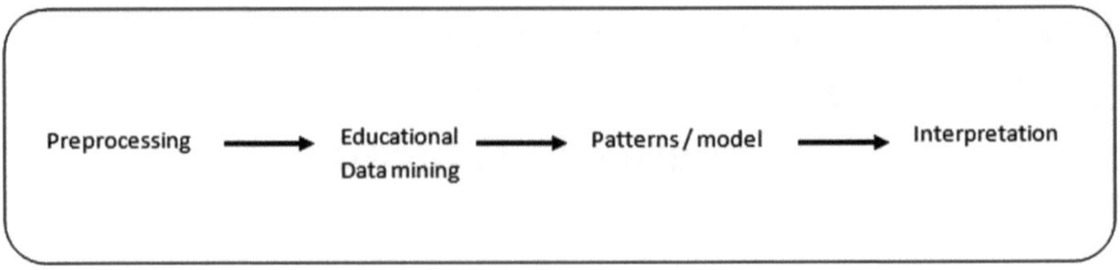

Fig. 1 Process of educational data mining (EDM)

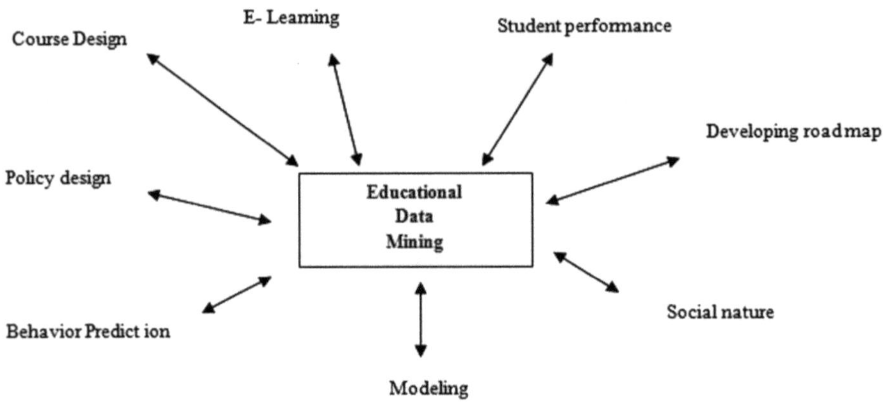

Fig. 2 Process of educational data mining (EDM)

to improve learning as well as strategies designing for administrative planning [1]. In this paper, we discuss machine learning algorithms applied on the student data set. These machine learning algorithms are capable to extract useful information from education-related data sets.

The process of educational data mining (EDM) is described in Fig. 1 [2, 3]. EDM contains mainly four stages. In the first phase, we discussed relationship between data, the second phase contains validation process, the third phase predicts result and the final phase is used for decision-making process. Based on the output of the third phase, the EDM is applied in some areas [4] described in Fig. 2.

The organization structure of the paper is as follows. Section 2 contains some of the interrelated work in the literature review. Section 3 discusses the machine learning algorithm used in this study. Section 4 describes the proposed work. Section 5 shows experiment and results of this paper. Section 6 concludes the work and final section contain references.

2 Literature Review

Many papers have been published in the area of educational data mining. Data mining in education field is very useful to extract useful information of educational outcomes that helps to administrators to make policies regarding learning and performance of students. The author of paper [5] discussed and reviewed various applications, tools, and concepts of EDM. The applications and tasks of educational data mining are discussed in detail in [6] and reviewed educational data mining tasks and decision support system to improve the performance of student. The research in the field of educational data mining (EDM) and educational learning analysis improved the overall education system and decision support system [7]. The hybrid classification approach [8] was described method of ensemble various machine learning algorithms to achieve better performance in term of precision, recall, and F measure on various data sets.

3 Machine Learning Algorithms

Machine learning is a method of producing algorithm to predict useful information from source data set. Machine learning algorithms can be applied in supervised or unsupervised way. The process of machine learning contains three basic steps—Initialization, learning and testing that provide efficient and accurate output data. This field can be applied on many application areas—face identification, weather prediction, speech reorganization, classification, fraud detection, spam filtering, and so on. In this paper, we discuss some important machine learning algorithms of data mining and implementation of these algorithms to estimate the result of students. The four machine learning technique—decision tree, Naïve Bayes, multilayer perception, and K-nearest neighbor algorithms have been used on student data set in this paper.

3.1 Decision Tree (J48)

Quinlan [9] developed decision tree algorithms ID3 and C4.5. In Weka, C4.5 is known as J48 algorithm. This tree method based on divide and conquer approach to generate decision tree by recursive calls. C4.5 algorithm is an advanced algorithm as compared with ID3 and overcomes all difficulties and limitations of ID3. This algorithm is used gain ration as feature selection of an attribute to create decision tree. The attribute with highest gain ratio holds the position of parent node and next values hold the position of children and so on till the completion of decision tree. There are two ways to create tree, pruned, or unpruned. The pruning method helps to improve the efficiency of decision tree by removing unwanted branches. Figure 3 shows the decision tree on student data set using J48 algorithm in Weka.

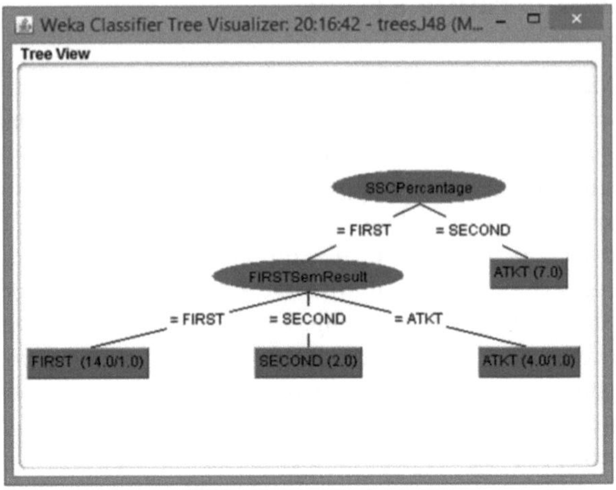

Fig. 3 Decision tree structure

3.2 K-nearest Neighbor Algorithm (IBK)

The K-nearest neighbor algorithm for classification in Weka is known as IBK algorithm comes under lazy learning algorithms. IBK algorithm uses Euclidean distance to measure K-nearest neighbor for classification data set.

3.3 Naïve Bayes (NB)

It is the popular classification technique based on conditional probabilities and Bayes theorem [10]. Naïve Bayes classification method is very useful for text classification. It takes less training data for classification of data.

3.4 Multilayer Perception (ANN)

In Weka, artificial neural network concept is offered by multilayer perception (MLP). This method contains an input layer, an output layer, and some hidden layers and used backpropagation approach for classification.

4 Proposed Method

The decision tree (J48), Naïve Bayes (NB), K-nearest neighbor (IBK) and multi-perception (ANN) machine learning algorithms are used to develop hybrid classification model using voting method. The detailed description of method is reflected in Algorithm 1. Here, all four algorithms mentioned above are grouped using voting approach to get better efficiency of prediction. A 10-fold cross-validation is used to predict the evaluation accuracy.

Algorithm1: Proposed work
I/P: Student data set for classification
O/P: Performance of classification in terms of accuracy.

Steps:
1. Input data set of students.
2. Apply preprocessing to remove unwanted information and attributes have low information gain.
3. Apply machine learning algorithms decision tree (J48), Naïve Bayes (NB), K- Nearest Neighbor (IBK), Multilayer perception (ANN) to student data set independently.
4. Development of hybrid classification model-
   (i) Find classification hypothesis of J48, IBK, ANN and NB using ensemble.
   (ii) Perform voting process by combining classifiers using average posterior probabilities rule.
5. Compare machine learning algorithm J48, NB, IBK and ANN with hybrid classification in terms of correctly classified instance, incorrectly classified instances and accuracy.
6. End.

5 Experiment and Result

This section contains implementation of machine learning algorithm on data set of students to analyze results. We used open-source Weka tools for the experiment purpose. Weka tool supports various data mining techniques and their algorithms to classify given data set. We used student data set of Department of Computer Science in csv format for classification and prediction. Data set can be easily transformed in arff format by ARFF tool provided by Weka. The student data set contains both academic and personal data. The attributes of data set are—Roll no, Address, Father's occupation, Mother's occupation, Father's Education, Gender, Caste, SSC Result, Graduation Result, First Semester Result, and Second Semester Result. The student data set contains around 27 instances and 11 attributes. The actual student data was in a continuous form that was converted into nominal form for experiment analysis. Accuracy rate is used to evaluate model of classification. The correct and incorrect classified attributes are defined by confusion matrix. In Confusion matrix, true positive and false positive rate are used to evaluate the classification model. The general confusion matrix is shown in Fig. 4, where rows indicate actual class and columns indicate prediction class. TP and TN represent the total number of true

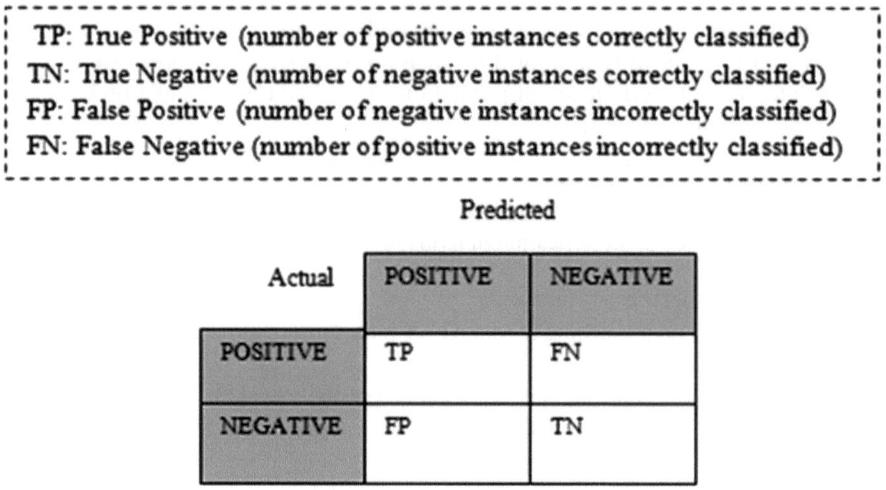

Fig. 4 Confusion matrix of binary class

positive correctly classified instance and the total number of true negative incorrectly classified instance. Accuracy of classifier is defined as the total number of correctly classified instances.

The overall performance [11] of classification can be measured by the following parameters:

Accuracy of classification is defined as the total number of instances that are correctly classified.

$$\text{Accuracy} = \frac{\text{TP} + \text{TN}}{\text{TP} + \text{TN} + \text{FP} + \text{FN}} \tag{1}$$

Table 1 show comparisons of classifiers efficiency in terms of correctly classified, incorrectly classified between proposed hybrid model and individual classifiers, i.e., J48, NB, IBK, and ANN. It is observed that the proposed hybrid method of classification achieved highest correctly classified instance over individual classifiers. Table 2 shows comparisons of accuracy or precision between proposed hybrid model and individual classifiers, i.e., J48, NB, IBK, and ANN.

It is observed that the proposed hybrid method of classification as shown in Fig. 5 is more accurate than individual classifiers, i.e., J48, NB, IBK, and ANN. This hybrid method achieved the highest accuracy of 92.59% and individual classifiers, i.e., J48, NB, IBK, ANN achieved an accuracy of 85.18, 81.48, 88.88, and 88.88% respectively.

Table 1 Comparisons of classifiers efficiency

Classifier	Execution time (s)	Efficiency (%)	Correctly classified instances	Incorrectly classified instances
Decision tree (J48)	0.41	85.18	23	4
K-nearest neighbor algorithm (IBK)	0	88.88	24	3
Naïve Bayes (NB)	0	81.48	22	5
Multilayer perception (ANN)	0.13	88.88	24	3
Proposed hybrid method	0.03	92.59	25	2

Table 2 Comparisons of classifiers accuracy

Classifier class	Decision tree (J48)	K-nearest neighbor algorithm (IBK)	Naïve Bayes (NB)	Multilayer perception (ANN)	Proposed hybrid method
First	0.867	0.875	0.813	0.875	0.933
Second	0.667	0.667	0.5	0.667	0.667
ATKT	0.889	1	0.889	1	1
Weighted avg.	0.861	0.91	0.82	0.91	0.941

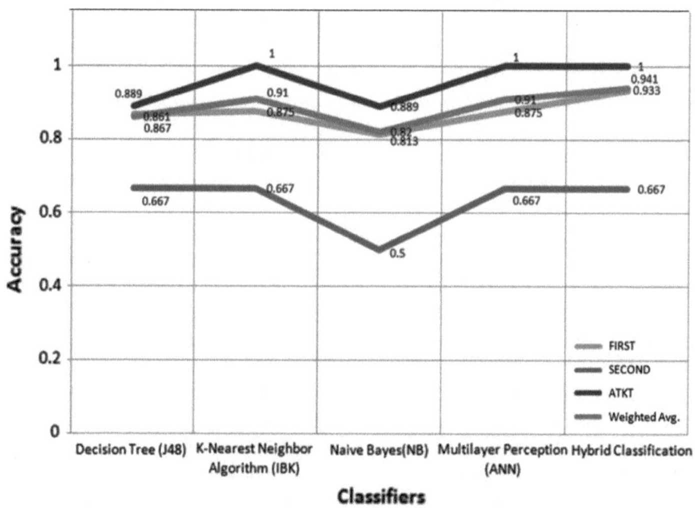

Fig. 5 Comparisons of classifiers accuracy

6 Conclusion

In this paper, the hybrid methodology of classification based on machine learning classifiers to predict performance evaluation is proposed and discussed. The four machine learning algorithms J48, NB, IBK, and ANN were used and ensembled through voting. The result of this paper indicated that the hybrid method of classification achieved better accuracy as compared to individual machine learning algorithms J48, NB, IBK, and ANN. This hybrid method achieved highest accuracy of 92.59% and individual classifiers, i.e., J48, NB, IBK, and ANN achieved an accuracy of 85.18, 81.48, 88.88, and 88.88%, respectively. Hence, the proposed hybrid method of classification is useful to predict the result of students and it can be used to develop strategies in education to improve performance. The model can also be applied in other domain of data mining and research can be further useful on other ensemble methods for classification of data sets.

References

1. B. Bakhshinategh, O.R. Zaiane, S. ElAtia et al., Educational data mining applications and tasks: a survey of the last 10 years. Educ. Inf. Technol. (2017)
2. C. Silva, J. Fonseca, Educational data mining: a literature review, in *Europe and MENA Cooperation Advances in Information and Communication Technologies*, vol. 520, Advances in Intelligent Systems and Computing, ed. by A. Rocha, M. Serrhini, C. Felgueiras (Springer, Cham, 2017)
3. R. Baker, Data mining for education, in *International Encyclopedia of Education*, vol. 7, 3rd edn., ed. by B. McGaw, P. Peterson, E. Baker (Elsevier, Oxford, UK, 2010), pp. 112–118
4. C. Romero, S. Ventura, Educational data mining: a review of the state-of-the-art. IEEE Trans. Syst. Man Cybern. Part C Appl. Rev. **40**(6), 601–618 (2010)
5. C. Romero, S. Ventura, in *Data Mining in Education, WIREs Data Mining Knowledge Discovery*, vol. 3 (Wiley, 2013), pp. 12–27. https://doi.org/10.1002/widm.1075
6. B. Behdad, R. Osmar, E. Samira, *Educational Data Mining Applications and Tasks: A Survey of the Last 10 Years* (Education Information Technology, Springer, 2017). https://doi.org/10.1007/s10639-017-9616z
7. C. Laura, A. Angel, Educational data mining and learning analytics: differences, similarities, and time evolution. RUSC **12**(3) (2015). Barcelona. July 2015. https://doi.org/10.7238/rusc.v1213.2515
8. T. Abinash, A. Abhishek, K. Santannu, Document-level sentiment classification using hybrid machine learning approach. J. Knowl. Inf. Syst. **53**, 805–831 (2017). https://doi.org/10.1007/s10115-017-1055-z
9. J.R. Quinlan, *C4.5: Programs for Machine Learning* (Morgan Kaufmann, San Francisco, 1993)
10. K. Chhogyal, A. Nayak (2016) An empirical study of a simple Naïve Bayes classifier based on ranking functions, in *AI 2016: Advances in Artificial Intelligence. AI 2016*, vol. 9992. Lecture Notes in Computer Science, ed. by B. Kang, Q. Bai (Springer, Cham, 2016)
11. D.M. Powers, Evaluation: from precision, recall and F-measure to ROC, in *Informedness, Markedness and Correlation* (2011)

A Novel Approach for the Classification of Leukemia Using Artificial Bee Colony Optimization Technique and Back-Propagation Neural Networks

Rudrani Sharma and Rakesh Kumar

Abstract This paper proposes a novel system of Leukemia detection based on Back-Propagation Neural Network (BPNN) classifier optimized by Principal Component Analysis (PCA) and Artificial Bee Colony (ABC) algorithm. PCA algorithm is used to extract the main characteristics of Leukemia data and to reduce the high-dimensional data in order to simplify the structure of the network and increase the training speed of BPNN. Then, ABC optimization algorithm is applied to select the most relevant subset of features. In the next step, the optimal feature set is used for training BPNN. The classification of Leukemia is finally implemented using the BPNN. PCA combined with ABC-BPNN reduces the amount of computation, increases the detection rate of the disease, and improves the accuracy of the system. Simulation results show that the PCA based ABC-BPNN approach gives better results than Genetic-based BPNN algorithm in terms of FAR, FRR, and accuracy.

Keywords Artificial Bee Colony · Artificial Neural Network · Back-Propagation Neural Network · Leukemia · Neural Network · Principal Component Analysis

1 Introduction

Leukemia is a form of a cancer that begins in the blood cells of the bone marrow. The blood cells that commonly contribute to Leukemia are white blood cells. Increased numbers of white blood cells act abnormally in the body causing low count of red blood cells and platelets that leads to the problems like anemia and blood clotting. Leukemia is the imbalance of white blood cells in the body. This disease can be treated if detected at an early stage. The motive of this paper is to develop an

R. Sharma (✉) · R. Kumar
Department of Computer Science and Engineering, National Institute of Technical Teachers Training and Research, Chandigarh, India
e-mail: Rudranisharma@gmail.com

R. Kumar
e-mail: Raakeshdhiman@gmail.com

C. R. Krishna et al. (eds.), *Proceedings of 2nd International Conference on Communication, Computing and Networking*, Lecture Notes in Networks and Systems 46, https://doi.org/10.1007/978-981-13-1217-5_68

efficient method for the detection of Leukemia that can improve the detection rate of the disease and give accurate results. This method is a combination of PCA, ABC technique and BPNN. The two main problems associated with BPNN include easily converging to local minima and slow convergence speed. In order to improve the performance of BPNN, PCA is applied in the first step to reduce the high-dimensional feature data. In the next step, ABC algorithm is applied on the data obtained from the first step to find the best features that most contribute to Leukemia. ABC is a global search technique that is used with BPNN to adjust weights of networks to avoid fall into local minima. Also, ABC has the fast heuristic learning and global convergence characteristics that when combined with excellent mapping capability of BPNN solves the problem of local minima and slow convergence. Feihu et al. in paper [1] presented a computer simulation of a system that is a combination of ABC and BPNN. Through experiments, they show that not only the combined systems solve the problem of slow convergence but also provide better mapping ability and come with the improvement in computation. Di Pan et al. [2] proposed a technique for classification using BPNN in combination with ABC optimization and found it suitable for adjusting threshold and weight on network. Through simulation, the authors show that the defined system overcomes the problem of low performance and slow convergence. Beatriz et al. [3] described a method for classification of DNA microarray using ABC and ANN where ANN is automatically designed by DE (Differential Evolution) algorithm and results show that the defined system is able to optimize transfer functions, the architecture and synaptic weights. The defined ABC algorithm for feature selection task involves binarization of each solution to select the best genes. Lan et al. in paper [4] designed a PCA based PSO-BPNN algorithm. The simulation results of the proposed method demonstrate that PCA when used with PSO-BPNN algorithm improves the efficiency of NN by simplifying the structure of NN and reducing the computation. Rama et al. [5] proposed a hybrid approach for the classification of gene expression microarray. The authors defined this hybrid approach as a combination of GA (Genetic algorithm), PCA and PNN (Probabilistic Neural Network) where GA and PCA are used for extraction of relevant features and PNN for classification. Experiments are performed using different datasets including Colon cancer, Large B-Cell Lymphoma, and Leukemia dataset. Numerical simulations show that PCA and GA, when combined together, are capable to obtained optimal number of features and efficiently optimized the structure of PNN. Roberto et al. [6] described a method that used ABC algorithm to train Spiking neuron for solving the pattern recognition problems. Results show that ABC, as a learning strategy, is useful in amending the synaptic weights of Spiking neuron structure in pattern classification task. Through comparison the authors of this paper show that ABC gives better results than DE and PSO algorithm on cancer datasets. Mustafa et al. [7] proposed a fusion approach that used ABC as feature selection algorithm and SVM as classifier. ABC optimization algorithm is used to reduce redundant and low-distinctive features which in turn increased the speed of the system. Classification accuracy of proposed system is calculated for different datasets and ten-fold cross-validation scheme is used to achieve a consistent performance of classifier. Experimental results show that the proposed system gives impressively good performance as compared to other outcomes attained.

2 Dimensionality Reduction Using PCA

PCA is a dimensionality reduction technique that transforms a high-dimensional dataset into a smaller dimensional subspace. This dimensionality reduction technique is often helpful to use prior to applying any machine learning algorithms because reducing the dimensionality of dataset increases the performance, running time of the algorithms and also reduces the risk of overfitting by reducing the degree of freedom of hypothesis. Apart from that dimensionality reduction, PCA can simplify the dataset, facilitating description, visualization, and insight. Basically, PCA tries to find the directions of most variation in the dataset.

2.1 PCA Technique

Suppose we have a data matrix X (n, m) with n samples and m features. In case of large m (number of features), visualizing the data becomes extraordinary difficult because it requires to do all kinds of convolution and lot of other steps to visualize the data. PCA allows us to reduce the high-dimensional data into something that can be explained in fewer dimensions so that we can better understand and visualize the data. Data can have some kind of correlated patterns or structures which may not be independent that we need to discover in order to apply PCA.

PCA is the eigendecomposition of covariance matrix $(X^{\mathrm{T}} \cdot X)$ which is a square matrix of size m. Eigendecomposition of matrix X is a set of eigenvectors W and set of eigenvalues Λ. We can use W and Λ to describe our data as follows;

$$T = X \cdot W \tag{1}$$

where X is a data matrix of size $n \times m$, W (also called "loadings") is the eigenvector of size $m \times m$ that is obtained by multiplying $(X^T \cdot X)$, T is the resultant matrix called "scores" of size $n \times m$. Each column of W is called "Principal Component". The columns of W are ordered by how large their corresponding eigenvalues are. So the principal component with largest eigenvalue will always be the first column and the principal component of next highest eigenvalue will be the second column and so on. Scores matrix T has exactly the same size as original data matrix X because it is just a transformed way of looking at the data matrix. In other terms, PCA just turns a high-dimensional cloud of points and describe it in a different basis vector. The idea is to change the projection of m features in W space. Since the columns of W are ordered by the corresponding eigenvalues Λ, we can choose to take first r number of components to best describe our data. This decomposition gives a very nice and principled way to pick r components. Suppose we pick first r principal components of W, W_r will represent the matrix containing the first r columns of W. Then scores for first r components can be defined as

$$T_r = X \cdot W_r \tag{2}$$

where W_r is the matrix of size $m \times r$ and T_r is the score matrix of size $n \times r$.

There is one more convenient and computationally efficient way to compute W called Singular Value Decomposition (SVD). SVD is usually implemented in a different way but mathematical identical to PCA. Mandeep et al. [8] implemented PCA with SVD technique for the extraction of features to efficiently determine the principal emotions for facial expression recognition. This method is computationally effective way of identifying features.

SVD can be implemented using original data matrix X of size $m\ X\ n$ that can be decomposed into product of three different matrices as follows:

$$X = U \cdot \Sigma \cdot V^* \tag{3}$$

where U denotes left singular vectors, conjugate transpose V^* denotes the right singular vectors, and Σ contains singular values on its diagonal. The diagonal of Σ matrix represents singular values while all the values other than diagonal are zero. If we will compare V with the matrix W, both are identical. Also, the diagonal singular values of Σ are proportional to eigenvalues of Λ. Other interesting properties can be represented by;

$$V^* \cdot V = 1 \tag{4}$$

Since W and V are identical, Scores T can also be represented in terms of U and Σ as;

$$T = X \cdot V \Rightarrow T = U \cdot \Sigma \cdot V^* \cdot V \Rightarrow T = U \cdot \Sigma$$

Also, if we take first r principal component, T_r can be defined as

$$T_r = U_r \cdot \Sigma_r \tag{5}$$

where U_r matrix contains first r components of matrix U and Σ_r is a matrix of size $r \times r$ containing first r block of Σ. Benefit of using SVD is that it is very fast and efficient means of computing PCA specially when the dataset is large. It is better to use SVD for computing PCA rather than calculating $(X^T \cdot X)$ with large number of features.

3 ABC Algorithm for Feature Selection

ABC algorithm is based on global optimization technique. In this research paper, ABC algorithm is exercised for feature selection. The proposed framework uses ABC algorithm to analyze the weights of BPNN. The ABC algorithm has significant characteristics of constructive greedy heuristic convergence, positive feedback and distributed computation [1].

3.1 ABC Concept

ABC meta-heuristic algorithm was first introduced by Karaboga in 2005. This swarm-based algorithm is a technique that is inspired by the sensible foraging behavior of honey bees [9].

ABC algorithm is based on the concept of sharing information among bees that helps in choosing feasible solution which can satisfy required criteria. There are three important components of bees for the selection of quality food. The first component is food source whose value is dependent on its quality or we can say concentration of its energy. The second component is employed bee. This type of bee is involved with a particular food source. They store all the information about a particular source, its direction, closeness from hive and quality. Employed bee shares this information with certain probability. The last component represents unemployed bees that are further classified into two classes namely scout bees and onlooker bees. The scout bees are always involved in process of exploration of new food source. The onlooker bees make decisions on the basis of the information shared by the employed bees and then accordingly choose the finest food source. Whenever the food source associated with an employed bee get exhausted, the employed bee turn into a scout bee to search a new food source again. Employed bees, through waggle dancing, share information regarding the concentration of food source in the dancing area of the hive. There is great probability of onlooker bees choosing the most profitable source as more information is shared about more profitable sources.

3.2 ABC Algorithm

ABC algorithm is simple and easy to implement and has a unique solution modification technique that results in quick convergence into optimal solution [10]. ABC optimization technique, when used with a classifier, is used to optimize the process of feature selection and yields the best optimal features to increases the accuracy of the classifier. The steps followed in ABC algorithm are described below:

Step 1 Each employed bees is assigned a food source.

Step 2 Employed bees move to the food sources, calculate the nectar amount of their sources, and then share information through waggle dance.
Step 3 Onlooker bees pick food sources for themselves by observing the waggle dance then evaluate the nectar amount of their selected food sources.
Step 4 Bees abandon food sources that are exhausted by them.
Step 5 Scout bees search for new food sources randomly and become employed bees after finding a new food source.
Step 6 Save best food sources and repeat Steps 2–5 till termination condition not met.

4 Back-Propagation Neural Networks for Classification

Back-propagation technique is one of the most accepted Neural Network algorithms. Back-propagation is a common approach used for training a Neural Network.

BPNN consists of at least three layers of units: the first layer is the input layer, the second layer consists of one or more hidden layers, and the last layer is the output layer. Units in BPNN are connected in a similar way to feedforward. The basic idea of back-propagation is to transfer errors from the units of output layer to the inputs of hidden layer so the problem of adjustments of the weights from input units to hidden units can be solved. The output of BPNN represents a classification decision. The efficiency of implementation of BPNN algorithm depends on the amount of input and output data given to the layers.

Back-propagation algorithm has the ability to learn by computing the errors encountered in output layer and then finding the errors in hidden layers. This back-propagating criterion works well on the issues where it is difficult to find any relationship between the output and inputs [11]. BPNN has been successfully implemented in various application areas due to its learning capabilities and flexibility.

5 Methodology

To validate the efficiency and accuracy of proposed Leukemia detection system using BPNN classification algorithm based on PCA and ABC optimization algorithm, we performed some experiments with the defined procedure on Leukemia data. The performance of the defined system is evaluated using some metrics such as False Rejection Rate, False Acceptance Rate, and Accuracy. The entire methodology of proposed work is given below:

Step 1 **Preprocessing of Data**: First, apply the PCA on the original data features for dimensionality reduction and store the extracted feature set. To optimize the extracted feature set, we apply the novel ABC optimization algorithm with the appropriate fitness function in order to get optimal features that mainly contribute to Leukemia.

Step 2 **Classification**: First, BPNN is used for the training of optimized dataset and to create a trained structure for the classification. When training got completed then moved towards classification process with trained Neural Network structure. In the classification, upload a test data and simulate with trained Neural Network structure to obtain classification results.

Step 3 **Computation of Performance Parameters**: FAR, FRR, and accuracy are computed to measure the performance of the proposed system. Accuracy is defined as the overall accurately classified features while FAR and FRR can be defined as follows:

FAR: False Acceptance Rate is the ratio of number of falsely accepted features to the total classified features.

FRR: False Rejection Rate is the ratio of number of falsely rejected features to the total classified features.

The flowchart of the proposed methodology is shown in Fig. 1.

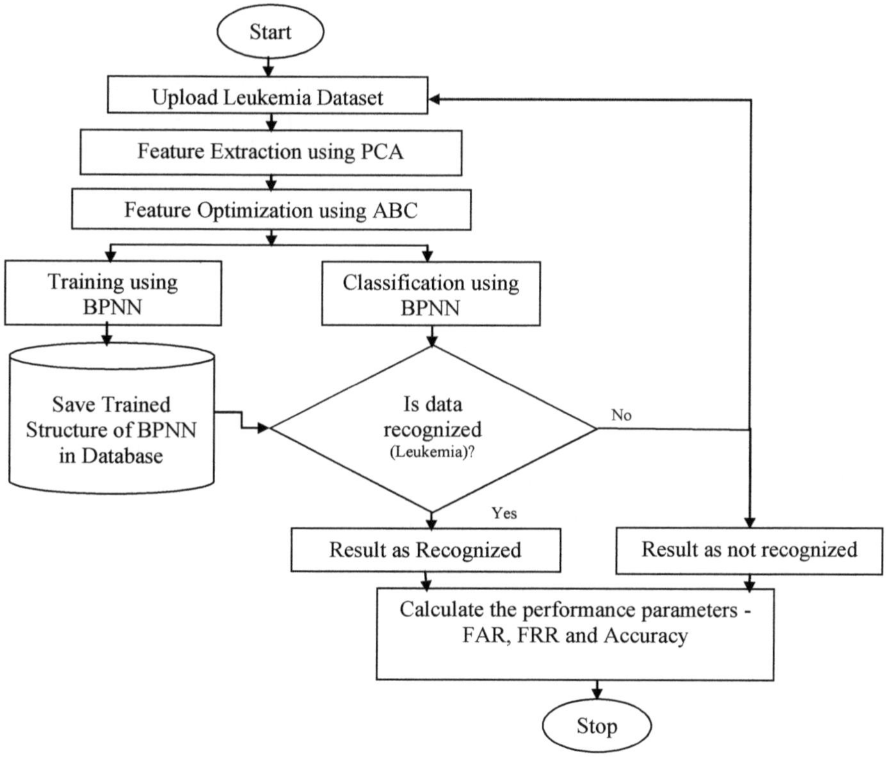

Fig. 1 Flowchart of the proposed methodology

6 Experimental Results

To check the efficiency of the system, the proposed model is programmed in MATLAB. This section of the paper describes the information regarding the dataset used, performance parameters of the proposed methodology as well as comparison of proposed work with another method.

6.1 Dataset

The proposed method has been evaluated through Leukemia dataset taken from mldata repository. Table 1 presents a description of the data set used, total number of samples, total number of features in the dataset and defined number of classes.

6.2 Performance Parameters of the Proposed Work

The performance of the defined method showed excellent results. Number of features selected in the first phase of the proposed method to train the Neural Network are 21 that made computation faster. The accuracy of the proposed system is shown in Table 2 along with FAR and FRR. It can be realized from the table that the classification accuracy is more than 97% in all trials of the experiments. The accuracy even reached 99.97% in its best trial.

Table 1 Data set

Dataset	Total samples	Total features	Number of classes
Leukemia	85	48	2

Table 2 Performance parameters

Trial No.	Accuracy (%)	FAR	FRR
1	98.48	0.563	0.404
2	98.97	0.583	0.377
3	98.57	0.479	0.521
4	97.97	0.35	0.475
5	99.02	0.542	0.431
6	98.47	0.457	0.416
7	99.45	0.485	0.454
8	99.97	0.634	0.578
9	98.67	0.547	0.278
10	97.58	0.478	0.398

Table 3 Comparison between ABC and GA

Iteration number	Accuracy (%)	
	ABC	GA
1	98.48	91.47
2	98.97	85.26
3	98.57	89.68
4	97.97	86.32
5	99.02	91.48
6	98.47	92.62

6.3 Comparison Against Other Method

In this section, the defined method is compared to a relevant swarm approach called genetic algorithm (GA). A comparison with GA is performed using the same Leukemia data after applying PCA. The comparison of the results on the basis of accuracy is given in Table 3. It is clear from comparison that BPNN model with proposed ABC based training is better than the BPNN with genetic based training.

7 Conclusion and Future Scope

This paper presented a technique of detecting Leukemia using ABC as a means to train BPNN. From the experimental outcomes, it can be determined that the proposed methodology of classifying Leukemia gave efficient results for the dataset used. Firstly, PCA is applied on the original Leukemia dataset for dimensionality reduction. Then ABC is used to successfully obtain the optimal feature set. Also, ABC algorithm works well with BPNN as ABC algorithm helps in global convergence. Classification results showed that average accuracy attained is 98.72% by using the proposed ABC-BPNN structure. The result showed that the performance of the PCA based ABC-BPNN system is better than PCA based GA-BPNN. Also, the proposed method reduces the amount of computation and improves the overall accuracy of the system. Future work can take account of the implementation of the ABC-BPNN algorithm with improved execution time and use different medical datasets.

References

1. F. Jin, G. Shu, Back propagation neural network based on artificial bee colony algorithm, in *Strategic Technology (IFOST), 2012 7th International Forum* (Tomsk, 2012), pp. 1–4
2. D. Pan, J. Cao, An improved neural network algorithm based on artificial bee colony algorithm and its application in sewage treatment, in *Behavioral. Economic and Socio-cultural Computing (BESC), 2015 International Conference*, (Nanjing, 2015), pp. 83–88

3. B.A. Garro, K. Rodrıguez, R.A. Vazquez, Designing artificial neural networks using differential evolution for classifying DNA microarrays. in *IEEE 2017 Congress on Evolutionary Computation (CEC)*, (San Sebastian, 2017), pp. 2767–2774
4. L. Shi, X. Tang, J. Lv, PCA-based PSO-BP neural network optimization algorithm. in *The 27th Chinese Control and Decision Conference (2015 CCDC)*, (Qingdao, 2015), pp. 1720–1725
5. R.S. Sreepada, S. Vipsita, P. Mohapatra, An efficient approach for classification of gene expression microarray data. in *2014 Fourth International Conference of Emerging Applications of Information Technology*, (Kolkata, 2014), pp. 344–348
6. R.A. Vazquez1, B.A. Garro, Training spiking neural models using artificial bee colony, Comput. Intell. Neurosci. **2015**(947098), 1–14 (2015)
7. M.S. Uzer, N. Yilmaz, O. Inan, Feature selection method based on artificial bee colony algorithm and support vector machines for medical datasets classification, Sci. World J. **2013**(419187), 1–10 (2013)
8. M. Kaur, R. Vashisht, N. Neeru, Recognition of facial expressions with principal component analysis and singular value decomposition. Int. J. Comput. Appl. (0975–8887), **9**(12), 36–40 (2010)
9. Kaswan, K.S., Choudhary, S., Sharma, K.: Applications of Artificial Bee Colony Optimization Technique: Survey. Computing for Sustainable Global Development (INDIACom), 2015 2nd International Conference, New Delhi, pp. 1660–1664 (2015)
10. B. Subanya, R.R. Rajalaxmi, Feature selection using artificial bee colony for cardiovascular disease classification. in *2014 International Conference on Electronics and Communication Systems (ICECS)*, (Coimbatore, 2014), pp. 1–6
11. M.A.Wani Saduf, Comparative study of back propagation learning algorithms for neural networks. Int. J. Adv. Res. Comput. Sci. Softw. Eng. **3**(12), 1151–1156 (2013)

Improving KFCM-F Algorithm Using Prototypes

K. Mrudula and E. Keshava Reddy

Abstract FCM algorithm is considered as the basis for all algorithms in fuzzy clustering. Due to its limitations in clustering the linearly inseparable and overlapping data, kernel-based fuzzy c-means was introduced. In literature, there are two kernel-based fuzzy clustering approaches, viz., KFCM-F and KFCM-K. In both kernel-based methods, the data items are implicitly mapped into a high-dimensional feature space, where the linearly inseparable clusters get well separable. In KFCM-F, the cluster centers are computed in the given input space, whereas in KFCM-K the cluster centers are present in the feature space and inverse mapping is used to compute these centers in the input space [3]. The time complexity of KFCM-F is $O(N)$, where N is the size of the dataset. KFCM-F becomes infeasible to work for large values of N. The objective of the present work is to reduce the time complexity of KFCM-F by selecting few prototypes *say*, M from the given data, where $M \ll N$. The key contribution of this work is that the memberships of a group of data items *say* S can be easily approximated using a single membership computation. Hence, our method requires only MC membership computations instead of NC, where C denotes the number of clusters. Experimentally, we proved that our proposed improvement over KFCM-F results in a great reduction in its running time.

Keywords Fuzzy clustering · Kernel FCM · KFCM-F · Prototypes

1 Introduction

Kernel-based clustering has been an important and interesting approach in fuzzy clustering. If the data is highly complex, nonspherical, overlapping, and linearly

K. Mrudula (✉)
Research Scholar, Jawaharlal Nehru Technological University Anantapur, Anantapur, AP, India
e-mail: mrudulasarma22207@gmail.com

E. Keshava Reddy
Department of Mathematics, Jawaharlal Nehru Technological University Anantapur, Anantapur, AP, India
e-mail: keshava.maths@jntua.ac.in

C. R. Krishna et al. (eds.), *Proceedings of 2nd International Conference on Communication, Computing and Networking*, Lecture Notes in Networks and Systems 46, https://doi.org/10.1007/978-981-13-1217-5_69

inseparable, then kernel-based FCM techniques are of greater importance. The basic idea of kernel-based clustering methods lies in applying a nonlinear transformation arbitrarily, say ϕ, on the given data in the original input space into a high-dimensional feature space or kernel space, and then apply the FCM algorithm in the kernel space. It is proved that by applying suitable transformation, the data items which are linearly inseparable in the input space will become separated linearly in the kernel space [7]. The distance between two different data items, say $\phi(p_i)$ and $\phi(p_j)$ in the kernel space, can be computed without knowing the explicit form of ϕ [2], using kernel trick [8]. Kernel FCM and its variants have potential applications in biomedical image analysis [4], remote sensing [11], document clustering [5], etc. [6].

There are two variants of kernel FCM algorithms presented by Graves et al. in [3], called, "KFCM-F (Kernel-based FCM with data items in Feature space and KFCM-K (Kernel-based FCM with data items in kernel space".

In KFCM-F, the data items are transformed into a high-dimensional kernel space using the function ϕ, and the cluster centers *say* v_is are computed in the input space, using the kernel trick. In KFCM-K, the cluster centers, *say* $\phi(v_i)$s, are initially computed in the Kernel space and finally its pre-image is computed to find the cluster center in the input space. The time complexities of KFCM-F and KFCM-K are $O(N)$ and $O(N^2)$, respectively. The current work focuses on KFCM-F and proposes a new approach to reduce the running time of KFCM-F to make it suitable to apply on large datasets.

The proposed improvement over KFCM-F is outlined as follows. Initially, a set of M prototypes are identified in the input space, such that each prototype represents a set of data items (for instance, S). In each iteration of KFCM-F, instead of computing the memberships of each one of the S data items, all their memberships are approximated using a single membership computation, that is, with the membership of its prototype. Hence, the proposed method computes only MC membership computations in each iteration, instead of NC computations, where C is number of clusters and $M \ll N$. Experimentally, it is proved that the proposed method is suitable for very large datasets.

The remaining paper is organized as follows. Section 2 explains the related work of the KFCM-F algorithm in detail. Section 3 introduces the proposed prototype-based KFCM-F algorithm. Empirical study is discussed in Sect. 4. Conclusions and future scope are outlined in Sect. 5.

2 KFCM-F

This section gives a detailed explanation of KFCM-F algorithm. KFCM-F algorithm is an extended version of kernel-based fuzzy c-means algorithm, produces better results when compared to FCM in case of nonspherical or linearly inseparable or overlapping clusters. For the sake of completeness of the paper, initially, we describe the FCM algorithm, in brief.

Let $D = (x_1, x_2, \ldots, x_N)$ be a dataset and C be the number of cluster centers. The FCM clustering algorithm determines a fuzzy clustering of D by minimizing the objective function given below.

$$J = \sum_{i=1}^{N} \sum_{j=1}^{C} \mu_{ij}^{m} ||x_i - v_j||^2, \tag{1}$$

where m is the fuzziness index, $m \in [1, \infty]$ and μ_{ij} represents the membership of ith data item in jth cluster.

After each iteration, the cluster centers are updated using the new membership values, computed as follows:

$$\mu_{ij} = \frac{1}{\sum_{k=1}^{C} \left(\frac{d(x_i, v_j)}{d(x_i, v_k)} \right)^{\frac{2}{m-1}}} \tag{2}$$

where $1 \leq i \leq N$ and $1 \leq j \leq C$.

$$v_i = \frac{\sum_{j=1}^{N} \mu_{ij}^{m} x_j}{\sum_{j=1}^{N} \mu_{ij}^{m}}; 1 \leq i \leq C \tag{3}$$

where m is the fuzziness index and μ_{ij} represents the membership of ith data item in jth cluster.

The KFCM-F algorithm aims to cluster the data by performing a nonlinear mapping ϕ on each data item from the original input space into high-dimensional feature space. It minimizes the following objective function subject to conditions as considered in FCM [3]:

$$J = \sum_{i=1}^{N} \sum_{j=1}^{C} \mu_{ij}^{m} ||\phi(x_i) - \phi(v_j)||^2 \tag{4}$$

In kernel space, the distance between the data items $\phi(x)$ and $\phi(y)$ are calculated using the kernel trick. Linear Kernel, polynomial kernel, and Gaussian kernel (RBF kernel) are some of the popularly used kernel functions [8, 9].

In case of RBF Kernel, the distance between $\phi(x_i)$ and $\phi(v_j)$ can be calculated as follows:

$$||\phi(x_i) - \phi(v_j)||^2 = 2(1 - K(x_i, v_j)) \tag{5}$$

In the iterative process of KFCM-F, the memberships and the cluster centers are updated as follows:

$$\mu_{ij} = \frac{1}{\sum_{k=1}^{C} \left(\frac{1-K(x_i, v_j)}{1-K(x_i, v_k)} \right)^{\frac{1}{(m-1)}}} \tag{6}$$

$$v_i = \frac{\sum_{k=1}^{N} \mu_{ik}^m K(x_k, v_i) x_k}{\sum_{k=1}^{N} \mu_{ik}^m K(x_k, v_i)} \tag{7}$$

In KFCM-F, the number of memberships to be computed in each iteration is NC. The scope of the current work is to reduce the number of such membership computations from NC to MC, where M is the number of prototypes and $M \ll N$. The following section explains the proposed method, called PKFCM-F, in detail.

3 Proposed Prototype-Based KFCM-F Algorithm

The proposed prototype-based KFCM-F algorithm, called PKFCM-F, works in two stages. In stage-1, few prototypes are identified in the dataset. Suppose that there are M number of prototypes, say $(p_1, p_2, \ldots, p_M)$. Each prototype p_j represents a few data items which lie within the distance ϵ. Each data item may belong to one or more prototypes depending on the memberships of the data item to the prototype, that is, a data item x_i may belong to one or more prototypes, i.e., p_js. However, the sum of memberships of a data item in all prototypes is equal to 1, i.e., $\sum_{j=1}^{M} \mu_{p_j x_i} = 1$.

The membership of the data item x_i to the prototype p_j is obtained by using the formula

$$\mu_{p_j x_i} = \frac{1}{\sum_{k=1}^{M} \left(\frac{1 - K(x_i, p_j)}{1 - K(x_i, p_k)} \right)^{\frac{1}{(m-1)}}} \tag{8}$$

where $1 \leq i \leq N$ and m is the fuzziness index.

Here, the number of prototypes M depends on ϵ. The greater the value of ϵ, the less the number of prototypes will be there and the smaller the value of ϵ gives more number of prototypes. After finding the prototypes, the memberships of the data items in each one of the prototypes are calculated. Then, a membership matrix M_1 of size $N \times M$ is obtained.

$$M_1 = \begin{bmatrix} \mu_{p_1 x_1} & \mu_{p_2 x_1} & \mu_{p_3 x_1} & \cdots & \mu_{p_M x_1} \\ \mu_{p_1 x_2} & \mu_{p_2 x_2} & \mu_{p_3 p_2} & \cdots & \mu_{p_M x_2} \\ \vdots & \vdots & \vdots & \ddots & \vdots \\ \mu_{p_1 x_N} & \mu_{p_2 x_N} & \mu_{p_3 x_N} & \cdots & \mu_{p_M x_N} \end{bmatrix}$$

In this matrix, the sum of elements in each row is equal to 1.

In stage-2, we apply KFCM-F algorithm by selecting (randomly) C cluster centers in the dataset D. Here, instead of taking entire N data items, consider only the above M prototypes, $(p_1, p_2, p_3, \ldots, p_M)$. Then, we calculate the memberships of these prototypes in each cluster. Here, each prototype may belong to one or more clusters. The membership of a prototype p_j in cluster c_k is obtained by using the formula

$$\mu_{c_k p_j} = \frac{1}{\sum_{i=1}^{C} \left(\frac{1-K(p_j,c_k)}{1-K(p_j,c_i)} \right)^{\frac{1}{(m-1)}}} \tag{9}$$

For each prototype p_j, the sum of its memberships in all clusters is equal to 1, i.e., $\sum_{k=1}^{C} \mu_{c_k p_j} = 1$.

After finding the memberships of all the prototypes in each cluster, a membership matrix M_2 of size $M \times C$ is formed.

$$M_2 = \begin{bmatrix} \mu_{c_1 p_1} & \mu_{c_2 p_1} & \mu_{c_3 p_1} & \cdots & \mu_{c_C p_1} \\ \mu_{c_1 p_2} & \mu_{c_2 p_2} & \mu_{c_3 p_2} & \cdots & \mu_{c_C p_2} \\ \vdots & \vdots & \vdots & \ddots & \vdots \\ \mu_{c_1 p_M} & \mu_{c_2 p_M} & \mu_{c_3 p_M} & \cdots & \mu_{c_C p_M} \end{bmatrix}$$

In each iteration of the proposed PKFCM-F, the membership of each data item x_i in the cluster c_k can be calculated by using the formula

$$\mu^*_{C_k x_i} = \frac{\mu_{p_j x_i} + \mu_{C_k p_j}}{2} \tag{10}$$

where p_j is the prototype, in which the data item x_i has greater membership among all prototypes.

It is easily seen that this membership $\mu^*_{C_k x_i}$ is very close to the actual membership of the data item x_i in the cluster c_k which is given by the formula

$$\mu_{c_k x_i} = \frac{1}{\sum_{j=1}^{C} \left(\frac{1-K(x_i,c_k)}{1-K(x_i,c_j)} \right)^{\frac{1}{(m-1)}}} \tag{11}$$

Hence, our proposed PKFCM-F computed only MC membership computations in each iteration, instead of NC computations, where $M \ll N$. Therefore, PKFCM-F takes lesser time compared to KFCM-F. However, the clustering quality may reduce. But it is negligible for an appropriate value of ϵ.

4 Empirical Study

To study the performance of the proposed PKFCM-F algorithm, experiments are conducted to compare the running times of KFCM-F and PKFCM-F on some benchmark datasets collected from [1]. The information about the datasets used in the study is mentioned in Table 1.

To compute the clustering quality (CQ), adjusted rand-index has been applied [10]. The clustering quality (CQ) and running time (RT) of KFCM-F and PKFCM-F are computed. The percentage of reduction in CQ and the percentage of reduction in RT are presented in Table 2.

Table 1 Datasets

Dataset	# of data items (N)	# of clusters (C)	# of dimensions (d)
OCR	10,003	10	192
Pendigits	10,992	10	16
LIR	20,000	26	16
Shuttle	58,000	7	9

Table 2 Reduction in CQ and RT values obtained using KFCM-F and PKFCM-F

Dataset	Reduction in clustering quality (%)	Reduction in running time (%)
OCR	0.53	81
Pendidits	0.21	62
LIR	0.45	79
Shuttle	0.82	84

5 Conclusions and Discussion

In this paper, we proposed a prototype-based KFCM-F algorithm, called PKFCM-F, to speed up KFCM-F algorithm. PKFCM-F algorithm runs by identifying some prototypes, where each prototype represents a small set of data items in the dataset. By calculating the memberships of the data items in the prototypes and also the memberships of these prototypes in each cluster, one can approximate the membership of the data item in each cluster. It is experimentally verified that the time complexity of our proposed PKFCM-F is reduced when compared with traditional KFCM-F, particularly in case of large datasets. For different values of ϵ, different number of prototypes can be obtained. The high clustering accuracy will be achieved by taking more number of prototypes.

References

1. A. Asuncion, D. Newman, UCI machine learning repository (2007)
2. R. Chitta, R. Jin, T.C. Havens, A.K. Jain, Scalable kernel clustering: approximate kernel k-means (2014), arXiv:1402.3849
3. D. Graves, W. Pedrycz, Kernel-based fuzzy clustering and fuzzy clustering: a comparative experimental study. Fuzzy Sets Syst. **161**(4), 522–543 (2010)
4. S.R. Kannan, R. Devi, S. Ramathilagam, T.P. Hong, A. Ravikumar, Effective kernel FCM: finding appropriate structure in cancer database. Int. J. Biomath. **9**(02), 1650018 (2016)
5. R. Kondadadi, R. Kozma, A modified fuzzy art for soft document clustering, in *Proceedings of the 2002 International Joint Conference on Neural Networks, 2002. IJCNN'02*, Vol. 3 (IEEE, 2002), pp. 2545–2549

6. K.-P. Lin, A novel evolutionary kernel intuitionistic fuzzy *c*-means clustering algorithm. IEEE Trans. Fuzzy Syst. **22**(5), 1074–1087 (2014)
7. J. Nayak, B. Naik, H.S. Behera, Fuzzy c-means (FCM) clustering algorithm: a decade review from 2000 to 2014, in *Computational Intelligence in Data Mining-Volume 2* (Springer, 2015), pp. 133–149
8. T.H. Sarma, P. Viswanath, Speeding-up the k-means clustering method: a prototype based approach, in *Proceedings of the 3rd International Conferences on Pattern Recognition and Machine Intelligence (PReMI)*. LNCS, vol. 5909 (Springer-Verlag, Berlin, Heidelberg, 2009), pp 56 –61
9. T.H. Sarma, P. Viswanath, A. Negi, Speeding-up the prototype based kernel k-means clustering method for large data sets, in *2016 International Joint Conference on Neural Networks (IJCNN)* (IEEE, 2016), pp. 1903–1910
10. D. Steinley, M.J. Brusco, L. Hubert, The variance of the adjusted rand index. Psychol. Methods **21**(2), 261 (2016)
11. C. Zhu, S. Yang, Q. Zhao, S. Cui, N. Wen, Robust semi-supervised kernel-fcm algorithm incorporating local spatial information for remote sensing image classification. J. Indian Soc. Remote Sens. **42**(1), 35–49 (2014)

A Time-Efficient Active Learning-Based Instance Matching System for Data Linking

Gulshakh Kaur, Shilpa Verma and Poonam Saini

Abstract Learning of link configuration can be classified into two subcategories, namely, *supervised* and *unsupervised*. The supervised learning of link configuration has an advantage over unsupervised learning in terms of high accuracy and low complexity. However, a common shortfall of supervised systems is the availability of training data. Here, *Active Learning* is a favorable semi-supervised learning technique in which a learning algorithm will actively query the user to label the most informative link candidate. This method is desirable in situation where data is abundant and manual labeling of links is expensive. Moreover, while dealing with large datasets, runtime efficiency of the instance matching system is also desirable. The paper proposes an extended framework to our previous work on a supervised learning-based instance matching system with runtime efficiency.

Keywords Linked data · Identity link · Instance matching · Link configuration
Supervised learning · Active learning

1 Introduction

As the web is growing exponentially in terms of information, end users expect an intelligent query processing. The task is more challenging in the presence of unstructured data which makes it difficult to extract semantics from the web. In order to improve a user's experience, we need to publish structured information with interlinking to other data sources. *Semantic web* is an extension to current web of docu-

G. Kaur (✉) · S. Verma · P. Saini
Department of Computer Science and Engineering, PEC University
of Technology, Chandigarh, India
e-mail: gulshakhkaur.pec@gmail.com

S. Verma
e-mail: shilpaverma.pec@gmail.com

P. Saini
e-mail: poonamsaini@pec.ac.in

C. R. Krishna et al. (eds.), *Proceedings of 2nd International Conference on Communication, Computing and Networking*, Lecture Notes in Networks and Systems 46, https://doi.org/10.1007/978-981-13-1217-5_70

ments into distributed web at the data level, i.e., web of data [1]. This contains open and interlinked data that can be shared and reused. A powerful mean to transform the web of documents into the web of interlinked data is *Linked Data*. Linked Data is machine-readable data format that interconnects structured data sources through the typed links, thereby creating global data space [2]. Linked Data is based on the set of principles proposed by Berners-Lee [3]:

- Use of URI (Uniform Resource Identifier) to name things on the web.
- Use of HTTP (HyperText Transfer Protocol) URI to name things so that Internet users can look up these names.
- HTTP URIs must return useful information when a user looks up these URIs.
- Include typed links to other URIs that specify the nature of connection.

1.1 Instance Matching

The essential component of Linked Data for linking related instances is *instance matching*. In Linked Data scenario, instance matching detects co-referent instances of data repositories for same object [4]. There are two basic techniques to perform the instance matching of data repositories, namely, *Configuration Learning* and *Binary Classification*. Instance matching step of the data linking workflow gives the collection of mappings between entities of two data sets. Instance matching is a technique that compares two object descriptions on the basis of metrics specified in the link configuration. Learning of link configuration can be classified into two categories, namely, supervised and unsupervised learning. Supervised learning of link configuration has an advantage over unsupervised learning in terms of high accuracy and low complexity. However, a common shortfall of supervised systems is the availability of training data. Active learning is a favorable strategy to reduce the human annotation burden. Active learning is a semi-supervised learning technique in which learning algorithm query the user in order to obtain the desired results. Learning algorithm will actively query the user to label the most informative link candidate. This method is desirable in situation where data is abundant and manual labeling of links is expensive. Active learning is querying for certain types of instances depending on the data that is seen so far, i.e., depending on the past experience. Moreover, while dealing with large datasets, runtime efficiency of the instance matching system is also desirable. The focus of most of the comparative analysis is on the effectiveness of the matching frameworks whereas runtime efficiency has not been considered.

2 Related Work

In existing literature, different frameworks and techniques have been designed to address the data linking problem for Linked Data on Semantic Web. A few techniques focus on time efficiency along with link configuration while other involves machine learning algorithms for classification. In 2009, *SILK* [5] was proposed that applies *Link Specification Language* for data linking where a user specifies settings for instance comparison. The framework has three components, namely, link discovery, link evaluation and link maintenance. Thereafter, a time-efficient framework was proposed in 2010 named as *LIMES* [6] to discover missing links between source and target dataset. It makes use of mathematical concept of *metric space* to filter out irrelevant pairs of instances that do not conform to the mapping. This, in turn, reduced the total number of comparisons required to compare the datasets. The *unsupervised learning* [7] technique provides fully automatic system used to learn link configuration without the labeled training data. The method employs genetic algorithm guided by the fitness criteria to evaluate the quality of decision rule and the corresponding derived identity links. However, unsupervised learning is impractical due to its low accuracy. On the other hand, supervised learning needs labeled data for training. Nevertheless, the scheme has high accuracy and low complexity. Other time-efficient and active supervised learning algorithms were proposed in 2012 called *EAGLE* [8] and *ActiveGenLink* [9] that combine genetic algorithm and active learning to interactively generate linkage rules in order to produce identity links. The algorithms take initial population as input and evolve it using genetic operators mutation and reproduction, to obtain efficient solution.

cLink [10] and *ScLink* [4] are the supervised instance matching systems that create optimal matching configuration automatically based on the heuristic search approach. *cLink* contains the supervised learning algorithm *cLearn* to learn link configuration. cLearn algorithm uses heuristic search to return the optimal combination of similarity function, similarity aggregator and the threshold parameter for the final filtering step. *ScLink* is the extension of *cLink* as it contains two supervised learning algorithms namely cLearn and minBlock. minBlock learning algorithm answers the scalability issue of large Linked Data repositories as it reduces the number of candidates that will be considered for the matching task. Besides link configuration methods, supervised instance matching can also be considered as a binary classification problem [11].

3 Assumptions and Proposed Approach

Most of the assumptions of the system model for the design of time-efficient and active learning-based instance matching are same as the previous proposed method [12] with additional points as follows:

1. Set of diverse and frequent properties are selected from both source and target repository.

2. The subset of source and target repository is labeled using the active learning technique-query by committee.
3. The heuristic search approach with active learning is used to learn the link configuration.
4. The concept of multithreading is used to parallelize the steps of the instance matching framework.

3.1 The Proposed System Model

Given two datasets Sc (source) and Tr (target), instance matching system determines all the pairs of instances (s, t) ϵ Sc × Tr, related via identity relation R. The proposed system employs the concept of active learning to get the desired link configuration which in turn is used to match the data instances. The list of data structures and functions used in the active learning-based link configuration algorithm is discussed as follows:

Data Structures

The data structures used in the algorithm 2 in addition to algorithm 1 [12] are as follows:

1. **Population (P)**: P represents the list of possible link configurations. The list evolves with the addition of new reference link until the termination condition is reached. The output of the algorithm is the optimal link configuration selected from P.
2. **Training Percentage (T)**: T contains the percentage of source data that will be used for training the learning algorithm.
3. **Unlabeled Pool (U)**: U is the list that contains all the instances which need to be labeled during the active learning in order to form the reference links.
4. **Reference Links (R)**: The list R is initially empty and every new link labeled by the user is added to R.

Functions

The list of functions used in the Algorithm 2 is as follows:

1. **Query**: The function includes the query strategy applied that will select the most informative link to be labeled by the user.
2. **Annotate**: The link candidate selected by the Query function is labeled as true or false using Annotate function.

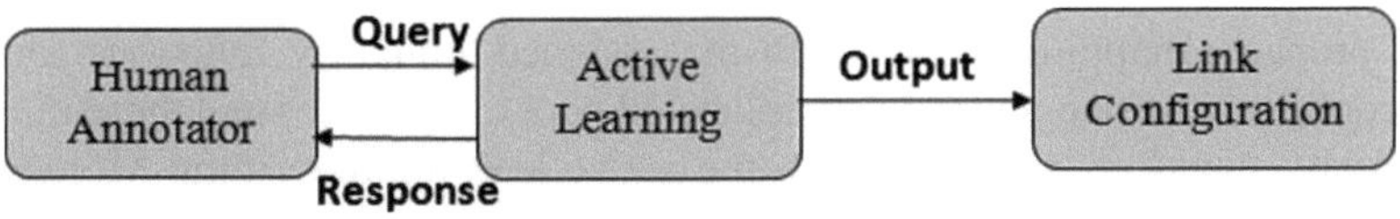

Fig. 1 General schema for active learning

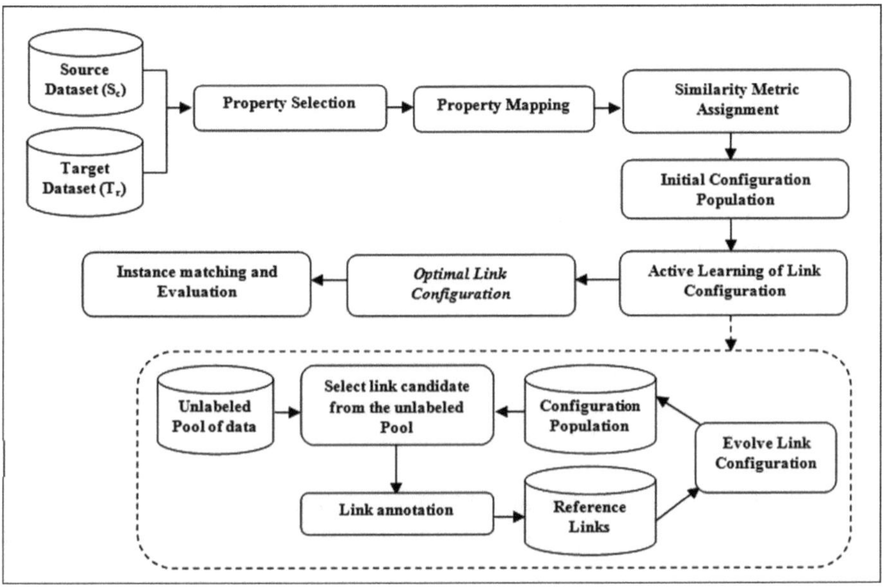

Fig. 2 Instance matching framework

3.2 *The Proposed Approach*

Active learning learns the link configuration by asking the user to confirm or reject the reference link candidates selected by the algorithm as shown in Fig. 1. This reduces the number of link candidates which need to be labeled by the user for training the learning algorithm. The main idea of adding the active learning is to generate the training data that provides the maximum information gain.

Given source and target datasets, the detailed system workflow is shown in Fig. 2. The workflow starts by selecting the list of diverse and frequent properties from both source and target dataset. The next step is the alignment of selected source and target properties depending on the similarity of the property values. The property mappings with the confidence or c score [12] in the threshold range 0.009 are selected. Afterwards, property mappings are assigned similarity metric depending on the data type of the property. Each metric computes the similarity score of two instances with reference to one property in the range [0, 1].

The property mappings along with the assigned similarity functions form the initial configuration population. The unlabeled pool of instances and these initial set of similarity functions act as input to the active learning algorithm. Moreover, in order to achieve the time efficiency, the concept of multithreading is applied. This involves parallel processing of all the steps using multithreading so as to enhance the speed of the system.

3.3 Active Learning of Link Configuration

The main idea behind using active learning combined with greedy approach is to evolve the population of link configuration while building the reference links. The algorithm starts with random linkage rules and the empty pool of reference links.

Every new iteration of the algorithm selects a link candidate to be labeled by the human annotator using the query-by-vote-entropy technique. In query-by-vote-entropy, the link candidate is determined by the voting of the members of the committee, i.e., the existing elements in the link configuration population. The link candidate for which current population is uncertain is selected by the query strategy. Algorithm 2 provides the pseudo code for the active learning procedure.

Algorithm 2: Active Learning of Link Configuration

Input: Initial population P, Unlabeled pool of instances U, R empty reference links, property alignments, similarity metrics M_{sim}, similarity aggregation functions S_{agg}, T percentage of training data.

1. *L* is empty set of reference links
2. *Lcounter* ← *Size(P)*
3. while *loop_count<=Lcounter && Precision<1 do*
4. *u* ← Query strategy selects link candidate from *U*
5. *r* ← *Annotate u*
6. *R* ← $R \cup r$
7. *P* ← Learn Link Configuration using Algorithm 1.
8. End
9. Return Optimal Link Configuration

At the start of the execution of the Algorithm 2, an initial population of link configuration is generated which includes the individual property alignments with the assigned similarity metrics. The population *P* is used to select the link candidates to be labeled using *Annotate* function. The labeled link candidates are passed to Algorithm 1 [12] where individual functions, as well as their combinations, are assigned the threshold value using these reference links. Only those combinations which have precision value better than the functions combined are added to the population *P* whereas rest is discarded. The evolved population is again used to select the reference links. The process continues until the loop_counter reaches the termination condition or the precision value for a particular element of P becomes 1. After the

Table 1 F1 score with variation of training data

Percentage of labeled data	F1 score	F1 score	F1 score	F1 score	F1 score
	D1	D2	D3	D4	D5
2	0.98	0.8281	0.9474	0.4012	0.3742
4	0.988	0.9471	0.9667	0.4123	0.3805
6	0.9984	0.9457	0.9714	0.45	0.3721
8	0.9991	0.944	0.971	0.468	0.39314
10	0.9987	0.9553	0.9704	0.50155	0.396
12	0.9987	0.9549	0.9697	0.5	0.39174
14	0.9987	0.9554	0.9703	0.5058	0.3992
16	0.99867	0.9553	0.9698	0.50439	0.3944
18	0.99865	0.9558	0.9695	0.504	0.3921
20	0.999	0.9559	0.9686	0.5056	0.3966

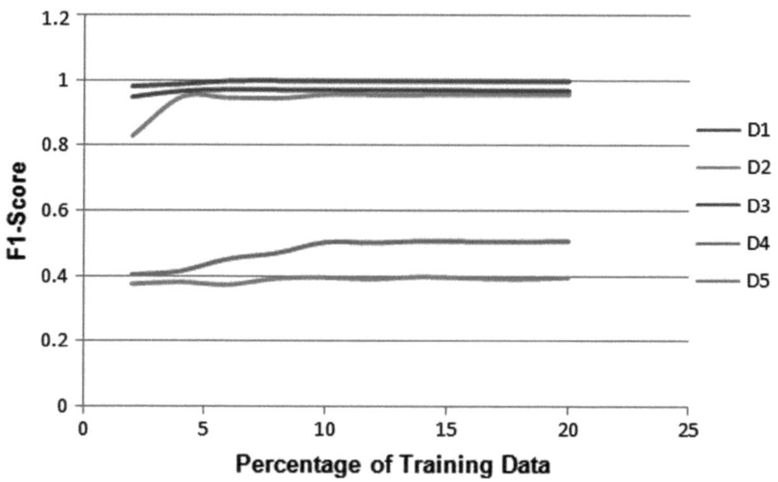

Fig. 3 Performance of the system with variation of training data size

execution of the loop at line 3, the link configuration with maximum precision score on the training data is returned as the output of the algorithm.

The trend of F1 score with variation in the size of training data has been analyzed. A small amount of training data is given to the configuration learning algorithm of the active learning-based instance matching system ranging from 2 to 20%. The F1 score after varying the training data is recorded in Table 1 and the corresponding graph is shown in Fig. 3.

4 Conclusion

Supervised learning of link configuration is more advantageous over unsupervised learning in terms of high accuracy and low complexity. The instance matching configuration is optimized by a learning algorithm that makes use of precision as a performance parameter. The result shows significant improvement of F1 score in comparison to other algorithms. In supervised learning approach, the major concern is the availability of labeled training data. Here, a semi-supervised learning technique called *Active Learning* can be used where a learning algorithm will actively query the user to label most informative link candidate. The paper proposes an extended framework to our previous work on supervised learning-based instance matching system with runtime efficiency. The proposed approach reduces the amount of labeled data required, by selecting the most informative candidates as the training data.

References

1. T. Berners-Lee, J. Hendler, O. Lassilia, The semantic web. Sci. Am. **284**(5), 34–44 (2001)
2. S. Rong, X. Niu, E.W. Xian et al., in *A Machine Learning Approach for Instance Matching Based on Similarity Metrics*. Proceedings of the 11th International Semantic Web Conference, Boston (2012), pp. 460–475
3. T. Heath, C. Bizer, Linked data: Evolving the web into a global data space, in *Synthesis Lectures on the Semantic Web: Theory and Technology*, 1st edn, ed. by J. Hendler, F. Harmelen (Morgan & Claypool Publishers, San Rafael, 2011)
4. K. Nguyen, R. Ichise, ScLink: Supervised instance matching system for heterogeneous repositories. J. Intell. Inf. Syst. 1–33 (2016)
5. J. Volz, C. Bizer, M. Gaedke, G. Kobilarov, in *Discovering and Maintaining Link on the Web of Data*. Proceedings of the 8th International Semantic Web Conference, Washington, DC (2009), pp. 650–665
6. A.C.N. Ngomo, S. Auer, in *LIMES—A Time-Efficient Approach for Large-Scale Link Discovery on the Web of Data*. Proceedings of the Twenty-Second International joint conference on Artificial Intelligence, Barcelona, Spain (2011), pp. 2312–2317
7. A. Nikolov, M. d'Aquin, E. Motta, in *Unsupervised Learning of Link Discovery Configuration*. Proceedings of the 9th Extended Semantic Web Conference, vol. 7295 (LNCS, Springer, Berlin, 2012), pp. 119–133
8. A.C.N. Ngomo, K. Lyko, in *Eagle: Efficient Active Learning of Link Specifications Using Genetic Programming*. Proceedings of the 9th Extended Semantic Web Conference, Greece (2012), pp. 149–163
9. R. Isele, C. Bizer, Active learning of expressive linkage rules using genetic programming. J. Web Semant. **23**, 2–15 (2013)
10. K. Nguyen, R. Ichise, in *Heuristic Based Configuration Learning for Linked Data Instance Matching*. Proceedings of 5th Joint International Semantic Technology Conference, Yichang, China (2015), pp. 56–72
11. A.C.N. Ngomo, J. Lehmann, S. Auer, K. Höffner, in *RAVEN- Active Learning of Link Specification*. Proceedings of the 6th International Workshop on Ontology Matching (2011), pp. 25–37
12. G. Kaur, P. Saini, S. Verma, in *Supervised Learning based Instance Matching System for Data Linking*. Proceedings of International Conference on Advances in Computing, Communication and Informatics (2017), pp. 38–44

Text Summarization Using WordNet Graph Based Sentence Ranking

Amita Jain, Sonakshi Vij and Devendra K. Tayal

Abstract Text summarization refers to the task of generating a summary from a given document that tries to replicate the most significant information of the original document. A number of techniques are available in the literature regarding the same and sentence ranking is one of them. This paper proposes a novel method for text summarization using WordNet graph based sentence ranking. The proposed method utilizes the degree, betweenness, and closeness centrality measures. This paper also analyzes the current research work going on in text summarization and sentence ranking. Web of Science (WoS) is used as the data source for the same. The co-occurrence of all the keywords in the research papers pertaining to sentence ranking is also visualized.

Keywords Sentence ranking · Text summarization · WordNet

1 Introduction

Text summarization refers to the process of generating a summary for a given document in such a way that most of the significant information of the original document is retained and the meaning is preserved. It is a key research area in the field of computational science and natural language processing. The trend of publication in text summarization from the year 2000 to 2016 is observed after extracting publication count from Web of Science (WoS) database. This is shown as in Fig. 1. It can be seen that 2016 accounts for the maximum record count in terms of number of publications in this field. In the coming years, the publication in this area is expected to grow further.

A. Jain (✉)
AIACTR (Ambedkar Institute of Advanced Communication Technologies & Research), New Delhi, India
e-mail: amita_jain_17@yahoo.com

S. Vij · D. K. Tayal
Indira Gandhi Delhi Technical University for Women, New Delhi, India

C. R. Krishna et al. (eds.), *Proceedings of 2nd International Conference on Communication, Computing and Networking*, Lecture Notes in Networks and Systems 46, https://doi.org/10.1007/978-981-13-1217-5_71

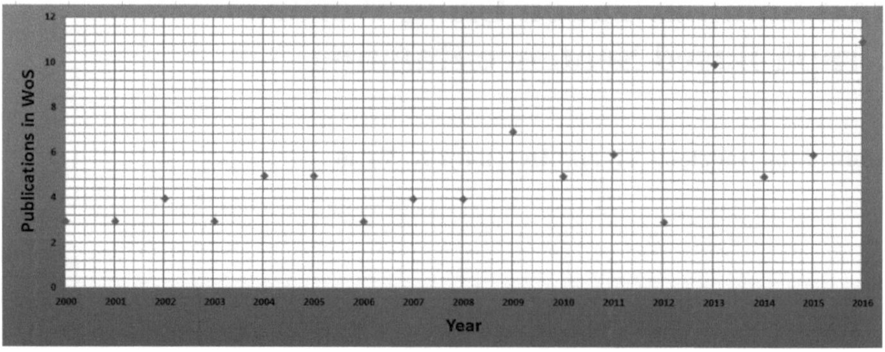

Fig. 1 Year-wise publication count

Sentence ranking, on the other hand, assigns a way to rank the sentences in a text, in order of the significance of information contained in them. Several methods exist which deal with sentence ranking algorithms to perform text summarization [1–3].

In this paper, a novel algorithm has been proposed to perform text summarization using WordNet graph based sentence scoring. The proposed method utilizes centrality measures degree, betweenness, and closeness. The rest of the paper is framed as follows: Sect. 2 highlights the background study in this field. Section 3 presents the proposed approach while Sect. 4 concludes the work.

2 Background Study

In order to get an in-depth insight of the research in sentence ranking, Web of Science (WoS) is chosen as the data source. The research papers extracted from it show that very little work has been done in this area [4–7]. These research papers are used to analyze the co-occurrence of the keywords, as shown in Fig. 2. These networks are visualized using VoSviewer.

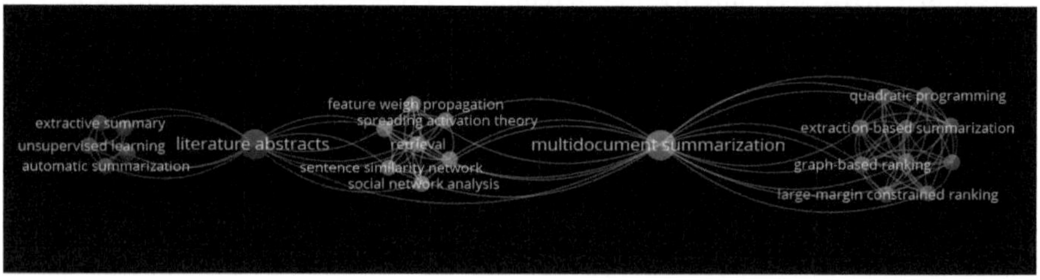

Fig. 2 Co-occurrence of the keywords

Table 1 Literature related to text summarization and sentence ranking

S.No.	Research paper	Concept	Pros	Cons
1	[1]	• Uses supervised learning for extractive text summarization • An "*averaged perceptron*" method is utilized	Query sensitivity is improved	Query expansion is not taken into consideration
2	[7]	This method obtains a set that consists of the "*topic-related documents*" in the form of a "*sentence similarity network*"	Node significance is determined	Measures of graph connectivity are not analyzed well
3	[2]	A semi-supervised learning based method is deployed for creating the text graph for summarization	Demonstrates how to generate a hypergraph for summarization	The amount of information contained in a graph can be measured using other graph connectivity measures
4	[4]	• Matrix notation is used for modeling CoRank method • Text redundancy is completely avoided, which helps in bringing better results	Better results than state-of-the-art methods	The method is not applicable for multi-document summarization
5	[5]	Takes into consideration, various factors for scoring, i.e., multiple decision criteria	Factors such as sentence length, sentence position etc. is taken into consideration	Results are not provided, i.e., just a proposal

The literature related to text summarization and sentence ranking is analyzed as in Table 1. The major research gap identified is that the graph-based text summarization approaches deployed till date; do not fully utilize the graph connectivity measures. Centrality measures like degree, betweenness, and closeness can be utilized to perform sentence based text summarization in an efficient and more effective manner. This paper tries to bridge this research gap.

Table 2 Proposed algorithm

Nomenclature :

i. Text to be summarized: T
ii. Total number of queries in T: N
iii. Degree Centrality Measure: D_{CM}
iv. Closeness Centrality Measure: C_{CM}
v. Betweenness Centrality Measure: B_{CM}
vi. Sum total of $D_{CM,}$ C_{CM} and B_{CM} : S
vii. Most Significant Sentence(s) : M_{SS}

(a) Input T
(b) Divide each sentence in T into queries
(c) For each query:
- Remove stop words & perform tokenization
- Generate WordNet graph (W_G) using each text token
- Calculate $D_{CM,}$ C_{CM} and B_{CM}
- Calculate S
- Top N/3 sentences with highest value of S: M_{SS}

(d) Show M_{SS}

Fig. 3 Process of proposed work

3 Proposed Approach

This section describes the proposed approach for performing text summarization using WordNet graph based approach. The algorithm for the same is given in Table 2 and the process is visualized as in Fig. 3.

4 Conclusions

This paper analyzes the current research work in the field of text summarization and sentence ranking using Web of Science (WoS) as the data source. The co-occurrence of all the keywords in the field of sentence ranking is visualized. This paper also proposes a method for text summarization that uses WordNet graph based sentence ranking. The proposed method utilizes the degree, betweenness, and closeness centrality measures so that the significance of information contained in the sentences is well analyzed. In the future, this proposed model will be implemented and the results will be compared to the state of the art.

References

1. S. Fisher, B. Roark, in *Query-Focused Summarization by Supervised Sentence Ranking and Skewed Word Distributions*. Proceedings of the Document Understanding Conference, DUC-2006, New York, USA (2006, June)
2. W. Wang, F. Wei, W. Li, S. Li, in *Hypersum: Hypergraph Based Semi-Supervised Sentence Ranking for Query-Oriented Summarization*. Proceedings of the 18th ACM Conference on Information and Knowledge Management (ACM, 2009, November), pp. 1855–1858
3. T.S.R. Raju, B. Allarpu, Text Summarization Using Sentence Scoring Method (2017)
4. C.J. Fang, D.J. Mu, Z.H. Deng, Z. Wu, Word-sentence co-ranking for automatic extractive text summarization. Expert Syst. Appl. **72**, 189 (2017). https://doi.org/10.1016/j.eswa.2016.12.021 (Pergamon-Elsevier Science Ltd.)
5. L.B. Yang, X.Y. Cai, Y. Zhang, P. Shi, Enhancing sentence-level clustering with ranking-based clustering framework for theme-based summarization. Inf. Sci. **260**, 37 (2014). https://doi.org/10.1016/j.ins.2013.11.026 (Elsevier Science Inc.)
6. X. Li, L. Du, Y.D. Shen, Update summarization via graph-based sentence ranking. IEEE Trans. Knowl. Data Eng. **25**(5), 1162 (2013). https://doi.org/10.1109/tkde.2012.42 (IEEE Computer Soc)
7. J.Y. Yeh, H.R. Ke, W.P. Yang, iSpreadRank: Ranking sentences for extraction-based summarization using feature weight propagation in the sentence similarity network. Expert Syst. Appl. **35**(3), 1451 (2008). https://doi.org/10.1016/j.eswa.2007.08.037 (Pergamon-Elsevier Science Ltd.)

Reckoning of Photosynthetic Pigments Using Remotely Sensed Spectral Responses of Vigna Radiata Crop for Surge Monitoring

Rupali R. Surase, Karbhari Kale, Amrsinh B. Varpe, Amol D. Vibhute, Hanumant Gite, Mahesh Solankar, Sandeep Gaikwad, Dhananjay Nalawade and Suresh Mehrotra

Abstract This paper outlines the intents to develop and assess remotely sensed spectral responses of Vigna radiata crop with an adaxial surface positioned at Aurangabad region by Latitude 19.897827 and Longitude 75.308666. Current exploration will be useful for crop surge monitoring centered on spectral features collected using

R. R. Surase (✉) · K. Kale · A. B. Varpe · A. D. Vibhute · H. Gite · M. Solankar
S. Gaikwad · D. Nalawade · S. Mehrotra
Geospatial Technology Research Laboratory, Department of Computer Science and IT,
Dr. Babasaheb Ambedkar Marathwada University, Aurangabad 431004, Maharashtra, India
e-mail: rupalisurase13@gmail.com

K. Kale
e-mail: kvkale91@gmail.com

A. B. Varpe
e-mail: varpeamarsinh@gmail.com

A. D. Vibhute
e-mail: amolvibhute2011@gmail.com

H. Gite
e-mail: hanumantgitecsit@gmail.com

M. Solankar
e-mail: mmsolankar13@gmail.com

S. Gaikwad
e-mail: sandeep.gaikwad22@gmail.com

D. Nalawade
e-mail: dhananjay.bamu@gmail.com

S. Mehrotra
e-mail: mehrotra.suresh15@gmail.com

C. R. Krishna et al. (eds.), *Proceedings of 2nd International Conference on Communication, Computing and Networking*, Lecture Notes in Networks and Systems 46, https://doi.org/10.1007/978-981-13-1217-5_72

ASD Field Spec 4 spectroradiometer intended for the estimation of photosynthetic pigments. The proposed approach resides preprocessing techniques followed via postprocessing methods instigated by python open source software. The preprocessing was prepared using parabolic correction technique with (.asd) files format. Spectral features were projected for detection of photosynthetic pigments using ten categories of indices. Among the diverse phonological patterns of crop progression, the respectable aggregate of a coefficient of correlations was found with fluctuating growth parameters at jointing phase of Vigna radiata. The spectral vegetation indices (SVI) desired for the investigation were composed at jointing and ripening crop phases. SVI was given superior outcomes with R^2 values speckled between 0.91 and 0.99, in addition, good amount of correlation was observed in between NDVI and PSSR-a. The NDVI index was found to be the appropriate parameter for healthy crops and ARI2 was found appropriate for detection of disease crops. Multiple regression equations were used by means of stepwise regression technique using open source software.

Keywords Spectral signature · Vegetation indices · Biochemical parameters Photosynthetic pigments · Crop analysis

1 Introduction

Reckoning of photosynthetic pigments from crops using non-imaging spectral signature is an emerging influential tool in remote sensing for crop surge monitoring era containing factors such as chlorophyll content, water contents, anthocyanin, xanthophyll, and carotenoids. The objective of the current study is to recognize biochemical parameters of Vigna radiata using several vegetation indices based on spectral features. The spectral response signifies the correlation between electromagnetic radiations (EMR) with the biological and chemical characteristics of objects, present on earth surface. These are the elementary aspects of data representation and fusion in the forms of passive remote sensing tool [1]. Latest sensing devices allow multiband frequencies for generation of spectral signatures for crop discrimination using SVI for the reckoning of photosynthetic pigments [2, 3]. The research study focuses on pulse crop because of its tremendous source of nutritive proteins. Vigna radiata is one of the needed pulses for sustaining proteins amongst populace in India. This crop type is grown up around 3.50 ha in India typically Maharashtra, Bihar, etc. [4]. The literature reveals that nondestructive technique provides crop type discrimination using spectral data collected using spectroradiometer of crops soybean, canola, wheat, and oat [5, 6]. The citrus plant types were identified using support vector machines, logistic regression, and K-nearest neighbor method with 0.3% error rate [7]. By considering the literature review, the research current study focuses towards collecting spectral responses of pulse crop to monitor growth.

The paper rumors on an experimental method to acquire reflectance signature of Vigna radiata and reckoning of their photosynthetic pigments for analysis of crop

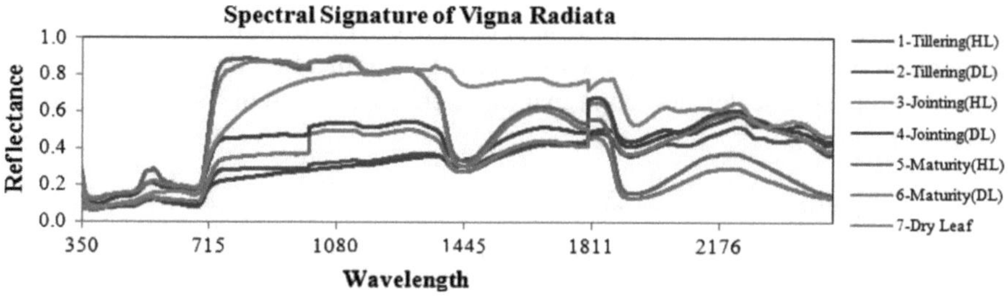

Fig. 1 Spectral responses of Vigna radiata crop surge phases

surge monitoring. This paper comprises four sections containing introduction with background knowledge of crops and their significance. Section 2 contains specifics about study sites, preprocessing techniques and algorithms selected for research. All comparative discussion about outcomes with the analysis is in Sect. 3. At last, conclusions are summarized followed by future scope.

2 Study Sites and Experimental Design

Vigna radiata is grown up on a black and red soil at Aurangabad region, Maharashtra, India. Crop samples from diverse agriculture fields were composed in four diverse stages of Tillering, Jointing, Maturity and dry growth phase. Shortly after anthesis, Vigna radiata leaves as per growth stages from 10 field locations were collected and transported in an airtight plastic bag for spectral measurement. Spectra response of leaves was measured using an ASD Field Spec 4 Spectroradiometer with a wavelength range of 350–2500 and 1.6 nm sampling recess with a resolution of 3.5 nm, respectively. Adaxial surfaces of each leaf were measured with 8 degrees Field Of View (FOV) in standard darkroom laboratory environment for pigment analysis along weather conditions dated July 2015–October 2015 between 11:00 a.m. to 2:20 p.m. as per the standard to maintain reflectance curve (Fig. 1).

A standard white reference panel was used to calibrate the gadget. Overall 10 shots of each sample were measured and total 70 spectral responses were transferred for further processing. The validated spectral indices provide good amount of results for crop growth monitoring.

2.1 Parabolic Correction

The spectral information of crops for the study is useful in the range of 350–2500 nm [8]. The spectral shift variations were occurred at the VNIR and the SWIR1 detector (1000 nm) and in between SWIR1 to SWIR2 (1800) spectral band. A parabolic correction was performed to remove stepped splice error towards .asd files format. Further analysis was done after parabolic correction to increase the accuracy of crop surge monitoring. The detector combinations were established in spectrum given [9] as Eqs. (1) and (2).

$$\text{VNIR} : \text{SWIR1} = (o(1001 - 1000))/(g(1001) - g(1000)) \tag{1}$$

$$\text{SWIR1} : \text{SWIR2} = (o(1801) - o(1800))/(g(1801) - g(1800)) \tag{2}$$

where g represents the gain of detector and o represents the offset of the detector.

2.2 Spectral Indices

Spectral Indices vary with their specific role for crop type identification using multispectral and hyperspectral imagery. Table 1 contains equations of all indices used in our study along with their indicators and respective references. R_i indicates reflectance at the ith wavelength.

The NDVI provides better prediction accuracy in photosynthetic activity because of near-infrared (NIR) and red band. The NDVI spectral profile has been correlated to many variables such as crop nutrient deficiency, crop health measurement, disease estimation, and long-term water stress due to atmospheric changes [10]. PSSR-a, PSSR-b, and PSSR-c indices provide spectral features for crop health assessment. A CRI2 and ARI2 index gives spectral information of crop disease severity. The CAI, NDLI, and X1 spectrum informs about crop capacity to make wood and for other uses. WI provides availability of water contents with respective spectral features.

2.3 Accuracy Assessment Techniques

Overall outcomes of spectral features were measured by stepwise regression analysis method. It is a form of investigating the relationship between dependent and independent variable. In the current study, dependent variables are NDVI spectral features correlated with other indices spectral features including PSSR-a, PSSR-b, PSSR-c, ARI2, CRI2, and WI, respectively. This method is based on adding or removing the iterations as per the data are given to the model [18] as given in Eq. (3),

$$\text{Percent change} = \left[\text{RMSE}_{\text{previous}} - \text{RMSE}_{\text{current}}/\text{RMSE}_{\text{current}}\right] * 100 \tag{3}$$

where $\text{RMSE}_{\text{previous}}$ denotes the squared error of the previous model iterated spectral features and $\text{RMSE}_{\text{current}}$ describes the squared error of the current model execution with spectral features.

Table 1 List of vegetation indices studies

Vegetation indices	Equations	Indicator	References
Normalized difference vegetation index	$NDVI = \frac{R_{800} - R_{670}}{R_{800} + R_{670}}$	Biomass, leaf area	Rouse et al. [10]
Pigment specific normalized difference chl a	$PSSR - b = R_{800}/R_{670}$	Chlorophyll a	Blackburn [11]
Pigment specific normalized difference chl b	$PSSR - b = R_{800}/R_{635}$	Chlorophyll b	Blackburn [11]
Pigment specific normalized difference C	$PSSR - c = R_{800}/R_{470}$	Carotenoid	Blackburn [11]
Cartenoid reflectance index 2	$CRI2 = 1/R_{510} - 1/R_{700}$	Cartenoid	Gitelson et al. [12]
Anthocyanin reflectance index 2	$ARI2 = R_{800}(1/R_{580} - 1/R_{700})$	Anthocyanin	Gitelson et al. [13]
Cellulose absorption index	$CAI = 0.5(R_{2000} + R_{2200}) - R_{2100}$	Cellulose	Daughtry et al. [14]
Normalized difference lignin index	$NDLI = \frac{\log(1/R_{1754}) - \log(1/R_{1680})}{\log(1/R_{1754}) + \log(1/R_{1680})}$	Lignin	Serrano et al. [15]
Water index	$WI = \frac{R_{900}}{R_{970}}$	Water content	Penuelas et al. [16]
Xanthophyll index	$X1 = \frac{(R_{528} - R_{567})}{(R_{528} + R_{567})}$	Xanthophyll	Gamon and Penuelas [17]

3 Results and Discussion

This section represents an analysis of result with assessment technique through spectral indices.

3.1 Spectral Feature Reckoning

Figure 2 shows graphical representations of vegetation indices with growth parameters. The symbol S_n represents nth samples of crop growth. The X-axis represents crop growth stages with reference to Fig. 1, and Y-axis represents observed spectral features with specific range respective to indices.

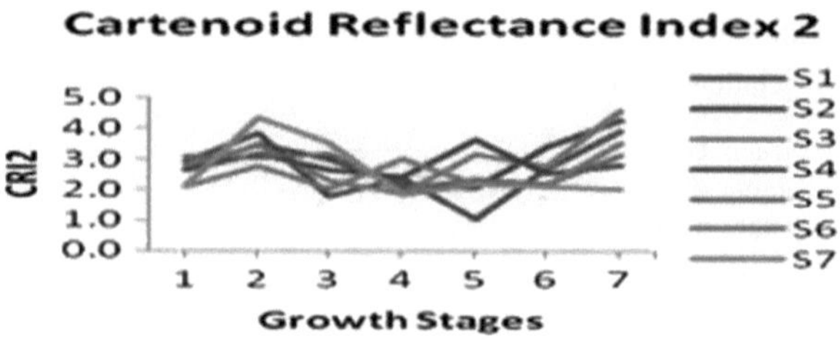

Fig. 2 Spectral feature estimation using CRI2 index

NDVI index varies between the accuracy 0.3–0.7 spectral range standards +1 to −1 vegetation indicator. CRI2 varies with the spectral range of 0.1–0.5 as an amount carotenoid pigments indicator which is graphically represented in Fig. 2. ARI2 is modified index for anthocyanin pigments identification which varies in an experiment from −0.50 to 1.50 which is also used to indicate the severity of disease of leaves than healthy leaves. CAI index varies from −0.03 to 0.0 that represents the amount cellulose contents present in leaf represented, NDLI represents the quantity of lignin pigment, furthermore total water content available in leaf was obtained using WI. There was significant difference measured in CRI2 at different levels for Vigna radiate crop. PSRI and MSI as reflectance were decreased from tillering and constant in between tillering to jointing stage due to the maintained level of nitrogen and water of the crops.

3.2 Stepwise Regression Analysis

The indices derived from visible near-infrared (VNIR) region like NDVI increased in the early vegetative period which is tillering up to full vegetative phases. It reached the peak around jointing stage then it moved towards maturity of a crop in all vegetative criteria. Table 2 represents the correlation between NDVI with CRI2, ARI2, and CAI spectral index. The stepwise regression between different growth parameters varies with healthy, disease, and dry stages. NDVI correlated with CRI2 at Jointing stage (HL) with highest accuracy 0.45, NDVI correlated with ARI2 at Tillering (DL) with highest accuracy 0.68 and NDVI with CAI at Maturity (DL) with accuracy 0.13 as per the given equation.

Table 3 represents the correlation between NDVI with pigment specific index including PSSR a with accuracy 0.99 for Jointing (HL), PSSR b with accuracy 0.92 for dry leaf and 0.66 with tillering (HL).

Linear regression interpretation between different vegetation indices with varying stages from healthy, diseased and dry samples with NDVI including same growth stages. The coefficient of correlation was successfully approved with SVI and respective crop health parameters collected at different growth stages.

Table 2 Regression analysis with growth parameters and SVI of crop species

Vegetation indices		CRI2		ARI2		CAI		NDLI		WI	
		R2	Regression model	R2	Regression model	R2	Regression model	R2	Regression model	R2	Regression model
NDVI	Tillering (HL)	0.06	$y = -2.6x + 4.17$	0.61	$y = -2.26x + 1.19$	0.08	$y = -0.04x + 0.05$	0.07	$y = 0.0x - 0.0$	0.14	$y = 0.0x + 0.9$
	Tillering (DL)	0.14	$y = -3.7x + 5.13$	0.68	$y = -2.4x + 1.35$	0.13	$y = 0.0x + 0.01$	0.02	$y = -0.0x + 0.0$	0.52	$y = 0.2x + 0.8$
	Jointing (HL)	0.45	$y = -25.6x + 16.8$	0.55	$y = -4.2x + 2.39$	0.06	$y = -0.0x + 0.01$	0.38	$y = 0.8x - 0.41$	0.51	$y = 0.27x + 0.8$
	Jointing (DL)	0.07	$y = -2.98x + 3.74$	0.38	$y = -2.4x + 1.43$	0.04	$y = 0.01x - 0.01$	0.08	$y = 0.0x + 0.03$	0.47	$y = 0.21x + 0.8$
	Maturity (HL)	0.06	$y = -6.69x + 7.23$	0.21	$y = -2.7x + 1.94$	0.02	$y = 0.00x - 0.02$	0.11	$y = 0.1x - 0.04$	0.06	$y = -0.04x + 1.0$
	Maturity (DL)	0.02	$y = -0.44x + 2.93$	0.33	$y = -5.9x + 4.51$	0.13	$y = 0.03x - 0.04$	0.13	$y = -0.0x + 0.1$	0.68	$y = 0.3x + 0.8$
	Dry leaf	0.12	$y = -6.76x + 7.21$	0.46	$y = -5.4x + 4.05$	0.11	$y = 0.03x - 0.01$	0.46	$y = -0.3x + 0.2$	0.27	$y = 0.08x + 0.8$

Table 3 Regression analysis with growth parameters and SVI of crop species

Vegetation indices		PSSR a		PSSR b		PSSR c		X1	
		R2	Regression model	R2	Regression model	R2	Regression model	R2	Regression model
NDVI	Tillering (HL)	0.98	$y = 7.94x - 0.96$	0.90	$y = 8.56x - 1.69$	0.66	$y = 3.82x + 1.29$	0.51	$y = 0.20x - 0.11$
	Tillering (DL)	0.99	$y = 6.73x - 0.38$	0.82	$y = 5.38x - 0.05$	0.15	$y = 1.36x + 2.44$	0.71	$y = 0.21x - 0.12$
	Jointing (HL)	0.99	$y = 10.71x - 2.48$	0.47	$y = 8.42x - 1.62$	0.21	$y = 4.30x + 1.36$	0.44	$y = 0.24x - 0.16$
	Jointing (DL)	0.98	$y = 7.89x - 0.96$	0.64	$y = 5.94x - 0.35$	0.35	$y = 2.58x + 2.21$	0.77	$y = 0.32x - 0.20$
	Maturity (HL)	0.93	$y = 24.06x - 11.21$	0.69	$y = 14.45x - 4.79$	0.02	$y = 4.85x + 3.14$	0.31	$y = 0.12x - 0.11$
	Maturity (DL)	0.91	$y = 20.38x - 8.51$	0.49	$y = 16.45x - 6.00$	0.48	$y = 13.19x - 2.70$	0.62	$y = 0.29x - 0.23$
	Dry leaf	0.97	$y = 12.55x - 3.53$	0.92	$y = 6.99x - 0.89$	0.28	$y = 4.01x + 2.89$	0.72	$y = 0.40x - 0.30$

4 Conclusion and Future Work

The research study concluded that VNIR region of the spectral signature is valuable intended for evaluation spectral features regard to crop patterns analysis. The respectable association was observed in jointing stage of crop based on SVI, viz., NDVI, PSSR-a, PSSR-b, PSSR-c, NDLI, WI, X_1, CAI, ARI2, and CRI2. This SVI was increasing differently as per crop criteria of leaves including healthy, diseased and dry from Tillering to dry stages of leaves. The R^2 varies within the range of 0.91–0.99, the given range describes the good accuracy of spectral features relationship. Finally, current research invention contributes in identification of crop surge monitoring system. Research concluded that NDVI with PSSRa index performs better for determination of different growth parameters and ARI2 classifies the disease severity details of leaves. This research will be used for yield forecasting and prediction of growth stages using spectral responses of crops. The current study was experimented by open source software (python-2.7.11). Future directions of current research will be to compare and extend the research for various types of the growth of the crops monitoring system.

Acknowledgements Authors would like to thank technical supports under UGC SAP (II), DST-NISA, DRS phase-II, and DST-FIST for partial financial assistance to the Department of CS & IT, Dr. Babasaheb Ambedkar Marathwada University Aurangabad, India and also thanks to UGC for providing financial support under UGC-BSR research fellowship for a research study.

References

1. K. Pfitzner, A. Bollhofer et al., A standard design for collecting vegetation reference spectraimplementation and implications for data sharing. J. Spat. Sci. **52** (2006)
2. K. Muller, U. Bottcher, H. Kage, Analysis of vegetation indices derived from hyperspectral reflection measurements for estimating crop canopy parameters of oilseed rape, Brassica napus L. Biosyst. Eng. **101**, 172–182 (2008)
3. E.I. Rabinowitch, Govindjee, *The Role of Chlorophyll in Photosynthesis* (Project Coordinator (MULLaRP), Indian Institute of Pulses Research)
4. A. Hueni, A. Bialek, Cause, effect, and correction of field spectroradiometer interchannel radiometric steps. IEEE J. Sel. Top. **10**(4) (2017)
5. J.H. Wilson, C. Zhang, J.M. Kovacs, Separating crop species in northeastern ontario using hyperspectral data. Remote Sens. ISSN 2072–4292 (2014)
6. S.M. Arafat, M.A. Aboelghar, I.F. Ahmed, Crop discrimination using field hyper spectral remotely sensed data. Adv. Remote Sens. 63–70 (2013)
7. A.R. Mishra, D. Karimi, R. Ehsani, W.S. Lee, Identification Of citrus greening (HLB) using a VIS-NIR spectroscopy technique. Am. Soc. Agric. Biol. Eng. **55**(2), 711–720 (2012). ISSN 2151-0032
8. B. Govaerts, N. Verhulst, *The Normalized Difference Vegetation Index Green Seeker TM Handheld Sensor: Toward the Integrated Evaluation of Crop Management* (International Maize and Wheat Improvement Center KatholiekeUniversity, Leuven, 2010)
9. A. Hueni, A. Bialek, Cause, effect, and correction of field spectroradiometer interchannel radiometric steps. IEEE J. **10**(4) (2017)

10. J.W. Rouse et al., Monitoring vegetation systems in the Great Plains with ERTS. Third Earth Resources Technology Satellite–1 Symposium, vol. I, NASA SP-35
11. G.A. Blackburn, Spectral indices for estimating photosynthetic pigment concentrations: A test using senescent tree leaves. Int. J. Remote Sens. **19** (1998)
12. A. Gitelson et al., Assessing carotenoid content in plant leaves with reflectance spectroscopy. Photochem. Photobiol. **75**, 272–281 (2002)
13. A. Gitelson, M. Merzlyak et al., Optical properties and nondestructive estimation of anthocyanin content in plant leaves. Photochem. Photobiol. 38–45 (2008)
14. Daughtry et al., Discriminating crop residues from soil by short-wave infrared reflectance. Agron. J. **93**, 125–131 (2001)
15. Serrano et.al., Remote sensing of nitrogen and lignin in mediterranean vegetation from AVIRIS data: decomposing biochemical from structural signals. Remote Sens. Environ. 355–364 (2001)
16. J. Penuelas, I. Filella, J. Gamon, Assessing the photosynthetic radiation-use efficiency of emergent aquatic vegetation from spectral reflectance. Aquat. Bot. **58**, 307–315 (1997)
17. J.A. Gamon, J. Penuelas, A narrow-waveband spectral index that tracks diurnal changes in photosynthetic efficiency. Remote Sens. Environ. **41**, 35–44 (1992)
18. Stepwise Regression, Chapter 311, NCSS Statistical software

Comparative Analysis of Machine Learning Algorithms for Breast Cancer Prognosis

Kashish Goyal, Prakriti Sodhi, Preeti Aggarwal and Mukesh Kumar

Abstract In the past few years, gynecological cancers have taken their toll on women's health. Breast cancer is the major cause of cancer death followed by ovarian cancer and others. In this paper, we aim at finding the cancer status of the patient, whether it is benign or malignant. Data is collected from WISCONSIN dataset of UCI machine learning Repository. The dataset includes the cases of patients who are at risk of developing the cancer or have redeveloped the cancer. Different attribute selection techniques are applied on the data set. Further, different classification algorithms are used to compare and analyze the results.

Keywords Breast cancer prognosis · Machine learning algorithms
Classification models · Feature selection techniques

1 Introduction

Cancer is one of the deadly diseases caused by unnatural cell growth. Our bodies are composed of millions of cells. The irregular growth of these cells lead to tumor called as primary tumor where the emergence of cancer starts. Approximately 1,688,780 new cases of cancer were estimated to be reported in 2017, out of which 600,920 were reported. A study in 2012–2014 shows that around 38.5% of men and women have more chances to be diagnosed with cancer [1, 2]. Among others, cancer stands

K. Goyal (✉) · P. Sodhi · P. Aggarwal · M. Kumar
Department of Computer Science & Engineering, University Institute of Engineering and Technology, Panjab University, Chandigarh, India
e-mail: kashishgoyal31@gmail.com

P. Sodhi
e-mail: prakritisodhi@gmail.com

P. Aggarwal
e-mail: pree_agg@pu.ac.in

M. Kumar
e-mail: mukesh_rai9@pu.ac.in

C. R. Krishna et al. (eds.), *Proceedings of 2nd International Conference on Communication, Computing and Networking*, Lecture Notes in Networks and Systems 46, https://doi.org/10.1007/978-981-13-1217-5_73

to be the major reason of deaths due to diseases around the globe [3, 4]. Nearly one in six deaths occur due to cancer. Roughly around 70% of cancer deaths occur in economically weak countries [5, 6]. Among other various reasons high body mass index, low nutritious diet, sedentary lifestyle, tobacco intake, infections, radiation, stress and consumption of alcohol are the most common ones for the deaths due to cancer [7, 8]. The most found cancers among women are breast cancer followed by ovarian and cervical cancers. These are the most common gynecological cancers affecting the health of women worldwide and in India as well [9–11]. Various machine learning algorithms are used to predict the risks factors causing cancer.

The leftover paper is organized in the following manner. Outline of data mining algorithms (like Adaboost, Random Forest Algorithm and Support Vector Machine) which are used for processing and getting the acute results are given in Sect. 2. Section 3 gives the experimental setup, i.e., the detail analysis of WISCONSIN Breast Cancer Dataset with different parameters and classifiers is discussed here. Section 4 includes Results and Sect. 5 includes Conclusion and future scope followed by References.

2 Data Mining Algorithms

2.1 Adaboost

Boosting is a method of creating an immensely accurate classifier by the combination of weak and inaccurate rules of thumb. In this, weak learners are combined to form a strong rule. Different distributions are applied to identify a weak rule. Since this being an iterative process a new tentative weak rule is predicted every time. A single strong prediction rule is obtained after the combination of weak rules. This rule is considered to be much more accurate than the others found previously [12–14]. Figure 1 shows the flowchart of the algorithm in steps followed by the equations used to calculate the rule.

Equations:

Eqn 1:

$$w_i = \left\{ \begin{array}{ll} \frac{1}{2l}, & \text{for positive} \\ \frac{1}{2m}, & \text{for neagtive} \end{array}, \quad i = 1 \ldots n \right\} \tag{1}$$

Eqn 2:

$$w_{t,i} = \frac{w_{t,i}}{\sum_{j=1}^{n} w_{t,j}}, \quad i = 1 \ldots n \tag{2}$$

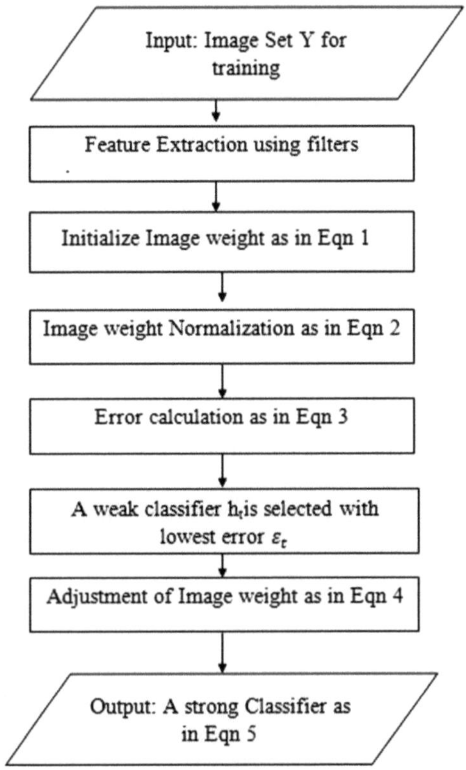

Fig. 1 Block diagram of Adaboost algorithm

Eqn 3:

$$\varepsilon_{i,j} = \sum_{k=1}^{n} w_{t,k}\left|h_{i,j}(x_k) - y_k\right|, \quad i = 1\ldots n, \quad j = 1\ldots N \tag{3}$$

Eqn 4:

$$w_{t+1,i} = \left\{ \begin{array}{ll} w_{t,i}\beta_t, & \text{if } x_i \text{ is classified correctly} \\ & \text{where } \beta_t = \frac{\varepsilon_t}{1-\varepsilon_t} \\ w_{t,i,} & \text{otherwise} \end{array} \right\} \tag{4}$$

Eqn 5:

$$H_T(x) = \sum_{t=1}^{T} \alpha_t h_t(x) \tag{5}$$

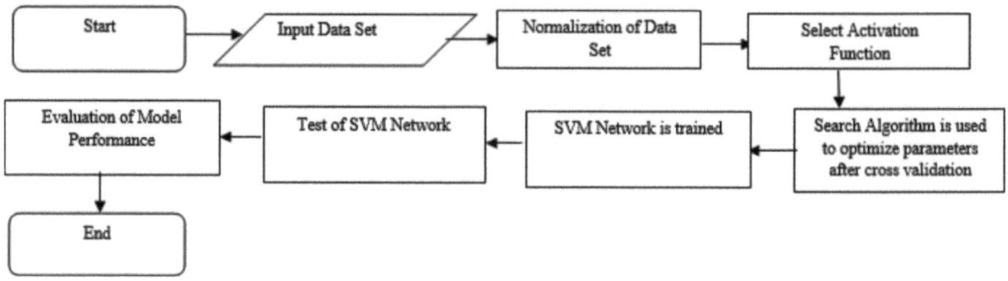

Fig. 2 Flowchart for SVM

2.2 Support Vector Machine

Support Vector Machine generally called as SVM are used for classification and regression purposes. They are used in image recognition, handwritten digit recognition, and in many more [15, 16]. Figure 2 illustrates the model and tells the different stages the data goes through.

2.3 Random Forest Algorithm

Random Forest algorithm is a supervised classification algorithm. It makes the forest with a combination of number of trees. The accuracy increases with the number of trees in the forest and vice versa [17]. Gini Index is measured on the contribution of variables on the nodes. With every split of a node Gini Index for the new child nodes are calculated which is then compared with the existing ones and the best ones are chosen [18, 19]. Figure 3 shows the flowchart to explain this algorithm in a detailed manner for better understanding.

3 Experimental Setup

The dataset has been taken from WISCONSIN Breast Cancer, UCI Machine Learning Repository available online [20]. It has 576 instances and 32 attributes including one class attribute of breast cancer being malignant and benign. The 32 attributes are *ID Number, Mean Radius, Mean Texture, Mean Perimeter, Mean Area, Mean Smoothness, Mean Compactness, Mean Concavity, Mean Concave Points, Mean Symmetry, Mean Fractal Dimension, Radius Standard Error, Texture Standard Error, Perimeter Standard Error, Area Standard Error, Smoothness Standard Error, Compactness Standard Error, Concavity Standard Error, Concave Points Standard Error, Symmetry Standard Error, Fractal Dimension Standard Error, Worst Radius, Worst Texture, Worst Perimeter, Worst Area, Worst Smoothness, Worst Compactness, Worst Concav-*

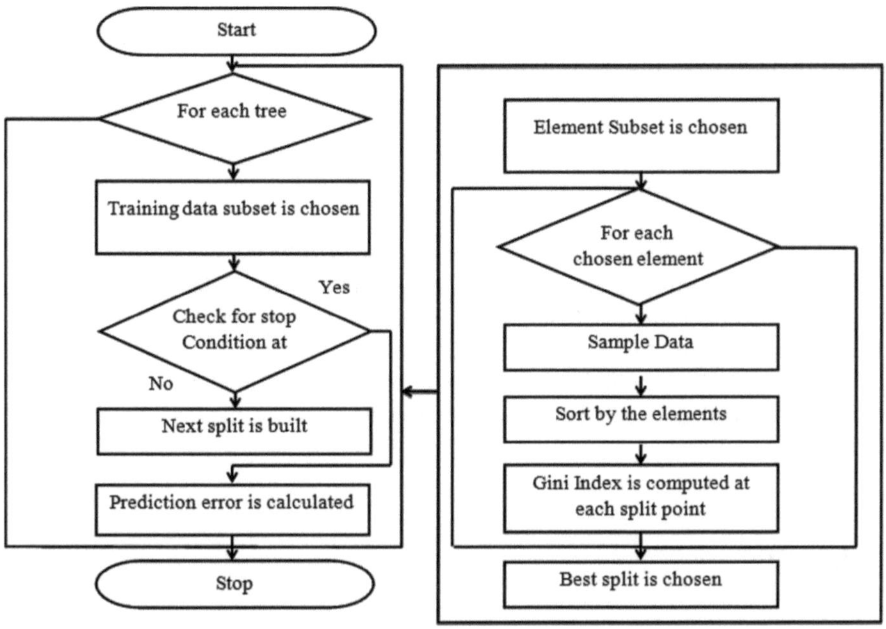

Fig. 3 Flowchart of random forest algorithms

ity, Worst Concave Points, Worst Symmetry, Worst Fractal Dimension and *Diagnosis* (the class with two values Benign and Malignant).

3.1 *Methodology*

In this experiment, the number of correctly classified instances is evaluated using different data mining classification techniques in WEKA 3.8. The techniques included are Naive Bayes, SVM, Decision Tree J48, Random Forest, Bagging, Ada Boost and Logistic Regression. These classifiers are applied using different test options such as cross validation and percentage split and the one that gives the better results are chosen. 66% data split and 10-fold cross validation have been used.

3.1.1 Data Preprocessing

There are no missing values in the data. The data has been normalized as the attributes had anonymous values and had different units and scales. Hence the data has been normalized on 0–1 scale and then data discretization is done using 10 bins and −1.0 as the weight of the instance per interval. In this method continuous data feature values are converted into finite set of intervals to minimize the information loss. Therefore, preprocessing is performed to make the comparison of different values efficiently.

3.1.2 Feature Selection

When the complete dataset is taken all 32 attributes are selected and then the accuracy is evaluated using different classifiers and in case of the reduced dataset the attributes are selected using Principal Component Analysis method (PCA). On performing PCA, 12 best features are selected out of 32 features. When Subset Evaluation using Best First method is applied, the number of attributes are reduced to 11 while in the case of Correlation Attribute Evaluation using Rankers method the number of features remain same.

4 Results

The results of correctly identified instances have been analyzed here. Figure 4 represents the accuracy for different classifiers using complete dataset as well as reduced dataset. PCA method is used for minimizing the features. In case of complete dataset, Random Forest gives the highest accuracy of 95.85% whereas the accuracy of 92.60% is given by AdaBoost which is worst when compared among other classification techniques. In the case of reduced dataset, Adaboost as well as Logistic Regression give the same accuracy of 97.92%. In some cases such as Naive Bayes and Random forest the accuracy is less as compared to when applied on complete dataset. When the results are analyzed using other techniques, the classifiers have performed better than when applied on entire dataset.

The accuracy results of different attribute selection methods measured along with different classifiers are presented in Fig. 5. It can be observed that, overall PCA method has outperformed the other techniques. Although, in some cases such as Naive Bayes and Random Forest, the Subset Evaluation using Best First has performed better than PCA. Ada Boost and Logistic Regression algorithms have performed better than the others in case of PCA with accuracy of 97.92%. The results given by Correlation Attribute Evaluation method are not as good when compared with other two methods.

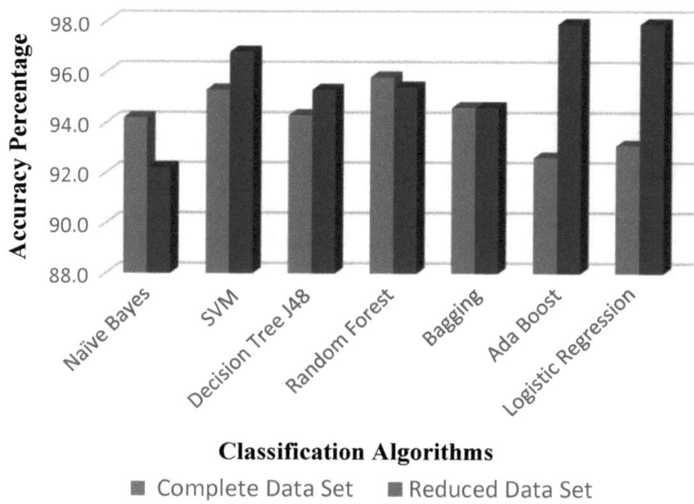

Fig. 4 Accuracy results for complete dataset and reduced dataset

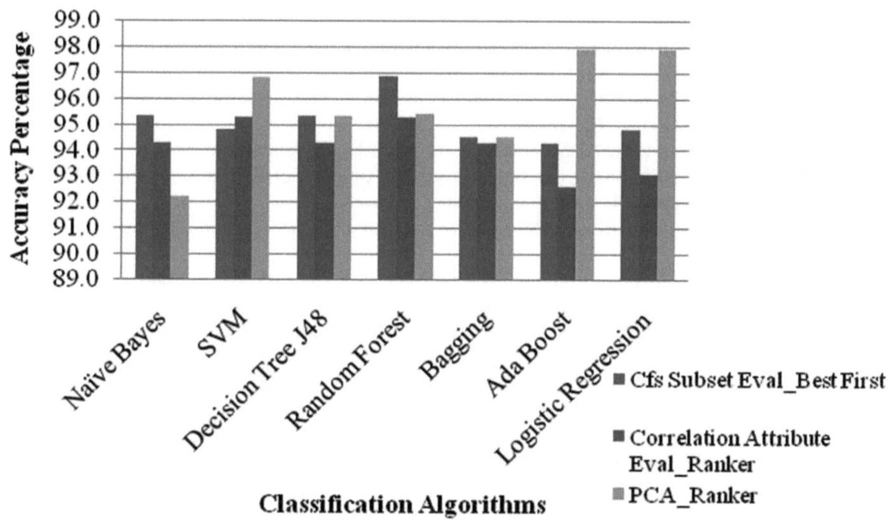

Fig. 5 Accuracy (in %) results using different feature selection technique

5 Conclusion and Future Scope

In this paper, best features for breast cancer prognosis are predicted. Different attribute selection methods along with different classification techniques are compared. In future, the accuracy can be increased by adding more features or by increasing the instances of the dataset. Also, the combination of existing classification techniques can be used to enhance the efficiency.

References

1. M. Baldonado, C.C.K. Chang, L. Gravano, A. Paepcke, The stanford digital library metadata architecture. Int. J. Digit. Libr. **1**, 108–121 (1997)
2. C. Paper, S. Ninkovic, K. Centar, Prediction models for estimation of survival rate and relapse for breast cancer patients, (March 2016) (2015)
3. K.B. Bruce, L. Cardelli, B.C. Pierce, Comparing object encodings, in *Theoretical Aspects of Computer Software. Lecture Notes in Computer Science*, vol. 1281, ed. by M. Abadi, T. Ito (Springer, Berlin, New York, 1997), pp. 415–438
4. K. Kourou, T.P. Exarchos, K.P. Exarchos, M.V. Karamouzis, D.I. Fotiadis, Machine learning applications in cancer prognosis and prediction. Comput. Struct. Biotechnol. J. (2015)
5. J. Van Leeuwen (ed.), *Computer Science Today. Recent Trends and Developments. Lecture Notes in Computer Science*, vol. 1000 (Springer, Berlin, 1995)
6. R.J. Kate, R. Nadig, Stage-specific predictive models for breast cancer survivability. Int. J. Med. Inform. **97**, 304–311 (2017)
7. Z. Michalewicz, *Genetic Algorithms+Data Structures=Evolution Programs*, 3rd edn. (Springer, Berlin, 1996)
8. K. Ahmed, Prediction of breast cancer risk level with risk factors in perspective to Bangladeshi women using data mining, **82**(November), 36–41 (2013)
9. A. Maheshwari, N. Kumar, U. Mahantshetty, Gynecological cancers: A summary of published Indian data. South Asian J. Cancer **5**(3), 112–120 (2017)
10. H. Asri, H. Mousannif, H. Al Moatassime, T. Noel, Using machine learning algorithms for breast cancer risk prediction and diagnosis. Procedia Comput. Sci. **83**(Fams), 1064–1069 (2016)
11. E. Gauthier, L. Brisson, P. Lenka, S. Ragusa, in *Breast Cancer Risk Score: A Data Mining Approach to Improve Readability*. The International Conference on Data Mining (2011), pp. 15–21
12. A. ToloieEshlaghy, A. Poorebrahimi, M. Ebrahimi, A.R. Razavi, L. Ghasem Ahmad, Using three machine learning techniques for predicting breast cancer recurrence (2013)
13. K. Park, A. Ali, D. Kim, Y. An, M. Kim, H. Shin, Robust predictive model for evaluating breast cancer survivability. Eng. Appl. Artif. Intell. **26**(9), 2194–2205 (2013)
14. J. Thongkam, G. Xu, Y. Zhang, in *AdaBoost Algorithm with Random Forests for Predicting Breast Cancer Survivability*. Proceedings of the International Joint Conference on Neural Networks (2008), pp. 3062–3069
15. J.A. Cruz, D.S. Wishart, Applications of machine learning in cancer prediction and prognosis. Cancer Inform. **2**, 59–77 (2006)
16. D.A. Aljawad, E. Alqahtani, N. Qamhan, N. Alghamdi, S. Alrashed, J. Alhiyafi et al., Using bayesian network and support vector. **8** (2017)
17. P.H. Abreu, M.S. Santos, M.H. Abreu, B. Andrade, D.C. Silva, Predicting breast cancer recurrence using machine learning techniques. ACM Comput. Surv. (2016)
18. J.M. Jerez-Aragone's, J.A. Gomez-Ruiz, G. Ramos-Jimenez, J. Munoz-Perez, E. Alba-Conejo, A combined neural network and decision trees model for prognosis of breast cancer relapse. Artif. Intell. Med. **27**, 45–63 (2003)
19. H.M. Zolbanin, D. Delen, A. Hassan Zadeh, Predicting overall survivability in comorbidity of cancers: A data mining approach. Decis. Support Syst. **74**, 150–161 (2015)
20. M. Lichman, UCI Machine Learning Repository (2013). http://archive.ics.uci.edu/ml. University of California, School of Information and Computer Science, Irvine, CA

Digital Assessment of Spatial Distribution of the Surface Soil Types Using Spatial (Texture) Features with MLC and SVM Approaches

Amol D. Vibhute, Karbhari V. Kale, Rajesh K. Dhumal, Ajay D. Nagne, Suresh C. Mehrotra, Amarsinh B. Varpe, Rupali R. Surase, Dhananjay B. Nalawade and Sandeep V. Gaikwad

Abstract In the present work, the effort has been made to identify and distribute of surface soil types using high spatial resolution multispectral (HSRM) image investigated in two ways. First, multispectral data is classified based on conventional approaches. Second, a method based on gray level co occurrence matrix (GLCM) as spatial objects extraction of the multispectral data is proposed. In this view, various texture parameters of the co-occurrence matrix method were used to highlight and extract the textures in the image. The method was computed on increasing matrix window size starting from original one. The Resourcesat-II Linear Imaging Self Scanning (LISS-IV) sensor multispectral image was used for testing the algorithms of the study area Phulambri Tehsil of Aurangabad region of Maharashtra state, India. The proposed approach was used as an input for Maximum Likelihood Classifier (MLC) and Support Vector Machine (SVM) approaches for identification and distribution of surface soil types and other patterns. The experimental outcomes of the present research were appraised on the basis of classification accuracy of methods. The overall accuracy of classification by MLC and SVM after spatial feature extraction was 92.82 and 97.32% with kappa value of 0.90 and 0.96 respectively. It was found that, the accuracy of the classification has increased after considering spatial features based on co-occurrence matrix. The results were promising to extract the mixed features for classification of soil type objects.

A. D. Vibhute (✉) · K. V. Kale · R. K. Dhumal · A. D. Nagne · S. C. Mehrotra · A. B. Varpe
R. R. Surase · D. B. Nalawade · S. V. Gaikwad
Geospatial Technology Research Laboratory, Department of Computer Science and IT,
Dr. Babasaheb Ambedkar Marathwada University, Aurangabad 431004, Maharashtra, India
e-mail: amolvibhute2011@gmail.com

C. R. Krishna et al. (eds.), *Proceedings of 2nd International Conference on Communication, Computing and Networking*, Lecture Notes in Networks and Systems 46, https://doi.org/10.1007/978-981-13-1217-5_74

Keywords Soil type classification · Multispectral data · Gray level co-occurrence matrix · Spatial features · Soil mapping · Maximum likelihood algorithm · Support vector machine

1 Introduction

Soils are very heterogeneous in nature which is one of the most imperative resources. Soils are assorted product of rock type, landform or topography, vegetation cover with climate. As a result, spectral features of soil or single landscape model cannot suffice to estimate soil features or soil boundary [1, 2]. Detailed catchment scale or farm scale maps of soil variability that offer high spatial resolution are requisite to estimate soil threats [3]. Moreover, the soil and land resource survey is vital in precision farming for better management and planning soil for better food production. However, traditional methods for analysis of soil and its mapping are time consuming, expensive and does not fulfill the spatiotemporal variability [4]. Furthermore, the topographic and cadastral maps as a reference were used by the conventional survey methods and the traditional methods are also time-consuming, formidable and subjective [5]. Consequently, soil features can be extracted from remote sensing datasets with its analysis and mapping. Nevertheless, the digital assessment of surface soil types, spatial distribution and its mapping is somewhat formidable task due to various soil attributes and assorted effect of numerous features of planet surface that can vary spectral and spatial features of soils and compose it non-consistent through the spectrum region [6].

The traditional methods of HSRM image classification consider solely spectral information while neglecting the spatial information. Under this constraint, we have tried to extract the spectral and spatial features of mixed land use and covered area for soil classification like settlements, hilly area, farmland, natural vegetations, etc., with HSRM images.

2 Materials and Methods

2.1 The Test Location, Used Satellite Data and Soil Sampling

The test location is geographically located at 19° 28′ 43.27″–20° 24′ 52.19″N latitude and 75° 13′ 10.75″–75° 30′ 14.87″E longitude which covers 72.70 km^2. The data was used for this study have been acquired from various sources like satellite data along with field data. Global Positioning System (GPS) was used to acquire the ground coordinate points. The base maps were developed using Survey of India (SOI) toposheet of 1:50,000 scales. The proposed approach was applied on Indian Remote Sensing (IRS) Resourcesat-II—P6 satellite imagery of LISS-IV sensor. The spatial resolution of LISS-IV image is 5.8 m with three spectral channels and 23.5 km swath. The imagery was geometrically and radiometrically corrected (orthorectified) by the provider [7]. Total 74 soil samples (0–20 cm depth) were collected and their GPS values were recorded with scenes. The soil sampling sites were selected based on spatial distribution. The field work for collecting soil samples and ground truth points were carried out during the period of February 10 to March 25, 2015 and in between 0800 and 1330 h on light days with clear environment. The soil samples were collected in air-tight container. Every specimen was air dried and passed through 2 mm sieve for laboratory analysis of some physicochemical soil properties. The soil properties were analyzed by standard laboratory methods at "MIT Soil and Water Testing Laboratory", Aurangabad, Maharashtra, India.

2.2 *Experimental Methodology*

In this research, image processing operations were performed through Environment for Visualizing Images (EVVI 5.1) image analysis software and ArcGIS 10 software. The methods are discussed as follows.

2.2.1 Spectral and Spatial Feature Information

A reliable approach to deal with pixel-based classification of satellite imagery is to consider its spectral information. The spectral features (information) of image is directly used as the input for the classification algorithms and processed as a feature vector. It is reliable for classification based on spectral reflectance of each object. Unfortunately, it does not consider spatial structures of various objects in the classification [8]. The texture-based spatial features were analyzed for the HSRM satellite imagery. The GLCM method is broadly used for texture analysis and pattern recognition also known as spatial co-occurrence matrix (SCM) which is more scientific depicter of texture and more accurate as compared to other methods. From the given spectral band or its subset, the GLCM method is implemented from the spatial stochastic properties [9, 10]. The co-occurrence matrix is two-dimensional matrix of joint probabilities between pairs of pixel values one with gray level value i and other with gray level value j, separated by a distance $d = (1,0)$ (i.e. neighboring pixels in the same row) at an angle of 0° (i.e., horizontally) in a given direction from left to right. The GLCM matrix is a symmetrical matrix which element $p\ (i, j \mid d, \theta)$ contains the second-order statistical probability values [9, 10]. The elements of the GLCM method were computed using Eq. (1).

$$p(i, j) = \sum_{i=1}^{Ng} \sum_{j=1}^{Ng} i \cdot p(i, j), \tag{1}$$

Table 1 GLCM textures and their equations

GLCM texture features	Equations	
Mean	$x = \sum_{i,j} i \cdot p(i, j)$	(2)
Variance	$f_4 = \sum_i \sum_j (i - u)^2 p(i, j)$	(3)
Contrast	$f_2 = \sum_{n=0}^{Ng-1} n^2 \left\{ \sum_{\substack{i=1 \\ \lvert i-j \rvert = n}}^{Ng} \sum_{j=1}^{Ng} p(i, j) \right\}$	(4)
Dissimilarity	$f = \sum_{n=1}^{Ng-1} n \left\{ \sum_{\substack{i=1 \\ \lvert i-j \rvert = n}}^{Ng} \sum_{j=1}^{Ng} p(i, j)^2 \right\}$	(5)

where, $p(i, j)$ is the GLCM, and Ng is the digit of gray levels.

The pairs of pixels of GLCM are computed in four angular directions such as 0° in horizontal, 45° in right diagonal, 90° in vertical and 135° in left diagonal within the instantly neighboring pixels of the image. Hence, to produce the texture image, the GLCM or the texture measures were usually calculated within a moving window. Haralick [9] has extracted 14 textural features from the GLCM which are the significant features. In this study, we have computed four texture measures based on co-occurrence matrix such as; mean, variance, contrast, and dissimilarity. These measures were computed from the co-occurrence matrix using Eqs. 2, 3, 4, and 5 drawn in Table 1.

2.2.2 Proposed Approach

The above texture measures with increasing window sizes of 3×3, 5×5, and 7×7 were considered. The x and y values were used to calculate the co-occurrence matrix. In the calculation of co-occurrence matrix by considering each pixel, 8-neighboring pixels were preferred. The four directions were used, i.e., 0°, 45°, 90°, and 135° separately for the calculation. The grayscale quantization level was used with 64-bit. The 64-bit value was beneficial when the grayscale values of the image are spread over a broad range. After the setting of grayscale values, co-occurrence matrix was computed in four directions. Mean, variance, contrast and dissimilarity features were extracted for further processing. In our experiments, four features were calculated; hence the feature set was four for each of the band. The used data has three bands; thus the final size of feature set was 12. The resulted textures of GLCM method is shown in Fig. 1 for window sizes of (a) 3×3, (b) 5×5, and (c) 7×7 respectively.

For each window size, these all twelve features were used and MLC as well as SVM methods were implemented over it. The window sizes were increased up to

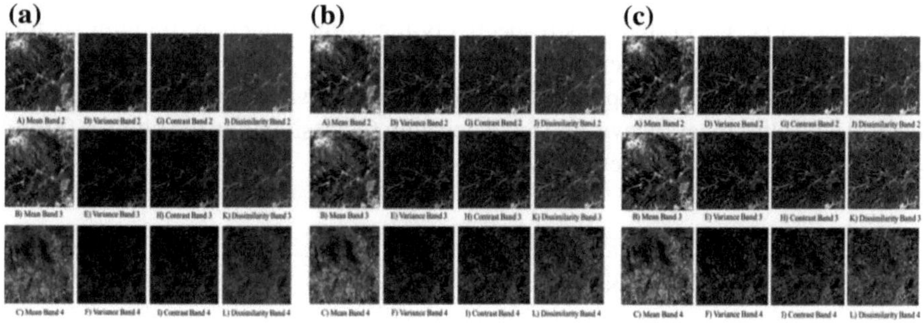

Fig. 1 Spatial features based on GLCM texture measures with window sizes of **a** 3×3, **b** 5×5 and **c** 7×7 respectively

7×7 due to the stability (similarity) of the textural features of window sizes 5×5 and 7×7. The GLCM feature set with increased window size was given as input for MLC and SVM algorithms. The LISS-IV image for MLC and SVM methods were trained accordingly field observations in the combinations with laboratory results, geolocated ground reference data with visual inspection. The ROIs were developed by said reference points which were the support vectors for SVM-based classification and also used for the MLC-based classification to train and test the HSRM image for classification. The available training and testing pixels were 920 and 2203 for said data.

The GLCM feature set with increased window size was given as input for MLC algorithm first. The data scale factor for MLC was one and probability threshold value was 0.1 for all training pixels. The MLC is parametric method based on the Bayes theorem [4, 10–12] is a widely acceptable by the researchers including remote sensing community. The MLC algorithm was implemented by the Eq. 6 [12].

$$p(X/Cj) = \frac{1}{(2\pi)^{n/2} \left|\sum j\right|^{0.5}} \times \exp\left[-\frac{1}{2} (\mathrm{DN} - \mu j)^T \sum_{j}^{-1} (\mathrm{DN} - \mu j) \right] \tag{6}$$

Additionally, SVM method was also considered for LISS-IV data for comparison of MLC results and for getting better results. The kernel of Gaussian Radial Basis Function (RBF) of SVM method was implemented using Eq. 7 for classification analysis. SVM non-parametric supervised machine learning algorithm [13] was chosen due to its high accuracy for classification of heterogeneous and noisy remote sensing data with less training pixels. The SVM method is originally formulated by Vapnik in 1995 based on statistical learning theory [10, 11, 13, 14].

Gaussian (RBF)

$$k(xi, xj) = \exp(-\gamma \cdot ||xi - xj||^2), \quad \gamma > 0, \tag{7}$$

where, $k(xi, xj)$, xi, xj and γ is respectively the kernel function, training vectors and kernel constant parameter.

The SVM method was implemented with gamma (γ) value 0.010, penalty parameter 100, and classification pyramid level one with reclassification threshold 0.90 and classification probability threshold 0.10.

3 Results and Discussion

According to the laboratory results of soil physicochemical properties and field data with visual inspection one major (black cotton soil or "Regur") and two minor (Lateritic soil and Sand dunes) soil types were identified and classified with other land features. The black cotton soil includes "vertisol", "inceptisol" and "entisol"; lateritic soil includes "alfisol" and sand dunes include "Typic Torripsamments". Five soil classes according to USDA soil taxonomy [15] were detected and classified on the basis of report generated by laboratory analysis of soil physicochemical properties and field investigations. The soil classes were "vertisol"; "inceptisol" and "entisol" of black cotton soil, "alfisol" of lateritic soil, "Typic Torripsamments" of sand dunes, and other land features such as vegetations, water bodies, settlements and boundaries with roads were also classified. The classification maps (Fig. 2 (a) for MLC method and (b) for SVM method) clearly indicate that, black cotton soils [16] have covered most area of the test site followed by vegetations, sand dunes, lateritic soil, settlements, roads and water bodies. As per the laboratory reports of soils, these black soils are deep or heavy and medium or lighter as per its physical properties. The textures of black soils are loamy to clayey with mixed carbonates (mostly $CaCo_3$) and are suitable for cotton cultivation. The organic carbon, organic matter and nitrogen found to be less in this soil and pH values are near about 7–9. The EC values vary from 0.25 and 0.46 d Sm-1 where values were less than 0.36 d Sm-1. The iron contents are good in black soils. The lateritic soil included only the "alfisol" in the studied areas as per USDA soil taxonomy which found to be hilly part and somewhat farming sectors of test site. The pH value of these soils is low and organic matter is high with fine texture. Sand dunes were observed to be more at riverside and hilly rocks due to the spectral structure of sand dunes and rocks. The texture of sand dunes was sandy. Electrical Conductivity (EC) values and organic matter contents are very low in sand dunes. Natural vegetations including agricultural crops were accurately classified and mapped.

First, the classification methods were computed on original preprocessed LISS-IV data (only spectral information) and outcome was evaluated with accuracy. The classified maps derived by MLC and SVM methods with spatial features are illustrated in Fig. 2a, b respectively. The soil classes were classified well with both the classifiers. As black cotton soil was found to be more followed by sand dunes and alfisol soils.

It was observed that, the accuracy was less with both methods on original image (Table 2). Accordingly, the spatial texture measures were pondered and accuracy

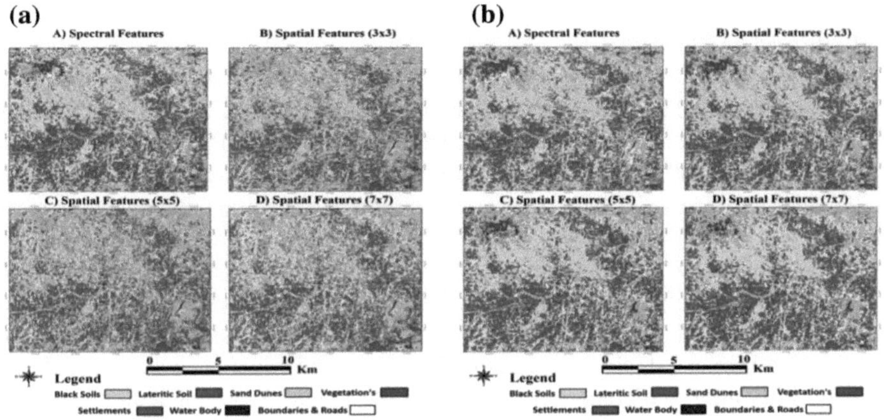

Fig. 2 Result of classification using (a) MLC and (b) SVM approaches respectively on **A** original data, **B** spatial information (3 × 3 window size), **C** spatial information (5 × 5 window size) and **D** spatial information (7 × 7 window size)

Table 2 Accuracy assessment of MLC and SVM algorithms

Methods/features	Spectral features		Spatial features (3 × 3)		Spatial features (5 × 5)		Spatial features (7 × 7)	
	OA	K	OA	K	OA	K	OA	K
MLC	83.38	0.79	89.83	0.87	92.23	0.90	92.82	0.90
SVM	84.52	0.80	91.51	0.89	94.55	0.93	97.32	0.96

Where, OA-Overall Accuracy and K-Kappa Value

was evaluated of each window sizes and found to be similarity in textures of 5 × 5 and 7 × 7 and hence window sizes were finalized up to 7 × 7. Producer's accuracy, user's accuracy, overall accuracy and kappa statistics were calculated for evaluation of the accuracy [10, 12]. The ground truth points were used for deriving the confusion matrix. The classified values against actual ground observation values at particular location were determined in the confusion matrix. The diagonal values of the confusion matrix depicts the correctly classified features, where as nondiagonal nonzero value demonstrates the misclassification between classified features from related view [17]. The resulted outcome of both MLC and SVM algorithms are drawn in Table 2 with overall accuracy and kappa values.

It was observed that, the spectral confusion were remarkable in between sand dunes and settlements with both the classification methods. The four texture measures were highlighted the individual objects, when increased the window size, because they are related to the size of the object. Larger the window size within the window, the object reflects higher gray values. Consequently, the objects with various sizes were separated each other. The classification methods were implemented on these four texture measures with increasing window sizes and achieved better accuracy

(Table 2) than the original one. The confusion was reduced with spatial information and achieved good results.

The accuracy has increased with increased window sizes. The soils and vegetations were classified more accurately than other classes. The window size 7×7 was given best results than other two window sizes such as 3×3 and 5×5 for four texture measures. In fact, all soil classes were classified accurately except sand dunes and settlements caused by similarity in spatial (texture) and spectral (signature) information.

The class-specific accuracies, black cotton soil, lateritic soil, and vegetations were classified with the higher accuracy excluding sand dunes which were misclassified with settlements. Our aim was to identify and classify soil classes accurately within the farming sectors which were classified well; however, we have also considered soils on hilly area, settlement area. It causes the low accuracy of settlement class within the test site. It can be concluded that, heterogeneous lad features can be correctly classified by using spatial information.

4 Conclusions

The present research highlighted the advantages of GLCM-based spatial features for the classification of five soil classes and other mixed land patterns. The research work introduces the new approach for soil taxonomy using spatial information along with spectral information. Only spectral information is not sufficient for classification of heterogeneous land features. Consequently, we have used HSRM satellite image of LISS-IV with 5.8 m of pixel size and three spectral bands. The MLC and SVM classification approaches were computed for spatial distribution of soil accordingly USDA soil taxonomy and four textures of GLCM were considered with three window sizes of 3×3, 5×5, and 7×7 matrix. It was observed that, sand dunes and settlement area, were not classified accurately when applied the two approaches on spectral as well as spatial information. Nevertheless, when included spatial features in the classification as an input with increased window sizes, the accuracy was also enhanced up to 10–13% more for MLC and SVM approaches respectively. According to the classification maps and accuracies, it is noticed that, spatial features are very imperative for mixed land features especially soils in different platforms. In the final conclusion, SVM method has given much better results than MLC method. The classification results can be useful for better utilization of soil for precision farming practices along with its planning and management.

Acknowledgements The authors would like to acknowledge the UGC-BSR fellowship; UGC-SAP(II)DRS Phase-I and Phase-II for providing infrastructure, DeitY, Government of India, under Visvesvaraya Ph.D. Scheme, DST and also extend our gratitude to DST-FIST program to Department of CSIT, Dr. BAM University, Aurangabad, M.S. India. We would also thankful to Prof. D. T. Bornare and his team for physiochemical analysis of soil specimens at "MIT Soil and Water Testing Laboratory, Aurangabad", Maharashtra, India.

References

1. K.S. Lee, G.B. Lee, E.J. Tyler, Determination of soil characteristics from thematic mapper data of a cropped organic-inorganic soil landscape. Soil Sci. Soc. Am. J. **52**(4), 1100–1104 (1988)
2. M.L. Manchanda, M. Kudrat, A.K. Tiwari, Soil survey and mapping using remote sensing. Trop. Ecol. **43**(1), 61–74 (2002)
3. F. Garfagnoli, A. Ciampalini, S. Moretti, L. Chiarantini, S. Vettori, Quantitative mapping of clay minerals using airborne imaging spectroscopy: New data on Mugello (Italy) from SIM-GA prototypal sensor. Eur. J. Remote Sens. **46**(1), 1–17 (2013)
4. A.D. Vibhute, B.W. Gawali, Analysis and modeling of agricultural land use using remote sensing and geographic information system: A review. Int. J. Eng. Res. Appl. (IJERA) **3**(3), 081–091 (2013)
5. S.S. Asadi, M.S. Kumar, M.V. Raju, B.V.T.V. Rao, Preparation of soil map using remote sensing and GIS: A case study from Andhra Pradesh. Int. J. Emerg. Trends Eng. Dev. **4**(2), 138–146 (2014)
6. E. Ben-Dor, K. Patkin, A. Banin, A. Karnieli, Mapping of several soil properties using DAIS-7915 hyperspectral scanner data-a case study over clayey soils in Israel. Int. J. Remote Sens. **23**(6), 1043–1062 (2002)
7. http://www.nrsa.gov.in/
8. M. Fauvel, J.A. Benediktsson, J. Chanussot, J.R. Sveinsson, Spectral and spatial classification of hyperspectral data using SVMs and morphological profiles. IEEE Trans. Geosci. Remote Sens. **46**(11), 3804–3814 (2008)
9. R.M. Haralick, K. Shanmugam, I. Dinstein, Textural features for image classification. IEEE Trans. Syst. Man Cybernet. **3**(6), 610–621 (1973)
10. J. Gao, *Digital Analysis of Remotely Sensed Imagery* (McGraw-Hill Professional, New York, 2008)
11. J.A. Richards, X. Jia, *Remote Sensing Digital Image Analysis An Introduction*, 4th edn. (Springer, Berlin, 2006)
12. A.D. Vibhute, R.K. Dhumal, A.D. Nagne, Y.D. Rajendra, K.V. Kale, S.C. Mehrotra, in *Analysis, Classification, and Estimation of Pattern for Land of Aurangabad Region Using High-Resolution Satellite Image*. Proceedings of the Second International Conference on Computer and Communication Technologies (Springer, New Delhi, 2016), pp. 413–427
13. V.N. Vapnik, An overview of statistical learning theory. IEEE Trans. Neural Netw. **10**(5), 988–999 (1999)
14. A.D. Vibhute, K.V. Kale, R.K. Dhumal, S.C. Mehrotra, in *Soil Type Classification and Mapping using Hyperspectral Remote Sensing Data*. IEEE, International Conference on Man and Machine Interfacing (MAMI) (2015), pp. 1–4
15. USDA, *Keys to Soil Taxonomy, United States Department of Agriculture* (Natural Resources Conservation Service, 2014)
16. T. Bhattacharyya, D.K. Pal, C. Mandal, P. Chandran, S.K. Ray, D. Sarkar, A.K. Sahoo et al., Soils of India: Historical perspective, classification and recent advances. Curr. Sci. 1308–1323 (2013)
17. A.D. Vibhute, A.D. Nagne, B.W. Gawali, S.C. Mehrotra, Comparative analysis of different supervised classification techniques for spatial land use/land cover pattern mapping using RS and GIS. Int. J. Sci. Eng. Res. **4**(7), 1938–1946 (2013)

Frequent Itemset Mining on Uncertain Database Using OWA Operator

Samar Wazir, M. M. Sufyan Beg and Tanvir Ahmad

Abstract In today's modern age, one of the most research focused technique for analyzing frequent pattern from the data is known as *Frequent Itemset Mining* (FIM). In the past three decades, many algorithms for data analysis have been proposed, and different categories of data came into existence. Now, various FIM algorithms are available to deal with certain, probabilistic or fuzzy data. In this paper, a novel solution named FuzzyApriori using OWA operator (FAOWA) algorithm is proposed to mine frequent items from the fuzzy uncertain transactional database. Earlier numerous techniques have been developed to calculate frequent items on the fuzzy uncertain transactional database by considering fuzzy min/max operator as a basis for minimum support. The fuzzy min/max operators consider only one value, i.e., minimum/maximum value of membership function to calculate support for an itemset, in place of considering all values of membership functions of all items in an itemset. Due to the lack of aggregating multicriteria to form a decision function, in this paper, fuzzy min/max operators are replaced by fuzzy OWA for calculating minimum support. Experiments are performed by using example dataset, and standard available dataset and performance are compared with probabilistic and fuzzy support based algorithms.

Keywords Certain and uncertain transactional database · Existential probability Expected support · Membership function · Fuzzy min/max · OWA

S. Wazir (✉) · T. Ahmad
Department of Computer Engineering, Jamia Millia Islamia, Delhi 110025, New Delhi, India
e-mail: samar.wazir786@gmail.com

T. Ahmad
e-mail: tahmad2@jmi.ac.in

M. M. Sufyan Beg
Department of Computer Engineering, Muslim University, Aligarh 202001, Uttar Pradesh, India
e-mail: mmsbeg@eecs.berkeley.edu

C. R. Krishna et al. (eds.), *Proceedings of 2nd International Conference on Communication, Computing and Networking*, Lecture Notes in Networks and Systems 46, https://doi.org/10.1007/978-981-13-1217-5_75

Table 1 Certain transactional database

TID	Items			
0	1	2	4	5
1	2	3	5	
2	1	2	4	5
3	1	2	3	5

Table 2 Uncertain transactional database

TID	Items				
0	1(0.25)	2(0.47)	3(0.0)	4(0.5)	5(0.5)
1	1(0.0)	2(0.33)	3(0.4)	4(0.0)	5(0.4)
2	1(0.25)	2(0.37)	3(0.0)	4(0.3)	5(0.3)
3	1(0.25)	2(0.53)	3(0.6)	4(0.0)	5(0.6)

1 Introduction

The technology used to monitor and analyze the generation pattern of data is known as *Frequent Itemset Mining* (FIM) [19]. The databases produced from different sources like records of *bank transactions, railway reservations,* and *market basket data* are generally *Certain Transactional DataBases* (CTDB as in Table 1). The presence of each item in every transaction is definite in such

databases, e.g., T_1 = *{Shirt, Trouser, Tie, Shoes}*

The transaction T_1 is an example of market basket data. We can say that in transaction T_1 when a customer purchased a "*Shirt*" then he/she also purchased "*Trouser*," so there is some association between "*Shirt*" and "*Trouser*." Extraction of such kind of patterns called *Association Rule Mining* (ARM) [19] and it can be said that FIM is a part of the ARM. If in transaction T_1 each item is associated with its chances of existence, then the database is called *Uncertain Transactional DataBase*(UTDB) [4, 5, 20, 22, 23] (as in Table 2), e.g.,

$$T_1 = \{Shirt : 0.5,\ Trouser : 0.7, Tie : 0.1,\ Shoes : 0.3\}$$

So, T_1 can be interpreted as if the chance of selling shirt is 50% then there are 70% chance of selling trouser, 10% for Tie and 30% for shoes. The uncertainty attached to each item can be probabilistic*(Existential Probability)* or fuzzy*(Membership Function)*. So, accordingly, the uncertain databases can be Probabilistic uncertain Transactional DataBase (PTDB) or Fuzzy uncertain Transactional DataBase (FTDB).

To handle large databases uncertainty was introduced [4–11] and in place of calculating the count of an item like in CTDB, the *Expected Support* (*ES*) of an item is calculated on UTDB. For example, in T_1 if the *ES* of the shirt, trouser, and tie is calculated then it will be 0.5 * 0.7 * 0.1 = 0.035. The limitation of this approach is when size increases *ES* approaches to zero. The positive thing with *ES* is that it considers *Existential Probability* [6, 7] of all items for calculating support, so in results association rules interestingness [19] exist. The limitation with *ES* can be overcome by using fuzzy min/max operator [24–26, 29–33] in FIM on FTDB [13–18]. For example, Fuzzy min always choose min value among the membership functions of items in a transaction, like the fuzzy support of shirt, trouser and tie in T_1 will be min(0.5, 0.7, 0.1) = 0.1, so it will never be zero if membership function of any item in itemset is greater than zero. The limitation of fuzzy min/max operator is that these operators are not capable of aggregating multicriteria or do not aggregate the association of all items in a transaction to generate an overall result. For example, if transaction looks like

$$T_1 = \{Shirt : 0.9,\ Trouser : 0.99,\ Tie : 0.1,\ Shoes : 0.999\}$$

In above case, fuzzy min results in 0.1 without any association with other items, so fuzzy min/max operators cannot generate results by considering all values and take only one value, i.e., min or max. Therefore, it can be said that with fuzzy min/max operator, FIM on FTDB can produce larger size itemset, but there is no association rules interestingness in results.

A new method is proposed in this paper for calculating frequent items on the fuzzy uncertain transactional database by using *Ordered Weighted Aggregation* (OWA) [26–28] operator to calculate fuzzy support. The algorithm follows the *Apriori* [1–3] principle for execution, run on the fuzzy uncertain transactional database, and use OWA operator to calculate support, so it is named as *FAOWA*(Fuzzy Apriori using OWA). The performance of FAOWA is compared by using example database and standard available database with probabilistic and fuzzy data mining algorithms in terms of time taken by algorithms and numbers of frequent items generated. So the proposed work has been completed in following stages:

1. *UApriori* is executed on uncertain transactional database.
2. *Fuzzy Data Mining Algorithm* by considering fuzzy min and max operator is executed on the uncertain transactional database.
3. *FAOWA* is executed on the uncertain transactional database.
4. Performance of *UApriori*, *Fuzzy Data Mining Algorithm,* and *FAOWA* is compared on the basis of time taken by algorithm and number of frequent items generated by each algorithm.

2 Related Works

2.1 Apriori [1]

The first algorithm proposed by Agrawal in 1994 is Apriori and used to calculate all items or itemsets whose count is greater than or equal to user-specified minimum support for a given CTDB. So it can be said that an item or itemset is frequent if

$$\text{count}(A\ in\ T_x) >= ms \tag{1}$$

Here A is any itemset; T_x is the transactional database and ms is the minimum support. All frequent items for the CTDB given in Table 1, for minimum support 1, can be calculated as

$$\text{Size - 1 FI} = \{1, 2, 3, 4, 5\},\ \text{Size - 2 FI} = \{12, 13, 14, 15, 23, 24, 25, 35, 45\}$$
$$\text{Size - 3 FI} = \{123, 124, 125, 135, 145, 235, 245\},\ \text{Size - 4 FI} = \{1235, 1245\}.$$

2.2 *UApriori [6]*

This algorithm was proposed by Chui in 2007, and it was the first algorithm used to calculate frequent items over UTDB based on *Expected Support*. The following formula is used for calculating expected support.

$$\text{Expected Support}(A) = \sum_{i=1}^{D} \sum_{j=1}^{K} P_{t_i}(a) \tag{2}$$

Here A is any itemset containing $a_1, a_2 \ldots a_{|K|}$ distinct items |D| is the number of transactions in UTDB. Let minimum support is 0.75 then ES-based frequent items on UTDB given in Table 2, can be calculated as

$$\text{Size - 1 FI} = \{1, 2, 3, 4, 5\},\ \text{Size - 2 FI} = \{25\}.$$

2.3 *Fuzzy Data Mining Algorithm(FDMA) [16] with Fuzzy Min/Max Operator*

The Fuzzy Data Mining Algorithm was proposed by T.P. Hong, which is used to calculate frequent items on FTDB by calculating fuzzy support based on fuzzy max operator. The following formula is used to calculate fuzzy support.

$$\text{FuzzySupport(A)} = \sum_{i=0}^{D} \max(MF(a1), MF(a2), MF(a3) \ldots .MF(a|K|)) \tag{3}$$

Here A is an itemset consist of $a_1, a_2, a_3, \ldots a_{|K|}$ distinct items,D is the number of transactions in FTDB. Let minimum support is 0.75 then fuzzy support based frequent items using a max operator on UTDB given in Table 2, can be calculated as

$$\text{Size - 1 FI} = \{1, 2, 3, 4, 5\},\ \text{Size - 2 FI} = \{12, 13, 14, 15, 23, 24, 25, 34, 35, 45\}$$

Size - 3 FI = {123, 124, 125, 134, 135, 145, 234, 235, 245, 345},
Size - 4 FI = {1234, 1235, 1245, 1345, 2345}, Size - 5 FI = {12345}

In another case, if fuzzy min operator is used to calculate fuzzy support then Eq. 3 can be written as

$$FuzzySupport(A) = \sum_{i=0}^{|D|} \min(MF(a1), MF(a2), MF(a3) \ldots MF(a|K|)) \quad (4)$$

So, according to above Eq. 4 fuzzy support based frequent items on UTDB in Table 2 by using fuzzy min operator can be calculated as:- Size-1 FI = {1, 2, 3, 4, 5}, Size-2 FI = {12, 15, 23, 24, 25, 35, 45}, Size-3 FI = {125, 235, 245}.

3 Problem Definition

In the process of FIM on UTDB, it is necessary to choose such approach which gives results by aggregating membership functions of all items in an itemset. The proposed algorithm FAOWA generates a number of frequent items more than UApriori, between FDMA min and FDMA max, keeps association rule interestingness. Fuzzy support using FAOWA can be calculated as

Let us take m as number of items in an itemset, so, according to Yager [26–28], relative quantifier$Q(r)$ for an itemset A = $\{x_1, x_2, \ldots x_m\}$ of size m can be calculated as

$$Q(r) = \begin{cases} 0 & \text{if } r < a \\ \frac{r-a}{b-a} & \text{if } a \leq r \geq b \\ 1 & \text{if } r > b \end{cases} \quad (5)$$

Here,a,b,r is between 0 and 1, and the values of$a$,b can be decided on the basis relative linguistic fuzzy quantifier, e.g., $a = 0$ and $b = 0.5$ for “At least half,” $a = 0.5$ and $b = 1$ for “As many as possible,” $a = 0.3$ and $b = 0.8$ for “most.”

Q(r) is used for calculating weights of OWA as

$$w_i = Q(i/m) - Q\left(\frac{i-1}{m}\right), \quad i = 1, 2, 3 \ldots m \quad \text{with } Q(0) = 0 \quad (6)$$

Here m is criteria or number of items in an itemset, $Q(i/m) = Q(r)$ and $\sum_{i=1}^{m} w_i = 1$

So, OWA for m different items, $x_1, x_2, \ldots x_m$ can be calculated as

$$\mathrm{OWA}(x_1, x_2, \ldots x_m) = \sum_{i=1}^{m} w_i y_i,$$

where y_i is the ith largest value amongst $x_1, x_2, \ldots x_m$

$$\text{Or} \quad \mathrm{OWA(A)} = \mathrm{OWA}(x_1, x_2, \ldots x_m) = \sum_{i=1}^{m} \left[Q\left(\frac{i}{m}\right) - Q\left(\frac{i-1}{m}\right) \right] y_i \qquad (7)$$

In the case of FAOWA, the "most" relative linguistic fuzzy quantifier is used for calculating fuzzy support as

$$FuzzySupport(A) = \sum_{i=1}^{D} most\, [\mathrm{OWA}(Ai)]_{\mathrm{a}=0.3,\mathrm{b}=0.8} \qquad (8)$$

4 FAOWA Algorithm

This algorithm is used to calculate frequent items on the fuzzy uncertain transactional database by using OWA operator. So the input of this algorithm is UTDB with D transactions and the user given minimum support *ms*. The algorithm executes as:-

FAOWA Algorithm

```
1. Input (UTDB D, ms)
2. L1 = {large-1 itemset};
3.  for (k=2; LK-1 ≠ ø; k++) do begin
4.          CK = apriori-gen (LK-1) // New Candidates
5.          for all transactions t ∈ D do begin
6.                  fs = fs + most[OWA(Ck)]a=0.3,b=0.8
7.           End
8.           LK = {c ∈ CK | fs >= ms}
9.    End
10. Answer = UKLK
```

In the above algorithm, L_1 is the large 1 itemset calculated by checking the total of membership function with minimum support. In line 3, 4 all possible candidates are generated by apriori-gen() function by joining and pruning of candidates from the previous iteration. In line 5, 6 the OWA for each candidate is calculated, and their total is compared with minimum support. If fuzzy support (fs) is greater than minimum support, then itemset is said to be frequent itemset.

Table 3 Weights calculation for $m = 2$

i	0	1	2
i/m	0	0.5	1
Q	0	$\frac{0.5-0.3}{0.8-0.3}$	1
	0	0.4	1
w	NR	0.4	0.6

Table 4 Weights calculation for $m = 3$

i	0	1	2	3
i/m	0	0.33	0.66	1
Q	0	$\frac{0.03}{0.5}$	$\frac{0.36}{0.5}$	1
	0	0.06	0.72	1
w	NR	0.06	0.66	0.28

5 FAOWA Example

Let us consider the UTDB in Table 2. If the minimum support is 0.75, then frequent items based on OWA operator can be calculated as

Size-1 FI

Size-1 frequent items can be calculated same as in FDMA min/max algorithm, e.g.,

$$FuzzySupport(2) = 0.47 + 0.33 + 0.37 + 0.53 \geq 0.75$$

so item 2 is frequent item, Therefore, all size-1 frequent items can be calculated as Size-1 FI = {1, 2, 3, 4, 5}

Size-2 FI

In this case according to Eq. 6, $m = 2$, $a = 0.3$, $b = 0.8$ so that w can be calculated as in Table 3. For example, fuzzy support for itemset (1,3) can be calculated as

$$\begin{aligned} FuzzySupport(1,3) &= 0.4 \times 0.25 + 0.6 \times 0.0 + 0.4 \times 0.4 + 0.6 \times 0.0 + 0.4 \times 0.25 \\ &+ 0.6 \times 0.0 + 0.4 \times 0.6 + 0.6 \times 0.25 = 0.75 \geq 0.75 \end{aligned}$$

So itemset (1, 3) is frequent here, Therefore, all size-2 frequent items can be calculated as:- Size-2 FI = {12,13,15,23,24,25,35,45}

Size-3 FI

In this case according to Eq. 6, $m = 3$, $a = 0.3$, $b = 0.8$ so that w can be calculated as in Table 4. For example, fuzzy support for itemset (1,2,3) can be calculated as

$$\begin{aligned} FuzzySupport(1,2,3) &= 0.06 \times 0.47 + 0.66 \times 0.25 + 0.28 \times 0.0 + 0.06 \times 0.4 \\ &+ 0.66 \times 0.33 + 0.28 \times 0.0 + 0.06 \times 0.37 + 0.66 \times 0.25 + 0.28 \times 0.0 \\ &+ 0.06 \times 0.6 + 0.66 \times 0.53 + 0.28 \times 0.25 = 1.078 \geq 0.75 \end{aligned}$$

Table 5 Weights calculation for $m = 4$

i	0	1	2	3	4
i/m	0	0.25	0.5	0.75	1
Q	0	0	$\frac{0.2}{0.5}$	$\frac{0.45}{0.5}$	1
	0	0	0.4	0.9	1
w	NR	0	0.4	0.5	0.1

So itemset (1,2, 3) is frequent here, Therefore, all size-3 frequent items can be calculated as:- Size-3 FI = {123,125,135,234,235,245}

Size-4 FI

In this case according to Eq. 6, $m = 4$, $a = 0.3$, $b = 0.8$, so w can be calculated as in

Table 5. For example fuzzy support for itemset (1,2,3,5) can be calculated as

$$\begin{aligned} FuzzySupport(1,2,3,5) &= 0.0 \times 0.5 + 0.4 \times 0.47 + 0.5 \times 0.25 + 0.1 \times 0.0 \\ &+ 0.0 \times 0.4 + 0.4 \times 0.4 + 0.5 \times 0.33 + 0.1 \times 0.0 + 0.0 \times 0.37 + 0.4 \times 0.3 + 0.5 \times 0.25 \\ &+ 0.1 \times 0.0 + 0.0 \times 0.6 + 0.4 \times 0.6 + 0.5 \times 0.53 + 0.1 \times 0.25 = 1.413 \geq 0.753 \end{aligned}$$

So itemset (1,2, 3, 5) is frequent here, Therefore, all size-4 frequent items can be calculated as:- Size-4 FI = {1235, 2345}.

6 Experiments and Result Analysis

The performance of the proposed FAOWA algorithm is analyzed by running the algorithm on example database given in Table 2 and Gazelle [12, 21] dataset. Gazelle uncertain transactional database has 59,602 total transactions and 497 distinct items. On the above databases total four algorithms (UApriori, FDMA min, FDMA max and FAOWA) are executed, and the performance is measured in terms of a number of frequent items generated and time taken by the algorithms. All the programs were implemented in Microsoft's Visual Studio 2015 in C language and executed on gcc compiler on Ubuntu version 15.10. The hardware specification of the machine is Intel(R) Core(TM) i5 2.2 GHz processor with 4 GB of RAM. The result of example dataset is given in Table 6 and Gazelle dataset in Table 7, Figs. 1, 2.

It is clear that FAOWA in the context of "number of frequent items generated" performs better than UApriori and FDMA min, however less than FDMA max. In UApriori, the expected support for an itemset tends to zero when the size of itemset increases due to the multiplication of probabilities. On the other hand, FDMA min considers the lowest value of the item in an itemset, so if items values are closer to zero than itemset fuzzy support become very low which results in less number of frequent items. The FDMA max considers the highest value of the item in an itemset, so if any of the items of itemset has a value closer to one than itemset fuzzy support become very high which results in more number of frequent items. In the

Table 6 FI by Example dataset in Table 2

Algorithms	Min. Sup. (ms = 0.75)
	No of FI
UApriori	6
FDMA min	15
FDMA max	31
FAOWA	21

Table 7 FI and time calculated on Gazelle dataset

Min. sup. (ms)	Algorithms							
	UApriori		FDMA min		FDMA max		FAOWA	
	Time	No. of FI	Time	No. of FI	Time	No. of FI	Time	No. of FI
0.2	0.472	1	0.290	1	0.473	1	0.469	1
0.1	1.892	3	1.194	3	1.182	3	1.197	3
0.02	91.852	25	67.178	25	70.429	25	79.431	25
0.01	1025.22	86	674.232	87	843.6	91	693.107	90
0.002	17157.7	900	17999.6	982	13684	2875	14287	1019
0.001	26856.9	3124	28721.6	3725	21799.9	4447	29278.2	4029

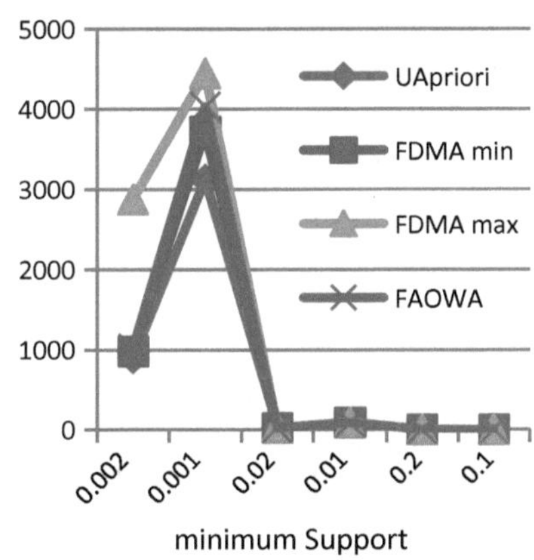

Fig. 1 Number of FI generated by UApriori, FDMA min, FDMA max, FAOWA on Gazelle

context of execution time taken by the algorithms, FAOWA generates items larger than UApriori, FDMA min and smaller than FDMA max but at some minimum support, it takes max time, e.g., ms = 0.02 and ms = 0.001 because FAOWA involves more computations compare to other algorithms.

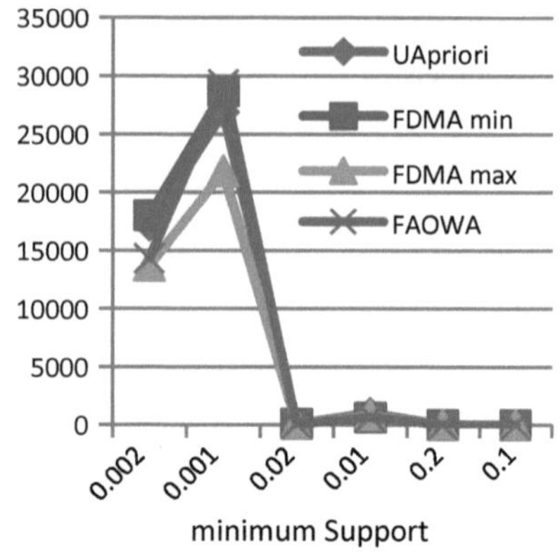

Fig. 2 Time taken(in Second) by UApriori, FDMA min, FDMA max, FAOWA on Gazelle

7 Conclusion and Future Directions

The uncertain databases were developed to fulfill the future requirement of customer or for the transactions who involves uncertainty. If an uncertain transactional database is developed for the symptom of different disease with their existential probability or membership function, e.g. diarrhea: 0.1, diabetes: 0.5 and cancer: 0.8. in this case, according to UApriori person having all symptoms should be less ill, according to FDMA min person must have diarrhea and according to FDMA max person must have cancer but practically person should be more ill, and this result is produced by FAOWA. So, it can be concluded that during the generation of frequent items association rule interestingness is also important.

References

1. R. Agrawal, R. Srikant, Fast algorithms for mining association rules, in *Proceeding of the 20th VLDB Conference* (Santiago, Chile, 1994), pp. 487–499
2. R. Agrawal, T. Imielinski, A.N. Swami, Mining association rules between sets of items in large databases, in *SIGMOD* (1993)
3. C.C. Aggarwal, Managing and mining uncertain data, (Kluwer Press, 2009)
4. T. Bernecker, R. Cheng, H.P. Kriegel, M. Renz, F. Verhein, A. Züfle, D.W. Cheung, S.D. Lee, W. Liang, Model-based probabilistic frequent itemset mining. Knowledge information system, vol. 37, (Springer), pp. 181–217
5. T. Bernecker, H.P. Kriegel, M. Renz, F. Verhein, A. Züfle, Probabilistic frequent itemset mining in uncertain databases, in *Proceedings of the 15th ACM Sigkdd Conference on Knowledge Discovery and Data Mining (Kdd'09)*, (Paris, France, 2009)
6. C.K. Chui, B. Kao, E. Hung, Mining frequent itemsets from uncertain data, in *11th Pacific-Asia Conference on Advances in Knowledge Discovery and Data Mining*, (PAKDD, China, 2007)
7. C.K. Chui, B. Kao, A decremental approach for mining frequent itemsets from uncertain data, (PAKDD, 2008), pp. 64–75
8. T. Calders, C. Garboni, B. Goethals, Approximation of frequentness probability of itemsets in uncertain data, in *ICDM* (2010)

9. T. Calders, C. Garboni, B. Goethals, Efficient pattern mining of uncertain data with sampling, in *PAKDD* (2010)
10. R. Cheng, D. Kalashnikov, S. Prabhakar Evaluating probabilistic queries over imprecise data, in *SIGMOD* (2003)
11. Y.L. Chen, K. Tang, R.J. Shen, Y.H. Hu Market basket analysis in a multiple store environment, Decis. Support Syst. **40**(2), (2005)
12. Frequent itemset mining implementations repository http://fimi.ua.ac.be/
13. T.P. Hong, P. Fournier-Viger, C.W.J. Lin, A fast algorithm for mining fuzzy frequent itemsets, J. Intell. Fuzzy Syst. (2015)
14. T.P. Hong, C.S. Kuo, S.C. Chi, Mining association rules from quantitative data. Intell. Data Anal. **3**, 363–376 (1999)
15. T.P. Hong, J.B. Chen, Finding relevant attributes and membership functions. Fuzzy Sets Syst. **103**, 389–404 (1999)
16. T.P. Hong, C.S. Kuo, S.L. Wang, A fuzzy AprioriTid mining algorithm with reduced computational time. Appl. Soft. Comput. **5**(1) (2004)
17. T.P. Hong, C.Y. Lee, Induction of fuzzy rules and membership functions from training examples. Fuzzy Set Syst. **84**(1) (1996)
18. T.P. Hong, K.Y. Lin, S.L. Wang, Fuzzy data mining for interesting generalized association rules. Fuzzy Set Syst. **138**(2) (2003)
19. J. Han, M. Kamber, Data mining concepts and techniques, 2nd edn, (Morgan Kaufmann Publishers, 2016)
20. J. Huang, MayBMS: a probabilistic database management system. in *SIGMOD* (2009)
21. SPMF an open-source data mining library. http://www.philippe-fournier-viger.com/spmf/index.php?link=datasets.php
22. Y. Tong, L. Chen, Y. Cheng, P.S. Yu, Mining frequent itemsets over uncertain databases, VLDB (2012)
23. Y. Tong, L. Chen, P.S. Yu UFIMT: an uncertain frequent itemset mining toolbox. KDD'12, (Beijing, China, 2012)
24. L.A. Zadeh, Fuzzy sets as a basis for a theory of possibility. Fuzzy Set Syst. **1**, 3–28 (1978)
25. L.A. Zadeh, Probability measures of fuzzy events. J. Math. Anal. Appl. **23**, 421–427 (1968)
26. R.R. Yager, Quantifiers in the formulation of multiple objective decision functions. Inform. Sci. **31**, 107–139 (1983)
27. R.R. Yager, On a general class of fuzzy connectives, Fuzzy Sets und Syst. **4** (1980)
28. R.R. Yager, On ordered weighted averaging aggregation operators in multicriteria decision-making. IEEE Trans. Syst. Man Cybern. **18** (1988)
29. L.A. Zadeh, Fuzzy sets and possibility distribution. StndFuzz **195**, 47–58 (2006)
30. L.A. Zadeh, Fuzzy Probabilities. Inf. Process. Manage. **20**, 363–372 (1984)
31. L.A. Zadeh Fuzzy sets, Inf Control **8** (1965)
32. Q. Zhang, F. Li, K. Yi, Finding frequent items in probabilistic data, in *SIGMOD* (2008)
33. S. Zhang, X. Wu, C. Zhang, J. Lu, Computing the minimum-support for mining frequent patterns, Knowl. Inf. Syst. **15**(2) (2008)

Recommender System Survey: Clustering to Nature Inspired Algorithm

Suraj Pal Singh and Shano Solanki

Abstract For e-commerce, besides increasing cross-selling by suggesting additional products for the customer to purchase, Recommender Systems (RS) are an effective way to increase customer satisfaction and consequently, customer loyalty. Thus personalization tools like RS is essential to support users to identify interesting products and services by considering the opinions of thousands if not millions of people all over the world. This article offers an outline of RS, their types as well as existing challenges. Further it also emphasizes on various evaluation metrices to be used for recommender system performance analysis. This paper surveys various existing algorithms that solve some problem of recommender system and the results of various nature inspired algorithms for recommender system are compared.

Keywords Recommender system · Content-based filtering · Collaborative filtering · Hybrid filtering · Knowledge-based · Data sparsity cold start · Gray sheep and black sheep · Cuckoo search · *k*-means · Gray wolf · Particle swarm optimization

1 Introduction

Internet, the fastest growing network the world has ever known, that has revolutionized the world by allowing millions of people to communicate with each other over vast distances, and across all kinds of boundaries. Nowadays, there are more than 3,726,100,580 Internet users all over the world [Internetlivestats.com, September, 2017], i.e., about 51% of the world's population now has access to the Internet, and this number is continuously increasing. These Internet users access more than 1,252,723,340 websites [internetlivestats.com, September, 2017] in the world. This

S. P. Singh · S. Solanki (✉)
Department of Computer Science and Engineering, NITTTR, Chandigarh, India
e-mail: s_solanki_2000@yahoo.com

S. P. Singh
e-mail: srjsingh21@gmail.com

C. R. Krishna et al. (eds.), *Proceedings of 2nd International Conference on Communication, Computing and Networking*, Lecture Notes in Networks and Systems 46, https://doi.org/10.1007/978-981-13-1217-5_76

huge number of Internet users accesses billions of documents (Web pages) that are interconnected by World Wide Web (Web for short) through the Internet.

A user who wants to search a relevant and reliable information on the web is not able to find it because of the overloaded of the information. To cope with these issues, among other solutions, a RS is used as information filtering tools that suggest items to a user and filter and sort information. The objective of RS is to support in decision-making process like browsing, entertainment, purchasing, reading, movies, jokes and so on. The information in the RS can be seized by two ways either explicitly or implicitly [1]. An explicit acquisition of user information typically involves collection of user's rating by different means and an implicit acquisition of user information typically involves monitoring of user's behavior such as downloaded applications, watched movie, purchase history, etc.

Following consideration are taken for generating RS [2].

1. Check what type of data is accessible in the database.
2. Apply filtering algorithm.
3. Choose model.
4. Consider employed techniques like Bayesian networks, probabilistic approaches, nearest neighbors' algorithms; bio-inspired algorithm like a particle swarm optimization(PSO), genetic algorithm, etc.
5. Check the desired scalability and sparsity level of the databases.
6. Check system performance that is memory and time consumption.
7. Check whether the objective is met.
8. Novelty, coverage, and precision etc. are checked to find whether it meets the desired quality.

The internal functions of the recommender system are categorized by the filtering algorithms. Mainly filtering algorithms are divided into five parts [2].

1.1 Collaborative Filtering

Collaborative filtering (CF) [3–6] predict or provide the ratings of a particular product for the individual user based on the previous ratings that are given by the users that have similar liking to the existing user. This basic idea of collaborative recommender is depicted in Fig. 1.

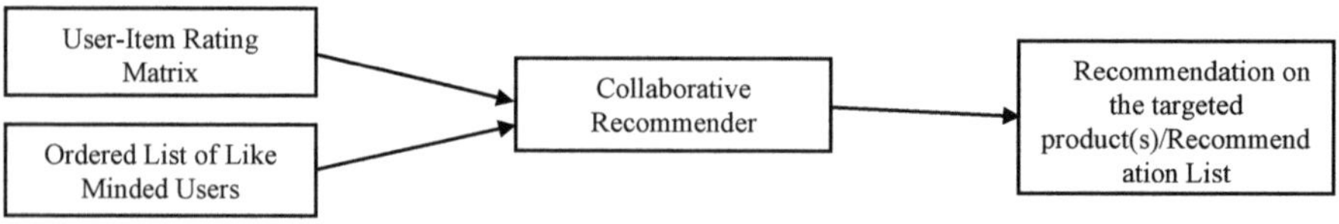

Fig. 1 Collaborative filtering based recommender system

There are two approaches for CF, that are User-User based CF and Item-Item based CF.

1.2 Content-Based Filtering

Content-based filtering (CBF) [7–9] work on the principle that it only build recommendations based on the person choice in the earlier, e.g., in an online book store if a person purchases few drama-based books then RS will probably recommend new drama book that comes to the market newly. CBF make recommendation using the content from items proposed for the recommendation, by this way a certain amount of content can be analyzed. This basic idea of content-based recommender is depicted in Fig. 2.

The pure CBF have mainly three shortcomings [10, 11].

- In few domains like videos, blogs, and music, it is a complex task to make the features for objects.
- Overspecialization problem occurs in CBF as it suggests only the similar type of items.
- It is difficult to obtain the more information from the user like feedback because in this type users typically did not rate the item and so it is difficult to check whether the recommendations are correct or not.

1.3 Demographic Filtering

Demographic Filtering [12–14] is based on the principal of person with common characteristics like country, age, sex, etc., will also have similar likings for the particular items. It is an effective and simple personalization way. Online websites can simply familiarize with the showing language based on, where a particular user live.

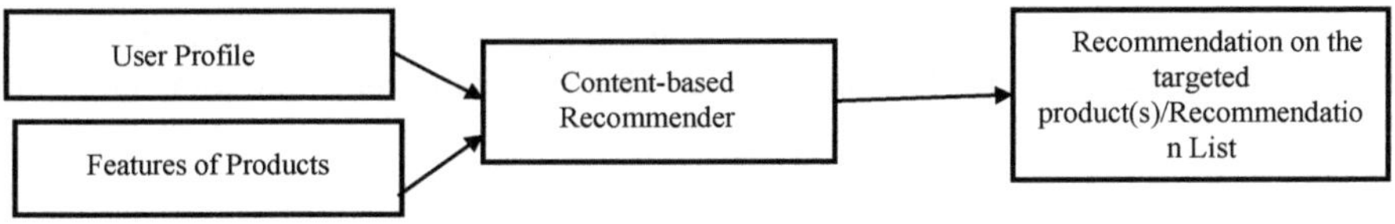

Fig. 2 Content-based filtering recommender system

1.4 Hybrid Filtering

Hybrid Filtering [13, 15] combines any of the two methods, generally by taking good features of each of the method, to eliminate the shortcomings of such RS. It commonly combines CF with demographic filtering [16] or CF with CBF.

Hybrid filtering is commonly centered on probabilistic or bio-inspired methods such as fuzzy genetic, Bayesian networks, clustering, genetic algorithm, neural networks, and latent features.

1.5 Knowledge-Based

Knowledge-based recommender systems make use of knowledge structure to make inference about the user needs and preferences. Such kind of recommender systems have knowledge about what kind of items are liked by a user, so, a relationship can be established between user needs and relevant recommendation for that user.

2 Challenges in Recommender System

Recommender system gave excellent results in many systems but there are few challenges that need to be dealt. Some of the challenges are [2]:

2.1 Data Sparsity

This type of problem arises when a number of recorded ratings are less than the number of ratings that should be predicted. In this situation, the user-item matrix will be very sparse and RS will not be capable to recommend correct opinion to the specific user. In any movie recommender system, there might be possibility that many movies are rated by only some person at that case these movies would recommend rarely, even if these movies had high rating. This may cause poor recommendation.

2.2 Scalability

When a number of users and item increase in large amount at that situation RS become inefficient to handle this huge amount of data because of computation resource beat practical level [2]. This problem increases the computation level for recommender system. For example, Amazon.com has 20 million customers and recommend 18 million items. Different method can be used to reduce scalability like reducing dimensionality, clustering and Bayesian Network. High dimensionality can be reduced by SVD.

2.3 Synonymy

When a number of items have similar attribute but the name is different than RS is not able to handle such type of situation and may generate recommendation list containing similar items and hence leads to degraded recommendation quality. For example, collaborative filtering memory based approach will treat "comedy film" and "comedy movie" differently but both have same meaning.

2.4 Cold Start Problem

When there is not enough previous rating for an item, cold start problem occurs, in this case, RS fails to create meaningful recommendation due to inadequate information. This problem can be categorized as cold start for new items and new users. When system does not have any information interrelated to former purchasing of a new user and user's taste is unknown to the system as the user might not rate any item, cold start for new users occurs. When there is not enough previous rating for an item, cold start problem for new items occurs. In these kinds of cases it is difficult for the recommender system to suggest particular items for the new users.

2.5 Gray Sheep and Black Sheep

Gray sheep is the small community of the group who do not agree or disagree with other people of the group so they are not beneficial to the CF. This increase error and sometimes failure may occur in the RS. Gray sheep users can be separated and identified from other normal users by applying k-mean clustering that is offline clustering technique. Using this way recommendation error is minimal and performance gets better.

In black sheep category of the user, there are a few people with whom they can correlate. It is very difficult to make recommendation for these users.

3 Evaluation Metrics of Recommender System

In the field of RS evaluation metrics and quality measures [17] are required to identify the eminence of the algorithms, techniques, and methods for prediction and recommendations. Evaluation metric [18] give many results for the similar type of problem and choice from various encouraging problems of research that can produce improved results.

Most commonly used evaluation metrics can be classified as [18].

3.1 Prediction Metrics

These metrics measures the closeness of predicted rating by the RS and the true ratings by users.

3.1.1 Mean Absolute Error (MAE)

It measures to what quantity a system can predict rating for users. MAE computes the average of the all absolute difference between true ratings and predictions. It can be given by Eq. (1)

$$\mathrm{MAE} = \sum \frac{|p_{i,j} - r_{i,j}|}{n}, \tag{1}$$

where $P_{i,j}$ is the predicted value of user i on item j, $R_{i,j}$ is the true rating, n is entire no of expected movies.

3.1.2 Root Mean Squared Error (RMSE)

RMSE is the deviation between the value observed from the environment that is being modeled and the values predicted by the model. RMSE can be given by Eq. (2)

$$\mathrm{RMSE} = \sqrt{\frac{\sum \left(p_{i,j} - r_{i,j}\right)^2}{n}}, \tag{2}$$

where $P_{i,j}$ is the predicted value of user i on item j, $R_{i,j}$ is the true rating, n is Total no. of user.

3.1.3 Normalized Mean Absolute Error (NMAE)

It normalizes MAE to make it self-determining from the rating measure. It can be given by Eq. (3)

$$\mathrm{NMAE} = \frac{\mathrm{MAE}}{\mathrm{r_{max}} - \mathrm{r_{min}}}, \tag{3}$$

where $r_{\min}$ and $r_{\max}$ are the lower and upper bounds of the ratings.

3.2 *Set Recommendation Metrics*

3.2.1 Precision

It provides how much proportion of recommended item is relevant to the total number of items. It can be calculated by the following Eq. (4):

$$\text{Precision} = \frac{\text{No. of corrected answer}}{\text{No. of item selected}} \tag{4}$$

3.2.2 Recall

It provides how many portions of recommended item are relevant to the total number of relevant items. It can be calculated by the following Eq. (5):

$$\text{Recall} = \frac{\text{No. of corrected answer}}{\text{Total no. of relevant item}} \tag{5}$$

3.2.3 F-Measure

It is a combination of Recall and Precision. It can be calculated by the following Eq. (6).

$$\mathrm{F-measure} = \frac{2 * \mathrm{Precision} \ * \mathrm{Recall}}{(\mathrm{Precision} + \mathrm{Recall})} \tag{6}$$

3.3 *Rank Recommendation Metrics*

If a number of recommended items are not small, then user generally gives the preference to the item that has occurred in the list first. This creates a serious mistake in recommending the item. There are two rank recommendation metric first is half- life [19] and second is discounted cumulative gain [19]. Half-life assumes that interest of the user exponentially decreases when user move from top to the bottom. Discounted cumulative gain assumes that interest of user decrease logarithmically when user moves from top to the bottom. Half-life and Discounted cumulative gain can be calculated by Eqs. (7) and (8) respectively.

$$\text{Half} - \text{life} = \frac{1}{U}\sum_{u \in U}\sum_{i=1}^{N}\frac{\max\left(r_{u.p_i} - d\right)}{2^{(i-1)/(\alpha-1)}} \tag{7}$$

$$\text{Discounted cumulative gain} = \frac{1}{U}\sum_{u \in U}\left(r_{u,p_i} + \sum_{i=2}^{k}\frac{r_{u,p_i}}{\log_2(i)}\right), \tag{8}$$

where U are the set of the users, $r_{u,pi}$ is the true rating of the user u for the item p_i, $p_1, \ldots, p_n$ represents recommendation list, k is the rank of the evaluated item, d is the default rating, α is the number of the item on the list.

4 Related Work

Recommendation can be done in many ways; one way is to use clustering approach. Clustering approach is widely used in generating the cluster as it is a powerful unsupervised learning method to correctly evaluate the large amount of data created by applications. The main purpose of clustering algorithm is to classify the data into cluster in a particular way so that data is assembled into the similar cluster if their attributes are same. Cluster formed should have minimum inter similarity and maximum intra similarity. If cluster formed are good, then recommendation will be better.

Many clustering algorithm like fuzzy c-mean (FCM), *k*-mean, *k*-medoids, self-organizing map (SOM), etc., are applied to any datasets like movielens (for movies), jester (for jokes), book-crossing (for books), last.fm (for music), amazon (for products), audio scrobbler (for music), nursing.home and hospital ratings (for healthcare), intitut fur informaticauniversital Freiburg (for book rating), etc. [20].

One can choose any clustering algorithm and any datasets for experiment, but widely used dataset is movielens data set as it has 26,000,000 ratings of 45,000 movies and 270,000 users [21]. In this paper we analyze various results that are applied on the movielens dataset contain 100,000 rating of 1682 movies by 943 users which is movielens 100 k dataset.

In [22] author proposed the three steps (proofing, inferring and predicting) for collaborative filtering recommendation methods while maintain prediction accuracy and computing speed. Author took case based reasoning (CBR) and self-organizing map (SOM) and changed an unsupervised learning method into the supervised user preference reasoning approach. Author took movielens dataset 100,000 rating from 943 users and 1682 movies which gave 0.1790 NMAE and value of ROC area as 0.8461.

In [23] authors ensemble the approach of SOM and k-means and applied on to the movielens 100 k dataset that gave better results when SOM and k-means applied individually. SOM of type-II gave 75.49% accuracy, 76.95% precision and 21.8% recall. k-means of type-II gave 74.94% accuracy, 77.46% precision and 25.66% recall. SOM ensembles by HGPA gave 76.41%. 78.03 and 20.54% accuracy, precision and recall respectively. k-means ensembles by HGPA gave 76.32, 7.54 and 19.60% accuracy precision and recall respectively. Naturally when authors used ensemble technique in both SOM and k-mean accuracy, precision and recall are better.

In [24] author provides various clustering algorithm that are applied onto the movielens 100 k dataset. Principle component analysis-Self organizing map (PCS-SOM) gave 0.98 MAE and 0.07 standard deviation (SD). As PCA is used to reduce the dimensionality of the user-rating matrix that is mostly unfilled at the starting [25–27]. Reducing dimensionality increase the value of MAE and standard deviation. If PCA is not applied and only SOM is applied the MAE was 0.75 and SD was 0.06 [24]. When k-means clustering method is applied on to the movielens 100 k dataset then MAE and SD were 0.69 and 0.10 respectively in this MAE is reduced but SD increased. If dimensionality of user-rating matrix is reduced by PCA and then apply k-mean clustering is applied it gave 0.93 MAE and 0.12 SD (both are increased due to PCA) [24]. So PCA solve the problem of dimensionality but increase the value of MAE and SD.

Problem of cold start can be solved by demographic filtering technique. In [28] authors worked on to the demographic information given in the movielens datasets then applied J48 classifier. This approach reduces the overall execution time of generating recommendation.

Results can be optimized by using the nature inspired algorithms. k-means fall into the local optima value because of the initial seed point which is generated randomly. Various authors applied clustering algorithm with nature-inspired algorithm so the search can be global optimum and convergence process can speed up.

Genetic algorithm is commonly use nature-inspired algorithm that is used in optimization problem. In [29] generate the movie recommendation on the movielenes 110 k dataset. Firstly authors dense the movie population by applying the PCA data reduction technique then apply genetic algorithm approach with k-means which gave 0.7821 MAE and 0.004747 standard deviation value. If only GAKM (genetic algorithm with k-means) is applied, then MAE were 0.8040 and SD were 0.002786.

In [30] authors used particle swarm optimization (PSO) [31–33] for optimizing the result of RS. Authors distributed ratings specified over the types that specific movie belongs using type division method [30]. They collaborated PSO, fuzzy c means and k-means. Initial seed point was found using PSO and k-means and then FCM [34] is used for finding the center for clustering. They worked on the movielens 100 k dataset which gave 0.7459 MAE and 0.006153 SD. These results are better than GAKM and other approaches. In [35] result is further optimized with the help of gray wolf optimizer techniques [36, 37] and FCM [34, 38]. Authors applied this

Table 1 Comparison of different methods

Method	Data set	MAE	SD
PCA-SOM	Movie-lens 100 k	0.98	0.07
SOM	Movie-lens 100 k	0.75	0.06
k-means	Movie-lens 100 k	0.69	0.10
PCA + k-means	Movie-lens 100 k	0.93	0.12
PCA + GAKM	Movie-lens 100 k	0.7821	0.004747
GAKM	Movie-lens 100 k	0.8040	0.002786
PSO + k-Means	Movie-lens 100 k	0.7459	0.006153
Gray wolf + FCM	Movie-lens 100 k	0.68	0.54
Cuckoo + k-means	Movie-lens 100 k	0.68	0.10

technique on movielens 100 k datasets which gave 0.68 MAE, 0.54 SD, 0.55 Precision and 0.49 recall. MAE of this technique is better than other previous approaches that were discussed in this paper but result of SD is not so good. In [39] authors apply cuttlefish optimization with k-means and experiment done on to the movielens dataset of 100 k which provide batter results.

A new approach cuckoo search [40, 41] with k-means has been used to optimize the result of RS in [24]. Authors used levy flight for the random walk which gave more accurate and precise results. When this approach is applied on to the movielens 100 k dataset, it gave 0.68 MAE and 0.10 SD. Value of MAE is better than all other approaches that are discussed previously in this paper.

5 Comparison of Various Recommender System Algorithms

MAE and SD are two important parameters when we evaluate recommender system. MAE is calculated by taking difference between predicted and true rating and then take average while SD how much rating is differ from the average value of rating. So less value of MAE and SD shows better result. Table 1 gives various methods that are applied on the movielens dataset 100 k and provide MAE and SD.

6 Conclusion

This paper presents various methods to generate recommender system. Mainly Recommender system is generated using clustering method that is an unsupervised learning method. But the results of evaluation metrics of RS are decreased when we try to solve the problem of recommender system like PCA that is used to decrease the problem of high dimensionality. Results can be optimized by various nature inspired algorithms that are integrated with clustering algorithms. A newly developed algorithm cuckoo search gives better result than other approaches.

References

1. D.H. Park, H.K. Kim, I.Y. Choi, J.K. Kim, A literature review and classification of recommender systems research. Expert Syst. Appl. **39**(11), 10059–10072 (2012)
2. J. Bobadilla, F. Ortega, A. Hernando, A. Gutiérrez, Recommender systems survey. Knowl. Based Syst. **46**, 109–132 (2013)
3. G. Adomavicius, A. Tuzhilin, Toward the next generation of recommender systems: A survey of the state-of-the-art and possible extensions. IEEE Trans. Knowl. Data Eng. **17**(6), 734–749 (2005)
4. A. Fahad et al., A survey of clustering algorithms for big data: Taxonomy and empirical analysis. IEEE Trans. Emerg. Top. Comput. **2**(3), 267–279 (2014)
5. J. Herlocker, J.A. Konstan, J. Riedl, An empirical analysis of design choices in neighborhood—based collaborative filtering. Inf. Retr. Boston. **5**(4), 287–310 (2002)
6. J.A. Konstan, J. Riedl, Recommender systems: From algorithms to user experience. User Model. User Adapt. Interact. **22**(1–2), 101–123 (2012)
7. J. Salter, N. Antonopoulos, CinemaScreen recommender agent: Collaborative and content-based filtering. Intell. Syst. **21**, 35–41 (2006)
8. K. Lang, in *NewsWeeder: Learning to Filter Netnews*. Proceedings of the 12th International Conference on Machine Learning (1995), pp. 331–339
9. R. Van Meteren, M. Van Someren, Using content-based filtering for recommendation. ECML/MLNET Workshop: Machine Learning in New Information Age (2000), pp. 47–56
10. Y. Shoham, Content-based, collaborative recommendation. Commun. ACM **40**(3) (1997)
11. M.J. Pazzani, D. Billsus, Content-based recommendation systems. Adapt. Web **4321**, 325–341 (2007)
12. M.J. Pazzani, A framework for collaborative, content-based and demographic filtering. Artif. Intell. Rev. **13**(5), 393–408 (1999)
13. C. Porcel, A. Tejeda-Lorente, M.A. Martínez, E. Herrera-Viedma, A hybrid recommender system for the selective dissemination of research resources in a technology transfer office. Inf. Sci. **184**(1), 1–19 (2012)
14. B. Krulwich, Lifestyle finder: Intelligent user profiling using large-scale demographic data. AI Mag. **18**(2), 37 (1997)
15. R. Burke, Hybrid web recommender systems. Adapt. Web, 377–408 (2007)
16. M.G. Vozalis, K.G. Margaritis, Using SVD and demographic data for the enhancement of generalized collaborative filtering. Inf. Sci. **177**(15), 3017–3037 (2007)
17. A. Gunawardana, G. Shani, A survey of accuracy evaluation metrics of recommendation tasks. J. Mach. Learn. Res. **10**, 2935–2962 (2009)
18. J.L. Herlocker, J.A. Konstan, L.G. Terveen, J.T. Riedl, Evaluating collaborative filtering recommender systems. ACM Trans. Inf. Syst. **22**(1), 5–53 (2004)
19. L. Baltrunas, T. Makcinskas, F. Ricci, in *Group Recommendations with Rank Aggregation and Collaborative Filtering*. Proceedings of the Fourth ACM Recommender Systems Conference (2010), pp. 119–126
20. https://gist.github.com/entaroadun/1653794
21. https://grouplens.org/datasets/movielens/
22. T. Hyup, K. Joo, I. Han, The collaborative filtering recommendation based on SOM cluster-indexing CBR. Expert Syst. Appl. **25**, 413–423 (2003)
23. C. Tsai, C. Hung, Cluster ensembles in collaborative filtering recommendation. Appl. Soft Comput. J. **12**(4), 1417–1425 (2012)

24. R. Katarya, O.P. Verma, An effective collaborative movie recommender system with cuckoo search. Egypt. Inform. J. (2016)
25. K. Goldberg, T. Roeder, D. Gupta, C. Perkins, Eigentaste: A constant time collaborative filtering algorithm. Inf. Retrieval **4**, 133–151 (2001)
26. K. Honda, N. Sugiura, H. Ichihashi, S. Araki, in Collaborative Filtering Using Principal Component Analysis and Fuzzy Clustering. Proceedings of the First Asia-Pacific Conference on Web Intelligence: Research and Development, Maebashi City, Japan (2001), pp. 394–402
27. A. Selamat, S. Omatu, Web page feature selection and classification using neural networks. Inf. Sci. **158**, 69–88 (2004)
28. S. Solanki, S. Batra, Recommender system using collaborative filtering and demographic characteristics of users. Int. J. Recent Innovative Trends Comput. Commun. (IJRITCC) **3**(7) (2015). ISSN 2321-8169
29. Z. Wang, X. Yu, N. Feng, Z. Wang, An improved collaborative movie recommendation system using computational intelligence. J. Vis. Lang. Comput. **25**(6), 667–675 (2014)
30. R. Katarya, O.P. Verma, A collaborative recommender system enhanced with particle swarm optimization technique. Multimed. Tools Appl. (2016)
31. J. Kennedy, R. Eberhart, in *Particle Swarm Optimization*. Proceedings of the IEEE International Conference on Neural Networks (1995), pp. 1942–1948
32. C. Chen, F. Ye, Particle swarm optimization algorithm and its application to clustering analysis. IEEE Netw. Sens. Control **2**, 789–794 (2004)
33. D. Chen, C. Zhao, Particle swarm optimization with adaptive population size and its application. Appl. Soft Comput. **9**, 39–48 (2008)
34. J.C. Bezdek, R. Ehrlich, W. Full, FCM: the fuzzy c-means clustering algorithm. Comput. Geo. sci. **10**, 191–203 (1984)
35. R. Katarya, Recommender system with grey wolf optimizer and FCM. Neural Comput. Appl. (2016)
36. S. Mirjalili, S. Saremi, S.M. Mirjalili, Multi- objective grey wolf optimizer: a novel algorithm for multi criterion optimization. Expert Syst. Appl. **47**, 106–119 (2016)
37. S. Mirjalili, S.M. Mirjalili, A. Lewis, Grey wolf optimizer. Adv. Eng. Softw. **69**, 46–61 (2014)
38. H. Koohi, K. Kiani, User based collaborative filtering using fuzzy c-means. Measurement, 134–139 (2016)
39. M. Senbagaraman, R. Senthilkumar, S. Subasankar, R. Indira, A movie recommendation system using collaborative approach and cuttlefish optimization (2017), pp. 95–99
40. X.S. Yang, S. Deb, Cuckoo Search via Levy flights. World Congress Nature and Biologically Inspired Computing (2009), pp. 210–214
41. R. Rajabioun, Cuckoo optimization algorithm. Appl. Soft Comput. J. **11**(8), 5508–5518 (2011)

Part VII
Electronics and Instrumentation

Systematic Review of Dependencies in Source Code of Software and Their Categorization

Mrinaal Malhotra and Jitender Kumar Chhabra

Abstract Dependencies represent that change in one of component will affect other component. This paper presents systematic study and review of dependencies in the source code. The paper also focuses on their computation methodologies and application of these dependencies. A state of art analysis of source code dependencies has been carried out in this paper and unaddressed research issues have been identified. Based on these gaps in the literature this paper proposes categorization of various source code dependencies, which can be useful for organizing dependencies based on the applications.

Keywords Code dependencies · Structural dependencies · Inter-connectivity

1 Introduction

Inter-connectivity between different parts of a software system is one of the most important characteristics and can be very helpful to developers which give rise to dependencies [1, 2]. Dependency among software component has been defined as "the quality or state of being dependent; especially: the quality or state of being influenced or determined by or subject to another" [3]. Source code Dependencies denote that change in one component is likely to affect or change the other component. Broadly, there are two types of dependencies: structural and non-structural [4]. Structural dependencies are a result of relationship between two or more compilation units at linking time or compile time. [3]. Non-structural/logical dependencies correspond to relationships based on identifiers and comments [5]. Logical dependencies can be used for improvement of modularization [2]. Syntax analysis is a prominent

M. Malhotra (✉) · J. K. Chhabra
Computer Engineering Department, National Institute of Technology,
Kurukshetra 136119, Haryana, India
e-mail: Mrinaalmalhotra@gmail.com

J. K. Chhabra
e-mail: jitenderchhabra@gmail.com

C. R. Krishna et al. (eds.), *Proceedings of 2nd International Conference on Communication, Computing and Networking*, Lecture Notes in Networks and Systems 46, https://doi.org/10.1007/978-981-13-1217-5_77

method for discovering dependencies [6]. But it is insufficient for module-related connections [7, 8] and many other techniques such as fan-in/fan-out analysis [9], requirement-patterns-identification [10], etc., are also used for the computation of various types of dependencies [11]. Dependencies are also useful for evaluating requirements of the software system [12]. This paper focuses on various dependencies, their computation, their applications and their categorization.

2 Systematic Review of Source Code Dependencies

2.1 Source of Information

In order to investigate the development in literature in the domain of source code dependencies various sources are studied for information. The primary sources include IEEEXplore, ACM, Springer and Elsevier. Total 111 papers were studied for literature survey.

Dependencies are studied by dividing them as black box and white box. Black box are mostly module-based dependencies in which only characteristics of modules are considered. Black box dependencies are: symbol dependencies, bad dependencies, acyclic dependencies, failure dependencies, service dependencies, SEA dependencies, feature dependencies and argument dependencies.

2.2 Literature Review of Black Box Dependencies

Lutellier et al. [13] in "Measuring the impact of code dependencies on software architecture recovery technique" proposed symbol dependencies. Symbol dependencies means if one module of file A is dependent on other module of file B. Symbol dependencies are considered as precise dependencies. Monitoring precise dependencies help a lot in maintaining the large software systems. Because of complex and fast evolving software, inter-method dependency becomes a major factor in debt in the software development process. In paper by Morgenthaler et al. [14], is managed by debt identification of bad dependencies.

Wang et al. [15] in "Generating precise dependencies for large software systems" proposed bad dependencies. There are two varieties of bad dependencies stated as, inconsistent and underutilized dependencies. Underutilized dependencies mean that if small portion of the module targeted is actually used in client module. But inconsistent dependencies are those that contradict with the design of the system. These dependencies should not form a cycle as described by parnas in acyclic dependency principle (ADP) [16]. Acyclic dependencies are explained in more detail below.

Zhoy et al. [17] in "A tree-based reliability model for composite web service with common cause failures" proposed acyclic dependency. Acyclic dependency means there is a parent–child relationship between different modules of a system. The relationships between various modules or functions should not from a cycle [18, 19]. Acyclic dependencies are used for checking reliability of services. Several other dependencies are also needed for checking reliability of services include failure and service dependencies which are discussed next.

Peng et al. [20] in "Reliability analysis of on-demand service based software systems considering failure dependencies" proposed failure and service dependencies. Consider a system which is responsible for providing different services to the user. Each of these services can be in two modes: operational and failure. And the state of each of these services is monitored with the help of random variable X, which can have only two values namely 1 or 0 indicating operational and failure modes respectively. Service dependency means that the operability of one service may depend on another

Some software systems, which provide different services, take advantage of reusability. These services come from same providers and lots of services can fail due to some common reasons. Failure dependency means the existence of these failures having one cause, called the root of the failure. Failure dependencies can be used for reusability, checking reliability of services and checking error propagation between different services.

There are several other acyclic dependencies such as SEA/SEB dependency and feature dependency described as follows.

Jasz et al. [21] in "Static execute after/before as a replacement of traditional software dependencies" proposed static execute after and static execute before dependencies (SEA/SEB). As the name implies these dependencies are executed statically. This technique is used for determination of dependencies among program methods with the consideration of execution order. Two procedures f and g are SEA dependent if g gets executed after f. Similarly two methods f and g are SEB dependent if g gets executed before f.

Lienhard et al. [22] in "Tracking objects to detect feature dependencies" proposed feature dependencies. A feature means one functional requirement or a behavioral unit of the system. If a feature F1 uses the objects which are created in feature F2 then feature F1 depends on feature F2. If the objects, which are used by F1, have any aliases, i.e., they are referred by a different name in any other feature (say F3) then F1 will depend on F3 as well. Feature dependencies are supposed to cross the boundaries of procedures as crossed boundaries remains undetected usually [23]. Addition of new features gives rise to more dependencies which need to be maintained [24]. Feature dependencies can be used for software maintenance activities [22]. Object aliasing is generally found in object oriented systems and it can give rise to argument dependencies. Argument dependencies are described by John Boyland very effortlessly.

Lienhard et al. [22], Boyland et al. [25] in "Capabilities for sharing_ A generalization of uniqueness and read-only" proposed argument dependencies. A collection of elements, a structure, is dependent on individual elements of the structure. If the argument of a collection is exposed, i.e., if any element of a collection/structure is modifiable from outside then changing that element from outside can disrupt the property of the entire structure of which it is a part. This is argument-dependency.

The dependencies which are computed by analyzing the complete code are called white box dependencies. White box dependencies are include dependencies, data and control dependencies, hidden dependencies, temporal dependencies, semantic dependencies, and circular dependencies.

2.3 Literature Survey of White Box Dependencies

Lutellier et al. [13] in "Measuring the impact of code dependencies on software architecture recovery technique" proposed include dependencies as well. Include dependencies means one file contains some another file example a.cpp include a.h. Include dependencies are used as an input for different techniques of architecture recovery [26]. Various other techniques are also presented for recovering architecture. These techniques use different compilation units (e.g., files and modules) which form a unit [27, 5, 28]. From [29], it can be seen that these techniques are not providing sufficiently accurate results. Hence, more information is needed for improving the accuracy like we can include some more dependencies such as data and control dependencies which are described next.

Alzamil [30] in" Redundant coupling detection using dynamic dependence analysis" proposed data and control dependencies. Data dependencies means one value is assigned to a variable in one statement and that value is used in another statement and control dependencies means that the evaluation of any particular statement depend on the execution of any other test node. Data and control dependencies are computed differently by different authors. ([11]) compute data dependencies statically by analyzing the source code.

Wu et al. [31] presented a different approach of computation of dependencies. In this paper dynamic evaluation of dependencies take place by analyzing the execution traces. Data and control dependencies have various applications which include fault localization [32], byte-code verification [33], service combination [34] code optimization by making a program dependence graph [35], coupling detection [30], concept localization [36] and change impact analysis (CIA) [37]. Change impact analysis means seeing the effects of modifications done in some parts of source code. In this paper [37], effects are monitored using data dependencies where they are used for making a state chart diagram. But some dependencies are left out. These dependencies are hidden dependencies which should be included during CIA.

Zhifeng et al. [38] in "Hidden dependencies in program comprehension and change propagation" proposed hidden dependencies. Several classes are included in code which gives rise to more dependencies [39]. Two classes have hidden dependency if both of the classes are not explicitly dependent on each other but a third class that has a dependency relation with both the classes which has data flow that happen between objects of both classes

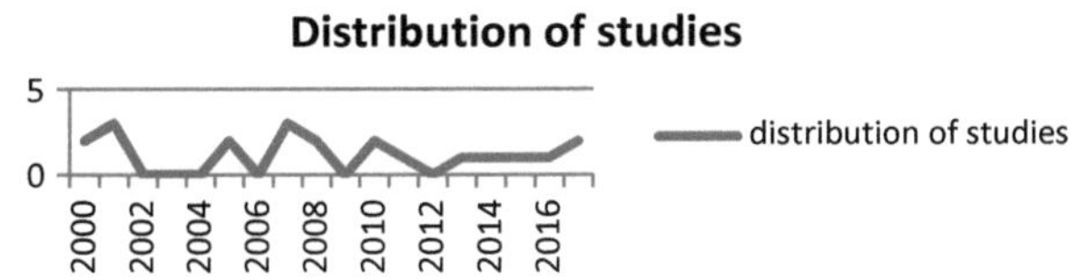

Fig. 1 Distribution of studies along different years of publications

In this paper [40] a new approach is exploited for computation of hidden dependency. Here, hidden dependencies are found out with the help of relations between different entities. It is established that most of studies include only hidden, data and control dependencies for CIA [41–43]. This results in imprecise results. Thus, more information is needed for CIA. Temporal, circular and Semantic dependencies should be added during the process of CIA which is discussed next.

Cataldo et al. [44], Garcia [45] in "Socio-technical congruence: A framework for assessing the impact of technical and work dependencies on software development" proposed temporal dependencies. If two or more modules are temporarily dependent on each other, one needs to execute before another. Semantic dependencies are explained as follows.

Podgurski and Clarke [46] in "A Formal Model of Program Dependences and Its Implications for Software Testing, Debugging, and Maintenance" proposed semantic dependencies. Semantic dependencies means the behavior of one statement is affected by the semantics of another statement present inside the code.

In this paper [47] semantic dependency is present in class level as well as procedural level. First one is class-level dependency. According to the class-level semantic dependency, consider classes A and B in which A is a sub-class of B. Then constructor of B impacts the object creation of class A. Second type is procedure level semantic dependency. According to procedure level semantic dependency if there are two overloaded methods and some modification in one of method's signature can affect the calling of both the overloaded methods.

Identification of circular dependencies is important which are discussed next. Al-Mutawa et al. [19] in "On the shape of circular dependencies in java programs" proposed circular dependencies. Consider a set of classes/packages (c1,c2,c3, … ,cn) if c1 depends on c2 and c2 depends on c3, c3 depend on c4 and so on and cn depend on c1 then this situation causes circular dependency among classes. Figure 1 illustrates distribution of studies over the last 17 years.

Based on this study it can be inferred that dependencies have applications in multiple fields. Our systematic study has helped us to identify various applications of different dependencies.

Table 1 Applicability of different dependencies

Dep.	Change impact analysis	Architecture recovery	Software defect removal.	Code optimization.	Software re-modularization	Reliability of services
Semantic	✓		✓			
Include		✓				
Symbol		✓				
Data	✓			✓	✓	
Control	✓			✓	✓	
Hidden	✓					
Circular	?					
Acyclic	*				✓	✓
SEA/SEB	*	*	✓			
Feature	*		✓		✓	
Failure						✓
Service						✓

Here ✓ =>used for specified purpose. ? =>used incorrectly. * =>should be used

3 Applicability of Dependencies

Table 1 gives an illustration of dependencies which are used in the industry.

According to our study, findings are

- Semantic dependency is used for CIA and software defect removal.
- Include and symbol dependency is used for architecture recovery.
- Data and control dependency is used for CIA, code optimization and software re-modularization.
- Hidden dependency is used for change impact analysis.
- Circular dependency is incorrectly used for change impact analysis.
- Acyclic dependency is used for software re-modularization and checking reliability of services and acyclic dependency should be used for change impact analysis.
- SEA/SEB dependency should be used for change impact analysis and architecture recovery and it is used, in the literature, for software defect removal.
- Feature and service dependency is used for checking reliability of services.

This applicability can be more systematic if we categorize them and identify different categories for different software activities. The work of categorization has not been attempted in the literature till now and to the best knowledge of authors it is first to propose such categorization along with their applicability.

4 Categorization of Dependencies

Categorization of dependencies is important as it helps in organization. After categorization, one can easily identify which group of dependencies is used for a particular field. Categorization can help in improving various fields as if, after recognition of group, some dependencies are not used, which belong to the group, then those can be added which will increase the accuracy and efficiency of the field investigated.

Code dependencies can be categorized on different basis. First, we can categorize code dependencies on the basis of hierarchy.. According to hierarchical classification there are five classes of code dependencies namely, file-level dependencies, procedural-level dependencies, class-level dependencies, statement-level dependencies and data-structure level dependencies.

Second, we can categorize code dependencies on the basis of design satisfaction. Design satisfaction means that the source code does not satisfy the design made, prior to source code. This is done in order to get a perfect software system. According to design, there are two classes namely, missing and spurious dependencies. Missing dependencies but are those which are present in the source code according to the design made but are not recorded anywhere. Missing dependencies include: Include and argument dependencies.

Spurious dependencies are those dependencies which should not be present in the source code according to the design but they are present because of any fault in coding or understanding the design of the system. Spurious dependencies include hidden dependencies and failure dependencies.

Thirdly, we can categorize dependencies on the basis of flow of control/data. These dependencies are flow-based dependencies. The flow-based dependencies are (i) Data and control dependencies. (ii) Temporal dependencies. (iii) SEA/SEB dependencies. (iv) Feature dependencies. (v) Acyclic dependencies.

And lastly, we can also categorize dependencies as static and dynamic.. Static dependencies are data and control dependencies, semantic dependencies, SEA/SEB dependencies and hidden dependencies Dynamic classification can also help in debugging [48]. Dynamic dependencies comprises of data, control and symbol dependencies. Figure 2 describes various sets of dependencies and there intersection. Dependencies can also be categorized on the basis of graph computation and matrix computation [49].

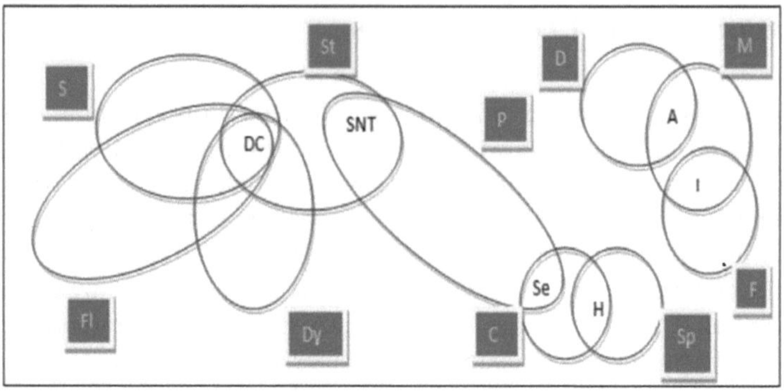

Where
S: Static dependency DC: Data and Control dependency Fl : Flow–based dependency
SNT: Semantic and temporal dependency Dy: Dynamic dependency M: Missing dependency
St: Statement-level dependency I: Include dependency D: Data-structure level dependency
H: Hidden dependency P: Procedural level dependency Se: Semantic dependency
C: Class-level dependency Sp: Spurious dependency F: File-level dependency

Fig. 2 Categorization of code dependencies

According to our study, we found out that

- Hierarchical classification is useful for complete recovering ground-truth architecture.
- Classification based on design should be performed when we need to check the reliability of services.
- Categorization based on flow of control or data is done for better understanding the impact of modifications done in some portion of the system, i.e., flow-based classification helps in change impact analysis.
- Dependencies are classified as static when familiar dependency graph is needed by developer.
- Dependencies are classified as dynamic when only one particular scenario analysis is required as during any execution approx 20% of the code is executed.

The primary motive of our study is to identify various groups of the dependencies and also investigate which of this group is useful for which of the activity such as change impact analysis, architecture recovery, code optimization, software re-modularization, etc.

5 Conclusion

This paper has presented literature survey of various source code dependencies in software system and review of previous works in this direction has been carried out systematically to identify various research issues in there computation and applications. Further the paper has proposed different categorization showing there overlapping as well as their applicability towards different activities such as change impact analysis, architecture recovery, code optimization, etc. Our study indicated that many dependencies have been incorrectly or incompletely used for different activities and proper categorization of these dependencies is highly desirable for better software management.

References

1. J. Amarjeet, K. Chhabra, FP-ABC: Fuzzy pareto-dominance driven artificial bee colony algorithm for many objective software clustering, Comput. Lang. Syst. Struct. **51**, 1–21 (2018)
2. J. Amarjeet, K. Chhabra, Harmony search based remodularization for object-oriented software systems, Comput. Lang. Syst. Struct. **47**(2), p 153–169 (2017)
3. G.A. Oliva, M.A. Gerosa, Experience report: How do structural dependencies influence change propagation? An empirical study, in *International Symposium on Software Reliability Engineering (ISSRE)* (2015), pp. 250–260
4. B.G. Ryder, F. Tip, Change impact analysis for object-oriented programs, in *Proceedings of the ACM SIGPLAN-SIGSOFT Workshop on Program Analysis for Software Tools and Engineering* (2001), pp. 46–53
5. A. Corazza, S. Di Martino, V. Maggio, G. Scanniello, Investigating the use of lexical information for software system clustering, in *15th European Conference on Software Maintenance and Reengineering (CSMR)* (2011), pp. 35–44
6. M. Konôpka, M. Bieliková, Software developer activity as a source for identifying hidden source code dependencies, in *International Conference on Current Trends in Theory and Practice of Informatics* (2015), pp. 449–495
7. A. Parashar, J.K. Chhabra, Package restructuring based on software change history, Nat. Acad. Sci. Lett. **40**(1), pp. 21–27 (2017)
8. M. Konopka, P. Navrat, M. Bielikova, Discovering code dependencies by harnessing developer's activity. in *Proceedings of the 37th International Conference on Software Engineering*, vol. 2 (2015), pp. 801–802
9. M. Marin, A.V. Deursen, L. Moonen, Identifying crosscutting concerns using Fan-In analysis. ACM Trans. Softw. Eng. Methodol. (TOSEM) **17**(1), pp. 3:1–3:37 (2007)
10. A. Ghabi, A. Egyed, Observations on the connectedness between requirements-to-code traces and calling relationships for trace validation, in *26 International Conference on Automated Software Engineering (ASE)* (Lawrence, Kansas, 2011), pp. 416–419
11. H. Kuang, P. Mader, H. Hu, A. Ghabi, L.G. Huang, L.V. Jian, A. Egyed, Do data dependencies in source code complement call dependencies for understanding requirements traceability? in *International Conference on Software Maintenance (ICSM)* (2012), pp. 181–190
12. A. Parashar, J.K. Chhabra, Mining software change data stream to predict changeability of classes of object-oriented software system. Evolving Syst. **7**(2), 117–128 (2016)
13. T. Lutellier, D. Chollak, J. Garciaz, L. Tan, D. Rayside, N. Medvidovi´cy, R. Kroeger, Measuring the impact of code dependencies on software architecture recovery techniques. IEEE Trans. Softw. Eng. (2017) pp.1–22
14. J.D. Morgenthaler, M. Gridnev, R. Sauciuc, S. Bhansali, Searching for build debt: Experiences managing technical debt at google. MTD'12
15. P. Wang, J. Yang, L. Tan, R. Kroeger, J.D. Morgenthaler, Generating precise dependencies for large software, in *4th International Workshop on Managing Technical Debt (MTD)* (2013), pp. 47–50
16. D.L. Parnas, Designing software for ease of extension and contraction, in *Proceedings ICSE'78* (1978), pp. 264–277

17. B. Zhou, K. Yin, S. Zhang, H. Jiang, A.J. Kavs, A tree-based reliability model for composite web service with common-cause failures, in *International Conference on Grid and Pervasive Computing* (Springer, Berlin, Heidelberg, 2010), pp. 418–429
18. R.C. Martin, Design principles and design patterns, in *Object Mentor* (2000), pp. 1–34
19. H.A. Al-Mutawa, J. Dietrich, S. Marsland, C. McCartin, On the shape of circular dependencies in java programs, in *23rd Australian Software Engineering Conference (ASWEC)* (2014), pp. 48–57
20. K.L. Peng, C.Y. Huang, Reliability analysis of on-demand service-based software systems considering failure dependencies. IEEE Trans. Serv. Comput. **10**(3), 423–435 (2017)
21. J. Jász, Á. Beszédes, T. Gyimóthy, V. Rajlich, Static execute after/before as a replacement of traditional software dependencies, in *International Conference on Software Maintenance ICSM* (2008), pp. 137–146
22. A. Lienhard, O. Greevy, O. Nierstrasz, Tracking objects to detect feature dependencies, in *International Conference on Program Comprehension ICPC* (2007), pp. 59–68
23. I. Rodrigues, M. Ribeiro, F. Medeiros, P. Borba, B. Fonseca, R. Gheyi, Assessing fine-grained feature dependencies. Inf. Softw. Technol. **78**, 27–52 (2016)
24. B.B.P. Cafeo, E. Cirilo, A. Garcia, F. Dantas, J. Lee, Feature dependencies as change propagators: An exploratory study of software product lines, Inf. Softw. Technol. **69**, 37–49 (2016)
25. J. Boyland, J. Noble, W. Retert, Capabilities for sharing: a generalization of uniqueness and sharing, in *European Conference for Object-Oriented Programming (ECOOP)* (2001)
26. T. Lutellier, D. Chollak, J. Garcia, L. Tan, D. Rayside, N. Medvidovic, R. Kroeger, Comparing software architecture recovery techniques using accurate dependencies. in *37th IEEE International Conference on Software engineering ICSE*, vol. 2 (1990), pp. 69–78
27. P. Andritsos, V. Tzerpos, Information-theoretic software clustering, IEEE Trans. Softw. Eng. **31**(2), 2005
28. K. Kobayashi, M. Kamimura, K. Kato, K. Yano, A. Matsuo, Feature-gathering Dependency-based Software Clustering using Dedication and Modularity, Proc. ICSM, **0**, 462–471 (2012)
29. J. Garcia, I. Ivkovic, N. Medvidovic, A comparative analysis of software architecture recovery techniques, in *Proceeding ASE* (2013), pp. 486–496
30. Z.A. Alzamil, Redundant coupling detection using dynamic dependence analysis, in *International Conference on Software engineering advances (ICSEA)* (2007), pp. 43–43
31. Y. Wu, J. Sun, Y. Liu, J.S. Dong, Automatically partition software into least privilege components using dynamic data dependency analysis, in *IProceedings of the 28th IEEE/ACM International Conference on Automated Software Engineering* (2013), pp. 323–333
32. W.E. Wong, Y. Qi, Effective program debugging based on execution slices and inter-block data dependency, J. Syst. Softw. **79**(7), 891–903 (2006)
33. C. Bernardeschi, G. Lettieri, L. Martini, P. Masci, Using control dependencies for space-aware bytecode verification. Comput. J. **49**(2), 224–234 (2006)
34. G. Monsieur, M. Snoeck, W. Lemahieu, Managing data dependencies in service compositions. J. Syst. Softw. **85**(11), 2604–2628 (2012)
35. J. Ferrante, K.J. Ottenstein,J.D. Warren, The program dependence graph and its use in optimization, ACM Trans. Program. Lang. Syst. (TOPLAS) **9**(3), 319–349 (1987)
36. M. Petrenko, V. Rajlich, Concept location using program dependencies and information retrieval (DepIR). Inf. Softw. Technol. **4**, 651–6559 (2013)
37. H. Ural, H. Yenigün, On capturing effects of modifications as data dependencies. in *36th Annual Computer Software and Applications Conference (COMPSAC)* (2012), pp. 350–351
38. Z. Yu, V. Rajlich, Hidden dependencies in program comprehension and change propagation, in *9th International Workshop on Program Comprehension IWPC* (2001), pp. 293–299
39. N. Ajienka, A. Capiluppi, Understanding the interplay between the logical and structural coupling of software classes. J. Syst. Softw. **134**, 120–137 (2017)
40. R. Vanciu, V. Rajlich, Hidden dependencies in software systems. in *International Conference on Software maintenance ICSM* (2010), pp. 1–10
41. M. Petrenko, V. Rajlich, Variable granularity for improving precision of impact analysis, in *IEEE 17th Intl Conference on Program Comprehension* (2009), pp. 10–19

42. M. Lee, A.J. Offutt, R.T. Alexander, Algorithmic analysis of the impacts of changes to object-oriented software, in *International Conference on Technology of Object-Oriented Languages and Systems, TOOLS* (2000), pp. 61–70
43. J.A. Chhabra, Improving modular structure of software system using structural and lexical dependency. Inf. Softw. Technol. **82**, 96–120 (2017)
44. M. Cataldo, J.D. Herbsleb, K.M. Carley, Socio-technical congruence: a framework for assessing the impact of technical and work dependencies on software development productivity, in *Proceedings of the Second ACM-IEEE International Symposium on Empirical Software Engineering and Measurement* (2008), pp. 2–11
45. A. Garcia, P. Greenwood, G. Heineman, R. Walker, Y. Cai, H.Y. Yang, E. Baniassad, C.V. Lopes, C. Schwanninger, J. Zhao, Assessment of contemporary modularization techniques-ACoM'07: workshop report, ACM SIGSOFT Softw. Eng. Notes **32**(5), 31–37 (2007)
46. A. Podgurski, L.A. Clarke, A formal model of program dependences and its implications for software testing, debugging, and maintenance, IEEE Trans. Softw. Eng. **16**(9), 965–979 (1990)
47. T. Sharma, G. Suryanarayana, Augur: incorporating hidden dependencies and variable granularity in change impact analysis, in *International Working Conference on source code analysis and manipulation (SCAM)* (2016), pp. 73–78
48. J. Malburg, A. Finder, G. Fey, Debugging hardware designs using dynamic dependency graphs. Microprocess. Microsyst. **47**, 347–359 (2016)
49. V. Jagtap, U. Shrawankar, Dependency analysis for secured code level parallelization. Procedia Comput. Sci. **78**, 831–837 (2016)

Efficacy of Softwares for Generation of Dental Aligners

Divya Bajaj, Aditi Rawat, Deepkamal Kaur Gill, Mamta Juneja and Prashant Jindal

Abstract Orthodontic treatment is a critical issue. The application of dental braces is both painful as well as displeasing to the eye. Hence, there has been a transition from the usage of dental braces to clear dental aligners. There are multiple software available in the market which provides a platform for manipulation of teeth and development of dental aligners. The demand for orthodontic treatment has resulted in the rapid evolution of software for proper treatment plans. The motive of this study is to (a) compare two different CAD/CAM software: Maestro 3D Ortho Studio and 3-Shape Ortho Analyzer® and (b) examine their methodology and respective approaches in the generation of clear dental aligners. The manipulation done using Maestro 3D is a cumbersome task as each stage needs to be manually generated whereas in 3-Shape Ortho Analyzer, only the final stage is generated manually and intermediate stages are automatically generated with appropriate manipulations in each stage. In case of 3-Shape Ortho Analyzer®, the aligners are fabricated using thermoforming while in Maestro 3D Ortho Studio, they are directly printed using the 3D printer with no involvement of making a 3D model of maxilla and mandible.

Keywords Mandible · Maxilla · Rapid prototyping · FDM (Fused deposition modeling) · CAD (Computer-aided design) · CAM (Computer-aided manufacturing) · 3D printing · Thermoforming

D. Bajaj (✉) · A. Rawat · D. K. Gill · M. Juneja · P. Jindal
University Institute of Engineering and Technology, Panjab University, Chandigarh 160014, India
e-mail: divyabajaj1997@gmail.com

A. Rawat
e-mail: aditiR2407@gmail.com

D. K. Gill
e-mail: deepkamal1597@gmail.com

M. Juneja
e-mail: mamtajuneja@pu.ac.in

P. Jindal
e-mail: jindalp@pu.ac.in

C. R. Krishna et al. (eds.), *Proceedings of 2nd International Conference on Communication, Computing and Networking*, Lecture Notes in Networks and Systems 46, https://doi.org/10.1007/978-981-13-1217-5_78

1 Introduction

The study of dental models has become a prerequisite for the orthodontic procedure to be carried out by the dentist for analyzing the details of the anatomic structure of the patient's teeth before and after treatment. The dental study requires the plaster models that are mounted properly with dental bases in appropriate occluded alignment for scanning. However, the generation of plaster models is a difficult, tedious, expensive, and time-consuming task as it needs to be cooled first and then needs removal of bubbles. An alternate method is to use optical 3D scanners to obtain the dental models directly in digital data format. An orthodontic model helps us to visualize the initial stage, treatment progress, and the final result. Various CAD/CAM Software are available through which modifications on the scans of dental models are performed to design the replicas dentition. Human dentition is the collection of both maxilla and mandible. Primary dentition is the first set of teeth which are 20 in number and also called or Milk teeth. Secondary dentition is the second set of teeth which are 32 in number and also called permanent teeth.

First, the scanner is used to collect the digital data (.stl files) which gives the topographical characteristics of teeth. Those STL files are input to CAD/CAM Software to carry out the required manipulation on digital data. Cupernus et al. [1] presented a study for validation and reproducibility for dental models using intraoral scanners. They initially scanned the dentition using chair oral scanner and then corrected them by computer programs and convert it to digital models by using software: Orthoproof. Finally, these files were converted to stereolithographic (STL) using the 3D printer with 3 M ESPE algorithm. The measurements of dentition and stereolithographic models were performed using a digital caliper and that of the digital models were performed using the DigiModel software. According to the analysis, differences were clinically insignificant. Ender et al. [2] evaluated the accuracy of new reference scanner. From the reference model, impressions were made and exported as digital models. By superimposing the digital model within each group, precision was measured. It was found that reference scanner delivered high accuracy with a precision of 1.6 ± 0.6 μm and a trueness of 5.3 ± 1.1 μm. Conventional impressions showed significantly higher precision 12.5 ± 2.5 μm and trueness values 20.4 ± 2.2 μm. Wiranto et al. [3] assessed the validity, reliability, and reproducibility of digital models obtained from the Oral scanner and CT scans of impressions. The intraoral scanner was used for scanning the maxilla and mandible. 22 sets of study models were taken for digital and manual measurements. On the plaster model, tooth widths were measured with a digital caliper and DigiModel software for measuring tooth widths on Digi models. The differences of tooth width measurements of the digital models and the intraoral scans never exceeded 1.5 mm which is clinically insignificant.

Hassan et al. 2016 [4] gave an analysis of the accuracy of measurements made with stone models and rapid prototyping. The stone models were scanned using a structured light scanner and exported as binary stereolithographic files. 10 sets of models for each category of crowding (mild, moderate, and severe) were printed using Zprinter 450. Digital calipers were used for measuring Stone and RP models. Various

3D printers were used to print 3D models. FDM (Fused Deposition Modeling) begins with a software process which processes an STL [5, 6] file by mathematically slicing and orienting the model for the build process. The designed object is fabricated as a three-dimensional part based solely on the precise deposition of thin layers of the extrudate. [7] Lee et al. [8] evaluated the accuracy of fabrication of teeth using FDM and Polyjet rapid prototyping technology. These 3D surface models were printed using FDM and Polyjet techniques. Geomagic software was used to perform mean deviation, linear, and volumetric measurements. The original crown width (in mm) was 11.441 and the same through FDM and Polyjet was 11.325 and 11.505, respectively. The original tooth height (in mm) was 17.226 and the same through FDM and Polyjet was 17.083 and 17.219, respectively. Ibrahim et al. [9] presented an analysis study of capacities of Selective Laser Sintering (SLS), 3DP, and Polyjet model to reproduce the anatomy of the mandible and find their dimensional errors. They started with the acquisition of CT images from dry mandible and then performed manipulations using Invesalius software. These files were printed using SLS, 3DP, and Polyjet printers. Results gave a dimensional error of 1.79% for SLS model, 2.14% for Polyjet, and 3.14% for 3DP. Hence, it was analyzed that SLS prototype had the greatest dimensional accuracy among three. But cost analyses showed that 3DP technique had the lowest final cost.

The accuracy of the 3D models has a direct relationship with the accuracy of the clear dental aligners in both the cases: 1. when aligners are made using thermoforming 2. when the aligners are directly printed using 3D printers. After printing of 3D models using FDM, Polyjet, and SLS—the rapid prototyping technique, aligners are directly printed for Maestro 3D Ortho Studio whereas aligners from these dental molds are created using the process called thermoforming. Thermoforming is a generic term of the art and science of forming commercial products by heating plastic sheet to a pliable state, pressing the sheet against a cool mold, holding the formed sheet against the mold until cool, and trimming the formed part from the web or skeleton surrounding it. Thermoforming is characterized as capable of forming many polymers, forming large surface area-to-wall thickness parts of forming parts against single-surface molds that can be made of any materials, and using moderate forming temperatures and pressures [10]. Clear transparent aligner is constructed by first forming an impression of a patient's upper or lower dentition and constructing a cast from the impression. The retainer is vacuum thermoformed over the cast using a sheet, plate, or disc of thermoformable plastic and a vacuum or pressure thermoforming machine [11].

2 Methodology Used

In this section, a detailed procedure for the designing of aligner from Maestro 3D Ortho Studio and 3Shape Ortho Analyzer has been presented which uses a series of steps, starting from the use of plaster model or direct scans using intraoral scanner

Table 1 Steps used for generation of aligners from Maestro 3D ortho studio

Step 1: Make plaster models 	Step 2: Take scans of plaster model using scanners 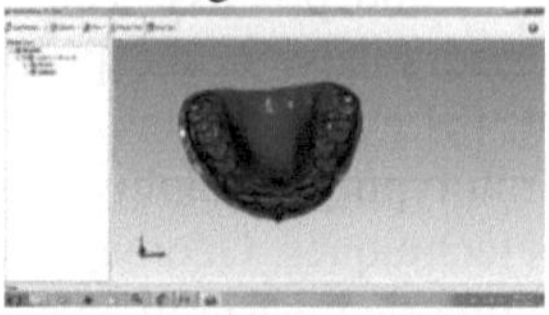	Step 3: Load scans as input (.stl) file 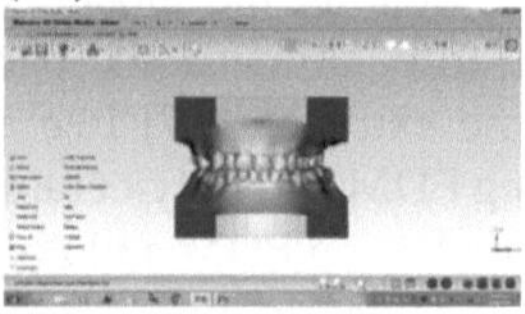
Step 4: Measurement of each tooth width 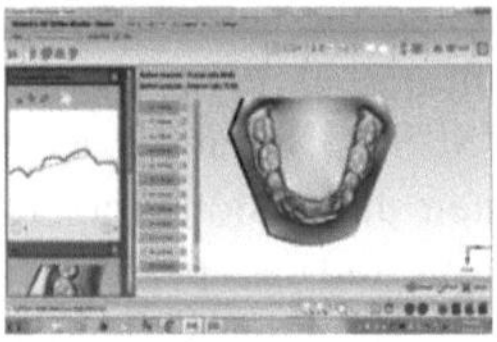	Step 5: Perform segmentation (define tooth boundaries) 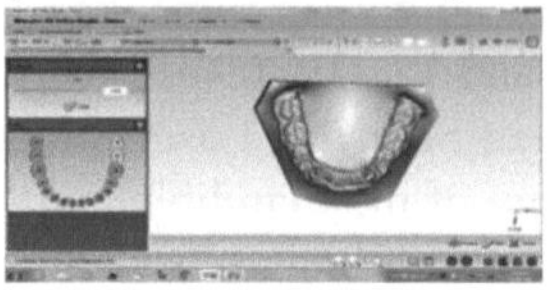	Step 6: Perform Tooth manipulation 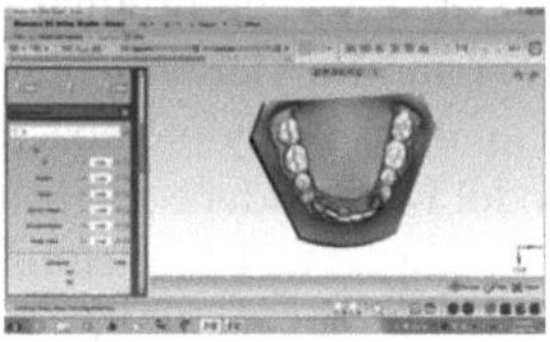
Step 7: Generate intermediate stages of models 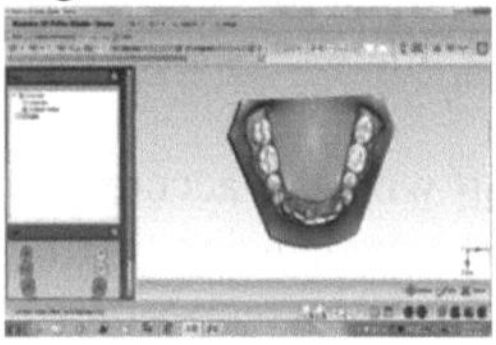	Step 8: Generate the aligner by defining boundary 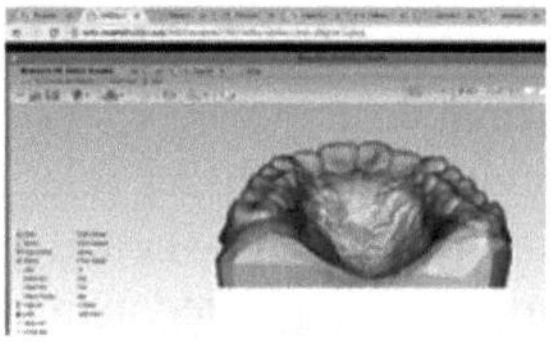	Step 9: Export The aligner to obtain aligner in .stl format 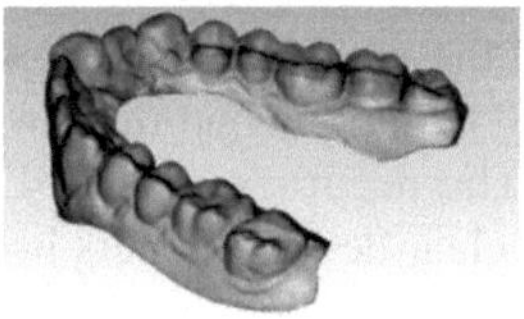

and ending with 3D printing of Clear Dental Aligners or the dental models which further undergo thermoforming to generate clear dental aligners.

First of all, there is the detailed methodology used by Maestro 3D Ortho Studio to generate clear dental aligners (Table 1):

Step 1: **Make plaster models**: Plaster models also called virtual dentition are generated using alginate powder and dental impression trays by taking negative impressions of hard and soft tissues of the oral-maxillofacial region [12, 13]. These impressions are then cooled down to obtain hard, stable dentition cast models that can be further used to be scanned.

Step 2: **Take scans of plaster model using scanner**: The plaster models are scanned using 3D scanners to obtain data (.stl) which is input to software for processing. An alternate method to decrease the overhead of making plaster models is the scanning of dentition using intraoral scanner.

Step 3: **Load scans in software**: The details of the patient including the name, date, age, sex, patient id, etc., are stored to keep the record of data. Then, the scans of mandible (lower) and maxilla (Upper) are imported in the software.

Step 4: **Measurement of each tooth width**: The width of each tooth is calculated using linear measurements and an ideal arch is also made to compare the sum of individual tooth length with the length of the ideal arch, according to which the planning of treatment is decided. The planning rectification of crowding can be done by Interproximal Reduction. [14]

Step 5: **Perform tooth segmentation**: In this step, we define the boundary of each tooth to make a distinction between the hard tissues (tooth) and soft tissues (gum). Dental contouring is also done in which odontoplasty, enameloplasty, tooth reshaping, stripping or sculpting, or the removal of small amounts of surface enamel from the tooth is carried out in order to get rid of minor imperfections.

Step 6: **Perform Tooth manipulation**: This step comes under the analysis and measures functionality of Maestro 3D in which the defective tooth is selected to do following manipulations to achieve alignment: (1) **Rotation**: A rotation is a circular movement of an object about a reference point. It can be negative (counterclockwise) as well as positive (clockwise). (2) **Buccal-lingual**: It is a forward and backward movement (in mm) of the tooth from its original position. It can be negative (forward) as well as positive (backward). (3) **Mesial-distal**: It is sideways movement (in mm) of the tooth to generate space for adjacent tooth or to remove the overlapping of two teeth. It can be negative (left) as well as positive (right). (4) **Torque**: It describes the angular movement of the tooth about the root. It can be inward as well as outward. (5) **Tip**: A tooth movement in which the angulations of the long axis of the tooth is altered. (6) **Extrusion/Intrusion**: It is an inward and outward movement of the tooth. It can be negative (inward) as well as positive (outward).

Step 7: **Generate the intermediate stages of models**: Before generating the intermediate stages, we need to perform the analysis concerning the degree of alignment required by the patient to have an ideal arch. Since the pressure we can apply to gums and tooth has a limit of about 0.0036 MPa [15]. Also, the biological limits of linear movements can be at the highest 0.5 mm and the angular rotation can be at the highest 3^0 as told by doctors. So at first, we need to do manipulations in order to achieve the final state from initial state, after that keeping in mind the biological constraints, we need to manually design each intermediate stage. One of the real-life cases, with intermediate stages' data, is as follows (Tables 2 and 3):

Similarly for making each intermediate stage till final stage, alignment values are manually given to software (Table 4).

Step 8: **Generate the aligner by defining boundary**: On each stage of maxilla and mandible, we define the outer boundary for aligner and under virtual setup, we generate the aligner and form a close covering over the model which is required to align the tooth.

Step 9: **Export The Aligner to obtain aligner in the .stl format**: The clear dental aligner files generated are then exported so that they can be printed using 3D printers and can be given to the patients to be applied over the tooth.

Below is the detailed methodology used by 3-Shape Ortho Analyzer to generate clear dental aligners: (Table 5).

Step 1: **Make Plaster Models**: Plaster models are generated using alginate powder and dental impression trays by taking negative impressions of hard and soft tissues of the oral-maxillofacial region [12, 13]. These impressions are then allowed to cool down to make stable dentition models that can be scanned properly.

Table 2 Manipulation from stage 0–stage 1

Tooth no.	T	R	TO	BL	E/I	M	Tooth no.	T	R	TO	BL	E/I	MD
18	0	0	0	0	0	0	48	0	0	0	0	0	0
17	0	0	0	0	0	0	47	0	0	0	0	0	0
16	0	0	0	0.49	0	0.5	46	0	0	0	0	0	0
15	0	0	0	0.5	0	0	45	0	1	0	0.046	0	0.18
14	0	0	0	−0.48	−0.41	0.4	44	0	0	0	−0.1	−0.5	0.48
13	0	-1	0	−0.28	0.42	0.11	43	0	0	0	0.5	−0.5	0.4
12	0	0	0	−0.49	−0.46	0.44	42	0	0	0	0	−0.5	0.41
11	0	0	0	−0.5	0	0.49	41	0	0	0	0	−0.47	0.4
21	0	0	0	−0.40	0	−0.2	31	0	0	0	0	−0.5	0
22	0	0	0	−0.5	0	−0.5	32	0	1	1	0.47	−0.49	−0.41
23	0	0	0	0	0	−0.3	33	0	0	0	−0.5	0.1	−0.5
24	0	0	0	0	0	−0.5	34	0	0	0	0	0	−0.4
25	0	0	0	0	0	−0.4	35	0	0	0	0	0.1	−0.1
26	0	0	0	−0.5	0	0	36	0	0	0	0	0.1	0
27	0	0	0	0	0	0	37	0	0	0	0	0	0
28	0	0	0	0	0	0	38	0	0	0	0	0	0

T Tip, *R* Rotation, *TO* Torque, *BL* Buccal-Lingual, *E* Extrusion, *I* Intrusion, *MD* Mesial-Distal

Table 3 Manipulations from stage 1–stage 2

Tooth no.	T	R	TO	BL	E/I	MD	Tooth no.	T	R	TO	BL	E/I	MD
18	0	0	0	−0.2	0	0	48	0	0	0	0	0	0
17	0	0	0	−0.5	0	0	47	0	0	0	0	0	0
16	0	0	0	1	0	0.5	46	0	0	0	0	0.2	0
15	0	0	0	0.49	0	0	45	0	2	2	−0.6	0.28	0.22
14	0	0	0	−1	−0.4	0.4	44	0	0	0	−0.1	−1	0.5
13	0	2	0	−0.29	0.41	−0.18	43	0	0	0	0.5	−1	0.4
12	0	0	0	−0.95	−0.47	0.94	42	0	0	0	0	−0.7	0.4
11	0	0	0	−0.9	0	0.5	41	0	0	0	0	−0.5	0.4
21	0	0	0	−1	0	−0.2	31	0	2	0	0	−0.5	0
22	0	0	0	−0.5	0	−0.5	32	0	2	2	0.9	−0.97	−0.21
23	0	0	0	0	0	−0.3	33	0	0	0	−1	0.1	−0.5
24	0	0	0	0	0	−0.5	34	0	0	0	0	0	−0.4
25	0	0	0	0	0	−0.4	35	0	0	0	0	0.1	−0.1
26	0	0	0	−0.5	0	0	36	0	0	0	0	0.2	0
27	0	0	0	0	0	0	37	0	0	0	0	0	0
28	0	0	0	0	0	0	38	0	0	0	0	0	0

Table 4 Initial stage and final stage after manipulations

Stage 0 (Maxilla)	Stage 0 (Mandible)	Final stage (Maxilla)	Final stage (Mandible)
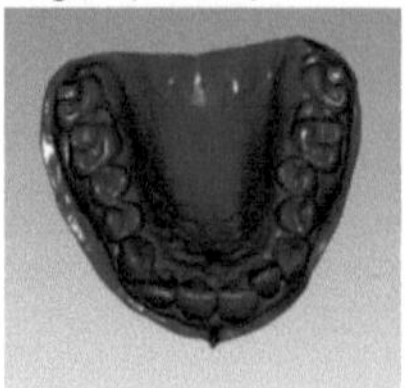	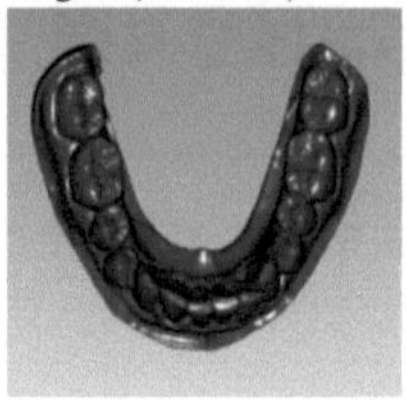	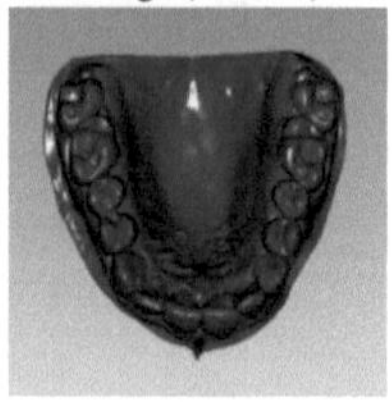	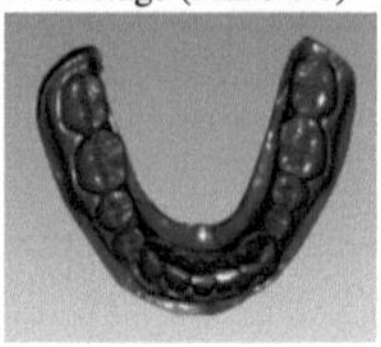

Table 5 Steps used for generation of aligners from 3-Shape ortho analyzer

Step 1: Make plaster models 	Step 2: Take scans of plaster model using scanners 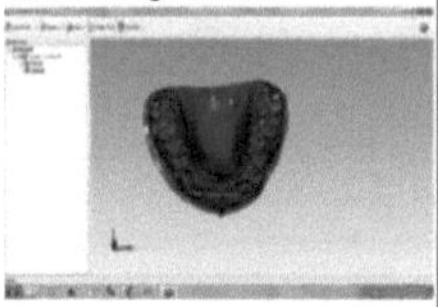	Step 3: Load scans and feed patient's details 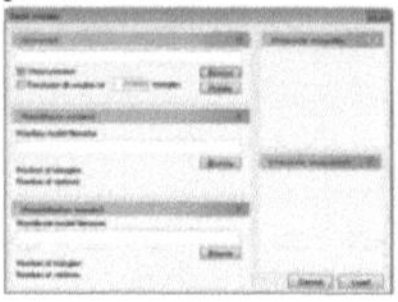	Step 4: Set occlusion & orientation plane of dentition 
Step 5: Measurement of tooth width 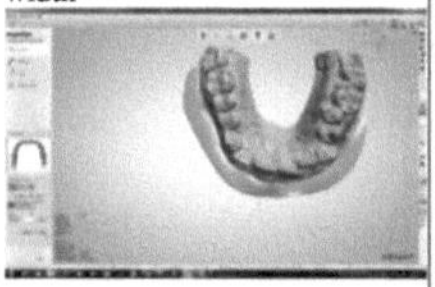	Step 6: Perform contouring (define tooth boundaries) 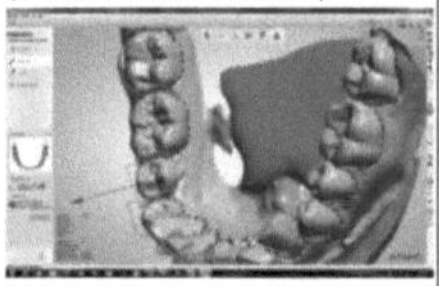	Step 7: Ideal dental arch analysis 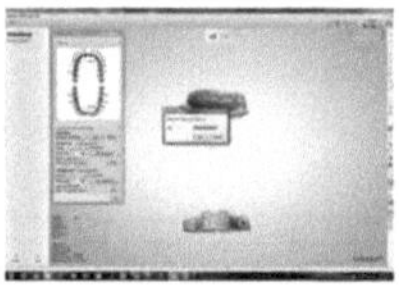	Step 8: Overbite/overjet problem analysis 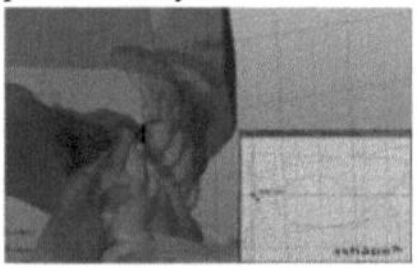
Step 9: Perform tooth manipulation 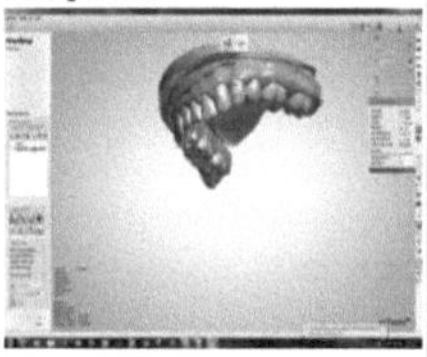	Step 10: Deciding & defining the number of intermediate stages 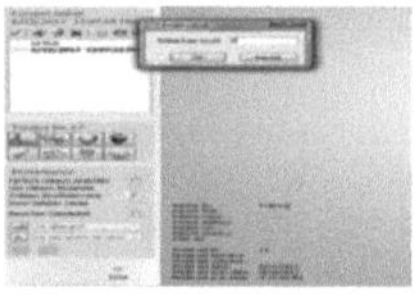	Step 11: Export digital models and print using 3D printers 	Step 12: Perform thermoforming on printed models 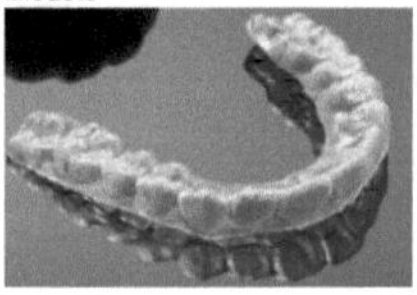

Step 2: **Take scans of Plaster Model using Scanner**: These models are scanned using 3D scanners to obtain digital data (.stl) which is given as an input to software for processing. An alternate to decrease the overhead of making plaster models is the recent technique of scanning the dentition with the intraoral scanner to provide the digital data (.stl) directly.

Step 3: **Load scans and feed patient's details**: The digital scans of the patient's tooth are loaded along with the details that will help in maintaining a database for future reference including Model Set ID, Comment, Operator, and Model Set Date.

Step 4: **Set occlusion and orientation plane of maxilla and mandible**: Occlusion of upper and lower jaw is set and the distance between the two virtual casts is defined such that they are moved forward or backward according to the orientation plane. It can be negative for moving up and positive for moving down.

Step 5: **Measurement of tooth width**: Individual tooth width is measured by the doctor using linear measurements tool to analyze the dentition so as to plan the treatment for alignment of teeth.

Step 6: **Perform contouring (Define tooth boundaries)**: The outer boundary of each tooth is defined so as to make a distinction between the gums and the tooth.

Step 7: **Ideal dental Arch analysis**: An ideal arch for both maxilla and mandible is made and the length of ideal arch is compared with the sum of individual tooth length and then the decision for alignment treatment takes place such as: (1) If the ideal arch length is greater than the sum of individual tooth, then there is spacing between the teeth which needs to be rectified. (2) If the ideal arch length is lesser than the sum of individual tooth, then tooth needs to be either cut or displaced for rectification. (3) If the ideal arch length is equal to the sum of individual tooth, then there is no need of rectification.

Step 8: **Overbite/Overjet problem analysis**: The maxilla and mandible are placed over each other as like human tooth are in the position of biting and analysis is made according to the point at which top teeth protrudes over the bottom ones.

Step 9: **Perform Tooth manipulation**: According to the required tooth alignment, the tooth manipulations are done which can be left/right, forward/backward or inclination/rotation/extrusion/intrusion/angulation movements. The movements are done keeping in mind the biological limits of applying pressure on the tooth, and linear movements at the highest 0.5 mm and rotation at the highest 3°.

Step 10: **Deciding and defining the number of subsets (intermediate stages)**: The final stage is created by performing required manipulations and then deciding the number of stages required for complete treatment according to biological factors. The maximum linear movement and angular movement are determined by comparing individual tooth movement and hence the deciding factor is calculated.

DECIDING FACTOR = max ((linear movements/0.5, (angular movements/3)) (Table 6).

Step 11: **Export digital models and print them through 3D printers**: All the stages created are saved and exported in .stl format which is then printed using various printing techniques like Polyjet, SLA and FDM and also using n various materials like Acrylonitrile, Butadiene, and Styrene (ABS), Polylactic Acid (PLA), etc. [16]

Table 6 Manipulation from initial stage to final stage

Tooth no.	I	R	A	L/R	E/I	F/B	Tooth no.	I	R	A	L/R	E/I	F/B
18	M	M	M	M	M	M	48	M	M	M	M	M	M
17	M	M	M	M	M	M	47	M	M	M	M	M	M
16	0	0	0	0	0	0	46	0	0	0	0	0	0
15	0	0	0	0	0	0	45	0	1	0	0.046	0	0.9
14	0	0	0	0	0	0	44	0	6.2	0	−0.1	−0.5	0.3
13	0	0	0	0	0	0	43	0	0	0	0.5	−0.5	0.4
12	0	−10.6	0	−0.49	−0.46	0.4	42	0	−5.2	0	0	−0.5	0.4
11	14.3	0	0	−0.5	0	0.4	41	0	10	0	1.2	−0.47	0.5
21	0	2.12	0	−0.40	0	−0.2	31	0	3.4	0	0	−0.5	0
22	0	0	0	−0.5	0	−0.5	32	0	7.2	1	0.47	−0.49	0
23	0	−5.36	0	0	0	−0.3	33	0	0	0	−0.5	0.1	−0.5
24	0	0	0	0	0	0	34	0	0	0	0	0	−0.4
25	0	0	0	0	0	−0.4	35	0	0	0	0	0.1	−0.1
26	0	0	0	0	0	0	36	0	0	0	0	0.1	0
27	0	0	0	0	0	0	37	0	0	0	0	0	0
28	M	M	M	M	M	M	38	0	0	0	0	0	0

Number of stages = 1(original stage) + deciding factor
I Inclination, *R* Rotation, *A* Angulation, *L/R* Left/Right, *E/I* Extrusion/Intrusion, *F/B* Forward/Backward, *M* Missing

Step 12: **Perform thermoforming on 3D printed models**: The 3D models undergo thermoforming to fabricate the aligners made of Duran sheet which is a biomedically approved material. This sheet can be used in various thicknesses like 0.5 mm, 0.625 mm, 0.75 mm, 1 mm, or 1.5 mm [17]. The selection of sheet thickness depends upon biological limits and treatment direction. Generally, it is estimated that an aligner needs to be applied for 20 h a day [18] and each stage requires 2–3 weeks [19] of application.

3 Comparison of Maestro and 3Shape Ortho Software

Product name	Maestro 3D Ortho studio	3-Shape ortho analyzer
Price	455570.76 Indian Rupee)	590,000 Indian Rupees
Input/output file format	STL\PLY\ORTHO	STL
Perform measurements of teeth	Yes	Yes
Output models	Aligner	Dentition (Mandible & Maxilla)
Automated generation of intermediate stages	No	Yes
Overhead of printing 3D models of Maxilla and Mandible	No	Yes
Comparison of two models simultaneously with reference models	No	Yes
Fixed constraints definition	No	Yes
Overbite/overjet problem analysis	No	Yes
Feature of articulator	No	Yes

4 Conclusion

This paper compared the methodology of developing clear dental aligners from both the CAD/CAM software: Maestro 3D Ortho Studio and 3-Shape Ortho Analyzer that follow different techniques and concludes that the aligners generated from thermoforming are more effective and have a greater practical use. This can help the practitioner to decide on which software to work on depending upon the requirements for making real-time clear transparent aligners.

Acknowledgements The authors are grateful to MHRD for funding this project of Design Innovation Center under the subtheme "Medical Devices & Restorative Technologies", "Director UIET, Panjab University, Chandigarh", "Director Dr. Harvansh Singh Judge Institute of Dental Sciences & Hospital" for their consistent support in completing this work.

References

1. A.M.R. Cuperus et al., Dental models made with an intraoral scanner: a validation study. Am. J. Orthod. Dentofac. Orthop. **142**(3), 308–313 (2012)
2. A. Ender, A. Mehl, Accuracy of complete-arch dental impressions: a new method of measuring trueness and precision. J. Prosthet. Dent. **109**(2), 121–128 (2013)
3. M.G. Wiranto et al., Validity, reliability, and reproducibility of linear measurements on digital models obtained from intraoral and cone-beam computed tomography scans of alginate impressions. Am. J. Orthod. Dentofac. Orthop. **143**(1), 140–147 (2013)
4. W.N. Hassan, Y.Y. Wan, N.A. Mardi, Comparison of reconstructed rapid prototyping models produced by 3-dimensional printing and conventional stone models with different degrees of crowding. Am. J. Orthod. Dentofac. Orthop. **151**(1), 209–218 (2017)
5. StereoLithography Interface Specification, 3D Systems, Inc. (July 1988)
6. StereoLithography Interface Specification, 3D Systems, Inc. (Oct 1989)
7. I. Zein et al., Fused deposition modeling of novel scaffold architectures for tissue engineering applications. Biomaterials **23**(4), 1169–1185 (2002)
8. K.Y. Lee et al., Accuracy of three-dimensional printing for manufacturing replica teeth. Korean J. Orthod. **45**(5), 217–225 (2015)
9. D. Ibrahim et al., Dimensional error of selective laser sintering, three-dimensional printing and PolyJet™ models in the reproduction of mandibular anatomy. J. Craniomaxillofac. Surg. **37**(3), 167–173 (2009)
10. J.L. Throne, *Thermoforming* (Wiley, Hoboken, NJ, 2003)
11. D.A. Schwartz, J.J. Sheridan, Thermoformed plastic dental retainer and method of construction. U.S. Patent No. 5,692,894, 2 Dec 1997
12. Dental Impression. Available from: https://www.en.wikipedia.org/wiki/Dentalimpression. Last accessed on 27 Dec 2016
13. G.E. Halverson, G.A. Nelson, Inventors, Excel Dental Studios, Inc., Assignee. Dental impression supply kit. U.S. Patent 4,763,791, 16 Aug 1988
14. E. Bhambri, J.P.S. Kalra, S. Ahuja, G. Bhambri, Evaluation of enamel surfaces following interproximal reduction and polishing with different methods: a scanning electron microscope study. Indian J. Dent. Sci. **9**(3), 153–159 (2017)
15. T. Drake, S.P. McGorray, C. Dolce, M. Nair, T.T. Wheeler, Orthodontic tooth movement with clear aligners. *ISRN Dentistry,* ID 657973
16. G. Budzik et al., Analysis of the accuracy of reconstructed two teeth models manufactured using the 3DP and FDM technologies. Strojniški vestnik-J. Mech. Eng. **62**(1), 11–20 (2016)
17. J.S. Kwon, Y.K. Lee, B.S. Lim, Y.K. Lim, Force delivery properties of thermoplastic orthodontic materials. Am. J. Orthod. Dentofac. Orthop. **133**(2), 228–234 (2008)
18. O. Malik, A. McMullin, D. Waring, Invisible orthodontics part 1: invisalign. Dental Update. **40**, 203–204, 207–210, 213–215. PMID 23767109 (April 2013)
19. J. Fogel, R. Janani, Intentions and behaviors to obtain Invisalign. J. Med. Mark. **10**(2), 135–145 (2010). ISSN 1745–7904

Craniofacial Model Generation Using CAD/CAM Software

Aditi Rawat, Deepkamal Kaur Gill, Divya Bajaj, Mamta Juneja, Anand Gupta and Prashant Jindal

Abstract Patients with skull and/or facial injuries resulting in bone damage require artificial models in order to replace the originally damaged bone portion. Traditionally, surgeons use the inaccurate method of estimating the size and shape of the bone to be replaced. The task of manual reconstruction is highly subjective in nature and may involve human errors and hence the same task when performed with the use of software provides highly accurate results. Eventually, there has been a transition from manual methods to CAD/CAM software and 3D printing technique to fabricate these models. These 3D printed models would provide assistance to surgeons performing aforementioned surgeries. The motive of this study is to: (a) propose a suitable methodology to fabricate the damaged portion of the craniofacial bone and (b) guide complex craniofacial surgery and implantology procedures.

Keywords CAD (Computer-aided design) · CAM (Computer-aided manufacturing) · Craniofacial

---

A. Rawat (✉) · D. K. Gill · D. Bajaj · M. Juneja · A. Gupta · P. Jindal
University Institute of Engineering and Technology, Panjab University, Chandigarh 160014, India
e-mail: aditiR2407@gmail.com

D. K. Gill
e-mail: deepkamal1597@gmail.com

D. Bajaj
e-mail: divyabajaj1997@gmail.com

M. Juneja
e-mail: mamtajuneja@pu.ac.in

A. Gupta
e-mail: dranandkgmc2@gmail.com

P. Jindal
e-mail: jindalp@pu.ac.in

C. R. Krishna et al. (eds.), *Proceedings of 2nd International Conference on Communication, Computing and Networking*, Lecture Notes in Networks and Systems 46, https://doi.org/10.1007/978-981-13-1217-5_79

1 Introduction

The need for craniofacial reconstruction arises when confronted by cases where patients have damaged their craniofacial bone beyond repair. There are numerous reasons contributing to this such as falls, sports-related accidents, interpersonal violence, etc. [1–3].

Several 3D manual methods as well as computerized/automated facial reconstruction methods have their origins in the nineteenth century. At that time, this technique was used by German anatomists to recreate the faces of important personalities [4–7]. In 1912, Wilder tried to reconstruct faces of Americans Indians [8]. In the late nineteenth century, reconstruction as well as recognition studies were carried out on cadavers [9, 10]. Thereafter anthropologists, anatomists, and odontologists began to study the relation between underlying bony structure of the skull and the soft tissues of the face. Using the needle depth probing method, the measurement of the facial soft tissues by Rhine and Campbell on Americans [11–13] as well as by Suzuki on Japanese [14] was implemented on cadavers. The results of these applications, however, show an innate error as the shrinkage of the tissues and extent of cadavers' dehydration cannot be restrained [11].

Other subsequent methods used in order to evaluate the thickness of facial tissue are lateral cranial radiographs [15]. This method of facial reconstruction involves following four steps: (1) a cephalometric study to evaluate individual skull and facial proportions; (2) the average thickness of soft tissues and angles plotted against the predesignated points in order to fix the average dimensions of the facial parts; (3) the linking of points in harmony with the known anthropomorphic data (race, gender, age); and (4) lastly, the profile is refined by adding hair patterns, age lines, skin tone, and other characteristics that can be found on the basis of an anatomic study of the skull [15]. Some 2D computer graphics approaches have been employed with meager success in the evaluation of frontal as well as lateral profiles collected by this method [9, 17, 18]. Another method employed was ultrasonic probing [16]. Also, methods such as computerized tomography (CT) scanning, magnetic resonance imaging (MRI) have been employed. Gerasimov, in 1924, developed the "Russian method" which held developing musculature of the skull and neck of fundamental importance [19, 20]. Thereafter, he developed a continual system of analysis and data collection of thicknesses of the heads' soft tissues. Eventually, enough concrete data had been collected which led to Gerasimov reconstructing more than 200 heads of primitive ancestors. He further worked on the facial anatomies of the primal known fossil men: Pithecanthropus as well as Neanderthaloid [27, 28].

In 1946, Krogmann, an American anatomist generated 5 basic principles to normalize the approach of recreating the soft tissues of the faces, which described the relationship of the shape of the nose's tip, of the eyeball to the socket, the position of the ear, the span of the mouth, and the length of the ear [23–26]. Caldwell, in 1981 worked on 3D reconstructions evaluating the equation between the details of the human face with the skull [13]. Richard Neave applied reconstruction technologies in an attempt to identify skeletal remains of possible murder victims and unrecognizable victims of natural disasters [21–23]. The beginning of the twentieth century saw the utilization of these reconstructions in museums such as the reconstruction of Cro-Magnon skulls [19]. The focus of traditional reconstruction for maxillofacial surgery has largely been on soft tissue reconstruction [10, 26] and nonvascularized bone deposits with limited successes in case of large defects. Also, such methods require multiple rounds of soft tissue rearrangements or bone grafting to obtain a satisfactory result [26] followed by 3D computer-aided reconstruction process. As of 1984, 3D reconstruction imaging techniques improved perception of critical facial defects. Applied to CT studies of convoluted abnormalities of craniofacial anatomy, this approach has aided surgical planning, improved objective evaluation post application of approach.

In 1984, leading aircraft design techniques aided by computer were adapted and applied to procedural planning of craniofacial surgery as well as study of surface contours acquired using CT scans [30]. A survey carried out in 1986 investigated three-dimensional reconstructions but was limited to three-dimensional reconstructions based upon an object's multiple parallel sections [31]. All these methods have a disadvantage of not allowing certain reconstructions [32].

In 1988, focus drifted toward the 3D perception of complex structures. Using this approach [33, 34], depth perspective was provided by Lozanoff by digitizing tissue outlines, and then superimposing contour lines. Thereafter in 1995, an algorithm was proposed by Mole for complex surfaces in order to aid oral surgeries [27]. They developed a new interpolation method called DSI (discrete smooth interpolation) for easy modeling of complex 3D surfaces. In 1999, geometric modeling was used to generate prototype that will be used as composite prosthesis for covering cranial defects [28]. Multiple imaging techniques were used to obtain 3D images of craniofacial structure [29].

In 2005, an approach taking input from CT(computerized tomography) imaging along with surface laser scanning was used in order to provide three-dimensional visualization of facial soft tissues [35]. As of then, facial soft tissue depth data was obtained using ultrasound, CT scans, and radiographies [36]. Until 2007, with the objective of checking the matching of facial landmarks and contours; methods focusing on representing a 3D model of the person in a 2D plane using either a laser scanner or two stereoscopic cameras were developed [37]. In 2011, Gordon et al. investigated accuracy and dependability of skull photo superimposition but the approach was found to be of limited use [38]. Later in 2013, Cummaudo et al. described the accurate definition of anatomical points and the significance of placing facial landmarks [39].

2 Methodology

Patient's data files are input in DICOM file format. These files are then stacked so as to form a 3D graphics object which would be manipulated upon. Then following a series of steps involving volume rendering, volume cropping, and thresholding, we obtain the resulting model which would be 3D printed.

Step 1: **Loading patient's data files**

It involves loading patient's data files in DICOM(Digital Imaging and Communications in Medicine) file format. A batch of DICOM files is imported. Each slice is a CT scan wherein the scans range from the top of the skull to the bottom of the skull (Fig. 1).

Step 2: **Superposition of DICOM files—Volume rendering**

Multiple DICOM files imported are then superimposed resulting in the formation of a 3D graphics object (Fig. 2).

Step 3: **Choice of display preset**

Among the multiple display presets whichever facilitated the visualization of the graphics object was selected (Fig. 3).

Step 4: **Cropping of required volume**

Since we may require only a portion of the craniofacial bone, we need to crop the representation such that only that fraction of the original object is selected which we wish to generate (Fig. 4).

Step 5: **Threshold Effect**

After we have sized down to the portion we require, we perform the threshold effect. In threshold effect, elements within the range in the source volume will acquire the selected label value. The labels are written into the label map. In our case, the label map is bone (Fig. 5).

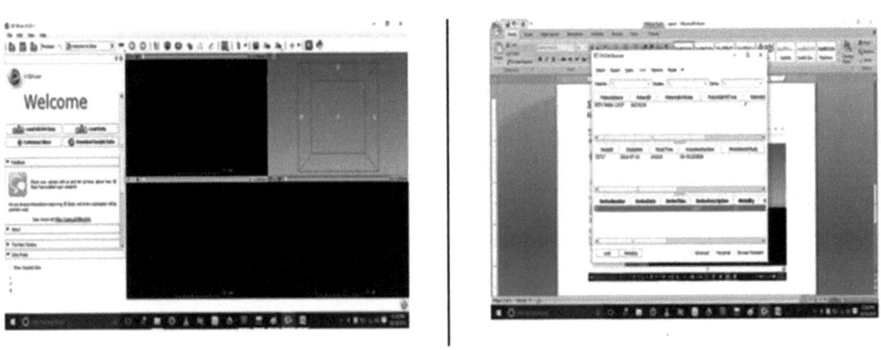

Fig. 1 Loading DICOM files

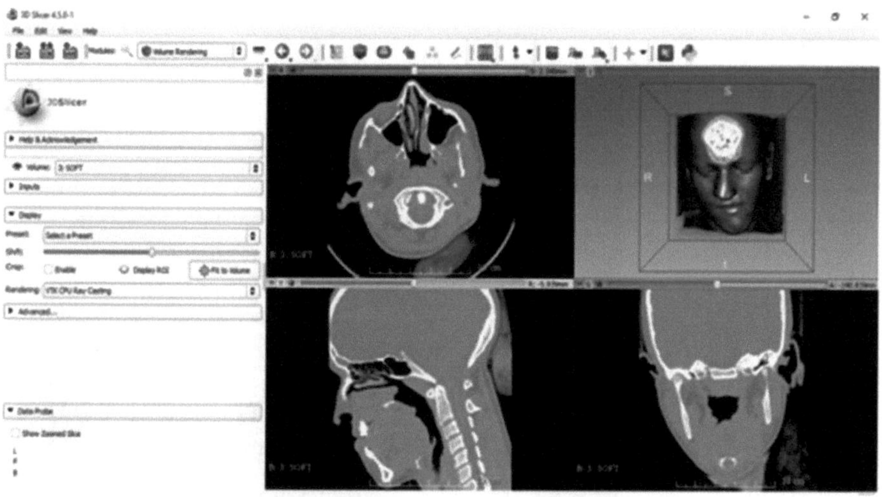

Fig. 2 Volume rendering

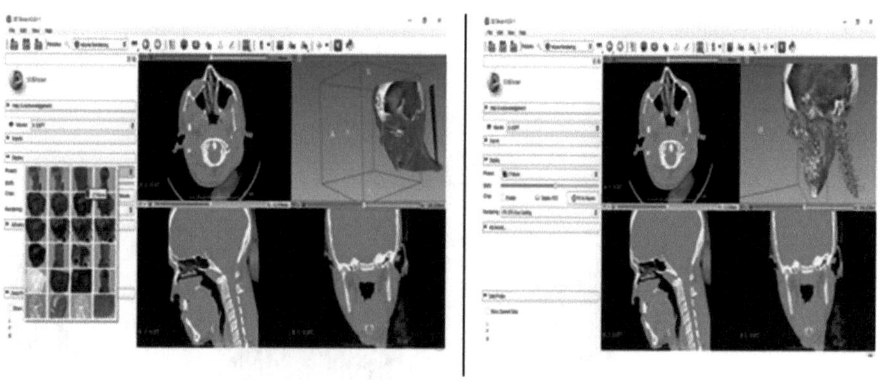

Fig. 3 Choosing a preset

Step 6: Exporting the models

After this, the resulting file is exported in a format suitable for 3D printing, i.e., STL. Then, the software generated STL file can be 3D printed which would aid the surgeon in case of such implantology procedures.

The resulting files can be viewed in any STL file viewing software (Fig. 6).

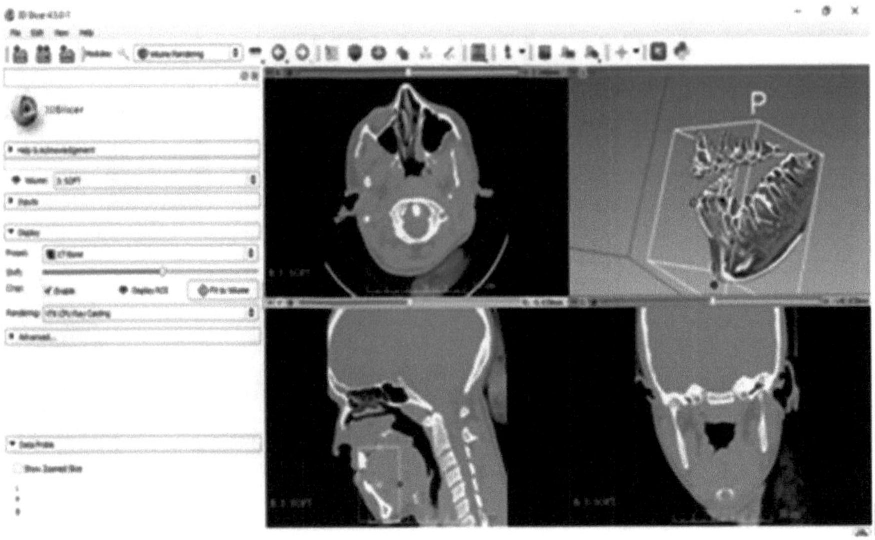

Fig. 4 Volume cropping

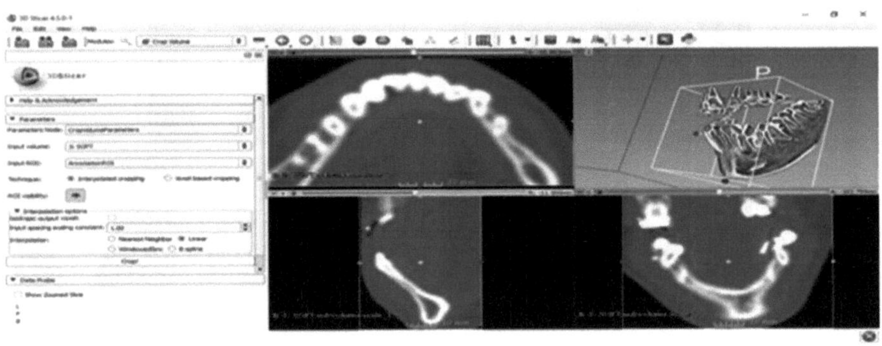

Fig. 5 Thresholding the selected portion

3 Conclusion

The manual development of craniofacial models is a highly inaccurate practice and also a cumbersome task for surgeons. This has rightfully led to the advent of computer-aided reconstruction methods so as to provide an accurate aid to the surgeons. The models generated by using CAD/CAM software and 3D printing can act as a template for the surgeons performing craniofacial implantology procedures. It would provide more accurate results as compared to the manual methods of reconstruction.

Fig. 6 Exporting the models

Acknowledgements The authors are grateful to MHRD for funding this project of Design Innovation Center under the subtheme "Medical Devices & Restorative Technologies" and "Director UIET, Panjab University, Chandigarh for their consistent support in completing this work.

References

1. K. Hussain, D.B. Wijetunge, S. Grubnic, I.T. Jackson, A comprehensive analysis of craniofacial trauma. J. Trauma Acute Care Surg. **36**(1), 34–47 (1994)
2. N.M. Wymann, A. Hölzle, Z. Zachariou, T. Iizuka, Pediatric craniofacial trauma. J. Oral Maxillofac. Surg. **66**(1), 58–64 (2008)
3. P.B. Rajendra, T.P. Mathew, A. Agrawal, G. Sabharawal, Characteristics of associated craniofacial trauma in patients with head injuries: an experience with 100 cases. J. Emergencies, Trauma Shock **2**(2), 89 (2009)
4. M. Vanezis, P. Vanezis, Cranio-facial reconstruction in forensic identification—historical development and a review of current practice. Med. Sci. Law **40**(3), 197–205 (2000)
5. W.A. Aulsebrook, M.Y. İşcan, J.H. Slabbert, P. Becker, Superimposition and reconstruction in forensic facial identification: a survey. Forensic Sci. Int. **75**(2–3), 101–120 (1995)
6. B.A. Fedosyutkin, J.V. Nainys, The relationship of skull morphology to facial features. Forensic Anal. Skull: Craniofac. Anal. Reconst. Ident. **8**, 119–213 (1993)
7. P.C. Caldwell, The relationship of the details of the human-face to the skull and its application in forensic anthropology. Am. J. Phys. Anthropol. **54**(2), 207–217 (1981)
8. H.H. Wilder, The phsysiognomy of the Indians of Southern New England. Am. Anthropol. **14**(3), 415–436 (1912)
9. P. Vanezis, R.W. Blowes, A.D. Linney, A.C. Tan, R. Richards, R. Neave, Application of 3-D computer graphics for facial reconstruction and comparison with sculpting techniques. Forensic Sci. Int. **42**(1–2), 69–84 (1989)
10. M. Domaracki, C.N. Stephan, Facial soft tissue thicknesses in Australian adult cadavers. J. Forensic Sci. **51**(1), 5–10 (2006)
11. V.M. Phillips, N.A. Smuts, Facial reconstruction: utilization of computerized tomography to measure facial tissue thickness in a mixed racial population. Forensic Sci. Int. **83**(1), 51–59 (1996)
12. J.S. Rhine, H.R. Campbell, Thickness of facial tissues in American blacks. J. Forensic Sci. **25**(4), 847–858 (1980)

13. J.S. Rhine, C.E. Moore, J.T. Weston, *Facial reproduction: tables of facial tissue thicknesses of American Caucasoids in forensic anthropology* (Maxwell Museum of Anthroplogy, The University of New Mexico, Albuquerque, 1982)
14. H. Suzuki, On the thickness of the soft parts of the Japanese face. J Anthropol Soc Nippon. **60**, 7–11 (1948)
15. R.M. George, The lateral craniographic method of facial reconstruction. J. Forensic Sci. **32**(5), 1305–1330 (1987)
16. V.M. Phillips, N.A. Smuts, Facial reconstruction: utilization of computerized tomography to measure facial tissue thickness in a mixed racial population. Forensic Sci. Int. **83**(1), 51–59 (1996)
17. W. Krogman, *The human skeleton in forensic medicine* (Charles C Thomas, Springfield, IL, 1962), p. 11
18. W.M. Krogman, M.J. McCue, *The Reconstruction of the Living Head from the Skull* (1946)
19. M.M. Gerasimov, *The face finder; translated from the German by Alan Houghton Brodrick* (Hutchinson & Co, London, UK, 1971)
20. M.M. Gerasimov, *Principles of Reconstruction of the Face on the Skull* (1949)
21. A. Hall, *Indelible Evidence*. (Blitz Editions, 1995), pp. 75–80; L. So, *An Introduction to Facial Reconstruction*
22. C. Wilkinson, Facial reconstruction–anatomical art or artistic anatomy? J. Anat. **216**(2), 235–250 (2010)
23. B. Lane, *The Encyclopaedia of Forensic Science*. (BCA, London, 1992), pp 167–169; B. Rai, J. Kaur, Craniofacial identification, in *Evidence-Based Forensic Dentistry* (Springer, Berlin, Heidelberg, 2013), pp. 65–71
24. G.F. Walker, The computer and the law: coordinate analysis of skull shape and possible methods of postmortem identification. J. Forensic Sci. **21**(2), 357–366 (1976)
25. J. Eisenfeld, D.J. Mishelevich, J.J. Dann III, W.H. Bell, Soft-hard tissue correlations and computer drawings for the frontal view. Angle Orthod. **45**(4), 267–272 (1975)
26. J.S. Vorrasi, A. Kolokythas, Controversies in traditional oral and maxillofacial reconstruction. Oral Maxillofac. Surg. Clin. **29**(4), 401–413 (2017)
27. C. Molé, H. Gérard, J.L. Mallet, J.F. Chassagne, N. Miller, A new three-dimensional treatment algorithm for complex surfaces: applications in surgery. J. Oral Maxillofac. Surg. **53**(2), 158–162 (1995)
28. B. Fallahi, M. Foroutan, S. Motavalli, M. Dujovny, S. Limaye, Computer-aided manufacturing of implants for the repair of large cranial defects: an improvement of the stereolithography technique. Neurol. Res. **21**(3), 281–286 (1999)
29. M.A. Papadopoulos, P.K. Christou, P.K. Christou, A.E. Athanasiou, P. Boettcher, H.F. Zeilhofer, R. Sader, N.A. Papadopulos, Three-dimensional craniofacial reconstruction imaging. Oral Surg. Oral Med. Oral Pathol. Oral Radiol. Endodontol. **93**(4), 382–393 (2002)
30. M.W. Vannier, J.L. Marsh, J.O. Warren, Three dimensional CT reconstruction images for craniofacial surgical planning and evaluation. Radiology **150**(1), 179–184 (1984)
31. D.P. Huijsmans, W.H. Lamers, J.A. Los, J. Strackee, Toward computerized morphometric facilities: a review of 58 software packages for computer-aided three-dimensional reconstruction, quantification, and picture generation from parallel serial sections. Anat. Rec. **216**(4), 449–470 (1986)
32. C. Mole, H. Gerard, P. Bouchet, R. Della Malva, S. Corbel, N. Miller, J. Penaud, Medical imaging holdings and perspectives 3-D modeling and stereolithographic plastic reconstruction. Odontostomatological News Encycl. Pract. **181**, 127–141 (1993)
33. S. Lozanoff, V.M. Diewert, B. Sinclair, K.Y. Wang, Surface modeling of craniofacial form in human embryos with a limited graphics terminal. J. Craniofac. Genet. Dev. Biol. **8**(2), 101–109 (1988)
34. S. Lozanoff, J.J. Deptuch, Implementing Boissonnat's method for generating surface models of craniofacial cartilages. Anat. Rec. **229**(4), 556–564 (1991)
35. L. Kovacs, A. Zimmermann, H. Wawrzyn, K. Schwenzer, H. Seitz, C. Tille, N.A. Papadopulos, R. Sader, H.F. Zeilhofer, E. Biemer, Computer aided surgical reconstruction after complex facial burn injuries–opportunities and limitations. Burns. **31**(1), 85–91 (2005)

36. S. De Greef, G. Willems, Three-dimensional craniofacial reconstruction in forensic identification: latest progress and new tendencies in the 21st century. J. Forensic Sci. **50**(1), JFS2004117-6 (2005)
37. C. Cattaneo, Forensic anthropology: developments of a classical discipline in the new millennium. Forensic Sci. Int. **165**(2), 185–193 (2007)
38. G.M. Gordon, *An investigation into the accuracy and reliability of skull-photo superimposition in a South African sample.* Doctoral dissertation, University of Pretoria
39. M. Cummaudo, M. Guerzoni, L. Marasciuolo, D. Gibelli, A. Cigada, Z. Obertovà, M. Ratnayake, P. Poppa, P. Gabriel, S. Ritz-Timme, C. Cattaneo, Pitfalls at the root of facial assessment on photographs: a quantitative study of accuracy in positioning facial landmarks. Int. J. Legal Med. **127**(3), 699–706 (2013)

Dimensional Accuracy of Surgical Guides Fabricated from Different Materials Using 3D Printer

Deepkamal Kaur Gill, Divya Bajaj, Aditi Rawat, Yash Gopal Mittal, Mamta Juneja and Prashant Jindal

Abstract *Purpose*: The recent advent of osseointegration has led to a major transformation in implantology and increased demand for methods for accurate placement of dental implants. Guided surgery using CAD/CAM (Computer-Aided Design/Computer-Aided Manufacturing) software is the most accurate way to accomplish this. *Materials and methods*: In this study, one edentulous test case was taken whose dentition was scanned using high-quality desktop scanner for Gypsum dental models. The STL (Standard Triangulation Language) file was generated and passed into 3 Shape Implant Studio® software where implant was planned and surgical guide model generated. The generated STL file was then printed by FDM (Fused Deposition Modelling) printer using different types of materials such as PLA (Polylactic Acid), ABS (Acrylonitrile Butadiene Styrene), PETG (Polyethylene Terephthalate with a Glycol) and Nylon. *Results*: The internal and external diameter and depth of all models were calculated and compared with STL file. Variation of each parameter from original was evaluated. PLA showed a deviation of −0.88, 0.83 and −1.37% for internal, external diameter and depth, respectively. For ABS, the values were 4.2, −2.28, −1.53%. Nylon showed greater deviations from original file with a variation of −10.8, 0.73 and −5.51% for the parameters. The values for PETG

D. K. Gill (✉) · D. Bajaj · A. Rawat · Y. G. Mittal · M. Juneja · P. Jindal
University Institute of Engineering and Technology, Panjab University, Chandigarh 160014, India
e-mail: deepkamal1597@gmail.com

D. Bajaj
e-mail: divyabajaj1997@gmail.com

A. Rawat
e-mail: aditiR2407@gmail.com

Y. G. Mittal
e-mail: ymymittal1@gmail.com

M. Juneja
e-mail: mamtajuneja@pu.ac.in

P. Jindal
e-mail: jindalp@pu.ac.in

C. R. Krishna et al. (eds.), *Proceedings of 2nd International Conference on Communication, Computing and Networking*, Lecture Notes in Networks and Systems 46, https://doi.org/10.1007/978-981-13-1217-5_80

were 5.09, 1.76 and −0.92% respectively. *Conclusion*: It was concluded that PLA is the most accurate material for printing surgical guides using FDM printer, closely followed by ABS and PETG. Nylon is the least suited.

Keywords ABS · CAD · CAM · FDM · Guided surgery · Implant · PLA PETG · STL

1 Introduction

The use of technology in dentistry dates back to the 1990s; the term referred to as Digital Dental Technology (DDT) [1]. The demand for osseointegration is large. While placing the implant, the surgeon has to take care of a number of factors such as correct angulation of implant determined by angulation of the bone, maintaining correct depth of the implant or avoiding any damage to the inferior alveolar nerve while performing implant on the mandible [2] to get the desired long-term outcome and prevent any infections resulting from the surgery. Earlier dentists used to place the implant at a place where the maximum amount of bone was present without caring about long-term outcome [3]. Thus, a computer-generated surgical guide helps by providing a specified pathway to the edentulous region with little or no scope for inaccuracies.

3D modelling involves data acquisition, software manipulation and display of a 3D virtual interactive model on computer. Special care must be taken while taking the scans. Delong et al. [4] in 2003 measured suitability of a system for creating 3D images of dental arches. Ten stone casts of a dental standard were made using vinyl polysiloxane impression materials and improved dental stone. They then scanned the impressions and casts using a Comet 100 optical scanner and generated computerised models from the scanned data. They took the average of the absolute differences between the computer models to calculate accuracy. Ten measures were taken repeatedly to check precision. Accuracy ± precision obtained for the casts and impressions were 0.024 ± 0.002 mm and 0.013 ± 0.003 mm. Sufficient accuracy for clinical uses was observed.

A procedure called Cone Beam Computed Tomography (CBCT) is generally used for getting scans. The use of Medical CT is increasing annually at an estimated rate of 15–20% [5]. Commonly used maxillofacial imaging methods include periapical radiography, panoramic radiography, and conventional tomography [6] which can produce only distorted 2D images. Hence, surgeons these days resort to CT for getting scans. Baumgaertel et al. [7] in 2007 scanned thirty human skulls with CBCT, and reconstructed the dentitions. They made ten measurements—overbite, overjet, maxillary and mandibular intermolar and intercanine widths, arch length available and desired-on the dentitions of models with a vernier calliper and on digital reconstructions of the skulls with some CAD software. They then accessed accuracy and reliability using intraclass correlation and paired Student *t* tests. Both CBCT and calliper results were found to be highly reliable with value of reliability greater than

0.9 and hence it was concluded that dental measurements from CBCT volumes can be used for quantitative analysis.

Although, digitisation of models is done both directly and indirectly—by scanning directly inside mouth using intraoral scanner or generating a plaster model first and then scanning it—but the latter has several advantages and hence generally preferred [8]. Even if scans get lost, the plaster moulds can be scanned again. Also being more tangible, they are easy to understand. Cupernus et al. [8] in 2012 presented a study for reproducibility for dental models using intraoral scanners. They scanned dentition using chair side oral scanner and checked for missing data by programs and improved it. These were converted into digital models using Orthoproof software. Finally, they converted these files to STL models. The measurements of dentition were done using digital calliper and those of digital models were done using DigiModel software. The measures used for analysis were tooth width, transversal distances, skulls segments and the dimensions of 3D and digital models. The differences were clinically insignificant. The standard values for analysis were mean measurements of skulls with cut-offs for segments of 0.2 mm, widths of mesiodistal of 0.1 mm, arch discrepancies and transversal distances of 1.0 mm and discrepancies in tooth size of 1.5 mm.

Reconstruction of the models from input scans can also be carried out using 3D printers by a method called Rapid Prototyping (RP) that adds layer over layer to generate 3D models [9]. Some common RP techniques include stereolithography, selective laser sintering(SLS) and ballistic particle manufacturing [10]. The most commonly used RP technology these days is FDM (Fused Deposition Modelling) which uses principle of fibre material deposition pushed through a heated extrusion nozzle thus melting the material [11]. The initial state of material can be solid, liquid or powder state. The solid state may be in wire or pellet form. FDM process was introduced commercially in the early 1990s by Stratasys Inc., USA [12].

Melnikova et al. [13] in 2014 analysed designs obtained by different textile or textile-based structures in a 3D graphics software-BlenderTM and repaired them with netfabb. Thereafter, slices of the STL files were made to get a layer model and fed to a 3D printer. Most of the tests were performed with FDM printer X400 produced by German RepRap with PLA (elongation at break about 4%, tensile modulus 1968 MPa [14, 15]) or soft PLA (softener added to PLA, elongation at break up to about 200%), BendLay (butadiene; impact strength 3 kJ/m^2, tensile modulus 1550 MPa with other materials). First tests on ABS showed insufficient values of these properties of models and hence it was excluded from the study. For comparing with FDM process, an SLS printer by Shapeways was used with Nylon and PLA. It was concluded that both hard PLA and nylon were too hard for use in textile industry. Soft PLA, combined with less flexible materials like BendLay, was best option to reproduce textile structures. However, 3D printed objects combined with fibrous materials cause a low adhesion force between the parts leading to delimitation of layers of materials [14]. Hence, bonding properties of parts of materials in contact must be enhanced in such cases.

2 Methodology

The main steps required to generate surgical guides include (1) getting the patient's dental scans, (2) using a CAD/CAM software to generate surgical guide, and (3) printing the models using a 3D printer with appropriate material. For ensuring best results, each step must be performed as gingerly as possible (Fig. 1).

The detailed outline of the steps followed is as stated below:

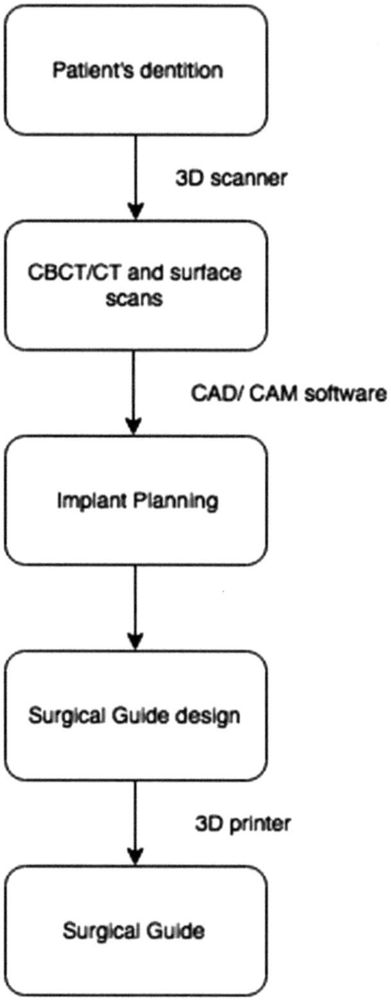

Fig. 1 General methodology for fabricating surgical guides

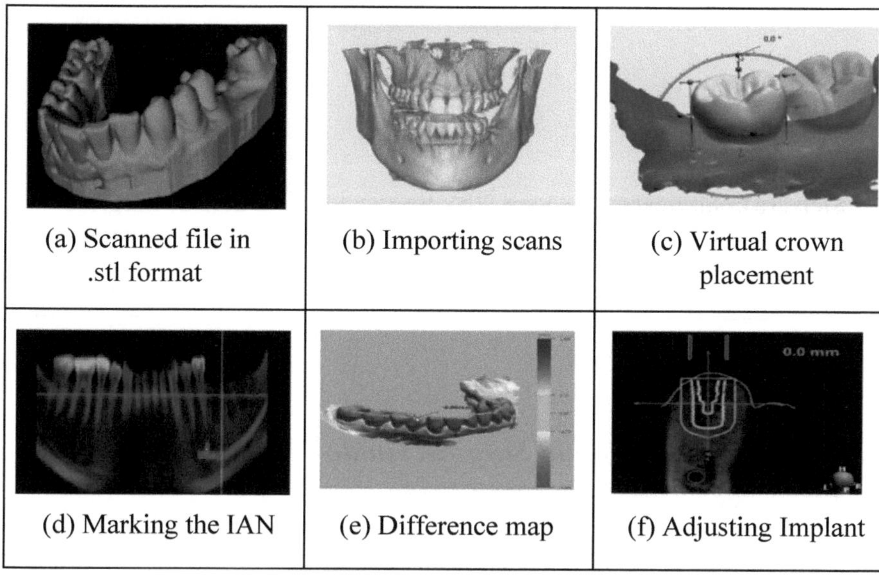

Fig. 2 Input scans and implant planning using CAD/CAM software

2.1 *Step 1: Input Scans and Implant Planning*

For this study, a patient's dentition was scanned using high-quality static desktop scanner for gypsum models and CBCT/CT and surface scans thus obtained were fed to 3Shape Implant Studio® software. The missing tooth/teeth was identified from the STL file and marked in software. A virtual crown similar to the actual crown to be placed in patient's edentulous part was planned for that tooth. Then panoramic curve was set to generate difference map showing complete overlap of CBCT/CT and surface scans. Next, inferior alveolar nerve (IAN) was marked. Identification of nerve canal is an essential step in case of maxillary implant since the implant must be at least 0.1 mm away from IAN. Also, the primary stability of implant body of first molar may be lower than that of second premolar, which should be carefully considered during dental implant surgery [16]. An implant too deep may rupture the nerve causing infections. So it must be identified with sufficient assistance from an orthodontist. The IAN is the most commonly injured nerve (64.4%), followed by the lingual nerve (28.8%) [14]. Too wide an implant will lead to damage to the neighbouring teeth. Hence, the depth and diameter of the implant must be set cautiously so as to generate the best results without any blight.

Keeping aforementioned factors in mind, the appropriate type of implant was chosen and depth and diameter of implant were adjusted with help from a practitioner. The detailed steps followed for implant planning are shown in Fig. 2.

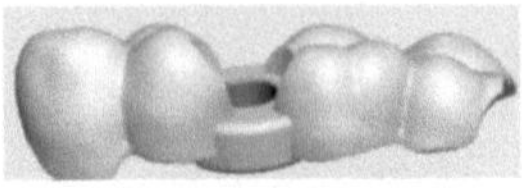	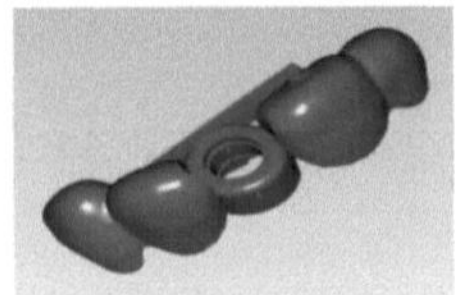	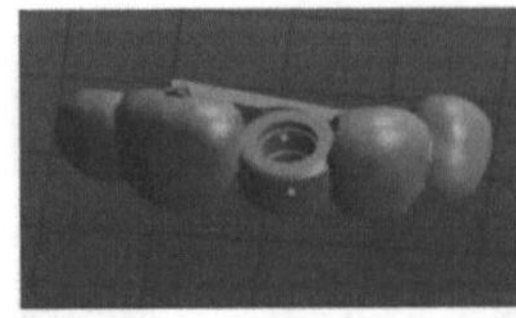
Step 2(a): Surgical Guide design generated	Step 2(b): Surgical guide STL	Step 3: Surgical guide GCODE

Fig. 3 Generation of surgical guide

2.2 *Step 2: Generation of Surgical Guide*

After planning the details of the implant in the software, the surgical guide was designed by marking the regions of the teeth nearest to the edentulous one on which the surgical guide would reside. Since the surgical guide may break due to pressure while drilling through it; appropriate bars and supports were added on it. Thereafter, the surgical guide was ready to be printed. Figure 3 illustrates these steps.

2.3 *Step 3: Printing*

For printing of the STL file, it was converted to Gcode format to be given as input to the FDM printer. The appropriate material was chosen one by one and print command given using compatible software. The materials used were white PLA, black ABS, Nylon and PETG. The printed surgical guides are as shown in Fig. 4.

2.4 *Step 4: Comparison*

Lastly, the surgical guides printed from the different materials were compared to check which material shows the best results with FDM printer.

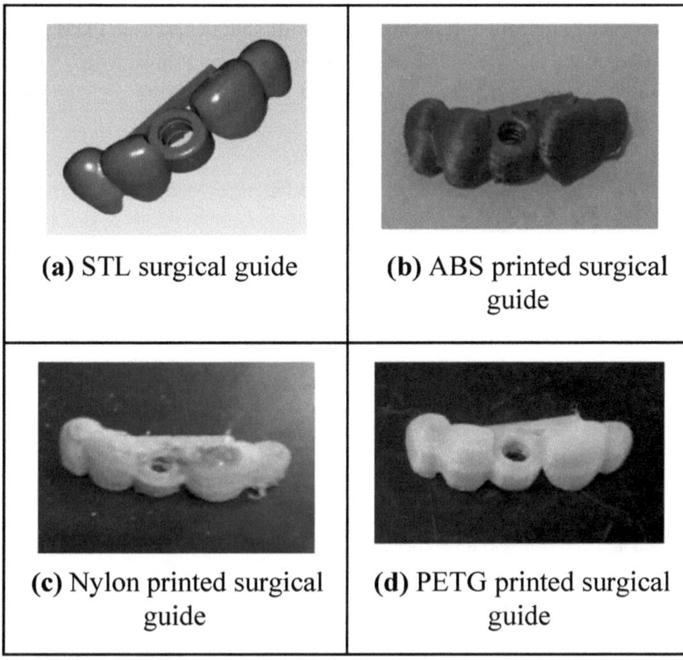

(**a**) STL surgical guide
(**b**) ABS printed surgical guide
(**c**) Nylon printed surgical guide
(**d**) PETG printed surgical guide

Fig. 4 The STL and printed surgical guides

3 Results and Discussions

The surgical guides were printed using PLA, ABS, PETG and Nylon one by one using the FDM technique. The depth and diameter (both internal and external) of each printed surgical guide were measured using Vernier Calliper and of the STL file using computer software. The differences of dimensions of each printed model from the STL file were calculated and based on that, the most suitable material for printing with FDM was established. The results for this are tabulated in Table 1.

It was observed that the PLA printed surgical guide showed minimum deviations from STL file for all three parameters. PETG and ABS also showed less differences and dimensions more close to STL file than Nylon; which showed high shrinkage. However, for best results with Nylon, the material should be preheated so that all moisture is lost. Hence, results show that PLA must be preferred for printing of surgical guides using FDM printer due to more closeness of surgical guides to STL file. However, PETG must be used preferably because it is biocompatible and suitable for placing in mouth while performing surgery for implant.

Table 1 Comparison of surgical guides printed using different materials in FDM printer

Material used	Original STL file	Printed with PLA	Printed with ABS	Printed with Nylon	Printed with PETG
Depth (in mm)	6.53	6.44	6.43	6.17	6.47
% age deviation from original	0%	−1.37%	−1.53%	−5.51%	−0.92%
Inner Diameter (in mm)	4.52	4.48	4.71	4.03	4.75
% age deviation from original	0%	−0.88%	4.2%	−10.8%	5.09%
Outer diameter (in mm)	9.64	9.72	9.42	9.71	9.81
% age deviation from original	0%	0.83%	−2.28%	0.73%	0.73%

4 Conclusion

An important concern for surgeons in implantology is accurate positioning of implant without injuring IAN or infections from surgery. Hence, guided surgery is best way to avoid any error. In this study, STL file of an edentulous patient's dentition was fed into 3Shape Implant Studio® software for designing surgical guide. The designed surgical guide was converted into Gcode format for printing with FDM printer. Different materials namely PLA, ABS, PETG and Nylon were used for printing. It was concluded that PLA shows best results with minimum deviation for depth, internal and external diameter from STL file of surgical guide. ABS also shows results close to STL file. While Nylon shows shrinkage after printing, PETG gives more accurate results than Nylon. Thus, PLA should be preferred for printing surgical guides requiring high precision since it shows least shrinkage after printing. But because of biocompatible nature, PETG is often recommended.

References

1. S.J. Lee, G.O. Gallucci, Digital versus conventional implant impressions: efficiency outcomes. Clin. Oral Implant Res. **24**(1), 111–115 (2013)
2. K. Misch, H.L. Wang, Implant surgery complications: etiology and treatment. Implant Dent. **17**(2), 159–168 (2008)

3. G. Widmann, R.J. Bale, Accuracy in computer-aided implant surgery–a review. Int. J. Oral Maxillofac. Implants **21**(2), 305–313 (2006)
4. R. DeLong, M. Heinzen, J.S. Hodges, C.C. Ko, W.H. Douglas, Accuracy of a system for creating 3D computer models of dental arches. J. Dent. Res. **82**(6), 438–442 (2003)
5. M.W. Vannier, Craniofacial computed tomography scanning: technology, applications and future trends. Orthod. Craniofac. Res. **6**(s1), 23–30 (2003)
6. P. Sukovic, Cone beam computed tomography in craniofacial imaging. Orthod. Craniofac. Res. **6**(s1), 31–36 (2003)
7. S. Baumgaertel, J.M. Palomo, L. Palomo, M.G. Hans, Reliability and accuracy of cone-beam computed tomography dental measurements. Am. J. Orthod. Dentofac. Orthop. **136**(1), 19–25 (2009)
8. A.M. Cuperus, M.C. Harms, F.A. Rangel, E.M. Bronkhorst, J.G. Schols, K.H. Breuning, Dental models made with an intraoral scanner: a validation study. Am. J. Orthod. Dentofac. Orthop. **142**(3), 308–313 (2012)
9. X. Yan, P.E. Gu, A review of rapid prototyping technologies and systems. Comput. Aided Des. **28**(4), 307–318 (1996 Apr 1)
10. J.P. Kruth, Material incress manufacturing by rapid prototyping techniques. CIRP Ann. Manuf. Technol. **40**(2), 603–614 (1991)
11. P. Dudek, FDM 3D printing technology in manufacturing composite elements. Arch. Metall. Mater. **58**(4), 1415–1418 (2013)
12. O.A. Mohamed, S.H. Masood, J.L. Bhowmik, Optimization of fused deposition modeling process parameters: a review of current research and future prospects. Adv. Manuf. **3**(1), 42–53 (2015)
13. R. Melnikova, A. Ehrmann, K. Finsterbusch, in *3D Printing of Textile-Based Structures by Fused Deposition Modelling (FDM) with Different Polymer Materials*. IOP Conference Series: Materials Science and Engineering, vol. 62, no. 1. (IOP Publishing, Bristol, 2014), p. 012018
14. C. Richter, S. Schmülling, A. Ehrmann, K. Finsterbusch, FDM printing of 3D forms with embedded fibrous materials, in *Design, Manufacturing and Mechatronics: Proceedings of the 2015 International Conference on Design, Manufacturing and Mechatronics* (*ICDMM2015*) (2016), pp. 961–969
15. A.B. Tay, J.R. Zuniga, Clinical characteristics of trigeminal nerve injury referrals to a university centre. Int. J. Oral Maxillofac. Surg. **36**(10), 922–927 (2007)
16. J.T. Hsu, H.L. Huang, L.J. Fuh, R.W. Li, J. Wu, M.T. Tsai, Y.W. Shen, M.G. Tu, Location of the mandibular canal and thickness of the occlusal cortical bone at dental implant sites in the lower second premolar and first molar. Comput. Math. Methods Med. **3**, 2013 (2013)

Reliability-Aware Design and Performance Analysis of QCA-Based Exclusive-OR Gate

Gurmohan Singh, R. K. Sarin and Balwinder Raj

Abstract Reliability-aware design of a new Quantum-dot Cellular Automata (QCA) based 2-input exclusive-OR (XOR) gate has been presented. The fault tolerance in proposed XOR gate has been implemented by introducing redundancy. The new design demonstrates fault tolerance capability against cell displacement, cell omission, misalignment, and extra cell deposition defects. The new QCA design of XOR gate successfully demonstrates fault tolerance capability of 85.7, 93.58, 79.68, and 56% against cell displacement/misalignment faults, extra cell deposition defects, single cell omission, and double cell omission, respectively. QCAPro tool has been utilized to compute energy dissipation results for proposed design. The functionality of new QCA design of XOR gate has been verified using QCA Designer version 2.0.3 tool.

Keywords QCA · XOR · Fault tolerant · Majority gate · Inverter · Wire-crossover

1 Introduction

Moore's law has guided the growth of CMOS technology in past 4 decades [1]. Due to many technological and fundamental issues, CMOS technology is approaching towards its end in very near future as predicted by International Technology Roadmaps for Semiconductors (ITRS 2015) [2]. So, new nanotechnologies are emerging to replace CMOS technology. A new QCA-based nanocomputing architecture has been emerging as potential candidate to replace the CMOS technology

G. Singh (✉)
Centre for Development of Advanced Computing, Mohali, Punjab 160071, India
e-mail: gurmohan@cdac.in

R. K. Sarin · B. Raj
Dr. B. R, Amberdkar National Instutute of Technology, Jalandhar, Punjab 144011, India
e-mail: sarinrk@nitj.ac.in

B. Raj
e-mail: rajb@nitj.ac.in

C. R. Krishna et al. (eds.), *Proceedings of 2nd International Conference on Communication, Computing and Networking*, Lecture Notes in Networks and Systems 46, https://doi.org/10.1007/978-981-13-1217-5_81

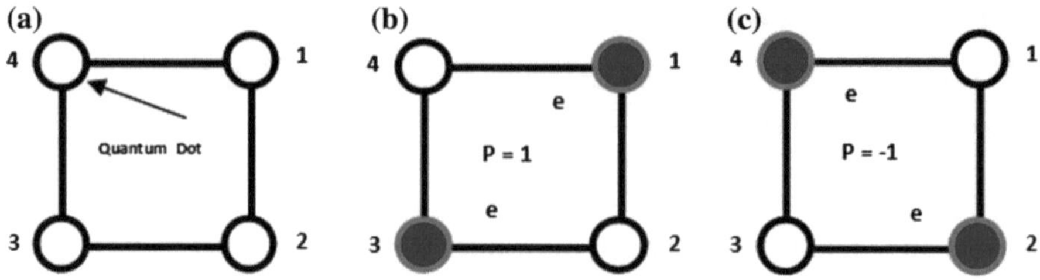

Fig. 1 **a** A typical 4-dot QCA cell, **b** logic '1' encoding, and **c** logic '0' encoding

Fig. 2 Binary wire realization in QCA logic

for binary computations [3]. The QCA technology offers many advantages such as extremely low power dissipation; clock-based pipeline type computing approach, high functional density, improved computing speed, and facilitates further miniaturization in nanoscale. A square nanostructure comprising of four quantum dots is called QCA cell and is the most basic computing entity in QCA nanotechnology [4]. Figure 1a depicts a typical QCA cell. Figure 1b, c represents QCA cell encoding for binary state '1' and '0', respectively.

The Coulombic interactions between nearby cells perform the meaningful computations and during this, no current flows between the cells. No electric current results in energy dissipation of the order of tens of milli electron-volts (meV). Figure 2 shows a QCA cell array used to implement a binary wire. The input cell polarization propagates towards output due to Columbic interactions among the cells demonstrating the functionality of a binary wire.

Figure 3a represents widely used QCA-based inverter implementation [5]. Figure 3b shows the most widely used decision-making gate used in QCA circuits called 3-input majority gate. The polarization of decision cell is decided by majority of inputs. By setting any of input at polarization level '−1' or '1', the majority gate may function as AND and OR gates. In QCA logic, any binary function could be realized using majority gates and inverters. Equation (1) describes the output of 3-input majority gate.

$$\mathrm{MAJ(A, B, C) = AB + AC + BC} \quad (1)$$

Only few attempts have been made to identify and characterize the QCA fabrication related defects/faults [6–9]. There are two phases during fabrication of QCA devices/circuits namely chemical synthesis phase and deposition phase. Each QCA cell is fabricated in chemical synthesis phase and this phase further comprises of two sub-steps called quantum-dot fabrication and free electrons injection into quantum

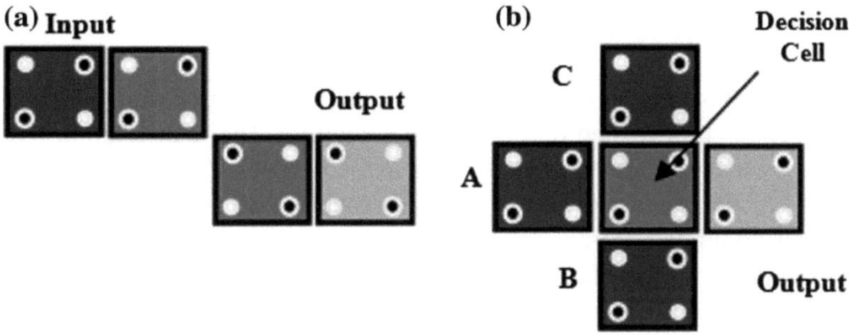

Fig. 3 **a** An inverter configuration, and **b** 3-input majority gate

dots in the QCA cells. There is probability of errors like an extra quantum dot may be produced or injection of extra free electron into a QCA cell. The fabricated QCA cells in chemical synthesis phase are attached on a substrate in deposition phase. The probability of occurrence of an error is significantly high when QCA cells are attached to the substrate. Three different types of errors that may occur during this phase are cell omission, extra cell deposition, and cell misplacement/misalignment. Researchers have observed that fabrication faults generated during deposition phase are more critical and may lead to severe errors in QCA devices/circuits.

Fault tolerance is defined as the ability of a device or circuit to maintain correct functionality in presence of defects/faults. A fault-tolerant device or circuit maintains its intended functionality may be at a reduced level, instead of failing completely when faults/defects are present. The fault tolerance generally results in hardware overhead and degradation in performance. The redundant components or devices added in binary computing circuits provide the fault tolerance. Triple Modular Redundancy (TMR) is widely used standard fault tolerance method. In TMR, three copies of a circuit are made to function in parallel and the final stage produces the result on majority-voting basis [10]. The fault tolerance techniques are required in mission critical applications like space and defense, etc. Von Newmann introduced Multiplexing redundancy algorithm (vN-MUX) [11] to improve the reliability of computing circuits. The algorithm is applicable for any arbitrary gate. Von Newmann illustrated this algorithm for NAND and majority gates in the presence of failures and for high redundancy factor.

The implementations of fault tolerance in QCA circuits have been investigated by only few researchers [12–17]. The XOR gates are widely used in digital circuits/systems and in arithmetic and logic units (ALUs). So, many researchers demonstrated QCA-based design of XOR gates in literature [18, 19]. Also, many researchers have demonstrated QCA-based design and simulation of digital computing circuits like adders, encoders, decoders, reversible gates, etc. in past two decades [20–22].

A new fault-tolerant design of QCA-based 2-input XOR gate is proposed and analyzed its performance in presence of deposition defects as well as in terms of generalized metrics. The redundant QCA cells are added in QCA wires, inverters, and majority gates in proposed fault-tolerant XOR design.

2 Proposed Fault-Tolerant Design

This paper demonstrates a QCA-based new 2-input XOR gate design which is fault-tolerant to deposition-related defects in QCA circuits. The fault-tolerant feature is introduced by adding redundant QCA cells in QCA wires and logic gates. The new QCA-based XOR gate is coplanar, structurally robust as no wire crossovers are used, and employs no rotated or translated (shifted) cells. The correctness of functionality of proposed design in absence and presence of deposition-related defects has been verified in QCADesigner version 2.0.3 simulation tool [23]. The bistable simulation engine parameters are adopted for all simulations.

Figure 4 describes schematic of QCA-based proposed 2-input XOR gate. It comprises of three majority gates and one inverter. The QCA-based layout of new fault-tolerant 2-input XOR gate is shown in Fig. 5. The fault-tolerant design of XOR gate layout utilizes only 70 QCA cells and consumes an area of 0.06 μm^2.

The simulation waveforms of proposed fault-tolerant QCA-based new 2-input XOR gate are shown in Fig. 6. A latency of 3 clock phases has been observed in output signal *'Out'*. The output signal polarization level varies between '0.985' and '−0.984' which are close to the ideal values of '1' and '−1'.

There is significant probability that a QCA cell in the layout may not be aligned properly with respect to nearby cells or may get misplaced within its original direction. The impact of displacement/misalignment defects in proposed QCA XOR gate has also been analyzed. It would be better if lesser number of QCA cells cause displacement or misalignment related defects. The standard dimensions of a four quantum-dot QCA cell are in nanometer scale (18 nm × 18 nm). So, the probability of occurrence of misalignment/displacement of few nanometer is significantly high during fabrication process. Hence, it becomes important to consider faults caused smaller displacements or alignments more critical. The radius of effect in bistable simulation engine set up in QCA Designer tool is set at 65 nm. In this work, we have

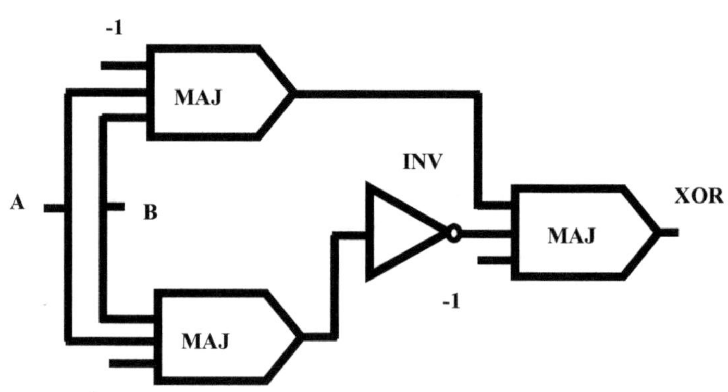

Fig. 4 Schematic of QCA-based new 2-input XOR gate

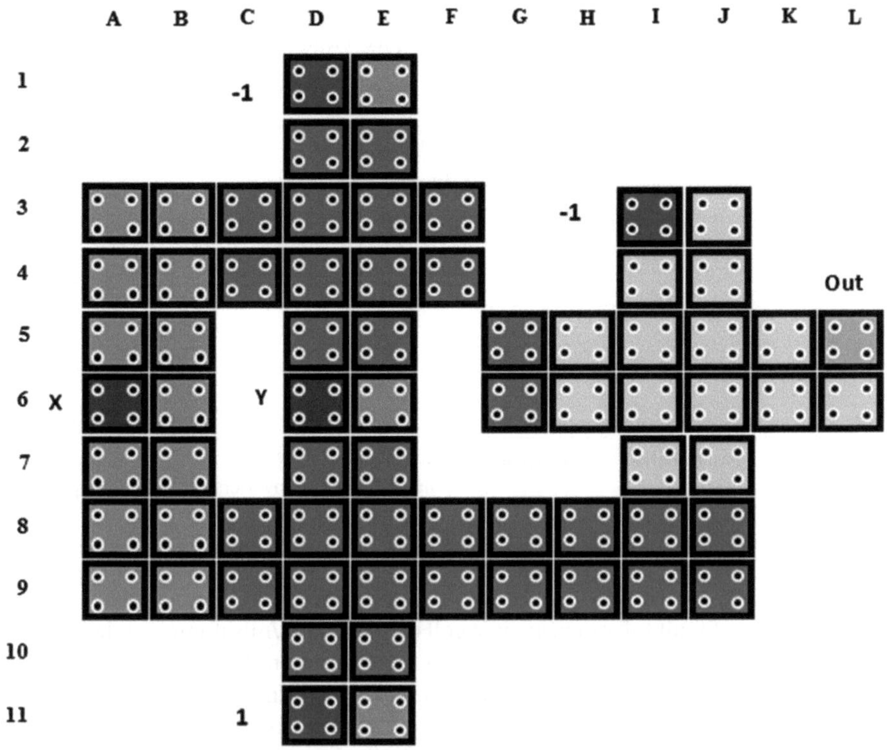

Fig. 5 QCA layout of proposed fault-tolerant 2-input XOR gate

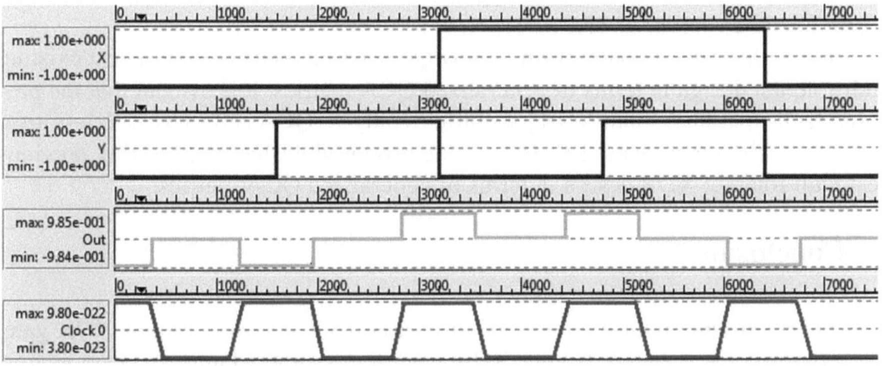

Fig. 6 Simulation waveforms for the proposed fault-tolerant XOR gate

ignored the instances of cell displacement/misalignment where it is more than 65 nm. However, the instances where the displaced/misaligned cell falls within 65 nm radius of nearby cells, its effect has been considered. So, it becomes vital to consider value of radius of effect appropriately as per requirement of simulation. A new metric to com-

Table 1 Cell displacement/misalignment defects of proposed fault-tolerant XOR gate

Cell	Cell displacement/misalignment (nm)			
	East	West	North	South
E5	≤18			
E6	≤18			
E7	≤36			
G4				≤6
G8			≤35	
H5		≤7	≤11	≤16
H7				≤12
K7	≤16			

Table 2 Comparison in terms of generalized QCA design metric for fault-tolerant XOR gates

XOR gate	No. of cells	Area (μm^2)	No. of majority gates	No. of inverters	Latency (clock phases)
[17]	85	0.078	3	2	5
[This Work]	70	0.06	3	1	3

pute the ability of fault tolerance against cell displacement/misalignment defects has been introduced. It is basically ratio of QCA cells count producing correct output to the total QCA cell count when subjected to displacement/misalignment. The higher the ratio, better is fault tolerance of the circuit against displacement/misalignment defects. Table 1 illustrates the instances of cell displacement/misalignment for new QCA XOR gate of Fig. 5. The instances of displacement/misalignment defect free cells in proposed fault-tolerant XOR gate are 48 out of total cases of 56, achieving a fault tolerance ability of 85.7%.

Table 2 illustrates a comparison of new QCA-based XOR gate with best existing fault-tolerant design in terms of generalized QCA metrics. It is evident that the proposed fault-tolerant 2-input XOR gate utilizes only 70 QCA cells, has 23% less area, and 40% less latency than [17]. Hence, the proposed design outperforms previous best fault-tolerant XOR design in terms of generalized QCA metrics.

3 Conclusion

This paper presents fault-tolerant design of a new QCA-based 2-input XOR gate. The proposed QCA-based 2-input XOR gate demonstrates fault tolerance against cell displacement/misalignment, extra cell deposition, and cell omission related faults. The new QCA design of XOR gate successfully demonstrates fault tolerance capability of 85.7, 93.58, 79.68, and 56% against cell displacement/misalignment faults, extra cell deposition defects, single cell omission, and double cell omission, respectively. Also, the new fault-tolerant QCA design of 2-input XOR gate utilizes only 70 QCA cells, has 23% less area, and 40% less latency than the previous best QCA-based fault-tolerant XOR gate.

References

1. G.E. Moore, Cramming more components onto integrated circuits. Electr. Mag. **38**(8), 114–117 (1965)
2. International Technology Roadmap for Semiconductors (ITRS), in *Emerging Research Devices (ERD)* (2015). Hyperlink: www.itrs2.net/itrs-reports.html
3. C.S. Lent, P.D. Taugaw, W. Porod, G.H. Berstein, Quantum cellular automata. Nanotechnology **4**(1), 49 (1993)
4. C.S. Lent, P.D. Taugaw, Lines of interacting quantum-dot cells: a binary wire. J. Appl. Phys. **74**, 6227–6233 (1993)
5. P.D. Taugaw, C.S. Lent, Logical devices implemented using quantum cellular automata. J. Appl. Phys. **75**, 1818–1825 (1994)
6. M.B. Tahoori, M. Momenzadeh, J. Huang, F. Lombardi, Testing of quantum cellular automata. IEEE Trans. On Nanotechnology **3**(4), 432–442 (2004)
7. T.J. Dysart, P.M. Kogge, C.S. Lent, M. Liu, An analysis of missing cell defects in quantum-dot cellular automata, vol. 3, in *IEEE International Workshop on Design and Test of Defect-Tolerant Nanoscale Architectures (NANOARCH)* (2005)
8. X. Ma. F. Lombardi, Fault-tolerant schemes for QCA systems, in *International Symposium on Defect and Fault-tolerance of VLSI Systems* (2008)
9. B. Sen, Y. Sahu, R. Mukherjee, R.K. Nath, B.K. Sikdar, On the reliability of majority logic structure in quantum-dot cellular automata. Microelectron. J. **47**, 7–18 (2016)
10. R.E. Lyons, W. Venderkulk, The use of triple modular redundancy to improve computer reliability. IBM J. **6**, 200–209 (1962)
11. Von Neumann, Probabilistic Logics and the Synthesis of Reliable Organisms from Unreliable Components. Automata Stud. Ann. Math. Stud. **34**, 43–98 (1956)
12. S. Roy, V. Beiu, majority multiplexing—economical redundant fault-tolerant designs for nanoarchitectures. IEEE Trans. Nanotechnol. **4**, 441–451 (2005)
13. A. Fijany, B. Toomarian, New design for quantum dots cellular automata to obtain fault-tolerant logic gates. Int. J. Nanop. Res. **3**(1), 27–37 (2001)
14. R. Roohi, F. DeMara, N. Khoshavi, Design and evaluation of an ultra-area-efficient fault-tolerant QCA full adder. Microelectron. J. **46**(6), 531–542 (2015)
15. R. Farazkish, F. Khodaparast, Design and characterization of a new fault-tolerant full-adder for quantum-dot cellular automata. Microp. Micros. **39**, 426–433 (2015)
16. V. Farazkish, A new quantum-dot cellular automata fault-tolerant full-adder. J. Comp. Elect. **14**(2), 506–514 (2015)
17. D. Kumar, D. Mitra, B.B. Bhattacharya, On fault-tolerant design of exclusive-OR gates in QCA. J. Comp. Elect. **16**(3), 896–906 (2016)
18. G. Singh, R.K. Sarin, B. Raj, A novel robust exclusive-or function implementation in QCA nanotechnology with energy dissipation analysis. J. Comp. Elect. **15**, 455–465 (2006)
19. A.N. Bahar, S. Waheed, N. Hossain, Md Asaduzzaman, A novel 3-input XOR function implementation in quantum dot-cellular automata with energy dissipation analysis. Alexandria Eng. J. (2017). https://doi.org/10.1016/j.aej.2017.01.022
20. W. Wang, K. Walus, G.A. Jullien, Quantum-dot cellular automata adders. IEEE Conf. Nanotechnol. **1**, 461–464 (2003)
21. G. Singh, B. Raj, R.K. Sarin, Design and performance analysis of a new efficient coplanar quantum-dot cellular automata adder. Indian J. Pure App. Phy. **55**, 97–103 (2017)
22. G. Singh, R.K. Sarin, B. Raj, A review of quantum-dot cellular automata based adders. Int. J. Hybrid Info. Tech. **10**, 41–58 (2017)
23. QCADesigner Tool Version 2.0.3. Hyperlink: http://www.mina.ubc.ca/qcadesigner_download

Design of Low Noise Amplifier for 802.16e

Makesh Iyer and T. Shanmuganantham

Abstract This work deals with the designing of low noise amplifier for WiMAX 802.16e application considering 3.4–3.6 GHz frequency band with 3.5 GHz as the center frequency. The amplifier is designed using Pseudomorphic HEMT ATF—34143 of Avago Technologies with different stabilization techniques to improve the stability of the potentially unstable device. It is observed that the noise immunity is more in source inductor degenerative stabilized HEMT than in other techniques with the help of noise figure which is obtained as 0.635 dB for HEMT. The comparative analysis of the LNA design is discussed in this paper.

Keywords WiMAX · LNA · HEMT · Power gain · Noise figure · VSWR

1 Introduction

Many inventions happened in the wireless technology and WiMAX i.e. Wireless Interoperability for Microwave Access has become the hot topic in today's scenario because it provides high data rate over a wide coverage area. There are many frequencies in which WiMAX operates like 2.3 GHz, 2.5 GHz, 3.3 GHz, 3.5 GHz, and 5.8 GHz in 2 to 11 GHz band [1]. G.-L. Ning, Z.-Y. Lei, et al have designed multiband low noise amplifiers with stepped impedance transformers and different active devices like CMOS, GaAs E-pHEMT for 3.5 GHz WiMAX using series and shunt feedback techniques and obtained the noise figures 4.03 dB and 1.2 dB respectively [2]. Ishaan Biswas, SC Bose, et al obtained noise figure of 1.102 dB for 2.3 GHz inductively loaded low noise amplifier with source inductive degeneration [3]. The lower frequency WiMAX is less susceptible to attenuation and hence the signal loss

M. Iyer · T. Shanmuganantham (✉)
Electronics Engineering Department, Pondicherry University,
Puducherry 605014, India
e-mail: shanmugananthamster@gmail.com

M. Iyer
e-mail: makwave.26791@gmail.com

C. R. Krishna et al. (eds.), *Proceedings of 2nd International Conference on Communication, Computing and Networking*, Lecture Notes in Networks and Systems 46, https://doi.org/10.1007/978-981-13-1217-5_82

is minimum which improves large coverage area but at the same time, its data rate is less and number of users per channel i.e. channel capacity is less. Abolfazl Zokaei, Amir Amirabadi designed a tunable multi-band LNA using the active post-distortion technique to improve LNA linearity and provide optimum noise figure and obtained noise figure of 3 dB with a gain of 12.95 dB [4]. Chang-Hsi Wu and Kuan-Lin Liu implemented the low noise amplifier using complementary current reuse topology and obtained noise figure of 4.6 dB and gain of 11 dB [5]. Hsuan-ling Kao, Bai-Hong Wei, et al used the AlGaN/GaN HEMT for designing the cascade topology LNA and obtained minimum noise figure of 3.3 dB [6]. Ameer Bansal, Anwar Jarndal designed a GaN based LNA for WiMAX applications and obtained a power gain of 14 dB and noise figure of 2.9 dB with input return loss of −2.44 dB [7]. Since there is a tradeoff between the gain, stability and noise figure of an LNA, a low noise amplifier of single stage with moderate gain and low noise figure is acceptable.

2 Design Aspects

The low noise amplifier is designed using pHEMT with different stability improvement techniques to obtain a better performing LNA in aspects of noise figure and gain of the amplifier. Low noise figure (NF) and high gain are required in an LNA for WiMAX application. The role of LNA is to filter noise and amplify weak signals and maintain high SNR of the system [8]. Advanced Design System (ADS) simulation tool is used for designing the low noise amplifier. There are two different types of device libraries available in the ADS software namely *S*—parameter library and RF Transistor library. *S*—Parameter library works on fixed bias, i.e., these parameters of the device are fixed for a particular bias point of the device. The RF transistor library consists of the normal transistor which can be biased toward using any technique and at any bias point. In this work, the *S*—parameter library devices are used.

2.1 Design Procedure

For designing a low noise amplifier, certain procedure has to be followed which are as follows:

- The proper active device should be selected, i.e., BJT/MOSFET/HBT/HEMT depending upon the frequency of operation and noise immunity requirement.
- Stability of the device should be checked using *S* parameters.
- If the device is unstable at the frequency of operation desired, proper techniques should be applied to stabilize the device.
- Proper biasing should be done based on the operating point required and the application of design.

- Matching circuits at input and output side of the amplifier should be an optimum design that can perform well in an amplifier.

In a low noise amplifier, the vital parameters to be considered are, the maximum gain provided by the active device which is termed as MAG (maximum available gain) and NF_{min} (minimum intrinsic noise) figure which in turn depends on the S parameters of the device. The S parameters determine the stability criteria of the device at various biasing points [9]. Theoretically, the stability of the device is checked using the $K - |\Delta|$ test where K is said to be Rollet's Stability factor which is given by

$$K = \frac{1 - |S_{11}|^2 - |S_{22}|^2 + |\Delta|^2}{2|S_{12}\,S_{21}|} \tag{1}$$

Also B is another parameter that is calculated for checking stability which should be positive for stable operation given by

$$B = 1 + |S_{11}|^2 - |S_{22}|^2 - |\Delta|^2 \tag{2}$$

And Δ is given by

$$\Delta = S_{11}S_{22} - S_{12}S_{21} \tag{3}$$

The condition for stability is that if $K > 1$, $|\Delta| < 1$ and $B = +\text{ve}$, then the device is unconditionally stable and if $K < 1$ then device is potentially unstable. This condition will tend the device to oscillate and the maximum available gain will be no more valid due to an unstable condition. Hence, the maximum gain produced by the device will now be said as MSG (maximum stable gain) which is mathematically expressed as

$$\text{MSG} = \frac{|S_{21}|}{|S_{12}|} \tag{4}$$

If the device is unconditionally stable, i.e., $K > 1$, then the gain obtained will be maximum available gain which is expressed as

$$\text{MAG} = \frac{|S_{21}|}{|S_{12}|}\left(K \pm \sqrt{K^2 - 1}\right) \tag{5}$$

2.2 *Different Techniques for Stability Improvement*

Different techniques are available to improve the stability of low noise amplifier [10]. But each technique has its own limitations and hence the optimum and efficient technique has to be determined from the analysis. These techniques are

- Connecting a series resistance to the gate of the active device.
- Connecting an inductor in series with the source terminal of the HEMT.
- Connecting a series resistor to the drain of the respective active device.

In this work, each technique is implemented with HEMT device and a comparative study is done to determine the most stable and best noise immune amplifier.

3 LNA Using HEMT

The device used in this work is a high range, low noise pseudomorphic HEMT ATF-34143 of Avago Technologies that is highly linear and provides excellent uniformity. The circuits of HEMT based LNAs are discussed further.

3.1 LNA with a Series Resistor in Gate Terminal of HEMT

A series resistor is connected to the gate terminal of HEMT to improve the stability of the device and convert it from a potentially unstable device to an unconditionally stable device. The circuit design of the low noise amplifier with gate resistor is shown in Fig. 1. The results of the gate resistor LNA design are shown below. Figure 2 shows the stability factor represented as stabfact1 parameter and delta (|Δ|) as stabmeas1 which is obtained as 1.844 and 0.913, respectively, which shows that the device is

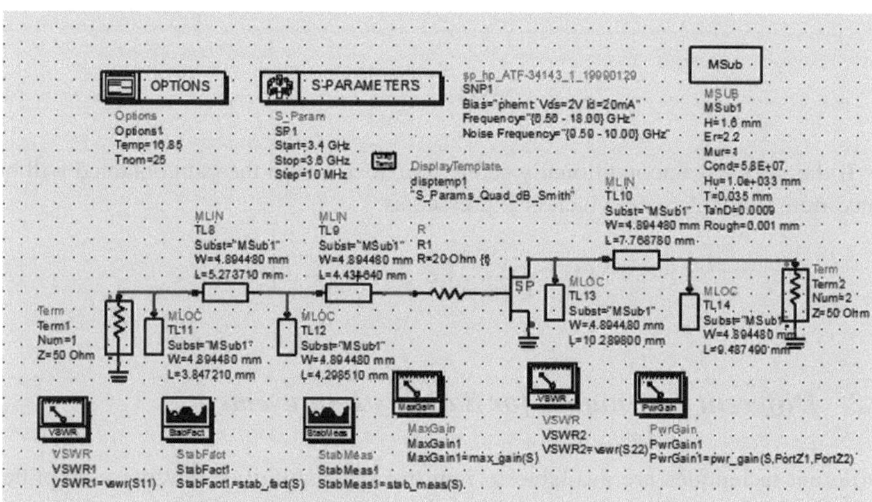

Fig. 1 LNA design with the gate resistor

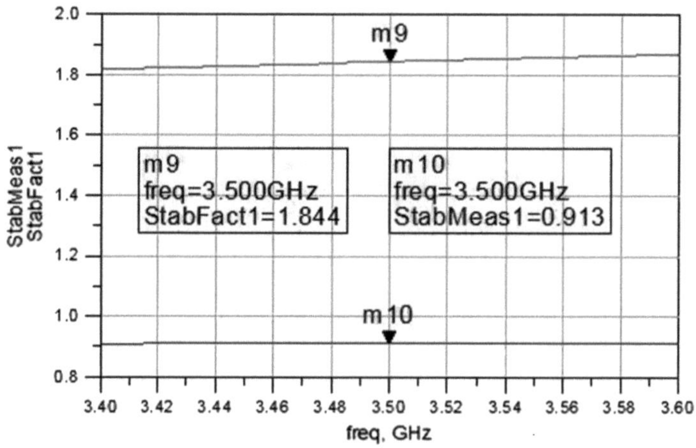

Fig. 2 Stability factor

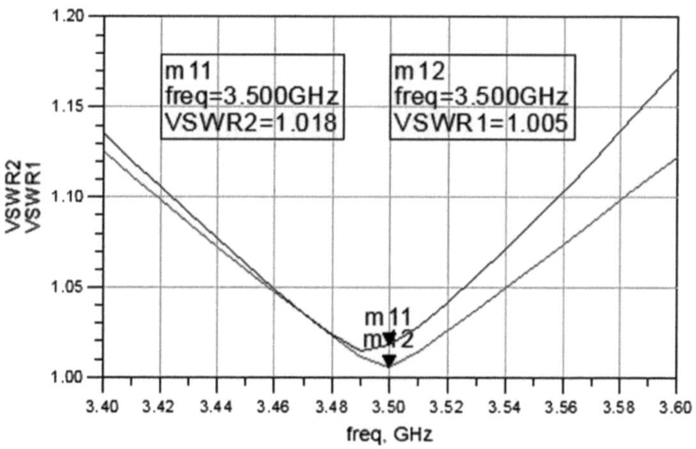

Fig. 3 VSWR

unconditionally stable. Figure 3 shows the VSWR, i.e., the voltage standing wave ratio which is obtained as 1.018 at the output side and 1.005 at the input side.

The *S* parameters of the amplifier are shown in Figs. 4 and 5 respectively which signifies S_{11} is −51.568 dB, S_{21} is 10.41 dB, S_{12} is −21.023 dB and S_{22} is −41.215 dB.

The noise figure obtained for the base resistor LNA is 3.066 dB and the power gain is 10.41 dB which is shown in Figs. 6 and 7, respectively.

From the above results, it is observed that the gain of the amplifier is high and also the noise figure is high. Hence, next technique is implemented where a series resistor is connected to the drain terminal of HEMT as shown in Fig. 8.

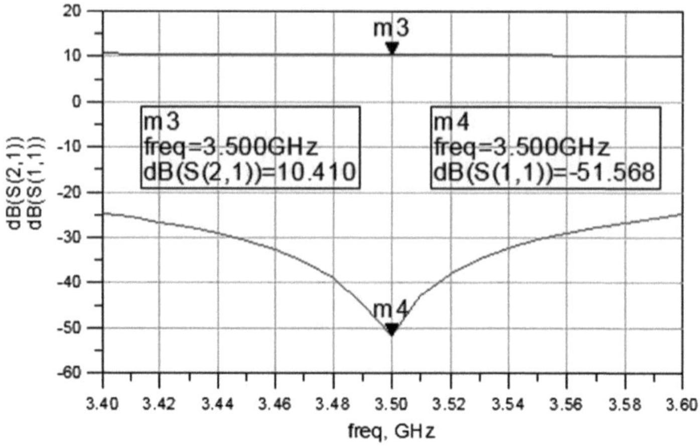

Fig. 4 S_{11} and S_{21}

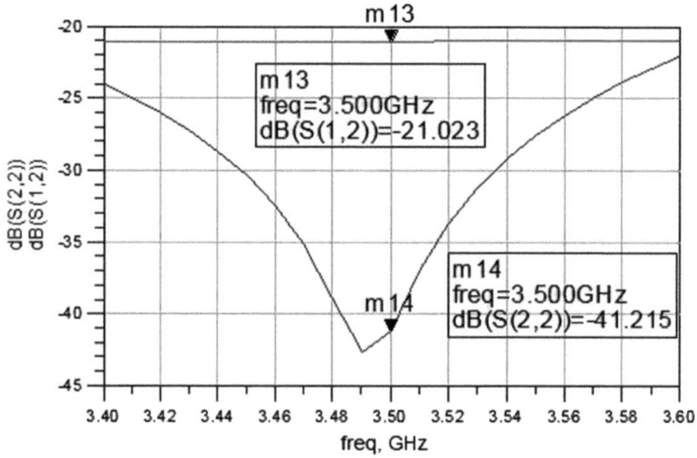

Fig. 5 S_{12} and S_{22}

3.2 LNA with a Series Resistor in Drain Terminal of HEMT

The results of the drain resistor LNA design are shown. Figure 9 shows the stability factor represented as stabfact1 parameter and delta ($|\Delta|$) as stabmeas1 which is obtained as 1.178 and 0.694, respectively, which manifests that the device is unconditionally stable. Figure 10 shows the VSWR, i.e., the voltage standing wave ratio which is obtained as 1.247 at the input and 2.434 at the output side of the amplifier. The *S* parameters of the amplifier are shown in Figs. 11 and 12 respectively which signifies S_{11} is -19.165 dB, S_{21} is 11.806 dB, S_{12} is -19.627 dB, and S_{22} is -7.585 dB.

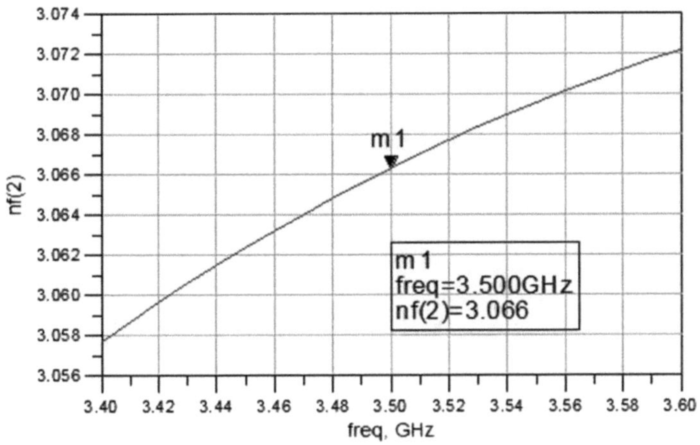

Fig. 6 Noise figure

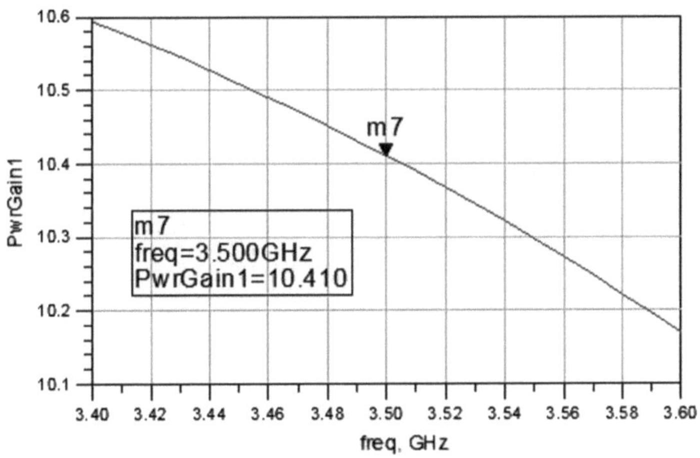

Fig. 7 Power gain

The noise figure obtained for the drain resistor LNA is 0.795 dB and the power gain is 11.806 dB which is shown in Figs. 13 and 14 respectively.

3.3 LNA with a Series Inductor in Source Terminal of HEMT

The results of the source inductor LNA designed in Fig. 15 are shown below. Figure 16 shows the stability factor represented as stabfact1 parameter and delta (|Δ|) as stabmeas1 which is obtained as 1.024 and 0.356 respectively. Figure 17 shows the VSWR,

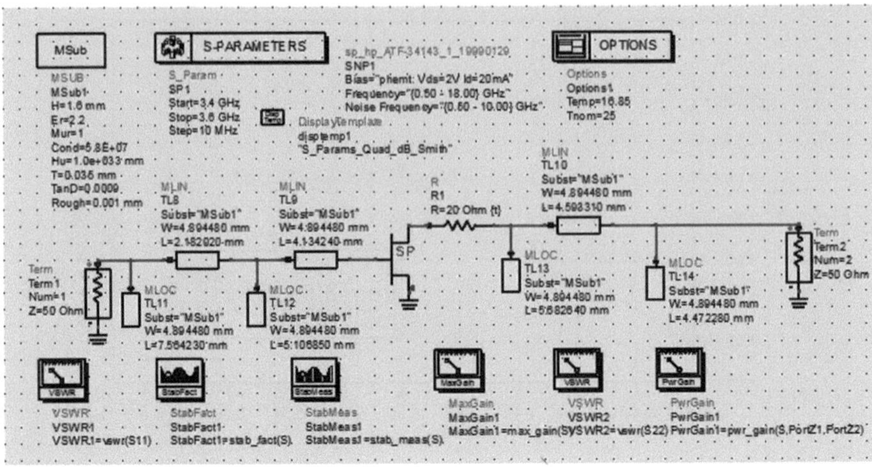

Fig. 8 LNA design with drain resistor

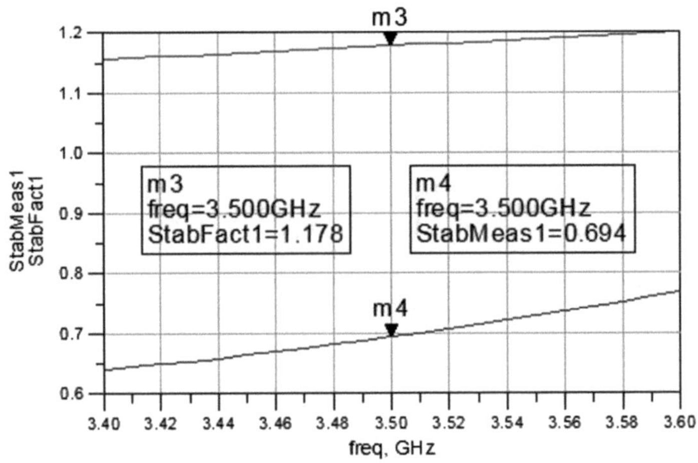

Fig. 9 Stability factor

i.e., the voltage standing wave ratio which is obtained as 1.017 both at the input and output side of the amplifier.

The S parameters of the amplifier are shown in Figs. 18 and 19 respectively which signifies S_{11} is −41.273 dB, S_{21} is 12.173 dB, S_{12} is −14.084 dB, and S_{22} is −41.936 dB. The noise figure obtained for the source inductor LNA is 0.635 dB and the power gain is 12.173 dB which is shown in Figs. 20 and 21 respectively. From the above results, it is observed that the gain of the amplifier has increased compared to both other techniques and the noise figure has decreased compared to other techniques which is shown in Table 1.

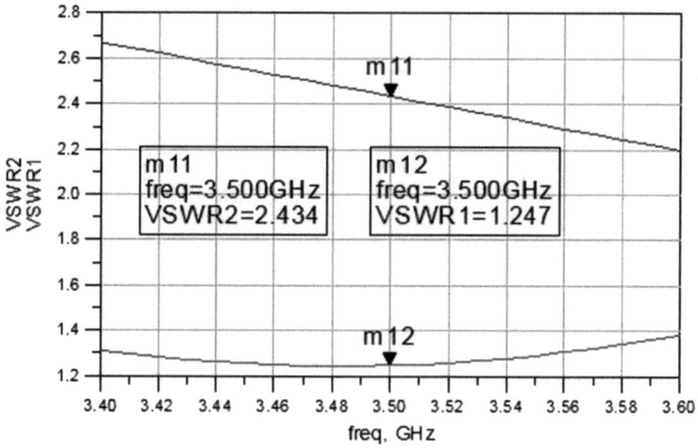

Fig. 10 VSWR

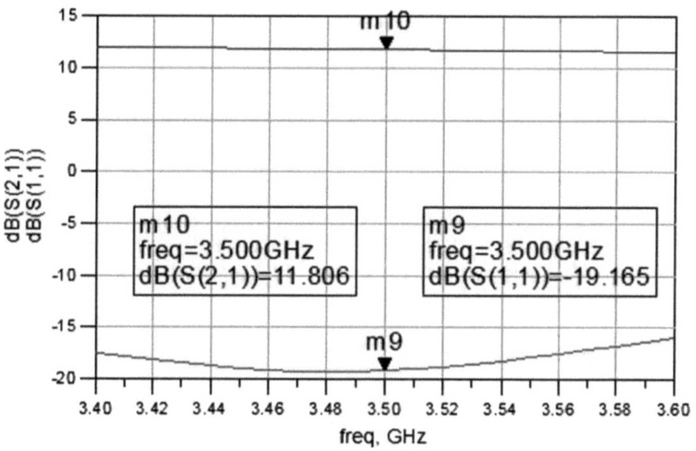

Fig. 11 S_{11} and S_{21}

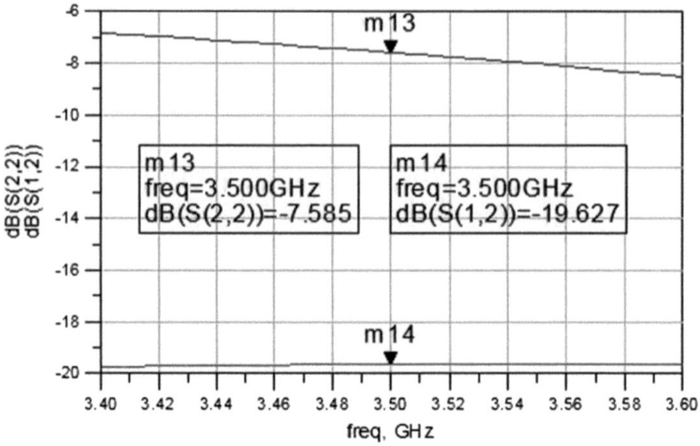

Fig. 12 S_{12} and S_{22}

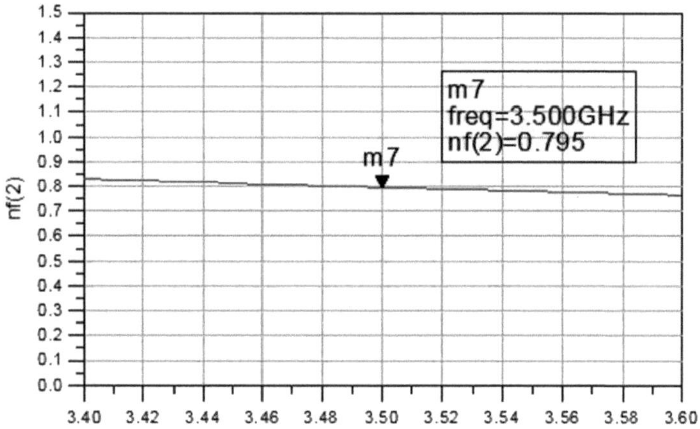

Fig. 13 Noise figure

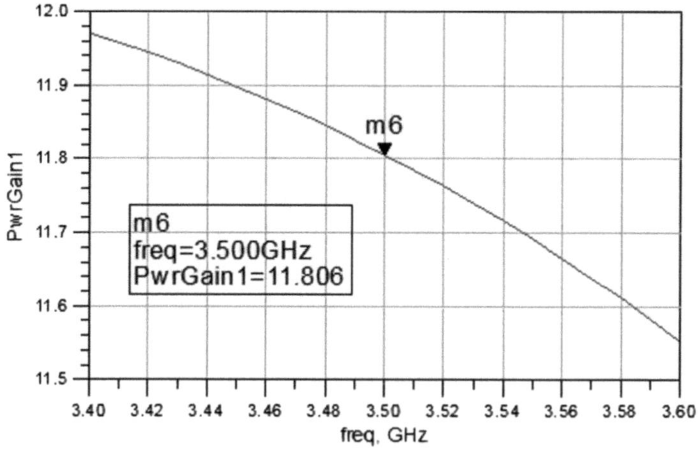

Fig. 14 Power gain

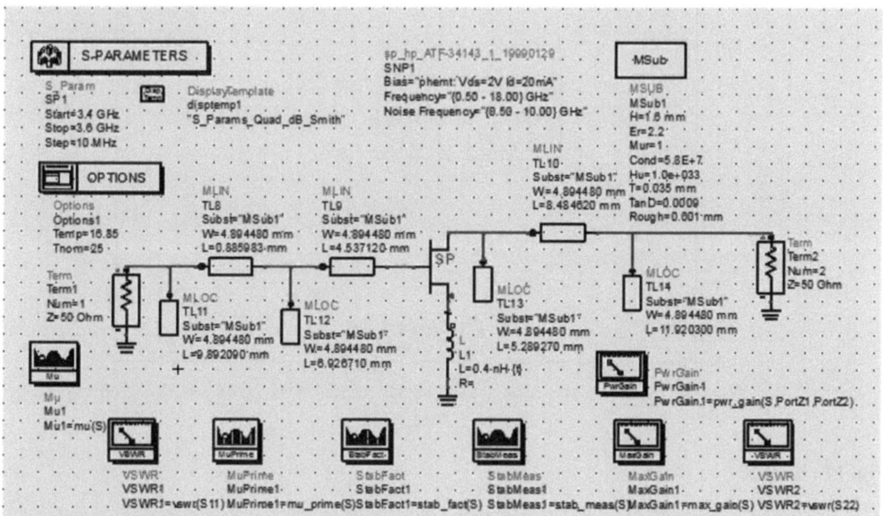

Fig. 15 LNA design with source inductor

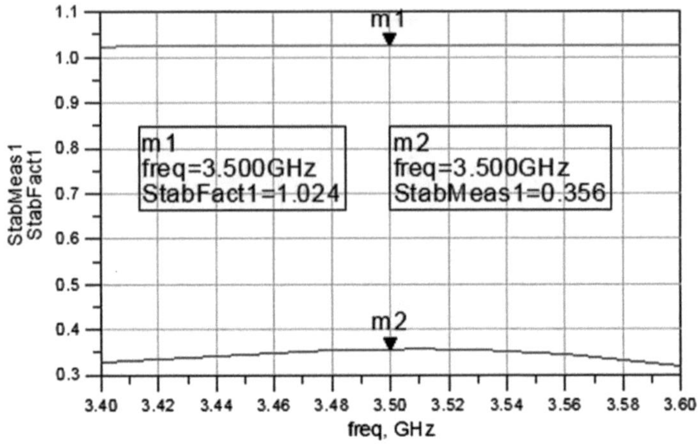

Fig. 16 Stability factor

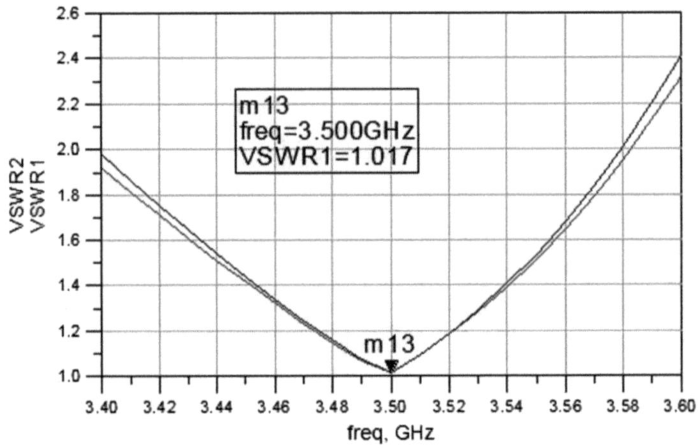

Fig. 17 VSWR

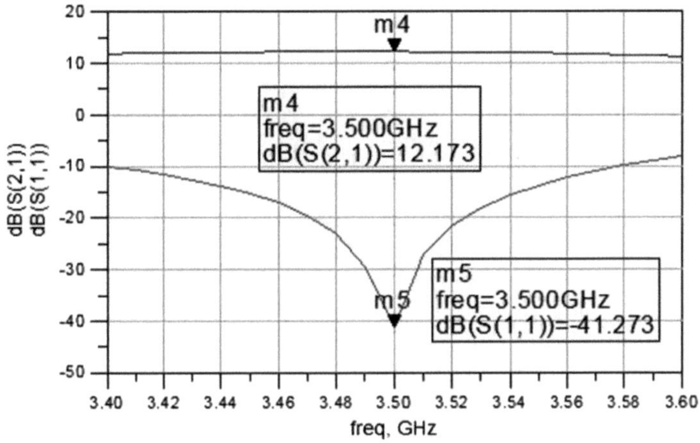

Fig. 18 S_{11} and S_{21}

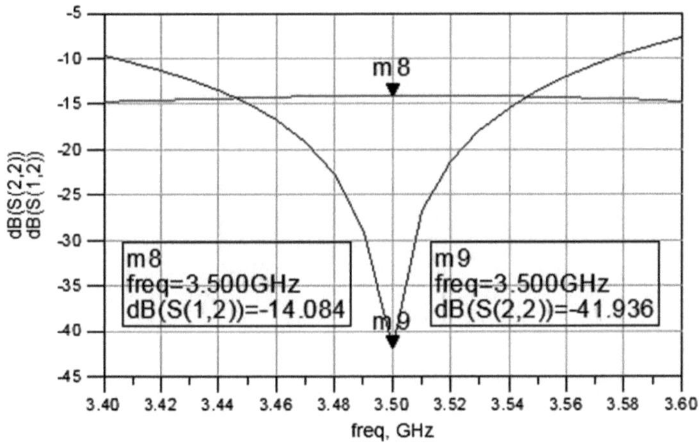

Fig. 19 S_{12} and S_{22}

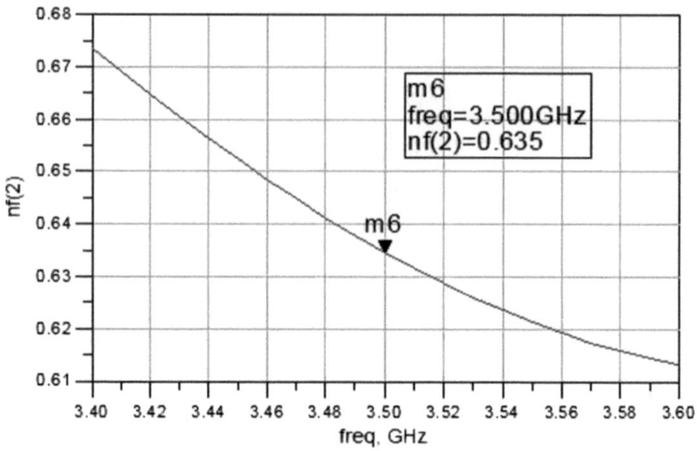

Fig. 20 Noise figure

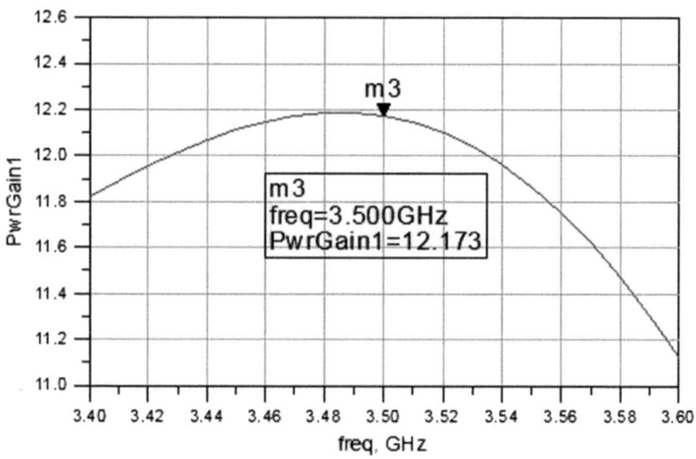

Fig. 21 Power gain

Table 1 Comparative results of LNA

Techniques	S_{21}/Power gain (dB)	Noise figure (dB)	S_{11} (dB)
[4] (cascade)	12.95	3	−10.3
[5] (cascade)	11	4.6	−12
[6] (cascade)	14.4	3.3	−15
[7] (cascade)	14	2.9	−2.44
Gate resistor LNA	10.41	3.066	−51.568
Drain resistor LNA	11.806	0.795	−19.165
Source inductor LNA	12.173	0.635	−41.273

4 Comparative Results

See Table 1.

5 Conclusion

The different techniques to improve the stability of the amplifier are discussed and analyzed to compare and determine the suitable design for the WiMAX applications. The results are obtained after implementing the designs in ADS software. The overall results are proving satisfactory and it can be concluded that the source degenerative inductor addition in series with the source terminal of HEMT gives optimum and efficient noise figure and gain. Further to improve the bandwidth of LNA and to reduce the noise figure cascading multistage low noise amplifiers can be designed.

References

1. https://www.fcc.gov/general/wimax-applications-public-safety
2. G.-L. Ning, Z.-Y. Lei et al., Design of concurrent low noise amplifier for multi–band applications. Prog. Electromagnet. Res. C **22**, 165–178 (2011)
3. I. Biswas, S.C. Bose et al., Design of a 2.3 GHz low noise amplifier for WIMAX application, in *International conference on Device, Circuits and Systems*, March (2012), pp. 105–109
4. A. Zokaei, A. Amirabadi, A low voltage low power 3.5/5.8 GHz dual-band common gate low noise amplifier, in *IEEE Regional Symposium on Micro and Nanoelectronics* (*RSM*) (2013)
5. C.-H. Wu, K.-L. Liu et al., Design of a low-noise amplifier using complementary and current-reused technology, in *International Conference on Microwave and Millimeter Wave Technology* (2012)
6. H.-L. Kao, B.-H. Wei et al., A 3.5 GHz low-noise amplifier in 0.35 μm GaN HEMT on Si-substrate for WiMAX applications. Int. J. Electron. Electr. Eng. **3**(1), 61–65 (2015)
7. A.M. Bassal, A.H. Jarndal, GaN low noise amplifier design for WiMAX applications, in *IEEE 16th European Miditerrean Microwave Symposium* (2016)

8. J.-T. Yang, H.-P. Fan et al., A RF CMOS low noise amplifier for WiMAX applications, in *Proceedings of the 13th WSEAS International Conference on CIRCUITS* (2004), pp. 123–127
9. M. Iyer, T. Shanmuganantham, Design of LNA for WLAN applications, in *IEEE International Conference of Circuits and Systems* (2017), pp. 319–323
10. M. Iyer, T. Shanmuganantham, Design of LNA for satellite uplink, in *Elsevier's SSRN Journal of Information Systems & eBusiness Network – ISSN: 1556-5068* (2017), pp. 255–264

The Effect of Various Parameters on the Nozzle Diameter and 3D Printed Product in Fused Deposition Modelling: An Approach

Jayant Giri, Ajit Patil and Harish Prabhu

Abstract Fused deposition Modelling (FDM) is the one of most striking and tremendously growing technology for manufacturing of 3D printed product. FDM technology was developed by stratasys. This technology was introduced in the era of 1980. FDM is additive manufacturing technology which fabricates products by extruding material technology through the nozzle in semi-molten state on the platform. Various thermoplastics such as PLA, ABS is widely used as a filament. But now a days modern FDM printers use abrasives such as aluminium, brass along with the PLA and ABS for printing to enhance the strength of printed specimen. This paper reviews effect of various parameters such as temperature, feed rate, print speed, orientation angle, infill density, nozzle size on thermal properties and mechanical properties of printed product. The temperature and feed rate are most important factors due to which wear of nozzle occurs.

Keywords Fused deposition modelling · Additive manufacturing · ABS · PLA Mechanical properties

1 Introduction

There are variousAs per the present scenario consumer demands wide range of products and in order to fulfil their requirements a new technology called rapid prototyping was introduced. 3D printing is a process of converting a digital file into a 3D object [1]. It is one of the best techniques and has become very popular due to high

J. Giri (✉) · A. Patil · H. Prabhu
Department of Mechanical Engineering, Yeshwantrao Chavan
College of Engineering, Nagpur, India
e-mail: jaytatnpgiri@gmail.com

A. Patil
e-mail: ajitpatilkarpo@gmail.com

H. Prabhu
e-mail: harishprabhu9898@gmail.com

C. R. Krishna et al. (eds.), *Proceedings of 2nd International Conference on Communication, Computing and Networking*, Lecture Notes in Networks and Systems 46, https://doi.org/10.1007/978-981-13-1217-5_83

degree of flexibility. In the coming years it is going to take this technology to a new world. There are various types of 3d printers used such as SLA (Stereolithography), Selective Laser Sintering (SLS) and Fused deposition modelling (FDM). FDM is one of the best technique which is widely used in for building complex products. There is minimal wastage of material. FDM is one of the most efficient process as compared to other rapid prototyping process. In Fused deposition modelling material is heated to a temperature till it gets melt and is extruded through the nozzle. Material is deposited through the nozzle layer by layer till it gets transformed into a complete product. The movement of the nozzle and the bed is controlled on the basis of G codes [2]. FDM has wide range of application in many areas such as medical industry, automobile industry, manufacturing industry and also for printing household products. Wear of the nozzle is mainly dependent upon the two factors such as temperature and volume feed rate [3]. FDM type is used for different type of thermoplastics such as ABS (Acrylonitrile-Butadiene-Styrene), PLA (Polylactic acid). Nozzle diameter plays a major role in determining the surface quality of the printed product, so, it very important to select an optimum nozzle diameter while printing a product [4]. The filament should be heated depending upon the type of filament used otherwise due to uneven heating nozzle clogging may occur and the temperature should be well enough for the adhesiveness between the layers. Mainly we apply adhesive glue to the bed because the first layer deposited should stick properly [5]. Temperature distribution plays a major role in determining the surface finish, temperature distribution is constant along the horizontal direction but in vertical direction it varies from point to point [6]. Mechanical properties of the printed product depend upon the temperature and should be heated depending upon the type of filament we are using otherwise due to uneven heating the properties of material can deteriorate [7]. L. M. Galantucci found that the mechanical properties of the printed product depend upon the layer height, volume feed rate and the print speed irrespective of the product printed from the open source printer or by an industrial system printer [8]. Infill density and layer thickness plays a vital role in determining the strength of the product and dimensional characteristics [9, 10]. The material can be deposited through the nozzle by varying the nozzle orientation angle from 0° to 90°. Product printed at 45° orientation angle and 100% infill density has higher tensile strength [11]. The tensile strength reduces by 50% if it is varied from 45° to 90° at 100% infill density [12]. Tensile strength along the vertical and horizontal direction is not evenly distributed, tensile strength in vertical direction is 5% more than in horizontal direction [13]. Dr. Tahseen Fadhil Abbas explained compressive strength is maximum at 80% infill density [14]. A printed specimen was evaluated to check the material uniformity along the tensile and axial direction, deflection is maximum in tensile direction as compared to that in axial direction [15]. Chieh Kung, carried out the experiment by printing a complex product such as screw, as he used a support structure while printing which was inherently designed in the cad software. Due to the support structure the gridded pattern remained on the product [16].

1.1 CAD CAM Interface

Initially at the very first stage we design a 3D object as per our required dimension in a cad software such as blender, solid works. Various types of commercial as well as open source software are available for 3D modelling. Blender is a software which is commonly used for creating a cad file. Every software has its own advantages and limitations. It is an open source software which is freely available. Blender has a special feature which tells that our 3D model is printable and exportable or not. Blender also tells us how the proper uniformity of material can be maintained. It is difficult to edit 3D Mesh surfaces in blender. Blender is occupied with numerous tools which can be used for designing complex models [17]. An online survey was conducted by I. materialise in which they found that Blender is very popularly software for 3D modelling as shown in the Fig. 1.

The cad file is converted to standard tessellation language (STL) which is inherently present in cad software. STL is also known by name of steriolithography or standard triangular language. There are various other file formats but STL is a standard format which is commonly used in 3D printing. STL cuts the object into various triangular layers and stores the information regarding the surface geometry of the object in ASCII and binary form. The main drawback of the STL file is that it doesn't give any information indication regarding the colour. It is very important to convert the cad model into STL format because the slicing software will only accept the STL file. In order do the further process STL file is been further sent to slicer software which cuts the object into numerous layers of same thickness. Various slicer software's are used such as cura, repetier host, Slic3r. In slicer software we can vary various parameters such as layer height, print speed, temperature and feed rate. Slicer software also specifies the amount of material required to print the object [1]. Cura is a special software where generates the program for the object we have to print. The program is totally based on the various set of G codes. The program is loaded

Fig. 1 Popularity of 3D modelling software used for 3D printing

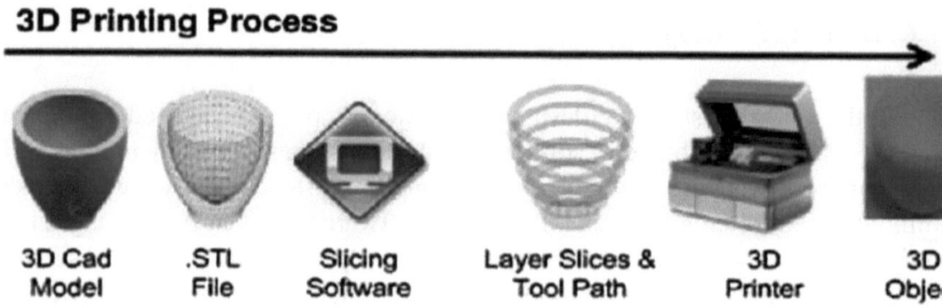

Fig. 2 Steps in 3D printing

into the controller to do the printing process as shown in Fig. 2 G-code is preparatory function it tells the controller to perform the task.

1.2 Working

The movement of the nozzle and the bed is totally governed by various G codes. Material is extruded through the nozzle layer by layer as shown in Fig. 3. The extruder assembly basically consists of two ends cold end and hot end. Cold end is present above the hot end. Material is stored in the spool and is pulled from spool through driving mechanism and is sent to hot end for further process as shown in. As the material enters into the hot zone it is melted in the heating chamber and is extruded through the nozzle as shown in Fig. 4. Finally, the melted material is deposited on the bed. The material should be heated at different temperature depending upon the type of filament used. Nichrome wire or cartridge heater is used for heating the filament. It is very important to control the temperature and the volume flow rate of material [2]. Thermistor is present to sense the temperature of the material the chamber and send feedback to controller. If we heat the material with excess temperature the material may either lose their properties and wearing of the nozzle can also occur. Due to wearing of the nozzle the surface quality of the printed specimen may effect. Nozzle is any important part of extruder assembly and it plays a vital role. The diameter of the nozzle should be properly selected as it affects the quality and surface finish of the product. The height of the layer should not be higher than the size of the nozzle because it may reduce the bond strength and overall quality will be affected. For, example if a 3D printer is using a nozzle of 0.4 mm, then the height of the layer should not be more than 0.3 mm. In FDM while using ABS as filament the melting temperature should be around 210 °C, and while using PLA it should be 160 and it varies as per the material. So it is very necessary to maintain required extrusion temperature to avoid jamming in the nozzle [5]. Jamming of the nozzle also occurs due to improper design of the extruder.

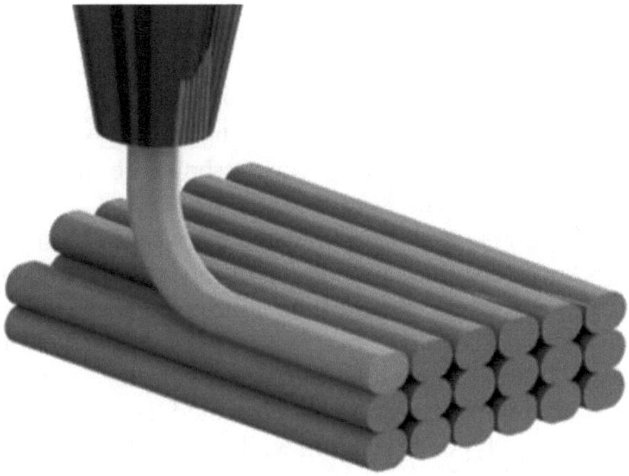

Fig. 3 Layer wise addition

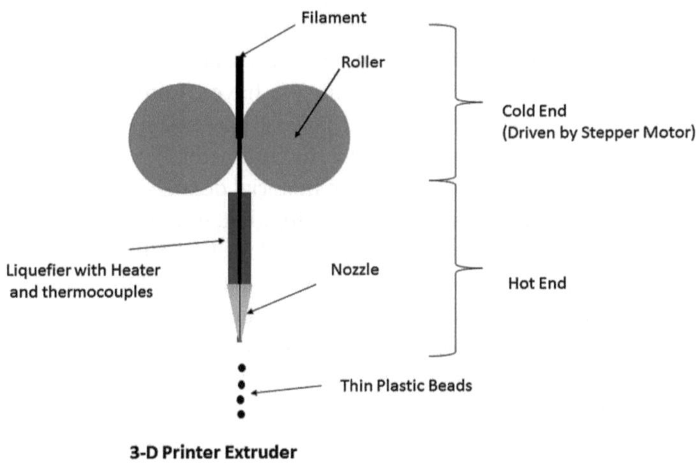

Fig. 4 Extruder

2 Literature Review

Yaman and Dolen et al. [1] explained how the cad cam interface is been generated in 3D printing. At the very first stage a cad file is designed of the specimen and then converted into STL format in order do the further process, STL file is been loaded into the slicer software. The slider software slices the object into various layers of some required thickness and it generates a G-code program for the product to be printed. G-code is a function which tells the controller to interpret and execute the task. The G code is fed into controller for the movement of the print head. Alabdullah et al. [2] explained the working of fused deposition modelling process. Filament enters into the cold end from the spool and is pushed in the hot end zone using the drive mechanism. The filament is heated with the help of nichrome wire till it gets melt in the hot end and a thermistor is also present to give feedback to the controller regarding the temperature. It is a process in which material is added layer by layer on the bed in semisolid state. The layer thickness should be smaller than the nozzle diameter for getting good results.

Pitayachaval and Masnok et al. [3] Evaluated the wear of the nozzle by varying various parameters such feed rate and volume of the material and nozzle diameter was maintained fixed at 0.4 mm. Wear of the nozzle was then evaluated by using a microscope. Finally, we can conclude that if we increase the feed rate the wear rate of the nozzle increases. There is linear relationship between the wear of the nozzle and feed rate. Sukindar and Ariffinetal et al. [4] presented the effect of diameter of the nozzle in fused deposition type 3D printing while using PLA as a filament. Various nozzle diameters are used such 0.2, 0.25, 0.3 and 0.4 mm to find effect on surface finish of the printed product. As here we have seen that pressure drop is maximum while using a nozzle of 0.2 mm diameter as pressure drop can influence the surface finish and the road width of the printed specimen. Road width affects the uniformity and accuracy of the printed product. So it very necessary to select optimum nozzle for extruding the filament at a constant rate. We have examined that the material extruded through 0.3 mm nozzle has good consistency and has similar road width for every layer as compared to nozzle of 0.2 mm diameter.

Kim and Espalin et al. [5] studied status of material deposition in fused deposition modelling. In FDM type of 3D printing nozzle clogging, uncertain substance invasion as well as Substrate deformation and object collapse are the common problems which can occur most of the time while depositing a material through the extruder. Nozzle clogging occurs due to inappropriate temperature and due to the presence dust particles on the filament. Due to which it may lead to irregular deposition. If the 1st layer has no proper adhesion with the further depositing layer from the nozzle, due to which the whole object can collapse so to avoid this the extruded material should property stick with the bed. Zhoua and Xiong et al. [6] studied the temperature analysis in fused deposition modelling. The bond strength between the layers depends upon the temperature distribution and it also affects the print quality. Hence temperature distribution plays a major role in determining the properties of the printed product. The temperature distribution along the length is non-linear. If we compare the temperature distribution in horizontal and vertical directions, then it is constant along the horizontal direction and in vertical direction there is a decreasing curve. Behzadnasab and Yousefi et al. [7] presented the effect of temperature on the physical and mechanical properties of the printed product. PLA is used as a filament. It is clearly evident that the due to increase in temperature of the nozzle the tensile strength and young's modulus of the printed product increases but there is no effect on the elongation. On raising the temperature adhesion properties between the layers will increase. But temperature should be in the range of 180–240. Strain break value has also been raised from 34 to 56 MPa. If the temperature crosses beyond 240 °C the material starts to deteriorate. Hence we can say that mechanical properties of the printed product depend on the temperature.

Galantucci and Bodi et al. [8] evaluated the dimension performance of the printed products by using 2 different FDM 3D printers. The two printers are industrial system and open source type. During the Experiment a rectangular specimen is printed using both the printers. ABS-P400 was used as a filament for printing the object. Some distortion could also occur while raising the volume feed rate through the nozzle. The specimen was evaluated using a microscope. As the product printed via both printers didn't have much distortion. The print quality does not depend on the printer but it fully depends on the layer height, volume feed rate and the print speed. Decuir and Phelan el at. [9] experimented to find the strength of two types of filament such as PLA and ABS. The strength of the filament such as PLA and ABS depends upon various parameters such as infill percentage, layer height, print orientation, extruding temperature, and build speed. It has been Observed that Tensile strength of the ABS filament is higher than PLA filament. While printing the extruding temperature was kept constant. From the tensile strength it has been seen that the strength is weakest at the infill percentage of 20%. As we raise the infill percentage of the material the strength of the printed object increases. The strength is maximum when the infill percentage is between 80 to 100%. As the infill capacity increases tensile strength also increases. Nuñez and Rivas et al. [10] carried out this experiment by varying infill density and the layer thickness of the specimen, in order to find the effect of dimensional characteristics on the printed product while using ABS as a filament. After evaluation it was found that best surface finish was obtained with a combination of layer thickness at 0.178 mm and at 100% infill density and there was minimum flatness error as compared to that with the layer thickness of 0.254 mm. The product with layer thickness at 0.254 mm and at 100% infill density has good dimensional characteristics but the deviation along the layer was maximum.

Afrose and Masood et al. [11] studied the experiment by using PLA as a filament. In fused deposition modelling type of 3D printing in order to find the fatigue tensile behaviour on the PLA printed parts. The specimen was deposited from the nozzle along both x and y at same feed rate and velocity. They have also deposited along 45° orientation. Tensile test was carried out in that they found the specimen printed in x direction has higher tensile strength as compared to that along the other two orientations. The tensile strength along the x axis was about 60–64% higher than along y axis. A specimen built in 45° orientation has the larger capacity to store the strain energy. Ahmed and Susmel el at. [12] studied the behaviour of strength and fatigue behaviour on the printed products by varying the deposition angle. The specimen was printed via 3D-printer Ultimaker 2 Extended+, by varying the deposition angle from 0° to 90°. As infill orientation angle is one of the major parameter that can affect the ultimate tensile strength. As we found that tensile strength reduces by 50% as the nozzle angle is varied from 0° to 90° along the horizontal axis. The tensile strength along the horizontal is 30% higher than along y axis. Yang and Zhao et al. [13] studied the mechanical properties of the specimen while using polymer parts as a specimen. The volume infill flow rate is varied from 10, 50–100% and then carried out the tensile test. It has been found that ultimate stress of Specimen of volume 100% filled capacity is approximately 20% larger than specimen of volume 10% filled capacity. Tensile test is carried out in both horizontal as well as vertical direction as along the vertical direction the tensile strength was 5% more than in horizontal direction.

Abbas and Othman et al. [14] studied the compression property of the printed specimen by varying the infill density of the melted material. PLA has been used to print the specimen. The infill density of the filament is varied from 0 to 100%. While setting infill density at 0% the printer prints only the outer layer as per the shell thickness and the inner part remains empty. As we increase the infill density the time required for printing also increases. The compressive was found maximum at 80% infill density. Baker and Mccoy el al. [15] studied the anisotropic properties of material been printed using various filaments such as PLA, ABS and electronically-conductive PLA (C-PLA). The properties of the printed products also vary as per the direction of the axis. The axial direction is along z axis, and transverse direction is along x and y axis. Experiments showed that PLA and ABS had anisotropic properties greater than 10% along both transverse and axial under compression. Strain rate was kept constant at 0.254 mm/min, and strain rate was measured using an extensometer. The deflection is maximum in tensile direction as compared to that in axial direction. ABS being more ductile has greater tendency to fail in transverse direction as compared to that in axial direction under compression. Kung and Lee et al. [16] performed the experiment by printing a screw. As screw is a complex structure, so a supporting material was used while printing. The screw is printed through the nozzle horizontally as well as vertically. The diameter of the nozzle is 0.4 mm and the heated bed temperature is 60 °C. The diameter of the filament is 3 mm. Support structure been used in both the cases which is been automatically designed in Cura software. Support structure is in contact with the specimen while printing, so due to which it leaves grid pattern on the product. Finally, we have seen that the vertical print is better than the horizontal print.

3 Conclusion

The present review shows that the mechanical properties, thermal properties and dimensional characteristics depends upon the various parameters such as temperature, layer thickness, infill density, path speed, feed rate, printing time and nozzle diameter. The optimum nozzle diameter for printing a product is 0.3 mm. Wear of the nozzle mainly depends upon two major factors such as temperature and volume feed rate. Tensile and compressive strength also depends upon the nozzle orientation angle and infill density. There is direct relationship between the parameters and surface quality of the printed product. So from the literature review we can conclude that for preparing a specimen of good quality we have select appropriate parameters.

References

1. U. Yaman, M. Dolen, *A Command Generation Approach for Desktop Fused Filament Fabrication 3D Printers* (IEEE, 2016) (978-1-5090-3474-1/16 2016)
2. F. Alabdullah, Fused deposition modeling (FDM) mechanism. Int. J. Sci. Eng. Res. **7**(5) (May 2016)
3. P. Pitayachaval, K. Masnok, Feed rate and volume of material effects in fused deposition modeling nozzle wear, in *2017 4th International Conference on Industrial Engineering And Application*
4. N.A. Sukindar, B.T. Ariffin, H.T. Baharudin, C.N.A. Jaafar, M.I.S. Ismail, Analysing the effect of nozzle diameter in fused deposition modeling for extruding polylactic acid using open source 3d printing. Jurnal Teknologi (Sciences and Engineering) **78**(10), 7–15 (2016)
5. C. Kim, D. Espalin, A. Cuaron, M.A. Perez, E. MacDonald, R.B. Wicker, A study to detect a material deposition status in fused deposition modeling technology, in *2015 IEEE International Conference on Advanced Intelligent Mechatronics (AIM)* (Busan, Korea, 7–11 July 2015)
6. Y. Zhoua, G. Xiong, T. Nyberg, D. Liue, Temperature analysis in the fused deposition modeling process, in *2016 3rd International Conference on Information Science and Control Engineering*
7. M. Behzadnasab, A.A. Yousefi, Effects of 3D printer nozzle head temperature on the physical and mechanical properties of PLA based product, in *12th International Seminar on Polymer Science and Technology* (Islamic Azad University, Tehran, Iran.L, 2–5 Nov 2016)
8. M. Galantucci, I. Bodi, J. Kacani, F. Lavecchia, *During the Exprimentan Rectangular Specimen Printed Using Both the Printers* (Published by Elsevier Procedia CIRP 28, 2015)
9. F. Decuir, K. Phelan, B. Hollins, Mechanical strength of 3-D printed filaments, in *2016 32nd Southern Biomedical Engineering Conference*
10. P.J. Nuñez, A. Rivas, E. García-Plaza, E. Beamud, A. Sanz-Lobera, Dimensional and surface texture characterization in fused deposition modelling (FDM) with ABS plus. Procedia Eng. **132**, 856–863 (2015)
11. M.F. Afrose, S.H. Masood, P. Iovenitti, M. Nikzad, I. Sbarsk, Effects of part build orientations on fatigue behaviour of FDM-processed PLA material. Prog. Addit. Manuf. **1**, 21–28 (2016). https://doi.org/10.1007/s40964-015-0002-3
12. A. Ahmed, L. Susmel, Additively manufactured PLA under static loading: strength/cracking behaviour vs. deposition angle. Struct. Integrity Procedia **3** (2017)
13. J.-H. Yang, Z.-J. Zhao, S.-H. Park, Evaluation of directional mechanical properties of 3D printed polymer parts, in *2015 15th International Conference on Control, Automation and Systems (ICCAS 2015)* (BEXCO, Busan, Korea, 13–16 Oct 2015)
14. T.F. Abbas, F.M. Othman, H.B. Ali, Effect of infill Parameter on compression property in FDM process. Int. J. Eng. Res. Appl. **7**(10), 16–19 (Oct 2017) (ISSN: 2248-9622) (Part -I)
15. A.M. Baker, J. Mccoy, B.S. Majumdar, Brinttanyrumley-Ouellette, J. Wahry, A.N. Marchi, J.D.B. Dusanpernjak, *Measurement and Modelling of Thermal and Mechanical Anisotropy of Parts Additively Manufactured Using Fused Deposition Modelling (FDM)* (2017 Research Gate)
16. C. Kung, T.-H. Lee, Selection of process parameters for the precision of screw-shaped objects on reprap 3-D printers, in *2016 International Symposium on Computer, Consumer and Control*
17. C. Nandi, A. Caspi, D. Grossman, Z. Tatlock, Programming language tools and techniques for 3D printing, in *2nd Summit on Advances in Programming Languages (SNAPL 2017)*

Computational Technique for Onset of P-Wave for Seismic Alert System

Satish Kumar, Pawan Kapur and Renu Vig

Abstract The disasters occurred during the major earthquakes result in destruction, losses of property and human life. In order to mitigate the damage impact, there is an urgent need for the design, development and installation of Seismic Alert System (SAS) for vital installations. The development of SAS requires real-time algorithms for fast determination of source parameters and their reliability estimation. Early warning can be generated by detecting the primary waves, which travel with higher velocity then the disaster waves. The computations of source parameters start after detecting the primary wave i.e. P-wave. The generated alert warning depends on the accurate detection of P-wave. The alert signal generated can be sent to the vital installations such as nuclear and thermal plants, fast running trains, hospitals etc. either for shutdown the activities or for resuming emergency facilities as per requirement. Therefore, for the development of SAS it becomes necessary to develop a computationally efficient method that can automatically determine the onset time of P-wave. This paper illustrates the computational technique for detecting the arrival of P-wave along with the analysis for early warning point of view by using the polarization attributes i.e. degree of polarization (DOP), Planarity (Planr) and Rectilinearity (Rect) of recorded seismic signal.

Keywords Alert system · Degree of polarization · Early warning · Earthquake Planarity · Rectilinearity · Seismic

S. Kumar (✉) · P. Kapur
CSIR-CSIO, Chandigarh, India
e-mail: satish@csio.res.in

R. Vig
University Institute of Engineering and Technology,
Panjab University, Chandigarh, India

C. R. Krishna et al. (eds.), *Proceedings of 2nd International Conference on Communication, Computing and Networking*, Lecture Notes in Networks and Systems 46, https://doi.org/10.1007/978-981-13-1217-5_84

1 Introduction

It is true that occurrence of earthquakes cannot be predicted accurately. Though, enormous scientific efforts are being made but even today, it is beyond our comprehension and also it gazes that it will be not possible to predict it in near future. To minimize the colossal losses which take place as a result of major earthquake, the design and configuration of SAS has become the urgent need of the day. The operational purpose of these early warning networks is to generate and send alert signals to the vital places at least few tens of seconds before the actual ground shaking starts at these places. The technical feasibility for early warning has significantly increased during the past years [1].

State-of-the-art instrumentation is required for the configuration SAS system, which not only include sophisticated sensing and processing modules but also require high computational software techniques for the computations of various earthquake parameters such as phase detection (P-wave), magnitude, location etc. The first arriving P-wave can be used to determine other significant parameters, and early warning can be generated by detecting the P-waves. The lead time i.e. amount of advance warning depends on the arrival time difference between the P-wave and the disaster wave, generally on the order of 10 s. Therefore, the effectiveness of SAS depends on accurate detection of the P-wave time and hence there is need for a more reliable robust and fast computational technique for the accurate P-wave detection from recorded seismic packet. This paper explained the developed technique for the identification of P-wave using sliding window technique with the polarization properties of the recorded three component seismic signal.

2 Polarization

One property of seismic phases is signal polarization and P-waves show a high degree of linear polarization [2]. Therefore, the important parameter for identifying the presence of P-wave energy is the polarization analysis of three components of seismic event. The characteristic function for the P-wave is obtained from the three component analysis of seismic signal, which allows to reconstruct the ground particle motion and to determine its polarization [3]. The particle motion can be represented by the attributes that show discrete features for seismic waves, which is used to identify the on-set of P-wave [4].

Polarization analysis of three component seismic signal is used to identify the peak of polarized energy that distinguishes the body-wave arrivals from seismic noise, which tends to be uncorrelated between the signal components. The signal covariance matrix $M = \sigma_{ij}$ is computed in a moving time window to get eigenvalues (λ) and corresponding eigenvectors for three components of seismic signal [5]. For propagation of seismic wave in three dimensions, there are three eigenvectors/eigenvalues. It has an eigenvector or polarization that is in the direction of the

P-wave [6]. Montalbetti and Kanasewich proposed a polarization filter by computing the rectilinearity and direction of particle motion from horizontal and vertical components of recorded seismic signal [2, 7].

Polarization attributes from three component seismic signal can be used to detect the onset of P-wave. The DOP is calculated from the total power that is carried by the different components of the seismic wave [8]. Measuring the degree of polarization helps in distinguishing the onset of seismic waves. Degree of polarization i.e. DOP is computed by using following equation:

$$\mathrm{DOP} = \frac{(\lambda_1 - \lambda_2)^2 + (\lambda_2 - \lambda_3)^2 + (\lambda_3 - \lambda_1)^2}{2(\lambda_1 + \lambda_2 + \lambda_3)^2}$$

Rectilinearity (*Rect*) is a property which states that waves propagate in straight lines. This property is a measure of how linear the wave field is polarized, yields values between zero and one [8]. Therefore, it measures the degree of linear polarization of three component seismic signal and attains its maximum value at P-wave arrival time. Rectilinearity is given by

$$\mathrm{Rect} = 1 - \frac{(\lambda_2 + \lambda_3)}{2\lambda_1}$$

In practice, seismic data shows that waves are polarized in a plane called main plane of polarization. Planarity (*Planr*) measures the distribution of energy and degree of polarization within the selected time window and has also been used to estimate the onset of P-Phase over above mentioned time windows [9]. Its range of values is between 0 and 1 for exactly linear polarization, and for an undetermined polarization [10]. Planarity is given by

$$\mathrm{Planr} = 1 - \frac{2\lambda_3}{\lambda_1 + \lambda_2}$$

3 Implementation

Seismic signals are a complicated mixture of source radiation, affects the spectral content and relative amplitude of generated seismic wave, their identification involves assessing their behaviour as a function of distance, measuring the type of ground motion they produce [11].

A signal-processing method has been developed for three-component seismic data that allows the accurate detection of P-wave using polarization property of seismic signal. Polarization characteristics of seismic signal has been analysed by computing the covariance of the three channels with each other. For detecting the P-wave particle motion analysis of ground motion amplitudes of three components has been performed by finding the covariance between different components. The particle motion of a three-component seismic event recording i.e. $S = (x, y, z)$ can be described using the three-component principal components method. The matrix (M) is the covariance matrix in a time window of length of N samples, computed as below [12, 13]:

$$M = \begin{bmatrix} \theta(x,x) & \theta(x,y) & \theta(x,z) \\ \theta(y,x) & \theta(y,y) & \theta(y,z) \\ \theta(z,x) & \theta(z,y) & \theta(z,z) \end{bmatrix}$$

where

$$\theta(x,y) = \frac{1}{N}\sum_{i=1}^{N}(x_i - x')(y_i - y')$$

The integer N is the number of points in the window. Parameters x, y and z represent amplitudes of horizontal and vertical components of seismograms. x' and y' are the average values of the seismic components. The matrix M is decomposed into eigenvectors e_1, e_2 and e_3 with corresponding Eigen values λ_1, λ_2 and λ_3, such that $\lambda_1 \geq \lambda_2 \geq \lambda_3$.

Polarization parameter is computed from the Eigen-values and Eigenvectors within the selected time window which describe the polarization direction (principal direction of oscillation) i.e. the degree of polarization (DOP), rectilinearity (rect), planarity (plan). The Eigen vector of the largest Eigen-value indicates the direction of a P-wave. The polarization attributes i.e. (a) degree of polarization, (b) rectilinearity, (c) planarity are computed using matrix M for different number of samples windows as- 10 (0.1 s), 20 (0.2 s), 50 (0.5 s), 100 (1.0 s) for a sample rate of recorded seismic event of 100 samples/s. The polarization attributes have been computed and analysed for both point-wise and without point-wise sliding window technique for these different sample windows as mentioned below.

3.1 Point-Wise Sliding Window Technique

In point-wise sliding window technique, value of covariance matrix is calculated at each point. It slides in a point-wise pattern by finding value of covariance matrix at every point. The new sample window is selected by discarding one sample and adding next sample. For example, if the window length is of 100 samples, then using this technique initially the value of covariance matrix from sample number 1–100 are computed, then window of 100 samples will be selected by discarding the 1st sample and adding 101th sample to it and its covariance matrix are computed and so on till the last sample has been considered. Figure 1 shows the window length in a point-wise translation.

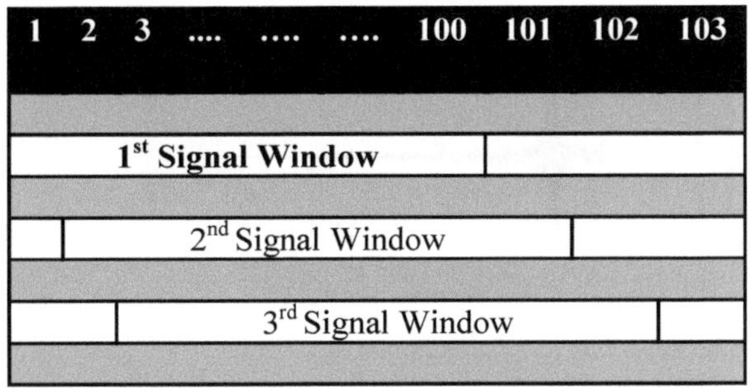

Fig. 1 Point-wise translation

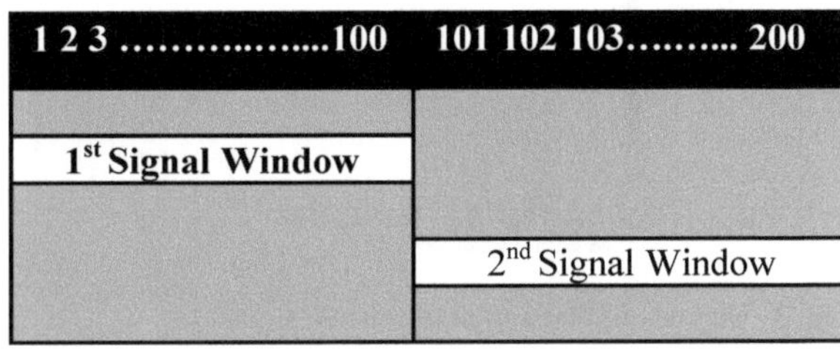

Fig. 2 Without point-wise translation

3.2 Without Point-Wise Sliding Window Technique

In without point-wise sliding window technique, the values of covariance matrix are not calculated at every point rather all old samples are discarded and only next samples are selected for computing the covariance matrix. For example, if the window length is taken as 100 samples, then as per this technique the value of covariance matrix are computed from sample number 1st to 100th, then from 101st to 200th and so on till the last sample. Figure 2 shows the window length in a without point-wise translation.

The selection of the time window is crucial mainly in reference to early warning application. Smaller time windows lead to instabilities and larger ones possibly average over phases and thus limiting the resolution of the analysis [10]. Therefore, the different time windows have been selected to optimize the algorithm to detect the P-wave in minimum possible time, as time is the critical factor for SAS to improve lead time.

4 Results and Analysis

For the estimation of the on-set of P-wave, polarization features—degree of polarization, planarity and rectilinearity have been computed from recorded seismic events. Total 37 seismic events of different magnitude and different locations recorded by Kyoshin Net (K-NET) network of National Research Institute for Earth Science and

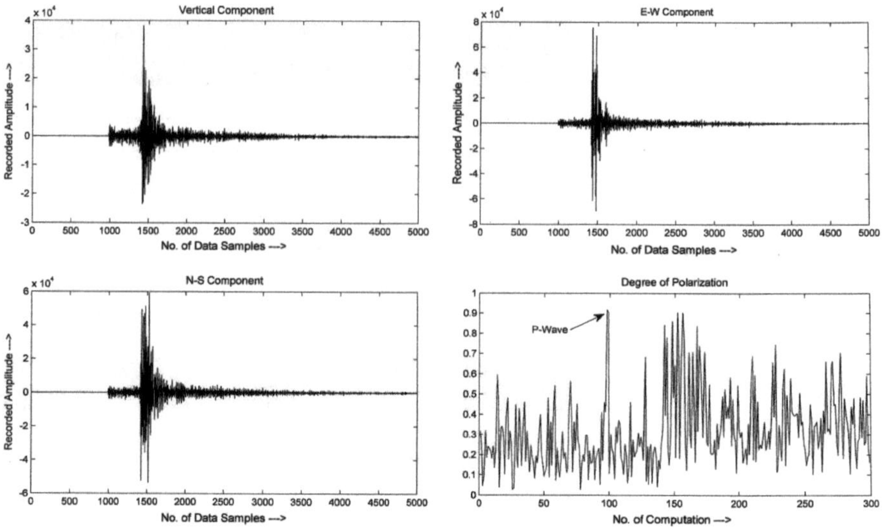

Fig. 3 Degree of polarization of seismic signal

Disaster Prevention, Japan are used for result and analysis for the identification of P-wave. K-NET is a system which records strong-motion data from different observatories deployed all over Japan [14].

The recorded data has been divided into 1, 2, 3, 4 and 5 s of data packets after triggering the event for extracting the polarization properties. The data packets are generated for the identification of P-wave mainly for SAS application. It is also required to conclude the required data size for accurate computations and to improve lead time of warning system. The seismic signal is processed for different window length of sample 10 i.e. 0.1 s, 20 i.e. 0.2 s, 50 i.e. 0.5 s and 100 i.e. 1.0 s for computing the polarization features. These features are computed for different seismic events using point-wise and without point wise sliding window translation.

Figure 3 shows the computed DOP from the seismic signal, the peak point indicated the on-set of P-wave in the selected window length i.e. 20 samples or 0.2 s. Figures 4 and 5 shows the computed planarity and rectilinearity features of recorded seismic signal of the selected window length, the peak point indicated the on-set of P-wave. Figure 6 shows the all the three features i.e. DOP, Planarity and Rectilinearity computed from the seismic signal, the dominating point indicated the on-set of P-wave.

Finally the error analysis has been computed for all the selected window length and parameters. It has been observed that to determine the on-set of P-wave from the 2 s of window length of seismic signal and Rectilinearity property of signal shows the best result with minimum error. Figures 7, 8 and 9 shows the error analysis of DOP, Planarity and Rectilinearity of the Japan earthquakes.

The comparative analysis of DOP, Plan and Rect has been made for the windows of 10, 20, 50 and 100 samples as shown in the following Table 1.

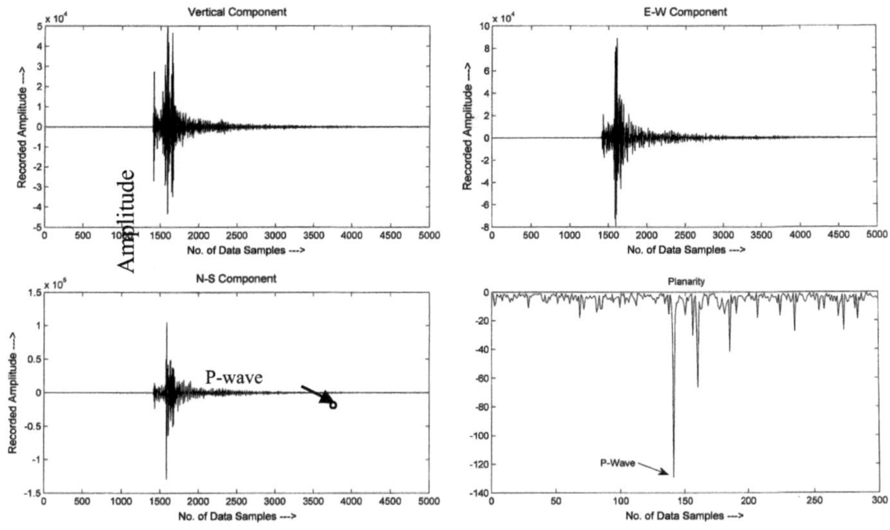

Fig. 4 Planarity property of seismic signal

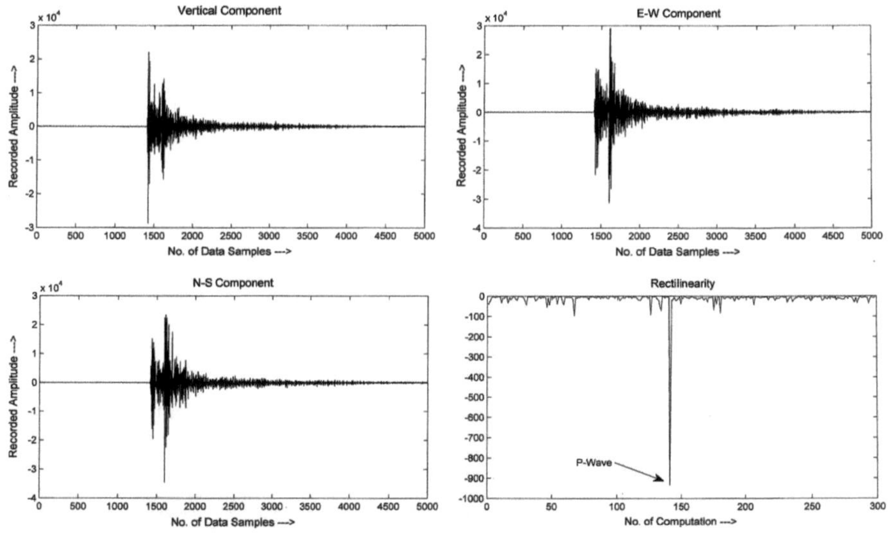

Fig. 5 Rectlinearity property of seismic signal

It has also been observed that the window size of 20 samples and polarization attributes Rectilinearity (Rect) have shown minimum error i.e. error between −0.54 and 1.34 s. Figure 10 shows the error analysis of DOP, Planarity and Rectilinearity of the Japan earthquakes for selected window of 20 samples. Figure 11 shows the error in picking of on-set of P-wave using Rect feature and indicated that the sample window of 0.2 s shows the best results.

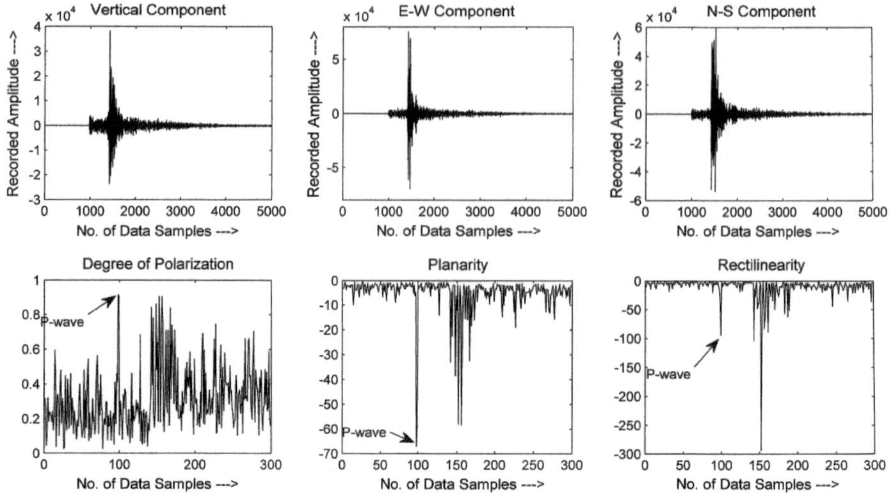

Fig. 6 DOP, planarity and rectilinearity property of seismic signal

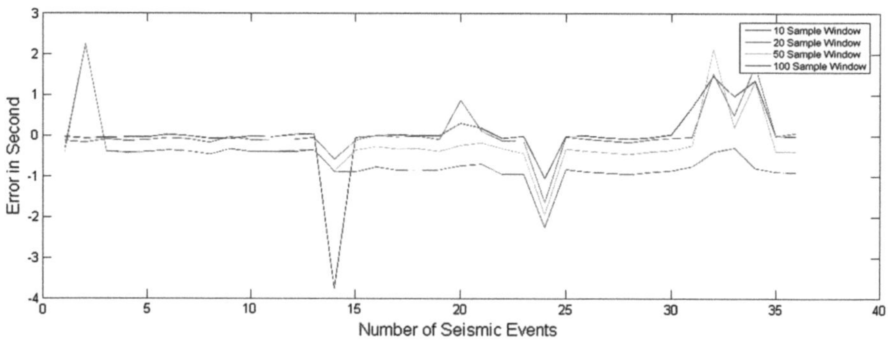

Fig. 7 Error analysis of degree of polarization

Table 1 Comparative error analysis of degree of polarization, planarity and rectilinearity (error in second)

Feature	0.1 s window	0.2 s window	0.5 s window	1.0 s window
DOP	−3.75 to 1.48	−1.63 to 1.73	−1.93 to 2.13	−2.24 to 2.25
Plan	−3.75 to 1.63	−1.63 to 1.87	−1.88 to 2.25	−2.24 to 0.23
Rect	−0.91 to 5.82	−0.54 to 1.34	−0.41 to 2.57	−1.36 to 1.98

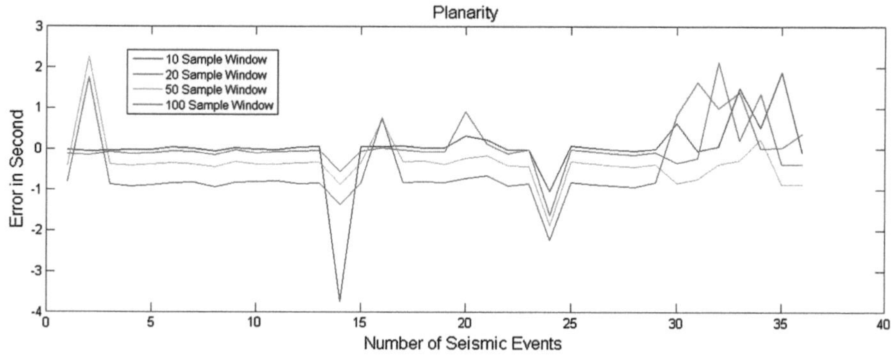

Fig. 8 Error analysis of planarity

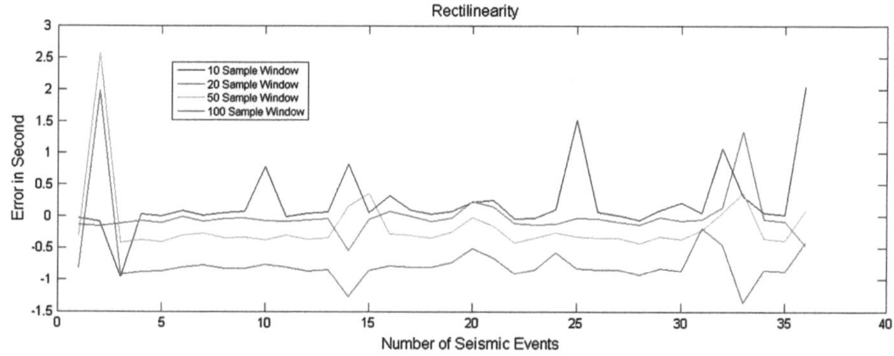

Fig. 9 Error analysis of rectilinearity

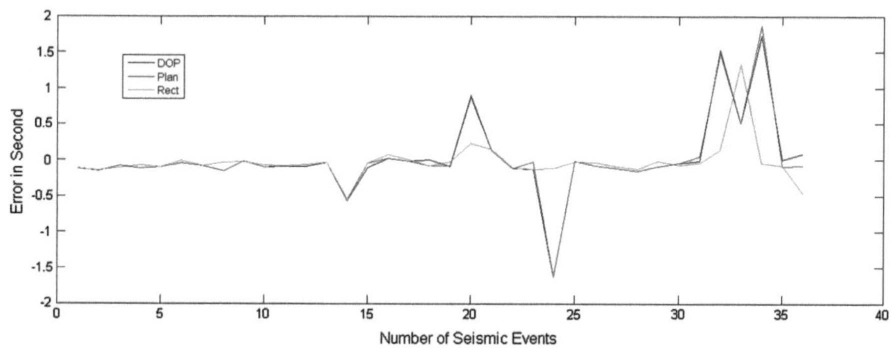

Fig. 10 Error analysis of DOP, planarity and rectilinearity

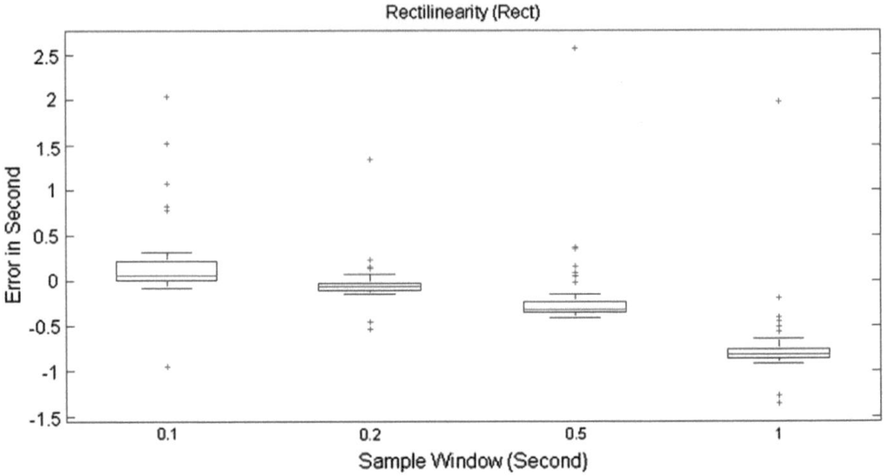

Fig. 11 Box plot of error analysis of rectilinearity with different Sample Windows

5 Conclusion

The importance of SAS has been felt around the world and lots of efforts are in progress in this direction. The important component of this system is the fast determination of parameters and their reliability after identification of P-wave. In view of this polarization features -degree of polarization, planarity and rectilinearity have been explored and a technique has been realized for the computations of these feature for the identification of P-wave and finally their comparative analysis have been made. The developed technique has been tested on the already recorded data from K-NET Japan. From the results it has been observed that for SAS point of view, a window size of 20 samples and polarization attributes rectilinearity (Rect) can be used for the identification of P-wave. The error in the result for rectilinearity feature has been observed from -0.54 to -1.34 s.

Acknowledgements We are thankful to Dr. Friedemann Wenzel, Professor, Geo Physical Institute (KIT), Karlsruhe, Germany for providing processed seismic data for research work in the area. Authors are thankful to National Research Institute for Earth Science and Disaster Prevention (NIED) for providing strong motion seismic data observed by K-NET Japan.

References

1. F. Wenzel, *An Earthquake Early Warning System for the Romanian Capital, Lecture Notes in Earth Sciences, Perspectives in Modern Seismology* (Springer, Berlin, Heidelberg, 2005), pp. 1–11
2. J.F. Montalbetti, E.R. Kanasewich, Geophy. J. Int. **21**, 119–129 (1970). https://doi.org/10.1111/j.1365-246X.1970.tb01771.x
3. P. Bormann, *IASPEI New Manual of Seismological Observatory Practice* (Germany: GFZ German Research Centre for Geosciences Potsdam, 1, 2002), pp. 1–70
4. M. Diallo, M. Kulesh, M. Holschneider, F. Scherbaum, F. Adler, Geophy **71**, 67–77 (2006). https://doi.org/10.1190/1.2194511
5. T. Fischer, A. Bouskova, L. Eisner, J.L. Calvez, *EAGE 69th Conference and Exibition* (London, 2007), pp. 11–14
6. G. Schubert, A. Dziewonski, B. Romanowicz, *Treatise on Geophysics Earthquake Seismology and Structure of the Earth* (Elsevier, 2015), pp. 1–91
7. Z. Du, G.R. Foulger, W. Mao, Geophy. J. Int. **141**, 820–828 (2000)
8. http://www.wikipedia.org. Accessed 6 Jan 2015
9. X. Li, E. Sze, D. Burns, N.M. Toksoz, *Technical Program Expanded Abstracts, Society of Exploration Geophysicists* (2004), pp. 151–154, doi:https://doi.org/10.1190/1.1851130
10. N. Maercklin, *Three-Component Processing and Analysis Tools for Seismic Data in SAC Format, DATASET* (2010), pp. 1–10. https://doi.org/10.13140/2.1.2222.0801
11. T. Lay, T.C. Wallace, *Seismology* (Academic Press, USA, 1995), pp. 35–171
12. O.K. Kedrov, V.E. Permyakova, Ann. Geophy. **37**(3), 255–266 (1994)
13. A. Jurkevics, Bull. Seis. Soc. Am. **78**, 1725–1743 (1998)
14. http://www.kyoshin.bosai.go.jp/. Accessed 3 Mar 2015

Object Oriented Dynamic Coupling and Cohesion Metrics: A Review

Navneet Kaur, Aradhana Negi and Hardeep Singh

Abstract The contribution of Software metrics in the evaluation of Computer applications has been tremendously increased during the past few decades. Several static and dynamic measures have been recommended in literature for assessing the different dimensions of Software. The static metrics can easily assess most of the features from the code itself, but to uncover the true dynamic behavior of the Software, the program needs to be measured during execution time. As the values for dynamic metrics can only be collected at run-time, therefore it is capable of capturing inheritance, dead code, dynamic binding, and polymorphism. This paper presents a review on the dynamic coupling and cohesion metrics for Object Oriented System (OOS). The system of measurement of coupling and cohesion gives an indication of high quality software which possesses attributes like reliability, maintainability, extendibility and usability. The observational perspective of review reveals two main facts: (1) little work on dynamic cohesion metrics, and (2) empirical validation of dynamic measures on a very small scale dataset, therefore practical applicability of these measures to industrial environment is questionable.

Keywords Cohesion · Coupling · Dynamic measures · Quality attributes

1 Introduction

The key reason behind the measurement of Software is to check its quality attributes like reliability, maintainability, extendibility, performance and usability. The quality attributes that necessitate for a particular product depends upon its type, e.g. in

N. Kaur (✉) · A. Negi · H. Singh
Department of Computer Science, Guru Nanak Dev University, Amritsar, India
e-mail: navneetsandhu02@gmail.com

A. Negi
e-mail: aradhana.csersh@gndu.ac.in

H. Singh
e-mail: hardeep.dcse@gndu.ac.in

C. R. Krishna et al. (eds.), *Proceedings of 2nd International Conference on Communication, Computing and Networking*, Lecture Notes in Networks and Systems 46, https://doi.org/10.1007/978-981-13-1217-5_85

incremental software development approach, the maintainability and extendibility are highly required quality attributes while in real time applications the performance is a crucial attribute. With the advent of Software coupling, cohesion and other design properties the desired quality attributes can be directly assessed from the design itself. The presence of low coupling and high cohesion in Software tells about the high quality of Software product, therefore it become vital to identify metrics for controlling the Software coupling and cohesion.

In object oriented paradigm (OOP) the coupling is defined as, "the degree of association among the objects" whereas cohesion is defined as, "the degree of conceptual consistency within an object" [1]. The inclination of a Software towards coupling leads to high complexity. To build reliable, maintainable and extendible Software, the high coupling can be proved as a serious impediment. For computing Software coupling, many researchers have made their contribution in the formulation of a variety of coupling metrics [2–10]. Similarly, high amount of work has been accomplished in the field of software cohesion measures [2, 11–14].

All the above discussed measures are static in nature, i.e. measurement of coupling and cohesion without executing the source code. It is pertinent to mention here that static measures are incapable of evaluating the dynamic behavior of the Software. To include the impact of OOP features in measurement, the concept of dynamic metrics has been introduced. This paper presents the review on Object Oriented dynamic coupling and cohesion metrics. The current paper is motivated by the work of Geetika and Singh [15], besides the dynamic coupling measures it also considers the dynamic cohesion measures. The rest of the review has been ordered in four Sections. In Sects. 2 and 3, various existing dynamic coupling and cohesion metrics are presented. Section 4 depicts the observational facts. Section 5 concludes the paper.

2 Object Oriented Dynamic Coupling Metrics

Major research work has been done in the field of coupling where low coupling is advised for high quality software. The researchers have claimed that dynamic coupling metrics are better than static coupling measures because the former measures are proficient in capturing advanced features like polymorphism, Inheritance, dynamic binding and dead code. The following subsection covers the Object Oriented Dynamic Coupling metrics proposed by different researchers.

2.1 *Yacoub et al. [16]*

The authors have suggested four measures to identify coupling between objects at an early design phase. The list of dynamic measures identified by the authors includes—*Export Object Coupling (EOC), Import Object Coupling (IOC), Object*

Request for Service (ORFS) and Object Response for Service (ORFS) (Table 1. Sr. No. 1). The authors have employed the Dynamic Modeling Diagrams to assess the runtime behavior of real time applications. The authors have claimed that the proposed measures are indicators of quality attributes such as understandability, error propagation, reusability, and maintainability. But no empirical validation has been provided by the authors to support the evaluation of suggested dynamic measures.

2.2 *Arisholm et al. [17]*

The authors have inspected the idea that dynamic coupling metrics are better predictor of change proneness as compared to static coupling metrics. The authors have proposed 12 metrics (listed in Table 1. Sr. No. 2) and validated them on 17 consecutive versions of Velocity, which is an open- source software system containing a huge number of application and library classes.

2.3 *Hassoun et al. [18–20]*

The authors have introduced a metric named as Dynamic Coupling Metric (DCM) to support the concept of Meta level architecture. The details of metric suggested by the authors are given in Table 1. Sr. No. 3.

2.4 *Mitchell et al. [21–23]*

Mitchell et al. have suggested several metrics based on tremendous research work conducted in the field of dynamic environment. The authors have ensured the validity of the metrics by correlating the measurement values with their focused quality attributes (complexity and maintainability). The authors have adopted the profile based tracing approach and used Java Platform Debug Architecture (JPDA) to collect the run-time information from Java programs [21, 22]. In another paper, the authors have also used Byte Code Engineering Library (BCEL) instrumentation to collect the dynamic measure traces at run time [23]. The list of measures proposed by the authors is included in Table 1. Sr. No. 4–6.

2.5 *Zaidman et al. [24]*

Maintainability requires the adequate understanding of the existing Software system by the programmer. In order to reduce the time required during the understandability

Table 1 Summary of object oriented dynamic coupling metrics

Sr. no.	Ref. no.	Theoretical validation	Dynamic analysis approach	Quality focus	Metric
1	[16]	–	Dynamic modeling diagrams	M, U, R, EP	EOC, IOC, ORFS and ORFS
2	[17]	Briand et al.	Profile based tracing	CP	IC_OD, IC_OM, IC_OC, IC_CD, IC_CM, IC_CC, EC_OD, EC_OM, EC_OC, EC_CD, EC_CM, and EC_CC
3	[18–20]	Braind et al. and Weyuker et al.	Aspect oriented tracing	R and M	Dynamic coupling metric
4	[21]	–	Profile based tracing	M, R, EP, T	Dynamic coupling between objects, degree of dynamic coupling between class A and class B, degree of dynamic coupling within a given set of classes
5	[22]	–	Profile based tracing	C and M	Run time import degree coupling (RD_I), Run time export degree coupling (RD_E), Run time import coupling between objects (R_I), and Run time export coupling(R_E)
6	[23]	–	BCEL	C and M	Run-time CBO, unique accesses to objects
7	[24]	–	Profile based tracing and aspect-oriented tracing	U	CRFS and IC_CC′
8	[25]	–	–	CIA	DFC
9	[26]	–	Profile based tracing	M, T, R, P	DCa, DKSC, DKCC, DKC

Note M Maintainability, *U* Understandability, *R* Reusability, *CP* Change proneness, *T* Testing, *EP* Error proneness, *C* Complexity, *P* Portability, *CIA* Change impact analysis]

of an existing Software, Zaidman and Demeyer have proposed two metrics-*Class Request for Service* (*CRFS*) and *IC_CC′*. The outcome of *CRFS* metric gives a clue to the programmer from where to start the program comprehension. The authors have validated their work against two object oriented systems, Apache Ant 1.6.1 and Jakarta JMeter 2.0.1 (Table 1. Sr. No. 7).

2.6 Beszedes et al. [25]

The authors have identified a metric named as *Dynamic Function Coupling* (*DFC*) to perform the impact analysis (Table 1. Sr. No. 8). Impact analysis is essential for evolving Software to incur the potential consequences of a change.

2.7 Sandhu and Singh [26]

The authors have introduced four metrics: Dynamic Afferent Coupling (DCa), Dynamic Key Server Class (DKSC), Dynamic Key Client Class (DKCC) and Dynamic Key Class (DKC) for measuring dynamic coupling of a Software (Table 1. Sr. No. 9).

3 Object Oriented Dynamic Cohesion Metrics

The number of static metrics suggested by researchers to measure the cohesion is high. But these static metrics are inefficient for considering the effects of dead code, dynamic binding, inheritance, and polymorphism. The following subsection discusses the key contribution of the authors in developing dynamic cohesion metrics.

3.1 Gupta and Chhabra [27]

The authors have proposed four dynamic cohesion metrics to identify four different types of association between the object elements. The measures given by the authors are: Dynamic Cohesion due to Read dependency of Methods on Attributes (DC_MA_X), Dynamic Cohesion due to Write dependency of Attributes on Methods (DC_AM_X), Dynamic Cohesion due to Call dependency between Methods (DC_MM_X), and Dynamic Cohesion due to Reference dependency between Attributes (DC_AA_X) (Table 2. Sr. No. 1). The authors have empirically validated the proposed measures and proved that the dynamic cohesion measures are an improved predictor of change-proneness rather than other cohesion measures.

Table 2 Summary of object oriented dynamic cohesion metrics

Sr. No.	Ref. no.	Theoretical validation	Dynamic analysis approach	Quality focus	Metric
1.	[20]	Briand	Aspect-oriented based tracing	CP	DC_AM_X, DC_MA_X, DC_MM_X, DC_AA_X
2.	[21]	–	Profiler based tracing	M, T, R, C, and EP	DLCOM, DCLCOM

Note CP Change proneness, *M* Maintainability, *T* Testing, *R* reusability, *C* Complexity, *EP* Error proneness

3.2 *Mitchell and Power [28]*

Mitchell and Power have made immense contribution in developing various dynamic metrics for evaluating the run time behavior of the Software. Along with dynamic coupling measures, the authors have also worked on the dynamic cohesion metrics. The static metric for cohesion: LCOM (outlined by Chidamber and Kemerer [2]) has been modified to work in the field of dynamic execution environment and two new measures have been introduced-*Dynamic Simple LCOM (DLCOM)* and *Dynamic Call-Weighted LCOM (DCLCOM)* (Table 2. Sr. No. 2).

4 Observational Results

One of the major problems in proving the validity of Software measures is the lack of clarity of measurement goals [29]. Briand et al. [29] has suggested that before formulating any metric, it is necessary to identify the objectives and measurement goals. The measurement goals of all metrics defined above is to evaluate the different quality attributes of a given Software. The authors of the current paper have noticed that many fault prediction models have been developed using static metrics. But, no such prediction model has been created using Dynamic measures. One of the significant reasons behind non-existence of such models is insufficient number of dynamic measures, focusing on fault proneness (Shown in Fig. 1). In Fig. 1, x-axis represents number of metrics and y-axis represents different quality focus.

Many authors have proposed framework containing a set of properties to validate the Software metrics theoretically, like properties outlined by Weyuker et al. [30] and Braind et al. [8]. The review of literature unveils that in most of the research work the theoretical validation of dynamic measures is missing (Shown in Fig. 2).

The authors have also observed that as compared to dynamic coupling measures the number of metrics formulated for dynamic cohesion measures are very low (Shown in Fig. 3).

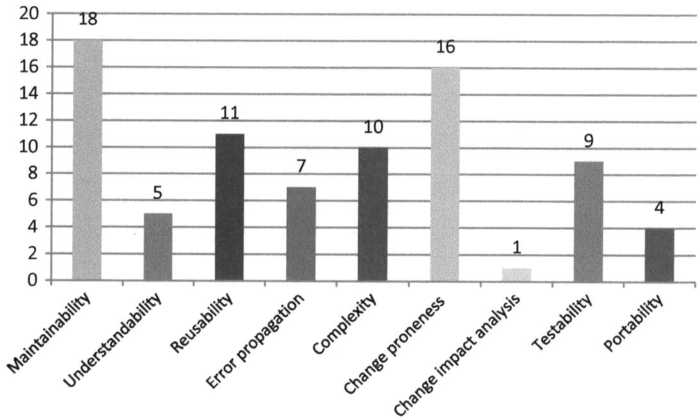

Fig. 1 Distribution of quality attributes focused by considered metrics

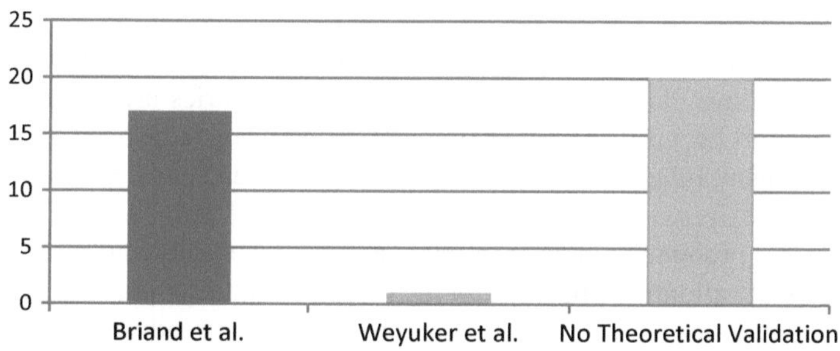

Fig. 2 Distribution of different theoretical validation properties adopted in coupling and cohesion metrics

Fig. 3 Percentage of dynamic coupling and cohesion metrics for OOS

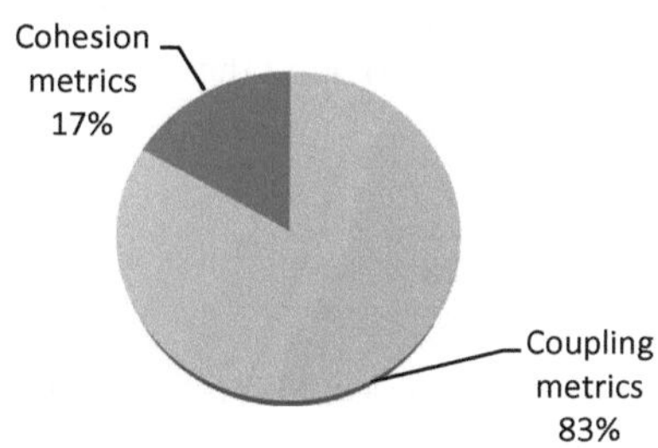

To measure the dynamic behavior of software, it is necessary to track and extract the execution events from the running software. Different dynamic analysis approaches as shown in Fig. 4, have been followed by researchers to track and record the runtime information such as method invocations and other events during program execution. The observations disclosed that the most frequently used Dynamic Analysis approach is Profile-Based Tracing [Note: In Figs. 1, 2, and 4 the x-axis represents

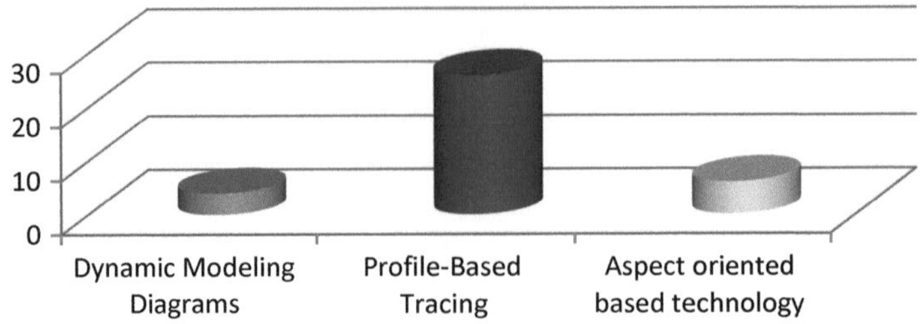

Fig. 4 Dynamic analysis approach followed by different dynamic coupling and cohesion metrics

the number of metrics and *y*-axis represents quality attributes, different theoretical validation properties, and dynamic analysis techniques respectively].

5 Conclusion

This paper presents the Object Oriented dynamic coupling and cohesion metrics. The researchers have claimed that dynamic coupling metrics are better than static coupling measures because of their proficiency in capturing advanced OOP features like, polymorphism, Inheritance, dynamic binding and dead code. Consequently, the dynamic measures provide improved results in terms of program comprehension, detecting change proneness, change impact analysis, testability and maintainability. But the empirical validation provided by the researchers, to support their conception is inadequate. In order to apply these dynamic measures in industrial environment, more empirical validation is essential. From the survey of existing literature some observational results have been revealed; (a) Work done in the field of cohesion metrics is not adequate. In order to realize the full potential, formulation of more dynamic cohesion measures are required, (b) In most of the included papers, the researchers have also not provided the theoretical validation of their respective measures.

References

1. R. Subramanyam, M. Krishnan, Empirical analysis of CK metrics for object-oriented design complexity: implications for software defects. IEEE Trans. Softw. Eng. **29**(4), 297–310 (2003)
2. S. Chidamber, C. Kemerer, A metrics suite for object oriented design. IEEE Trans. Softw. Eng. **20**(6), 476–493 (1994)
3. W. Li, S. Henry, Object-oriented metrics that predict maintainability. J. Syst. Softw. **23**(2), 111–122 (1993)
4. R. Harrison, S. Counsell, R. Nithi, An evaluation of the MOOD set of object-oriented software metrics. IEEE Trans. Softw. Eng. **24**(6), 491–496 (1998)
5. Y. Lee, Measuring the coupling and cohesion of an object-oriented program based on information flow, in *International Conference on Software Quality* (Maribor, Slovenia, 1995)
6. M. Hitz, B. Montazeri, Measuring coupling and cohesion in object-oriented systems, in *International Symposium on Applied Corporate Computing* (Monterrey Mexico, 1995)

7. B. Henderson-Sellers, *Object-Oriented Metrics: Measures of Complexity* (ACM, NJ, USA, 1996)
8. L. Briand, J. Daly, J. Wust, A unified framework for coupling measurement in object-oriented systems. IEEE Trans. Softw. Eng. **25**(1), pp. 91–121 (Jan/Feb 1999)
9. F.B. Abreu, R. Carapuça, Object-oriented software engineering: measuring and controlling the development process, in IEEE *International Conference on Software Quality* (McLean, VA, USA, 1994)
10. J. Eder, G. Kappel, M. Schrefl, *Coupling and Cohesion in Object Oriented Systems* (1994)
11. J.M. Bieman, B.K. Kang, Cohesion and reuse in an object-oriented system, in *SSR '95 Proceedings of the 1995 Symposium on Software reusability* (Seattle, Washington, USA, 1995)
12. L. Briand, J. Daly, J. Wust, A unified framework for cohesion measurement in object-oriented systems, in *Fourth International Software Metrics Symposium* (1997)
13. H.S. Chae, Y.R. Kwon, A cohesion measure for classes in object-oriented systems, in *Fifth International Software Metrics Symposium (Metrics 1998)* (Bethesda, MD, USA, USA, 1998)
14. Y. Zhou, B. Xu, J. Zhao, H. Yang, ICBMC: an improved cohesion measure for classes, in *International Conference on Software Maintenance* (Montreal, Quebec, Canada, Canada, 2002)
15. R. Geetika, P. Singh, Dynamic coupling metrics for object oriented software systems: a survey. ACM SIGSOFT Soft. Eng. Notes **39**(2), 1–8 (2014)
16. S. Yacoub, H. Ammar, T. Robinson, Dynamic metrics for object oriented designs, in *Sixth International Software Metrics Symposium* (1999)
17. E. Arisholm, L. Briand, A. Foyen, Dynamic coupling measurement for object-oriented software. IEEE Trans. Softw. Eng. **30**(8), 491–506 (2004)
18. Y. Hassoun, R. Johnson, S. Counsell, A dynamic runtime coupling metric for meta-level architectures, in *Eighth European Conference on Software Maintenance and Reengineering* (2004)
19. Y. Hassoun, R. Johnson, S. Counsell, *Empirical Validation of a Dynamic Coupling Metric* (2004)
20. Y. Hassoun, S. Counsell, R. Johnson, Dynamic coupling metric: proof of concept. IEE Proc. Softw. **152**(6), 273–279 (2005)
21. A. Mitchell, J. Power, Run-time coupling metrics for the analysis of java programs-preliminary results from the SPEC and grande suites (2003)
22. Á. Mitchell, J.F. Power, An empirical investigation into the dimensions of run-time coupling in Java programs, in *Proceedings of the 3rd international symposium on Principles and practice of programming in Java* (Las Vegas, Nevada, USA, 2004)
23. A. Mitchell, J.F. Power, Using object-level run-time metrics to study coupling between objects, in *Proceedings of ACM symposium on Applied computing* (Santa Fe, New Mexico, 2005)
24. A. Zaidman, S. Demeyer, Analyzing large event traces with the help of coupling metrics, in *Proceedings of the Fourth International Workshop on OO Reengineering* (University Antwerpen, 2004)
25. A. Beszedes, T. Gergely, S. Farago, T. Gyimothy, F. Fischer, The dynamic function coupling metric and its use in software evolution, in *11th European Conference on Software Maintenance and Reengineering* (Amsterdam, Netherlands, 2007)
26. P. Singh, H. Singh, Class-level dynamic coupling metrics for static and dynamic analysis of object-oriented systems. Int. J. Inf. Telecommun. Technol. **1**(1), 16–28 (2010)
27. V. Gupta, J.K. Chhabra, Dynamic cohesion measures for object-oriented software. J. Syst. Architect. **57**(4), 452–462 (2011)
28. A. Mitchell, J.F. Power, *Run-Time Cohesion Metrics for the Analysis of Java Programs—Preliminary Results from the SPEC and Grande Suites* (2003)
29. L. Briand, S. Morasca, V. Basili, An operational process for goal-driven definition of measures. IEEE Trans. Softw. Eng. **28**(12), 1106–1125 (2002)
30. E. Weyuker, Evaluating software complexity measures. IEEE Trans. Softw. Eng. **14**(1), 1357–1365 (1998)

IFSOM: A Two-Phase Framework for COTS Evaluation and Selection

Vikram Bali, Shivani Bali and Sushila Madan

Abstract Software users demand cheaper software, faster deliveries, and reliable products, whereas software developers seek to minimize the development cost and maximize the profit margin by simultaneously meeting the competitive requirements. This calls for a trade-off between the conflicting objectives of reliability and cost. Component-based software system (CBSS) approach of software development can reduce the development time and thereby provides a reliable system within the limited budget. In CBSS approach, different software components are integrated to form a complete system. These software components can be either purchased from the software vendors or can also be built within the organization. This work aims at proposing a framework for evaluation and selection of software components. The proposed framework is termed as "IFSOM" (intuitionistic fuzzy set and optimization model).

Keywords COTS · Intuitionistic fuzzy set · Optimization model · CBSS Evaluation and selection

1 Introduction

Software development has gone through a number of rapid changes, since its inception. The traditional methods of software development are no longer applicable to the software development industry. Component-based software system (CBSS) has taken

V. Bali (✉)
J.S.S. Academy of Technical Education, Sector-62, Noida, Uttar Pradesh, India
e-mail: vikramgcet@gmail.com

S. Bali
Lal Bahadur Shastri Institute Of Management, Delhi, India
e-mail: lbsshivani@gmail.com

S. Madan
Lady Shri Ram College for Women, University of Delhi, Delhi, India
e-mail: sushila_lsr@yahoo.com

C. R. Krishna et al. (eds.), *Proceedings of 2nd International Conference on Communication, Computing and Networking*, Lecture Notes in Networks and Systems 46, https://doi.org/10.1007/978-981-13-1217-5_86

over the classical software development process. In the very beginning, the software were developed from scratch. In today's competitive world, the reuse of components plays a major role to satisfy customer demand. The motivation behind using CBSS is to reduce the cost of development and to compete with the fast-changing technology. In CBSS approach, different software components are integrated to develop software system. Developers can procure software components either from different software vendors or from the software market, or can also be built within the organization. These ready-to-use software components are called commercial off-the-shelf (COTS) components. Different parameters need to be considered while selecting the right COTS components. This gave birth to the multi-criteria decision-making problem, where we must evaluate the different COTS products. "Build-or-buy strategy" for component selection is also the focus of this study. When a variety of components are available, a complete framework is required to systematically evaluate, rank, and select the best available option.

In this work, two-phase methodology is adopted for the selection of a best mix of components for development of CBSS. The proposed method takes into consideration the quantitative criteria along with the qualitative ones. In the first phase "IFS (intuitionistic fuzzy set method)" is used along with "TOPSIS (technique for order preference by similarity to an ideal solution)". Different COTS vendors are evaluated based on multiple qualitative criteria by more than one decision-maker. Based on the evaluation, the vendors are then assigned ranks. In the second phase, an optimization model is developed with vendor ranking as the objective function which is to be maximized. The constraints with the model are reliability, cost, and delivery time which are expressed in quantitative terms. The information regarding these parameters is obtained either from the COTS vendors in case of COTS products or estimated by software development team in case of in-house built component. This methodology helps in choosing the best components from those vendors who were assigned high ranks and at the same time selects them under the constraints of reliability, cost, and delivery time. Since the COTS selection is done using two approaches, intuitionistic fuzzy set (IFS) and optimization model, therefore, the proposed framework is termed as intuitionistic fuzzy set and optimization model (IFSOM).

A lot of works have been done by researchers in COTS selection methods [1–7]. COTS evaluation is the core activity during COTS selection process as at this stage the fitness-of-use of COTS product is determined. Different strategies have been proposed in the literature to evaluate the COTS product [8–16]. Many authors have attempted to propose the mathematical models for selection of COTS components. At the outset, Berman and Ashrafi [17] proposed an optimization model for component selection for development of modular software system. Build-or-buy strategy for selection of COTS components was proposed by Cortellessa et al. [18, 19]. A lot of authors have proposed multi-objective optimization model for COTS selection for development of CBSS [20–25]. Due to the vagueness and impreciseness of information, fuzzy optimization models were proposed for COTS selection [20, 21, 26–28].

This paper is organized as follows: Sect. 2 presents a two-phase methodology that takes care of evaluation and selection of software components for the development of CBSS. In Sect. 3, a case of medium-sized software building company is illustrated to describe the projected methodology. Conclusion is presented in Sect. 4.

2 Evaluation and Selection Method—IFSOM

Component selection is a challenging task and involves an intelligent mechanism for choosing the best mix of components. Therefore, two-phase methodology is proposed that takes care of evaluating the components based on qualitative as well as quantitative criteria.

2.1 *Phase I: IFS Method for COTS Vendor Selection*

The component selection process involves assigning rating to the software vendors based on the expert evaluation. Let $A = \{A_1, A_2, \ldots, A_m\}$ denote set of alternatives and $X = \{X_1, X_2, \ldots, X_n\}$ denote set of criteria. Below are steps of intuitionistic fuzzy TOPSIS approach (adopted from [29]).

Step I Assigning weights to decision-makers

Suppose there are l experts or decision-makers, their importance is stated as linguistic variable and is measured as intuitionistic fuzzy number (IFN). $E_k = [\mu_k, \nu_k, \pi_k]$ denotes an IFN for assigning rating to kth expert. Therefore, weight of kth expert can be stated as [29, 30]

$$\lambda_k = \frac{\left(\mu_k + \pi_k\left(\frac{\mu_k}{\mu_k+\nu_k}\right)\right)}{\sum_{k=1}^{l}\left(\mu_k + \pi_k\left(\frac{\mu_k}{\mu_k+\nu_k}\right)\right)}, \quad \sum_{k=1}^{l} \lambda_k = 1$$

Step II Develop aggregated intuitionistic fuzzy decision matrix

For each expert, let $R^{(k)} = (r_{ij}^{k})_{m\times n}$ be an intuitionistic fuzzy decision matrix. An aggregated intuitionistic fuzzy decision matrix is created to fuse all individual expert views into a group view. For the same, IFWA operator is used [31].

$r_{ij}R = (r_{ij})_{m\times n};$ *where*

$$= IFWA_\lambda(r_{ij}^{(1)}, r_{ij}^{(2)}, \ldots, r_{ij}^{(l)}) = \lambda_1 r_{ij}^{(1)} \oplus \lambda_2 r_{ij}^{(2)} \oplus \cdots \oplus \lambda_l r_{ij}^{(l)}$$

$$= \left[1 - \prod_{k=1}^{l}\left(1 - \mu_{ij}^{(k)}\right)^{\lambda_k}, \prod_{k=1}^{l}\left(\nu_{ij}^{(k)}\right)^{\lambda_k}, \prod_{k=1}^{l}\left(1 - \mu_{ij}^{(k)}\right)^{\lambda_k} - \prod_{k=1}^{l}\left(\nu_{ij}^{(k)}\right)^{\lambda_k}\right]$$

where

$$r_{ij} = (\mu_{Ai}(x_j), \nu_{Ai}(x_j), \pi_{Ai}(x_j)) \quad \text{for all } (i = 1, 2, \ldots, m; j = 1, \ldots, n)$$

The aggregated intuitionistic fuzzy decision matrix is given below:

$$R = \begin{bmatrix} r_{11} & r_{12} & r_{13} & \cdots & r_{1n} \\ r_{21} & r_{22} & r_{23} & \cdots & r_{2n} \\ r_{31} & r_{32} & r_{33} & \cdots & r_{3n} \\ \vdots & \vdots & \vdots & \vdots & \vdots \\ r_{m1} & r_{m2} & r_{m3} & \cdots & r_{mn} \end{bmatrix}$$

$$= \begin{bmatrix} (\mu_{A1}(x_1), \nu_{A1}(x_1), \pi_{A1}(x_1)) & (\mu_{A1}(x_2), \nu_{A1}(x_2), \pi_{A1}(x_2)) & \cdots & (\mu_{A1}(x_n), \nu_{A1}(x_n), \pi_{A1}(x_n)) \\ (\mu_{A1}(x_1), \nu_{A1}(x_1), \pi_{A1}(x_1)) & (\mu_{A1}(x_2), \nu_{A1}(x_2), \pi_{A1}(x_2)) & \cdots & (\mu_{A1}(x_n), \nu_{A1}(x_n), \pi_{A1}(x_n)) \\ \vdots & \vdots & \vdots & \vdots \\ (\mu_{A1}(x_1), \nu_{A1}(x_1), \pi_{A1}(x_1)) & (\mu_{A1}(x_2), \nu_{A1}(x_2), \pi_{A1}(x_2)) & \cdots & (\mu_{A1}(x_n), \nu_{A1}(x_n), \pi_{A1}(x_n)) \end{bmatrix}$$

Step III Determination of criteria weights

Let $w_j^{(k)} = \left[\mu_j^{(k)}, \nu_j^{(k)}, \pi_j^{(k)}\right]$ denotes an IFN for criterion X_j provided by the kth expert. By using IFWA operator, criteria weights can be calculated as follows:

$$w_j = \mathrm{IFWA}_\lambda(w_j^{(1)}, w_j^{(2)}, \ldots, w_j^{(l)}) = \lambda_1 w_j^{(1)} \oplus \lambda_2 w_j^{(2)} \oplus \cdots \oplus \lambda_l w_j^{(l)}$$
$$= \left[1 - \prod_{k=1}^{l}\left(1 - \mu_j^{(k)}\right)^{\lambda_k}, \prod_{k=1}^{l}\left(\nu_j^{(k)}\right)^{\lambda_k}, \prod_{k=1}^{l}\left(1 - \mu_j^{(k)}\right)^{\lambda_k} - \prod_{k=1}^{l}\left(\nu_j^{(k)}\right)^{\lambda_k}\right]$$

$$W = \left[w_1, w_2, w_3, \ldots, w_j\right], \text{ where } w_j = (\mu_j, \nu_j, \pi_j); j = 1, 2, \ldots, n$$

Step IV Construction of aggregated weighted intuitionistic fuzzy decision matrix

The following expression is used [32]:

$$R \otimes W = \{\langle x, \mu_{Ai}(x), \mu_w(x), \nu_{Ai}(x) + \nu_w(x) - \nu_{Ai}(x), \nu_w(x)\rangle / x \in X\}$$

and

$$\pi_{Ai}, w(x) = 1 - \nu_{Ai}(x) - \nu_w(x) - \mu_{Ai}(x), \mu_w(x) + \nu_{Ai}(x), \nu_w(x)$$

Therefore, aggregated intuitionistic fuzzy decision matrix is expressed below:

$$R' = \begin{bmatrix} r'_{11} & r'_{12} & r'_{13} & \cdots & r'_{1n} \\ r'_{21} & r'_{22} & r'_{23} & \cdots & r'_{2n} \\ r'_{31} & r'_{32} & r'_{33} & \cdots & r'_{3n} \\ \vdots & \vdots & \vdots & \vdots & \vdots \\ r'_{m1} & r'_{m2} & r'_{m3} & \cdots & r' \end{bmatrix}$$

$$= \begin{bmatrix} (\mu_{A1W}(x_1), \nu_{A1W}(x_1), \pi_{A1W}(x_1)) & (\mu_{A1W}(x_2), \nu_{A1W}(x_2), \pi_{A1W}(x_2)) & \ldots & (\mu_{A1W}(x_n), \nu_{A1W}(x_n), \pi_{A1W}(x_n)) \\ (\mu_{A1W}(x_1), \nu_{A1W}(x_1), \pi_{A1W}(x_1)) & (\mu_{A1W}(x_2), \nu_{A1W}(x_2), \pi_{A1W}(x_2)) & \ldots & (\mu_{A1W}(x_n), \nu_{A1W}(x_n), \pi_{A1W}(x_n)) \\ \vdots & \vdots & \vdots & \vdots \\ (\mu_{A1W}(x_1), \nu_{A1W}(x_1), \pi_{A1W}(x_1)) & (\mu_{A1W}(x_2), \nu_{A1W}(x_2), \pi_{A1W}(x_2)) & \ldots & (\mu_{A1W}(x_n), \nu_{A1W}(x_n), \pi_{A1W}(x_n)) \end{bmatrix}$$

where

$$\begin{aligned} r'_{ij} &= (\mu'_{Ai}(x_j), \nu'_{Ai}(x_j), \pi'_{Ai}(x_j)) \\ &= (\mu'_{AiW}(x_j), \nu'_{AiW}(x_j), \pi'_{AiW}(x_j)) \quad (i = 1, 2, \ldots, m; j = 1, 2, \ldots, n) \end{aligned}$$

Step V Get IFPIS (Intuitionistic fuzzy positive ideal solution) and IFNIS (intuitionistic fuzzy negative ideal solution)

In TOPSIS method, the assessment criteria are classified into two classes, benefit and cost. Assume G_1 be the benefit criteria and G_2 the cost criteria. A^+ is an IFPIS and A^- is an IFNIS. A^+ and A^- are expressed as follows:

$$\begin{aligned} A^+ = \Bigg[\Bigg\{ x_j, \Bigg\langle \Big(\max_i \mu_{Ai,W}(x_j)/g \in G_1 \Big), \Big(\min_i \mu_{Ai,W}(x_j)/g \in G_2 \Big) \Bigg\rangle, \\ \Bigg\langle \Big(\min_i \nu_{Ai,W}(x_j)/g \in G_1 \Big), \Big(\max_i \nu_{Ai,W}(x_j)/g \in G_2 \Big) \Bigg\rangle \Bigg\} | i = 1, 2, \ldots, m \Bigg] \\ A^- = \Bigg[\Bigg\{ x_j, \Bigg\langle \Big(\min_i \mu_{Ai,W}(x_j)/g \in G_1 \Big), \Big(\max_i \mu_{Ai,W}(x_j)/g \in G_2 \Big) \Bigg\rangle, \\ \Bigg\langle \Big(\max_i \nu_{Ai,W}(x_j)/g \in G_1 \Big), \Big(\min_i \nu_{Ai,W}(x_j)/g \in G_2 \Big) \Bigg\rangle \Bigg\} | i = 1, 2, \ldots, m \Bigg] \end{aligned}$$

Step VI Distance calculation from IFPIS and IFNIS

For measuring distance of each alternative from IFPIS and IFNIS by using normal Euclidean distance given by Szmidt and Kacprzyk [33], the intuitionistic separation measures are expressed as follows:

$$S^+ = \sqrt{\frac{1}{2n} \sum_{j=1}^{n} \begin{bmatrix} (\mu_{Ai,W}(x_j) - \mu_{A^+,W}(x_j))^2 \\ +(\nu_{Ai,W}(x_j) - \nu_{A^+,W}(x_j))^2 + (\pi_{Ai,W}(x_j) - \pi_{A^+,W}(x_j))^2 \end{bmatrix}}$$

$$S^- = \sqrt{\frac{1}{2n} \sum_{j=1}^{n} \begin{bmatrix} (\mu_{Ai,W}(x_j) - \mu_{A^-,W}(x_j))^2 \\ +(\nu_{Ai,W}(x_j) - \nu_{A^-,W}(x_j))^2 + (\pi_{Ai,W}(x_j) - \pi_{A^-,W}(x_j))^2 \end{bmatrix}}$$

Step VII Determine CC_i (relative closeness coefficient), and ranking of alternatives CC_i of each alternative based on intuitionistic fuzzy ideal solutions can be estimated with the help of the expression given below:

$$CC_i = \frac{S_{i^-}}{S_{i^+} + S_{i^-}} \quad \text{where } 0 \le CC_i \le 1$$

Greater value of CC_i states an alternative stand closer to IFPIS and farther from IFNIS. Next, CC_i of each alternative are ranked in decreasing order. The best ideal alternative is the one having the maximum value.

2.2 Phase II: Optimization Model for Selection of Software Components

In phase II of the proposed methodology, an optimization model is developed that helps in optimal selection of software components that are either ready-to-use COTS components or in-house built components. Objective function for an optimization problem is the closeness coefficient (CC_i) of each software vendor and is an outcome of phase I. The model aims at selecting the best software components by maximizing (CC_i) under the constraints of budget, reliability, and delivery time. The proposed model follows the same assumptions based on the work of Bali [26] (Table 1).

Table 1 Notations

x_{ij}	$\begin{cases} \text{if } j\text{th COTS alternative is selected for } i\text{th module} \\ 0, \text{ otherwise} \end{cases}$
y_i	$\begin{cases} \text{if } i\text{th module is in - house developed} \\ 0, \text{ otherwise} \end{cases}$
CC_{ij}	Closeness coefficient of the "jth COTS alternative for ith module"
CC_i	Closeness coefficient of the in-house developed component for module i
c_{ij}	Cost of "alternative j for module i" of a COTS component
C_i	Unitary development cost of in-house developed component of module i
B	Overall budget
t_i	Estimated development time for in-house developed component for module i
ω_i	Average time required to perform a test case for an in-house developed component for module i
π_i	Probability that a single execution of software fails on a test case taken from some definite input distribution
R_i	Reliability of module i
r_{ij}	Reliability of "jth COTS alternative of module i"
R	System quality measure
d_{ij}	Delivery time of alternative j of module i of a COTS component

Optimization Models

$$\text{Maximize}\, Z = \sum_{i=1}^{m}\left(\sum_{j=1}^{n} CC_{ij}x_{ij} + CC_i y_i\right)$$

Subject to

$$y_i + \sum_{j=1}^{n} x_{ij} \geq 1 \tag{1}$$

$$\sum_{i=1}^{m}\left(C_i\left(t_i + \omega_i N_i^{\text{tot}}\right)y_i\right) + \sum_{j=1}^{n} c_{ij}x_{ij} \leq B \tag{2}$$

$$N_i^{\text{suc}} = (1 - \pi_i)N_i^{\text{tot}} \tag{3}$$

$$\rho_i = \frac{1 - \pi_i}{(1 - \pi_i) + \pi_i(1 - \pi_i)^{N_i^{\text{suc}}}} \tag{4}$$

$$R_i = \rho_i y_i + \left(1 - \prod_{j=1}^{n}(1 - r_{ij})^{x_{ij}}\right) \tag{5}$$

$$\prod_{i=1}^{n} R_i \geq R \tag{6}$$

$$DT_i = \left(t_i + \omega_i N_i^{\text{tot}}\right)y_i + \sum_{j=1}^{n} d_{ij}x_{ij} \tag{7}$$

$$\underset{i=1,2,\ldots,m}{\text{Max}}\,(DT_i) \leq T \tag{8}$$

The objective function maximizes the closeness coefficient (CC_i) of each alternative of a module. It integrates COTS as well as in-house built components. Constraint (1) denotes "build-or-buy" constraint. For module i, an alternative can be either bought (i.e., $x_{ij} = 1$) or is in-house built (i.e., $y_i = 1$) or both can be selected. Redundancy can be created by selecting more than one COTS component. Constraint (2) is a budget constraint. Constraint (3) is count of failure-free tests done on the jth alternative of a module. Constraint (4) denotes the probability that in-house built alternative is failure-free on a single execution provided N_i^{suc} test cases were successfully performed. Constraint (5) is the reliability expression of the module i. Constraint (6) is the overall reliability of the software. Constraints (7) and (8) are constraints related to delivery time which incorporates delivery time in case of COTS components and development time of in-house built components.

Table 2 Linguistic terms for rating the importance of criteria and the experts

Linguistic terms	IFNs
Very important	(0.90, 0.10)
Important	(0.75, 0.20)
Medium	(0.50, 0.45)
Unimportant	(0.35, 0.60)
Very unimportant	(0.10, 0.90)

3 Case Study

A local software development company signed a contract to build a software system for a retail organization. As per the functional requirements of the retailer, the software development company has identified three modules, viz, front office, finance, and back office. The company plans to develop software using CBSS approach. For software, components are either ready-to-integrate COTS components or components that can be built in-house. Based on a preliminary study, four COTS vendors (A1–A4) were identified, from where the components can be procured. If the component is not procured from any of these vendors, it is developed within the organization. In-house developed component is termed as (A5). A two-phase methodology is adopted to select best components, to develop a reliable system with optimized cost and delivery time. In the first phase, IFS is used to evaluate and rank the software vendors based on the qualitative criteria and is presented in Sect. 3.

3.1 Solution Methodology of Intuitionistic Fuzzy Set (Phase I)

- Step I Linguistic weighting variables were used by decision-makers to assign ratings to the criteria and are shown in Table 2. Table 4 presents importance of the experts on group decision.
- Step II The experts employ linguistic terms (see Table 3) to assess the rating of alternatives assigned to each criterion. For ratings provided by decision-makers to five alternatives for each module, refer Table 5.
- Step III Table 6 presents criteria importance denoted in linguistic terms. Aggregated intuitionistic fuzzy decision matrixes along with criteria weights are given in Table 7.
- Step IV Aggregated weighted intuitionistic fuzzy decision matrices is created (refer Table 8).
- Step V Determine IFPIS and IFNIS with the help of Table 8. Functionality, compatibility, and vendor reputation are the benefit criteria $G_1 = \{X_1, X_2, X_3, X_4\}$, whereas pricing strategy is the cost criteria $G_2 = \{X_5\}$. Then, IFPIS and IFNIS are obtained (refer Table 9).

Table 3 Linguistic terms: alternatives' rating

Linguistic terms	IFNs
Extremely good (EG)/extremely high (EH)	(1.00, 0.00)
Very very good (VVG)/very very high (VVH)	(0.90, 0.10)
Very good (VG)/very high (VH)	(0.80, 0.10)
Good (G)/high (H)	(0.70, 0.20)
Medium good (MG)/medium high (MH)	(0.60, 0.30)
Fair (F)/medium (M)	(0.50, 0.40)
Medium bad (MB)/medium low (ML)	(0.40, 0.50)
Bad (B)/low (L)	(0.25, 0.60)
Very bad (VB)/very low (VL)	(0.10, 0.75)
Very very bad (VVB)/very very low (VVL)	(0.10, 0.90)

Source Boran et al. [19]

Table 4 Expert's importance and their weights

	Expert 1	Expert 2	Expert 3
Linguistic terms	Important	Medium	Very important
Weight	0.356	0.238	0.406

Step VI Estimate the distance measure from IFPIS and IFNIS. Relative closeness coefficient (CC_i) is calculated and alternatives are ranked in decreasing order of (CC_i). See Table 10 and 11.

Solution of Phase I: Finally, ranks are assigned to the alternatives of each module based on the decreasing order of (CCi) and are given in Table 11.

3.2 *Solution Methodology of Phase II*

In phase I, ranks are assigned to the vendors based on their performance on qualitative parameters. In the second phase, values of (CC_i) are the input to the objective function of the optimization model which is maximized. The constraints to the model are of reliability, cost, and delivery time and are expressed in quantitative terms. The model helps in selecting the best components that are integrated to build the software system. As discussed earlier, software is also developed by assembling three modules, namely, front office, finance, and back office. For each module, there are five alternatives present. Out of them, four are COTS components and are denoted by (x_{ij}) and the fifth one is the in-house built component (y_i). The data set for development of optimization model is given in Tables 12 and 13. Structure of software is given in Fig. 1.

Table 5 The ratings of the alternatives for the three modules

Criteria	Supplier	Module 1: Front office			Module 2: Finance			Module 3: Back office		
		Expert 1	Expert 2	Expert 3	Expert 1	Expert 2	Expert 3	Expert 1	Expert 2	Expert 3
X1 functionality	A1	G	VG	G	VG	G	VG	VVG	VG	VG
	A2	VVG	VG	VG	MG	G	F	G	G	G
	A3	MG	G	F	VVG	VG	VG	VG	G	VVG
	A4	MG	G	G	MG	G	G	F	MG	MG
	A5	VG	VVG	VG	F	MG	MG	MG	G	G
X2 compatibility	A1	VG	G	VG	G	VG	G	VG	F	VG
	A2	F	MG	G	F	MG	G	F	MG	G
	A3	F	MG	G	G	G	VG	VG	G	VG
	A4	F	F	MG	F	F	MG	MB	F	F
	A5	G	VG	VG	B	F	F	F	F	MG
X3 vendor reputation	A1	VG	G	VG	VG	VG	VG	G	G	MG
	A2	VG	VG	G	MG	G	G	VG	VG	G
	A3	G	MG	MG	G	VG	VG	VG	MG	G
	A4	VG	G	G	VG	VG	G	VG	G	G
	A5	G	G	G	G	G	G	G	G	G
X4 pricing strategy	A1	H	H	H	H	H	H	MH	M	MH
	A2	MH	M	MH	MH	M	MH	M	M	MH
	A3	MH	M	MH	MH	M	MH	H	H	H
	A4	H	MH	MH	H	MH	MH	H	MH	MH
	A5	H	H	VH	MH	M	M	M	MH	M

Table 6 The important weight of the criteria

Criteria	Module 1: Front Office			Module 2: Finance			Module 3: Back Office		
	Expert 1	Expert 2	Expert 3	Expert 1	Expert 2	Expert 3	Expert 1	Expert 2	Expert 3
Functionality	VI	VI	I	VI	VI	I	VI	VI	I
Compatibility	I	I	M	I	I	M	I	I	M
Vendor reputation	I	I	I	I	I	I	I	I	I
Pricing strategy	M	I	M	M	I	M	M	I	M

Table 7 Aggregated intuitionistic fuzzy decision matrix along with criteria weight

Alternative	Module 1: Front office			
	X1	X2	X3	X4
A1	(0.728, 0.170, 0.102)	(0.780, 0.118, 0.102)	(0.780, 0.118, 0.102)	(0.700, 0.200, 0.100)
A2	(0.849, 0.100, 0.051)	(0.605, 0.292, 0.103)	(0.769, 0.128, 0.103)	(0.578, 0.321, 0.101)
A3	(0.769, 0.128, 0.103)	(0.605, 0.292, 0.103)	(0.644, 0.254, 0.102)	(0.578, 0.321, 0.101)
A4	(0.663, 0.236, 0.101)	(0.538, 0.361, 0.101)	(0.746, 0.151, 0.103)	(0.644, 0.254, 0.102)
A5	(0.830, 0.100, 0.070)	(0.764, 0.133, 0.103)	(0.800, 0.100, 0.100)	(0.740, 0.156, 0.104)
Weight	(0.861, 0.128, 0.011)	(0.750, 0.200, 0.050)	(0.680, 0.267, 0.053)	(0.576, 0.371, 0.053)
	Module 2: Finance			
A1	(0.780, 0.118, 0.102)	(0.728, 0.170, 0.102)	(0.800, 0.100, 0.100)	(0.700, 0.200, 0.100)
A2	(0.596, 0.302, 0.102)	(0.605, 0.292, 0.103)	(0.663, 0.236, 0.101)	(0.578, 0.321, 0.101)
A3	(0.849, 0.100, 0.051)	(0.740, 0.156, 0.104)	(0.764, 0.133, 0.103)	(0.578, 0.321, 0.101)
A4	(0.663, 0.236, 0.101)	(0.538, 0.361, 0.101)	(0.769, 0.128, 0.103)	(0.644, 0.254, 0.102)
A5	(0.562, 0.337, 0.101)	(0.410, 0.472, 0.118)	(0.700, 0.200, 0.100)	(0.543, 0.356, 0.101)
Weight	(0.861, 0.128, 0.011)	(0.750, 0.200, 0.050)	(0.680, 0.267, 0.053)	(0.576, 0.371, 0.053)
	Module 3: Back office			
A1	(0.849, 0.100, 0.051)	(0.751, 0.139, 0.110)	(0.668, 0.231, 0.101)	(0.578, 0.321, 0.101)
A2	(0.700, 0.200, 0.100)	(0.605, 0.292, 0.103)	(0.769, 0.128, 0.103)	(0.538, 0.361, 0.101)
A3	(0.769, 0.128, 0.103)	(0.780, 0.118, 0.102)	(0.728, 0.166, 0.106)	(0.700, 0.200, 0.100)
A4	(0.562, 0.337, 0.101)	(0.462, 0.438, 0.100)	(0.746, 0.151, 0.103)	(0.644, 0.254, 0.102)
A5	(0.663, 0.236, 0.101)	(0.538, 0.361, 0.10)	(0.700, 0.200, 0.100)	(0.526, 0.374, 0.100)
Weight	(0.861, 0.128, 0.011)	(0.750, 0.200, 0.050)	(0.680, 0.267, 0.053)	(0.576, 0.371, 0.053)

Table 8 Aggregated weighted intuitionistic fuzzy decision matrix

Alternative	Module 1: Front office			
	X1	X2	X3	X4
A1	(0.627, 0.276, 0.097)	(0.585, 0.294, 0.121)	(0.530, 0.353, 0.117)	(0.403, 0.497, 0.100)
A2	(0.731, 0.215, 0.054)	(0.454, 0.433, 0.113)	(0.523, 0.361, 0.116)	(0.333, 0.573, 0.094)
A3	(0.662, 0.240, 0.098)	(0.454, 0.433, 0.113)	(0.438, 0.454, 0.108)	(0.333, 0.573, 0.094)
A4	(0.571, 0.334, 0.095)	(0.404, 0.489, 0.107)	(0.507, 0.378, 0.115)	(0.371, 0.531, 0.098)
A5	(0.715, 0.215, 0.070)	(0.573, 0.306, 0.121)	(0.544, 0.340, 0.116)	(0.426, 0.469, 0.105)
	Module 2: Finance			
A1	(0.672, 0.231, 0.097)	(0.546, 0.336, 0.118)	(0.545, 0.340, 0.116)	(0.403, 0.497, 0.100)
A2	(0.513, 0.391, 0.096)	(0.454, 0.433, 0.113)	(0.451, 0.440, 0.109)	(0.333, 0.573, 0.094)
A3	(0.731, 0.215, 0.054)	(0.555, 0.325, 0.120)	(0.520, 0.364, 0.116)	(0.333, 0.573, 0.094)
A4	(0.571, 0.334, 0.095)	(0.404, 0.489, 0.107)	(0.523, 0.361, 0.116)	(0.371, 0.531, 0.098)
A5	(0.484, 0.422, 0.094)	(0.308, 0.577, 0.115)	(0.476, 0.414, 0.110)	(0.313, 0.595, 0.092)
	Module 3: Back Office			
A1	(0.731, 0.215, 0.054)	(0.564, 0.311, 0.125)	(0.454, 0.436, 0.110)	(0.333, 0.573, 0.094)
A2	(0.603, 0.302, 0.095)	(0.454, 0.433, 0.113)	(0.523, 0.361, 0.116)	(0.310, 0.598, 0.092)
A3	(0.662, 0.240, 0.098)	(0.585, 0.294, 0.121)	(0.435, 0.389, 0.116)	(0.403, 0.497, 0.100)
A4	(0.484, 0.422, 0.094)	(0.346, 0.550, 0.104)	(0.507, 0.378, 0.115)	(0.371, 0.531, 0.098)
A5	(0.571, 0.334, 0.095)	(0.404, 0.489, 0.107)	(0.476, 0.414, 0.110)	(0.303, 0.606, 0.091)

Solution of Phase II: LINGO software package was used to solve the model [34]. Sensitivity analysis is also performed by changing the total budget.

Case 1: When Cost = 100 units and Delivery Time = 7 weeks

Since total budget is low, redundancy is created only in module 2, and hence the overall reliability is low. Also, the in-house component is selected for the first module and for module 2 and 3 COTS alternative got selected (Table 14).

Table 9 Intuitionistic fuzzy positive ideal solution (IFPIS) and intuitionistic fuzzy negative ideal solution (IFNIS)

	Module 1: Front office			
A^+	(0.731, 0.215, 0.054)	(0.585, 0.294, 0.121)	(0.530, 0.353, 0.116)	(0.333, 0.573, 0.094)
A^-	(0.571, 0.334, 0.095)	(0.404, 0.489, 0.107)	(0.438, 0.454, 0.108)	(0.426, 0.469, 0.105)
	Module 2: Finance			
A^+	(0.731, 0.215, 0.054)	(0.555, 0.325, 0.120)	(0.544, 0.340, 0.116)	(0.313, 0.595, 0.092)
A^-	(0.484, 0.422, 0.094)	(0.308, 0.577, 0.115)	(0.451, 0.440, 0.109)	(0.403, 0.497, 0.100)
	Module 3: Back office			
A^+	(0.731, 0.215, 0.054)	(0.585, 0.294, 0.121)	(0.523, 0.361, 0.116)	(0.303, 0.606, 0.091)
A^-	(0.484, 0.422, 0.094)	(0.346, 0.550, 0.104)	(0.454, 0.436, 0.110)	(0.403, 0.497, 0.100)

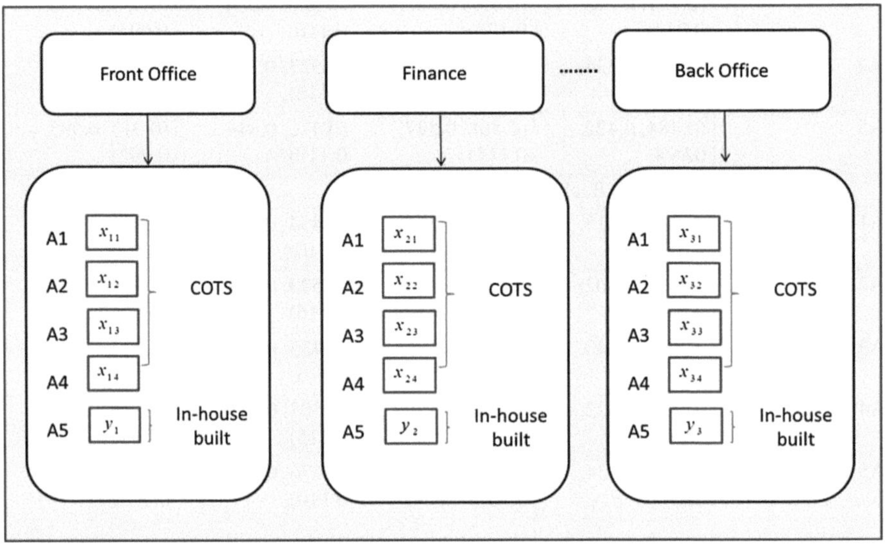

Fig. 1 Software structure

Case 2: When Cost = 150 units and Delivery Time = 7 weeks

In this case, the budget is increased that resulted in redundancy to appear in the two modules, viz, module 1 and module 3. The system reliability is significantly improved in this case and approaches to 0.99. Here, only in case of module 1, both COTS and in-house components are selected (Table 15).

Table 10 Closeness coefficient of the alternatives

Alternatives	Module 1: Front office			Module 2: Finance			Module 3: Back office		
	S^+	S^-	*CCi*	S^+	S^-	*CCi*	S^+	S^-	*CCi*
A1	0.058	0.11	0.654	0.054	0.16	0.746	0.041	0.166	0.803
A2	0.068	0.102	0.6	0.124	0.083	0.4	0.088	0.102	0.536
A3	0.088	0.073	0.451	0.016	0.177	0.917	0.062	0.155	0.713
A4	0.121	0.047	0.279	0.112	0.076	0.405	0.173	0.033	0.159
A5	0.058	0.112	0.659	0.173	0.049	0.22	0.121	0.075	0.384

Table 11 Ranking of the alternatives

Module 1: Front office		Module 2: Finance		Module 3: Back office	
Ranks	Alternative	Ranks	Alternative	Ranks	Alternative
1	A5	1	A3	1	A1
2	A1	2	A1	2	A3
3	A2	3	A4	3	A2
4	A3	4	A2	4	A5
5	A4	5	A5	5	A4

Table 12 Data set for COTS components

Module	Alternative	Cost (c_{ij})	Reliability (r_{ij})	Delivery time (d_{ij})	Closeness coefficient (CC_{ij})
Front office	A1	20	0.95	5	0.654
	A2	50	0.96	7	0.600
	A3	35	0.98	3	0.451
	A4	70	0.94	4	0.279
Finance	A1	25	0.92	4	0.746
	A2	30	0.84	5	0.400
	A3	15	0.96	2	0.917
	A4	42	0.87	5	0.405
Back office	A1	42	0.89	3	0.803
	A2	54	0.88	1	0.536
	A3	34	0.92	1	0.713
	S4	37	0.83	2	0.159

Table 13 Data set for in-house built component

Module	Development time (t_i)	Testing time (ω_i)	Unitary development cost (C_i)	Probability of testability (π_i)	Closeness coefficient (CC_i)
Front office	4	0.0005	5	0.0002	0.659
Finance	4	0.0005	12	0.0002	0.220
Back office	8	0.0005	10	0.0002	0.384

Table 14 Solution of case 1

Total cost	System reliability	Component selected		
		Module 1	Module 2	Module 3
99	0.91	$y_1 = 1$	$x_{22} = x_{23} = 1$	$x_{33} = 1$

Table 15 Solution of case 2

Total cost	System reliability	Component selected		
		Module 1	Module 2	Module 3
146	0.99	$x_{13} = y_1 = 1$	$x_{23} = 1$	$x_{31} = x_{33} = 1$

4 Conclusion

An attempt has been made to select the components for software development using CBSS strategy. Two-phase methodology was proposed. In the first phase, the IFS method was employed to select best COTS vendor. Different COTS vendors are evaluated based on multiple qualitative criteria by more than one decision-maker. On the basis of the evaluation, the vendors are then assigned ranks. In the second phase, an optimization model is developed with vendor ranking as the objective function which is to be maximized. The constraints to the model are reliability, cost, and delivery time, and are expressed in quantitative terms. Build-or-buy strategy was also incorporated in the optimization model.

References

1. L. Chung, K. Cooper, S. Courtney, COTS aware requirements engineering: the CARE process. in *Proceedings of the 2nd International Workshop on Requirements Engineering for COTS Components (RECOTS 2004)*, Kyoto, Japan, 7 September 2004
2. G. Grau, J.P. Carvallo, X. Franch, C. Quer, DesCOTS: A software system for selecting COTS components. in *Proceedings of the 30th IEEE Euromicro Conference on (EUROMICRO'04)*, 2004
3. J. Kontio, A COTS selection method and experiences of its use. in *Twentieth Annual Software Engineering Workshop, NASA Goddard Space Flight Center* (Greenbelt, Maryland, 1995)
4. G. Kotonya, J. Hutchinson, Viewpoints for specifying component-based systems. Lect. Notes Comput. Sci. **3054**, 114–121 (2004)
5. K.R.P.H. Leung, H.K.N. Leung, On the efficiency of domain-based COTS product selection method. Inf. Softw. Technol. **44**, 703–715 (2002)
6. C. Rolland, Requirement engineering for COTS based systems. Inf. Softw. Technol. **41**, 985–990 (1999)
7. R.C. Wang, T.F. Liang, Applying possibilistic linear programming to aggregate production planning. Int. J. Prod. Econ. **98**, 328–341 (2005)
8. J. Bhuta, B. Boehm, *A method for compatible COTS component selection* (ICCBSS, LNCS (Springer), 2005), p. 3412
9. A. Cechich, M. Piattini, in *Filtering COTS Components Through an Improvement-Based Process* (ICCBSS, LNCS, Springer, 2005), p. 3412
10. C. Couts, P. Gerdes, Integrating COTS software: lessons from a large healthcare organisation. IT Prof. **12**, 50–58 (2010)
11. S. Mahmood, The impact of acceptance tests on analyzing component-based systems specifications: an experimental evaluation. in *10th IEEE International Conference on Computer and Information Technology* (CIT, 2010)
12. N.R. Mead, *An Evaluation of A-SQUARE for COTS Acquisition*. A Technical report (Carnegie Mellon University, SEI, 2014)

13. A. Mohamed, G. Ruhe, A. Eberlein, *Decision Support for Handling Mismatches between COTS Products and System Requirements* (ICCBSS'07, Banff, Canada, 2007)
14. J. Sheng, B. Wang, Evaluating COTS components using gap analysis. in *The 9th International Conference for Young Computer Scientists, IEEE*, 2008, https://doi.org/10.1109/icycs.2008.472
15. H.J. Shyur, COTS evaluation using modified TOPOSIS and ANP. J. Appl. Math. Comput. **177**, 251–259 (2006)
16. F. Tarawneh, F. Baharom, J.H. Yahaya, F. Ahmad, Evaluation and selection COTS software process: the state of art. Int. J. New Comput. Archit. Their Appl. (IJNCAA) **1**(2), 344–357
17. O. Berman, A. Ashrafi, Optimization models for reliability of modular software systems. IEEE Trans. Softw. Eng. **19**(11), 1119–1123 (1993)
18. V. Cortellessa, F. Marinelli, P. Potena, Automated selection of software components based on cost/reliability trade-off. Lect. Notes Comput. Sci. **4344**, 66–81 (2006)
19. V. Cortellessa, F. Marinelli, P. Potena, An optimization framework for "build-or-buy" decisions in software architecture. J. Comput. Oper. Res. **35**, 3090–3106 (2008)
20. P.C. Jha, S. Bali, P.K. Kapur, Fuzzy approach for selecting optimal COTS based software products under consensus recovery block scheme. BVICAM's Int. J. Inf. Technol. (BIJIT) **3**(1), (ISSN No. 0973-5658)
21. P.C. Jha, S. Bali, U. Kumar, A fuzzy approach for optimal selection of COTS components for modular software system under consensus recovery block scheme incorporating execution time. Turkish J. Fuzzy Syst. **2**(1), 45–63 (2011)
22. P.C. Jha, P.K. Kapur, S. Bali, U.D. Kumar, Optimal component selection of COTS based software system under consensus recovery block scheme incorporating execution time. Int. J. Reliab. Qual. Saf. Eng. (IJRQSE) **17**(3), 209–222 (2010)
23. U.D. Kumar, Reliability analysis of fault tolerant recovery blocks. J. Oper. Res. Soc. India **35**(4), 281–294 (1998)
24. C.K. Kwong, J.F. Tang Mu, X.G. Luo, Optimization of software components selection for component-based software system development. Comput. Ind. Eng. **58**, 618–624
25. T. Neubauer, C. Stummer, Interactive decision support for multi-objective COTS selection. in *Proceedings of the 49th Hawaii International Conference on System Sciences*, 2007, pp. 1–7
26. S. Bali, P.C. Jha, U.D. Kumar, H. Pham, Fuzzy multi-objective build-or-buy approach for component selection of fault tolerant software system under consensus recovery block scheme with mandatory redundancy in critical modules. Int. J. Artif. Intell. Soft Comput. **4**(2/3), 98–119 (2014)
27. P. Gupta, H. Pham, M.K. Mehlawat, S. Verma, A fuzzy optimization framework for COTS products selection of modular software systems. Int. J. Fuzzy Syst. **15**(2), 91–109 (2013)
28. P. Gupta, S. Verma, M.K. Mehlawat, A membership function approach for cost-reliability trade-off of COTS selection in fuzzy environment. Int. J. Reliab. Qual. Saf. Eng. **18**(6), 573–595 (2011)
29. F.E. Boran, S. Gença, M. Kurt, D. Akay, A multi-criteria intuitionistic fuzzy group decision making for supplier selection with TOPSIS method. J. Expert Syst. Appl. Elsevier **36**, 11363–11368 (2009)
30. B.D. Rouyendegh, Developing an Integrated AHP and intuitionistic fuzzy TOPSIS methodology. Tehničkivjesnik **21**(6), 1313–1319 (2014)
31. Z.S. Xu, Intuitionistic fuzzy aggregation operators. IEEE Trans. Fuzzy Syst. **15**(6), 1179–1187 (2007)
32. K.T. Atanassov, Intuitionistic fuzzy sets. Fuzzy Sets Syst. **20**, 87–96 (1986)
33. E. Szmidt, J. Kacprzyk, Distances between intuitionistic fuzzy sets. Fuzzy Sets Syst. **114**(3), 505–518 (2000)
34. H. Thiriez, OR software LINGO. Eur. J. Oper. Res. **124**, 655–656 (2000)

HL-7 Based Middleware Standard for Healthcare Information System: FHIR

Meenakshi Sharma and Himanshu Aggarwal

Abstract Fast health Interoperability Resources (FHIR), a advanced proposed emerging standard of Health Level 7 (HL7) that inheritance various advantageous of HL7- v2 and v3 for providing health interoperability. HL7 messaging Standard has been widely implemented and adopted by healthcare domain internationally from last few decades. Among hospital HL7-V2 (version 'v2') is preferred choice as compared to standard v3 to exchange healthcare information like electronic health records (EHR) among local hospitals. HL7-V3 was successor of the HL7-V2 that inherits various features and overcome various shortcomings of the V2. HL7-V3 standard had been highly criticized by the healthcare industry due to various shortcomings like complex documentation, implementation and maintenance cost high along with stalled system. HL7 standards has been introduced new approach FHIR standard which yet under experimental stage. FHIR has various attractive features like user friendly features, various built in modules and widely compatible with existing web standards. This research paper will provide substantiation evolution of the HL-7 standards pattern messaging, prologue related to the HL7 FHIR and comparison among HL7 standards.

Keywords RESTful · Interoperability · Electronic health records · HL7 · FHIR Agile · Standards

M. Sharma (✉)
Computer Science and Engineering Department, GIMET, Amritsar 143501, India
e-mail: sharma.minaxi@gmail.com

H. Aggarwal
Computer Engineering Department, Punjabi University, Patiala 147002, India
e-mail: himanshu.pup@yahoo.com

C. R. Krishna et al. (eds.), *Proceedings of 2nd International Conference on Communication, Computing and Networking*, Lecture Notes in Networks and Systems 46, https://doi.org/10.1007/978-981-13-1217-5_87

1 Introduction

In today's era, key requirement to enhance quality of the care in the domain of healthcare is Interoperability that has been acknowledged widely. With the ease of Interoperability among HIS (healthcare information system) will provides the economic based value support to the nation [1]. Public safety improved drastically with appropriate deploying interoperability standards as patient accurate information can be accessible at anytime from anywhere while handling patients. Interoperability can be attained by parts with effective utilization of available standards, that can provide the semantic and synthetic significance of the right information. Standards deployment will reduce risks factor, provide the consistency in care, improve the health care and will drastically reduce the cost factors. Many common industries like financial etc. have attained a superior level in the field of integration, automation and interoperability among organisational domain with support of standards like SWIFT [2]. In field of healthcare, the HL7 [3] standards facilitated various interoperability standards to provide the integration in the health care information system but yet it has been not widely adopted. Around the globe tremendous investment regarding resources has been made by industry along with various guidelines imposed by jurisdictional yet the adaptability of integration is elusive in domain of healthcare. HL7 standards has recently introduced new FHIR standard [4] known as 'Fast Health Interoperability Standard' to overcome shortcomings of HL7 v2 and v3 [3]. FHIR standard considered both processes and product for the messaging, has iterative incremental approach to handle complex real world scenario barriers. Scalability, compatibility and reusability are attractive measured features of FHIR standard. FHIR has various built in terminologies for administrator and clinical purpose along with the its based upon web standards like JSON, Atom, http and xml. These unique features of this standard considered the cause of getting popularity and attraction among the healthcare industry. In the rest of the paper, we provide the chronicle review about the development of the messaging of HL7 standards from the v2, v3 and about the newly proposed FHIR. Also present the introduction about FHIR standards along with relevant technologies and the comparison of existing messaging standards.

2 Literature Review

On the basis of the semantic interoperability along the aspects of the model driven number of authors reports about the advantages of the HL7v3 over v2. Many authors published about complexity of V3 message implementation and high cost of implementation. Author Mead [5] provide differentiate between v3 and v2 on basis of data specification, cross information domain model and approaches of the development model. Vernadat [6] projected enterprise interoperable system which provide services for process implementation and handling was supposed to be main focal point.

Authors also enlighten about semantic interoperability at organisational and technical level and necessity about the standard to provide the semantic interoperability.

In year 2002, trends related to the integration of legacy health care system with the standards was on boom. With utilization of emerging technologies and architecture like web-services, SOA (service oriented architecture) and ESB (Enterprise Service Bus) proposed models [7, 8]. A novel architecture proposed by Liu et al. [8] that provide the interoperability between PACS(picture archiving and communication system) using DICOM and healthcare information system. They provide the gateway between information system and HLv3. WebReach Inc. [9] developed a healthcare messaging system Mirth for interoperability. Mirth client server based architecture used enterprise service bus that consist of connectors, filters and various transforming modules that perform various roles like receive/send, parse and transform messages from HL v2 to legacy format and vice versa. Sujansky et al. [10] provide guide report how to integrate messages among EHR (electronic health records) and HL7 v2 standard. Author uses the principle of optionality for the sake of interoperability for HL7 standards and various modules of hospital information system to reduce complexity of HL7 v2. Many authors worked in distributed environment framework to scale integration among various hospital information system [11–13]. Configurations and structural framework vary from project to project. Sartipi et al. [14] provide various barriers in HL7 v3 message to provide integration with legacy clinical reporting system. Semantic level integration to provide the interoperability of clinical decision support system with specialist of EMR. Jayaratna et al. [15] developed a semantic web based tool known as TAMMP to categorize HL7 v3 massages that perform integration among health information system. This process provide new significance to HL7 messages that has been extracted from hierarchy of existing HL7 messages. Dehmoobad [16] proposed cross domain architecture by utilizing HL7 messages to provide interoperability among healthcare system and insurance corporation. Variety of HL7 v3 based tools have been developed to provide assistance to stakeholders, physicians, developers and researchers. These tools can perform number of different tasks. Few of the popular tools are: R-MIM Designer graphics based tools that used to design static models of the HL7 information. V3-Generator take input inform of XML expression of HMD and provide various HL7-artifacts like schema of static and interaction, various HTML based table-views, data types, repository, static models for MIF-files [2]. Mapping of v2 and v3 files for transformation of data along with semantic-mapping can be performed with support of the Eclipse tool. Extension of HL7 v2 and v3 being used. It also performs testing, mapping and translation into other formats. One of translator named TAMMP that can translate level of the healthcare into the v3 messages of HL7 [15].

3 Existing Interoperability Healthcare Standards: Health Level 7 (HL7)

HL7 is a non-profitable standard organisation that had been established in year of 1987, main goal was to provide standards for HIS. Presently, HL7 is an well known international level organization of technocrats experts that developed health care interoperability standards to exchange information along with resources among various hospital information-system.

HL7-Version 2

Version 2 of HL7 standard was developed in year of 1989 on ad hoc mode to provide integration among various sections of hospitals like administration and clinical information system. Earlier hospital various departments like the laboratory, billing system and discharge section can't communicate each other and execute their working independently. To automate each department individual software installed and used. v2 provide all features, well adopted by clinician along with local communities of hospitals. In north America, was widely adopted by the vendors of the HIS. Its ad hoc nature based standard, has limitation of scalability, can't support multiple large scale environment information system as well as can't globally support inherit enterprise identifiers along with it has highly reliant on local communities with ease of the "Z-segments". Interfaces of the HL7 v2 has been designed in such a way that 80% designed by the specification wise and rest of 20% can be designed as per local customization requirements. Another major drawback is lack of identifying the appropriate ontology that is capable enough to identifying concepts for exchanging of messaging and interfaces.

HL7 Version 3

To overcome shortcomings of v2, v3 was introduced in year of 1995. HL7-v3 was not improved version of v2, a novel standard that introduced development process framework called "HDF" (HL7 Development Framework) and model for central information known as "RIM"(Reference- Information-Model). All lexical and semantic structural representation of elements concerned defined by RIM. For HDF architecture, v3 utilized top down approach for all models related to clinical information system along with automate generation of messaging among interface. Process models decoupled from implementation along agnostic platform. It supports the XML schema that has been bundled along with standards of HL7 during distribution. XML schema partially useful for verification of various messaging but can't be useful for rest of other task like implementation to automate generation of classes of software. Implementation of v3 for clinical models as per customization purpose complex transformation compiler model required to make model platform specific but no such specific tools available in v3. Comprehensive indulgent of the RIM model is required for proper implementation of HL7-v3. Structure of RIM consist of information in form of entities, roles, acts etc. Development model of HL7-v3 is designed with process of constraint of RIM, which is not a normal practice in field of software industry and has complex kind of transmission of wire format along with messaging format. SHALL, kind of informational logical model that is foundation key for v3 and SAIF has been derived from RIM(Reference information Model). This standard unify structural contents, provide representation in consistence form for model and made implementation simpler that has been derived from numerous issue. RIM is type of class model in which it includes the attributes, associated state-machines and association related to those classes. Universal edition of Version 3 has been designed in such a way that it can't be directly implementable, it consists of various kind of initial template that has been utilized for implementation region and local wise. It's a tedious process for implementer. Addition to that the it can't support the compatibility features with real world scenario as well as not provide consistence solution. Top down model approach required the complex application roles during the implementation and higher level description for the language translation required. Many of developed countries like Canada, UK, Germany, Australia, China, Japan, Denmark etc. adopted by providing integration with the EHR, but not to be considered in the United State due to act of HITECH. Due to this it was main reason for failure in investment project and not able to attract vendors due to implementation complexities as well as many data-acquisition modules was missing like food, drug-administration Addition to implementation complexities, v3 and v2 are not compatible with each other along with does support the interoperability among them.

4 Fast Healthcare Interoperability Resources (FHIR)

Due to failure of HL7-v3 and lack of funding support due to ineligible for US HITECH in month of January 2011, the task force of Body of Governance has again taken the initiated to improve HL7 messaging standard. With inspiration the group of software engineers architect a novel approach for interoperability initially named as "RHF" (resources for Health) that latterly known as "FHIR" [2] (Fast Healthcare Interoperability Resources). This novel approach based on the architecture of the RESTful approach that has been introduced by fielding [17]. FHIR standards are robust, scalable, simpler, easily adoptable and in many services as open source standards also. Format of the standard quite simple so no need of complex tools, FHIR standard easily adaptable in nature and various test servers and implementation examples are available. Learning of the FHIR standard can be go through with the support of HL7-v3. FHIR basically support four paradigms, that have distinct approach to make usage of these paradigms to accommodate various workflow of system. Four paradigm of FHIR and when to utilize them are as: (i) REST: Its small and light weight exchange preferred when the coupling among the system is low. (ii) Messages: communicate with number of resources in single exchange. Architecture for the message exchange shown in Fig. 1. (iii) Documents: When want to span the data on multiple resources and the focal point is persistence. (iv) Services: Make the usage of the customize service while the services of the other paradigm not appropriate as per the requirement.

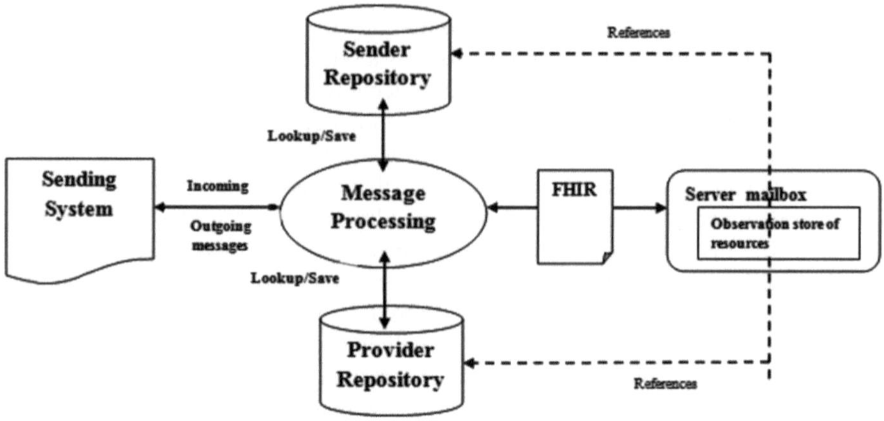

Fig. 1 FHIR message system architecture

Resources of the FHIR are quite similar to like v2 and v3. Resources basically small and logical define unit for exchange and define behaviour, meaning, location, that can be one unit of transaction etc. Limit of resources in FHIR around 100–150. As per requirement, number of resources can be increased. Some of examples of resources are as: (i) Administrative: Patient, Organization, Location. (ii) Clinical: Allergy, Family History, Care Plan. (iii) Infrastructure. Document, Message Profile, Conformance. In spite of paradigm content still have dependency on resources. On basis of paradigm resources are bundled in different data sets. FHIR has potential to be used in amount of workflows like from small mobile device to multi-speciality Hospitals-Information System as shown in Fig. 1. Also not limited to patient engagement related workflows but also provide traditional communication among the applications. Timeline diagram of the FHIR workflow shown in Fig. 2. General workflows that has been included in the FHIR are as: Application to application interoperability among the four-walls. External connectivity: HIE/ACO's), for National level exchanges, in the area of social-web, for the mobile based applications, in usage of home related health devices.

Earlier there was requirement of interface engine like brokers' application that behave as middleware among existing standards and application. FHIR standard remove this barrier. Key objective of the FHIR standards are the Patient-rendezvous. With support of REST API (light weight) FHIR provide the leverage of providing alters and data along make the secure access of patient data repositories to authentic users where and when its required. Implementers of the FHIR written coding as well as prepared interface by keeping users in their mind.

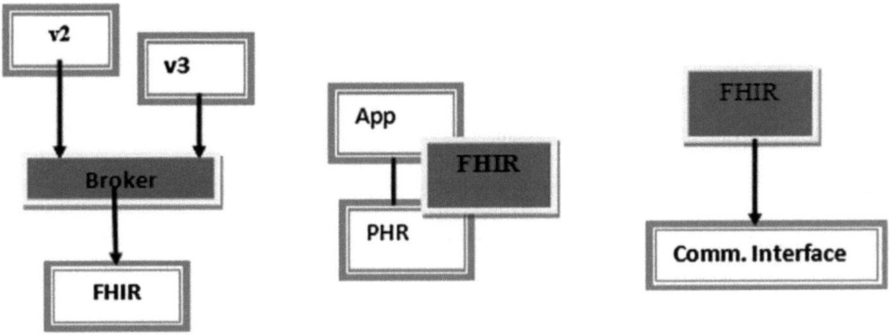

Fig. 2 Timeline workflow of FHIR

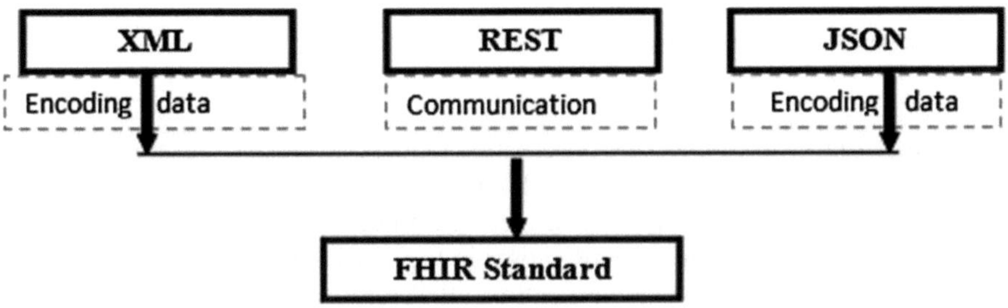

Fig. 3 Emerging standards of the FHIR

Availability provide to servers after passing through robust testing phases. Its supporter API's accessible through number of programming languages like Java, Java-script, C#, C, JASON etc. Emerging standards of the FHIR shown in the Fig. 3.

HL7 has developed various flavour for the interoperability like HL7, V2 and V3, Development process of the HL7 v2 utilized the ad hoc approach while the V3 used the top-down approach that known as HDF.

FHIR standard used the incremental and iterative approach for the development process. Interoperatiability among HIS provided with the support of the special message formant of FHIR.

In server side repository, data stored in FHIR format for exchanging this information among various HIS. Interfaces of HL7 used for data encoding of ORU (observation-result) messages format.

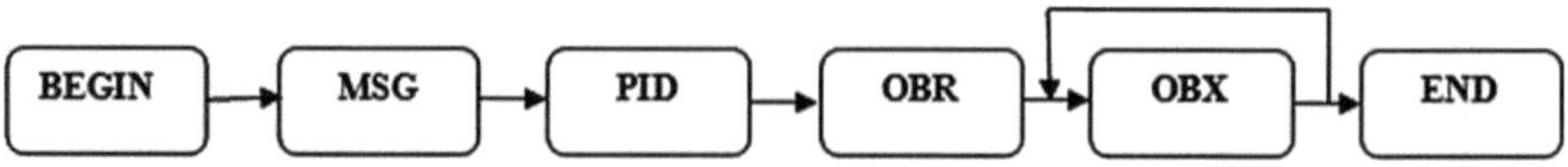

Fig. 4 Flow diagram of HL7-FHIR message

```
MSH|^~\&|GHH LAB|ELAB-3|GHH OE|BLDG4|200202150930||ORU^R01|CNTRL-3456|P|2.4<cr>

PID|||555-44-4444||EVERYWOMAN^EVE^E^^^^L|JONES|19620320|F|||153 FERNWOOD DR.^
^STATESVILLE^OH^35292||(206)3345232|(206)752-121||||AC555444444||67-A4335^OH^20030520<cr>

OBR|1|845439^GHH OE|1045813^GHH LAB|1554-5^GLUCOSE|||200202150730||||||||555-55-
5555^PRIMARY^PATRICIA P^^^^MD^^|||||||||F||||||444-44-4444^HIPPOCRATES^HOWARDH^^^^MD<cr>

OBX|1|SN|1554-5^GLUCOSE^POST 12H CFST:MCNC:PT:SER/PLAS:QN||^182|mg/dl|70  105|H|||F<cr>
```

Fig. 5 Message format example of HL7-FHIR

Foremost segments of ORU are observation request (OBR) and observation result (OBX). Segment OBR(Observation read) carried all the information regarding date of order placed, time etc. [17]. Flow of message process is shown in Fig. 4.

While the format pattern of FHIR shown in Fig. 5 where MSG used for message details, PID for patient unique identification, OBR for observation request and OBX for purpose of observation result.

5 Fast Healthcare Interoperability Resources (FHIR)-Artifacts

FHIR, advanced standard of HL7 in which entities are used to define resources that are used for exchange of health care information. Each entity distinctly identifiable for each of resources. Specification of FHIR standards defined various specification for resource attributes which are as follows: Boundary of the resources should be clearly define that similar to one or more transaction logical scopes. -Resources name can be differ from each other but the usage and processing should be similar. -Normal identification of the resources. -Commonly used resources should be preferred that can be easily useable in the industry transactions. -Resource can use the other resource, every resource should provide a novel content. -all the resources are organized in the logical framework on the foundation of the cohesion of other resources and provision of the link associated to resources. -Size of the resources should be large that can be used for industry meaningful purpose.

Number of resources more than 150 has been defined in FHIR standard like device, patient, staff and document etc. FHIR team also claimed that it has maintained rigour of existing standard HL7-v3 but representation and processing is quite simple. Commonly reusable structures in HL7 v3 are known as CMETs (Common Message Element Types), that are compatible to FHIR data structures. In year of 2010, number of CIMET was 194 in v2 while after development of FHIR (subsequent to 17 months) it was around 35. This signify of radical lessening in amount of concepts that are required for implementation.

6 Criticism About FHIR Standard

FHIR provide the resource oriented environment, that provide simplicity in implementation, diligence and better distribution of information. Major issue is that lack of documentation and the guidance regarding how to utilized the iterative and incremental approach for enhance the relationships. CRUD operations are defined in detailed but not the detailed about workflow and dynamic behaviour. This divergence may lead towards the scarcity of interoperability. Association among the resource aggregations and documentation is the grey area that need to be defined in detail. In the base standard the resources and their inputs are defined but their association among resources and how they will behave is not described.

7 Conclusion

Through this paper we have present the chronicle study of the existing interoperability standards of the HL7 family that are used in the healthcare system. Various attributes of the HL7 v2, v3 and FHIR standard has been studied with the analytical premise. Table 1 provide the comparison among the salient features of the these standards which reflects the strength and weakness of these standards. FHIR standard is under development phase, it got quite popularity among the stakeholders due to usage of RIM model, quite simple to implement and above all based on iterative and incremental model. Still some ambiguities are there that how FHIR can utilized the strengths of its predecessors. Developers are keep on working how to make the drastically improvement in the processing of the FHIR. Future work includes how to implement the FHIR standard for the IOS platform. Further considered the FHIR standard to be utilized for the clinical decision support system for the prediction of the chronic disease.

Table 1 Comparison of FHIR with existing HL7 standards

Features	HL7-v2	HL7-v3	HL7-FHIR
Introduced of year	1987	1997	2011
Methodology approach favoured for development	Bottom–up approach/adhoc approach	Top–down approach, MDA	Iterative along incremental approach
Semantic for ontology	No	Yes	Yes
Acquaintance overhead	Duration of weeks	Duration of months	Duration of weeks
Explicit tools essential?	Yes—parser	Yes—model compiler	No
Instantly consume-able?	Yes	No	Yes
Documentation size specification	Volume of pages in hundreds	Volume of pages in thousands	Volume of pages in hundreds
Specific examples for implementation	Yes	Minimal	Yes
Availability of references from HL7	No	No	Yes
Support form Industry and society	Strong	Weak	N/A—quite new
Compatibility with smart phones	No	No	Yes
Quantity of various category of messages	Evidently not specific	Around 450	Around 30
Ration of adaptability	Quite high	Quite low	Quite-new
Kind of information model	Adhoc	Constrained-RIM	Agile
Internationally character level support	No (ASCII)	Conceptually yes	Yes (UTF8)
Internationally support for message format	Particular global level standard	Restricted by dominion	Particular global level standard

Acknowledgements The corresponding author indebted to Dr. Sandeep Sharma (Prof. and Head GNDU) and Dr. Raman Sharma (Prof. Govt. Medical College) for their throughout guidance and motivation in my research journey.

References

1. C. Bertolini, Z. Liu, M. Schaf, V. Stolz, Towards a formal integrated model of collaborative healthcare workflows. Found. Heal. Inf. Eng. Syst. **7151**, 57–74 (2012)
2. SWET, The global provider of secure financial messaging services (2016). http://www.swift.com/standards

3. HL7, (2016). http://www.hl7.org/
4. FHIR: Fast healthcare interoperability resources (2016). http://hl7.org/implement/standards/fhir/
5. C.N. Mead, Data interchange standards in healthcare it computable semantic interoperability: now possible but still difficult, do we really need a better mousetrap? J. Heal. Inf. Manage. **20**(1), 71–78 (2006)
6. V. François, D. Chen, Architectures for enterprise integration and interoperability: past, present and future. Comput. Ind. **59**(7), 647– 659 (2008)
7. T.H.Y. et al., A scalable multi-tier architecture for the national taiwan university hospital information system based on HL7 standard. in *Symposium on Computer-Based Medical Systems* (2006), pp. 99–104
8. B. Liu, X. Li, Z. Liu, Q. Yuan, X. Yin, Design and implementation of information exchange between HIS and PACS based on HL7 standard. in *Information Technology and Applications in Biomedicine* (2008), pp. 522–555
9. S. Hsieh et al. Middleware based inpatient healthcare information system. in *Bioinformatics and Bioengineering* (2007), pp. 1230–1234
10. W.V. Sujansky et al., The development of a highly constrained health level 7 implementation guide to facilitate electronic laboratory reporting to ambulatory electronic health record systems. J. Am. Med. Inform. Assoc. **16**(3), 285–290 (2009)
11. S. Hsieh et al., Middleware based inpatient healthcare information system. in *Bioinformatics and Bioengineering* (2007), pp. 1230–1234
12. W. Wang, M. Wang, S. Zhu, Healthcare information system integration: a service oriented approach. in *Services Systems and Services Management* (2005), pp. 1475–1480
13. L.F. Ko et al., HL7 middleware framework for healthcare information system. in *e-Health Networking, Applications and Services* (2006), pp. 152–156
14. K. Sartipi, M. Yarmand, Standard-based data and service interoperability in e-health systems. in *IEEE International Conference on Software Maintenance (ICSM)* (2008), pp. 187–196
15. P. Jayaratna, K. Sartipi, HL7 v3 message extraction using semantic web techniques. Int. J. Knowl. Eng. Data Min. **2**(1), 89–115 (2012)
16. Dehmoobad, K. Sartipi. Cross-domain information and service interoperability. in *IEEE International Conference on Information Integration and Web-based Applications and Services* (2008), pp. 25–32
17. J. Choi, S. Yoo, H. Park, J. Chun, MobileMed: a PDA-based mobile clinical information system. IEEE Trans. Inf. Technol. Biomed. **10**(3), 627–635 (2006)

An Artificial Fuzzy Logic Intelligent Controller Based MPPT for PV Grid Utility

Neeraj Priyadarshi, Farooque Azam, Akash Kumar Bhoi and S. Alam

Abstract An intelligent Fuzzy Logic Control (FLC) based trackers are employed for optimum power generation from photovoltaic (PV) module. Proposed model evaluates its performance will result in obtaining a more accurate response from the system. The characteristics of photovoltaic systems are inherently nonlinear and are functions of environmental parameters such as light, ambient temperature and its attaching charge. Therefore, we can receive the maximum power from PV array by choosing appropriate work point of PV when the amount of light and temperature are constant. With changes in environmental conditions (light and temperature) the work point of array will be changed and consequently by using different algorithms of maximum power point tracking (MPPT) it can be always kept the received power amount from array at its maximum value, in other words, maximum power point can be tracked. The advantage of fuzzy controllers is in working with imprecise and nonlinear inputs, no need to accurate mathematical model and fast convergence and the lowest fluctuations in MPPT.

Keywords FLC · MPPT · MATLAB · PV

N. Priyadarshi (✉) · F. Azam · S. Alam
Millia Institute of Technology, Purnea 854301, India
e-mail: neerajrjd@gmail.com

F. Azam
e-mail: farooque53786@gmail.com

S. Alam
e-mail: ershamsher.alam@gmail.com

A. K. Bhoi
Department of EEE, SMIT, SMU, Rangpo, Sikkim 737136, India
e-mail: akash730@gmail.com

C. R. Krishna et al. (eds.), *Proceedings of 2nd International Conference on Communication, Computing and Networking*, Lecture Notes in Networks and Systems 46, https://doi.org/10.1007/978-981-13-1217-5_88

1 Introduction

Due to running out of energy resources including fossil fuels, it is required now to use new energies, especially solar energy and thus to identify the ways to receive maximum power and the best output of existing converters. The necessity of using new energies rather than fossil fuels, has led human to use renewable energies such as solar energy. Meanwhile, the systems to convert solar energy to electrical energy as a new energy source and free from environmental pollution are so important [1–3]. But what is important in a meantime is to find methods in order to receive maximum power from converters of these energies. One of the works done in this field is to present various methods to track maximum power point of solar modules. Because these modules have nonlinear current-voltage characteristic and produce maximum power just in one particular operating point and this point varies with changes in temperature and light intensity. In this study, using power processor in order to achieve maximum power point which depends on inference rule base have been presented to optimize the use of these modules and to reduce the costs of solar systems tracking [4–7]. In this paper, it is recommended in proposed method to control the duty cycle of boost converter switching attached to solar module in a way to achieve maximum power point of module under various conditions of temperature and light intensity. The obtained important advantage in compared to previous methods is the independency of control method from module type used and no need to reference signal or need to any kind of physical information about the nature of module. The advantage of fuzzy controllers is in working with imprecise and nonlinear inputs, no need to accurate mathematical model and fast convergence and the lowest fluctuations in MPPT [8, 9]. The capability of fuzzy systems, online tracking of maximum power, robustness against light and temperature changes and no need to external sensors are to measure temperatures and light intensity, this means that there is no need to have temperature and light intensity sensors in this method and fuzzy controller can track the maximum power point in unforeseen circumstances. Solar arrays are constant in various methods, therefore the received power from PV arrays will reduce with the movement of sun and he changes of light angle on the surface of solar arrays but again in this case, extraction PV power is important in various environmental conditions. Using controlling system of maximum power, the system is adjusted in a way to have maximum power independent from environmental conditions or charge conditions. Inverter control strategy has been implemented using FLC based control [3].

2 Achieving Maximum Power Point

Photovoltaic systems are appropriate choices to supply electrical energy of homes and industrial sites due to the lack of environmental pollution and long life. However, since the cost of these systems is high, a FLC based MPPT can achieve better tracked

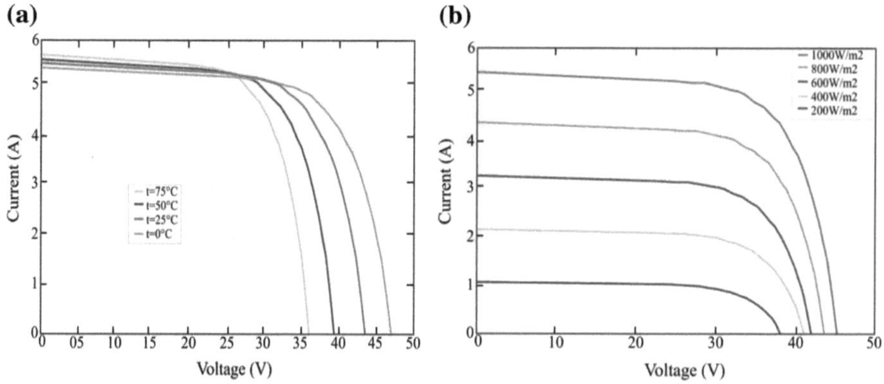

Fig. 1 I–V characteristics of PV panel **a** at variable irradiance **b** at variable temperature

Table 1 180 W SUNFORCE PV panel's specifications

Parameter	Value
Rated power	180 W
Max. voltage	35.2 V
Max. current	5.10 A
Open circuit voltage	43.2 V
Short circuit current	5.5 A

efficiency with boost converter control. For this purpose, we should have a mechanical system that places solar panels on the side of direct sunlight, at any time. Nevertheless, there is a need to have an electronic system to place the output of solar panels in proper operating point that has the maximum transfer power. Studies of solar modules show that current-voltage characteristic of modules is highly nonlinear and function of the temperature of the cell, light intensity, longevity and charge characteristic. In addition, there is only one operating point for a certain temperature and light intensity that results in maximum power.

The I–V curves of PV module using MATLAB/Simulink is presented at variable irradiance level and constant temperature (25 °C) in Fig. 1a. Also Fig. 1b depicts the solar I–V curve at varying temperature and constant insolation level (1000 W/m^2) using MATLAB/Simulink. Table 1 presents the design specification of 180 W Sunforce PV panel.

2.1 Multiplier DC/DC BOOST Converter

Output voltage of PV arrays with series-parallel connection is relatively low. So it is required to have multiplier DC/DC converters with high efficiency in order to convert

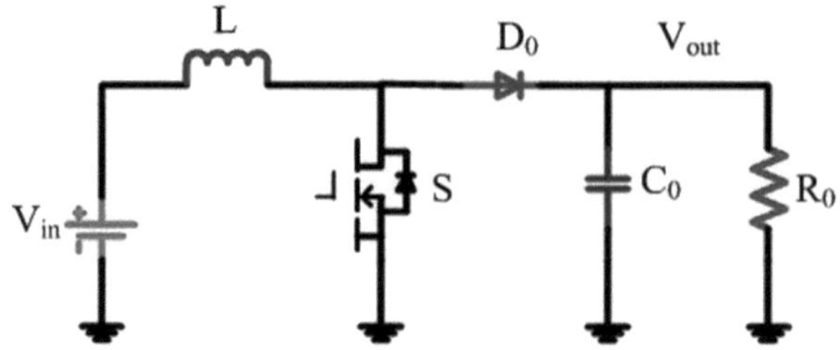

Fig. 2 Simple boost converter

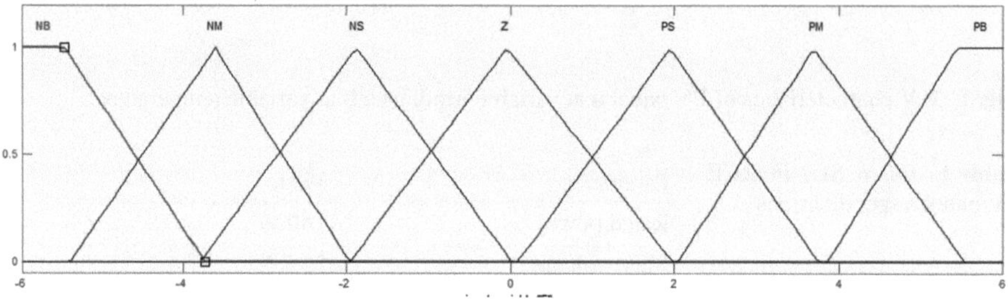

Fig. 3 Input and output membership functions

low voltage of PV arrays to higher level output. Figure 2 depicts the equivalent boost converter employed for renewable energy application due to simple circuit structure.

2.2 *Designing MPPT System Based on Fuzzy Logic Control*

The block diagram shows a fuzzy system to control solar cell system with two inputs and one output.

The stage of fuzzification to convert input variable to linguistic variable has been described on the basis of membership function presented in Fig. 3. Here, seven fuzzy stages including NB (negative big), NM (negative medium), NS (negative small), ZE (zero), PS (positive small), PM (positive medium) and PB (positive big).

The E error and dE error changes of inputs of fuzzy logic are based on MPPT controller. Since dP/dV in MPPT obtained is nearly one. Alternatively, the signal error can be calculated as output fuzzy controller converts a linguistic to a numeric parameter by adjusting membership functions. In Fuzzification process controller converts analog to digital signal which regulates the PV power capability. Output of fuzzification values are measured and simulated by fuzzy rules. The efficiency of fuzzy rules based on table is change requirement in DC. In the stage of defuzzification, numeric value of ΔD is determined through converting linguistic values. Finally, the necessary switching to convert power to MPPT is applied through an analog to digital converter and a signal gate drive. Under different weather conditions, fuzzy

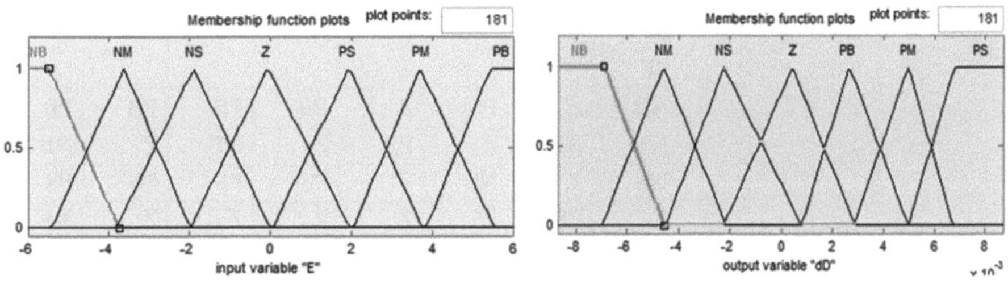

Fig. 4 Input and output membership function

logic controller shows good performance with MPPT regulation. Moreover, FLC performances are dependent on measurement of error accuracy and developed rules of basic table by user. For better efficiency, functions and rules of the basic table can be continuously updated or can be adjusted by achieving optimal performance similar to the adaptive fuzzy logic controller. Thus the strong convergence to MPPT and minimal fluctuations of MPPT can be obtained. As stated, the applied FLC comprises two inputs and one output based on MPPT.

In order to convert the fuzzy controller inputs from numeric variables to linguistic variables, we use fuzzification. Input and output variables are defined by linguistic variables set: including: NB (negative big), NM (negative medium), NS (negative small), ZE (zero), PS (positive small), PM (positive medium) and PB (positive big), which should be selected for each of the appropriate fuzzy membership functions [19]. The initial choice of membership functions for linguistic variables is done with respect to the empirical knowledge of the photovoltaic system. In order to simulate the fuzzy controller, we use fuzzy logic software in toolbox of MTLAB. First, we define a new environment to produce fuzzy controller in which we save FIS editor in the name of solar tracker. However, we should define any inputs of system that we name two inputs of input1 and input2 as E and dE. Now we have to define each inputs of E and dE according to existing membership functions in software that are compatible to figures for every two inputs, we use trimf and trapmffunctions. As in Fig. 4, we use five membership functions of NB, NM, NS, Z, PS, PM & PB like inputs to define dD output.

The input variables after fuzzification are provided to fuzzy inference machine in order to make fuzzy decision by fuzzy rules to define dD variable i.e. the amount of duty cycle change of PWM. To design fuzzy controller and selection of fuzzy rules, we should have a complete understanding of the behavior of a photovoltaic system. The ultimate goal of maximum power point tracking of solar array is under various conditions of temperature, light intensity, charge and other factors.

After assigning fuzzy rule inference shown in Table 2 and membership functions of the input variables, we should propose fuzzy inference machine to produce ΔD variable. For this purpose, Mamdani's inference method in order to make fuzzy decisions that are more famous in control engineering sector and are used more than

Table 2 Fuzzy inference rules

dI / dP	NB	NM	NS	Z	PS	PM	PB
NB	Z	PM	Z	PB	PB	PB	PB
NM	NB	Z	PS	Z	NB	Z	NB
NS	Z	NS	Z	PS	PS	PS	PS
Z	PS	PS	Z	Z	Z	Z	NS
PS	NS	NS	NS	NS	Z	Z	Z
PM	NS	NS	Z	Z	NS	NM	NS
PB	NB	NB	NB	NB	Z	NB	Z

other inference methods and Max–Min Mamdani method in order to combine fuzzy rules have been used.

3 Simulation Results

The MATLAB/Simulink simulation environment has been applied to validate performance of proposed PV integrated system. The developed design has been examined under varying operating states as steady state operation and dynamic operation.

3.1 Steady State Operation

Figure 5 interprets effective control of grid connected PV system under steady state operation. The PV system is operating with 1000 W/m^2 irradiance level and 25° ambient temperatures. In case of steady state operation, the maximum power is extracted from the PV panel with small oscillation in relatively short time. Figure 5d describes synchronized grid voltage and grid current. Generated current injected to the utility grid is sinusoidal in nature and gives unity power factor and low total harmonic distortion (THD).

3.2 Dynamic Operation of the PV System

Simulation is performed for variable irradiance level to validate the effectiveness of the proposed control system. Figure 6 shows the simulated response of the proposed system with variable irradiance level from 500 to 1000 W/m^2. Under dynamic condition, the controller tracks peak PV power. The simulated responses reveal that the proposed controller tracks maximum power point (MPP) and injects sinusoidal

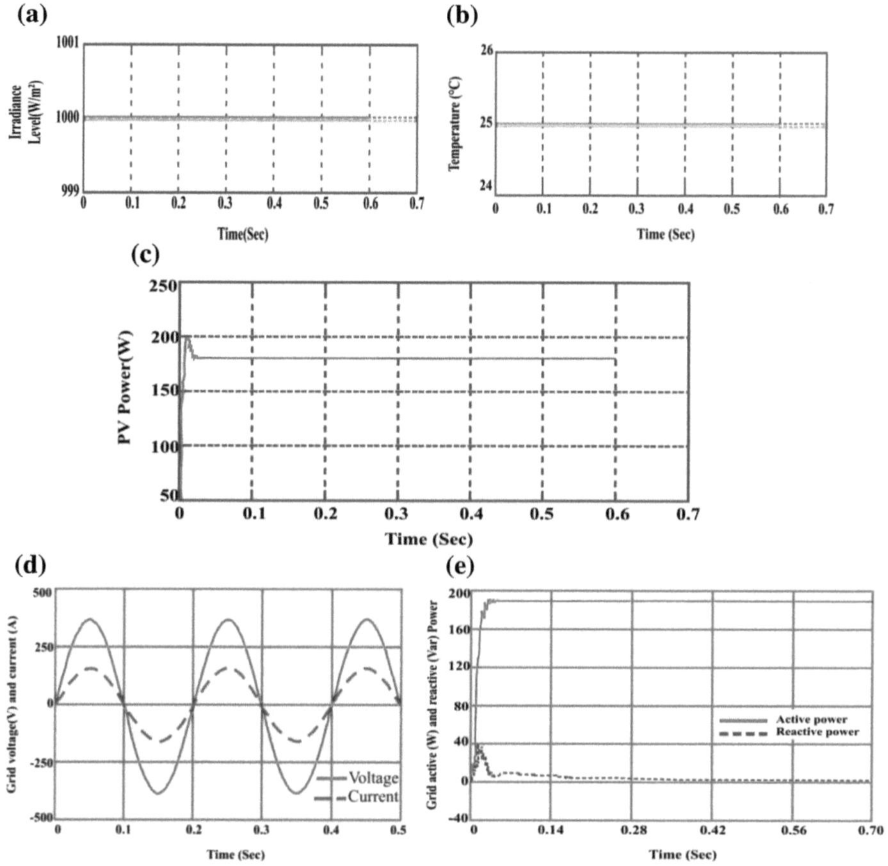

Fig. 5 Simulated responses of **a** irradiance level **b** temperature **c** PV power **d** grid voltage **e** active and reactive power

current to the grid to ensure better tracking efficiency, less oscillation and unity p.f in dynamic conditions.

3.3 *Partial Shading Condition*

The performance of the proposed system is tested under partial shading condition. Figure 7 shows the P–V characteristics with local and global maxima under partial shading condition. The simulated response depicts that in partial shadow condition, where global and local points present, the proposed MPPT controller tracks MPP under abrupt atmospheric conditions. The proposed MPPT controller resets the value of stored global maxima and the process is repeated to get new global maximum.

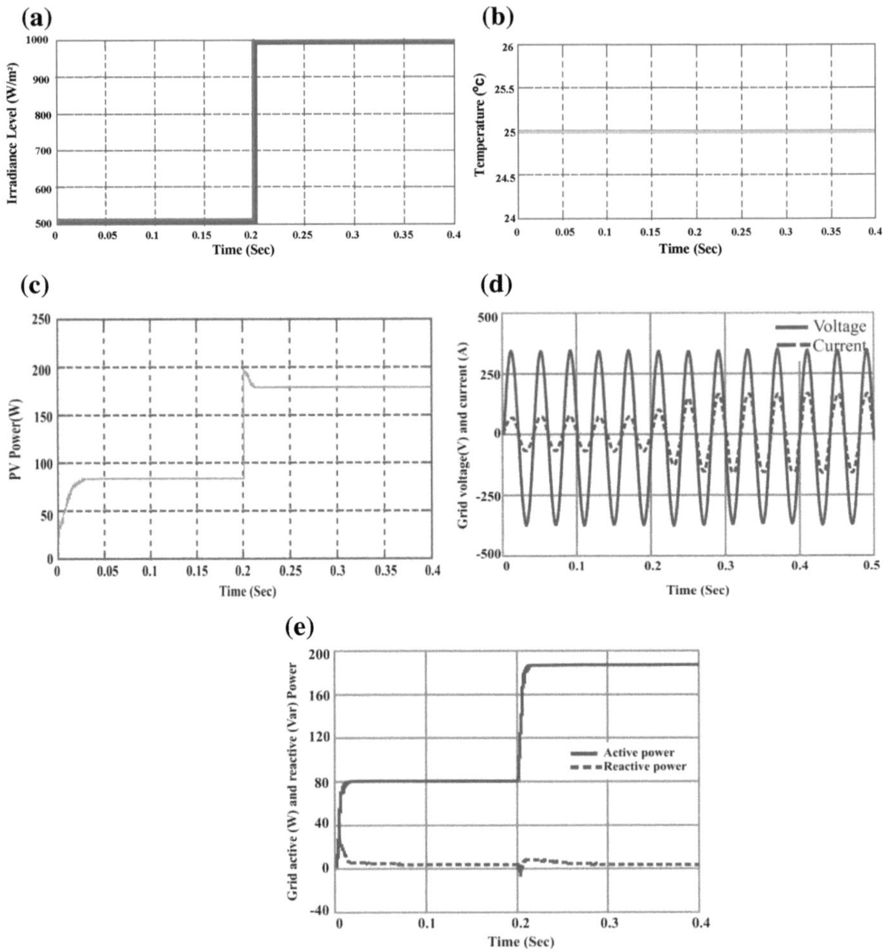

Fig. 6 Simulated responses of **a** irradiance level **b** temperature **c** PV power **d** grid voltage **e** active and reactive power

4 Conclusions

In this paper, in order to optimize the use of these modules and further reduction of tracking costs of solar systems, it was recommended to use power processor to achieve maximum power point using FLC MPPT of proposed method, switching duty cycle of power processor attached to solar module was controlled in a way to obtain maximum power point of module under various conditions of temperature and light intensity. An important advantage obtained compared to previous methods is the control method independency of module type used and no need to reference signal or need to any kind of physical information about nature of module. The advantage

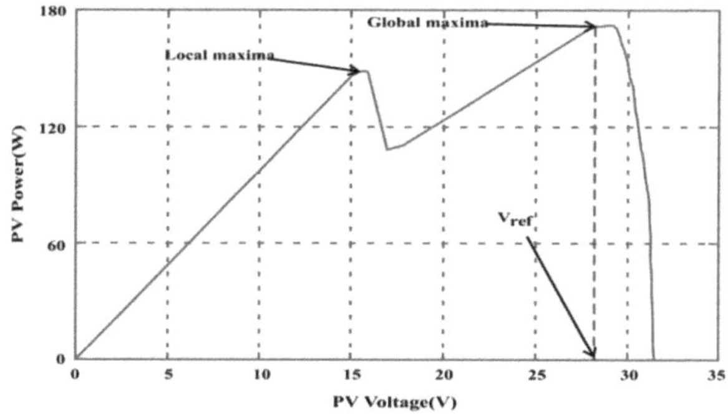

Fig. 7 Simulated P-V curves for proposed system in partial shadow condition

of fuzzy controllers is in working with imprecise and nonlinear inputs, no need to accurate mathematical model and fast convergence and the lowest fluctuations in MPPT.

References

1. N. Priyadarshi, A. Anand, A.K. Sharma, F. Azam, V.K. Singh, R.K. Sinha, An experimental implementation and testing of GA based maximum power point tracking for PV system under varying ambient conditions using dSPACE DS 1104 controller. Int. J. Renew. Energy Res. **7**(1), 255–265 (2017)
2. N. Priyadarshi, V. Kumar, K. Yadav, M. Vardia, An experimental study on zeta buck-boost converter for application in PV system. in *Handbook of Distributed Generation* (Springer). https://doi.org/10.1007/978-3-319-51343-0_13
3. N. Priyadarshi, A.K. Sharma, S. Priyam, in *An Experimental Realization of Grid-Connected PV System with MPPT Using dSPACE DS 1104 Control Board*. Advances in Smart Grid and Renewable Energy. Lecture Notes in Electrical Engineering, vol. 435 (Springer, Singapore, 2018)
4. N. Priyadarshi, A.K. Sharma, F. Azam, A hybrid firefly-asymmetrical fuzzy logic controller based MPPT for PV-wind-fuel grid integration. in Int. J. Renew. Energy Res. **7**(4) (2017)
5. N. Kumar, I. Hussain, B. Singh, B.K. Panigrahi, Maximum power peak detection of partially shaded PV panel by using intelligent monkey king evolution algorithm. IEEE Trans. Ind. Appl. 1–10 (2017)
6. N. Kumar, I. Hussain, B. Singh, B. K. Panigrahi, MPPT in dynamic condition of partially shaded PV system by using WODE technique. IEEE Trans. Sustain. Energy 1–10 (2017)
7. N. Kumar, I. Hussain, B. Singh, B.K. Panigrahi, Rapid MPPT for uniformly and partial shaded PV system by using JayaDE algorithm in highly fluctuating atmospheric conditions. IEEE Trans. Ind. Inf. 1–10 (2017)
8. A. Lashab, D. Sera, J.M. Guerrero, L. Mathe, A. Bouzid, Discrete model predictive control-based maximum power point tracking for PV systems: overview and evaluation. IEEE Trans. Power Electr. 1–14 (2017)
9. A.K. Mishra, A. Kumar, B. Singh, Solar photovoltaic array dependent dual output converter based water pumping using switched reluctance motor drive. IEEE Trans. Ind. Appl. 1–8 (2017)

Identification and Classification of Water Stressed Crops Using Hyperspectral Data: A Case Study of Paithan Tehsil

Sandeep V. Gaikwad, Amol D. Vibhute, K. V. Kale, S. C. Mehrotra, Rajesh K. Dhumal, Amarsinh B. Varpe and Rupali R. Surase

Abstract Globally, agricultural drought is the heterogeneous issue which causes the reduction of food production. The conventional methods have many limitations. Moreover, the use of multispectral remote sensing in drought condition monitoring possesses a limited spectral resolution which is insignificant for an understanding of water stress in the vegetation. In this regard, the study has been examined the agricultural droughts using ground observation, meteorological data and hyperspectral remote sensing (HRS) for assessment of crop water stress. The objective of this research was to: (a) examine the meteorological and hyperspectral data set for drought assessment (b) examine the agricultural stress tool for agricultural crop stress classification. The experimental results were evaluated and validated. The overall accuracy was obtained 86.66% with kappa coefficient 0.80. The research study has investigated the severe drought in the study area due to scanty rainfall during the Kharif season of year 2014. The present work is beneficial for identifying and monitoring the agricultural drought for better planning and management of crops.

Keywords QUAC · Hyperspectral processing · Spectral indices · Crop stress Agricultural stress tool

1 Introduction

The rainfall deficiency over the long term period forms the drought episode. The drought is primarily classified into three types viz Meteorological, Hydrological, and Agricultural. Whereas agricultural drought can be cause to the degradation of

S. V. Gaikwad (✉) · A. D. Vibhute · K. V. Kale · S. C. Mehrotra · R. K. Dhumal · A. B. Varpe
R. R. Surase
Geospatial Technology Research Laboratory, Department of Computer Science and IT,
Dr. Babasaheb Ambedkar Marathwada University, Aurangabad, Maharashtra 431004, India
e-mail: sandeep.gaikwad22@gmail.com

A. D. Vibhute
e-mail: amolvibhute2011@gmail.com

C. R. Krishna et al. (eds.), *Proceedings of 2nd International Conference on Communication, Computing and Networking*, Lecture Notes in Networks and Systems 46, https://doi.org/10.1007/978-981-13-1217-5_89

crop health which is necessary for future food production to sustain the humans. Agricultural sectors are significantly affected by climate changes which are highly imbalance by the gap between demand and supply of food. The economies of various countries are depending upon agriculture and associated businesses. The India is a witness of 25 major drought episodes since 1871–2015. The drought of 1987 was the one of the worst drought among the many droughts; it has 19% of the overall rainfall deficiency, which are affected 59–60% of the normal cropped area and a population of 285 million. The similar drought episode repeated in the year 2002 when the overall rainfall deficiency was 19%. Over 300 million people across the 18 states, and 150 million cattle were badly affected by severe drought. Moreover, in 2009, the overall rainfall deficiency in the India was 22%, which were caused to reduction of yield by 16 million tons. India has been faced drought disaster in various states in the 2013–2016, so that the suicidal rate of farmer was increased [1–4]. The better understanding of regional climate and topography plays an important role in the monitoring of droughts and its assessment. The drought is complex in nature and its vulnerability is diverse as per the region. The conventional approach of drought monitoring and assessment is based on rainfall, temperature, evapotranspiration, extends to the sown area, and crop condition. It has certain limitation like subjective observation, non-spatial nature, insufficient coverage, and distinct ground observational data [5]. To overcome the limitations of conventional drought monitoring systems we have tried to cover the impact of agricultural droughts using remote sensing approach. The approach is a utilization of remote sensing based parameter for drought monitoring and impact assessment. Similar studies were carried out by the authors [6], [7] using remote sensing technology in drought classification and its analysis.

Recently, the HRS data provides significant spectral information of the spatial objects for various applications [8]. The various studies have concluded the use of HRS data in the analyzing of vegetations instead of conventional multispectral remote sensing approach. Moreover, the spectral indices are significantly used to study the interaction of light and object at a small number of wavelengths [9, 10]. The HRS has enabled to study the spectral response of the crop in 400–2500 mm spectrum. The scientific community has developed various narrowband vegetation indices for the analysis of biophysical and biochemical properties of vegetation, which include leaf area index, pigment content (e.g., anthocyanin, carotenoid, and chlorophyll), biomass, crop stress (Nitrogen, water stress), crop damages, and biochemical properties (e.g. cellulose, lignin) etc. [11, 12].

The hyperspectral vegetation indices can be divided into three main categories: (1) Plant physiology/stress; (2) Biochemistry; and (3) Structure. The Hyperspectral Vegetation Indices (HVI) can be used as the potential parameter for crop stress analysis. Consequently, the physiological and stress indices are used to measure the subtle changes in the state of xanthophyll's, chlorophyll, fluorescence, and water moisture in leaf due to abiotic stress [12]. The studied literature has suggested that, the Greenness/leaf pigment indices, light use efficiency indices, and leaf water indices are significantly used for analyzing the crop stress. The objectives of the present research study are (1) identification and classification of crop stress, (2) spectral analysis of the study area, (3) meteorological analysis of the study region.

2 Study Area, Used Datasets and Proposed Methodology

For the current research study, Bidkin, Islampur, Gidhada, Bangala Tanda, and Banni Tanda villages of Paithan Tehsil were chosen which is located at latitude 19°42′ 14.29″N and longitude 75°18′ 0.54″E of Aurangabad district of Maharashtra state of India. The average rainfall of Paithan is 734 mm and the minimum temperature is 4.6 °C whereas maximum temperature is 45.9 °C. The meteorological data, such as rainfall and temperature has acquired from Maharain and Weather-underground weather services. The EO-1 Hyperion satellite image was acquired on October 15, 2014 from USGS earthexplorer. The Hyperion data were provided in standard HDF Version 4.1 (release 5), written as a band-interleaved-by-line (BIL) files stored in 16-bit signed integer radiance values. The EO-1 Hyperion level 1Gst product of 242 bands, which is radiometric corrected and resembled for geometric correction. The acquired image was registered using geographic map projection WGS84 with UTM zone 43 N. The product was orthorectified using digital elevation model (DEM) data and stored in 16-bit signed integer radiance values. The preprocessing operations such as bad band removal, bad column removal, radiometric calibration, atmospheric correction were performed in ENVI software. The Hyperion image consists visible-near-infrared (VNIR) spectrum with 70 bands, and the shortwave-infrared (SWIR) spectrum with 172 bands. We have removed 1–7, 58–76, 121–126, 167–180, 222–242 bands which were zero or no values using bad band removal algorithm [13]. The total 175 among the 242 bands were extracted for vegetation analysis. It is essential to convert radiance image into reflectance image by applying an atmospheric correction algorithm like Fast Line-of-sight Atmospheric Analysis of Hypercubes (FLAASH), QUick Atmospheric Correction (QUAC) and Atmospheric CORrection (ATCOR). In the present study, QUAC algorithm [14] was used for atmospheric correction and to determine the atmospheric correction parameter directly from the observed pixel spectra in a scene without metadata information. It performs a more approximate atmospheric correction than FLAASH algorithm, generally producing reflectance spectra within the range of approximately 10% of ground control points [13, 14]. The QUAC rectifies the hyperspectral images by removing cloud end members with 940–1020 nm water absorption band.

2.1 The Agricultural Stress Tool and Used Indices

The agricultural stress tool was used to classify crop water stress with the help of hyperspectral narrowband indices. It was used to create spatial maps based on distribution of crop stress. The tool analyzes green leaf area, photosynthesis and canopy water content for interpretation of crop stress. The vegetation indices were used for identification and classification of crop stress using Eqs. (1)–(3). The Agricultural Stress Tool classifies input image into nine classes. The higher value shows high stress and lower values shows lower stress [15].

In this research study, the Normalized Difference Vegetation Index (NDVI) with minimum valid greenness value 0 is used for masking out the non-vegetative classes. The NDVI is a worldwide popular index, which is used for drought monitoring and vegetation health assessment. The NDVI values range between −1 and +1, whereas lower values shows non vegetation classes and higher value shows dense vegetation.

$$\mathrm{NDVI} = \frac{\mathrm{NIR} - \mathrm{RED}}{\mathrm{NIR} + \mathrm{RED}} \quad (1)$$

The canopy water content VIs is based on amount water present in foliage canopy. The healthier canopy has a high percentage of water which is a vital element for crop growth and fire resistance. The canopy water content VIS uses reflectance spectra of NIR and SWIR region to determine the absorption feature of water present in the canopy, and it measures the total water content [15]. The Water Band index is very sensitive to changes in water content in the canopy. It is also useful for fire hazard condition analysis, cropland management, yield analysis, crop bio characteristics analysis and vegetation health analysis. The WBI shows green vegetation ranges 0.8–1.2.

$$\mathrm{WBI} = \frac{\rho_{970}}{\rho_{900}} \quad (2)$$

The light use efficiency indices are based on the properties of light utilizes for the photosynthesis process. The Photochemical Reflectance Index (PRI) measures the changes in carotenoid pigment in live foliage. The carotenoid pigment is very sensitive to the rate of carbon dioxide uptake by foliage per unit light energy absorbed. It is used for analysis of crop stress, yield estimation, vegetation health assessment and forest health monitoring. The value of PRI ranges from −1 to 1. The range of green vegetation is −0.2 to 0.2 [16, 17].

$$\mathrm{PRI} = \frac{\rho_{531} - \rho_{570}}{\rho_{531} + \rho_{570}} \quad (3)$$

3 Results and Discussion

The present research study investigates the assessment of crop health in the Kharif season 2014. The monsoon arrives in Maharashtra in the first week of June month. The sowing activity commence when enough rainfall is received. The first week of June 2014 (Fig. 1a) has received 9.1 mm rainfall and 7 mm in the third week, which is not sufficient for sowing operations. In July, satisfactory rainfall has received in a first 3 weeks, which help to resume sowing activity. Weekly rainfall is required to sustain the health of the crop. The August month has received highest rainfall 141.9 mm in 3rd and 4th week of the month. The August and September month is an important for the flowering process of many crops, it is also known as the maturity period of the plant. The September month has received total 59.1 mm rainfall, which was not enough to sustain the health of the crop. September is an important month for flowering of cotton. The irregularity and deficiency of rainfall cause water stress in plant which result drooping of leaves and boll, yellowing, browning and finally death of the plants. The October month is known as fruiting of plant, the October month has received lowest rainfall during the season. The high humidity and cloudy days accompanied by low temperature have been reported to increase the incidence of Reddening or Lalya Disease. Figure 1b shows, the temperature start decreasing at the beginning of October month. It was clearly noticed that, rainfall and temperature are inversely proportional to each other. The soil moisture decreases due to prolonged onset rainfall and raising temperatures.

In the present research study, the historical Google map images for visual interpretation and ground observation data were used for analysis of vegetation health conditions. We have carried out field visit along with interview of farmer who have faced severe drought.

Figure 2 depicted the spectral profile of the classified stress level of agricultural crop. The algorithm has been used to classify the drought into 0–9 classes, which consist 0 classes belong to unclassified and 1–9 indicate the health condition from healthy to severe stress respectively. The unclassified group has included non-vegetation classes such as settlement, water body, and hill with rocks. We have to merge spectrally relative classes into the group of 2 classes each.

Figure 3a illustrate the result of the classification algorithm which is divided into 6 groups such as unclassified, healthy, moderate, stressed, highly stressed, and fallow land. Where, the class 1 and 2 are merged into healthy vegetation class. The class 3 and 4 are merged into normal vegetation class. The class 5 and 6 are merged into stressed vegetation class. The class 7 and 8 are merged into highly stressed vegetation class. Class 9 shows the non-agricultural crop like a grass or harvested crop. So, we have skipped non-vegetation class 0 and 9 in the analysis because the scope of this research study is limited to the agricultural crops. Figure 3b illustrates the classification result of agricultural stress tool. The healthy class area is 745.87 ha which is lowest in the group due to deficiency of rainfall and surface water. The normal vegetation in the study area was a 1350.44 ha, which was less than stressed

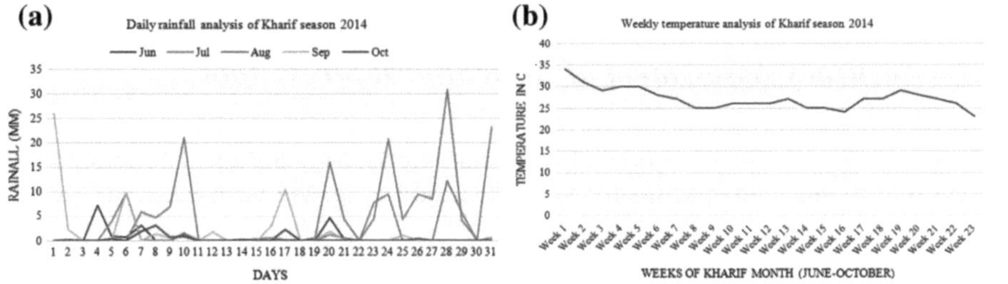

Fig. 1 **a** Daily rainfall analysis of Kharif season 2014 and **b** weekly temperature analysis of Kharif season 2014

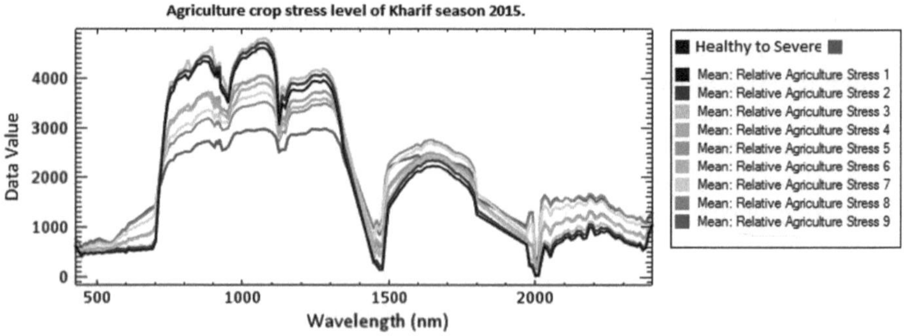

Fig. 2 Spectral signatures of water stressed crop

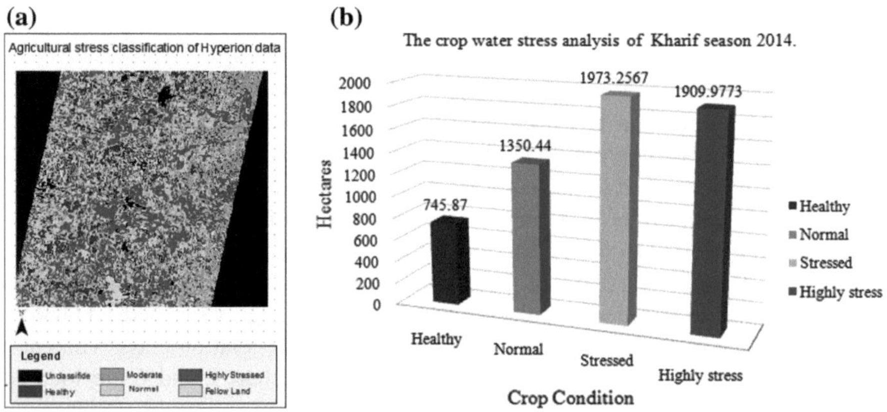

Fig. 3 **a** Classified image using agricultural stress tool and **b** the crop water stress analysis of Kharif season 2014

vegetation. The stressed vegetation area was 1973.25 ha and the highly stressed area was 1909.97. It is clearly noticed that, actual rainfall received from the Kharif season was not enough to sustain the healthy condition of the crop.

3.1 Accuracy Assessment of Agricultural Stress Tool

The accuracy assessment is an essential for classification algorithms. The agricultural stress tool is intended specifically for agricultural crop. We have taken 40 ground truth using ODK Collect android app.

The 20 GCP has taken to for vegetation and other 20 GCPs were received of other land covers. The classified image with historical image of year 2014 has used for overlay operation in Google earth software which is helpful for visual interpretation and analysis of experimental results. Accuracy estimation (Table 1) was evaluated

Table 1 Error matrix resulting from classifying training sets of pixels

Error matrix					
Classes	Vegetation	Settlement	Barren land	Water body	Total
Unclassified	0	0	0	0	0
Vegetation	18	0	2	0	20
Settlement	0	10	1	0	11
Barren land	1	1	4	0	6
Water body	0	0	0	3	3
Total	19	11	7	3	40
PA (%)	94.74	91	57.14	100	
UA (%)	90	90	66.66	100	

PA producers accuracy, *UA* users accuracy, Overall accuracy −86.66, kappa value −0.80

in terms of producer's accuracy, user's accuracy, overall accuracy and kappa coefficient after generating confusion matrix for agriculture stress classification. The non-diagonal values (Table 1) indicate the error between classified classes from corresponding ground observation [18, 19]. According to accuracy assessment results of agriculture stress tool, the overall accuracy of error matrix was 86.66% and Kappa Coefficient was 0.80.

4 Conclusions

The current research study utilizes meteorological data and hyperspectral data to investigate the impact of drought over the Kharif season. The study area has faced drought episode since 2013. The impact of agricultural drought differs from crop to crop and one crop species to another. The understanding of anatomy of crop stress will remain the challenge due to our limited understanding of water demand and supply for crop growth stages. It was observed that, understanding of soil-water relationship, soil nutritional components, seed variety, regional climate and regional hydrological condition are required for assessment of agricultural droughts. Also, it was observed that, farmers could not sow in June month due to unavailability of sufficient soil moisture. The late arrival of monsoon has affected the production and nutritional ingredient of the food crop. Moreover, weekly change in temperature along with rainfall could not sustain the health of the crop, which is having a severe impact on flowering process, drooping of leaves and boll. Such severe impacts lead to substantial decrease in yield. Therefore, improving the understanding of agricultural drought is necessary to develop a drought simulation model by considering daily changing climatic condition, obsolete ground reference data, farming practices, soil-water relationship, plant biophysical and biochemical properties.

Acknowledgements The authors would like to acknowledge and thanks to UGC, India for granting UGC SAP (II) DRS Phase-I and Phase-II F. No. 3-42/2009 and 4-15/2015/DRS-II for Laboratory facility to Department of CS and IT, Dr. BAM University, Aurangabad, Maharashtra, India and financial assistance under UGC BSR Fellowship for this work. The author is thankful Vaijapur Tehsil office and the Agricultural office for providing meteorological and sown area data. The author is also thankful to USGS for providing all the satellite images.

References

1. NRAA, in *Contingency and Compensatory Agriculture Plans for Droughts and Floods in India 2012*. Position paper No. 6. National Rainfed Area Authority, NASC Complex, DPS Marg, New Delhi, 110012, India, 2013, 87 p
2. DAC&FW, in *Crisis Management Plan: Drought (National)* (Department of Agriculture, Cooperation and Farmers Welfare, Government of India, 2016)
3. DAC, in *Manual of Drought Management, Department of Agriculture and Cooperation, Government of India* (2016), 192 p
4. A.D. Vibhute, B.W. Gawali, Analysis and modeling of agricultural land use using remote sensing and geographic information system: a review. Int. J. Eng. Res. Appl. (IJERA) **3**(3), 081–091 (2013)
5. S.V. Gaikwad, K.V. Kale, Agricultural drought assessment of post monsoon season of Vaijapur Taluka Using Landsat8. Int. J. Res. Eng. Tec. **4**(04), 405–412 (2015)
6. S.V. Gaikwad, K.V. Kale, S.B. Kulkarni, A.B. Varpe, G.N. Pathare, Agricultural drought severity assessment using remotely sensed data: a review. Int. J. Adv. Remote Sens. GIS **4**(1), 1195 (2015)
7. S.V. Gaikwad, K.V. Kale, R.K. Dhumal, A.D. Vibhute, Analysis of TCI index using Landsat8 TIRS sensor data of Vaijapur region. Int. J. Comput. Sci. Eng. **3**(8), 60–64 (2015)
8. A.D. Vibhute, K.V. Kale, R.K. Dhumal, S.C. Mehrotra, Soil type classification and mapping using hyperspectral remote sensing data. in *2015 International Conference on Man and Machine Interfacing (MAMI)* (IEEE, 2015), pp. 1–4
9. G.A. Carter, Reflectance bands and indices for remote estimation of photosynthesis and stomatal conductance in pine canopies. Remote Sens. Environ. **63**, 61–72 (1998)
10. P.S. Thenkabail, J.G. Lyon, A. Huete, in *Hyperspectral Remote Sensing of Vegetation and Agricultural Crops: Knowledge Gain and Knowledge Gap After 40 years of Research: Chapter 28* (CRC Press/Taylor and Francis Group, Boca Raton, London, New York, 2011)
11. S. Ullah, M. Schlerf, A.K. Skidmore, C. Hecker, Identifying plant species using mid-wave infrared (2.5–6 m) and thermal infrared (8–14 m) emissivity spectra. Remote Sens. Environ. **118**, 95–102 (2012)
12. S.E. El-Hendawy, W.M. Hassan, N.A. Al-Suhaibani, U. Schmidhalter, Spectral assessment of drought tolerance indices and grain yield in advanced spring wheat lines grown under full and limited water irrigation. Agric. Water Manag. **182**, 1–12 (2017)
13. A.D. Vibhute, K.V. Vibhute, R.K. Dhumal, S.C. Mehrotra, Hyperspectral imaging data atmospheric correction challenges and solutions using QUAC and FLAASH algorithms. in *IEEE, International Conference on Man and Machine Interfacing (MAMI)*, 2015, pp. 1–6
14. L.S. Bernstein, X. Jin, B. Gregor, S.M. Adler-Golden, Quick atmospheric correction code: algorithm description and recent upgrades. Opt. Eng. **51**(11), 111719 (2012)
15. H. Geospatial, Envi 5.3 manual, https://www.harrisgeospatial.com/docs/AgriculturalStressTool.html. Accessed 10 Dec 2017
16. J. Penuelas et al., The reflectance at the 950–970 region as an indicator of plant water status. Int. J. Remote Sens. **14**, 1887–1905 (1995)
17. C. Champagne, et al., Mapping crop water stress: issues of scale in the detection of plant water status using hyperspectral indices. In *Physical Measurements and Signatures in Remote Sensing. International symposium*, 2001, pp. 79–84

18. A.D. Vibhute, R.K. Dhumal, A.D. Nagne, Y.D. Rajendra, K.V. Kale, S.C. Mehrotra, Analysis, classification, and estimation of pattern for land of Aurangabad region using high-resolution satellite image. In *Proceedings of the Second International Conference on Computer and Communication Technologies*, Springer, India, 2016, pp. 413–427
19. A.D. Vibhute, A.D. Nagne, B.W. Gawali, S.C. Mehrotra, Comparative analysis of different supervised classification techniques for spatial land use/land cover pattern mapping using RS and GIS. Int. J. Sci. Eng. Res. **4**(7), 1938–1946 (2013)

Part VIII
Computing Technologies

Social Spider Foraging Based Resource Placement Policies in Cloud Environment

Preeti Abrol and Savita Gupta

Abstract Expansion in the cloud infrastructure leads to the challenge of resource placement. The existing resource placement techniques are not sufficiently effective. In this paper, the mathematical model of social spider cloud web algorithm is presented that targets the improvement in the utilization and focuses on the overall cloud performance. A new novel nature-inspired algorithm, social spider cloud web algorithm, helps in resource placement and load balancing of the cloud. It works on the foraging behavior of social spider and sorts the tasks and allocates the resources which leads to the efficient cloud performance.

Keywords Cloud computing · Cloud architecture · Resource placement module Social Spider Cloud Web Algorithm (SSCWA)

1 Introduction

The optimal resource utilization is very important in the cloud management. Due to the scalable nature of cloud, complexity issues creep into the resource placement, and hence a very efficient resource placement technique is required to be designed. A hierarchal multilayer Cloud Architectural Framework (CAF) is proposed that can handle the requested task sets for the available resources, tagged as the resource management problem in PaaS layer of cloud, and the utilization devoted to the execution of each task at each resource, tagged as the resource placement problem in IaaS layer of cloud [1, 2]. The resource placement can be followed subsequently after resource management according to the user's requirement to locate the resources. Then, resource placement is performed by mapping the resources with those tasks. The framework executes the requests as follows:

P. Abrol (✉)
STD, CDAC, Mohali, Punjab, India
e-mail: Karanpreet@cdac.in

S. Gupta
CSE, UIET, Chandigarh, India

C. R. Krishna et al. (eds.), *Proceedings of 2nd International Conference on Communication, Computing and Networking*, Lecture Notes in Networks and Systems 46, https://doi.org/10.1007/978-981-13-1217-5_90

- The client sends the request dynamically with the help of request generator in order to execute it on a certain number of resources.
- Further, the request is handled by the admission control module where the tasks are lined up according to the traffic in the network.
- The cloud scheduler is a decision-making body that implements allocation and prioritization policies while optimizing the execution of the task set for high efficiency and performance. It assigns the task set to the Cloud Service Providers (CSP) which is having appropriate information about the resources as per the client's demand.
- The CSP is having all the information about the resources available in the hosts of a datacenter as it updates the resource status by communicating with datacenter controller.
- RPM is a core module that implements SSCWA on the IaaS layer of cloud, for the purpose of resource placement. Now, the provisioned and scheduled resources are mapped with the tasks in the RPM.

In the related work, we can recall few studies that are close to the proposed approach. Spider algorithm is implemented for the search engine optimization by Whitehouse and Lublin in 1999 [3, 4], Levin [5, 6]. A behavioral study of the foraging strategy of the colonial spider [7, 8, 9] is conducted by offering different prey sizes and recorded their interactions. The foraging behavior of social spider is proposed in [10] where authors emphasized the population-based algorithm called Social Spider Optimization (SSO). An another paper [11] specifies the framework that is based on the foraging strategy of social spiders, dependent on the concept of the vibrations for communication over the spider web to determine the position of prey for the purpose of global optimization. The structure of the paper is as follows: Sect. 2 enlightens contributions. Metaheuristic SSCWA is specified in Sect. 3. Mathematical model is discussed in Sect. 4. Section 5 presents conclusion and future scope of the paper.

2 Contributions

Cloud Architectural Framework (CAF) model has been defined for cloud computing in which a Resource Management Module (RMM) in PaaS layer of cloud and Resource Placement Module (RPM) in IaaS layer of cloud have been incorporated. The first step handled by the RMM would ascertain a set of resources which meets the requirements of the user. The RPM verifies resource placement policies and hence the provisioned set of resources is listed out of the total available resources in the CAF using the Social Spider Cloud Web Algorithm (SSCWA). The contribution of the paper is manifold as follows:

- The hierarchical multilayer Cloud Architectural Framework (CAF) that guarantees the stability of system and constraints satisfaction.

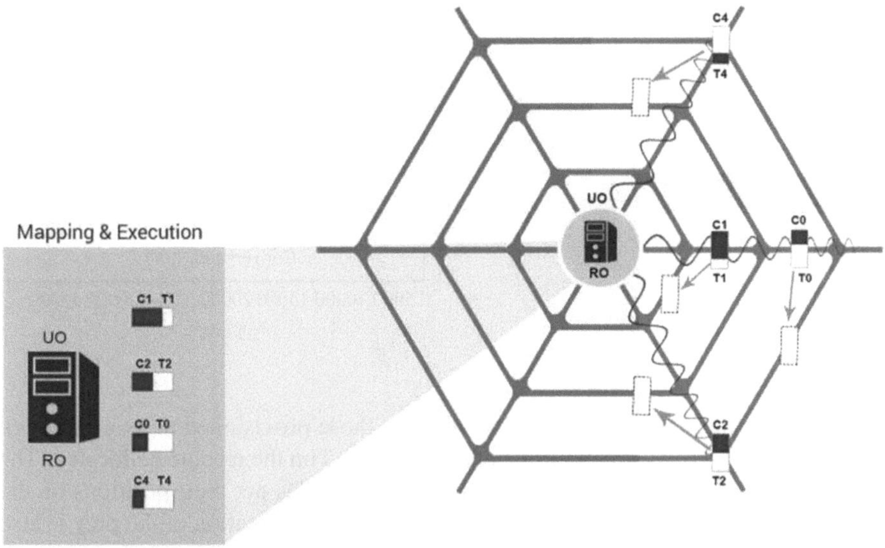

Fig. 1 SSCWA functionality

- The Resource Placement Module (RPM) which captures the dynamic behavior of the underlying infrastructure in PaaS layer by implementing various existing nature-inspired algorithms on the framework.
- Optimized technique for analysis and mathematical model is presented.
- The determination of the set of feasible task set confronts the complexity of the global optimization problem, thereby avoiding the convergence issue (Fig. 1).

3 Social Spider Cloud Web Algorithm—SSCWA

Many researchers developed algorithms based on optimization with the idea of ants, bees, bacteria, and fireflies behavior. A collection of similar kind of animals' exhibit swarm behavior. Most of the spiders are of the type solitaire. The intensity of vibration generated by each spider is equal to fitness, i.e., larger for the maximization and smaller for the minimization problem. The fitness evaluation is performed by itself. If a prey is caught in the web, then it will put efforts to free it from the web and hence make vibrations on the web. This vibration on the spider web is sensed by the various spiders present on the web as shown in Fig. 2. Here, *Ro* is the resource/prey and $C0$, $C1$, $C2$, $C3$, and $C4$ are the spiders or task.

Social Spider Cloud Web Algorithm (SSCWA) mimics the nature of social spider, implemented at the IaaS layer of the CAF model in order to implement the RPM policies for the purpose of mapping of resources with the task. Initially, all the provisioned tasks and resources information and status are passed to the resource placement mod-

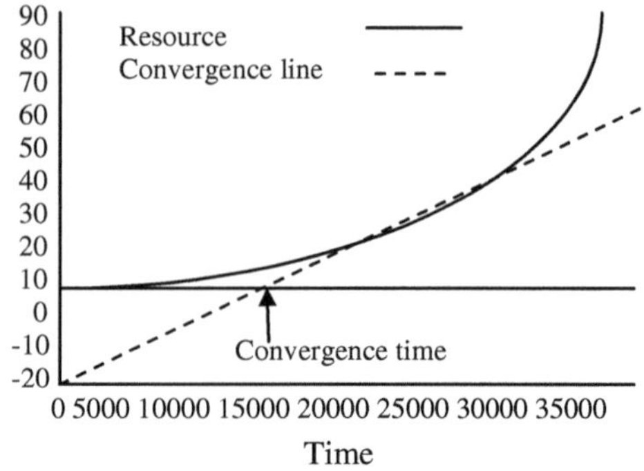

Fig. 2 Convergence of SSCWA w.r.t. resources

ule. SSCWA places the tasks to the resource as these provisioned tasks go through the position change event and ultimately get executed on the resource allocated. The fitness values of each task and resource are calculated as per their positions on the research space and its utilization. Similarly, the fitness evaluation of the prey is also performed on the basis of its capacity. Now each spider *s* will search the strongest intensity of vibration as V_{best} among vib and will sort all the vibration intensity as per attenuation in its memory, leading to it with its own vibration intensity named as the target vibration V_{tar} at position P_{tar} which is stored in its memory. *s* will store V_{best} as V_{tar} if the intensity of *v* best *s* is larger; otherwise, the previous V_{tar} is retained and each time the number of iterations is incremented. The algorithm then sorts the spiders such that the position change is performed as per the increasing value of vib.

3.1 Position Change

Each spider has a parameter DIM, representing the dimension of the spider web that can be multidimensional. Initially, DIM is zero. Now the probability of the spider to change its position is $(1 - P_C)$ where P_C is a control parameter such that $P_C \; \varepsilon \; (0, 1)$

where $P_C = \frac{r_j}{e^{-\frac{D_{x,p}}{D_{\max}}}}$

such that $D_{x,p}$ =Euclidean distance, x=position of the spider x at time t, p = position of prey p at time t, and r_j is a position change rate parameter. After the dimension, DIM is assigned to the spider; a new position P_o is allotted to the spider based on the DIM, where r is a random number [12]. Now, there are two constraints faced by the spider in the event of position change as discussed follows.

The P_o calculated must not be computed out of the boundary of the spider web, i.e., the spider web must have boundary constraints and the new position of the spider must not be blotted out of those boundaries.

Some other spider might not be already placed on P_o as no two spiders can take one position on the web.

As a summary of different boundary conditions, a brief description of each is given below:

Absorbing: When a particle moves out of the solution boundary of the search space in the given particular dimensions, its position is relocated on the boundary in the dimension of the search space, and the velocity component is set to zero.

Reflecting: When an individual jumps out of the solution space in one of the dimensions, its position is relocated on the boundary in the dimension of the search space and the sign of the velocity component in that dimension is changed.

Damping: When an individual jumps out of the solution space in one of the dimensions, its position is relocated on the boundary in the dimension of the search space; velocity component in that dimension is changed in the opposite direction with a random factor between 0 and 1.

Periodic: This strategy maps an infeasible location, for each variable violating the boundary to a feasible location such that an assumption is made that the search space is infinite. This is done by placing repeated copies of original search space along the dimension of interest. The periodic method handles all the variables separately and allows the infeasible solution to re-enter the search space from an end which is opposite to where it left the search space. For the problems with optima at the center of the boundary, the periodic approach may be useful.

Here, we will use the periodic method for handling boundary constraints. In the position change event, we can calculate the new position P_o of those spiders that have probability to change their positions.

$$P_o = \begin{cases} (U - \Delta P) * \gamma & \text{where } \Delta P = L - P_{\text{tar}}, L < P_{\text{tar}} \\ (L + \Delta P) * \gamma & \text{where } \Delta P = P_{\text{tar}} - U, U > P_{\text{tar}} \end{cases}$$

P_{tar} Depicts the position of the spider with source vibration v_{tar}. γ is the random number generator. The walk performed by the spider in the spider web toward the prey can be calculated as follows:

$$P_s(t) = \text{Rand} \odot (P_o - P_s(t)) + \gamma(P_s(t) - P_s(t-1))$$

where $\odot$ is the element-wise multiplication such that $P_s(t-1)$ is the position of the spider s at time $(t-1)$, $P_s(t)$ is the position of the spider s at time t, and P_o is the position of the spider s at time $(t+1)$.

4 Mathematical Models

Assume that CAF is requested with T independent tasks such that each task is denoted by T_i where $0 \leq i \leq m$, where m is the maximum number of tasks requested by the client to get executed. The available resources are denoted by R such that each resource is denoted by R_j where $0 \leq j \leq n$ where n is the maximum number of resources required. Optimization function $\left(O^t\right)$ can be defined as a simple weighted sum function of makespan and cost. Makespan is used to indicate the general performance of the cloud. Makespan is the execution time ET $\left(R_j\right)$ of a task set on the resource. Execution Time (ET) $\left(R_j\right)$ can be defined as the time taken by the resource in the completion of the execution of all the tasks assigned along with the task T_i on resource R_j where ET $\left(R_j, T_i\right)$ is the execution time. It can be calculated as $\text{ET}\left(R_j\right) = \text{AT}\left(R_j\right) + \text{BT}\left(R_j, T\right)_i$ where AT $\left(R_j\right)$ is the available time and BT $\left(R_j, T_i\right)$ is the time required by each task T_i to get executed on the resource R_j, such that $R_j * T_i$ represents the size of the matrix. Lesser makespan specifies that the tasks are handled effectively and efficiently. The makespan of the resource depends upon the fitness of the resource; hence cost, C_r, is also dependent on the fitness of the resource. If the fitness of the resource is more, lesser will be the makespan of the resource, lesser will be C_r. The optimization function is calculated as the sum of makespan and cost. Makespan can be defined as the total finishing time that each task needs before presenting the next job.

The intensity of vibration generated by the prey can be computed as

$$I_{x,p}(t) = \begin{cases} \left(\frac{1}{U_{\max} - f(P_p)}\right)^{\text{pop}} & \text{for maximum Capacity} \\ \log\left(\frac{1}{f(P_p) - U_{\min}} * \text{pop}\right) & \text{for minimum Capacity} \end{cases} \tag{1}$$

where pop = population size of the spider/prey, $U_{\max}$ = maximum resource utilization constant, $U_{\min}$ = minimum resource utilization constant, $f\left(P_p\right)$ = fitness function of prey P at any position p in the web.

$$f\left(P_p\right) = \sum_{x=1}^{n} \text{Capacity of } R_j \tag{2}$$

where $\text{Task}_{\text{Length}i}$ is the length of task.

Mathematically, optimization function $\left(O^t\right)$ targets to spot a better optimization function O^{t+1} iteratively at time t. For the purpose of finding the optimal values of $f\left(O^t\right)$, the Newton–Raphson method is used to spot the critical points leading to $f'\left(O^t\right) = 0$ in the solution space.

$$O^{t+1} = O^t - f\left(O^t\right)/f'\left(O^t\right) = Z\left(O^t\right) \tag{3}$$

Usually, in the evolutionary computational algorithms, once a better solution is found, the entire population converges to a single solution, even if that solution is not of high quality as expected. Hence, the entire population got "stuck". This state is called convergence. In SSCWA, the concept of attenuation in the intensity of vibration has helped in avoiding convergence prematurely. Attenuation is inversely proportional to the convergence. In this graph, x-axis represents resource and y-axis specifies time. Convergence can be seen in the resource curve that is represented experimentally for one of the simulations. This curve is exploited to measure a convergence rate.

The linear part of the resource curve signifies convergence. Hence, from the resource curve, we can see that initially the convergence is less and increase gradually with the increase in time, linearly. The convergence rate is the intersection of the dotted line and the time axis, and the rate of convergence is the inverse of the time of convergence.

To avoid the convergence, rate of convergence must be quadratic in nature. So, the above equation is required to be modified by adding a parameter q as follows:

$$O^{t+1} = O^t - qf(O^t)/f'(O^t) = Z'(O^t, q) \tag{4}$$

such that $q = \frac{1}{1-Z'(O^*)}$

where O^* is the optimal solution of the iterative process.

As per Taylor's theorem, any function $f(x)$ having continuous second derivative can be specified with expansion to a point which is close to the root of $f(x)$. Then, the expansion of $f(\Upsilon)$ around O^t can be

$$f(\Upsilon) = f(O^t) + f'(O^t)(\Upsilon - O^t) + R_1 \tag{5}$$

where $R_1 = \frac{1}{2!}f''(x)(\Upsilon - O^t)^2$ such that x lies in between Υ and O^t.

$$f(\Upsilon) = f(O^t) + f'(O^t)(\Upsilon - O^t) + \frac{1}{2!}f''(x)(\Upsilon - O^t)^2 \tag{6}$$

When Υ is root, use value of R_1.

Now this equation is divided by $f'(O^t)$

$$1 = \frac{f(O^t)}{f'(O^t)} + (\Upsilon - O^t) + \frac{1}{2!}\frac{f''(x)}{f'(O^t)}(\Upsilon - O^t)^2$$

$$-\frac{f''(x)}{2f'(O^t)}(\Upsilon - O^t)^2 = \frac{f(O^t)}{f'(O^t)} + \Upsilon - O^t \tag{7}$$

$$-\frac{f''(x)}{2f'(O^t)}(\Upsilon - O^t)^2 = \Upsilon - O^{t+1} \tag{8}$$

This proves that the rate of convergence is quadratic in nature. In addition, there are often a set of p parameters in an algorithm such as in particle swarm optimization, there are four parameters (two learning parameters, one inertia weight, and the population size). We can write Eq. (10) with k parameters and r random variables as

$$O^{t+1} = Z(O^t, r(t), p(t)) \tag{9}$$

In above, the equation satisfies the single agent of the system. For the swarm of pop agents or solutions, the equation can be extended as the following formula:

$$\begin{pmatrix} O_1 \\ O_2 \\ \vdots \\ O_n \end{pmatrix}^{t+1} = Z\Big(\big(O_1^t\ O_2^t \ldots O_n^t\big)\big(r_1\ r_2 \ldots r_n\big)\big(p_1\ p_2 \ldots p_n\big)\Big)\begin{pmatrix} O_1 \\ O_2 \\ \vdots \\ O_n \end{pmatrix}^{t} \tag{10}$$

where $p_1\ p_2 \ldots p_n$ are n algorithm-dependent parameters and $O_1\ O_2 \ldots O_m$ are n random variables. Hence, we prove that the rate of convergence is quadratic in nature.

5 Conclusion

The challenging problem of resource placement is focused in this paper, where SSCWA is presented mathematically. Different test scenarios are implemented to the RMM in CAF for the performance evaluation in CloudSim. When SSCWA implemented on IaaS layer in RPM module, SSCWA outstands in makespan and cost analysis, in comparison to other algorithms, therefore avoids premature convergence, whereas other metaheuristics fails to do so. It proves that the rate of convergence for the SSCWA is quadratic in nature. In future, the social spider foraging behavior must be mimicked for the purpose of resource provisioning in the cloud framework.

Acknowledgements We want to thank Mr. Sukhwinder Singh for his guidance.

References

1. P. Abrol, S. Gupta, K. Kaur, Social spider cloud web algorithm (SSCWA): a new meta-heuristic for avoiding premature convergence in cloud. Int. J. Innov. Res. Comput. Commun. Eng. **3**(6), 5698–5704 (2015). ISSN (Online): 2320-9801, ISSN (Print): 2320-9798. . https://doi.org/10.15680/ijircce.2015.0306113
2. P. Abrol, S. Gupta, K. Kaur, in *Analysis of resource management and placement policies using a new nature inspired meta heuristic SSCWA avoiding premature convergence in cloud*. International Conference on Computational Techniques in Information and Communication Technologies (ICCTICT) (2016), pp. 127–132. ISSN 978-1-5090-0082-1/16/$31.00 ©2016 IEEE
3. L. Bater, in *Incredible insects: answers to questions about miniature marvels*. Vero Beach: Rourke Publishing LLC. Post Office Box 3328 (2007). ISBN 978-1-60044-348-0
4. C. Eric, K.S. Yip, in *Cooperative capture of large prey solves scaling challenge faced by spider societies*. Proceedings of the National Academy of Sciences of the United States of America (vol. 105, Issue 33, 2008), pp. 11818–11822

5. S. Levin, in *Encyclopedia of biodiversity* (Academic Press, Elsevier Inc, London, 2013a, 2013b, 2013c, 2013d, 2013e, 2013f). ISBN 978-0-12-384719-5
6. T.B. Lubin, in *The Evolution of Sociality in Spiders*, ed. by H.J. Brockmann. Advances in the Study of Behavior, (vol. 37, 2007), pp. 83–145
7. F.F. Campon, Group foraging in the colonial spider parawixia bistariata (Araneidae): effect of resource level and prey size. Animal Behav. Elsevier (2007), http://dx.doi.org/10.1016/j.anbehav.2007.02.030
8. A. Tchernykh, U. Schwiegelsohn, V. Alexandrov, E. Talbi, Towards understanding uncertainty in cloud computing resource provisioning. Procedia Comput. Sci. **51**, 1772–1781 (2015). https://doi.org/10.1016/j.procs.2015.05.387
9. C.E. Klein, E.H.V. Segundo, V.C. Mariani, L.D.S. Coelho, Modified social-spider optimization algorithm applied to electromagnetic optimization. IEEE Trans. Mag. **52**(3), (2016) https://doi.org/10.1109/tmag.2015.2483059
10. E. Cuevas, M. Cienfuegos, D. Zaldivar, M. Perez-Cisneros, A swarm optimization algorithm inspired in the behavior of the social-spider. Expert Syst. Appl. **40**(16), 6374–6384 (2013)
11. J.Q. Yu James, O.K. Li Victor, A social spider algorithm for global optimization. J. Appl. Soft Comput **30**(C), 614–627. Elsevier Science (2015)
12. E. Cuevas, M. Cienfuegos, R. Rojas, A. Padilla, in *A Computational Intelligence Optimization Algorithm Based on the Behavior of the Social-Spider*. Computational Intelligence Applications in Modeling and Control, Studies in Computational Intelligence (Springer, 2015) pp. 123–146. https://doi.org/10.1007/978-3-319-11017-2_6
13. C. Erik, C. Miguel, Z. Daniel, P.-C. Marco, A swarm optimization algorithm inspired in the behavior of the social-spider. Expert Syst. Appl. **40**, 6374–684 (2013)

Distributed Mutual Exclusion Algorithm with Improved Performance

Amit Kumar Maurya, Arvind Kumar, Akhilesh Kumar Mishra, Bharat Kumar and Priyanshu Vishwakarma

Abstract This paper presents a non-token-based algorithm for mutual exclusion in distributed systems in which sites communicate with other sites by message passing. The Ricart–Agrawala algorithm is one of the mutual exclusion algorithms for a distributed system. This algorithm uses message passing concept to decide which site will execute the Critical Section. This algorithm is a modified version of Lamport mutual exclusion algorithm that takes $3(n - 1)$ numbers of messages to enter into the CS for any site, where n is the number of sites. Ricart–Agrawala algorithm requires $2(n - 1)$ messages per Critical Section execution. In our proposed algorithm, the number of messages required for any site to enter into the CS will always be less than the number of messages required in Ricart–Agrawala algorithm. We have reduced the number of messages by cleverly removing the number of reply messages for a site that has already executed the CS.

Keywords Distributed system (DS) · Mutual exclusion · Critical Section (CS)

A. K. Maurya (✉) · A. Kumar · A. K. Mishra (✉) · B. Kumar · P. Vishwakarma
Ashoka Institute of Technology & Management, Varanasi, Uttar Pradesh, India
e-mail: amit.maurya.mit@gmail.com

A. K. Mishra
e-mail: mailforakhil29@gmail.com

A. Kumar
e-mail: omarvinda1812@gmail.com

B. Kumar
e-mail: bharatkumar093@gmail.com

P. Vishwakarma
e-mail: priyu.vishwakarma@gmail.com

C. R. Krishna et al. (eds.), *Proceedings of 2nd International Conference on Communication, Computing and Networking*, Lecture Notes in Networks and Systems 46, https://doi.org/10.1007/978-981-13-1217-5_91

1 Introduction

Mutual exclusion in a distributed system is an important concept in computer science. The problem of mutual exclusion states that only one process can enter into the Critical Section at a time. Once a process completed its execution of Critical Section, it simply releases the CS and other processes may execute their Critical Sections. In a distributed system, mutual exclusion is widely studied where execution of Critical Section for a process is done by passing asynchronous messages. Algorithms to maintain mutual exclusion is basically divided into two parts, i.e., Token-based mutual exclusion algorithm [1–3] and Non-token-based mutual exclusion algorithm [3]. Following are few examples of Token-based mutual exclusion algorithms like Singhal's heuristic algorithm, Suzuki–Kasami's algorithm [3], Raymond's tree-based algorithm [2], Naimi et al.'s algorithm, and Yan et al.'s algorithm [4]. Examples of Non-token-based algorithm are Lamport's algorithm, Ricart–Agrawala's algorithm [3, 5], Carvalho–Roucairlo's description of the Ricart–Agrawala algorithm [3, 5], Maekawa's algorithm, and Singhal's algorithm. In a distributed system, there is no concept of shared memory (except virtual shared memory [3, 6]) and a global clock. So if there is no concept of shared memory then each process needs to maintain their own queue to handle all the requests generated by all other processes, and each process maintains a clock value with a starting value is equal to 0 (zero). Every moment when a process wants to enter to the Critical Section, it sends a timestamp value with a request which is equal to sender's clock value (if there are multiple processes). There are some criteria on which a mutual exclusion algorithm will be treated as good mutual exclusion algorithm [1, 7]. In which of them the main criteria are fairness, and only the non-token-based algorithms like Ricart–Agrawala's and Lamport's mutual exclusion algorithm are fair [3, 6]. If we talk about Singhal's heuristic algorithm, that ensures some degree of fairness.

In Sect. 2 already existing Ricart–Agrawala algorithm is explained in a very simple way. Section 3 of this paper explains the proposed algorithm, while Sect. 4 gives the idea that how the proposed algorithm works with a suitable example. At last, that is in Sect. 5 we have concluded the proposed idea.

2 The Ricart–Agrawala's Algorithm

Each step in the algorithm is executed automatically. The following steps give a general idea of Ricart–Agrawala's algorithm [5].

- The reply message sent by a process is blocked only by processes that are requesting to enter into the Critical Section with greater priority.
  So, when a process sends a reply message to all the delayed requests, the process with the next highest priority request receives the required reply message and enters the Critical Section.

- The execution of Critical Section requests in this algorithm is always in the order of their nonincreasing priority.

For per Critical Section execution, there are exactly $2(n - 1)$ messages that are $(n - 1)$ requests and $(n - 1)$ replies messages, where n is the total number of processes.

3 Proposed Algorithm

3.1 Description and Basic Idea

The proposed algorithm is based on the concept of Ricart–Agrawala's algorithm for mutual exclusion, in which the underlying network channels follow the FIFO strategy. The new concept will reduce the number of messages needed per Critical Section as compared to Ricart—Agrawala's algorithm [3]. As in the previous algorithm, there are local request queues maintained by all the processes, which contain all the requests of processes that want to execute the Critical Section. This algorithm uses three stages: (a) Requesting the Critical Section, (b) Executing the Critical Section, and (c) Releasing the Critical Section.

Similar to RA algorithm, all requests are ordered by priority. Using the *local_request_queue*, one process can simply identify which process will execute the CS first.

1. **Requesting the Critical Section**

Any process which wants to enter into the Critical Section, need to send the request message to all the other processes. If P_i is a process which wants to execute the Critical Section, then it will send the request message to all other processes and maintains the *local_request_queue* same as we already did in RA algorithm. Requesting and maintaining *local_request_queue* is done by all the processes same as RA algorithm. A process, which does not want to execute the Critical Section will also maintain the same *local_request_queue* to set the order of execution of CS.

- Suppose a site-I, let us say S_i, wants to execute the Critical Section; it sends a time stamp request message to all the sites in the request sets.
- When a receiver site-J, let us say S_j, gets a request message from the sender site, i.e., S_i, it sends a reply message to the sender site S_i.
  If any of the given conditions are satisfied.
  (a) The receiver site, i.e., site S_j is not requesting to enter into the Critical Section and also it is not executing the Critical Section or,
  (b) If the receiver site S_j is requesting for the Critical Section execution and the sender site, i.e., Si's request is lesser than S_j's own request timestamp.
- The request is differed otherwise.

2. **Execution of the Critical Section**

Site S_i execute the CS after it has received reply message from all the sites in its request set. Once the request set is maintained by all the processes or say sites in *local_request_queue* then they will look up for the process having highest priority and let it to execute the CS first.

3. **Releasing the Critical Section**

As in RA algorithm, once a process or site completed the execution of Critical Section, it simply sends the release message to all the processes, by which other process can execute the CS. But in the proposed algorithm, instead of sending the release message to every other process, it simply sends a release message to the next site or process that have second highest priority by looking in the *local_request_queue.*

For example, the *local_request_queue* have the requests on the basis of priority which are P_i, P_j, *and Pk*. When process P_i will execute the CS, then it simply sends the release message to only P_j. Once the next highest priority process gets the release message it simply removes all the entries before the timestamp of itself from the *local_request_queue*. This will reduce the number of release messages for a process or site.

4 Example and Illustration of Algorithm

Let us take an example of three sites namely, S1, S2, S3. All the three sites are requesting to execute the CS and maintain the *local_request_queue.*

The given example is to compare with RA algorithm. Let us say *local_request_queue as* LRQS and V vectors at different instants of time. Following are the conditions throughout the process:

- If no process has sent out the request:

1. Vt1 = [0, 0, 0] and LRQS1 = ()
2. Vt2 = [0, 0, 0] and LRQS2 = ()
3. Vt3 = [0, 0, 0] and LRQS3 = ()

- If all sites have sent the request message but none of them have received any reply message from other sites or processes.

1. Vt1 = [1, 0, 0] and LRQS1 = ((1, 1))
2. Vt2 = [0, 1, 0] and LRQS2 = ((2, 2))
3. Vt3 = [0, 0, 1] and LRQS3 = ((3, 3))

- If all the sites have got the reply message of all requests from other sites then

1. Vt1 = [1] and LRQS1 = ((1, 1), (2, 2), (3, 3))
2. Vt2 = [1] and LRQS2 = ((1, 1), (2, 2), (3, 3))
3. Vt3 = [1] and LRQS3 = ((1, 1), (2, 2), (3, 3))

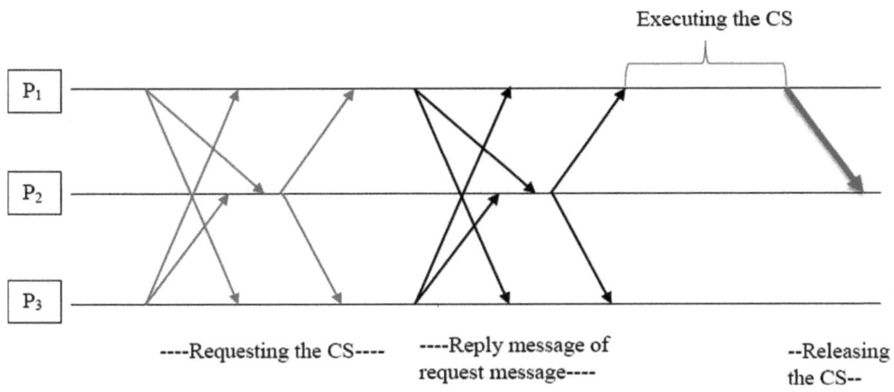

Fig. 1 Communication between processes to execute the Critical Section

Here, we have taken the two elements (process id, priority), which are the combination of process id and priority, respectively, in the timestamp. Now all the sites will look into their *local_request_queue* and starts executing the CS.

- Let us say, all the sites have got the reply message of all requests from all the other processes or sites

1. Vt1 = [1] and LRQS1 = ((1, 1), (2, 2), (3, 3))
2. Vt2 = [1] and LRQS2 = ((1, 1), (2, 2), (3, 3))
3. Vt3 = [1] and LRQS3 = ((1, 1), (2, 2), (3, 3))

In the above example, we have taken that priority of process 1 is higher than others so it will start executing the Critical Section. Once it completes the execution of Critical Sections, it simply removes the entry of itself from *local_request_queue* of the site and then sends a release message to the next process, which contains second higher priority in the request set.

4.1 P1 Finishes the CS

1. Vt2 = [1] and LRQS2 = ((2, 2), (3, 3))
2. Vt3 = [1] and LRQS3 = ((1, 1), (2, 2), (3, 3))

The same process will be executed until the *local_request_queue* of the last site is not empty, and the last process or site does not require to send any release message to any other site. All these processes definitely reduces the number of messages to execute the CS for a process or site (Fig. 1).

5 Conclusion

As we can see in the introduction part of this paper, there are various algorithms that sent the number of messages to communicate to each other and to execute the Critical Section. As explained in RA algorithm, the number of transferred messages will be at least $2(N - 1)$. But as in the proposed new algorithm, because it is an improved version of RA algorithm, it will take less than $2(N - 1)$ messages to execute the Critical Section for more than two processes always.

References

1. Guohong Cao, A delay-optimal quorum-based mutual exclusion algorithm for distributed system. Parallel Distrib. Syst. **12**(12), 1256–1267 (2001)
2. K. Raymond, A tree-based algorithm for distributed mutual exclusion. ACM Trans. Comput. Syst. **7**, 61–77 (1989)
3. M. Singhal, N. Shivaratri, in *Advanced Concepts in Operating Systems: Distributed, Database and Multiprocessor Operating Systems* (Tata McGraw Hill, 2001)
4. N. Naimi, M. Trehel, A. Arnold, A log(n) distributed mutual exclusion algorithm based on path reversal. J. Parallel Distrib. Comput. **34**, 1–13 (1996)
5. G. Ricart, A.K. Agrawala, An optimal algorithm for mutual exclusion in computer networks. Commun. ACM **24**(1), 9–17 (1981)
6. S. Lodha, A. Kshemkalyani, A fair distributed mutual exclusion algorithm. Parallel Distrib. Syst. **11**(6), 537–548 (2000)
7. M. Naimi, M. Trehel, An improvement of the log n distributed algorithm for mutual exclusion. ICDS 371–377 (1987)
8. J.-M. Helary, A. Mostefaoui, A o(log2 n) fault-tolerant distributed mutual exclusion algorithm based on open-cube structure. ICDCS 89–96 (1994)

Big Data's Biggest Problem: Load Imbalancing

Kanak Meena and Devendra K. Tayal

Abstract Uneven distribution of data between the nodes causes the data skewness problem. Due to this problem, various problems occur during the processing. So, this paper presents the brief analysis of the existing techniques related to load imbalancing with their pros and cons. Also, types of data skewness have been discussed in this paper.

Keywords Data skewness · Load imbalancing · Big data · Hadoop

1 Introduction

Technology has developed hugely within a previous couple of years. Additionally, the quantity of individuals utilizing innovation has expanded exponentially. As indicated by a source, every 2 days we make as much data as we did from the earliest starting point of time until 2003 [1–5]. More than 90% of the information on the world was made in the previous 2 years alone. Furthermore, it is expected that by 2020 the measure of digital data in the present would have grown from 3.2 zettabytes today to 40 zettabytes. There are various sources for generation of this big amount of data namely:

1. **People**
   As indicated by a source [2], the quantity of cell phone clients for 2016 were 4.61 billion and it is relied upon to cross 5 billion marks in 2019. Furthermore, with this expanding number, so is the information produced from these gadgets expanding in a humongous rate.

K. Meena (✉) · D. K. Tayal
Indira Gandhi Delhi Technical University for Women, New Delhi, India
e-mail: Kanak556@gmail.com

D. K. Tayal
e-mail: dev_tayal2001@yahoo.com

C. R. Krishna et al. (eds.), *Proceedings of 2nd International Conference on Communication, Computing and Networking*, Lecture Notes in Networks and Systems 46, https://doi.org/10.1007/978-981-13-1217-5_92

2. **Organization**
   Information created by associations is getting huge and complicated. Numerous associations now manage unstructured data also [3]. Henceforth, they are a major supporter of this huge information. In addition, associations require some answer for managing this enormous information produced to take better choices, make more benefits and satisfy more customers [1–7].
3. **IoT-generated data**
   With everything getting associated with the Internet, IoT creates a great deal of information. Gradually and consistently the things around us are showing signs of improvement [4]. The Internet of Things (IoT) has exploded over these previous 3 years, and as indicated by Gartner research, 25 billion associated "things" will be associated with the Internet by 2020 [1–7].

From the above introduction, it can be concluded that a lot of data has been generated and will be generated in future. These data may be structured and well defined or it may be unstructured/semi-structured as well as. Big data is a combination of different properties which makes it very complex, composite and tangled. Whereas, our conventional database systems fail to process such humongous and unstructured data. To overcome these problems and to handle big data, Apache Hadoop [8] becomes the trademark nowadays [6, 7].

MapReduce framework is now popular to handle a large amount of data due to its remarkable features scalability, fault tolerance, and simplicity. Management of this huge amount of data is really difficult which causes various problems viz. Data skewness and load imbalancing (uneven distribution of data causes data skewness problem). Due to the complexity of data, it increases the runtime, it also increases the cluster throughput and the slowest job takes time to complete the task [8–11].

Parallel data management systems try to limit the runtime of a complicated processing task master via adopting the method partitioning, where it partitions the data and distribute the load evenly among the present machines. Unless the data are partitioned in a reasonable manner, the runtime of the slowest machine will effortlessly rule the aggregate runtime of the program. For a given machine, its runtime is totally depended upon various factors to different parameters (including the speed of the processor and the extent of the memory), however, our primary concern is on the effect actuated by data assignment. The important issue is dividing the information genuinely, with the end goal that every machine is allotted an equi-sized data [11].

For real data sets, the regularly experienced problem is data skewness, thus plain partitioning schemes that are not enough to handle, viz., MapReduce platform, which easily falls short. More critically, when the data is similarly divided to the accessible machines, approach runtime cannot be ensured. The reason for this the situation is some machines may require complex information of the data, while others may essentially contain information where the processing of the query is not required. Providing advanced load balancing mechanisms appointing equal shares of useful work to the capable time that aim to increase the efficiency of query handling [9–11].

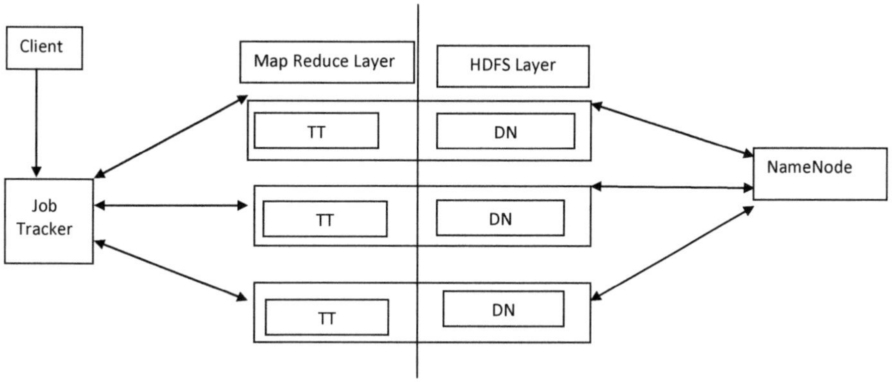

Fig. 1 Hadoop architecture

Table 1 Issues of Hadoop during the large-scale query processing

Shortcomings	Methods can be used to handle
Load balancing	Pre-processing, repartitioning
Lack of interactive real-time processing	Streaming, pre-computation, and pipeline
Communication cost is high between map and reduce	Partitioning
Access in input data is very difficult	Indexing and different data structures
Re-computation	Materialization
Long throughput	Sorting and sampling
Lack of iteration	Pipeline, recursion, and loop-aware processing

1.1 Hadoop Framework

Hadoop is a platform which is open source (MapReduce) to manage the big data, which is widely used in various organizations for the management of the data. Hadoop consists of two major parts for the processing of the data namely: HDFS (Hadoop-Distributed File System) and MapReduce. Hadoop architecture has been explained diagrammatically in Fig. 1 and issues of Hadoop platform during the analysis of large-scale processing have been explained in Table 1.

HDFS [8–14]: They are designed to store large files and to manage them with streaming. It is further divided into two sections Name-Node and Data-Node (DN). To improve the reliability of the systems, HDFS (data blocks) replicates and stored on three different machines, for the availability of these data they store one of the replicas in a different rack. Name-Node is used to maintain the metadata file. These metadata include the mapping from file to block and location of the block. Periodically, for the configuration of the data, the Name-Node communicates its metadata to a secondary Name-Node [8–14].

MapReduce [9–14]: It also works in two sections namely: Job-Tracker and Task-Tracker. The access point for clients is Job-Tracker and it also taking care of fair and efficient scheduling of incoming MapReduce jobs. It also assigns tasks to the Task-Tracker (TT). Execution of the given task is a responsibility of Task-Tracker and it can run a number of tasks simultaneously but it depends on the available resources (e.g., two map tasks and two reduce tasks). The small size of each task compared with a large number of tasks in total to ensure the load balancing. While map task depends on the size of input and in reduce task the number of particular jobs is defined by the user [9–14].

There are some issues related to large and parallel scale processing in MapReduce framework during the management of big data as shown in the given Table 1[11].

This paper has been organized in the following manner: Sect. 2 focuses on the types of data skewness existing in the literature and Sect. 3 discussed the literature review of the existing techniques with respect to load imbalancing, Sect. 4 outlines the comparative study and Sect. 5 presents the conclusion of this paper.

2 Types of Data Skewness

This section provides the types of data skewness: when the computational load in imbalanced the situation arises of lower cluster throughput and long job execution time so the problem arises data skewness. It occurs in both sides map & reduces side so it is known as map-skew and reduce-skew. Further, we discuss the types of skewness under and reduce side (map = 3 type and reduce = 2 type).

2.1 Types of Skewness at Map Side

1. Expensive Record [12–14]: Map assignments normally process a collection of records as a key-value pair, one-by-one. Ideally, from record to record, the handling time does not change basically. However, more CPU and memory has been required for some records as it totally depends upon the application. These costly records may fundamentally be bigger than other records, or the MapReduce runtime may rely upon the record's value.

   For example: Page rank algorithm [15]

2. Heterogeneous Maps [12–14]: it is a unary operator, be that as it may, can be utilized to copy a n-array operation by logically connecting numerous datasets as an isolated input. For task runtime, every dataset may require leading a multi-modular distribution and diverse processing.

For example: Cloudburst [16]

3. Non-Homomorphic Map [12–15]: In MapReduce framework, the clients can run discretionary code as soon it confirms to the MapReduce interface. It commonly initiates and cleanup, this kind of adaptability enables clients to push, when important, the limits of the map and reduce stages have been designed to do: every map output can be depended upon a group of data records (inputs).

For example: Sort–merge algorithm [12] and friends of friends (fof) [12].

2.2 *Types of Skewness at Reduce Side*

1. Expensive Input [12–15]: In MapReduce, reduce tasks perform a sequence of pairs as follows: (key, set of values). Expensive records prepared by a map which can cause skew during the runtime of the task at reducer side. Since reduce handles the group of the key in place of an individual key. The costly input issue can be more articulated, particularly at the point when reduce is an integrated operation.
   For example: Cloudburst [16].
2. Partitioning skew [12–17]: In MapReduce, the contribution of map task is appropriately distributed among reduce task by the standard method namely hashes-partitioning or some user customized partitioning method. The default hash-partitioning is generally, uniformly dispersing the information. But still, reduced skew occurs.
   For example: Page rank algorithm [15].

3 Literature Review

This section presents the literature survey on data skewness problem. All the existing work has been discussed here, with respect to all those techniques who handle the data skewness problem in Hadoop Framework only. It occurs due to load imbalancing which can occur on either side: map or reduce namely map-skew and reduce-skew. Skew-reduce [12] is a system which has been designed to handle skewness manually, data partitioning is defined by user. Reduce-skew studied by Ibrahim et al. and Gufler et al. [13]. In both of the approaches, the sequence of output as input cannot be assured. Skewtune [13, 14] handles the skew at both side map side and reduce side as well as it has been designed to handle data automatically. It also minimizes the skew mitigation side effects by preserving the order of input.

Adaptive MapReduce using situation-aware mappers [17] was proposed to handle the skewness at map side only although it is very flexible and adaptive technique. It is user-friendly and also it is having the continuous monitoring features to modify its feature. It has adaptive combiner and partitioner. Locality-aware and fairness-aware key partitioning [LEEN] [18, 19] has been designed to handle map-skew and it works on locality-aware and fairness-aware schemes. It also saves the communication bandwidth during shuffling phase. It also keeps track of buffered nodes which are used for hosting the data nodes.

Handling Data Skew in MapReduce Cluster by Using Partition Tuning [PTSH] [20], it has been designed to handle skewness by providing the virtual partitioning at map side and repartitioning at reduce side. Spongefiles [21] it has been developed to handle skewness problem in distributed memory. It uses the splitting of large datasets into small chunks. It handles the skewness during the high memory use at reduce side.

Dynamic Resource Allocation for MapReduce with Data Skew [DREAMS] [22] has been used to handle the data skewness by adjusting the time of task during the runtime. As well as, it eliminates overhead during repartitioning. Lightweight Data Skew Mitigation in MapReduce [LIBRA] [23], it has been designed to handle reduce-skew. It is the best solution for the huge datasets and does not support much for the small datasets. During the presence of data skewness, the overhead is negligible and there is no overhead in the absence of data skewness.

4 Comparative Study

This section provides the conclusive study of existing literature in the one direction, how much work has been done on big data to handle load imbalancing. A number of methods have been used to develop the best partitioning method. To save communication time and communication cost during the shuffling time and mapper to reducer communication. Here, we present the benefits and disadvantage of the type of skewness they can handle during the processing of the existing techniques in a chronicle order (Table 2).

5 Conclusion

In this paper, we have done the literature review of various techniques related to the load imbalancing. We have done the comparative analysis of various existing techniques. Different techniques have different benefits and shortcomings in different situations. So, every technique is suitable for different scenarios. According to our survey, we have concluded that LIBRA and DREAMS are the best techniques to handle the data skewness in big data on Hadoop platform.

Table 2 Shows the comparative analysis of different techniques

Technique	Advantage	Disadvantage	Skew type
Skew-reduce [12–14]	• User-defined partitioning function	• User needs the knowledge about the domain • Limited to specific application	• Map-skew
Skewtune [13, 14]	• Automatic • Cost-effective	• Slow • Not user-friendly	• Map-skew and reduce-skew
Adaptive MapReduce using situation-aware mappers [17]	• Adaptively resplit • Continuous monitoring	• Cannot handle the skew at reduce side	• Map-skew
LEEN [18, 19]	• Save bandwidth of network during the shuffle phase • Full control of key distributions	• Static approach • Do not support small keys	• Map-skew
PTSH [20]	• Synchronous • Virtual partitioning at map side	• Cannot handle small datasets • Slow	• Reduce-skew
Spongefiles [21]	• Distributed memory • Flexible	• Nonuniform in nature • Failure of any node during the process caused the job failure • High vulnerability	• Map and reduce-skew
DREAMS [22]	• Minimize the overhead over repartitioning • Automatic • Cost-effective	• Cannot handle the skewness at map side • Not flexible	• Reduce-skew
LIBRA [23]	• Lightweight strategy • Minimal overhead • Transparent for users • Supports heterogeneous environment	• Do not support homogenous environment • Not a suitable solution for the small datasets	• Reduce-skew

If we consider the data mining techniques to handle data skewness problem, there are very less in existing in the survey. Whereas, when we consider the database joins, they are able to handle the data skewness problem. Where, PTSH is very effective during the analysis of medical data. But there is a need to develop a solution to handle data skewness effectively.

References

1. https://www.slideshare.net/BernardMarr/big-data-25-facts/4-It_is_expectedthat_by_2020
2. https://www.statista.com/statistics/274774/forecast-of-mobile-phone-users-worldwide/
3. http://wersm.com/how-much-data-is-generated-every-minute-on-social-media/
4. https://en.wikipedia.org/wiki/Big_data
5. https://www.versatek.com/blog/how-much-data-will-the-internet-of-things-IoT-generate-by-2020/
6. https://www.cisco.com/c/en/us/solutions/collateral/service-provider/visual-networking-index-vni/vni-hyperconnectivity-wp.html
7. http://searchcloudcomputing.techtarget.com/definition/Hadoop
8. J. Dean, S. Ghemawat, MapReduce: simplified data processing on large clusters. in Proceedings of the 6th OSDI Symposium, 2004
9. Apache Hadoop Project, Powered By Hadoop, http://wiki.apache.org/Hadoop/PoweredBy/ (2011)
10. "Hadoop," http://Hadoop.apache.org/
11. C. Doulkeridis, K. Nørvåg, A survey of large-scale analytical query processing in MapReduce. VLDB J. **23**(3), 355–380 (2014)
12. Y. Kwon, M. Balazinska, B. Howe, J. Rolia, A study of skew in MapReduce applications. Open Cirrus Summit **11**
13. Y. Kwon, M. Balazinska, B. Howe, J. Rolia, Skewtune in action: mitigating skew in MapReduce applications. Proc. VLDB Endowment **5**(12), 1934–1937 (2012)
14. Y. Kwon, K. Ren, M. Balazinska, B. Howe, J. Rolia, Managing Skew in Hadoop. IEEE Data Eng. Bull. **36**(1), 24–33 (2013)
15. S. Brin, L. Page, The anatomy of a large-scale hypertextual web search engine. in Proceedings of the 7th WWW Conference, 1998, pp. 107–117
16. M.C. Schatz, Cloudburst. Bioinformatics **25**, 1363–1369 (2009)
17. R. Vernica, A. Balmin, K.S. Beyer, V. Ercegovac. Adaptive MapReduce using situation-aware mappers. in Proceedings of the 15th EDBT Conference, 2012, pp. 420–431
18. S. Ibrahim, H. Jin, L. Lu, S. Wu, B. He, L. Qi, Leen: locality/fairness-aware key partitioning for MapReduce in the cloud. in 2010 IEEE Second International Conference on Cloud Computing Technology and Science (CloudCom). IEEE, 2010, pp. 17–24
19. S. Ibrahim, H. Jin, L. Lu, B. He, G. Antoniu, S. Wu, Handling partitioning skew in MapReduce using leen. Peer-to-Peer Netw. Appl. **6**(4), 409–424 (2013)
20. Y. Gao, Y. Zhou, B. Zhou, L. Shi, J. Zhang, Handling data skew in MapReduce cluster by using partition tuning. J. Healthc. Eng. (2017)
21. Z. Liu, Q. Zhang, M.F. Zhani, R. Boutaba, Y. Liu, Z. Gong, Dreams: dynamic resource allocation for MapReduce with data skew. in *2015 IFIP/IEEE International Symposium on Integrated Network Management (IM)* (IEEE, 2015), pp. 18–26
22. K. Elmeleegy, C. Olston, B. Reed, Spongefiles: mitigating data skew in MapReduce using distributed memory. in *Proceedings of the 2014 ACM SIGMOD International Conference on Management of Data* (ACM, 2014), pp. 551–562
23. Q. Chen, J. Yao, Z. Xiao, Libra: lightweight data skew mitigation in MapReduce. IEEE Trans. Parallel Distrib. Syst. **26**(9), 2520–2533 (2015)

Recent Trends in Green Cloud Computing

Shatakshi Kaushal, Dweep Gogia and Bishwesh Kumar

Abstract Cloud computing is a ubiquitous technology which is spreading its roots in every sphere of modern computing. The benefits of these services are top-notch, but the data centers that power these services consume a huge amount of energy and pose serious threats to the environment due to the increase in carbon footprint. This stems the need to shift to green cloud computing, a crucial area of research these days. Green cloud computing provides a methodology for energy management, recycling, efficient cooling, load balancing, and virtualization of servers. We have reviewed potential domains to tackle the problems that the growth of cloud computing brings along, including underutilized resources like traditional DBMS servers and high energy consumption by processors, servers, and cooling infrastructure. Along with the advantages, we have mentioned some of the disadvantages of the techniques as well. However, these disadvantages are not a major concern for the large-scale implementation of these methods. Once implemented, they are bound to alleviate the problem of hefty energy consumption and growing carbon footprint of the cloud data centers. We have also discussed various parameters which can be used to compute the power consumption by the data centers and also to quantify the green energy coefficient of the cloud services.

Keywords Cloud computing · Data center · Energy efficiency · Green cloud computing · Sustainable computing · Reducing energy consumption

S. Kaushal (✉) · D. Gogia · B. Kumar
University Institute of Engineering and Technology, Panjab University, Chandigarh, India
e-mail: shatakshi.kaushal@gmail.com

D. Gogia
e-mail: dweepgogia@gmail.com

B. Kumar
e-mail: bishwesh211@gmail.com

C. R. Krishna et al. (eds.), *Proceedings of 2nd International Conference on Communication, Computing and Networking*, Lecture Notes in Networks and Systems 46, https://doi.org/10.1007/978-981-13-1217-5_93

1 Introduction

Cloud computing, a rapidly growing Internet based technology, is defined as the "On Demand" service which allows the sharing of resources and handling of data over internet rather than on local computers. As stated by the National Institute of Standards and Technology [1], "Cloud computing is a model for providing suitable, cost effective and energy efficient network access to the shared system resources which can be rapidly delivered with least management effort or client and service provider interaction." The cloud technology encompasses three main services including Software as a Service (SaaS), Platform as a Service (PaaS), and Infrastructure as a Service (IaaS).

Cloud computing services offer surplus of advantages which provides the clients an edge over the conventional computing. Some of them are little management toil, less focus on infrastructure, faster and cost-effective application development facilities, scaling the services according to the user requirement. It also provides easy access to services which are readily available over the internet round the clock. It reduces the total cost of ownership because instead of buying and managing dedicated on-premise infrastructure, the user can avail the cloud services, conveniently on-demand. Moreover, cloud services provide security and allow the users to create backup of their data.

As a result of these benefits, various large scales as well as small scale enterprises have largely shifted to cloud computing in the past few years. Forrester Research made an estimate that cloud computing services expenditure will increase from $40.7 billion in 2011 to $236 billion in 2020 [2].

The rising deployment of cloud computing has extremely increased the amount of energy spent at data centers used for managing and distributing large chunks of data. The vast network of data centers utilizes colossal amount of electricity for running servers, processors, web peripherals, and for cooling down the heat produced by processor chips. According to United States Data Center Energy Usage Report, the overall power usage by US data centers in 2014 was approximately 70 billion kilowatt-hours which is equivalent to the amount of energy utilized by 6.4 million American homes in 2014 and is an increase of 4% from 2010 to 2014 [3]. The increasing energy usage in data centers is not only a threat to the environment, but also leads to inflated energy expenses. The multinational companies like Google, Amazon, Microsoft and Yahoo contribute half (50%) of the three-year entire cost incurred by data centers to power consumption [4].

The massive power consumption by data centers releases excessive carbon gases which are detrimental for the environment. The carbon emissions by data centers contributed to the 60% of global greenhouse gas emissions in 2011 [5]. The CO_2 emissions increase the carbon footprint which affects the global climate, increases pollution, leads to global warming and has various health impacts. If the carbon footprint goes on increasing unchecked, it will have calamitous effects on the environment.

These factors make it necessary to implement Green Cloud Computing to reduce carbon dissipation and save energy. Though some companies have taken steps to resolve the issue but still further steps need to be taken to make IT sector environment friendly on a larger scale. The increasing carbon footprint arises alarm which makes it necessary to give priority to green cloud computing and discover varied methodologies to implement it.

The paper is structured as follows: we examine the existing work done in this field under the Sect. 2. Under the Sect. 3, we discuss about various parameters which can be used for measuring power utilization by data centers. In Sect. 4, we present various potential domains that can foster green cloud computing. We discuss the future work in Sect. 5 followed by the conclusion in Sect. 6.

2 Related Work

The increased demand of cloud services among the clients has led to an increase in the data center's energy consumption and carbon emission. This has made green cloud computing trending and popular topic for research. A variety of power conserving approaches have been proposed like turning the computers off or in Sleep or Hibernate mode when not in use, using LCD monitors that use less energy, recycling the waste from old computers, using equipment with energy star label, integrated memory controller, using techniques like quad core designs which increases performance per watt [6].

Apart from these, virtualization, a cost effective and energy efficient characteristic of cloud computing, enables the processing of numerous requests and managing data by creating multiple virtual environments on a single computer system. Kaur et al. [7] discuss in their paper how virtualization reduces the hardware costs and leads to less energy utilization through VM migration and reducing CPU idle time.

Furthermore, various algorithms have been proposed to optimize the task scheduling and resource allocation strategies on cloud computing environments. Improved Differential Evolution Algorithm (IDEA) [8] is one such example. It is a combination of two algorithms namely Differential Evolution Algorithm (DEA) and the Taguchi method. The DEA algorithm has a robust exploration capability, whereas the Taguchi method provides it an excellent exploitation capability, thus making IDEA algorithm a balanced combination of exploration and exploitation capabilities. The paper [8] showed that IDEA algorithm can significantly improve the resource allocation and task scheduling process, thus making an optimum use of the server's power consumption.

Zhao et al. [9] implemented another solution to decrease the data centers energy consumption by improving energy proportionality, a relationship between amount of power consumed and performance of computer, through dynamic provisioning and CPU power scaling. Dynamic provisioning uses methods like Wake-On-LAN and VM migration that help to remotely turn on/off the servers. However, these approaches are less effective as the chances to go into sleep mode are less and waking up the servers and VM migration induce extra costs. CPU power scaling makes use of Dynamic Voltage/Frequency Scaling (DVFS) which lowers CPU frequency as the work load drops to maximize energy conservation. But, this technique doesn't deliver adequate results because even if CPU energy utilization is proportionate with system load, other peripherals like motherboard, storage devices, and disk utilize the same amount of power.

Load balancing is one of the key components for distributing workload across different nodes to prevent overburdening of a single node. Kansal et al. [10] discussed the various current load balancing approaches and then provided a comparison based on different metrics used in those techniques like resource utilization, throughput, performance etc. They also examined these methods according to the amount of energy utilization and carbon release. Also, Jayant Baliga et al. performed inclusive analysis of energy utilization of public as well as private cloud and of the three services of cloud. They analyzed the energy usage in transport, storage, processing and showed how cloud computing can employ energy efficient techniques. Also, according to the paper, cloud computing can sometimes make use of more energy than traditional computing [11].

3 Parameters to Measure Data Center Power Consumption

There are several parameters that act as a metric and can be used to quantify the power consumption of the data centers. These parameters also help us to determine the sustainability as well as the green energy coefficient of the data centers. The five most prominent parameters are discussed below:

Power Usage Effectiveness (PUE) [12] [13]. PUE was presented in 2006 and advanced by the Green Grid in 2007. PUE is the ratio of the total power consumed by the data center to that of the power consumed only by its computing infrastructure. A PUE value approaching 1.0 means approximately all the energy consumed is utilized for computing.

$$\text{PUE} = (\text{Total Facility Power})/(\text{IT Equipment Power}) \tag{1}$$

Data Center Energy Productivity (DCeP) [14]. DCeP evaluates the 'valuable work' that a data center produces in view of the measure of energy it expends. Most intriguing is that DCeP permits each data center to characterize "valuable work" as it applies to its own business.

$$\text{DCeP} = (\text{Useful Work Produced})/\left(\text{T}_{\text{power}}\right) \tag{2}$$

Energy Reuse Factor (ERF) [15]. The ERF value of a data center indicates the amount of energy which is re-used for operations concentrated outside of a data center. The primary purpose of ERF parameter is to boost the reuse of energy for some other operations instead of simply its dismissal. The value of ERF lies between

0.0 and 1.0. A value to approaching 0 indicates that no or minimal energy is reused whereas a value closer to 1 signifies that most of the energy is reused for some operations outside of the data center.

$$\text{ERF} = (\text{Energy Reused})/(\text{Total Facility Power}) \tag{3}$$

Green Energy Coefficient (GEC) [15]. GEC parameter allows an organization to measure the amount of renewable energy used by its data center as part of the total energy consumed. Green energy sources can be termed as any type of energy that is renewable and which gets replenished naturally over time. The GEC estimation of Energy devoured is measured in kWh. GEC has a maximum value of 1.0, which would indicate that approximately all the energy used by the data center is green.

$$\text{GEC} = (\text{Green Energy Consumed})/(\text{Total Energy Consumed}) \tag{4}$$

Carbon Usage Effectiveness (CUE) [15]. CUE parameter can be used to evaluate the total emanations of the various Greenhouse Gases like Carbon dioxide, Methane, Chlorofluorocarbons etc. from the data center with respect to its total IT energy consumption. The perfect value of CUE is 0.0, demonstrating that no carbon usage is related with the data center's operations. The unit of CUE is Kg CO_2 eq/kWh.

$$\text{CUE} = (\text{Total CO}_2\text{ Emission by the Data Center})/(\text{Total Facility Power}) \tag{5}$$

4 Potential Domains Fostering Green Cloud Computing

To the extent of our knowledge, energy consumption by the data centers majorly falls under two categories: energy consumed by IT equipments like network, storage, and servers and by infrastructure requirements like cooling. As per the study by the Infotech group shown in Fig. 1, a substantial amount of energy is utilized by the cooling infrastructure, followed by the server and storage requirements.

Therefore, the below-mentioned methods, that we have proposed, tackle both these major concerns.

4.1 Shifting Datacenter Workload on Microservers

Low-power embedded microservers are more energy efficient alternatives to the conventional servers and can make data center infrastructure more environment friendly in the long run. These low energy substitutes depend on Intel low power central processing units (CPUs). They can be more productive than larger monolithic systems, given the right workloads. Large clusters of these microservers powering the data center workloads can considerably improve work done per joule, thus, achieving

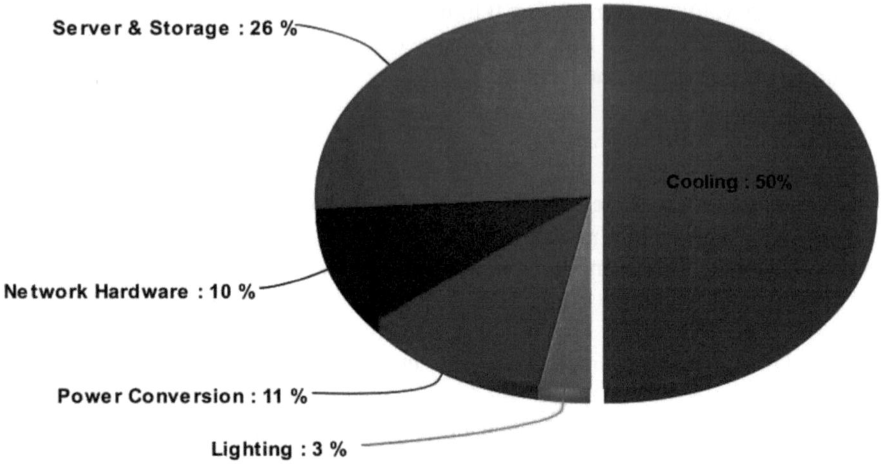

Fig. 1 Power consumption breakdown of a datacenter [16]

higher energy efficiency. Often, microservers are wrongly attributed with degraded performance. However, large swarm of these microservers can provide near-native performance of conventional servers, especially during non-peak hours [17]. Shifting data center workloads on microservers, therefore, is a viable option, particularly for managing huge chunks of archived data which is infrequently accessed.

Further, there are an abundance of microservers available in the form of defective mobile phones. This is especially true in India, which is one of the largest Smartphone markets in the world. The processors of defective mobile phones can be utilized in making large clusters of microservers. These clusters can be used to manage the data center workloads. This step will not only reap the aforementioned advantages of the microservers, but will also help towards managing piles of defective smartphones, which currently, are a threat to the environment.

4.2 *Improving Datacenter Design*

Uninterrupted functioning of cloud services makes cooling mechanisms to work endlessly. It causes cooling to absorb most part of the data centers overall energy. Therefore, effective methods need to be employed to cut cooling expenses by using efficient cooling methods or by decreasing the heat released. The locations of data centers can play a major role in amount of heat released and in the activity of cooling. The data centers can be shifted to geographical areas with abundant renewable sources of energy which can be utilized for cooling. Alternatively, the data centers can be moved to colder regions for downsizing the cooling costs. Moreover, instead of dissipating the extra heat produced into the environment, it can be deployed to maintain room temperature for data centers at regions with negative climates like the polar regions. Besides, the orientation and structure of data centers racks can also improve the cooling process up to a fair amount. Increasing the height of data centers ceiling and presence of proper ventilation system in the data centers can also help in decreasing the cooling costs. All these methods can help to make the cooling process cost-effective.

4.3 Utilizing a Combination of Wimpy and Traditional Servers

Traditional DBMS servers in the data centers are largely energy-inefficient due to the fact that they waste a lot of energy while underutilized. These traditional high end servers do not support flexibilities to scale down when workload is low. A lot of energy can be saved if some part of the Data Center Workload, preferably those which require only database accessing, is shifted to clusters of small (wimpy) servers, where the number of nodes forming the cluster can dynamically increase or decrease as per the workload. Although, these wimpy servers have challenges of their own, they are excellent for database accessing applications.

Using wimpy servers in data centers also foster the concept of microservices, which is an architecture in which various functionalities of an application are loosely coupled independent services that communicate with each other to deliver the business goals. Due to a surfeit of benefits like independent and expedite production and easy updating of application components without affecting other microservices, businesses have started converting their monolithic applications into blocks of microservices. To improve the fault tolerance of systems, replicas of these microservices should ideally run on different servers. So, running huge number of microservices (sometimes thousands) over traditional servers hampers the energy-efficiency as well as increases the overall infrastructure cost to a great extent. Essentially, using wimpy servers, especially for microservices, is helpful both for the environment and the cloud providers.

4.4 Increasing Transparency

Higher transparency regarding carbon and energy footprint among data companies will provide right statistics and help to tackle the problem of high carbon emissions and energy consumption better. After the plea was made by GreenPeace [18], various multinational companies like Google, Facebook and Apple became crystal clear about their energy and carbon footprint and also decided to go 100% renewable in coming years. However, there are other companies who remain ambiguous about their data. These data center operators should break the silence as increasing transparency will help to take better measures to curb the impending danger. It will also increase cooperation in the IT sector. The governments of various countries should also take measures to address the increasing carbon footprint problem by setting transparency standards and enforcing environment related laws. A governing committee or a consortium can also be founded to oversee the problem of transparency and keep in check the increasing power consumption by data centers. Increasing transparency seems an insubstantial solution but any measure in this field, however small, can pave the way for an environment healthy IT sector.

The suggested methods can be used to resolve the problem and foster the implementation of green cloud computing. These practices should be executed on a large scale to get the desired results.

5 Future Work

There are several lines of future work arising from this research paper that can be pursued. We have discussed various methods that can be implemented to make cloud services greener. Shifting to microservers can pose a problem when the workload is very high. Thus, the capability of microservers to handle greater traffic is an area of future scope. Moreover, new methods can be explored so that the cost of revamping the existing infrastructure to accommodate low power embedded micro servers is minimized.

Also, there exists a trade-off between performance and energy utilization while using Wimpy Servers. One such disadvantage of using Wimpy servers is that clusters suffer from "friction losses" and thus they may not be able to quickly adapt to changing workload environments, unlike brawny servers which deliver performance instantaneously. So, further research can be done to devise algorithms which can enhance the adaptability of wimpy servers to dynamic workload, reduce the performance lag and provide a mechanism which allows improved communication between different clusters of wimpy servers even when the workload increases substantially.

Furthermore, an effective cooling method is to find proper balance between turning off the idle servers (Power Management) and minimizing the heat generation (Thermal Management). Power management techniques focus on turning off idle servers to save power which highly increase the temperature of busy servers, consequently, increasing the amount of cooling. More research can be conducted on this balancing technique which can be a significant solution to the problem of cooling.

These areas of improvements are very crucial lines of future research work, which can make the proposed methods an optimal solution for green cloud computing. All these methods can be studied and tested extensively at a larger scale to eliminate the slight performance lag, keeping the power consumption in check.

6 Conclusion

Evidently, cloud computing is a ground-breaking advancement in the field of computer science but it poses a great threat to the environment. This raises the concern and makes it obligatory to shift to green cloud computing. This paper demonstrates deployable techniques for sharpening the energy performance of cloud computing service providers. We explored the existing work done in the field of green cloud. We discussed various parameters for evaluating power intake of data centers. We also proposed different solutions for decreasing energy utilization and carbon emanation of these data centers. The proposed methods should be adopted on a larger scale not only by giant technical companies but also by small service providers. Measures should be taken to make IT sector more ecofriendly by regulating the inflating energy utilization and CO_2 emissions. It has become the need of the hour to shift to green cloud computing to modulate the expanding danger.

References

1. P. Mel, T. Grance, in *The NIST Definition of Cloud Computing*, National Institute of Standards and Technology, SP 800–145, September 2011
2. The Public Cloud Services Market Will Grow Rapidly To $236 Billion in 2020. https://www.forrester.com/report/The+Public+Cloud+Services+Market+Will+Grow+Rapidly+To+236+Billion+In+2020/-/E-RES132004
3. A. Shehabi, S. Smith, D. Sartor, R. Brown, M. Herrlin, J. Koomey, E. Masanet, N. Horner, I. Azevedo, W. Lintner, United States Data Center Energy Usage Report, Lawrence Berkeley National Laboratory, Berkeley, California, Tech. Rep. LBNL-1005775, June 2016
4. V. Vasudevan, D. Andersen, M. Kaminsky, L. Tan, J. Franklin, I. Moraru, Energy-efficient cluster computing with fawn: Workloads and implications, in *Proceedings of the 1st International Conference on Energy-Efficient Computing and Networking* (ACM, 2010), pp. 195–204
5. D. Bouley, Estimating a Data Center's Electrical Carbon Footprint, in *White Paper*, vol. 66 (Schneider Electric—Data Center Science Center, 2011)
6. A. Jain, M. Mishra, S. Kumar, N. Jain, Energy Efficient Computing—Green Cloud Computing, in *International Conference on Energy Efficient Technologies for Sustainability (ICEETS)* (2013), pp. 978–982
7. T. Kaur, I. Chana, Energy efficiency techniques in cloud computing: a survey and taxonomy. ACM Comput. Surv. (CSUR) 48(2) (2015)
8. J. Tsai, J. Fang, J. Chou, Optimized task scheduling and resource allocation on cloud computing environment using improved differential evolution algorithm. Comput. Oper. Res. **40**(12), 3045–3055 (2013)
9. Y. Zhao, S. Li, S. Hu, H. Wang, S. Yao, H. Shao, T. Abdelzaher, An experimental evaluation of datacenter workloads on low-power embedded micro servers, in *Proceedings of the VLDB Endowment*, vol. 9 (2016), pp. 696–707
10. N.J. Kansal, I. Chana, Cloud load balancing techniques: a step towards green computing. IJCSI Int. J. Comput. Sci. Issues **9**(1), 238–246 (2012). ISSN (Online): 1694-0814
11. J. Baliga, R. Ayre, K. Hinton, R.S. Tucker, Green cloud computing: balancing energy in processing, storage and transport. Proc. IEEE **99**(1), 149–167 (2010)
12. A.Y. Zomaya, Y.C. Lee, in *Energy Efficient Distributed Computing Systems* (Wiley, 2012)
13. Efficiency: How we do it—Data Centers—Google. https://www.google.com/about/datacenters/efficiency/internal/

14. M.S. Obaidat, A. Anpalagan, I. Woungang, in *Handbook of green information and communication systems*, 1st edn. (Academic Press Inc., 2012)
15. E. Ariwa, in *Green Technology Applications for Enterprise and Academic Innovation* (Information Science Reference Publication, 2014)
16. Info-Tech, *Top 10 energy-saving tips for a greener data center* (Info-Tech Research Group, London, ON, Canada, 2010). http://static.infotech.com/downloads/samples/070411_premium_oo_greendc_top_10.pdf
17. Y. Zhao, S. Li, S. Hu, H. Wang, S. Yao, H. Shao, T. Abdelzaher, An experimental evaluation of datacenter workloads on low-power embedded micro servers. Proc. VLDB Endowment **9**(9), 696–707 (2016)
18. G. Cook, J. Lee, T. Tsai, A. Kong, J. Deans, B. Johnson, E. Jardim, *Clicking Clean: Who Is Winning The Race To Build A Green Internet?* (NW United States, Greenpeace Inc., 2017), pp. 1–102. http://www.clickclean.org/downloads/ClickClean2016%20HiRes.pdf

A FOG Computing Based Battery Swapping Model for Next Generation Transport

Md. Muzakkir Hussain, Mohammad Saad Alam and M. M. Sufyan Beg

Abstract It has been a consensus persuasion from automotive industries, policy-makers, R&Ds and vehicle vendors that electric vehicle is the powertrain archetype for future transport. The current Electric, plug in electric and plug in hybrid electric vehicles (xEVs) no longer remain only a means of commute, but can act as prime actors to have active business participation with various markets in the power system such as V2G, demand side management (DSM) etc. The modern development in the information and communication technology (ICT) evolves such vehicle into intelligent vehicle (IV) and augments their utility to provide diverse services for Intelligent Transportation (ITS) infrastructure. However, due to lack of viable charging infrastructures the contemporary power system fails to accommodate the incoming xEV flux. The inability is manifested in the form poor quality of service, which causes customer dissatisfaction and ultimately lower adoption of xEVs. This work proposes an energy efficient battery swapping topology (BSS) adopting the notion of Internet of Things (IoT). The work introduced the innovative notion of integrating internet of things (IoT) into smart charging infrastructures and proposed a data driven IoT-BSS model whose operation is regulated through Fog computing and Big Data analytics. Further, a four layer fog computing execution stack is developed to set up the service oriented architecture (SOA) for an efficient and real-time decision making framework for next generation intelligent transportation. The work also highlights the data science prospects and challenges that can elucidate in course of realization the proposed infrastructure.

Keywords Battery swap station (BSS) · Big data analytics · Cloud computing
Fog computing · Internet of things (IoT) · Intelligent transportation system (ITS)

Md. M. Hussain (✉) · M. M. Sufyan Beg
Department of Computer Engineering, ZHCET, AMU, Aligarh, India
e-mail: md.muzakkir@zhcet.ac.in

M. M. Sufyan Beg
e-mail: mmsbeg@cs.berkeley.edu

M. S. Alam
Department of Electrical Engineering, ZHCET, AMU, Aligarh, India
e-mail: saad.alam@zhcet.ac.in

C. R. Krishna et al. (eds.), *Proceedings of 2nd International Conference on Communication, Computing and Networking*, Lecture Notes in Networks and Systems 46, https://doi.org/10.1007/978-981-13-1217-5_94

1 Introduction

The era of urbanism in and everyone is in rush to gather slots in condos. Smart megacities and related projects are rapidly sprouting even in nations like India, Malaysia, and Africa etc. However, due to rise in environmental pollution, degrading ambient air quality, escalating number of IC engine vehicles and rise in the carbon footprint from the conventional vehicles in such cities, the emphasis of the entire research community is towards implementing sustainable technologies in the field of automotive transportation. Such concerns compel developed as well as developing nations to anticipate heavy investment for electric vehicles (EV) uptake in the forthcoming years. The rollout of Electric (EV), plug-in electric vehicles (PEV), and plug in hybrid electric vehicles (PHEV), collectively termed as xEV technologies is a welcomed research and development thrust towards transportation electrification that became a center of commotion in xEV automotive industry. Preliminary research threads estimate that by 2020, the size of xEV population will range to five million in China while by that time it will constitute at least one tenth of US vehicle fleet [1]. It had also been anticipated that by the end of the decade, the xEV will contribute to 25% of market share in these nations [2]. A properly planned xEV fleet is multitenant to act as a temporary energy source for smart grid in the demand side and will enhance the efficiency of smart grid frequency control [3]. The transportation as a service (TaaS) paradigm will establish a secure and hassle free commuting avenues for the urban folks.

Despite the evidence of its positives, there are still some downsides of such inventions that the stakeholders need to contemplate on, before they adjudicate to invest in electric car rollouts. The pitfalls may arise due to scarce availability of recharging points/sockets, coercion on back-end energy grid, longer recharging delay, and high upfront costs and above all is the innate Electric Vehicle Range Anxiety (EVRA) syndrome that exists in a considerable percentage of xEV users and manifests as the major cause for ceasing xEV adoption [3]. It has been reflected by literatures and surveys that the length of nearly a half of xEV user's routine trip is greater than their full-electric-range [1]. Inspired by the trend and need to adopt eco-friendly modes of transport, developed as well as many developing countries are in motion to adopt and promote methodologies for sustainable energy and carbon footprint reduction. The possible and prominent charging technologies for the electric vehicle include the conductive charging (transferring the power by means of connector and receiver), inductive charging (charging the vehicle wirelessly by means of the induction effect) and the battery swapping technique (involving the replacement of the discharged battery with a fully charged battery). Among all, battery swapping seems to be the exemplary recourse for the contemporary scenario. While there are expedited efforts in the development of the conductive and inductive charging, the third one i.e. battery swapping technologies has still not achieved the state of commercially viable deployment [4]. Contrast to conductive charging through charge points in battery charging stations (BCS), the battery swapping offers one major advantage i.e. prompt charging. It's an effortless undertaking where the xEV driver just drives to a swapping station, steers on a platform, battery is autonomously swapped, transaction done and drives away in a span of minutes. The whole swapping job in a BSS gets done in less than five minutes through a robotic shuttle that unplugs the depleted battery from rear/front/underside of the car [5].

In the recent days it has been felt by the research and development enterprises that if intelligently designed the battery swap station (BSS) architecture has the potential to provide a reliable platform where the massive fleet of electric and hybrid electric vehicles (xEVs) could be successfully installed [6]. The battery swapping station subsystem, where the depleted batteries gets replaced by fully or partially charged ones incurring a time shift of a few minutes, may calibrate EV adoption by executing similar role to that of contemporary gasoline stations. Battery Swapping Station (BSS) strategies are arising as a promising alternative to the traditional Battery Charging Station (BCS) approach since they provide a wider set of business opportunities to the devoted stakeholders. The anticipated electrification of transportation via BSS infrastructure can serve as demand-side regulation resources and integrate with automatic generation control (AGC), thereby stabilizing the dynamic performance of grid frequency control.

In the age of intelligent transportation (ITS), the electric vehicles that are being released into the current fleet are blessed with intelligent sensors, devices and smart service applications and utilities. The integration of electric vehicles into advanced information and communication technologies (ICT) have now allowed it to act not just as a standalone passive nodes but has now evolved into an active entity. New xEV versions are smart, secure and a potential source of Big-Data analytics. Moreover, such vehicle fleet needs efficient infrastructures where its monitoring and coordination can be under autonomous control. The fleet requires data driven real-time decision making framework where the dynamism of the infrastructure can be engrained and enable the xEVs to confront every adversary that emanates on its way.

Motivated by the benefits of efficient swapping paradigms towards viable realization of green electromobility and the need of integrating intelligent vehicles to the ICT services, this work aims to develop an optimization framework for smart charging and swapping management of electric vehicles. The underlying principle lays in injection of intelligence into the BSS network through the notion internet of things (IoT) and fog computing for decision making from real-time analytics of Big Data [7]. The work focuses on multidimensional aspects of an ideal BSS deployment, considering customer, the BSS owner and the smart grids. This research emerges to be novel of its kind, addresses and overcomes the shortcomings of previous innovations and will establish cornerstones for next generation sustainable transport. The model proposed here is a novel approach that endeavors to simulate a practical realization platform that congregates the power of cloud and fog computing paradigm and notion of internet of things into a multi-objective business model that can accommodate and facilitate the adoption of a giant xEV fleet. The perception of "smartness" is introduced into the infrastructure by devising UNIQUE identification addressing schemes for the involved entities, installing on board sensors (OBS) into the vehicles, actuators, RFIDs, CANs and miscellaneous intelligent devices into the physical backbone. Further, the data collection, storage, analytics, routing and coordination mechanism of the whole infrastructure is administered by utility aware fog computing technologies and data analytics modules installed at cloudlets and cloud data centers. The significant contributions of this work can be highlighted as:

1. Introduced the concept of Internet of Things (IoT) for contemporary power system infrastructures, discussed its significance in launching economically viable swap station networks and also proposed a BSS topology that works under IoT paradigm.
2. Discussed how internet of things (IoT) paradigms shall be emerged as the supreme technology in providing utilities to intelligent transport specifically for smart charging of xEVs.
3. Proposed a multilayer layer fog computing architecture for intelligent transportation and explained its significance in designing smart electric vehicle battery swapping stations (BSS).
4. Highlighted the data science prospects and challenges that may evolve during installation and running of an IoT aware smart BSS infrastructure.

2 IoT for Battery Swap Station Networks (BSSN)

The technology of battery swapping involves the replacement of the depleted or nearly depleted battery with a fresh one to support the electric drive. The history of swapping dates back to early 19th century when the exchangeable battery service was first proposed with an aim to overcome the limited operating range of electric cars and trucks [1], where the swapping process was completed employing the human workforce. Since then, there had been expedited efforts and evolutions in the Battery Swap Station research and deployment. But due to excruciating capital investment, lack of intelligent business models, lack of viable charging infrastructure and poor coordination between manufacturers and policy makers, the xEVs under BSS plan couldn't achieve the desired height of penetration and ultimately lead to the liquidation of companies. But, since latter half of the past decade somewhere in 2008, when BSS was used by China during Beijing summer Olympics where approximately 500 buses were made to run through switched batteries [8], the BSS paradigm got a competitive focus both by research community as well as automobile vendors. This invention opens a doorway to BSS deployment in other countries like Australia, Denmark, Israel, Japan etc. Study of achievements in contemporary transportation depicts that works on BSS paradigm is still in premature phase.

A BSS warehouse can acts as the power supplier not only for the electric vehicles but can also actively participate in mitigating miscellaneous power demands and regulation services thereby maintaining load equilibrium in the whole power system. The BSS has dedicated interfaces that ensure seamless exchange of energy to and from power sources such as smart grid, micro gird, xEVs etc. Thus BSS model is proves multi-tenant in having active role in diverse modes of power trading such as V2B, B2 V, B2B, B2G, G2B, R2B, B2R etc. where letters V, B, G

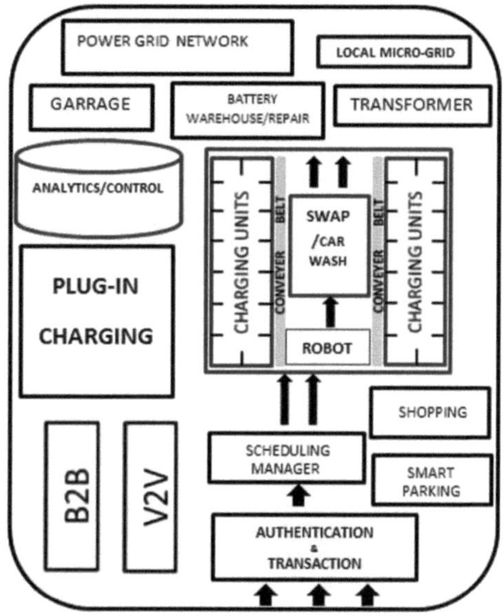

Fig. 1 Topology of a typical Battery Swap Station (BSS) its interaction with various power sources and its participation in varying markets

and R is acronym for xEV, BSS, smart grid and renewables respectively. Figure (1) shows the topology of a typical BSS, its interaction with various power sources and its participation in varying markets. The proposed model adopted a paradigm shift and advocated to devise a xEV battery swap/charge monitoring system based on internet of things (IoT). The work discloses a xEV battery charging/swapping and monitoring under IoT paradigm. The framework comprises of terminal data gathering module, monitoring sub-system, management module for coordinating the operation of a cluster of BSS collocated in the same neighbour. The terminal data collecting module carries out real-time tracking, identification, state information collection and dynamic orientation of xEV fleets on a terminal device through the interconnected mesh of sensors, actuators, RSUs and other transportation telematics, transmits it to the battery charge/swap station monitoring subsystem through dedicates wired links. It synthesizes and transfers the status of the operating state and orientation to the regional monitoring device through wireless IoT sensor network. The analytics modules deployed at the BSS/BCS site acts on such data to carry out metering and billing procedures. Further, it also offloads the transaction data to the regional data centers for historical and bulk analytics. The regional monitoring cloud monitors the terminal devices and the battery charge/swap stations in its jurisdiction area, and prepares statistics on the xEV routing behavior, metering and billing transactions, driving profiles etc. Such cloud level analytics enables the system to learn and evolve according to the fleet penetration, thus ensuring a reliable and effective management of whole infrastructure. Figure 2 embodies a structural layout and control flow of an IoT aware Smart xEV battery swapping/charging framework.

The smart IoT aware battery swapping management framework prototyped in the previous section is primarily surveilled by dedicated micro-datacenters deployed at

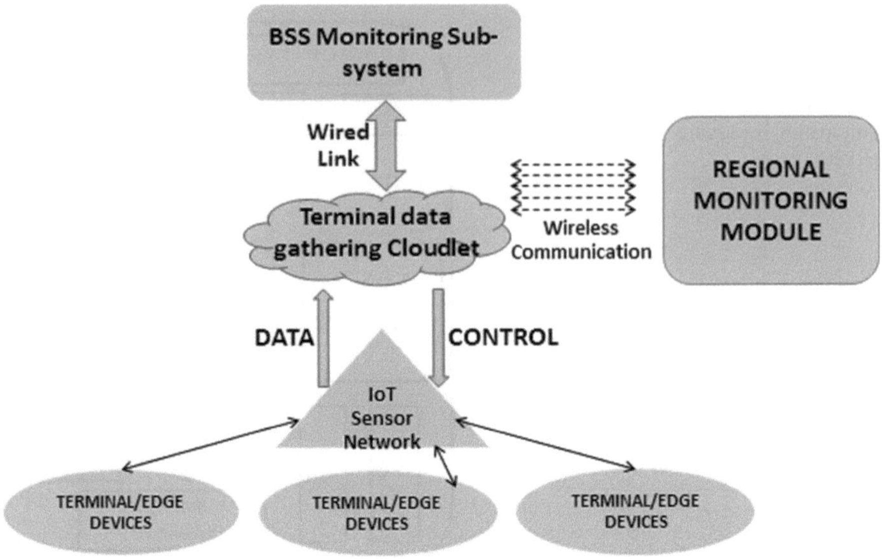

Fig. 2 Architecture of a typical IoT aware Battery Swap Station (BSS)

appropriate spots. The deployment of fog nodes or micro-datacentres geodistributed edges and cloud data centres for centralized analytics sets the basis for hierarchical cloud-fog architecture. The former does time critical processing and computation on the generated data to ensure prompt responses while it offloads the less critical or historical data analysis tasks to the mega data centres or cloud.

3 Fog Computing for BSS

In this section we elaborate how to integrate Cloud-Fog computing technologies into a Battery swap Station and what functionalities ad protocols need to be defined for a viable deployment of proposed model. As shown in Fig. 3, there is a granularization and distribution of tasks among the fog nodes as well as the data centers. Each of these micro centres will execute the real-time, critical and light weight tasks and will offload bulky and analytics activities to the global or hybrid data centers. This will ensure a smooth transport of data and control traffic across the whole network. However, robust protocols and standards need to be defined to synchronize the flow of control and information across dedicated layers. The data that needs historical analysis and long term storage are dispatched to global data centres via fog-cloud WAN communication interfaces. Fog nodes, implemented at the edge of network, provide highly virtualized platform bestowed with compute, storage, and networking services between end points and mega datacenters. In contrast to the centralized cloud services, fog nodes are targeted to serve the consumers with widely geo-distributed

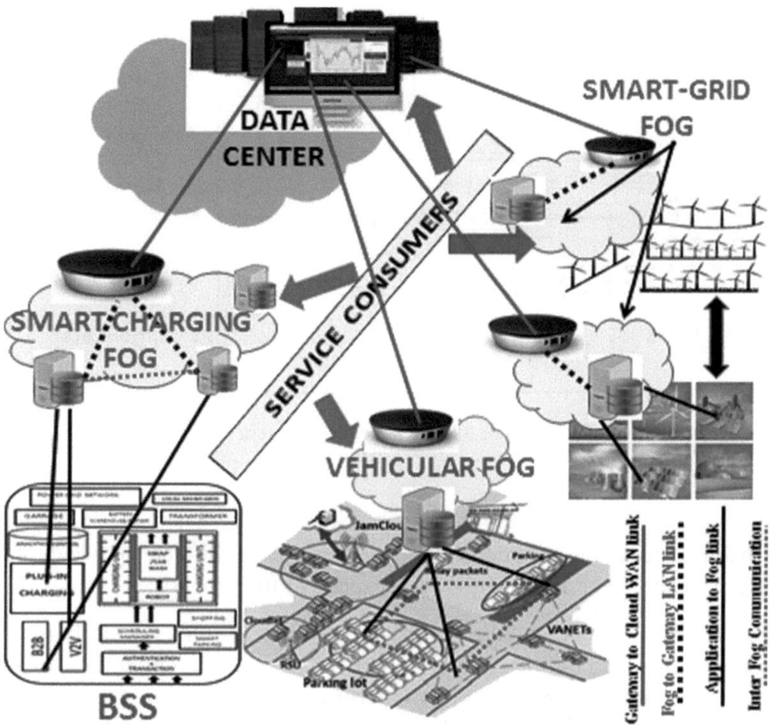

Fig. 3 Cloud-IoT architecture for intelligent Battery Swap Station (BSS)

deployments. The unique attributes of fog services such as low latency, consistency, context and location awareness makes is adaptable for mission critical infrastructures like vehicular network, Smart Grid, Smart Charging, Smart healthcare etc.

Figure 4 depicts the service oriented architecture for an integrated Cloud-Fog model supposed for a smart swap services. The proposed four layer architectural prototype for smart charging/swapping of electric vehicle batteries is a comprehensive cloud-fog architecture control stack along with defined protocols, standards & services at corresponding layers. The model inherits and parallels properties of prevalent five layer architecture for internet of things (IoT). These micro data centres also termed as fog nodes specifically Vehicular Cloud for the case of intelligent vehicles in Fig. 4 are closest to the network edge and they ingest the real-time data from IoT devices [9]. The applications installed in these nodes also direct varying types of data to upper cloud layer for aggregation & analysis. Such design possesses a range of advantages; both form computational as well as business perspectives. The computational benefits may be in the form load balancing, bandwidth conservation, real-time decision making, switching to semi-distributed control etc. whereas advantages from perspective of latter will be greater agility, enhanced privacy & security,

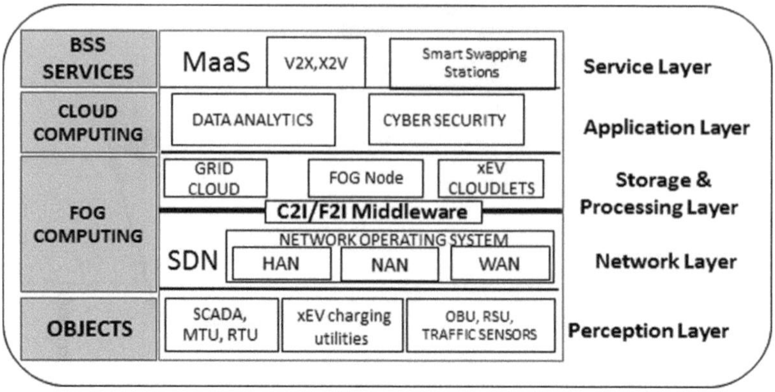

Fig. 4 Service Oriented Architecture of a Cloud-Fog system for IoT aware battery swapping/charging station

deeper insight, reduction in operational expenses etc. In this section we explain the processing of each layer along with their respective functionalities.

A. ***OBJECT LAYER:*** Also called physical layer, constitutes primarily real world devices, things and entities that are contributing to IoT architecture. In analogy to five layer architecture this layer is termed as perception layer but compared to latter, the objects layer not only encompasses smart meters, physical sensors, actuators and smart embedded systems but the layer includes the whole universe of active objects in the power system. In context of smart charging such entities may be the xEVs, the smart roads, streets, micro-nano grids, power grids, road side units (RSU), on board units (OBU) and smart people etc. This layer only defines data not information, which means the semantics of generated data are not defined in this layer. The sensors, data accumulators, smart meters, RSUs and actuators sense and absorb diverse parameters such as location; SOCs of vehicles, customer driving behavior etc. and digitize the data to transfer to upper layers to act on via secure channels. Moreover, the formats, schema and abstractions for the data are defined in this layer. This layer not only defines the means and standards for transmitting raw data, but also defines the syntax and representation standards for efficiently understanding the data.
B. ***FOG COMPUTING LAYER***: This layer extends the cloud to be in proximity to the entities that produce and act on internet of things (IoT) data. The means and protocol for networking and communication are defined here. This layer articulates the inception of computation and analytics in the IoT architecture. The micro data centers also called fog nodes collect and act on the real time data, execute edge computing paradigms wherever necessary. This layer also delineates the data based on its significance in respect of decision parameters like response time, latencies, bandwidth etc. It offloads the less sensitive data (that demand low responses, more & longer storage) to the cloud layer for further analytics and performs instant and real time analytics to the critical and sensi-

tive data [10]. For transferring the data to the cloud, there may be numerous technologies at hand such as RFID, 3G, GSM, DSRC, UMTS, WiFi, Bluetooth, infrared, ZigBee etc. Fog applications are as diverse as the IoT itself & what they have in common is monitoring or analyzing real-time data from network-connected things and then initiating an action in real time. There can be unlimited possibilities of actions which may further involve machine-to-machine (M2 M) communications or human-machine interaction (HMI).

C. ***CLOUD COMPUTING LAYER***: This layer resides above the fog computing layer and performs the bulk of analytics. This layer resides above the fog computing layer and performs the bulk of analytics. For smart charging purpose it is the house for analytics having geographically global coverage. It has seamless interaction with the fog nodes and is the center for Big Data analytics. It defines the middleware and service management strategies and enables IoT application developers to play with heterogeneous objects independent of any hardware platforms. The cloud processes the data to make decisions, and delivers the requirements of the service layer. The distinctive analytics at cloud computing level provide various functionalities and utility business operations to multiple entities in the power system. The Big Data processing framework can be utilized to improve the planning methodologies of both short and long term distribution system. The impacts of big data analytics to utility business operations include state estimation, visualization, short term, long term and renewable forecasting, failure prediction and recovery, impact analysis, frequency regulation, theft detection, usage and behavioral modeling, spatial and temporal demand forecasting etc.

D. ***SERVICE LAYER***: As name suggests, it forms the interfaces and abstractions that are primarily linked to users, such that the end users have seamless interaction through queries and devices. It inherits the both application and business layer characteristics from five layer architecture of IoT. For smart charging, the service layer provides a platform to build consistent representation of analysis performed at lower layers. The service and business outputs may be in the form of reports, plots, models, flowcharts etc. which reflects the outputs of design, analysis, monitoring, implementation and evaluation algorithm applied on IoT data.

4 Challenges and Issues

The notion of IoT-BSS architecture is still in concept phase and due to lack of viable technologies, challenges may need to be confronted while realizing it successfully. The need for assessment and evaluation of such challenges is continuously felt since the inception of the design and modeling process. IoT based vehicular fog nodes and data clouds needs to be designed intelligently to enable wide scale deployment. However contemporary implementation and design models, algorithms and mechanisms are still far from the vision of IoT and are confronted with many challenges.

Fundamental challenges that need to be addressed include availability, reliability, security, efficiency, interoperability etc. [7]. The research for viable realization of next generation swapping stations is in progress but all are trapped in local loop. As is reflected from the failure and bankruptcy of BetterPlace, there persist plenty of challenges to confront. We discuss some major challenges that may arise in the IoT enabled BSS infrastructures.

A. ***Management:*** In the proposed IoT model, there is a seamless trading of energy between each pair from ay of power grid, smart charging stations or aggregators, distributed sources (micro-nano grids and other non-conventional sources), forming range of markets namely G2 V, V2G, V2S, S2 V, R2 V, V2R, V2 V, G2B, B2G, B2G etc. Further, in smart cities the domain of renewables is extended to integrate smart homes, smart parking lots etc. The power exchange or flow can be conceptualized as a complete graph, where nodes represent entities while edges denote the bidirectional power flow. Integration of IoT technologies into such infrastructure further complicates the set-up as entry of billions of smart sensing devices demand robust and efficient communication infrastructure. Concurrent management of bi-level (energy and data) transport thus poses daunting issues for the configuration managers as well as service providers. Light weight remote management protocols needs to be standardized to build service agnostic schemes for a viable realization of IoT-BSS deployment. Comprehensive attention into research projects and investments from governments, policy makers, academia and industries are strongly welcomed into such disruptive innovation. Though some labs are involved conducting IoT research studies, integrating them with transportation frameworks is still in its infancy and is restricted within simulation phase only. Demonstrating the feasibility of prepared prototypes needs real world applications and test beds, but are still in concept phase.
B. ***Synchronization and QoS guarantee:*** The infrastructure unifies multiple dimensions and entities such as power grid, renewables, xEVs etc. into a common platform and the processing and analytics also ranges to multiple magnitudes. So there is need of robust synchronization mechanism to enable long and smooth running and operation of infrastructure. Evaluating the performance of transport service offered by the proposed infrastructure is again a cumbersome task. Real world datasets need to be generated to assess whether the QoS metrics are guaranteed. Evaluation metrics or QoS parameters from smart charging and power system perspective are service loss probability, average busy chargers, blocking rate and mean waiting time for swapping service [11] while from IoT viewpoint are throughput, jitter, bandwidth, overhead etc. [12]. The extent to which the metrics are fulfilled also depend how they are defined. Thus, proper QoS metrics need to defined taking perspectives from both power system as well as contemporary IoT architectures.
C. ***Interoperability:*** Enabling the population of heterogeneous IoT entities belonging to multiple platforms to be interoperable is another limitation to contemporary architectures. The products and standards delivered from applications and IoT OEM's end should be interoperable to ensure the delivery of services to vary-

ing range of customers irrespective of hardware specifications. Furthermore, the objects connected through IoT demand a unified platform for delivering diversified services. Such platforms can also be provided as transport as a services (TaaS) for IoT applications. The charging/swapping services employed should work efficiently for all xEVs irrespective of specification of batteries. Multiple swapping protocols, variations, ambiguities in interpretation of IoT-BSS standards may pose hindrance to interoperability.

D. ***Reliability and electromobility:*** A reliable swapping service is the result of sequence of executions performed at multiple schema formed by smart grid, aggregators, edge networks etc. the overall performance is decided by the cumulative operation characteristics across all levels. The objective function is to maximize the success rate of power delivery which is manifested in the form of power availability, economic power, minimum queuing delay, power quality etc. Further, both the hardware and software modules of the energy and data transport network should be resilient to threats, thefts, intrusions and failures as networks under these circumstances will cause unreliable perceptions and wrong decisions leading to disastrous consequences. As the routing and trip behavior of xEVs is random and uncertain, relieving the drivers out of EVRA while on the move is the prime premise of the proposed architecture.

E. ***Scalability, Security and Privacy:*** The BSS infrastructure should be scalable i.e. provide provisions for augmenting new devices, services and functionalities, to allow more and more xEV penetration without any significant degradation in power quality and power delivery. The swapping platform must be modeled from the ground up to enable extendibility in services and operations. Presenting the data analytics daemons with automation and intelligence will guarantee scalability in such substructures. The programmers and developers for both hardware and software components should design platform independent and compatible applications to allow addition of functionalities and integrate with other technologies and standards. Since the underlying core of IoT-BSS is based on trading of energy and transfer of data between galaxies of objects, lack of common security and privacy standard is a notable challenge for BSS installation. The integration of IoT will increase the transparency and ease the access control procedures, thus removing the possible intrusions in the power system. More research efforts are welcomed in this trend for it to mature. Efficient location routing of intelligent objects, sensors, charging stations, context aware services through secure networking paradigms remains a nascent research thrust for the research and automotive community.

5 Conclusion

This work proposes a novel xEV battery swapping infrastructure for smart and efficient charging management. The proposed work incorporates the concept of Internet of Things (IoT) into swap station dynamics to ensure duplex exchange and trade of energy between entities. To ensure a realistic integration of information and communication technologies into electrified transportation domain, the work applied the internet of things (IoT) fundamentals into battery swapping mechanics to develop a novel and intelligent IoT aware BSS architecture. Further, for management of the data driven storage, processing and computational aspects, the work introduced the notion of fog computing paradigms and suggested a modular service oriented control and processing stack defining the complete concept map for transforming the IoT data into real-time analytic and decision making utilities. Finally, the research and development challenges as well as future data science and business prospects that will evolve in realization of the whole infrastructure are also outlined.

References

1. I.S. Bayram, G. Michailidis, I. Papapanagiotou, M. Devetsikiotis, Decentralized control of electric vehicles in a network of fast charging stations, in *Globecom 2013-Symposium Selected Areas Communications (GC13 SAC)* (2013), pp. 2785–2790
2. S. Yang, J. Yao, T. Kang, X. Zhu, Dynamic operation model of the battery swapping station for EV (electric vehicle) in electricity market. Energy **65**, 544–549 (2014)
3. Q. Wang, X. Liu, J. Du, F. Kong, Smart charging for electric vehicles: a survey from the algorithmic perspective. IEEE Commun. Surv. Tutorials **18**(2), 1500–1517 (2016)
4. M.R. Sarker, H. Pandžić, M.A. Ortega-Vazquez, Optimal operation and services scheduling for an electric vehicle battery swapping station. IEEE Trans. Power Syst. **30**(2), 901–910 (2015)
5. M. Borup, Electric mobility case study for Denmark; Case study : better place—an effort of creating new actor roles and infrastructure for electric car mobility (Top-NEST Proj. Number RD 2011-42, no. April, 2014), pp. 1–34
6. C. Chen, G. Hua, A new model for optimal deployment of electric vehicle charging and battery swapping stations. Int. J. Control Autom. **7**(5), 247–258 (2014)
7. A. Al-Fuqaha, M. Guizani, M. Mohammadi, M. Aledhari, M. Ayyash, Internet of things: a survey on enabling technologies, protocols, and applications. IEEE Commun. Surv. Tutorials **17**(4), 2347–2376 (2015)
8. R. Earley, L. Green-Weiskel, L. Kang, F. An, in *Electric Vehicles in the Context of Sustainable Development in China*, no. 9 (2011)
9. E.-K. Lee, M. Gerla, G. Pau, U. Lee, J.-H. Lim, Internet of vehicles: from intelligent grid to autonomous cars and vehicular fogs. Int. J. Distrib. Sens. Networks **12**(9), 241–246 (2016)
10. B. Alohali, M. Merabti, K. Kifayat, A secure scheme for a smart house based on Cloud of Things (CoT), in *2014 6th Computer Science and Electronic Engineering Conference (CEEC)* (2014), pp. 115–120
11. B. Sun, X. Tan, D.H.K. Tsang, Optimal charging operation of battery swapping stations with QoS guarantee, in *2014 IEEE International Conference on Smart Grid Communication (SmartGridComm)*, no. 619312 (2015), pp. 13–18
12. A. Nieto, J. Lopez, Analysis and taxonomy of security/QoS tradeoff solutions for the future internet. Secur. Commun. Networks **7**(12), 2778–2803 (2014)

A Modified FIFO with Distributed and Pipelining Scheduling in Hadoop

Sunil Kumar and Maninder Singh

Abstract Hadoop MapReduce is an open-source platform for the processing of large-scale data. For the high performance, Hadoop scheduler must avoid the unused data transmission by reducing the time processing. Scheduling refers to the allocation of resources in the best possible way among multiple tasks and arrangement of tasks in queue. First in first out is the default scheduler present in Hadoop, which is operated without pipeline, i.e., the preceding task has to wait until the ongoing task has finished its execution. This makes the CPU idle for some time until the next task comes for execution, which results in inefficient CPU usage. In this paper, the concept of pipelining is introduced with FIFO scheduler, so as to decrease the job completion time and make CPU usage efficient. In our experiments modified scheduler is tested with five jobs of different sizes and then compared with the default scheduler. On experimental analysis, it is found that the modified scheduler successfully reduces the job completion time, thereby making the CPU usage efficient.

Keywords Hadoop · MapReduce · FIFO · Scheduler

1 Introduction

MapReduce is a programming model suitable for parallel handling of data and its implementation under Hadoop environment [1]. MapReduce is used for commercial applications as well as scientific applications. MapReduce paradigm can be developed in several languages, but in the current scenario, Hadoop MapReduce is a prominent Java-based framework for easily developing the applications. It has the capability to handle a large amount of data (multi-terabyte datasets) with the abil-

S. Kumar (✉)
Directorate Livestock Farms, Guru Angad Dev Veterinary and Animal Sciences University, Ludhiana, India
e-mail: sunilkapoorldh@gmail.com

M. Singh
Department of Computer Science, Punjabi University, Patiala, India

C. R. Krishna et al. (eds.), *Proceedings of 2nd International Conference on Communication, Computing and Networking*, Lecture Notes in Networks and Systems 46, https://doi.org/10.1007/978-981-13-1217-5_95

ity of parallel on the large clusters, i.e., number of nodes, commodity software and hardware in a reliable and fault tolerant manner [2]. In Hadoop, the processing of jobs can be executed in a parallel manner. While there are a number of jobs in the queue then the load increases in the cluster. As the cluster has a number of jobs and needs time for the execution of the jobs. Scheduling of the jobs improves the overall performance of the cluster.

The default Hadoop scheduler uses FIFO queue for execution. The jobs are first partitioned into separate tasks, and after loading them into the queue-free slots are assigned to these jobs. The jobs in the queue would have to wait for their turn in case of default FIFO scheduling [3]. It results in inefficient CPU usage, as the CPU would also have to wait for the next upcoming tasks. The already working schedulers are also not able to make the CPU usage efficiently. So, a modified FIFO with distributed and pipelining scheduling algorithm in Hadoop is proposed in which the concept of pipelining is used in the distribution of tasks to increase the CPU throughput, i.e., to make CPU usage efficient. Pipelining increases the task throughput of CPU—the number of tasks finished per unit time. The increase in task throughput means that a job runs with a faster speed and, therefore, has lesser total execution time. Until 2008, only a single scheduler was supported with Hadoop that was in built with the JobTracker logic [4]. However, after this, the scenario changed to pluggable schedulers which could be implemented easily. The use of these pluggable schedulers enables the use of new scheduling algorithms in helping to optimize the jobs that have particular characteristics.

1.1 Pipelining Scheduler

In order to reduce the job completion time and make CPU usage efficient, the pipelining scheduler is used with FIFO scheduler. This modifies the FIFO scheduler with pipelining processing instead of sequential processing. Pipelining scheduler helps in arranging the waiting tasks in the form of pipes. As the job to be executed it divided into various tasks by MapReduce. These tasks get processed one by one in a single-user environment of Hadoop. So a large time gets wasted in waiting for these tasks to be executed. So, the concept of pipelining is introduced which allows the tasks to change its stage each time the succeeding task is getting processed (Fig. 1). This could be better explained diagrammatically as follows.

The processing of parallel execution is splitting a given task into multiple subtasks. Now, the tasks do not have to wait for their stage to get changed. As the succeeding tasks move forward, the task takes its place, and this process goes on as all tasks have finished their part. The major benefit of using parallel execution is that it reduces the execution time for a particular job to complete. Henceforth, this helps in having efficient CPU usage as the throughput of CPU instruction cycle have increased.

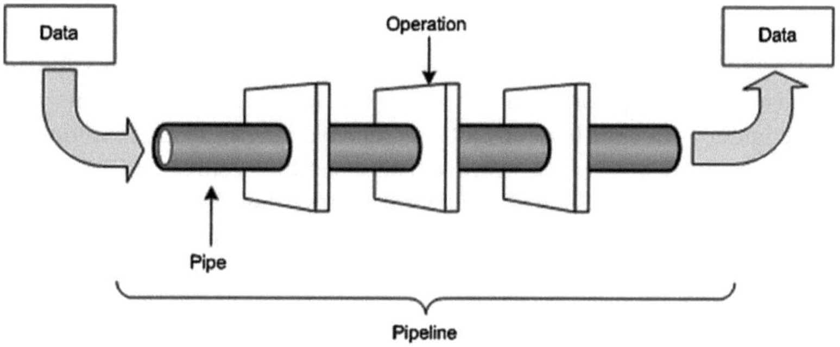

Fig. 1 Pipelining workflow

2 Literature Review

Dean et al. [4] gave the detailed description about the functioning and implementation of MapReduce through various examples. MapReduce is a programming model used for generating and processing a large cluster of datasets. Two functions—the map function and the reduce function are specified by the users. MapReduce library is responsible for fragmentation of the input data, set the scheduling process of program's execution across a cluster of various nodes, handle fault tolerance, and manage the required communication among the machines. Zaharia et al. [5] proposed an algorithm having name delay scheduling in order to address the conflict between data locality and fairness. The analysis of its performance shows that optimal data locality is achieved through delay scheduling in a different of workloads and can high the throughput by up to 2x while preserving fairness. Rao et al. [6] focused on Hadoop and various current schedulers used in its MapReduce programming. Parallel processing of Google's MapReduce is implemented through Apache Hadoop. Hadoop is responsible for performing the parallel tasks while hiding its details. It also distributes data to the various processing nodes, restarts the failed tasks and consolidates the results after computation. Their pros and cons are being overviewed. Scheduler improvements through LATE (Longest Approximate Time to End), Delay scheduling, Dynamic Priority scheduling, Deadline Constraint scheduling and Resource-Aware scheduling are also summarized. However, all the schedulers discussed by them assume homogeneous clusters rather than heterogeneous clusters.

Chen et al. [7] developed a new Matchmaking algorithm of MapReduce scheduling in order to enhance the data locality rate of map tasks and also to reduce the average response time of clusters. This technique is integrated into Hadoop fair scheduler and Hadoop default FIFO scheduler. The result shows that the lowest response time and the highest data locality is achieved through this technique. Kumar et al. [8] presented CASH, i.e., Context-Aware Scheduler for Hadoop which knows the resource characteristics and context, i.e., job characteristics of the nodes in clusters. CASH algorithm is implemented both in Hadoop and MapReduce cluster. The experimental result shows that overall execution time of various jobs is significantly improved as compared to the default FIFO Hadoop scheduler. Phuong et al. [9] proposed an algorithm based on dynamic priority known as Hybrid Scheduler (HybS). This algorithm reduced the latency for simultaneous jobs of variable length, while retaining the data locality. The result shows that the average response time is enhanced 2.1x faster than the Hadoop Fair scheduler. Jisha et al. [10] introduced Task Tracker-Aware Scheduling for Hadoop MapReduce framework. This scheduler addresses the issues related to jobs that depend on the external devices like Web service or database that can cause failure of tasks due to overloading.

3 Proposed Modified FIFO Scheduler Algorithm

The proposed algorithm is designed for a single node using Hadoop ecosystem. This algorithm, first of all, initializes all the parameters and then resources are added. resTopreempt() function is used to add the resources. If resources required are greater than resources available then running containers are collected using Get LeafQueue() function. In the same way, one more running container is collected. In order to send the task for execution, it makes the container and puts the task in it.

Simultaneously, another container is collected and preceding task is put into it. These two containers are then added using add() function in order to make a pipeline. Now, if there is some task left whose time spent and still wait time before killing is less than the present time, then the status of container to be preempted is created using container.getContainerId() function so as the whole process is rescheduled and the task is recalculated. In this way by making a pipelining, various tasks are scheduled.

The modified FIFO algorithm Pipelining Scheduler in Hadoop is as follows (Fig. 2).

4 Experimental Setup and Result Analysis

In order to analyze our proposed algorithm, we have compared it with the default FIFO scheduler. We have the cluster of one head node and 5 slave nodes for the experiments. We have the version of Hadoop-0.28 and modify the default FIFO scheduler. The block size of the cluster is configured with 128 MB. The performance of our proposed algorithm is evaluated using best, average and worst cases by testing each job 10 times. 5 jobs are submitted at a fixed particular time. The algorithm is evaluated by using job completion time as a parameter. The experimental analysis is carried out using a MapReduce WordCount job as a benchmark. The WordCount example counts the rate of appearances of every word which is given in a file. We are submitting a single job at a time which is split into the tasks by the MapReduce. The starting and ending time are noted and hence the job completion time is obtained as:

Fig. 2 Algorithm code of modified FIFO

```
Initialize from Config file
Minimum allocation, maximum allocation, node locality
Initialize queues
For (Csleaf Q: queueManager)
 {          Resource. add (resTopreempt)
 If (Resource greater than Resource Calculator)
 Preempt + resource (quemagr. Get LeafQueue)
    // collect running container
 for (sched: scheds)
{           If (Resource. greater than Resource Calculator)
{       for (appsched: sched.getAppSchedulable)
{       for (RMContainer: getLiveContainer)
{       Running container. add(c)
                Apps.put (c, appsched.getApp ()
                Queues. put (c, sched)
}       } }           //kill container
If (time! = null)
{       If (time + waitTime BeforeKill < clock.getTime ())
      Create preempted container Status (container .getContainerId ())
   Rescheduled the manager ()
   Initialize the queueManager () {
   Task recalculated ()}
CompletedContainer (containers, status, RMcontainer.KILL)
}           }
```

Table 1 Shows the result of Job1

Testing of jobs with the proposed scheduler		Testing of jobs with the default scheduler	
Job J1	Total time taken (in s)	Job J1	Total time taken (in s)
1	00:43	1	00:48
2	00:39	2	00:45
3	00:41	3	00:42
4	00:39	4	00:41
5	00:39	5	00:51
6	00:39	6	00:47
7	00:36	7	00:49
8	00:43	8	00:47
9	00:40	9	00:42
10	00:40	10	00:44
Best case: 00:36		Best case: 00:41	
Worst case: 00:43		Worst case: 00:51	
Average case: 00:39		Average case: 00:47	

Job completion time = Ending time of job—Starting time of the job (Tables 1, 2, 3, 4 and 5).

The modified FIFO scheduler is processed with the five jobs. The performance of our modified FIFO is analyzed by comparing it with the default FIFO scheduler who is present in Hadoop, by taking best, average, and worst cases. Figures 3, 4 and 5, shows the results obtained after completion of the jobs. Modified FIFO improves the job execution time with workload. The same tasks are assigned to the main node

Table 2 Shows the result of Job2

Testing of jobs with the proposed scheduler		Testing of jobs with the default scheduler	
Job J2	Total time taken (s)	Job J2	Total time taken (s)
1	00:40	1	00:48
2	00:46	2	00:48
3	00:49	3	00:46
4	00:49	4	00:48
5	00:50	5	00:50
6	00:46	6	00:47
7	00:48	7	00:48
8	00:48	8	00:43
9	00:46	9	00:48
10	00:46	10	00:49
Best case: 00:40		Best case: 00:43	
Worst case: 00:50		Worst case: 00:50	
Average case: 00:46		Average case: 00:48	

Table 3 Shows the result of Job3

Testing of jobs with the proposed scheduler		Testing of jobs with the default scheduler	
Job J3	Total time taken (in s)	Job J3	Total time taken (in s)
1	00:57	1	00:56
2	00:56	2	00:59
3	00:57	3	00:58
4	00:55	4	00:58
5	00:56	5	00:54
6	00:58	6	00:59
7	00:56	7	00:57
8	00:51	8	00:55
9	00:55	9	01:19
10	00:55	10	00:57
Best case: 00:51		Best case: 00:54	
Worst case: 00:58		Worst case: 1:19	
Average case: 00:56		Average case: 00:59	

without aware of a feature of job and without specifying the workload on the node. Both schedulers have the same hardware configuration, software configuration, and power management. From the experimental analysis, it is clear the modified FIFO is better than default Hadoop's FIFO in terms of best case, average case, and worst case. MapReduce performance parameters automatically manage jobs which require large memory for the processing. It also observed average time with modified FIFO

Table 4 Shows the result of Job4

Testing of jobs with the proposed scheduler		Testing of jobs with the default scheduler	
Job J4	Total time taken (in s)	Job J4	Total time taken (in s)
1	01:04	1	01:00
2	01:03	2	01:01
3	01:04	3	00:59
4	01:01	4	00:59
5	01:02	5	01:02
6	01:03	6	01:04
7	01:03	7	01:03
8	00:53	8	00:57
9	01:02	9	01:07
10	01:01	10	01:01
Best case: 1:01		Best case: 00:57	
Worst case: 1:04		Worst case: 1:07	
Average case: 1:02		Average case: 1:03	

Table 5 Shows the result of Job5

Testing of jobs with the proposed scheduler		Testing of jobs with the default scheduler	
Job J5	Total time taken (in s)	Job J5	Total time taken (in s)
1	01:18	1	01:14
2	01:09	2	01:09
3	01:14	3	01:11
4	01:09	4	01:08
5	01:10	5	01:09
6	01:07	6	01:11
7	01:10	7	01:10
8	01:08	8	01:10
9	01:02	9	01:07
10	01:10	10	01:09
Best case: 1:02		Best case: 1:07	
Worst case: 1:18		Worst case: 1:14	
Average case: 1:09		Average case: 1:10	

is reduced as compared to Hadoop's FIFO. This is because the workload is under consideration while scheduling the job, which reduces the average time also.

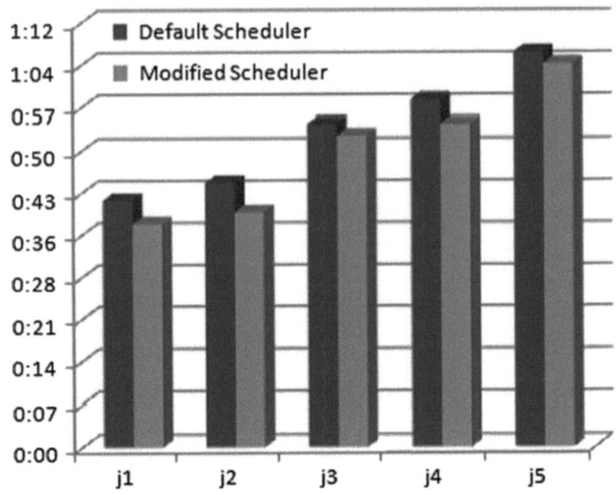

Fig. 3 Analysis of best case

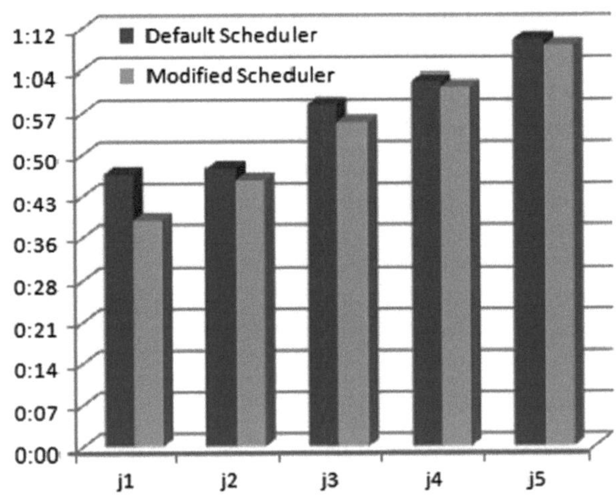

Fig. 4 Analysis of average case

5 Conclusion

On studying various schedulers, we come to know that the CPU usage is not taken into consideration while developing different schedulers. So there was a need to focus on this issue too. Hence a novel distribution algorithm is proposed which implements the pipelining concept in order to have efficient CPU usage by decreasing the job execution time. From the experimental analysis, we could conclude that the job completion time in case of the proposed scheduler is lesser than the default scheduler. This is because of pipelining scheduling which we have used in the proposed algorithm. This enables the faster execution of various tasks, thereby making the CPU usage efficient as the CPU does not have to waste its time in waiting for the split tasks. In last, we found that the proposed scheduler or the modified scheduler has

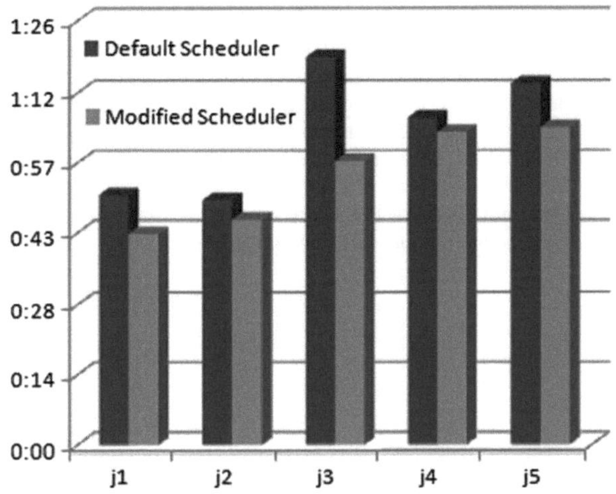

Fig. 5 Analysis of worst case

lesser job completion time than the default scheduler in each best case, average case, and worst case. We have applied the pipeline scheduler in a single node Hadoop so in future It could be further used for a multiple node Hadoop, i.e., for a Hadoop cluster also.

References

1. C. Philip, C. Zhang, Data-intensive applications, challenges, techniques and technologies: a survey on Big Data. Elsevier J. Inf. Sci. **275**, 314–347 (2014)
2. B. Sarladevi, N. Pazhaniraja, Big Data and Hadoop—A study in security perspective, in *Elsevier 2nd International Symposium on Big Data and Cloud Computing, Procedia Computer Science*, vol 50 (2015), pp. 596–601
3. D.P. Augustine, Leveraging big data analytics and Hadoop in developing India's healthcare services. Int. J. Comput. Appl. **89**, 44–50 (2014)
4. J. Dean, S. Ghemawat, MapReduce: simplified data processing on large clusters, in P*roceeding of the 6th Symposium on Operating systems Design and Implementation* (*OSDI*) (USENIX Association, December 2004), pp. 137–150
5. M. Zaharia, D. Borthankur, J. Sarma, K. Elmellegy, S. Shenker, I. Stoica, Delay scheduling: a simple technique for achieving locality and fairness in cluster scheduling, in *Proceedings of the 5th European conference on Computer systems, ACM* (2010), pp. 265–278
6. B.T. Rao, L.S.S. Reddy, Survey on improved scheduling in Hadoop MapReduce in cloud environments. Int. J. Comput. Appl. **1**, 229–232 (2011)
7. C. He, Y. Lu, D. Swanson, Matchmaking: a new mapreduce scheduling technique, in *IEEE Third International Conference on Cloud Computing Technology and Science* (*CloudCom*) (December 2011), pp. 40–47

8. K.A.Kumar, V.K. Konishetty, K.Voruganti, G.V. Rao, *CASH: context aware scheduler for Hadoop*, in *Proceedings of the International Conference on Advances in Computing, Communications and Informatics*, New York, USA, (Aug 2012), pp. 52–61
9. N.Phuong, T.Simon, M.Halem, D.Chapman, Q.Le, A Hybrid scheduling algorithm for data intensive workloads in a mapreduce environment, in *Proceedings of the 2012 IEEE/ACM Fifth International Conference on Utility and Cloud Computing*, Washington DC, USA (Nov 2012), pp. 161–167
10. J.S.Manjaly, V.S.Chooralil, TaskTracker aware scheduling for Hadoop MapReduce, in *Third International Conference on Advances in Computing and Communications (ICACC)* (2013)

Understanding Concepts, Future Trends, and Case Studies of Big Data Technologies

Khyati Ahlawat and Amit Prakash Singh

Abstract In this paper, several promising big data technologies and their case studies are discussed in the context of their application in a real-world scenario and their potential to meet growing demands of mainstream big data customers. The prevailing big data technologies include Predictive Analytics, NoSQL Databases, Real-Time Analytics, Data Virtualization, and Knowledge Discovery in Big Data. Each of these requires one or more different big data handing platforms and tools. A brief study of big data preparation and analytics platforms for the above-mentioned technologies is also given in the paper. Big Data Analytics is rapidly becoming popular, spawning the need to comprehend its associated technologies. This paper will provide a baseline study for many researchers who wish to gain knowledge on current behavior and future trends of big data technologies. Currently, numerous platforms are available, both in the online and offline mode to handle big data, so a detailed review of all of such technologies is out of the scope of this paper. Hence, it deals with ones which are highly accepted and popular.

Keywords Big data analytics · NoSQL databases · Real-time analytics
Predictive analytics · MapReduce

1 Introduction

Any large sized and complex data, which limits the capabilities of traditional data processing applications and tools to handle it, can be termed as Big Data. It has certain characteristics that are defined as several Vs by various researchers [1], however, significant ones are Volume, Variety, and Velocity. Big data can comprise of all of

K. Ahlawat (✉) · A. P. Singh
University School of Information Communication and Technology, Guru Gobind Singh Indraprastha University, Sector 16C, Dwarka 110078, Delhi, India
e-mail: khyatiahlawat@ipu.ac.in

A. P. Singh
e-mail: amit@ipu.ac.in

C. R. Krishna et al. (eds.), *Proceedings of 2nd International Conference on Communication, Computing and Networking*, Lecture Notes in Networks and Systems 46, https://doi.org/10.1007/978-981-13-1217-5_96

these characteristics or any one of them. Volume defines substantial amount of data that challenges storage and processing capacity of currently used tools and processors. Variety refers to different types of data present for analysis like multimedia data (audio, video) and textual data, images, social media data [2], etc. Velocity on the other hand, deals with the pace at which big data is getting generated and propagated for analysis. Streaming data analysis is one example of high-velocity big data where the challenge lies in the rapid analysis of constantly generating streams of data via different sources. Similarly, researchers in [3] have introduced HACE theorem suggesting major characteristics of big data as heterogeneous and diverse data sources, autonomous with distributed control, complex, and evolving in knowledge.

Analysis of big data by incorporating its characteristics to discover and predict valuable information from it is termed as Big Data Analytics. The basic aim of analysis is to extrapolate big data, interpret it, and assist in prediction, diagnosis, and decision-making. It is one of the major research areas in parallel and distributed systems in future generation due to its wide application areas like health welfare, government and public sector, e-commerce, social networking, and natural processes [4] etc.

KDD, Knowledge Discovery, and Data Mining, is a process comprising of basic three steps, i.e. input, analysis, and output. Currently, numerous techniques are available for normal-sized data analysis. However, new techniques and algorithms are required for big data analysis to match highly complex nature of big data. Currently available tools for big data concentrate on two types of analysis, batch processing and streaming data analysis [5]. Most established batch processing tool is Apache Hadoop platform. Apache Hadoop consists of two basic components, HDFS and MapReduce Paradigm. HDFS is Hadoop-Distributed File System and is used for storing voluminous big data in master–slave fashion. MapReduce Paradigm is based on the parallel computation of big data and taking into consideration computation time, parallel processing is the most important future trend [6]. Other tools for batch processing include Apache Mahout, Dryad, Tableau, etc.

On the other hand, streaming data tools include Apache Spark, Storm, S4, Kafka, etc. It is also known as real-time processing in which ongoing data processing requires a very low latency rate for efficient analysis. Apache Storm is mostly used tool for this purpose. It is scalable, fault tolerant, and easy to use and operate. Other tools include S4, Splunk, Kafka, SAP Hana, etc. [5] has very well demonstrated the differences between batch processing and real-time tools.

One needs to decide which platform is most suitable for which type of data according to the application area it is being generated from. For this purpose, various factors are important to consider like data size, throughput optimization, training of model, scalability, etc. [7]. Rest of the paper is as follows: Sect. 2 describes various technologies in big data domain covering almost all available techniques and areas along with a discussion of some case studies. Section 3 discusses trends and forecasts for big data technologies according to Forbes. Section 4 is conclusion and future scope of this chapter.

2 Technologies in Big Data Domain

Big data, consisting of some unique features, gets generated from diverse sources in various formats like batch and streaming. It is observed that most of the big data that is generated is unstructured. Challenge lies in gaining insights into such data directly or by first converting it to structured data and then analyzing it. Storage and analysis of structured or unstructured data are always a challenge when size of data increases enormously. With the evolution of big data, techniques related to it are also developing to efficiently analyze it and extract important information from it. This section discuses top-level technologies in Fig. 1 that are currently being worked on in the big data domain. These include both for storage and analysis purpose.

Each type of big data technologies described in Fig. 1 such as Predictive Analytics, NoSQL Databases, Real-Time Analytics, Data Virtualization, and Cloud Computing are discussed next.

2.1 *Predictive Analytics*

Predictive analytics is an advanced level of analytics in which predictions are made for future events based upon previously available data. It is also known as Classification. For this purpose, machine learning algorithms are used to train the machine based on historic data and then it is tested for the quality of predictions it can make. Many practitioners are working toward applying existing machine learning techniques in big data domain or scaling them to map big data level.

There are multiple platforms for machine learning in big data scenario as discussed in [8] like Mahout, H2O, SAMOA, Apache Spark, etc. Mahout is the most well-known platform with a wide range of machine learning algorithms that are basically focused on classification, clustering, and collaborative filtering. It has some shortcomings also like its inefficient runtimes due to slow MapReduce principle and difficulty to set it on a Hadoop cluster. On the other hand, Apache Spark is a recent

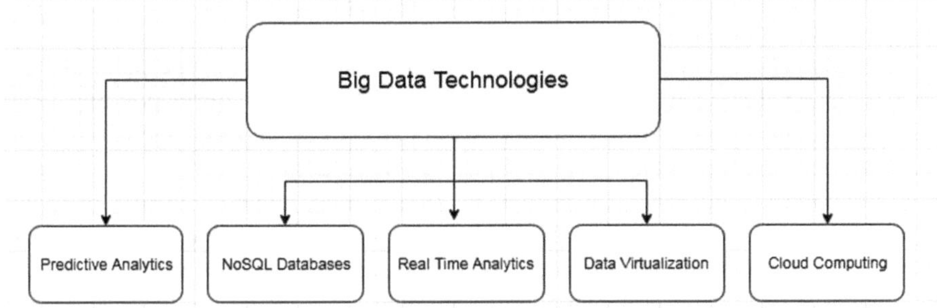

Fig. 1 Big data technologies

platform which is based upon MLlib, machine learning library. MLlib implements all algorithms of Mahout along with some regression models, feature extraction, etc., which is the lacking factor for Mahout.

H_2O is an enterprise edition with a notable feature of GUI, graphical user interface, and deep neural network toolkits. Users can work on this platform via programming or web-based interface. SAMOA stands for Scalable Advanced Massive Online Analysis and is a machine learning platform for streaming data. It is highly scalable and extensible but with low coverage of algorithms.

Apart from this, one more promising platform for machine learning in big data is R [9]. Though R provides more than 70 packages but they lack scaling to big data. Therefore, R has provided specialized packages for big data like bigrf, biglm, etc. Bigrf provides Random Forest functionality on big data while biglm deals with linear regression for data that is too large to fit in memory. R also provides an interface with Hadoop known as Radoop, which works in the form of map reduce jobs on the data stored in HDFS. Researchers have discussed and compared various machine learning platforms on big data in detail which is in [9].

In addition to this, a general-purpose framework, Petuum is introduced in [10] that systematically addresses data and works efficiently in comparatively less time.

Case Study: For instance, started as an online bookstore, Amazon has led its journey to top-level online retail store. Product recommendation is used based on customers profile and their past experiences in amazon which serves as a major big data challenge. It uses the concepts of content-based and collaborative filtering to group customers and analyzes their preferences.

2.2 NoSQL Databases

In this era of big data analytics, a variety of data in different formats are available and traditional relational databases are becoming obsolete to manage them. In their place, NoSQL databases are providing promising and efficient data storage and analysis platform to handle big data. Some popular NoSQL databases include Redis, MongoDB, Couchbase, Cassandra, HBase, etc. These can be grouped based on their characteristics in four categories as Key-value Store, Document Store, Column Family Store and Graph databases [11].

Key-value Store is where data is stored in the form of key-value pairs as hash tables where values can be easily accessed with the help of corresponding keys. In document store, semi-structured documents are stored and managed in formats like XML or JSON. Column Family Store is where data is stored in the form of rows and columns and based on similarities of columns, related columns formulate column family. Graph databases store information in the form of graph data structure and make use of graph algorithms to analyze them. It is mainly used to represent recommendation system or social networking data.

Many researchers are working in this field to analyze the capabilities and performance of NoSQL databases to handle big data. In [12], it is concluded that Redis which is a key value data store has performed better in terms of efficiency and is found well suited for loading and executing workloads. Apart from this, practitioners have also proposed a fuzzy-based methodology in [13] to store consistent fuzzy geospatial data based on certain validations and constraints for semantics. Researchers in [14] have addressed the problem of data migration in heterogeneous NoSQL databases by providing a fault-tolerant approach to migrate data. In some applications, a hybrid architecture of relational and NoSQL databases can prove useful. In [15], a data adapter system to support hybrid database model supporting both types of databases with three modes for query approach.

Case Study: EHR system as discussed in [16] replaced existing system to NoSQL database for primary data store and to prepare a local cache at each site to improve request latency and availability.

2.3 Real-Time Analytics

Real-time big data analytics is when the timely prediction of streaming data is provided. This is a very challenging task in big data scenario as storage and analysis of big data at the same time requires large in-memory and correct platform. Its applications include majorly sensor related data and social web data. Examples of real-time analytics tools are S4, Storm, and Apache Spark.

In recent times, the role of streaming analytics has been increased in the field of sentiment analysis. A distributed system based on Vertical Hoeffding Tree, a decision tree classifier for same has been presented in [17]. Apache Samoa, a tool for distributed streaming classification along with evaluation criteria to measure the performance of streaming classifiers is depicted in paper [18].

Case Study: A case study on finding the effectiveness of Storm to analyze streaming data in real time from two applications to predict trending news is given in [19].

2.4 Data Visualization

Visualization is the concept of representing the results and analysis part of big data in a graphical manner. It is becoming a concerned technology because traditional visualization approaches do not map well to big data scenario. In case of a large-scale data visualization, processing methods such as feature extraction and geometric modeling are also used. In the current scenario, researchers are actively working on searching for new visualization platforms to scale them for big data. Here, augmented reality techniques are proving to be next milestone in this scenario. In [20], practitioners have discussed utilizing virtual reality and augmented reality techniques for big data visualization. A new algorithm for visualization of big data from high dimension to a two-dimensional space is presented in [21] and found to perform noticeably well.

Similarly, in [22], high dimensional big data is visualized in three dimension using unsupervised dimension reduction techniques.

Case Study: To maintain the consistency and integrity of constantly increasing data in Qualcomm, a data virtualization solution was produced to make data available fast and in the virtual view.

2.5 Cloud Computing

Distributed computing along with Internet in an extended form can be conceptualized as cloud computing. Here, cloud refers to a network which can be accessed via the Internet anytime and anywhere to use it as any service. With the evolution of big data, storing and analyzing big data in clouds seems to be perfect solution to almost all data storage problems. Still, there exist some issues in this solution out of which security is most important. In [23], an efficient approach to handle secure storage of big data in a cloud environment is presented which is based on compressing and decompressing of data.

One more problem associated with distributed cloud computing is heavy data traffic among data centers. This leads to high communication costs in query evaluation. An online query evaluation system to deal with this problem of big data in distributed clouds is discussed in [24].

Case Study: Several case studies in this research area are discussed in [25] namely SwiftKey, Nokia, redBus, etc. SwiftKey is a language technology that uses Apache Hadoop on Amazon Cloud for scalable and multilayered model along which applies artificial intelligence component to manage constantly increasing data in prediction problems. redBus is responsible for internet bus ticketing system in India and has implemented its system using Google Query for large data analysis. On the other hand, Nokia makes use of Hadoop Distributed File System to store multi-structured data in cloud environment.

3 Trends and Forecasts for Big Data Technologies

Big data market is constantly expanding in mainstream business and there is a need to analyze which big data technologies are in demand. According to Forbes, an analysis of growing requirements of big data technologies is done and a trajectory of more than 20 techniques in big data scenario is presented as Fig. 2.

As shown in this figure, a wide range of big data technologies are shown as per their business value in coming time. Predictive analytics is the technology that is highly popular in today's business needs and the only one with maximum time for next phase. Apart from it, NoSQL databases, Stream analytics, search and knowledge discovery are some other technologies that are currently in demand. No big data technology is achieving the minimum success line on graph showing the promising

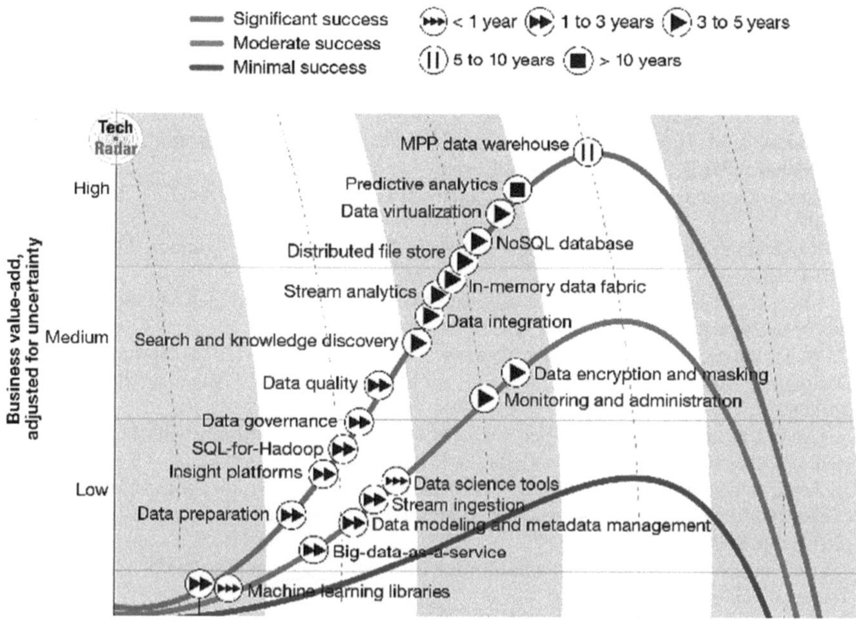

Fig. 2 Big data technologies described by Forbes

nature of the technologies in this area. Very few technologies namely data science tools and machine learning libraries are achieving moderate success and are likely to be popular in less than 1 year. Based on TechRadar methodology of Forrester, above-described technologies in this paper are most popular and likely to remain persistent for more than 5 years [26].

4 Conclusion

Big data technologies are trending in almost all present business requirements and their understanding has become a burning need in this field. In this paper, a brief study on some popular big data technologies like predictive analytics, NoSQL databases, Real-time analytics, cloud computing, and data virtualization is discussed. Prediction for future trends of popular techniques and their lifespan is also presented as described by Forbes. Sample case studies for each discussed technology are also given to understand the scope and purpose of this study.

References

1. H. Hu, Y. Wen, T.S. Chua, X. Li, Toward scalable systems for big data analytics: a technology tutorial. IEEE J. Mag. **2**, 652–687 (2014)
2. A. Gandomi, M. Haider, Beyond the hype: big data concepts, methods and analytics. Int. J. Inf. Manage. **35**(2), 137–144 (2015). Elsevier
3. X. Wu, X. Zhu et al., Data mining with big data. IEEE Trans. Knowl. Data Eng. **26**(1), 97–107 (2014)
4. K. Kambatla et al., Trends in big data analytics. J. Parallel Distrib. Comput. **74**, 2561–2573 (2014). Elsevier
5. C. Chen, C. Zhang, Data-intensive applications, challenges, techniques and technologies: a survey on big data. Inf. Sci. **275**, 314–347 (2014). Elsevier
6. C.-W. Tsai et al., Big data analytics: a survey. J. Big Data **2**(1), 21 (2015). Springer
7. D. Singh, C. Reddy, A survey on platforms for big data analytics. J. Big Data **2**(1), 8 (2014). Springer
8. S. Landset et al., A survey of open source tools for machine learning with big data in the Hadoop ecosystem. J. Big Data **2**(1), 24 (2015). Springer
9. J. Zheng, A. Dagnino, An initial study of predictive machine learning analytics on large volumes of historical data for power system applications, in *International Conference on Big Data, IEEE* (2014), pp. 952–959
10. P. Eric Xing et al., Petuum: a new platform for distributed machine learning on big data. IEEE Trans. Big Data **1**, 49–67 (2015)
11. R. Zafar et al., Big data: the NoSQL and RDBMS review, in *International Conference on Information and Communication Technology, IEEE* (2016), pp. 120–126
12. E. Tang, Y. Fan, Performance comparison between five NSQL databases, in *International Conference on Cloud Computing and Big Data, IEEE* (2016), pp. 105–109
13. B. Khalfi et al., A new methodology for storing consistent fuzzy geospatial data in big data environment, vol. 3, in *IEEE Transactions on Big Data* (2017)
14. M. Scavuzzo et al., Providing Big data applications with fault-tolerant data migration across heterogeneous NoSQL databases, in *International Workshop on BIG Data Software Engineering, ACM* (2016), pp. 26–32
15. Y.T. Liao et al., Data adapter for querying and transformation between SQL and NoSQL database. Future Gener. Comput. Sys. **65**, 111–121 (2016). Elsevier
16. J. Klein et al., Performance evaluation of NoSQL databases: a case study, in *International Workshop on Performance Analysis of Big Data Systems, ACM* (2015), pp. 5–10
17. A. Hossein, A. Rahnama, Distributed real-time sentiment analysis for big data social streams, in *International Conferences on IEEE* (2014), pp. 789–794
18. A. Bifet et al., Efficient online evaluation of big data stream classifiers, in *Conference on Knowledge Discovery and Data Mining, ACM* (2015), pp. 59–68
19. T. Chardonnens et al., Big data analytics on high velocity streams: a case study, in *International Conference on Big Data, IEEE* (2013), pp. 784–787
20. E. Olshannikova et al., Visualizing big data with augmented and virtual reality: challenges and research agenda. J. Big Data **2**(1), 22 (2015). Springer
21. B. Wu, B.M. Wilamowski, An algorithm for visualization of big data in a two-dimensional space, in *41st Annual conference of the IEEE Industrial Electronics Society, IEEE* (2015), pp. 53–58
22. Y. Xie et al., Visualization of big high dimensional data in a three dimensional space, in *3rd International Conference on Big Data Computing, Applications and Technologies, ACM* (2016), pp. 61–66

23. A. Kuma et al., Efficient and secure cloud storage for handling big data, in *International Conferences on IEEE* (2012), pp. 162–166
24. Q. Xia et al., Data locality-aware query evaluation for big data analytics in distributed clouds, in *Second International Conference on Advanced Cloud and Big Data, IEEE* (2014), pp. 1–8
25. I. Abaker et al., The rise of "big data" on cloud computing: review and open research issues. Inf. Syst. **47**, 98–115 (2015). Elsevier
26. G. Press, in *Top 10 Hot Big Data Technologies* [Online] (2013, Mar 14). Available: https://www.forbes.com/sites/gilpress/2016/03/14/top-10-hot-big-data-technologies/#1643589c65d7

Issues in Software-Defined Networking

Amit Nayyer, Aman Kumar Sharma and Lalit K. Awasthi

Abstract Software-Defined Network is a nascent emerging architecture that provides solutions to most of the challenges in traditional networks. It is a model for heterogeneous integration of different network devices. It separates control logic from the network devices, leaving the devices with only data-forwarding responsibilities. It makes configuration and management of the network simpler, manageable, and provides flexibility for various innovative network designs. Conclusively, it increases the overall performance of the network. This article provides a comprehensive survey of different issues in SDN. We discuss various research areas in SDN. Our target is to introduce a newbie to this emerging field of SDN and make him aware of various issues available for research in this field.

Keywords Software-defined network · Scalability · Security · Congestion
Flow control mechanism

1 Introduction

Software-defined networking (SDN) makes configuration and management of the network simpler and manageable. It brings into picture the concept of separating the control plane and the forwarding plane in the network. The control plane is responsible of controlling the overall behavior of the network, whereas the forwarding plane just send and receive the data as well as instruction from other devices in the network. This approach reduces the number of existing challenges in the network

A. Nayyer (✉) · A. K. Sharma
HPU, Shimla, India
e-mail: nayyer.amit@gmail.com

A. K. Sharma
e-mail: sharmaas1@gmail.com

L. K. Awasthi
NIT, Jalandhar, India
e-mail: lalitdec@yahoo.com

C. R. Krishna et al. (eds.), *Proceedings of 2nd International Conference on Communication, Computing and Networking*, Lecture Notes in Networks and Systems 46, https://doi.org/10.1007/978-981-13-1217-5_97

and introduces efficiency and sustainability. It also removes the vendor dependencies of the devices as they are now controlled by a specific controller [1]. The protocols are defined and imposed by the controller in such network. Issues like security and reliability can now be handled in a better way using this new technological paradigm. Dynamic updating becomes more feasible and impactful. Using central controller, the decision can be implemented energy efficiently, and hence increase the lifetime of the network. Most of the energy-consuming functions are now with controller and it has more power resources as compared to other nodes in the network. OpenFlow emerged as de facto SDN protocol.

2 Architecture and Behavior

The SDN divides the network into two major parts. The control panel: It is where all the decision-making capabilities reside. It is responsible for network configuration and managements and it is the intelligent part of the architecture. The data plane: It is where only data-forwarding responsibilities reside. It accepts the policies as decided by the control plane and not intelligent in any way. The data plane stores the policies that contain the instructions from the controller to handle any message when the device receives them [2]. The architecture further divides the APIs (Application interfaces) into northbound APIs and southbound APIs. These APIs are named just due to their interaction toward north and the south directions of the control plane. The APIs provide an interface between different layers in the architecture. Figure 1 represents the architecture of SDN, representing different planes and APIs.

The APIs shown as 2 and 4 in the Fig. 1 represent Northbound and Southbound APIs. The control plane interacts with these APIs for any communication between the lower and the upper planes of the architecture. The data plane shown in the Fig. 1 consists of heterogeneous types of nodes. Some architectures discussed in other research articles also named it as physical plane as this plane deals with hardware associated with the network. The nodes forward its data to different routers in the network, which further routes the data to the central controller for decision-making.

3 SDN: Issues

The concept of SDN is to implement the architecture based upon central controller. This architecture no doubt provided solution to many existing problems, still had some limitations in other areas. Some of the research issues in this paradigm are as follows.

Fig. 1 Architecture of SDN 1: Application Plane 2: Northbound API 3: Control plane 4: Southbound API 5: Data Plane

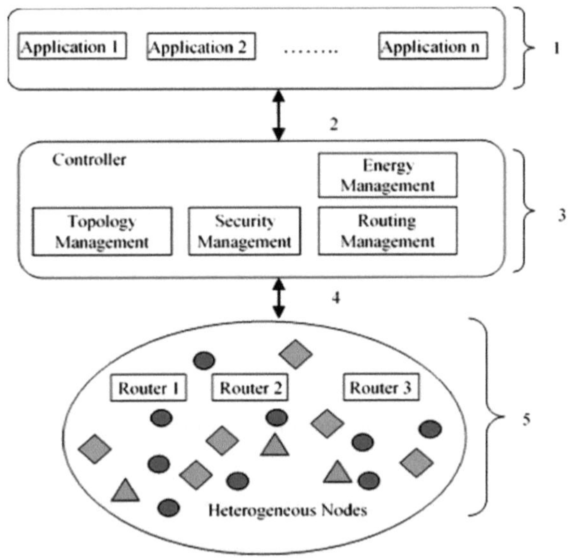

1:Application Plane 2:Northbound API 3: Control plane
4: Southbound API 5: Data Plane

3.1 Scalability

In Software-Defined Networks (SDN), scalability is also one of the major issues. Many researchers are concerned with scalability in SDN due to its centralized control architecture [3]. Scalability can be related to either controller's scalability (processing power) or with respect to communication capabilities of control plane–data plane. The capability of controller may be hampered due to the increase in a number of elements in the network as it grows with time. The Controller may fail to provide the required services to all the elements depended on it and, therefore, restricting the number of elements connected to it. Scalability is also a concern in SDN due to its architecture, which allows decoupling of data plane and control plane. The centralized control plane can become a bottleneck in case of addition of elements and thereof requests to it. No doubt, that moving all the control functionality to centralized controller brings network-wide view at a single point and controlling policies become easier. In fact, SDN adds a level of flexibility and management for scalability [4]. Different factors that can be used to measure the scalability of any SDN are the following:

(a) Functions: The capability of SDN to allow the addition of new functionalities at minimum efforts.
(b) Administration: The ability of network to allow increased number of users without losing control over them.

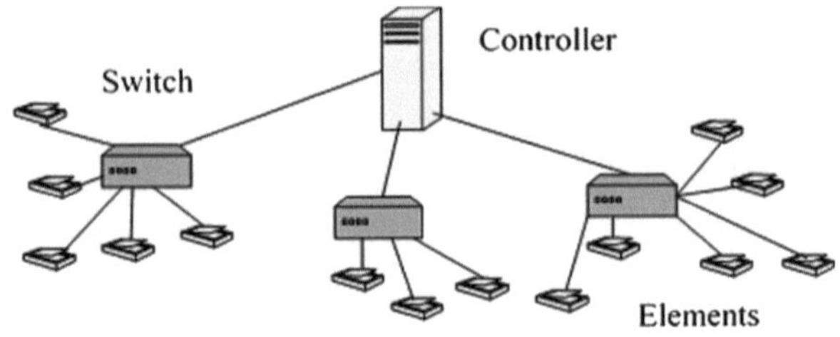

Fig. 2 Illustration of controller, switch, and element placement in SDN

(c) Geographic: To maintain performance regardless of location of the expansion either local or distributed geographically.
(d) Load Scalability: The capability of increasing and decreasing its resource pool to accommodate the number of elements in the network.

Increase of load at the controller is the main factor that affects scalability in SDN. There are different other reasons that contribute to the issue of scalability in SDN:

(a) Capability of Switches: The switches in the SDN are proposed to have very limited processing capabilities and data storing capacity. The switches are not able to hold more flow of table entries, which results in sending a repeated request to the controller for new entry and, therefore, increases load and finally affecting scalability [5]. The less capability of switches turns the SDN model into a reactive approach.
(b) Link failures: Failure of any link in the network is one of the natural phenomena. In SDN, also link failure is a factor which further increases the load on the controller. Whenever there is a failure of link detected by any switch in the network, it notifies the controller about the event [6]. The controller computes the new route information and sends back the update to switches for their forwarding table in Fig. 2.

No doubt, scalability is one of the major issues in SDN, but due to network-wide view capability of controller and flexibility of the management it can be easily eradicated. Multiple controller installations are an effective way to deal with this challenge.

3.2 *Security*

Security is one of the benefits of SDN, as it provides global visibility of the network. Due to the reason that the centralized controller is responsible for managing the whole network, it is more prone to attacks and once affected it affects the whole network. The availability of central controller opens other challenges in security including attacks such as man in middle attack, DoS attacks, and saturation attack [7]. The updating of any security policy can be done from controller rather than going to each forwarding hardware and updating their relative software. The security lapses in controller path communication can lead to illegitimate access and the usage of network resources. The security challenges in SDN can be divided into three main different planes as follows [8]:

(a) *Application plane*: Most of the functions are implemented with the applications in the SDN. If any fraud application is spread across the network, it can hamper the activity of the network. There is absence of authentication and authorization mechanism for application and there is a threat to security in case of large number of third-party applications in the network [9]. The fraud applications can generate different flow rules and it will become difficult to check if and application is compromised or not. Access control and accountability on third-party applications is also difficult to implement.
(b) *Control plane*: Due to the more visibility of this plane in the network, it is more prone to DoS attacks. Also, in this plane there is no compelling mechanism for enforcing access control on the application. Scalability and availability challenges are other types of threat for such centralized architectures.
(c) *Data plane*: The data plane is most static part of the network and fraudulent flow rules are one of the major attacks on such part of the network. The tables called flow table stores limited number of flow rules. As data plane is dependent on the control plane, so controller hijacking or compromised controller are other attacks [10]. TCP-level attacks and man in middle attacks are also targeted at this plane.

3.3 Fault Tolerance

Fault tolerance is the major topic of concern to any network. The same behavior is expected in SDN. Fault tolerance is a challenging task in SDN due to the presence of multiple heterogeneous elements in the network [11]. Google's SDN named as B4 has shown the resilience architecture up to some extent [12]. The resilience of an SDN network depends on fault tolerance at two planes as follows:

(a) Control plane: SDN is prone to single point of failure due to centralized controller model. The solution for making fault-tolerant control plane consists of two methods. First one is primary backup replication in which only one primary controller exists and performs all the operations. It updates the backup replicas periodically. In case of failure of primary controller, one of the backup replicas takes control and acts as primary controller. Second one is state machine replica model, in which client issues commands to all replicas that execute them in a deterministic way. Implementing replicated state machines are not providing the desired performance in case of controller failure. Efforts are being made to increase the resilience of network by control plane split architecture. Distributed controller architectures increase resilience in the network but increased other tradeoffs.

(b) Data plane: The requirement of fault tolerance in this plane arises when a switch detects a link failure and updates the controller regarding the failure. The controller updates the flow table of the switch based on some intra-calculations. The approach is reactive and needs large restoration time for updating of flow tables at the switch. Another strategy can be worked out to provide an efficient solution to the problem of link failure in the network.

FatTire [13] presents a new language for writing fault-tolerant programs in SDN. New programming construct is proposed based on regular expressions. The set of paths is defined by the programmer and the packets may take that path in the network. The algorithm was proposed for compiling FatTire programs to Open Flow switch configuration. The model only deals with link-level failure. It can be extended such that the controller can be updated regarding the link failures. Ravana [14] offers the abstraction of a fault-free centralized controller to control applications. In this model, the control message is processed transitionally and exactly one time. The replicated state machine is extended with light weighted switch side mechanism for correctness. It is a distributed protocol that enables unmodified controller application to execute in fault-tolerant way. The model can be extended so that multiple threaded control applications can also be supported. Slick Flow [15] provides the idea of using packet to carry extra path to provide resilient routing. If there is a path failure, alternative path can be chosen randomly. This method eliminates the need of contacting the controller in case of a link failure is detected by the switch. Byzantine controllers [16] are another type of SDN controllers that can tolerate Byzantine faults in control and data planes. Byzantine fault tolerance is a characteristic of distributed system that tolerates the Byzantine Generals problem described by Leslie Lamport in his research paper. Reactive controller logic is considered in the model for updating flow tables from control to data plane. The Byzantine controller model does not perform well in case of large-scale network; therefore, it can be modified to work for larger networks.

3.4 Congestion

Congestion is the reduction of quality of service that occurs when a link is carrying more data than it can handle due to whatsoever the reason is. Another factor is insufficient memory space at the switches to store the data. The capability of the processor is also a reason for congestion due to heavy load at some moments. As discussed in the last section, the control plane and data plane are affected due to congestion in the network. The controller itself can be congested due to different reasons. Moreover, the centralized concept of the model itself is prone to the congestion. The links between data and controller are also congested due to the same reason.

SDTCP [17] provides a SDN-based TCP congestion control mechanism. The controller selects a long-lived flow to reduce sending rate by adjusting TCP receive window of acknowledgement packets. This behavior of controller is defined when

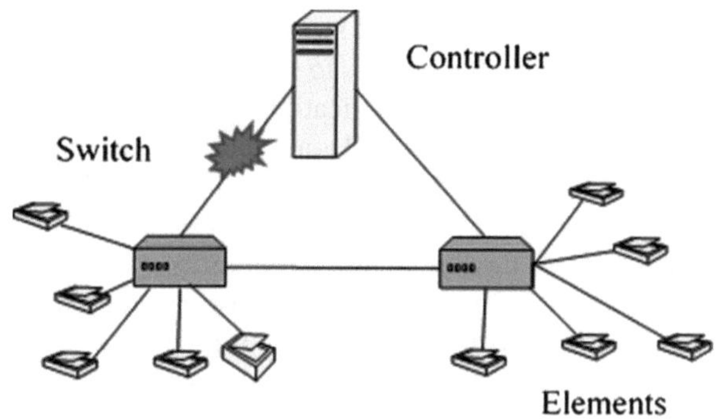

Fig. 3 SDN-based congestion control

there is a congestion message to controller by the switch. This model provides a benefit of accurately decelerate the rate of long-lived flow to ensure the performance of other flows in the network. It provides a guarantee output to high-priority flows. The congestion in the link between controller and switch is detected by examining the ports of the OpenFlow-enabled switches [18]. The controller keeps communication with switches on different events to know the statistics going on there. Flows from the congested ports are rerouted from such affected ports to some other less congested ports. This method decongests the affected port but increases the overhead of the controller. Moreover, new research can be done to include the concept of machine learning methods for decongestion improvements. The centralized controller receives large volumes of traffic due to single point of administration in SDN. As a consequence, congestion can occur frequently in such network. The article [19] provides a congestion-free routing strategy, suited to the concept of global view by SDN controller. A time slot allocation is done for the incoming packets and then the routing paths are computed for each packet. The heuristic approach is adapted for time slot allocation algorithm whereas bin packing problem is modeled for path selection. The proposed method performs well in all factors like throughput, queues and end-to-end round trip time, but the performance of protocol for large and scalable network is yet to be established. In Fig. 3, the congestion in SDN is depicted.

3.5 *Flow Management*

SDN controller acts as a brain of the network. It satisfies your versatile needs with an open architecture which provide flexibility and risk-free change management. Applications on the controller provide control on the flow of the network by defining policies. These policies are required to be stored in any flow table, which requires storage space. Both controller and switches face the difficulty of storing a large number of such flow tables in their respective storage spaces. Proactive flow management allows a switch to respond to the first packet received according to the flow rules installed in the flow table of the switch itself. This method is not a feasible solution as the limited availability of storage space at the switches. The reactive approach is much better in such scenarios to save the storage space at switches but it suffers from round-trip delay of communication with the controller.

The TimeoutX [20] provides a solution to such problems by means of adaptive flow table management. Adaptive timeouts are set for the entries to be in the flow table after which the entries in the table can be replaced by other entries. The method combines flow types and flow table utilization ratio to decide the time out of each entry. The method uses the flow table to its best use and supports more flows. Domain Flow [21] is another method for providing the fine-grained flow control of Ethernet switching. The network is split into sections and exact matches are used to enable practical flow management using OpenFlow for commodity switches. In the flow model, a flow is controlled by bit masking and wild-carded matching rules in different domains. This method can be extended for construction of workload-aware high-performance networks.

4 Conclusion

SDN provides a flexibility of innovative solution in any network, thus making it easily manageable and automated policy control. In this paper, we discuss the major issues in SDN and in this context examine different-related solutions. The extension work to already proposed solutions is provided, to the best of our knowledge. Most of the problems discussed here can be avoided by implementing multiple controllers in the SDN paradigm. The implementation of multiple controllers require synchronization among them and it will remain as one of the major issues in such networks. Synchronization studied in other networks can be applied with some modifications. As a future extension to this article, we would like to discuss all such issues in multiple controller platforms in SDN. The issues provided in this article are not meant to be exhaustive.

References

1. W. Xia, Y. Wen, C.H. Foh, H. Xie, A survey on software-defined networking, IEEE Commun. Surv. Tutorials **17**(1) (2015)
2. D. Kreutz, F.M.V. Ramos, P.E. Verissimo, C.E. Rothenberg, S. Azodolmolky, S. Uhlig, Software defined networking: a comprehensive survey. Proc IEEE **103**(1), 14–76 (2015)
3. S. Yeganeh, A. Tootoonchian, Y. Ganjali, On scalability of software-defined networking. IEEE Commun. Mag. **51**(2), 136–141 (2013)
4. D. Erickson, The beacon open flow controller, in *Proceeding 2nd ACM SIGCOMM Workshop Hot Topics Software Defined Net*work (2013)
5. A. Dixit, F. Hao, S. Mukherjee, T. Lakshman, R. Kompella, Elasticon: an elastic distributed SDN controller, in *Proceeding 10th ACM/IEEE Symposium Architecture Network Communication System* (2014)

6. M.F. Bari, A.R. Roy, S.R. Chowdhury, Q. Zhang, M.F. Zhani, R. Ahmed, R. Bautaba, Dynamic controller provisioning in software defined networks, in *Proceeding 9th International Conference Network Service Manager* (2013)
7. D. Kreutz, F.M. Ramos, P. Verissimo, Towards secure and dependable software-defined networks, in *Proceeding 2nd ACM SIGCOMM Workshop Hot Topics Software Defined Network* (2013)
8. S. Shin, Y. Song, T. Lee, S. Lee, J. Chung, P. Porras, V. Yegneswardn, J. Noh, B.B. Kang, Rosemary: a robust, secure, high-performance network operating system, in *Proceeding 21st ACM Conference Computer Communication Security*, (Scottsdale, AZ, USA, 2014)
9. S. Sezer, S. Scott-Hayward, P.K. Chouhan, Are we ready for SDN? Implementation challenges for software-defined networks. IEEE Commun. Mag. **51**(7), 36–43 (2013)
10. S. Shin, G. Gu, Attacking software-defined networks: a first feasibility study, in *Proceeding 2nd Workshop Hot Topics Software Defined Networks* (2013)
11. M. Desai, T. Nandagopal, Coping with link failures in centralized control plane architectures, in *Proceeding 2nd International Conference Communication System Network* (2010)
12. S. Jain, A. Kumar, S. Mandal, J. Ong, L. Poutievski, A. Singh, S. Venkatta, J. Wanderer, M. Zhu, J. Zholla, U Holzle, S. Stuart and A. Vahdat: B4: experience with a globally-deployed software defined wan, in *Proceeding ACM SIGCOMM Conference* (2013)
13. M. Reitblatt, M. Canini, A. Guha, N. Foster, FatTire: declarative fault tolerance for software defined networks, in *Proceeding 2nd Workshop Hot Topics Software Defined Network* (2013)
14. N. Katta, H. Zhang, M. Freedman, J. Rexford: Ravana: controller fault-tolerance in software-Defined networking, ACM (2015)
15. R. Ramos, M. Martinello, C. Esteve Rothenberg, SlickFlow: resilient source routing in data center networks unlocked by openflow, in *Proceeding IEEE 38th Conference Local Computer Network* (2013)
16. K. ElDefrawy, T. Kaczmarek, Byzantine fault tolerant software-defined networking (SDN) Controller, Computer Software and Applications Conference (COMPSAC), (2016)
17. J.R. Santos, Y. Turner, G. Janakiraman, End-to-end congestion control for infiniband in INFO-COM. in *Twenty-Second Annual Joint Conference of the IEEE Computer and Communications. IEEE Societies*, vol 2 (2003)
18. M. Al-Fares, A. Loukissas, A. Vahdat, A scalable, commodity data center network architecture, in *ACM SIGCOMM Computer Communication Review*, vol 38, no. 4 (2008)
19. Y. Zhang, N. Ansari, On architecture design, congestion notification, TCP incast and power consumption in data centers. Commun. Surv. Tutorials IEEE **15**(1), 39–64 (2013)
20. Y. Liu, B. Tang, D. Yuan, J. Ran, H. Hu, A dynamic adaptive timeout approach for SDN switch, Computer and Communications (ICCC), (2016)
21. Bolla Raffaele, R. Bruschi, F. Davoli, C. Lombardo, Fine-grained energy-efficient consolidation in SDN networks and devices. IEEE Trans. Netw. Serv. Manage. **12**(2), 132–145 (2015)

Feasibility of Fog Computing in Smart Grid Architectures

Md. Muzakkir Hussain, Mohammad Saad Alam and M. M. Sufyan Beg

Abstract Contemporary Smart Grid (SG) systems are enticed by smart devices and entities due to unfolded developments in both the IT sectors viz. Intelligent Transportation and Information Technology. The intelligent transportation infrastructure elements when bestowed with Internet of Things (IoT) and sensor network of latter IT (Information Technology), makes every object active and brings them online. In such scenario, the traditional cloud deployment perishes to meet the analytics and computational exigencies for such dynamic cum resource-time critical subsystems. Starting with highlighting the mission-critical requirements of an idealized SG infrastructure, this work proposes an edge-centered FOG (From cOre to edGe) computing model primarily focused to realize the processing and computational objectives of SG. The objective of this work is to comprehend the applicability of FOG computing algorithms to interplay with the core-centered cloud computing support, thus enabling to come up with a new breed of real-time and latency free utilities. Further, for demonstrating the feasibility of the proposed framework, the SG use case is considered and an exemplary FOG Service-Oriented Architecture (SOA) is depicted. Finally, the potential adoption challenges elucidated in the realization of the proposed framework are highlighted along with nascent research domains that call for efforts and investments in successfully guiding the FOG approaches into a pinnacle.

Keywords Smart grid (SG) · Internet of things (IoT) · Fog computing
Cloud computing · Advanced metering infrastructures (AMI)
Software-defined networking (SDN)

Md. Muzakkir Hussain (✉) · M. M. Sufyan Beg
Department of Computer Engineering, ZHCET, AMU, Aligarh, India
e-mail: md.muzakkir@zhcet.ac.in

M. M. Sufyan Beg
e-mail: mmsbeg@cs.berkeley.edu

M. S. Alam
Department of Electrical Engineering, ZHCET, AMU, Aligarh, India
e-mail: saad.alam@zhcet.ac.in

C. R. Krishna et al. (eds.), *Proceedings of 2nd International Conference on Communication, Computing and Networking*, Lecture Notes in Networks and Systems 46, https://doi.org/10.1007/978-981-13-1217-5_98

1 Introduction

There are relentless economic as well as environmental arguments in the academia, industries, R&Ds, and legislative bodies for the overhaul of the contemporary power grid comprehended by a full Smart Grid rollout [1]. The latter integrates green cum renewable energy production utilities, robust power monitoring schemes, adapts, and evolves with the consumption behavior and requirements. However, the unique feature that overlays on the heap of a SG amenities is connectivity and real-time analytics [2]. The recent advancements in information and communication infrastructures in general and Internet of Things (IoT) utilities in specific redefine the notion of "SMART" in current SG architectures. This work outlines the FOG computing paradigm and examines its primacy over the cloud computing counterpart that became ubiquitous in fulfilling the computational and analytics needs of a reliable, robust, resilient, and sustainable SG. The notion of smartness has been introduced into the contemporary SG architectures where the local nodes will be leveraged with computational capabilities. They will no longer remains a "thing" rather will be transformed into active computing nodes or "objects". Every component of SG network whether it is at the generation, transmission, or service level, will act as active nodes in the entire transportation web. They are now called object in the sense that they will be having attributes, gateways, states, etc. The whole transportation system can be encapsulated as a network of active nodes having deterministic state transitions. This is achieved through the notion of Internet of Things (IoT). IoT will ensure real-time transport of information to and from the utilities, smart grid system, and other components in the power system and charging infrastructure in a way that the current as well as future needs of these entities along with dedicated business engagement can be determined [3]. Such technologies are enabled by the recent developments in RFID, smart sensors, communication standards, and Internet protocols [4]. The basic principle is to have an environment where smart sensors collaborate directly with the "objects" without human involvement aiming to deliver multitudes of applications and services [5, 6]. It is a consensus belief by industries as well as research giants that down the line, in near future IoT will emerge as a technology enabler for smart transport, X2X (where X may be any of but not necessarily same from entities like vehicles, grids, homes, Microgrid etc.) data and energy exchange topologies, optimal renewable integration, and intelligent charging infrastructures [7].

However, it is obvious that in the IoT architecture, the population of connected entities will overshoot the current growth drift and will jeopardize the normal configurations [8]. This will in turn cause an exponential escalation in data generation, handling of which is key task to ensure viable implementation of the infrastructure. Connecting the objects through edge networks and technologies such as Wireless Sensor Networks (WSN), Zigbee, Bluetooth, RFID, Wi-Fi, 3G, and 4G, etc., will increase the complexity of underlying communication architectures. Efficient and robust data analytics setup that can establish a real-time cum intelligent decision-making atmosphere at every edge services becomes the need of hour. An exhaustive review of existing control paradigms reveal the presence centralized coordination strategies such as cloud computing, grid computing, etc. [8, 9]. However, the service demands of IoT architecture IoT-based transportation calls for storage and computation strategies that can execute locally at the edge itself. The prevalent cloud models are not intended to handle the seven unprecedented V's (Volume, Velocity, Variety, Variability, Veracity, Visualization, and Value) in the data generated by IoT architectures and coupling the whole universe of "things" or "objects" directly to the cloud is nearly unfeasible [10]. FOG computing approaches seem to be the preeminent preference for computations at the extreme edges such as vehicles, roadways, charging station, etc. [10–13].

However, installation of fogs (mini data centers) everywhere across the edges of the networks and entities may not be cost productive. The infrastructure demands varying levels of services which in turn have specified QoS requirements. Transporting tera–peta bytes of data from millions of edge devices to the central cloud in real-time is quite infeasible and even unessential as a significant percentage of data are passive and do not contribute to any decision-making process. Furthermore, there exist several tasks that do not even entail storage, processing, and analytics at cloud scale. Such requirements motivate the need of a hybrid control architecture where the mining and analytics activities are intelligently dispersed.

The Smart Grid applications require location-aware geo-distributed intelligence in services such as metering information updates, power thefts, distribution outages, network intrusions, etc., and require prompt and reflex actions to evolve and organize according to the adversaries. However, the current Smart Grid is under immense pressure owing to its sullen response to the abovementioned computational demands. Also, due to its fragility, concerns in SG control and coordination subsystems, repercussions of power outages, resiliency, and reliability issues are growing ever more serious. Upgrading to a computationally smarter, reliable, and resilient grid has escalated from being a desirable vision, to an urgent imperative. Here, we itemize few but not the least, of some of the mission-critical requirements of an ideal SG infrastructure plus the somber experiences encountered while going for pure cloud computing deployment.

1.1 Support for Scalable Real-Time Services

The need of real-time analytics and decisions is being emerged as the need for the hour to carry up the timing requirements of mission-critical SG utilities [14]. Even if some servers' fiascos occur, the system should heal itself with just graceful degradation in latency services. The current cloud models support for SGs can provide rapid response mechanisms but adversaries still pose threats to responsiveness.

1.2 Support for Scalable, Consistency Guaranteed, Fault-Tolerant Services

Consistency for cloud-hosted utilities is a broad term associated with ACID (Atomicity, Consistency, Isolation, and Durability) guarantees, support for state machine replication, virtual synchrony, and support for only limited count of node failures [1]. Today's Smart Grid cloud infrastructures often "embrace inconsistency", thus implementing consistency preserving computational structures constitute a nascent thrust domain for the research and development sector.

1.3 Privacy and Security

The woeful protection services of current cloud deployments often stimulate the cloud vendors to recapitulate their security management folks to "not be evil". Stern efforts are in progress across the power system and transportation communities to come up with SG cloud utilities and platforms leveraged with robust protective contrivances where the stakeholders could entrust the storage of sensitive and critical data even under concurrent share and access architectures [15, 16].

1.4 Highly Assured Connectivity

Added with power outages, the Smart Grid consumers also experience intermittence in data connectivity. Projects for establishing mechanisms dedicated to support secured multipath data routing from user edges to cloud services are on headway. Critical components of the future Smart Grid applications demand better Quality of Service (QoS) and Quality of Experience (QoE) from the data routing backbone that underlie the cloud-hosted utilities.

Motivated by the abovementioned mission-critical Smart Grid requirements, the pitfalls associated with current cloud computing infrastructures to meet such needs, and having the assumption that the Smart Grid community is not in a position to reinvent a remotely owned Internet infrastructure or to develop computing platforms and elements from scratch, this work presents a FOG computing framework whose principle underlie on offloading the time and resource critical operations From cOre to edGe (FOG). The argument here is not to cannibalize the existing cloud support for SG, but to comprehend the applicability of FOG computing algorithms to interplay with the core-centered cloud computing support leveraged with a new breed of real-time and latency free utilities.

2 FOG Architecture for Smart Grid

This work presents a three schema computing architecture where the significant portions of Smart Grid control and computations are nontrivially hybridized alongside the cloud computing support. The objective is to overcome the disruption caused by the development of IoT utilities where the control, storage, networking, and computational needs are actively proliferated across the edges or end-points. The lowermost schema, namely physical schema primarily comprises of a wide range of smart IoT enabled devices which come within the SG domain. For simplicity, the entities are abstracted into logical clusters of applications, directly, or indirectly influenced by the expediency of SG operations.

The first cluster (C1) represents vehicular applications where the intelligent vehicles are arranged to form vehicular fogs. The existing transportation telematics support such as cellular telephony, on-board sensors (OBS), road-side units (RSU), and smart wearable devices will uncover the computational as well as networking capabilities latent in the underutilized vehicular resources. The notion is to employ the underutilized vehicular resources into communicational and analytics use, where a collaborative multitude of end-user clients or near-user edge devices carry out communication and computation, based on better utilization of individual storage, communication and computational resources of each vehicle [5]. Similarly, the similar presence of clusters (C2) could also be traced in smart home networks that have a noteworthy contribution in consistent operations of the backend SG support. The intelligent IoT equipped home gadgets such as washing machines, AC, freezes, parking lots, CC camera, etc., are also potentially active to provide storage, analysis, and computational support for satisfying the prompt and local decision-making services. The third but not the least, cluster **C3** depicts similar structure that can be constituted by utilities involved at the extreme ends of an SG infrastructure viz. micro-nano grid, PLCs, automated circuit breakers, and other entities associated to diverse range of SG generation, transmission, and distribution services. The smart nodes within such clusters sense and cultivate the heterogeneous physical attributes and transmit it to the upper layers through dedicated edge gateways. However, the whole or a portion of data generated within these physical clusters are accumulated at the interim across access points such as Global Positioning (GPS), GIS, Road-Side Units (RSUs), Remote Terminal Units (RTUs), Intelligent Electronic Devices (IEDs), Phasor Data Concentrators (PDC), and other field arrays (Fig. 1).

The next tier constitutes the FOG computing layer comprising of intelligent fog devices such as SCADA, smart meters, routers, switches high-end proxy servers, intelligent agent and commodity hardware, etc., having peculiar ability of storage, computation and packet routing. The Software-Defined Networking (SDN) assembles the physical clusters to form virtualized Inter-Cluster Private Networks (ICPNs) that route the generated data to the fog devices spanned across the FOG computing layer. The fog devices and its corresponding utilities form geographically distributed virtual computing snapshots or instances that are mapped to lower layer devices in order to serve the processing and computing demands of corresponding physi-

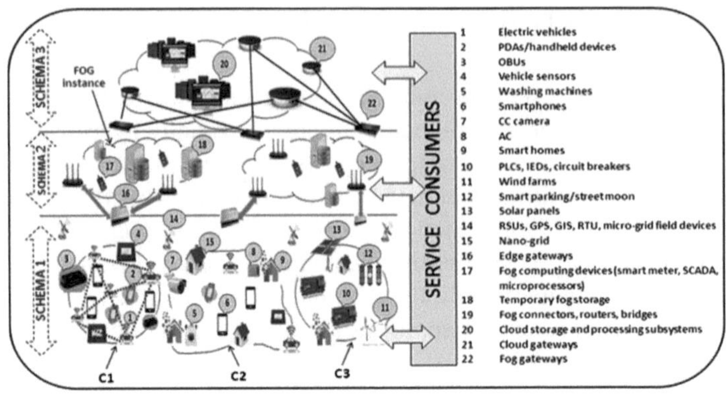

Fig. 1 Topology of FOG computing paradigm in a smart grid

cal entities or logical clusters. However, the location dynamics or mobility patterns of intelligent entities such as smart vehicles, OBUs, and other Smart Grid sensors make the FOG instances inflexible, heterogeneous, and dynamic. The FOG instances undertake the processing and analytics tasks in framing decision-making standards of when and what data to be offloaded to the upper cloud computing tier [14]. Generally, the bulky and resource exhaustive tasks are redirected to the cloud data centers for distributed processing. The cloud schema is primarily responsible for permanent and voluminous data chunks for high-performance computing and bulky analysis activities [17].

3 Service-Oriented Architecture for FOG Computing Layer

The SG use cases and requirements discussed in above sections nail down to devise a Service-Oriented Software Architecture (SOA) for defining the networking, storage and analytics activities of FOG elements defined in schema 2. An ideal SOA should meet the following key goals corresponding to FOG computing services.

3.1 Support for Heterogeneity and Interoperability

It should facilitate seamless resource management of heterogeneous FOG nodes deployed in diverse environments and under computational landscapes such as core, edge, RAM support, storage structures, end terminals, etc. It should provide ubiquitous support for Smart Grid applications from different programmers and developers to be compatible interoperable across heterogeneous platforms. Moreover, the dynamic FOG instances in Smart Grid infrastructures formed from heterogeneous network and computing elements such as sensors, actuators, and Machine-to-Machine (M2M) architectures demand high-bandwidth communication backbone. The communication links and protocols should provide resiliency support and access mechanisms for data transfer to and from enterprise data centers or clouds.

3.2 Support for Abstraction and Virtualization

The architecture should expose generic APIs that will hide the platform heterogeneity and provide a uniform and programmable interface for seamless resource management of multitudes of FOG computing applications across various SG domains and verticals such as microgrid, electric vehicle infrastructures, smart homes, etc. The APIs should be capable of monitoring, provisioning and controlling physical resources such as CPU, memory, network, and energy. In order to support multi-tenancy in the FOG computing paradigms and to enable improve resource utilization by incorporating the capabilities to run multiple operating systems or service containers on same physical machine, the architecture must have robust virtualization support. The software architecture will also have provisions for guaranteeing the security, privacy, consistency, and isolation policies for service execution among the tenants.

3.3 Support for Service Orchestration

The FOG software framework should provide efficient policy-based service orchestration utilities in order to undertake scalable and reliable management and execution of individual SG subsystems. Intelligently programmed software architecture will potentially assure distributed orchestration services that are primarily motivated to behold the dynamic, policy-based lifecycle management of FOG properties as well as resiliency of Smart Grid services. Moreover, a coherent lifecycle management platform will offer abilities for composing, configuring, dispatching, activating and deactivating, adding and removing, and updating the SG applications.

4 Illustrative Use-Case-the Smart Grid

A smart grid probably offers the richest use case to demonstrate the FOG computing scenario. The intent of this paper is to abstract the major FOG computing requirements that would get exposed in the emerging IoT-aware SG infrastructures. The SG manifests itself as a computationally full-fledged system envisioned by numerous subsystems such as Connected Rail Transportation (CRT), microgrid, and elec-

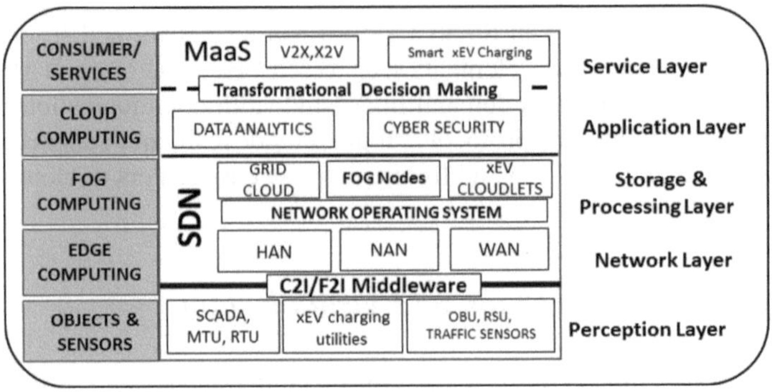

Fig. 2 Service-Oriented Architecture (SOA) of FOG computing model in a Smart Grid-based transportation system

trically driven Intelligent Vehicular Systems (IVSs), etc., that calls for potentially enough requirements for compute and analytics platform at the edge. The major goals of an IoT-aware connected rail network that FOG computing paradigms are purposed for include real-time location updates, accident prevention, steady flow of rail traffic, historical analytics utilities for continuous evaluation and feedback, etc. While the first two cases are location-constrained services and necessitate uncompromised real-time response, the third one requires a pseudo real-time. The historical analysis part relates to result obtained after collection and processing of global data over the desired lapse of time. For such instances, where the infrastructure requires a noble mix of offline/online, real-time/aggregated analytics, and decisions, FOG computing paradigms complemented with cloud computational backends will be the best recourse. The CRT FOG standards support multi-tenancy to anticipate an integrated hardware architecture and software platform for streamlining deployment of intelligent and situation-aware services.

Similarly, the microgrid subsystems expose characteristics and requirements shared by multitudes of Internet of Things/everything (IoT/IoE) deployments viz. interplay between real-time and batch analytics, stringent interaction between sensors and actuators and so in closed control loops, wider geo-distributed deployments of autonomous microgrid utilities, etc. Recent utility-scale developments have enabled installation of microgrid equipment such as large and flexible turbine structures, massive solar panels equipped with amenities for effectively harnessing wind and solar power, improving solar generation density, reducing structural loading of wind turbines, etc. A solar plant may consist of series of individual panels or a wind farm may include thousands of peer wind turbines, whose performance can be enhanced by integrating intelligent microcontrollers, sensors, and actuators deployed at the desired sites. Such devices can also be utilized to realize a global coordination framework whose services are smart and optimal. Such infrastructure requires a robust communication network and synchrony among the smart grid subsystems (FOG), between the system components and the internet at whole (cloud) (Fig. 2).

A Smart Grid or microgrid microcontroller having a global scope will utilize data fed from the assigned subsystems to determine and implement desired decision policy for each domain. Depending on the geo-distribution of each utility, such assignments may be global or local. Both variants require low latency compute, storage, and analytics services at edges of the utility network. Further, the IoT-aware SG systems generate voluminous datasets which are majorly actionable in real-time. Beyond such time constrained requirements, a significant portion of data in the warehouse demand analytics over aged (days, months, or years), recent or a mix of both categories of datasets and over wider scenarios such as solar panels, wind farms, or other forecasting or historical data. For such computational requirements, high-performance computationally intensive algorithms need to be implemented in cloud data centers for batch processing and analytics. Such infrastructural setups require middleware that mediates between FOG applications at the edges while cloud computing at the core.

5 Adoption Challenges and Future Prospects

The proposed SG FOG computing architecture due to lack of viable technologies will face difficulties in realizing its vision successfully. IoT-based vehicular FOG nodes and data clouds need to be designed intelligently to enable wide scale deployment. However, contemporary implementation and design models, algorithms and mechanisms are still far from the vision of IoT and are confronted with many challenges. Fundamental challenges that need to be addressed include availability, reliability, security, efficiency, interoperability, etc. [5, 12]. The research for viable realization of next generation Smart Grid utilities is in progress but all are trapped in local loop. Some key challenges that may arise while executing FOG algorithms in the IoT-enabled SG infrastructures are given here.

5.1 Management

In the proposed computing model, there is a seamless trading of energy between each pair from any of power grid, smart charging stations, or aggregators, distributed sources (micro-nano grids and other nonconventional sources), smart vehicles, Advanced Metering Infrastructures (AMIs), etc. Further, in smart cities, the domain of renewables is extended to integrate smart homes, smart parking lots, etc. The power exchange or flow can be conceptualized as a complete graph, where nodes represent entities while edges denote the bidirectional power flow. Integration of IoT technologies into such infrastructure further complicates the setup, as entry of billions of smart sensing devices elucidates the need for robust and efficient communication frameworks. Concurrent management of bi-level (energy and data) transport, thus poses daunting issues for the configuration managers as well as service providers.

Lightweight remote management protocols needs to be standardized to build service agnostic schemes for a viable realization of low-level processing and storage deployment. Comprehensive attention into research projects and investments from governments, policymakers, academia, and industries are strongly welcomed into such disruptive innovation. Though some labs are involved conducting FOG research studies, integrating them with transportation frameworks is still in its infancy and is restricted within simulation phase only. Demonstrating the feasibility of prepared prototypes needs real-world applications and test beds, that are not executed on large scale.

5.2 Stakeholder's Reliance and Confidence

It is one of the most dominating factors that decide the frequency of adoption of FOG paradigms. For that purpose, robust business models as proposed in this work or even more improved and efficient prototypes need to be developed. It calls for future attention towards simulation and demonstration through real-world datasets; as such analysis will be more realistic to the FOG stakeholder in IoT aware environments.

5.3 Synchronization and QoS Guarantee

The IoT-enabled SG infrastructure unifies multiple dimensions and entities such as power grid, renewables, xEVs, etc., into a common platform whose processing and analytics demand also ranges across multiple magnitudes. Thus, there is need of robust synchronization mechanism to enable long and smooth running and operation of infrastructure. Evaluating the performance of transport service offered by the proposed SG infrastructure is again a cumbersome task. Real-world datasets need to be generated to assess whether the QoS metrics are guaranteed. Evaluation metrics or QoS parameters from smart charging and power system perspective are service loss probability, power tariffs; Mean Time to Failure (MTTF), expense in security subsystems, etc. from SG power delivery perspectives while from IoT viewpoint throughput, jitter, bandwidth, overhead, etc., are key parameters that capture the QoS and Quality of Experience (QoE). The extent to which the metrics can be realized also depends on how they are defined.

5.4 Interoperability

Enabling the population of heterogeneous IoT entities belonging to multiple platforms to be interoperable is another limitation to contemporary SG architectures. The products and standards delivered from applications and IoT, OEMs end should be interoperable to ensure the delivery of services to varying range of customers irrespective of hardware specifications. Furthermore, the objects connected through IoT demand a unified platform for delivering diversified services. Such platforms can also be provided as Transport as a Services (TaaS) for IoT applications.

5.5 *Reliability and Electromobility*

A reliable FOG service is the result of a sequence of executions performed at multiple schemas formed by smart grid, aggregators, edge networks, etc., the overall performance is decided by the cumulative operation characteristics across all levels. The objective function is to maximize the success rate of power delivery which is manifested in the form of power availability, economic power, minimum queuing delay, power quality, etc. Further, both the hardware and software modules of the energy and data transport network should be resilient to threats, thefts, intrusions, and failures as networks under these circumstances will cause unreliable perceptions and wrong decisions leading to disastrous consequences. As the behavioral dynamics of SG utilities random and uncertain, relieving the stakeholders out of inconsistencies while on the move is the prime premise of the proposed architecture.

5.6 *Scalability, Security, and Privacy*

The FOG computing set up in IoT aware SG infrastructure should be scalable, i.e., provide provisions for augmenting new devices, services, and functionalities, to allow more and more consumer penetration without any significant degradation in power quality and power delivery [12]. The FOG computing platforms must be modeled from the ground up to enable extendibility in services and operations. Presenting the data analytics daemons with automation and intelligence guarantees scalability in such substructures. The programmers and developers for both hardware and software components should design platform independent and compatible applications to allow the addition of functionalities and integrate with other technologies and standards. Since the underlying core of IoT-SG is based on trading of energy and transfer of data between galaxies of objects, lack of common security and privacy standard is a notable challenge for FOG installation. The integration of IoT will increase the transparency and ease the access control procedures, thus removing the possible intrusions in the power system. More research efforts are welcomed in this trend for it to mature. Efficient location routing of intelligent objects, sensors, charging stations, context-aware services through secure networking paradigms remains a nascent research thrust for the research community.

6 Conclusion

In this paper, an edge-centered FOG computing model is proposed for Smart Grids infrastructures. The work first exposes the inadequacies of pure cloud-based SG subsystem to meet the processing and computational requirements and proposes a model that ensures proper distribution and offloading of computations across the cores (cloud) as well as edges. For assessing the viability of the proposed framework, an SG use case is explored. The work also proposes a FOG computing Service-Oriented Architecture (SOA) for IoT-aware intelligent transportation systems running on SG backbone. Finally, the latent implementation challenges encountered while the realization of the proposed framework is identified along with budding research thrusts that call for efforts and investments in viable deployment of the FOG paradigms.

References

1. K. Birman, L. Ganesh, R. van Renesse, Running smart grid control software on cloud computing architectures 1. introduction : the evolving power grid, in *Computational Needs for the Next Generation Electric Grid* (2010), pp. 1–28
2. X. Fang, S. Misra, G. Xue, D. Yang, Smart grid—the new and improved power grid. IEEE Commun. Surv. Tutorials **14**(4), 944–980 (2012)
3. R. Emilia, R. Emilia, P.G.V. Naranjo, M. Shojafar, L. Vaca-cardenas, Big data over smartgrid-a fog computing perspective big data over smartgrid—a fog computing perspective, no. (2016)
4. A. Al-Fuqaha, M. Guizani, M. Mohammadi, M. Aledhari, M. Ayyash, Internet of things: a survey on enabling technologies, protocols, and applications. IEEE Commun. Surv. Tutorials **17**(4), 2347–2376 (2015)
5. X. Hou, Y. Li, M. Chen, D. Wu, D. Jin, S. Chen, Vehicular fog computing: a viewpoint of vehicles as the infrastructures. IEEE Trans. Veh. Technol. **65**(6), 3860–3873 (2016)
6. F.Y. Okay, S. Ozdemir, A fog computing based smart grid model **10**(6), 1–6 (2016)
7. F. Bonomi, R. Milito, J. Zhu, S. Addepalli, Fog computing and its role in the internet of things, in *Proceedings of the First Edition of the MCC Workshop on Mobile Cloud Computing* (2012), pp. 13–16
8. E. van der Zee, H. Scholten, Big data and internet of things: a roadmap for smart environments. Stud. Comput. Intell. **546**, 137–168 (2014)
9. Z. Xiao, Y. Xiao, S. Member, Security and privacy in cloud computing **15**(2), 843–859 (2013)
10. B. Davis, The 7 pillars of big data, *Pet. Rev.* no. 34–36 (2015)
11. S. Sarkar, S. Chatterjee, S. Misra, Assessment of the suitability of fog computing in the context of internet of things. IEEE Trans. Cloud Comput. **PP**(99), 1–14 (2015)
12. I. Stojmenovic, S. Wen, The fog computing paradigm: scenarios and security issues. Proc. 2014 Fed. Conf. Comput. Sci. Inf. Syst. **2**, 1–8 (2014)
13. B. Varghese, N. Wang, D.S. Nikolopoulos, R. Buyya, Feasibility of fog computing (2017)
14. Cisco Systems, Fog computing and the internet of things: extend the cloud to where the things are, Www.Cisco.Com, (2016), p. 6
15. Y. Wang, T. Uehara, R. Sasaki, Fog computing: Issues and challenges in security and forensics, Proc. Int. Comput. Softw. Appl. Conf. **3**, 53–59 (2015)
16. K. Lee, D. Kim, D. Ha, U. Rajput, H. Oh, On security and privacy issues of fog computing supported Internet of Things environment, *2015 Int. Conf. Netw. Futur. NOF 2015* (2015)
17. A.V. Dastjerdi, H. Gupta, R.N. Calheiros, S.K. Ghosh, Fog computing : principles, architectures, and applications, pp. 1–26

A Hybrid Approach for Energy-Efficient Task Scheduling in Cloud

Sunil Kumar and Mala Kalra

Abstract Task scheduling is a big problem in the cloud computing. As large number of users simultaneously request for the resources, the tasks must be allocated to resources rapidly and in an optimized manner. Energy consumption by data centers worldwide has increased tremendously. This leads to focus on developing eco-friendly and energy-efficient scheduling algorithms. A lot of techniques for energy-efficient task scheduling have already been proposed, yet there is a lot of room for improvements. We are presenting a hybrid of Genetic Algorithm and Artificial Bee Colony-based approach along with DVFS to achieve energy-efficient task scheduling. Empirical analysis of the proposed approach proves its better performance over Modified Genetic Algorithm with respect to makespan and energy consumption.

Keywords Artificial bee colony (ABC) · Cloud · Dynamic voltage and frequency scaling (DVFS) · Genetic algorithm (GA) · Task scheduling

1 Introduction

Cloud computing can be described as on-demand delivery of IT resources and IT applications through Internet using pay-as-you-go pricing. The National Institute of Standards and Technology describes cloud computing as a "model for enabling ubiquitous, convenient, on-demand network access to a shared pool of configurable computing resources (e.g. networks, servers, storage, applications and services) that can be rapidly provisioned and released with minimal management effort or service provider interaction" [1]. The cloud abstracts details of software and hardware implementation from the users and makes available unlimited resources through virtualization. Due to increase in demand of cloud infrastructure, there has been a

S. Kumar (✉) · M. Kalra
Department of Computer Science and Engineering, NITTTR, Chandigarh, India
e-mail: suniel.vns@gmail.com

M. Kalra
e-mail: malakalra2004@yahoo.co.in

C. R. Krishna et al. (eds.), *Proceedings of 2nd International Conference on Communication, Computing and Networking*, Lecture Notes in Networks and Systems 46, https://doi.org/10.1007/978-981-13-1217-5_99

lot of increase in consumption of energy by cloud data centers. More energy consumption leads to more operating cost as well as more carbon emissions. Carbon dioxide emissions of the Information and Communication Technology industries are presently estimated to be two percent of the global emissions [2]. On an average rate, a data center requires energy as equivalent to 25,000 households [3]. Therefore, green computing comes into the light because it reduces operational costs as well as minimizes environmental impacts. The processor requires much more energy when compared to other system resources. Along with this, idle server also consumes a lot of energy; an idle server on average requires roughly 70 percent of power required by server running at maximum speed of CPU. Energy required by data centers has grown by 56% from 2005 to 2010 worldwide, and in 2010, is estimated to be between 1.1 and 1.5% of the total energy consumption [4]. All these lead to focus on developing eco-friendly and energy-efficient cloud computing systems.

The rest portion of the paper is arranged as follows; Sect. 1 gives Genetic Algorithm introduction and Artificial Bee Colony introduction, Sect. 2 gives related works, Sect. 3 details proposed approach, Sect. 4 discusses experimental setup, whereas Sect. 5 analyzes the achieved results. Section 6 concludes the paper with scope for future work.

1.1 Genetic Algorithm (GA)

The GA uses the biological idea of population creation. Important entities used in GA are population size, fitness function, chromosomes selection, chromosomes crossover, and mutation. The population size can be defined as the total search space of individuals which are known as chromosomes. From initial population size, some chromosomes are selected based on some criteria and mating takes place between them. The fitness function measures the quality of chromosomes based on the objective parameter of the scheduling. Selection of any individual will take place according to its fitness value. All chromosomes are ranked by their fitness value. Crossover is the process in which two parent chromosomes produces child chromosomes which could have better fitness value as compared to their parents. Mutation is the process which increases the uniqueness of individuals in the population.

1.2 Artificial Bee Colony (ABC)

In ABC, there are three categories of artificial bees: first, are employed bees assigned with food sources, second, are onlooker bees monitor the employed bees for choosing a food source, and third, are scout bees looking for food sources. The location of foods in ABC shows the probable solution to the problem. Initially, all locations of the foods are initialized by scout bees. Employee bees calculate the fitness of neighbor food sources. Employee bees tell about their food sources to onlooker bees, and onlooker bees take their decisions based on received information. Scout bees are those who choose their food sources randomly. The employee bees whose results do not improve after certain attempts becomes scout bees and their results are destroyed. The new scout bees begin to look for new food sources randomly. This process repeatedly takes place for all tasks to be allocated to virtual machines until criterion for termination is achieved [5].

2 Related Work

Singh et al. [6] proposed a Modified Genetic Algorithm in which starting population of chromosomes is produced using an improved max–min algorithm. The authors have taken makespan as the fitness function. The authors have tested their algorithm taking two scenarios. In the first scenario, virtual machines number is kept four and the number of tasks is varied. In the second scenario, virtual machines number is kept eight and the number of tasks is varied. The authors have compared their approach with enhanced max–min, IGA, and GA-LCFP. Kumar et al. [7] discussed assigning tasks to the resources using improved GA. The authors have reported that initial population in GA should be generated using min–min and max–min approach. Authors have compared their approach with the standard genetic algorithm. The authors have also reported that their approach is producing better resource utilization. Sindhu et al. [8] reported reducing makespan and increase of processor utilization. The authors have generated the initial population using four different methods like random, LCFP, SCFP, and MCT. The authors concluded that GA-LCFP provides an optimal solution over other approaches by simulating in CloudSim. Shojafor et al. [9] presented a combination of metaheuristics approaches for optimally balancing load considering cost and execution time. In this paper, fuzzy logic is applied along with GA for allocation of jobs on the resources. Beloglazov et al. [10] presented an environment-friendly scheduling algorithm. They used dynamic consolidation of VMs for reducing energy consumption during resource allocation. Using simulation outputs, they proved their method dominates static resource allocation techniques in reducing energy consumption in clouds data centers. Kaur et al. [11] gives a modified method of GA with the combination of GA with LCFP and SCFP for scheduling task. The authors compared their algorithm with the standard genetic algorithm based on makespan and execution cost.

3 The Proposed Algorithm

The proposed hybrid approach is being discussed here. Start allocation process with GA, where the population is the number of tasks.

1. Initialize population of GA by randomly allocating VMs to tasks.
2. Evaluate candidates.
3. Repeat until termination condition occurs.

   a. Take best chromosomes.
   b. Perform linear crossover over selected chromosomes.
   c. Mutate the resulting chromosomes.
   d. Evaluate new individuals.
   e. Make selection of candidates for next generation.

4. Start working with ABC to find the best VM for tasks to be allocated. The output of GA will be input for ABC algorithm.
5. Find distance between tasks and VMs and start working with employee bees, scout bees, and onlooker bees.

   5.1 Determine new food positions for employee bees.
   5.2 If onlooker bees are not distributed, then select from neighbor table.
   5.3 Memorize position for onlooker bees.
   5.4 If Food is not searched for the first time
      a. Scout_bee = Employee_bee;
      b. Employee_bee = Onlooker_bee;
      c. Onlooker_bee = TaskIndex;
   5.5 Do until all tasks are allocated to virtual machines.

6. Initialize DVFS energy-aware model.
7. Find the makespan and energy consumption of schedule.
8. Output the list of tasks allocated to VMs.

The DVFS energy consumption model is as given in [12]. In DVFS, the computation of power for any processing element can be obtained using the below expression:

$$P(V, f) = aCV^2 f \tag{1}$$

Here, V stands for voltage, f stands for frequency and "a", C are frequency multipliers. From the above expression, it is clear that the decrease in frequency and voltage implies that less computation power is consumed. $P_{\text{idle}}(V, f)$ is power consumption in an idle state of CPU and $p_{\text{busy}}(V, f)$ is the power consumption when CPU is in a busy state.

$$p_{\text{idle}}(V, f) = aCV_{\text{idle}}^2 f_{\text{idle}} \tag{2}$$

$$p_{\text{busy}}(V, f) = aCV_{\text{busy}}^2 f_{\text{busy}} \tag{3}$$

where V_{idle} represents voltage and f_{idle} represents frequency in the idle state of the CPU, V_{busy} represents voltage, and f_{busy} represents frequency in busy state of the CPU.

E_{busy} is the energy consumed by the busy processor; an expression for E_{busy} is given by:

$$E_{\text{busy}} = \sum_{i=1}^{M} p_{i,\text{busy}}(V, f) * t(i) \tag{4}$$

where $t(i)$ is time duration for which the ith VM is in busy state and M represents number of VMs in the schedule. E_{idle} is the energy consumed by the idle processor; an expression for E_{idle} is given by:

$$E_{\text{idle}} = \sum_{i=1}^{M} p_{i,\text{idle}}(V, f) * t'(i) \tag{5}$$

where $t'(i)$ is time duration for which the ith VM is in an idle state.

4 Experimental Setup

The proposed work is executed in a cloud environment produced using CloudSim [13]. The execution time of tasks is calculated using VMs processing capacity and length (instruction count) of cloudlets. Table 1 gives the GA parameter settings, Table 2 gives the ABC parameter settings, Table 3 gives CloudSim parameter settings, and Table 4 provides parameter settings for DVFS in CloudSim.

Table 1 Parameter settings for GA

Parameters	Values
Population count	Total nummber of jobs
Maximum generation	40
Rate of crossover	0.5
Rate of mutation	0.1

Table 2 Parameter settings for ABC

Colony size	Total number of jobs
Onlooker bee number (n_o), Employed bee number (n_e)	50% of the colony
Maximum cycle number (MCN)	40
Limit	$n_e * 2$

Table 3 Parameter settings of CloudSim simulator

Type	Parameter	Value
Datacenters	Number of data centers	1
	Number of hosts	5
	Types of manager	Time shared
VM	Number of VMs	10–40
	Processing capacity	250–2200 MIPS
	VM memory (RAM)	128 MB
	Bandwidth	2500
	Policy	Time shared
Tasks	Number of tasks	10–40
	Length of tasks	200–4000

Table 4 Parameter settings for DVFS in CloudSim [14]

Frequency (in GHz)	1.6	1.867	2.133	2.40	2.67
Base frequency (2.67 GHz)	59.925	69.93	79.89	89.89	100
CloudSim (in MIPS)	1498	1748	1997	2247	2500
Power (in W) 0% (p_{idle})	82.7	82.85	82.95	83.10	83.25
Power(in W)100%(p_{busy})	88.77	92	95.5	99.45	103

5 Results Analysis

We are comparing our proposed hybrid approach with the Modified Genetic Algorithm (MGA) [6]. In the MGA, the authors have generated initial population using modified approach. The individuals have been evaluated based on makespan.

Case 1: **Tasks fixed (40) and varying number of VMs (Based on Makespan)**
In this case, we are comparing MGA and hybrid approach by keeping number of tasks as 40 and varying number of VMs from 10 to 40. The numbers of VMs are marked on X-axis and makespan values are marked on Y-axis.
From Fig. 1, we can say that as number of VMs increases, makespan reduces rapidly in case of MGA but its value is always greater than the hybrid approach's makespan value, if the number of tasks is fixed. The highest difference of makespan between MGA and hybrid approach is 84.00% when there are 10 VMs and 40 tasks.

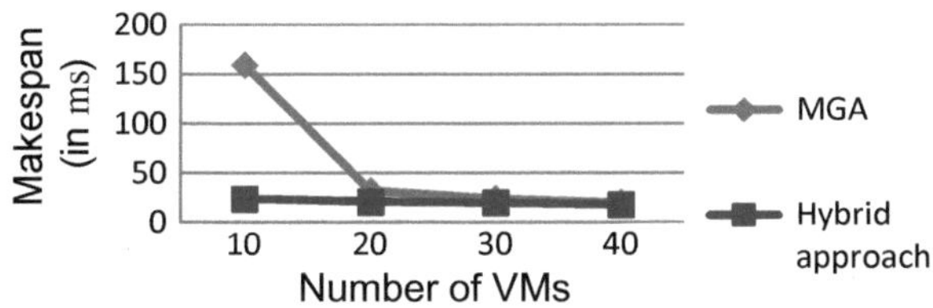

Fig. 1 Makespan for fixed tasks and varying VMs

Fig. 2 Makespan for fixed VMs and varying tasks

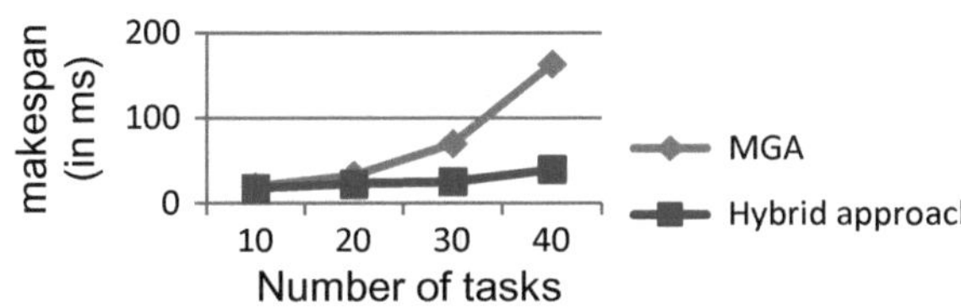

Case 2: **Number of VMs fixed (10) and varying number of tasks (Based on Makespan)**

In this case, we are comparing MGA and hybrid approach by taking 10 numbers of VMs and changing number of tasks from 10 to 40. The numbers of tasks are marked on X-axis and makespan values are marked on Y-axis.

From Fig. 2, we can conclude that as number of tasks increases, makespan also increases but rate of increase in hybrid approach is slower than MGA if number of VMs are fixed. The highest difference of makespan between MGA and hybrid approach is 75.5% when number VMs are 10 and number of tasks are 40. The lowest difference of makespan between MGA and hybrid approach is 1.00% when there are 10 VMs and 10 tasks.

Case 1: **Number of VMs fixed (10) and varying no. of tasks (Based on Energy)**

In this case, we are comparing MGA and hybrid approach by taking 10 numbers of VMs and changing number of tasks from 10 to 40. The numbers of tasks are marked on X-axis and total energy consumption values are marked on Y-axis.

From Fig. 3, we can conclude that as number of tasks increases, total energy consumption also increases in both the approaches but the increase in MGA is more than in hybrid approach. The highest difference of total energy consumption between MGA and hybrid approach is 81.14% when number of VMs are 10 and number of tasks are 40. The lowest difference total energy consumption between MGA and the hybrid approach is 56.61% when numbers of VMs are 10 and number of tasks are 10.

Case 2: **Tasks fixed (40) and varying number of VMs (Based on Energy)**

In this case, we are comparing MGA and the hybrid approach by taking 40 numbers of tasks and changing number of VMs from 10 to 40. The number of VMs is marked on X-axis and total energy consumption values are marked on Y-axis.

From the Fig. 4, we can conclude that as number of VMs increases, energy consumption reduces in MGA and in the hybrid approach, total energy consumption

Fig. 3 Energy consumption for fixed VMs and varying tasks

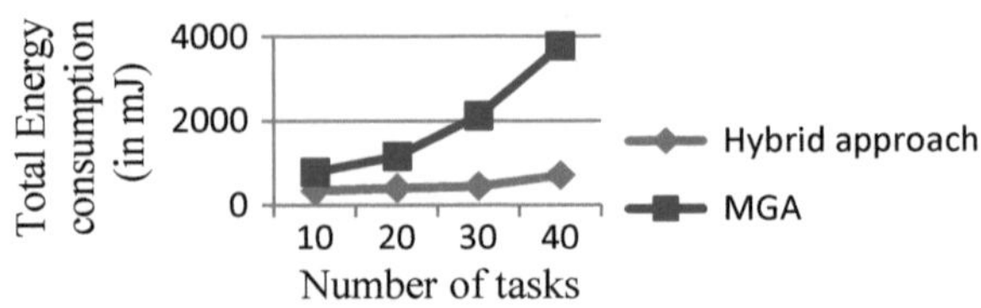

Fig. 4 Energy consumption for fixed tasks and varying VMs

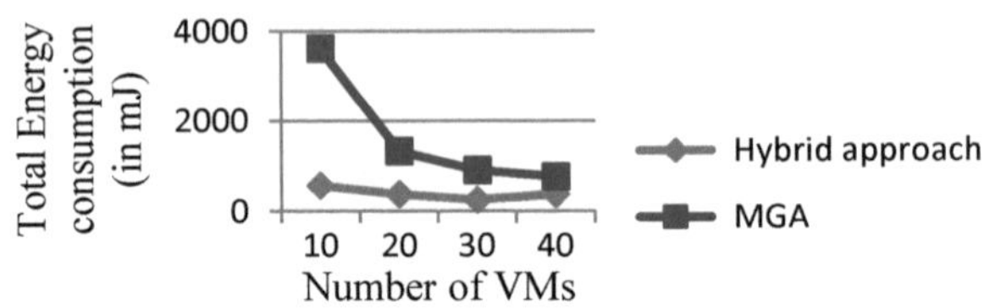

first decreases and then increases but its total energy consumption is always lesser than the MGA. The highest difference of total energy consumption between MGA and the hybrid approach is 84.14% when the numbers of VMs are 10 and numbers of tasks are 40. The lowest difference of energy consumption between MGA and the hybrid approach is 51.29%.

6 Conclusion and Future Scope

This work presented a novice approach which combines GA and Artificial Bee Colony algorithm and focuses on reducing makespan and energy consumption. The DVFS power model has been used for the calculation of total energy consumption by the resources. We have evaluated our proposed hybrid approach performance with Modified Genetic Algorithm based on makespan and total energy consumption. The experimental outputs show that our hybrid approach outperforms modified genetic algorithm. This work can be extended by considering execution cost in addition to existing objectives.

References

1. P. Mell, T. Grance, *The NIST Definition of Cloud Computing* (NIST, Gaithersburg, 2011)
2. J. Kaplan, W. Forrest, N. kindler, in *Revolutionizing Data Center Energy Efficiency*. Technical Report (Mckinsey & Company, July 2008)
3. Gartner Inc., in *Gartner Estimates ICT Industry Accounts for 2 Percent of Global CO_2 Emissions* (2007)
4. J. Koomey, *Growth in Data Center Electricity Use 2005 to 2010* (Analytical Press, Berkeley, 2011)
5. D. Karaboga, Artificial bee colony algorithm. Scholarpedia **5**(3), 6915 (2010)

6. S. Singh, M. Kalra, Scheduling of independent tasks in cloud computing using modified genetic algorithm, in *6th IEEE proceedings of International Conference on Computational Intelligence and Communication Networks* (2014), pp. 565–569
7. S. Sindhu, S. Mukherjee, A genetic algorithm based scheduler for cloud environment, in *4th IEEE proceedings of International Conference on Computer and Communication Technology* (2013), pp. 23–27
8. P. Kumar, A. Verma, Scheduling using improved GA in cloud computing for independent tasks, in *The proceedings of the International Conference on Advances in Computing, Communications and Informatics,* August 3–5 (2012), pp. 137–142
9. M. Shojafar, S. Javanmardi, S. Abolfazli, N. Cordeschi, FUGE: A joint meta-heuristic approach to cloud job scheduling algorithm using fuzzy theory and a genetic method. Cluster Comput. **18**(2), 829–844 (2015). Springer
10. A. Beloglazov, J. Abawajy, R. Buyya, Energy-aware resource allocation heuristics for efficient management of data centers for Cloud computing. Future Gener. Comput. Syst. **28**(5), 755–768 (2012). Elsevier
11. A. Kaur, A. Verma, An efficient approach to genetic algorithm for task scheduling in cloud computing environment. Int. J. Inf. Technol. Comput. Sci. **10**, 74–79 (2012)
12. I.T.C. Ruiz, R.P. Prado, S.G. Galan, J.E. Muñoz-Expósito, N.R. Reyes, Dynamic voltage frequency scaling simulator for real workflows energy-aware management in green cloud computing. Public Libr. Sci. **12**(1), e0169803 (2017)
13. R.N. Calheiros, R. Ranjan, A. Beloglazov, C.A.F. De Rose, R. Buyya, CloudSim: a toolkit for modeling and simulation of cloud computing environments & evaluation of resource provisioning algorithms. Softw. Pract. Experience **41**(1), 23–50 (2011)
14. T. Guérout, T. Monteil, G.D. Costa, R.N. Calheiros, R. Buyya, M. Alexandru, Energy-aware simulation with DVFS. Simul. Model. Pract. Theory **39**, 76–91 (2013)

Improving the Performance of Pre-copy Virtual Machine Migration Technique

Aditya Bhardwaj and C. Rama Krishna

Abstract Virtualization is the key technology running behind the cloud servers to deploy users services efficiently. So, when a user requests for cloud services, resources are allocated in the form of VM instance. With the tremendous growth in cloud services, the number of applications running over cloud servers is increasing. Therefore, the chances of server overloading and maintenance are quite often. This scenario is handled by live VM migration, that allows migration of VMs from one host to another with minimum disruption to cloud user services. Cloud hypervisors use pre-copy technique to perform the VM migration task. However, it suffers from page-level content redundancy problem, where complete VM memory page is transferred instead of modified contents of the page. Therefore, a technique is proposed which transfers only modified contents of VM memory page. The performance of proposed technique is evaluated with VMs running read-and write-intensive workloads configured on QEMU-KVM virtualized environment. The results obtained demonstrate that for read-intensive workload, compared to existing pre-copy VM migration scheme, our proposed technique minimize downtime by 69.4%, migration time by 73.5%, and reduces network traffic generated by 74.21%. Similarly, for write-intensive workload, downtime, migration time, and network traffic generated are reduced by 47.23, 49.16, and 51.37%. Thus, the proposed technique effectively improves the performance of VM migration scheme.

Keywords Cloud computing · Virtualization · Pre-copy VM migration · Content redundancy · Downtime · Migration time · Network traffic

---

A. Bhardwaj (✉) · C. Rama Krishna
Department of Computer Science & Engineering, NITTTR, Chandigarh, India
e-mail: aditya.cse@nitttrchd.ac.in

C. Rama Krishna
e-mail: ramakrishna.challa@gmail.com

C. R. Krishna et al. (eds.), *Proceedings of 2nd International Conference on Communication, Computing and Networking*, Lecture Notes in Networks and Systems 46, https://doi.org/10.1007/978-981-13-1217-5_100

1 Introduction

In the last decade, cloud computing came out as one of the popular computing paradigm by offering storage, computing power, and software services based on utility computing. Virtualization at the cloud data center has gained a lot of attention to deploy cloud services efficiently. This is achieved with the help of a hypervisor which abstracts the underlying physical resources. Virtualization enables cloud service providers to create multiple operating systems on a single physical machine. Virtualization offers advantages such as server consolidation, dynamic resource management, hardware optimization, heterogeneous system operation, faster provisioning time, and dynamic load balancing. Popular hypervisor used by cloud service providers are kernel-based Virtual Machine (KVM) [1], Citrix XenServer [2], Microsoft Hyper-V [3], and VMware ESXi [4]. KVM is an open-source hypervisor and it is widely adopted by the research communities. Therefore, in our study, we setup virtualization environment using KVM hypervisor.

Cloud computing inherits one of the key features provided by server virtualization called Virtual Machine (VM) migration. It is defined as, "a process of transferring VMs from one server to another to achieve load balancing, fault tolerance, server consolidation and server maintenance". Cloud admin triggers VM migration from one data center to another if there is any chance of server overloading, fault in server, or maintenance of data center is required. Pre-copy is a widely used VM migration technique adopted by cloud service providers. In this mechanism, memory pages are transmitted from source to destination host in a iterative manner. During the iterative process, because VM is in running state at source server, so its memory pages gets dirtied before their transfer to the destination host is completed. To maintain the consistency between source and destination VM, the pages modified during the previous iteration are retransmitted. This process is repeated until any of the conditions are met: (i) number of iterations exceeds to 30 or (ii) number of dirtied pages reaches to predefined threshold. After this, the VM is stopped at source server and migrated to destination with minimum downtime. In the existing pre-copy VM migration technique, complete VM memory page is transferred instead of modified contents of the page. This leads to increase in migration time, downtime, and network traffic generated during the migration. To address these issues, we investigated page-level content redundancy to improve the VM migration performance.

The main contribution of this paper is as follows:

- A technique is proposed to reduce the contents of VM memory pages transferred.
- Testbed cloud virtualization platform is designed and implemented using QEMU-KVM open-source hypervisor technology.
- Experiments are performed using real-time workloads to test the VM memory pages dirtied rate.
- To eliminate the overhead of storage synchronization, Network File System (NFS) server is configured.

- Finally, the performance of proposed technique is evaluated with existing pre-copy VM migration scheme in terms of migration time, downtime, and network traffic generated.

This paper is organized as follows: Section 2 provides recent "Related Work". Section 3 presents "Background on virtual machine migration. Section 4 provides "Proposed VM migration technique". "Cloud VM migration testbed setup" is described in Sect. 5. "Comparison and results with respect to existing pre-copy scheme," are highlighted in Sect. 6, and finally in Sect. 7, "Concluding remarks" are presented.

2 Related Work

Although, recent studies have primarily focused on applying migration technique to balance the load in cloud computing data centers, there has been little attention on the performance improvement of VM migration process while taking into account when VM pages are transferred [5].

In [6], a CloudNet-based bandwidth optimization VM migration scheme is proposed to trigger migration across the Wide Area Network (WAN) links. CloudNet uses Content-Based Redundancy (CBR) scheme to fragment memory blocks into fixed-size regions. To calculate hash value, the author proposed scheme uses SuperFastHash mechanism. Furthermore, the delta-based method is used to ensure transfer of only single instance of memory page. Network connection is managed by Virtual Private Network (VPN). Storage consistency is ensured using synchronous and asynchronous migration policy. However, CloudNet-based approach has the limitation of large migration storage and WAN latencies.

Hirofuchi et al. [7] proposed SimGrid, an open-source simulation framework to simulate virtual machine migration environment. This simulator supports resource allocation to each computational task. It enables experiments with a number of VMs launched on their simulation framework and controls these VMs like in real-time framework. VM support is provided by adding virtual workstation module which is responsible for resource management in a Physical Machine (PM) or VM layer. This simulator also supports simulation of VM migration model with some limited features. However, this simulation toolkit lacks the consideration to support simulation mechanism of memory update pattern in order to correctly simulate migration behavior of virtual machines. Furthermore, bandwidth allocation for a VM request is not considered.

In Zhang et al. [8], explored the problem of frequently varied bandwidth requirement by the widely used pre-copy VM migration method. To solve this issue, the authors theoretically calculate the amount of bandwidth that should be allocated to guarantee the VM migration completion time. A novel transport protocol rSAB (Rate-Aware Bandwidth Sharing by Allocating Switch Buffer) has been developed that satisfy the bandwidth requirement of VM migration process. From this study, it is observed that VM migration traffic contends for maximum bandwidth but allocating maximum bandwidth leads to bandwidth wastage. However, this study lacks the consideration of page dirtying rate pattern in order to overcome the performance degradation of existing VM migration technique.

In [9], the authors proposed web-based system to manage KVM-based cloud virtualization environment. In this system, libvirt API has been used that supports communication with cloud management hypervisors. This system has been developed by using Java Server Pages (JSP) and Servlet technologies, JSP enables the view of web pages and Servlets acts as a controller for the developed system. To store the disks of VMs, share storage is used that allows VMs migration across the network. This study considered only graphical web-based mechanism for cloud virtualization environment but it lacks the consideration of VM migration implementation on shared storage.

In [10], the authors proposed remote backup approach named LiveRB to save the running state of VMs. This approach is based on the idea of virtual disk device which cache I/O operations in memory and then save virtual machine disk data to the remote server. When any application write VM memory pages, LiveRB records these memory pages and then transfer the dirty pages to remote server. This work solved issues regarding VM running state backup, but the limitation of this approach is that still dirty pages are sent repeatedly resulting in increase in network traffic.

3 Background on Virtual Machine Migration

In cloud computing, there are two types of migration techniques that are implemented in hypervisor and plays an important role for migrating VMs from one data center to another. These techniques are categorized as "pre-copy", and "post-copy". In post-copy approach, the first execution is transferred and then memory, i.e., VMs CPU states are transmitted to the destination machine and after that VM is resumed on destination. If the memory pages requested by VM context are not found at destination host, then page fault is triggered to obtain the pages from source host. Repetition in page faults at different time intervals during execution of VM is a main drawback of post-copy VM migration approach. So, pre-copy is widely used for cloud data center virtualization due to its reliability and less downtime [11]. Therefore, in this work, we also considered pre-copy approach.

3.1 Pre-copy VM Migration Technique

Pre-copy VM migration is executed mainly in three phases: (i) memory copy phase, (ii) iteration phase, and (iii) stop-and-copy phase [12]. In this technique, the memory pages allocated to the VM are iteratively copied to the target Physical Machine (PM) while its execution at the source PM continues. In the first phase, complete

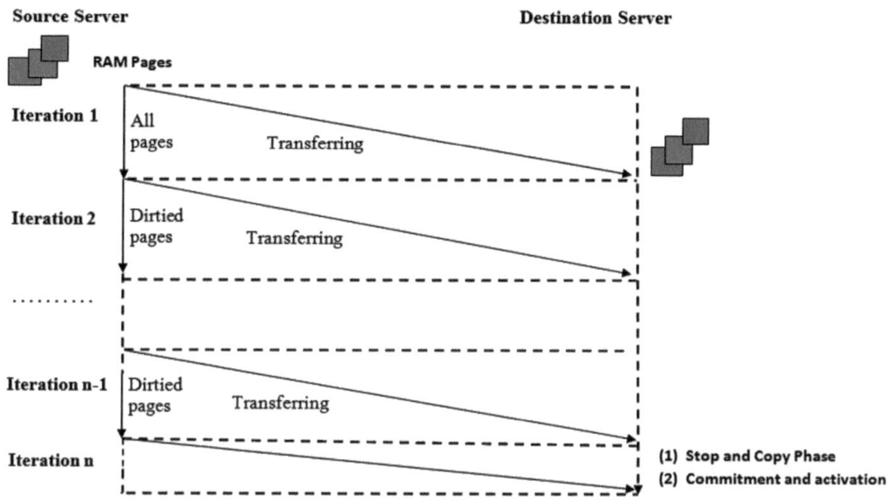

Fig. 1 Iterative rounds in pre-copy VM migration technique

memory pages allocated to VM are transferred from source to the target PM, while its execution at the source PM continues [13]. During the transfer of memory pages in memory copy phase, some portion of the VM memory pages gets modified. The modified memory pages are repeatedly sent to the target PM through several iteration phases [14]. When the migration iteration count reaches maximum limit, the execution of the VM at the source PM is suspended and remaining dirtied pages and VM's CPU states copied to the target server in stop-and-copy phase. The migrated VM then relinquishes all the consumed resources at source and starts its execution at the target PM. The performance of this approach is measured using migration time and downtime. The logical steps for VM migration process using pre-copy technique are shown in Fig. 1.

4 Proposed VM Migration Technique

Existing pre-copy VM migration technique suffers from page-level content redundancy problem [15, 16]. In other words, during the iterative phase of pre-copy VM migration, the virtual machine memory pages are transferred over the network from source to the destination host. As the VM continues to run at the source host, VM memory pages gets modified before its transfer to the destination host is completed. For read-intensive workloads, usually only a couple of VM page's contents are modified. In the existing pre-copy VM migration technique, complete VM memory page is transmitted instead of modified contents of the page. This leads to increase in downtime, migration time, and network traffic generated during the migration.

The migration time (T_m) shown in Eq. 1, can be calculated as summation of all the push phase duration time:

$$T_m = \sum_{k=1}^{N} \frac{V_{m,k}}{b_{0,k}} \left(1 + \sum_{i=1}^{n_{max-1}} \prod_{l=1}^{i} \frac{d_{l-1,k}}{b_{l,k}} \right) \quad (1)$$

and downtime (T_d) can be calculated as shown in Eq. 2:

$$T_d = \sum_{k=1}^{N} \frac{V_{m,k}}{b_{0,k}} \left(\prod_{l=1}^{n_{max}} \frac{d_{l-1,k}}{b_{l,k}} \right) \quad (2)$$

4.1 Reducing Contents of VM Memory Pages Transferred

The idea of our proposed technique is that instead of transferring complete VM memory page only modified contents of the page should be transferred from source to target server. This is because during the iterative rounds only fraction part of the page is dirtied. At the source server, modified contents of VM page is computed by applying logical operator XOR for the old and current version of VM memory page. At the destination, the original page is reconstructed by performing XOR operation between the received contents and page stored in system cache.

Systematic steps of proposed technique implementation as a modification to QEMU-KVM hypervisor source code is depicted in Fig. 2.

5 Cloud VM Migration Testbed Setup Using QEMU-KVM Hypervisor

To perform live VM migration, we conducted experiment using *virt-manager* managed cloud hypervisor testbed experimental setup which comprises of three HP servers running Linux Redhat 6.5 with kernel version 2.6.32-431, QEMU-KVM v 1.7 hypervisor, and libvirt 2.0.0 [17] on a Intel Core i7 (3rd Gen) 3770/3.4 GHz processor, with 8 GB RAM storage. Physical servers are connected to the Ethernet Local Area Network (LAN) interface for operating VM migration, whereas VMs are connected using virtual bridge interface to manage the networking complexity. To eliminate the overhead of storage synchronization during the migration process, Network File System (NFS) server is installed on one server which is used for sharing the migrated VM's disk image as depicted in Fig. 3. The other two servers are installed with VMs. The migrated VM is configured with Ubuntu 16.04 Linux kernel v3.8.0-29, one logical processor and 4 GB of RAM. These VMs are running Idle

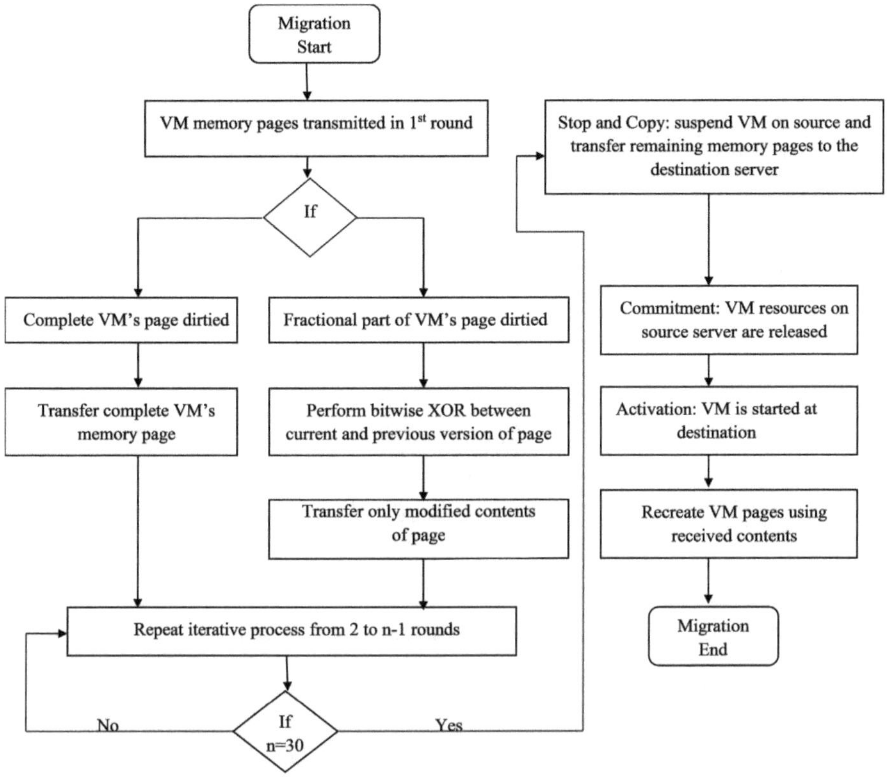

Fig. 2 Flowchart for reducing contents of VM memory pages transferred

VM OS pages, LMbench, SPECjvm2008, and kernel compile workload to generate data traffic and simulate cloud user application benchmarks.

5.1 Steps to Setup QEMU-KVM VM Migration Environment

In order to set up QEMU-KVM VM migration environment, we designed the following methodologies:

- Step1: A secure SSH link is setup between source and target server.
- Step2: VM is started at the source host, and channel bandwidth parameter is configured for the QEMU-KVM hypervisor.
- Step3 : Cloud workload benchmarks are run on the VMs. Transfer all VM pages in the first round. VM instance is still active during first iteration, so VM pages gets updated during the first iteration. Transfer these updated VM pages iteratively until threshold criteria such as maximum number of iterations are reached.

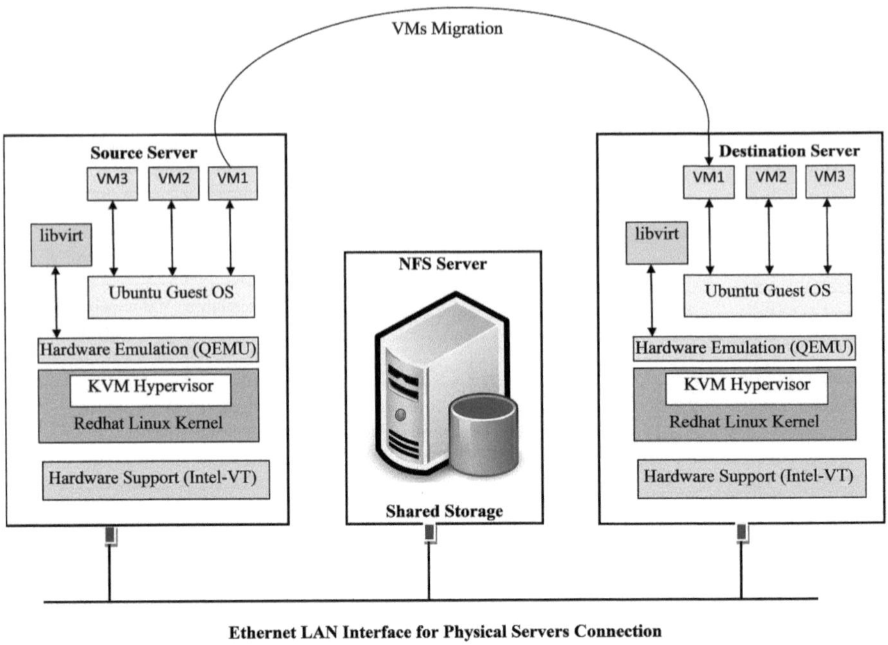

Fig. 3 VM migration experimental setup using QEMU-KVM Hypervisor

- Step4: When the hypervisor triggers for stop-and-copy condition (e.g., number of updated pages reached to a predefined value), then VM is suspended and transfer remaining pages to the target server.
- Step5: Now, VM connection is terminated from the source server.
- Step6: Finally, resume VM at the target server and migration is completed successfully.

The command to migrate VMs is triggered from the source server. Once this command is received by the hypervisor, a socket connection is set up between the source and target server. After this, when the reply is received from the target server VM migration is started.

6 Performance Evaluation with Existing Pre-copy VM Migration Technique

To evaluate the performance with existing pre-copy VM migration scheme [18], we executed four types of test workloads, namely Idle, LMbench, SPECjvm, and kernel compile. "Idle" indicates the test case when no application is running at the VM. "LMbench" is a UNIX-based benchmark that represents high dirty rate. It is designed to produce database workloads. "SPECjvm" represents the class of applications running on cloud data centers that belongs to high compute-intensive workload. This type of workloads aims to investigate the system performance with emphasis on processing capacity of hardware system. Finally, "kernel-compile" workload represents write-intensive category. In this benchmark, kernel source code is compiled at the VM.

For read-intensive workload, i.e., when VM is running low dirty rate benchmark like "Idle", page's content modification rate is less. As a result, pre-copy scheme performs poorly. This is because unmodified contents are also transmitted from source to target server. For write-intensive workloads although page modification rate is high, but VM page's bits which are not dirtied is also transferred. To resolve both of these issues, our proposed technique transfers only modified contents of VM memory page.

6.1 Downtime-Based Performance Evaluation

Figure 4 presents downtime-based performance comparison between pre-copy and proposed VM migration technique. Downtime represents unresponsive time for the VM. Downtime is calculated using ping requests from source to target server. Figure 4a, b shows that downtime is minimum for our proposed technique because by transferring only the modified contents of the page, the VM pages in the last iteration will be migrated with shorter duration.

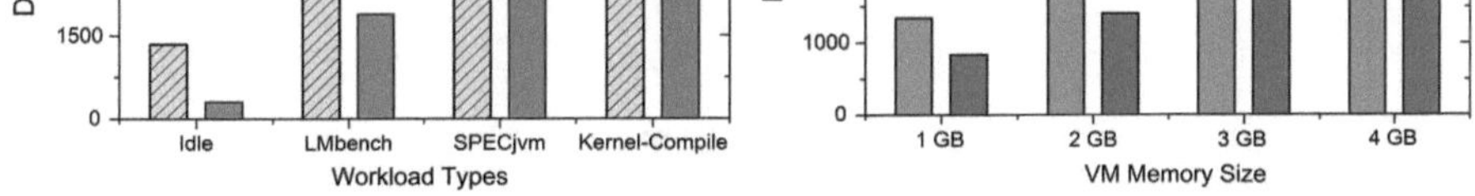

(a) T_d Downtime for diverse workloads

(b) T_d Downtime for diverse VM memory size

Fig. 4 Downtime results for pre-copy and proposed VM migration technique

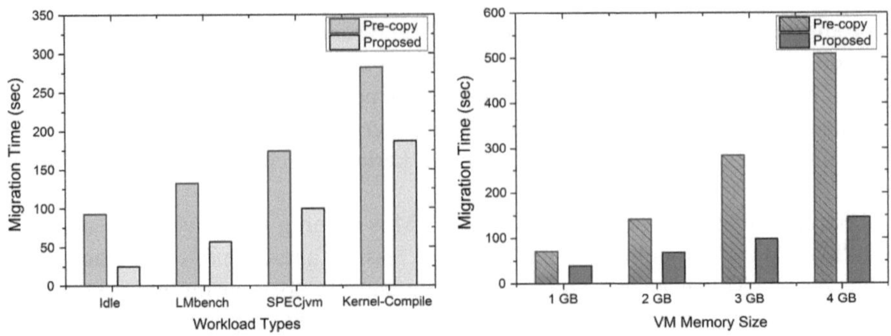

(a) T_m Migration time for diverse workloads

(b) T_d Migration time for diverse VM memory size

Fig. 5 Migration time results for pre-copy and proposed VM migration technique

6.2 Migration Time-Based Performance Evaluation

Figure 5 presents migration time-based performance comparison between pre-copy and proposed VM migration technique. Migration time depends upon the amount of data transmitted during iterative process. Figure 5a, b shows that migration time is minimum for our proposed technique because by eliminating unmodified data transfers, the migration process is completed with shorter duration.

6.3 Network Traffic-Based Performance Evaluation

Figure 6 presents performance evaluation in terms of network traffic generated by both of these techniques. Network traffic is caused due to the amount of data transmit-

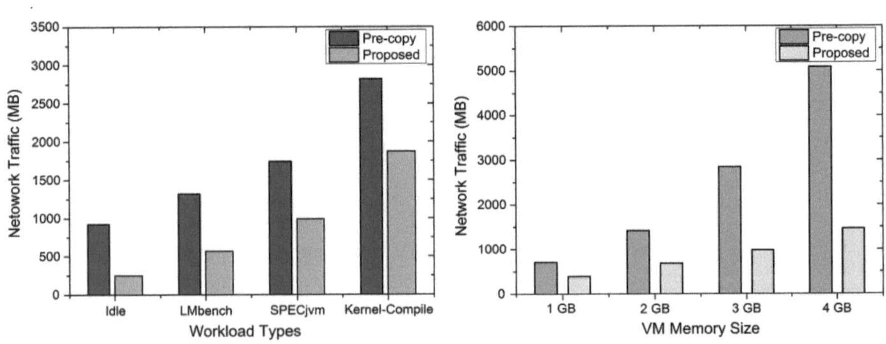

(a) Network traffic for diverse workloads

(b) Network traffic for diverse VM memory size

Fig. 6 Network traffic results for pre-copy and proposed VM migration technique ted during migration mechanism. Figure 6a, b shows that network traffic is minimum for our proposed technique because by eliminating unmodified data during VM memory pages transfer, our approach reduces the data transmitted from source to target server. Thus, it results in less network traffic generated.

7 Conclusion

VM migration is a fundamental operation for cloud virtualization environment. However, its performance is limited by page-level content redundancy problem where complete VM memory page is transferred instead of modified contents of the page. Therefore, in this paper, a technique is proposed to identify modified contents of VM memory page and transfers only the dirtied contents. First, we presented cloud virtualization testbed environment using QEMU-KVM hypervisor. Then, evaluated the performance of proposed technique compared to existing pre-copy VM migration scheme. The experiments are demonstrated by running different types of workloads benchmarks. From the results obtained, it is found that compared to existing pre-copy VM migration scheme, our proposed technique minimize downtime by 69.4%, for read-intensive workload and 47.23% for write-intensive benchmark. Similarly, migration time is reduced to 73.5% for low dirty rate and 49.16% for high dirty rate applications. Also, the amount of network traffic generated is reduced to 74.21% for read and 51.37% for write-intensive workloads. Therefore, the results obtained can help cloud service providers to meet the required goal of efficient deployment of cloud services.

References

1. Kernel Virtual Machine (2017), https://www.linux-kvm.org/page/Downloads//. Accessed 10 April 2017
2. XEN Hypervisor (2017), https://www.xenproject.org/developers/teams/hypervisor.html//. Accessed 26 Sep 2017
3. Microsoft Hyper-V (2017), https://technet.microsoft.com/en-us/hyper-v-server-docs/hyper-v-server-2017//. Accessed 26 Oct 2017
4. VMware (2017), https://www.vmware.com/products/esxi-and-esx.html//. Accessed 15 Oct 2017
5. Z. Zhang, L. Xiao, M. Zhu, L. Ruan, Mvmotion: a metadata based virtual machine migration in cloud. Cluster Comput. **17**(2), 441–452 (2014)
6. T. Wood, K.K. Ramakrishnan, P. Shenoy, J. Van der Merwe, J. Hwang, G. Liu, L. Chaufournier, CloudNet: dynamic pooling of cloud resources by live WAN migration of virtual machines. IEEE/ACM Trans. Netw. **23**(5), 1568–1583 (2014)
7. T. Hirofuchi, A. Lebre, L. Pouilloux, SimGrid VM: virtual machine support for a simulation framework of distributed systems. IEEE Trans. Cloud Comput. **6**(1), 1–14 (2015)
8. J. Zhang, F. Ren, R. Shu, T. Huang, Y. Liu, Guaranteeing delay of live virtual machine migration by determining and provisioning appropriate bandwidth. IEEE Trans. Comput. **65**(9), 2910–2917 (2016)

9. M.R. Ms, N. Patel, K. Sharma, Distributed virtualization manager for KVM based cluster. Procedia Comput. Sci. **79**, 182–189 (2016)
10. Z. Wang, J. Zeng, T. Lv, B. Shi, B. Li, A remote backup approach for virtual machine images, in *2016 IEEE 3rd International Conference on Cyber Security and Cloud Computing (CSCloud)* (IEEE, 2016), pp. 252–255
11. F. Xu, F. Liu, H. Jin, A.V. Vasilakos, Managing performance overhead of virtual machines in cloud computing: a survey, state of the art, and future directions. Proc. IEEE **102** (2014). https://doi.org/10.1109/JPROC.2013.2287711
12. H. Liu, H. Jin, X. Liao, C. Yu, C.Z. Xu, Live virtual machine migration via asynchronous replication and state synchronization. IEEE Trans. Parallel Distrib. Syst. **22**(12), 1986–1999 (2011)
13. R. Tomar, A. Khanna, A. Bansal, V. Fore, An architectural view towards autonomic cloud computing, in *Data Engineering and Intelligent Computing* (Springer, 2018), pp. 573–582
14. J. Thaman, M. Singh, SLA conscious VM migration for host consolidation in cloud framework. Int. J. Commun. Netw. Distrib. Syst. (Inderscience Publishers) **19**(1), 46–64 (2017)
15. J. Liu, Y. Li, D. Jin, SDN-based live VM migration across datacenters, in *ACM SIGCOMM Computer Communication Review*, vol. 44, no. 4 (ACM, 2014), pp. 583–584
16. H. Maziku, S. Shetty, Network aware VM migration in cloud data centers, in *Research and Educational Experiment Workshop (GREE), 2014 Third GENI*. (IEEE, 2014), pp. 25–28
17. libvirt Virtualization API (2017), https://libvirt.org. Accessed 06 Aug 2017
18. W. Fan, Z. Zhang, T. Wang, B. Hu, S. Qing, D. Sun, Research on security algorithm of virtual machine live migration for KVM virtualization system, in *International Conference on Information and Communications Security* (Springer, 2016), pp. 54–70

Improved Active Monitoring Load-Balancing Algorithm in Cloud Computing

Pawan Kumar and Rakesh Kumar

Abstract The primary job of load balancing in cloud is to allocate millions of incoming user requests on available distributed resources to increase cloud performance. With increasing number of user requests and demand for computing resources, it becomes a big challenge for load balancer to manage all tasks properly to maintain both Quality of Service (QoS) and performance in cloud. Improper distribution of tasks may decrease the utilization of resources as well as performance of cloud and situation of underutilization or overutilization may occur. Load balancing helps to overcome performance degradation problems and to increase resource utilization. So, it becomes essential to use an efficient load balancing algorithm in data center to select virtual machines for tasks execution. In our work, we have presented a load-balancing algorithm which allocates the tasks on suitable virtual machines based on load and availability of virtual machine. We implemented and evaluated proposed algorithm on cloud analyst simulator. We analyzed the results and found that the proposed algorithm provides higher resource load balancing and performance.

Keywords Load balancer · Cloud analyst · Cloud data center · Userbase · Virtual machine

1 Introduction

In modern paradigm of technology, cloud provides various services which are available on-demand in powerful data centers. Due to pay- as-you-go nature of cloud, it is attracting companies and IT professionals to use cloud services and to shift their business on cloud. According to NIST, "cloud computing is a model that allows us to use shared pool of resources (such as network, applications, storage, etc.) in a

P. Kumar (✉) · R. Kumar
NITTTR, Chandigarh, India
e-mail: pawanyadav235@gmail.com

R. Kumar
e-mail: raakeshdhiman@gmail.com

C. R. Krishna et al. (eds.), *Proceedings of 2nd International Conference on Communication, Computing and Networking*, Lecture Notes in Networks and Systems 46, https://doi.org/10.1007/978-981-13-1217-5_101

convenient way and also allows to increase or decrease with minimal interaction of service providers [1]". Cloud provides the following basic services: Infrastructure as a Service (IaaS), Platform as a service (PaaS), and Software as a Service (SaaS) which are enabled in data centers in virtual form [2, 3].

Cloud provides computing resources using concept of virtualization in the form of virtual machines (VMs) running on physical servers [4, 5]. With the help of virtualization, a user request can be distributed on different servers to provide QoS and lower response rate. A data center consists of large number of VMs to process user requests, and it becomes necessary for service providers to allocate these tasks on suitable VM to increase resource utilization.

Load balancer in cloud provides the facility to manage incoming user requests on available resources [6–8]. Load balancer helps to minimize overutilization or underutilization situation for resources by allocating the user requests to best suitable VMs available in data center which leads to improvement in performance of cloud. Many researchers have developed various load-balancing algorithms like Round Robin, Active Monitoring, and Throttled load balancing algorithms, optimal resource allocation [9] which have different performance criteria.

In this paper, we proposed an improved active monitoring load-balancing approach which allocates resources to user requests based on current state and load of VMs. Response time is considered as main evaluation parameter for the proposed algorithm as well as in earlier developed techniques. The proposed technique improves the existing active monitoring load-balancing algorithm in cloud analyst [10] and gives better results as compared to other existing load-balancing algorithms.

In rest of the paper: Sect. 2, presents research work corresponds to load-balancing techniques. In Sect. 3, we discussed the proposed load-balancing approach and flowchart of load- balancing algorithm. Section 4 describes experimental setup and the evaluation of the proposed algorithm. Conclusion and future work are presented in Sect. 5.

2 Related Work

A plenty of work has been done in the field of load balancing. Here, we discuss some research literature on load balancing that has been done for improving cloud performance.

Domanal et al. [11] presented an optimal load- balancing algorithm on the availability of VM in data center based on round-robin technique to distribute the workload on all available VMs. When a new request comes, load balancer checks next available VM in data center rather than the load on a VM which may lead to underutilization of VMs and more power consumption also. We need to invent some load-balancing technique that also considers remaining load of VMs along with VMs having high processing power to increase the performance and resource utilization in cloud.

Soni et al. [12] proposed central load balancer which selects VM from data center based on state and preference of VMs. According to authors, the technique will calculate the priority of machine depending on CPU speed & memory availability and arrange them in priority order. It selects a VM from same data center and if most

VMs are busy, then it will transfer the request to low priority VM. It will lead to high response time and lower the performance of cloud. So, we need to develop an algorithm which allows to execute tasks on another data center also.

Ajit et al. [13] proposed a Signature-based load- balancing technique that minimizes response time to the user request. They described using simulation on cloud analyst tool that their proposed technique outperformed the existing Round Robin and Throttled algorithms in terms of response time. But the algorithm has not evaluated and compared other load-balancing techniques such as active monitoring load balancer, optimal resource allocator which provides better result compared to existing techniques in simulator.

In [14], researchers proposed a Stochastic hill climbing technique for distribution of tasks and to maximize the optimization of resources on cloud computing. The authors developed the algorithm based on soft computing load-balancing concept. They evaluated the algorithm on simulator and the results stated that response time in the proposed algorithm is minimum as compared to existing techniques such as First-Come-First-Serve and Round Robin.

In [15], the authors developed an algorithm, namely Enhanced Equally Load-Balancing Algorithm (EEDLBA) that dynamically allocates the load on virtual machines and they showed with the help of simulation that their algorithm gives better response time in comparison with other algorithms as Round Robin, server-based load for Internet-distributed services, Scheduling techniques for Load distribution on VMs, and task scheduling algorithm.

Sharma et al. [16] discussed basic load-balancing approaches based on different metrics. This research work also gives the direction to design a new algorithm on the basis of different metrics by analyzing the behavior of various existing algorithms. The Author has analyzed various static load-balancing techniques, which are not useful in cloud environment where estimation of workload in advance is not possible due to continuous request from users. They have also discussed some dynamic load-balancing algorithms.

Behal et al. [17] presented a load-balancing algorithm for performance evaluation heterogeneous cloud environment. They have evaluated Round Robin and Throttled in combination with service broker policies in cloud analyst simulator. Based on experiments, the authors analyzed that Throttled algorithm with optimized response time gives the best performance in heterogeneous cloud computing. They have not implemented these algorithms in real- time environment and evaluated only existing techniques in simulator. In [18], researchers compared various static and dynamic algorithms present in cloud computing. They presented the comparison based on the challenges present in cloud computing. They found that in many algorithms, Load-Balancing Min-Min (LBMM) uses less response time and it is more effective in terms of resource utilization parameters.

3 Proposed Load-Balancing Approach

In the proposed algorithm, load balancer allocates incoming user requests on least loaded VMs in data center. Service broker continuously detects user requests coming from Internet Cloudlet for execution. Service broker selects best suitable data center according to user requirements and forward requests to corresponding data center controller (DCC). DCC detects incoming user requests from service broker and process them initially for execution. Proposed load balancer preserves an index table of VMs consisting of virtual machine ID (VMid), State of VM, and current allocated load on VM in data center along with priority of VM to execute the task. As described in load-balancing architecture Fig. 1, when load balancer detects a request form DCC, it checks index table and identifies VM with least load and higher priority. If more than one VM is detected, it selects the VM with lower memory utilization. After selecting a VM, it sends corresponding VM id to DCC which allocates VM to the task for execution and load balancer updates index table accordingly. After completion of task, load balancer accepts request from DCC to vacate VM and update index table.

The proposed load-balancing algorithm is described below:

Step 1: Information of available virtual machine ID (VMid), state of VM, and number of currently allocated requests to VM are stored in an index table.

Step 2: Data center receives new user requests and send to data center controller (DCC).

Step 3: DCC queries improved load balancer for next user request allocation.

Step 4: Load balancer go through the VM index table and identifies VMs with least load along with available state.

Step 5: If load balancer identifies list of more than one VM.

- (i.) Improved load balancer identifies VM with highest priority along with memory utilization of each VM.
- (ii.) Load balancer returns VMid of VM with highest priority and least load to DCC.
- (iii.) DCC notifies load balancer for VM allocation
- (iv.) Load balancer updates index table accordingly.

Step 6: Load balancer returns VMid to DCC allocated to task.

Step 7: DCC transfers user request to selected VM for execution.

Step 8: Once task execution is complete, DCC sens request to load balancer to de-allocate VM.

Step 9: Load balancer updates index table accordingly.

Step 10: If more requests, go to step 3.

4 Results and Discussion

Here, we used different configurations for evaluating the proposed policy and existing algorithms using various number of data center and user bases. In the below tables, five user bases and six data centers are taken for evaluation. In the below tables, configuration for user base in Table 1, data centers in Table 2, and other configurations in Table 3 are shown. These configurations are used to evaluate the experiment and results.

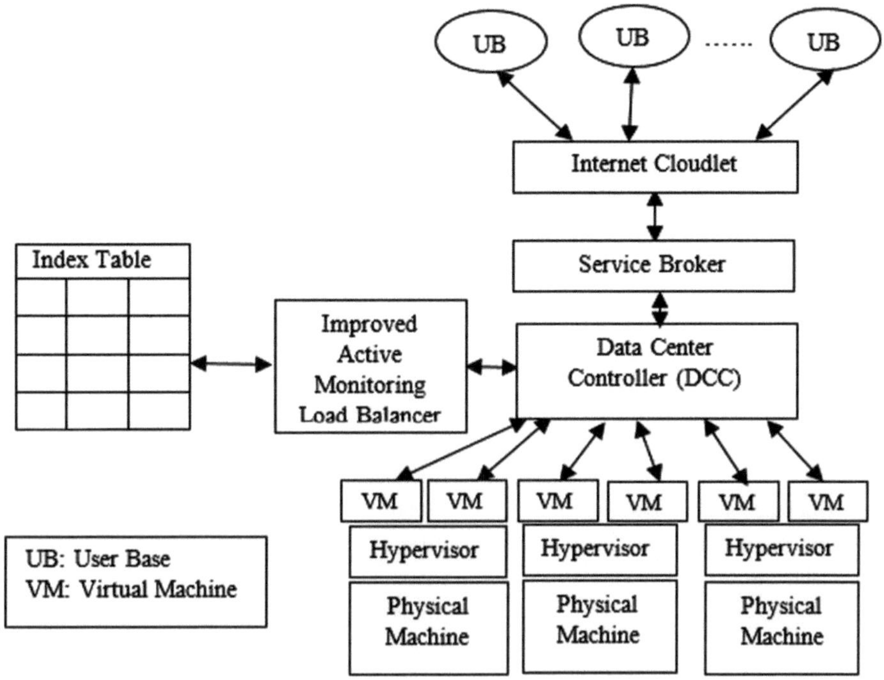

Fig. 1 Architecture of Improved Load-Balancing Approach

Table 1 User base configuration

Userbase name	Region	No. of per user request per hour	Data size per request (Bytes)	Peak hours	Average users on-peak	Average users off-peak
UB1	0	80	200	3:00–6:00	500,000	100,000
UB2	1	80	200	3:00–6:00	300,000	80,000
UB3	2	80	200	5:00–8:00	400,000	75,000
UB4	3	80	200	8:00–11:00	200,000	40,000
UB5	4	80	200	9:00–12:00	100,000	10,000

4.1 Experimental Setup

In experimental setup, we configured five userbases from different regions along with separate off-peak and on-peak hours as shown in Table 1 in cloud analyst simulator. We have configured six data centers having various configuration of VMs in Table 2. Some other parameters used in data center configuration are shown in Table 3.

Table 2 Data center configuration

DC name	Region	Cost/VM ($/h)	Data transfer cost ($/GB)	Speed (MIPS)
DC1	0	0.05	0.05	20,000
DC2	0	0.1	0.1	10,000
DC3	0	0.12	0.12	15,000
DC4	0	0.15	0.15	20,000
DC5	0	0.20	0.20	17,000
DC6	0	0.18	0.18	16,000

Table 3 Other parameters used

Parameter	Value used
Service broker algorithm	Optimize response time
Simulation duration	24 hour
User grouping factor in userbase	1000
Request grouping factor	100
Executable Instruction Length	500
VM image size	10,000
VM bandwidth	1000
Data center architecture	×86
Data center—Memory per machine	2048 MB
Data center—Processor speed	10,000 MIPS
Data center processor	4

4.2 Result Analysis

For evaluation of the proposed algorithm, we integrated it in cloud analyst with existing load-balancing techniques. To evaluate performance of developed technique, we have considered two different cases. In the first case, we evaluated techniques with constant data per request and varying number of VMs in data center as shown in Fig. 2. Then in the second case, we evaluated the technique by increasing the size of data size per request as shown in Fig. 3.

As shown in the results, we observed that the proposed improved active monitoring load-balancing algorithm provides reduced response time as compared to other existing techniques.

5 Conclusion and Future Work

In cloud data center condition of overloading and under loading of VMs may occur, if the proper load distribution between VMs is not done, then it may lower the performance of cloud. In the proposed algorithm, we attempted to evade the condition of resource underutilization. Improved load-balancing algorithm distributes the tasks among VMs in an effective way based on VMs current load, priority, memory utilization, and state. Consequently, the proposed algorithm effectively distributes the user requests between different available VMs which increased both resource utilization and performance of cloud.

In the future, we need to advance the proposed technique to distribute user requests among the VMs according to their reliability and processor utilization and also

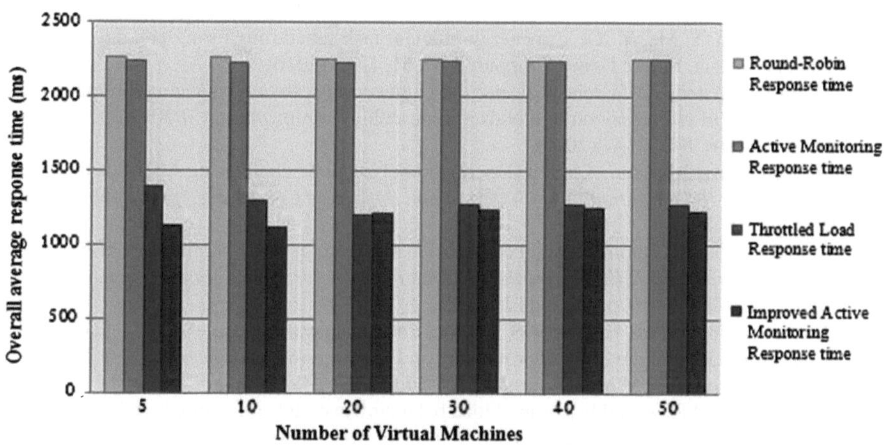

Fig. 2 Comparison of load-balancing algorithm when load is constant

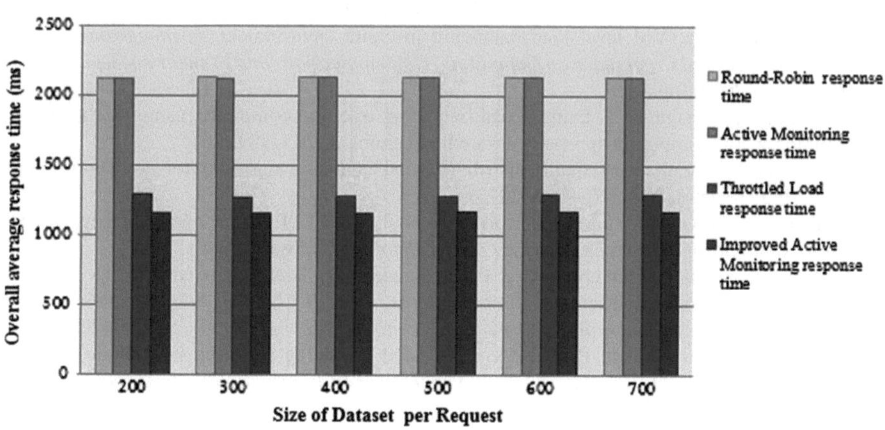

Fig. 3 Comparison of load-balancing algorithm when number of VMs is constant

consider some more parameters such as execution cost of VM and data transfer cost to increase the performance as well as minimize usage cost for cloud users.

References

1. T. Mell, T. Grance, Future of Health Insurance, vol. 145 (National Institute of Standards and Technology, Information Technology Laboratory, 2011), pp. 7
2. M. Armbrust, A. Fox, R. Griffith, A.D. Joseph, R. Katz, A. Konwinski, G. Lee, D. Patterson, A. Rabkin, I. Stoica et al., A view of cloud computing. Commun. ACM **53**(4), 50–58 (2010)
3. R. Buyya, J. Broberg, M.A. Goscinski, *Cloud Computing: principles and paradigms*, vol. 87 (Wiley, 2010)
4. Y, Xing, Y, Zhan, Virtualization and cloud computing. Future Wirel. Netw. Inf. Syst. 305–312 (2012)
5. M. Forsman, A. Glad, L. Lundberg, D. Ilie D, PT US CR. J. Syst. Softw. (2014)
6. T. Liu, F. Chen, Y. Ma, Y. Xie, An energy-efficient task scheduling for mobile devices based on cloud assistant. Future Gener. Comput. Syst. **61**, 1–12 (2016)
7. E. Pacini, C. Mateos, C. Garcia, Advances in engineering software balancing throughput and response time in online scientific clouds via ant colony optimization (SP2013/2013/00006). Adv. Eng. Softw. **84**, 31–47 (2015)
8. M. Suresh, S. Karthik et al., A load balancing model in public cloud using ANFIS and GSO, in *proceedings of IEEE International Conference on Intelligent Computing Applications (ICICA)* (2014), pp. 85–89
9. P. Kumar, Kumar, R, Optimal resource allocation approach in cloud computing environment," in *proceedings of 2nd IEEE International Conference on Next Generation Computing Technologies (NGCT)* (2016), pp. 112–117
10. B. Wickremasinghe, N.R. Calheiros, R. Buyya, Cloudanalyst: a cloudsim-based visual modeller for analysing cloud computing environments and applications, in *Proceedings of 24th IEEE Advanced Information Networking and Applications (AINA)* (2010), pp. 446–452
11. G.S. Domanal, M.R.G. Reddy, Optimal load balancing in cloud computing by efficient utilization of virtual machines, in *proceedings of 6th IEEE International conference on Communication Systems and Networks (COMSNETS)* (2014), pp. 1–4
12. G. Soni, M. Kalra, A novel approach for load balancing in cloud data center, in *proceedings of IEEE International Advance Computing Conference (IACC)* (2014), pp. 807–812
13. M. Ajit, G. Vidya, VM level load balancing in cloud environment, in *proceedings of 4th IEEE International Conference on Computing, Communications and Networking Technologies (ICCCNT)* (2013), pp. 1–5
14. B. Mondal, K. Dasgupta, P. Dutta, Load balancing in cloud computing using stochastic hill climbing-a soft computing approach. Procedia Technol. **4**, 783–789 (2012)
15. S. Mulay, S. Jain, Enhanced equally distributed load balancing algorithm for cloud computing. Int. J. Eng. Technol. **2**(6), 976–980 (2013)
16. E. Sharma, S. Singh, M. Sharma, Performance analysis of load balancing algorithms, in *38th World Academy of Science Engineering and Technology* (Citeseer, 2008)
17. V. Behal, A. Kumar, Cloud computing: Performance analysis of load balancing algorithms in cloud heterogeneous environment, in *proceedings of 5th IEEE International Conference on Confluence the Next Generation Information Technology Summit* (2014), pp. 200–205
18. B. Ghutke, U. Shrawankar, Pros and cons of load balancing algorithms for cloud computing, in *Proceedings of IEEE International Conference on Information Systems and Computer Networks (ISCON)* (2014), pp. 123–127

Analyzing Privacy Issues in Cloud Computing Using Trust Model

Parmod Kalia, Sanjeev Sofat and Divya Bansal

Abstract In the ever-increasing usage of Internet, a new paradigm of pervasive and ubiquitous cloud computing has evolved for providing computing delivery services. This has infused an increased interest amongst researchers and industry/user due to its ease in scalability and provisioning of computing resources on demand. It has caused concerns amongst its users on security, privacy, and trust-related issues. Therefore, it is essential to address these concerns, especially the privacy issues. In this paper, privacy issues concerning the cloud user are analyzed using trust model for taking effective measures in protecting the privacy of cloud users.

1 Introduction

Cloud computing has emerged as an increasingly convenient and transparent method of utilizing computing resources according to necessity/requirement. Its salient features such as scalability, accessibility, and availability of procuring/obtaining IT Services comprising of software-based, platform-based, and infrastructure enables individual as well as organizations to adopt it [7]. However, many organizations are apprehensive about its usage with the fear of losing confidentiality and privacy. The lack of trust in cloud computing prevents its wider acceptance by the users. Since users do not have direct control over the data, therefore, evaluating trust over the security and privacy issues would be significant and crucial concern of these organizations. The secure access of personal data and processing of this data on the cloud causes a huge challenge. It is essential that the identity of user and his/her sensitive data residing in the cloud must be protected as the data and information came

P. Kalia (✉) · S. Sofat · D. Bansal
Punjab Engineering College, Chandigarh, India
e-mail: parmod_kalia@hotmail.com

S. Sofat
e-mail: sanjeevsofat@pec.ac.in

D. Bansal ·
e-mail: divya@pec.edu.in

C. R. Krishna et al. (eds.), *Proceedings of 2nd International Conference on Communication, Computing and Networking*, Lecture Notes in Networks and Systems 46, https://doi.org/10.1007/978-981-13-1217-5_102

from different and diverse sources. The exchange of data/information over the cloud should be processed only by the authorized user authenticated to avail and access the information considering the prevailing privacy regulations [1, 17].

2 Privacy Challenges and Issues in Cloud Computing

Privacy has been made as a fundamental right by the United Nations Universal Declaration of Human Rights (1948), which was in due course adopted in the European Convention on Human Rights and UK Human Rights Act 1998 [2, 11]. In cloud computing, the nature of data/information comes from different and diverse sources. Security and privacy of data/information becomes an inevitable and massive challenge. The key challenges depicted in the Table 1 Ruiter and Warnier [18] and Buyya et al. [5] are:

Table 1 Key challenges in cloud computing

Type of challenge	Understanding
Security and privacy	It is a common fear among the users that the hosting of personal sensitive data on cloud maintains and controlled by the Cloud Service Provider (CSP) entails lack of privacy
Interoperability and portability	Migration of data from and into the cloud and changeover providers whenever they require, with no lock-in period
Reliability and availability	Data transfer, i.e., data movement/transfer between countries having different privacy procedures and regulations is a big challenge privacy
Performance and quality of service assurance	Sufficient bandwidth is required for effectively delivering intensive and complex data over the network, which is a costly affair these days
Governance and control	Risk of losing control or governance by organizations using Cloud computing services is high
Complexity	Establishing cloud services by the user is another challenging task
Lack of resource/expertise	Lack of expertise and resources makes protection of user privacy in the shared cloud environment more challenging
Compliance and disclosure	Government surveillance and access of databases, unwarranted mandatory disclosure of privacy databases causes huge concern to its users

3 Universal Privacy Principles

On the basis of privacy laws as applicable, the US developed fair information practices [11] in 1970 for the protection of the privacy of an individual, business/organization. Subsequently, these practices were as later adopted by the Organization for Economic Cooperation and Development (OECD) and the Council of Europe [8, 12] and declared as Universal Privacy Principles. The classification of these principles is as follows:

1. **Limitation of data collection**: defines that the data/information must be collected/assembled legally but restricted as per need with consent of the data subject.
2. **Quality of data**: means that data gathered should be relevant and pertinent to the need.
3. **Purpose detailing**: specifies that the purpose for which the data is collected should be clearly well defined.
4. **Use limitation**: defines that the personal and sensitive information should be used and limited only for the purpose for which it is collected and not for other purpose unless consent of the data subject is obtained.
5. **Security**: signifies that the personal and sensitive data should be well guarded and secured.
6. **Openness**: defines that the data subject should be able to find out the personal sensitive details comprising the data and to what extent it is used.
7. **Individual participation**: ensures that data collected by the data controller can be viewed and challenged if it is found to be incorrect.
8. **Accountability**: signifies that it is the accountability of data controller to comply with these principles.

4 Related Works

In recent times, researchers have explored the concept of trust model in cloud computing environment. They have carried out studies usually on basis of mutual agreement between stakeholders, use of signature certificate/secret key, feedback, etc. Each trust model has its advantages and disadvantages. Some of the related work done about security, privacy, and trust in the cloud are reviewed below in the segment.

Marsh [15] is one of the first to present concepts on trust and an implementable mechanism for trust and introduced computational trust model to enable user to make trust-based decisions. Beth et al. [4] also proposed a trust model for distributed networks and formulated algorithms to compute trust values derived from direct trust. Josang and Ismail [13] presented a trust model based on Bayesian network model which accumulates positive and negative feedback about a member using the beta probability density function. Ahmad et al. [3] proposed a three turns trust model in which at the first turn, the user satisfies itself with the previous experience of cloud service provider and the subsequent turn, the user gets the knowledge about the service level agreement once the first two turns are satisfying for the user, then the user organization can trust on reliability of the cloud service provider. Shen et al. [19] used trusted computing patform to provide authentication, data protection, and rule-based access in the cloud computing. A trust model provided by Canedo et al. [6] ensured safe file exchange among the nodes in a private cloud computing environment. On the basis of intercommunication between the nodes, values recognized as weights and ranging between (0–1) are established. Operation indicators such as node stored space, operating system, network bandwidth, and process capacity is assessed and based on these the value of trust are determined. Li et al. [14] introduced a computing model based on multi-tenancy trust for the IaaS layer to ensure that the user and cloud service provider collaborate with each other to build and maintain a trustworthy cloud computing environment. A collaborative trust model for firewalls in cloud computing was proposed by Yang et al. [20] in which a trust value is ascertained based on the context and historical behavior of an entity. A system of domain trust and intra and internodes is utilized to determine their relation. Also interdomain trust and weights value of the domain nodes is gaged. Goyal et al. [10] developed a framework for establishing priority of the quality of service in terms of service reliability, accountability, cost performance, security and privacy in cloud environment on the basis of ranking mechanism. To "improve the quality of service" in cloud infrastructure, Garg et al. [9] presented a trust management model based on an efficient cost algorithm as service. The trust values obtained and separated are based on the calculation of the trust value. The reliable and unreliable data center parameters including starting time, cost, processing speed, bandwidth, and failure rate are utilized to calculate the trust values. Naseer et al. [16], offered a model to secure dependable and efficient providers of cloud service based on the performance/capabilities of supplier, caliber of service, downtime, response time, uptime, etc. Evaluation of customers is also taken into consideration. A choice is provided to the user to assess cloud service provider based on their requirements.

5 Proposed Trust Model for Privacy Issues Analysis

Trust model is considered to be an important and significant approach for making decision in cloud computing where contrasting and distinctive applications are utilized by the user. Large numbers of user in a cloud computing environment embraced trust-based security structure to safeguard operations of the applications. Though most researchers have unanimously concluded that a trust model should have three different components [11], i.e., 1. Subjectivity, i.e., individual preferences may vary for different entities with same view of things. 2. The expected probability, i.e., the

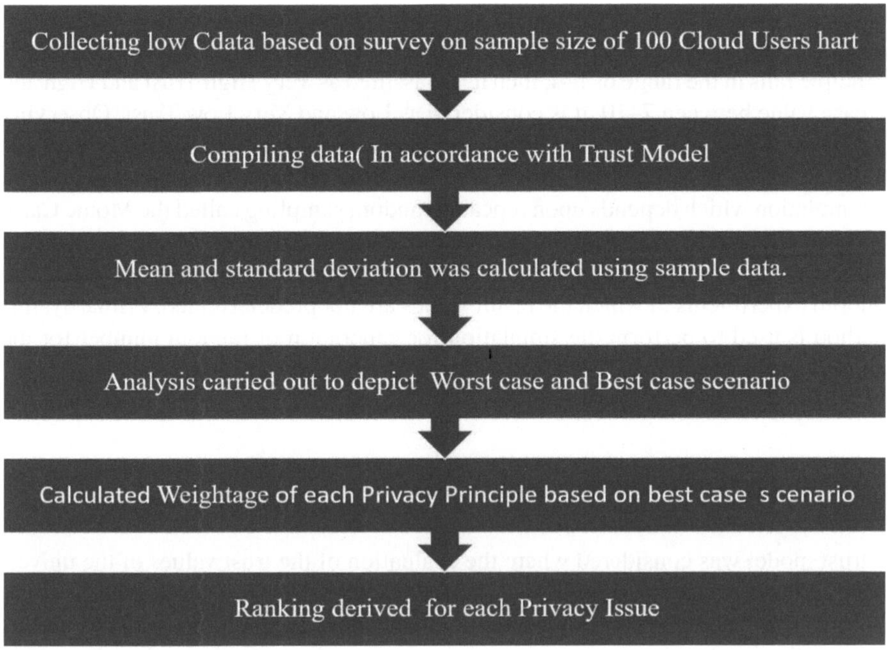

Fig. 1 Proposed methodology flow chart

expected value of trust obtained from the probability-weighted average of all possible values of trust. 3. Relevance, i.e., certainty and ability of specific content. A dependable cloud computing environment is provided by the trust model to tackle issues which include issues concerning privacy and preservation. The trust model is expected to provide higher degree of availability, reliability, integrity, and security.

In this paper, we propose a trust model to gain an all-inclusive appraisal of the trust information and establish a ranking of trustworthy privacy principles to secure privacy among users in a cloud. In this model as depicted in Fig. 1, value of the trust determines whether the privacy principle is trustworthy or not. Analysis has been performed on privacy issues such as accountability, security, compliance, data collection, usage of data, etc. Considering the privacy principles as previously described, the trust value of privacy issues is established based on the response of the queries sent to the cloud users. Each privacy principle maintains two tables, i.e., Direct Table and the Reference Table. In order to determine the direct trust tables, user feedback on the questionnaire pertaining to each privacy principle is assigned a value. In the process to normalize the data, ambiguous data has been removed to gain consistency for analysis purpose. The obtained values against each principle are tabulated as a Direct Table. The calculated value in the Direct Table is compared with the reference values for trust table to access the status of each privacy principle. In accordance to the Reference Trust Table 2, if the calculated value in the Direct Table of a privacy principle falls in the range of 1–4, then it is classified as Very High Trust and High and in case value between 7–10, it is considered as Low and Very Low Trust. Observing the Direct Table and Reference Table, the decision to trust or not largely depends upon the value obtained against each principle from these simulations. Also, a type of simulation which depends upon repeated random sampling called the Monte Carlo has been effectively used to investigate the privacy concerns of cloud service decision users and estimate the results. This method of simulation is very closely related to random experiments in which the result values are not predetermined. Primarily, this method is used to perform the simulation for generation of random number for the ten universal privacy principles.

6 Experimental Design and Analysis

A trust model was considered where the evaluation of the trust values of the universally adopted privacy principles determines whether the privacy principle is trustworthy. Trust model provides an analytical viewpoint for the privacy issues and its preservation in the cloud computing environment. Monte Carlo method is used to perform the simulation for generation of random numbers for validation of the analysis. Analysis can be performed based on accountability, security, compliance, data collection, usage of data, etc. The trust value is built upon the response of the queries sent to the cloud users on privacy issues considering the privacy principles. Each privacy principle maintains two tables, i.e., Direct Table and the Reference Table. The environment computing public cloud and its users were selected because with a specific context of analyzing privacy issues based on privacy principles.

In order to access the importance of privacy, questions were raised to approximately 100 users on the privacy principles to access their inputs for the analysis purpose. The questions were asked about their usage of the cloud service providers such as Google, Amazon, etc., thereby accessing the importance of privacy issues. Based on the input given by them in the questionnaire, the data was compiled and taken into consideration in one sample survey where in the average rating on the ranks given by the users was calculated on each privacy principle. As explained above, the feedback from 100 cloud service users was obtained in the form of questionnaire comprising of 72 questions on 10 privacy principles of significance, and importance to the users and analysis using all privacy principles as variables were done.

Step1. First survey is conducted on sample size of 100 cloud service users. Questions were raised to them on the privacy principles to access their inputs for the analysis purpose. The questions were asked about their usage of the cloud service providers such as Google, Amazon etc., thereby accessing the importance of privacy issues.

Table 2 Reference values of trust

Very high trust	High 1	Average trust	Low trust	Very low trust
1–2	3–4	5–6	7–8	9–10

Step 2. Based on the input given by them in the questionnaire, the data was compiled and taken into consideration in one sample survey where in the average rating on the ranks given by the users was calculated on each privacy principle as shown in Table 3.

Step 3. Further, the survey sample was expanded to sample size of 100 cloud users to drive out the impact analysis based on the average rating of all users.

Step 4. To apply the average rating over a sample size of 1000 cloud users, Monte Carlo method was used, thereby calculating the mean and standard deviation from the sample survey considered in Step 3 as shown in Table 4.

Step 5. In the Monte Carlo method, random value was generated for each privacy principle by applying random function. And thereafter, normal inverse function was applied using mean and standard deviation on the random function on the random value generated earlier to get each privacy principle average rating.

Step 6. On the basis of Step 4, calculations were made to access worst-case scenario and best-case scenario for each privacy principle. ±5% cases were considered for evaluating the best-case and worst-case scenarios which are depicted as below. Further, the survey sample was expanded to sample size of 100 cloud users to drive out the impact analysis based on the average rating of all users. Monte Carlo method was used to ascertain user rating of each privacy principle by calculating the mean and standard deviation from the sample survey. As depicted in Table 5, the weighted average value for each privacy principle was calculated in both the cases, i.e., worst-case scenario and best-case scenario as in Fig. 2.

Step 7. Each privacy principle was assigned ranking based on the best-case scenario.

It is very evident from Table 6 that the high-ranking privacy principles carry a higher degree of trust as compared to other privacy principles. Final analysis is presented by taking the entire variable set together to find out the ranking of all the privacy principles selected for analysis. The ranking of privacy issues is based on the value of trust calculated using the trust model and applying Monte Carlo method. Table 6 presents the analysis taking all the privacy principles for the analysis, comparing the values of privacy principle from Direct table with the reference values. Security issues have the highest levels of concern as compared to other issues as observed by the value of Direct table and Reference table. The second highest value is for accountability which poses the highest concern after security. The value for openness, access, and accuracy of privacy issues are low which associate these variables with low levels of concern as compared to other variables as in Fig. 3.

Table 3 User rating of each privacy principle

Sr No	Accountability	Security	Compliance	Disclosure	Usage	Data collection	Consent	Openness	Access	Accuracy
1	4	2.5	4	3.8	5.8	8	8.3	9.4	7.9	8.8
2	3.9	2.3	4	3.8	5.8	8	8.3	9.3	7.9	8.8
3	3.9	2.3	4	3.8	5.7	7.9	8	9.3	7.9	8.8
4	3.9	2.3	3.8	3.8	5.7	7.9	7.8	9.3	7.9	8.8
5	3.9	2.3	3.8	3.8	5.7	7.9	6.9	9.3	7.9	8.8
...	...	...	...	...	...	...	...	...	...	...

Table 4 Mean and standard deviation value of each privacy principles

Privacy aspects	Mean value	Standard deviation
Accountability	2.5	0.82
Security	1.1	0.41
Compliance	2.6	0.97
Disclosure	2.5	0.79
Usage	4.8	0.49
Data collection	5.9	1.20
Consent	7.6	0.77
Openness	7.8	0.81
Access	7	0.49
Accuracy	7.6	0.76

Table 5 Weighted average value of each privacy principles

Privacy aspects	Worst case	Best case	Weightage	Weightage (%)
Accountability	3.87	1.19	0.217	21.7
Security	1.92	0.73	0.354	35.4
Compliance	2.14	2.87	0.089	9
Disclosure	3.83	2.53	0.102	10.2
Usage	4.07	4.6	0.056	5.6
Data collection	5.67	6.27	0.041	4.1
Consent	7.54	8	0.032	3.2
Openness	6.95	7.82	0.033	3.3
Access	7.49	7.11	0.036	3.6
Accuracy	7.57	6.82	0.0378	3.8

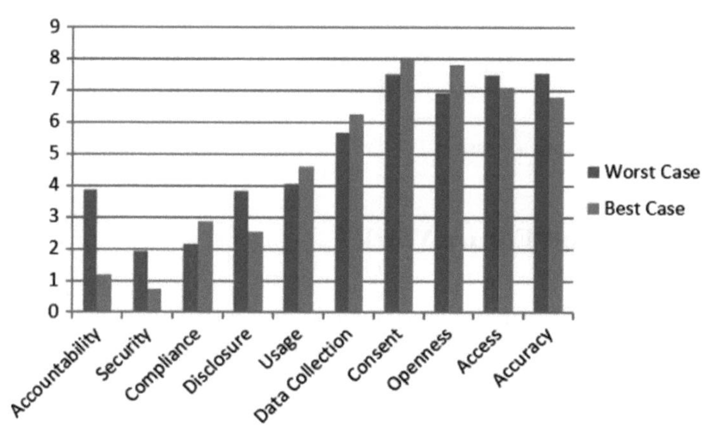

Fig. 2 Privacy issues weighted avg. value

Table 6 Ranking table on privacy issues

RANK	Privacy issue	Weightage (%)
1	Security	35.4
2	Accountability	21.7
3	Disclosure	10.2
4	Compliance	9.0
5	Data usage	5.6
6	Data collection	4.1
7	Consent	3.8
8	Openness	3.6
9	Access	3.3
10	Accuracy	3.2

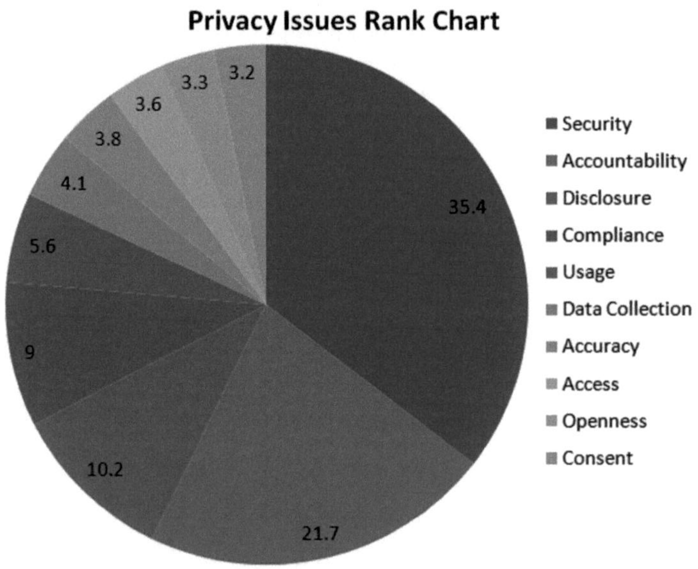

Fig. 3 Ranking graph on privacy issues

7 Conclusion and Future Work

In this paper, we described that the trust model achieved a comprehensive assessment of the trust information and could ensure a trustworthy cloud computing environment to the user. From this trust model using Monte Carlo method, it is concluded that the high-ranking principle is more significant and important in designing any sort of privacy framework for cloud computing environment. Privacy issues which are more sensitive and which possess high levels of trust of the individual are to be considered first by the cloud service provider. The inclusion of clauses on the protection of privacy and personal information in the agreement or in the regulations is of paramount importance. Various risks and challenges concerning the privacy of the cloud user that may not have been considered in the past by the cloud service provider are required to be analyzed and evaluated in future.

References

1. Key cloud privacy concerns in 2012, http://techgenix.com/key-cloud-privacy-concerns-2012/
2. Justice and fundamental rights, https://ec.europa.eu/info/strategy/justice-and-fundamental-rights_en
3. S. Ahmad, B. Ahmad, S.M. Saqib, R.M. Khattak, Trust model: Clouds provider and clouds user. Int. J. Adv. Sci. Technol. **44**, 69–80 (2012)
4. T. Beth, M. Borcherding, B. Klein, Valuation of trust in open networks, in *Computer Security ESORICS*, vol. 94 (1994), pp. 1–18
5. R. Buyya, J. Broberg, A.M. Goscinski. *Cloud Computing: principles and paradigms*, vol. 87 (Wiley, 2010)
6. E.D. Canedo, R.T. de_Sousa Junior, R. de_Oliveira Albuquerque, F.L.L. de Mendonça, File exchange in a private cloud supported by a trust model, in *2012 International Conference on Cyber-Enabled Distributed Computing and Knowledge Discovery (CyberC)* (IEEE, 2012), pp. 89–96
7. U.S.P.P.S, Commission. Personal privacy in an information society: the report of the Privacy Protection Study Commission, vol. 1. The Commission: for sale by the Supt. of Docs., US Govt. Print. Off., 1977
8. Organisation for Economic Co-operation and Development. Guidelines on the Protection of Privacy and Transborder Flows of Personal Data. OECD, 1981
9. S.K. Garg, S. Versteeg, R. Buyya, A framework for ranking of cloud computing services. Future Gener. Comput. Syst. **29**(4), 1012–1023 (2013)
10. M.K. Goyal, A. Aggarwal, P. Gupta, P. Kumar, Qos based trust management model for cloud IaaS, in *2012 2nd IEEE International Conference on Parallel Distributed and Grid Computing (PDGC)* (IEEE, 2012), pp. 843–847
11. W. House, Consumer data privacy in a networked world: a framework for protecting privacy and promoting innovation in the global digital economy. Feb 2012
12. W. Jansen, T. Grance, *Sp 800–144, Guidelines on Security and Privacy in Public Cloud Computing* (National Institute of Standards & Technology, Gaithersburg, MD, 2011)
13. A. Josang, R. Ismail, The beta reputation system, in *Proceedings of the 15th Bled Electronic Commerce Conference*, vol. 5 (2002), pp. 2502–2511
14. X. Li, M. R. Lyu, J. Liu, A trust model based routing protocol for secure ad hoc networks, in *2004 Proceedings of 2004 IEEE Conference on Aerospace*, vol. 2 (IEEE, 2004), pp. 1286–1295
15. S.P. Marsh, *Formalising Trust as a Computational Concept* (1994)
16. M.K. Naseer, S. Jabbar, I. Zafar, A novel trust model for selection of cloud service provider, in*2014 World Symposium on Computer Applications & Research (WSCAR)* (IEEE, 2014), pp. 1–6
17. S. Pearson, Privacy, security and trust in cloud computing, in *Privacy and Security for Cloud Computing* (Springer, 2013), pp. 3–42
18. J. Ruiter, M. Warnier, Privacy regulations for cloud computing: compliance and implementation in theory and practice, in *Computers, privacy and data protection: an element of choice* (Springer, 2011), pp. 361–376

19. Z. Shen, L. Li, F. Yan, and X. Wu. Cloud computing system based on trusted computing platform. In *2010 International Conference on Intelligent Computation Technology and Automation (ICICTA)*, vol. 1 (IEEE, 2010), pp. 942–945
20. Z. Yang, L. Qiao, C. Liu, C. Yang, G. Wan, A collaborative trust model of firewall-through based on cloud computing, in *2010 14th international conference on Computer supported cooperative work in design (CSCWD)* (IEEE, 2010), pp. 329–334

Author Index

A
Aakanksha Mahajan, 393
Aanshi Bhardwaj, 3
Aastha Maheshwari, 153
Aditi Rawat, 783, 795, 805
Aditya Bhardwaj, 1021
Ajay D. Nagne, 735
Ajit Patil, 839
Akashdeep, 375
Akash Kumar Bhoi, 901
Akhilesh Kumar Mishra, 933
Alam, S., 901
Amandeep Kaur, 439
Aman Kumar Sharma, 989
Amarendra Goap, 335
Amarsinh B. Varpe, 735, 911
Amita Jain, 661, 711
Amit Doegar, 469
Amit Kumar Garg, 345
Amit Kumar Maurya, 933
Amit Nayyer, 989
Amit Prakash Singh, 979
Amol D. Vibhute, 717, 735, 911
Amol P. Bhondekar, 335
Amrsinh B. Varpe, 717
Anand Gupta, 795
Ankushe, K., 83
Anurag Jagetiya, 15
Anureet Singh, 271
Aradhana Negi, 861
Arindam Biswas, 383
Arushi Arora, 233
Arvind Kumar, 933
Ashwani Kumar Vashishtha, 53
Atul Sharma, 3

B
Baljit Singh Saini, 25
Balwinder Raj, 815
Bhable, S., 83
Bharat Kumar, 933
Birmohan Singh, 637
Bishwesh Kumar, 947

C
Charu Gupta, 661
Charu Nimesh, 353
Charu Sharma, 165

D
Damanjeet Kaur, 323
Deepak Sharma, 335
Deepali Gupta, 217
Deepkamal Kaur Gill, 783, 795, 805
Devendra K. Tayal, 711, 939
Dhananjay B. Nalawade, 735
Dhananjay Nalawade, 445, 717
Dinesh Kumar, 407
Dipanwita Thakur, 173
Divya Bajaj, 783, 795, 805
Divya Bansal, 669, 1041
Divya Mamgai, 653
Dweep Gogia, 947

F
Farooque Azam, 901

G
Garima Saini, 313
Gaurav Kumar, 469
Girisha Kumar, 313

C. R. Krishna et al. (eds.), *Proceedings of 2nd International Conference on Communication, Computing and Networking*, Lecture Notes in Networks and Systems 46, https://doi.org/10.1007/978-981-13-1217-5

Gulshakh Kaur, 703
Gurjeet Kaur, 35
Gurleen Kaur, 117
Gurmohan Singh, 815
Gurneet Kaur, 353
Gurpreet kaur, 427

H
Hanumant Gite, 717
Hanumant R. Gite, 445
Hardeep Singh, 861
Hardeep Singh Saini, 271
Harish Kumar, 293, 539
Harish Prabhu, 839
Harkiran Kaur, 551, 559
Harkirat Singh, 599
Harpreet Kaur, 479
Himanshu Aggarwal, 889
Hitender Kumar Tyagi, 407

J
Jadhav, V. S., 83
Jaspal Kaur Saini, 669
Jayant Giri, 839
Jitender Kumar Chhabra, 771
Jitendra N. Dongre, 83

K
Kale, K. V., 83, 911
Kamaljeet Kaur, 479
Kamaljit Singh Bhatia, 25
Kanak Meena, 939
Kannan, R., 205
Karbhari Kale, 717
Karbhari V. Kale, 445, 735
Kashish Goyal, 727
Kawaljeet Singh, 551, 559
Keshava Reddy, 695
Keshav Singh Rawat, 677
Khyati Ahlawat, 979
Krishan Kumar, 3
Kulwinder Singh Mann, 303, 439
Kumud Arora, 627

L
Lalit K. Awasthi, 989

M
Mahesh M. Solankar, 445
Mahesh Solankar, 717
Maher, S., 83
Mahreen Saleem, 93
Malhan, I. V., 677
Maitreyee Dutta, 469
Makesh Iyer, 823
Mala Kalra, 1011
Mamta Juneja, 783, 795, 805
Manas Khatua, 173
Maninder Singh, 969
Manisha Singla, 615
Mankirat Kaur, 577
Manoj Duhan, 359
Manpreet Kaur, 637
Mehrotra, S. C., 911
Md. Muzakkir Hussain, 957, 999
Meenakshi Sharma, 889
Mohammad Ahsan Chishti, 93
Mohammad Saad Alam, 957, 999
Mohit Dua, 653
Mohit Mittal, 135
Mrinaal Malhotra, 771
Mrudula, K, 695
Mukesh Kumar, 599, 727
Musheer Ahmad, 63

N
Nagsen Bansod, 83
Naman Bansal, 375
Namita Kathpal, 345
Narottam Chand, 153
Navdeep Kaur, 25
Naveen Aggarwal, 43, 375, 577
Naveen Kumar, 271
Navneet kaur, 861
Navneet Kaur Sehgal, 183
Nazia Parveen, 489
Neelam Saleem Khan, 93
Neeraj Priyadarshi, 901
Neeti Taneja, 479
Nisheeth Joshi, 523, 661

O
Ochin Sharma, 523
Oishila Bandyopadhyay, 383

P
Pankaj Kumar, 63
Pardeep Kumar, 261
Parminder Singh, 415
Parmod Kalia, 1041
Partha Pratim Bhattacharya, 145
Pawan Kapur, 849
Pawan Kumar, 1033
Payal Kamboj, 127
Poonam Garg, 627
Poonam Jaglan, 359

Poonam Saini, 703
Prakriti Sodhi, 727
Prasad Reddy Padala, V. G. D, 195
Prashant Jindal, 783, 795, 805
Preeti Abrol, 923
Preeti Aggarwal, 599, 727
Preethi Padala, 195
Priyanshu Vishwakarma, 933

R
Rajarshi Mahapatra, 283
Rajesh K. Dhumal, 445, 735, 911
Rajeshwar Dass, 359
Rajiv Chechi, 53
Rakesh Kumar, 685, 1033
Rama Krishna Challa, 15, 53, 127, 183, 513, 1021
Ranbir Singh, 303
Ranjana Thalore, 145
Rashid Ali, 589
Rashid Mustafa, 241, 567
Rashmi Vishraj, 457
Raunak Monir, 145
Ravi Kant, 607
Ravneet Kaur, 539
Reddy, S. R. N., 127, 217
Reena Rathee Jaglan, 241, 567
Reeta Devi, 407
Rekh Ram Janghel, 645
Renu Vig, 3, 427, 849
Rigzin Angmo, 43
Rishu Chhabra, 513
Riya Goyal, 293
Rohit Bansal, 393
Rohit Vaid, 165
Rohit Yadav, 653
Rudrani Sharma, 685
Rupali R. Surase, 445, 717, 735, 911

S
Saeed, S. H, 489
Sachin Lalar, 73
Sakshi Kaushal, 293
Samar Wazir, 745
Sandeep Gaikwad, 717
Sandeep V. Gaikwad, 735, 911
Sanjeev Sofat, 1041
Sarbjeet Singh, 539, 577
Sarin, R. K., 815
Saroj Kumar Pandey, 645
Satendra Kumar, 135
Satish Kumar, 849
Savita Gupta, 353, 457, 923
Seema Verma, 261, 513
Senthil Kumaran, M., 205
Shanmuganantham, T., 823
Shano Solanki, 757
Shashi Bhushan, 73
Shatakshi Kaushal, 947
Shende, K. V., 83
Shilpa Verma, 703
Shivani Bali, 871
Shukla, A. K., 335
Shukla, K. K., 615
Siddharth B. Dabhade, 83
Sirat Kaur, 323
Sobin, C.C., 107
Solai Manohar, S., 205
Sonakshi Vij, 711
Sonali Brodiya, 653
Sudesh Rani, 251
Sufyan Beg, M. M., 745, 957, 999
Sukhpreet Kaur Gill, 415
Sukhwinder Singh, 323, 427, 457
Suman Balhwan, 217
Sumandeep Kaur, 251
Sumit Kr Yadav, 233
Sunil Agrawal, 35, 117, 241, 567
Sunil Kumar, 969, 1011
Suniti Dutt, 117
Suraj Pal Singh, 757
Surender, 73
Suresh C. Mehrotra, 735
Suresh Koduru, 195
Suresh Kumar, 523
Suresh Mehrotra, 717
Sushila Madan, 871
Sushil Kumar, 393

T
Tamanna Goyal, 669
Tamanna Puri, 183
Tanu Goyal, 293
Tanvir Ahmad, 745
Tawseef Ayoub Shaikh, 589
Tejinder Kaur, 551, 559
Tharewal, S., 83
Thorat, S., 83
Trilok C. Aseri, 251

U
Ujjwal Kumar Kamila, 383

V
Veenu Mangat, 3, 43
Vikram Bali, 871
Vinay Kumar, 607
Vinod Kumar, 63

Y
Yash Gopal Mittal, 805
Yousuf Haider, 15